T0318817

Developments in Marine Geology

Volume 7

Earth and Life Processes Discovered from Subseafloor Environments

Developments in Marine Geology

Volume 7

Earth and Life Processes Discovered from Subseafloor Environments

A Decade of Science Achieved by the
Integrated Ocean Drilling Program (IODP)

Edited by

Ruediger Stein
Division of Geosciences, Alfred Wegener Institute for
Polar and Marine Research, Bremerhaven, Germany

Donna K. Blackman
Scripps Institution of Oceanography, University of California San Diego,
La Jolla, CA, USA

Fumio Inagaki
Kochi Institute for Core Sample Research, Japan Agency for Marine-Earth
Science & Technology (JAMSTEC), Nankoku, Kochi, Japan

Hans-Christian Larsen

AMSTERDAM • BOSTON • HEIDELBERG • LONDON
NEW YORK • OXFORD • PARIS • SAN DIEGO
SAN FRANCISCO • SINGAPORE • SYDNEY • TOKYO

ELSEVIER

Elsevier
Radarweg 29, PO Box 211, 1000 AE Amsterdam, Netherlands
The Boulevard, Langford Lane, Kidlington, Oxford OX5 1GB, UK
225 Wyman Street, Waltham, MA 02451, USA

First edition 2014

Notices
Knowledge and best practice in this field are constantly changing. As new research
and experience broaden our understanding, changes in research methods, professional
practices, or medical treatment may become necessary.

Practitioners and researchers must always rely on their own experience and knowledge
in evaluating and using any information, methods, compounds, or experiments
described herein. In using such information or methods they should be mindful of
their own safety and the safety of others, including parties for whom they have a
professional responsibility.

To the fullest extent of the law, neither the Publisher nor the authors, contributors, or
editors, assume any liability for any injury and/or damage to persons or property as a
matter of products liability, negligence or otherwise, or from any use or operation of
any methods, products, instructions, or ideas contained in the material herein.

ISBN: 978-0-444-62617-2
ISSN: 1572-5480

For information on all Elsevier publications
visit our web site at http://store.elsevier.com

Contents

2.2.1. Biomass, Diversity, and Metabolic Functions of Subseafloor Life

Yuki Morono and Jens Kallmeyer

2.2.2. Genetic Evidence of Subseafloor Microbial Communities

Andreas Teske, Jennifer F. Biddle and Mark A. Lever

2.3. The Underground Economy (Energetic Constraints of Subseafloor Life)

Steven D'Hondt, Guizhi Wang and Arthur J. Spivack

2.4. Life at Subseafloor Extremes

Ken Takai, Kentaro Nakamura, Douglas LaRowe and Jan P. Amend

3.   Environmental Change, Processes and Effects

3.1. Introduction: Environmental Change, Processes and
      Effects—New Insights From Integrated Ocean Drilling
      Program (2003–2013)

Ruediger Stein

3.2. Cenozoic Arctic Ocean Climate History: Some
      Highlights from the Integrated Ocean Drilling Program
      Arctic Coring Expedition

*Ruediger Stein, Petra Weller, Jan Backman, Henk Brinkhuis,
Kate Moran and Heiko Pälike*

3.3. From Greenhouse to Icehouse at the Wilkes Land
      Antarctic Margin: IODP Expedition 318 Synthesis
      of Results

Carlota Escutia, Henk Brinkhuis & the Expedition 318 Scientists

4.   Solid Earth Cycles and Geodynamics

4.1.   Introduction

Donna K. Blackman

4.2.1. Formation and Evolution of Oceanic Lithosphere: New Insights on Crustal Structure and Igneous Geochemistry from ODP/IODP Sites 1256, U1309, and U1415

Benoit Ildefonse, Natsue Abe, Marguerite Godard, Antony Morris, Damon A.H. Teagle and Susumu Umino

4.4.1. Subduction Zones: Structure and Deformation History

Harold Tobin, Pierre Henry, Paola Vannucchi and Elizabeth Screaton

4.4.2. Seismogenic Processes Revealed Through the Nankai Trough Seismogenic Zone Experiments: Core, Log, Geophysics, and Observatory Measurements

Masataka Kinoshita, Gaku Kimura and Saneatsu Saito

xiv Contents

Contributors

Natsue Abe Institute for Research on Earth Evolution, Japan Agency for Marine-Earth Science and, Technology (JAMSTEC), Yokosuka, Japan

Jeffrey Alt Department of Earth and Environmental Sciences, University of Michigan, Ann Arbor, MI, USA

Jan P. Amend Department of Earth Sciences, and Department of Biological Sciences, University of Southern California, Los Angeles, CA, USA

Wolfgang Bach Department of Geosciences, Center for Marine Environmental Sciences (MARUM), University of Bremen, Bremen, Germany

Jan Backman Department of Geology and Geochemistry, Stockholm University, Stockholm, Sweden

Keir Becker Department of Marine Geosciences, Rosenstiel School of Marine and Atmospheric Science, University of Miami, Coral Gables, FL, USA

Jennifer F. Biddle School of Marine Science and Policy, University of Delaware, Lewes, DE, USA

Donna K. Blackman Scripps Institution of Oceanography, University of California San Diego, La Jolla, CA, USA

Henk Brinkhuis Royal Netherlands Institute for Sea Research NIOZ, Netherlands; and Instituto Andaluz de Ciencias de la Tierra, CSIC-Univ. Granada, Granada, Spain

Gilbert Camoin CEREGE, UMR 7330 CNRS, Europôle Méditerranéen de l'Arbois, Aix-en-Provence, France

James E.T. Channell Department of Geological Sciences, University of Florida, Gainesville, FL, USA

Steven D'Hondt Graduate School of Oceanography, University of Rhode Island, RI, USA

Katrina J. Edwards Department of Earth Sciences, and Department of Biological Sciences, University of Southern California, Los Angeles, CA, USA

Bert Engelen Institut für Chemie und Biologie des Meeres (ICBM), Carl von Ossietzky Universität Oldenburg, Postfach, Oldenburg, Germany

Carlota Escutia Instituto Andaluz de Ciencias de la Tierra, CSIC-Univ. Granada, Granada, Spain

Timothy G. Ferdelman Department of Biogeochemistry, Max Planck Institute for Marine Microbiology, Bremen, Germany

Andrew T. Fisher Earth and Planetary Sciences Department, University of California, Santa Cruz, CA, USA

Marguerite Godard Géosciences Montpellier, Université Montpellier, Montpellier, France

Robert N. Harris Oregon State University, Corvallis, OR, USA

Pierre Henry CEREGE, Aix-Marseille Université, Marseille, France

David A. Hodell Godwin Laboratory for Palaeoclimate Research, Department of Earth Sciences, University of Cambridge, Cambridge, UK

Benoit Ildefonse Géosciences Montpellier, Université Montpellier, Montpellier, France

Hiroyuki Imachi Department of Subsurface Geobiological Analysis and Research, Japan Agency for Marine-Earth Science & Technology (JAMSTEC), Yokosuka, Japan

Fumio Inagaki Kochi Institute for Core Sample Research, Japan Agency for Marine-Earth Science & Technology (JAMSTEC), Nankoku, Kochi, Japan

Jens Kallmeyer Deutsches GeoForschungsZentrum GFZ, Section 4.5 Geomicrobiology, Telegrafenberg, Potsdam, Germany

Miriam Kastner Scripps Institution of Oceanography, La Jolla, CA, USA

Gaku Kimura Department of Earth and Planetary Science, Graduate School of Science, University of Tokyo, Japan

Masataka Kinoshita Kochi Institute for Core Sample Research, Japan Agency for Marine-Earth Science & Technology (JAMSTEC), Nankoku, Kochi, Japan

Anthony A.P. Koppers College of Earth, Ocean & Atmospheric Sciences, Oregon State University, Corvallis, OR, USA

Douglas LaRowe Department of Earth Sciences, University of Southern California, Los Angeles, CA, USA

Mark A. Lever Department of Environmental Systems Science, ETH Zurich, Zurich, Switzerland

Jeannette Lezius Geosciences Department, Alfred Wegener Institute for Polar and Marine Research, Bremerhaven, Germany

Mitchell W. Lyle College of Geosciences, Texas A&M University, TX, USA

Kate Moran Neptune Ocean Networks Canada, University of Victoria, Victoria, BC, Canada

Yuki Morono Kochi Institute for Core Sample Research, Japan Agency for Marine-Earth Science & Technology (JAMSTEC), Nankoku, Kochi, Japan

Antony Morris School of Geography, Earth and Environmental Sciences, Plymouth University, Plymouth, UK

Kentaro Nakamura Laboratory of Ocean-Earth Life Evolution Research (OELE), Japan Agency for Marine-Earth Science & Technology (JAMSTEC), Yokosuka, Japan; Submarine Hydrothermal System Research Group, Japan Agency for Marine-Earth Science & Technology (JAMSTEC), Yokosuka, Japan; Department of Systems Innovation, School of Engineering, University of Tokyo, Tokyo, Japan

Hiroshi Nishi The Center for Academic Resources and Archives, Tohoku University Museum, Tohoku University, Sendai, Japan

Beth N. Orcutt Bigelow Laboratory for Ocean Sciences, East Boothbay, ME, USA

Victoria Orphan Division of Geological and Planetary Sciences, California Institute of Technology, Pasadena, CA, USA

Heiko Pälike MARUM – Center for Marine Environmental Sciences, University of Bremen, Leobener Strasse, Bremen, Germany

Isabella Raffi Dipartimento di Ingegneria e Geologia (InGeo)–CeRSGeo, Università "G. d'Annunzio" di Chieti-Pescara, Chieti Scalo, Italy

William W. Sager Department of Earth and Atmospheric Sciences, University of Houston, Houston, TX, USA

Saneatsu Saito R&D Center for Ocean Drilling Sciences, JAMSTEC, Japan

Elizabeth Screaton Department of Geological Science, University of Florida, Gainesville, FL, USA

Evan A. Solomon School of Oceanography, University of Washington, Seattle, WA, USA

Arthur J. Spivack Graduate School of Oceanography, University of Rhode Island, RI, USA

Ruediger Stein Geosciences Department, Alfred Wegener Institute for Polar and Marine Research, Bremerhaven, Germany

Ken Takai Department of Subsurface Geobiological Analysis and Research, Japan Agency for Marine-Earth Science & Technology (JAMSTEC), Yokosuka, Japan; Laboratory of Ocean-Earth Life Evolution Research (OELE), Japan Agency for Marine-Earth Science & Technology (JAMSTEC), Yokosuka, Japan; Submarine Hydrothermal System Research Group, Japan Agency for Marine-Earth Science & Technology (JAMSTEC), Yokosuka, Japan; Earth-Life Science Institute (ELSI), Tokyo Institute of Technology, Tokyo, Japan

Damon A.H. Teagle National Oceanography Centre Southampton, University of Southampton, Southampton, UK

Andreas Teske Department of Marine Sciences, University of North Carolina at Chapel Hill, Chapel Hill, NC, USA

Harold Tobin Department of Geoscience, University of Wisconsin–Madison, Madison, WI, USA

Marta E. Torres Oregon State University, Corvallis, OR, USA

Susumu Umino Department of Earth Sciences, Kanazawa University, Kanazawa, Japan

Paola Vannucchi Department of Earth Sciences, University of London, London, UK

Guizhi Wang State Key Laboratory of Marine Environmental Science, Xiamen University, Xiamen, China

Jody Webster Geocoastal Research Group, School of Geosciences, University of Sydney, NSW, Australia

Laura M. Wehrmann School of Marine and Atmospheric Sciences, Stony Brook University, Stony Brook, NY, USA

Petra Weller Geosciences Department, Alfred Wegener Institute for Polar and Marine Research, Bremerhaven, Germany

Preface

Since 1968, the Deep Sea Drilling Project (DSDP: 1968–1983), Ocean Drilling Program (ODP: 1983–2003) and Integrated Ocean Drilling Program (IODP: 2003–2013) have provided crucial records of past and present processes and interactions within and between the biosphere, cryosphere, atmosphere, hydrosphere, and geosphere. The early DSDP–ODP exploratory phase resulted in the confirmation of the unifying theory of plate tectonics, shortly followed by the development of new fields of research such as paleoceanography, astronomically tuned geochronology, structure and geodynamics of the ocean crust, mineral resources in oceanic hydrothermal systems, marine gas-hydrate reservoirs, subseafloor life and biogeochemical cycles, and others. Research within the IODP encompasses a wide range of fundamental and applied topics that affect society, such as global change, evolution and diversity of life, natural hazards such as earthquakes, submarine landslides and volcanism, and the internal structure and dynamics of our planet (Figure 1). Unlike the previous, single vessel drilling programmes, the IODP operated with three different drilling vessels: the riser-equipped drilling vessel D/V *Chikyu* provided by Japan, the non-riser drilling vessel *JOIDES Resolution* provided by the United States, and ad-hoc chartered drilling vessels tailored to specific missions such as the high Arctic or carbonate reef drilling. These mission-specific platforms were provided by the European Consortium for Scientific Ocean Drilling (ECORD). With this improved and more diversified drilling capability, the IODP focused on three broad scientific themes laid out in its guiding science plan: (1) The Deep Biosphere and the Subseafloor Ocean, (2) Environmental Change, Processes and Effects, and (3) Solid Earth Cycles and Geodynamics.

This book is the first comprehensive compilation of synthesis papers presenting key results of cutting-edge research carried out within the IODP during the decade 2003–2013 of IODP operations. The different specialty chapters are written by internationally well-known and accepted experts in their fields of research.

The book is divided into five parts. Chapter 1 summarizes the major highlights obtained during the one decade of ocean science research within the IODP. The following Chapters 2 to 4, the "heart of the book," present in more detail the main IODP results devoted to the three main IODP themes: *The Deep Biosphere and the Subseafloor Ocean*, *Environmental Change, Processes and Effects*, and *Solid Earth Cycles and Geodynamics*. In Chapter 5, an appendix,

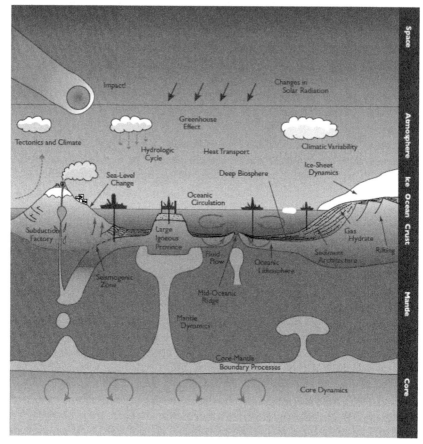

FIGURE 1 Earth system components, processes, and phenomena to be studied by scientific ocean drilling. *Figure from the IODP Initial Science Plan;* www.iodp.org/initial-science-plan.

background information about each expedition is shortly presented in one-page summaries.

This book is suitable for lecturers, graduate students as well as scientists interested in all types of Earth system studies, with special emphasis on the deep biosphere, paleoclimate, and solid earth dynamics.

Acknowledgments

We are very grateful to each of the Reviewers, listed below, for numerous suggestions and comments for improvement of the manuscripts. Furthermore, special thanks to Derek Coleman and Susan Dennis from the Elsevier staff for their support during all stages of this project.

Ruediger Stein, Donna Blackman, Fumio Inagaki
and Hans-Christian Larsen (Eds.)
August 2014

List of Reviewers

Wolfgang Bach, University of Bremen, Germany

Jan Backmann, Stockholm University, Sweden

Ian Bailey, Exeter University, UK

Peter Barrett, Victoria University of Wellington, New Zealand

Gail Christeson, University of Texas, Austin, TX, USA

Mike Coffin, University of Tasmania, Australia

Thomas M. Cronin, US Geological Survey, Reston, VA, USA

André Droxler, Rice University, Houston, TX, USA

Christian Dullo, GEOMAR Kiel, Germany

Patrick Fulton, University of California, Santa Cruz, USA

Joris Gieskes, Scripps Institution of Oceanography, San Diego, CA, USA

Eberhard Gischler, Frankfurt University, Germany

Rob Harris, Oregon State University, OR, USA

Roland von Huene, USGS Emeritus Scientist, Menlo Park, CA, USA

Eystein Jansen, University of Bergen, Norway

Anna H. Kaksonen, CSIRO, Floreat Park, WA, Australia

Mike Kaminski, King Fahd University of Petroleum and Minerals, Dhahran, Saudi Arabia

Dick Kroon, University of Edinburgh, UK

Mark A. Lever, ETH, Zurich, Switzerland

Karen G. Lloyd, University of Tennessee, TN, USA

Andrew McCaig, Leeds University, UK

Lisa McNeil, University of Southampton, UK

Heath Mills, University of Houston, Clear Lake, TX, USA

Juli Morgan, Rice University, USA

Satoshi Nakagawa, Kyoto University, Japan

David Naafs, Bristol University, UK

Clive Neal, University Notre Dame, IN, USA

Geoff Wheat, University of Alaska, USA

Laura de Santis, OGS, Sgonico, Italy
Axel Schippers, BGR, Hannover, Germany
Stefan Sievert, Woods Hole Oceanographic Institution, MA, USA
John Sinton, University of Hawaii, HI, USA
David C. Smith, University of Rhode Island, RI, USA
Glenn Spinelli, New Mexico Tech, NM, USA
Alfred Spormann, Stanford University, CA, USA
Yohey Suzuki, University of Tokyo, Japan
Roy Wilkens, University of Hawaii at Manoa, HI, USA

Chapter 1

Major Scientific Achievements of the Integrated Ocean Drilling Program: Overview and Highlights

Keir Becker

Department of Marine Geosciences, Rosenstiel School of Marine and Atmospheric Science, University of Miami, Coral Gables, FL, USA

E-mail: kbecker@rsmas.miami.edu

1.1 INTRODUCTION

The Integrated Ocean Drilling Program (IODP, 2003–2013) was built on the rich heritage of the Ocean Drilling Program (ODP, 1983–2003) and its predecessor the Deep Sea Drilling Project (DSDP, 1968–1983) to renew and expand one of the most successful international programs in all of science. DSDP and ODP each provided access to a single scientific drillship that could be utilized by a wide community of geoscientists to address a number of important scientific problems in typesettings beneath the seafloor. With a wider international funding base, IODP provided expanded access for a larger international community to three types of scientific drilling platforms that could address an even greater range of scientifically and societally important themes beneath the seafloor. No other programs have provided such scientific access beneath nearly 70% of Earth's surface; the result is a huge legacy of important results from scientific ocean drilling (e.g., National Research Council, 2011).

IODP was planned around an ambitious and challenging 10-year Initial Science Plan (ISP; IODP, 2001). The ISP was based partly on extending the performance and ODP legacy of the U.S.-provided riserless drillship *JOIDES Resolution*, with the addition of two major new elements (Figure 1.1) reflecting the unique technological capabilities of the Japanese riser drillship *Chikyu* and mission-specific platforms (MSPs) provided by ECORD (the European Consortium for Ocean Research Drilling). Like this entire volume, this chapter is organized in parallel with the three main themes and eight initiatives identified in the ISP, and in that sense it represents not just an overview of IODP achievements

Developments in Marine Geology, Volume 7. http://dx.doi.org/10.1016/B978-0-444-62617-2.00001-3
1

FIGURE 1.1 IODP drilling vessels, left to right: *JOIDES Resolution* (USIO), *Vidar Viking* (ECORD charter for Expedition 302), *Chikyu* (CDEX).

but also a broad initial assessment of the program's success in addressing its original plan.

Note that IODP formally started as an international program in 2003, but IODP drilling operations did not begin until July 2004. Note also that the operating time realized for all IODP drilling vessels was considerably less than initially envisioned, mainly because of (1) shipyard delays affecting delivery of *Chikyu* and refitting of *JOIDES Resolution* and (2) a program-wide mismatch between operating expenses and financial resources. Partly because of the former, only 15 IODP expeditions were completed during the first 5 years of the program, whereas over two-thirds of all IODP expeditions were conducted between 2009 and 2013.

This chapter presents an overview and selected highlights of published results from IODP drilling, but it is not intended as a comprehensive review of all 51 IODP expeditions. As it typically takes a few years after expeditions for comprehensive scientific results to be produced, the results that have emerged to date from the last two-thirds of IODP expeditions during remain incomplete and are unavoidably underrepresented in this chapter, last revised in July 2014. Locations of the IODP expeditions and sites discussed in this chapter are shown in Figure 1.2. (Note that the Appendix, chapter 5 of this volume, includes both a map and table that include all IODP expeditions.) In most cases, later chapters in this book explore these contributions in much greater detail from the perspectives of experts in the respective subjects. While the focus here is on IODP results, it is impossible to separate the strong links to late stages of ODP. For example, 20 of the 51 IODP expeditions were based on proposals submitted originally to ODP. Also, the last 2 years of ODP drilling produced major scientific papers during the IODP era that were significant contributions toward IODP initiatives and so are included in those sections below.

Despite the underrepresentation of later IODP expeditions, it is already evident that excellent, sometimes breakthrough scientific progress was made on all three main themes and a majority of the eight initiatives identified in the ISP, and several IODP expeditions had extraordinary impact in the public media. More modest progress was made on a couple of the ISP initiatives, and

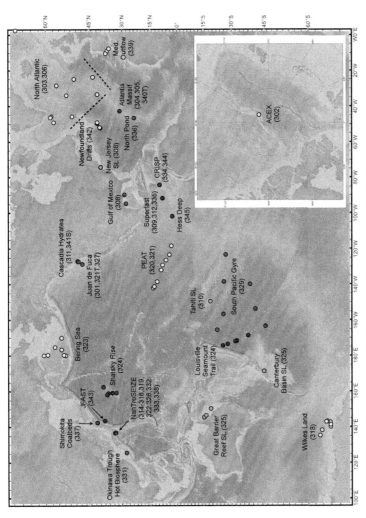

FIGURE 1.2 Locations of IODP Expeditions discussed in this chapter. Color code: purple (dark gray in print versions) = deep biosphere and subseafloor ocean; yellow (white in print versions) = extreme climates and rapid climate change; orange (light gray in print versions) = sea level change; blue (black in print versions) = solid earth cycles and geodynamics. Topographic basemap produced using GeoMapApp (Ryan et al., 2009).

for various reasons one was basically not attempted. None were completely addressed, which is not surprising given the complexity of the Earth system. In many cases, important IODP results led to identifying equally important remaining questions that provide compelling opportunities and challenges for a 10-year renewal of the program—with a different programmatic organization as the International Ocean Discovery Program. Although the post-2013 program shares the same acronym and platforms, in this chapter "IODP" is used exclusively to refer the 2003–2013 IODP.

1.2 THE DEEP BIOSPHERE AND THE SUBSEAFLOOR OCEAN (INITIATIVES IN DEEP BIOSPHERE AND GAS HYDRATES)

The importance of using scientific ocean drilling to understand fluid flow and hydrology beneath the subseafloor was recognized about midway through DSDP around the time of the discovery of hydrothermal circulation in the oceanic crust. This became a strong theme during ODP, particularly in several type examples of subduction zones and thickly sedimented young oceanic crust. It also motivated development of ODP long-term "CORK" (Circulation Obviation Retrofit Kit) hydrological observatories (Becker & Davis, 2005; Davis, Becker, Pettigrew, Carson, & MacDonald, 1992) that were highlighted in the ISP as an important technological approach for IODP. The ISP theme described efforts to understand "the subseafloor ocean in various geological settings" including mid-ocean ridges, ridge flanks, old ocean basins, large igneous provinces (LIPs), subduction zones, passive margins, and carbonate platforms. IODP expanded the strong ODP efforts in ridge flank and subduction zone hydrogeology, as described in chapters by Fisher et al. (chapter 4.2.2 of this volume) and Kastner et al. (chapter 4.4 of this volume). IODP also made pioneering efforts in the overpressured Gulf of Mexico passive margin and the Okinawa Trough sediment-covered back-arc spreading center. The subduction zone contributions were largely in the context of (1) the Vancouver margin gas hydrates province described in this section under the gas hydrates initiative and (2) the Nankai Trough and Costa Rica seismogenesis programs (NanTroSEIZE and CRISP, respectively) that are described in Section 1.4 of this chapter (Solid Earth Cycles and Geodynamics).

IODP ridge flank hydrogeological contributions centered on long-term monitoring and active experiments in two classical type locations previously investigated during DSDP and ODP: the thickly sedimented eastern flank of the Juan de Fuca Ridge (Fisher, Urabe, Klaus, & the IODP Expedition 301 scientists, 2005; Fisher et al., 2012) and the isolated "North Pond" sediment pond on the western flank of the Mid-Atlantic Ridge (Edwards, Bach, Klaus, & the IODP Expedition 336 Scientists, 2014). In both cases, the IODP hydrological contributions involved dedicated long-term monitoring efforts and a focus on biosphere studies, the latter described in this section under the "Deep Biosphere" initiative (see also Orcutt and Edwards, chapter 2.5 of this volume). In both settings, ODP experiments and long-term monitoring had demonstrated that there must be very large lateral fluid fluxes in highly permeable, sediment-covered upper oceanic basement, but that very small pressure differentials were associated with the fluid fluxes (e.g., Davis & Becker, 1998; Davis, Wang, Becker, & Thomson, 2000). ODP had not resolved the actual flow pathways, nor the depths to which such flow extends within basement. IODP returned to these locations and established arrays of borehole observatories that penetrate up to a few hundred meters into oceanic basement (e.g., Figure 1.3). These were installed in spatial patterns designed to address hypotheses based on detailed survey data for primary flow directions in a ridge-parallel direction as opposed to traditional two-dimensional cartoons showing schematic flow along ridge-perpendicular transects. On the Juan de Fuca flank, the borehole arrays provided an assessment of large-scale upper crustal permeability based on transmission of drilling signals from one site to a monitoring hole over 2 km away (Fisher, Davis, & Becker, 2008). They have also allowed for the very first planned hole-to-hole tracer experiments in the sub-seafloor to resolve actual flow pathways (Fisher et al., 2012, Figure 1.3). These field experiments were to be concluded in 2014, so results are not yet available, but already they represent an important scientific and technological innovation. As the North Pond observatories were installed in late 2011, full results are also

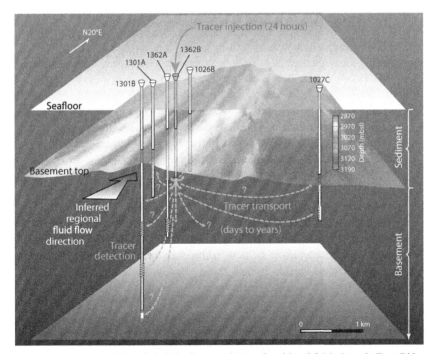

FIGURE 1.3 Array of IODP borehole observatories emplaced in ~3.3 Ma Juan de Fuca Ridge flank during Expeditions 301 and 327, illustrating the first planned hole-to-hole tracer experiments ever conducted in oceanic crust. *From ISP, after Fisher et al. (2012).*

not yet available but initial data demonstrated good quality of the installations (Edwards et al., 2014).

In the Gulf of Mexico, Expedition 308 coring and downhole measurements documented high overpressures in thick and quickly accumulating sediments of the Ursa Basin and showed how they are related to slope instabilities (e.g., Behrmann, Flemings, John, & the IODP Expedition 308 Scientists, 2006; Flemings et al., 2008). The original vision for this program included installation of borehole observatories to measure the magnitude of the overpressure and monitor its variation over time, but those have never been completed owing to a combination of financial and safety constraints.

In the Okinawa Trough, Expedition 331 cored within the Iheya North active hydrothermal field for hydrothermal and microbiological objectives, and established three artificial vent observatories for future hydrothermal and microbiological studies (Takai, Mottl, Nielsen, & the IODP Expedition 335 scientists, 2012). Notably, the core includes the first sphalerite-rich Kuroko-type black ore yet recovered from the modern seafloor, providing a modern analog for formation of economically important Kuroko-type deposits. Postdrilling submersible surveys and experiments documented a significant impact of drilling on the natural hydrothermal system (Kawagucci et al., 2013).

1.2.1 Initiative: Deep Biosphere

ODP made important pioneering efforts in a deep biosphere "pilot project," most notably the dedicated Leg 201 to sedimentary environments in the eastern Pacific (D'Hondt et al., 2004). This and earlier ODP successes (e.g., Parkes, Cragg, & Wellsbury, 2000) motivated ISP definition of a Deep Biosphere Initiative that described two main IODP approaches: (1) a global mapping of the subseafloor extent of the microbial biosphere and its boundaries as defined by variables such as temperature, pressure, pH, redox potential, and lithology; and (2) focused, process-oriented studies to define the biogeochemical impacts of the subseafloor biota. To address the first objective, most IODP expeditions have included a biosphere component in analyzing the cores, and considerable effort has been devoted to establishing sampling protocols. With respect to the second objective, the latter part of IODP witnessed a blossoming of focused deep biosphere investigations, with four dedicated biosphere expeditions in 2010–2012 and development of new techniques including in situ borehole observatory instrumentation. Two of these expeditions were to regions with very low organic fluxes to the seafloor (South Pacific Gyre and North Pond), whereas the other two were in areas of high nutrient potential (Okinawa Trough and Shimokita Coal Beds). While results from the four dedicated expeditions are still in progress, the following IODP highlights have already emerged. (See also sub chapters in chapter 2 of this volume.)

The global set of ODP and IODP samples has provided crucial materials for IODP-era genomic analyses that to date are demonstrating that a significant fraction of subseafloor archaeal and bacterial species is distinct from the subaerial microbial world (Figure 1.4) (D'Hondt et al., 2007; Inagaki, 2010). ODP and IODP samples have also revealed a significant presence of fungi (Orsi, Edgcomb, Christman, & Biddle, 2013) and viruses (Engelhardt, Kallmeyer, Cypionka, & Engelen, 2014) in subseafloor sedimentary ecosystems. Finally, the global sample set has also led to continually refined estimates of the global microbial biomass in seafloor sediments. A good example is the reduced sedimentary biomass estimate by Kallmeyer, Pockalny, Adhikari, Smith, and D'Hondt (2012) based on accounting for lower cell counts in remote oceanic areas away from highly productive coastal zones where previous sampling may have been concentrated. This was reinforced by IODP results from Expedition 329 in the remote South Pacific Gyre that indicate extremely low cell counts and microbial respiration levels throughout the sediment column in this very low nutrient environment (D'Hondt, Inagaki, Alvarez-Zarikian, & the IODP Expedition 329 Scientific Party, 2013; D'Hondt and Spivack, chapter 2.3 of this volume). The global

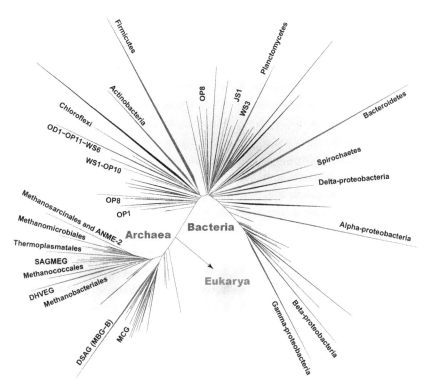

FIGURE 1.4 Phylogenetic tree of the three domains of life on Earth. Red (light gray in print versions) and blue (dark gray in print versions) branches represent archaeal and bacterial groups that are frequently detected in subseafloor sediment samples recovered by IODP and are mostly distinct from subaerial branches shown in black. *From ISP after Inagaki (2010).*

sample set has also been used as the primary constraint in simulations of the global biogeochemical importance of microbial sulfate reduction metabolism in subseafloor sediments (Bowles, Mogollón, Kasten, Zabel, & Hinrichs, 2014).

Likewise, IODP results are central to demonstrating the existence of a microbial biosphere within the upper basaltic crust beneath the sediments and estimating its potential biomass (e.g., Edwards, Becker, & Colwell, 2012). Recently, Lever et al. (2013) reported clear existence of sulfur- and methane-cycling microbes in the basaltic crust cored on the Juan de Fuca flank by Expedition 301. In situ cultivation experiments in long-term borehole observatories at the same Juan de Fuca site also provided evidence for a basaltic subseafloor microbial community (Orcutt et al., 2011; Smith et al., 2011).

Two recent *Chikyu* expeditions with a deep biosphere focus were located in settings with high subseafloor nutrient availability: the Okinawa Trough hydrothermal site and the carbon-rich deeply buried coalbeds offshore Shimokita. Owing to very high subseafloor temperatures, the Okinawa Trough hydrothermal settings returned very low cell counts except in the coolest upper 30 m (Takai et al., 2012). Shimokita coalbed results are not yet available, but Kawai et al. (2014) recently reported metagenomic evidence for a significant level of anaerobic organohalide respiration in shallow sediments cored at this location during *Chikyu* shakedown operations in 2006.

1.2.2 Initiative: Gas Hydrates

ODP successfully conducted pioneering gas hydrates studies on the Blake Ridge and at the Chile and Cascadia subduction margins. The ISP set out a strategy of process-related drilling studies to document gas hydrate generation, migration, and accumulation and their relationship to bottom-simulating reflectors (BSRs), as well as sampling gas hydrates in a variety of geological type environments. IODP contributions mainly involved three return expeditions for process-oriented studies to the section of the Cascadia Margin originally visited by ODP offshore Vancouver Island, where there are a regionally extensive BSR as well as focused hydrate discharge sites. The goal of sampling a greater variety of gas hydrate settings was not achieved by IODP per se, although it should be noted that during the 2003–2013 time period both IODP drillships were used for non-IODP, charter work investigating gas hydrates and their resource potential at several locations.

On the Vancouver Margin, Expedition 311 cored and logged a trench-normal transect of sites that verified that virtually all the gas hydrate there is microbially produced by reduction of CO_2 within the subseafloor gas hydrate stability zone (Riedel, Collett, & Malone, 2010). Results led to a refinement of an earlier model (Hyndman & Davis, 1992) for generation, migration, and accumulation of hydrates based partly on spatial variability of fluid expulsion rates within the accretionary prism above the Cascadia subduction zone (Figure 1.5), with permeable sand layers within the hydrate stability zone being favored for

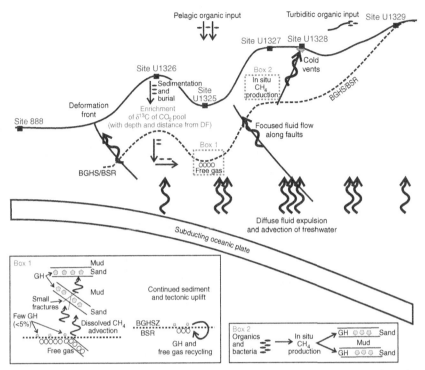

FIGURE 1.5 Schematic cartoon (not to scale) of the model of fluid flow and gas hydrate formation for the Vancouver Cascadia margin, *(from Riedel et al. (2010))*. DF = deformation front, GH = gas hydrate, BSR = bottom-simulating reflector, BGHS = base of gas hydrate stability, BGHSZ = base of gas hydrate stability zone.

accumulation. A site at an active hydrate vent with massive hydrates close to the seafloor revealed that more deeply generated hydrates (also of microbial origin) are being transported there by focused fluid expulsion, probably guided by faults within the accretionary prism.

Borehole observatories were originally proposed for several of the Expedition 311 sites, to better define the gas hydrate stability zone and to monitor temporal variability associated with fluid flow. However, these have not yet been completed owing mainly to financial and scheduling constraints. Nevertheless, IODP has made a credible start at this effort with successful installations of two simple borehole observatories near the original ODP Site 889: a basic pressure-monitoring Advanced CORK (ACORK) in 2010 (Expedition 328, Davis, Heesemann, & the IODP Expedition 328 Scientists and Engineers, 2012) and a prototype "SCIMPI" ("Simple Cabled Instrument to Measure Parameters In situ") observatory during 2013 (Lado-Insua, Moran, Kulin, Farrington, & Newman, 2013). The ACORK is planned for connection to the Ocean Networks Canada/NEPTUNE Canada cabled observatory in 2015.

Finally, it has been hypothesized that massive releases of gas hydrates once sequestered in seafloor sediments may have played a significant role in triggering major "hyperthermal" warming events in Earth's past (e.g., Zachos et al., 2005; review by; Dickens, 2011). Focused IODP coring described in the next section targeting the Paleogene Period, when several such hyperthermal events occurred, may provide additional evidence to clarify this debate.

1.3 ENVIRONMENTAL CHANGE, PROCESSES, AND EFFECTS (INITIATIVES IN EXTREME CLIMATES AND RAPID CLIMATE CHANGE)

Marine sediments contain the world's most complete archives of past climate change, and DSDP and ODP sediment coring and logging efforts made fundamental contributions that underpin much of the science of paleoceanography. Building on this foundation, the ISP defined a comprehensive program to understand past climate variability and its consequences at various timescales ranging from tens of millions of years to centuries or even shorter. The ISP highlighted the importance of understanding the effects of "internal" (Earth-based) factors like tectonically induced changes (mountain building, plate reorganizations, LIPs), "external" factors like periodic Milankovitch orbital variations and unpredictable bolide impacts, relationship to glaciations and sea level variations, and the complex interactions that produce millennial- and shorter-scale climate variations. Two initiatives were defined: Extreme Climates, referring mainly to the differences between the warm greenhouse world of the Cretaceous and Paleogene Periods and our current icehouse world, and Rapid Climate Change, referring to abrupt climate changes mainly in our current icehouse world but also in the past greenhouse world. The technological approach was to involve a continued emphasis on paleoceanography by *JOIDES Resolution* with a significant new element made possible by MSPs in shallow water and Arctic regions inaccessible to the two IODP drillships. (In addition, the ISP envisioned using *Chikyu* for recovering deeply buried sections of Mesozoic greenhouse sediments, but that was not attempted owing to factors described in Section 1.4 of this chapter.)

Since the ISP was written, public recognition of the importance of understanding global climate change has increased dramatically, and the essential contributions from scientific ocean drilling were highlighted in the two most recent IPCC reports (Solomon et al., 2007; Stocker et al., 2013). Thus, this objective is one of the most visible aspects of IODP, and it is noteworthy, if not IODP's greatest success, that IODP made major new contributions spanning most of the objectives for this theme outlined in the ISP. Although the overall IODP effort in this theme was assembled primarily from individual proposals, the sum includes strong interexpedition links that will eventually produce integrated global perspectives. Good examples of such linkages include (1) multiple expeditions to investigate the Paleogene greenhouse world and its effects in the

Arctic Ocean, equatorial Pacific, and North Atlantic; (2) complementary programs offshore Antarctica, in the Arctic Ocean, Bering Sea, and North Atlantic that investigated the transition to the icehouse world and development of polar ice sheets; and (3) multiple expeditions to assess the global sea level record during the Neogene and Quaternary, including coral reef sites in the tropical south Pacific and siliciclastic records on the New Jersey and New Zealand margins. Integrated results are discussed in several of the sub chapters in chapter 3 of this volume. For the more recent expeditions, much work is ongoing on many of the samples recovered, so only initial results are available to date. Nevertheless, the highlights are many; they include, but will not be limited to, the following.

1.3.1 Initiative: Extreme Climates

This initiative has particularly strong roots in the final years of ODP, when the report of the Extreme Climates Program Planning Group sets the stage for a coordinated global effort to investigate the greenhouse world of the Cretaceous and Paleogene Periods. This started with four legs during the final 2 years of ODP (two each in the Pacific and Atlantic oceans), with several major papers published early in the IODP period that have had significant influence on planning subsequent IODP expeditions to investigate potential societally important effects of global warming, rapid hyperthermal events under warm conditions, and ocean acidification. For example, the depth transect of Walvis Ridge sites cored during ODP Leg 208 provided clear evidence of a geologically rapid shoaling of the calcite compensation depth (CCD) at the time of the Paleocene-Eocene Thermal Maximum (PETM), consistent with a massive release of carbon from methane hydrates and resultant ocean acidification (Zachos et al., 2005). Cores from that leg plus the preceding Leg 207 to Demerara Rise and Leg 198 to Shatsky Rise also provided evidence for smaller-magnitude Eocene hyperthermal events that were apparently paced by orbital eccentricity cycles (Lourens et al., 2005) and showed more rapid recoveries than the PETM probably involving carbon resequestration within the oceans (Sexton et al., 2011). Those same cores also provided globally important records of the Chicxulub asteroid impact and mass extinctions at the Cretaceous-Paleogene boundary (e.g., Schulte et al., 2010).

1.3.1.1 Arctic

The first MSP expedition in IODP, the Arctic Coring Expedition (ACEX) to the Lomonosov Ridge near the North Pole, produced a huge public impact as well as breakthrough scientific results in understanding the Cenozoic climatic history of the Arctic (Moran et al., 2006; Backman & Moran, 2009; Stein et al., chapter 3.2 of this volume). This was achieved despite encountering an unexpectedly condensed section from ~44 to 18 Ma (Moran et al., 2006). This was initially interpreted as a hiatus (Backman et al., 2008) using limited biostratigraphic controls with large uncertainties, but with the addition of Osmium isotope dating has since been reinterpreted as a time of very slow marine sediment accumulation (Poirier & Hillaire-Marcel, 2011) (Figure 1.6). Organic-rich Paleocene and Eocene sediments below this section were deposited in a neritic environment in a restricted Arctic Ocean (Moran et al., 2006; Stein, Boucsein, & Meyer, 2006). They reveal: (1) summer Arctic sea surface temperatures much higher than previously estimated at the time of the PETM (Moran et al., 2006), (2) the influence of mid-Eocene periodic pulses of freshwater input to the Arctic (Brinkhuis et al., 2006), and (3) initial occurrence of seasonal sea ice in the late Eocene, much earlier than previously hypothesized (Stickley et al., 2009). Above the condensed section, marine sediments accumulated at higher rates in a ventilated Arctic Ocean connected to the North Atlantic by opening of the Fram Strait (Jakobsson et al., 2007) and document a synchroneity of Arctic cooling and expansion of Greenland ice. Even with the uncertainty in interpreting the condensed section, these results have fundamental implications for the Arctic's

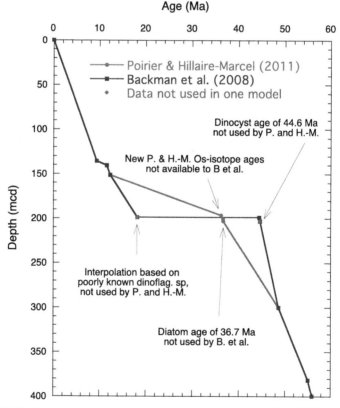

FIGURE 1.6 Contrasting stratigraphic age models for the Arctic based on ACEX core results (Backman et al., 2008) and later osmium isotope dates (Poirier & Hillaire-Marcel, 2011). See also complementary figure and further explanation in Stein et al. (chapter 3.2 of this volume).

role in early Cenozoic climate; many of these implications are explored in the wealth of publications since 2006 (see, e.g., Backman & Moran, 2008, 2009; Stein et al., chapter 3.2 of this volume). Not surprisingly, the results have also raised many important new questions, so returning to complementary sites in the Arctic has been identified as a high priority for MSP operations during the first 5 years of the post-2013 program.

1.3.1.2 Antarctic

In 2010, Expedition 318 used JOIDES Resolution to core Eocene to Holocene sediments offshore Wilkes Land, providing a record of the evolution of the Antarctic from greenhouse conditions to current icehouse conditions. Specific objectives included documenting the onset of Cenozoic glaciation in Antarctica, Miocene history of the East Antarctic Ice Sheet, and Neogene glacial history. Initial results are already quite significant in terms of documenting: (1) mild Antarctic climate conditions during the warm Eocene (Pross et al., 2012), (2) an association of the onset of global cooling after the early Eocene climatic optimum with initial opening of the Tasmanian gateway (Bijl et al., 2013), (3) major changes in the Southern Ocean planktonic ecosystems associated with early Oligocene onset of continental-scale Antarctic glaciation (Houben et al., 2013), and (4) the active response of the East Antarctic ice sheet during warm intervals of the Pliocene that had similar temperature and atmospheric CO_2 levels to those predicted for the end of this century (Cook et al., 2013). Other objectives include linking the development of the East Antarctic Ice Sheet and evolution of the circum-Antarctic current system and providing a high-resolution record of the last deglaciation (Escutia, Brinkhuis, Klaus, & the IODP Expedition 318 scientists, 2011). (The last will be a strong contribution to the Rapid Climate Change Initiative.) In addition, the Wilkes Land cores have provided direct evidence to constrain glacial-hydro-isostatic models of relative sea level rise on the Antarctic margin as extensive East Antarctic ice sheets developed in the Oligocene accompanied by a global eustatic sea level drop of ~60 m (Stochhi, et al., 2013). Results from ACEX and Wilkes Land drilling suggest a greater degree of bipolar synchroneity between development of northern and southern hemisphere ice sheets in the Cenozoic than previously thought, as described in more detail by Stein et al. (chapter 3.2 of this volume) and Escutia et al. (chapter 3.3 of this volume).

1.3.1.3 Pacific Equatorial Age Transect

The role of the tropics during Cenozoic climate evolution from greenhouse to icehouse was documented by Expeditions 320/321 in the equatorial Pacific region. The strategy took advantage of the northwestward motion of the Pacific plate through the equatorial high-productivity region, to select eight sites from which high-resolution records of different intervals could be spliced together to form a nearly continuous high-resolution sequence from Eocene to present. A particularly important objective was successfully achieved: to document the evolution of the Pacific CCD (carbonate compensation depth) since the mid-Eocene warm period and its relationship to the change in global climate from greenhouse to icehouse conditions. The results (Figure 1.7) show a deepening of the CCD from 3–3.5 to 4.6 km as the global climate changed, consistent with hypotheses of a corresponding increase in weathering and decrease in atmospheric CO_2 during the same interval (Pälike et al., 2012; Pälike, chapter 3.4 of this volume). The mid-to-late Eocene interval was characterized by large fluctuations in CCD that document rapid climate change in the greenhouse world. This was followed by a deepening of the CCD during the Eocene–Oligocene transition from greenhouse conditions to the icehouse world that has persisted to the present. Pälike et al. (2012) suggest that these fluctuations were caused by changes in continental weathering and mode of organic-carbon deposition.

1.3.1.4 North Atlantic Paleogene

In a highly complementary effort, Expedition 342 in 2012 cored high-resolution North Atlantic Paleogene drift deposits with a primary objective of documenting evolution of the CCD in the North Atlantic during a similar time interval

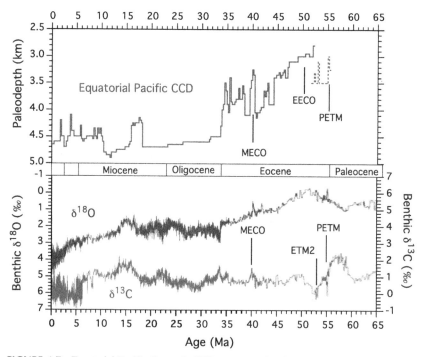

FIGURE 1.7 Equatorial Pacific Cenozoic CCD reconstruction from Pälike et al. (2012), compared with benthic foraminiferal $\partial^{18}O$ and $\partial^{13}C$ compilation of Zachos et al. (2008). EECO = Early Eocene Climatic Optimum, MECO = Mid Eocene Climatic Optimum, PETM = Paleocene-Eocene Thermal Maximum.

as Pacific Equatorial Age Transect (PEAT). Shipboard results (Norris, Wilson, Blum, & the Expedition 342 Scientists, 2014) indicate at least 2 km variation in the North Atlantic CCD during this interval, consistent with the large variations in South Atlantic CCD at Walvis Ridge (Zachos et al., 2005). The cored sections spanned older intervals than PEAT and also provide preliminary evidence of relatively rapid fluctuations in CCD through the Cretaceous/Paleogene greenhouse world and into the middle Oligocene icehouse. Suggested causes include rapid climate change in the greenhouse world and/or rebalancing of ocean alkalinity after mass extinction events.

1.3.2 Initiative: Rapid Climate Change

As noted above, rapid climate change in the Paleogene greenhouse world was investigated during several expeditions grouped under the Extreme Climates initiative. Rapid climate change in the Pliocene and Quaternary icehouse world was a strong focus of high-latitude IODP expeditions in three regions: the North Atlantic (Expeditions 303/306 in 2004–2005), Bering Sea (Expedition 323 in 2009), and the Wilkes Land Expedition 318 in 2010 described above. In addition, Expedition 339 investigated the imprint of Mediterranean outflow on evolution of North Atlantic circulation during the same interval.

1.3.2.1 North Atlantic

The North Atlantic results are more mature and have produced significant records of millennial-scale variations that highlight the relationship between northern hemisphere glacial oscillations and strength of the thermohaline circulation system (Channell et al., 2006; Channell and Hodell, chapter 3.5 of this volume). By targeting regions of high sediment accumulation in the "Ruddiman IRD (ice-rafted debris) belt" off the Hudson Strait, the expeditions provided excellent core to refine the isotope and magnetostratigraphy since the Pliocene (e.g., Channell, Xuan, & Hodell, 2009). Results have documented a history of millennial-scale variability throughout the Pleistocene, with Heinrich events identified as early as the mid-Pleistocene transition from a 41-kya cyclicity to a 100-kya cyclicity since then (Hodell, Channell, Curtis, Romero, & Röhl, 2008). These results indicate a strong link between Hudson Strait ice discharge and weakening of the Atlantic Meridional Overturn Circulation (AMOC), supporting hypotheses of an ice volume threshold mechanism for rapid climate change in the current icehouse world.

1.3.2.2 Bering Sea

The 2009 Bering Sea expedition explored the paleoceanographic history recorded in Bering Sea sediments over a similar Pliocene-Quaternary interval (Takahashi, Ravelo, Alvarez Zarikian, & the IODP Expedition 323 scientists, 2011), and hence should provide a complementary North Pacific marginal sea perspective. Sedimentology of the cores indicates that productivity in the Bering Sea was higher during interglacials than during glacials, and also higher during the Pliocene warm period (Aiello & Ravelo, 2012). A site on the Bowers Ridge documents millennial-scale events over the past 90 ka and possible links to changes in circulation of North Pacific Intermediate Water (Schlung et al., 2013). In addition, since nutrient-rich surface waters of the Bering Sea currently flow across the 50-m deep Bering Strait into the Arctic Ocean, the Bering Sea cores should provide important constraints on the effects of transport across the Bering Strait in Arctic paleoceanography as polar ice sheets waxed and waned and sea level rose or fell. As noted above, Wilkes Land Antarctic cores should provide a complementary southern hemisphere record over the Pliocene and Quaternary; in combination, these expeditions should better constrain the degree of bipolar synchronicity in millennial-scale variations.

1.3.2.3 Mediterranean Outflow

This winter 2011–2012 expedition (Hernández-Molina, Stow, Alvarez-Zarikian, & Expedition IODP 339 Scientists, 2013) investigated the influence on North Atlantic thermohaline circulation of tectonically controlled Pliocene-Quaternary variations in Mediterranean outflow of warm, salty water as an important source for North Atlantic Deep Water. A high-profile initial summary paper (Hernández-Molina et al., 2014; see also commentary by; Filippelli, 2014) documented (1) intensifications of Mediterranean outflow during the mid-Pliocene warm period and late Pliocene that may be associated with development of the present AMOC patterns, and (2) a further intensification at the time of the mid-Pleistocene transition to the current pattern of eccentricity-dominated 100,000 glacial–interglacial cycles. This expedition also cored a high-resolution Iberian margin site for a continuous record of millennial-scale variability over the past 1.5 my (Hodell, Lourens, Stow, Hernández-Molina, Alvarez-Zarikian, 2013) that will commemorate the contributions of Sir Nicholas Shackleton, one of the pioneers of paleoceanography and its central role in scientific ocean drilling.

1.3.3 Sea Level Change

Although it was not identified explicitly in the ISP as an initiative, with hindsight the case can be made that IODP will have effectively mounted a de facto initiative on understanding eustatic sea level change during the Neogene icehouse world. This effort has involved three MSP expeditions (310, 313, 325) and one JOIDES RESOLUTION expedition (317). Of these, two cored southwest Pacific corals (310 in Tahiti, 325 at the Australian Great Barrier Reef) to document sea level rise since the last glacial maximum in areas relatively far from polar ice sheets and where tectonic subsidence is well understood (Camoin, Iryu, McInroy, & the IODP Expedition 310 Scientists, 2007; Yokoyama et al., 2011; Camoin and Webster, chapter 3.6 of this volume). The other two cored Miocene-Pleistocene sequences in siliciclastic margins in the North Atlantic

(New Jersey, Expedition 313) and South Pacific (Canterbury Basin, Expedition 317) to refine timing, amplitudes, and mechanisms of eustatic change and their effects on continental margin sequence stratigraphy through the icehouse world of the Neogene Period (Fulthorpe, Hoyanagi, Blum, & the IODP Expedition 317 scientists, 2011; Mountain, Proust, & the Expedition 313 scientists, 2010).

Results from the Tahiti program were noteworthy for capturing a nearly complete record of sea level change of the last deglaciation between 16 and 8 kya (Camoin et al., 2012; Deschamps et al., 2012). This record (Figure 1.8) shows a generally smooth rise of sea level punctuated by a sharp 14–18 m rise of sea level at the time of meltwater pulse 1A, but no evidence of reef-drowning events caused by exceptionally rapid sea level rises previously inferred at the times of meltwater pulses. Equally important, they accurately document the duration

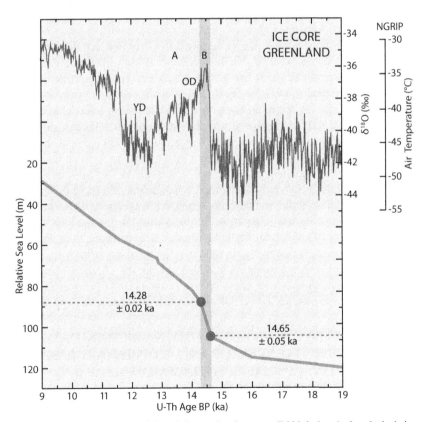

FIGURE 1.8 Reconstruction of the relative sea level curve at Tahiti during the last deglaciation. Note the coincidence of the largest discrete rise of 14–18 m (dark blue (light gray in print versions) shading) with the Bolling warm period (B) as recorded in the Greenland ice cores, followed by a smooth rise of sea level since 14.28 ka. A = Allerod warm period, OD = Older Dryas, YD = Younger Dryas *From ISP, after Deschamps et al. (2012).*

and timing of meltwater pulse 1A as lasting no more than 340 years coinciding with the Bölling warming. Preliminary results from Great Barrier Reef drilling indicate that older corals were sampled, so that the combined record can be extended back through 30 kya (Camoin and Webster, chapter 3.6 of this volume). In addition, radiometric dating of even older Tahiti samples provided important constraints on the timing of the penultimate deglaciation starting around 140 ka, and show a contrasting relationship to extrema in northern hemisphere insolation than observed for the last deglaciation (Thomas et al., 2009).

In siliciclastic margin settings, results from the 2010 Canterbury Basin expedition are still preliminary, but it appears that it recovered sufficient core and log information to achieve objectives to document the history of eustasy and effects on sedimentary margin architecture through the Neogene and even back into the Oligocene (Fulthorpe et al., 2011). The 2009 New Jersey expedition successfully targeted inner shelf Miocene icehouse sequences that were poorly sampled during previous ODP legs to the outer shelf and slope or onshore drilling (Mountain et al., 2010). Integrated analysis of core, seismic, and log data from the three-site transect across inner margin sequences (1) verified that the sequence stratigraphic boundaries can be documented as seismic impedance contrasts (Miller, Browning, et al., 2013) and (2) produced an important refinement of sequence stratigraphic models for development of the clinothem units (Miller, Mountain, et al., 2013). With biostratigraphic and strontium-isotopic age control, Browning et al. (2013) showed that the timing of the Miocene sequence boundaries correlates with the deep-sea oxygen isotopic record and inferred orbital controls, verifying glacioisostasy as a major factor in forming continental margin sequence boundaries (as opposed to tectonic factors associated with basin development). They also documented a correlation of a large increase in sedimentation rate with the mid-Miocene climate changes associated with the development of a permanent East Antarctic Ice Sheet. Further analysis should lead to refinement of estimates of the magnitudes of eustatic sea level changes associated with orbitally controlled waxing and waning of ice sheets through the Miocene. The New Jersey expedition was also notable for a financial contribution from the International Continental Drilling Program (ICDP), hopefully paving the way for even better ICDP–IODP cooperation in the post-2013 programs.

1.4 SOLID EARTH CYCLES AND GEODYNAMICS (INITIATIVES IN CONTINENTAL BREAKUP AND SEDIMENTARY BASIN FORMATION, LIPS, 21ST CENTURY MOHOLE, AND SEISMOGENIC ZONE)

This broad ISP theme encompasses IODP objectives to better understand the complex relationships among core and mantle convection, plate tectonics, and formation of oceanic and continental crust. These subjects have fundamental scientific interest, and they also have societal implications for volcanic and earthquake hazards as well as the global carbon cycle and long-term (millions

of years and longer) climate change on Earth. Two subthemes were defined:
(1) Formation of Rifted Continental Margins, Oceanic LIPs, and Oceanic Crust
and (2) Recycling of Oceanic Lithosphere into the Mantle and Formation of
Continental Crust. Within the first subtheme, three special initiatives were highlighted: Continental Breakup and Sedimentary Basin Formation, LIPs, and the
21st Century Mohole. The second subtheme highlighted a Seismogenic Zone
initiative in subduction settings that was a major thrust of IODP drilling. As is
described further in this section, IODP made essentially no progress on the first
of the four initiatives, credible progress on the second and third, and substantial
but still incomplete progress on the last.

All four initiatives included very deep drilling, so the implementation plan in
the ISP envisioned focused efforts by *Chikyu* to address them all. However, fiscal realities and resultant limitations on *Chikyu* IODP operations time precluded
fulfilling the vision to use *Chikyu* during the 10 years of IODP to address more
than one of these initiatives (or to core deeply buried, thick Mesozoic sedimentary sections that could address the Extreme Climates initiative). Hence those
unfulfilled deep drilling objectives remain prime opportunities for the post-2013
International Ocean Discovery Program. In this section we start with the second subtheme and Seismogenic Zone initiative, partly because it was the main
focus for *Chikyu* operations in IODP, but also because it defined a new mode of
planning and implementation that sets further context for understanding IODP
progress (or lack thereof) in the other three initiatives within this theme.

1.4.1 Initiative: Seismogenic Zone

As defined in the ISP, the Seismogenic Zone initiative focuses on the plate boundary fault in subduction systems, specifically the section landward of the trench
where Earth's most destructive earthquakes originate, where "the subducting
and overriding plates are coupled to some degree so that elastic strain accumulates and is eventually released as an earthquake." There is obvious societal relevance in terms of understanding earthquake hazards at subduction zones, where
about 90% of Earth's seismic energy is released. The Seismogenic Zone is typically many kilometers to tens of kilometers deep, but there are a few subduction
zones where the updip limit is within reach of the deep drilling capabilities of
Chikyu. Scientific objectives include sampling the fault materials to understand
how they behave under stress and during earthquakes, refining our understanding of the factors that control the updip limit of the Seismogenic Zone, and
long-term monitoring of the fault system to better understand temporal behavior,
processes of earthquake nucleation, and potential links to fluid flow.

In recognition of the technological challenges and extensive time and financial commitments required to address such an objective, a new planning mode
was defined more than 2 years before IODP began. This was the "complex drilling project" (CDP) that could involve multiple phases over periods of years,
possibly with more than one drilling platform. CDP proposals were submitted

for Seismogenic Zone drilling and instrumentation in two complementary sub-duction systems where the updip limit was in reach of *Chikyu*. These were the accretionary Nankai Trough system (NanTroSEIZE), where sediments from the incoming plate are accreted to the overriding plate and the erosional Costa Rica subduction system (CRISP), where all the sediments on the incoming plate are currently subducted and the forearc is eroded during subduction. During the 10 years of IODP, NanTroSEIZE has been the principal focus of *Chikyu* opera-tions, both nonriser and riser, and JOIDES Resolution completed preparatory nonriser drilling for CRISP. Highlights of IODP progress on both programs are summarized in this section, but it should be noted that the prime long-term objectives of both NanTroSEIZE and CRISP remain to be completed; in par-ticular, their ultimate long-term monitoring objectives remain opportunities and technological challenges for the post-2013 program. Finally, this section closes with a summary of two highly successful 2012 "rapid-response" expeditions with *Chikyu* in deepwater riserless mode to the site of the destructive, tsunami-genic 2011 Tohoku-Oki earthquake—a major scientific success with high pub-lic impact and implications for future response efforts in the post-2013 program.

But before summarizing scientific highlights, it should also be noted that the NanTroSEIZE implementation experience helped define a new mode of plan-ning for complicated IODP programs and initiatives, using a long-lived "proj-ect management team" (PMT) involving lead scientists, drilling operators, and IODP-MI. This mode of planning helped to maintain scientific and operational continuity through the still-unfinished lifetime of NanTroSEIZE. It was also a model for a management effort 2–4 years into IODP to define a new "mission" mode of planning and implementation for thematically integrated programs that would require coordinated technological development and multiple drilling expeditions. One call for mission proposals was issued in 2007, and the IODP scientific community responded with three proposals, for an Asian Monsoon mission, a Continental Breakup and Birth of Oceans mission, and a Mission Moho. For a number of reasons, the review by the Science Advisory Struc-ture (SAS) resulted in none of these being adopted formally as IODP missions. One interpretation is that this review occurred at nearly the same time when it became clear to the whole IODP community that financial resources would not be sufficient to allow full-time IODP operation of either *Chikyu* or JOIDES Resolution through the end of IODP in 2013, so for SAS to commit to a mission then would have entailed an unacceptable financial impact on all other IODP objectives. As two of the three mission proposals were for two of the four Solid Earth initiatives, this had a significant effect on IODP progress toward address-ing those initiatives, as described below.

1.4.1.1 NanTroSEIZE

The NanTroSEIZE program focused on a trench-normal transect to the south-east of the Kii peninsula of southwest Honshu Island, above and directly off-shore the rupture zone of the 1944 Tonankai earthquake, where the historical recurrence interval for large earthquakes is on the order of a century. It built on ODP results in comparable transects in the Ashizuri and Muroto regions offshore Shikoku Island farther to the southwest. However, at those locations, the updip limit of the Seismogenic Zone is out of reach, whereas offshore Kii it is within reach of *Chikyu*. As originally proposed, the CDP had four main components: (1) coring and logging at reference sites on the incoming Shikoku Basin plate and at the toe of the accretionary prism to document inputs into the Seismogenic Zone; (2) coring and logging into (a) the mega-splay fault system that breaks off from the plate boundary fault may accommodate a considerable amount of plate motion, and intersects seafloor high up on the prism, as well as (b) a site farther landward in the Kumano forearc basin; (3) a deep riser site to the plate boundary fault, with long-term observatory, landward of the mega-splay sites; and (4) further observatories at the seaward reference sites cored in the first component.

Figure 1.9 illustrates the transect and actual drilling progress to date. Early on, the PMT organized the program into a series of four operational stages that did not correspond to the four main components described above. Stage I included essentially all of the shallow coring and logging that could be done with riserless drilling, and was basically completed using *Chikyu* in riserless mode. Stage II was planned to include (1) initial riser drilling fairly deep to the mega-splay fault (C0001) along with an initial observatory there, (2) a shallower riserless observatory site (C00010) above the mega-splay seaward of (1), (3) basement coring and observatory installation at the incoming plate reference sites, and (4) a riserless observatory site above the rupture zone in the landward section of the Kumano Basin. Stage III was to include completion of the deep riser site to the plate boundary fault (C0002) with an initial observatory, plus potential deepening of the site crossing the toe of the accretionary prism and an additional shallower hole to the mega-splay fault if necessary. Stage IV was envisioned for installation of the ultimate long-term monitoring systems in the two deep riser holes to replace initial observatories installed during Stages II and III.

The overall plan has evolved with time and experience in using *Chikyu*, and in recognition of the fiscal realities that limited *Chikyu* IODP operations time. The primary conceptual modification to date has been to scale back the plans for deep riser holes and observatories; instead of two instrumented sites, one deep to the mega-splay fault and the other penetrating even deeper all the way to the plate boundary, the major goal is now one deep riser/observatory site (C0002) across the mega-splay fault (Figure 1.9). Also, a shallow site was added (C0018) to establish a Quaternary history of mass transport deposits in the accretionary prism and constrain tsunamigenic potential of landslides in the prism. When IODP finished, it was in the midst of addressing the objectives originally defined for Stages II and III. The final *Chikyu* IODP expedition encountered difficult drilling conditions shallower than expected in its attempt to install casing to enable the deep mega-splay objectives be achieved in the future. Thus, the

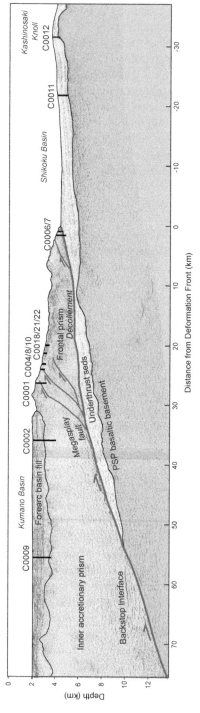

FIGURE 1.9 Cross-section of the Nankai Trough offshore Kii showing IODP progress to date on the NanTroSEIZE transect. PSP = Philippine Sea Plate. *Modified after Moore et al. (2013).*

ultimate goals of the overall program to drill to and instrument the plate boundary at its mega-splay fault remain to be addressed in the post-2013 program.

In that sense, the IODP NanTroSEIZE work mainly sets the stage for future breakthrough science originally envisioned for the overall project. Nevertheless, IODP accomplished nearly all the original riserless objectives, with the exception of observatories at the seaward reference sites. This has produced important successes in their own right including the following:

- Strasser et al. (2009) used IODP core from the shallow portion of the mega-splay fault and seismic data to constrain the initiation of the mega-splay faulting at about 1.95 Ma, followed by a short period of low activity and then uplift and reactivation at about 1.55 Ma.
- Coring and logging through the sediments and into basement at two contrasting reference sites, one near the peak of the sediment buried Kashinosaki volcanic Knoll (Site C0011) and the other in more normal crust closer to the trench (Site C0012). Together the sites provide a good geochemical and geophysical description of the typical Shikoku Basin inputs into the subduction system, including good thermal constraints (Henry, Kanamatsu, Moe, Strasser, & the IODP Expedition 333 Scientific Party, 2012; Underwood, Saito, Kubo, & the IODP Expedition 322 scientists, 2010).
- Documentation from borehole breakouts of the state of stress at four sites (Tobin et al., 2009). Three in the upper prism indicate maximum compressive stress consistent with but slightly offset from the subduction convergence direction. The landward fourth site in the Kumano Basin indicates maximum compressive stress nearly perpendicular to the convergence, which is consistent with the normal faulting observed within the forearc basin.
- Successful installation of prototype observatories, including temporary autonomous "smart plug" and "genius plug" instruments in Site C0010 (Hammerschmidt, Davis, & Kopf, 2013), as well as the first long-term borehole monitoring system (LTBMS) in Site C0002, including broadband seismometer, strainmeter, and thermistor string and now hooked up to the DONET (Dense Oceanfloor Network System for Earthquakes and Tsunamis) submarine cable (Toczko, Kopf, Araki, & the IODP Expedition 332 Scientific Party, 2012).

1.4.1.2 Costa Rica Seismogenesis Project

As noted above, CRISP shares fundamental goals with NanTroSEIZE but is sited in a contrasting, erosional subduction zone. The location offshore of the Osa peninsula of Costa Rica, where the Cocos Ridge is being subducted, is currently the only known erosive subduction system at which the updip limit of the Seismogenic Zone is within reach of *Chikyu*. Although NanTroSEIZE was the main focus of *Chikyu* drilling during IODP, two IODP expeditions (334 and 344) used the JOIDES Resolution to set the stage for eventual *Chikyu* deep drilling by successfully coring and logging a transect from the incoming plate to the toe of the forearc prism, up the forearc slope, and into the forearc basin above the updip limit of the Seismogenic Zone. These two expeditions were relatively recent (2011 and 2012) but it is clear from even preliminary reports that significant results were achieved such that the riser phase of CRISP drilling is ready for potential *Chikyu* operations in the post-2013 program.

As summarized by Vannucchi, Ujiie, Stroncik, and the IODP Expedition 334 Scientific Party (2013) and Harris, Sakaguchi, Petronotis, and the Expedition 344 Scientists (2013), possibly the most important result was documenting the nature of the forearc basement. This is especially important in an erosional subduction system because eroded forearc material is input into the Seismogenic Zone in addition to sediments and oceanic basement of the downgoing plate. Recovered cores showed that the forearc basement consists of lithified forearc basin sediments, not a mélange of previously accreted oceanic materials or an extension of the Caribbean LIP as had previously been hypothesized. Vannucchi, Sak, Morgan, Ohkushi, and Hujii (2013) interpreted the core to show that tectonic processes associated with subduction erosion led to the creation of forearc basins that trapped much of the arc-derived sediment supply from reaching the trench itself. In addition, the two expeditions have provided: a suitable reference section in the downgoing oceanic plate; structural information that will allow resolution of the state of stress within the forearc; geochemical evidence that will provide constraints on the role of fluid flow in the subduction system and seismogenesis; and recovery of a number of tephras that will provide constraints on the effects of Cocos Ridge subduction on arc volcanism.

1.4.1.3 Japan Trench Fast Drilling Project

About 2.5 years before the formal end of IODP, the Japan Trench was the site of the tsunamigenic Tohoku-Oki earthquake in which the plate boundary broke all the way to the floor of the trench at a depth of about 7000 m, with a total slip of about 30–60 m—the largest slip ever accurately documented in an earthquake. IODP responded with a highly successful "rapid response" drilling program (JFAST, Japan Trench Fast Drilling Project) a little more than a year later (Chester, Mori, Toczko, Eguchi, & the Expedition 343/343T scientists, 2012), using *Chikyu* in deepwater nonriser mode to sample the fault material and emplace a thermal observatory across the fault zone to record the frictional heating effects of the 2011 earthquake; the latter was recovered in January 2013.

This program produced four 2013 papers in Science and a very high public impact. As described by Chester et al. (2013), the spot-coring program successfully sampled the relatively thin (5 m) main fault zone, which is localized in subducting pelagic clays and accommodated nearly all of the total slip. Lin et al. (2013) analyzed postearthquake borehole breakouts to show that the earthquake probably involved a nearly full stress release consistent with the large slip all the way to the seafloor. Ujiie et al. (2013) conducted laboratory experiments on fault zone samples to simulate conditions during the earthquake and showed that the clays become much weaker once slip starts ("velocity-weakening" behavior), facilitating the large slip and stress release. Finally, Fulton et al. (2013) reported that the fault motion produced a smaller frictional heating anomaly than predicted using reasonable static frictional coefficients, consistent with the conclusion of velocity-weakening fault properties. The implication is that the plate boundary fault at the Japan Trench is much weaker than expected, as was also concluded at the San Andreas Fault (Zoback, Hickman, Ellsworth, & the SAFOD science team, 2011). As noted by Ujiie et al. (2013) and Chester et al., (2013), the high amount of pelagic clay in the subducting sediments at Japan Trench may explain the higher risk of large-slip tsunamigenic earthquakes there, compared to other subduction zones like Nankai Trough.

It should also be noted that the Tohoku-Oki earthquake produced significant pressure signals at two ACORK observatories installed late in ODP in the Nankai Trough offshore Shikoku Island. The timing and nature of the pressure responses suggest that the stress release between Pacific and Okhotsk plates at the Japan Trench also produced slow slip and a readjustment of the stress field between Philippine Sea Plate and Eurasian Plate nearly 1000 km away at Nankai Trough (Davis et al., 2013).

1.4.2 Initiative: Continental Breakup and Sedimentary Basin Formation

ODP had made considerable progress in understanding the processes of continental rifting and breakup on North Atlantic margins, with eight dedicated legs in response to strong proposal pressure and the integration of the overall effort by a North Atlantic Rifted Margins Detailed Planning Group (NARM-DPG). The NARM-DPG sets out a strategy of coring basement on the conjugate margins of both volcanic and nonvolcanic North Atlantic margins to constrain the processes of initial rifting and transition to true seafloor spreading. This plan and ODP results set the stage to define a focused initiative in IODP, particularly given the expectation that *Chikyu*'s deep-penetration capabilities would allow reaching basement beneath thick continental margin sediments. However, this initiative was not addressed in IODP, even though it was the focus of one of the three "mission proposals" submitted but not adopted in 2007. One interpretation is that by then it had become clear that the mismatch of operating funds and costs would keep *Chikyu* dedicated mainly to NanTroSEIZE and otherwise close to Japan through the end of IODP. Regular proposal pressure was already relatively weak for this initiative, and it dropped even more after the decision about the mission proposals.

1.4.3 Initiative: LIPs

ODP had verified that oceanic LIPs were produced by voluminous, episodic magmatic events, but they probably remain the least understood of volcanic processes on Earth. The ISP defined an LIP initiative with two lines of emphasis: understanding underlying mantle processes associated with LIP eruption and understanding potential links between LIP emplacement and long-term environmental change (climate, ocean chemistry, etc.). The envisioned strategy involved drilling in both Mesozoic and Cenozoic LIPs, documenting correlations with nearby sedimentary records of climate change, and coordination with the broadband seismology community in International Ocean Network (ION) broadband seismic observatories to image mantle convection processes associated with LIPs. Unfortunately, while ODP had drilled selected holes for ION seismic observatories, less than half were actually instrumented as observatories, so the last vision remains largely unfulfilled. Also, as noted above, the dedication of the majority of *Chikyu* IODP drilling time to NanTroSEIZE meant its deep drilling capabilities were not applied to the LIP initiative during IODP. Nevertheless, one JOIDES Resolution IODP expedition was quite successful in coring an important Mesozoic LIP example: Shatsky Rise in the western Pacific. Another JOIDES Resolution expedition successfully cored a hotspot trail, the Louisville Seamount hotspot trail in the southwestern Pacific, which presumably resulted from mantle processes similar to those giving rise to LIPs.

These expeditions were relatively recent (2009 and 2011, respectively), but both have already produced high-profile publications (see also Sager and Koppers, chapter 4.3 of this volume). The Shatsky Rise expedition (Sager, Sano, Geldmacher, & the IODP Expedition 324 scientists, 2011) was designed to understand the formation of a relatively large LIP in a well-constrained tectonic setting at a ridge triple junction during the latest Jurassic and earliest Cretaceous, and in particular to test the mantle "plume-head" hypothesis for LIP emplacement. As reported by Sager et al. (2013) in a paper that garnered considerable media interest, the cores and survey data demonstrate that the largest and oldest section of Shatsky Rise, the TAMU Massif, is a "supervolcano" comparable in size to Olympus Mons on Mars and possibly the largest volcano on Earth. The recovered core appears to be sufficient to test the plume-head hypothesis, but definitive result is still to come. The Louisville Seamount expedition (Koppers, Yamazaki, Geldmacher, & the IODP Expedition 330 Scientific Party, 2013) was designed as a complement to the late ODP leg to the Hawaii-Emperor Seamount Chain that demonstrated that the Hawaiian hotspot did not remain fixed in a mantle reference frame but migrated ~15°S between 80 and 50 Mya (Tarduno et al., 2003). In contrast, the Louisville Seamount cores show that its hotspot remained relatively fixed during the same time period, demonstrating that the Pacific hotspots can move independently, and are not constrained to consistent motion by a Pacific-wide "mantle wind" (Koppers et al., 2012).

1.4.4 Initiative: 21st Century Mohole

This ISP initiative revived the long-standing scientific goal to recover a complete section (order of 5-km long) through intact oceanic crust and into uppermost mantle, to better understand the formation of oceanic crust and geochemical mass balances between mantle and crust. In fact, this was the primary motivation behind the first important effort at scientific ocean drilling, Project Mohole (Bascom, 1961), but it was out of reach of the technological capabilities of the nonriser drilling vessels of DSDP and ODP. The ISP strategy involved deep drilling capabilities of *Chikyu* pending development of 4-km riser technology. That development was not feasible during IODP, and also *Chikyu* could not be applied to this initiative during this period for reasons discussed above. Nevertheless, JOIDES Resolution nonriser drilling, especially during the first and last two years of IODP operations, made significant progress in a staged approach defined in the ISP as a long-term strategy toward the 21st Century Mohole. As summarized by Ildefonse et al. (chapter 4.2.1 of this volume), the results produced fundamental advances in understanding oceanic crustal structure and formation in both fast- and slow-spread crust.

In crust formed at fast spreading rates under conditions of vigorous magma supply, IODP produced two important breakthrough achievements: (1) the first complete penetration though upper crustal basalts and into the transition to underlying gabbros (Wilson et al., 2006), and (2) the first subseafloor recovery of cumulate gabbros formed by magmatic processes deep in the lower oceanic crust (Gillis et al., 2014). The first was achieved by one ODP and three IODP expeditions to the "Superfast" Hole 1256D in 15 my-old eastern Pacific crust formed at very high spreading rate. Here two 2005 expeditions (309 and 312) reached uppermost gabbros at the relatively shallow depth predicted for the very high spreading rate, based on accretion models and global compilation of seismically determined depths to axial melt lenses (Wilson et al., 2006). Although the first gabbros were reached, their composition is mainly basaltic so they cannot yet represent cumulate residues from a magma lens or mush zone; thus, the hole penetrates only the very top of the transition to more primitive cumulate gabbros below. The transition between seismic layers 2 and 3 is deeper than the hole extends, but it is possible the seismic transition corresponds more closely to the full transition to the primitive cumulates predicted to lie deeper in the section. The recovered gabbros crystallized at very high temperatures from a melt lens, and adjacent dikes were exposed to corresponding high-temperature contact metamorphism (granulite facies). A return expedition in 2011 generated considerable media interest, but hole conditions and the toughness of the metamorphosed formation precluded significant advancement even if exciting large-diameter samples were recovered from the dike–gabbro transition zone (Teagle, Ildefonse, Blum, & the IODP Expedition 335 scientists, 2012).

The second breakthrough—the first core recovery of predicted layered cumulate gabbros from the lower crust formed at fast spreading rate—was achieved during Expedition 345 at a tectonic window that exposes the lower crust in Hess Deep in the equatorial Pacific (Gillis et al., 2014). Results from the >200 m drilled sections in the cumulates, together with previous sampling of other sections of the Hess Deep oceanic crust by dredging, submersibles, and ODP drilling, provides the most complete composite section of fast-spread crust to date. They suggest that the crust here is formed from a mantle mix of compositionally diverse magmas, and provide important constraints on flow processes in the residual melts that form the lower crust after extraction of magmas that form the basaltic upper crust.

In crust formed at slow spreading rates with discontinuous magma delivery and a strong tectonic imprint, two 2004–2005 IODP expeditions (304 and 305) targeted the Atlantis Massif oceanic core complex, with particular success at Hole U1309D in the footwall that tectonically exposes lower crustal and mantle rocks on the seafloor. With extensive exposures of mantle lithologies on the seafloor, it was somewhat unexpected that the >1400 m deep hole penetrated a section of predominantly gabbros. The cored gabbros display relatively low deformation and alteration, and consist of distinct magmatic units showing the body was built by an extended history of multiple magmatic pulses. Structural and paleomagnetic data show a bulk rotation that supports the "rolling hinge" flexural unloading model for the detachment fault footwall (Morris et al., 2009). ODP had also recovered unexpectedly thick sections of gabbros at other tectonic exposures of slow-spread crust where mantle peridotites were predicted to be cored. The results from Hole U1309D inspired an important refinement of the Cannat (1996) "plum pudding" model for accretion at slow spreading rate to explain the thick sections of gabbro unexpectedly recovered by drilling at several oceanic core complexes. In that refined model (Ildefonse et al., 2007; Blackman et al., 2011), formation of core complexes may be associated with large gabbroic intrusions that are less dense but stronger than surrounding mantle peridotites. As a result, the gabbros are preferentially brought close to the seafloor and survive better as more coherent bodies than surrounding mantle peridotites in the low-angle normal-faulting processes that tectonically expose lower crust and mantle at core complexes.

1.5 BOREHOLE OBSERVATORY ACCOMPLISHMENTS

A few unifying comments about IODP accomplishments in borehole observatory science are in order, partly because the descriptions of those accomplishments are scattered among several initiatives above, but also because they contributed significantly toward definition of a fourth main theme ("Earth in Motion") in the new Science Plan for the post-2013 program. Although long-term monitoring had been identified as an initiative in the final years of ODP, borehole observatory science was not explicitly called out as an IODP initiative. Nevertheless, the ISP highlighted borehole observatory science as an important IODP technological approach and one of six "principles of implementation." The same financial constraints that limited IODP platform operations also imposed serious limits on IODP observatory installations, especially when compounded by the additional financial commitments required for postinstallation servicing with submersibles. As a result, a few IODP drilling programs noted above were implemented without their proposed observatory components, and IODP had to adopt a considerably more conservative approach toward observatories than originally envisioned. Nevertheless, the sum of the IODP observatory effort is impressive and must be counted as a success. This includes: the two arrays of ridge-flank hydrological/microbiological CORKs described in Section 1.2; artificial hydrothermal vent wellheads installed in the Okinawa Trough; a sub-seafloor thermistor array deployed on the Norwegian margin to record the signal of bottom water temperature variations in the last century (Harris and the IODP Expedition 306 scientists, 2006); the simple ACORK and prototype SCIMPI in the Vancouver margin hydrates area; the Smart Plug, Genius Plug, and proto-type LTBMS installed in NanTroSEIZE; and the detailed temperature monitoring string successfully installed and then recovered at the JFAST rapid-response site. While the ridge-flank CORKs shared many common design elements, nearly every other installation had a unique design. That they all succeeded can be attributed to the strong teamwork among IODP scientists, engineers, and funding managers, setting a good model for success in the Earth in Motion theme in the International Ocean Discovery Program even if funding pressures continue into the post-2013 program.

REFERENCES

Aiello, I., & Ravelo, A. C. (2012). Evolution of marine sedimentation in the Bering Sea since the Pliocene. *Geosphere*, *8*, 1231–1253. http://dx.doi.org/10.1130/GES00710.1.

Backman, J., & Moran, K. (2008). Introduction to special section on cenozoic paleoceanography of the central Arctic Ocean. *Paleoceanography*, *23*. http://dx.doi.org/10.1029/2007PA001516 PA1S01.

Backman, J., & Moran, K. (2009). Expanding the cenozoic paleoceanographic record in the central Arctic Ocean: IODP Expedition 302 synthesis. *Central European Journal of Geosciences*, *1*, 157–175. http://dx.doi.org/10.2478/v10085-009-0015-6.

Backman, J., Jakobsson, M., Frank, M., Sangiorgi, F., Brinkhuis, H., Stickley, C., et al. (2008). Age model and core-seismic integration for the cenozoic Arctic coring expedition from the Lomonosov Ridge. *Paleoceanography*, *12*. http://dx.doi.org/10.1029/2007PA001476 PA1S03.

Bascom, W. (1961). *A hole in the bottom of the sea: The story of the mohole project*. New York: Doubleday and Company. 352 pp.

Becker, K., & Davis, E. E. (2005). A review of CORK designs and operations during the Ocean Drilling Program. *Proc. IODP* (Vol. 301) College Station, TX: IODP-MI. http://dx.doi.org/10.2204/iodp.proc.301.104.2005.

Behrmann, J. H., Flemings, P. B., John, C. M., & the IODP Expedition 308 Scientists. (2006). Rapid sedimentation, overpressure, and focused fluid flow, Gulf of Mexico continental margin. *Scientific Drilling*, *3*, 12–17. http://dx.doi.org/10.2204/iodp.sd.3.03.2006.

Bijl, P. K., Bendle, J. A. P., Bohaty, S. M., Pross, J., Schouten, S., Tauxe, L., & Expedition 318 Scientists., et al. (2013). Eocene cooling linked to early flow across the Tasmanian Gateway. *Proceedings of the National Academy of Sciences*, *110*, 9645–9650. http://dx.doi.org/10.1073/pnas.1330872110.

Blackman, D. K., Ildefonse, B., John, B. E., Ohara, Y., Miller, D. J., Abe, N., et al. (2011). Drilling constraints on lithospheric accretion and evolution at Atlantis Massif, Mid-Atlantic Ridge 30° N. *Journal of Geophysical Research*, *116*, B07103. http://dx.doi.org/10.1029/2010JB007931.

Bowles, M. W., Mogollón, J. M., Kasten, S., Zabel, M., & Hinrichs, K.-U. (2014). Global rates of marine sulfate reduction and implications for sub-sea-floor metabolic activities. *Science, 344,* 889–891. http://dx.doi.org/10.1126/science.1249213.

Brinkhuis, H., Schouten, S., Collinson, M. E., Sluijs, A., Sinninghe Damsté, J. S., Dickens, G. R., & the Expedition 302 Scientists., et al. (2006). Episodic fresh surface waters in the Eocene Arctic Ocean. *Nature, 441,* 606–609. http://dx.doi.org/10.1038/nature04692.

Browning, J. V., Miller, K. G., Sugarman, P. J., Barron, J., McCarthy, F. M. G., Kulhanek, D., et al. (2013). Chronology of Eocene-Miocene sequences on the New Jersey shallow shelf: implications for regional, interregional, and global correlations. *Geosphere, 9,* 1434–1456. http://dx.doi.org/10.1130/GES00857.1.

Camoin, G. F., Iryu, Y., McInroy, D. B., & the IODP Expedition 310 Scientists. (2007). IODP Expedition 310 reconstructs, sea level, climatic, and environmental changes in the South Pacific during the last deglaciation. *Scientific Drilling, 5,* 4–12. http://dx.doi.org/10.2204/iodp.sd.5.01.2007.

Camoin, G. F., Seard, C., Deschamps, P., Webster, J. M., Abbey, E., Braga, J. C., et al. (2012). Reef response to sea-level and environmental changes during the last deglaciation: Integrated Ocean Drilling Program Expedition 310, Tahiti Sea Level. *Geology, 40,* 643–646. http://dx.doi.org/10.1130/G32057.1.

Cannat, M. (1996). How thick is the magmatic crust and slow spreading oceanic ridges? *Journal of Geophysical Research, 101,* 2847–2857. http://dx.doi.org/10.1029/95JB03116.

Channell, J. E. T., Sato, Y., Kanamatsu, T., Stein, R., Malone, M., Alvarez-Zarikian, C., & the IODP Expeditions 303 and 306 scientists. (2006). IODP Expeditions 303 and 306 monitor Miocene-Quaternary climate in the North Atlantic. *Scientific Drilling, 2,* 4–10. http://dx.doi.org/10.2204/iodp.sd.2.01.2006.

Channell, J. E. T., Xuan, C., & Hodell, D. A. (2009). Stacking paleointensity and oxygen isotope data for the past 1.5 Myrs (PISO-1500). *Earth and Planetary Science Letters, 283,* 14–23. http://dx.doi.org/10.1016/j.epsl.2009.03.012.

Chester, F. M., Mori, J. J., Toczko, S., Eguchi, N., & the Expedition 343/343T scientists. (2012). Japan trench fast drilling project (JFAST). *IODP Prel Rept, 343/343T.* http://dx.doi.org/10.2204/iodp.pr.343343T.2012.

Chester, F. M., Rowe, C., Ujiie, H., Kirkpatrick, J., Regalla, C., Remitte, F., & Expedition 343 and 343T Scientists., et al. (2013). Structure and composition of the plate-boundary slip zone for the 2011 Tohoku-Oki earthquake. *Science, 342,* 1208–1211. http://dx.doi.org/10.1126/science.1243719.

Cook, C. P., van de Flierdt, T., Williams, T., Hemming, S. R., Iwai, M., Kobayashi, M., & IODP Expedition 318 Scientists., et al. (2013). Dynamic behavious of the East Antarctica ice sheet during Pliocene warmth. *Nature Geoscience, 6,* 765–769. http://dx.doi.org/10.1038/ngeo1889.

Davis, E. E., & Becker, K. (1998). Borehole observatories record driving forces for hydrothermal circulation in young oceanic crust. *EOS, Transactions American Geophysical Union, 79*(369), 377–378.

Davis, E., Heesemann, M., & the IODP Expedition 328 Scientists and Engineers. (2012). IODP Expedition 328: early results of Cascadia subduction zone ACORK observatory. *Scientific Drilling, 13,* 12–18. http://dx.doi.org/10.2204/iodp.sd.13.02.2012.

Davis, E. E., Wang, K., Becker, K., & Thomson, R. (2000). Formation-scale hydraulic and mechanical properties of oceanic crust inferred from pore-pressure response to periodic seafloor loading. *Journal of Geophysical Research, 105,* 13,423–13,435.

Davis, E., Becker, K., Pettigrew, T., Carson, B., & MacDonald, R. (1992). CORK: a hydrologic seal and downhole observatory for deep ocean boreholes. In E. Davis, M. Mottl, et al. (Eds.), *Proc. ODP. Init. Repts.* (Vol. 139) (pp. 43–53).

Davis, E., Kinoshita, M., Becker, K., Wang, K., Asano, Y., & Ito, Y. (2013). Episodic deformation and inferred slow slip at the Nankai subduction zone during the first decade of CORK borehole pressure and VLFE monitoring. *Earth and Planetary Science Letters, 368*, 110–118. http://dx.doi.org/10.1016/j.epsl.2013.03.009.

Deschamps, P., Durand, N., Bard, E., Hamelin, B., Camoin, G., Thomas, A., et al. (2012). Ice sheet collapse and sea-level rise at the Bølling warming, 14,600 yr ago. *Nature, 483*, 559–564. http://dx.doi.org/10.1038/nature10902.

D'Hondt, S., Jørgensen, B. B., Miller, D. J., Batzke, A., Blake, R., Cragg, B. A., et al. (2004). Distributions of microbial activities in deep subseafloor sediments. *Science, 306*, 2216–2221. http://dx.doi.org/10.1126/science.1101155.

D'Hondt, S., Inagaki, F., Ferdelman, T., Jorgensen, B. B., Kato, K., Kemp, P., et al. (2007). Exploring subseafloor life with the integrated ocean drilling program. *Scientific Drilling, 5*, 26–37. http://dx.doi.org/10.2204/iodp.sd.5.03.2007.

D'Hondt, S., Inagaki, F., Alvarez-Zarikian, C., & the IODP Expedition 329 Scientific Party (2013). IODP Expedition 329: life and habitability beneath the seafloor of the South Pacific Gyre. *Scientific Drilling, 15*, 4–10. http://dx.doi.org/10.2204/iodp.sd.15.01.2013.

Dickens, G. R. (2011). Down the rabbit hole: toward appropriate discussion of methane release from gas hydrate systems during the Paleocene-Eocene thermal maximum and other past hyperthermal events. *Climate of the Past, 7*, 831–846. http://dx.doi.org/10.5194/cp-7-831-2011.

Edwards, K. J., Becker, K., & Colwell, F. (2012). The deep, dark energy biosphere: intraterrestrial life on Earth. *Annual Review of Earth and Planetary Sciences, 40*, 551–568. http://dx.doi.org/10.1146/annurev-earth-042711-105500.

Edwards, K. J., Bach, W., Klaus, A., & the IODP Expedition 336 Scientists. (2014). Mid-Atlantic Ridge microbiology: inititiation of long-term coupled microbiological, geochemical, and hydrological experimentation within the seafloor at North Pond, western flank of the Mid-Atlantic Ridge. *Scientific Drilling, 17*, 13–18. http://dx.doi.org/10.5194/sd-17-13-2014.

Engelhardt, T., Kallmeyer, J., Cypionka, H., & Engelen, B. (2014). High virus-to-cell ratios indicate ongoing production of viruses in deep subsurface sediments. *The ISME Journal, 8*, 1503–1509. http://dx.doi.org/10.1038/ismej.2013.245.

Escutia, C., Brinkhuis, H., Klaus, A., & the IODP Expedition 318 scientists. (2011). IODP Expedition 318: from greenhouse to icehouse at the Wilkes Land Antarctic Margin. *Scientific Drilling, 12*, 15–23. http://dx.doi.org/10.2204/iodp.sd.12.02.2011.

Filippelli, G. (2014). A salty start to modern ocean circulation. *Science, 344*, 1228–1229. http://dx.doi.org/10.1126/science.1255553.

Fisher, A. T., Davis, E. E., & Becker, K. (2008). Borehole-to-borehole hydrologic response across 2.4 km in the upper oceanic crust: implications for crustal-scale properties. *Journal of Geophysical Research, 113*, B07106. http://dx.doi.org/10.1029/2007JB005447.

Fisher, A. T., Urabe, T., Klaus, A., & the IODP Expedition 301 scientists. (2005). IODP Expedition 301 installs three borehole crustal observatories, prepares for three-dimensional, crosshole experiments in the northeastern Pacific Ocean. *Scientific Drilling, 1*, 6–11. http://dx.doi.org/10.2204/iodp.sd.1.01.2005.

Fisher, A. T., Tsuji, T., Petronotis, K., Wheat, C. G., Becker, K., Clark, J. F., & the IODP Expedition 327 and Atlantis Expedition AT18-07 Scientific Parties., et al. (2012). IODP Expedition 327 and Atlantis Expedition AT 18-07: observatories and experiments on the eastern flank of the Juan de Fuca Ridge. *Scientific Drilling, 15*, 4–11. http://dx.doi.org/10.2204/iodp.sd.13.01.2012.

Flemings, P. B., Long, H., Dugan, B., Germaine, J., John, C. M., Behrmann, J. H., & IODP Expedition 308 Scientists., et al. (2008). Pore pressure penetrometers document high overpressure on the seafloor where multiple landslides have occurred on the continental slope, offshore Loiusiana, Gulf of Mexico. *Earth and Planetary Science Letters*, *269*, 309–325. http://dx.doi.org/10.1016/j.epsl.2007/12.005.

Fulthorpe, C. S., Hoyanagi, K., Blum, P., & the IODP Expedition 317 scientists. (2011). IODP Expedition 317: exploring the record of sea-level change off New Zealand. *Scientific Drilling*, *12*, 4–14. http://dx.doi.org/10.2204/iodp.sd.12.01.2011.

Fulton, P. M., Brodsky, E. E., Kano, Y., Mori, J., Chester, F., Ishikawa, T., & Expedition 343, 343T, and KR-13-08 Scientists., et al. (2013). Low coseismic friction on the Tohoku-Oki fault determined from temperature measurements. *Science*, *342*, 1214–1217. http://dx.doi.org/10.1126/science.1243641.

Gillis, K. M., Snow, J. E., Klaus, A., Abe, N., Adrião, A. B., Akizawa, N., et al. (2014). Primitive layered gabbros from fast-spreading lower oceanic crust. *Nature*, *505*, 204–207. http://dx.doi.org/10.1038/nature12778.

Hammerschmidt, S., Davis, E., & Kopf, A. (2013). Fluid pressure and temperature transients detected at the Nankai Trough megasplay fault: results from the SmartPlug borehole observatory. *Tectonophysics*, *600*, 116–133. http://dx.doi.org/10.1016/j.tecto.2013.02.010.

Harris, R. N., & the IODP Expedition 306 scientists. (2006). Borehole observatory installations on IODP Expedition 306 reconstruct bottom-water temperature changes in the Norwegian Sea. *Scientific Drilling*, *2*, 28–31. http://dx.doi.org/10.2204/iodp.sd.2.03.2006.

Harris, R. N., Sakaguchi, A., Petronotis, K., & the Expedition 344 Scientists. (2013). In *Proc. IODP* (344). College Station, TX: Integrated Ocean Drilling Program. http://dx.doi.org/10.2204/iodp.proc.344.2013.

Henry, P., Kanamatsu, T., Moe, K. T., Strasser, M., & the IODP Expedition 333 Scientific Party. (2012). IODP Expedition 333: return to Nankai Trough subduction inputs sites and coring of mass transport deposits. *Scientific Drilling*, *14*, 4–17. http://dx.doi.org/10.2204/iodp.sd.14.01.2012.

Hernández-Molina, F. J., Stow, D. A. V., Alvarez-Zarikian, C., & Expedition IODP 339 Scientists. (2013). IODP Expedition 339 in the Gulf of Cadiz and off West Iberia: decoding the environmental significance of the Mediterranean outflow water and its global influence. *Scientific Drilling*, *16*, 1–11. http://dx.doi.org/10.5194/sd-16-1-2013.

Hernández-Molina, F. J., Stow, D. A. V., Alvarez-Zarikian, C. A., Acton, G., Bahr, A., Balestra, B., et al. (2014). Onset of Mediterranean outflow into the North Atlantic. *Science*, *344*, 1244–1249. http://dx.doi.org/10.1126/science.1251306.

Hodell, D. A., Channell, J. E. T., Curtis, J. H., Romero, O. E., & Röhl, U. (2008). Onset of "Hudson Strait" Heinrich events in the eastern North Atlantic and the end of the middle Pleistocene transition (~640 ka)? *Paleoceanography*, *23*, PA4218. http://dx.doi.org/10.1029/2008PA001591.

Hodell, D. A., Lourens, L., Stow, D. A. V., Hernández-Molina, F. J., Alvarez-Zarikian, C. A., & the Shackleton Site Project Members. (2013). The "Shackleton site" (IODP site U1385) on the iberian margin. *Scientific Drilling*, *16*, 13–19. http://dx.doi.org/10.5594/sd-16-13-2013.

Houben, A. J. P., Bijl, P. K., Pross, J., Bohaty, S. M., Passchier, S., Stickley, C. E., & the Expedition 318 Scientists., et al. (2013). Reorganization of Southern Ocean plakton ecosystem at the onset of Antarctic glaciation. *Science*, *340*, 341–344. http://dx.doi.org/10.1126/science.1223646.

Hyndman, R. D., & Davis, E. E. (1992). A mechanism for the formation of methane hydrate and seafloor bottom-simulating reflectors by vertical fluid expulsion. *Journal of Geophysical Research*, *97*, 7025–7041. http://dx.doi.org/10.1029/91JB03061.

Ildefonse, B., Blackman, D., John, B. E., Ohara, Y., Miller, D. J., MacLeod, C. J., & IODP Expeditions 304/305 Science Party. (2007). Oceanic core complexes and crustal accretion at slow-spreading ridge. *Geology, 35*, 623–626. http://dx.doi.org/10.1130/G23531A.1.

Inagaki, F. (2010). Deep subseafloor microbial communities. In *Encyclopedia of life sciences.* Chichester, UK: John Wiley & Sons, Ltd. http://dx.doi.org/10.1002/9780470015902.a0021894.

IODP. (2001). *Earth, oceans, and Life: Scientific investigations of the earth system using multiple drilling platforms and new Technologies, integrated ocean drilling program initial science plan 2003-2013.* Washington DC: International Working Group Support Office. 110 pp.

Jakobbson, M., Backman, J., Rudels, B., Nycander, J., Frank, M., Mayer, L., et al. (2007). The early Miocene onset of a ventilated circulation regime in the Arctic Ocean. *Nature, 447*, 986–990.

Kallmeyer, J., Pockalny, R., Adhikari, R. R., Smith, D. C., & D'Hondt, S. (2012). Global distribution of microbial abundance and biomass in subseafloor sediment. *Proceedings of the National Academy of Sciences, 109*, 16213–16216. http://dx.doi.org/10.1073/pnas.1203849109.

Kawagucci, S., Miyazaki, J., Nakajima, T., Takaya, Y., Kato, Y., Shibuya, T., et al. (2013). Post-drilling changes in fluid discharge pattern, mineral deposition, and fluid chemistry in the Iheya North hydrothermal field, Okinawa Trough. *Geochemistry, Geophysics, Geosystem, 14*, 4774–4790. http://dx.doi.org/10.1002/2013GC004895.

Kawai, M., Futagami, T., Toyoda, A., Takaki, Y., Nishi, S., Hori, S., et al. (2014). High frequency of phylogenetically diverse reducative dehalogenase-homologous genes in deep subseafloor sedimentary metagenomes. *Frontiers in Microbiology, 5*, 1–15. http://dx.doi.org/10.3389/fmicb.2014.00080.

Koppers, A. A. P., Yamazaki, T., Geldmacher, J., & the IODP Expedition 330 Scientific Party. (2013). IODP Expedition 330: drilling the Louisville Seamount trail in the SW Pacific. *Scientific Drilling, 15*, 11–22. http://dx.doi.org/10.2204/iodp.sd.15.02.2013.

Koppers, A. A. P., Yamazaki, T., Geldmacher, J., Gee, J. S., Pressling, N., Hoshi, H., et al. (2012). Limited latitudinal mantle plume motion for the Louisville hotspot. *Nature Geoscience, 5*, 911–917. http://dx.doi.org/10.1038/ngeo1638.

Lado-Insua, T., Moran, K., Kulin, I., Farrington, S., & Newman, J. B. (2013). SCIMPI: a new borehole observatory. *Scientific Drilling, 16*, 57–61. http://dx.doi.org/10.5194/sd-16-57-2013.

Lever, M. A., Rouxel, O., Alt, J. C., Shimizu, N., Ono, S., Coggon, R., et al. (2013). Evidence for microbial carbon and sulfur cycling in deeply buried ridge flank basalt. *Science, 339*, 1305–1308. http://dx.doi.org/10.1126/science.1229240.

Lin, W., Conin, M., Moore, J. C., Chester, F. M., Nakamura, Y., Mori, J. J., & Expedition 343 Scientists., et al. (2013). Stress state in the largest displacement area of the 2011 Tohoku-Oki earthquake. *Science, 339*, 687–690. http://dx.doi.org/10.1126/science.1229379.

Lourens, L. J., Sluijs, A., Droon, D., Zachos, J. C., Thomas, E., Röhl, U., et al. (2005). Astronomical pacing of late Palaeocene to Eocene global warming events. *Nature, 435*, 1083–1087. http://dx.doi.org/10.1038/nature03814.

Miller, K. G., Browning, J. V., Mountain, G. S., Bassetti, M. A., Monteverde, D., Katz, M. E., et al. (2013). Sequence boundaries are impedance contrasts: core-seismic-log integration of Oligocene-Miocene sequences, New Jersey shallow shelf. *Geosphere, 9*, 1257–1285. http://dx.doi.org/10.1130/GES00858.1.

Miller, K. G., Mountain, G. S., Browning, J. V., Katz, J. E., Monteverde, D., Sugarman, P. J., et al. (2013). Testing sequence stratigraphic models by drilling Miocene foresets on the New Jersey shallow shelf. *Geosphere, 9*, 1236–1256. http://dx.doi.org/10.1130/GES00884.1.

Moran, K., Backman, J., Brinkhuis, H., Clemens, S. C., Cronin, T., Dickens, G. R., et al. (2006). The cenozoic palaeoenvironment of the arctic ocean. *Nature, 441*, 601–605. http://dx.doi.org/10.1038/nature04800.

Moore, G. F., Kanagawa, K., Strasser, M., Dugan, B., Maeda, L., Toczko, S., & the IODP Expedition 338 Scientific Party. (2013). IODP Expedition 338: NanTroSEIZE Stage 3; NanTroSEIZE plate boundary deep riser 2. *Scientific Drilling, 17*, 1–12. http://dx.doi.org/10.5194/sd-17-1-2014.

Morris, A., Gee, J. S., Pressling, N., John, B., MacLeod, C. J., Grimes, C. B., et al. (2009). Footwall rotation in an oceanic core complex quantified using reoriented Integrated Ocean Drilling Program core samples. *Earth and Planetary Science Letters, 287*, 217–228. http://dx.doi.org/10.10.1016/j.epsl.2009.08.007.

Mountain, G., Proust, J.-N., & the Expedition 313 scientists. (2010). The New Jersey margin scientific drilling project (IODP Expedition 313): untangling the record of global and local sea-level changes. *Scientific Drilling, 10*, 26–34. http://dx.doi.org/10.2204/iodp.sd.10.03.2010.

National Research Council (Committee on the review of the scientific achievements and assessment of the potential for transformational discoveries with U.S.-supported scientific ocean drilling). (2011). *Scientific ocean drilling: Accomplishments and challenges*. Washington, DC: National Academies Press.

Norris, R. D., Wilson, P. A., Blum, P., & the Expedition 342 Scientists. (2014). In *Proc. IODP, 342*. College Station, TX: Integrated Ocean Drilling Program. http://dx.doi.org/10.2204/iodp.proc.342.2014.

Orcutt, B. N., Bach, W., Becker, K., Fisher, A. T., Hentscher, M., Toner, B. M., et al. (2011). Colonization of subsurface microbial observatories deployed in young ocean crust. *ISME J, 5*, 692–703.

Orsi, W. D., Edgcomb, V. P., Christman, G. D., & Biddle, J. F. (2013). Gene expression in the deep biosphere. *Nature, 499*, 205–208. http://dx.doi.org/10.1038/nature12230.

Pälike, H., Lyle, M. W., Nishi, H., Raffi, I., Ridgwell, A., Gamage, K., et al. (2012). A Cenozoic record of the equatorial Pacific carbonate compensation depth. *Nature, 488*, 609–614. http://dx.doi.org/10.1038/nature11360.

Parkes, R. J., Cragg, B. A., & Wellsbury, P. (2000). Recent studies on bacterial populations and processes in subseafloor sediments: a review. *Hydrogeology Journal, 8*, 11–28. http://dx.doi.org/10.1007/PL00010971.

Poirier, A., & Hillaire-Marcel, C. (2011). Improved Os-isotope stratigraphy of the Arctic Ocean. *Geophysical Research Letters, 38*, L14607. http://dx.doi.org/10.1029/2011GL047953.

Pross, J., Contreras, L., Bijl, P. K., Greenwood, D. R., Bohaty, S. M., Schouten, S., & Integrated Ocean Drilling Program Expedition 318 scientists., et al. (2012). Persistent near-tropical warmth on the Antarctic continent during the early Eocene epoch. *Nature, 488*, 73–77. http://dx.doi.org/10.1038/nature11300.

Riedel, M., Collett, T. S., & Malone, M. (2010). Expedition 311 synthesis: scientific findings. In J. Riedel, T. S. Collett, M. J. Malone, & the Expedition 311 scientists (Eds.), *Proc. IODP* (Vol. 311). Washington, DC: Integrated Ocean Drilling Program Management International, Inc. http://dx.doi.org/10.2204/iodp.proc.311.214.2010.

Ryan, W. B. F., Carbotte, S. M., Coplan, J. O., O'Hara, S., Melkonian, A., Arko, R., et al. (2009). Global multi-resolution topography synthesis. *Geochemical, Geophysical, Geosystems, 10*, Q03014. http://dx.doi.org/10.1029/02008GC002332.

Sager, W. W., Sano, T., Geldmacher, J., & the IODP Expedition 324 scientists. (2011). IODP Expedition 324: ocean drilling at Shatsky Rise gives clues about oceanic plateau formation. *Scientific Drilling, 12*, 24–31. http://dx.doi.org/10.2204/iodp.sd.12.03.2011.

Sager, W. W., Zhang, J., Korenaga, J., Sano, T., Koppers, A. A. P., Koppers, M., et al. (2013). An immense shield volcano within the Shatsky Rise oceanic plateau, northwest Pacific Ocean. *Nat Geosci, 6*, 976–981. http://dx.doi.org/10.1038/ngeo1934.

Schlung, S. A., Ravelo, A. C., Aiello, I. W., Andreasen, D. H., Cook, M. S., Drake, M., et al. (2013). Millenial-scale climate change and intermediate water circulation in the Bering Sea from 90 ka: a high-resolution record form IOPD Site U1340. *Paleoceanography, 28*, 54–67. http://dx.doi. org/10.1029/2012PA002365.

Schulte, P., Alegret, L., Arenillas, I., Arz, J. A., Barton, P. J., Bown, P. R., et al. (2010). The Chicxulub asteroid impact and mass extinction at the Cretaceous-Paleogene boundary. *Science, 327*, 1214–1218. http://dx.doi.org/10.1126/science.1177265.

Sexton, P. F., Norris, R. D., Wilson, P. A., Pälike, H., Westerhold, T., Röhl, U., et al. (2011). Eocene global warming events driven by ventilation of oceanic dissolved organic carbon. *Nature, 471*, 349–353. http://dx.doi.org/10.1038/nature09826.

Smith, A., Popa, R., Fisk, M., Nielsen, M., Wheat, C. G., Jannasch, H. W., et al. (2011). In situ enrichment of ocean crust microbes on igneous minerals and glasses using an osmotic flow-through device. *Geochemical, Geophysical, Geosystems, 12*, Q06007. http://dx.doi.org/10.1029/2010GC003424.

Solomon, S., Qin, D., Manning, M., Chen, Z., Marquis, M., Averyt, K. B., et al. (Eds.). (2007). *Contribution of working group I to the fourth assessment report of the intergovernmental panel on climate change, the physical science basis.* Cambridge UK: Cambridge University Press.

Stein, R., Boucsein, B., & Meyer, H. (2006). Anoxia and high primary productivity in the Paleogene central Arctic Ocean: first detailed records from Lomonosov Ridge. *Geophysical Research Letters, 33*, L18606. http://dx.doi.org/10.1029/2006GL026776.

Stickley, C. E., St John, K., Koç, N., Jordan, R. W., Passchier, S., Pearce, R. B., et al. (2009). Evidence for middle Eocene Arctic sea ice from diatoms and ice-rafted debris. *Nature, 460*, 376–379. http://dx.doi.org/10.1038/nature08163.

Stocchi, P., Escutia, C., Houben, A. J. P., Vermeersen, B. L. A., Bijl, P. K., Brinkhuis, H., & IODP Expedition 318 scientists., et al. (2013). Relative sea-level rise around East Antarctica during Oligocene glaciation. *Nature Geoscience, 6*, 380–384. http://dx.doi.org/10.1038/ngeo1783.

Stocker, T. F., Qin, D., Plattner, G.-K., Tignor, M., Allen, S. K., Boschung, J., et al. (Eds.). (2013). *IPCC 2013: Climate Change 2013: The Physical science basis. Contribution of working group I to the fifth assessment report of the intergovernmental panel on climate change.* Cambridge UK: Cambridge University Press.

Strasser, M., Moore, G. F., Kimura, G., Kitamura, Y., Kopf, A. J., Lallemant, S., et al. (2009). Origin and evolution of a splay fault in the Nankai accretionary wedge. *Nature Geoscience, 2*, 648–652. http://dx.doi.org/10.1038/ngeo609.

Takahashi, K., Ravelo, A. C., Alvarez Zarikian, C., & the IODP Expedition 323 scientists. (2011). IODP Expedition 323 – Pliocene and Pleistocene paleoceanographic changes in the Bering Sea. *Scientific Drilling, 11*, 4–13. http://dx.doi.org/10.2204/iodp.sd.11.01.2011.

Takai, K., Mottl, M. J., Nielsen, S. H. H., & the IODP Expedition 335 scientists. (2012). IODP Expedition 331: strong and expansive subseafloor hydrothermal activities in the Okinawa Trough. *Scientific Drilling, 15*, 19–27. http://dx.doi.org/10.2204/iodp.sd.13.03.2012.

Tarduno, J. A., Duncan, R. A., Scholl, D. W., Cottrell, R. D., Steinberger, B., Thordarson, T., Kerr, B. C., Neal, C. R., Frey, F. A., Torii, M., & Carvallo, C., (2003). The Emperor Seamounts: southward motion of the Hawaiian hotspot plume in Earth's mantle. *Science, 301*, 1064–1069. http://dx.doi.org/10.1126/science.1086442.

Teagle, D. A. H., Ildefonse, B., Blum, P., & the IODP Expedition 335 scientists. (2012). IODP Expedition 335: deep sampling in ODP Hole 1256D. *Scientific Drilling, 13*, 28–34. http:// dx.doi.org/10.2204/iodp.sd.13.04.2012.

Thomas, A. L., Henderson, G. M., Deschamps, P., Yokoyama, Y., Mason, A. J., Bard, E., et al. (2009). Penultimate deglacial sea-level timing from uranium/thorium dating of Tahitian corals. *Science, 324*, 1186–1189. http://dx.doi.org/10.1126/science.1168754.

Tobin, H., Kinoshita, M., Ashi, J., Lallemant, S., Kimura, G., Screaton, E., & IODP Expeditions 314/315/316 Scientific Party., et al. (2009). NanTroSEIZE stage 1 expeditions 314, 315, and 316; first drilling program of the nankai trough seismogenic zone Experiment. *Scientific Drilling*, *8*, 4–17. http://dx.doi.org/10.2204/iodp.sd.8.01.2009.

Toczko, S. T., Kopf, A. J., Araki, E., & the IODP Expedition 332 Scientific Party. (2012). The IODP Expedition 332: eyes on the prism, the NanTroSEIZE observatories. *Scientific Drilling*, *14*, 34–38. http://dx.doi.org/10.2204/iodp.sd.14.04.2012.

Ujiie, H., Tanaka, H., Saito, T., Tsutsumi, A., Mori, J. J., Kameda, J., & Expedition 343 and 344T Scientists., et al. (2013). Low coseismic shear stress on the Tohoku-Oki megathrust determined from laboratory experiments. *Science*, *342*, 1211–1214. http://dx.doi.org/10.1126/science.1243485.

Underwood, M. B., Saito, S., Kubo, Y., & the IODP Expedition 322 scientists. (2010). IODP Expedition 322 drills two sites to document inputs to the Nankai subduction zone. *Scientific Drilling*, *10*, 14–25. http://dx.doi.org/10.2204/iodp.sd.10.02.2010.

Vannucchi, P., Ujiie, K., Stroncik, N., & the IODP Expedition 334 Scientific Party. (2013). IODP Expedition 334: an investigation of the sedimentary record, fluid flow, and state of stress on top of the seismogenic zone of an erosive subduction margin. *Scientific Drilling*, *15*, 23–30. http://dx.doi.org/10.2204/iodp.sd.15.03.2013.

Vannucchi, P., Sak, P. B., Morgan, J. P., Ohkushi, K., Ujiie, K., & the IODP Expedition 334 Shipboard Scientists. (2013). Rapid pulses of uplift, subsidence, and subuction erosion offshore Central America: implications for building the rock record of convergent margins. *Geology*, *41*, 995–998. http://dx.doi.org/10.1130/G34355.1.

Wilson, D. S., Teagle, D. A. H., Alt, J. C., Banerjee, N. R., Umino, S., Miyashita, S., et al. (2006). Drilling to gabbro in intact ocean crust. *Science*, *312*, 1016–1020. http://dx.doi.org/10.1126/science.1126090.

Yokoyama, Y., Webster, J., Cotterill, C., Braga, J. C., Jovane, L., Mills, H., & the IODP Expedition 325 scientists., et al. (2011). IODP Expedition 325: Great Barrier Reefs reveals past sea-level, climate and environmental changes since the last ice age. *Scientific Drilling*, *12*, 32–45. http://dx.doi.org/10.2204/iodp.sd.12.04.2011.

Zachos, J. C., Röhl, U., Schellenberg, S., Sluijs, A., Hodell, D. A., Kelly, D. C., et al. (2005). Rapid acidification of the ocean during the Palecene-Eocene Thermal Maximum. *Science*, *308*, 1611–1615. http://dx.doi.org/10.1126/science.1109004.

Zachos, J. C., Dickens, G. R., & Zeebe, R. E., (2008). An early Cenozoic perspective on greenhouse warming and carbon-cycle dynamics. *Nature*, *451*, 279–283. http://dx.doi.org/10.1038/nature06588.

Zoback, M., Hickman, S., Ellsworth, W., & the SAFOD science team. (2011). Scientific drilling into the San Andreas Fault Zone – an overview of SAFOD's first five years. *Scientific Drilling*, *11*, 14–28. http://dx.doi.org/10.2204/iodp.sd.11.02.2011.

Chapter 2

New Frontier of Subseafloor Life
and the Biosphere

Chapter 2.1

Exploration of Subseafloor Life and the Biosphere Through IODP (2003–2013)

Fumio Inagaki[1],* and Victoria Orphan[2]
[1]*Kochi Institute for Core Sample Research, Japan Agency for Marine-Earth Science & Technology (JAMSTEC), Nankoku, Kochi, Japan;* [2]*Division of Geological and Planetary Sciences, California Institute of Technology, Pasadena, CA, USA*
Corresponding author: E-mail: inagaki@jamstec.go.jp

2.1.1 BACKGROUND: THE DEEP SUBSEAFLOOR BIOSPHERE

Deep subseafloor environments have long been considered an inanimate world harboring only fossil records of geologic material. Pioneering work by Morita and ZoBell in the 1950s demonstrated the presence of approximately 1000 cells per gram of pelagic sediments cored down to 8 m below the seafloor (mbsf) and pointed to a subseafloor microbial community defined by slow, continuous activity (Morita & ZoBell, 1955). Forty years later, pilot microbiological studies of marine sediments recovered by the Ocean Drilling Program (ODP) expanded these numbers by over two orders of magnitude, reporting $>10^5$ microbial cells per cubic centimeters of Pacific margin sediment down to approximately 800 mbsf (Parkes et al., 1994, Parkes, Cragg, & Wellsbury, 2000). The cell numbers are generally logarithmically decreasing with depth, suggesting correlations to geophysical and nutrient conditions in the deep subseafloor microbial habitat. The unexpected abundance of microbial biomass in subseafloor sediments was estimated to be approximately one-third of total living biomass carbon on Earth (Whitman, Coleman, & Wiebe, 1998); however, critical factors that convert cell numbers to biomass carbon, as well as the coverage of oceanographic settings (e.g., open ocean subsurface), were poorly constrained (Hinrichs & Inagaki, 2012).

ODP Leg 201 in 2002 was the first microbiology- and biogeochemistry-dedicated scientific drilling expedition that was conducted by the US drilling research vessel *JOIDES Resolution* at the eastern Equatorial Pacific and the Peru margin (D'Hondt et al., 2004; Jørgensen, D'Hondt, & Miller, 2006); the results from Leg 201 have significantly expanded our previous knowledge of deep

39

subseafloor life and the biosphere. During Leg 201, potential contamination of drill fluids was continuously monitored using both perfluorocarbon (PFC) tracer and fluorescent microsphere beads, providing quality assurance of uncontaminated sediment cores for microbiological studies down to approximately 420 mbsf (House, Cragg, Teske, & the Leg 201 Scientific Party, 2003). Geochemical measurement of pore waters revealed that potential electron acceptors for adenosine triphosphate (ATP)-yielding anaerobic microbial respirations (e.g., nitrate, sulfate, and metal oxides) are supplied either from the overlying seawater or the crustal fluids underlying sediments (D'Hondt, Rutherford, & Spivack et al., 2002; D'Hondt et al., 2004). Microbial cells were evaluated onboard by microscopic direct counts using the nucleic acid stain acridine orange (AODC), revealing abundant but generally decreasing microbial numbers with depth. Some sites displayed interesting geochemical profiles and variable cell numbers in response to changing geochemistry. For example, at Site 1229 on the Peruvian shelf, sulfate reduction was observed in the deepest horizons of the sediment core overlying the basement rock, and corresponded with remarkably high numbers of microbial cells ($\sim 10^{10}$ cells cm^{-3}) in sediments associated with this second, deeper sulfate–methane transition zone (SMTZ) (Parkes et al., 2005). Diagenetic calculations of pore water chemical constituents suggested that microbial activities represented by sulfate reduction and methanogenesis are relatively high in near surface and organic-rich Peru margin sediments, whereas relatively organic-poor sediments in the eastern Equatorial Pacific sites appeared to be characterized by lower activity and oxidative processes (D'Hondt et al., 2004). This has been attributed to a very low flux of nutrient and energy substrates from photosynthetic primary organic production in the overlying seawater.

During the ODP period, the diversity of deep subseafloor microbes was investigated by culture-independent molecular ecological techniques: the environmental genomic DNA (or RNA) was extracted from the innermost part of sediment core, and then 16S ribosomal RNA (rRNA) genes from the domains Bacteria and Archaea were amplified using polymerase chain reaction (PCR). Phylogenetic analysis of PCR-amplified 16S rRNA gene clone libraries revealed that subseafloor microbial communities are mainly composed of previously uncultured hence unidentified archaea and bacteria. Some of the common groups recovered included members of Deep-Sea Archaeal Group (DSAG: alternatively designated as Marine Benthic Group-B), Miscellaneous Crenarchaeotic Group (MCG) and South African Gold Mine Euryarchaeotic Group for Archaea, and JS1 Candidate Division, *Chloroflexi*, *Planctomycetes* for Bacteria (e.g., Inagaki et al., 2003, Inagaki et al., 2006; Teske, 2006; Webster et al., 2006; see also Chapter 2.2a). At Site 1227 on the Peru Margin, notably high RNA concentrations were recovered from sediments at the SMTZ, in which DSAG archaea were predominantly detected by sequencing the reverse transcribed 16S rRNA (cDNA), suggesting the presence of active archaeal communities at geochemical interfaces within the deep subseafloor biosphere (Sørensen & Teske, 2006).

Recent progress in high-throughput DNA sequencing with next-generation sequencers (NGS) has enabled the retrieval of metagenomic information from natural habitats, even for relatively low biomass or low DNA-recovery samples. Small amounts of DNA have been successfully recovered from deep subseafloor biosphere samples and coupled with NGS-sequencing techniques, including targeted gene sequencing of "tagged" fragments of 16S rRNA genes or whole genomes/metagenomes. Using ODP Leg 201 samples from the Peru Margin, environmental DNA was extracted from four sediment horizons down to depths of 50 mbsf at Site 1229, and then whole genomic DNA was amplified with multiple displacement amplification using a virus-derived phi29 polymerase, yielding approximately 62 Mbp of genetic information (Biddle, Fitz-Gibbon, Schuster, Brenchley, & House, 2008). The first metagenomic results of the microbial assemblage revealed that only 3–8% of sequences from subseafloor sediments showed homologies to functionally known sequences in the databases, suggesting that >90% of metagenomic sequences in the deep subseafloor biosphere remain functionally unknown.

If the activity of microbes at SMTZs is supported by the anaerobic oxidation of methane (AOM), the carbon isotopic compositions (δ ^{13}C) of cellular biomolecules should correspond to isotopically light carbon (i.e., ^{12}C-enriched carbon) derived from methane as the carbon source. Using secondary ion mass spectrometry (SIMS), natural abundance carbon isotopic compositions of intact archaeal cells detected by fluorescence in situ hybridization (FISH) were measured and compared with the δ ^{13}C values of archaeal intact polar lipids extracted from the same sediment samples (Biddle et al., 2006). The δ ^{13}C values obtained by SIMS and intact polar lipid analysis were in good agreement and suggested that most metabolically active archaeal components at SMTZs were not assimilating methane into biomass, but rather might be mixotrophs, assimilating buried organic matter as a carbon source, while gaining energy (making ATP) from sulfate-dependent AOM (Biddle et al., 2006). The microbes mediating the AOM reaction in the deep subsurface may be using a different metabolic strategy than their counterparts inhabiting near seafloor advective methane seeps. The diversity and ecological strategies of methane-consuming microorganisms in the deep subseafloor biosphere are a topic requiring further study. In addition, based on the δ ^{13}C values of pore water-dissolved inorganic carbon, acetate, and hydrocarbons in Leg 201 samples, it has been suggested that other novel microbial metabolisms may also contribute to the production of ethane and propane using hydrogen and acetate (Hinrichs et al., 2006).

It has long been unknown if the deeply buried subseafloor microbes are alive, just surviving, or dead. Activity measurements using radiotracers such as ^{35}S-labeled sulfate for sulfate reduction and ^{14}C-labeled bicarbonate or acetate for methanogenesis have indicated that microbial activities in deep subseafloor sediments are extremely low, with mean generation times of up to thousands of years (Hoehler & Jørgensen, 2013; Schippers et al., 2005). The proportion of presumably living microbial cells in the deep subseafloor was independently assessed using RNA-based molecular techniques on ODP Leg 201 samples. Using a fluorescence microscopy method known as catalyzed reporter deposition fluorescence in situ hybridization, Shippers et al. (2005) reported a large fraction of ribosomal RNA-containing bacterial cells, suggesting that a significant component of the microbial assemblage may be viable. Cultivation efforts also resulted in the successful isolation of a number of (mostly facultative) anaerobic bacteria from deep sediments (D'Hondt et al., 2004), although the predominant microbial components (identified with cultivation-independent molecular methods) were highly resistant to growth in conventional batch-type cultivation (Chapter 2.6). Recent studies of subseafloor sedimentary microbial cells tackled the issues regarding the physiological state and strategies of deep subseafloor sedimentary microbes for long-term survival. Using nanoscale secondary ion mass spectrometry (NanoSIMS), Morono et al. (2011) studied the number of live cells that show incorporation of stable isotope-labeled substrates (e.g., ^{13}C-glucose, ^{13}C-acetate, ^{13}C-bicarbonate, ^{15}N-ammonia, and $^{13}C^{15}N$-amino acids) with a 450,000-year-old sediment sample recovered from 219 mbsf, indicating that at least >75% of total cells that can be stained with SYBR Green I are "live cells" as physiologically stand-by in the deep and old sedimentary habitat (Jørgensen, 2011). In this study, nitrogen assimilation rates were found to be generally higher than carbon assimilation rates except for amino acids, suggesting a hypothesis that subseafloor microbes may suppress energy-consuming nitrogen assimilation in situ despite the presence of ammonia in sediment pore water. Instead, deep subseafloor microbes may invest their limited energy for sustaining the essential metabolic demands required for long-term survival.

Using ODP Leg 201 core samples, Lomstein et al. calculated microbial turnover using a molecular clock proxy from amino acid racemization, indicating that the deep subseafloor microbial community turns over at exceedingly slow rates of 200–2000 years (Lomstein, Langerhuus, D'Hondt, Jørgensen, & Spivack, 2012). The concentration of dipicolinic acid in sediment, the biomarker of bacterial endospores, as well as of muramic acid suggested that endospores in deep subseafloor sediments are as abundant as vegetative cells that have been estimated by AODC or other cell enumeration methods (Lomstein et al., 2012; Langerhuus, et al., 2012; see also a review by Hoehler & Jørgensen, 2013). These studies also indicated that, in deep and old subseafloor sediments, large fractions of amino acids (and possibly other biomolecules) were produced in situ by dead biomass (i.e., necromass). These findings are in good agreement with recently described physiological and metabolic characteristics of subseafloor sedimentary archaea: Takano et al. (2010) studied the membrane lipid turnover of marine benthic archaea by in situ stable isotope probing and revealed that benthic archaea generally recycle lipid components without new lipid synthesis via the energy-consuming isoprenoid-synthesis pathway. Lloyd et al. (2013a) studied single cell-derived genome fragments from the MCG and Marine Benthic Group-D and found that these cells encoded extracellular protein-degrading enzymes such as gingipain and clostripain, suggesting that subseafloor sedimentary archaea have the potential to degrade proteins derived from necromass for their long-term survival under the energy-limited condition.

Taken together, these data indicated that the deep subseafloor sedimentary biosphere consists of a mixture of live, dormant (or just surviving), and dead archaeal and bacterial cells, although details of in situ physiological state as well as the strategies and mechanisms of long-term growth and/or survival have remained largely unknown. While the majority of deep subseafloor biosphere research has focused on archaea and bacteria, there is recent evidence for the occurrence of eukaryotes such as fungi (Orsi et al., 2013a,b), as well as viruses (Engelhardt, Kallmeyer, Cypionka, & Engelen, 2014; Yanagawa et al., 2014a; Yoshida, Takaki, Eitoku, Nunoura, & Takai, 2013), are present in the subseafloor sedimentary habitats examined from ODP Leg 201 and Integrated Ocean Drilling Program (IODP). However, functional and evolutional characteristics of those deeply buried eukaryotes and viruses, as well as the ecological roles in the subseafloor microbial ecosystems, are largely unknown.

To verify the true nature of the deep subseafloor microbial community, especially in low biomass habitats close to the limit of the habitable zone, aseptic sampling, sample processing, and storage/subsampling techniques of uncontaminated fresh samples and careful quality assurance and quality control (QA/QC) are required. In addition, potential methodological biases should be carefully considered since the deep subseafloor microbial life may be very different from that of the surface biosphere: Teske and Sørensen (2008) pointed out the possible bias in PCR-dependent molecular ecological surveys caused by mismatches of PCR-primer sequences, especially for subseafloor sedimentary archaea. To address this issue, primer-independent metagenomic and metatranscriptomic approaches (e.g., Biddle, White, Teske, & House, 2011, Biddle et al., 2008; Kawai et al., 2014; Orsi et al., 2013a) as well as poly (A)-tailing reverse transcription PCR (Hoshino & Inagaki, 2013) are a useful complement for the study of the subseafloor microbial community. In addition to the issue of primer- or probe-sequence mismatches, bias caused by DNA extraction and purification must also be assessed and is critical for accurate gene quantification. In fact, a large fraction of deeply buried subseafloor sedimentary microorganisms were found to be recalcitrant to cell lysis using conventional enzymatic approaches, resulting in only approximately 30% lysis of total cells. Customized protocols for subseafloor sediment samples using chemical and/or physical treatments have significantly improved the efficiency of cell lysis (Lipp, Morono, Inagaki, & Hinrichs, 2008; Morono, Terada, Hoshino, & Inagaki, 2014). PCR inhibitors and/or substances that affect fluorescence intensity coeluting with extracted DNA will also impact results from fluorescence-based conventional real-time PCR. In this regard, newly developed technologies such as digital PCR coupled with microfluidics may be a useful alternative for the accurate quantification of genes in the deep subseafloor biosphere

(Hoshino & Inagaki, 2012). So far, quantification of the specific cells and/ or genes has been performed using a variety of conventional techniques in different laboratories (Lloyd et al., 2013b). Given the issues of potential biases and data accuracy with these challenging samples, and rapid progress on technological developments, the geographical distribution of subseafloor life needs to be systematically studied at the regional to global scale using consistent techniques for molecular quantification and sequencing with appropriate sample processing that minimizes potential sample biases.

Within the subseafloor biosphere, the deep aquifer in the permeable upper oceanic crust—so-called "Subseafloor Ocean"—represents a vast environment that is potentially habitable for microbial life (Bach & Edwards, 2003; Staudigel et al., 2008; Thorseth, Torsvik, Furnes, & Muehlenbachs, 1995). Microscopic observations of basaltic core samples recovered by ODP showed that cell-like fabrics are present in the volcanic glass and fractures (Fisk, Giovannoni, & Thorseth, 1998). Based on the volume of glass, the total global biomass of basalt-hosted microbes was estimated to be 0.6 Tg of cellular carbon, representing only a small fraction of total living biomass on Earth. However, the real abundance of microbial cells in the rock aquifer remains unknown. During the ODP Leg 186, an experimental borehole seal, or Circulation Obviation Retrofit Kit aka "CORK" (Chapter 2.5), was deployed at Site 1026B in the eastern flank of the Juan de Fuca ridge, where 247 m of sedimentary column and 48 m of basaltic ocean crust were drilled. Although the CORK at Site 1026B did not completely seal the borehole and may have introduced ambient seawater, molecular ecological analysis of the borehole fluids collected from a Biocolumn device installed on the CORK recovered diverse bacterial and archaeal sequences related to thermophilic microorganisms that are potentially derived from the crustal aquifer (Cowen et al., 2003). Thermodynamic and bioenergetic calculations also suggest that water–rock reactions such as iron oxidation and hydrogen production of relatively young basaltic crusts may provide energetically habitable conditions for chemolithotrophic subseafloor life (Bach & Edwards, 2003; Edwards, Bach, & McCollom, 2005).

2.1.2 IODP EXPEDITIONS RELATIVE TO THE DEEP-BIOSPHERE RESEARCH

The successful pilot studies from the ODP highlighted the abundance and potential metabolic diversity of life in the subseafloor and paved the way for the deep subseafloor biosphere being selected as one of the top scientific objectives for the IODP (International Ocean Drilling Program, 2001). Starting from the first IODP Expedition 301 at the Juan de Fuca Ridge off Oregon in 2004, microbiologists and (bio)geochemists often participated in the IODP expeditions, including several comprehensive microbiology- and biogeochemistry-dedicated drilling expeditions (Figure 2.1.1).

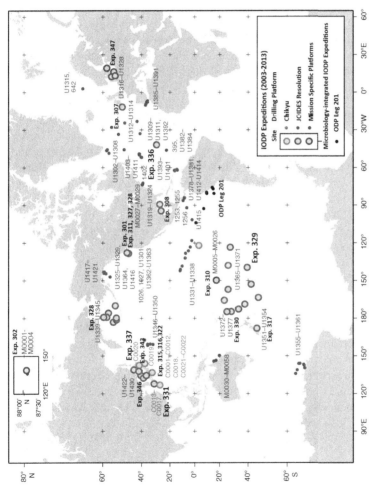

FIGURE 2.1.1 A global map showing drilling sites during the IODP 2003–2013. Yellow (gray in print versions) dots represent sites of expeditions that microbiology and/or biogeochemistry were/was partly integrated into the shipboard scientific program. Expeditions 329, 331, 336, and 337 were the microbiology- and biogeochemistry-dedicated IODP cruises. Drilling sites of the first microbiology- and biogeochemistry-dedicated scientific ocean drilling, ODP Leg 201 off Peru and the Eastern Equatorial Pacific, are also shown by black (black in print versions) dots.

2.1.2.1 Expedition 329: South Pacific Gyre Subseafloor Life

Until Expedition 329, the nature of subseafloor life in sediment beneath mid-ocean gyres, which covers most of the area of the open ocean, has been poorly understood. Almost all sites where subseafloor life has been studied are on ocean margins or in the equatorial ocean. In 2010, using the *JOIDES Resolution*, Expedition 329 explored deep-water sediments (water depth >5000 m) down to the basement at seven sites in the ultraoligotrophic South Pacific Gyre (SPG). SPG is the largest of the ocean gyres where chlorophyll concentrations and primary photosynthetic productivity in the seawater are significantly lower than in other regions of the ocean, and hence, the total sedimentary organic carbon here, as well as the burial rate, is the lowest in the global ocean. Pilot studies in shallow gyre sediments demonstrated that the microbial assemblage and its net metabolic activities are substantially lower than those recorded in Pacific margin sediment (D'Hondt et al., 2009; Kallmeyer, Pockalny, Adhikari, Smith, & D'Hondt, 2012; Røy et al., 2012). During Expedition 329, a total of seven sites (Sites U1365-1370) located along two transects crossing the heart of the oligotrophic gyre were drilled. Because of the low sedimentation rate, sediment core was recovered down to the sediment–basement interface at all sites and even penetrated into the uppermost basaltic basement at three locations (U1365, U1367, and U1368). Preliminary shipboard and shore-based data showed that dissolved oxygen and nitrate are present throughout the entire SPG sediment column, and alteration of the basement basalt occurs continuously on the timescale of formation fluid replacement, even within Mesozoic-aged basaltic basement (i.e., 80–120 Ma) (Expedition 329 Scientists, 2011). In addition, preliminary shipboard cell counts showed that microbial abundances are generally three or more orders of magnitude lower than at the same sediment depths in all sites previously cored by scientific ocean drilling. However, because of the extremely low microbial biomass in the SPG sedimentary habitat, accurate cell counts as well as careful molecular characterizations of the microbial diversity associated with these aerobic and ultraoligotrophic subseafloor environments require further study, and implementation of improved, ultraclean cell count methods and molecular ecological techniques (e.g., Morono, Terada, Kallmeyer, & Inagaki, 2013; also see Chapter 2.1a).

2.1.2.2 Expedition 331: Deep Hot Biosphere

Over the past several decades, global surveys of deep-sea hydrothermal vents have demonstrated the widespread occurrence of diverse and highly productive chemosynthetic ecosystems, which thrive in the steep redox gradients created during the mixing of hydrothermal vent fluids and seawater. Geochemical and microbial investigations of vent fluids suggest the occurrence of microbial activity in the subseafloor system (e.g., Huber, Stotters, Cheminee, Richnow, & Stetter, 1990; Deming & Barros, 1993; Huber, Butterfield, & Baross, 2003; Kawagucchi et al., 2011; Nakagawa et al., 2005; Takai et al., 2004). However, direct evidence for the existence of a subseafloor biosphere beneath active hydrothermal vents and fluid discharge zones has long been debated due to the difficulty of direct sampling by scientific ocean drilling. In 2010, using the *Chikyu*, Expedition 331 explored the Iheya North hydrothermal field in the middle Okinawa Trough, an active spreading backarc basin. Multiple holes were drilled at five sites (Sites C0013–C0017), sampling down to the maximum penetration depth of 151 mbsf at Site C0017 (Expedition 331 Scientists, 2010). During Expedition 331, the first Kuroko-type blackish ore (massive sulfide) deposits were successfully recovered from subseafloor environment of the Iheya North hydrothermal field. Onboard, geochemical analyses of interstitial water and headspace gas showed complex patterns with depth and lateral distance of only a few meters, but clearly indicated the occurrence of phase separation and stagnation, secondary mineral deposition and alterations, and potential evidence of biogenic respiration/production of carbon compounds. Shipboard microbiological analyses demonstrated the occurrence of microbial communities in relatively low-temperature sediment near the seafloor, but little evidence of life in the deeper hydrothermally altered zones. Yet, shore-based molecular ecological analysis revealed DNA signatures of archaea belonging to the Hot Water Crenarchaeotic Group IV within the deepest sedimentary section at approximately 90 °C (Yanagawa et al., 2014b). Three of the holes were transformed into artificial hydrothermal vents after drilling and were later cased and caped with a steel mesh platform, providing valuable postcruise opportunities for microbiological and geochemical sampling of fluids and mineral deposits by a manned submersible or remotely operated vehicle (Takai et al., 2010).

2.1.2.3 Expedition 336: Mid-Atlantic Ridge Flank Microbiology

Oceanic crust is one of the least explored subseafloor environments that potentially harbor a significant fraction of Earth's subseafloor microbial biomass. The extent to which microbes colonize, alter, and evolve in the crustal habitat remains largely unknown (Edwards et al., 2005). In 2011, using the *JOIDES Resolution*, Expedition 336 explored the North Pond of the Mid-Atlantic Ridge, an 8- × 15-km large sediment pond, approximately 300-m-thick overlying eight million-year-old basaltic basement rock where vigorous circulation of seawater occurs. Expedition 336 drilled multiple holes at three North Pond sites (U1382–U1384) and recovered the sediment and basaltic cores down to a maximum depth of 331 mbsf (Expedition 336 Scientists, 2012). The recovered upper basement basalts are fresh to moderately altered, aphyric to highly plagioclase–olivine–phyric tholeiites that fall on a liquid line of descent controlled by olivine fractionation. Dissolved oxygen in the sediment pore water showed an upward supply of oxygen from fluids circulating within the crust (Orcutt et al., 2013). Based on the modeled fluid flow rates, the calculated rates of oxygen consumption are ≤ 1 nmol cm^{-3}ROCK d^{-1} in young, cool basaltic crust. In addition to conventional downhole logging, a newly developed deep ultraviolet (UV) fluorescence-based logging tool called the Deep Exploration Biosphere Investigative tool, was also deployed for the first time on Expedition 336 to detect in situ microbial signatures. Two of the holes (U1382A and U1383C) were cased with a new fiberglass casing, and CORKs were installed as a long-term geophysical, geochemical, and microbiological observatory. Postcruise research from these borehole observatories will provide useful information to understand (1) the extent and activity of microbial life in basalt and its relation to basalt alteration by circulating seawater, and (2) the mechanism of microbial inoculation of an isolated sediment pond (Edwards, Bach, Klaus, & the IODP Expedition 336 Scientific Party, 2014; also see Chapter 2.5).

2.1.2.4 Expedition 337: Deep Coalbed Biosphere Off Shimokita

The limits and extent of deep subseafloor life and the biogeochemical carbon cycle associated with hydrocarbon reservoirs in continental margin remain largely unknown because of the absence of opportunities to conduct scientific ocean drilling initiatives. Expedition 337 in 2012 was the first expedition dedicated to subseafloor microbiology that used riser-drilling technology. Primary objectives during Expedition 337 were to study the relationship between the deep microbial biosphere and the subseafloor coalbed and to explore the limits of life in horizons deeper than ever probed before by scientific ocean drilling. Among the questions that guided this research were "Do deeply buried hydrocarbon reservoirs such as coalbeds act as geobiological reactors that sustain subsurface life by releasing nutrients and carbon substrates?" Using the *Chikyu*, a 2466 m-deep sedimentary sequence was drilled at Site C0020 off the Shimokita Peninsula, Japan, including a series of coal layers at approximately 2 km below the seafloor (Inagaki, Hinrichs, Kubo, & the Expedition 337 Scientists, 2012). The use of riser-drilling technology in very deep sediment created both unique opportunities and new challenges in the study of subseafloor life. Onboard sedimentological and paleontological analyses showed the record of dynamically changing depositional environments in the former forearc basin off the Shimokita Peninsula during the late Oligocene to Miocene. New shipboard facilities, such as the mud-gas monitoring laboratory and the radioisotope laboratory, were successfully implemented and strongly contributed to the success of Expedition 337. Real-time measurement of mud-gas chemistry and isotopic compositions provides the first indication of the existence of a subseafloor biosphere in deep horizons associated with the coalbed. In addition, for the first time in scientific ocean drilling, almost full downhole logging operations including in situ fluid sampling and analysis using a QuickSilver probe yielded data of unprecedented quality that provide a comprehensive view of sediment properties at Site C0020. Preliminary microbiological data showed that deep sediments recovered from Expedition 337 harbor very small microbial cells close to the quantification limit ($\sim 10^2$ cells cm^{-3}). To study very low numbers of "indigenous" deep subseafloor life with ultrasensitive molecular ecological and isotopic analyses, a rigorous program dedicated to

QA/QC of the cored materials and data has been conducted, for example, x-ray computed tomography scan for selecting undisturbed parts of core for microbiology, chemical tracers to monitor the levels of drilling mud-derived contaminants, and study of various control samples for nonindigenous contaminants including drill muds, chemical regents, commercial kits, and even air in laboratories.

2.1.2.5 Other Microbiology-Integrated IODP Expeditions

In addition to those microbiology- and biogeochemistry-dedicated IODP expeditions, deep-biosphere research has been systematically integrated into various multidisciplinary drilling expeditions (Figure 2.1.2). For example, during the first IODP Expeditions 301 at the eastern Juan de Fuca Ridge flank, which primarily aimed to hydrogeology and CORK recovery and installation, Lever et al. (2013) reported molecular and isotopic evidence for subseafloor microbial activities associated with biogeochemical carbon and sulfur cycles within the upper basaltic basement underlying the ridge flank sediment. Potential contamination of cored samples (sediment and basalt) was routinely monitored using PFC chemical tracers. Cell counts and activity measurements indicated that the upward flux of circulating hydrothermal fluids might support and stimulate

FIGURE 2.1.2   Onboard microbiological sample processing. (A) Subsampling of minicore sediments using sterilized tip-cut syringes from the core-cutting end on the catwalk deck of the *JOIDES Resolution* (Expedition 329). (B) Microsensor measurement of dissolved O_2 in the cold room of the *JOIDES Resolution* (Expedition 329). (C) Molecular works in the UV-lamina flow clean bench of the *Chikyu* (Expedition 337). (D) Anaerobic sample preparation of outside-scraped whole round core with N_2 flush in the microbiology laboratory of the *Chikyu* (Expedition 337). *Photograph: F. Inagaki.*

microbial communities in sediments above the oceanic crust (Engelen et al., 2008). During Expedition 301 in 2004, shipboard microbiologists collected samples of black rust on the CORK recovered from Hole 1026B, approximately 8 years after installation during ODP Leg 168. Although the CORK at 1026B was not completely sealed, hot crustal fluid measuring temperatures of 64 °C was flowing continuously. Nakagawa et al. (2006) studied black rusts formed on the recovered CORK using cultivation and cultivation-independent molecular approaches, showing diverse microbial communities including thermophilic methanogens and sulfate-reducing archaea and bacteria. One sulfate-reducing archaeon, *Archaeoglobus sulfaticallidus*, was isolated from the black rust and identified as new species (Steinsbu et al., 2010). The extensive microbiological studies associated with the reinstalled CORK at Site U1301 reported the in situ colonization of iron-oxidizing microbes (Orcutt, Wheat, & Edwards, 2010, Orcutt et al., 2011; also see Chapter 2.5).

In 2004, the second IODP Expedition 302 using mission-specific platforms traveled to the Lomnosov Ridge in the Central Arctic. Expedition 302 was dedicated to the study of climate and paleoenvironmental evolutionary records and also included some microbiological studies. Microbial assemblages associated with three sediment samples down to 242 mbsf were studied, showing the presence of a relatively high number of cells and diverse bacterial communities that are similar to those typically found in organic-rich sediments from continental margins (Forschner, Sheffer, Rowley, & Smith, 2009; Kallmeyer, Smith, Spivack, & D'Hondt, 2008).

During Expedition 307 in 2005, the Challenger Mound in the Porcupine Basin off the southwestern continental shelf of Ireland was drilled down to the basement (Ferdelman, Kano, Williams, & the IODP Expedition 307 Scientists, 2006; Kano, Ferdelman, & Williams, 2010). The diversity of microbial assemblages and metabolic genes associated with sediment and coral reef fossils was characterized by culture-independent molecular approaches. Results revealed that sedimentological characteristics are important factors affecting subseafloor microbial diversity and community structure (Hoshino et al., 2011; Webster et al., 2009). Studying these cold-water coral reefs and associated sedimentary environments also provided some insight into how paleoenvironmental change and sedimentary processes corelate with the diversity and biogeochemical processes of the extant deep subseafloor biosphere (Chapter 2.7).

The timing and course of the last deglaciation are an important component for understanding the dynamics of glacial ice sheets and the effect on the global sea level and climate change (Chapter 3). As witnessed with Expedition 307 in 2005, paleocoral reefs are excellent indicators for the dating of glacial–interglacial cycles and sea temperature. During Expedition 310, coral reef terraces in Tahiti were drilled down to 79 mbsf using a mission-specific platform DP *Hunter* at various water depths ranging from 41.6 to 161 m. In these relatively shallow subseafloor coral deposits, microbiologists observed dense biofilms in ancient reef cavities, which were associated with brownish iron and manganese deposits (Expedition 310 Scientists, 2007). Quantification of ATP showed that the highest levels of ATP activity (20,600 RFU) occurred within the subseafloor biofilms, suggesting this dense microbial biomass was made up of active microorganisms.

Expedition 308 in 2005 was dedicated to study the overpressure fluid flow system in the continental slope of the Gulf of Mexico. Abundance, diversity, and metabolic potential of subseafloor microorganisms in the Brazos-Trinity Basin IV and the Mars-Ursa Basin were studied by molecular ecological analyses (Nunoura et al., 2009) and metagenomic approaches (Biddle et al., 2011). For sediment samples collected during Expeditions 302, 307, and 308, hydrogenase activities were measured using a tritium-based radiotracer method, showing a strong correlation with sediment porosity (Soffientino, Spivack, Smith, & DHondt, 2009; Nunoura et al., 2009). The positive correlation between hydrogenase activity and porosity suggests that sedimentological characteristics such as pore-throat size and the related geohydrological flux of dissolved nutrients are important factors for controlling in situ metabolic activity and habitability in the deep sedimentary biosphere.

Specific IODP targets have included gas hydrates along continental margins to enhance understanding of formation/accumulation mechanisms of hydrocarbon reservoirs, and associated microbial processes such as methanogenesis and methane consumption. Together with the deep-biosphere initiative, the exploration of gas hydrates on the continental margins is also considered as one of the important IODP implementations to expand our knowledge of the formation/accumulation mechanisms of hydrocarbon reservoirs, including microbiological associations such as methanogenesis and methane consumption. In 2005, Expedition 311 was focused on methane hydrate-bearing sediments of the Hydrate Ridge on the Cascadia Margin. Using the "*JOIDES Resolution*," a high-pressure coring equipped with microbiological sampling device "DeepIsoBUG" was performed at near in situ pressure (~14 MPa), which resulted in successful enrichment cultivation of some anaerobic bacteria such as the genera *Clostridium* and *Acetobacterium* (Parkes et al., 2009). From sediment cores obtained during Expedition 311, Briggs et al. (2011) discovered pink biofilms at the SMTZ, which consisted of ANME-1 archaea and sulfate-reducing bacteria that mediate the AOM. Sediment core samples from Hydrate Ridge (ODP Leg 204) and cultivation of hydrogenotrophic methanogenic archaea (*Methanoculleus submarinus*) indicate that the abundance of methanogens at this site is generally <1000 cells g^{-1} sediment, and support low rates of methanogenic activity, at 4 fmol methane per gram sediment per day (Colwell et al., 2008). This value is close to the rate derived from geochemical modeling, but was a few orders of magnitude lower than the rates measured by radiotracer experiments with sediment obtained from Expedition 311 (Yoshioka et al., 2010). To better understand the biogeochemical processes occurring in Hydrate Ridge and other subseafloor organic-rich sedimentary ecosystems, Heuer, Pohlman, Torres, Elvert, & Hinrichs, (2009) analyzed the stable carbon isotope values of acetate and other carbon-based metabolites in sediment pore waters. Results from this work indicate that hydrogen levels are particularly significant for controlling the thermodynamically most favorable electron accepting processes such as acetogenesis and hydrogenotrophic methanogenesis (Heuer, Pohlman, Torres, Elvert, & Hinrichs, 2009; also see, Lever, 2012).

Using the *Chikyu*, geologically active, seismogenic subduction zones within and around the Nankai Trough and Japan Trench were intensively investigated by IODP over a 10-year-period (2003–2013) (Chapter 4.4). Using sediment samples during Expeditions 315 in 2007 and 316 in 2008, as well as other drilling sites, the potential for anaerobic respiration coupled to organohalide compounds (i.e., dehalogenation), including the conversion of tribromophenol to phenol, was documented in enrichment cultivation experiments (Futagami, Morono, Terada, Kaksonen, & Inagaki, 2009). Remarkably diverse dehalogenase gene homologs were discovered from subseafloor sediments using compatible molecular ecological approaches including PCR-based cloning and sequencing, substrate induced gene expression, and metagenomic analysis (Futagami, Morono, Terada, Kaksonen, & Inagaki, 2012, Futagami et al., 2009; Kawai et al., 2014). These molecular observations indicate that the buried organohalide compounds, which are mostly derived from the burial of bioproducts in the overlying water column or seabed such as algae or sponges, play significant roles in supporting deep subseafloor microbial respiration in sediments where availability of dissolved electron acceptors is severely limited.

The plate-subduction zones explored by drilling with the *Chikyu*, such as the Nankai Trough (NanTroSEIZE project) and Japan Trench (i.e., Expedition 343 J-FAST), provided information that implicate possible microbial reactions associated with this active plate-subduction system (e.g., Toki et al., 2012). For example, at the frontal thrust region of the Nankai subduction zone (e.g., Site C0007), pore water geochemistry and sulfur and oxygen isotopic analyses indicate the occurrence of microbial sulfur and iron cycling associated with the downward- and upward-diffusing sulfate (Riedinger et al., 2010). At the slope apron sediment associated with a branch of mega-splay faults at Site C0004, Mills et al. (2012) demonstrated the metabolically active bacterial populations in the sediment above, within, and below the SMTZ using RNA-based molecular ecological approaches, showing the occurrence of low diversity bacterial populations in deep sediment samples, including members of *Firmicutes* and *Betaproteobacteria* as well as some lower abundance bacterial phylotypes within the *Actinobacteria* and *Fusobacteria*.

The importance of hydrogen as the central electron donor in anaerobic ecosystems is well recognized. Abiotic hydrogen production through geological processes is therefore considered to be a fundamental factor that may support subseafloor microbial life. These processes include serpentinization, Fischer–Tropsch-type reactions, radiolysis, and faulting in the deep subsurface (e.g., McCollom, 2007; Ménez, Pasini, & Brunelli, 2012; Nealson, Inagaki, & Takai, 2005; Sleep & Zoback, 2007). In fact, Hirose, Kawagucci, and Suzuki (2011) demonstrated through friction experiments using different sedimentary rock types and crusts that significant amounts of hydrogen can be produced by mechanoradical reactions on fault surfaces, similar to those produced by earthquake faulting. On the other hand, it is important to carefully assess whether the hydrogen detected is indigenous or an artifact produced during drilling in highly fractured zones through water–rock reactions. Long-term observatory of geochemical signatures as well as in situ microbiological experimentations through IODP boreholes cased by carbon-reinforced plastic or other types of anticorrosion glass fiber pipes will open a new window to study the geologically active zones that are difficult to obtain high-quality core samples for microbiology and geochemistry, including the biosphere within the oceanic crust (see a review by Orcutt and Edwards in Chapter 2.5).

In 2009, Expedition 323 was conducted at Bering Sea. The primary scientific objective of Expedition 323 was to study the dynamics of the climate system associated with the recent warming of the high latitudes in the Northern Hemisphere but also the expedition provided an excellent opportunity to study the deep sedimentary biosphere in a region of high surface productivity and deep-water nutrient concentrations (Expedition 323 Scientists, 2010). Using microbiological samples from Expedition 323, the number of microbial cells and viruses in microbiological sediment samples were evaluated, indicating the occurrence of organic-rich subseafloor sedimentary biosphere (Kallmeyer et al., 2012; Engelhardt et al., 2014; also see Chapter 2.2a). In addition, based on the geochemical, sedimentological, and microbiological results from Expedition 323, the biogeochemical consequences of the modern and past sedimentary microbial activity, including paleoceanographic settings and methane flux from the subseafloor biosphere, were systematically studied (Wehrmann, Arndt, März, Ferdelman, & Brunner, 2013, Wehrmann et al., in press; also see Chapter 2.7).

In 2010, during Expedition 317, a 1,927-m-deep Holocene to late Eocene sedimentary sequence (water depth: 344 m) was cored by the *JOIDES Resolution* at Site U1352 in the Canterbury Basin off New Zealand. The distribution and changes in the microbial community with depth were assessed using 454 pyrosequencing of PCR-amplified 16S and 18S rRNA genes in addition to targeted metabolic gene analysis (e.g., *mcrA*; alpha subunit of the methyl coenzyme M reductase gene for methanogenesis, *dsrA*; alpha subunit of the sulfite reductase gene for sulfate reduction, *cbbL*; large subunit of the ribulose-1,5-bisphosphate carboxylase/oxygenase gene for autotrophic CO_2 fixation via RubisCO) (Ciobanu et al., 2014). Cell counts of 13 samples were performed using SYBR Green I staining, revealing a relatively high number of microbial-like cells ($>10^4$ cells cm^{-3}) throughout the sediment column. Although contamination of drill fluids was not monitored during the expedition (e.g., PFC tracer), the results extended the depth of molecular ecological data obtained from previous scientific ocean drilling down to 1740 mbsf for eukaryotic 18S rRNA genes (i.e., fungi), and to 1922 mbsf for bacterial 16S rRNA genes (Ciobanu et al., 2014).

In 2013, IODP Expeditions 346 (Asian Monsoon: Tada, Murray, & Alvarez Zarikian, 2013) and 347 (Baltic Sea Paleoenvironment: Andrén, Jørgensen, & Cotterill, 2012) were conducted at the Japan Sea and Baltic Sea using the *JOIDES Resolution* and a mission-specific platform (the DV *Greatship Manisha*), respectively. The primary scientific objectives of those expeditions were to understand the past climate and environmental change, but the research of deep subseafloor biosphere has been systematically integrated as a part of these projects. Questions addressed during these multidisciplinary expeditions included "How has the modern deep subseafloor sedimentary biosphere responded to the paleoenvironmental changes (e.g., glacial–interglacial cycles)?"; "How have past depositional processes that include transformation of organic matter from the surface world (i.e., terrestrial and seawater environments including ice coversheets) affected the indigenous subseafloor microbial diversity, activity, and functional characteristics?"; and "Is there evidence for new genetic or biomarker proxies that allow better resolution of past environmental conditions?" Postcruise research associated with Expeditions 346 and 347, as well as Expedition 323 at the Bering Sea and others, is still ongoing and multidisciplinary programs, which integrate paleontology, sedimentology, geochemistry, and microbiology, offer exciting potential to address fundamental questions regarding Earth's systems.

During the 2003–2013 IODP program, intensive cultivation efforts have resulted in the successful enrichment and isolation of diverse subseafloor microorganisms (Chapter 2.6.). In addition to conventional batch-type cultivation efforts, the use of downflow hanging sponge bioreactors yielded successful enrichment and isolation of anaerobic subseafloor sedimentary microorganisms including multiple species of methanogens and heterotrophic bacteria belonging to the phylum *Chloroflexi* (i.e., *Pelolinea submarina*; see Imachi et al., 2011, 2014). More detailed studies of successfully isolated indigenous microbes from the subseafloor sediments will significantly expand our current knowledge of physiological and functional characteristics required for survival and adaptation to energy-limited conditions.

2.1.3 SAMPLE STORAGE FOR THE FUTURE DEEP-BIOSPHERE RESEARCH

During 2003–2013, or earlier, there has been a long history of task forces being formed to make recommendations to the ODP/IODP concerning sampling and procedures for archiving microbiological samples (e.g., D'Hondt et al., 2007). A reservoir of archived, deep-frozen samples for the future shore-based microbiological and biogeochemical studies have tremendous value to the scientific community, especially for more comprehensive studies of global biomass and molecular surveys across the spectrum of oceanographic conditions and depths. During IODP, a report from the microbiology task force with recommendations for long-term biological archives was produced by the science technology panel (STP), and then approved by the IODP-Management International. Based on the documented community recommendation, routine microbiology sampling (RMS) and the use of chemical tracers to monitor potential contamination have been employed during multiple expeditions. Typically, microbiological sampling consists of a whole round core sample collected directly adjacent to or nearby interstitial water (IW) chemistry samples during all sedimentary drilling expeditions and immediately stored at −80 °C (or lower). For long-term storage of deep-frozen samples at core repositories, an aseptic subsampling procedure is also required, but working with this frozen material causes some technological challenges in terms of the QA/QC. To overcome some of these challenges, Masui, Morono, and Inagaki (2009) developed a semiaseptic diamond-saw system for subsampling of cores in a frozen state without thawing. In addition, recent progress of snap-freezing technique using an alternating magnetic field under supercooled liquid phase (i.e., cell alive system) minimizes cell destruction by providing fast, uniform freezing of samples with minimal ice crystal formation (Morono et al., in press). Although the STP recommended RMS strategy has not yet been fully employed (i.e., the processing of multiple smaller volume samples e.g., paraformaldehyde-fixed slurries and anaerobic fresh core samples in addition to deep frozen whole round cores), the framework for deep-frozen core samples has been an important first step for advancing future microbiological research in deep subseafloor environments. Following the recommendation for collecting biological samples for long-tem storage from the IODP community, whole round cores or syringe-plugged sediment samples (Figure 2.1.2) have been routinely requested from each IODP expedition from the biocurator of core repository. For example, the Kochi Core Center in Japan, one of the three IODP core repositories, established "The DeepBIOS" to store deep-frozen core samples at either −80 °C deep freezers or liquid nitrogen tanks and has been working to develop the QA/QC and aseptic subsampling techniques (see, www.kochi-core.jp/DeepBIOS) for the international scientific community.

2.1.4 CONCLUSION AND PERSPECTIVES

In conclusion, IODP expeditions during 2003–2013 have resulted in a wealth of new information regarding the abundance, diversity, and activity of the deep subseafloor biosphere. The global subseafloor biomass was reevaluated with additional core samples from open ocean and other oceanographic settings, resulting in 2.9×10^{29} cells and 4 Gt of biomass carbon (Chapter 2.2a). The number of subseafloor microbial cells in the organic-rich, anaerobic subseafloor biosphere on the Pacific margins is generally several orders of magnitude higher than that in the organic-poor, aerobic subseafloor biosphere on the open Pacific gyres (D'Hondt et al., 2009; Kallmeyer et al., 2012). In addition, the geographical distribution of subseafloor microbial cells is consistent to that of the occurrence of microbial sulfate reduction in sediment: globally, 11.3 Tmol of sulfate is reduced yearly, accounting for the oxidation of 12–29% of the organic carbon flux from the seawater to the seafloor (Bowles, Mogollón, Kasten, Zabel, & Hinrichs, 2014). The global geographical studies of subseafloor life generally indicate that the subseafloor sedimentary biosphere is a "geobiofilm" that covers Earth's oceanic crusts, playing important functions in the global elemental cycles. Along with this growing appreciation and interest in life within the ocean sediments and crust, researchers have also begun to recognize the need for improvements in the scientific approach and methods used to study these challenging ecosystems. In particular, when the environment approaches the limit of habitable zone, the so-called "biotic fringe," strict control of QA/QC and application of ultrasensitive sampling and analytical techniques under the ultraclean condition will be necessary in the future. Remarkable advancements in molecular technology in recent years has now made it possible to extract genomic DNA from single cell and to measure the elemental and isotopic composition. Combining these state-of-the-art analytical techniques, cultivation or stimulation of live cells in the laboratory or perhaps even in situ will add valuable information for expanding our understanding of the physiological and metabolic capabilities of the largely uncharacterized subseafloor biosphere.

In "*Chapter 2. New Frontiers of Subseafloor Life and the Biosphere,*" microbiological discoveries associated with IODP 2003–2013 are summarized by experts who contributed to the research of deep biosphere. The chapter includes theories regarding the energetic and thermodynamic constraints for habitability of subseafloor life under various geological, geophysical, and geochemical conditions (Chapters 2.3 and 2.4; also see a review of Hoehler & Jørgensen, 2013). The research of subseafloor life and deep biosphere is still ongoing with technological developments and a number of new discoveries ranging from information at the level of single cells to the global scale can be expected from previous ODP/IODP expeditions into the next phase of scientific ocean drilling. The expanding evidence and knowledge for the deep sedimentary and crustal subseafloor biosphere have stimulated a number of major knowledge gaps. Assessing fundamental questions such as "What are the environmental constraints for habitability of subseafloor life?"; "How and why can they live in such deep and extremely energy-limited conditions?"; "To what extent do subsurface microbial communities impact the transformation and cycling of carbon and other elements?" and "What is the genetic diversity, longevity, and metabolic function of deeply buried microbial communities?" Beginning to understand these questions and others will be central to ongoing efforts by the international scientific ocean drilling community.

ACKNOWLEDGMENTS

The authors thank the crews and scientists who participated in the IODP Expeditions; the members of IODP Microbiology Thematic Review Committee provided useful discussions for reviewing the progress of the deep-biosphere research during the IODP 2003–2013 and the perspectives.

REFERENCES

Andrén, T., Jørgensen, B. B., & Cotterill, C. (2012). Baltic sea basin paleoenvironment: paleoenvironmental evolution of the Baltic sea basin through the last glacial cycle. *IODP Scientific Prospectus, 347*. http://dx.doi.org/10.2204/iodp.sp.347.2012.

Bach, W., & Edwards, K. J. (2003). Iron and sulfide oxidation within the basaltic ocean crust: Implications for chemolithoautotrophic microbial biomass production. *Geochimica et Cosmochimica Acta, 67*, 3871–3887.

Biddle, J. F., Fitz-Gibbon, S., Schuster, S. C., Brenchley, J. E., & House, C. H. (2008). Metagenomic signatures of the Peru Margin subseafloor biosphere show a genetically distinct environment. *Proceedings of the National Academy of Sciences of the United States of America, 105*, 10583–10588.

Biddle, J. F., Lipp, J. S., Lever, M. A., Lloyd, K. G., Sørensen, K. B., Anderson, R., et al. (2006). Heterotrophic archaea dominate sedimentary subsurface ecosystems off Peru. *Proceedings of the National Academy of Sciences of the United States of America, 103*, 3846–3851.

Biddle, J. F., White, J. R., Teske, A. P., & House, C. H. (2011). Metagenomics of the subsurface Brazos-Trinity Basin (IODP site 1320): comparison with other sediment and pyrosequenced metagenomes. *ISME Journal, 5*, 1038–1047.

Bowles, M. W., Mogollón, J. M., Kasten, S., Zabel, M., & Hinrichs, K.-U. (2014). Global rates of marine sulfate reduction and implications for sub-sea-floor metabolic activities. *Science, 344*, 889–891.

Briggs, B. R., Pohlman, J. W., Torres, M., Riedel, M., Brodie, E. L., & Colwell, F. S. (2011). Macroscopic biofilms in fracture-dominated sediment that anaerobically oxidize methane. *Applied and Environmental Microbiology, 77*, 6780–6787.

Ciobanu, M.-C., Burgaud, G. E. T., Dufresne, A., Breuker, A., dou, V. R. E., Ben Maamar, S., et al. (2014). Microorganisms persist at record depths in the subseafloor of the Canterbury Basin. *ISME Journal, 8*, 1370–1380.

Colwell, F. S., Boyd, S., Delwiche, M. E., Reed, D. W., Phelps, T. J., & Newby, D. T. (2008). Estimate of biogenic methane production rates in deep marine sediments at Hydrate Ridge, Cascadia Margin. *Applied and Environmental Microbiology, 74*, 3444–3452.

Cowen, J. P., Giovannoni, S. J., Kenig, F., Johnson, H. P., Butterfield, D., Rappé, M. S., et al. (2003). Fluids from aging ocean crust that support microbial life. *Science, 299*, 120–123.

Deming, J. W., & Baross, J. A. (1993). Deep-sea smokers: Windows to a subsurface biosphere? *Geochimica et Cosmochimica Acta, 57*, 3219–3230.

D'Hondt, S., Inagaki, F., Ferdelman, T. G., Jørgensen, B. B., Kato, K., Kemp, P., et al. (2007). Exploring subseafloor life with the integrated ocean drilling program. *Scientific Drilling, 5*, 26–37.

D'Hondt, S., Jørgensen, B. B., Miller, D. J., Batzke, A., Blake, R., Cragg, B. A., et al. (2004). Distributions of microbial activities in deep subseafloor sediments. *Science, 306*, 2216–2221.

D'Hondt, S., Rutherford, S., & Spivack, A. J. (2002). Metabolic activity of subsurface life in deep-sea sediments. *Science, 295*, 2067–2070.

D'Hondt, S., Spivack, A. J., Pockalny, R., Ferdelman, T. G., Fischer, J. P., Kallmeyer, J., et al. (2009). Subseafloor sedimentary life in the South Pacific Gyre. *Proceedings of the National Academy of Sciences of the United States of America, 106*, 11651–11656.

Edwards, K., Bach, W., Klaus, A., & the IODP Expedition 336 Scientific Party. (2014). IODP Expedition 336: initiation of long-term coupled microbiological, geochemical, and hydrological experimentation within the seafloor at North Pond, western flank of the Mid-Atlantic Ridge. *Scientific Drilling, 17*, 13–18.

Edwards, K. J., Bach, W., & McCollom, T. M. (2005). Geomicrobiology in oceanography: microbe-mineral interactions at and below the seafloor. *Trends in Microbiology*, *13*, 449–456.

Engelen, B., Ziegelmüller, K., Wolf, L., Köpke, B., Gittel, A., Cypionka, H., et al. (2008). Fluids from the oceanic crust support microbial activities within the deep biosphere. *Geomicrobiology Journal*, *25*, 56–66.

Engelhardt, T., Kallmeyer, J., Cypionka, H., & Engelen, B. (2014). High virus-to-cell ratios indicate ongoing production of viruses in deep subsurface sediments. *ISME Journal*, *8*, 1503–1509.

Expedition 310 Scientists. (2007). Tahiti sea level. *IODP Preliminary Report*, *310*. http://dx.doi.org/10.2204/iodp.pr.310.2007.

Expedition 323 Scientists. (2010). Bering Sea paleoceanography: Pliocene–Pleistocene paleocean-ography and climate history of the Bering Sea. *IODP Preliminary Report*, *323*. http://dx.doi.org/10.2204/iodp.pr.323.2010.

Expedition 329 Scientists. (2011). South Pacific Gyre subseafloor life. *IODP Preliminary Report*, *329*. http://dx.doi.org/10.2204/iodp.pr.329.2011.

Expedition 331 Scientists. (2010). Deep hot biosphere. *IODP Preliminary Report*, *331*. http://dx.doi.org/10.2204/iodp.pr.331.3010.

Expedition 336 Scientists. (2012). Mid-Atlantic Ridge microbiology: initiation of long-term cou-pled microbiological, geochemical, and hydrological experimentation within the seafloor at North Pond, western flank of the Mid-Atlantic Ridge. *IODP Preliminary Report*, *336*. http://dx.doi.org/10.2204/iodp.pr.336.3012.

Ferdelman, T. G., Kano, A., Williams, T., & the IODP Expedition 307 Scientists. (2006). IODP Expedition 307 drills cold-water coral mound along the Irish continental margin. *Scientific Drilling*, *2*, 11–16.

Fisk, M. R., Giovannoni, S. J., & Thorseth, I. H. (1998). Alteration of oceanic volcanic glass: tex-tural evidence of microbial activity. *Science*, *281*, 978–980.

Forschner, S., Sheffer, R., Rowley, D., & Smith, D. C. (2009). Microbial diversity in Cenozoic sediments recovered from the Lomonosov Ridge in the Central Arctic basin. *Environmental Microbiology*, *11*, 630–639.

Futagami, T., Morono, Y., Terada, T., Kaksonen, A. H., & Inagaki, F. (2009). Dehalogenation activi-ties and distribution of reductive dehalogenase homologous genes in marine subsurface sedi-ments. *Applied and Environmental Microbiology*, *75*, 6905–6909.

Futagami, T., Morono, Y., Terada, T., Kaksonen, A. H., & Inagaki, F. (2012). Distribution of deha-logenation activity in subseafloor sediments of the Nankai Trough subduction zone. *Philo-sophical Transactions of the Royal Society of London B: Biological Sciences*, *368*, 20120249. http://dx.doi.org/10.1098/rstb.2012.0249 .

Heuer, V. B., Pohlman, J. W., Torres, M. E., Elvert, M., & Hinrichs, K.-U. (2009). The stable carbon isotope biogeochemistry of acetate and other dissolved carbon species in deep sub-seafloor sediments at the northern Cascadia Margin. *Geochimica et Cosmochimica Acta*, *73*, 3323–3336.

Hinrichs, K.-U., Hayes, J. M., Bach, W., Spivack, A. J., Hmelo, L. R., Holm, N. G., et al. (2006). Biological formation of ethane and propane in the deep marine subsurface. *Proceedings of the National Academy of Sciences of the United States of America*, *103*, 14684–14689.

Hinrichs, K.-U., & Inagaki, F. (2012). Downsizing the deep biosphere. *Science*, *338*, 204–205.

Hirose, T., Kawagucci, S., & Suzuki, K. (2011). Mechanoradical H_2 generation during simulated faulting: Implications for an earthquake-driven subsurface biosphere. *Geophysical Research Letters*, *38*, L17302. http://dx.doi.org/10.1029/2011GL048850.

Hoehler, T. M., & Jørgensen, B. B. (2013). Microbial life under extreme energy limitation. *National Reviews Microbiology*, *11*, 83–94.

Hoshino, T., & Inagaki, F. (2012). Molecular quantification of environmental DNA using microfluidics and digital PCR. *Systematic and Applied Microbiology, 35*, 390–395.

Hoshino, T., & Inagaki, F. (2013). A comparative study of microbial diversity and community structure in marine sediments using poly (A) tailing and reverse transcription PCR. *Frontiers in Microbiology, 4*, 160. http://dx.doi.org/10.3389/fmicb.2013.00160.

Hoshino, T., Morono, Y., Terada, T., Imachi, H., Ferdelman, T. G., & Inagaki, F. (2011). Comparative study of subseafloor microbial community structures in deeply buried coral fossils and sediment matrices from the Challenger Mound in the Porcupine Seabight. *Frontiers in Microbiology, 2*, 231. http://dx.doi.org/10.3389/fmicb.2011.00231.

House, C. H., Cragg, B. A., Teske, A., & the Leg 201 Scientific Party. (2003). Drilling contamination tests during ODP leg 201 using chemical and particulate tracers. In S. L. D'Hondt, B. B. Jørgensen, D. J. Miller, et al. (Eds.), *Proc ODP int repts* (201) (pp. 1–19).

Huber, J. A., Butterfield, D. A., & Baross, J. A. (2003). Bacterial diversity in a subseafloor habitat following a deep–sea volcanic eruption. *FEMS Microbiology Ecology, 43*, 393–409.

Huber, R., Stotters, P., Cheminee, J. L., Richnow, H. H., & Stetter, K. O. (1990). Hyperthermophilic archaebacteria within the crater and open-sea plume of erupting Macdonald Seamount. *Nature, 345*, 179–182.

Imachi, H., Aoi, K., Tasumi, E., Saito, Y., Yamanaka, Y., Saito, Y., et al. (2011). Cultivation of methanogenic community from subseafloor sediments using a continuous-flow bioreactor. *ISME Journal, 5*, 1913–1925.

Imachi, H., Sakai, S., Lipp, J. S., Miyazaki, M., Saito, Y., Yamanaka, Y., et al. (2014). *Pelolinea submarina* gen. nov., sp. nov., an anaerobic, filamentous bacterium of the phylum *Chloroflexi* isolated from subseafloor sediment. *International Journal of Systematic and Evolutionary Microbiology, 64*, 812–818.

Inagaki, F., Hinrichs, K.-U., Kubo, Y., & the Expedition 337 Scientists. (2012). Deep coalbed biosphere off Shimokita: microbial processes and hydrocarbon system associated with deeply buried coalbed in the ocean. *IODP Preliminary Report, 337*. http://dx.doi.org/10.2204/iodp.pr.337.2012.

Inagaki, F., Nunoura, T., Nakagawa, S., Teske, A., Lever, M., Lauer, A., et al. (2006). Biogeographical distribution and diversity of microbes in methane hydrate-bearing deep marine sediments on the Pacific Ocean Margin. *Proceedings of the National Academy of Sciences of the United States of America, 103*, 2815–2820.

Inagaki, F., Suzuki, M., Takai, K., Oida, H., Sakamoto, T., Aoki, K., et al. (2003). Microbial communities associated with geological horizons in coastal subseafloor sediments from the Sea of Okhotsk. *Applied and Environmental Microbiology, 69*, 7224–7235.

Integrated Ocean Drilling Program. (2001). Earth, Oceans and Life: scientific investigation of the Earth system using multiple drilling platforms and new technologies. *Integrated Ocean Drilling Program Initial Science Plan, 2003–2013*, 1–109.

Jørgensen, B. B. (2011). Deep subseafloor microbial cells on physiological standby. *Proceedings of the National Academy of Sciences of the United States of America, 108*, 18193–18194.

Jørgensen, B. B., D'Hondt, S. L., & Miller, D. J. (2006). Leg 201 synthesis: Controls on microbial communities in deeply buried sediments. In B. B. Jørgensen, S. L. D'Hondt, & D. J. Miller (Eds.), *Proc ODP sci results* (201) (pp. 1–45).

Kallmeyer, J., Pockalny, R., Adhikari, R. R., Smith, D. C., & D'Hondt, S. (2012). Global distribution of microbial abundance and biomass in subseafloor sediment. *Proceedings of the National Academy of Sciences of the United States of America, 109*, 16213–16216.

Kallmeyer, J., Smith, D. C., Spivack, A. J., & D'Hondt, S. (2008). New cell extraction procedure applied to deep subsurface sediments. *Limnology and Oceanography:Methods, 6*, 236–245.

Kano, A., Ferdelman, T. G., & Williams, T. (2010). The Pleistocene cooling built challenger Mound, a deep-cold water coral mound in the NE Atlantic: synthesis from IODP Expedition 307. *Sedimentary Records*, *8*, 4–9.

Kawagucchi, S., Chiba, H., Ishibashi, J., Yamanaka, T., Toki, T., Muramatsu, Y., et al. (2011). Hydrothermal fluid geochemistry at the Iheya North field in the mid-Okinawa Trough: Implication for origin of methane in subseafloor fluid circulation systems. *Geochemical Journal*, *45*, 109–124.

Kawai, M., Futagami, T., Toyoda, A., Takaki, Y., Nishi, S., Hori, S., et al. (2014). High frequency of phylogenetically diverse reductive dehalogenase-homologous genes in deep subseafloor sedimentary metagenomes. *Frontiers in Microbiology*, *5*, 80. http://dx.doi.org/10.3389/fmicb.2014.00080.

Langerhuus, A. T., Røy, H., Lever, M. A., Morono, Y., Inagaki, F., Jørgensen, B. B., et al. (2012). Endospore abundance and D: L-amino acid modeling of bacterial turnover in Holocene marine sediment (Aarhus Bay). *Geochimica et Cosmochimica Acta*, *99*, 87–99.

Lever, M. A. (2012). Acetogenesis in the energy-starved deep biosphere—a paradox? *Frontiers in Microbiology*, *2*, 284. http://dx.doi.org/10.3389/fmicb.2011.00284.

Lever, M. A., Alperin, M., Engelen, B., Inagaki, F., Nakagawa, S., Steinsbu, B. R. O., et al. (2006). Trends in basalt and sediment core contamination during IODP Expedition 301. *Geomicrobiology Journal*, *23*, 517–530.

Lever, M. A., Rouxel, O., Alt, J. C., Shimizu, N., Ono, S., Coggon, R. M., et al. (2013). Evidence for microbial carbon and sulfur cycling in deeply buried ridge flank basalt. *Science*, *339*, 1305–1308.

Lipp, J. S., Morono, Y., Inagaki, F., & Hinrichs, K.-U. (2008). Significant contribution of Archaea to extant biomass in marine subsurface sediments. *Nature*, *454*, 991–994.

Lloyd, K. G., Schreiber, L., Petersen, D. G., Kjeldsen, K. U., Lever, M. A., Steen, A. D., et al. (2013a). Predominant archaea in marine sediments degrade detrital proteins. *Nature*, *496*, 215–218.

Lloyd, K. G., May, M. K., Kevorkian, R. T., & Steen, A. D. (2013b). Meta-analysis of quantification methods shows that Archaea and Bacteria have similar abundances in the subseafloor. *Applied and Environmental Microbiology*, *79*, 7790–7799.

Lomstein, B. A., Langerhuus, A. T., D'Hondt, S., Jørgensen, B. B., & Spivack, A. J. (2012). Endospore abundance, microbial growth and necromass turnover in deep sub-seafloor sediment. *Nature*, *484*, 101–104.

Masui, N., Morono, Y., & Inagaki, F. (2009). Bio-archive core storage and subsampling procedure for subseafloor molecular biological research. *Scientific Drilling*, *8*, 35–37.

McCollom, T. M. (2007). Geochemical constraints on sources of metabolic energy for chemolithoautotrophy in ultramafic-hosted deep-sea hydrothermal systems. *Astrobiology*, *7*, 933–950.

Ménez, B., Pasini, V., & Brunelli, D. (2012). Life in the hydrated suboceanic mantle. *Nature Geoscience*, *5*, 1–5.

Mills, H. J., Mills, H. J., Reese, B. K., Shepard, A. K., Riedinger, N., Dowd, S. E., et al. (2012). Characterization of metabolically active bacterial populations in subseafloor Nankai Trough sediments above, within, and below the sulfate–methane transition zone. *Frontiers in Microbiology*, *3*, 113. http://dx.doi.org/10.3389/fmicb.2012.00113.

Morita, R. Y., & ZoBell, C. L. (1955). Occurrence of bacteria in pelagic sediments collected during the Mid-Pacific expedition. *Deep-Sea Research*, *3*, 66–73.

Morono, Y., Terada, T., Hoshino, T., & Inagaki, F. (2014). Hot-alkaline DNA extraction method for deep-subseafloor archaeal communities. *Applied and Environmental Microbiology*, *80*, 1985–1994.

Morono, Y., Terada, T., Kallmeyer, J., & Inagaki, F. (2013). An improved cell separation technique for marine subsurface sediments: applications for high-throughput analysis using flow cytometry and cell sorting. *Environmental Microbiology*, *15*, 2841–2849.

Morono, Y., Terada, T., Nishizawa, M., Ito, M., Hillion, F., Takahata, N., et al. (2011). Carbon and nitrogen assimilation in deep subseafloor microbial cells. *Proceedings of the National Academy of Sciences of the United States of America, 108*, 18295–18300.

Morono Y., Terada T., Yamamoto Y., Xiao N., Hirose T., Sugeno M., Ohwada N., & Inagaki F. Intact preservation of environmental samples by freezing under alternating magnetic field. *Environmental Microbiology Reports,* in press.

Nakagawa, S., Inagaki, F., Suzuki, Y., Steinsbu, B. O., Lever, M. A., Takai, K., et al. (2006). Microbial community in black rust exposed to hot ridge flank crustal fluids. *Applied and Environmental Microbiology, 72*, 6789–6799.

Nakagawa, S., Takai, K., Inagaki, F., Chiba, H., Ishibashi, J.-I., Kataoka, S., et al. (2005). Variability in microbial community and venting chemistry in a sediment-hosted backarc hydrothermal system: Impacts of subseafloor phase-separation. *FEMS Microbiology Ecology, 54*, 141–155.

Nealson, K. H., Inagaki, F., & Takai, K. (2005). Hydrogen-driven subsurface lithoautotrophic microbial ecosystems (SLiMEs): do they exist and why should we care? *Trends in Microbiology, 13*, 405–410.

Nunoura, T., Soffientino, B., Blazejak, A., Kakuta, J., Oida, H., Schippers, A., & Takai, K. (2009). Subseafloor microbial communities associated with rapid turbidite deposition in the Gulf of Mexico continental slope (IODP Expedition 308). *FEMS Microbiology Ecology, 69*, 410–424.

Orcutt, B. N., Bach, W., Becker, K., Fisher, A. T., Hentscher, M., Toner, B. M., et al. (2011). Colonization of subsurface microbial observatories deployed in young ocean crust. *ISME Journal, 5*, 692–703.

Orcutt, B., Wheat, C. G., & Edwards, K. J. (2010). Subseafloor ocean crust microbial observatories: development of FLOCS (FLow-through osmo colonization system) and evaluation of borehole construction materials. *Geomicrobiology Journal, 27*, 143–157.

Orcutt, B. N., Wheat, C. G., Rouxel, O., Hulme, S., Edwards, K. J., & Bach, W. (2013). Oxygen consumption rates in subseafloor basaltic crust derived from a reaction transport model. *Nature Communications, 4*, 1–8.

Orsi, W., Biddle, J. F., & Edgcomb, V. (2013a). Deep sequencing of subseafloor eukaryotic rRNA reveals active fungi across marine subsurface provinces. *PLoS One, 8*, e56335.

Orsi, W. D., Edgcomb, V. P., Christman, G. D., & Biddle, J. F. (2013b). Gene expression in the deep biosphere. *Nature, 499*, 205–208.

Parkes, R. J., Cragg, B. A., Bale, S. J., Getliff, J. M., Goodman, K., Rochelle, P. A., et al. (1994). Deep bacterial biosphere in Pacific Ocean sediments. *Nature, 371*, 410–413.

Parkes, R. J., Cragg, B. A., & Wellsbury, P. (2000). Recent studies on bacterial populations and processes in subseafloor sediments: a review. *Hydrogeology Journal, 8*, 11–28.

Parkes, R., Sellek, G., Webster, G., Martin, D., Anders, E., Weightman, A., et al. (2009). Culturable prokaryotic diversity of deep, gas hydrate sediments: first use of a continuous high-pressure, anaerobic, enrichment and isolation system for subseafloor sediments (DeepIsoBUG). *Environmental Microbiology, 11*, 3140–3153.

Parkes, R. J., Webster, G., Cragg, B. A., Weightman, A. J., Newberry, C. J., Ferdelman, T. G., et al. (2005). Deep sub-seafloor prokaryotes stimulated at interfaces over geological time. *Nature, 436*, 390–394.

Riedinger, N., Brunner, B., Formolo, M. J., Solomon, E., Kasten, S., Stasser, M., et al. (2010). Oxidative sulfur cycling in the deep biosphere of the Nankai Trough, Japan. *Geology, 38*, 851–854.

Røy, H., Kallmeyer, J., Adhikari, R. R., Pockalny, R., Jørgensen, B. B., & D'Hondt, S. (2012). Aerobic microbial respiration in 86-million-year-old deep-sea red clay. *Science, 336*, 922–925.

Schippers, A., Neretin, L. N., Kallmeyer, J., Ferdelman, T. G., Cragg, B. A., John Parkes, R., et al. (2005). Prokaryotic cells of the deep sub-seafloor biosphere identified as living bacteria. *Nature, 433*, 861–864.

Sleep, N. H., & Zoback, M. D. (2007). Did earthquakes keep the early crust habitable? *Astrobiology*, *7*, 1023–1032.

Soffientino, B., Spivack, A. J., Smith, D. C., & DHondt, S. (2009). Hydrogenase activity in deeply buried sediments of the Arctic and North Atlantic Oceans. *Geomicrobiology Journal*, *26*, 537–545.

Sørensen, K. B., & Teske, A. (2006). Stratified communities of active archaea in deep marine subsurface sediments. *Applied and Environmental Microbiology*, *72*, 4596–4603.

Staudigel, H., Furnes, H., McLoughlin, N., Banerjee, N. R., Connell, L. B., & Templeton, A. (2008). 3.5 billion years of glass bioalteration: volcanic rocks as a basis for microbial life? *Earth-Science Reviews*, *89*, 156–176.

Steinsbu, B. O., Thorseth, I. H., Nakagawa, S., Inagaki, F., Lever, M. A., Engelen, B., et al. (2010). *Archaeoglobus sulfaticallidus* sp. nov., a thermophilic and facultatively lithoautotrophic sulfate-reducer isolated from black rust exposed to hot ridge flank crustal fluids. *International Journal of Systemic and Evolutionary Microbiology*, *60*, 2745–2752.

Tada, R., Murray, R. W., & Alvarez Zarikian, C. A. (2013). Asian Monsoon: onset and evolution of millennial-scale variability of Asian monsoon and its possible relation with Himalaya and Tibetan plateau uplift. *IODP Scientific Prospectus*, *346*. http://dx.doi.org/10.2204/iodp.sp.346.2013.

Takai, K., Gamo, T., Tsunogai, U., Nakayama, N., Hirayama, H., Nealson, K. H., et al. (2004). Geochemical and microbiological evidence for a hydrogen-based, hyperthermophilic subsurface lithoautotrophic microbial ecosystem (HyperSLiME) beneath an active deep-sea hydrothermal field. *Extremophiles*, *8*, 269–282.

Takai, K., Mottle, M. J., Nielsen, S. H. H., & the IODP Expedition 331 Scientists. (2010). IODP Expedition 331: Strong and expansive subseafloor hydrothermal activities in the Okinawa Trough. *Scientific Drilling*, *13*, 19–27.

Takano, Y., Chikaraishi, Y., Ogawa, N. O., Nomaki, H., Morono, Y., Inagaki, F., et al. (2010). Sedimentary membrane lipids recycled by deep-sea benthic archaea. *Nature Geoscience*, *3*, 858–861.

Teske, A. (2006). Microbial communities of deep Marine subsurface sediments: molecular and cultivation surveys. *Geomicrobiology Journal*, *23*, 357–368.

Teske, A., & Sørensen, K. B. (2008). Uncultured archaea in deep marine subsurface sediments: have we caught them all? *ISME Journal*, *2*, 3–18.

Thorseth, I. H., Torsvik, T., Furnes, H., & Muehlenbachs, K. (1995). Microbes play an important role in the alteration of oceanic crust. *Chemical Geology*, *126*, 137–146.

Toki, T., Uehara, Y., Kinjo, K., Ijiri, A., Tsunogai, U., Tomaru, H., et al. (2012). Methane production and accumulation in the Nankai accretionary prism: results from IODP Expeditions 315 and 316. *Geochemical Journal*, *46*, 89–106.

Webster, G., Blazejak, A., Cragg, B. A., Schippers, A., Sass, H., Rinna, J., et al. (2009). Subsurface microbiology and biogeochemistry of a deep, cold-water carbonate mound from the Porcupine Seabight (IODP Expedition 307). *Environmental Microbiology*, *11*, 239–257.

Webster, G., John Parkes, R., Cragg, B. A., Newberry, C. J., Weightman, A. J., & Fry, J. C. (2006). Prokaryotic community composition and biogeochemical processes in deep subseafloor sediments from the Peru Margin. *FEMS Microbiology Ecology*, *58*, 65–85.

Wehrmann, L. M., Arndt, S., März, C., Ferdelman, T. G., & Brunner, B. (2013). The evolution of early diagenetic signals in Bering Sea subseafloor sediments in response to varying organic carbon deposition over the last 4.3 Ma. *Geochimica et Cosmochimica Acta*, *109*, 175–196.

Wehrmann L.M., Ockert C., Mix A.C., Gussone N., Teichert B.M.A., & Meister P. Repeated occurrences of methanogenic zones, diagenetic dolomite formation and linked silicate alteration in southern Bering Sea sediments (Bowers Ridge, IODP Exp. 323 Site U1341). *Deep-Sea Research Part*, in press. http://dx.doi.org/10.1016/j.dsr2.2013.09.008

Whitman, W. B., Coleman, D. C., & Wiebe, W. J. (1998). Prokaryotes: the unseen majority. *Proceedings of the National Academy of Sciences of the United States of America, 95,* 6578–6583.

Yanagawa, K., Morono, Y., Yoshida-Takashima, Y., Eitoku, M., Sunamura, M., Inagaki, F., et al. (2014a). Variability of subseafloor viral abundance at the geographically and geologically distinct continental margins. *FEMS Microbiology Ecology, 88,* 60–68.

Yanagawa, K., Breuker, A., Schippers, A., Nishizawa, M., Ijiri, A., Hirai, M., et al. (2014b). Microbial community stratification controlled by the subseafloor fluid flow and geothermal gradient at the Iheya North hydrothermal field in the Mid-Okinawa Trough (IODP Expedition 331). *Applied and Envionmental Microbiology, 80,* 6126–6135.

Yoshida, M., Takaki, Y., Eitoku, M., Nunoura, T., & Takai, K. (2013). Metagenomic analysis of viral communities in (Hado)Pelagic sediments. *PLoS One, 8,* e57271.

Yoshioka, H., Maruyama, A., Nakamura, T., Higashi, Y., Fuse, H., Sakata, S., et al. (2010). Activities and distribution of methanogenic and methane-oxidizing microbes in marine sediments from the Cascadia Margin. *Geobiology, 8,* 223–233.

Chapter 2.2.1

Biomass, Diversity, and Metabolic Functions of Subseafloor Life

Detection and Enumeration of Microbial Cells in Subseafloor Sediment

Yuki Morono[1,*] and Jens Kallmeyer[2]

[1]Kochi Institute for Core Sample Research, Japan Agency for Marine-Earth Science Technology (JAMSTEC), Nankoku, Kochi, Japan; [2]Deutsches GeoForschungsZentrum GFZ, Section 4.5 Geomicrobiology, Telegrafenberg, Potsdam, Germany
*Corresponding author: E-mail: morono@jamstec.go.jp

2.2.1.1 THE HISTORY OF DETECTION AND ENUMERATION OF MICROBIAL CELLS IN DEEP SUBSEAFLOOR SEDIMENT

Although Morita and ZoBell did not use the term "deep subseafloor biosphere," they were the first to study microbial life in deep marine sediments by attempting to quantify the abundance and distribution of microbes in pelagic Pacific Ocean sediments (Morita & ZoBell, 1955). Due to their inability to culture bacteria from greater depths, they defined the depth limit of the marine biosphere as occurring at 7.47 m below the seafloor (mbsf). Despite the fact that their definition of the depth limit of the biosphere was quite inaccurate, a number of conclusions and questions that arose from their findings remain the subject of intense study and debate today (1) The central Pacific Ocean has a very low sedimentation rate (2–3 mm/ky), and therefore, microbes found in the upper few meters of the seafloor sediment must have been viable for hundreds of thousands to millions of years; (2) the decrease in the number of cells with depth reflects the decreasing availability of organic carbon. For over two decades, Morita and ZoBell's conclusion that the depth limit of the biosphere occurs only a few meters below the sediment surface was not challenged despite the fact that even Morita and ZoBell already stated in their paper that they most probably underestimate abundance due to the limitations of the culture media. However, they did not consider the inherent limitations of culturing to affect their interpretation of the lower limits of the biosphere. The first indications that microbes are active in deeper sediment layers came not from microbiological studies but from geochemical data obtained from drill cores recovered by the first ocean-drilling program, the Deep-Sea Drilling Program (DSDP). Hathaway et al. (1979) studied sulfate and methane pore water profiles down to 300 mbsf. Upon finding striking similarities to profiles reported for near-shore locations (Martens & Berner, 1974), they postulated that microbial activity could occur in 20-My-old sediments. Oremland et al. (1982) reported methanogenic microbial activity down to about 20 mbsf in DSDP cores. Through inhibition experiments, they clearly showed that methane is generated biologically. Whelan et al. (1986) conducted experiments in which they amended sediment from DSDP cores with radiolabeled substrates and detected the formation of radiolabeled products down to a depth of 167 mbsf. The turnover rates in the deep cores were several orders of magnitude lower than in surface sediments. They detected sulfate reduction, methanogenesis, and fermentation. Reduction of CO_2 by hydrogen was suggested as an important pathway of methane production in deep sediments. Despite the geochemical evidence for an active microbial biosphere in marine sediments, there were no quantitative data regarding its biomass, extent, and limits.

Based on mostly theoretical considerations and observations from hydrocarbon reservoirs and deep terrestrial mines, Gold (1992) postulated that microbial life is widespread at depths in the crust of the Earth. This life was supposed to be independent of the surface, as it does not rely on solar energy and photosynthesis for its primary energy supply but instead depends upon chemical sources derived from fluids that migrate upward from deeper levels. Gold's publication introduced the term "deep biosphere" as a general term for subsurface life.

Beginning in the late 1980s, the group of John Parkes began quantifying microbial biomass and activity in deep Ocean Drilling Program (ODP) sediment cores, using Acridine Orange Direct Counting (AODC) for biomass estimates and radiotracer incubations for turnover rate measurements (review in Parkes, Cragg et al., 2000). Initially, all cell counts of ODP core samples were performed by Parkes' research group, mainly by Barry Cragg (Cragg et al., 1990, 1992, 1995a, 1995b, 1997, 1998, 1999, 2000, 2002; Cragg & Parkes, 1993, 1994; Cragg, 1994; Cragg & Kemp, 1995; Roussel et al., 2008; Wellsbury, et al., 2000). The fact that all measurements were carried out by a single research group (and primarily by the same person) created a data set of almost unrivaled consistency. This work culminated in a seminal paper that proved the existence of a widespread and diverse microbial community in deep subseafloor sediments (Parkes et al., 1994). This publication can be considered as the starting point for systematic microbiological subseafloor exploration.

Gold (1992) speculated about the fraction of Earth's total living biomass in subsurface microbes. Based on the bacterial counts obtained from ODP cores, Parkes et al. (1994) estimated that although the biomass of the entire subseafloor biosphere accounts for only 0.004% of global sedimentary organic carbon, it accounts for approximately 10% of the living biomass in the global biosphere. The profiles of bacterial distribution were consistent between cores collected from different locations in the Pacific Ocean. This database was later extended to other oceans (see Parkes et al., 2000 for a review and Parkes et al., 2014 for an update).

Whitman et al. (1998) calculated the first global estimate of living biomass. Based on ODP cell count data from the Parkes group, they estimated that the subseafloor sedimentary microbial abundance was $35.5 \cdot 10^{29}$ cells, comprising 55–86% of Earth's prokaryotic biomass and 27–33% of Earth's living biomass. Recent advances in cell counting (Kallmeyer et al., 2008; Morono et al., 2013) and exploration of new sampling sites (D'Hondt et al., 2009) led to a reassessment of the global microbial subseafloor distribution. Compared to previous studies, this new estimate (Kallmeyer et al., 2012) reduced the total number of microbes and the total living biomass by 50–78% and 10–45%, respectively. Although the biomass and the number of subseafloor microbes were significantly reduced by this new estimate, Jørgensen stated that "…the importance of these deeply buried communities for driving carbon and nutrient cycling and catalyzing a multitude of reactions between rocks, sediments, and fluids is not challenged (by the new biomass assessment). Neither is the persisting enigma of slow life beneath the surface of Earth. The deep biosphere is still alive" (Jørgensen, 2012).

A major issue when counting microbial cells in deep subsurface sediments is the minimum quantification limit (MQL), which is defined by statistically meaningful range of the quantity data, of the counting method used. Most early counts were done with AODC, and this method had an MQL of around 10^5–10^6 cells/cm^3. In several earlier studies, for example, on ODP Leg 190, Site 1173 (Moore et al., 2001) cell counts fell below the MQL. Although cells were still present at these depths, the data could not be used for global estimates. By excluding counts that fell below the MQL, the data population becomes biased toward higher cell abundances. When calculating the rate at which cell abundance decreases with depth, the trend will indicate a smaller decrease with depth. Even with new counting techniques that have a much lower (but still have a) MQL, this problem is not completely remedied, but rather shifted toward lower values. Integrated Ocean Drilling Program (IODP) Exp. 329 to the South Pacific Gyre provides a good example. The MQL for on-board counts was 10^3 cells/cm^3, but at most sites, cell abundance fell below this limit at greater depths (Expedition 329 Scientists, 2011). Figure 2.2.1.1 gives an overview of subseafloor cell counts from ODP/IODP cores and other expeditions. The majority of ODP/IODP counts done with AODC (black circles) lie above the MQL of 10^5–10^6 cells/cm^3. Cell counts from the Bering Sea (IODP

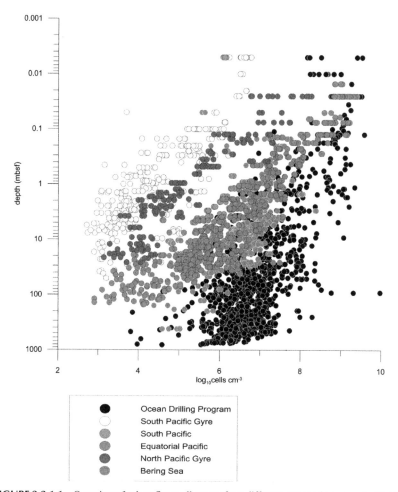

FIGURE 2.2.1.1 Overview of subseafloor cell counts from different areas and processed with different techniques. All Ocean Drilling Program counts (black circles) were done with AODC, which has an minimum quantification limit around 10^5–10^6 cells/cm^3. All other counts were done with SYBR-I, and most counts employed a cell separation prior to counting, which has an MQL around 10^3 cells/cm^3. See the text for details.

Exp. 323, gray circles in Figure 2.2.1.1) were done using a cell extraction and therefore have a lower MQL. They follow the general trend of the ODP/IODP data in the upper part but continue toward lower cell abundances. The South Pacific Gyre data (Figure 2.2.1.1, yellow circles) are sharply cut off at 10^3 cells/cm^3, reflecting the on-board MQL. Calculating a rate of decrease by using only the counts above the MQL would be different in comparison to a trend that includes also the counts below the MQL. The competition between using only statistically valid numbers on the one side and providing a more realistic, albeit statistically less safe estimate on the other can only be solved by improving counting techniques. Recent developments (see section "Lowering the quantification limit" in this chapter) might provide a solution for this problem.

2.2.1.2 TECHNICAL CHALLENGES IN ESTIMATING BIOMASS AND MICROBIAL DIVERSITY IN SUBSEAFLOOR ENVIRONMENTS

One of the major difficulties in finding life in the subseafloor biosphere is that organisms are very closely associated with the geological matrix (i.e., sediment grains or rocks). Even assuming that 1 cc of sediment may harbor 10^9 spherical and/or slightly rod-shaped cells with a size of around 0.5 μm (resulting in a biovolume of $6.5–9.8 \times 10^{-2}$ μm^3), the microbial biomass would account for <0.01% of the total sediment volume. Finding and enumerating microbial cells representing such a tiny proportion of an overwhelmingly large volume of sediment matrix is not an easy task. However, through proper sample treatment (subsampling, fixing, staining, etc.), microbial cells can be detected and made visible within the proverbial "forest" of matrix grains (Kepner & Pratt, 1994).

As already mentioned above, the conventional and most common method for counting cells is to observe samples under a fluorescence microscope after staining with a DNA-specific dye, the most common ones being 4′,6-diamidino-2-phenylindole (DAPI, Porter & Feig, 1980), acridine orange (AO, Daley & Hobbie, 1975), and SYBR Green I (SYBR-I, Noble & Fuhrman, 1998). Fluorescence microscopy has been successfully implemented in the analysis of cells from aquatic environments (e.g., Patel et al., 2007); however, studies of geological samples (i.e., sediment and rock) have long been hampered by difficulties in discriminating cell-derived signals due to nonspecific background fluorescence associated with the geological matrix. The natural background native fluorescence from minerals prevents specific recognition of microbial cells in geological samples. A wide range of minerals fluoresce under ultraviolet (UV) excitation (Robbins, 1994). DAPI, one of the most commonly used cell stains, also fluoresces under UV excitation, and therefore causes problems in detecting sediment-associated microbial cells. For this reason, DAPI is not commonly used for such analyses.

Another type of background fluorescence is caused by nonspecific binding of dyes to the geological matrix. SYBR-I is now considered to be a more effective fluorescent dye than AO for cell enumeration in marine sediments due to its higher fluorescence intensity (Engelen et al., 2008; Lunau et al., 2005). However, many particles are found to emit bright fluorescence after SYBR-I staining, even in heat-combusted samples (Morono et al., 2009). SYBR-I is known to exhibit a high fluorescence quantum yield (Johnson & Spence, 2010) upon forming a complex with DNA molecules, and its background fluorescence is quite low. However, a high accumulation of SYBR-I dye molecules could increase the fluorescence intensity of SYBR-stainable particulate matter (SYBR-SPAM), such that these particles become visible upon microscopic observation (Morono et al., 2013). Signals associated with SYBR-SPAM are generally difficult to distinguish from cell-derived SYBR-I fluorescence signals, which can result in a serious overestimation of cell abundance or inconsistencies in data generated by different investigators. However, Morono et al. (2009) found that the fluorescence spectrum of SYBR-SPAM shifts toward longer wavelengths (Figure 2.2.1.2). Based on images taken from two different fluorescence wavelength regions using an automated microscope system and dividing the green fluorescence signal by the red signal, they could highlight only signals from dsDNA-SYBR complexes and thereby discriminatively enumerate living cells (Morono et al., 2009; Morono & Inagaki, 2010).

To prepare samples suitable for microscopic observation, fixed samples (slurries) are placed on membrane filters with pore sizes smaller than microbial cells (usually 0.2 μm). For enumeration of viruses, a much smaller pore size

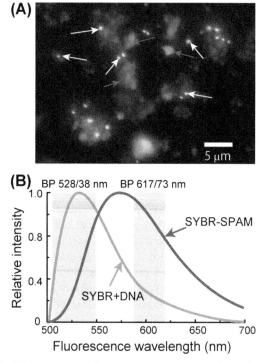

FIGURE 2.2.1.2 (A) Examples of fluorescence-producing cellular and noncellular objects stained with SYBR-I. Red (gray in print versions) arrows = yellowish (grayish in print versions) SYBR-SPAM. White arrows = green (white in print versions) *Escherichia coli* cells. (B) Spectrum patterns show "red shift" of SYBR-I fluorescence. When SYBR-I binds to SPAM, the fluorescence spectra shift to longer wavelengths overall (red line (dark-gray line in print versions)), whereas the fluorescence spectra of cell-derived SYBR-I (solid green line (light-gray line in print versions)) show little or no shift relative to the original SYBR-I spectrum.

must be used (usually 0.02 μm; Noble & Fuhrman, 1998). Due to their extremely small size, virus enumeration requires highly specialized sample preparation. Despite recent advances in quantifying viral abundance in marine sediment (Danovaro et al., 2002; Engelhardt et al., 2014; Yanagawa et al., 2014), reports are still scarce due to the technical challenges involved. We therefore focus only on microbial (i.e., >0.2 μm in size) cell counting in this chapter. Since microbes can only be observed through direct visualization (i.e., they cannot be observed when obstructed by other particles), any cell behind a matrix particle will remain uncounted. Therefore, during sample analysis, all cells on sediment grains (Figure 2.2.1.1) should be detached and preferably separated from the mineral matrix prior to placing them on the filter. There are, however, statistical methods to account for particle-associated cells that are shielded from view (Fry, 1988).

Over the last several decades, many methods for separating microbial cells from soil matrix or shallow sediment particles for various downstream applications have been published (Barra Caracciolo et al., 2005; Berry et al., 2003; Bertaux et al., 2007; Delmont et al., 2011; Frischer et al., 2000; Gabor et al., 2003; Katayama et al., 1998; Lindahl, 1996; Maron et al., 2006; Prieme et al., 1996; Unge et al., 1999; Zaporozhenko et al., 2006). Kallmeyer et al. (2008) introduced a cell extraction technique specifically optimized for subseafloor biosphere samples. Their method uses a combined chemical and mechanical cell detachment approach followed by separation of cells from the sediment matrix by density centrifugation. After refining of the method, Morono et al. (2013) were able to place two orders of magnitude more sediment on the membrane and bring the MQL down to 10^3 cells/cm^3.

From the author's experience, cells on sandy matrix particles are difficult to detach and require a longer treatment time. Lappé and Kallmeyer (2011) recently reported a modified sample treatment protocol for oily sediments that involves washing samples with organic solvent prior to processing for cell counting. Silica-depositing microbes (Inagaki et al., 2003) or microbes residing in rock (Lever et al., 2013) are strongly associated with the matrix and are thus difficult to detach. However, we should keep in mind that overly strong chemical or mechanical treatments can cause lysis of cells. Therefore, although the cell detachment procedure is important, it may need to be optimized depending on the type of sample.

2.2.1.3 COUNTING STATISTICS

For cell enumeration based on microscopic counting, counting statistics are the next critical factor when enumerating low-abundance subseafloor microbes. Kirchman et al., (1982) were the first to discuss the statistical problems associated with counting low cell abundances. Although their paper focuses on samples from the water column with much higher cell numbers than of deep subseafloor sediment, the fundamental problems are already mentioned.

With the aforementioned filter protocol, the MQL of DNA-containing particles (i.e., archaea and bacteria, and viruses) is constrained by the volume of sediment that can be loaded onto the filter and the time required for counting a certain area of the filter. For the commonly used filter membranes (25 mm in diameter, roughly $2.0\,cm^2$ of filterable surface area), $3.5 \times 10^{-3}\,cm^3$ of sediment was found to be the maximum load (Morono & Inagaki, 2010), although this number may vary depending on the composition and grain size of the sample. Loading more sediment onto the membrane results in stacking of sediment particles, which shields other particles from view and causes a loss of focus throughout the microscopic field because the particles are distributed beyond the depth of field. Assuming 200 microscopic images are used in a cell counting experiment, this would correspond to observing $1.3 \times 10^{-4}\,cm^3$ of sediment. According to Kallmeyer et al. (2008), to detect one cell in such an analysis with a probability of 95%, 83 cells must be on the membrane, which corresponds to $2.4 \times 10^4\,cells/cm^3$. To enumerate cells at a lower abundance (i.e., at or below MQL), the volume of sediment analyzed must be increased, and therefore, an increase in the number of imaging fields (i.e., the field of view) is necessary. To reach an MQL of $1 \times 10^2\,cells/cm^3$, 4.5×10^4 imaging fields must be counted, equivalent to seven 25-mm-diameter membranes. From a practical standpoint, a reasonable range for this simple filtration approach would be $10^4-10^5\,cells/cm^3$ (Kallmeyer, 2011).

2.2.1.4 OVERCOMING THE LIMITATIONS

The development of methods for separating cells from geological matrix particles enabled the detection and enumeration of microbes using flow cytometry (FCM). FCM is a powerful tool for identifying and enumerating fluorescent-labeled cells on the basis of size, fluorescence intensity, and wavelength. FCM is commonly used in the medical sciences and has been used to study the ecology of microbial communities in a variety of aquatic environments (Irvine-Fynn et al., 2012; Miteva & Brenchley, 2005; Monger & Landry, 1993; Porter et al., 1995; Wang et al., 2010). However, due to analytical interferences associated with nonbiological particles such as mineral grains, until recently, it was not possible to analyze sediment and soil samples using FCM. Morono et al. (2013) developed an improved cell separation technique involving multiple density layers to achieve quantitative recovery of microbial cells from sediment samples for enumeration using FCM. Although some noncellular particles remain after cell separation using this approach, the green-fluorescing cells are clearly recognizable and can be enumerated, with results that do not significantly differ from microscopic count data (Figure 2.2.1.3). The advantage of using FCM is that it is a flow-through type of measurement. Although the current protocol involves filter trapping at the SYBR-I staining step (Morono et al., 2013) because sufficient staining cannot be achieved in suspension, the method still provides a time advantage in terms of the data acquisition capacity. FCM

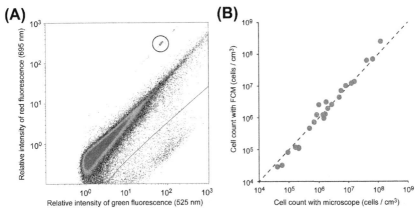

FIGURE 2.2.1.3   Flow cytometry (FCM) analysis of density-separated sediment samples. (A) FCM cytogram (scatter plot) of a sample collected off the Shimokita Peninsula, Japan (Site C9001, core 1H-1) following separation and staining with SYBR-I. The lower right portion below the solid line shows the region of cell-derived fluorescence signals. Signals surrounded by a solid circle are from volume calibration beads. (B) The number of microbial cells as determined by microscopic counting and FCM. The dotted line shows the 1:1 line for the counts determined by microscopy and FCM. *Images are reproduced from Morono et al. (2013).*

can acquire 10,000–100,000 events (i.e., particle data points) per second, which is orders of magnitude faster than even the most advanced automated micros-copy-based systems. Although complete discrimination of the background sig-nals remains problematic, the high-throughput capability of FCM gives it great potential for future application.

It should be noted that in all the protocols for detecting and enumerating microbial cells in geological samples, detachment of the cells from matrix par-ticles is of the utmost importance. With sufficient detachment from the sediment matrix, Morono et al. (2013) were able to enumerate cells using FCM without cell separation, although the MQL remained high, at 10^5 cells/cm^3.

2.2.1.5 COMBATING CONTAMINATION

Obtaining pristine samples without contamination is of primary importance. Methodological details with practical explanations can be found in Kallmeyer (2011). For sediments, a small volume of sample (usually 2 or 3 cm^3) is taken from the drilled (cored) material with sterile tip-cut syringes as soon as possible. On the drill ship, this is usually done right after retrieval of the drilled core from the center of the freshly cut core sections. This procedure is well established and has the advantage of allowing the exact determina-tion of the sample volume, enabling proper calculation of cell number per unit volume, even on moving ships in heavy seas. The sample is then chemi-cally deactivated (fixed) by suspending it in a solution of the same salinity (usually sterile-filtered seawater) containing an aldehyde at a concentration of a few single percent to prevent sample deterioration or growth of the microbes during sample storage (Kallmeyer, 2011). These steps must be carried out quickly (i.e., directly on site, immediately after cutting the core) to preserve the in situ microbial biomass.

In cases where a particle tracer like fluorescent microspheres is used to monitor contamination (Griffin et al., 1997), extra care must be taken to avoid spreading the microspheres through the entire laboratory and causing false positives in the blank samples. It is unavoidable that there will be microsphere-containing drill mud in the small gap between the outside of the core and the inside of the liner. When sectioning a core, this drill mud will almost certainly seep out and thereby contaminate surfaces, tools gloves, etc. On a normal laboratory bench, the outside of the core can be shaved off with disposable tools relatively easily and disposed. However, when sectioning inside a glove box or glove bag to avoid any contact with atmospheric oxygen, things look quite different. Even with the greatest care possible, some contamination will occur almost certainly. In the closely confined space of a glove box, drying mud can liberate microspheres into the atmosphere, from where they can land on any surface (or sample). A good way to ensure minimum crosscontamination inside a glove box is to install a standard household air purifier with an HEPA (high efficiency particulate airfilter) filter. Purifiers are easily available in many shapes and sizes, which makes it easy to fit them inside a glove box. Experience shows that the number of airborne particles inside a glove box can be reduced significantly (Kallmeyer et al., 2006).

2.2.1.6 LOWERING THE QUANTIFICATION LIMIT

How can we lower the MQL further? Obtaining a clean laboratory environment is an important but often overlooked issue. The MQL is determined in part by the mean number (and standard deviation) of microbial cells found in negative controls containing no sample (usually the sample is replaced with clean buffer or heat-combusted sediment) but is otherwise treated the same as the actual samples from the environment. Count values below the MQL are not statistically reliable because of the chance of counting airborne contaminants. Typical laboratory air contains >20,000 0.5-μm particles per cubic foot. According to US Federal Standard 209D (US GSA 1988), such an environment corresponds to class 100,000. Even if researchers pay careful attention to ensure preparation of clean samples, particles floating in the laboratory air can get into the samples, resulting in a high MQL. To obtain a clean working environment, there are several options, such as the use of a clean bench, clean booth, and/or clean room. Placing a clean bench within a clean booth (Figure 2.2.1.4) is an effective way to obtain an extraclean space with minimal investment; no particles (even 0.3-μm in size) were detected with a commercial dust monitor inside the clean bench shown in the figure.

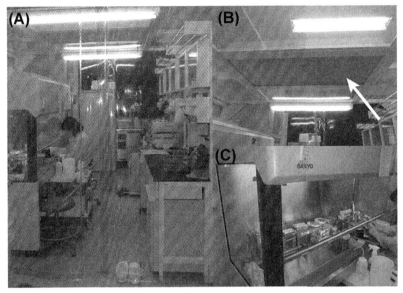

FIGURE 2.2.1.4   (A) Clean experimental booth installed at the laboratory of the Kochi Institute for Core Sample Research, Japan Agency for Marine-Earth Science and Technology (JAMSTEC). (B) High Efficiency Particulate Airfilter (HEPA) filter (white arrow) for the clean booth, (C) Clean bench located inside the clean experimental booth, where the level of floating dust is extremely low.

Another point to consider is the need to eliminate static electricity on experimental materials. Although the use of presterilized plastic laboratory ware is important to reduce sample contamination, such laboratory ware is easily electrostatically charged and thus may attract airborne particles. Therefore, placing a static electricity remover in the working environment is another effective measure to reduce the MQL. Using the aforementioned approach, the MQL in the JAMSTEC Kochi laboratory was reduced to 10^2 cells/cm^3.

To date, the record depth for finding microbial life in the subseafloor environment is 1922 mbsf (Ciobanu et al., 2014) or much deeper realms (Inagaki et al., 2013). How deep can microbial life persist or survive? From previous studies, we know that cells become increasingly rare with depth. It is therefore of paramount importance to have methods at hand that are sensitive enough to accurately detect even the faintest traces of life. Eventually, some methods might reach their physical or chemical limits (e.g., the number of fluorophores that bind to a given molecule ultimately controls the strength of a fluorescence signal). So far, our findings regarding the limit at which living organisms can be detected in the subseafloor environment reflect the limits of our ability to detect life rather than the true biological limits. Therefore, we will have to continue optimizing existing tools and keep our eyes open for new ones.

2.2.1.7 POTENTIAL ALTERNATIVES FOR DETECTING LIFE IN SUBSURFACE ENVIRONMENTS

Despite major advances in molecular biology over the last twenty years, fluorescence-based cell counting remains the gold standard for estimating total biomass. However, such counts only provide information regarding the total number of cells that contain nucleic acids; they do not provide information regarding the presence of biomolecules other than DNA or information pertaining to metabolism, the physiological state of the organisms, or their relationship to a specific phylogenetic group.

Fluorescent in situ hybridization (FISH) (Amann et al., 1990a, 1990b, 1995) is one of the most commonly used techniques for investigating microbial phylogeny at the cellular level in environmental samples. However, subseafloor microbes contain too little RNA to be detected using FISH. Schippers et al. (2005) overcame this limitation by applying catalyzed reporter deposition (CARD)-FISH (Teira et al., 2004) and were able to identify rRNA-containing microbes in subseafloor sediments.

FISH, CARD-FISH, and quantitative polymerase chain reaction (qPCR) are now the most commonly used techniques for the general characterization of microbial diversity in subsurface environments. However, the many studies that have utilized these techniques have reported vastly different results, even when examining the very same samples. For some archaeal lineages, it is known that the oligonucleotide probes for FISH staining cannot pass through the microbial cell wall (Nakamura et al., 2006). In a large metadata study, Lloyd et al. (2013) showed that many studies suffer from inadequate sample preparation. For example, estimates of archaeal abundance might vary by more than an order of magnitude, depending on which permeabilization agent is chosen. Also, certain general archaeal qPCR primers appear to be unsuitable for analyses of subsurface environments or at least do not adequately capture the total archaeal diversity (Teske & Sørensen, 2008).

Several other techniques have been used to detect microbial life, mainly in aquatic samples. However, not all these methods have been applied to sedimentary and/or subsurface environments. Zubkov et al. (1999) stained proteins associated with planktonic bacteria with SYPRO dye and detected the cells using FCM. They found that the intensity of SYPRO-protein fluorescence of these bacteria strongly correlated with their total protein content.

Another approach exploits the endoenzymatic activity of microbial cells. Fluorescein diacetate and related compounds have been used to evaluate the viability of various types of cells (Morono et al., 2004). Hydrophobic compounds such as fluorescein diacetate can diffuse into the cytoplasm, where they are hydrolyzed by intracellular nonspecific esterases, producing a green-fluorescent, hydrophilic product, carboxyfluorescein. Thus, the enzymatic activity associated with living cells can be detected using fluorescence. Although as yet there are no reports describing the application of the two above-mentioned approaches to subseafloor samples, it is anticipated that these methods could be used complementarily to provide important information regarding subseafloor life.

Lomstein et al. (2012) recently reported that subseafloor sediment contains a higher number of endospores than AO-stainable microbial cells, and the authors speculated that endospores cannot be stained using fluorescent DNA-specific dyes because of their extremely tight cell walls. Therefore, the use of stain-free methods will be necessary to obtain more precise estimations of subseafloor life. Bhartia et al. (2010) reported the use of deep UV (<250 nm) illumination for detecting native fluorescent signals associated with biomolecules while avoiding background fluorescence associated with minerals. They detected specific signals from microbial cells on gypsum particles (which fluoresce upon UV excitation) using a customized microscope. In addition, they found a shifted fluorescence spectrum associated with spores. Using this system, the authors could detect microbes on basalt chips incubated in a deep-sea environment. Igisu et al. (2012) reported a label-free micro-Fourier transform infrared spectroscopic approach for identifying microbial cells. They identified differences in the chemical structure of membrane lipids and used the method to identify microbes at the domain level in a label-free manner by calculating aliphatic CH_3/CH_2 absorbance ratios.

Several bulk approaches are interesting with regard to the exploration of life signatures. In an analysis of subseafloor sediment, Takano et al. (2013) recently reported the sensitive detection of coenzyme factor 430 (F430), a prosthetic group of methyl coenzyme M reductase, which is involved in archaeal methanogenesis. Due to labile character of F430, the authors suggested that it may be possible to predict the methanogenic or methanotrophic potential and activity of living methanogens in sedimentary environments. Intact polar lipids (IPLs), another class of biomolecules associated with the cellular membrane, can be used as an indicator of living biomass in the subseafloor biosphere (Lipp et al., 2008). However, the same research group recently published new results showing that the rate of IPL degradation differs in bacteria and archaea, indicating that IPLs are limited to serving as proxies for the abundance of archaeal cells (Xie et al., 2013).

The subseafloor biosphere is an energy-poor environment in which biological activity is restricted (Hoehler & Jørgensen, 2013), and the net rate of in situ metabolic activity is estimated to be significantly lower than that of the surface biosphere. Accordingly, biomolecules, which are known to be labile in the surface biosphere, should have a longer half-life or slower decay rate in the subseafloor biosphere. Therefore, research discussions based on the quantification of biomolecules in the subseafloor biosphere should consider the existence of geomolecules, which are molecules no longer associated with living organisms. Distinguishing biomolecules and geomolecules in the deep sedimentary biosphere based on decay rates should be taken into consideration when interpreting data.

2.2.1.8 CONCLUDING REMARKS

Through the relatively brief history of the deep biosphere research, there was a dramatic development in our understanding of the distribution, abundance, and limits of life in subseafloor sediment, starting with Morita and Zobell's pioneering study over Cragg's studies on many ODP sites, on which Parkes, Whitman, and Lipp built their estimates, to Kallmeyer's recent revision. Although none of the models can fully predict microbial abundance for every site on the world's ocean floor, new data will hopefully help to design even better models in the future.

This progress is driven by technology, from agar plates to AO to SYBR Green to FCM. It will be exciting to see what new technologies will arise and how the new results will impact our understanding of subsurface microbiology. Continued progress should result in the lowering of the MQL and help us to finish what Morita and ZoBell started 60 years ago, defining the lower limit of life. The addition of new samples will increase the geographic range as well as the range of geochemical settings being studied.

The distribution of organisms is a central tenet of ecology. Only by having an accurate estimate of the distribution of biomass can we hope to determine the environmental factors that control their distribution. We have made great progress in quantifying subseafloor microbes, but there are still many challenges ahead of us.

REFERENCES

Amann, R. I., Binder, B. J., Olson, R. J., Chisholm, S. W., Devereux, R., & Stahl, D. A. (1990a). Combination of 16S rRNA-targeted oligonucleotide probes with flow cytometry for analyzing mixed microbial populations. *Applied and Environmental Microbiology, 56*, 1919–1925.

Amann, R. I., Krumholz, L., & Stahl, D. A. (1990b). Fluorescent-oligonucleotide probing of whole cells for determinative, phylogenetic, and environmental studies in microbiology. *Journal of Bacteriology, 172*, 762–770.

Amann, R. I., Ludwig, W., & Schleifer, K. H. (1995). Phylogenetic identification and in situ detection of individual microbial cells without cultivation. *Microbiological Reviews, 59*, 143–169.

Barra Caracciolo, A., Grenni, P., Cupo, C., & Rossetti, S. (2005). In situ analysis of native microbial communities in complex samples with high particulate loads. *FEMS Microbiology Letters, 253*, 55–58.

Berry, A. E., Chiocchini, C., Selby, T., Sosio, M., & Wellington, E. M. H. (2003). Isolation of high molecular weight DNA from soil for cloning into BAC vectors. *FEMS Microbiology Letters, 223*, 15–20.

Bertaux, J., Gloger, U., Schmid, M., Hartmann, A., & Scheu, S. (2007). Routine fluorescence in situ hybridization in soil. *Journal of Microbiological Methods, 69*, 451–460.

Bhartia, R., Salas, E. C., Hug, W. F., Reid, R. D., Lane, A. L., Edwards, K. J., et al. (2010). Label-free bacterial imaging with deep-UV-laser-induced native fluorescence. *Applied and Environmental Microbiology, 76*, 7231–7237.

Ciobanu, M. C., Burgaud, G., Dufresne, A., Breuker, A., Redou, V., Ben Maamar, S. J., et al. (2014). Microorganisms persist at record depths in the subseafloor of the Canterbury Basin. *The ISME Journal, 8*, 1370–1380.

Cragg, B. A. (1994). Bacterial profiles in deep sediment layers from the Lau Basin (Site 834). In J. Hawkins, L. Parsons, & J. Allan (Eds.), *Proceedings of the ocean drilling program, scientific results* (Vol. 135) (pp. 147–150). College Station, TX: Ocean Drilling Program.

Cragg, B. A., Harvey, S. M., Fry, J. C., Herbert, R. A., & Parkes, R. J. (1992). Bacterial biomass and activity in the deep sediment layers of the Japan Sea, Hole 798b. In K. A. Pisciotto, J. C. J. Ingle, M. T. von Breymann, & J. Barron (Eds.), *Proceedings of the ocean drilling program, scientific results* (Vol. 127/128) (pp. 761–776). College Station, TX, USA: Ocean Drilling Program.

Cragg, B. A., & Kemp, A. E. S. (1995). Bacterial profiles in deep sediment layers from the Eastern Equatorial Pacific, Site 851. In *Proceedings of the ocean drilling program, scientific results* (Vol. 138) (pp. 599–604). College Station, TC, USA: Ocean Drilling Program.

Cragg, B. A., Law, K. M., Cramp, A., & Parkes, R. J. (1997). Bacterial profiles in Amazon Fan sediments, Sites 934 and 940. In R. D. Flood, D. J. W. Piper, A. Klaus, & L. C. Peterson (Eds.), *Proceedings of the ocean drilling program - scientific results* (Vol. 155) (pp. 565–572). College Station, TX: Ocean Drilling Program.

Cragg, B. A., Law, K. M., Cramp, A., & Parkes, R. J. (1998). The response of bacterial populations to sapropels in deep sediments of the Eastern Mediterranean. In A. H. F. Robertson, K.-C. Emeis, C. Richter, & A. Camerlenghi (Eds.), *Proceedings of the ocean drilling program - scientific results* (Vol. 160) (pp. 303–308). College Station, TX: Ocean Drilling Program.

Cragg, B. A., Law, K. M., O'Sullivan, G. M., & Parkes, R. J. (1999). Bacterial profiles in deep sediments of the Alboran Sea, Western Mediterranean, Sites 976–978. In R. Zahn, M. C. Comas, & A. Klaus (Eds.), *Proceedings of the ocean drilling program - scientific results* (Vol. 161) (pp. 433–438). College Station, TX: Ocean Drilling Program.

Cragg, B. A., & Parkes, R. J. (1993). 29. Bacterial profiles in hydrothermally active deep sediment layers from Middle Valley (NE Pacific), sites 857 and 858. In M. J. Mottl, E. E. Davis, A. T. Fisher, & J. F. Slack (Eds.), *Proceedings of the ocean drilling program - scientific results* (Vol. 139). College Station, TX: Ocean Drilling Program. http://dx.doi.org/10.2973/odp.proc.sr.2139.1994.

Cragg, B. A., & Parkes, R. J. (1994). Bacterial profiles in hydrothermally active deep sediment layers from Middle Valley (NE Pacific), Sites 857 and 858. In *Proceedings of the ocean drilling program, scientific results* (Vol. 139) (pp. 509–516). College Station, TX, USA: Ocean Drilling Program.

Cragg, B. A., Parkes, R. J., Fry, J. C., Herbert, R. A., Wimpenny, J. W. T., & Getliff, J. M. (1990). Bacterial biomass and activity profiles within deep sediment layers. In E. Suess, & R. von Huene (Eds.), *Proceedings of the ocean drilling program, scientific results* (Vol. 112) (pp. 607–619). College Station, TX: Ocean Drilling Program.

Cragg, B. A., Parkes, R. J., Fry, J. C., Weightman, A. J., Maxwell, J. R., Kastner, M., et al. (1995a). Bacterial profiles in deep sediments of the Santa Barbara Basin, Site 893. In J. P. Kennett, J. G. Baldauf, & M. Lyle (Eds.), *Proceedings of the ocean drilling program, scientific results* (Vol. 146). College Station, TX, USA: Ocean Drilling Program. Pt 2.

Cragg, B. A., Parkes, R. J., Fry, J. C., Weightman, A. J., Rochelle, P. A., Maxwell, J. R., et al. (1995b). 27. The impact of fluid and Gas venting on bacterial populations and processes in sediments from the Cascadia margin Accretionary system (Sites 888–892) and the geochemical consequences. In B. Carson, G. K. Westbrook, R. J. Musgrave, & E. Suess (Eds.), *Proceedings of the ocean drilling program - scientific results* (Vol. 146) (pp. 399–413). College Station, TX: Ocean Drilling Program.

Cragg, B. A., Summit, M., & Parkes, R. J. (2000). Bacterial profiles in a sulfide mound (Site 1035) and an area of active fluid venting (Site 1036) in hot hydrothermal sediments from Middle Valley (Northeast Pacific). In R. A. Zierenberg, Y. Fouquet, D. J. Miller, & W. R. Normark (Eds.), *Proceedings of the ocean drilling program, scientific results* (Vol. 169). College Station, TX, USA: Ocean Drilling Program.

Cragg, B. A., Wellsbury, P., Murray, R. W., & Parkes, R. J. (Eds.). (2002). *Bacterial populations in deepwater low-sedimentation-rate marine sediments and evidence for subsurface bacterial manganese reduction (ODP site 1149, Izu-Bonin Trench. Proceedings of the ocean drilling program - scientific results.* College Station, TX: Ocean Drilling Program.

D'Hondt, S., Spivack, A. J., Pockalny, R., Fischer, J., Kallmeyer, J., Ferdelman, T. G., et al. (2009). Subseafloor sedimentary life in the South Pacific Gyre. In *Proceedings of the national academy of sciences of the United States of America* (Vol. 106) (pp. 11651–11656).

Daley, R. J., & Hobbie, J. E. (1975). Direct counts of aquatic bacteria by a modified epifluorescence technique. *Limnology and Oceanography, 20,* 875–882.

Danovaro, R., Manini, E., & Dell'anno, A. (2002). Higher abundance of bacteria than of viruses in deep Mediterranean sediments. *Applied and Environmental Microbiology, 68,* 1468–1472.

Delmont, T. O., Robe, P., Cecillon, S., Clark, I. M., Constancias, F., Simonet, P., et al. (2011). Accessing the soil Metagenome for studies of microbial diversity. *Applied and Environmental Microbiology, 77,* 1315–1324.

Engelen, B., Ziegelmüller, K., Wolf, L., Köpke, B., Gittel, A., Cypionka, H., et al. (2008). Fluids from the oceanic crust support microbial activities within the deep biosphere. *Geomicrobiology Journal, 25,* 56–66.

Engelhardt, T., Kallmeyer, J., Cypionka, H., & Engelen, B. (2014). High virus-to-cell ratios indicate ongoing production of viruses in deep subsurface sediments. *ISME J, 8,* 1503–1509.

Expedition 329 Scientists. (2011). *South Pacific Gyre Subseafloor Life.* http://dx.doi.org/10.2204/iodp.pr.329.2011 IODP Preliminary Report, 329.

Frischer, M. E., Danforth, J. M., Healy, M. a. N., & Saunders, F. M. (2000). Whole-cell versus total RNA extraction for analysis of microbial community structure with 16S rRNA-Targeted oligonucleotide probes in salt marsh sediments. *Applied and Environmental Microbiology, 66,* 3037–3043.

Fry, J. (1988). Determination of biomass. In *Methods in aquatic bacteriology* (pp. 27–72). Wiley.

Gabor, E. M., Vries, E. J., & Janssen, D. B. (2003). Efficient recovery of environmental DNA for expression cloning by indirect extraction methods. *FEMS Microbiology Ecology, 44,* 153–163.

Gold, T. (1992). The deep, hot biosphere. *Proceedings of the National Academy of Sciences of the United States of America, 89,* 6045–6049.

Griffin, W., Phelps, T., Colwell, F., & Fredrickson, J. (1997). Methods for obtaining deep subsurface microbiological samples by drilling. In *The microbiology of the terrestrial deep subsurface* (pp. 23–44). Boca Raton, FL: CRC Press.

Hathaway, J., Poag, C., Valentine, P., Manheim, F., Kohout, F., Bothner, M., et al. (1979). U.S. geological survey core drilling on the Atlantic shelf. *Science, 206,* 515–527.

Hoehler, T., & Jørgensen, B. (2013). Microbial life under extreme energy limitation. *Nature Reviews in Microbiology, 11,* 83–94.

Igisu, M., Takai, K., Ueno, Y., Nishizawa, M., Nunoura, T., Hirai, M., et al. (2012). Domain-level identification and quantification of relative prokaryotic cell abundance in microbial communities by micro-FTIR spectroscopy. *Environmental Microbiology Reports, 4,* 42–49.

Inagaki, F., Motomura, Y., & Ogata, S. (2003). Microbial silica deposition in geothermal hot waters. *Applied Microbiology and Biotechnology, 60,* 605–611.

Inagaki, F., Hinrichs, K.-U., Kubo, Y., & the Expedition 337 Scientists (2013). Proc. IODP, 337: Tokyo (Integrated Ocean Drilling Program Management International, Inc.). http://dx.doi.org/10.2204/iodp.proc.337.2013.

Irvine-Fynn, T., Edwards, A., Newton, S., Langford, H., Rassner, S., Telling, J., et al. (2012). Microbial cell budgets of an Arctic glacier surface quantified using flow cytometry. *Environmental Microbiology, 14,* 2998–3012.

Jørgensen, B. B. (2012). Shrinking majority of the deep biosphere. *Proceedings of the National Academy of Sciences, 109*, 15976–15977.

Johnson, I., & Spence, M. T. Z. (2010). *Molecular probes handbook, a guide to fluorescent probes and labeling technologies.* Life Technologies.

Kallmeyer, J. (2011). Detection and quantification of microbial cells in subsurface sediments. In S. S. Allen, I. Laskin, & M. G. Geoffrey (Eds.), *Advances in applied microbiology* (Vol. 76) (pp. 79–103). Academic Press.

Kallmeyer, J., Mangelsdorf, K., Cragg, B., & Horsfield, B. (2006). Techniques for contamination assessment during drilling for terrestrial subsurface sediments. *Geomicrobiology Journal, 23*, 227–239.

Kallmeyer, J., Pockalny, R., Adhikari, R. R., Smith, D. C., & D'Hondt, S. (2012). Global distribution of microbial abundance and biomass in subseafloor sediment. *Proceedings of the National Academy of Sciences of the United States of America, 109*, 16213–16216.

Kallmeyer, J., Smith, D. C., D'Hondt, S. L., & Spivack, A. J. (2008). New cell extraction procedure applied to deep subsurface sediments. *Limnology and Oceanography: Methods, 6*, 236–245.

Katayama, A., Kai, K., & Fujie, K. (1998). Extraction efficiency, size distribution, colony formation and [H-3]-thymidine incorporation of bacteria directly extracted from soil. *Soil Science and Plant Nutrition, 44*, 245–252.

Kepner, R., & Pratt, J. R. (1994). Use of fluorochromes for direct enumeration of total bacteria in environmental samples: past and present. *Microbiological Reviews, 58*, 603–615.

Kirchman, D., Sigda, J., Kapuscinski, R., & Mitchell, R. (1982). Statistical analysis of the direct count method for enumerating bacteria. *Applied and Environmental Microbiology, 44*, 376–382.

Lappé, M., & Kallmeyer, J. (2011). A cell extraction method for oily sediments. *Frontiers in Microbiology, 2*.

Lever, M., Rouxel, O., Alt, J., Shimizu, N., Ono, S., Coggon, R., et al. (2013). Evidence for microbial carbon and sulfur cycling in deeply buried ridge flank basalt. *Science, 339*, 1305–1308.

Lindahl, V. (1996). Improved soil dispersion procedures for total bacterial counts, extraction of indigenous bacteria and cell survival. *Journal of Microbiological Methods, 25*, 279–286.

Lipp, J. S., Morono, Y., Inagaki, F., & Hinrichs, K.-U. (2008). Significant contribution of Archaea to extant biomass in marine subsurface sediments. *Nature, 454*, 991–994.

Lloyd, K., Schreiber, L., Petersen, D., Kjeldsen, K., Lever, M., Steen, A., et al. (2013). Predominant archaea in marine sediments degrade detrital proteins. *Nature, 496*, 215–218.

Lomstein, B., Langerhuus, A., D'Hondt, S., Jørgensen, B., & Spivack, A. (2012). Endospore abundance, microbial growth and necromass turnover in deep sub-seafloor sediment. *Nature, 484*, 101–104.

Lunau, M., Lemke, A., Walther, K., Martens-Habbena, W., & Simon, M. (2005). An improved method for counting bacteria from sediments and turbid environments by epifluorescence microscopy. *Environmental Microbiology, 7*, 961–968.

Maron, P.-A., Schimann, H., Ranjard, L., Brothier, E., Domenach, A.-M., Lensi, R., et al. (2006). Evaluation of quantitative and qualitative recovery of bacterial communities from different soil types by density gradient centrifugation. *European Journal of Soil Biology, 42*, 65–73.

Martens, C. S., & Berner, R. A. (1974). Methane production in the interstitial waters of sulfate-depleted marine sediments. *Science, 185*, 1167–1169.

Miteva, V., & Brenchley, J. (2005). Detection and isolation of ultrasmall microorganisms from a 120,000-year-old Greenland glacier ice core. *Applied and Environmental Microbiology, 71*, 7806–7818.

Monger, B. C., & Landry, M. R. (1993). Flow cytometric analysis of marine bacteria with Hoechst 33342. *Applied and Environmental Microbiology, 59*, 905–911.

Moore, G. F., Taira, A., & Klaus, A. (2001). Initial report Leg 190. In *Proceedings of the ocean drilling program.* College Station, TX: Ocean Drilling Program.

Morita, R. Y., & ZoBell, C. E. (1955). Occurrence of bacteria in pelagic sediments collected during the Mid-Pacific Expedition. *Deep Sea Research (1953), 3*, 66–73.

Morono, Y., & Inagaki, F. (2010). Automatic slide-loader fluorescence microscope for discriminative enumeration of subseafloor life. *Scientific Drilling, 9*, 32–36.

Morono, Y., Takano, S., Miyanaga, K., Tanji, Y., Unno, H., & Hori, K. (2004). Application of glutaraldehyde for the staining of esterase-active cells with carboxyfluorescein diacetate. *Biotechnology Letters, 26*, 379–383.

Morono, Y., Terada, T., Kallmeyer, J., & Inagaki, F. (2013). An improved cell separation technique for marine subsurface sediments: applications for high-throughput analysis using flow cytometry and cell sorting. *Environmental Microbiology, 15*(10), 2841–2849. http://dx.doi.org/10.1111/1462-2920.12153.

Morono, Y., Terada, T., Masui, N., & Inagaki, F. (2009). Discriminative detection and enumeration of microbial life in marine subsurface sediments. *ISME Journal, 3*, 503–511.

Nakamura, K., Terada, T., Sekiguchi, Y., Shinzato, N., Meng, X.-Y., Enoki, M., et al. (2006). Application of pseudomurein endoisopeptidase to fluorescence in situ hybridization of methanogens within the family Methanobacteriaceae. *Applied and Environmental Microbiology, 72*, 6907–6913.

Noble, R. T., & Fuhrman, J. A. (1998). Use of SYBR Green I for rapid epifluorescence counts of marine viruses and bacteria. *Aquatic Microbial Ecology, 14*, 113–118.

Oremland, R., Culbertson, C., & Simoneit, B. (1982). Methanogenic activity in sediment from Leg-64, Gulf of California. *Initial Reports of the Deep Sea Drilling Project, 64*, 759–762.

Parkes, R. J., Cragg, B. A., Bale, S. J., Getliff, J. M., Goodman, K., Rochelle, P. A., et al. (1994). Deep bacterial biosphere in Pacific Ocean sediments. *Nature, 371*, 410–413.

Parkes, R. J., Cragg, B. A., & Wellsbury, P. (2000). Recent studies on bacterial populations and processes in subseafloor sediments: a review. *Hydrogeology Journal, 8*, 11–28.

Parkes, R. J., Sass, H., Cragg, B., Webster, G., Roussel, E., & Weightman, A. (2014). Studies on prokaryotic populations and processes in subseafloor sediments-an update. In J. Kallmeyer, & D. Wagner (Eds.), *Microbial life of the deep biosphere* (Vol. 1) (pp. 1–27). Berlin, Boston: deGruyter.

Patel, A., Noble, R. T., Steele, J. A., Schwalbach, M. S., Hewson, I., & Fuhrman, J. A. (2007). Virus and prokaryote enumeration from planktonic aquatic environments by epifluorescence microscopy with SYBR Green I. *Nature Protocols, 2*, 269–276.

Porter, J., Diaper, J., Edwards, C., & Pickup, R. (1995). Direct measurements of natural planktonic bacterial community viability by flow cytometry. *Applied and Environmental Microbiology, 61*, 2783–2786.

Porter, K. G., & Feig, Y. S. (1980). The use of DAPI for identifying and counting aquatic microflora. *Limnology and Oceanography, 25*, 943–948.

Prieme, A., Bonilla Sitaula, J. I., Klemedtsson, A. K., & Bakken, L. R. (1996). Extraction of methane-oxidizing bacteria from soil particles. *FEMS Microbiology Ecology, 21*, 59–68.

Robbins, M. (1994). *Fluorescence: gems and minerals under ultraviolet light.* Geoscience Press.

Roussel, E. G., Bonavita, M.-a. C., Querellou, J., Cragg, B. A., Webster, G., Prieur, D., et al. (2008). Extending the sub-sea-floor biosphere. *Science, 320*, 1046.

Schippers, A., Neretin, L. N., Kallmeyer, J., Ferdelman, T. G., Cragg, B. A., Parkes, R. J., et al. (2005). Prokaryotic cells of the deep sub-seafloor biosphere identified as living bacteria. *Nature, 433*, 861–864.

Takano, Y., Kaneko, M., Kahnt, J., Imachi, H., Shima, S., & Ohkouchi, N. (2013). Detection of coenzyme F430 in deep sea sediments: a key molecule for biological methanogenesis. *Organic Geochemistry, 58*, 137–140.

Teira, E., Reinthaler, T., Pernthaler, A., Pernthaler, J., & Herndl, G. J. (2004). Combining catalyzed reporter deposition-fluorescence in situ hybridization and microautoradiography to detect substrate utilization by bacteria and archaea in the deep ocean. *Applied and Environmental Microbiology, 70*, 4411–4414.

Teske, A., & Sørensen, K. B. (2008). Uncultured archaea in deep marine subsurface sediments: have we caught them all? *ISME Journal, 2*, 3–18.

Unge, A., Tombolini, R., Molbak, L., & Jansson, J. K. (1999). Simultaneous monitoring of cell number and metabolic activity of specific bacterial populations with a dual gfp-luxAB marker system. *Applied and Environmental Microbiology, 65*, 813–821.

Wang, Y., Hammes, F., De Roy, K., Verstraete, W., & Boon, N. (2010). Past, present and future applications of flow cytometry in aquatic microbiology. *Trends in Biotechnology, 28*, 416–424.

Wellsbury, P., Goodman, K., Cragg, B. A., & Parkes, R. J. (2000). The geomicrobiology of deep marine sediments from Blake Ridge containing methane hydrate (Sites 994, 995, and 997). In C. K. Paull, R. Matsumoto, P. J. Wallace, & W. P. Dillon (Eds.), *Proceedings of the ocean drilling program - scientific results* (Vol. 164) (pp. 379–391). College Station, TX: Ocean Drilling Program.

Whelan, J. K., Oremland, R., Tarafa, M., Smith, R., Howarth, R., & Lee, C. (1986). Evidence for sulfate-reducing and methane-producing microorganisms in sediments from sites 618, 619, and 622. *Initial Reports Deep Sea Drilling Project Leg, 96*, 767–775.

Whitman, W. B., Coleman, D. C., & Wiebe, W. J. (1998). Prokaryotes: the unseen majority. *Proceedings of the National Academy of Sciences of the United States of America, 95*, 6578–6583.

Xie, S., Lipp, J. S., Wegener, G., Ferdelman, T. G., & Hinrichs, K.-U. (2013). Turnover of microbial lipids in the deep biosphere and growth of benthic archaeal populations. *Proceedings of the National Academy of Sciences of the United States of America, 110*, 6010–6014.

Yanagawa, K., Morono, Y., Yoshida-Takashima, Y., Eitoku, M., Sunamura, M., Inagaki, F., et al. (2014). Viability of subseafloor viral abundance at the geographically and geologically distinct continental margins. *FEMS Microbiology Ecology, 88*(1), 60–68.

Zaporozhenko, E., Slobodova, N., Boulygina, E., Kravchenko, I., & Kuznetsov, B. (2006). Method for rapid DNA extraction from bacterial communities of different soils. *Microbiology, 75*, 105–111.

Zubkov, M. V., Fuchs, B. M., Eilers, H., Burkill, P. H., & Amann, R. (1999). Determination of total protein content of bacterial cells by SYPRO staining and flow cytometry. *Applied and Environmental Microbiology, 65*, 3251–3257.

Chapter 2.2.2

Genetic Evidence of Subseafloor Microbial Communities

Andreas Teske,[1,*] Jennifer F. Biddle[2] and Mark A. Lever[3]

[1]Department of Marine Sciences, University of North Carolina at Chapel Hill, Chapel Hill, NC, USA; [2]School of Marine Science and Policy, University of Delaware, Lewes, DE, USA; [3]Department of Environmental Systems Science, ETH Zurich, Zurich, Switzerland
*Corresponding author: E-mail: teske@email.unc.edu

2.2.2.1 RIBOSOMAL RNA AS PHYLOGENETIC MARKER

The commonly adopted "gold standard" for gene-based identification of microorganisms is the small subunit ribosomal ribonucleic acid (SSU rRNA) gene. This gene codes for rRNAs that are integral parts of the ribosome, the multienzyme complex in all living cells that translates genetic information into proteins of multiple physiological functions. The SSU rRNAs serve as scaffolding molecules that maintain the complex three-dimensional structure of the ribosome; they occur in two size classes, 16S rRNA in Bacteria and Archaea with c.1500 nucleotides length, and 18S rRNA in Eukaryotes with c.2400 nucleotides length. Very short rRNAs, the 5S rRNA, and the longer large subunit rRNA (23S rRNA in Bacteria and Archaea, 28S rRNA in Eukaryotes) are also integral structural components of the ribosome. Due to extremely strong functional and structural constraints that conserve all components of the fine-tuned ribosomal machinery, rRNA molecules—and the genes that are coding for them—have evolved sufficiently slowly to reveal, by degree of nucleotide sequence similarity, the evolutionary paths and phylogenetic relationships of all life forms on Earth (Woese, 1987). Ribosomal RNA molecules preserve a memory of deep evolutionary divergence that reaches in principle back to the inferred ancestral form of cellular life with coherent, inheritable genomes, the last universal common ancestor or progenote (Woese, 1998). Milestones in the recognition and development of SSU rRNA as a universal phylogenetic marker include the initial discovery of the deep evolutionary divergence between the two prokaryotic domains of life, the Bacteria and the Archaea (Woese & Fox, 1977); the first universal rRNA-based tree of life (Fox et al., 1980); the designation of the Bacteria, Archaea, and Eukaryotes as the highest-order taxonomic divisions (domains) corresponding to the actual evolutionary divergence of life (Woese, Kandler, & Wheelis,

Developments in Marine Geology, Volume 7. http://dx.doi.org/10.1016/B978-0-444-62617-2.00004-9

1990); and the validation of rRNA phylogenies by complete genome sequencing that supported the deep genomic and evolutionary divergences between the bacterial, archaeal, and eukaryotic domains (Bult et al., 1996).

With increasingly fine-scaled phylogenies corresponding to biochemically and physiologically coherent groups of microorganisms, rRNA phylogenies became the molecular road map of the microbial world. They provide the current basis for standard references, such as Bergey's Manual for Determinative Bacteriology (2nd edition) and The Prokaryotes (3rd edition). With improved sequencing techniques, rRNA genes became indispensible in mapping microbial diversity directly in the environment, and demonstrated the extent of previously unrecognized microbial diversity in nature that has so far eluded cultivation efforts (Pace, 1997). The tally of originally 12 bacterial phyla (the highest-order lineages within the bacterial domain) before the onset of environmental rRNA sequencing (Woese, 1987) grew quickly to 36 (Hugenholtz et al., 1998) before increasing further to 52 (Rappé & Giovannoni, 2003). More recently, 15 additional new bacterial phyla were added from deep rRNA sequencing of a single marine microbial mat (Rey et al., 2006). The most recent comprehensive update includes novel lineages that were uncovered with genomic and metagenomic approaches, and counts 80 mutually exclusive phylum-level bacterial lineages, of which 38 remain uncultured; the archaeal count—although on a more finely resolved taxonomic scale—yields a total of 72 lineages, of which 52 are uncultured (Baker & Dick, 2013).

The archaeal domain consisted originally of two subdomain-level clusters, the *Crenarchaeota*, represented by hyperthermophilic, sulfur-dependent archaea from hot springs and hydrothermal vents, and the *Euryarcheota*, containing the methanogens and extreme halophiles (Woese, 1987). While 18 separately branching archaeal lineages were recognized in 2002 (Hugenholtz, 2002), it became increasingly difficult to fit them into the crenarchaeotal/euryarchaeal divide. The "misfits" that defy categorization (Schleper, Jurgens, & Jonuscheit, 2005) include the *Korarchaeota* (Barns, Delwiche, Palmer, & Pace, 1996), the Ancient Archaeal Group lineage (Takai & Horikoshi, 1999), the *Thaumarchaeota* (Brochier-Armanet, Boussau, Gribaldo, & Forterre, 2008), the *Aigarchaeota* (Nunoura et al., 2011), the obligate symbiont *Nanoarchaeum equitans* (Huber et al., 2002), and the major subsurface lineages Marine Benthic Group B (MBG-B) (Vetriani, Jannasch, MacGregor, Stahl, & Reysenbach, 1999) and Miscellaneous Crenarchaeotal Group (MCG) (Inagaki et al., 2003), also proposed as a new archaeal phylum-level lineage "Bathyarchaoeta" (Meng et al., 2014). Current efforts for a genomic census of the widest possible range of bacterial and archaeal lineages (Rinke et al., 2013) recognize the increasingly complex archaeal phylogenetic landscape, and propose bacterial and archaeal "superphyla," for example the archaeal "TACK" superphylum (Guy & Ettema, 2011), consisting of phylum-level lineages united by common ancestry.

The fundamental biological discoveries through rRNA sequencing surveys—in scope and extent only comparable to the golden age of zoological and botanical discoveries in the late eighteenth and early nineteenth century—have started to reveal the taxonomic and evolutionary diversity of microbial life in the deep subsurface biosphere, and the extent of microbially colonized deep subsurface environments. While exhaustive, lineage-by-lineage surveys of microbial diversity in the deep subsurface are provided in specialized reviews (Durbin & Teske, 2012; Fry, Parkes, Cragg, Weightman, & Webster, 2008; Orcutt, Sylvan, Knab, & Edwards, 2011; Teske & Sørensen, 2008), the following sections discuss some general characteristics of the deep subsurface biosphere in the light of rRNA sequencing.

2.2.2.1.1 Ribosomal RNA Surveys in the Deep Sedimentary Biosphere

The sedimentary marine subsurface is permeated by microbial life. Recently, global quantifications have shown that the total contribution of the marine sedimentary biosphere amounts to c.0.6% of Earth's living biomass (Kallmeyer, Pockalny, Adhikari, Smith, & D'Hondt, 2012). Even highly oligotrophic open-ocean sediments—oxygenated and depleted of organic carbon—contain active microbial populations that consume oxygen and presumably remineralize buried organic substrates at very slow rates (D'Hondt et al., 2009; Røy et al., 2012). The subsurface biosphere is therefore an omnipresent mediator between biological and geological elemental cycles in the subsurface (Hinrichs & Inagaki, 2012).

Persistent microbiological surveys within the Ocean Drilling Program (ODP) and its successor programs (Integrated Ocean Drilling Program, IODP; renewed as International Ocean Discovery Program in 2013) have uncovered the gene-based evidence for this subsurface biosphere. The first 16S rRNA gene survey of subsurface sediments, from Hole 798B in the Japan Sea drilled during ODP Leg 128, detected a new bacterial lineage that could not be subsumed under any other taxonomic category within the 16S rRNA phylogeny of the bacterial world (Rochelle, Cragg, Fry, Parkes, & Weightman, 1994). Since Leg 128, many more deep sediments recovered by ODP and IODP drilling have been examined by SSU rRNA sequencing: ODP 190 site 1173 in the Nankai Trench Accretionary wedge (Newberry, Webster, Weightman, & Fry, 2004) and site 1176 overlying the Nankai subduction zone (Kormas, Smith, Edgcomb, & Teske, 2003); ODP leg 201 Eastern Equatorial Pacific site 1225 (Inagaki et al., 2006; Teske, 2006); Peru Margin and Peru Trench sites 1227 to 1230 of ODP leg 201 (Biddle et al., 2006; Edgcomb, Beaudoin, Gast, Biddle, & Teske, 2011; Inagaki et al., 2006; Parkes et al., 2005; Sørensen & Teske, 2006) and Peru Basin site 1231 of ODP leg 201 (Sørensen et al., 2004); ODP leg 204 sites 1244 and 1251 on the Cascadia Margin (Inagaki et al., 2006; Nunoura, Inagaki, Delwiche, Colwell, & Takai, 2008); and ODP leg 210 site 1276 on the Newfoundland Margin (Roussel et al., 2008). Within IODP, 16S rRNA gene sequencing started immediately on sediments (Lever, 2008), borehole crusts (Nakagawa et al., 2006), and seafloor basalt crust (Lever et al., 2013) obtained at the Juan de Fuca Ridge flanks during Expedition 301, and has since then continued unabated. Major studies include IODP expedition 302 on the Lomonosov Ridge in the Arctic Ocean (Forschner, Sheffer, Rowley, & Smith, 2009), IODP expedition 307 to the Porcupine Seabight (Hoshino et al., 2011; Webster et al., 2007), and IODP expedition 308 to the Brazos-Trinity and Mars-Ursa Basins on the Gulf of Mexico continental slope (Nunoura et al., 2009). Recently, sequencing surveys of sediments from the site survey cruise of IODP Expedition 329 have examined bacterial and archaeal communities of oligotrophic South Pacific sediments (Durbin & Teske, 2010, 2011). Currently, 16S rRNA gene sequencing surveys are not only performed for their own sake, but also as a baseline for increasingly sophisticated functional gene analyses and high-throughput metagenome and metatranscriptome surveys (see later sections in this chapter). For the vast range of subsurface and deep-ocean microbial habitats, 16S rRNA genes provide the essential taxonomic "grid" for mapping microbial biodiversity in the subsurface; 16S rRNA datasets ensure consistent comparisons and meta-analyses of ever-growing datasets, and they accommodate novel microbial lineages that continue to be discovered in the subsurface (reviewed in Fry et al., 2008; Teske & Sørensen, 2008; Orcutt et al., 2010; Durbin & Teske, 2012).

2.2.2.1.2 Bacterial Lineages

Cloning and sequencing of bacterial 16S rRNA genes from marine subsurface sediments have consistently detected several bacterial phylum-level lineages, notably the JS-1 group, *Chloroflexi*, *Planctomycetes*, and the *Proteobacteria*. These phyla have become extensively populated with bacterial rRNA sequences, phylotypes, and sequence clusters of subsurface origin (Blazejak & Schippers, 2010; Durbin & Teske, 2011; Inagaki et al., 2006), and are introduced in the following four paragraphs.

2.2.2.1.2.1 Japan Sea Group I

The first subsurface lineage of bacteria discovered during ODP Leg 128 (Rochelle et al., 1994) was designated Japan Sea Group I (JS-1) after the location of its discovery, although it also occurs in surface sediments (Webster, Parkes, Fry, & Weightman, 2004). This group accounted for approximately a quarter of all bacterial sequences recovered from Pacific continental margin subsurface sediments (Fry et al., 2008). Members of the JS-1 and of the *Chloroflexi* subphylum I have been enriched from marine sediments under anaerobic long-term incubations, using artificial seawater media amended with glucose, acetate, and sulfate (Webster et al., 2011). In a study of shallow intertidal subsurface sediments of the German Wadden Sea, JS-1 bacteria were frequently found in organic-rich, sulfate-depleted sediment layers, and appeared only rarely in clone libraries from organic-poor intertidal sediments (Webster et al., 2007); they represented the most frequently found bacterial group in clone library surveys of organic-rich, highly reducing sediments of the Peru Trench and on the Cascadia Margin (Inagaki et al., 2006). These results are compatible with a heterotrophic, anaerobic metabolism for JS-1 bacteria.

2.2.2.1.2.2 Chloroflexi

Clone library sequencing surveys of shallow subsurface sediments have shown that the *Chloroflexi* increase in detection frequency within a few meters down-core (Durbin & Teske, 2011; Wilms et al., 2006) accounts for a substantial proportion (on average c.20%) of all bacterial clones recovered from Pacific continental margin subsurface sediments (Fry et al., 2008). Previously, the subsurface abundance of *Chloroflexi* has been linked to their dehalogenating capabilities. For example, species and strains of the genera *Dehalogenimonas* and *Dehalococcoides* are anaerobic H_2-oxidizers that respire with halogenated hydrocarbons (Moe, Yan, Fernanda Nobre, da Costa, & Rainey, 2009). Growth on organohalides in the subsurface is supported by the detection of reductive dehalogenase homologous (*rdhA*) genes in marine subsurface sediments (Futagami, Morono, Terada, Kaksonen, & Inagaki, 2009; Futagami, Morono, Terada, Kaksonen, & Inagaki, 2013; also see next section). Interestingly, substrate-induced gene expression screening of actively dehalogenating enrichment samples from Nankai Trough subsurface sediments yielded *Dehalococcoides* phylotypes but no detectable *rdhA* genes, suggesting that the *rdhA* functional gene assay needs to be redesigned to be phylogenetically more inclusive (Futagami et al., 2013). Since the subsurface phylotypes of *Chloroflexi* form separate phylogenetic lineages that are only distantly related to cultured dehalogenating strains and species (Durbin & Teske, 2011; Inagaki et al., 2006), other metabolic modes besides or in addition to halorespiration are likely to support subsurface *Chloroflexi*. Cultivations and genome analyses of new *Chloroflexi* indicate that representatives of this phylum are metabolically versatile: several new genera and species of filamentous, mostly anaerobic, and fermentative *Chloroflexi* have been isolated from organic-rich, reducing habitats such as anaerobic bioreactors and wastewater processing plants (Yamada et al., 2006, 2007). Recently, the new species and genus *Pelolinea submarina*, growing anaerobically on different carbohydrates amended with yeast extract, was isolated from marine subsurface sediments offshore the Shimokita Peninsula in northeast Japan, and thus represents the first cultured subseafloor *Chloroflexi* isolate (Imachi et al., 2014). In molecular assays combining microsphere adhesion to cells with enzyme-labeled fluorescence, several uncultured *Chloroflexi* populations were shown to express chitinase, esterase, galactosidase, and glucuronidase activity under aerobic conditions (Kragelund et al., 2007). The recently published single-cell genome of a *Chloroflexi* cell from shallow marine sediments contains aromatics- and fatty acid-degrading pathways and the reductive acetyl-CoA pathway, but no indication of dehalogenation (Wasmund et al., 2014). Three *Chloroflexi* genomes reconstructed from metagenomes in aquifer sediments encode aerobic and/or anaerobic carbohydrate degradation pathways and acetogenesis via the acetyl-CoA pathway (Hug et al., 2013).

2.2.2.1.2.3 Planctomycetes

The phylum *Planctomycetes* has been detected intermittently in organic-rich, reducing subsurface sediments of the Peru Margin and the Nankai Trough (reviewed in Fry et al., 2008); it is one of the most frequently found phyla within and below the nitrate-reducing zone in organic-lean marine subsurface sediments of the South Pacific (Durbin & Teske, 2011). The redox regime of this habitat resembles that of the suboxic water column below the Black Sea chemocline, where *Planctomycetes* occur in increased abundance as well (Kirkpatrick et al., 2006). Many genera of *Planctomycetes* isolated from soil, freshwater, and peat bogs are chemoheterotrophic, and are capable of degrading various complex polysaccharides, whereas a distinct, autotrophic lineage within this phylum harbors bacteria capable of anaerobic ammonia oxidation (anammox) (Fuerst & Sagulenko, 2011). All currently cultured members of the *Planctomycetes* are characterized by unusual intracellular compartmentalization and peculiar cell walls lacking peptidoglycan (Fuerst & Sagulenko, 2011).

The *Proteobacteria* defy succinct summaries. Cumulatively, they account for almost 40% of all bacterial sequences in Pacific continental margin sediments (Fry et al., 2008). Phylogenetically, they are divided into at least six subphyla (alpha, beta, gamma, delta, epsilon, and zetaproteobacteria), and resemble in this regard the "superphyla" that have been proposed for mono-phyletic associations of other major bacterial phyla that appear to be linked by shared common ancestry (Rinke et al., 2013; Wagner & Horn, 2006). The *Deltaproteobacteria* are of major interest, since they include numerous cultured sulfate-, sulfur-, and iron-reducing species and genera that are in many cases isolated from marine sediments (Lonergan et al., 1996; Rabus, Hansen, & Widdel, 2006). Sulfate-reducing *Deltaproteobacteria* are tracked in marine subsurface sediments with functional gene approaches, as introduced in Section 2.2.2.2 Functional Genes of this chapter In contrast to the uncultured JS-1 bacteria and the sparsely cultured *Chloroflexi* and *Planctomycetes*, the *Proteobacteria* contain the majority of cultivated bacterial species that are known today, with a wide range of heterotrophic, autotrophic, aerobic, and anaerobic metabolisms. Extensive cultivation surveys of deep subsurface bacteria have succeeded in isolating predominantly *Proteobacteria*, together with *Firmicutes* (D'Hondt et al., 2004; Batzke, Engelen, Sass, & Cypionka, 2007; Parkes et al., 2009). So far, the best example of a cultured subsurface bacterium that contributes substantially to the molecular tally of in situ subsurface populations is the widespread species *Rhizobium radiobacter* within the *Alphaproteobacteria* (Young, Kuykendall, Martinez-Romero, Kerr, & Sawada, 2001). *Rhizobium radiobacter* has been documented in Mediterranean sapropel sediments (Suess, Engelen, Cypionka, & Sass, 2004; Suess et al., 2006) and in Peru Margin subsurface sediments; at the latter location it is has been investigated as a host of subsurface virus populations (Engelhardt, Sahlberg, Cypionka, & Engelen, 2013).

2.2.2.1.3 Archaeal Lineages

Within the archaeal domain, the most commonly encountered subsurface phylum-level lineages are the MCG; discovered in deep marine sediments (Inagaki et al., 2003), the MBG-A to MBG-E first found in surficial open-ocean sediments (Vetriani et al., 1999), the Deep-Sea Hydrothermal Vent Euryarchaeotal groups 1 to 7 (DHVEG-1 to 7), originally detected in hydrothermal vent habitats (Takai & Horikoshi, 1999), and South African Goldmine Euryarchaeotal Group (SAGMEG); first found in terrestrial deep goldmines (Takai, Moser, DeFlaun, Onstott, & Fredrickson, 2001) but also widespread in marine subsurface sediments (Figure 2.2.2.1). Surprisingly, cultured methanogens are only a minor component of archaeal communities as detected in 16S rRNA gene surveys of marine subsurface sediments, and do not appear at all in many published datasets (reviewed in Fry et al., 2008; also see next section). Different lineages are found in oligotrophic marine sediments (Deep-Sea Euryarchaeotal Groups 1 to 4, DSEG 1 to 4; Durbin & Teske, 2011); some of these oligotrophic sediment clones appear also within the DHVEG clades that were originally found at hydrothermal vents (Takai & Horikoshi, 1999).

The 16S rRNA phylogeny provides the taxonomic skeleton for a complex archaeal subsurface biosphere, and suggests linkages between the sedimentary and the hydrothermal biospheres. The names or acronyms of these archaeal lineages reflect the absence of ecophysiological clues and hypothesis-generating phylogenetic affiliations. Genomic and cultivation approaches will be required to elucidate the physiologies and ecological functions of these poorly known lineages. Recently, the MCG, one of the most frequently detected archaeal subsurface lineages, has been investigated by fosmid sequencing (Li et al., 2012; Meng et al., 2009), molecular quantification in multiple environments (Kubo et al., 2012), and single-cell genome sequencing (Lloyd et al., 2013). Members of this group have been enriched from marine and estuarine sediments in stable isotope probing experiments with ^{13}C-labeled acetate (Webster et al., 2010) and other ^{13}C-labeled substrates including glycine, urea, algal lipid extract, and complex growth medium (Seyler, McGuinness, & Kerkhof, 2014). General heterotrophic and anaerobic media have allowed significant MCG enrichment but not pure culture isolation, suggesting that more specific growth requirements have to be met to allow pure culture isolation (Gagen, Huber, Meador, Hinrichs, & Thomm, 2013). Single-cell genome sequencing of MCG and MBG-D cells from surficial sediments of Aarhus Bay has uncovered genomic pathways for the degradation of detrital proteins. The suite of genes includes extracellular peptidases, oligopeptide transporters for transfer into the cell,

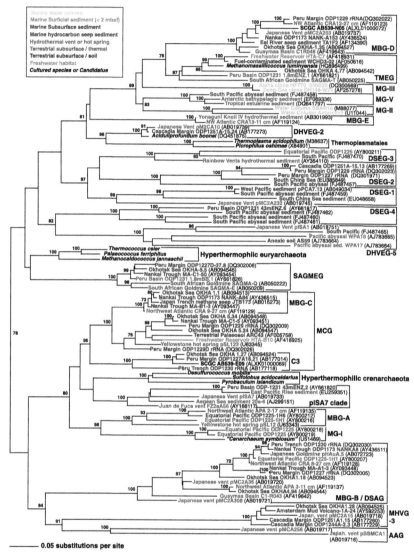

FIGURE 2.2.2.1 Archaeal 16S rRNA phylogeny, based on distance analysis of 16S rRNA nucleotide positions 24 to 906 (distance measure HKY85, optimality criterion=minimum evolution, gamma distribution factor of 0.5) performed with PAUP4.0 (Swofford, 2000). The tree topology was checked by 200 bootstrap reruns. References for archaeal lineages: Marine Benthic Group (MBG) (A to E): Vetriani et al. (1999); Marine Group I and II: DeLong (1992); Marine Group III: López-García, Moreira, Lopez-Lopez, and Rodriguez-Valera (2001); Deep-Sea Hydrothermal Vent Euryarchaeotal Groups (DHVEG): Takai and Horikoshi (1999); Miscellaneous Crenarchaeotal Group (MCG): Inagaki et al. (2003); SAGMEG: Takai et al., (2001); Deep-Sea Euryarchaeotal Groups (DSEG) 1–4 and pSIA17 clade: Durbin and Teske (2011); C3 archaea: Inagaki et al. (2006); Marine Hydrothermal Vent Group (MHVG) III and Ancient Archaeal Group (AAG): Takai and Horikoshi (1999). The C3 archaeal group is congruent with MCG lineage 15 (Kubo et al., 2012). DSEG-2 is synonymous with the VAL III clade, originally found in freshwater lakes (Jurgens et al., 2000); DSEG-1 is synonymous with Deep Sea Euryarchaeotal Group, and DHVEG-5 matches the Miscellaneous Euryarchaeotic Group, MEG (both Takai et al., 2001). The 16S rRNA sequences SCGC AB539-E09 and SCGC AB540-N05 are from contigs of single cell genomes (Lloyd et al., 2013).

intracellular peptidases, aminotransferases that deaminate amino acids to 2-keto acids, and oxidation of the 2-keto acids to acyl-CoA and organic acids (Lloyd et al., 2013). These genome-based inferences were substantiated by enzymatic measurements of extracellular peptidases (the types found in the MCG and MBG-D genomes) in the same sediments (Lloyd et al., 2013).

2.2.2.1.4 Eukaryotic Lineages

The third domain of life, the eukaryotes, was initially neglected in SSU rRNA sequencing surveys of the marine subsurface. The first comprehensive survey focused on the Peru Margin and Peru Trench (ODP sites 1227, 1229 and 1230) and detected mostly fungi in the subsurface (Edgcomb et al., 2011). The subsurface fungal phylotypes were derived from rRNA gene and also from rRNA sequences; most of these subsurface fungal phylotypes were related to single-celled yeasts within the *Basidiomycota* (Edgcomb et al., 2011). Metagenome sequence reads associated with *Ascomycota* and *Basiomycota* (Biddle, Fitz-Gibbon, Schuster, Brenchley, & House, 2008) and live counts of fungal colonies per gram sediment (Biddle, House, & Brenchley, 2005a,b), both from Peru Margin site 1229, further support the notion of a eukaryotic subsurface biosphere dominated by small, single-cell fungi that assimilate buried organic compounds (Edgcomb & Biddle, 2011). A recent 18S rRNA survey showed unique phylogenetic profiles of fungi from subsurface sediments of continental margins and open-ocean locations of the Atlantic and Pacific oceans (Hydrate Ridge offshore Oregon, the Benguela Upwelling system, the Eastern Equatorial Pacific, the Peru Margin, and sedimented North Pond near the Mid-Atlantic Ridge), and coastal surface sediments near Woods Hole, Massachusetts (Orsi et al., 2013a). These subsurface fungal signatures were significantly correlated to geochemical parameters such as total organic carbon, nitrate, sulfide, and dissolved inorganic carbon, suggesting that specific geochemical regimes select for specific fungal (and other eukaryotic) communities (Orsi et al., 2013a).

2.2.2.1.5 Distinct Subsurface Habitats and Their Microbial Communities

The 16S rRNA gene sequences of different bacterial and archaeal lineages are not randomly distributed in the marine sedimentary subsurface, but show evidence of biogeographical structure most likely controlled by in situ chemical regime. For example, sediment-hosted methane hydrates (Inagaki et al., 2006), volcanic ash layers embedded in marine sediments (Inagaki et al., 2003), methane/sulfate transition zones (Sørensen & Teske, 2006), oxygen/nitrate porewater gradients (Durbin & Teske, 2011), and organic carbon content and redox status of subsurface sediments (Durbin & Teske, 2012) appear to select in favor of phylogenetically distinct bacterial and archaeal lineages.

Of all subsurface geochemical interfaces, the methane/sulfate transition zone has received the most sustained attention; microbial communities that oxidize methane with sulfate as the electron donor were predicted to be a major component of the subsurface biosphere (D'Hondt, Rutherford, & Spivack, 2002). Interestingly, 16S rRNA surveys of the methane/sulfate transition zones in deep subsurface sediments yielded quite consistent microbial community signatures within and around these sediment horizons. The archaeal lineages MCG, MBG-B, SAGMEG, and MBG-D were repeatedly detected based on reverse-transcribed 16S rRNA (Biddle et al., 2006; Sørensen & Teske, 2006) and also based on sequencing of 16S rRNA genes (Nunoura et al., 2009; Parkes et al., 2005; Webster et al., 2006). With few exceptions (Briggs et al., 2011; Roussel et al., 2008), anaerobic, sulfate-dependent methanotrophic archaea (ANME) remained below detection limit with general archaeal 16S rRNA gene primers; their amplification and successful detection required redesigned functional gene primers (Lever, 2008, 2013; see also next section). Obviously, sulfate-dependent methane oxidation is not the only process that sustains subsurface microbial cells in the sulfate/methane transition zone (SMTZ); heterotrophic archaea that assimilate buried organic matter predominate to such an extent that their 16S rRNA sequences account for all PCR amplicons if general archaeal primers are used (Biddle et al., 2006). In contrast to the deep sedimentary subsurface, ANME archaea are easily detectable by general archaeal 16S rRNA sequencing in shallow methane/sulfate transitions zones of organic-rich, coastal surface sediments (Webster et al., 2011; Lloyd, Alperin, & Teske, 2011; Knittel & Boetius, 2009), and in surficial sediments of methane seeps (Lloyd, Lapham, & Teske, 2006; Lloyd et al., 2010).

Subseafloor sediments in the open ocean are more oxidized and typically do not harbor distinct methane/sulfate transition zones. Recent sequencing surveys are examining drilling sites and marine sedimentary environments representing different trophic regimes, from thoroughly oxidized and extremely organic-lean sediments that accumulate with very slow sedimentation rates in the ultraoligotrophic South Pacific Gyre (D'Hondt et al., 2009), to reduced, organic-rich, sulfidic, or methanogenic sediments on highly productive continental margins, such as the Peru Margin or the Cascadia Margin (D'Hondt et al., 2004). This ultraoligotrophic to eutrophic spectrum should influence the composition and activity of subsurface microbial communities, in response to different redox regime, organic carbon content, and available substrate spectra. Indeed, a comparison of contrasting sites shows that archaeal lineages change systematically over a spectrum of organic-lean, oxidized marine sediments from abyssal plains to organic-rich, reduced sediments on continental margins (Figure 2.2.2.2) (Durbin & Teske, 2012).

The correlation between organic carbon availability and redox status and microbial community structure in marine subsurface sediments, as reflected in 16S rRNA gene diversity, indicates a linkage between geochemical regime and 16S rRNA gene-based microbial community structure, even in the metabolically slow sedimentary subsurface. Decoupling mechanisms, for example, accumulations of

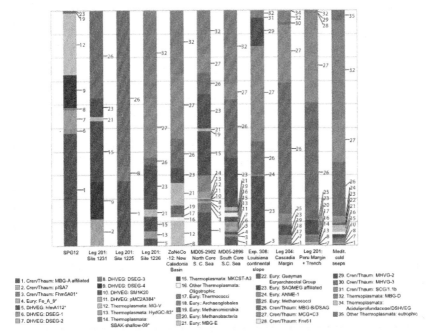

FIGURE 2.2.2.2   Relative abundance of uncultured archaeal lineages (Durbin & Teske, 2012) in 16S rRNA gene clone libraries for all oligotrophic sediment sites for which public relative-abundance data and geochemical information is available. Clades 1–16, colored blue, were either shared with eutrophic end-member sites but comprised less than 1.5% of total clones in any eutrophic site, or were entirely absent from eutrophic end-members. Selected eutrophic sediment sites include the Peru Margin, Peru Trench, and Cascadia Margin; Mediterranean mud volcanoes are included as a surface expression of subsurface processes. Sediment sampling sites are arranged as increasingly oligotrophic from right to left, as suggested by available geochemical parameters (details in Durbin & Teske, 2012).

fossil and/or inactive microbial cells, remnant populations from past geological and geochemical regimes, or dispersed cells from distant source habitats (Inagaki et al., 2001) have to be carefully considered in individual case studies and specific localities, but are unlikely to overwrite and invalidate global distribution and abundance patterns of active in situ microbial populations and living cells with intact rRNA and rRNA genes. By the same token, comprehensive meta-analyses of the microbial community composition and abundance in the deep subsurface biosphere, and their correlations to geochemical habitat characteristics and controls should be a high research priority for the immediate future.

So far, SSU rRNA sequencing surveys in all three domains of life are consistent with the notion of a heterotrophic subsurface biosphere in marine sediments that ultimately depends on buried organic matter of planktonic, photosynthetic origin as carbon and energy source (Biddle et al., 2006). However, a chemosynthetic contribution to life in subsurface sediments cannot be ruled out a priori, and is certainly apparent in the deep basalt biosphere (Lever et al., 2013). The Marine Group I Archaea, whose few cultured representatives are autotrophic ammonia oxidizers (Pester, Schleper, & Wagner, 2011), are abundant in the marine water column, but occur also in oxidized marine sediments as long as oxygen or nitrate is present (Durbin & Teske, 2010, 2011). The ANME-1 Archaea, who preferentially assimilate dissolved inorganic carbon (DIC) for biosynthetic purposes, constitute another candidate group of subsurface autotrophs (Kellermann et al., 2012).

The working hypothesis of a heterotrophic subsurface biosphere has to be qualified by the persistent sampling bias toward organic-rich, reducing continental margin sediments where buried organic biomass is omnipresent, and by the general paucity of physiological knowledge on uncultured microorganisms in the subsurface. Inferring physiological and geochemical key characteristics of deep subsurface microorganisms requires different approaches, such as functional gene sequencing and metagenomic surveys that are discussed in the following sections.

2.2.2.2 FUNCTIONAL GENES

Protein-coding (functional) genes have been used as phylogenetic markers across a wide range of microorganisms since the late 1980s, when Amann, Ludwig, and Schleifer (1988) demonstrated that bacterial phylogenetic trees based on gene sequences of the β-subunit of ATP-synthase were congruent with ones based on 16S rRNA genes. The study confirmed 16S rRNA gene-based interpretations of evolution in Bacteria and demonstrated that specific, highly conserved functional genes are suitable for the study of evolutionary relationships. Functional genes thus offer a tool to link the genotype to the phenotype, and provide a record of how genes are vertically and laterally transferred between organisms over time. The study of genes and gene transcripts involved in known metabolic, biosynthetic, maintenance, transport, and other cellular processes can even provide insights to the identity and activity of unknown or uncultivated organisms in the environment (reviewed in Lever, 2013). As outlined previously, functional genes used for process and phylogenetic studies should have the following characteristics: (1) the gene encodes an enzyme used for one specific reaction; (2) a large gene sequence database and alignment is available to infer the phylogenetic identity of environmental sequences; (3) sufficiently conserved sequence sections are available so that genes can be comprehensively targeted by general or group-specific probes, such as PCR primers; and (4) the evolutionary history is well constrained, so organisms which have obtained the gene through lateral transfer can be identified (Lever, 2013).

In practice, functional genes rarely meet all these criteria. Highly conserved genes may encode different versions of enzymes that catalyze key reactions within a pathway in different directions. For example, the gene encoding the alpha subunit of methyl coenzyme M reductase (*mcrA*) encodes the enzyme that reduces the coenzyme M-bound methyl group to methane in methanogens, but also performs the reverse reaction in anaerobic methanotrophs (Hallam et al., 2004; Scheller,

Goenrich, Boecher, Thauer, & Jaun, 2010; Shima et al., 2012; Thauer, 2011). The dissimilatory sulfite reductase gene (*dsrAB*) also codes for two versions of this enzyme, which operate in reductive or oxidative direction, depending on whether it is found in sulfate (or sulfite)-reducing or sulfide-oxidizing microbes (Loy et al., 2009). Although the *dsrAB* gene has been laterally transferred multiple times in its evolutionary history, it is nonetheless a suitable marker for the study of microbial sulfur cycling due to a comprehensive phylogenetic framework: sulfate reducers and sulfide oxidizers fall into distinct genetic clusters, and lateral gene transfer events are precisely identified by comparison against 16S rRNA phylogenies (Klein et al., 2001; Loy, Duller, & Wagner, 2008; Zverlov et al., 2005). The clear physiological and process-specific role of genes such as *mcrA* and *dsrAB* cannot always be taken for granted. For example, the gene for formyl-tetrahydrofolate synthetase (*fhs*), an enzyme that catalyzes the ATP-dependent activation of formate, and is used as a marker gene of acetogenesis (Leaphart, Friez, & Lovell, 2003), also occurs in aerobic heterotrophic, fermentative, and sulfate-reducing organisms (Lever et al., 2010). To verify metabolic function, functional gene assays can be complemented with process rate measurements, concentrations, and stable isotopic compositions of metabolic educts and products, and thermodynamic model calculations (Lever, 2012; Lever et al., 2010).

To our knowledge, published analyses of samples collected during ODP and IODP expeditions have focused on four different functional genes: the aforementioned *mcrA*, *dsrAB*, and *fhs*, as well as the gene for reductive dehalogenase (*rdhA*), an enzyme which catalyzes the reductive removal of halogen groups in halogenated organic compounds (Futagami et al., 2009, 2013). All publications have used PCR-based methods, both to obtain sufficient amounts of gene product for DNA sequencing followed by phylogenetic analyses, as well as for quantitative analyses via real-time PCR. The phylogenetic and metabolic diversity of organisms detected so far, as well as the sites where they were found, and information on their closest cultured relatives will be presented in the following sections.

2.2.2.2.1 Methanogenesis and Anaerobic Methane Oxidation

The community composition of methane-cycling Archaea has been characterized at several ocean drilling locations, mostly on the Pacific Rim. These locations include the Nankai Trough (Newberry et al., 2004), the Peru Trench (Inagaki et al., 2006), the Peru Margin (Webster et al., 2006), the Cascadia Margin (Yoshioka et al., 2010), the Shimokita Peninsula on the northeast of Japan (Imachi et al., 2011), and the Eastern flank of the Juan de Fuca Ridge (Lever et al., 2013). In addition, *mcrA* genes have been quantified via real-time PCR in cores from Cascadia Margin (Colwell et al., 2008), and Porcupine Seabight in the Atlantic Ocean (Webster et al., 2009). At most of these sites, *mcrA* genes were only reliably detected and analyzed at few sampling depths. Although the community composition of methanogens and anaerobic methanotrophs should in principle correlate with environmental variables, such as geochemical gradients or lithostratigraphy, these controls remain to be explored in depth. Since *mcrA* and 16S rRNA gene sequences of known methanogens or methanotrophs have turned out hard to detect, even at sites with high biogenic methane concentrations and high microbial biomass, current methods or commonly used PCR primers need to be improved. In addition, methanogen populations may not be as dominant as expected, and small methanogen stocks have over geologic time produced the vast biogenic methane reservoir found in subseafloor sediments (Inagaki et al., 2006).

The published *mcrA* sequences show a considerable diversity of methane-cycling Archaea in subseafloor sediments (Figure 2.2.2.3(A)), with genera belonging to the orders Methanosarcinales (*Methanosarcina*, *Methanococcoides*, *Methanosaeta*, ANME-2), Methanocellales (*Methanocella*), Methanomicrobiales (*Methanoculleus*), Methanobacteriales (*Methanobrevibacter*, *Methanobacterium*), Methanococcales (*Methanococcus*), ANME-1, and the novel order *Methanomassiliicoccales* (Borrel et al., 2013). Members of the genera *Methanosarcina*, *Methanobrevibacter*, and *Methanobacterium* (Imachi et al., 2011; Newberry et al., 2004; Webster et al., 2006; Yoshioka et al., 2010) have been found in several locations; the ANME-1 lineage and an uncultivated group of *Methanosarcinales* have, moreover, been detected in deep subsurface basalt (Lever et al., 2013). In some cases the degree of *mcrA* sequence similarity between sites is striking, with nearly identical phylotypes related to *Methanosarcina barkeri* detected in the Nankai Trough (Newberry et al., 2004) and the Peru Margin (Webster et al., 2006), and equally similar *mcrA* sequences related to *Methanobrevibacter arboriphilus* detected in the Nankai Trough (Newberry et al., 2004), Peru Margin (Webster et al., 2006), and off the Shimokita Peninsula (Imachi et al., 2011). All other genera have only been detected in single locations, indicating locally variable subsurface communities of methane-cycling Archaea.

Interestingly, none of the *mcrA* groups detected during drilling expeditions are unique to the deep biosphere. Several groups are, however, characteristic of marine environments. These include the putatively methanotrophic ANME-1 and ANME-2 Archaea, and the methanogenic genera *Methanococcoides* and *Methanococcus*. ANME-1 and ANME-2 Archaea have mainly been found in sulfate-rich habitats or within/near SMTZs, from estuarine and coastal sediments to deep-sea cold seep and hydrothermal sediments (e.g., Boetius et al., 2000; Michaelis et al., 2002; Thomsen, Finster, & Ramsing, 2001; Teske et al., 2002; Lloyd et al., 2011; Yanagawa et al., 2011; Biddle et al., 2012; for review, see Knittel & Boetius, 2009). Members of the genus *Methanococcoides*, obligately methylotrophic methanogens within the *Methanosarcinales*, are predominantly detected in and repeatedly isolated from marine sediment samples (Sowers & Ferry, 1983; Dhillon et al., 2005; Singh, Kendall, Liu, & Boone, 2005; Roussel et al., 2009; Parkes et al., 2012; reviewed in; Whitman, Bowen, & Boone, 2006). The H_2/ CO_2 and formate-utilizing genus *Methanococcus* has mainly been detected and isolated in salt marsh, estuarine, and coastal sediment (Whitman, Shieh, Sohn, Caras, & Premachandran, 1986; Franklin, Wiebe, & Whitman, 1988; Ward, Smith, & Boone, 1989; Kendall et al., 2006; reviewed in; Whitman et al.,

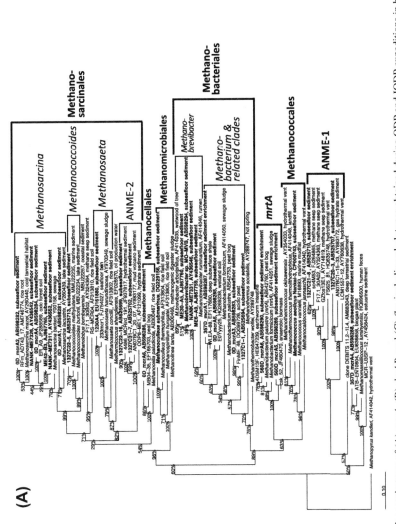

FIGURE 2.2.2.3 Phylogenetic trees of (A) *mcrA*, (B) *dsrAB*, and (C) *flis*, including phylotypes obtained during ODP and IODP expeditions in bold. All trees are based on nucleotide sequences and were produced using the neighbor-joining function with Jukes-Cantor Correction in the software program ARB (Ludwig et al., 2004). The recently published *mcrA* and *dsrAB* sequences from deep Juan de Fuca basalt (Lever et al., 2013) group near clone RS-MCR04 (AF313810) within the *Methanosarcinales*, within the ANME-1 lineage (Figure 2.2.2.3(A)) and within Cluster IV (Figure 2.2.2.3(B)), respectively.

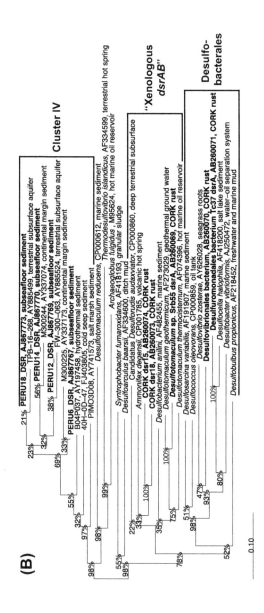

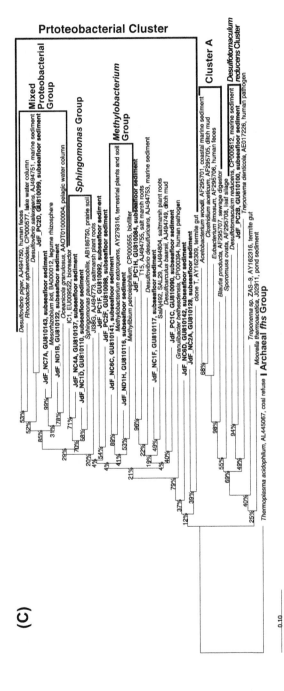

FIGURE 2.2.2.3 *Continued*

2006). Other genera, such as *Methanosarcina, Methanosaeta, Methanocella,* and *Methanoculleus,* cannot be considered 'typically marine,' as they are found not only in estuarine and marine sediment, but also across a wide range of non-marine anoxic habitats, including animal feces and sewage sludge, landfills, lake sediment, and freshwater wetlands (e.g., Lanoil, Sassen, La Duc, Sweet, & Nealson, 2001; Lueders, Chin, Conrad, & Friedrich, 2001; Luton, Wayne, Sharp, & Riley, 2001; Castro, Ogram, & Reddy, 2004; Dhillon et al., 2005; Liu & Whitman, 2008; Nunoura et al., 2008; Sakai et al., 2008; Zhang et al., 2008; Roussel et al., 2009; Parkes et al., 2012; reviewed in; Whitman et al., 2006; Liu & Whitman, 2008). Perhaps the most surprising finding is the detection of the genera *Methanobacterium* and *Methanobrevibacter* in marine subsurface sediments. Most known *Methanobacterium* species are sensitive to salinity, with culture media [NaCl] in excess of 0.2 M (~40% of typical seawater salinity) inhibiting cell growth (Whitman et al., 2006). The genus *Methanobrevibacter* is a typical inhabitant of animal intestines, sewage sludge, and decaying plant matter on land (Miller & Wolin, 1986; Ufnar et al., 2006; Whitman et al., 2006; Zeikus & Henning, 1975)—environments that differ strikingly from the energy-depleted deep subsurface biosphere.

According to thermodynamic competition theory, methanogenic Archaea using electron donors such as H_2, formate, and acetate are outcompeted by organisms performing energetically more favorable reactions, such as sulfate reduction, as long as the necessary electron acceptors are available (e.g., Cappenberg, 1974; Cord-Ruwisch, Seitz, & Conrad, 1988; Lovley & Goodwin, 1988); microbial communities in coastal marine sediments are no exception (Hoehler, Alperin, Albert, & Martens, 1998). Hence, only methanogens using noncompetitive substrates, i.e., C1 compounds such as methylamines, methyl sulfides, and methanol that are not used by most sulfate reducers, should be able to coexist with sulfate reducers in sulfate-rich sediment. Based on closest cultured relatives, methanogens in the deep subsurface consume the full spectrum of known methanogenic substrates (H_2/CO_2, formate, acetate, C1 compounds). Their distribution only partially reflects the zonation observed elsewhere, however. Consistent with the notion of thermodynamics-driven competition, *mcrA* genes are below detection or occur in low numbers in sulfate-reducing sediment of the Peru Trench and Cascadia Margin, respectively, and increase within the methanogenesis zone (Inagaki et al., 2006; Yoshioka et al., 2010); but they also occur at peak abundance in sulfate-rich sediment above the SMTZ at IODP Sites 1244 and 1245 in the Cascadia Margin (Colwell et al., 2008). Surprisingly, *mcrA* genes with high sequence similarity to hydrogenotrophic *Methanobrevibacter arboriphilus* appear in sulfate-rich surface sediments of ODP Site 1174 (Newberry et al., 2004). The highest *mcrA* abundances detected so far in the deep subseafloor are from sulfate-rich sediments near the Porcupine Seabight Challenger Mound (IODP U1318), in-depth layers with concomitantly high rates of hydrogenotrophic and aceti-clastic methanogenesis (Webster et al., 2009).

2.2.2.2.2 Sulfate Reduction

Sulfate-reducing bacteria are essential players in the terminal remineralization of organic matter in surficial and subsurface marine sediments (Jørgensen, 1982; D'Hondt et al., 2002, 2004), and this geochemical role motivates their detection, identification, and quantification. Subsurface sulfate-reducing bacteria and archaea are targeted with functional gene surveys based on PCR, cloning, and sequencing of two functionally conserved and phylogenetically informative key genes, dissimilatory sulfite reductase alpha and beta subunit (*dsrA* and *dsrB*), and adenosine-5′-phosphosulfate reductase alpha subunit (*aprA*). These genes allow the selective detection and phylogenetic identification of sulfate-reducing and sulfite-reducing prokaryotes against considerably more abundant background populations (Wagner et al., 1998; Klein et al., 2001; Friedrich, 2002; Zverlov et al., 2005).

Gene copy numbers of *dsrA* have been quantified via real-time PCR in sediments of the Peru Margin (ODP Sites 1227, 1229, 1230; Schippers & Neretin, 2006), Porcupine Seabight (IODP Sites 1316-18; Webster et al., 2009), the Gulf of Mexico continental slope (IODP Sites 1319-20, 1322, 1324; Nunoura et al., 2009), the Black Sea, and the Benguela upwelling system (Blazejak & Schippers, 2012; Schippers, Kock, Höft, Köweker, & Siegert, 2012). So far, all published real-time PCR quantifications on *dsrA* in subseafloor sediments have been obtained by the same primer pair (Kondo, Nedwell, Purdy, & de Queiroz Silva, 2004). In qPCR quantifications of Peru Margin site 1227, high *dsrA and aprA* gene copy numbers in the range of 10^7–10^8 gene copies per gram sediment were found in near-surface sediments; these numbers decrease toward 10^5–10^6 gene copies per gram sediment in the shallow subsurface near 1 m depth and gradually toward 10^3–10^4 gene copies per gram sediment at 10 m sediment depth, and sink below detection threshold at 121 and 42 mbsf (meters below seafloor), respectively (Blazejak & Schippers, 2012). A similar decline from 10^7 to 10^8 gene copies per gram sediment near the surface to below 10^3 gene copies per gram sediment was observed over the top 6 m of sediments from the central basin of the Black Sea (Blazejak & Schippers, 2012; Schippers et al., 2012). In sediments from the highly productive Benguela upwelling area offshore Namibia, the initially high *dsrA* and *aprA* gene copy numbers at the sediment surface (quite variable, 10^6–10^9) stabilized at higher values (10^4–10^7 down to 5 m depth) than in the Black Sea and the Peru Margin (Schippers et al., 2012). The *dsrA* and *aprA* gene copy numbers remained mostly two to three orders of magnitude below 16S-rRNA gene-based qPCR quantifications at the Peru Margin, in the Black Sea, and offshore Namibia. Thus, sulfate reducers generally represent on the order of 1% or less of the overall bacterial population in reducing subsurface sediments of the Black Sea and the Peru Margin (Blazejak & Schippers, 2012; Schippers & Neretin, 2006). Similar trends were found independently in Eastern North Atlantic sediments of the Porcupine Basin, where relatively low *dsrA* gene copy numbers in the range of 10^3–10^5 gene copies per milliliter remain mostly three orders of magnitude below bacterial 16S rRNA gene counts (Webster et al., 2009). In organic-poor deepwater turbidite sediments, *dsrA* gene detection becomes spotty, and quantification remains in the low range of 10^2–10^4 gene copies per milliliter (Nunoura et al., 2009). In coastal, organic-rich sediments with high sulfate-reducing activity (Aarhus Bay, Denmark), the proportion of sulfate reducers determined by *dsrA* gene qPCR increased to over 10% of the total bacterial count (Leloup et al., 2009). Viewed together, the *dsrA* and *aprA* gene copy trends are consistent with the geochemical prognosis for abundance and activity of sulfate reducers: the highest numbers are found in surficial sediments where fresh organic matter from the water column is supplied; gene copy numbers decrease downcore, and from organic-rich sediments in shallow coastal waters to organic-lean or recalcitrant sediments in deepwater locations. (Yet, the lower fraction of total microbial populations that is accounted for by sulfate reducers in organic-lean or recalcitrant offshore subsurface sediments compared to organic-rich coastal subsurface sediments remains unexplained.)

So far, phylogenetic identifications of sulfate (sulfite)-reducing microbes have been published from subseafloor sediment of the Peru Margin (ODP site 1227, Blazejak & Schippers, 2012; ODP Site 1228, Webster et al., 2006), from rust deposits of a circulation obviation retrofit kit at the seafloor on the Juan de Fuca Ridge Flank (ODP Site 1026; Nakagawa et al., 2006), from deep subsurface basalt in the Juan de Fuca Ridge flanks (IODP site 1036B; Lever et al., 2013), from the Black Sea, and from the Benguela Upwelling system offshore Namibia (Blazejak & Schippers, 2012; Schippers et al., 2012).

The *dsrA* and *aprA* genes recovered from the Peru Margin and Black Sea subsurface sediments belonged to diverse clusters and lineages of *Deltaproteobacteria* (families *Desulfobacteriaceae*, *Desulfovibrionaceae*, *Desulfobulbaceae*, *Desulfo-arculaceae*, *Synthrophobacteraceae*), *Firmicutes* (genus *Desulfotomaculum*), and clusters without cultured representatives (*dsrA* A and B) (Blazejak & Schippers, 2012). Some of the *dsrA* gene phylotypes recovered from deep subsurface sediments were nearly identical to those of cultured sulfate-reducing bacteria (*Desulfovibrio acrylicus*, Peru Margin). Moreover, *dsrA* phylotypes of *Desulfobacterium autotrophicum* were recovered both from Peru Margin and Black Sea sediments (Blazejak & Schippers, 2012). Independent *dsrAB* clone library surveys of sediment cores from the Peru Margin (Webster et al., 2006), the Black Sea, and the coastal Baltic Sea (Leloup et al., 2007; Leloup et al., 2009) have recovered members of the *Desulfobacteraceae*, the *Firmicutes* and numerous uncultured groups. Thus, the marine sedimentary subsurface is not dominated by a single group or by a phylogenetically restricted subgroup of sulfate reducers, but harbors multiple lineages that could differ in substrate spectra and geochemical niche.

One of the unsolved mysteries of deep subsurface microbiology is the identity of Cluster IV, a deeply branching *dsrAB* lineage that was initially reported from hydrothermal sediment of the Guaymas Basin (Dhillon, Teske, Dillon,

Stahl, & Sogin, 2003; Figure 2.2.2.3(B)), and has yet to be linked to a 16S rRNA lineage. This group has been detected in subseafloor sediments of the Peru Margin (Webster et al., 2006) and deeply buried basalt of the Juan de Fuca Ridge flank (Lever et al., 2013), in addition to—under various monikers— a wide range of estuarine to deep-sea marine sediments and cold seeps (e.g., Bahr et al., 2005; Jiang et al., 2009; Kaneko, Hayashi, Tanahashi, & Naganuma, 2007; Leloup et al., 2009; Ye et al., 2009), as well as sulfidic springs, subsurface aquifers, and freshwater sediment on land (Bagwell, Liu, Wu, & Zhou, 2005; Elsahed et al., 2003; Pester, Bittner, Pinsurang, Wagner, & Loy, 2010). The repeated detection of Cluster IV within sulfate-depleted sediments (Harrison, Zhang, Berelson, & Orphan, 2009; Leloup et al., 2009; Pester et al., 2010) suggests that it is capable of growth under low sulfate concentrations.

2.2.2.2.3 Acetogenesis

Phylogenetic characterizations of *fhs* are so far limited to one site, IODP Site 1301 on the Juan de Fuca Ridge Flank, a location with strong geochemical potential for acetogenesis (Heuer, Pohlman, Torres, Elvert, & Hinrichs, 2009; Lever et al., 2010). Using a previously published degenerate primer pair (Leaphart et al., 2003) and a less degenerate primer pair that amplifies a smaller gene fragment (Lever et al., 2010), a high diversity of *fhs* was detected across eight depths spanning the upper sulfate reduction zone, the upper and lower SMTZ, and one depth in the methanogenesis zone. While no sequences of the classic acetogens (Cluster A) were found and many of the phylotypes detected have no close cultured relatives, the considerable similarity of several sequences to known *Proteobacteria* is noteworthy (Figure 2.2.2.3(C)). Proteobacterial relatives include sulfate reducers within the *Deltaproteobacteria* (*Desulfovibrio salexigens, Dsv. desulfuricans*), as well as several methylotrophic (*Methylobacterium extorquens, Methylibium petroleiphilum*) and fermentative (*Mesorhizobium loti*) *Alphaproteobacteria*. Two further sequences have sulfate-reducing *Deltaproteobacteria* and *Firmicutes* as their closest cultured relatives (*Desulfoarculus baarsii, Desulfotomaculum reducens*). Since *Desulfoarculus baarsii* and several *Desulfotomaculum* species are facultative acetogens, their relatives detected at IODP Site 1301 may (at least partially) account for the ^{13}C-depleted acetate that indicates in situ acetogenesis at this site (Lever et al., 2010). Given that *fhs* occurs widely in nature and is not unique to acetogens, reliable interpretations regarding the organisms involved in acetogenesis in subsurface sediments are currently not possible. This may change as complementary genetic information is gained from the sequencing of whole genomes of uncultivated subsurface Bacteria and Archaea, and as innovative isotope labeling experiments that combine measurements of process rates and isotopic incorporation into biomass with genetic analyses are further applied to samples from the deep subsurface.

2.2.2.2.4 Reductive Dehalogenation

Phylotypes affiliated with the *Chloroflexi* have been frequently found in sub-seafloor sediments (Reed et al., 2002; Inagaki et al., 2003; Kormas et al., 2003; Parkes et al., 2005; Inagaki et al., 2006; also see review by Teske, 2006), raising questions regarding potential energy sources utilized by this phylum. The 16S rRNA sequence similarity to members of the *Dehalococcoides*, a genus comprised of obligate dehalorespiring bacteria, has led to the profiling and intercomparison of *rdhA* gene phylogenetic composition across six drilling sites within the Pacific Ocean (Futagami et al., 2009). These include ODP Site 1226 in the eastern equatorial Pacific, Site 1227 on the Peru Margin, Site 1230 in the Peru Trench, IODP Site 1301 on the Juan de Fuca Ridge Flank, as well as sites C9001 off Shimokita and C0002 in the Nankai Trough Forearc Basin (Futagami et al., 2009). Several primer pairs targeting *rdhA* were tested; however, only one, designed to specifically target *Dehalococcoides* (Krajmalnik-Brown et al., 2004), has led to successful detections. Sequences obtained uniformly share *Dehalococcoides ethenogenes* as their closest cultured relative. While several sequences are fairly closely related to terrestrial *Dehalococcoides*, two of the clusters (Subseafloor *rdhA* clusters I and II) are highly divergent from any previously detected *Dehalococcoides* sequences. The diversity and quantity of *rdhA* clusters and *rdhA* homologues were significantly extended with a primer-independent, metagenomic survey of subsurface sediments from offshore Shimokita (northwestern Japan), and with a reanalysis of previously published metagenome sequences from the Peru Margin, suggesting dehalogenation as a widespread metabolic mode in the marine subsurface (Kawai et al., 2014).

2.2.2.2.5 Future Perspectives of Functional Gene Analysis

Given that only a small number of functional genes have been explored in sub-seafloor habitats from a few locations, our ability to link biogenic processes that are evident from geochemical gradients (e.g., D'Hondt et al., 2002, 2004; 2009), isotopic compositions of metabolic educts and products (e.g., Böttcher et al., 2006; Heuer et al., 2009; Lever et al., 2010), and measured process rates (e.g., Jørgensen, D'Hondt, & Miller, 2006; Parkes et al., 2005) to the organisms responsible remains marginal. The following paragraphs cover possible future foci of functional gene research in the subseafloor. Topics discussed include technical ones related to detection issues, contamination risks, and the hunt for unstudied genes, as well as the integration of functional gene surveys with single-cell genomic, metagenomic, and experimental data.

The detection of genes related to biogenic processes that are evident from geochemical gradients is a common problem of functional gene surveys in the deep subseafloor (Nunoura et al., 2009; Schippers & Neretin, 2006; Webster et al., 2006, 2009). One plausible reason is that organisms responsible for these processes in subseafloor habitats are too divergent from their relatives in the surface world to be detected with the same methods. The sequencing of single-cell genomes of uncultivated phyla holds great promise in this regard (Stepanauskas, 2012), as it may lead to the discovery of divergent sequences of known functional genes that cannot be targeted with existing primers due to sequence mismatches. A further problem may lie in the high degree of degeneracy or limited phylogenetic coverage of existing primers. The following examples will demonstrate this problem: as previously mentioned, *mcrA* genes could only be detected at one depth at ODP Site 1230 (Inagaki et al., 2006). This was using a highly degenerate general *mcrA* primer pair with broad phylogenetic coverage (Springer, Sachs, Woese, & Boone, 1995). A less degenerate general primer pair with lower phylogenetic coverage (Hales et al., 1996) yielded no successful *mcrA* amplifications (Inagaki et al., 2006). By removing redundant degeneracies in the highly degenerate primer pair this problem has been solved and *mcrA* genes detected throughout the methanogenesis zone (Lever, 2008). Moreover, since ANME-1 Archaea were found to have highly divergent genetic sequences in the target loci of these general primer pairs, an ANME-1-specific *mcrA* primer pair was designed. This then led to the detection of ANME-1 *mcrA* genes throughout the SMTZ and upper 10 m of methanogenic sediment at ODP Site 1230 (Lever, 2008). A similar case occurred with *fhs* genes in the sediment column of IODP Site 1301. While a highly degenerate published primer pair (Leaphart et al., 2003) allowed *fhs* detection at four depths, *fhs* genes could be detected at four additional depths using a new, less degenerate primer pair that targets a smaller gene fragment (Lever et al., 2010). All three cases illustrate that unsuccessful functional gene detections may often be due to low PCR amplification efficiency resulting from high primer degeneracy or mismatches. These problems can be solved by careful in silico analyses, followed by redesign of existing primers, or design of altogether new primers.

Contamination of samples obtained by drilling is a major concern in any studies on the subseafloor biosphere, due to the use of drilling fluid (surface seawater, drilling mud), with significant populations of microbes. An important factor is the recirculation of drilling mud in a riser system, where an outer casing surrounds the drill pipe to provide return-circulation of drilling fluid to maintain pressure balance within the borehole (http://www.iodp.org/riser-vessel). The extent of sediment contamination during riserless drilling on the R/V *JOIDES Resolution* was controlled and monitored with fluorocarbon tracers and microspheres based on published protocols (Lever et al., 2006; Smith et al., 2000; House et al., 2003) and showed that seawater contamination affected mostly the outer centimeter layer of sediment cores. The detection limit on Leg 201 was 0.01 μl seawater per gram sediment, and the typical range of contamination was 0.01–1 μl seawater per gram sediment; 1 μl seawater corresponds to c.50 prokaryotic cells, or 500.000 cells per milliliter in open-ocean seawater (House et al., 2003). This estimate depends ultimately on the microbial cell content of seawater, and the assumption that remains unchanged during the drilling process. Contamination tests with perfluorocarbon tracers during riser drilling on R/V *Chikyu* during IODP Expedition 337 (Shimokita coalbed) showed similar patterns, from below 0.1 μl drilling fluid per gram sediment in the center of a core to almost 10 μl drilling fluid per gram sediment at the periphery. Due to the high cell density of at least 10^8 cells per milliliter in drilling mud, this range of contamination volumes corresponds to an equivalent of $<10^4–10^6$ nonindigenous cells per gram sediment (Inagaki, Hinrichs, Kubo, & the Expedition 337 Scientists, 2012); thus, contamination potential is significantly elevated compared to nonriser drilling. Strategies to constrain contamination levels include using offshore seawater to prepare drilling fluid, rather than water from polluted harbors. The baseline of drilling-induced contamination could be identified by microbial population analyses in wastewater and drilling mud onboard, in surface seawater, and circulating fluids, before and after riser drilling. The development of genetic assays by which indicator organisms, e.g., *Escherichia coli* of human sewage, or *Xanthomonas* of drilling mud (Masui, Morono, & Inagaki, 2008), are monitored in potential contamination sources and in samples, deserve consideration.

The default assumption should be that contamination risk is omnipresent, as shown by the following examples. During sampling of subseafloor basalt, contamination is evident from high concentrations of contamination tracer on the exterior of basalt rock (Lever et al., 2006), as well as from 16S rRNA gene inventories of rock surfaces that are dominated by sequences typically associated with the human microbiome (Santelli, Banerjee, Bach, & Edwards, 2010). The detection of *E. coli* and chloroplast DNA in sediments that are hundred thousands of years old suggests that even subseafloor sediments obtained by riserless drilling are at risk of being contaminated (Kormas et al., 2003). Similarly, reports of functional genes of *Methanobrevibacter* (Imachi et al., 2011; Newberry et al., 2004; Webster et al., 2006), a genus otherwise known from human/animal intestines, activated sludge, and decaying wood (Whitman et al., 2006) and used as a tracer to identify sewage pollution (Ufnar et al., 2006), require further confirmation. The same is true for other methanogenic genera, e.g., *Methanobacterium*, *Methanoculleus*, *Methanosaeta*, and *Methanosarcina*, which are also common inhabitants of human wastewater (Liu & Whitman, 2008). Contamination may occur if sewage dumping occurs during drilling operations. Sewage released to surface waters may be entrained into drilling fluid or give rise to microbial blooms in water surrounding the ship (Santelli et al., 2010).

Functional gene surveys have so far focused on a few genes that are indicative of anaerobic metabolism—with emphasis on terminal remineralization pathways. In comparison, genes that are diagnostic of intermediate processes, e.g., fermentation, and autotrophy have been relatively neglected (Lever, 2013). As a result, little is known about the organisms involved in these critical processes in the deep biosphere. This could change with the development of PCR primers targeting key genes of organisms involved in the breakdown of larger molecules and complex organic matter. A recent study on bioreactors, in which glycoside hydrolase genes of cellulose degradation (*cel48*, *cel5*) and iron hydrogenase genes indicative of fermentation (*hydA*) were successfully targeted using newly developed PCR primers, illustrates that organisms involved in the degradation of complex organic matter can also be studied using the functional gene approach (Pereyra, Hiibel, Prieto Riquelme, Reardon, & Pruden, 2010). The functional gene approach can also investigate chemoautotrophy (e.g., Campbell & Cary, 2004; Elsaied & Naganuma, 2001; Gagen et al., 2010; Hügler et al., 2010) and nitrogen fixation (Zehr, Jenkins, Short, & Steward, 2003), which may play key roles in sustaining microbial ecosystems in subseafloor basalts (Lever et al., 2013). Furthermore, given that much of the global seafloor is located underneath subtropical ocean gyres with very low primary production, oxygen penetrates deep into sediments and underlying basalt (D'Hondt et al., 2009; Expedition 329 Scientists, 2011; Ziebis et al., 2012; Roy et al., 2012). Therefore, functional genes of oxygen-dependent pathways, such as anammox or aerobic methylotrophy, could provide important clues to the in situ energy sources of microbes inhabiting these energy-starved environments.

In the age of whole-genome and high-throughput sequencing, the functional gene approach may seem dated, given that entire genomes of single cells can be assembled and annotated within days. Yet, there are several factors that make functional genes attractive targets for process-based studies of microbial populations. These include the fact that microbes involved in certain processes can be specifically targeted via PCR. Ecologically important minority groups such as sulfate reducers and methanogens, which may only represent a small portion of total microbial populations in the subseafloor, can be phylogenetically characterized and quantified in a cost-effective way (e.g., Pester et al., 2010), and without the bioinformatics challenges of large genomic datasets. Combined with analyses of functional gene transcripts, active use of a given gene and quantitative importance of the associated reaction can be inferred (Friaz-Lopez et al., 2008). To subseafloor microbial ecologists interested in linking microbial identity to processes across a large number of samples, sites, habitats, and experimental treatments, this can be a considerable advantage over other approaches. Yet, the reliance on genomic data to link functional gene to 16S rRNA identity and pure-culture isolation and/or isotope-labeling experiments to confirm (newly) invoked gene functions, shows the crucial importance of these approaches as a background to successful functional gene applications.

2.2.2.3 METAGENOMIC INVESTIGATIONS OF COMPLEX SUBSEAFLOOR COMMUNITIES

While individual ribosomal and functional genes have advanced the study of subsurface communities, the ability to analyze all genes in a community regardless of their origin, a metagenomics approach, has also enriched our understanding of subseafloor microbes. During the first microbiology-dedicated ocean drilling expedition, ODP Leg 201 in 2002 (D'Hondt et al., 2003), the technologies required to explore the genomics of subseafloor communities were becoming feasible, building on developments of the previous years (Ronaghi, Uhlen, & Nyren, 1998). The low yields from DNA extractions on subseafloor sediments hampered progress since many extractions yielded only a few nanograms of DNA (Webster, Newberry, Fry, & Weightman, 2003), and most genomic techniques required microgram amounts. This roadblock was overcome by the development of whole-genome amplification methods, also known as multiple-displacement amplification (MDA), which allowed the amplification of entire genomes without use of gene-specific primers and production of micrograms of DNA from picograms of starting material (Dean et al., 2002). Another prerequisite for the metagenomic analysis of the subsurface has been the establishment of second-generation sequencing technologies, which now enable the routine sequencing of millions of DNA sequences in single sequencing runs.

The first metagenomic analysis from deeply buried sediments was performed on sediments from ODP Leg 201 Site 1229 on the Peru Margin (Biddle et al., 2008). Site 1229 was of significant interest due to cell count maxima in SMTZs (D'Hondt et al., 2004; Biddle et al., 2006), which indicated that cells were active and responding to in situ conditions at these biogeochemical interfaces (Biddle et al., 2006; Parkes et al., 2005). These initial investigations at this site warranted a fuller exploration of the functional potential of subsurface microbes by metagenomics. Samples of site 1229 were analyzed at 1, 16, 32, and 50 mbsf and amplified using MDA. Metagenomes were created using a GS20 pyrosequencer from 454 Life Sciences (now Roche). Over 60 million base pairs of sequence data were generated, yet less than 15% of these were identifiable based on gene homologies with previously published and archived genes (Biddle et al., 2008). This could have been due to artifacts created by the MDA technique, the short sequence read (~100 base pairs) of the GS20 sequencer, or the novelty of the subsurface genomes. The bacterial phyla *Proteobacteria, Firmicutes,* and *Chloroflexi* and the archaeal phylum *Euryarchaeota* dominated the phylogenetic composition of this subsurface metagenome (Biddle et al., 2008). Despite the prominent geochemical profiles of sulfate and methane at this site, few functional genes for methanogenesis, sulfate-dependent methane oxidation, and sulfate reduction were found. Moreover, genes linked to these processes did not increase with depth, indicating that microbes involved in sulfate and methane geochemistry are not numerically dominant in the subsurface. This result is congruent with qPCR surveys of sulfate-reducing and methane-cycling key genes (see Section 2 of this chapter), with the relative scarcity of sulfate-reducing key genes (*dsrA, dsrB*) based on average coverage ratio in comparison to universally conserved single-copy genes in metagenomic surveys (Kawai et al., 2014), and with the $\delta^{13}C$-isotopic composition of prokaryotic cells and intact polar lipids in marine subsurface sediments, which indicate that methane and DIC are not significant carbon substrates for subsurface cells (Biddle et al., 2006). Thus, several strands of evidence suggest that subsurface microbial communities subsist heterotrophically on buried organic carbon of photosynthetic origin (Biddle et al., 2006; Lipp, Morono, Inagaki, & Hinrichs, 2008; Morono et al., 2011).

The Peru Margin metagenome was complemented with the metagenome of site U1320 in the Brazos-Trinity Basin in the Gulf of Mexico, drilled on IODP Expedition 308 (Flemings, Behrmann, John, & the Expedition 308 Scientists, 2006). This study again utilized MDA technology; however, a newer version of next-generation sequencing was used, the GS FLX system from 454/Roche, Inc. (Biddle, White, Teske, & House, 2011). The advanced sequencing technology analyzed over 100 million base pairs from a single depth (8 mbsf) of IODP Site 1320. Sequence reads were over 200 base pairs long, and 24% of the data were identifiable, suggesting that shorter reads had previously confounded the identification and annotation of a mostly novel metagenome. Genes with high-sequence similarities to those of Site 1229 were found, with similarly low numbers of identifiable genes for specific geochemical processes (Biddle et al., 2011). Cluster analysis showed that the sites could be discriminated by functional genes. However, phylogenetic associations showed distinct clusters of shallow and deep samples. Correlations to geochemistry were examined using available data from shipboard ODP and IODP analyses: ammonium concentrations showed a strong positive correlation with relative abundances of genes related to posttranslational modification and energy production/conversion categories, and negative correlations for genes related to cell motility and membrane biosynthesis (Biddle et al., 2011), indicative of distinct microbial responses to variable nitrogen availability within consistently organic-rich, reducing, and sulfidic sediments at Site 1229.

Due to the low recovery of DNA in DNA extractions from the subsurface, advances in amplification and sequencing technology have drastically increased our ability to analyze subsurface metagenomes. However, the use of MDA is associated with biases that are difficult to overcome, including uneven amplification and chimera formation (Binga, Lasken, & Neufeld, 2008). Thus, new methods are being investigated for the generation of sediment metagenomes. One such method is to skip the MDA step in sequencing of low-biomass samples and instead amplify the small quantity of DNA using random primers attached to sequencing pipeline adapters. This technique, termed Random Amplification Metagenomic PCR (Martino et al., 2012), was tested on both control genomes and the Peru Margin Site 1229 samples analyzed previously (Biddle et al., 2008). This technique shows promise in that it is able to generate similar data with fewer concerns about artifacts introduced by MDA. This study also showed that an improved sequencing technology that produced sequence reads over 400 base pairs long further increased the ability to identify sequence, up to 65% of the dataset (Martino et al., 2012). Fortunately, despite the further increase in identification, the interpretation of subsurface genomes did not change drastically relative to previous studies.

Important advances in sequencing technology are impacting subsurface microbiology. Next-generation sequencing technologies to analyze PCR-amplified ribosomal and functional genes are resulting in nearly 1000 times more data per sample than the classic clone library approach. For example, pyrosequencing of 16S rRNA gene fragments has opened up investigations of the microbial "rare biosphere"—microorganisms that occur in low densities and are therefore overlooked in low-throughput clone library analyses, but collectively make up a substantial proportion of all recovered gene signatures in pyrosequencing surveys (Sogin et al., 2006). The subsurface rare biosphere was examined as part of the International Census of Marine Microbes effort, with most sites from ODP Leg 201 being examined. Interesting new datasets come from IODP Expedition 316 to the Nankai Trough and Expedition 325 to the Great Barrier Reef (Mills et al., 2012; Mills, Reese, & Peter, 2012). Bacterial rRNA profiles of sediment samples from 1, 19, and 57 mbsf in the Nankai Trough showed a drastic decrease in diversity between 1 mbsf and the deeper samples (Mills et al., 2012). Similar to the metagenomic analyses, the bacterial phyla of *Proteobacteria* and *Chloroflexi* are well represented (>50% of the dataset across all depths). Although the ribosomal gene does not indicate functional potential, genera sampled are known to participate in sulfur cycling, and made up 5–15% of the dataset. The most abundant potential function was nitrogen cycling, comprising 29–50% of the sampled genera. Lineages associated with fermentation accounted for 14–28% of the sequence dataset. With over 12,020 sample sequences representing over 1000 operational taxonomic units, this amplicon sequencing approach drastically increases our ability to interrogate the deep biosphere. However, such analyses are critically dependent on sample integrity and preservation: The bacterial 16S rRNA pyrosequencing signatures in sediment core samples from 2, 20, and 40 mbsf at the Great Barrier Reef that had been stored at 4 °C for 3 months differed strongly from controls that had been frozen immediately after recovery. Extended live storage selected against the originally dominant fermentative bacterial families, and replaced them by an enrichment of diverse aerobic, nitrate-reducing, metal-reducing, and sulfate-reducing families—with implicit consequences for geochemical changes in the sediment samples (Mills, Reese, & Peter, 2012). Thus, samples for microbiological and biogeochemical analyses need to be stored under fixed or frozen conditions, as previously suggested (Lin et al., 2010), or they have to be examined immediately after recovery.

The volume of sequencing surveys is increasing rapidly, even between different revisions of this chapter as it is being written. For an extensive metatranscriptome of the Peru Margin subsurface, over 1 billion cDNA sequences were analyzed (Orsi et al., 2013b); the results defy a brief summary, but indicate that anaerobic metabolism of amino acids, carbohydrates, and lipids is the dominant metabolic process, consistent with the notion of the heterotrophic subsurface biosphere (Orsi et al., 2013b). Interestingly, the mRNA profiles are taxonomically at odds with previous 16S rRNA gene and 16S rRNA transcript analyses of the same Peru Margin site (ODP 1229); gamma-proteobacterial mRNA transcripts abound especially in deeper sediment layers, whereas transcripts derived from *Archaea* or the *Chloroflexi* (major community components at site 1229) appear in reduced relative proportions. Such discrepancies might be attributed to divergent mRNA expression levels across different microbial phyla, to the effects of lateral gene transfer, or to difficulties in assigning taxonomic positions to transcripts with low homologies to reference genes—a problem that could be solved by limiting the phylogenetic assignments to transcripts that satisfy stringent taxonomic criteria.

We conclude this chapter on gene sequence-based exploration of the deep subsurface biosphere with the caveat that, if cultivation efforts could keep up with the rapid development of sequencing technologies, molecular signatures would be easier to interpret. Most major groups of subsurface microbes remain for the time being uncultured. Continued cultivation efforts, metagenomic surveys, and single-cell genome sequencing of novel lineages will provide more insight into the genomic potential, physiological capabilities, biochemical pathways, and biogeochemical function of subsurface microbial life.

REFERENCES

Amann, R., Ludwig, W., & Schleifer, K. H. (1988). β-Subunit of ATP-synthase: a useful marker for studying the phylogenetic relationship of Eubacteria. *Journal of General Microbiology, 134*, 2815–2821.

Bagwell, C. E., Liu, X., Wu, L., & Zhou, J. (2005). Effects of legacy nuclear waste on the compositional diversity and distributions of sulfate-reducing bacteria in a terrestrial subsurface aquifer. *FEMS Microbiology Ecology, 55*, 424–431.

Bahr, M., Crump, B., Klepac-Ceraj, V., Teske, A., Sogin, M., & Hobbie, J. (2005). Molecular characterization of sulfate-reducing bacteria in a New England salt marsh. *Environmental Microbiology, 7*, 1175–1185.

Baker, B. J., & Dick, G. J. (2013). Omic approaches in microbial ecology: charting the unknown. *ASM News, 9*, 353–360.

Barns, S. M., Delwiche, C. F., Palmer, J. D., & Pace, N. R. (1996). Perspectives on archaeal diversity, thermophily and monophyly from environmental rRNA sequences. *Proceedings of the National Academy of Sciences USA, 93*, 9188–9193.

Batzke, A., Engelen, B., Sass, H., & Cypionka, H. (2007). Phylogenetic and physiological diversity of cultured deep-biosphere bacteria from equatorial Pacific Ocean and Peru margin sediments. *Geomicrobiology Journal, 24*, 261–273.

Biddle, J. F., Cardman, Z., Mendlovitz, H., Albert, D. B., Lloyd, K. G., Boetius, A., & Teske, A. (2012). Anaerobic oxidation of methane at different temperature regimes in Guaymas Basin hydrothermal sediments. *ISME J, 6*, 1018–1031.

Biddle, J. F., House, C. H., & Brenchley, J. E. (2005a). Enrichment and cultivation of microorganisms from sediment from the slope of the Peru Trench (ODP site 1230). In B. B. Jørgensen, S. L. D'Hondt, & D. J. Miller (Eds.), *Proc. ODP, sci. results* (Vol. 201). [Online]. Available from World Wide Web: http://www-odp.tamu.edu/publications201_SR/107107.htm.

Biddle, J. F., House, C. H., & Brenchley, J. E. (2005b). Microbial stratification in deeply-buried marine sediment reflects changes in sulfate/methane profiles. *Geobiology, 3*, 287–295.

Biddle, J. F., Lipp, J. S., Lever, M. A., Lloyd, K. G., Sørensen, K. B., Anderson, R., et al. (2006). Heterotrophic archaea dominate sedimentary subsurface ecosystems off Peru. *Proceedings of the National Academy of Sciences USA, 103*, 3846–3851.

Biddle, J. F., Fitz-Gibbon, S., Schuster, S. C., Brenchley, J. E., & House, C. H. (2008). Metagenomic signatures of the Peru Margin subseafloor biosphere show a genetically distinct environment. *Proceedings of the National Academy of Sciences USA, 105*, 10583–10588.

Biddle, J. F., White, J. R., Teske, A., & House, C. H. (2011). Metagenomics of the subsurface Brazos-Trinity Basin (IODP Site 1320): comparison with other sediment and pyrosequenced metagenomes. *The ISME Journal, 5*, 1038–1047.

Binga, E. K., Lasken, R. S., & Neufeld, J. D. (2008). Something from (almost) nothing: the impact of multiple displacement amplification on microbial ecology. *The ISME Journal, 2*, 233–241.

Blazejak, A., & Schippers, A. (2010). High abundance of JS-1- and chloroflexi-related bacteria in deeply buried marine sediments revealed by quantitative, real-time PCR. *FEMS Microbiology Ecology, 72*, 198–207.

Blazejak, A., & Schippers, A. (2012). Real-time quantification and diversity analysis of the functional genes *aprA* and *dsrA* of sulfate-reducing prokaryotes in marine sediments of the Peru continental Margin and the Black Sea. *Front Microbiology, 2*, 253. http://dx.doi.org/10.3389/jmicb.2011.00253.

Boetius, A., Ravenschlag, K., Schubert, C. J., Rickert, D., Widdel, F., Gieseke, A., et al. (2000). A marine microbial consortium apparently mediating anaerobic oxidation of methane. *Nature, 407*, 623–626.

Borrel, G., O'Toole, P. W., Harris, H. M. B., Peyret, P., Brugere, J.-F., & Gribaldo, S. (2013). Phylogenomic data support a seventh order of methylotrophic methanogens and provide insights into the evolution of methanogenesis. *Genome Biology and Evolution, 5*, 1769–1780.

Böttcher, M. E., Ferdelman, T. G., Jørgensen, B. B., Blake, R. E., Surkov, A. V., & Claypool, G. E. (2006). 6. Sulfur isotope fractionation by the deep biosphere within sediments of the Eastern Equatorial Pacific and Peru Margin. In B. B. Jørgensen, S. L. D'Hondt, & D. J. Miller (Eds.). *Proc. ODP, sci. res: Vol. 201.* (pp. 1–45). College Station, TX: ODP.

Briggs, B. R., Pohlmann, J. W., Torres, M., Riedel, M., Brodie, E. L., & Colwell, F. S. (2011). Macroscopic biofilms in fracture-dominated sediment that anaerobically oxidize methane. *Applied Environmental Microbiology, 77*, 6780–6787.

Brochier-Armanet, C., Boussau, B., Gribaldo, S., & Forterre, P. (2008). Mesophilic crenarchaeota: proposal for a third archaeal phylum, the thaumarchaeota. *Nature Reviews Microbiology, 6*, 245–252.

Bult, C. J., White, O., Olsen, G. J., Zhou, L., Fleischmann, R. D., Sutton, G. G., et al. (1996). Complete genome sequence of the methanogenic archaeon, *Methanococcus jannaschii*. *Science, 273*, 1058–1073.

Campbell, B. J., & Cary, S. C. (2004). Abundance of reverse tricarboxylic acid cycle genes in free-living microorganisms at deep-sea hydrothermal vents. *Applied Environmental Microbiology, 70*, 6282–6289.

Cappenberg, T. E. (1974). Inter-relations between sulfate-reducing and methane-producing bacteria in bottom deposits of a fresh-water lake. 2.Inhibition experiments. *Antonie Van Leeuwenhoek, 56*, 1247–1258.

Castro, H., Ogram, A., & Reddy, K. R. (2004). Phylogenetic characterization of methanogenic assemblages in eutrophic and oligotrophic areas of the Florida Everglades. *Applied Environmental Microbiology, 70*, 6559–6568.

Colwell, F. S., Boyd, S., Delwiche, M. E., Reed, D. W., Phelps, T. J., & Newby, D. T. (2008). Estimates of biogenic methane production rates in deep marine sediments at Hydrate Ridge, Cascadia Margin. *Applied Environmental Microbiology, 74*, 3444–3452.

Cord-Ruwisch, R., Seitz, H.-J., & Conrad, R. (1988). The capacity of hydrogenotrophic anaerobic bacteria to compete for traces of hydrogen depends on the redox potential of the terminal electron acceptor. *Archives of Microbiology, 149*, 350–357.

Dean, F. B., Hosono, S., Fang, L., Wu, X., Faruqi, A. F., Bray-Ward, P., et al. (2002). Comprehensive human genome amplification using multiple displacement amplification. *Proceedings of the National Academy of Sciences USA, 99*, 5261–5266.

D'Hondt, S., Rutherford, S., & Spivack, A. J. (2002). Metabolic activity of subsurface life in deep-sea sediments. *Science, 295*, 2067–2070.

D'Hondt, S. L., Jørgensen, B. B., Miller, D. J., et al. (Eds.). (2003). *Proc. ODP, init. repts* (Vol. 201). College Station, TX: Ocean Drilling Program. http://dx.doi.org/10.2973/odp.proc.ir.201.2003.

D'Hondt, S., Jørgensen, B. B., Miller, D. J., Batzke, A., Blake, R., Cragg, B. A., et al. (2004). Distribution of microbial activities in deep subseafloor sediments. *Science, 306*, 2216–2221.

D'Hondt, S., Spivack, A. J., Pockalny, R., Ferdelman, T. G., Fischer, J. P., Kallmeyer, J., et al. (2009). Subseafloor sedimentary life in the South Pacific Gyre. *Proceedings of the National Academy of Sciences USA, 106*, 11651–11656.

DeLong, E. F. (1992). Archaea in coastal marine environments. *Proc Natl Acad Sci USA, 89*, 5685–5689.

Dhillon, A., Teske, A., Dillon, J., Stahl, D. A., & Sogin, M. L. (2003). Molecular characterization of sulfate-reducing bacteria in the Guaymas Basin. *Applied Environmental Microbiology, 69*, 2765–2772.

Dhillon, A., Lever, M., Lloyd, K. G., Albert, D. B., Sogin, M. L., & Teske, A. (2005). Methanogen diversity evidenced by molecular characterization of methyl coenzyme M reductase A (*mcrA*) genes in hydrothermal sediments of the Guaymas Basin. *Applied Environmental Microbiology, 71*, 4592–4601.

Durbin, A. M., & Teske, A. (2010). Sediment-associated microdiversity within the marine group I crenarchaeota. *Environmental Microbiology Reports, 2*, 693–703.

Durbin, A. M., & Teske, A. (2011). Microbial diversity and stratification of South Pacific abyssal marine sediments. *Environmental Microbiology, 13*, 3219–3234.

Durbin, A. M., & Teske, A. (2012). Archaea in organic-lean and organic-rich marine subsurface sediments: an environmental gradient reflected in distinct phylogenetic lineages. *Frontiers in Microbiology, 3*, 168. http://dx.doi.org/10.3389/fmicb.2012.00168.

Edgcomb, V. P., Beaudoin, D., Gast, R., Biddle, J. F., & Teske, A. (2011). Marine subsurface eukaryotes: the fungal majority. *Environmental Microbiology, 13*, 172–183.

Edgcomb, V. P., & Biddle, J. F. (2011). Microbial eukaryotes in the marine subsurface. In A. V. Altenbach, et al. (Ed.), *Anoxia: evidence for eukaryote survival and paleontological strategies, cellular origin, life in extreme habitats and astrobiology* (Vol. 21) (pp. 479–493). http://dx.doi.org/10.1007/978-94-007-1896-8_25.

Elsahed, M. S., Senko, J. M., Najar, F. Z., Kenton, S. M., Roe, B. A., Dewers, T. A., et al. (2003). Bacterial diversity and sulfur cycling in a mesophilic sulfide-rich spring. *Applied Environmental Microbiology, 69*, 5609–5621.

Elsaied, H., & Naganuma, T. (2001). Phylogenetic diversity of ribulose-1,5-bisphosphate carboxylase/oxygenase large-subunit genes from deep-sea microorganisms. *Applied Environmental Microbiology, 67*, 1751–1765.

Engelhardt, T., Sahlberg, M., Cypionka, H., & Engelen, B. (2013). Biogeography of *Rhizobium radiobacter* and distribution of associated temperate phages in deep subseafloor sediments. *The ISME Journal, 7*, 199–209.

Expedition 329 Scientists. (2011). South Pacific Gyre subseafloor life. *IODP Preliminary Report, 329.* . http://dx.doi.org/10.2204/iodp.pr.329.2011.

Flemings, P. B., Behrmann, J. H., John, C. M., & the Expedition 308 Scientists (Eds.). (2006). *Proc. IODP* (Vol. 308). College Station TX: Integrated Ocean Drilling Program Management International, Inc. http://dx.doi.org/10.2204/iodp.proc.308.101.2006.

Forschner, S. R., Sheffer, R., Rowley, D. C., & Smith, D. C. (2009). Microbial diversity in cenozoic sediments from the Lomonosov Ridge in the central Arctic Ocean. *Environmental Microbiology, 11*, 630–639.

Fox, G. E., Stackebrandt, E., Hespell, R. B., Gibson, J., Maniloff, J., Dyer, T. A., et al. (1980). *Science, 209*, 457–463.

Franklin, M. J., Wiebe, W. J., & Whitman, W. B. (1988). Populations of methanogenic bacteria in a Georgia salt marsh. *Applied Environmental Microbiology, 54*, 1151–1157.

Frias-Lopez, J., Shi, Y., Tyson, G. W., Coleman, M. L., Schuster, S. C., Chisholm, S. W., et al. (2008). Microbial community gene expression in ocean surface waters. *Proceedings of the National Academy of Sciences USA, 105*, 3805–3810.

Friedrich, M. W. (2002). Phylogenetic analysis reveals multiple lateral transfers of adenosine-5'-phosphosulfate reductase genes among sulfate-reducing microorganisms. *Journal of Bacteriology, 184*, 278–289.

Fry, J. C., Parkes, R. J., Cragg, B. A., Weightman, A. J., & Webster, G. (2008). Prokaryotic biodiversity and activity in the deep subseafloor biosphere. *FEMS Microbiology Ecology, 66*, 181–196.

Fuerst, J. A., & Sagulenko, E. (2011). Beyond the bacterium: planctomycetes challenge our concepts of microbial structure and function. *Nature Reviews Microbiology, 9*, 403–413.

Futagami, T., Morono, Y., Terada, T., Kaksonen, A. H., & Inagaki, F. (2009). Dehalogenation activities and distribution of reductive dehalogenase homologous genes in marine subsurface sediments. *Applied Environmental Microbiology, 75*, 6905–6909.

Futagami, T., Morono, Y., Terada, T., Kaksonen, A. H., & Inagaki, F. (2013). Distribution of dehalogenation activity in subseafloor sediments of the Nankai Trough subduction zone. *Philosophical Transactions of the Royal Society B, 368*, 20120249.

Gagen, E. J., Denman, S. E., Padmanabha, J., Zadbuke, S., Jassim, R. A., Morrison, M., et al. (2010). Functional gene analysis suggests different acetogen populations in the bovine rumen and Tammar Wallaby forestomach. *Applied Environmental Microbiology, 76*, 7785–7795.

Gagen, E. J., Huber, H., Meador, T., Hinrichs, K.-U., & Thomm, M. (2013). A novel cultivation-based approach for understanding the miscellaneous crenarchaeotic group (MCG) archaea from sedimentary ecosystems. *Applied Environmental Microbiology, 79*, 6400–6406.

Guy, L., & Ettema, T. J. (2011). The archaeal 'TACK' superphylum and the origin of eukaryotes. *Trends in Microbiology, 19*, 580–587.

Hales, B. A., Edwards, C., Ritchie, D. A., Hall, D., Pickup, R. W., & Saunders, J. R. (1996). Isolation and identification of methanogen-specific DNA from blanket bog peat by PCR amplification and sequence analysis. *Applied Environmental Microbiology, 62*, 668–675.

Hallam, S. J., Putnam, N., Preston, C. M., Detter, J. C., Rokhsar, D., Richardson, P. M., et al. (2004). Reverse methanogenesis: testing the hypothesis with environmental genomics. *Science, 305*, 1457–1462.

Harrison, B. K., Zhang, H., Berelson, W., & Orphan, V. J. (2009). Variations in archaeal and bacterial diversity associated with the sulfate-methane transition zone in continental margin sediments (Santa Barbara Basin, California). *Applied Environmental Microbiology, 75*, 1487–1499.

Heuer, V. B., Pohlman, J. W., Torres, M. E., Elvert, M., & Hinrichs, K.-U. (2009). The stable carbon isotope biogeochemistry of acetate and other dissolved carbon species in deep subseafloor sediments at the northern Cascadia Margin. *Geochim Cosmochim Acta, 73*, 3323–3336.

Hinrichs, K.-U., & Inagaki, F. (2012). Downsizing the deep biosphere. *Science, 338*, 204–205.

Hoehler, T. M., Alperin, M. J., Albert, D. B., & Martens, C. S. (1998). Thermodynamic control on hydrogen concentrations in anoxic sediments. *Geochim Cosmochim Acta, 62*, 1745–1756.

Hoshino, T., Morono, Y., Terada, T., Imachi, H., Ferdelman, T. G., & Inagaki, F. (2011). Comparative study of subseafloor microbial community structures in deeply buried coral fossils and sediment matrices from the Challenger Mound in the Porcupine Seabight. *Frontiers in Microbiology, 2*, 231. http://dx.doi.org/10.3389/fmicb.2011.00231.

House, C. H., Cragg, B. A., Teske, A., & Leg 201 Scientific Party (2003). 2. Drilling contamination tests during ODP Leg 201 using chemical and particulate tracers. *Proc Ocean Drilling Prog Init Rept, 201*, 1–18.

Huber, R., Rossnagel, P., Woese, C. R., Rachel, R., Langworth, T. A., & Stetter, K. O. (1996). Formation of ammonium from nitrate during chemolithoautotrophic growth of the extremely thermophilic bacterium *Ammonifex degensii* gen. nov. sp. nov.. *Systematic and Applied Microbiology, 19*, 40–49.

Huber, H., Hohn, M. J., Rachel, R., Fuchs, T., Wimmer, V. C., & Stetter, K. O. (2002). A new phylum of archaea represented by a nanosized hyperthermophilic symbiont. *Nature, 417*, 63–67.

Hug, L. A., Castelle, C. J., Wrighton, K. C., Thomas, B. C., Sharon, I., Frischkorn, K. R., et al. (2013). Community genomic analyses constrain the distribution of metabolic traits across the *Chloroflexi* phylum and indicate roles in sediment carbon cycling. *Microbiome, 1*, 22. http://dx.doi.org/10.1186/2049-2618-1-22.

Hügler, M., Gärtner, A., & Imhoff, J. F. (2010). Functional genes as markers of sulfur cycling and CO_2 fixation in microbial communities of hydrothermal vents of the Logatchev field. *FEMS Microbiology Ecology, 73*, 526–537.

Hugenholtz, P., Pitulle, C., Hershberger, K. L., & Pace, N. R. (1998). Novel division level bacterial diversity in a Yellowstone hot spring. *Journal of Bacteriology, 180*, 366–376.

Hugenholtz, P. (2002). Exploring prokaryotic diversity in the genomic era. *Genome Biology, 3*, 1–8.

Imachi, H., Aoi, K., Tasumi, E., Saito, Y., Yamanaka, Y., Saito, Y., et al. (2011). Cultivation of methanogenic community from subseafloor sediments using a continuous-flow bioreactor. *ISME Journal, 5*, 1913–1925.

Imachi, H., Sakai, S., Lipp, J. S., Miyzaki, M., Saito, Y., Yamanaka, Y., et al. (2014). *Pelolinea submarina* gen.nov., sp. nov., an anaerobic, filamentous bacterium of the phylum *Chloroflexi* isolated from subseafloor sediment. *International Journal of Systematic and Evolutionary Microbiology, 64*, 812–818.

Inagaki, F., Takai, K., Komatsu, T., Kanamatsu, T., Fujiioka, K., & Horikoshi, K. (2001). Archaeology of archaea: geomicrobiological record of pleistocene thermal events concealed in a deep-sea subseafloor environment. *Extremophiles, 5*, 385–392.

Inagaki, F., Suzuki, M., Takai, K., Oida, H., Sakamoto, T., Aoki, K., et al. (2003). Microbial communities associated with geological horizons in coastal subseafloor sediments from the Sea of Okhotsk. *Applied Environmental Microbiology, 69*, 7224–7235.

Inagaki, F., Nunoura, T., Nakagawa, S., Teske, A., Lever, M. A., Lauer, A., et al. (2006). Biogeographical distribution and diversity of microbes in methane hydrate-bearing deep marine sediments on the Pacific Ocean Margin. *Proceedings of the National Academy of Sciences USA, 103*, 2815–2820.

Inagaki, F., Hinrichs, K.-U., Kubo, Y., & the Expedition 337 Scientists (2012). Deep coalbed biosphere off shimokita: microbial processes and hydrocarbon system associated with deeply buried coalbed in the ocean. *IODP Preliminary Report, 337*. http://dx.doi.org/10.2204/iodp.pr.337.2012.

Jiang, L., Zheng, Y., Peng, X., Zhou, H., Zhang, C., Xiao, X., et al. (2009). Vertical distribution and diversity of sulfate-reducing prokaryotes in the Pearl River estuarine sediments, Southern China. *FEMS Microbiology Ecology*, *70*, 249–262.

Jørgensen, B. B. (1982). Mineralization of organic matter in the sea bed - the role of sulphate reduction. *Nature*, *296*, 643–645.

Jørgensen, B. B., D'Hondt, S. L., & Miller, D. J. (2006). Leg 201 synthesis: controls on microbial communities in deeply buried sediments. In B. B. Jørgensen, S. L. D'Hondt, & D. J. Miller (Eds.), *Proc. ODP, sci. res* (Vol. 201) (pp. 1–45). College Station, TX: ODP.

Jurgens, G., Glöckner, F. O., Amann, R., Saano, A., Montonen, L., Likolammi, M., et al. (2000). Identification of novel archaea in bacterioplankton of a boreal forest lake by phylogenetic analysis and fluorescent in situ hybridization. *FEMS Microbiology Ecology*, *34*, 45–56.

Kallmeyer, J., Pockalny, R., Adhikari, R. R., Smith, D. C., & D'Hondt, S. (2012). Global distribution of microbial abundance and biomass in subseafloor sediment. *Proceedings of the National Academy of Sciences USA*, *109*, 16213–16216.

Kaneko, R., Hayashi, T., Tanahashi, M., & Naganuma, T. (2007). Phylogenetic diversity and distribution of dissimilatory sulfite reductase genes from deep-sea sediment cores. *Marine Biotechnology*, *9*, 429–436.

Kawai, M., Futagami, T., Toyoda, A., Takaki, Y., Nishi, S., Hori, S., et al. (2014). High frequency of phylogenetically diverse reductive dehalogenase-homologius genes in deep subsurface sedmentary metagenomes. *Frontiers in Microbiology*, *5*, 80. http://dx.doi.org/10.3389/fmic.201400080.

Kellermann, M., Wegener, G., Elvert, M., Yoshinaga, M. Y., Lin, Yu-S., Holler, T., et al. (2012). Autotrophy as a predominant mode of carbon fixation in anaerobic methane-oxidizing microbial communities. *Proceedings of the National Academy of Sciences USA*, *109*, 19321–19326.

Kendall, M. M., Liu, Y., Sieprawska-Lupa, M., Stetter, K. O., Whitman, W. B., & Boone, D. R. (2006). *Methanococcus aeolicus* sp. nov., a mesophilic, methanogenic archaeon from shallow and deep marine sediments. *International Journal of Systematic and Evolutionary Microbiology*, *56*, 1525–1529.

Kirkpatrick, J., Oakley, B., Fuchsman, C., Srinivasan, S., Staley, J. T., & Murray, J. W. (2006). Diversity and distribution of *Planktomycetes* and related bacteria in the suboxic zone of the Black Sea. *Applied Environmental Microbiology*, *72*, 3079–3083.

Klein, M., Friedrich, M., Roger, A. J., Hugenholtz, P., Fishbain, S., Abicht, H., et al. (2001). Multiple lateral transfers of dissimilatory sulfite reductase genes between major lineages of sulfate-reducing prokaryotes. *Journal of Bacteriology*, *183*, 6028–6035.

Kondo, R., Nedwell, D. B., Purdy, K. J., & de Queiroz Silva, S. (2004). Detection and enumeration of sulphate-reducing bacteria in estuarine sediments by competitive PCR. *Geomicrobiology Journal*, *21*, 145–157.

Kormas, A. K., Smith, D. C., Edgcomb, V., & Teske, A. (2003). Molecular analysis of deep subsurface microbial communities in Nankai Trough sediments (ODP Leg 190, Site 1176). *FEMS Microbiology Ecology*, *45*, 115–125.

Kubo, K., Lloyd, K. G., Biddle, J. F., Amann, R., Teske, A., & Knittel, K. (2012). Archaea of the miscellaneous crenarchaeotal group (MCG) are abundant, diverse and widespread in marine sediments. *The ISME Journal*, *6*, 1949–1965.

Knittel, K., & Boetius, A. (2009). Anaerobic oxidation of methane: progress with an unknown process. *Annual Review of Microbiology*, *63*, 311–334.

Kragelund, C., Levantesi, C., Borger, A., Thelen, K., Eikelboom, D., Tandoi, V., et al. (2007). Identity, abundance and ecosphysiology of filamentous *Chloroflexi* species present in activated sludge treatment plants. *FEMS Microbiology Ecology*, *59*, 671–682.

Krajmalnik-Brown, R., Hölscher, T., Thomson, I. N., Saunders, F. M., Ritalahti, K. M., & Löffler, F. E. (2004). Genetic identification of a putative vinyl chloride reductase in *Dehalococcoides* sp. strain BAV1. *Applied Environmental Microbiology, 70*, 6347–6351.

Lanoil, B. D., Sassen, R., La Duc, M. T., Sweet, S. T., & Nealson, K. H. (2001). Bacteria and archaea physically associated with Gulf of Mexico gas hydrates. *Applied Environmental Microbiology, 67*, 5143–5153.

Leaphart, A. B., Friez, M. J., & Lovell, C. R. (2003). Formyltetrahydrofolate synthetase sequences from salt marsh plant roots reveal a diversity of acetogenic bacteria and other bacterial functional groups. *Applied Environmental Microbiology, 69*, 693–696.

Leloup, J., Fossing, H., Kohls, K., Holmkvist, L., Borowski, C., & Jørgensen, B. B. (2009). Sulfate-reducing bacteria in marine sediment (Aarhus Bay, Denmark): abundance and diversity related to geochemical zonation. *Environmental Microbiology, 11*, 1278–1291.

Leloup, J., Loy, A., Knab, N. J., Borowski, C., Wagner, M., & Jørgensen, B. B. (2007). Diversity and abundance of sulfate-reducing microorganisms in the sulfate and methane zones of a marine sediment, Black Sea. *Environ Microbiol, 9*, 131–142.

Lever, M. A., Alperin, M. J., Engelen, B., Inagaki, F., Nakagawa, S., Steinsbu, B., et al. (2006). Trends in basalt and sediment core contamination during IODP Expedition 301. *Geomicrobiology Journal, 23*, 517–530.

Lever, M.A. (2008). Anaerobic carbon cycling pathways in the subseafloor investigated via functional genes, chemical gradients, stable carbon isotopes, and thermodynamic calculations. (PhD thesis) The University of North Carolina at Chapel Hill, Dept. of Marine Sciences.

Lever, M. A. (2012). Acetogenesis in the energy-starved deep biosphere – a paradox? *Frontiers in Microbiology, 2*, 1–18.

Lever, M. A. (2013). Functional gene surveys from ocean drilling expeditions - a review and perspective. *FEMS Microbiol Rev, 84*, 1–23.

Lever, M. A., Heuer, V. B., Morono, Y., Masui, N., Schmidt, F., Alperin, M. J., et al. (2010). Acetogenesis in deep subseafloor sediments of the Juan de Fuca Ridge Flank: a synthesis of geochemical, thermodynamic, and gene-based evidence. *Geomicrobiology Journal, 27*, 183–211.

Lever, M. A., Rouxel, O., Alt, J. C., Shimizu, N., Ono, S., Coggon, R. M., et al. (2013). Evidence for microbial carbon and sulfur cycling in deeply buried Ridge Flank Basalt. *Science, 339*, 1305–1308.

Li, P.-Y., Xie, B. B., Zhang, X. Y., Qin, Q-L., Dang, H. Y., Wang, X. M., et al. (2012). Genetic structure of three fosmid-fragments encoding 16S rRNA genes of the miscellaneous crenarchaeotic group (MCG): implications for physiology and evolution of marine sedimentary archaea. *Environmental Microbiology, 14*, 467–479.

Lin, Y.-S., Biddle, J. F., Lipp, J. S., Orcutt, B., Holler, T., Teske, A., et al. (2010). Effect of storage conditions on archaeal and bacterial communities in subsurface marine sediments. *Geomicrobiology Journal, 27*, 261–272.

Lipp, J. S., Morono, Y., Inagaki, F., & Hinrichs, K.-U. (2008). Significant contribution of archaea to extant biomass in marine subsurface sediments. *Nature, 454*, 991–994.

Liu, Y., & Whitman, W. B. (2008). Metabolic, phylogenetic, and ecological diversity of the methanogenic Archaea. *Annals of the New York Academy of Sciences, 1125*, 171–189.

Lloyd, K. G., Alperin, M., & Teske, A. (2011). Environmental evidence for net methane production and oxidation in putative anaerobic methanotrophic (ANME) archaea. *Environmental Microbiology, 13*, 2548–2564.

Lloyd, K. G., Lapham, L., & Teske, A. (2006). An anaerobic methane-oxidizing community of ANME-1 archaea in hypersaline Gulf of Mexico sediments. *Applied Environmental Microbiology, 72*, 7218–7230.

Lloyd, K. G., Albert, D., Biddle, J. F., Chanton, L., Pizarro, O., & Teske, A. (2010). Spatial structure and activity of sedimentary microbial communities underlying a *Beggiatoa* spp. mat in a Gulf of Mexico hydrocarbon seep. *PLoS One, 5*(1), e8738. http://dx.doi.org/10.1371/journal.pone.0008738.

Lloyd, K. G., Schreiber, L., Petersen, D. G., Kjeldsen, K. U., Lever, M. A., Steen, A. D., et al. (2013). Predominant archaea in marine sediments degrade detrital proteins. *Nature, 496*, 215–218.

Lonergan, D. J., Jenter, H. L., Coates, J. D., Phillips, E. J. P., Schmidt, T. M., & Lovley, D. E. (1996). Phylogenetic analysis of dissimilatory Fe(III)-reducing bacteria. *Journal of Bacteriology, 178*, 2402–2408.

López-García, P., Moreira, D., Lopez-Lopez, A., & Rodriguez-Valera, F. (2001). A novel haloarchaeal-related lineage is widely distributed in deep oceanic regions. *Environmental Microbiology, 3*, 72–78.

Lovley, D. R., & Goodwin, S. (1988). Hydrogen concentrations as an indicator of the terminal electron-accepting reactions in aquatic sediments. *Geochim Cosmochim Acta, 52*, 2993–3003.

Loy, A., Duller, S., & Wagner, M. (2008). Evolution and ecology of microbes dissimilating sulfur compounds: insights from siroheme sulfite reductases. In C. Dahl, & C. G. Friedrich (Eds.), *Microbial sulfur metabolism* (pp. 46–59). Berlin: Springer.

Loy, A., Duller, S., Baranyi, C., Mussmann, M., Ott, J., Sharon, I., et al. (2009). Reverse dissimilatory sulfite reductase as phylogenetic marker for a subgroup of sulfur-oxidizing prokaryotes. *Environmental Microbiology, 11*, 289–299.

Ludwig, W., Strunk, O., Westram, R., Richter, L., Meier, H., Yadhukumar, A., et al. (2004). ARB: a software environment for sequence data. *Nucleic Acids Research, 32*, 1363–1371.

Lueders, T., Chin, K. J., Conrad, R., & Friedrich, M. (2001). Molecular analyses of methyl-coenzyme M reductase α-subunit (*mcrA*) genes in rice field soil and enrichment cultures reveal the methanogenic phenotype of a novel archaeal lineage. *Environmental Microbiology, 3*, 194–204.

Luton, P. E., Wayne, J. M., Sharp, R. J., & Riley, P. W. (2001). The *mcrA* gene as an alternative to 16S rRNA in the phylogenetic analysis of methanogen populations in landfill. *Microbiology, 148*, 3521–3530.

Martino, A. J., Rhodes, M. E., Biddle, J. F., Brandt, L. D., Tomsho, L. P., & House, C. H. (2012). Novel degenerate PCR method for whole-genome amplification applied to Peru Margin (ODP Leg 201) subsurface samples. *Frontiers in Microbiology, 3*, 17. http://dx.doi.org/10.3389/fmicb.2012.00017.

Masui, N., Morono, Y., & Inagaki, F. (2008). Microbiological assessment of circulation mud fluids during the first operation of riser drilling by the deep-earth research vessel Chikyu. *Geomicrobiology Journal, 25*, 274–282.

Meng, J., Wang, F., Zheng, Y., Peng, X., Zhou, H., & Xiao, X. (2009). An uncultivated crenarchaeota contains functional bacteriochlorophyll a synthase. *The ISME Journal, 3*, 106–116.

Meng, J., Xu, J., Qin, D., He, Y., Xiao, X., & Wang, F. (2014). Genetic and functional properties of uncultivated MCG archaea assessed by metagenomic expression analysis. *The ISME Journal, 8*, 650–659.

Michaelis, W., Seifert, R., Nauhaus, K., Treude, T., Thiel, V., Blumenberg, M., et al. (2002). Microbial reefs in the Black Sea fueled by anaerobic oxidation of methane. *Science, 297*, 1013–1015.

Miller, T. L., & Wolin, M. J. (1986). Methanogens in human and animal intestinal tracts. *Systematic and Applied Microbiology, 7*, 223–229.

Mills, H. J., Reese, B. K., Shepard, A. K., Riedinger, N., Dowd, S. E., Morono, Y., et al. (2012). Characterization of metabolically active bacterial populations in subseafloor nankai trough sediments above, within, and below the sulfate–methane transition zone. *Frontiers in Microbiology, 3*, 113. http://dx.doi.org/10.3389/fmicb.2012.00113.

Mills, H. J., Reese, B. K., & Peter, C. S. (2012). Characterization of microbial population shifts during sample storage. *Frontiers in Microbiology*, *3*, 49. http://dx.doi.org/10.3389/fmicb.2012.00049.

Moe, W. M., Yan, J., Fernanda Nobre, M., da Costa, M. S., & Rainey, F. A. (2009). *Dehalogenimonas lykanthroporepellens* gen. nov., sp. nov., a reductively dehalogenating bacterium isolated from chlorinated solvent-contaminated groundwater. *International Journal of Systematic and Evolutionary Microbiology*, *59*, 2692–2697.

Morono, Y., Terada, T., Nishizawa, M., Ito, M., Hillion, F., Takahata, N., et al. (2011). Carbon and nitrogen assimilation in deep subseafloor microbial cells. *Proceedings of the National Academy of Sciences USA*, *108*, 18295–18300.

Nakagawa, S., Inagaki, F., Suzuki, Y., Steinsbu, B. O., Lever, M. A., Takai, K., & Integrated Ocean Drilling Program Expedition 301 Scientists., et al. (2006). Microbial community in black rust exposed to hot ridge flank crustal fluids. *Applied Environmental Microbiology*, *72*, 6789–6799.

Newberry, C. J., Webster, G., Weightman, A. J., & Fry, J. C. (2004). Diversity of prokaryotes and methanogenesis in deep subsurface sediments from the Nankai Trough, ocean drilling program Leg 190. *Environmental Microbiology*, *6*, 274–287.

Nunoura, T., Inagaki, F., Delwiche, M. E., Colwell, F. S., & Takai, K. (2008). Subseafloor microbial communities in methane-hydrate bearing sediments at two distinct locations (ODP Leg 204) in the Cascadia Margin. *Microbe Environments*, *23*, 317–325.

Nunoura, T., Soffientino, B., Blazejak, A., Kakuta, J., Oida, H., Schippers, A., et al. (2009). Subseafloor microbial communities associated with rapid turbidite deposition in the Gulf of Mexico continental slope (IODP Expedition 308). *FEMS Microbiology Ecology*, *69*, 410–424.

Nunoura, T., Takai, Y., Kakuta, J., Nishi, S., Sugahara, J., Kazama, H., et al. (2011). Insights into the evolution of archaea and eukaryotic protein modifier systems revealed by the genome of a novel archaeal group. *Nucleic Acids Research*, *39*, 3204–3223.

Orcutt, B. N., Bach, W., Becker, K., Fisher, A. T., Hentscher, M., Toner, B. M., et al. (2010). Colonization of subsurface microbial observatories deployed in young ocean crust. *The ISME Journal*, *5*, 692–703.

Orcutt, B. N., Sylvan, J. B., Knab, N. J., & Edwards, K. J. (2011). Microbial ecology of the dark ocean above, at, and below the seafloor. *Microbiology and Molecular Biology Reviews*, *75*, 361–422.

Orsi, W., Biddle, J. F., & Edgcomb, V. (2013a). Deep sequencing of subseafloor eukaryotic rRNA reveals active fungi across marine subsurface provinces. *PloS One*, *8*, e56335. http://dx.doi.org/10.1371/journal.pne.0056335.

Orsi, W., Edgcomb, V., Christman, G. D., & Biddle, J. F. (2013b). Gene expression in the deep biosphere. *Nature*, *499*, 205–208.

Pace, N. R. (1997). A molecular view of microbial diversity and the biosphere. *Science*, *276*, 734–740.

Parkes, R. J., Webster, G., Cragg, B. A., Weightman, A. J., Newberry, C. J., Ferdelman, T. G., et al. (2005). Deep sub-seafloor prokaryotes stimulated at interfaces over geological time. *Nature*, *436*, 390–394.

Parkes, R. J., Sellek, G., Webster, G., Martin, D., Anders, E., Weightman, A. J., et al. (2009). Culturable prokaryotic diversity of deep, gas hydrate sediments: first use of a continuous high-pressure, anaerobic, enrichment and isolation system for subseafloor sediments (DeepIsoBUG). *Environmental Microbiology*, *11*, 3140–3153.

Parkes, R. J., Brock, F., Banning, N., Hornibrook, E. R. C., Rousel, E. G., Weightman, A. J., et al. (2012). Changes in methanogenic substrate utilization and communities with depth in a saltmarsh, creek sediment in southern England. *Estuarine Coastal and Shelf Science*, *96*, 170–178.

Pereyra, L. P., Hiibel, S. R., Prieto Riquelme, M. V., Reardon, K. F., & Pruden, A. (2010). Detection and quantification of functional genes of cellulose-degrading, fermentative, and sulfate-reducing bacteria and methagenic archaea. *Applied Environmental Microbiology*, *76*, 2192–2202.

Pester, M., Bittner, N., Pinsurang, D., Wagner, M., & Loy, A. (2010). A 'rare biosphere' microorganism contributes to sulfate reduction in a peatland. *ISME Journal*, *4*, 1591–1602.

Pester, M., Schleper, C., & Wagner, M. (2011). The Thaumarchaeota: an emerging view of their phylogeny and ecophysiology. *Current Opinion in Microbiology*, *14*, 300–306.

Rabus, R., Hansen, T. A., & Widdel, F. (2006). Dissimilatory sulfate- and sulfur-reducing prokaryotes. Chapter 1.22. In (3rd ed.) M. Dworkin, & K.-H. Schleifer (Eds.), *The prokaryotes* (Vol. 2) , pp. 659–768).

Rappé, M. S., & Giovannoni, S. J. (2003). The uncultured microbial majority. *Annual Review of Microbiology*, *57*, 369–394.

Reed, D. W., Fujita, Y., Delwiche, D. E., Blackwelder, D. B., Sheridan, P. P., Uchida, T., & Colwell, F. S. (2002). Microbial communities from methane hydrate-bearing deep marine sediments in a Forearch Basin. *Appl Environ Microbiol*, *68*, 3759–3770.

Rinke, C., Schwientek, P., Sczyrba, A., Ivanova, N.N., Anderson, I.J., Cheng, J-F., et al. (2013). Insights into the phylogeny and coding potential of microbial dark matter. *Nature*, *499*, 431–437.

Rochelle, P. A., Cragg, B. A., Fry, J. C., Parkes, R. J., & Weightman, A. J. (1994). Effect of sample handling on estimation of bacterial diversity in marine sediments by 16S rRNA gene sequence analysis. *FEMS Microbiology Ecology*, *19*, 215–226.

Ronaghi, M., Uhlen, M., & Nyren, P. (1998). A sequencing method based on real-time pyrophosphate. *Science*, *281*, 363–365.

Roy, H., Kallmeyer, J., Adhikari, R. R., Pockalny, R., Jørgensen, B. B., & D'Hondt, S. (2012). Aerobic microbial respiration in 86-million-year-old deep-sea red clay. *Science*, *336*, 922–925.

Roussel, E. G., Bonavita, M.-A. C., Querellou, J., Cragg, B. A., Webster, G., Prieur, D., et al. (2008). Extending the sub-sea-floor biosphere. *Science*, *320*, 1046.

Roussel, E. G,, Sauvadet, A.-L., Chaduteau, C., Fouquet, Y., Charlou, J.-L., Prieur, D., et al. (2009). Archaeal communities associated with shallow to deep subseafloor sediments of New Caledonia Basin. *Environmental Microbiology*, *11*, 2446–2462.

Rey, R. E., Harris, J. K., Wilcox, J., Spear, J. R., Miller, S. R., Bebout, B. M., et al. (2006). Unexpected diversity and complexity of the Guerrero Negro hypersaline microbial mat. *Applied Environmental Microbiology*, *72*, 3685–3695.

Sakai, S., Imachi, H., Hanada, S., Ohashi, A., Harada, H., & Kamagata, Y. (2008). *Methanocella paludicola* gen. nov., sp. nov., a methane-producing archaeon, the first isolate of the lineage 'Rice Cluster I', and proposal of the new archaeal order *Methanocellales* ord. nov. *International Journal of Systematic and Evolutionary Microbiology*, *58*, 929–936.

Santelli, C. M., Banerjee, N., Bach, W., & Edwards, K. J. (2010). Tapping the subsurface ocean crust biosphere: low biomass and drilling-related contamination calls for improved quality controls. *Geomicrobiology Journal*, *27*, 158–169.

Scheller, S., Goenrich, M., Boecher, R., Thauer, R., & Jaun, B. (2010). The key nickel enzyme of methanogenesis catalyzes the anaerobic oxidation of methane. *Nature*, *465*, 606–609.

Schippers, A., & Neretin, L. N. (2006). Quantification of microbial communities in near-surface and deeply buried marine sediments on the Peru continental margin using real-time PCR. *Environmental Microbiology*, *8*, 1251–1260.

Schippers, A., Kock, D., Höft, C., Köweker, G., & Siegert, M. (2012). Quantification of microbial communities in subsurface sediments of the Black Sea and off Namibia. *Frontiers in Microbiology*, *3*, 16. http://dx.doi.org/10.3389/fmicb.2012.00016.

Schleper, C., Jurgens, G., & Jonuscheit, M. (2005). Genomic studies of uncultivated archaea. *Nature Reviews Microbiology*, *3*, 479–488.

Seyler, L. M., McGuinness, L. M., & Kerkhof, L. J. (2014). Crenarchaeal heterotrophy in salt marsh sediments. *ISME J*, *8*, 1534–1543.

Shima, S., Krüger, M., Weinert, T., Demmer, U., Kahnt, J., Thauer, R. K., et al. (2012). Structure of a methylcoenzyme M reductase from Black Sea mats that oxidize methane anaerobically. *Nature, 481*, 98–101.

Singh, N., Kendall, M. M., Liu, Y., & Boone, D. R. (2005). Isolation and characterization of methylotrophic methanogens from anoxic marine sediments in Skan Bay, Alaska: description of *Methanococcoides alaskense* sp. nov., and emended description of *Methanosarcina baltica*. *International Journal of Systematic and Evolutionary Microbiology, 55*, 2531–2538.

Smith, D. C., Spivack, A. J., Fisk, M. R., Haveman, S. A., & Staudigel, H. (2000). Tracer-based estimates of drilling-induced microbial contamination of deep sea crust. *Geomicrobiology Journal, 17*, 207–219.

Sogin, M. L., Morrison, H. G., Huber, J. A., Welch, D. M., Huse, S. M., Neal, P. R., et al. (2006). Microbial diversity in the deep sea and the underexplored "rare biosphere". *Proceedings of the National Academy of Sciences USA, 103*, 12115–12120.

Sørensen, K. B., Lauer, A., & Teske, A. (2004). Archaeal phylotypes in a metal-rich and low-activity deep subsurface sediment of the Peru Basin, ODP Leg 201, Site 1231. *Geobiology, 2*, 151–161.

Sørensen, K. B., & Teske, A. (2006). Stratified communities of active archaea in deep marine subsurface sediments. *Applied Environmental Microbiology, 72*, 4596–4603.

Sowers, K. R., & Ferry, J. G. (1983). Isolation and characterization of a methylotrophic methanogen, *Methanococcoides methylutens* gen. nov., sp. nov. *Applied Environmental Microbiology, 45*, 684–690.

Springer, E., Sachs, M. S., Woese, C. R., & Boone, D. R. (1995). Partial gene sequences for the A subunit of methyl-coenzyme M reductase (*mcrI*) as a phylogenetic tool for the family Methanosarcinaceae. *International Journal of Systematic Bacteriology, 45*, 554–559.

Suess, J., Engelen, B., Cypionka, H., & Sass, H. (2004). Quantitative analysis of bacterial communities from mediterranean sapropels based on cultivation-dependent methods. *FEMS Microbiology Ecology, 51*, 109–121.

Stepanauskas, R. (2012). Single cell genomics: an individual look at microbes. *Current Opinion in Microbiology, 15*, 613–620.

Suess, J., Schubert, K., Sass, H., Cypionka, H., Overmann, J., & Engelen, B. (2006). Widespread distribution and high abundance of *Rhizobium radiobacter* within mediterranean subsurface sediments. *Environmental Microbiology, 8*, 1753–1763.

Swofford, D. L. (2000). *PAUP*. Phylogenetic analysis using parsimony (and other methods), version 4*. Sunderland, Massachusetts: Sinauer Associates.

Takai, K., & Horikoshi, K. (1999). Genetic diversity of archaea in deep-sea hydrothermal vent environments. *Genetics, 152*, 1284–1297.

Takai, K., Moser, D. P., DeFlaun, M., Onstott, T. C., & Fredrickson, J. K. (2001). Archaeal diversity in waters from deep South African gold mines. *Applied Environmental Microbiology, 67*, 5750–5760.

Teske, A., Hinrichs, K.-U., Edgcomb, V., Gomez, A. D., Kysela, D., Sylva, S. P., et al. (2002). Microbial diversity of hydrothermal sediments in the guaymas basin: evidence for anaerobic methanotrophic communities. *Applied Environmental Microbiology, 68*, 1994–2007.

Teske, A. P. (2006). Microbial communities of deep marine subsurface sediments: molecular and cultivation surveys. *Geomicrobiology Journal, 23*, 357–368.

Teske, A., & Sørensen, K. B. (2008). Uncultured archaea in deep marine subsurface sediments: have we caught them all? *The ISME Journal, 2*, 3–18.

Thauer, R. K. (2011). Anaerobic oxidation of methane with sulfate: on the reversibility of the reactions that are catalyzed by enzymes also involved in methanogenesis from CO_2. *Current Opinion in Microbiology, 14*, 292–299.

Thomsen, T. R., Finster, K., & Ramsing, N. B. (2001). Biogeochemical and molecular signatures of anaerobic methane oxidation in a marine sediment. *Applied Environmental Microbiology, 67*, 1646–1656.

Ufnar, J. A., Wang, S. Y., Christiansen, J. M., Yampara-Iquise, H., Carson, C. A., & Ellender, R. D. (2006). Detection of the *nifH* gene of *Methanobrevibacter smithii*: a potential tool to identify sewage pollution in recreational waters. *Journal of Applied Microbiol, 101*, 44–52.

Vetriani, C., Jannasch, H. W., MacGregor, B. J., Stahl, D. A., & Reysenbach, A. L. (1999). Population structure and phylogenetic characterization of marine benthic archaea in deep-sea sediments. *Applied Environmental Microbiology, 65*, 4375–4384.

Wagner, M., & Horn, M. (2006). The *Planctomycetes, Verrucomicrobia, Chlamydiae* and sister phyla comprise a superphylum with biotechnological and medical relevance. *Current Opinion in Biotechnology, 17*, 241–249.

Wagner, M., Roger, A. J., Flax, J. L., Brusseau, G. A., & Stahl, D. A. (1998). Phylogeny of dissimilatory sulfite reductases supports an early origin of sulfate reduction. *J Bacteriol, 180*, 2975–2982.

Ward, J. M., Smith, P. H., & Boone, D. R. (1989). Emended description of strain PS (=OGC 70 =ATCC 33273 =DSM 1537), the type strain of *Methanococcus voltae*. *International Journal of Systematic Bacteriology, 39*, 493–494.

Wasmund, K., Schreiber, L., Lloyd, K. G., Petersen, D. G., Schramm, A., Stepanauskas, R., et al. (2014). Genome sequencing of a single cell of the widely distributed marine subsurface *Dehalococcoidia*, phylum *Chloroflexi*. *The ISME Journal, 8*, 383–397.

Webster, G., Newberry, C. J., Fry, J. C., & Weightman, A. J. (2003). Assessment of bacterial community structure in the deep sub-seafloor biosphere by 16S rDNA-based techniques: a cautionary tale. *Journal of Microbiological Methods, 55*, 155–164.

Webster, G., Parkes, R. J., Fry, J. C., & Weightman, A. J. (2004). Widespread occurrence of a novel division of bacteria identified by 16S rRNA gene sequences originally found in deep marine sediments. *Applied Environmental Microbiology, 70*, 5708–5713.

Webster, G., Parkes, R. J., Cragg, B. A., Newberry, C. J., Weightman, A. J., & Fry, J. C. (2006). Prokaryotic community composition and biogeochemical processes in deep subseafloor sediments from the Peru Margin. *FEMS Microbiology Ecology, 58*, 65–85.

Webster, G., Yarram, L., Freese, E., öster, J. K., Sass, J. H., Parkes, R. J., et al. (2007). Distribution of candidate division JS1 and other bacteria in tidal sediments of the German Wadden Sea using targeted 16S rRNA gene PCR-DGGE. *FEMS Microbiology Ecology, 62*, 78–89.

Webster, G., Blazejak, A., Cragg, B. A., Schippers, A., Sass, H., Rinna, J., et al. (2009). Subsurface microbiology and biogeochemistry of a deep, cold-water carbonate mound from the Porcupine Seabight (IODP Expedition 307). *Environmental Microbiology, 11*, 239–257.

Webster, G., Rinna, J., Roussel, E. G., Fry, J. C., Weightman, A. J., & Parkes, R. J. (2010). Prokaryotic functional diversity in different biogeochemical depth zones in todal sediments of the Severn Estuary, UK, revealed by stable-isotope probing. *FEMS Microbiology Ecology, 72*, 179–197.

Webster, G., Sass, H., Cragg, B. A., Gorra, R., Knab, N. J., Green, C. J., et al. (2011). Enrichment and cultivation of prokaryotes associated with the sulphate-methane transition zone of diffusion-controlled sediments of Aarhus Bay, Denmark, under heterotrophic conditions. *FEMS Microbiology Ecology, 77*, 248–263.

Whitman, W. B., Shieh, J., Sohn, S., Caras, D. S., & Premachandran, U. (1986). Isolation and characterization of 22 mesophilic methanococci. *Systematic Applied Microbiology, 7*, 235–240.

Whitman, W. B., Bowen, T. L., & Boone, D. R. (2006). The methanogenic bacteria. In M. Dworkin, S. Falkow, E. Rosenberg, K.-H. Schleifer, & E. Stackebrandt (Eds.), *The prokaryotes: an evolving electronic resource for the microbiological community* (3) (pp. 165–207). New York, NY: Springer.

Wilms, R., Köpke, B., Sass, H., Chang, T. S., Cypionka, H., & Engelen, B. (2006). Deep biosphere-related bacteria within the subsurface of tidal flat sediments. *Environmental Microbiology, 8,* 709–719.

Woese, C. R. (1987). Bacterial evolution. *Microbiological Review, 51,* 221–271.

Woese, C. R., & Fox, G. E. (1977). The phylogenetic structure of the prokaryotic domains. *Proceedings of the National Academy of Sciences USA, 74,* 5088–5090.

Woese, C. R., Kandler, O., & Wheelis, M. L. (1990). Towards a natural system of organisms: proposal for the domains archaea, bacteria and eucarya. *Proceedings of the National Academy of Sciences USA, 87,* 4576–4579.

Woese, C. R. (1998). The universal ancestor. *Proceedings of the National Academy of Sciences USA, 95,* 6854–6859.

Yamada, T., Sekiguchi, Y., Hanada, S., Imachi, H., Ohashi, A., Harada, H., et al. (2006). *Anaerolinea thermolimosa* sp. nov., *Levilinea saccarolytica* gen. nov., sp. nov. and *Leptolinea tardivitalis* gen. nov., sp. nov., novel filamentous anaerobes, and description of the new classes *Anaerolineae* classis nov. and *Caldineae* classis nov. in the bacterial phylum *Chloroflexi*. *International Journal of Systematic and Evolutionary Microbiology, 56,* 1331–1340.

Yamada, T., Imachi, H., Ohashi, A., Harada, H., Hanada, S., Kamagata, Y., et al. (2007). *Bellilinea caldifistulae* gen. nov., sp. nov. and *Longilinea arvoryzae* gen. nov., sp. nov., strictly anaerobic, filamentous bacteria of the phylum *Chloroflexi* isolated from methanogenic propionate-degrading consortia. *International Journal of Systematic and Evolutionary Microbiology, 57,* 2299–2306.

Yanagawa, K., Sunamura, M., Lever, M. A., Morono, Y., Hiruta, A., Ishizaki, O., et al. (2011). Niche separation of methanotrophic archaea (ANME-1 and -2) in methane-seep sediments of the Eastern Japan Sea offshore Joetsu. *Geomicrobiology Journal, 28,* 118–129.

Ye, G., Wang, S., Jiang, L., Xiao, X., Wang, F., Noakes, J., et al. (2009). Distribution and diversity of bacteria and archaea in marine sediments affected by gas hydrates at Mississippi Canyon in the Gulf of Mexico. *Geomicrobiology Journal, 26,* 370–381.

Yoshioka, H., Maruyama, A., Nakamura, T., Higashi, Y., Fuse, H., Sakata, S., et al. (2010). Activities and distribution of methanogenic and methane-oxidizing microbes in marine sediments from the Cascadia Margin. *Geobiology, 8,* 223–233.

Young, J. M., Kuykendall, L. D., Martinez-Romero, E., Kerr, A., & Sawada, H. (2001). A revision of *Rhizobium* Frank 1889, with an emended description of the genus, and the inclusion of all species of *Agrobacterium* Conn 1942 and *Allorhizobium undicola* de Lajudie et al. 1998 as new combinations: *Rhizobium radiobacter, R. rhizogenes, R. rubi, R. undicola* and *R. vitis. International Journal of Systematic and Evolutionary Microbiology, 51,* 89–103.

Zehr, P. J., Jenkins, B. D., Short, S. M., & Steward, G. F. (2003). Nitrogenase gene diversity and microbial community structure: a cross-system comparison. *Environmental Microbiology, 5,* 539–554.

Zeikus, J. G., & Henning, D. L. (1975). *Methanobacterium arboriphilum* sp. nov. An obligate anaerobe isolated from wetwood of living trees. *Antonie van Leeuwenhoek, 41,* 543–552.

Ziebis, W., McManus, J., Ferdelman, T., Schmidt-Schierhorn, F., Bach, W., Muratl, J., et al. (2012). Interstitial fluid chemistry of sediments underlying the North Atlantic Gyre and the influence of subsurface fluid flow. *Earth and Planetary Science Letters, 323–324,* 79–91.

Zhang, G., Tian, J., Jiang, N., Guo, X., Wang, Y., & Dong, X. (2008). Methanogen community in Zoige wetland of Tibetan plateau and phenotypic characterization of a dominant uncultured methanogen cluster ZC-I. *Environmental Microbiology, 10,* 1850–1860.

Zverlov, V., Klein, M., Lücker, S., Friedrich, M. W., Kellermann, J., Stahl, D. A., et al. (2005). Lateral gene transfer of dissimilatory (bi)sulfite reductase revisited. *Journal of Bacteriology, 187,* 2203–2208.

Chapter 2.3

The Underground Economy (Energetic Constraints of Subseafloor Life)

Steven D'Hondt,[1]* Guizhi Wang[2] and Arthur J. Spivack[1]

[1]*Graduate School of Oceanography, University of Rhode Island, RI, USA;* [2]*State Key Laboratory of Marine Environmental Science, Xiamen University, Xiamen, China*
Corresponding author: E-mail: dhondt@mail.uri.edu

2.3.1 INTRODUCTION

A major result of scientific ocean drilling is the discovery that microbes in subseafloor sediment persist at extraordinarily slow rates of metabolic activity. Per-cell rates of microbial respiration calculated from geochemical profiles in marine sediment appear to be three to six orders of magnitude slower than rates in shallow marine sediment and microbial cultures, respectively (D'Hondt, Rutherford, & Spivack, 2002; Price & Sowers, 2004; Jørgensen & D'Hondt, 2006; Hoehler & Jørgensen, 2013). The mean turnover time of microbial biomass in subseafloor sediment appears to be in the range of hundreds to thousands of years (Lomstein, Langerhuus, D'Hondt, Jørgensen, & Spivack, 2012).

Given this discovery, scientific ocean drilling provides a tremendous opportunity to advance the understanding of microbial survival, community interactions, and natural selection under conditions of extremely low energy flux.

In this paper, we briefly discuss (i) what little is known about microbial energetics in subseafloor sediment and (ii) examples of what may be learned from the study of this extraordinary ecosystem, about energetic limits to life, microbial communities, and natural selection.

2.3.2 ENERGY-CONSERVING ACTIVITIES IN MARINE SEDIMENT

All known biological energy-conserving pathways involve oxidation–reduction (redox) reactions. In chemoorganotrophic systems, two types of redox reactions are linked, fermentation and respiration.

Developments in Marine Geology, Volume 7. http://dx.doi.org/10.1016/B978-0-444-62617-2.00005-0
127

In fermentation, a single chemical species is split to form an oxidized product and a reduced product. For example, consider acetoclastic methane production:

$$C_2H_3O_2^- + H_2O \rightarrow CH_4 + HCO_3^-$$

In this reaction, the acetate ($C_2H_3O_2^-$) is split to form reduced methane (CH_4) and oxidized bicarbonate (HCO_3^-). The splitting of a single chemical into an oxidized product and a reduced product is termed disproportionation.

In respiration, an oxidized chemical (an electron acceptor) is used to oxidize a separate reduced chemical (an electron donor). For example, consider the following reactions:

$$O_2 + CH_2O \rightarrow HCO_3^- + H^+$$

$$SO_4^{2-} + 2CH_2O \rightarrow HS^- + 2HCO_3^- + H^+$$

In the first reaction, oxygen (O_2) is the electron acceptor, and CH_2O stoichiometrically represents generic organic matter with the oxidation state of a simple carbohydrate. In the second reaction, sulfate (SO_4^{2-}) is the electron acceptor, and CH_2O is the electron donor.

Products of respiration or disproportionation reactions may serve as electron donors or electron acceptors for other metabolic reactions. For example, the methane created by acetoclastic methane production can be oxidized with an external electron acceptor, such as O_2 or SO_4^{2-}:

$$SO_4^{2-} + CH_4 \rightarrow HS^- + H_2O + HCO_3^-$$

2.3.2.1 Electron Acceptors and Electron Donors in Marine Sediment

Electron acceptors in marine sediment include dissolved chemical species {e.g., O_2, nitrate (NO_3^-), and SO_4^{2-}} and oxidized elements that reside in solid phases {e.g., minerals that include oxidized iron [Fe(III)] or oxidized manganese [(Mn(IV)]}. Electron donors in marine sediment also include solid compounds {e.g., buried organic matter and minerals with reduced metal, e.g., reduced iron [Fe(II)] and reduced manganese [Mn(II)]} and dissolved chemicals {e.g., hydrogen (H_2) from in situ water radiolysis (Blair, D'Hondt, Spivack, & Kingsley, 2007) or mineral weathering}. In most subseafloor sediment studied to date, the principal electron donor is organic matter that originated in the surface photosynthetic world (e.g., D'Hondt et al., 2004). However, in the highly oxidized and carbon-poor sediment of oligotrophic ocean regions, dissolved H_2 from in situ water radiolysis may be a significant electron donor (D'Hondt et al., 2009).

2.3.2.2 Vertical Distribution of Electron-Accepting Activities

In the standard model of sedimentary organic oxidation, electron acceptors are used in the following sequence with increasing sediment depth (Jørgensen, 2006):

$$O_2 \rightarrow NO_3^- \rightarrow Mn\,(IV) \rightarrow Fe\,(III) \rightarrow SO_4^{2-} \rightarrow CO_2$$

This sequence of successive electron -acceptor zones broadly matches pore-water evidence of net respiration in shallow sediment with high rates of organic oxidation (Froelich et al., 1979); dissolved O_2 disappears from the sediment at shallower depths than does dissolved NO_3^-, followed by the appearance of dissolved manganese [Mn(II)] and dissolved iron [Fe(II)] at a greater depth, with SO_4^{2-} disappearing at a still greater depth and abundant CH_4 produced below the depth of SO_4^{2-} exhaustion. In continental margin sediment, where organic oxidation rates are high, the O_2 and NO_3^- reduction zones are typically limited to the first few millimeters or centimeters below the seafloor (Glud, 2008; Revsbech, Jørgensen, & Blackburn, 1980; Shaw, Gieskes, & Jahnke, 1990), the Mn(IV) reduction zone is limited to centimeters or decimeters below seafloor (Froelich et al., 1979; Shaw et al., 1990), and SO_4^{2-} is exhausted in meters to tens of meters below seafloor (Jørgensen & Parkes, 2010; Kastner, Torres, Solomon, & Spivack, 2008).

This vertical succession of electron-accepting zones is generally interpreted as linked to the redox potential of the oxidant, with successively lower energy yields associated with the electron acceptors used in successively deeper zones (e.g., O_2 reduction yields more energy per mole of organic matter than does NO_3^- reduction per mole of the same organic compound) (Froelich et al., 1979; Jørgensen, 2006).

Scientific ocean drilling expeditions have discovered many variations on this standard model. In the simplest variation, organic oxidation is relatively low, and successive zones are drawn out over great depths. For example, at Peru Basin Ocean Drilling Program (ODP) Site 1231, the zone of Mn(IV) reduction extends 60 m below the seafloor (mbsf) (Shipboard Scientific Party, 2003a) (Figure 2.3.1(A)), and SO_4^{2-} is not reduced anywhere in the sediment column (Böttcher et al., 2006). The most extreme example of this situation occurs in the abyssal clay of the South Pacific Gyre, where the O_2-reducing zone spans the entire sediment column (D'Hondt, Inagaki, Alvarez Zarikian, & the Expedition 329 Scientists, 2011).

A second common variation occurs at open-ocean sites where reversal of the standard model occurs in the lower sediment column, sustained by upward diffusion of oxic water from the underlying aquifer (Figure 2.3.1) (D'Hondt et al., 2004). A similar pattern occurs at the Peru Margin, where upward diffusion of SO_4^{2-} from deeply buried brine causes a CH_4/SO_4^{2-} interface at depths, mirroring the standard SO_4^{2-}/CH_4 interface in the overlying sediment (Shipboard Scientific Party, 1988; D'Hondt et al., 2004).

In a more complicated variation, concentration peaks of dissolved manganese and/or iron occur deep below the sulfate-oxidizing zone, but above the reversed electron-accepting sequence at depths (Figure 2.3.1(C)). Paleoceanographic variation in the relative accumulation rates of oxidized metals and organic matter provides a potential explanation of this pattern. For example, at equatorial Pacific ODP Site 1226, high burial of MnO_2 relative to labile organic matter in the late Miocene may sustain present-day microbial reduction of the MnO_2 beneath an active SO_4^{2-} zone (which spans sediments with higher concentrations of biodegradable organic matter deposited during a later interval of time) (D'Hondt et al., 2004).

The most striking deviation from the standard model is perhaps best described as a violation of the model. In short, porewater chemical data from several scientific drilling expeditions demonstrate that hypothetically competing electron-accepting processes commonly co-occur. Several studies have shown that CH_4 production is not limited to sediment beneath the depth of SO_4^{2-} exhaustion, but consistently occurs within the SO_4^{2-}-reducing zone (Mitterer et al., 2001; Bralower et al., 2002; D'Hondt et al., 2004, 2002; Wang, Spivack, Rutherford, Manor, & D'Hondt, 2008). Others have shown that Fe(III) reduction also occurs in the SO_4^{2-}-reducing zone (Wang et al., 2008). These competing processes co-occur for hundreds of meters in subseafloor sediment deposited over millions of years. These co-occurrences are so counter to the standard model that we describe evidence for their existence in some detail below.

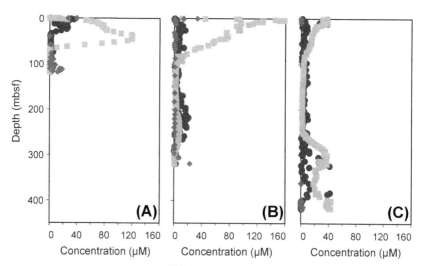

FIGURE 2.3.1 Profiles of dissolved NO_3^- (red (dark gray in print versions) diamonds), manganese (light blue (light gray in print versions) squares), and iron (dark blue (black in print versions) circles) at ODP Sites (A) 1230, (B) 1225, and (C) 1226. The dissolved manganese and iron are inferred to result from the reduction of Mn(IV) and Fe(III), respectively. *Figure modified from D'Hondt et al. (2004).*

2.3.2.3 Co-occurrence of Electron-Accepting Processes

Methane is commonly present in subseafloor sediment, even at sites where SO_4^{2-} is present throughout the entire sediment column (Mitterer et al., 2001; Bralower et al., 2002; D'Hondt et al., 2002, 2004). Careful examination of dissolved SO_4^{2-} and CH_4 profiles shows that they commonly exhibit opposite curvature, with concave SO_4^{2-} profiles paired with convex CH_4 profiles (Figure 2.3.2(A)); because the curvature (the second derivative) of each concentration profile results from in situ reactions, this opposite curvature indicates that CH_4 is created in the same sedimentary intervals where SO_4^{2-} is reduced. From such relationships, Wang et al. (2008) calculated net rates of co-occurring SO_4^{2-} reduction and CH_4 production at equatorial Pacific ODP Site 1226 (Figure 2.3.2(B)–(C)).

The evidence for the co-occurrence of Fe(III) reduction and SO_4^{2-} reduction in deep subseafloor sediment is more subtle, but quite solid. In short, quantification of chemical reactions at ODP Site 1226 indicates that alkalinity is created in SO_4^{2-}-reducing sediment at much higher rates than can result from in situ rates of net SO_4^{2-} reduction (Wang, 2006). Because alkalinity is created by metal reduction at much higher stoichiometric ratios than is created by SO_4^{2-} reduction, in situ Fe(III) reduction provides the simplest explanation of the excess alkalinity production. The alkalinity production rate at Site 1226 indicates that the rate of Fe(III) reduction is comparable to the rate of SO_4^{2-}

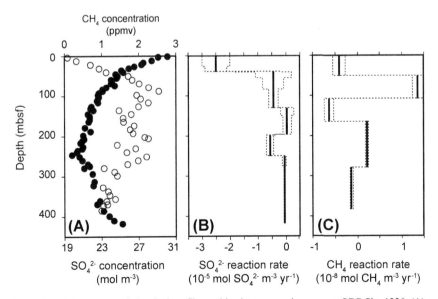

FIGURE 2.3.2 Dissolved chemical profiles and in situ net reaction rates at ODP Site 1226: (A) dissolved SO_4^{2-} and CH_4 concentration profiles, (B) net SO_4^{2-} reduction rates, and (C) net CH_4 production rates. Chemical data are from Shipboard Scientific Party (2003b), and reaction rates are from Wang et al. (2008). *Figure modified from Wang et al. (2008).*

reduction, but Fe(III) reduction only accounts for 6–16% of the carbon oxidized (Wang, 2006) (Figure 2.3.3).

These discoveries are consistent with those of studies of other subsurface environments. Kirk et al. (2004) inferred iron-reducing activity in both the methanogenic zones and sulfate-reducing zones of the Mahomet aquifer in central Illinois. Holmkvist, Ferdelman, and Jørgensen (2011) identified a cryptic sulfur cycle beneath the SO_4^{2-}/CH_4 interface in marine sediment, with reduced sulfur oxidized by iron reduction and the newly oxidized sulfur in turn used to oxidize organic matter.

In our marine examples (Holmkvist et al., 2011; Wang, Spivack, & D'Hondt, 2010), SO_4^{2-} reduction, Fe(III) reduction, and methanogenesis co-occur. As described above, Kirk et al. (2004) found Fe(III) reducers to co-occur with SO_4^{2-} reducers in some regions of the Mahomet aquifer and with methanogens in other regions of the aquifer, but hypothesized co-occurring SO_4^{2-} reducers and Fe(III) reducers to competitively exclude methanogens. Subsequent modeling experiments by Bethke, Ding, Jin, and Sanford (2008) predict that under some conditions in freshwater aquifers, SO_4^{2-} reducers and Fe(III) reducers can coexist, and under other conditions, SO_4^{2-} reducers and methanogens can coexist. Unlike the marine studies, none of the freshwater aquifer studies found all three processes [SO_4^{2-} reduction, Fe(III) reduction, and CH_4 production] to co-occur.

At least three different hypotheses might explain the co-occurrence of two or more electron-accepting pathways.

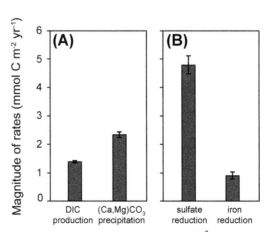

FIGURE 2.3.3 Depth-integrated rates of carbon oxidation, SO_4^{2-} reduction, and Fe(III) reduction at ODP Site 1226. For ease of comparison, all rates are given in oxidized carbon equivalents. (A) Rates of Dissolved Inorganic Carbon (DIC) production and $(Ca,Mg)CO_3$ precipitation (total carbon oxidation is the sum of both processes). (B) Rates of SO_4^{2-} reduction and Fe(III) reduction. The Fe(III) reduction rate estimate is the rate required to balance alkalinity production and precipitate the $(Ca,Mg)CO_3$, given the SO_4^{2-} reduction rate. *Figure modified from Wang (2006).*

First, some studies have proposed that co-occurring electron-accepting pathways do not actually compete, because they utilize different electron donors. For example, Mitterer et al. (2001) suggested that co-occurring methanogens and sulfate reducers in subseafloor sediment of the Great Australian Bight utilize different electron donors. This explanation also applies to the scenario of Holmkvist et al. (2011), which identified sulfur as the electron donor for Fe(III) reduction and organic matter as the electron donor for SO_4^{2-} reduction. In a sense, however, the sulfur in this scenario is an intermediate in the oxidation of (trace amounts of) organic matter by reduction of iron.

A second possible explanation is that the different electron-accepting processes occur in different microenvironments. However, this explanation is difficult to square with the relatively long turnover times of dissolved reactants and products (e.g., SO_4^{2-}, CH_4, and dissolved inorganic carbon) and the very long time available to equilibrate dissolved chemical concentrations in deeply buried marine sediment.

The third possible explanation is thermodynamic cooperation between electron-accepting pathways that compete for electron donors (Bethke, Ding, Jin & Sanford, 2008; Kirk et al., 2004; Wang et al., 2010). Accumulation of reaction products reduces the energy produced by chemical reactions (Gibbs energies of reaction). For example, a high concentration of dissolved sulfide diminishes the Gibbs energy of SO_4^{2-} reduction. Consequently, co-occurring Fe(III) reduction and SO_4^{2-} reduction can cooperate by co-precipitating the dissolved Fe(II) and HS- that result from—and would otherwise hamper—these respective redox activities (Bethke et al., 2008; Wang et al., 2010). Similarly, SO_4^{2-}-reducing or Fe(III)-reducing methanotrophy can aid methanogens by removing the CH_4 that the methanogens produce.

These three explanations make very different predictions. If the coexisting pathways do not actually compete, they must be incapable of utilizing the same electron donors at in situ concentrations (e.g., if the microbes do not have the right biochemical pathways). A microenvironmental explanation requires energetically significant differences in metabolite concentrations on very short length scales. Thermodynamic cooperation requires (i) clear evidence of how the co-occurring pathways cooperate and (ii) Gibbs energies of reaction that are negative enough for the cooperating pathways to simultaneously proceed.

2.3.2.4 Vertical Distribution of Organic-Fueled Respiration

The rate of organic degradation in marine sediment is generally modeled as declining very rapidly with increasing sediment age. Arndt et al. (2013) provide a thorough review of the primary models. These models typically assume relatively constant rates and composition of organic burial.

Scientific ocean drilling expeditions have helped to enrich the understanding of subseafloor organic oxidation in two significant ways.

First, they have demonstrated a strong local influence of dissolved chemical transport on the vertical distribution of organic-fueled respiration. Such transport can decouple the vertical distribution of organic-fueled respiration from the vertical distribution of particulate organic degradation. The clearest example of such decoupling is provided by sulfate-reducing methane oxidation zones, where abundant CH_4 (usually from deeper in the sediment column) comes into contact with abundant SO_4^{2-} (usually from the overlying ocean). These zones sustain unusually high rates of sulfate reduction (Borowski, Paull, & Ussler, 1996; D'Hondt et al., 2002) and unusually high cell abundances (D'Hondt et al., 2004; Parkes et al., 2005) relative to the overlying and underlying sediment (Figure 2.3.4).

Second, they have demonstrated that the ocean history of organic deposition influences present-day vertical distributions of organic-fueled respiration (Arndt, Brumsack, & Wirtz, 2006; Wehrmann, Arndt, März, Ferdelman, & Brunner, 2013) and cell abundance (Aiello & Bekins, 2010). For example, Arndt et al.'s (2006) study of Demerara Rise (ODP Leg 207) sites showed that Cretaceous black shale dominates the SO_4^{2-}, CH_4, and barium (Ba^{2+}) cycling of the entire sediment sequence nearly 100 myrs after shale deposition. Study of Mediterranean sapropels provides further evidence of how ancient histories of organic deposition affect the composition and activity of present-day subseafloor communities (Coolen, Cypionka, Sass, Sass, & Overmann, 2002).

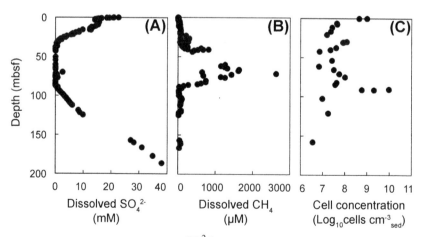

FIGURE 2.3.4 Profiles of (A) dissolved SO_4^{2-}, (B) dissolved CH_4, and (C) total cell abundance at ODP Site 1229. *Figure modified from D'Hondt et al. (2004).*

2.3.3 LIFE UNDER EXTREME ENERGY LIMITATION

2.3.3.1 Respiration Rates in Subseafloor Sediment

Reaction-transport modeling of porewater chemical profiles and microbial abundances at ODP and integrated ODP (IODP) sites shows that mean per-cell respiration in deep subseafloor sediment is many orders of magnitude slower than mean per-cell respiration in surface marine sedimentary communities or pure cultures (D'Hondt et al., 2002; Hoehler & Jørgensen, 2013).

In most cases, per-cell rates of activity are calculated by dividing the rates of activity by the total number of cells calculated from cell counts of representative samples. The standard cell count has been based on microscopic observations of cells dyed with DNA-binding stains that fluoresce under specific wavelengths of light (e.g., Kallmeyer, Pockalny, Adhikari, Smith, & D'Hondt, 2012; Parkes et al., 1994). More recently, studies have begun to use automated counting of stained cells separated from their sedimentary matrix (Morono, Terada, Kallmeyer, & Inagaki, 2013).

For SO_4^{2-} reduction rates, different studies have sometimes reported per-cell rates that differ by an order of magnitude. This difference results from the use of two different denominators. For example, D'Hondt et al. (2002) divided net SO_4^{2-} reduction by the total number of cells, on the basis of SO_4^{2-} being the terminal electron acceptor for the entire ecosystem and organic matter being the terminal electron donor for the entire ecosystem, for example,

$$3SO_4^{2-} + C_6H_{12}O_6 \rightarrow 3HS^- + 6HCO_3^- + 3H^+$$

In this case, the depth-integrated rate of SO_4^{2-} reduction is a direct measure of total dissimilatory activity (D'Hondt et al. 2002). From this perspective, the net catabolic activity of the entire ecosystem is represented by organic-oxidizing SO_4^{2-} reduction coupled to fermentation. In contrast, Hoehler and Jørgensen (2013) divided SO_4^{2-} reduction by 10% of the total cell number, because (i) only a fraction of the total cells directly engage in SO_4^{2-} reduction, and (ii) a previous study found SO_4^{2-} reducers to constitute approximately 10% of the total cells in shallow marine sediment (Ravenschlag, Sahm, Knoblauch, Jørgensen, & Amann, 2000). This perspective accounts for the SO_4^{2-}-reducing oxidation of fermentation products, but not the total pathway of SO_4^{2-} reduction coupled to fermentation.

The mean per-cell rate of carbon oxidation by deep subseafloor anaerobes at eastern equatorial Pacific Site 1226 is 2×10^{-17} mol e^-/cell/year (assuming that four electrons were transferred per mole of carbon oxidized) (Wang, 2006). Because some of the carbon oxidized by subseafloor microbes is precipitated as $(Ca,Mg)CO_3$ in the sediment, Wang (2006) calculated the carbon oxidation rate by adding net dissolved inorganic carbon production to the absolute value of net Ca^{2+} and Mg^{2+} loss (to account for carbonate precipitation). The mean per-cell rate is relatively constant throughout the 420-m sediment column (Wang, 2006).

It is four to six orders of magnitude slower than mean per-cell rates in pure cultures of sulfate-reducing bacteria (Jørgensen, 1978; Knoblauch, Jørgensen, & Harder, 1999) (Figure 2.3.5).

Because very little CH_4 and no fermentation products other than dissolved inorganic carbon escape the Site 1226 sediment column (Shipboard Scientific Party, 2003b), the total rate of carbon oxidation in the Site 1226 sediment column effectively sums the microbial activity of the entire organic-consuming community within the sediment.

The depth-integrated rate of carbon oxidation at Site 1226 [6×10^{-7} mol C/cm^2/year = 2.4×10^{-6} mol e$^-$/cm^2/year (Wang, 2006)] approximates the combined depth-integrated rates of net SO_4^{2-} reduction [1.8×10^{-7} mol SO_4^{2-}/cm^2/year (Wang et al., 2008) = 1.4×10^{-6} mol e$^-$/cm^2/year] and Fe(III) reduction [4×10^{-7} mol Fe/cm^2/year (Wang, 2006) = 4×10^{-7} mol e$^-$/cm^2/year], totaling 1.8×10^{-6} mol e$^-$/cm^2/year. The relatively close match between these rates suggests that (i) SO_4^{2-} is the principal terminal electron acceptor in this organic-fueled subseafloor sedimentary ecosystem, and (ii) the net rate of SO_4^{2-} reduction is close to the gross rate.

Mean per-cell respiration rates in oxic deep subseafloor sediment are also very low. The rate reported for deep subseafloor aerobes from piston cores in the North Pacific Gyre is 1.5×10^{-15} mol e$^-$/cell/year (assuming four electrons were transferred by reduction of each O_2 molecule) (Røy et al., 2012).

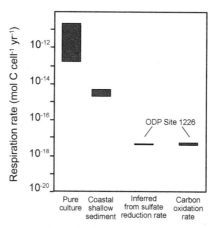

FIGURE 2.3.5 Mean per-cell carbon oxidation rates in pure cultures of sulfate reducers, coastal shallow sediment, and subseafloor sediment of ODP Site 1226. Rates for pure cultures and shallow marine sediment are from Jørgensen (1978) and Knoblauch et al. (1999). Rates for ODP Site 1226 are derived from shipboard cell counts (Shipboard Scientific Party, 2003b) and reaction rates integrated throughout the sediment column. Sulfate reduction rates (D'Hondt et al., 2004; Wang et al., 2008) are used to infer the per-cell carbon oxidation rate in the third column. Net carbon oxidation rates (sum of dissolved inorganic carbon ($HCO_3^- + CO_3^{2-} + CO_2$) production and (Ca,Mg)CO$_3$ precipitation) are used to calculate the per-cell carbon oxidation rate in the fourth column (Wang, 2006). All rates are normalized to total cell abundance. *Figure modified from Wang (2006).*

In situ rates of per-cell respiration are often much lower than potential rates measured in isotope tracer experiments with deep subseafloor sediment (Hoehler & Jørgensen, 2013). This large discrepancy between in situ rates and potential rates suggests that subseafloor sedimentary microbial communities are capable of respiring at much higher rates than subseafloor conditions allow.

Although the total depth-integrated rates of respiration in these examples are generally well constrained, per-cell rates of subseafloor respiration must be interpreted with caution. Populational variation in the respiration rate of subseafloor microbes is not yet known. For example, it is not clear if all the counted cells are equally active and respiring at the same slow rate—or if a very small fraction of the counted cells respires at a much faster rate than the remainder. Further, the fraction of the population that derives energy from fermentation rather than anaerobic respiration is not yet known.

2.3.3.2 Energy per Reaction (The Invisible Hand of Thermodynamics)

At ODP Site 1226, in situ Gibbs energies of reaction for acetoclastic methanogenesis, iron-reducing acetate oxidation, sulfate-reducing methane oxidation, and sulfate-reducing acetate oxidation are relatively constant for 275 m of sediment deposited over millions of years (Wang et al., 2010) (Figure 2.3.6).

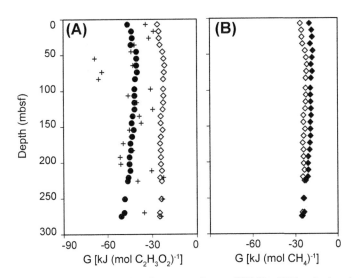

FIGURE 2.3.6 In situ Gibbs energies of some reactions at ODP Site 1226 at depths where dissolved sulfide is detectable: (A) in situ Gibbs energies per mole of acetate for sulfate-reducing acetate oxidation (black circles), iron-reducing acetate oxidation (crosses), and acetoclastic methane production (white diamonds); (B) in situ Gibbs energies per mole of methane for sulfate-reducing CH_4 oxidation (black diamonds) and acetoclastic methane production (white diamonds). *Figure modified from Wang et al. (2010).*

The in situ Gibbs energies of these reactions are not only constant throughout the sedimentary sequence, but are nearly identical in mean value: the Gibbs energies per mole of acetate are 42.9 ± 2.7 kJ for acetate-oxidizing sulfate reduction, 41.6 ± 12.4 kJ for acetate-oxidizing iron reduction, and 42.9 ± 2.4 kJ for the combination of acetoclastic methanogenesis (23.7 ± 1.2 kJ) and sulfate-reducing methanotrophy (19.2 ± 2.0 kJ).

These in situ energies of reaction are very different from the standard-state Gibbs energies of the same reactions (Wang et al., 2010). However, the in situ energies match the Gibbs energies of reactions for the same reactions in a variety of surface environments and cultivation experiments (Wang et al., 2010). They also closely match the minimum Gibbs energy of the reaction believed necessary for an organism to sustain an energy-conserving reaction (10–20 kJ/mol C) [compare Wang et al. (2010) to Thauer and Morris (1984), Schink (1997), and Hoehler, Alperin, Albert, and Martens (1998)]. Gibbs energies of reaction for sulfate-reducing ammonium oxidation at Indian National Gas Hydrate Program Sites 14A and 18A also match the biological minimum Gibbs energy (Schrum, Spivack, Kastner, & D'Hondt, 2009).

These results suggest that the microbes utilizing these reactions in this deep subseafloor sediment are literally working for minimum wage.

This circumstance helps to explain the persistent co-occurrence of competing electron-accepting reactions in subseafloor sediment (Wang et al., 2010). At such low energy yields, even a small increase in the concentration of a reaction product [dissolved sulfide, Fe(II) or CH_4] will drive the Gibbs energy of reaction below the minimum energy needed for an organism to sustain the reaction. Mutualistic removal of reaction products [e.g., coprecipitation of dissolved Fe(II) and dissolved sulfide and/or oxidation of CH_4 by SO_4^{2-} reduction or Fe(III) reduction] will help return the Gibbs energies of multiple reactions to the minimum value needed by the organisms.

2.3.3.3 Biomass Turnover and Energy Use

We can use the rates of microbial respiration, Gibbs energies of reaction, and an independent estimate of biomass turnover to estimate the relative importance of (i) de novo biomolecule production and (ii) building biomolecules from existing organic components. It is energetically cheaper to repair existing amino acids or recycle them from necromass into vegetative cells than to produce amino acids de novo. The minimum power required for amino acid recycling and repair can be estimated from the energy required for peptide bond formation and the rate of amino acid turnover.

The first step in this estimation is to determine the mean Gibbs energy per respiration reaction. As discussed previously, the principal energy-conserving activity in most subseafloor sediment is the degradation of organic matter. Sulfate is the principal terminal electron-accepting process in the Site 1226 sediment, with Fe(III) oxidation playing a secondary role.

[The alkalinity mass balance indicates that Fe(III) oxidation accounts for ~16% of the carbon oxidized (Wang, 2006).] Because SO_4^{2-} reduction and Fe(III) reduction yield nearly identical energies per mole of carbon oxidized [1/2 of 43 kJ/mol C for SO_4^{2-} reduction and 1/2 of 42 kJ mol C for Fe(III) reduction], the energy yield per mole of carbon released by organic oxidation is approximately 22 kJ/mol C.

During fermentation, energy is also conserved. Using the Emden–Meyerhoff glycolytic pathway as a model, two adenosine triphosphate (ATP) molecules are produced for every six carbon atoms metabolized (e.g., Madigan, Martinko, Dunlap, & Clark, 2010). The canonical in vivo Gibbs energy of reaction for ATP formation is 50 kJ/mol (Thauer, Jungermann, & Decker, 1977). Thus, approximately 17 kJ is conserved per mole of carbon. This gives a total of approximately 39 kJ/mol C for the complete oxidation of a carbohydrate to CO_2 (22 + 17 kJ/mol C).

The next step is to calculate energy turnover (kilojoules per-cell per year), by multiplying the mean rate of carbon oxidation per-cell by the Gibbs energy per mole of carbon oxidized:

$$\text{Energy turnover (kJ/cell/yr)} = [C \text{ oxidation/cell/yr}]^{*}$$
$$[\text{Gibbs energy yield (kJ)/mole } C].$$

Multiplication of the mean per-cell rate of carbon oxidation in subseafloor sediment of ODP Site 1226 (Wang, 2006) (0.5×10^{-17} mol C/cell/yr) by the Gibbs energy per mole of carbon oxidized (39 kJ/mol C) yields approximately 2.0×10^{-16} kJ/cell/year. If all the counted cells are alive, this rate can be viewed as the average energy flux needed to sustain this microbial population.

We can calculate the energy cost (actually power, since it is energy per time) of adding amino acids to growing peptide chains, $E_{Peptide}$ during amino acid recycling and repair from (i) an independent estimate of amino acid turnover time ($\tau_{cell, ATP}$) and (ii) the Gibbs energy required per peptide bond (Q).

$$E_{peptide} = {}^{1}\!/_{\tau_{cell}} \times AA_{cell} \times Q,$$

where AA_{cell} is the average amino acid content of a cell.

Lomstein et al. (2012) estimate the turnover time of living biomass in subseafloor sediment as 200–4000 years, using ratios of aspartic-acid stereoisomers and visual cell counts from three Peru Margin sites (ODP Sites 1227, 1229, and 1230). In this calculation, we use an intermediate value of this range (1000 years). Q is calculated based on number of moles of ATP consumed to create a peptide bond and an in vivo Gibbs energy of ATP formation of approximately 50 kJ/mol (Thauer et al., 1977). During peptide bond synthesis, two guanosine triphosphate (GTP) molecules are converted to guanosine diphosphate (GDP), and one ATP molecule is converted to adenosine monophosphate (AMP) per peptide bond. This conversion is equivalent to consumption of four

ATP molecules since it takes the equivalent of two ATP to convert AMP to ATP, and GTP and ATP are enzymatically equilibrated by nucleoside-diphosphate kinases. This then translates to approximately 200 kJ/mol of peptide bonds. The calculated value of $E_{Peptide}$ is thus 2.1×10^{-16} kJ/cell/year. This value is slightly >100% of the per-cell energy consumption estimated at Site 1226 based on diffusion/reaction analysis and measured in situ Gibbs energies (see preceding subsections).

This near balance implies that most of the energy available to microbes in anoxic subseafloor sediment is used for rebuilding proteins from preexisting amino acids scavenged from the surrounding environment. This does not leave much energy for RNA synthesis, which is energy intensive. This limited availability of energy for RNA synthesis implies that mRNA and tRNA molecules are used many times.

Because these implications are fundamental to the understanding of biomass turnover in the subseafloor sedimentary ecosystem, their underlying assumptions must be tested carefully. For example, the above calculation assumes that biomass turnover time is the same at equatorial Pacific Site 1226 as at Peru Margin Sites 1227, 1229, and 1230. Further, the turnover time calculation of Lomstein et al. (2012) relies on cell counts to quantitatively separate living biomass from necromass.

2.3.4 DISCUSSION

2.3.4.1 What Do We Know?

The previous subsections laid out much of what is known about the energetic constraints on subseafloor life in marine sediment.

In summary, most microbes in subseafloor sediment rely primarily on buried organic matter that originated in the surface world. Both dissolved and solid electron acceptors are used in subseafloor sediment; the oxidizing power of these electron acceptors also generally originated in the surface world. However, dissolved H_2 from in situ water radiolysis may be a principal electron donor in very organic-poor sediment (D'Hondt et al., 2009).

The standard model of successive electron acceptor zones is often modified in subseafloor sediment (D'Hondt et al., 2004). In our example anoxic sedimentary sequence (ODP Site 1226), where energy yields of multiple redox reactions are near a biologically harvestable minimum (e.g., Thauer & Morris, 1984), reaction pathways that may compete for electron donors likely coexist for millions of years (Wang et al., 2010). Thermodynamic cooperation, where coexisting reactions aid each other by removing each other's reaction products, provides the simplest explanation of this long-term coexistence.

Mean per-cell rates of catabolic activity (moles e−/cell/yr) (e.g., D'Hondt et al., 2002; Hoehler & Jørgensen, 2013), energy flux (kilojoules per cell per year), and biomass turnover (moles C fixed/cell/yr) (e.g., Lomstein et al., 2012) are all orders of magnitude slower in subseafloor sediment than in the surface world.

Comparison of the total organic-fueled energy flux at Site 1226 (Wang, 2006) to an independent estimate of biomass turnover time at Sites 1227, 1229, and 1230 (Lomstein et al., 2012) suggests that most anabolic activity at Site 1226 is focused on building of biomass from organic matter scavenged from the surrounding environment.

2.3.4.2 What Do We Not Know?

The present knowledge of energetic constraints on subseafloor microbes points the way to several fundamental mysteries. Some of these mysteries are centered on microbial communities. To what extent do counted cells in subseafloor sediment constitute a deep microbial necrosphere? How do different kinds of microbes interact to sustain their mean activity at low average rates for millions of years? Other mysteries relate to individual cells. How slowly can a cell metabolize? How long can a cell survive at such low rates of activity? What properties allow microbes to be sustained by low fluxes of energy? In what ways do subseafloor organisms balance the benefit(s) of maximizing energy recovery with their need to minimize biochemical cost(s) of energy recovery?

2.3.4.2.1 Deep Biosphere or Deep Necrosphere?

Do most of the counted cells in subseafloor sediment constitute a deep microbial "necrosphere" or a real portion of Earth's biosphere? Although the populational distribution of activity is unknown, two lines of evidence suggest that most of the counted cells are in some sense alive. First, fluorescence-in-situ-hybridization (FISH) studies have shown that a large fraction of the cells counted in sediment from many drill sites contain identifiable RNA (Biddle et al., 2006; Mauclaire, Zepp, Meister, & McKenzie, 2004; Schippers et al., 2005). Although these FISH results do not demonstrate that the cells are alive, they do demonstrate that the cells were active recently enough to still contain a relatively transient molecule (RNA). More definitive proof is provided by the second line of evidence—activity experiments with isotope-labeled substrates. In such experiments on 460,000-year-old sediment from IODP Site C9001, most of the counted cells incorporated isotope-labeled substrates (Morono et al., 2011). Although this result does not demonstrate that most of the counted cells from this sample are active in situ, it does demonstrate that most of the cells in this sample are alive and capable of taking up substrates when given the opportunity. Similar results from sediment samples that span a diverse range of sediment ages and habitat types will be needed to set strong constraints on the relative abundances of living organisms and potentially dead organisms in cell counts from subseafloor sediment.

2.3.4.2.2 *How Do Different Kinds of Microbes Interact to Sustain Their Individual Activity at Low Average Rates for Millions of Years?*

The nature of energy flow in subseafloor communities is almost entirely unknown. Very little is known about the details of metabolic coexistence in deep subseafloor sediment. The full extent of thermodynamic cooperation and the conditions of its onset in deep subseafloor sediment remain to be determined. The role, if any, played by other redox processes (e.g., fermentation, Mn(IV) reduction, NO_3^- reduction) in subseafloor thermodynamic cooperation is little explored.

The distribution of catabolic processes within subseafloor sedimentary communities is largely unexamined. The relative abundance and metabolic diversity of fermentative organisms are essentially unknown. The relationship between community structure and thermodynamic cooperation is also unknown. To what extent does thermodynamic cooperation take place between organisms? To what extent does it take place within organisms (with single organisms undertaking two or more metabolic processes (e.g., Fe(III) reduction and SO_4^{2-} reduction))?

Bacteria and Archaea are both abundant in deep subseafloor sediment (Biddle et al., 2006; Inagaki et al., 2006; Schippers et al., 2005). However, their respective roles in subseafloor communities are largely unknown. The trophic structure of deep subseafloor sedimentary communities is largely unexplored. Viruses are abundant (Engelhardt, Sahlberg, Cypionka, & Engelen, 2013), but their influence on community composition and energy cycling is unknown. The presence and community consequences (or absence) of predatory bacteria in deep subseafloor sediment are unexamined. Recent studies have reported evidence of eukaryotes (fungi) from samples of deep subseafloor sediment (Edgcomb, Beaudoin, Gast, Biddle, & Teske, 2010; Orsi, Biddle, & Edgcomb, 2013); however, it is not yet known if any eukaryotes play a significant role in the underground economy.

2.3.4.2.3 *How Slowly Can a Cell Live?*

At the most basic level, we do not yet know the minimum flux of energy necessary to sustain an individual cell in subseafloor sediment. Our present understanding is based on calculations of mean rates averaged over large numbers of cells. The relative activity of individual cells within the counted population is essentially unknown. We cannot yet distinguish between two extreme models. In the first model, most or all of the cells are equally active and respire equally slowly. In the second model, a very small fraction of the cells respire as rapidly as cells in the surface world, and the vast remainder are inactive. It is possible that both these models characterize the same subseafloor communities, but on very different timescales; a very small fraction of the population may be active at any single point in time, but the mean activity of most or all of the cells may be relatively constant over long stretches of time.

2.3.4.2.4 What Properties Allow Microbes to be Sustained by Low Fluxes of Energy?

The mechanisms that allow subseafloor cells to survive at extraordinarily slow energy fluxes are largely unexplored. Several mechanisms are conceivable. We give just a few examples here.

In the first example, microbes may survive as inactive spores for geologically long periods of time. Spores are not included in typical cell counts (because the fluorescent stain does not penetrate the spore wall) (Schippers et al., 2005). However, abundant dipicolinic acid and muramic acid show bacterial spores to be roughly as abundant as counted cells in million-year-old sediment of the Peru Margin (dipicolinic acid is limited to bacterial spores and muramic acid is much more abundant in spores than in vegetative bacteria) (Lomstein et al., 2012). The role of spores in maintaining subseafloor populations on timescales of many millions of years is unknown (Hoehler & Jørgensen, 2013).

A second example is selection for energy-conserving properties, such as (i) the use of sodium ions for energy storage, rather than proton motive force and (ii) the use of a less permeable cell membrane to sustain ionic gradients (Hoehler & Jørgensen, 2013). Both of these properties (reliance on Na^+ channels and formation of a relatively impermeable membrane) have been hypothesized to help build and sustain archeal dominance relative to bacteria in energy-limited ecosystems (e.g., Hoehler & Jørgensen, 2013; Valentine, 2007). However, the relative abundances of archaea and bacteria, and the properties that sustain their abundances, in subseafloor sediment are subjects of active debate (e.g., Schouten, Middelburg, Hopmans, & Sinninghe Damsté, 2010; Xie, Lipp, Wegener, Ferdelman, & Hinrichs, 2013).

Our third example is a "Frankencell" strategy, in which cells incorporate organic molecules from the surrounding sediment, which might allow cells to build or repair biomass more cheaply than cells that build biomass de novo. Takano et al. (2010) showed the use of this strategy by archaea in nearshore marine sediment, which incorporated membrane lipids from the surrounding sediment. This strategy is consistent with the result of the peptide bond calculations that we introduced in a previous subsection. However, the full extent to which this specific strategy is used by deep subseafloor microbes is not yet known.

A fourth example is the preferential loss of genes that code for unused biochemical pathways and the proteins used in those pathways. Maintenance and expression of a small genome is energetically cheaper than maintenance and expression of a large genome. From a natural selection perspective, preferential loss from subseafloor populations of genes that code for unused pathways could be as straightforward as preferential survival of microbes whose genetic repertoire generally lacks the unused pathways.

Our final example is the prospect of individual cells surviving for many millions of years. Because molecular repair, or even replacement of single molecules, is energetically much cheaper than the wholesale synthesis of new cells, the ongoing birth and death of cells is energetically much costlier than ongoing repair of existing cells (e.g., Hoehler & Jørgensen, 2013). Taken to an extreme, energy-starved cells in subseafloor sediment may survive for many millions of years without reproducing, but simply repairing or replacing their molecular machinery as their energy flux allows.

The extent to which any of the above mechanisms, or other mechanisms, sustain microbes at their extraordinarily low per-cell rates of respiration in deep subseafloor sediment remains to be discovered.

2.3.5 CONCLUSIONS

Scientific ocean drilling has greatly advanced the understanding of subseafloor sedimentary habitability and energetic constraints on subseafloor sedimentary life.

Rates of catabolic activity, energy flux, and biomass turnover per counted cell are orders of magnitude slower in subseafloor sediment than in the surface world.

Potentially competing metabolic pathways co-occur for hundreds of meters in subseafloor sediment deposited over millions of years.

Comparison of energy flux estimates from Site 1226 with biomass turnover rates from other sites suggests that most of the energy flux to subseafloor sedimentary microbes may be used for building biomolecules from existing components, rather than for de novo biosynthesis.

Given these discoveries, scientific ocean drilling provides an extraordinary opportunity to examine the nature of microbial survival, community structure, and natural selection under extreme energy limitation.

ACKNOWLEDGMENTS

We thank Joris Gieskes and Alfred M. Spormann, whose thoughtful reviews improved the manuscript. We thank Victoria M. Fulfer, Maureen J. Hayden, and Justine F. Sauvage for assistance with figures. This study was founded by the US National Science Foundation (NSF) through the Center for Dark Energy Biosphere Investigations (C-DEBI) and by the US National Aeronautics and Space Administration (NASA) through the NASA Astrobiology Institute (NAI). This is C-DEBI contribution 230.

REFERENCES

Aiello, I. W., & Bekins, B. A. (2010). Milankovitch-scale correlations between deeply buried microbial populations and biogenic ooze lithology. *Geology, 38,* 79–82.

Arndt, S., Brumsack, H.-J., & Wirtz, K. (2006). Cretaceous black shales as active bioreactors: a biogeochemical model for the deep biosphere encountered during ODP Leg 207 (Demerara Rise). *Geochimica et Cosmochimica Acta, 70,* 408–425.

Arndt, S., Jorgensen, B. B., LaRowe, D. E., Middelburg, J. J., Pancost, R. D., & Regnier, P. (2013). Quantification of organic matter degradation in marine sediments: a synthesis and review. *Earth-Science Reviews, 123,* 53–86.

Bethke, C. M., Ding, D., Jin, Q., & Sanford, R. A. (2008). Origin of microbiological zoning in groundwater flows. *Geology, 36*, 739–742.

Biddle, J. F., Lipp, J. S., Lever, M. A., Lloyd, K. G., Sorenson, K. B., Anderson, R., et al. (2006). Heterotrophic Archaea dominate sedimentary subsurface ecosystems off Peru. *Proceedings of the National Academy of Sciences of the United States of America, 103*, 3846–3851. http://dx.doi.org/10.1073/pnas.0600035103.

Blair, C. C., D'Hondt, S., Spivack, A. J., & Kingsley, R. H. (2007). Potential of radiolytic hydrogen for microbial respiration in subseafloor sediments. *Astrobiology, 7*(6), 951–970.

Borowski, W. S., Paull, C. K., & Ussler, W., III (1996). Marine pore-water sulfate profiles indicate in situ methane flux from underlying gas hydrate. *Geology, 24*, 655–658.

Böttcher, M. E., Ferdelman, T. G., Jørgensen, B. B., Blake, R. E., Surkov, A. V., & Claypool, G. E. (2006). Sulfur isotope fractionation by the deep biosphere within sediments of the eastern equatorial Pacific and Peru margin. In B. B. Jørgensen, S. L. D'Hondt, & D. J. Miller (Eds.), *Proc. ODP, Sci. Results, 201* (pp. 1–21). College Station, TX: Ocean Drilling Program. http://dx.doi.org/10.2973/odp.proc.sr.201.109.2006.

Bralower, T. J., Premoli Silva, I., Malone, M. J., et al. (2002). In *Proc. ODP, Init. Repts* (198). http://www-odp.tamu.edu/publications/198_IR/198ir.htm. Accessed on 08.08.06.

Coolen, M., Cypionka, H., Sass, A., Sass, H., & Overmann, J. (2002). Ongoing modification of Mediterranean pleistocene sapropels mediated by prokaryotes. *Science, 296*, 2407–2410.

D'Hondt, S., Jørgensen, B. B., Miller, D. J., Batzke, A., Blake, R., Cragg, B. A., et al. (2004). Distributions of metabolic activities in deep subseafloor sediments. *Science, 306*, 2216–2221. http://dx.doi.org/10.1126/science.1101155.

D'Hondt, S., Inagaki, F., Alvarez Zarikian, C. A., & the Expedition 329 Scientists (2011). South Pacific Gyre subseafloor life. In *Proc. IODP, 329*. Tokyo: Integrated Ocean Drilling Program Management International, Inc. http://dx.doi.org/10.2204/iodp.proc.329.2011.

D'Hondt, S., Rutherford, S., & Spivack, A. J. (2002). Metabolic activity of subsurface life in deep-sea sediments. *Science, 295*, 2067–2070.

D'Hondt, S., Spivack, A., Pockalny, R., Ferdelman, T., Fischer, J., et al. (2009). Subseafloor sedimentary life in the South Pacific Gyre. *Proceedings of the National Academy of Sciences, 106*(28), 11651–11656.

Edgcomb, V. P., Beaudoin, D., Gast, R., Biddle, J., & Teske, A. (2010). Marine subsurface eukaryotes: the fungal majority. *Environmental Microbiology, 13*(1), 172–183.

Engelhardt, T., Sahlberg, M., Cypionka, H., & Engelen, B. (2013). Biogeography of *Rhizobium radiobacter* and distribution of associated temperate phages in deep subseafloor sediments. *The ISME Journal, 7*, 199–209. http://dx.doi.org/10.1038/ismej.2012.92.

Froelich, P. N., Klinkhammer, G. P., Bender, M. L., Luedtke, N. A., Heath, G. R., Cullen, D., et al. (1979). Early oxidation of organic matter in pelagic sediments of the eastern Equatorial Atlantic: suboxic diagenesis. *Geochimica et Cosmochimica Acta, 43*, 1075–1090.

Glud, R. N. (2008). Oxygen dynamics of marine sediments. *Marine Biology Research, 4*, 243–289.

Hoehler, T. M., Alperin, M. J., Albert, D. B., & Martens, C. S. (1998). Thermodynamic control on hydrogen concentrations in anoxic sediments. *Geochimica et Cosmochimica Acta, 62*, 1745–1756.

Hoehler, T. M., & Jørgensen, B. B. (2013). Microbial life under extreme energy limitation, Nature Reviews. *Microbiology, 11*(02), 83–94.

Holmkvist, L., Ferdelman, T. G., & Jørgensen, B. B. (2011). A cryptic sulfur cycle driven by iron in the methane zone of marine sediment (Aarhus Bay, Denmark). *Geochimica et Cosmochimica Acta, 75*, 3581–3599.

Inagaki, F., Nunoura, T., Nakagawa, S., Teske, A., Lever, M., Lauer, A., et al. (2006). Biogeographical distribution and diversity of microbes in methane hydrate-bearing deep marine sediments on the Pacific Ocean Margin. *Proceedings of the National Academy of Sciences of the United States America, 103*(8), 2815–2820.

Jørgensen, B. B. (1978). Comparison of Methods for the quantification of bacterial sulfate reduction in coastal Marine-Sediments. *Geomicrobiology Journal, 1,* 11–64.

Jørgensen, B. B. (2006). Bacteria and marine biogeochemistry. In H. D. Schulz, & M. Zabel (Eds.), *Marine Geochemistry* (2nd ed.) (pp. 173–207). Berlin: Springer.

Jørgensen, B. B., & D'Hondt, S. (2006). A starving majority deep beneath the seafloor. *Science, 314,* 932–934.

Jørgensen, B. B., & Parkes, R. J. (2010). Role of sulfate reduction and methane for anaerobic carbon cycling in eutrophic fjord sediments (Limfjorden, Denmark). *Limnology and Oceanography, 55,* 1338–1352.

Kallmeyer, J., Pockalny, R., Adhikari, R., Smith, D. C., & D'Hondt, S. (2012). Global distribution of subseafloor sedimentary biomass. *Proceedings of the National Academy of Sciences, 109*(40), 16213–16216.

Kastner, M., Torres, M., Solomon, E., & Spivack, A. J. (2008). Marine pore fluid profiles of dissolved sulfate; do they reflect in situ methane fluxes? *Fire Ice, Summer 2008,* 6–8.

Kirk, M. F., Holm, T. R., Park, J., Jin, Q., Sanford, R. A., Fouke, B. W., et al. (2004). Bacterial sulfate reduction limits natural arsenic contamination in groundwater. *Geology, 32,* 953–956.

Knoblauch, C., Jørgensen, B. B., & Harder, J. (1999). Community size and metabolic rates of psychrophilic sulfate-reducing bacteria in Arctic marine sediments. *Applied Environmental Microbiology, 65,* 4230–4233.

Lomstein, B. A., Langerhuus, A. T., D'Hondt, S., Jørgensen, B. B., & Spivack, A. J. (2012). Spore abundance, microbial growth and necromass turnover in deep subseafloor sediment. *Nature, 484,* 101–104. http://dx.doi.org/10.1038/nature10905.

Madigan, M. T., Martinko, J. M., Dunlap, P. V., & Clark, D. P. (2010). *Brock biology of microorganisms* (12th ed.). Menlo Park, CA: Benjamin Cummings.

Mauclaire, L., Zepp, K., Meister, P., & McKenzie, J. A. (2004). Direct in situ detection of cells in deep-sea sediment cores from the Peru Margin (ODP Leg 201, Site 1229). *Geobiology, 2,* 217–223.

Mitterer, R. M., Malone, M. J., Goodfriend, G. A., Swart, P. K., Wortmann, U. G., Logan, G. A., et al. (2001). Co-generation of hydrogen sulfide and methane in marine carbonate sediments. *Geophysical Research Letters, 28*(20), 3931–3934.

Morono, Y., Terada, T., Kallmeyer, J., & Inagaki, F. (2013). An improved cell separation technique for marine subsurface sediments: applications for high–throughput analysis using flow cytometry and cell sorting. *Environmental Microbiology, 15,* 2841–2849.

Morono, Y., Terada, T., Nishizawa, M., Ito, M., Hillion, F., Takahata, N., et al. (2011). Carbon and nitrogen assimilation in deep subseafloor microbial cells. *Proceedings of the National Academy of Sciences of the United States America, 108,* 18295–18300.

Orsi, W., Biddle, J., & Edgcomb, V. (2013). Deep sequencing of subseafloor eukaryotic rRNA reveals active Fungi across multiple subseafloor provinces. *PLoS ONE, 8*(2), e56335.

Parkes, R. J., Cragg, B. A., Bale, S. J., Getliff, J. M., Goodman, K., Rochelle, P. A., et al. (1994). Deep bacterial biosphere in Pacific Ocean sediments. *Nature, 371,* 410–413.

Parkes, R. J., Webster, G., Cragg, B. A., Weightman, A. J., Newberry, C. J., Ferdelman, T. G., et al. (2005). Deep sub-seafloor prokaryotes stimulated at interfaces over geological time. *Nature, 436,* 390–394. http://dx.doi.org/10.1038/nature03796.

Price, P. B., & Sowers, T. (2004). Temperature dependence of metabolic rates for microbial growth, maintenance, and survival. *PNAS, 101,* 4631–4636.

Ravenschlag, K., Sahm, K., Knoblauch, C., Jørgensen, B. B., & Amann, R. (2000). Community structure, cellular rRNA content, and activity of sulfate-reducing bacteria in marine Arctic sediments. *Applied Environmental Microbiology*, 66, 3592–3602.

Revsbech, N. P., Jørgensen, B. B., & Blackburn, T. H. (1980). Oxygen in the sea bottom measured with a microelectrode. *Science*, 207, 1355–1356.

Røy, H., Kallmeyer, J., Adhikari, R. R., Pockalny, R., ørgensen, B. B. J., & D'Hondt, S. (2012). Aerobic microbial respiration in 86-million-year-old deep-sea red clay. *Science*, 336(6083), 922–925. http://dx.doi.org/10.1126/science.1219424 Erratum (figure correction) in *Science* 336 (6088), 1506, DOI: 10.1126/science.336.6088.1506.

Schink, B. (1997). Energetics of syntrophic cooperation in methanogenic degradation. *Microbiology and Molecular Biology Reviews*, 61, 262–280.

Schippers, A., Neretin, L. N., Kallmeyer, J., Ferdelman, T. G., Cragg, B. A., Parkes, R. J., et al. (2005). Prokaryotic cells of the deep sub-seafloor biosphere identified as living bacteria. *Nature*, 433, 861–864. http://dx.doi.org/10.1038/nature03302.

Schouten, S., Middelburg, J. J., Hopmans, E. C., & Sinninghe Damsté, J. S. (2010). Fossilization and degradation of intact polar lipids in deep subsurface sediments: a theoretical approach. *Geochimica et Cosmochimica Acta*, 74, 3806–3814.

Schrum, H. N., Spivack, A. J., Kastner, M., & D'Hondt, S. (2009). Sulfate-reducing denitrification: a thermodynamically feasible metabolic pathway in subseafloor sediments. *Geology*, 37(10), 939–942.

Shaw, T. J., Gieskes, J. M., & Jahnke, R. A. (1990). Early diagenesis in differing depositional environments—the response of transition metals in pore water. *Geochimica et Cosmochimica Acta*, 54, 1233–1246.

Shipboard Scientific Party. (1988). Introduction, objectives, and principal results, Leg 112, Peru continental margin. In E. Suess, R. von Huene, et al. (Eds.), *Proc. ODP, Init. Repts* (112) (pp. 5–23). College Station, TX: Ocean Drilling Program. http://dx.doi.org/10.2973/odp.proc.ir.112.102.1988.

Shipboard Scientific Party. (2003a). Site 1231. In S. L. D'Hondt, B. B. Jørgensen, D. J. Miller, et al. (Eds.), *Proc. ODP, Init. Repts., 201* (pp. 1–64). College Station, TX: Ocean Drilling Program. http://dx.doi.org/10.2973/odp.proc.ir.201.112.2003.

Shipboard Scientific Party. (2003b). Site 1226. In S. L. D'Hondt, B. B. Jørgensen, D. J. Miller, et al. (Eds.), *Proc. ODP, Init. Repts., 201* (pp. 1–96). College Station, TX: Ocean Drilling Program. http://dx.doi.org/10.2973/odp.proc.ir.201.107.2003.

Takano, Y., Chikaraishi, Y., Ogawa, N. O., Nomaki, H., Morono, Y., Inagaki, F., et al. (2010). Sedimentary membrane lipids recycled by deep-sea benthic archaea. *Nature Geoscience*, 3(12), 858–861. http://dx.doi.org/10.1038/NGEO983.

Thauer, R. K., Jungermann, K., & Decker, K. (1977). Energy conservation in chemotrophic anaerobic bacteria. *Bacteriological Reviews*, 41, 100–180.

Thauer, R. K., & Morris, J. G. (1984). Metabolism of chemotrophic anaerobes: old views and new aspects. *A Symposium of the Society for General Microbiology*, 36, 123–168.

Valentine, D. L. (2007). Adaptations to energy stress dictate the ecology and evolution of the Archaea. *Nature Reviews Microbiology*, 5, 316–323.

Wang, G., (2006). *Metabolic activities in deep subseafloor sediments*. University of Rhode Island. PhD Dissertation. http://digitalcommons.uri.edu/dissertations/AAI3248245

Wang, G., Spivack, A. J., & D'Hondt, S. (2010). Gibbs energies of reaction and microbial mutualism in anaerobic deep subseafloor sediments of ODP Site 1226. *Geochimica et Cosmochimica Acta*, 74, 3938–3947.

Wang, G., Spivack, A. J., Rutherford, S., Manor, U., & D'Hondt, S. (2008). Quantification of co-occurring reaction rates in deep subseafloor sediments. *Geochimica et Cosmochimica Acta*, 72, 3479–3488.

Wehrmann, L. M., Arndt, S., März, C., Ferdelman, T. G., & Brunner, B. (2013). The evolution of early diagenetic signals in Bering Sea subseafloor sediments in response to varying organic carbon deposition over the last 4.3 Ma. *Geochimica et Cosmochimica Acta, 109*, 175–196.

Xie, S., Lipp, J. S., Wegener, G., Ferdelman, T. G., & Hinrichs, K.-U. (2013). Turnover of microbial lipids in the deep biosphere and growth of benthic archaeal populations. *Proceedings of the National Academy of Sciences of the United States America, 110*, 6010–6014.

Chapter 2.4

Life at Subseafloor Extremes

Ken Takai,[1,2,3,4,*] Kentaro Nakamura,[2,3,5] Douglas LaRowe[6] and Jan P. Amend[6,7]

[1]Department of Subsurface Geobiological Analysis and Research, Japan Agency for Marine-Earth Science & Technology (JAMSTEC), Yokosuka, Japan; [2]Laboratory of Ocean-Earth Life Evolution Research (OELE), Japan Agency for Marine-Earth Science & Technology (JAMSTEC), Yokosuka, Japan; [3]Submarine Hydrothermal System Research Group, Japan Agency for Marine-Earth Science & Technology (JAMSTEC), Yokosuka, Japan; [4]Earth-Life Science Institute (ELSI), Tokyo Institute of Technology, Tokyo, Japan; [5]Department of Systems Innovation, School of Engineering, University of Tokyo, Tokyo, Japan; [6]Department of Earth Sciences, University of Southern California, Los Angeles, CA, USA; [7]Department of Biological Sciences, University of Southern California, Los Angeles, CA, USA

*Corresponding author: E-mail: kent@jamstec.go.jp

2.4.1 INTRODUCTION

Oceanic and terrestrial subsurface environments have been recognized to host biotopes that are potentially near the boundary between habitable and uninhabitable. Although the concept of habitability has many aspects, it is widely accepted that the most challenging terrains for life exist in deep subsurface environments (Takai, 2011). Potential physical and chemical constraints dictating the limits of life in the deep subsurface include temperature, pressure, physical space, liquid water, and sufficient access to nutrients and energy such as carbon and nitrogen and electron donors and acceptors. Microbial communities in deep subsurface biospheres may have specific adaptations and/or live in consortia for very long time periods in order to survive in environments that may be characterized by such constraints. Thus, exploration of the extent of Earth's biosphere and the elucidation of the evolutionary adaptations of the extremophiles that inhabit oceanic and terrestrial subsurface environments are key to determining the conditions that gave rise to and enabled the early evolution of life on Earth and its distribution beyond this planet.

In this chapter, we first discuss the physical and chemical characteristics of subseafloor environments with respect to the known extremes of laboratory-based microbial growth experiments. Although many of these pure-culture microbial experiments only explore one particular environmental extreme at a time (e.g., temperature), in real systems, there are often multiple physical,

Developments in Marine Geology, Volume 7. http://dx.doi.org/10.1016/B978-0-444-62617-2.00006-2
149

chemical, and biological factors simultaneously shaping the habitability of a particular setting. In particular, microbial symbioses and synergetic functions are difficult to capture in a controlled laboratory experiment. Indeed, in the Ocean Drilling Program (ODP) and the following Integrated Ocean Drilling Program (presently International Ocean Discovery Program) (IODP), several expeditions have sought to investigate the microbial communities and their functions in subseafloor fringe environments. Examples include the fluids emerging near subseafloor hydrothermal vent systems (Cragg & Parkes, 1994; Cragg, Summit, & Parkes, 2000; Kimura, Asada, Masta, & Naganuma, 2003; Reysenbach, Holm, Hershberger, Prieur, & Jeanthon, 1998; Takai, Mottl, Nielsen, & the Expedition 331 Scientists, 2011), highly alkaline serpentinized mud environments (Salisbury, Shinohara, Richter, & Shipboard Scientific Party, 2002), deep crustal rocky habitats (Blackman et al., 2006), and midocean gyre sediments containing low concentrations of organic compounds and nutrients (D'Hondt, Inagaki, Alvarez Zarikian, & the Expedition 329 Scientists, 2011). Although the nature, distribution, and size of the fringe biosphere remains to be fully resolved, the ODP- and IODP-expedition-based microbiological investigations have indicated that indigenous microbial communities in parts of the subseafloor live near the limits (e.g., temperature, pH, and energy) of life. Using the limited data that scientific drilling expeditions have gathered, we summarize what is known about the biotic fringe, knowing that it is insufficient to characterize the limit to life. Furthermore, we discuss predicted patterns in the fringe microbial communities in subseafloor hydrothermal fluids (HFs) based on thermodynamic estimation of abundance and composition of microbial catabolic and anabolic potentials.

2.4.2 POSSIBLE PHYSICAL AND CHEMICAL CONSTRAINTS ON LIFE IN SUBSEAFLOOR ENVIRONMENTS

In subseafloor environments, many physical and chemical parameters can limit microbial activity. One of the most studied is temperature. In surface environments, liquid water boils at $100\,^\circ$C, but with the pressure that accompanies increasing depth, liquid water can exist up to $373\,^\circ$C for pure water and $407\,^\circ$C for seawater (SW) (critical points) (Bischoff & Rosenbauer, 1988). Indeed, the highest temperature record of liquid water ($407\,^\circ$C) was found in a deep-sea hydrothermal vent of the Mid-Atlantic Ridge (Table 2.4.1) (Koschinsky et al., 2008). Thus, since the first discovery of high-temperature hydrothermal vents in 1979 at 21°N on the East Pacific Rise (Spiess & Rise Group, 1980), microbiologists have been interested in empirically determining the upper temperature limit (UTL) for life. At present, the highest temperature at which an organism has been grown is $122\,^\circ$C under 20 and 40 MPa of hydrostatic pressure (*Methanopyrus kandleri* strain 116; see Table 2.4.2) (Takai, Nakamura et al., 2008). This record of UTL for life shows one actual physical constraint limiting microbial growth but it is a transient one. If new methodological and technological

TABLE 2.4.1 Physical and Chemical Properties of Subseafloor Environments Reported in Studies Supported by the Ocean Drilling Program (ODP) and Integrated Ocean Drilling Program (IODP)

Properties	Representative Subseafloor Environment Expected	Representative Subseafloor Environment Explored
Lowest temperature	>0 °C	>0 °C
	Abyssal seawater	Abyssal seawater
Highest temperature	>407 °C	312 °C
	Subseafloor high-temperature hydrothermal fluid regimes	Subseafloor hydrothermal fluid regimes of the PACMANUS field (ODP Leg 193)
	(Koschinsky et al., 2008)	(Kimura et al., 2003)
Lowest pressure	>0.1 MPa	>0.1 MPa
Highest pressure	>110 MPa	>78 MPa
	Subseafloor environments in the Challenger Deep of the Mariana Trench	830 m below seafloor at a water depth of 6985 m in the Japan trench (IODP Exp 343)
	(Kato, Li, Tamaoka, & Horikoshi, 1997)	(Chester et al., 2013)
Lowest pH	pH < 1	pH 6.1
	Subseafloor hydrothermal fluid regimes in arc-backarc submarine volcanoes	Subseafloor hydrothermal fluid regimes of the Iheya North field (IODP Exp 331)
	(Resing et al., 2007)	(Takai et al., 2011)
Highest pH	>pH 12.5	pH 12.5
	Subseafloor environments in serpentinization-driven fluid regimes	Subseafloor serpentinite mud in the south Chamorro Seamount (ODP Leg 195)
	(Salisbury et al., 2002)	(Salisbury et al., 2002)
Highest salinity	Saturated	5.5 M to saturated
	Subseafloor salt evaporite and brine fluids	Subseafloor sediments and evaporites (ODP Leg 160)
	(Swallow & Crease, 1965)	(Emeis et al., 1996)

TABLE 2.4.2 Temperature, Pressure, pH, and Salinity Limits for Microbial Growth for all Microorganisms on Earth and for Those Found in Deep-Sea and Subseafloor Environments

Physical and Chemical Property	Of all the Microorganisms	Of Deep-Sea and Subseafloor Microorganisms
Lowest temperature limit for growth	<0 °C	<0 °C
	Many psychrophiles	Many deep-sea psychrophiles
	(Rainy & Oren, 2006)	(Rainy & Oren, 2006)
Highest temperature limit for growth	122 °C at 20–40 MPa	122 °C at 20–40 MPa
	Methanopyrus kandleri strain 116	*Methanopyrus kandleri* strain 116
	(Takai, Nakamura et al., 2008)	(Takai, Nakamura et al., 2008)
Highest pressure limit for growth	130 MPa at 2 °C strain MT41	130 MPa at 2 °C strain MT41
	(Yayanos, 1986)	(Yayanos, 1986)
Lowest pH limit for growth	pH 0	pH 3.3
	Picrophilus oshimae and *Picrophilus torridus*	*Aciduliprofundum boonei*
	(Schleper, Pühler, Kühlmorgen, & Zillig, 1995)	(Reysenbach et al., 2006)
Highest pH limit for growth	pH 12.4	pH 11.4
	Alkaliphilus transvaalensis	*Marinobacter alkaliphilus*
	(Takai, Moser et al., 2001)	(Takai et al., 2005)
Highest salinity limit for growth	Saturated	5.1 M at 45 °C
	Many extreme halophiles	*Halorhabdus tiamatea*
	(Rainy & Oren, 2006)	(Antunes et al., 2008)

innovations are applied to the cultivation of currently unknown hyperthermophiles, the UTL could be extended further. On a related note, the highest survival temperature of deep-sea hydrothermal vent microorganisms in a laboratory for a limited period of time is several hours at around 130 °C (Takai, Nakamura et al., 2008). However, the molecular signals of microorganisms and even living hyperthermophiles have been retrieved from samples of high-temperature (>250 °C) HFs or materials (e.g., Takai, Gamo et al., 2004; Takai, Nakamura et al., 2008; Takai, Nunoura et al., 2008). These molecular signals and living microorganisms are likely derived from a tiny fraction of the microbial populations that originally thrived at much lower temperatures, but have survived with exposure to the sometimes extraordinary temperatures of HF for relatively short time periods (Takai, Gamo et al., 2004; Takai, Nunoura et al., 2008). To our knowledge, no experiments have quantitatively shown that hyperthermophilic microorganisms can survive in natural high-temperature HFs exceeding 130 °C.

Although deep-sea hydrothermal vents are observed on the seafloor, the network of fissures, cracks, and permeable rock layers supporting the flux of HFs under chimneys can be extensive (e.g., Wilcock & Fisher, 2004). Thus, the subseafloor environments that are widespread around mid-ocean ridges (MOR), arc-backarc (ABA) volcanoes and spreading centers, and hot-spot volcanoes and their flank regions can host fluids that span the range of liquid water (up to 407 °C) and even higher (supercritical fluid) (Shock, 1992) (Table 2.4.1). In contrast, most of the subseafloor environments that are characterized by typical bottom water temperatures (1–4 °C) are not at all fatal to most living forms (Tables 2.4.1 and 2.4.2). Because such great ranges of temperature and pressure can be found in submarine hydrothermal systems, it seems likely that the UTL for life would be found in these environments.

Elevated hydrostatic and lithostatic pressure, common in the subseafloor environments, is another important physical parameter constraining the extent of life. The deepest habitat that has been explored is the Challenger Deep in the Mariana Trench, at a water depth of ~10,900 m (Glud et al., 2013; Kato et al., 1997), which corresponds to 110 MPa of hydrostatic pressure (Table 2.4.1). The current laboratory-based upper pressure limit for microbial growth is 130 MPa for a deep-sea psychrophilic heterotroph (strain MT41) isolated from Challenger Deep (Table 2.4.2) (Yayanos, 1986). In fact, a great diversity of phylogenetic (based on rRNA gene sequences), physiological, and highly active microbial metabolic functions have also been identified from the Challenger Deep sediments of the Mariana Trench (Glud et al., 2013; Kato et al., 1997; Takai, Inoue, & Horikoshi, 1999; Takami, Inoue, Fuji, & Horikoshi, 1997). Even many macrofauna thrive there (Bellaev & Brueggeman, 1989; Kobayashi, Hatada, Tsubouchi, Nagahama, & Takami, 2012). Thus, it is evident that even the greatest pressure conditions that have been explored in subseafloor environments by ocean drilling expeditions (e.g., in IODP Leg 343, core samples at around 830 m below seafloor at a water depth of 6985 m were recovered) (Chester et al., 2013, 2012; Lin et al., 2013) are not enough pressure to prevent life (Table 2.4.1).

Salinity and pH extremes are also found in subseafloor environments. Volcanic activity often leads to extremely acidic waters in which pH values can drop close to zero (e.g., Schleper et al., 1995). This extreme acidity is caused by inputs of sulfuric acid and hydrochloric acid originally provided from magmatic volatiles associated with volcanic activity. The lowest pH value ever reported from a hydrothermal vent fluid is 1.6 from the TOTO caldera field in the Mariana Arc (Nakagawa et al., 2006), while several more acidic hydrothermal vent fluids (pH < 1) are now being observed in western Pacific submarine volcanoes

(Table 2.4.1) (Butterfield et al., 2011; Resing et al., 2007). As far as we know, the lowest pore water pH value ever recorded in the ODP/IODP investigations is pH 6.1 in the subseafloor sediments associated with HFs in the Iheya North field of the Okinawa Trough (Takai et al., 2011). It is predicted that the extremely acidic habitats for subseafloor microbial communities are widespread beneath the hydrothermally active seafloor in such submarine volcanoes (Table 2.4.1). The acidic pH limit for growth and survival, irrespective of environment, is well established: extremely acidophilic *Archaea* belonging to the phylum *Thermoplasmata*, such as *Picrophilus* and *Ferroplasma*, can grow at pH 0 (Edwards, Bond, Gihring, & Banfield, 2000; Schleper et al., 1995), The most acidophilic microorganism from a submarine hydrothermal environment, "*Aciduliprofundum boonei*," can grow at pH 3.3 (Table 2.4.2) (Reysenbach et al., 2006). Many 16S rRNA gene phylotypes related to this thermophilic archaeon have also been identified in potentially acidic habitats of various deep-sea hydrothermal environments (Nakagawa, Takai, Suzuki, Hirayama, & Konno, 2006; Reysenbach, Longnecker, & Kirshtein, 2001; Takai & Horikoshi, 1999; Takai, Komatsu, Inagaki, & Horikoshi, 2001). Thus, previously identified but uncultivated deep-sea *Archaea* within the phylum *Thermoplasmata* may be able to grow near pH 0. However, it is unlikely that such low values of pH can be found in high-temperature systems due to the solubility of acids under hydrothermal conditions (Seyfried, Ding, & Berndt, 1991). As a result, it is likely that the lowest pH limit for the growth of "*A. boonei*" (pH 3.3) may be equivalent to the lowest in situ pH condition in the subseafloor hydrothermal environments.

On the other end of the pH spectrum, alkaline environments are also generated in deep subseafloor hydrothermal (water–rock) processes. It is well known that the serpentinization reaction of water and mafic minerals, which are common in rocks present in oceanic crust along ultraslow to intermediate spreading centers, generates highly alkaline waters (McCollom & Bach, 2009). Indeed, the highest pH value ever recorded in subseafloor environments is 12.5, in the serpentinite mud pore water of the South Chamorro Seamount in the Mariana Forearc (Table 2.4.1) (Mottl, Komor, Fryer, & Moyer, 2003; Salisbury et al., 2002). This is close to the highest pH ever reported (pH 12.9) in the extremely alkaline underground water in the Maqarin "bituminous marl formation" in Jordan (Pedersen, Nilsson, Arlinger, Hallbeck, & O'Neill, 2004). Highly alkaline hydrothermal systems driven by serpentinization could potentially be widespread in off-axis environments far from magmatic activity, such as the Lost City hydrothermal field (Kelley et al., 2001, 2005).

Alkaliphilic bacteria are known to be ubiquitous in nonalkaline habitats such as soil, freshwater, and ocean environments and are usually able to grow up to pH 10–11 (Horikoshi, 1999). This pH range is almost equivalent to the alkaline pH limit for life (pH 10–12) that has been often described in the literature (Rainy & Oren, 2006; Rothschild & Mancinelli, 2001). The most alkaliphilic microorganism known is *Alkaliphilus transvaalensis*, which was isolated from an ultradeep South African gold mine (3.2 km deep below land surface) and grown at pH 12.4 (Table 2.4.2) (Takai, Moser et al., 2001). Since *A. trans-vaalensis* is a gram-positive, spore-forming bacterium, the cells and spores would be able to survive under more alkaline pH conditions than the alkaline pH limit for growth. Nevertheless, the cellular and spore survival was not yet characterized. This bacterium has never been found in deep-sea and subseafloor environments. In the subseafloor serpentinite mud environment of the South Chamorro Seamount, the pH of pore waters is as high as 12.5 (Table 2.4.1) (Salisbury et al., 2002). From this hyperalkaline environment, an approximately 30-m-deep core sample was obtained on ODP Leg#195 (Salisbury et al., 2002; Takai et al., 2005). Down to a depth of 1.5 m below the seafloor (mbsf) (up to a pH of 11), a living alkaliphilic heterotroph *Marinobacter alkaliphilus* was detected (Takai et al., 2005). However, from deeper parts of the samples (pH > 11), no living microorganisms were identified (Takai et al., 2005). From the same core samples, considerable amounts of bacterial and archaeal lipids were obtained (Mottl et al., 2003). Thus, living microorganisms may be present in the hyperalkaline subseafloor environments. The pH limit for growth of *M. alkaliphilus* (pH 11.4) is probably the present highest pH value ever reported for growth of subseafloor microorganisms (Table 2.4.2).

Subseafloor environments can host microbial habitats with a variety of salinities from almost fresh to hypersaline, and potentially even in salt deposits. In subseafloor HF systems, the rapid decompression of upwelling high-temperature HFs can induce phase separation and partition of the fluid into vapor- and brine-rich phases (Bischoff & Pitzer, 1989). Phase-separation-influenced hydrothermal systems are well known, and both highly brine-enriched and vapor-dominating fluids have been identified (Von Damm, 1995). It is thought that the brine-dominated fluids contribute to high-saline subseafloor HF flows. Furthermore, many deep-sea brine pools and deep subseafloor salt deposits have been found (Table 2.4.1) (e.g., Krijgsman, Hilgen, Raffi, Sierro, & Wilson, 1999; Swallow & Crease, 1965). These environments result from past evaporative events of SW induced by sea level change and tectonic events. In contrast, environments with very low salt concentrations can be generated in subseafloor environments by the inputs of terrestrial freshwater outflows near coasts, the condensation of vapor-phase HFs and magmatic volatiles, and the dissociation water of gas (CO_2 and CH_4) hydrates that exclude the dissolved salts during hydrate formation. The natural range of salinities on Earth is within the habitable zone for microbial growth. Many freshwater microorganisms can grow in distilled water only supplemented with complex organic substrates, while extreme halophiles grow in NaCl-saturated media and can survive in salt crystals over geologic time (Vreeland, Rosenzweig, & Powers, 2000). The only cultivated organism from deep-sea environments is the extremely halophilic archaeon *Halorhabdus tiamatea*, which was isolated from deep-sea brine pool sediment in the Red Sea (Table 2.4.2) (Antunes et al., 2008), although many halophilic bacterial strains and prokaryotic 16S rRNA gene sequences have been identified in deep-sea hydrothermal vent chimneys and in deep-sea brine pools (Antunes, Ngugi, & Stingl, 2011; Eder, Ludwig, & Huber, 1999; Takai, Komatsu et al., 2001).

Other than temperature, pressure, pH, and salinity, microbial activity and survival in subseafloor environments can be limited by a lack of energy and essential elements (Valentine, 2007). However, based on recent exploration of subseafloor environments, microorganisms are always present in even the most energetically barren settings (D'Hondt et al., 2009; Morono et al., 2012; other chapters in this book). This is likely due to the fact that microorganisms, collectively, have tremendous metabolic potentials and can gain energy for growth, maintenance, and survival from numerous chemical reactions. It is likely that the amount of energy and essential elements present in the subseafloor limits cell densities and microbial activity levels. Furthermore, the type of catabolic reactions that are thermodynamically feasible, and the extent to which they are favored, will impact the abundance, functional diversity, and metabolic composition of subseafloor microbial communities. While there is no known example of habitats where the energy and elemental fluxes are below that needed to sustain a microbial community, it has recently been shown that the amount of energy available in marine sediments correlates with the amount of biomass in them (LaRowe & Amend, 2014). The energetic impact on the abundance and function of subseafloor microbial communities is discussed in detail later.

2.4.3 CHALLENGE FOR LIMITS OF BIOSPHERE IN OCEAN DRILLING EXPEDITIONS OF ODP AND IODP

In the history of the ODP and its successor program, the IODP, several expeditions were conducted to obtain core samples from extreme subseafloor environments with certain physical and chemical constraints that might determine the boundary between habitable regions (biosphere) and uninhabitable regions. In this section, we review several ODP and IODP expeditions, potentially to address the limits of the biosphere in subseafloor environments.

Although it has not been their primary scientific objective, several ODP/IODP expeditions have probed the UTL for life by investigating subseafloor hydrothermal vent systems (Cragg & Parkes, 1994; Cragg et al., 2000; Kimura et al., 2003; Reysenbach et al., 1998; Takai et al., 2011) and other areas with relatively large heat flow (Blackman et al., 2006). For example, ODP Leg 158 was conducted in the proximity of high-temperature hydrothermal discharges at the TAG field in the Mid-Atlantic Ridge (Humphris, Herzig, Miller, & Shipboard Scientific Party, 1996), ODP Legs 139 and 169 investigated the hydrothermal vent and flank regions of the sediment-covered hydrothermal system at the Middle Valley field on the Juan de Fuca Ridge (Davis, Mottl, Fisher, & Shipboard Scientific Party, 1992; Fouquet, Zierenberg, Miller, & Shipboard Scientific Party, 1998), ODP Leg 193 explored the sulfide deposits close to the active hydrothermal vents at the PACMANUS field in the Manus Basin (Binns, Barriga, Miller, & Shipboard Scientific Party, 2002), and IODP expedition 331 observed the hydrothermal activity center near the Iheya North field in the Okinawa Trough (Takai et al., 2011). IODP expedition 304/305 obtained plutonic rock samples from the crest of Southern Ridge in the Atlantis Massif (Blackman et al., 2006), which hosts the serpentinization-driven Lost City hydrothermal vent field (about 4 km south from Site U1309) (Kelley et al., 2001; 2005).

A number of microbiological techniques were used to search for organisms in core samples taken at various depths (0–52.1 mbsf) during ODP Leg 158 (Humphris et al., 1996), Microscopy, attempts at cultivation, and DNA extractions all failed to find evidence of a biosphere or viable microbial communities in the subsurface near the TAG hydrothermal field (Reysenbach et al., 1998). It is not clear whether the environment beneath the TAG field is inhospitable for life or if low microbial populations were undetectable due to the technical and methodological limitations inherent in the study.

In the adjacent and flank regions of the Middle Valley hydrothermal field on the Juan de Fuca Ridge, microbial cell counts were carried out on sediment core samples that were influenced by subseafloor HF flow (Cragg & Parkes, 1994; Cragg et al., 2000). The vertical distribution of microbial communities in areas characterized by steep thermal gradients (Site 858) was estimated to a maximum depth of 67 mbsf (Cragg & Parkes, 1994). Estimates of the temperature constraints suggested that potentially viable microbial communities were restricted to relatively shallow subseafloor environments that were lower than 76 °C (Cragg & Parkes, 1994). However, in several deeper subseafloor horizons, probably where HF circulates, detectable microbial populations were observed at a temperature range of 155–185 °C (Cragg & Parkes, 1994; Cragg et al., 2000). Similar depth and temperature profiles of microbial cell abundance were obtained from sediments in the hydrothermal flank regions (Site 1036) (Cragg et al., 2000). However, it has not been determined if these cells were viable, what their functions are, and whether or not they are indigenous since they are simply counted based on acridine orange stain. As a result, the data from this ODP Leg suggest that an active subseafloor biosphere faces a thermal maximum lower than that of intact cells.

The ocean drilling expedition in the PACMANUS hydrothermal field (Leg 193) was the first one conducted in an active ABA hydrothermal system (Binns et al., 2002). Coring operations extended down to 386.7 mbsf, and samples were subjected to microbiological characterization (Kimura et al., 2003). Microbial cell populations were detected in the core samples at depths down to about 69 mbsf (Site 1188) and about 80 mbsf (Site 1189), while the ATP (adenosine triphosphate) concentrations indicative of viable microbial populations were quantitatively significant down to 48.8 mbsf (Site 1188A) and 39.1 mbsf (Site 1189) (Kimura et al., 2003). Although the bottom temperatures of the drilled holes were determined by using wire-line logging tools several days after the drilling operation (312 °C at a depth of 386.7 mbsf at Hole 1188F and 68 °C at a depth of 206.0 mbsf at Hole 1189B), the temperature ranges of detectable microbial cell populations and potentially viable microbial populations were uncertain only based on the bottom temperature measurements (Kimura et al., 2003). Perhaps, potentially viable microbial communities are more extensive at relatively low temperatures (shallower zones) little affected by subseafloor high-temperature HF flow. In addition, successful enrichments of thermophiles were obtained from the core samples, some of which were even deeper than the depth limit of detectable microbial cell populations (Kimura et al., 2003). It is still uncertain whether these thermophiles derived from the indigenous viable or survived microbial communities or from drilling fluids that were contaminated at shallower depths. However, these results suggest that viable microbial communities may be distributed in a complex patchwork of habitable zones whose geometry is determined by the complex hydrogeologic structure of submarine high-temperature hydrothermal systems.

These ODP expedition-based microbiological investigations for the temperature limits of the biosphere have provided some useful guidelines for future investigations seeking the limits to life: sediment-associated hydrothermal systems are operationally and technically better targets for the drilling-science-based approach than the sediment-starved systems and sedimentary environments with moderate temperature gradients (e.g., drilling and core recovery spanning a temperature range of about 50–200 °C) should be extensively studied using a multidisciplinary approach: hydrogeology, temperature and physical properties, mineralogy, pore water chemistry and multiple microbiological techniques. Using these guidelines, a biogeochemistry- and microbiology-dedicated IODP expedition (IODP Exp 331) was conducted in the Iheya North hydrothermal field in the middle Okinawa Trough (Takai et al., 2011; Yanagawa et al., 2013). The onboard microbial cell counts revealed that the microbial cell populations were likely distributed in the hydrothermally active subseafloor sediments down to depths of about 20 mbsf (for Site C0014) and about 70 mbsf (for Site C0017) (Takai et al., 2011; Yanagawa et al., 2013). In these sediments, the likely in situ temperatures were estimated to be up to 70 °C (for Site C0014) and 25 °C (for Site C0017) (Takai et al., 2011; Yanagawa et al., 2013). Although multiple lines of evidence for the possible distribution of viable microbial communities are now being obtained from various biogeochemical and microbiological characterizations, such as stable isotope analyses of various energy, carbon, nitrogen, and sulfur sources, prokaryotic 16S rRNA and functional gene detection and quantification, metabolic activity measurements, and cultivation tests (Aoyama, Nishizawa, Takai, & Ueno, 2014), the onboard microbial cell counts suggest that the possible boundaries between the habitable and the uninhabitable regions are present at particular depths of the subseafloor sedimentary environment and that the estimated temperature range of the habitable region (<70 °C for Site C0014) is far below the experimentally determined UTL for growth.

Although it was only conducted near a subseafloor HF flow regime, IODP expedition 304/305 explored the plutonic rock environments with relatively steep thermal gradients in the Atlantis Massif near the MAR (Blackman et al.,

2006). One of the drilling holes (Hole U1309D) penetrated down to 1415.5 mbsf through very thin sediments and massive gabbroic rocks and obtained core samples down to 1395 mbsf (Blackman et al., 2006). After completion of drilling Hole U1309D, a temperature measurement was performed throughout the drilled depth, and the bottom of the hole at 1415 mbsf was found to be 119 °C (Blackman et al., 2006). For the purpose of microbiological investigation, 26 sediment and rock samples were subsampled from the cores at depths of 0.45–1391 mbsf in order to carry out microbial cell counts and polymerase chain reaction (PCR)-based 16S rRNA gene analysis (Mason et al., 2010). Except for the uppermost carbonate sediments (0.5 mbsf), no evidence of microbial cellular population was detected. However, 16S rRNA genes were amplified from DNA extracted from the gabbroic rocks (Mason et al., 2010). The maximal temperature for the bacterial 16S rRNA gene detection was 79 °C at a depth of 1313 mbsf (Mason et al., 2010). However, the bacterial phylotypes obtained from ~80 °C of rock habitats were closely related with mesophilic *Ralstonia* spp., indicating that microbial contamination was likely (Mason et al., 2010). Thus, it seems unlikely that most of the bacterial 16S rRNA gene phylotypes obtained from the deep subseafloor gabbroic rock habitats represent indigenous microbial populations near the observed temperature limit.

As described above, microbiological explorations associated with ocean drilling expeditions in deep-sea hydrothermal systems and surrounding subseafloor environments have provided preliminary insights into the possible temperature limits for the subseafloor biosphere. In several sediment-associated hydrothermal systems, the UTLs of possible subseafloor microbial cell populations are found to be similar (e.g., 76 °C in Site 858 of the Middle Valley field and about 70 °C in Site C0014 of the Iheya North field). Interestingly, this temperature range is similar to that of putative indigenous microbial cell populations and prokaryotic 16S rRNA gene sequences that were taken from deep continental margin sediments (60–100 °C) during ODP and IODP expeditions (Ciobanu et al., 2014; Roussel et al., 2008). However, the UTLs of possible subseafloor microbial communities are estimated based on the preliminary microbiological characterizations such as microbial cell counts and PCR-based 16S rRNA gene sequencings using very limited numbers of samples. Ongoing multidisciplinary investigation of the subseafloor sediment samples in IODP Exp 331 that are characterized by the moderate temperature gradients (Sites C0014 and C0017) will provide more definitive evidence of a possible temperature limit for the subseafloor biosphere.

Discoveries made during ODP Leg 195 represent another possible limit for the subseafloor biosphere, extremely alkaline pHs (Salisbury et al., 2002). At Site 1200, the subseafloor near cold seepages in a serpentinizing system (the South Chamorro Seamount) was drilled down to a depth of 128.3 mbsf (Salisbury et al., 2002). The pore water pH increased steeply with increasing depth just beneath the sediment–SW interface to above 12 at about 1 mbsf (Salisbury et al., 2002). The highest pore water pH found, pH 12.5, is higher than the known upper pH limit for microbial growth (pH 12.4) (Takai, Moser et al., 2001). As a result, studies were carried out to find and identify alkaliphilic and/or alkalitolerant organisms (Mottl et al., 2003; Takai et al., 2005). Based on the depth profile of dissolved sulfate, methane, ethane, and ammonium concentrations in pore waters, microbiological sulfate reduction coupled to methane- and/or ammonium oxidation accounts for the pore water chemistry down to a depth of about 25 mbsf (Hole 1200E) (Mottl et al., 2003). In addition, the potentially viable biomass of subseafloor bacterial and archaeal populations were estimated by quantification of phospholipid fatty acid and phospholipid-derived diphytanyl diether content (Mottl et al., 2003). The detectable bacterial lipids (equivalent to 10^4–10^6 cells/g dry weight of sediment) were only found at one sample taken just beneath the seafloor (0.45 mbsf and pH 8.5), while the archaeal lipids were detected at a depth range of 0–4.4 mbsf (pH range of 8.5–12.4) and a depth range of 9–13.4 mbsf (pH range of 12.2–12.4) (Mottl et al., 2003). The microscopic cell counts revealed that about 5×10^7 to 2×10^5 cells/cc were found in the subseafloor down to a depth of 29.2 mbsf (up to a pH value of 12.4), and cell abundance decreased with increasing depth and pH values (Takai et al., 2005). The viable bacterial population counts were verified by quantitative cultivation tests and found to occur at a depth range of 0–1.5 mbsf (pH range of 8.4–9.5) (Takai et al., 2005). These results strongly suggest that the possible boundary between the habitable and the uninhabitable regions is present in a particular depth range (10–100 mbsf) in the subseafloor serpentinite mud environment constrained by high values of pH. Although only one example of an extremely alkaline subseafloor environment has been explored so far and the microbial community composition and function remain uncertain, ODP expedition-based microbiological investigations suggest that the possible alkaline limit for subseafloor biosphere is present around pH 12.5, quite similar to the upper pH limit for microbial growth known so far.

2.4.4 THERMODYNAMIC ESTIMATION OF ABUNDANCE AND COMPOSITION OF MICROBIAL METABOLISMS IN SUBSEAFLOOR BOUNDARY BIOSPHERE

Previous and ongoing biogeochemical and microbiological explorations associated with ODP and IODP expeditions have been used to establish the possible temperature and pH limits for subseafloor life in high-temperature hydrothermal systems and serpentinization-driven alkaline fluids. However, the temperature and pH boundaries between the habitable and unhabitable zones in the subseafloor can only be resolved as finely as the detection limits for microbial communities. Previous and ongoing investigations have encountered various operational, technical, and analytical difficulties, and have not yet provided significant insights into the abundances, composition, and function of the microbial communities straddling the limits of the biosphere. However, using the physical and chemical properties of such boundary habitats, the abundance, composition, and function of fringe microbial communities can be inferred and estimated based on thermodynamic quantification of potential catabolic reactions and the energetics of anabolism (Amend, LaRowe, McCollom, & Shock, 2013; Amend, McCollom, Hentscher, & Bach, 2011; LaRowe & Amend, 2014; LaRowe, Dale, & Regnier, 2008; McCollom, 2007; McCollom & Shock, 1997; Nakamura & Takai, 2014; Takai & Nakamura, 2010; Takai & Nakamura, 2011; Shock & Holland, 2004). In the previous ODP- and IODP-related investigations of fringe biospheres, however, the data sets of physical and chemical properties are not complete. In particular, important data such as temperature and inorganic gas concentrations are largely absent.

In the case of subseafloor high-temperature hydrothermal environments, the energetic potential of plausible catabolic pathways that arise due to the mixing of HFs (end-member HFs) with SW can be calculated if the physical and chemical properties of the mixed fluids are well characterized. For the deep-sea hydrothermal systems that have been explored by ODP and IODP expeditions (the TAG field, the Middle Valley field, the PACMANUS field, and the Iheya North field), detailed physical and chemical properties of end-member HFs have been characterized (Table 2.4.3) (Butterfield, McDuff, Franklin, & Wheat, 1994; Charlou, Donval, Jean-Baptiste, Dapoigny, & Rona, 1996; Edmonds et al., 1996; Kawagucci et al., 2011; Reeves et al., 2011). These are used in the

TABLE 2.4.3 Physical and Chemical Properties of End-Member Hydrothermal Fluids Sampled from the TAG, Middle Valley, PACMANUS, and Iheya North Fields

Field	PACMANUS	TAG	Middle Valley	Iheya North
Vent Site	Fenway		Dead Dog 1990	
Vent	F3	1993BS	Site 858	NBC
Temperature (°C)	358	362.5	276.0	309.0
Depth (m)	1710	3630	2424.5	980
pH	2.7	3.22	5.5	5
H_2 (mM)	0.306	0.205	2.5	0.229
H_2S (mM)	18.8	3.5	3	4.5
CH_4 (mM)	0.042	0.155	19.3	3.7
CO_2 (mM)	56.1	3.435	10.8	227

Continued

TABLE 2.4.3 Physical and Chemical Properties of End-Member Hydrothermal Fluids Sampled from the TAG, Middle Valley, PACMANUS, and Iheya North Fields—Cont'd

Field	PACMANUS	TAG	Middle Valley	Iheya North
Vent Site	Fenway		Dead Dog 1990	
Vent	F3	1993BS	Site 858	NBC
SO$_4$ (mM)	0	0	0	0
Na (mM)	397	546.5	398	407
Cl (mM)	562	646.8	578.0	557.0
Ca (mM)	22.3	29	81	21.9
Mg (mM)	0	0	0	0
K (mM)	76.1	18	18.7	72.4
Fe (mM)	11.8	5.175	0.015	0.16
Mn (mM)	3.8	0.6995	0.063	0.658
Si (mM)	12.2	19.675	10.15	12.3
References	Reeves et al. (2011)	Edmonds et al. (1996), Charlou et al. (1996)	Butterfield, et al. (1994), Von Damm et al. (2005)	Kawagucci et al. (2011)

current study to compute the energetic potential of particular catabolic activities and the bioenergetic cost of synthesizing biomolecules, which are in turn used to delineate the abundance and composition of microbial populations in the subseafloor biosphere.

2.4.4.1 Catabolic Reaction Energetics

In this section, reaction energetics for the seven net chemolithotrophic catabolic processes listed in Table 2.4.4 were evaluated for four hydrothermal systems (TAG, Middle Valley, PACMANUS, and Iheya North fields). The calculations for the reaction energetics were performed following the method described in McCollom (2007). In this calculation, a simple mixing between HF and SW was assumed, and all redox reactions including the oxidation of H_2 were presumed to be kinetically inhibited at the timescale of the fluid mixing. Previous studies showed that the simple mixing model can provide a good first-order approximation of physical and chemical conditions in mixing environments that

TABLE 2.4.4 Chemolithotrophic Catabolic Reactions Considered in This Study

Chemolithotrophic Energy Metabolism	Overall Chemical Reaction
Aerobic methanotrophy	$CH_4 + 2O_2 = CO_2 + 2H_2O$
Hydrogenotrophic O_2-reduction	$H_2 + 1/2O_2 = H_2O$
Thiotrophic (H_2S-oxidizing) O_2 reduction	$H_2S + 2O_2 = SO_4^{2-} + 2H^+$
Fe(II)-oxidizing O_2 reduction	$Fe^{2+} + 1/4O_2 + H^+ = Fe^{3+} + 1/2H_2O$
Hydrogenotrophic methanogenesis	$H_2 + 1/4CO_2 = 1/4CH_4 + 1/2H_2O$
Hydrogenotrophic SO_4 reduction	$H_2 + 1/4SO_4^{2-} + 1/2H^+ = 1/4H_2S + H_2O$
Anoxic methanotrophy with SO_4 reduction	$CH_4 + SO_4^{2-} = HCO_3^- + HS^- + H_2O$

control quantitative and spatial distribution of chemolithoautotrophic microbial habitat in various types of seafloor hydrothermal systems (Amend et al., 2011; McCollom, 2007; Takai & Nakamura, 2010; Takai & Nakaruma, 2011; Nakamura & Takai, 2014).

As shown in Figure 2.4.1, the total available energy for these chemolithotrophic metabolisms in all four systems is greatest at low temperatures (especially at <20 °C) and decreases rapidly with increasing temperatures. Potential energy yields by aerobic reactions (e.g., H_2S oxidation, H_2 oxidation, CH_4 oxidation, and Fe(II) oxidation) are responsible for the trends of the total available energy in all the hydrothermal systems. This estimation is supported by the fact that dense populations of microbial communities (e.g., microbial mats) and chemosynthetic macrofaunal communities, in symbiosis with chemolithotrophic microorganisms, develop at the low-temperature mixing zones near the seafloor of all the hydrothermal vent systems considered here. On the other hand, at higher temperatures, aerobic reactions can produce only tiny amounts of catabolic energy, less than ~0.1 kJ/kg HF (Figure 2.4.1). The total available energy under high-temperature conditions, thus, depends largely on potential energy yields by anaerobic reactions (e.g., methanogenesis, sulfate reduction, and anaerobic methane oxidation), which are in turn controlled mainly by concentrations of H_2 or CH_4 in the end-member HFs.

TAG field: In the subseafloor mixing zones of the basalt-hosted TAG hydrothermal system, H_2S oxidation is predicted to be the most energetically favorable catabolism over the entire temperature range (Figure 2.4.1(A)), although aerobic CH_4 oxidation seems to compete with H_2S oxidation at the temperature range higher than ~90 °C. Fe(II) oxidation can also yield moderate amounts of energy at low temperatures (<40 °C) accounting for up to ~20% of the total

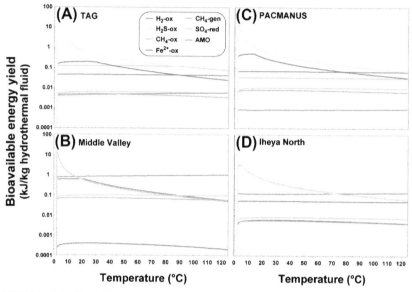

FIGURE 2.4.1   Catabolic energy available to communities of microorganisms inhabiting the mixing zones of submarine hydrothermal environments in (A) the TAG field, (B) Middle Valley field, (C) the PACMANUS field, and (D) the Iheya North field.

potential energy at 25 °C. Aerobic H_2 oxidation yields relatively small amounts of energy in the temperature range considered, although it contributes considerably to the total potential energy yield (up to ~20%) at higher temperatures (Figure 2.4.1(A)). On the other hand, all of the anaerobic catabolisms yield little energy for the subseafloor microorganisms in the TAG field.

Middle Valley field: The subseafloor mixing environments in the sediment-associated Middle Valley hydrothermal system is characterized by abundant energy yields from both aerobic and anaerobic CH_4 oxidation. The aerobic and anaerobic oxidation of CH_4 are likely the predominant metabolisms for the subseafloor microbial communities living in low-temperature conditions of <7 °C (up to ~80% of total potential energy at ~2 °C) and high-temperature conditions of >15 °C (up to 75% of total potential energy at 125 °C) (Figure 2.4.1(B)). H_2S oxidation outcompetes aerobic CH_4 oxidation from ~7 °C to ~15 °C (Figure 2.4.1(B)). Aerobic H_2 oxidation can also yield a moderate proportion of the total potential energy over the entire temperature range, accounting for up to ~20%. For temperatures higher than ~20 °C, the potential energy yield from aerobic H_2 oxidation outcompetes those from aerobic CH_4- and H_2S oxidation (Figure 2.4.1(B)). The relatively high energy yield from aerobic H_2 oxidation is due to the relatively high concentrations of H_2 in the end-member HFs (Table 2.4.3). Due to the high H_2 concentration in the end-member HF, methanogenesis and sulfate reduction yield significant amounts of energy over the entire temperature range (Figure 2.4.1(B)), commonly one order of magnitude higher than in the subseafloor environments of other hydrothermal fields. In particular, anaerobic methane oxidation dominates the total potential energy available for chemolithotrophy (up to ~80%) at the high-temperature range (Figure 2.4.1(B)). Energy yield from Fe(II) oxidation is essentially negligible in the subseafloor mixing environments of the Middle Valley field (Figure 2.4.1(B)).

PACMANUS field: The subseafloor mixing environment of the PACMANUS field is characterized by a high energy yield from H_2S oxidation (Figure 2.4.1(C)), especially at lower temperature than ~15 °C (up to ~95%). In addition, the relative contribution of Fe(II) oxidation at temperatures less than ~80 °C (up to ~30%) is also a characteristic feature of the subseafloor environment here (Figure 2.4.1(C)), while the energy yield from aerobic H_2 oxidation outcompetes Fe(II) oxidation as a catabolic strategy at temperatures above ~80 °C (Figure 2.4.1(C)). As compared to aerobic H_2S- and Fe(II) oxidation, all the anaerobic metabolisms and aerobic CH_4 oxidation yield far less energy at all temperatures (Figure 2.4.1(C)).

Iheya North field: In the subseafloor mixing environments of Iheya North field, a sediment-associated CH_4-rich hydrothermal system, both aerobic CH_4- and H_2S oxidation dominate the total potential energy below ~7 °C (Figure 2.4.1(D)). For temperatures higher than ~75 °C, however, anaerobic CH_4 oxidation becomes the most energetically favorable catabolism (Figure 2.4.1(D)). This is similar to the energetic state for the subseafloor microbial productivity in the Middle Valley field, which also represents a sediment-associated CH_4-rich hydrothermal system. Aerobic H_2 oxidation and all of the anaerobic catabolisms other than anaerobic CH_4 oxidation provide very little energy over the entire temperature range (Figure 2.4.1(D)) due to the relatively low concentrations of H_2 in the end-member HF (Table 2.4.3).

2.4.4.2 Anabolic Reaction Energetics

As with catabolism, the amount of energy required to synthesize biomass (anabolism) is a function of the temperature, pressure, and chemical composition describing the environment in which these reactions occur. However, an accurate accounting of the energy required to produce new biomass under extreme conditions requires knowledge of the type and amount of biomolecules needed by the organisms in situ. Although this level of detail is rarely available, the molecular composition of biomass can be approximated by dividing cellular biomass into its constituent monomers—amino acids, nucleotides, fatty acids, saccharides, and amines (Morowitz, 1968)—and by stipulating that, to the first order, the proportion of biomolecules in a model prokaryotic organism (*Escherichia coli*) is similar to that for other microorganisms. With this information in hand, along with the composition, temperature, and pressure in hydrothermal environments, the Gibbs energy (ΔG_r) required to synthesize biomass in any environment can be computed.

Thus far, the Gibbs energies for the formation of biomonomers from inorganic precursors (HCO_3^-, NH_4^+, H_2S, and H_2) have been evaluated at a range of extreme conditions. For example, the energetics of anabolism were calculated under microoxic and anoxic conditions (McCollom & Amend, 2005) as well as under hydrothermal conditions (Amend & McCollom, 2009; Amend et al., 2011). Based on the composition of *E. coli*, McCollom and Amend (2005) computed that 18,435 J are required to synthesize all of the amino acids, ribo- and deoxyribo-nucleotides, fatty acids, saccharides, amines, and other compounds commonly found in a dry gram of prokaryotic cells at 25 °C and 1 bar under relatively oxidizing conditions (Eh = 77 mV). Under more reducing conditions (Eh = −27 mV), the quantity decreases substantially to 1434 J/(g dry cell mass). Furthermore, it was noted that if the sources of nitrogen and sulfur are NO_3^- and SO_4^{2-} (rather than NH_4^+ and H_2S), then an additional 3170 J/(g dry cell mass) is required. This is consistent with the previously recognized higher biomass yield per unit energy in anaerobic autotrophs compared with their aerobic counterparts (Heijnen & van Dijken, 1992).

The energetics of biomass synthesis have also been determined in 12 deep-sea hydrothermal systems at different SW to HF ratios (Amend et al., 2011). The host rocks characterizing these systems include basalt (Edmond, Endeavor, EPR 21°N, Lucky Strike, TAG, Menez Gwen), peridotite (Rainbow, Logatchev, Lost City), felsic rock (Brothers, Mariner), and a troctolite–basalt hybrid (Kairei). Owing to different rock types, the concentrations of key reactants vary widely: pH (2.7–9), H_2 (0.04–16 mM), H_2S (0.1–9.7 mM), NH_4^+ (0.1–503 mM), and CH_4 (0.007–2.5 mM), among others. Consequently, the energetics of biomonomer synthesis show a wide range among the different systems. Predominantly because of the high H_2 levels, the formation of biomass yielded energy in the peridotite and troctolite–basalt hybrid systems, up to approximately 900 J/(g dry cell mass). In the basalt-hosted and felsic rock-hosted systems, the energetics were far less favorable even at the optimum conditions considered, with values ranging from −400 to +275 J/(g dry cell mass).

Amend et al. (2011) also calculated the effect of these compositional differences on biomass synthesis as a function of temperature in the above-stated 12 hydrothermal systems. It was shown that the Gibbs energies for the formation of total cell biomass as a function of the ratio of SW to HF mixing minimizes between approximately 10 °C and 50 °C and an SW:HF ratio of approximately 50–5 (Figure 2.4.2). In 7 of the 12 systems investigated (Rainbow, Logatchev, Kairei, Lost City, Endeavor, EPR 21°N, and Lucky Strike), this minimum is at $\Delta G_r < 0$; in other words, the synthesis of cellular biomonomers is exergonic at these conditions. Among the four hydrothermal systems (TAG, Middle Valley, PACMANUS, and Iheya North fields) that the previous ODP and IODP expeditions have explored, Middle Valley and Iheya North are most similar to Endeavor while the PACMANUS system is most similar to the Mariner field (Figure 2.4.2). If the energetics of biomass synthesis indeed correspond to the composition of the suggested subseafloor hydrothermal mixing environments

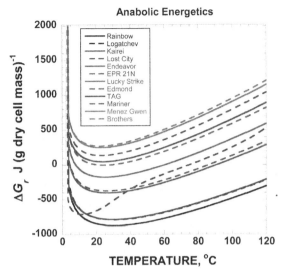

FIGURE 2.4.2   Gibbs energies (joules per gram dry cell mass) of anabolic reactions that represent the energy required to synthesize the biomonomers that constitute prokaryotic cells as a function of temperature in 12 deep-sea hydrothermal systems. *Redrawn after Amend et al. (2013).*

in sediment-associated hydrothermal systems, environments like the Middle Valley and the Iheya North fields should support a greater biomass of indigenous microbial communities than hydrothermal settings that are more oxidizing. Consequently, the spatial and temporal extent of the biosphere would exceed that of areas where the demand for anabolic energy is larger. It should be emphasized, however, that these calculations consider only the net energetics of reaction from inorganic compounds to biomonomers; other possible energy costs are not included so that the total anabolic process may well have a positive Gibbs energy. It should also be noted that the energetics differed demonstrably among the different biomolecule families. Generally, amino acid and fatty acid synthesis reactions were thermodynamically the most favorable. The formation of amines, saccharides (both with $\Delta G_r \approx 0 \, J/(g$ dry cell mass)), and nucleotides ($\Delta G_r > 0$) are energetically much less favorable. In fact, nucleotide synthesis was endergonic in each system and at all conditions considered, perhaps reflecting the structural complexity (e.g., double-bonded carbon–nitrogen rings) and the relatively high carbon redox state of these compounds (see LaRowe & Van Cappellen, 2011).

So far, we have discussed here only the progress that has been made in determining how chemical and physical variables affect the formation energetics of relatively simple biomolecules. However, microorganisms are largely composed of biomacromolecules, such as RNA, DNA, proteins, lipids, and polysaccharides, polymeric versions of the aforementioned biomonomers. Recently, the energetics of amino acid polymerization into polypeptides were determined as a function of temperature and pressure (Amend et al., 2013), and thus extreme environments. In that study, it was concluded that the energy required to polymerize all the protein in a gram of dry prokaryotic (*E. coli*) cells at 25 °C and 0.1 MPa is 191 J/(g dry cell mass). Although the concentrations of protein and amino acids inside cells will affect this value, the exterior cell chemistry does not, since the chemical reaction describing peptide formation is simply a dehydration reaction (i.e., no change in oxidation state). At low (saturation pressure) and high (50 MPa) pressures, the standard state Gibbs energies $\left(\Delta G_r^0 \right)$ of amino acid polymerization increase with increasing temperature, reaching maxima at approximately 120–150 °C; with a further increase in temperature, ΔG_r^0 decreases. It is worth noting that the pressure effect (independent of temperature) accounts for only approximately 0.5 kJ per mole of peptide bond formed, and the temperature effect from 0 °C to 150 °C (independent of pressure) accounts for up to 1.5 kJ per mole of peptide bond.

2.4.5 CONCLUDING REMARKS AND PERSPECTIVES

Of the many potential limits to life in the subseafloor, we have reviewed those that have been addressed, however marginally, by research made possible by ODP/IODP expeditions: high temperature, high pH, and energy. Because both temperature and composition (e.g., pH) influence the energetics of metabolism, thermodynamic calculations have been carried out to determine in which environments particular catabolic strategies are favored and how expensive biomass synthesis is under a variety of conditions. This type of analysis permits prediction of the dominant metabolic activities in a specific setting and the proportion of the microbial community engaged in them. Furthermore, the cost of growth in different extreme environments can be compared by computing the energetics of biomolecular synthesis and polymerization.

Based on these kinds of thermodynamic calculations, observation about the extent of the biosphere in the deep subsurface made possible by ODP/IODP sponsored work can be rationalized. For instance, it seems that, although microorganisms can be grown in the laboratory at 122 °C, there is not much evidence of subseafloor life above ~80 °C. This can be reconciled with the thermodynamic calculations carried out and reviewed in this communication (Figures 2.4.1 and 2.4.2): catabolic reactions occurring at lower temperatures are more exergonic than those at higher temperatures and anabolism becomes more expensive at higher temperatures. Both of these energetic realities should at the very least limit the size of microbial communities in hot subseafloor environments.

The energetics of catabolism in the sites considered in this study are largely due to the variable geologic settings of these environments. For example, fluids in habitats with temperatures >70 °C from sediment-associated hydrothermal systems, such as the Middle Valley and the Iheya North fields, are able to provide more abundant catabolic energy than fluids from sediments-starved hydrothermal systems along MOR, such as the TAG field (Figure 2.4.1). In addition, the energetics of biomass synthesis (anabolism) also point that the subsea-floor hydrothermal mixing habitats in the sediments-associated hydrothermal systems have a potential to host greater indigenous biomass than those in the sediments-starved systems (Figure 2.4.2). As already mentioned, the previous ODP- and IODP-expedition-based microbiological investigations have shown that the sediment-associated hydrothermal systems provide clearer signatures for the occurrence of indigenous microbial communities and therefore better established limits for the boundary conditions of life in subseafloor environments. Thus, further multidisciplinary ocean-drilling-based investigation of sediment-associated hydrothermal systems will provide more significant evidence of realistic limits for the subseafloor biosphere in the future. In addition, the thermodynamic estimation of energy states of potential microbial catabolic and anabolic metabolisms provides important targets for the drilling-science-based approach to find a temperature limit of subseafloor biospheres. In the long history of the ODP and IODP, no expedition has focused on ultramafic rock-associated hydrothermal systems. The energetics of both catabolic and anabolic metabolisms strongly suggest the favorable energetics in such hydrothermal systems, including the Rainbow, Logatchev, Kairei, and Lost City fields (Amend et al., 2011; McCollom, 2007; Takai & Nakamura, 2011). Thus, future IODP expeditions based on the microbiological exploration of ultramafic-hosted submarine hydrothermal systems may help expand the boundaries of the observable biosphere.

REFERENCES

Amend, J. P., LaRowe, D. E., McCollom, T. M., & Shock, E. L. (2013). The energetics of organic synthesis inside and outside the cell. *Philosophical Transaction of the Royal Society Biological Sciences*, *368*, 20120255.

Amend, J. P., & McCollom, T. M. (2009). Energetics of biomolecule synthesis on early Earth. In L. Zaikowski, J. M. Friedrich, & S. R. Seidel (Eds.), *Chemical evolution II: From the origins of life to modern society American Chemical Society symposium series: Vol. 1025.* (pp. 63–94). American Chemical Society.

Amend, J. P., McCollom, T. M., Hentscher, M., & Bach, W. (2011). Catabolic and anabolic energy for chemolithoautotrophs in deep-sea hydrothermal systems hosted in different rock types. *Geochimica et Cosmochimica Acta*, *75*, 5736–5748.

Antunes, A., Ngugi, D. K., & Stingl, U. (2011). Microbiology of the Red Sea (and other) deep-sea anoxic brine lakes. *Environmental Microbiology Reports*, *3*, 416–433.

Antunes, A., Taborda, M., Huber, R., Moissl, C., Nobre, M. F., & da Costa, M. S. (2008). *Halorhabdus tiamatea* sp. nov., a non-pigmented, extremely halophilic archaeon from a deep-sea, hypersaline anoxic basin of the Red Sea, and emended description of the genus *Halorhabdus*. *International Journal of Systematic and Evolutionary Microbiology*, *58*, 215–220.

Aoyama, S., Nishizawa, M., Takai, K., & Ueno, Y. (2014). Microbial sulfate reduction within the Iheya North subseafloor hydrothermal system constrained by quadruple sulfur isotopes. *Earth and Planetary Science Letters*, *398*, 113–126.

Bellaev, G. M., & Brueggeman, P. L. (1989). *Deep sea ocean trenches and their fauna*. Scripps Institution of Oceanography Technical Report. Scripps Institution of Oceanography, UC San Diego.

Binns, R. A., Barriga, F. J. A.S., Miller, D. J., & Shipboard Scientific Party. (2002). In *Proceedings of the ocean drilling program, initial reports* (Vol. 193). http://dx.doi.org/10.2973/odp.proc.ir.193.2002.

Bischoff, J. L., & Pitzer, K. S. (1989). Liquid-vapor relations for the system NaCl–H_2O; summary of the P–T–x surface from 300 degrees to 500 degrees C. *American Journal of Science, 289*, 217–248.

Bischoff, J. L., & Rosenbauer, R. J. (1988). Liquid-vapor relations in the critical region of the system NaCl–H_2O from 380 to 415 °C: a refined determination of the critical point and two-phase boundary of seawater. *Geochimica et Cosmochimica Acta, 52*, 2121–2126.

Blackman, D. K., Ildefonse, B., John, B. E., Ohara, Y., Miller, D. J., MacLeod, C. J., et al. (2006). In *Proceedings of the integrated ocean drilling program* (Vol. 304/305). http://dx.doi.org/10.2204/iodp.proc.304305.101.2006.

Butterfield, D. A., McDuff, R. E., Franklin, J., & Wheat, C. G. (1994). Geochemistry of hydrothermal vent fluids from Middle Valley, Juan de Fuca Ridge. In *Proceedings of the ocean drilling program, scientific results* (Vol. 139) (pp. 395–410).

Butterfield, D. A., Nakamura, K., Takano, B., Lilley, M. D., Lupton, J. E., Resing, J. A., et al. (2011). High SO_2 flux, sulfur accumulation, and gas fractionation at an erupting submarine volcano. *Geology, 39*, 803–806.

Charlou, J. L., Donval, J. P., Jean-Baptiste, P., Dapoigny, A., & Rona, P. A. (1996). Gases and helium isotopes in high temperature solutions sampled before and after ODP Leg 158 drilling at TAG hydrothermal field (26°N, MAR). *Geophysical Research Letter, 23*, 2491–3494.

Chester, F. M., Rowe, C., Ujiie, K., Kirkpatrick, J., Regalla, C., Remitti, F., et al. (2013). Structure and composition of the plate-boundary slip zone for the 2011 Tohoku-Oki Earthquake. *Science, 342*, 1208–1211.

Ciobanu, M.-C., Burgaud, G., Dufresne, A., Breuker, A., Redou, V., Maamar, S. B., et al. (2014). Microorganisms persist at record depths in the subseafloor of the Canterbury Basin. *ISME Journal, 8*, 1370–1380.

Cragg, B. A., & Parkes, R. J. (1994). Bacterial profiles in hydrothermally active deep sediment layers from Middle Valley (NE Pacific), Sites 857 and 858. In *Proceedings of the ocean drilling program, scientific results* (Vol. 139) (pp. 509–516).

Cragg, B. A., Summit, M., & Parkes, R. J. (2000). Bacterial profiles in a sulfide mound (Site 1035) and an area of active fluid venting (Site 1036) in hot hydrothermal sediments from Middle Valley (Northeast Pacific). In *Proceedings of the ocean drilling program, scientific results* (Vol. 169). http://dx.doi.org/10.2973/odp.proc.sr.169.105.2000.

Davis, E. E., Mottl, M. J., Fisher, A. T., & Shipboard Scientific Party. (1992). In *Proceedings of the ocean drilling program, initial reports* (Vol. 139). http://dx.doi.org/10.2973/odp.proc.ir.139.1992.

D'Hondt, S., Inagaki, F., Alvarez Zarikian, C. A., & the Expedition 329 Scientists. (2011). In *Proceedings of the integrated ocean drilling program* (Vol. 329). http://dx.doi.org/10.2204/iodp.proc.329.101.2011.

D'Hondt, S., Spivack, A. J., Pockalny, R., Ferdelman, T. G., Fischer, J. P., Kallmeyer, J., et al. (2009). Subseafloor sedimentary life in the South Pacific Gyre. *Proceedings of National Academy of Sciences in USA, 106*, 11651–11656.

Eder, W., Ludwig, W., & Huber, R. (1999). Novel 16S rRNA gene sequences retrieved from highly saline brine sediments of Kebrit Deep, Red Sea. *Archives of Microbiology, 172*, 213–218.

Edmonds, H. N., German, C. R., Green, D. R. H., Huh, Y., Gamo, T., & Edmond, J. M. (1996). Continuation of the hydrothermal fluid chemistry time series at TAG, and the effects of ODP drilling. *Geophysical Research Letter, 23*, 3487–3489.

Edwards, K. J., Bond, P. L., Gihring, T. M., & Banfield, J. F. (2000). An archaeal iron-oxidizing extreme acidophile important in acid mine drainage. *Science, 287*, 1796–1799.

Emeis, K.-C., & Shipboard Scientific Party. (1996). 2. Paleoceanography and Sapropel Introduction. In *Proceedings of the ocean drilling program, initial reports* (Vol. 160). http://dx.doi.org/10.2973/odp.proc.ir.160.102.1996.

Fouquet, Y., Zierenberg, R. A., Miller, D. J., & Shipboard Scientific Party. (1998). In *Proceedings of the ocean drilling program, initial reports* (Vol. 169). http://dx.doi.org/10.2973/odp.proc.ir.169.1998.

Glud, R. N., Wenzhöfer, F., Middelboe, M., Oguri, K., Turnewitsch, R., Canfield, D. E., et al. (2013). High rates of microbial carbon turnover in sediments in the deepest oceanic trench on Earth. *Nature Geoscience, 6*, 284–288.

Heijnen, J. J., & van Dijken, J. P. (1992). In search of a thermodynamic description of biomass yield for the chemotrophic growth of microorganisms. *Biotechnology and Bioengineering, 39*, 833–858.

Horikoshi, K. (1999). Alkaliphiles: some applications of their products for biotechnology. *Microbiology and Molecular Biology Reviews, 63*, 735–750.

Humphris, S. E., Herzig, P. M., Miller, D. J., & Shipboard Scientific Party. (1996). In *Proceedings of the ocean drilling program, initial reports* (Vol. 158). http://dx.doi.org/10.2973/odp.proc.ir.158.1996.

Kato, C., Li, L., Tamaoka, J., & Horikoshi, K. (1997). Molecular analyses of the sediment of the 11,000-m deep Mariana Trench. *Extremophiles, 1*, 117–123.

Kawagucci, S., Chiba, H., Ishibashi, J., Yamanaka, T., Toki, T., Muramatsu, Y., et al. (2011). Hydrothermal fluid geochemistry at the Iheya North field in the mid-Okinawa Trough: implication for origin of methane in subseafloor fluid circulation systems. *Geochemical Journal, 45*, 109–124.

Kelley, D. S., Karson, J. A., Blackman, D. K., Fruh-Green, G. L., Butterfield, D. A., Lilley, M. D., et al. (2001). An off-axis hydrothermal vent field near the Mid-Atlantic Ridge at 30 degrees N. *Nature, 412*, 145–149.

Kelley, D. S., Karson, J. A., Fruh-Green, G. L., Yoerger, D. R., Shank, T. M., Butterfield, D. A., et al. (2005). A serpentinite-hosted ecosystem: the lost city hydrothermal field. *Science, 307*, 1428–1434.

Kimura, H., Asada, R., Masta, A., & Naganuma, T. (2003). Distribution of microorganisms in the subsurface of the manus basin hydrothermal vent field in Papua New Guinea. *Applied and Environmental Microbiology, 69*, 644–648.

Kobayashi, H., Hatada, Y., Tsubouchi, T., Nagahama, T., & Takami, H. (2012). The hadal amphipod *Hirondellea gigas* possessing a unique cellulase for digesting wooden debris buried in the deepest seafloor. *PLoS One, 7*, e42727. http://dx.doi.org/10.1371/journal.pone.0042727.

Koschinsky, A., Garbe-Schönberg, D., Sander, S., Schmidt, K., Gennerich, H.-H., & Strauss, H. (2008). Hydrothermal venting at pressure-temperature conditions above the critical point of seawater, 5°S on the Mid-Atlantic Ridge. *Geology, 36*, 615–618.

Krijgsman, W., Hilgen, F. J., Raffi, I., Sierro, F. J., & Wilson, D. S. (1999). Chronology, causes and progression of the Messinian salinity crisis. *Nature, 400*, 652–655.

LaRowe, D. E., & Amend, J. P. (2014). Energetic constraints on life in marine deep sediments. In J. Kallmeyer, & K. Wagner (Eds.), *Life in extreme environments: Microbial life in the deep biosphere* (pp. 279–302). De Gruyter Publishing.

LaRowe, D. E., Dale, A. W., & Regnier, P. (2008). A thermodynamic analysis of the anaerobic oxidation of methane in marine sediments. *Geobiology*, *6*, 436–449.

LaRowe, D. E., & Van Cappellen, P. (2011). Degradation of natural organic matter: a thermodynamic analysis. *Geochimica et Cosmochimica Acta*, *75*, 2030–2042.

Lin, W., Conin, M., Moore, J. C., Chester, F. M., Nakamura, Y., Mori, J. J., et al. (2013). Stress state in the largest displacement area of the 2011 Tohoku-Oki Earthquake. *Science*, *339*, 687–690.

McCollom, T. M. (2007). Geochemical constraints on sources of metabolic energy for chemolithoautotrophy in ultramafic-hosted deep-sea hydrothermal systems. *Astrobiology*, *7*, 933–950.

McCollom, T. M., & Amend, J. P. (2005). A thermodynamic assessment of energy requirements for biomass synthesis by chemolithoautotrophic micro-organisms in oxic and anoxic environments. *Geobiology*, *3*, 135–144.

McCollom, T. M., & Bach, W. (2009). Thermodynamic constraints on hydrogen generation during serpentinization of ultramafic rocks. *Geochimica et Cosmochimica Acta*, *73*, 856–875.

McCollom, T. M., & Shock, E. L. (1997). Geochemical constraints on chemolithoautotrophic metabolism by microorganisms in seafloor hydrothermal systems. *Geochimica et Cosmochimica Acta*, *61*, 4375–4391.

Mason, O. U., Nakagawa, T., Rosner, M., Van Nostrand, J. D., Zhou, J., Maruyama, A., et al. (2010). First investigation of the microbiology of the deepest layer of ocean crust. *PLoS One*, *5*, e15399.

Morono, Y., Terada, T., Nishizawa, M., Ito, M., Hillion, F., Takahata, N., et al. (2012). Carbon and nitrogen assimilation in deep subseafloor microbial cells. *Proceedings of National Academy of Sciences in USA*, *108*, 18295–18300.

Morowitz, H. J. (1968). *Energy flow in biology: Biological organization as a problem in thermal physics*. New York: Academic Press.

Mottl, M. J., Komor, S. C., Fryer, P., & Moyer, C. L. (2003). Deep-slab fluids fuel extremophilic Archaea on a Mariana forearc serpentinite mud volcano: Ocean Drilling Program Leg 195. *Geochemistry Geophysics Geosystem*, *4*. http://dx.doi.org/10.1029/2003GC000588.

Nakagawa, T., Takai, K., Suzuki, Y., Hirayama, H., Konno, U., Tsunogai, U., et al. (2006). Geomicrobiological exploration and characterization of a novel deep-sea hydrothermal system at the TOTO caldera in the mariana Volcanic Arc. *Environmental Microbiology*, *8*, 37–49.

Nakamura, K., & Takai, K. (2014). Theoretical constraints of physical and chemical properties of hydrothermal fluids on variations in chemolithotrophic microbial communities in seafloor hydrothermal systems. *Progress in Earth and Planetary Science*, *1*. (5). http://dx.doi.org/10.1186/2197-4284-1-5.

Pedersen, K., Nilsson, E., Arlinger, J., Hallbeck, L., & O'Neill, A. (2004). Distribution, diversity and activity of microorganisms in the hyper-alkaline spring waters of Maqarin in Jordan. *Extremophiles*, *8*, 151–164.

Rainy, F. A., & Oren, A. (2006). Extremophile microorganisms and the method to handle them. Extremophiles. In *Methods in microbiology* (Vol. 35)London, UK: Elsevier.

Reeves, E. P., Seewald, J. S., Saccocia, P., Bach, W., Craddock, P. R., Shanks, W. C., et al. (2011). Geochemistry of hydrothermal fluids from the PACMANUS, Northeast Pual and Vienna Woods hydrothermal fields, Manus Basin, Papua New Guinea. *Geochimica et Cosmochimica Acta*, *75*, 1088–1123.

Resing, J. A., Lebon, G., Baker, E. T., Lupton, J. E., Embley, R. W., Massoth, G. J., et al. (2007). Venting of acid-sulfate fluids in a high-sulfidation setting at NW rota-1 submarine volcano on the Mariana Arc. *Economic Geology*, *102*, 1047–1061.

Reysenbach, A.-L., Holm, N. G., Hershberger, K., Prieur, D., & Jeanthon, C. (1998). In search of a subsurface biosphere at a slow-spreading ridge. In *Proceedings of the ocean drilling program, scientific results* (Vol. 158) (pp. 355–365).

Reysenbach, A.-L., Liu, Y., Banta, A. B., Beveridge, T. J., Kirshtein, J. D., Schouten, S., et al. (2006). A ubiquitous thermoacidophilic archaeon from deep-sea hydrothermal vents. *Nature, 442*, 444–447.

Reysenbach, A.-L., Longnecker, K., & Kirshtein, J. (2001). Novel bacterial and archaeal lineages from an in situ growth chamber deployed at a Mid-Atlantic Ridge hydrothermal vent. *Applied and Environmental Microbiology, 66*, 3798–3806.

Rothschild, L. J., & Mancinelli, R. L. (2001). Life in extreme environments. *Nature, 409*, 1092–1101.

Roussel, E. G., Cambon Bonavita, M.-A., Querellou, J., Cragg, B. A., Webster, G., Prieur, D., et al. (2008). Extending the sub-sea-floor biosphere. *Science, 320*, 1046.

Salisbury, M. H., Shinohara, M., Richter, C., & Shipboard Scientific Party. (2002). In *Proceedings of ocean drilling program initial reports* (Vol. 195)College Station, TX: Ocean Drilling Program. http://dx.doi.org/10.2973/odp.proc.ir.195.2002.

Schleper, C., Pühler, G., Kühlmorgen, B., & Zillig, W. (1995). Life at extremely low pH. *Nature, 375*, 741–742.

Seyfried, W. E., Jr., Ding, K., & Berndt, M. E. (1991). Phase equilibria constraints on the chemistry of hot spring fluids at mid-ocean ridges. *Geochimica et Cosmochimica Acta, 55*, 3559–3580.

Shock, E. L. (1992). Chemical environments of submarine hydrothermal systems. *Origin of Life and Evolution of Biosphere, 22*, 67–107.

Shock, E. L., & Holland, M. E. (2004). Geochemical energy sources that support the subsurface biosphere. In *The subseafloor biosphere at mid-ocean ridges Geophysical Monograph* (Vol. 144) (pp. 153–165).

Spiess, F. N., & RISE Group. (1980). East pacific rise; hot springs and geophysical experiments. *Science, 297*, 1421–1433.

Swallow, J. C., & Crease, J. (1965). Hot salty water at the bottom of the Red Sea. *Nature, 205*, 165–166.

Takai, K. (2011). Limits of life and the biosphere: lessons from the detection of microorganisms in the deep-sea and deep subsurface of the Earth. In M. Gargaud, P. Lopez-Garcia, & H. Martin (Eds.), *Origins and evolution of life—An astrobiological perspective* (pp. 469–486). Cambridge University Press.

Takai, K., Gamo, T., Tsunogai, U., Nakayama, N., Hirayama, H., Nealson, K. H., et al. (2004). Geochemical and microbiological evidence for a hydrogen-based, hyperthermophilic subsurface lithoautotrophic microbial ecosystem (HyperSLiME) beneath an active deep-sea hydrothermal field. *Extremophiles, 8*, 269–282.

Takai, K., & Horikoshi, K. (1999). Genetic diversity of archaea in deep-sea hydrothermal vent environments. *Genetics, 152*, 1285–1297.

Takai, K., Inoue, A., & Horikoshi, K. (1999). *Thermaerobacter marianensis* gen. nov., sp. nov., an aerobic extremely thermophilic marine bacterium from the 11,000 m deep Mariana Trench. *International Journal of Systematic Bacteriology, 49*, 619–628.

Takai, K., Komatsu, T., Inagaki, F., & Horikoshi, K. (2001). Distribution and colonization of archaea in a black smoker chimney structure. *Applied and Environmental Microbiology, 67*, 3618–3629.

Takai, K., Moser, D. P., Onstott, T. C., Spoelstra, N., Pfiffner, S. M., Dohnalkova, A., et al. (2001). *Alkaliphilus transvaalensis* gen. nov., sp. nov., an extremely alkaliphilic bacterium isolated from a deep South African gold mine. *International Journal of Systematic and Evolutionary Microbiology, 51*, 1245–1256.

Takai, K., Mottl, M. J., Nielsen, S. H., & the Expedition 331 Scientists. (2011). In *Proceedings of integrated ocean drilling program expedition 331*. Tokyo: Integrated Ocean Drilling Program Management International, Inc. http://dx.doi.org/10.2204/iodp.proc.331.2011.

Takai, K., Moyer, C. L., Miyazaki, M., Nogi, Y., Hirayama, H., Nealson, K. H., et al. (2005). *Marinobacter alkaliphilus* sp. nov., a novel alkaliphilic bacterium isolated from subseafloor alkaline serpentine mud from Ocean Drilling Program (ODP) Site 1200 at South Chamorro Seamount, Mariana Forearc. *Extremophiles, 9*, 17–27.

Takai, K., & Nakamura, K. (2011). Archaeal diversity and community development in deep-sea hydrothermal vents. *Current Opinion in Microbiology, 14*, 282–291.

Takai, K., Nakamura, K., Toki, T., Tsunogai, T., Miyazaki, M., Miyazaki, J., et al. (2008). Cell proliferation at 122 °C and isotopically heavy CH_4 production by a hyperthermophilic methanogen under high pressures cultivation. *Proceedings of National Academy of Sciences in USA, 105*, 10949–10954.

Takai, K., Nunoura, T., Ishibashi, J., Lupton, J., Suzuki, R., Hamasaki, H., et al. (2008). Variability in the microbial communities and hydrothermal fluid chemistry at the newly discovered Mariner hydrothermal field, southern Lau Basin. *Journal of Geophysical Research, 113*, G02031. http://dx.doi.org/10.1029/2007JG000636.

Takami, H., Inoue, A., Fuji, F., & Horikoshi, K. (1997). Microbial flora in the deepest sea mud of the Mariana Trench. *FEMS Microbiology Letters, 152*, 279–285.

Valentine, D. L. (2007). Adaptations to energy stress dictate the ecology and evolution of the Archea. *Nature Reviews Microbiology, 5*, 316–323.

Von Damm, K. L. (1995). Controls on the chemistry and temporal variability of seafloor hydrothermal fluids. In *Seafloor hydrothermal systems: physical, chemical, biological, and geological interactions Geophysical Monograph* (Vol. 91) (pp. 222–247).

Von Damm, K. L., Parker, C. M., Zierenberg, R. A., Lilley, M. D., Olson, E. J., Clague, D. A., et al. (2005). The Escanaba Trough, Gorda Ridge hydrothermal system: temporal stability and subseafloor complexity. *Geochimica et Cosmochimica Acta, 69*, 4971–4984.

Vreeland, R. H., Rosenzweig, W. D., & Powers, D. W. (2000). Isolation of a 250 million-year-old halotolerant bacterium from a primary salt crystal. *Nature, 407*, 897–900.

Wilcock, W. S. D., & Fisher, A. T. (2004). Geophysical constraints on the subseafloor environment near mid-ocean ridge. In *The subseafloor biosphere at mid-ocean ridges Geophysical Monograph* (Vol. 144) (pp. 51–74).

Yanagawa, K., Nunoura, T., McAllister, S. M., Hirai, M., Breuker, A., Brandt, L., et al. (2013). The first microbiological contamination assessment by deep-sea drilling and coring by the D/V Chikyu at the Iheya North hydrothermal field in the Mid-Okinawa Trough (IODP Expedition 331). *Frontiers in Microbiology, 4*, 327.

Yayanos, A. A. (1986). Evolutional and ecological implications of the properties of deep-sea barophilic bacteria. *Proceedings of National Academy of Sciences in USA, 83*, 9542–9546.

Chapter 2.5

Life in the Ocean Crust: Lessons from Subseafloor Laboratories

Beth N. Orcutt,[1,*] Katrina J. Edwards[2]
[1]*Bigelow Laboratory for Ocean Sciences, East Boothbay, ME, USA;* [2]*Department of Earth Sciences, and Department of Biological Sciences, University of Southern California, Los Angeles, CA, USA*
Corresponding author: E-mail: borcutt@bigelow.org

2.5.1 INTRODUCTION

One of the major discoveries of the scientific ocean drilling program over the past three decades is the confirmation of active and abundant life "buried alive" in the marine "deep biosphere," including deeply buried sediment and igneous oceanic crust (IODP, 2001; IODP, 2011). Following pioneering studies using deep marine sediment that confirmed the biological production of methane (Hoehler, Borowski, Alperin, Rodrigiuez, & Paull, 2000; Oremland, Culbertson, & Simoneit, 1982) and visualized microbial cells (Cragg et al., 1990; Cragg, Harvey, Fry, Herbert, & Parkes, 1992; Cragg & Kemp, 1995; Parkes et al., 1994), subsequent dedicated ocean drilling research confirmed the widespread existence of a marine deep biosphere (D'Hondt et al., 2004; D'Hondt, Rutherford, & Spivack, 2002; Inagaki et al., 2006). These studies lead to a revolutionary idea that the sedimentary marine deep biosphere might contain the majority of Earth's microorganisms and up to a third of all life on Earth (Parkes, Cragg, & Wellsbury, 2000; Whitman, Coleman, & Wiebe, 1998); however, these early estimates have since been revised downward by about an order of magnitude following increased sampling of a larger diversity of sedimentary systems, including the largely oligotrophic sedimentary deep biosphere (Kallmeyer, Pockalny, Adhikari, Smith, & D'Hondt, 2012). Though diminished in scope, the novel function and genetic diversity of the sedimentary marine deep biosphere are still significant (Hinrichs & Inagaki, 2012; Jørgensen, 2012).

By comparison, the existence and extent of a deep biosphere hosted in the igneous oceanic crust—the hard rock that resides beneath sediment or exposed near mid-ocean ridge spreading centers—are considerably less understood.

Developments in Marine Geology, Volume 7. http://dx.doi.org/10.1016/B978-0-444-62617-2.00007-4

Igneous oceanic crust is hydrologically active, with the entire volume of the oceans circulating through the rocky, porous, fractured crust on the order of hundreds of thousands of years (Wheat, McManus, Mottl, & Giambalvo, 2003). The crustal reservoir is also roughly 10 times larger than the sedimentary deep biosphere (Orcutt, Sylvan, Knab, & Edwards, 2011)—suggesting that this realm may harbor a larger abundance of microorganisms than sediment. Early molecular studies implicated microbial life in crustal samples (e.g., Giovannoni, Fisk, Mullins, & Furnes, 1996), but confirmation of an extensive biosphere in deep oceanic crust has been a challenging and long process. Unlike oceanic crust, deep sediment is readily sampled though the ocean drilling program, and previous studies have demonstrated that sediment can be recovered relatively uncontaminated and used for molecular biological surveys of life (Smith, Spivack, Fisk, Haveman, & Staudigel, 2000). There are significant challenges to sampling and study of igneous oceanic crust, in contrast. Coring in upper basement is notoriously difficult with the current drilling technologies used on the RV *JOIDES Resolution*, the drilling vessel most commonly used to sample igneous oceanic crust. Sample recovery during oceanic drilling is also low by comparison to sediment coring, with generally less than 25% of drilled material being recovered in cores (Edwards et al., 2012). Also, the sample material recovered generally skews to more massive rock that is easier to drill than highly fractured, altered, porous, and permeable basalts, although the latter is the more likely location for microbial colonization due to fractures allowing fluid and cell transport. Finally, drilling contamination can be a major concern in the rocks that are recovered (Lever et al., 2006), potentially obfuscating native rock-hosted microbial life.

Persistence by the scientific community to overcome these challenges has resulted in a substantial body of evidence supporting the existence of a diverse and active deep biosphere hosted in igneous oceanic crust—the subject of this chapter. Here, we synthesize recent molecular biological and biogeochemical studies conducted on samples of igneous oceanic crust collected through the Integrated Ocean Drilling Program (IODP) from various locations around the globe (Figure 2.5.1), as well as from seafloor-exposed outcrops and hydrothermal deposits. We also highlight recent novel microbial experimental techniques using subseafloor observatory technologies that started during the recent phase of the ocean drilling program. Such microbial observatories present alternative methods for overcoming the challenges associated with drilling and coring in oceanic crust, opening a window into the hard-to-access deep igneous oceanic crust for microbiological analyses.

2.5.2 GENERAL OVERVIEW OF THE DIVERSITY, ACTIVITY, AND ABUNDANCE OF MICROBIAL LIFE IN IGNEOUS OCEANIC CRUST

At the start of IODP back in 2003, little was then known about the existence of microbial life in oceanic crust. The Initial Science Plan for IODP (IODP, 2001) hinted at the existence of microbial life in oceanic crust, based on pioneering

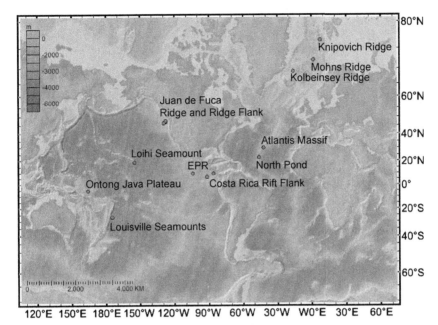

FIGURE 2.5.1 Global map of locations where seafloor and/or subsurface crustal rocks have been collected for microbiological investigation (see text for more details). East Pacific Rise (EPR). Inset in upper left indicates water depth in meters, and scale bar in bottom left indicates horizontal distance in kilometers. *Map was created using the default Global Multi-Resolution Topography Synthesis (Ryan et al., 2009) basemap in* GeoMapApp *version 3.2.1* (www.geomapapp.org).

studies that identified structures seen in oceanic crust rock thin sections that were attributed to microbial activity (Furnes et al., 2001; Fisk, Giovannoni, & Thoreth, 1998; Furnes & Staudigel, 1999; Giovannoni et al., 1996; Torsvik, Furnes, Muehlenbacks, Thorseth, & Tumyr, 1998). Other work also documented the existence of conspicuous stalked filaments on seafloor and subsurface basalts (Thorseth et al., 2001; Thorseth, Pedersen, & Christie, 2003), which are only known to be produced by iron-oxidizing bacteria (Chan, Fakra, Emerson, Fleming, & Edwards, 2010; Emerson et al., 2007; Emerson & Moyer, 2002). Studies to identify the identity and function of the microorganisms that colonized the rock surfaces indicated that various groups of Proteobacteria—a diverse phylum of the Domain Bacteria—were involved in iron oxidation on basalt surfaces through unknown biochemical mechanisms (Edwards, McCollom, Konishi, & Buseck, 2003; Rogers, Santelli, & Edwards, 2003; Thorseth et al., 2001).

These early studies prompted a wave of research into the composition of microbial communities inhabiting seafloor-exposed and subsurface basalts (Figure 2.5.2), buoyed by the concurrent findings mentioned above of a vast deep biosphere hosted in deep marine sediment. A series of papers documented the relatively high diversity of bacterial communities supported on seafloor-exposed basalts as compared to deep seawater and other deep-sea habitats (Mason et al.,

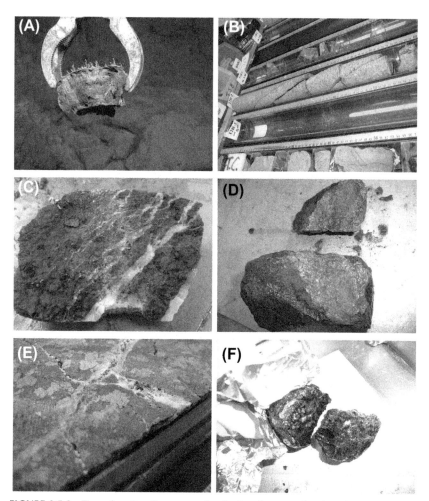

FIGURE 2.5.2 Examples of seafloor-exposed and subsurface crustal rocks collected for micro-biological investigation. (A) Collection of a seafloor-exposed basalt covered in iron-oxidizing bacteria from the Loihi Seamount by the ROV Jason II. (B) Example subsurface drill cores of fractured massive basalt from the Juan de Fuca Ridge flank. (C) Close-up of serpentinized breccia collected from the subsurface of the western Mid-Atlantic Ridge flank at North Pond. (D) Close-up of secondary mineral formation in veins of basalt collected from North Pond. (E) Close-up of vein-filling secondary minerals in serpentinized breccia collected from North Pond. (F) Oxidized glassy basalt recovered from North Pond. *Photograph in A courtesy of Woods Hole Oceanographic Institution; photographs in B–F copyright Beth Orcutt.*

2008; Santelli et al., 2008; Santelli, Edgcomb, Bach, & Edwards, 2009; Toner et al., 2013). These works also indicated that bacterial community diversity increased as the rock surfaces became progressively more altered and oxidized (Santelli et al., 2008, 2009), although another study did not observe an increase in bacterial abundance with age of basalts (Einen, Thorseth, & Ovreas, 2008).

The dominant members of the basalt hosted microbial communities grouped with uncultured members of the Proteobacteria phylum, and in particular in the Gamma- and Alpha-Proteobacteria classes (Mason et al., 2008; Santelli et al., 2009). Archaea were also detected on seafloor-exposed basalts, although in relatively low abundance and diversity (Mason et al., 2008; Santelli et al., 2008, 2009). The function of the Archaeal members in the rock-hosted communities is unclear, however, as the phylogenetic affiliation of the observed members was predominantly to uncultured representatives with unknown function. These seafloor-exposed basalts have been estimated to harbor 10^5–10^9 cells g^{-1} of rock (Santelli et al., 2008), and there was some indication that genes for carbon fixation and methane and nitrogen cycling were expressed in these microbial communities (Mason et al., 2008).

Concurrently, efforts were underway to document microbial life in deeper oceanic crust samples (Figure 2.5.2). Geochemical methods including stable isotope analysis of mineral deposits in crustal samples—not as susceptible to contamination issues as microbiological analysis—had already indicated microbial sulfate reduction in upper basement and deep mantle-type rocks (Alt et al., 2007; Alt et al., 2003), although the timing of the microbial activity was unclear as stable isotopic signatures can be a record of past activity. New methods to collect deep basalt samples suitable for molecular biological investigation were developed in 2004 during IODP Expedition 301 (Lever et al., 2006). Earlier attempts to document microbial diversity in crustal samples were stymied by contamination (Santelli, Bach, Banerjee, & Edwards, 2010), although another study with gabbroic rocks from the Atlantis Massif was not as impacted by contamination issues (Mason et al., 2010). These stringent methods have paid off with the recent documentation of functional genes for methane and sulfur cycling—indicating recent microbial activity—coregistered with stable isotopic evidence of carbon and sulfur cycling in mineral deposits in subseafloor basalts collected from the warm and chemically reducing upper basement of the eastern flank of the Juan de Fuca Ridge (Lever et al., 2013). These recent results document a relatively low diversity of microbial members involved in methane and sulfur cycling in this system, with only two groups of methane cyclers and one group of sulfate reducers present (Lever et al., 2013). Interestingly, although sulfate reduction and methane cycling were confirmed metabolic strategies in these samples, the regional fluid chemistry of the Juan de Fuca Ridge flank suggests that microbial sulfate reduction rates are very low in this chemically reducing environment (Hulme & Wheat, submitted for publication; Wheat et al., 2003b; Wheat & Mottl, 2000).

More recently, basement samples have been collected for microbiological analysis during several recent IODP expeditions to a variety of settings (Figure 2.5.1). Building off previous work (Lever et al., 2013), a return expedition to the eastern flank of the Juan de Fuca Ridge allowed collection of more upper basement basalts (Expedition 327 Scientists, 2011). Upper basement samples were collected from aged oceanic crust during IODP Expedition 329 to the South Pacific Gyre (Expedition 329 Scientists, 2011), and seamount basalts were also recently collected from the Louisville Seamounts during IODP Expedition 330 (Expedition 330 Scientists, 2011). Hard rock samples from hydrothermal vent stock work sampling during IODP Expedition 331 are also being analyzed for hydrothermal end-members in the deep biosphere (Expedition 331 Scientists, 2010). In 2011, a major IODP expedition to document life in oceanic crust focused on upper basement of the western flank of the Mid-Atlantic Ridge for microbiological investigation (Expedition 336 Scientists, 2012). This site has been the focus of several expeditions over the history of the ocean drilling program, as it represents a relatively young (~8 Ma) and cool (<25 °C) region of upper basement (Bartetzko, Pezard, Goldberg, Sun, & Becker, 2001; Becker, 1990; Becker, Bartetzko, & Davis, 2001; Becker, Langseth, & Hyndman, 1984; Lawrence, Drever, & Kastner, 1979; McDuff, 1984; Ziebis et al., 2012). Some of the first results from this recent expedition indicate oxygen consumption in upper basement, likely a product of microbial activity as abiotic reactions would be sluggish at these cooler temperatures (Orcutt et al., 2013b), confirming projections made from shallow sediment gravity cores in the area that indicated upward flow of oxygen from upper basement into basal sediments (Ziebis et al., 2012). Studies are currently underway to identify and quantify microbial groups in the upper basement rocks from this site (Expedition 336 Scientists, 2012). At the time of this review, analyses of these recently collected samples are in full swing but results are not available yet.

Two important improvements have been critical in enabling these studies within IODP—careful sample collection to minimize contamination (with routine contamination control), and modified DNA extraction and amplification protocols to maximize yield from low-biomass samples. A more extensive review of these improvements and their importance to the next phase of IODP has recently been published (Orcutt et al., 2013a). The recent demonstration of the ability to extract and amplify actively expressed RNA molecules from deep marine sediment (Orsi, Edgcomb, Christman, & Biddle, 2013) offers a tantalizing prospect for soon being able to do the same with deep igneous oceanic crust.

2.5.3 SUBSEAFLOOR OBSERVATORIES: ANOTHER TOOL FOR STUDYING LIFE IN OCEANIC CRUST

To overcome the limitations in hard rock sample collection from deep oceanic crust, microbiologists began collaborating around the turn of the century with geophysical, geochemical, and hydrogeological scientists using subseafloor observatories set in upper basaltic basement (Figures 2.5.3 and 2.5.4). The use of subseafloor observatories began in the 1990s, with the first Circulation Obviation Retrofit Kit (CORK) installed on the eastern flank of the Juan de Fuca Ridge during Ocean Drilling Program Leg 139 (Davis, Becker, Pettigrew, Carson, & Macdonald, 1992). In brief, a CORK is similar to a well used on land

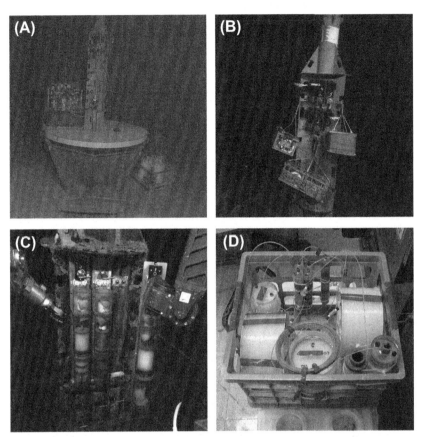

FIGURE 2.5.3 Examples of seafloor platforms of Circulation Obviation Retrofit Kit (CORK) subseafloor observatories used for microbiological research. (A) View of the Hole U1383C North Pond CORK observatory instrumented with a GeoMICROBE sled (unit on left-hand side of image) connected to several fluid sampling lines, and with a OsmoSampler package also connected to fluid sampling lines (unit on lower right-hand side of image). The ball-valve sampling unit is located in the center of the image, and pressure monitoring equipment is mounted on the back of the CORK platform in this configuration. (B) View of the Hole U1362A Juan de Fuca CORK observatory instrumented with OsmoSampler packages (in the green milk crates) being connected to the fluid sampling valves with the ROV Jason-II. (C) View of OsmoSampler packages being mounted on the Hole U1301A CORK observatory by the ROV Jason-II. (D) Close-up view of various OsmoSampler units for microbiological investigation housed inside a sampling milk crate, including several spools of Teflon tubing, osmotic pumps, and flow-through colonization experiments packaged with various mineral substrates. *Photographs in A–C courtesy of the Woods Hole Oceanographic Institution; photograph in D copyright of Beth Orcutt.*

to access freshwater in an underground aquifer, except that CORKs are designed to access hydrological units in upper oceanic crust, where modified seawater circulates beneath the seafloor. CORKs are installed from a drilling ship like the RV *JOIDES Resolution* following drilling and/or coring of the hole as well as placement of a seafloor Re-Entry Cone and Casing and a Borehole Instrument

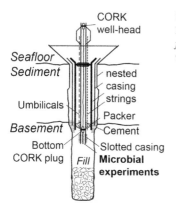

FIGURE 2.5.4 Schematic of Circulation Obviation Retrofit Kit subseafloor borehole observatories. *Modified from Orcutt et al. (2011) ISME Journal and reprinted here with permission.*

Hanger for latching the CORK hardware into place (Graber, Pollard, Jonasson, & Shulte, 2002). CORK hardware can reach several hundreds of meters below the seafloor using a series of nested pipes, often referred to as casings, and the pipes can be slotted or perforated to allow fluid movement into and out of the pipe. Once installed, a CORK can then serve as a long-term monitoring and access point into the subsurface.

CORK designs (Figures 2.5.3 and 2.5.4) have evolved over the decades, ranging from relatively simple open-pipe designs in the early stages for thermal and pressure characterization of subseafloor hydrology (Davis et al., 1992) to designs with subseafloor seals to isolate different hydrological units (Becker & Davis, 2005; Edwards et al., 2012; Fisher et al., 2005, 2011). Similarly, the construction materials used to fabricate CORKs have evolved from generic steel, which corrodes easily and can confound microbiological experiments, to corrosion-resistant fiberglass and epoxy-coated steel (Edwards et al., 2012; Fisher et al., 2011; Orcutt, Barco, Joye, & Edwards, 2012). Most CORKs feature a seafloor installation, often referred to as a "Christmas tree," where various valves and sensors can be accessed with a remotely operated vehicle or submersible (Figures 2.5.3 and 2.5.4). The valves and sensors at the seafloor are connected to small-bore tubing (roughly 3–12 mm inner diameter) that reaches to different depths into the oceanic crust on the outside of the CORK casing, allowing sampling of fluids. Traditionally, stainless steel tubing has been used for these "umbilicals"; however, problems with corrosion and biofouling have lead to the usage of Teflon-coated umbilicals as well (Edwards et al., 2012; Fisher et al., 2005, 2011). For sampling, valves can be opened to allow free flow of overpressured fluids, or pumps can be used to pull fluids up from depth (Cowen et al., 2011; Wheat et al., 2010, 2011). More recently, modified "lateral CORKs" or "L-CORKs" are designed with a ball valve on the seafloor installation to allow easier access to fluids sourced from depth (Edwards et al., 2012; Fisher et al., 2011). Another sampling and monitoring mechanism associated with CORKs is the deployment of instrument strings on the inside of the CORK (Edwards et al.,

2012; Fisher et al., 2005, 2011). Here, narrow-bore (less than 10 cm diameter) sensors, samplers, and experiments are connected together on a long Spectra® cable that hangs from the top of the CORK installation, with instruments then suspended at various depths within the CORK casings.

Monitoring, sampling, and experimentation for deep biosphere investigations are available both at the seafloor and at depth within the CORK (Figures 2.5.3 and 2.5.4). In addition to the battery-powered temperature and pressure sensors located at the wellheads and at depth on instrument strings (Edwards et al., 2012; Fisher et al., 2005), which provide important contextual information about the nature of fluid flow in the subsurface, other sensors can also be deployed to measure fluid chemistry. For example, in situ electrochemistry analyzers (Luther et al., 2008) have been deployed on CORK wellheads connected to fluid sampling valves to record the redox chemistry (i.e., iron, sulfur, and oxygen concentrations) of CORK fluids (Cowen et al., 2011; Wheat et al., 2011). More recently, in situ oxygen sensors have been deployed on CORK instrument strings to monitor oxygen concentrations over time (Edwards et al., 2012). Fluid chemistry within upper basement can also be documented using nonbattery-powered fluid sampling systems called "OsmoSamplers" (Jannasch, Wheat, Plant, Kastner, & Stakes, 2004) that can be deployed both at the wellheads and at depth for extended periods of time (i.e., years; Kastner et al., 2006; Wheat, Elderfield, Mottl, & Monnin, 2000; Wheat et al., 2011). With OsmoSamplers, fluid samples are continuously collected at rates of roughly $0.5\,ml\,d^{-1}$ into small-bore ($\sim 1\,mm$ inner diameter) tubing, with the pumping rate dependent on the surface area of membranes contained with the osmotic pumps that "power" the samplers (Jannasch et al., 2004). Major ion concentrations can be determined from fluids collected into acid-washed Teflon coils, while minor and trace element concentrations can be determined from fluids fixed in situ with acid in Teflon tubing coils using an acid-addition OsmoSampler (Wheat et al., 2011). Gas concentrations can be determined on fluids contained within copper tubing, which prevents outgassing and loss of pressure (Jannasch et al., 2004; Lapham et al., 2008; Wheat et al., 2011). These fluid sampling instruments provide important chemical information about the fluids circulating within basement, allowing constraint of the redox reactions occurring that might support a deep biosphere within oceanic crust.

Most excitingly for deep biosphere research, sampling and in situ experimentation for microbiological investigations are also possible with CORKs—recent microbiological research using such experimentation is described in more detail in the next section. In situ fluid sampling, filtering and concentration have occurred using BioColumns (Cowen et al., 2003) and in situ pumping systems called GeoMICROBE sleds and mobile pumping units (Cowen et al., 2011; Jungbluth, Grote, Lin, Cowen, & Rappe, 2012; Jungbluth, Lin, Cowen, Glazer, & Rappe, 2014; Lin, Cowen, Olson, Amend, & Lilley, 2012). These sample collection systems allow the collection at snapshots in time of discrete volumes of fluids, up to several liters at a time, for shipboard and shore-based investigations of fluid chemistry and cell abundance, diversity, and activity.

More recently, fluid sampling has also occurred with customized syringe samplers via wellhead ball valves and fluid sampling lines. Continuous and long-term fluid collection for cell abundance and diversity studies is also possible through a Biological OsmoSampler (BOSS) that introduces a preservative into the sampling stream, fixing the sample for DNA- and RNA-based laboratory investigation (Robidart et al., 2013). BOSS systems can be deployed both at CORK wellheads and at depth within a CORK on instrument strings (Edwards et al., 2012; Wheat et al., 2011).

Increasingly, OsmoSampler systems are also being used to pump in situ fluids through defined colonization substrates to study microbe-mineral interactions in deep oceanic crust (Edwards et al., 2012; Fisher et al., 2005, 2011; Orcutt et al., 2011; Smith et al., 2011; Wheat et al., 2011) (Figure 2.5.5). Here, sterilized substrates such as basalts, pyrites, and other solid materials are deployed in flow-through columns connected to osmotic pumps (Figures 2.5.3 and 2.5.4), allowing in situ microorganisms to colonize the materials (Orcutt et al., 2011; Smith et al., 2011). Recently, OsmoSamplers have also been designed to "feed" enrichment substrates into the flow-through columns to encourage microbial growth (Edwards et al., 2012; Fisher et al., 2011; Orcutt et al., 2011; Wheat et al., 2011). As with other OsmoSampler systems, these microbiology OsmoSamplers can be deployed both at CORK wellheads and at depth on the instrument string for years at a time. Combined, these technologies represent avenues for sampling the deep biosphere hosted in deep oceanic crust, although care must be taken to interpret colonization results in the context of the experimental conditions, since flow-through observatory experiments have different water–rock ratios than the native crustal environment.

2.5.4 RECENT DEEP BIOSPHERE DISCOVERIES FROM SUBSEAFLOOR OBSERVATORIES

The earliest deep biosphere studies using CORK observatories were conducted in the late 1990s at an early-generation CORK installed at ODP Hole 1026B on the eastern flank of the Juan de Fuca Ridge (Cowen et al., 2003). A BioColumn filtering unit was attached to the wellhead fluid outflow valve, designed to scavenge particles and organics from the escaping warm (54–64 °C) basement fluids sourced from the upper 50 m of basaltic basement underlying roughly 250 m of sediment. In these early studies, the cell density in the basement fluids was determined to be $8.5 \pm 4.1 \times 10^4$ cells ml^{-1}, and the dominant microorganisms based on a sequencing survey of small-subunit rRNA genes from environmental DNA extracts grouped near the *Desulfotomaculum* genus of the Firmicutes phylum and the *Archaeoglobus* genus of the Euryarchaeota (Cowen et al., 2003). This pioneering study documented the existence of a moderately diverse microbial community sourced from upper basaltic basement, although the degree of influence

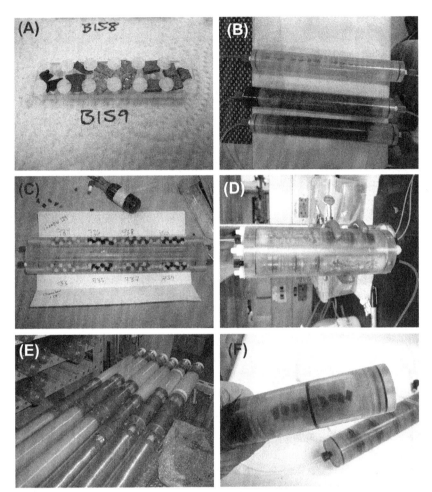

FIGURE 2.5.5 Overview of flow-through microbial colonization experiments (FLOCS) used in subsurface observatories. (A) Close-up of mineral coupons mounted on a plastic grid, including (from L to R) two types of basalt, chalcopyrite, pyrite, pyrrhotite, and magnetite. (B) View of flow-through "FLOCS" columns packed with (from bottom to top) basalt chips, pyrite chips, and various mineral coupons suspended in a glass wool matrix, in preparation for deployment on a Circulation Obviation Retrofit Kit (CORK) wellhead. (C) Mineral coupons mounted on the outside of FLOCS columns in preparation for deployment on a CORK instrument string. (D) A different downhole FLOCS design with internal cassettes packed with mineral substrates and glass beads. (E) Assembled downhole OsmoSampler packages in preparation for deployment in North Pond CORKs, with the osmotic pumps (top right) connected to sampling coils of Teflon (left) and copper (right). (F) Close-up view of FLOCS colonization experiment recovered from the Hole 1026B wellhead after 1 year of deployment. *All photographs copyright Beth Orcutt.*

that the corroding CORK hardware had on the microbial ecology was not previously possible to distinguish. These microbial groups were thought to be involved in sulfur and/or nitrogen cycling in this system.

In 2004, the early-generation CORK at Hole 1026B was upgraded to a second-generation CORK system, and new CORKs were also installed in the same area at IODP Holes U1301A and U1301B during IODP Expedition 301 (Fisher et al., 2005). For the first time, instrument strings were deployed in these CORKs containing mineral colonization experiments along with a variety of OsmoSamplers for measuring fluid chemistry over time (Fisher et al., 2005) (Figure 2.5.5). In 2008, these instrument strings were recovered after a dynamic recovery of the basement aquifer to warm and reducing "predrilling" conditions following up to 3 years of cool conditions due to seawater inflow into the borehole through a bad seal (Wheat et al., 2010). Mineral coupons of basalt and pyrite deployed at depth in the CORKs in Holes U1301A and 1026B were colonized during the cool (and presumably oxic) period of seawater inflow by iron oxide stalk-forming microorganisms (Figure 2.5.6)—likely iron-oxidizing bacteria such as *Mariprofundus* (Emerson et al., 2007; Emerson & Moyer, 2002) as evidenced by "fossil" stalks observed in biofilms on the mineral surfaces (Orcutt et al., 2011). Sequencing of small-subunit rRNA genes in DNA extracts from these chips revealed a community similar in composition to that observed in the Cowen et al. (2003) study—Firmicutes-group bacteria and *Archaeoglobus* Archaea being the dominant clones (Orcutt et al., 2011)—documenting roughly similar microbial community compositions under consistent thermal and chemical conditions in the upper basaltic basement at this location. In contrast, multiple years of fluid sampling at Hole U1301A following the recovery of the instrument strings revealed a fluctuating composition of the basement fluid microbial communities, transitioning from a Firmicutes-dominated system in 2008 to a Gammaproteobacteria-dominated system in 2010 (Jungbluth, Grote, Lin, Cowen, & Rappe, 2012), and more recent sampling at the Hole 1026B observatory also reveals a potential shift in microbial community structure (Jungbluth et al., 2014). It is likely that the repeated recovery and deployment of CORK instrument strings during those years caused shifts in the fluid sources to the basement aquifer, overprinting on the microbial community composition observed. These fluids had cell densities of roughly 1×10^4 cells ml^{-1} (Jungbluth et al., 2012), which is lower than that observed in the earlier Bio-Column studies (Cowen et al., 2003). Flow-through columns packed with various mineral substrates and connected to osmotic pumps in these same CORK intervals were also colonized by cells, with cell densities ranging from 2.3 to 39×10^7 cells g^{-1}, with the highest densities observed on iron-bearing olivines (Smith et al., 2011). Heterotrophic enrichment cultures inoculated with the colonized minerals yielded colonies phylogenetically related to known marine halophiles such as *Alcanivorax*, *Halomonas*, and *Marinobacter* (Smith et al., 2011), although these phylotypes were rare or absent in other studies (Cowen et al., 2003; Jungbluth et al., 2012; Orcutt et al., 2011).

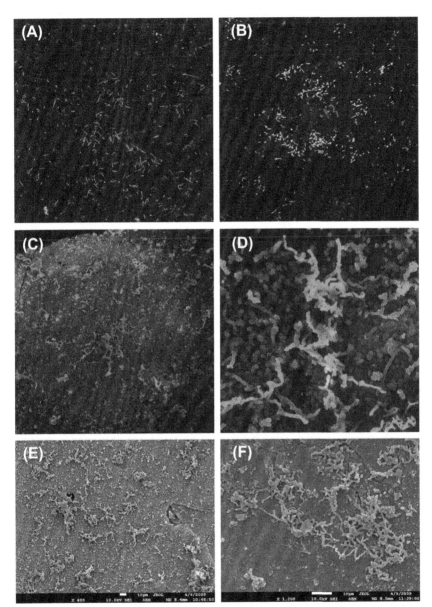

FIGURE 2.5.6 Microbial cells and iron-oxide stalks observed on FLOCS colonization substrates deployed at the Juan de Fuca CORKs. (A and B) Microbial cells colonizing two different basalt chips deployed at the Hole 1026B CORK wellhead for 1 year. Cells were stained with propidium iodide and SYBR Green DNA stains as described in Orcutt, Wheat, and Edwards (2010). (C–F) Stalks formed from iron-oxidizing bacteria on chips of basalt (C and E) and pyrite (D and F) after 4 years of deployment in Hole U1301A CORK observed by epifluorescence microscopy (C and D) and scanning electron microscopy (E and F) as described in Orcutt et al. (2011). *Images copyright of Beth Orcutt (A–D) and Katrina Edwards (E and F).*

This same area of the Juan de Fuca Ridge flank was again targeted with newer-generation CORKs during IODP Expedition 327 in 2010 (Fisher et al., 2011). Unlike earlier CORKs in this region, the new CORKs installed at IODP Holes U1362A and U1362B were constructed with epoxy-coated steel casing to minimize impacts from steel corrosion on microbial community composition and activity (Orcutt, Barco, Joye, & Edwards, 2012). Flow-through, osmotic pump-driven, colonization experiments called "FLOCS" (Orcutt, Wheat, & Edwards, 2010) (Figure 2.5.5) were deployed on the instrument strings in these Holes (Fisher et al., 2011; J.P. Baquiran, G. Ramirez, B. N. Orcutt, K. J. Edwards, unpublished data) as well as on the CORK wellheads. Similar flow-through colonization experiments recently deployed for 1 year on the Hole U1301A CORK wellhead revealed microbial communities that were more similar in composition to seafloor-exposed basalts and sulfides than to the earlier studies of the Hole U1301A subsurface, suggesting an important role for temperature in controlling microbial community structure in the basaltic deep biosphere, as the wellhead experiments were conducted at bottom seawater temperatures (J.P. Baquiran, G. Ramirez, B. N. Orcutt, K. J. Edwards, unpublished data).

Although the vast majority of subsurface observatory research has been conducted on the eastern flank of the Juan de Fuca Ridge, other deep crustal systems also yield important clues to the nature of a crustal deep biosphere. A borehole observatory system deployed in Hole 896A on the Costa Rica Rift flank produced fluids that were recently sampled for microbial community composition analysis (Nigro et al., 2012). In many ways, the Hole 896A environment is similar to that of the Juan de Fuca Ridge flank, including similar rock types in upper basement of roughly the same age and similar thermal structure and fluid composition, although possible seawater leakage into the observatory cannot be excluded (Nigro et al., 2012). Based on a sequencing survey of small-subunit rRNA genes in environmental DNA extracts, the dominant microbial groups in the Hole 896A samples were related to the *Thiomicrospira* genus, with sulfur-oxidizing cultivated members, although other forms of metabolism may also be possible. The community structure was more similar to seafloor-exposed basalts and hydrothermal sulfides than to the microbial communities observed in the Juan de Fuca Ridge flank subsurface, though, suggesting an important role for seawater mixing with deeply sourced hydrothermal fluids in controlling microbial community composition in the crustal subsurface (Nigro et al., 2012). More recently, a series of next-generation CORKs constructed of fiberglass and epoxy-coated steel was deployed into young (<8 Ma) and cool (<25 °C) ridge flank crust on the western flank of the Mid-Atlantic Ridge at "North Pond" during IODP Expedition 336 in 2011 (Edwards, Bach, & Klaus, 2010; Edwards et al., 2012). These new CORKs are replete with downhole and wellhead fluid samplers and colonization experiments to examine the nature of microbial communities resident in this type of crust (Edwards et al., 2010).

2.5.5 THE FUTURE OF SUBSEAFLOOR LABORATORIES FOR DEEP BIOSPHERE RESEARCH

The utility of subseafloor laboratories for deep biosphere research is expected to continue in the new drilling program, as they offer unprecedented access to the most difficult to reach habitat on Earth. The CORK network on the Juan de Fuca Ridge flank represents a model study for collaboration between geophysicists, geochemists, hydrologists, and microbiologists in understanding the understudied upper basement environment, allowing key questions about crustal permeability and fluid flow (Fisher & Becker, 2000) and global chemical cycling (Wheat, Jannasch, Kastner, Plant, & DeCarlo, 2003a; Wheat et al., 2004) to be answered in addition to opening a window into the crustal subsurface biosphere. This network will undoubtedly be targeted for continued investigation, although it represents only one end-member (i.e., warm and reducing chemical environment) of the global crustal system. On the another end of the spectrum, the recently installed CORK network on the Mid-Atlantic Ridge flank (Edwards et al., 2012) is poised to yield new insights into the nature of microbial communities harbored in young, cool, and oxic upper basaltic crust—an end-member of crustal conditions that has not received much attention until recently.

Other CORKs and subseafloor observatories are also being tapped for microbiological research. As was the case for the Juan de Fuca CORKs, these globally distributed observatories were established for nonmicrobiological research purposes, but microbiological experimentation techniques are being adapted to integrate into established research programs. As mentioned above, fluid samples from the observatory currently in Hole 896A allowed comparison of the microbial communities hosted in another warm and chemically reducing basement environment with the Juan de Fuca Ridge flank system (Nigro et al., 2012). Microbial colonization experiments were also recently (2009) deployed in CORK observatories positioned across the Costa Rica convergent margin at ODP Leg 205 Holes 1253 and 1255; some of the samples were only recently recovered in December 2013. Ongoing research in the Nankai Trough has seen the incorporation of downhole fluid sampling and microbial colonization experiments associated with the GeniusPlug retrievable observatory systems (Kopf et al., 2011). Elsewhere in the Nankai system, a CORK system installed in ODP Hole 808I was recently revisited in late 2011 for fluid collection for microbiological analysis, allowing investigation of the microbial communities present in deep sediment above the Nankai subduction zone décollement (J.P. Baquiran, B. N. Orcutt, K. J. Edwards, S. Hulme, G. Wheat, unpublished data). During IODP Expedition 331 to the Okinawa Trough, a new wireline in situ borehole fluid sampler was also used to access hydrothermal fluids made accessible through drilling (IODP, 2011). Finally, recent results from other, older-generation observatories elsewhere on the Juan de Fuca Ridge flank offer a unique look at the possible evolution of microbial communities in deep oceanic crust under different thermal and age conditions (Jungbluth et al., 2014).

New subseafloor observatory networks and designs are in the pipeline, as well. A CORK network is planned for the Cascadia Margin gas hydrate system, which incorporates the new SCIMPI (Simple Cabled Instrument for Measuring Properties In Situ) observatory design for simplified geophysical monitoring (Moran et al., 2006). The first SCIMPI was recently deployed in 2013 during IODP Expedition 341S. SCIMPI is not designed for microbiological research; however, other CORKs in the network are expected to incorporate some microbiological experimentation including BOSS OsmoSampler systems (Robidart et al., 2013; Wheat et al., 2011). Another new observatory design recently on the scene is CORK-Lite, an observatory system designed to be placed into previously drilled and cased "legacy" boreholes to enable geophysical, geochemical, and microbiological monitoring (Wheat et al., 2012). The first CORK-Lite deployment occurred in 2012 in the legacy borehole at IODP Hole U1382B, making it part of the observatory network at North Pond. The successful deployment of CORK-Lite in this system opens up the opportunity for turning other legacy boreholes around the globe into observatory systems, too (Edwards et al., 2012).

2.5.6 THE SIZE OF THE DEEP BIOSPHERE HOSTED IN IGNEOUS OCEANIC CRUST

Given the appreciable size of the igneous oceanic crust as a reservoir for life on Earth coupled with the Deep Biosphere initiative of IODP, it is tempting to speculate on the quantity of biomass in the crustal deep biosphere. Is igneous oceanic crust a significant reservoir of life, and if so, what are the impacts on global models of elemental cycling? A model-driven estimate of the size of the crustal deep biosphere—based on assumptions on the volume or igneous oceanic crust, the percentage of pore space occupied by cells, and assumed cell sizes—suggests that roughly 200 Pg of biomass carbon is stored in the crust (Heberling, Lowell, Liu, & Fisk, 2010). At the time of this model, this amount was comparable to the amount of biomass predicted to be harbored in deep marine sediments (Lipp, Morono, Inagaki, & Hinrichs, 2008; Whitman et al., 1998); however, more recent estimates of the sedimentary deep biosphere at about 4 Pg carbon (Kallmeyer, Pockalny, Adhikari, Smith, & D'Hondt, 2012) would indicate a larger biosphere in the crustal realm, if these estimates are confirmed. Furthermore, if such estimates are accurate even within an order of magnitude, then the size of the crustal deep biosphere would make it a significant contributor to global biomass on Earth (Whitman et al., 1998). It should be emphasized, though, that these estimates were made in a vacuum of data due to a lack of suitable samples for ground-truthing estimates. It is hoped that the recent underway efforts to analyze upper basement materials from a variety of settings (Figure 2.5.1) will enable more robust estimates to be made.

2.5.7 CONCLUSIONS

The last decade of deep biosphere research within IODP and associated programs has confirmed the existence of an active deep biosphere hosted in igneous oceanic crust, making this the largest habitat for life on Earth if it is inhabited throughout. News studies are now needed in a range of settings to explore the extent of life in the crustal ecosystem, as there is a basic lack of understanding about the limits of life in the deep biosphere. Is the extent of life limited by availability of carbon or other energy sources or even nutrients? By the age of the crust? By depth or pressure or temperature? These are fundamental questions that need to be addressed in the next phase of ocean drilling (IODP, 2011). The upcoming ship track for the RV *JOIDES Resolution* in the new drilling program (Humpris & Koppers, 2013) passes by a number of sites ripe with possibilities for exploring deep life in different oceanic crust settings (Edwards, Fisher, & Wheat, 2012). Recent advances in DNA sequencing techniques, sample collection, and handling have made possible recent analyses of igneous oceanic crust, revealing the diversity of microbial groups resident in oceanic crust under different conditions. Pioneering studies are now needed to push the analytical boundary forward to understand the biochemical mechanisms used by deep crustal microorganisms to sustain life "on the rocks" and to understand the genetic connection of deep crustal microbial ecosystems to the rest of the Earth environment.

REFERENCES

Alt, J. C., et al. (2007). Hydrothermal alteration and microbial sulfate reduction in peridotite and gabbro exposed by detachment faulting at the Mid-Atlantic Ridge, 15°20′N (ODP Leg 209): a sulfur and oxygen isotope study. *Geochemistry, Geophysics, Geosystems, 8*(8), Q08002.

Alt, J. C., & Teagle, D. A. (2003). Hydrothermal alteration of upper oceanic crust formed at a fast-spreading ridge: mineral, chemical, and isotopic evidence from ODP Site 801. *Chemical Geology, 201*, 191–211.

Bartetzko, A., Pezard, P., Goldberg, D. S., Sun, Y.-F., & Becker, K. (2001). Volcanic stratigraphy of DSDP/ODP Hole 395A: an interpretation using well-logging data. *Marine Geophysical Research, 22*, 111–127.

Becker, K. (1990). *Measurements of the permeability of the upper oceanic crust at Hole 395A, ODP Leg 109.* College Station, TX: Ocean Drilling Program.

Leg 174B synopsis: revisiting hole 395A for logging and long-term monitoring of off-axis hydrothermal processes in young oceanic crust. In K. Becker, A. Bartetzko, & E. E. Davis (Eds.), *Proceedings of the ocean drilling program, scientific results,* (174B). (2001). College Station, TX: Ocean Drilling Program.

Becker, K., & Davis, E. E. (2005). *A review of CORK designs and operations during the ocean drilling program.* In A. T. Fisher, T. Urabe, & A. Klavs (Eds.), The Expedition 301 Scientists Proc. IODP, 301. College Station TX (Integrated Ocean Drilling Program Management International, Inc.). http://dx.doi.org/10.2204/iodp.proc.301.104.2005.

Becker, K., Langseth, M. G., & Hyndman, R. D. (1984). Temperature measurements in Hole 395A, Leg 78B. In R. D. Hyndman, & M. H. Salisbury (Eds.), *Initial reports, DSDP* (Vol. 78B) (pp. 689–698). U.S. Government Printing Office.

Chan, C. S., Fakra, S. C., Emerson, D., Fleming, E. J., & Edwards, K. J. (2010). Lithotrophic iron-oxidizing bacteria produce organic stalks to control iron mineral growth; implications for biosignature formation. *The ISME Journal, 5*, 717–727.

Cowen, J., et al. (2003). Fluids from aging ocean crust that support microbial life. *Science, 299,* 120–123.

Cowen, J. P., et al. (2011). Advanced instrument system for real-time and time-series microbial geo-chemical sampling of the deep (basaltic) crustal biosphere. *Deep Sea Research Part I, 61,* 43–56.

Cragg, B. A., et al. (1990). Bacterial biomass and activity profiles within deep sediment layers. In E. Suess, & R. von Huene (Eds.), *Proceedings of the ocean drilling program.* College Station, TX: Scientific Results, Leg 112, Book 112. Ocean Drilling Program.

Cragg, B. A., Harvey, S. M., Fry, J. C., Herbert, R. A., & Parkes, R. J. (1992). Bacterial biomass and activity in the deep sediment layers of the Japan Sea, Hole 798B. In O. D. Program (Ed.), *Proceedings of the ocean drilling program.* College Station, TX: Scientific Results. Ocean Drilling Program pp. 127/128:761-776.

Cragg, B. A., & Kemp, A. E. S. (1995). Bacterial profiles in deep sediment layers from the eastern equatorial Pacific Ocean, Site 851. In *Proceedings of the ocean drilling program* (pp. 599–604). College Station, TX: Scientific Results. Ocean Drilling Program.

D'Hondt, S., et al. (2004). Distributions of microbial activities in deep subseafloor sediments. *Science, 306,* 2216–2221.

D'Hondt, S., Rutherford, S., & Spivack, A. J. (2002). Metabolic activity of subsurface life in deep-sea sediments. *Science, 295,* 2067–2070.

Davis, E. E., Becker, K., Pettigrew, T. L., Carson, B., & Macdonald, R. (1992). CORK: a hydro-logical sea and downhole observatory for deep-ocean boreholes. *Proc ODP Init Repts, 139,* 43–53.

Edwards, K. J., et al. (2012). Design and deployment of borehole observatories and experiments during IODP Expedition 336 Mid-Atlantic Ridge flank at North Pond. In K. J. Edwards, W. Bach, A. Klaus, & Expedition 336 Scientists (Eds.), *Proc. IODP* (336). Tokyo: Integrated Ocean Drilling Program Management International, Inc.

Edwards, K. J., Bach, W., & Klaus, A. (2010). Mid-Atlantic ridge flank microbiology: Initiation of long-term coupled microbiological, geochemical, and hydrological experimentation within the seafloor at North Pond, western flank of the Mid-Atlantic ridge. *IODP Scientific Prospectus, 336,* 62 Integrated Ocean Drilling Program, College Station, TX.

Edwards, K. J., Becker, K., & Colwell, F. (2012). The deep, dark biosphere: Intraterrestrial life on Earth. *Annual Review of Earth Planetary Sciences, 40,* 551–568.

Edwards, K. J., Fisher, A. T., & Wheat, C. G. (2012). The deep subsurface biosphere in igneous ocean crust: frontier habitats for microbiological exploration. *Frontiers in Microbiology, 3* Article 8.

Edwards, K. J., McCollom, T. M., Konishi, H., & Buseck, P. R. (2003). Seafloor bioalteration of sulfide minerals: results from in-situ incubation studies. *Geochimca et Cosmochimca Acta, 67*(15), 2843–2856.

Einen, J., Thorseth, I. H., & Ovreas, L. (2008). Enumeration of archaea and bacteria in seafloor basalt using real-time quantitative PCR and fluorescence microscopy. *FEMS Microbiology Letters, 282*(2), 182–187.

Emerson, D., et al. (2007). A novel lineage of proteobacteria involved in formation of marine Fe-oxidizing microbial mat communities. *PLoS One, 2*(7), e677.

Emerson, D., & Moyer, C. L. (2002). Neutrophilic Fe-oxidizing bacteria are abundant at the Loihi seamount hydrothermal vents and play a major role in Fe oxide deposition. *Applied and Environmental Microbiology, 68*(6), 3085–3093.

Expedition 327 Scientists. (2011). Site U1362. In A. T. Fisher, T. Tsuji, K. Petronotis, & Expedition 327 Scientists (Eds.), *Proceedings of IODP* (Vol. 327). Tokyo: Integrated Ocean Drilling Program Management International, Inc.

Expedition 329 Scientists. (2011). Methods. In S. D'Hondt, F. Inagaki, & C. A. Alvarez Zarikian (Eds.), *Proc. IODP* (Vol. 329). Tokyo: Integrated Ocean Drilling Program. http://dx.doi. org/10.2204/iodp.proc.329.102.2011.

Expedition 330 Scientists. (2011). Louisville seamount trail: implications for geodynamic mantle flow models and the geochemical evolution of primary hotspots. *IODP Preliminary Report, 330.*

Expedition 331 Scientists. (2010). Deep hot biosphere. *IODP Preliminary Report, 331.*

Expedition 336 Scientists (2012). Mid-Altantic Ridge microbiology: Initiation of long-term coupled microbiological, geochemical, and hydrological experimentation within the seafloor at North Pond, western flank of the Mid-Atlantic Ridge. *IODP Preliminary Reportt, 336.* http:// dx.doi.org/10.2204/iodp.pr.336.2012.

Fisher, A. T., et al. (2005). Scientific and technical design and deployment of long-term, subseafloor observatories for hydrogeologic and related experiments, IODP Expedition 301, eastern flank of Juan de Fuca Ridge. In A. T. Fisher, T. Urabe, A. Klaus, et al. (Eds.), *Proceedings IODP, expedition 301.* College Station, TX: Integrated Ocean Drilling Program.

Fisher, A. T., et al. (2011). Design, deployment, and status of borehole observatory systems used for single-hole and cross-hole experiments, IODP Expedition 327, eastern flank of Juan de Fuca Ridge. In A. T. Fisher, T. Tsuji, K. Petronotis, & Expedition 327 Scientists (Eds.), *Proc. IODP* (327). Tokyo: Integrated Ocean Drilling Program Management International, Inc.

Fisher, A. T., & Becker, K. (2000). Channelized fluid flow in oceanic crust reconciles heat-flow and permeability data. *Nature, 403,* 71–74.

Fisk, M. R., Giovannoni, S. J., & Thoreth, I. H. (1998). Alteration of oceanic volcanic glass: textural evidence of microbial activity. *Science, 281,* 978–980.

Furnes, H., et al. (2001). Bioalteration of basaltic glass in the oceanic crust. *Geochemistry, Geophysics, Geosystems, 2* 2000GC000150.

Furnes, H., & Staudigel, H. (1999). Biological mediation in ocean crust alteration: how deep is the deep biosphere? *Earth and Planetary Science Letters, 166,* 97–103.

Giovannoni, S. J., Fisk, M. R., Mullins, T., & Furnes, H. (1996). Genetic evidence for endolithic microbial life colonizing basaltic glass/seawater interfaces. In *Proceedings of the ocean drilling program.* College Station, TX: Scientific Results. Ocean Drilling Program.

Graber, K. K., Pollard, E., Jonasson, B., & Shulte, E. (2002). *Overview of ocean drilling program engineering tools and hardware.* ODP Technical Note. Book 31.

Heberling, C., Lowell, R., Liu, L., & Fisk, M. R. (2010). Extent of the microbial biosphere in the oceanic crust. *Geochemistry, Geophysics, Geosystems, 11*(8), Q08003.

Hinrichs, K. U., & Inagaki, F. (2012). Downsizing the deep biosphere. *Science, 338,* 204–205.

Hoehler, T. M., Borowski, W. S., Alperin, M. J., Rodrigiuez, N. M., & Paull, C. K. (2000). Model, stable isotope, and radiotracer characterization of anaerobic oxidation in gas hydrate-bearing sediments of the blank ridge. In O. D. Program (Ed.), *Proceedings of the ocean drilling program.* College Station, TX: Scientific Results.

Hulme, S., Wheat, C. G. Quantifying geochemical and fluid fluxes across the rough basement transect of the western Juan de Fuca Ridge flank. *Geochemistry Geophysics Geosystems* (submitted for publication).

Humphris, S. E., & Koppers, A. P. (2013). Planning for the future ocean drilling with the *Joides Resolution. EOS, 94*(26), 229–230.

Inagaki, F., et al. (2006). Biogeographical distribution and diversity of microbes in methane hydrate-bearing marine sediments on the Pacific Ocean Margin. *Proceedings of the National Academy of Sciences USA, 103,* 2815–2820.

IODP. (2001). *Earth, oceans, and life – scientific investigations of the Earth using multiple drilling platforms and new technologies.* Washington, DC: International Working Group Support Office. IODP.

IODP. (2011). *Science plan for 2013–2023: Illuminating Earth's past, present, and future.* Washington, DC: Integrated Ocean Drilling Program Management International. IODP.

Jannasch, H. W., Wheat, C. G., Plant, J. N., Kastner, M., & Stakes, D. S. (2004). Continuous chemical monitoring with osmotically pumped water samplers: OsmoSampler design and applications. *Limnology and Oceanography, 2,* 2843–2856.

Jørgensen, B. B. (2012). Shrinking majority of the deep biosphere. *Proceedings of the National Academy of Sciences USA, 109,* 15976–15977.

Jungbluth, S. P., Grote, J., Lin, H.-T., Cowen, J. P., & Rappe, M. S. (2012). Microbial diversity within basement fluids of the sediment-buried Juan de Fuca Ridge flank. *The ISME Journal, 7*(1), 161–172.

Jungbluth, S., Lin, H.-T., Cowen, J., Glazer, B. T., & Rappe, M. S. (2014). Phylogenetic diversity of microorganisms in subseafloor crustal fluids from boreholes 1025C and 1026B along the Juan de Fuca Ridge flank. *Front Microbiol, 5,* 119. http://dx.doi.org/10.3389/fmicb.2014.00119.

Kallmeyer, J., Pockalny, R., Adhikari, R. R., Smith, D. C., & D'Hondt, S. (2012). Global distribution of microbial abundance and biomass in subseafloor sediment. *Proceedings of the National Academy of Sciences USA, 109,* 16213–16216.

Kastner, M., et al. (2006). New insights into the hydrogeology of the oceanic crust through long-term monitoring. *Oceanography, 19*(4), 46–57.

Kopf, A., et al. (2011). The SmartPlug and GeniusPlug: simple retrievable observatory systems for NanTroSEIZE borehole monitoring. In A. Kopf, E. Araki, S. Toczko, & Expedition 332 Scientists (Eds.), *Proceedings of the integrated ocean drilling program* Vol.332. Tokyo: Integrated Ocean Drilling Program Management International.

Lapham, L. L., et al. (2008). Measuring temporal variability in pore-fluid chemistry to assess gas hydrate stability: development of a continuous pore-fluid array. *Environmental Science and Technology, 42,* 7368–7373.

Lawrence, J. R., Drever, J. J., & Kastner, M. (1979). Low temperature alteration of basalts predominates at Site 395. In *Initial reports of the deep sea drilling project* (Vol. 45) (pp. 609–612). Washington, D.C: U.S. Government Printing Office.

Lever, M., et al. (2006). Trends in basalt and sediment core contamination during IODP Expedition 301. *Geomicrobiology Journal, 23,* 517–530.

Lever, M. A., et al. (2013). Evidence for microbial carbon and sulfur cycling in deeply buried ridge flank basalt. *Science, 399,* 1305–1308.

Lin, H.-T., Cowen, J. P., Olson, E. J., Amend, J. P., & Lilley, M. D. (2012). Inorganic chemistry, gas compositions, and dissolved organic carbon in fluids from sedimented young basaltic crust on the Juan de Fuca Ridge flanks. *Geochemica et Cosmochimca Acta, 85,* 213–227.

Lipp, J. S., Morono, Y., Inagaki, F., & Hinrichs, K.-U. (2008). Significant contribution of archaea to extant biomass in marine subsurface sediments. *Nature, 454,* 991–994.

Luther, G. W., et al. (2008). Use of voltammetric solid-state (micro)electrodes for studying biogeochemical processes: laboratory measurements to real time measurements with an *in situ* electrochemical analyzer (ISEA). *Marine Chemistry, 108,* 221–235.

McDuff, R. E. (1984). The chemistry of interstitial waters from the upper oceanic crust, Site 395, Deep Sea Drilling Project Leg 78B. In R. D. Hyndman, & M. H. Salisbury (Eds.), *Initial reports of the deep sea drilling project, 78B* (pp. 795–799). Washington, D.C: U.S. Government Printing Office.

Mason, O. U., et al. (2008). The phylogeny of endolithic microbes associated with marine basalts. *Environmental Microbiology, 9*(10), 2539–2550.

Mason, O. U., et al. (2010). First investigation of the microbiology of the deepest layer of ocean crust. *PLoS One, 5,* e15399.

Moran, K., Farrington, S., Massion, E., Paull, C. K., Stephen, R., Trehu, A. et al. (2006). SCIMPI: A New Seafloor Observatory System, OCEANS. Boston, MA, pp.1–6.

Nigro, L. M., et al. (2012). Microbial communities at the borehole observatories on the Costa Rica Rift flank (Ocean Drilling Program Hole 896A). *Frontiers in Microbiology, 3,* 232.

Orcutt, B. N., et al. (2011). Colonization of subsurface microbial observatories deployed in young ocean crust. *The ISME Journal, 5,* 692–703.

Orcutt, B. N., et al. (2013a). Microbial activity in the marine deep biosphere: progress and prospects. *Frontiers in Microbiology, 11*(4), 189. http://dx.doi.org/10.3389/fmicb.2013.00189.

Orcutt, B. N., et al. (2013b). Oxygen consumption in subseafloor basaltic crust. *Nature Communications, 4,* 2539. http://dx.doi.org/10.1038/ncomms3539.

Orcutt, B. N., Barco, R. A., Joye, S. B., & Edwards, K. J. (2012). Summary of carbon, nitrogen, and iron leaching characteristics and fluorescence properties of materials considered for subseafloor observatory assembly. In K. J. Edwards, W. Bach, A. Klaus, & Expedition 336 Scientists (Eds.), *Proc. IODP* (Vol. 336). Tokyo: Integrated Ocean Drilling Program Management International, Inc.

Orcutt, B. N., Sylvan, J. B., Knab, N. J., & Edwards, K. J. (2011). Microbial ecology of the dark ocean above, at and below the seafloor. *Microbiology and Molecular Biology Reviews, 75*(2), 361–422.

Orcutt, B., Wheat, C. G., & Edwards, K. J. (2010). Subseafloor ocean crust microbial observatories: development of FLOCS (fLow-through osmo colonization system) and evaluation of borehole construction methods. *Geomicrobiology Journal, 27,* 143–157.

Oremland, R. S., Culbertson, C., & Simoneit, B. (1982). *Methanogenic activity in sediments from leg 64, Gulf of California, initial reports deep sea drilling project, Leg 64.* Washington, DC: U.S. Govt. Printing Office.

Orsi, W. D., Edgcomb, V. P., Christman, G. D., & Biddle, J. F. (2013). Gene expression in the deep biosphere. *Nature, 499,* 205–208.

Parkes, R. J., et al. (1994). Deep bacterial biosphere in Pacific Ocean sediments. *Nature, 371,* 410–413.

Parkes, R. J., Cragg, B. A., & Wellsbury, P. (2000). Recent studies on bacterial populations and processes in sub seafloor sediments: a review. *Hydrogeology Journal, 8,* 11–28.

Robidart, J., Callster, S. J., Song, P., Nicora, C. D., Wheat, C. G., & Girguis, P. R., (2013). Characterizing microbial community and geochemical dynamics at hydrothermal vents using osmotically driven continuous fluid samplers. *Environmental Science and Technology, 47,* 4399–4407.

Rogers, D. R., Santelli, C. M., & Edwards, K. J. (2003). Geomicrobiology of deep-sea deposits: estimating community diversity from low-temperature seafloor rocks and minerals. *Geobiology, 1,* 109–117.

Ryan, W. B. F., et al. (2009). Global multi-resolution topography synthesis. *Geochemistry, Geophysics, Geosystems, 10,* Q30114.

Santelli, C. M., et al. (2008). Abundance and diversity of microbial life in ocean crust. *Nature, 453,* 653–657.

Santelli, C. M., Bach, W., Banerjee, N. R., & Edwards, K. J. (2010). Tapping the subsurface ocean crust biosphere: low biomass and drilling-related contamination calls for improved quality controls. *Geomicrobiology Journal, 27,* 158–169.

Santelli, C. M., Edgcomb, V. P., Bach, W., & Edwards, K. J. (2009). The diversity and abundance of bacteria inhabiting seafloor lavas positively correlate with rock alteration. *Environmental Microbiology, 11*(1), 86–98.

Smith, A., et al. (2011). In situ enrichment of ocean crust microbes on igneous minerals and glasses using osmotic flow-through device. *Geochemistry, Geophysics, Geosystems, 12,* (6) Q06007, http://dx.doi.org/10.1029/2010GC003424.

Smith, D. C., Spivack, A. J., Fisk, M. R., Haveman, S. A., & Staudigel, H. (2000). Tracer-based estimates of drilling-induced microbial contamination of deep sea crust. *Geomicrobiology Journal*,
17, 207–217.

Thorseth, I. H., et al. (2001). Diversity of life in ocean floor basalt. *Earth and Planetary Science
Letters*, *194*, 31–37.

Thorseth, I. H., Pedersen, K., & Christie, D. M. (2003). Microbial alteration of 0–30-Ma seafloor
and sub-seafloor basaltic glasses from the Australian Antarctic Discordance. *Earth and Planetary Science Letters*, *215*, 237–247.

Toner, B. M., et al. (2013). Mineralogy drives bacterial biogeography at hydrothermally-inactive
seafloor sulfide deposits. *Geomicrobiology Journal*, *30*, 313–326.

Torsvik, T., Furnes, H., Muehlenbacks, K., Thorseth, I. H., & Tumyr, O. (1998). Evidence for microbiol activity at the glass-alteration interface in oceanic basalts. *Earth and Planetary Science
Letters*, *162*, 165–176.

Wheat, C. G., et al. (2004). Venting formation fluids from deep-sea boreholes in a ridge flank setting: ODP sites 1025 and 1026. *Geochemistry, Geophysics, Geosystems*, *5*, Q08007. http://
dx.doi.org/10.1029/2004GC000710.

Wheat, C. G., et al. (2010). Subseafloor seawater-basalt-microbe reactions: continuous sampling
of borehole fluids in a ridge flank environment. *Geochemistry, Geophysics, Geosystems*, *11*,
Q07011.

Wheat, C. G., et al. (2011). Fluid sampling from oceanic borehole observatories: design and methods for CORK activities (1990–2010). In A. T. Fisher, T. Tsuji, K. Petronotis, & Expedition
327 Scientists (Eds.), *Proc. IODP* (Vol. 327). Tokyo: Integrated Ocean Drilling Program Management International, Inc.

Wheat, C. G., et al. (2012). CORK-Lite: bringing legacy boreholes back to life. *Science Drilling*,
14, 39–43.

Wheat, C. G., Elderfield, H., Mottl, M. J., & Monnin, C. (2000). Chemical composition of basement
fluids within an oceanic ridge flank: Implications for along-strike and across-strike hydrothermal circulation. *Journal of Geophysical Research*, *105*(B6), 13437–13447.

Wheat, C. G., Jannasch, H. W., Kastner, M., Plant, J. N., & DeCarlo, E. (2003a). Seawater transport
in the upper oceanic basement: chemical data from continuous monitoring of sealed boreholes
in a ridge flank environment. *Earth and Planetary Science Letters*, *216*, 549–564.

Wheat, C. G., McManus, J., Mottl, M. J., & Giambalvo, E. (2003b). Oceanic phosphorous
imbalence: the magnitude of the ridge-flank hydrothermal sink. *Geophysical Research Letters*,
30, OCE5.1–OCE5.4.

Wheat, C. G., & Mottl, M. J. (2000). Composition of pore and spring waters from baby bare: global
implications of geochemical fluxes from a ridge flank hydrothermal system. *Geochimca et Cosmochimca Acta*, *64*, 629–642.

Whitman, W. B., Coleman, D. C., & Wiebe, W. J. (1998). Prokaryotes: the unseen majority. *Proceedings of the National Academy of Science USA*, *95*, 6578–6583.

Ziebis, W., et al. (2012). Interstitial fluid chemistry of sdiments underlying the North Atlantic
Gyre and the influence of subsurface fluid flow. *Earth and Planetary Science Letters*,
323-324, 79–91.

Chapter 2.6

Cultivation of Subseafloor Prokaryotic Life

Bert Engelen[1,*] and Hiroyuki Imachi[2]
[1]Institut für Chemie und Biologie des Meeres (ICBM), Carl von Ossietzky Universität Oldenburg, Postfach, Oldenburg, Germany; [2]Department of Subsurface Geobiological Analysis and Research, Japan Agency for Marine-Earth Science Technology (JAMSTEC), Yokosuka, Japan
*Corresponding author: E-mail: bert.engelen@uni-oldenburg.de

2.6.1 THE NECESSITY OF CULTURING SUBSEAFLOOR PROKARYOTES

The cultivation of microorganisms from a given environment is still the gold standard to identify their physiological capabilities and specific adaptations and to understand their role in biological interactions or elemental cycling. Even in the "omics" age, cultured strains are essential to test hypotheses that derive from predicted gene functions (Giovannoni & Stingl, 2007). Typically, about 40% of all proteins that are found in bacterial proteomes cannot be assigned to certain catalytic reactions (Nichols, 2007). These molecules represent a reservoir of bioactive compounds that might be exploited in future biotechnological applications. On the other hand, genes identified to be indicative for so far hidden metabolic pathways can be used to design genomically guided cultivation strategies.

However, all cultivation studies suffer from the "great plate count anomaly" that describes the discrepancy between the number of cells to be detectable in a sample and the number of colonies that grow on conventional media (Staley & Konopka, 1985). This fact results in generally low cultivation efficiencies for most habitats and especially for deep-sea sediments (Table 2.6.1). Furthermore, the vast microbial diversity in an environmental sample that was first displayed by Torsvik, Goksøyr, and Daae (1990) becomes more and more obvious from high-throughput sequencing (Sogin et al., 2006). Thus, from molecular investigations it was postulated that more than 99% of all bacteria from natural environments are "unculturable" (Amann, Ludwig, & Schleifer, 1995), which is now established in the literature as a "well-known fact." However, this term reflects much more the challenge to mimic various environmental settings

Developments in Marine Geology, Volume 7. http://dx.doi.org/10.1016/B978-0-444-62617-2.00008-6
197

TABLE 2.6.1 Cultivation Success of Subseafloor Sedimentary Microbes Compared with Other Environments

Habitat	Cultivation Success (%)	References
Seawater	0.001–0.15	Ferguson, Buckley, and Palumbo (1984), Gram, Melchiorsen, and Bruhn (2010), Kogure, Simidu, and Taga, (1979, 1980)
Freshwater	0.25	Jones (1977)
Mesotrophic lake	0.1–1	Staley and Konopka (1985)
Unpolluted estuary	0.1–3	Ferguson et al. (1984)
Activated sludge	1–15	Wagner, Amann, Lemmer, and Schleifer (1993, 1994)
Sediments	0.25	Jones (1977)
Soil	0.3–14	Janssen, Yates, Grinton, Taylor, and Sait (2002), Torsvik et al. (1990)
Alfresco air	0.5–3	Lighthart and Tong (1998)
Deep-sea sediments	0.00001–1	D'Hondt et al. (2004), Inagaki et al. (2003), Parkes et al. (2000)

Cultivation success was determined as a percentage of cultivated bacteria (colony-forming units) in comparison to total cell counts.
The table was compiled after Amann et al. (1995).

in the laboratory than the fact that these organisms refuse to grow under controlled conditions (Cypionka, 2005). Nevertheless, using nanoscale secondary ion mass spectrometry, over 70% of sedimentary microbes showed assimilation activity of ^{13}C- and/or ^{15}N-labeled substrate with very slow incorporation rates, suggesting that subseafloor microbes are indeed alive and physiologically standby (Jørgensen, 2011; Morono et al., 2011).

In general, cultivation attempts should be accompanied by molecular techniques (Stevenson, Eichorst, Wertz, Schmidt, & Breznak, 2004; Toffin, Webster, Weightman, Fry, & Prieur, 2004). The discrepancy between phylotypes detected in initial enrichment cultures and those that are present in culture collection (Fichtel, Mathes, Könneke, Cypionka, & Engelen, 2012) suggests that not only the cultivation techniques but also the isolation procedure itself needs to be improved. If the phylogenetic position of a species is consistent with its culture requirements, the design of specific culture conditions could eventually lead to a more successful isolation. The isolation of *Nitrosopumilus maritimus*, for instance, is a great example from pelagic studies that shows what can be expected if highly abundant but not-yet cultured organisms such as the Marine Group I *Thaumarchaeota* are brought into culture (Könneke et al., 2005). This organism accounts for a high percentage of all prokaryotic 16S rRNA genes in deep ocean waters, exhibits an autotrophic lifestyle, and has the capability to oxidize ammonium. The cultivation was a breakthrough for various disciplines working with archeal genomes (Brochier-Armanet, Boussau, Gribaldo, & Forterre, 2008), the marine nitrogen cycle (Wuchter et al., 2006), or using crenarchaeol as a paleotemperature proxy (Schouten, Forster, Panoto, & Damste, 2007). A similar outcome for understanding prokaryotic life in the subsurface will probably derive from a successful isolation of members of the *Chloroflexi* phylum (Imachi et al., 2011, 2014), which is one of the predominant bacterial phylotypes in clone libraries from deep subsurface sediments.

One basic requirement for the isolation of pure cultures is the ability to form colonies in agar-solidified media. Agar, in turn, has inhibitory effects on the growth of some bacteria (Hara, Isoda, Tahvanainen, & Hashidoko, 2012; Liesack, Janssen, Rainey, Ward-Rainey, & Stackebrandt, 1997). The use of Gellan gum or Gelrite as gelling agents might overcome this bias (Tamaki, Hanada, Sekiguchi, Tanaka, & Kamagata, 2009). It can be imagined that several phylotypes present in initial enrichments are unable to form colonies (Simu & Hagström, 2004) and thus get lost during isolation. For such species, single-cell manipulations would be appropriate (Ashida et al., 2010; Fröhlich & König, 2000). On the other hand, many microbes do not grow independently from syntrophic partners. Alternative strategies such as cocultures (Kaeberlein, Lewis, & Epstein, 2002; Sakai et al., 2007) or dialysis cultures might be required for their isolation.

Especially for prokaryotic subseafloor life, the described difficulties lead to the consequence that most diversity studies are so far based on cultivation-independent approaches (Inagaki et al., 2001, 2003, 2006; Marchesi, Weightman, Cragg, Parkes, & Fry, 2001). These surveys indicate the presence of highly diverse archeal communities within deep subseafloor sediments that in most cases are not related to any cultivated representatives (Teske & Sorensen, 2008). Speculations on their metabolic capacities are rather difficult to make, especially as their distribution does not seem to be linked to concentration gradients of electron donors and acceptors (Teske, 2006). Culture collections, in turn, are often reviewed quite skeptical in terms of being representative for the respective habitat (DeLong, 2004). Whatever cultivation attempt is performed one is confronted with the question whether the obtained isolates represent relevant members of indigenous communities. These collections frequently contain high numbers of gram-positive bacteria (Batzke, Engelen, Sass, & Cypionka, 2007). It cannot be ruled out that these strains might have derived from formerly deposited spores (Lomstein, Langerhuus, D'Hondt, Jorgensen, & Spivack, 2012) that have germinated during the cultivation process. However, many isolates with comparable phylogenetic affiliation were enriched from various subseafloor habitats (D'Hondt et al., 2004; Süß, Engelen, Cypionka, & Sass, 2004; Toffin et al., 2004) or match environmental sequences from the subsurface. Others have close relatives at the seafloor or in terrestrial habitats. It was seen unlikely that such microorganisms represent indigenous members of the deep subseafloor biosphere (DeLong, 2004). However, molecular quantification of *Rhizobium radiobacter*, for instance, the most frequently isolated phylotype from subseafloor sediments (Batzke et al., 2007; Süß et al., 2006), has proven its presence at various sites and depths with up to 5.6% of all bacterial 16S rRNA gene sequences (Engelhardt, Sahlberg, Cypionka, & Engelen, 2013).

2.6.2 THE SPECIFIC CHALLENGES TO CULTIVATE PROKARYOTIC LIFE FROM THE SUBSEAFLOOR

The subseafloor environment is generally energy limited and characterized by low energy fluxes (Jørgensen & D'Hondt, 2006). The principal food source for subseafloor prokaryotes is generated in the surface world (D'Hondt et al., 2004). However, even at ocean margins, only complex organic matter reaches the seafloor (Suess, 1980) as most remineralization is already taking place during sedimentation throughout the water column. With increasing sediment depth, organic matter becomes more recalcitrant and thus less bioavailable. Accordingly, microbial respiration leads to a consecutive depletion of electron acceptors with depth following the redox tower (Froelich et al., 1979). Thus, microbial activities are governed by the availability of electron acceptors and the quality of electron donors (Engelen & Cypionka, 2009). The diminished substrate availability leads to a selection for many indigenous subsurface microorganisms to be capable to degrade a large variety of carbon sources. In recent investigations, enrichment media were designed to contain a complex suite of single substrates to recover a broad diversity of cultivable bacteria from the subsurface (Batzke et al., 2007; Fichtel et al., 2012; Süß et al., 2004; Teske & Sorensen, 2008). The obtained subsurface isolates were identified as generalists.

Due to the general energy limitation, subseafloor prokaryotes exhibit extremely slow growth. Generation times were estimated from the rate of organic carbon supply and turned out to be in the range of one or two thousand years (Whitman, Coleman, & Wiebe, 1998). If this feature is obligatory for some indigenous prokaryotes, those will never grow under normal laboratory settings. An adaptation to low organic carbon flux might in part be mimicked in enrichment cultures by applying submillimolar concentrations, gradients (Figure 2.6.1) of diffusive substrates (Köpke, Wilms, Engelen, Cypionka, & Sass, 2005), or continuous flow reactors (Imachi et al., 2011). However, a substrate shock (Straskrabova, 1983) might not be circumvented in all cases. This and other environmental factors, such as low temperature, low porosity, and high-pressure conditions makes subseafloor prokaryotes extremely difficult to

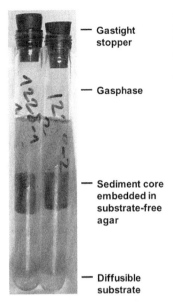

Gastight stopper

Gasphase

Sediment core embedded in substrate-free agar

Diffusible substrate

FIGURE 2.6.1 Enrichment of microorganisms in undisturbed sediment samples. Substrates slowly diffuse into the sediment subcores to allow indigenous microorganisms to adapt to an increasing substrate supply. Growth is stimulated in the presence of the original sediment matrix. Displayed are examples of two sediment gradient cultures from ODP Leg 201, Site 1228. *Photo: Heribert Cypionka.*

cultivate. While several isolates have been obtained from porous ash or sand layers, highly condensed hemi-pelagic clay samples turned out to be difficult to cultivate indigenous microbes by the conventional batch culture and/or colony formation methods (Inagaki et al., 2003; Kobayashi et al., 2008). Especially the combination of low temperature and high pressure on prokaryotic cell membranes causes problems when decompressed samples are incubated in the laboratory. One so far disregarded aspect for a diminished cultivation success from the subsurface might derive from the high percentage of prophages detected in deep-subseafloor bacteria (Engelhardt, Sahlberg, Cypionka, & Engelen, 2011). These enter the lytic stage upon a certain trigger when environmental conditions change. The setup of enrichment cultures might represent such a trigger and thus results in cell lysis and death.

2.6.3 CULTIVATION ATTEMPTS USING CONVENTIONAL BATCH-TYPE CULTIVATION

Historically, the concept of cultivation using enrichments from the environment derives from the work of Beijerinck (1851–1931) and Winogradsky (1856–1953) (Overmann, 2006). The specific design of culture conditions favors the growth of microorganisms that are capable of certain metabolic reactions. In principle, culture media provide the typical elements (C, O, H, N, S, P, K, Mg, Fe, and Ca) that are essential to build up prokaryotic cells and constitute their dry mass. Additionally, trace elements and vitamins are required as cofactors for most enzymes. The physiological function of the single elements is nicely compiled in Overmann (2006). Other growth factors are electron donors (and acceptors) as well as suitable physicochemical conditions such as temperature, pH, redox potential, osmolarity, and pressure. The medium composition can either be provided by complex substrates or specific combinations of the single components.

Following this concept, media for the enrichment of sediment microorganisms were adapted to the conditions determined for this habitat. This in turn lead to the directed isolation of those organisms that are involved in establishing biogeochemical gradients within the sediments such as the methanogenesis zone (*Methanococcus* sp. (Jones, Leigh, Mayer, Woese, & Wolfe, 1983)) or the sulfate-reduction zone (*Desulfovibrio* sp. (Bale et al., 1997; Fichtel et al., 2012)). Except these examples, a set of fermenting *Gammaproteobacteria* (e.g., several *Shewanella* species (Wang, Wang, Chen, & Xiao, 2004) or *Photobacterium* species (Nogi, Masui, & Kato, 1998; Süß et al., 2008)) were isolated. A compilation of isolates that were previously retrieved from subsurface sediments and other deep subsurface habitats is given in Orcutt, Sylvan, Knab, and Edwards (2011).

One of the most common cultivation approach is the most probable number (MPN) or dilution-to-extinction technique (Ferris, Ruff-Roberts, Kopczynski, Bateson, & Ward, 1996; Sekiguchi, Takahashi, Kamagata, Ohashi, & Harada, 2001). For MPN counts, a series of tenfold dilutions is prepared from initial slurries and incubated, e.g., in 96 deep-well microtiter plates (Figure 2.6.2). The incubation in microtiter plates is advantageous over classical enrichments in glass tubes. It saves space, which is always limited on ships, and as

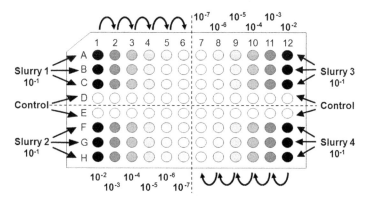

FIGURE 2.6.2 Dilution-to-extinction cultures performed in a deep-well microtiter plate. Triplicates of four different sediment slurries (dilution 10^{-1}) are diluted down to 10^{-7} and incubated simultaneously in one plate. For each sample, one row remains uninoculated as control. The wells are sealed with a capmat and the plate is stored either under oxic conditions or in an anaerobic bag with an oxygen-consuming catalyst and an oxygen indicator. After incubation, growth can be monitored by an SYBR Green assay using a microtiter plate reader (Martens-Habbena & Sass, 2006). The results are used for calculating the most probable numbers (MPN) of microorganisms. The highest dilutions harbor abundant microorganisms and serve for their isolation.

a multichannel pipette can be used for inoculation and subsampling, it also saves time. Using this technique, a sufficient number of dilutions is performed in three parallels until a concentration is reached where growth cannot further be detected. Based on the number of dilutions that still show growth, cell concentrations within the original sample can be calculated. The number of viable microorganisms is estimated by the statistical maximum-likelihood method (de Man, 1977). As MPN counts are strongly dependent on the chosen cultivation conditions, the quantitative information is only limited. However, isolation of microorganisms from the highest dilutions is an effective tool to obtain cultures of species that are numerically abundant in the original sediment sample. Less abundant, but fast-growing microorganisms that can cope better with the given cultivation conditions can be eliminated. So far, various physiological types of microorganisms have been enriched and isolated from marine subsurface sediments using this classical cultivation technique (DeLong, 2004; Inagaki et al., 2003; Parkes, Cragg, & Wellsbury, 2000). As this technique has a quantitative component, relative numbers of these isolates can be determined.

2.6.4 METABOLIC CAPABILITIES OF AVAILABLE ISOLATES FROM SUBSEAFLOOR SEDIMENTARY ENVIRONMENTS

Over the last years, reports on fully characterized isolates from subseafloor sediments have gradually increased. Although the major microbial components have not been isolated yet, cultivation attempts using traditional techniques resulted in the successful isolation of novel prokaryotes in pure culture. Detailed characterization revealed several physiological properties that suit the strains for living in deep-sea sedimentary environments. In this review, we focus on well-characterized strains isolated from deep subseafloor sediments.

Generally, the organoheterotrophic metabolism has been considered to play a major ecological role in the Earth's carbon cycle. Microbial biomass in subseafloor sediments is highly dependent on organic matter produced by photosynthetic organisms in the overlaying seawater or introduced from terrestrial origins (Biddle et al., 2006; D'Hondt, Rutherford, & Spivack, 2002; D'Hondt et al., 2009; Inagaki et al., 2006; Kallmeyer, Pockalny, Adhikari, Smith, & D'Hondt, 2012; Lipp, Morono, Inagaki, & Hinrichs, 2008). Indeed, cultivation-dependent characterization using conventional cultivation techniques such as MPN has yielded a diverse set of aerobic and anaerobic organoheterotrophic bacteria. Most of the organoheterotrophs are members of the *Alpha-* and *Gamma-proteobacteria*, *Firmicutes*, *Actinobacteria*, and *Bacteroidetes* (Batzke et al., 2007; D'Hondt et al., 2004; Kobayashi et al., 2008). Of those deep subseafloor organotrophic isolates, several strains have been fully characterized and display the unique physiological characteristics to fit their natural settings.

Recently, organoheterotrophs isolated from marine subsurface sediments off the Shimokita Peninsula of Japan have been characterized and proposed as new taxa (Miyazaki et al., 2012; Takai et al., 2013; Tsubouchi et al., 2012; 2013a,b).

Sunxiuqinia faeciviva, a facultative anaerobe, as well as *Desulfovibrio profundus*, showed piezotolerant growth under in situ pressure (Takai et al., 2013). The biostratigraphic age models for the site off the Shimokita Peninsula indicate very high sedimentation rates (ranging from 54 to 95 cm/ky) (Aoike, 2007; Aoike et al., 2010; Domitsu et al., 2010). These high sedimentation rates probably support an active organoheterotrophic metabolism and thus many organoheterotsrophs were cultivated from the sediments.

An alkaliphilic bacterium, *Marinobacter alkaliphilus*, has been obtained from subseafloor alkaline mud from Ocean Drilling Program (ODP) Hole 1220D at a serpentine mud volcano, the South Chamorro Seamount in the Mariana Forearc (Takai et al., 2005). After *D. profundus*, this bacterium is the second species taxonomically fully characterized from core samples recovered in the frame of ODP. *M. alkaliphilus* displays physiological properties that suit its natural settings. It can grow up to a pH of 11.4 and its optimal pH for growth is 9.0, which is equivalent to the pore water pH of the original sediment (pH 9.5).

Three thermophiles have been isolated from marine subsurface sediments. *Thermosediminibacter oceani* and *Thermosediminibacter litoriperuensis* have been obtained from deep-sea sediments collected at the Peru Margin during ODP Leg 201 (Lee et al., 2005). *Thermoaerobacter marianensis* has been isolated from surface sediment of the Mariana Trench Challenger Deep at a depth of 10,897 m (Takai, Inoue, & Horikoshi, 1999). All strains were isolated from a low-temperature environment (in situ temperature was 12 °C in the case of *Thermosediminibacter* species). Actually, there is no evidence how these thermophiles survive in situ and where they originally might come from.

Piezophilic growth, which defines optimal growth at pressures higher than 0.1 MPa and the requirement for increased pressure for growth, is one of the distinguished features in the deep marine subsurface. *D. profundus* is the first piezophilic species that was isolated from deep subseafloor sediments collected at Japan Sea (ODP Leg 128). This sulfate-reducing bacterium (SRB) can grow at up to 40 MPa, with optimum activities occurring between 10 and 15 MPa (Bale et al., 1997). Interestingly, the optimum sulfide production was found at the corresponding in situ pressure, strongly indicating the indigenous origin of this SRB.

Photobacterium profundum, isolated from sediments of the Ryukyu Trench, Japan, at a water depth of 5110 m, is another example for a piezophilic organoheterotroph (Nogi et al., 1998). Interestingly, gene expression analyses of *P. profundum* indicated that specific metabolic pathways, such as fermentation of amino acids or the degradation of biopolymers, might be expressed only at elevated hydrostatic pressure (Vezzi et al., 2005). In addition, Süß et al. (2008) showed remarkable phylogenetic and physiological heterogeneities at the subspecies level within members of *Photobacterium* genus using 22 *Photobacterium* strains isolated from Mediterranean subseafloor sediments.

Even though sulfate reduction is supposed to be a relevant mineralization process in subseafloor sediments, only *D. profundus* has been fully characterized (Bale et al., 1997). To extend the amount of available isolates that derive from subseafloor sediments, Fichtel et al. (2012) attempted the cultivation and isolation of sulfate-reducing bacteria (SRB) from subseafloor sediments of the Juan de Fuca Ridge, Northeast Pacific (IODP Site U1301). At this site, sulfate-containing crustal fluids diffuse into the sediments from the underlying basaltic aquifer. As a result, three SRB strains have successfully been isolated from a depth of 260 m below seafloor (mbsf) that are characterized to grow chemolithotrophically with hydrogen as energy source.

Methanogenic archaea play an important biogeochemical role in the subseafloor biosphere. Most of the large quantities of methane stored in the subseafloor (500–2500 Gt of methane-bound carbon) are considered to be of microbiological origin (Milkov, 2004). In spite of the importance of methanogenesis, only three subseafloor methanogens have been fully characterized so far: *Methanoculleus submarinus* (Mikucki, Liu, Delwiche, Colwell, & Boone, 2003), *Methanococcus aeolicus* (Kendall et al., 2006), and *Methanosarcina baltica* (von Klein, Arab, Völker, & Thomm, 2002). *M. submarinus* is the first methanogenic isolated from methane hydrate-bearing deep subseafloor sediments of the Nankai Trough. Interestingly, this methanogen could grow in a wide range of salinities with optimum growth at 0.1–0.4 M Na$^+$. This physiological characteristic suits methane hydrate-bearing sediments as these are considered to be low-salinity environments. This is especially the case around sediment layers that contain methane hydrate. Recently, Watkins, Roussel, Webster, Parkes, and Sass (2012, 2013) reported that *Methanococcoides* strains from subseafloor sediments are able to utilize choline, *N,N*-dimethylethanolamine, and glycine betaine for growth and methane production. As these compounds have not been recognized as methanogenic substrates, this finding expands our general knowledge on methanogenesis.

In addition to methanogenesis and sulfate reduction, acetogenesis must be considered as another relevant microbiological process in marine subsurface sediments (Lever, 2012). Toffin et al. (2004) reported the successful isolation of an acetogen (strain LT18), which is affiliated to the genus *Acetobacterium*, but the detailed physiological properties of the strain have not been reported yet.

2.6.5 NOVEL TECHNIQUES FOR THE CULTIVATION OF SUBSEAFLOOR PROKARYOTIC LIFE

Traditional batch-type cultivation efforts have always encountered the strong resistance of numerically and/or functionally relevant microbial components to be enriched and isolated from subseafloor sediments (D'Hondt et al., 2004; Fry, Parkes, Cragg, Weightman, & Webster, 2008; Inagaki, 2010; Toffin et al., 2004). In addition, a large discrepancy appears between the major prokaryotic phylotypes detected by 16S rRNA gene-based analysis and isolated strains. In almost all cases, conventional cultivation techniques only allowed to grow bacterial strains from distinctive phylogenetic groups that already dominate public microbial culture collections (Hugenholtz, 2002). These so-called big four belong to the *Proteobacteria* (particularly the classes of alpha, gamma, and delta), *Firmicutes,*

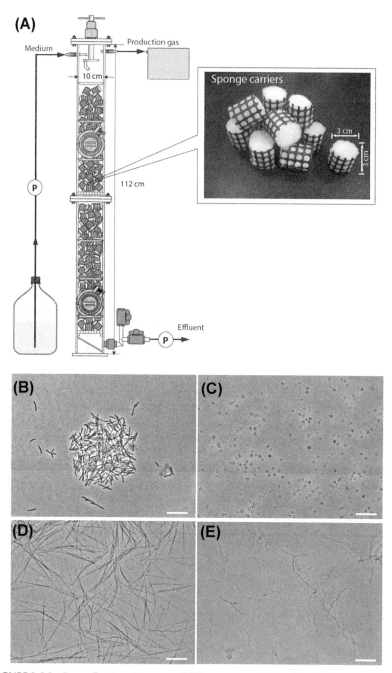

FIGURE 2.6.3 Down-flow hanging sponge (DHS) reactor used in Imachi et al. (2011) and micro-photographs of anaerobic microorganisms isolated from the DHS reactor. (A) Schematic diagram of the DHS reactor. The sponge cubes, which represent a microbial habitat, are encased in plastic

Actinobacteria, and *Bacteroidetes*. To move beyond our present microbiological understanding that is dominated by these four phyla and to expand our knowledge concerning subseafloor sedimentary microbes, the development of new strategies for the isolation of subseafloor prokaryotes is needed. So far, there are only a few reports available on the application of novel cultivation techniques for deep subseafloor biomes, but the number of such reports will probably increase in the future. In the following, we will introduce some examples for novel approaches to cultivate subseafloor sedimentary microorganisms.

The first example for a novel cultivation approach is the high-pressure anaerobic enrichment system DeepIsoBUG, which was described by Parkes et al. (2009b). Until this report, there was no system available to sample and handle subseafloor sediments without depressurization except for one previous investigation using the DEEP-BATH system (Yanagibayashi, Nogi, Li, & Kato, 1999). So far, all deep subseafloor sediments used for high-pressure experiments were depressurized once before starting the experiments. The DeepIsoBUG is a combination of the HYACINTH pressure-retaining drilling and core storage system and the PRESS core cutting and processing system. Both have been originally developed for collecting gas hydrates from subseafloor environments (Parkes et al., 2009, Schultheiss, Holland, & Humphrey, 2009). The most distinguished feature is that subseafloor sediments can be collected and handled under in situ pressure. Moreover, the system even enables the subsequent enrichment and isolation of prokaryotes at pressures of up to 100 MPa without any depressurization after sediment recovery from the subseafloor. Parkes et al. (2009) applied the DeepIsoBUG to gas-hydrate sediments collected from three distinct geographical sites (the Indian Continental Shelf, Cascadia Margin, and Gulf of Mexico). Considerable cell proliferation was observed after three months of incubation and several isolates were successfully obtained from the system. However, again all isolates were affiliated to "the big four" and closely related to previously isolated environmental strains (>96% 16S rRNA gene sequence identity). Additionally, phylotypes belonging to the phylum *Chloroflexi*, one of the dominant bacterial groups in subseafloor environments, appeared during the incubation. In addition, no archaea were enriched, although they have been suggested to be the dominant prokaryotes in subseafloor sediments (Lipp et al.,

---

nets to prevent crushing of the sponges. The medium is pumped into the reactor from the top inlet port, flows down by gravity, passes the sponge carriers that are randomly packed into the bioreactor column and is finally pumped out of the reactor. The sponge carriers are freely placed in the reactor, allowing an effective exchange of the medium fluid not only on the surface of but also inside the sponges. This setup leads to a greater yield of biomass in the DHS reactor in comparison to other continuous-flow systems (Onodera et al., 2013). (B) Strain MO-MB1, a methanogenic archaeon affiliated to the genus *Methanobacterium*. (C) Strain MO-MCD, a methanogenic archaeon belonging to the *Methanococcoides*. (D) *Pelolinea submarina* strain MO-CFX1, a chemoorganotrophic bacterium affiliated to the phylum *Chloroflexi* (Imachi et al., 2014). (E) Strain MO-SPC1, a chemoorganotrophic *Spirochaete*. Scale bars in (B) to (E) are 10 μm.

2008). Even though high-pressure is one of the key factors for culturing subseafloor sedimentary prokaryotes, the effect of depressurization remains unclear. Recently, Vossmeyer, Deusner, Kato, Inagaki, and Ferdelman (2012) reported that high hydrostatic pressure sustains microbial metabolic activities of sulfate reducers during long-term storage of the sediment samples. This strongly suggests that high-pressure should be considered for studying subseafloor prokaryotes even for sample storage.

The next example is a cultivation approach using a continuous-flow bioreactor. This technique is not new in microbiology and was already highlighted in previous reviews as an alternative cultivation strategy (Málek & Řičica, 1965; Zhang, Arends, Van de Wiele, & Boon, 2011). It was successfully applied to enrich fastidious microbes that usually escaped from conventional cultivation strategies such as anaerobic ammonium-oxidizing bacteria or anaerobic methanotrophs that live in syntrophic relation with nitrite/nitrate reducers (Haroon et al., 2013; Raghoebarsing et al., 2006; Strous et al., 1999). Applied for deep subseafloor sediments, Imachi et al. (2011) successfully cultured and isolated a diverse collection of anaerobic heterotrophs and methanogenic archaea from up to 106.7 mbsf deep sediments off Shimokita Peninsula, Japan, using this type of reactor. The so-called down-flow hanging sponge (DHS) reactor (Figure 2.6.3) has originally been developed for a low-cost treatment of municipal sewage in developing countries (Agrawal et al., 1997; Uemura & Harada, 2010). Special features of the DHS reactor are polyurethane sponges that provide enlarged surface areas for microbial colonization and the extension of cell residence times. The DHS has relatively long sludge retention times (=cell residence times) compared to other bioreactors developed in the field of wastewater treatment (Onodera et al., 2013). Thus, Imachi et al. selected the DHS to enrich slow-glowing subseafloor sedimentary microbes with low metabolic properties at a flow-through cultivation mode. An advantage is that substrates can be provided at low concentrations similar to those found in the natural environment. Additionally, the continuous-cultivation mode allows the outflow of metabolic products that might be accumulated and may inhibit microbial growth. Thus, the use of continuous-flow reactors offers a great opportunity to enrich subseafloor microorganisms in a controlled manner and serves as a source for further isolation attempts.

Recently, the combination of both high-pressure and continuous-flow cultivation techniques has been reported (Ohtomo et al., 2013; Sauer, Glombitza, & Kallmeyer, 2012; Zhang, Henriet, Bursens, & Boon, 2010). Using the high-pressure reactor cultivation technique, microbial communities capable of anaerobic oxidation of methane coupled with sulfate reduction (AOM-SR) were successfully enriched from subseafloor sediments. The AOM-SR reaction in marine sediments is a significant component of the global carbon cycle (Knittel & Boetius, 2009; Reeburgh, 2007), but the microorganisms involved have yet to be brought into pure culture. Gaseous substrates such as methane exhibit a generally low solubility in seawater. Thus, high pressure increases methane solubility and the thermodynamic feasibility of the AOM-SR reaction. In addition, the elimination of hydrogen sulfide during the continuous-flow cultivation reduces the toxicity of this by-product of the AOM-SR reaction for microbial cells. Recently, Ohtomo et al. (2013) developed a continuous-flow high-pressure bioreactor to examine potential geophysical, geochemical, and microbial impacts associated with CO_2 sequestration into subsurface oil, gas, or coal formations. Using this system, homoacetogenic CO_2 conversion to acetate was observed in a reaction column filled with coal and sand. The continuous high-pressure bioreactor method will be a promising cultivation technique to cultivate subseafloor sedimentary microorganisms.

In conclusion, the specific challenges of the subseafloor environment require novel cultivation techniques. In combination with molecular techniques and modern systems biology, the analysis of cultured microbial key players will help to elucidate their metabolic adaptations and to identify their role in ecosystem functioning.

REFERENCES

Agrawal, L. K., Ohashi, Y., Mochida, E., Okui, H., Ueki, Y., Harada, H., et al. (1997). Treatment of raw sewage in a temperate climate using a UASB reactor and the hanging sponge cubes process. *Water Science and Technology, 36*, 433–440.

Amann, R., Ludwig, L., & Schleifer, K. H. (1995). Phylogenetic identification and in situ detection of individual microbial cells without cultivation. *Microbiological Reviews, 59*, 143–169.

Aoike, K. E. (2007). *CK06–06 D/V Chikyu shakedown cruise offshore Shimokita, laboratory operation report.* Yokohama, Japan: CDEX-JAMSTEC. http://www.godac.jamstec.go.jp/catalog/data/doc_catalog/media/CK06-06_902_all.pdf.

Aoike, K., Nishi, H., Sakamoto, T., Iijima, K., Tsuchiya, M., Taira, A., et al. (2010). Paleoceanographic history of offshore Shimokita Peninsula for the past 800,000 years based on primary analyses on cores recovered by D/V Chikyu during the shakedown cruises. *Fossils, 87*, 65–81.

Ashida, N., Ishii, S., Hayano, S., Tago, K., Tsuji, T., Yoshimura, Y., et al. (2010). Isolation of functional single cells from environments using a micromanipulator: application to study denitrifying bacteria. *Applied Microbiology and Biotechnology, 85*, 1211–1217.

Bale, S. J., Goodman, K., Rochelle, P. A., Marchesi, J. R., Fry, J. C., Weightman, A. J., et al. (1997). *Desulfovibrio profundus* sp. nov., a novel barophilic sulfate-reducing bacterium from deep sediment layers in the Japan Sea. *International Journal of Systematic Bacteriology, 47*, 515–521.

Batzke, A., Engelen, B., Sass, H., & Cypionka, H. (2007). Phylogenetic and physiological diversity of cultured deep-biosphere bacteria from equatorial Pacific Ocean and Peru Margin sediments. *Geomicrobiology Journal, 24*, 261–273.

Biddle, J. F., Lipp, J. S., Lever, M. A., Lloyd, K. G., Sørensen, K. B., Anderson, R., et al. (2006). Heterotrophic Archaea dominate sedimentary subsurface ecosystems off Peru. *Proceedings of the National Academy of Sciences of the United States of America, 103*, 3846–3851.

Brochier-Armanet, C., Boussau, B., Gribaldo, S., & Forterre, P. (2008). Mesophilic crenarchaeota: proposal for a third archaeal phylum, the Thaumarchaeota. *Nature Reviews Microbiology, 6*, 245–252.

Cypionka, H. (2005). The physiological challenge. *Environmental Microbiology, 7*, 472.

D'Hondt, S., Jørgensen, B. B., Miller, D. J., Batzke, A., Blake, R., Cragg, B. A., et al. (2004). Distributions of microbial activities in deep subseafloor sediments. *Science, 306*, 2216–2221.

D'Hondt, S., Rutherford, S., & Spivack, A. J. (2002). Metabolic activity of subsurface life in deep-sea sediments. *Science, 295*, 2067–2070.

D'Hondt, S., Spivack, A. J., Pockalny, R., Ferdelman, T. G., Fischer, J. P., Kallmeyer, J., et al. (2009). Subseafloor sedimentary life in the South Pacific Gyre. *Proceedings of the National Academy of Sciences of the United States of America, 106,* 11651–11656.

de Man, J. C. (1977). MPN tables for more than one test. *European Journal of Applied Microbiology, 4,* 307–310.

DeLong, E. F. (2004). Microbial life breathes deep. *Science, 306,* 2198–2200.

Domitsu, H., Nishi, H., Uchida, J., Oda, M., Ogane, K., Taira, A., & Group, t.S.M.R., et al. (2010). Age model of core sediments taken by D/V Chikyu during the shakedown cruises off Shimokita Peninsula. *Fossils, 87,* 47–64.

Engelen, B., & Cypionka, H. (2009). The subsurface of tidal-flat sediments as a model for the deep biosphere. *Ocean Dynamics, 59,* 385–391.

Engelhardt, T., Sahlberg, M., Cypionka, H., & Engelen, B. (2011). Induction of prophages from deep-subseafloor bacteria. *Environmental Microbiology Reports, 3,* 459–465.

Engelhardt, T., Sahlberg, M., Cypionka, H., & Engelen, B. (2013). Biogeography of *Rhizobium radiobacter* and distribution of associated temperate phages in deep subseafloor sediments. *ISME Journal, 7,* 199–209.

Ferguson, R. L., Buckley, E. N., & Palumbo, A. V. (1984). Response of marine bacterioplankton to differential filtration and confinement. *Applied and Environmental Microbiology, 47,* 49–55.

Ferris, M. J., Ruff-Roberts, A. L., Kopczynski, E. D., Bateson, M. M., & Ward, D. M. (1996). Enrichment culture and microscopy conceal diverse thermophilic *Synechococcus* populations in a single hot spring microbial mat habitat. *Applied and Environmental Microbiology, 62,* 1045–1050.

Fichtel, K., Mathes, F., Könneke, M., Cypionka, H., & Engelen, B. (2012). Isolation of sulfate-reducing bacteria from sediments above the deep-subseafloor aquifer. *Frontiers in Microbiology, 3,* 65.

Froelich, P. N., Klinkhammer, G. P., Bender, M. L., Luedtke, N. A., Hea, G. R., Cullen, D., et al. (1979). Early oxidation of organic matter in pelagic sediments of the eastern equatorial atlantic: suboxic diagenesis. *Geochimica et Cosmochimica Acta, 43,* 1075–1090.

Fröhlich, J., & König, H. (2000). New techniques for isolation of single prokaryotic cells. *FEMS Microbiology Reviews, 24,* 567–572.

Fry, J. C., Parkes, R. J., Cragg, B. A., Weightman, A. J., & Webster, G. (2008). Prokaryotic biodiversity and activity in the deep subseafloor biosphere. *FEMS Microbiology Ecology, 66,* 181–196.

Giovannoni, S., & Stingl, U. (2007). The importance of culturing bacterioplankton in the 'omics' age. *Nature Reviews Microbiology, 5,* 820–826.

Gram, L., Melchiorsen, J., & Bruhn, J. B. (2010). Antibacterial activity of marine culturable bacteria collected from a global sampling of ocean surface waters and surface swabs of marine organisms. *Marine Biotechnology, 12,* 439–451.

Hara, S., Isoda, R., Tahvanainen, T., & Hashidoko, Y. (2012). Trace amounts of furan-2-carboxylic acids determine the quality of solid agar plates for bacterial culture. *PLoS ONE, 7,* e41142.

Haroon, M. F., Hu, S., Shi, Y., Imelfort, M., Keller, J., Hugenholtz, P., et al. (2013). Anaerobic oxidation of methane coupled to nitrate reduction in a novel archaeal lineage. *Nature, 500,* 567–570.

Hugenholtz, P. (2002). Exploring prokaryotic diversity in the genomic era. *Genome Biology, 3.*

Imachi, H., Aoi, K., Tasumi, E., Saito, Y., Yamanaka, Y., Saito, Y., et al. (2011). Cultivation of methanogenic community from subseafloor sediments using a continuous-flow bioreactor. *ISME Journal, 5,* 1913–1925.

Imachi, H., Sakai, S., Lipp, J. S., Miyazaki, M., Saito, Y., Yamanaka, Y., et al. (2014). *Pelolinea submarina* gen. nov., an anaerobic filamentous bacterium of the phylum *Chloroflexi* isolated from subseafloor sediment offshore Shimokita, Japan. *International Journal of Systematic and Evolutionary Microbiology, 64,* 812–818.

Inagaki, F. (2010). *Deep Subseafloor Microbial Communities*. In: eLS. John Wiley & Sons Ltd. Chichester. http://dx.doi.org/10.1002/9780470015902.a0021894. http://www.els.net.

Inagaki, F., Nunoura, T., Nakagawa, S., Teske, A., Lever, M., Lauer, A., et al. (2006). Biogeographical distribution and diversity of microbes in methane hydrate-bearing deep marine sediments, on the Pacific Ocean Margin. *Proceedings of the National Academy of Sciences of the United States of America, 103*, 2815–2820.

Inagaki, F., Suzuki, M., Takai, K., Oida, H., Sakamoto, T., Aoki, K., et al. (2003). Microbial communities associated with geological horizons in coastal subseafloor sediments from the Sea of Okhotsk. *Applied and Environmental Microbiology, 69*, 7224–7235.

Inagaki, F., Takai, K., Komatsu, T., Kanamatsu, T., Fujioka, K., & Horikoshi, K. (2001). Archaeology of Archaea: geomicrobiological record of Pleistocene thermal events concealed in a deep-sea subseafloor environment. *Extremophiles, 5*, 385–392.

Janssen, P. H., Yates, P. S., Grinton, B. E., Taylor, P. M., & Sait, M. (2002). Improved culturability of soil bacteria and isolation in pure culture of novel members of the divisions *Acidobacteria, Actinobacteria, Proteobacteria*, and *Verrucomicrobia*. *Applied and Environmental Microbiology, 68*, 2391–2396.

Jones, J. G. (1977). The effect of environmental factors on estimated viable and total populations of planktonic bacteria in lakes and experimental enclosures. *Freshwater Biol, 7*, 67–91.

Jones, W. J., Leigh, J. A., Mayer, F., Woese, C. R., & Wolfe, R. S. (1983). *Methanococcus jannaschii* sp nov, an extremely thermophilic methanogen from a submarine hydrothermal vent. *Archives of Microbiology, 136*, 254–261.

Jørgensen, B. B. (2011). Deep subseafloor microbial cells on physiological standby. *Proceedings of the National Academy of Sciences of the United States of America, 108*, 18193–18194.

Jørgensen, B. B., & D'Hondt, S. (2006). Ecology - a starving majority deep beneath the seafloor. *Science, 314*, 932–934.

Kaeberlein, T., Lewis, K., & Epstein, S. S. (2002). Isolating "uncultivable" microorganisms in pure culture in a simulated natural environment. *Science, 296*, 1127–1129.

Kallmeyer, J., Pockalny, R., Adhikari, R. R., Smith, D. C., & D'Hondt, S. (2012). Global distribution of microbial abundance and biomass in subseafloor sediment. *Proceedings of the National Academy of Sciences of the United States of America, 109*, 16213–16216.

Kendall, M. M., Liu, Y., Sieprawska-Lupa, M., Stetter, K. O., Whitman, W. B., & Boone, D. R. (2006). *Methanococcus aeolicus* sp. nov., a mesophilic, methanogenic archaeon from shallow and deep marine sediments. *International Journal of Systematic and Evolutionary Microbiology, 56*, 1525–1529.

Knittel, K., & Boetius, A. (2009). Anaerobic oxidation of methane: progress with an unknown process. *Annual Review of Microbiology, 63*, 311–334.

Kobayashi, T., Koide, O., Mori, K., Shimamura, S., Matsuura, T., Miura, T., et al. (2008). Phylogenetic and enzymatic diversity of deep subseafloor aerobic microorganisms in organics- and methane-rich sediments off Shimokita Peninsula. *Extremophiles, 12*, 519–527.

Kogure, K., Simidu, U., & Taga, N. (1979). A tentative direct microscopic method for counting living marine bacteria. *Canadian Journal of Microbiology, 25*(3), 415–420.

Kogure, K., Simidu, U., & Taga, N. (1980). Distribution of viable marine bacteria in neritic seawater around Japan. *Canadian Journal of Microbiology, 26*(3), 318–323.

Könneke, M., Bernhard, A. E., de la Torre, J. R., Walker, C. B., Waterbury, J. B., & Stahl, D. A. (2005). Isolation of an autotrophic ammonia-oxidizing marine archaeon. *Nature, 437*, 543–546.

Köpke, B., Wilms, R., Engelen, B., Cypionka, H., & Sass, H. (2005). Microbial diversity in coastal subsurface sediments: a cultivation approach using various electron acceptors and substrate gradients. *Applied and Environmental Microbiology, 71*, 7819–7830.

Lee, Y.-J., Wagner, I. D., Brice, M. E., Kevbrin, V. V., Mills, G. L., Romanek, C. S., et al. (2005). *Thermosediminibacter oceani* gen. nov., sp. nov. and *Thermosediminibacter litoriperuensis* sp. nov., new anaerobic thermophilic bacteria isolated from Peru Margin. *Extremophiles, 9*, 375–383.

Lever, M. A. (2012). Acetogenesis in the energy-starved deep biosphere - a paradox? *Frontiers in Microbiology, 2*, 284.

Liesack, W., Janssen, P. H., Rainey, F. A., Ward-Rainey, N. L., & Stackebrandt, E. (1997). Microbial diversity in soil: the need for a combined approach using molecular and cultivation techniques. In J. D. van Elsas, J. T. Trevors, & E. M. H. Wellington (Eds.), *Modern soil microbiology* (pp. 375–439). New York, N.Y: Marcel Dekker.

Lighthart, B., & Tong, Y. (1998). Measurements of total and culturable bacteria in the alfresco atmosphere using a wet-cyclone sampler. *Aerobiologia, 14*, 325–332.

Lipp, J. S., Morono, Y., Inagaki, F., & Hinrichs, K.-U. (2008). Significant contribution of Archaea to extant biomass in marine subsurface sediments. *Nature, 454*, 991–994.

Lomstein, B. A., Langerhuus, A. T., D'Hondt, S., Jørgensen, B. B., & Spivack, A. J. (2012). Endospore abundance, microbial growth and necromass turnover in deep sub-seafloor sediment. *Nature, 484*, 101–104.

Málek, I., & Řičica, J. (1965). Continuous cultivation of microorganisms. *Folia Microbiologica, 10*, 302–323.

Marchesi, J. R., Weightman, A. J., Cragg, B. A., Parkes, R. J., & Fry, J. C. (2001). Methanogen and bacterial diversity and distribution in deep gas hydrate sediments from the Cascadia Margin as revealed by 16S rRNA molecular analysis. *FEMS Microbiology Ecology, 34*, 221–228.

Martens-Habbena, W., & Sass, H. (2006). Sensitive determination of microbial growth by nucleic acid staining in aqueous suspension. *Applied and Environmental Microbiology, 72*, 87–95.

Mikucki, J. A., Liu, Y., Delwiche, M., Colwell, F. S., & Boone, D. R. (2003). Isolation of a methanogen from deep marine sediments that contain methane hydrates, and description of *Methanoculleus submarinus* sp. nov. *Applied and Environmental Microbiology, 69*, 3311–3316.

Milkov, A. V. (2004). Global estimates of hydrate-bound gas in marine sediments: how much is really out there? *Earth Science Reviews, 66*, 183–197.

Miyazaki, M., Koide, O., Kobayashi, T., Mori, K., Shimamura, S., Nunoura, T., et al. (2012). *Geofilum rubicundum* gen. nov., sp. nov., isolated from deep subseafloor sediment. *International Journal of Systematic and Evolutionary Microbiology, 62*, 1075–1080.

Morono, Y., Terada, T., Nishizawa, M., Hillion, F., Ito, M., Takahata, N., et al. (2011). Carbon and nitrogen assimilation of deep subseafloor microbial cells. *Proceedings of the National Academy of Sciences of the United States of America, 108*, 18295–18300.

Nichols, D. (2007). Cultivation gives context to the microbial ecologist. *FEMS Microbiology Ecology, 60*, 351–357.

Nogi, Y., Masui, N., & Kato, C. (1998). *Photobacterium profundum* sp. nov., a new, moderately barophilic bacterial species isolated from a deep-sea sediment. *Extremophiles, 2*, 1–7.

Ohtomo, Y., Ijiri, A., Ikegawa, Y., Tsutsumi, M., Imachi, H., Uramoto, G., et al. (2013). Biological CO2 conversion to acetate in subsurface coal-sand formation using a high-pressure reactor system. *Frontiers in Microbiology, 4* Article no. 361.

Onodera, T., Matsunaga, K., Kubota, K., Taniguchi, R., Harada, H., Syutsubo, K., et al. (2013). Characterization of the retained sludge in a down-flow hanging sponge (DHS) reactor with emphasis on its low excess sludge production. *Bioresource Technology, 136*, 169–175.

Orcutt, B. N., Sylvan, J. B., Knab, N. J., & Edwards, K. J. (2011). Microbial ecology of the dark ocean above, at, and below the seafloor. *Microbiology and Molecular Biology Reviews, 75*, 361–422.

Overmann, J. (2006). Principles of enrichment, isolation, cultivation and preservation of prokaryotes. In M. Dworkin, S. Falkow, E. Rosenberg, K.-H. Schleifer, & E. Stackebrandt (Eds.), *The prokaryotes* (Third edn) (pp. 80–136). New York: Springer.

Parkes, R. J., Cragg, B. A., & Wellsbury, P. (2000). Recent studies on bacterial populations and processes in subseafloor sediments: a review. *Hydrogeology Journal, 8*, 11–28.

Parkes, R. J., Martin, D., Amann, H., Anders, E., Holland, M., Collett, T., et al. (2009a). Technology for high-pressure sampling and analysis of deep-sea sediments, associated gas hydrates, and deep-biosphere processes. In T. Collett, A. Johnson, C. Knapp, and R. Boswell, eds., *Natural Gas hydrates - energy resource potential and associated geologic hazards AAPG Memoir* (Vol. 89) (pp. 672–683). http://dx.doi.org/10.1306/13201131M893362.

Parkes, R. J., Sellek, G., Webster, G., Martin, D., Anders, E., Weightman, A. J., et al. (2009b). Culturable prokaryotic diversity of deep, gas hydrate sediments: first use of a continuous high-pressure, anaerobic, enrichment and isolation system for subseafloor sediments (DeepIsoBUG). *Environmental Microbiology, 11*, 3140–3153.

Raghoebarsing, A. A., Pol, A., van de Pas-Schoonen, K. T., Smolders, A. J. P., Ettwig, K. F., Rijpstra, W. I. C., et al. (2006). A microbial consortium couples anaerobic methane oxidation to denitrification. *Nature, 440*, 918–921.

Reeburgh, W. (2007). Oceanic methane biogeochemistry. *Chemical Reviews, 38*.

Sakai, S., Imachi, H., Sekiguchi, Y., Ohashi, A., Harada, H., & Kamagata, Y. (2007). Isolation of key methanogens for global methane emission from rice paddy fields: a novel isolate affiliated with the clone cluster rice cluster I. *Applied and Environmental Microbiology, 73*, 4326–4331.

Sauer, P., Glombitza, C., & Kallmeyer, J. (2012). A system for incubations at high gas partial pressure. *Frontiers in Microbiology, 3*, 25.

Schouten, S., Forster, A., Panoto, F. E., & Damste, J. S. S. (2007). Towards calibration of the TEX86 palaeothermometer for tropical sea surface temperatures in ancient greenhouse worlds. *Organic Geochemistry, 38*, 1537–1546.

Schultheiss, P., Holland, M., & Humphrey, G. (2009). Wireline coring and analysis under pressure: recent use and future developments of the HYACINTH system. *Scientific Drilling, 7*, 44–50.

Sekiguchi, Y., Takahashi, H., Kamagata, Y., Ohashi, A., & Harada, H. (2001). In situ detection, isolation, and physiological properties of a thin filamentous microorganism abundant in methanogenic granular sludges: a novel isolate affiliated with a clone cluster, the green non-sulfur bacteria, subdivision I. *Applied and Environmental Microbiology, 67*, 5740–5749.

Simu, K., & Hagström, Å. (2004). Oligotrophic bacterioplankton with a novel single-cell life strategy. *Applied and Environmental Microbiology, 70*, 2445–2451.

Sogin, M. L., Morrison, H. G., Huber, J. A., Mark Welch, D., Huse, S. M., Neal, P. R., et al. (2006). Microbial diversity in the deep sea and the underexplored "rare biosphere". *Proceedings of the National Academy of Sciences of the United States of America, 103*, 12115–12120.

Staley, J. T., & Konopka, A. (1985). Measurement of in situ activities of nonphotosynthetic microorganisms in aquatic and terrestrial habitats. *Annual Review of Microbiology, 39*, 321–346.

Stevenson, B. S., Eichorst, S. A., Wertz, J. T., Schmidt, T. M., & Breznak, J. A. (2004). New strategies for cultivation and detection of previously uncultured microbes. *Applied and Environmental Microbiology, 70*, 4748–4755.

Straskrabova, V. (1983). The effect of substrate shock on populations of starving aquatic bacteria. *Journal of Applied Bacteriology, 54*, 217–224.

Strous, M., Fuerst, J. A., Kramer, E. H., Logemann, S., Muyzer, G., van de Pas-Schoonen, K. T., et al. (1999). Missing lithotroph identified as new planctomycete. *Nature, 400*, 446–449.

Suess, E. (1980). Particulate carbon flux in the oceans - surface productivity and oxygen utilization. *Nature, 288,* 260–263.

Süß, J., Engelen, B., Cypionka, H., & Sass, H. (2004). Quantitative analysis of bacterial communities from Mediterranean sapropels based on cultivation-dependent methods. *FEMS Microbiology Ecology, 51,* 109–121.

Süß, J., Herrmann, K., Seidel, M., Cypionka, H., Engelen, B., & Sass, H. (2008). Two distinct *Photobacterium* populations thrive in ancient mediterranean sapropels. *Microbial Ecology, 55,* 371–383.

Süß, J., Schubert, K., Sass, H., Cypionka, H., Overmann, J., & Engelen, B. (2006). Widespread distribution and high abundance of *Rhizobium radiobacter* within mediterranean subsurface sediments. *Environmental Microbiology, 8,* 1753–1763.

Takai, K., Abe, M., Miyazaki, M., Koide, O., Nunoura, T., Imachi, H., et al. (2013). *Sunxiuqinia faeciviva* sp. nov., a novel facultatively anaerobic, organoheterotrophic bacterium within the *Bacteroidetes* isolated from deep subseafloor sediment offshore Shimokita, Japan. *International Journal of Systematic and Evolutionary Microbiology, 63,* 1602–1609.

Takai, K., Inoue, A., & Horikoshi, K. (1999). *Thermaerobacter marianensis* gen. nov., sp. nov., an aerobic extremely thermophilic marine bacterium from the 11,000 m deep Mariana Trench. *International Journal of Systematic Bacteriology, 49*(Pt 2), 619–628.

Takai, K., Moyer, C. L., Miyazaki, M., Nogi, Y., Hirayama, H., Nealson, K. H., et al. (2005). *Marinobacter alkaliphilus* sp. nov., a novel alkaliphilic bacterium isolated from subseafloor alkaline serpentine mud from Ocean Drilling Program Site 1200 at South Chamorro Seamount, Mariana Forearc. *Extremophiles, 9,* 12–27.

Tamaki, H., Hanada, S., Sekiguchi, Y., Tanaka, Y., & Kamagata, Y. (2009). Effect of gelling agent on colony formation in solid cultivation of microbial community in lake sediment. *Environ Microbiol, 11,* 1827–1834.

Teske, A. P. (2006). Microbial communities of deep marine subsurface sediments: molecular and cultivation surveys. *Geomicrobiology Journal, 23,* 357–368.

Teske, A., & Sørensen, K. B. (2008). Uncultured archaea in deep marine subsurface sediments: have we caught them all? *ISME Journal, 2,* 3–18.

Toffin, L., Webster, G., Weightman, A. J., Fry, J. C., & Prieur, D. (2004). Molecular monitoring of culturable bacteria from deep-sea sediment of the Nankai Trough, Leg 190 Ocean Drilling Program. *FEMS Microbiology Ecology, 48,* 357–367.

Torsvik, V., Goksøyr, J., & Daae, F. L. (1990). High diversity in DNA of soil bacteria. *Applied and Environmental Microbiology, 56,* 782–787.

Tsubouchi, T., Shimane, Y., Mori, K., Miyazaki, M., Tame, A., Uematsu, K., et al. (2013a). *Loktanella cinnabarina* sp. nov., isolated from a deep subseafloor sediment, and emended description of the genus *Loktanella. International Journal of Systematic and Evolutionary Microbiology, 63,* 1390–1395.

Tsubouchi, T., Shimane, Y., Mori, K., Usui, K., Hiraki, T., Tame, A., et al. (2012). *Polycladomyces abyssicola* gen. nov., sp. nov., a thermophilic filamentous bacterium isolated from hemipelagic sediment. *International Journal of Systematic and Evolutionary Microbiology, 63,* 1972–1981.

Tsubouchi, T., Shimane, Y., Usui, K., Shimamura, S., Mori, K., Hiraki, T., et al. (2013b). *Brevundimonas abyssalis* sp. nov., a dimorphic prosthecate bacterium isolated from deep-subseafloor sediment. *International Journal of Systematic and Evolutionary Microbiology, 63,* 1987–1994.

Uemura, S., & Harada, H. (2010). Application of UASB technology for sewage treatment with a novel post-treatment process. In H. H. P. Fang (Ed.), *Environmental aanaerobic technology* (pp. 91–112). London, United Kingdom: Imperial College Press.

Vezzi, A., Campanaro, S., D'Angelo, M., Simonato, F., Vitulo, N., Lauro, F. M., et al. (2005). Life at depth: *Photobacterium profundum* genome sequence and expression analysis. *Science, 307*, 1459–1461.

von Klein, D., Arab, H., Völker, H., & Thomm, M. (2002). *Methanosarcina baltica*, sp. nov., a novel methanogen isolated from the Gotland Deep of the Baltic Sea. *Extremophiles, 6*, 103–110.

Vossmeyer, A., Deusner, C., Kato, C., Inagaki, F., & Ferdelman, T. G. (2012). Substrate-specific pressure-dependence of microbial sulfate reduction in deep-sea cold seep sediments of the Japan Trench. *Frontiers in Microbiology, 3*, 1–12.

Wagner, M., Amann, R., Lemmer, H., & Schleifer, K. H. (1993). Probing activated sludge with oligo-nucleotides specific for proteobacteria: inadequacy of culture-dependent methods for describing microbial community structure. *Applied and Environmental Microbiology, 59*, 1520–1525.

Wagner, M., Erhart, R., Manz, W., Amann, R., Lemmer, H., Wedi, D., et al. (1994). Development of an rRNA-targeted oligonucleotide probe specific for the genus *Acinetobacter* and its application for in situ monitoring in activated sludge. *Applied and Environmental Microbiology, 60*, 792–800.

Wang, F. P., Wang, P., Chen, M. X., & Xiao, X. (2004). Isolation of extremophiles with the detection and retrieval of *Shewanella* strains in deep-sea sediments from the west Pacific. *Extremophiles, 8*, 165–168.

Watkins, A. J., Roussel, E. G., Parkes, R. J., & Sass, H. (2013). Glycine betaine as a direct substrate for methanogens (*Methanococcoides* spp.). *Applied and Environmental Microbiology, 80*(1), 289–293.

Watkins, A. J., Roussel, E. G., Webster, G., Parkes, R. J., & Sass, H. (2012). Choline and *N,N*-dimethylethanolamine as direct substrates for methanogens. *Applied and Environmental Microbiology, 78*, 8298–8303.

Whitman, W. B., Coleman, D. C., & Wiebe, W. J. (1998). Prokaryotes: the unseen majority. *Proceedings of the National Academy of Sciences of the United States of America, 95*, 6578–6583.

Wuchter, C., Abbas, B., Coolen, M. J. L., Herfort, L., van Bleijswijk, J., Timmers, P., et al. (2006). Archaeal nitrification in the ocean. *Proceedings of the National Academy of Sciences of the United States of America, 103*, 12317–12322.

Yanagibayashi, M., Nogi, Y., Li, L., & Kato, C. (1999). Changes in the microbial community in Japan Trench sediment from a depth of 6292 m during cultivation without decompression. *FEMS Microbiology Letters, 170*, 271–279.

Zhang, Y., Arends, J. B. A., Van de Wiele, T., & Boon, N. (2011). Bioreactor technology in marine microbiology: from design to future application. *Biotechnology Advances, 29*, 312–321.

Zhang, Y., Henriet, J.-P., Bursens, J., & Boon, N. (2010). Stimulation of in vitro anaerobic oxidation of methane rate in a continuous high-pressure bioreactor. *Bioresource Technology, 101*, 3132–3138.

Chapter 2.7

Biogeochemical Consequences of the Sedimentary Subseafloor Biosphere

Laura M. Wehrmann[1] and Timothy G. Ferdelman[2,*]
[1]*School of Marine and Atmospheric Sciences, Stony Brook University, Stony Brook, NY, USA;*
[2]*Department of Biogeochemistry, Max Planck Institute for Marine Microbiology, Bremen, Germany*
Corresponding author: E-mail: tferdelm@mpi-bremen.de

2.7.1 INTRODUCTION

The subseafloor sedimentary ocean comprises a substantial fraction of the Earth's aqueous marine environment. When considering the total volume of sediments (Kennett, 1982, p. 321) and an average porosity of 0.5, the subsurface ocean contains $0.53 \times 10^{18} \, m^3$ of interstitial water (Figure 2.7.1), or 28% of the total water volume in the combined sedimentary and pelagic ($1.37 \times 10^{18} \, m^3$) ocean. Based on sheer volume alone, chemical and biological processes occurring in the deep subsurface ocean will expectedly have consequences for the evolution of the Earth's ocean and chemistry. Much effort in the past decade of deep biosphere research and scientific drilling within the framework of Ocean Drilling Program (ODP) and Integrated Ocean Drilling Program (IODP) has been placed on understanding this habitat, particularly in exploring the abundance, diversity, and physiology of the microbial biocenosis. A stated goal of recent scientific drilling and deep subseafloor biosphere research has been to evaluate the biogeochemical impacts of subseafloor biota. This deep subseafloor biosphere exhibits a strong geological time component that is tightly coupled to variations in climate and ocean circulation. In response, microbial processes leave their imprint on the overall geochemistry of deeply buried sediments and eventually on the global redox cycle. Here, we specifically ask what we have learned about the biogeochemical consequences of microbial activity in deeply buried sediments, on the geochemistry of those sediments, and the overall biogeochemistry of the ocean.

Although the subseafloor sedimentary ocean forms an integral part of the Earth's ocean, important characteristics distinguish it from the overlying pelagic ocean. The pelagic ocean comprises one connected system that mixes on the order of 1000 years (e.g., Libes, 2009), whereas sedimentary systems

Developments in Marine Geology, Volume 7. http://dx.doi.org/10.1016/B978-0-444-62617-2.00009-8
217

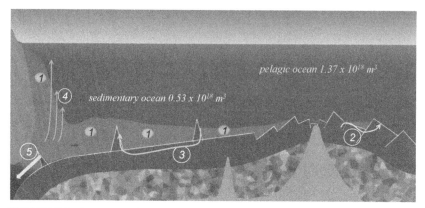

FIGURE 2.7.1 Important environments of the sedimentary subseafloor ocean under consideration in this chapter: (1) diffusive environments, (2) sediment ponds on mid-ocean ridge flanks, (3) subsediment flow through crustal aquifers as per Bekins, Spivack, Davis, and Mayer (2007), (4) fluid flow driven by tectonics and sediment accretion at continental margins, and (5) subduction of ocean sediment and removal of oxidants and reductants from the Earth's ocean-atmosphere system.

comprise water constrained in a dense particle matrix, whose varying chemistries depend greatly on the depositional histories of the often isolated basins. With the exception of fluid flow along constrained flow paths in some defined environments, molecular diffusion dominates physical transport in the sedimentary subseafloor ocean (Figure 2.7.1). The average time required for a dissolved compound to diffuse through to any given depth scales with the square of the diffusion path length (for an excellent discussion of this theme see Jørgensen, 2006). This has serious consequences for the location and rates of reactions. A medium diffusion time for a dissolved ion, e.g., Cl^-, through the middle of the thinly blanketed South Pacific Gyre sediments (about 20 m thick) is around 4000 years. Thicker sediment packages of 100 m and more, which are typical for the continental margins and high-productivity regions, yield diffusion times of ~400,000 years. Glacial–interglacial changes in ocean chlorinity and $\delta^{18}O$–H_2O composition can be observed as the diffusion front propagates through sediments (e.g., Adkins & Schrag, 2001; Adkins, Mcintyre, & Schrag, 2002; Insua, Spivack, Graham, D'hondt, & Moran, 2014; Mcduff, 1984; Schrag, Hampt, & Murray, 1996; Schrag et al., 2002). Likewise, at depth scales of 1–50 m the microbial communities can react to changing ocean chemistry and fluxes to the surface on glacial–interglacial timescales, as demonstrated by Contreras et al. (2013) for changes in the sulfate–methane transition (SMT) zone in sediments of ODP Leg 201 Site 1229 on the Peru Margin. At the other extreme, deeply buried layers in the absence of tectonically induced flow through permeable layers are essentially cut off from communication with the overlying pelagic ocean by long diffusion path lengths and extremely low porosities. An example is the Shimokita Oligocene-age coal-bearing sediments (IODP Expedition 337; Inagaki, Hinrichs, Kubo, & the Expedition 337 Scientists, 2013), which contain an active biosphere at depths of 2046–2466 mbsf. The consequences of biogeochemical processes within these sediments will be more likely expressed over geological time as the sediments eventually enter the rock record and are weathered.

2.7.2 BIOGEOCHEMICAL ZONATION IN SUBSEAFLOOR SEDIMENTS

The remineralization of deposited organic matter, predominately originating from photosynthetic activity in ocean surface waters and on land, is the main driver for biogeochemical processes in subseafloor sediments (D'Hondt, Rutherford, & Spivack, 2002; D'Hondt et al., 2004). Organic matter oxidation proceeds via a sequence of terminal electron accepting processes (TEAPs), encompassing O_2, NO_3^-, Mn(IV) and Fe(III) oxides, and SO_4^{2-}, followed by methanogenesis as a function of free energy gain (Froelich et al., 1979; Stumm & Morgan, 1996; Table 2.7.1(a)). In the subseafloor sediment, this progression typically leads to the installation of distinct, yet overlapping biogeochemical zones where the respective oxidation–reduction processes occur. The thicknesses of the different zones are, among other things, controlled by the rate of organic carbon oxidation, electron acceptor availability, and sediment accumulation rates (see Arndt et al., 2013, for review). Consequently, regional and global trends in the distribution of diagenetic processes are identified and continuously updated as studies of previously little-explored ocean regions are being published, integrated into diagenetic models and turned into regional and global estimates of deep sediment respiratory activity (Bowles, Mogollón, Kasten, Zabel, & Hinrichs, 2014; D'Hondt et al., 2004, 2002). For example, the upwelling areas of eastern boundary systems are delineated as high-productivity regions characterized by elevated organic carbon turnover rates. Here, oxygen, nitrate, and reactive metal oxides are consumed during TEAPs or related diagenetic reactions in the upper few centimeters of the sediment, and organic carbon mineralization in the remainder of the sediment column is proceeding via sulfate reduction and methanogenesis (Figure 2.7.2(a)). Sediments underlying the Benguela Upwelling System along the Southwest African margin and the Peru Margin Upwelling System, two of the world's most productive ocean areas, were drilled during ODP Legs 175 and Leg 112 (revisited during Leg 201), respectively. Alkalinity values exceeding 170 mM (Leg 175, Site 1085; Murray, Wigley, & Shipboard Scientific Party, 1998) and 260 mM (Leg 112, Site 688; Kastner et al., 1990), and ammonium concentrations greater than 50 mM, are a direct consequence of the elevated rates of primary productivity in the surface waters reflected in the very high supply and mineralization of organic carbon matter in the underlying sediments. In contrast, central gyres, such as the South Pacific Gyre, drilled during IODP Exp. 329, are extremely oligotrophic regions, and oxygen and nitrate penetrate tens of meters into the sediment (Figure 2.7.2(b); D'Hondt, Inagaki, Alvarez Zarikian, & the IODP Expedition 329 Science Party, 2013; Fischer, Ferdelman, D'Hondt, Røy, & Wenzhöfer, 2009; Røy et al., 2012).

TABLE 2.7.1 (a) Metabolic Reactions, Including Reactions of Organic Matter Oxidation in Typical Redfield Stoichiometry (C/N/P) Where x, y, and z are Taken as 106, 16, and 1 (Modified From Tromp, Van Cappellen, & Key, 1995; Aller, 2014) and (b) Important Secondary Reactions Occurring in Subseafloor Sediments

(a) Metabolic Reactions

Organic Matter Oxidation Reactions

Aerobic respiration	$(CH_2O)_x(NH_3)y(H_3PO_4)z + (x+2y)\textbf{O}_2 \rightarrow xCO_2 + (x+y)H_2O + yHNO_3 + zH_3PO_4$
Nitrate reduction	$5(CH_2O)_x(NH_3)y(H_3PO_4)z + 4x\textbf{NO}_3^- \rightarrow xCO_2 + 3xH_2O + 4xHCO_3^- + 2xN_2 + 5yNH_3 + 5zH_3PO_4$
Dissimilatory manganese oxide reduction	$(CH_2O)_x(NH_3)y(H_3PO_4)z + 2x\textbf{MnO}_2 + 3xCO_2 + xH_2O \rightarrow 2xMn^{2+} + 4xHCO_3^- + yNH_3 + zH_3PO_4$
Dissimilatory iron oxide reduction	$(CH_2O)_x(NH_3)y(H_3PO_4)z + 4x\textbf{Fe(OH)}_3 + 7xCO_2 \rightarrow 4xFe^{2+} + 8xHCO_3^- + 3xH_2O + yNH_3 + zH_3PO_4$
Organoclastic sulfate reduction	$2(CH_2O)_x(NH_3)y(H_3PO_4)z + x\textbf{SO}_4^{2-} \rightarrow xH_2S + 2xHCO_3^- + 2yNH_3 + 2zH_3PO_4$
Hydrogenotrophic methanogenesis	$CO_2 + 4H_2 \rightarrow \textbf{CH}_4 + 2H_2O$
Acetoclastic methanogenesis	$CH_3COOH \rightarrow CO_2 + \textbf{CH}_4$
Anaerobic oxidation of methane coupled to sulfate reduction	$CH_4 + \textbf{SO}_4^{2-} \rightarrow HS^- + HCO_3^- + H_2O$
Anaerobic oxidation of methane coupled to iron reduction	$CH_4 + 8\textbf{Fe(OH)}_3 + 15H^+ \rightarrow HCO_3^- + 8Fe^{2+} + 21H_2O$
Anammox	$NH_4^+ + NO_2^- \rightarrow N_2 + 2H_2O$ $NH_4^+ + 3/_2MnO_2 + 2H^+ \rightarrow 3/_2Mn^{2+} + 1/_2N_2 + 3H_2O$

(b) Secondary Reactions

Reaction type	Reactions
Oxidation of aqueous metal species	$Mn^{2+} + \tfrac{1}{4}O_2 + \tfrac{3}{2}H_2O \rightarrow MnOOH + 2H^+$ $Mn^{2+} + \tfrac{1}{2}O_2 + H_2O \rightarrow MnO_2 + 2H^+$ $\mathbf{Fe^{2+}} + \tfrac{1}{4}O_2 + \tfrac{5}{2}H_2O \rightarrow Fe(OH)_3 + 2H^+$ $MnO_2 + 2\mathbf{Fe^{2+}} + 4H_2O \rightarrow Mn^{2+} + 2Fe(OH)_3 + 2H^+$ $5\mathbf{Fe^{2+}} + \mathbf{NO_3^-} + 12H_2O \rightarrow 5Fe(OH)_3 + \tfrac{1}{2}N_2 + 4H^+$ $\tfrac{5}{2}\mathbf{Mn^{2+}} + \mathbf{NO_3^-} + 2H_2O \rightarrow \tfrac{5}{2}MnO_2 + \tfrac{1}{2}N_2 + 4H^+$
Iron monosulfide and pyrite formation	$Fe^{2+} + H_2S \rightarrow \mathbf{FeS} + 2H^+$ $FeS + S_x^{2-} \rightarrow \mathbf{FeS_2} + (x-1)S_x^{2-}$ $FeS + H_2S \rightarrow \mathbf{FeS_2} + H_2$
Manganese oxide reduction	$4H_2S + 4\mathbf{MnO_2} \rightarrow 4S^0 + 4Mn^{2+} + 8OH^-$ $NH_4^+ + 4\mathbf{MnO_2} + 6H^+ \rightarrow 4Mn^{2+} + \mathbf{NO_3^-} + 5H_2O$
Sulfur oxidation and disproportionation	$2FeOOH + 3H_2S \rightarrow 2\mathbf{FeS} + S^0 + 4H_2O$ $2\mathbf{H_2S} + 8FeOOH + 14H^+ \rightarrow S_2O_3^{2-} + 8Fe^{2+} + 13H_2O$ $4S^0 + 4H_2O \rightarrow 3H_2S + SO_4^{2-} + 2H^+$ $8\mathbf{HS_2^-} + 4H_2O \rightarrow SO_4^{2-} + 7HS^- + 7H^+$ $\mathbf{S_2O_3^{2-}} + H_2O \rightarrow H_2S + SO_4^{2-}$
Alkalinity generation via sulfate reduction	$3SO_4^{2-} + 6CH_2O + 2FeOOH \rightarrow 6\mathbf{HCO_3^-} + FeS_2 + FeS + 4H_2O$ $3SO_4^{2-} + 3CH_4 + 2FeOOH \rightarrow 3\mathbf{CO_3^{2-}} + FeS_2 + FeS + 4H_2O$

Bold indicates eponymous reactant.

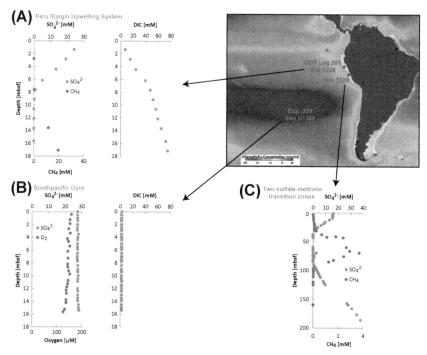

FIGURE 2.7.2   (A) Interstitial water concentration profiles of sulfate (SO_4^{2-}) and dissolved inorganic carbon (DIC), and methane (CH_4) concentration profiles from sediments of the Peru Margin Upwelling System (ODP Leg 201 Site 1230; D'Hondt, Jørgensen, Miller et al., 2003); (B) SO_4^{2-}, oxygen, and DIC concentrations in South Pacific Gyre sediments (IODP Exp. 329 Site U1368; D'Hondt, Inagaki, Alvarez Zarikian et al., 2011), and (C) SO_4^{2-} and CH_4 concentration profiles from ODP Leg 201 Site 1229 sediment, which features two sulfate–methane transition zones. Also shown is the average ocean chlorophyll concentration measured by SeaWiFS (SeaWiFS Project, NASA Goddard Space Flight Center).

A curious biogeochemical feature discovered at many sites drilled during the legacy of Deep Sea Drilling Project (DSDP), ODP, and IODP is the inversion of the sequence of electron acceptor activities near the basement, sometimes even expressed in the occurrence of two SMT zones, e.g., at Site 1229 drilled during ODP Leg 201 (Figure 2.7.2(c); D'Hondt et al., 2004).

2.7.3 SECONDARY BIOGEOCHEMICAL REACTIONS

Microbially mediated diagenetic processes driven by carbon mineralization leave a fundamental imprint on the pore water composition and lithogenic and biogenic solid-phase components of marine sediments (Table 2.7.1; Figure 2.7.3). For instance, the metabolic by-products of the subseafloor microbial carbon cycle, such as hydrogen sulfide and dissolved inorganic carbon (DIC), react with pore water and/or solid-phase components of the sediments. An additional, vast

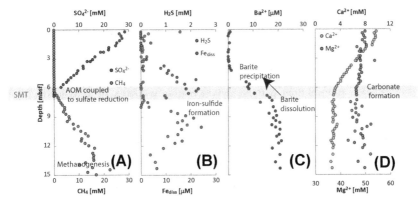

FIGURE 2.7.3 Interstitial water concentration profiles of (A) sulfate SO_4^{2-} and methane (CH_4), (B) hydrogen sulfide (H_2S) and dissolved iron (Fe_{diss}), (C) barium (Ba^{2+}), and (D) calcium (Ca^{2+}) and magnesium (Mg^{2+}) from IODP Expedition 323 Site U1345. *Expedition 323 Scientists, 2011.*

array of diagenetic processes, both inorganic and microbially mediated, occur as a consequence of the initial microbial reactions (Table 2.7.1(b)). These reactions often create a range of short-lived, intermediate compounds that react further via a network of oxidation–reduction reactions. These diagenetic reactions represent important recycling mechanisms that link the carbon, sulfur, phosphorus, silica, and metal cycles.

In the upper sediment column and in sediments above the ocean crust, reactions involving oxygen, nitrogen compounds, and dissolved and particulate metal species often play a prevalent role (Table 2.7.1(b)), for example, the oxidation of (up- or downward diffusing) aqueous iron and manganese to new metal oxide phases, and the oxidation of aqueous iron via manganese oxides phases. At depths where sulfate reduction is occurring, processes involving intermediate sulfur species and hydrogen sulfide are important, such as the reaction of these sulfur compounds with dissolved iron and solid-phase iron (oxyhydr)oxides to form elemental sulfur, iron monosulfide, and pyrite (Figure 2.7.3(b); Berner, 1970, 1984). Throughout the sediment column, the production of DIC during organic carbon mineralization and increasing alkalinity affects the saturation of carbonate mineral phases as outlined below.

2.7.4 INTERACTION OF BIOGEOCHEMICAL PROCESSES AND THE SEDIMENT

The diagenetic processes in subseafloor sediments lead to the dissolution, oxidation, reduction, and/or recrystallization of sedimentary deposits, and the formation of new authigenic precipitates. The genesis of authigenic mineral phases, such as carbonates (Figure 2.7.3(d)), can change sediment properties, e.g., by facilitating lithification and thus reducing sediment permeability. The occurrence and elemental and isotopic composition of these mineral phases often allow reconstruction of the microbially mediated processes and associated early diagenetic reactions in the past and present, and can reveal changes of the environmental conditions in the overlying water column over time, including oxygenation conditions (Figure 2.7.4; e.g., Goldhaber, 2003; Lyons & Severmann, 2006; Raiswell, Buckley, Berner, & Anderson, 1988; Raiswell & Canfield, 1998;). The alteration of primary sediment components by diagenetic processes, however, can also affect paleoproxies commonly used in paleoceanographic studies, e.g., by dissolution of skeletal fossils and alteration of initial biogenic constituents such as barite (Brumsack & Gieskes, 1983; McManus et al., 1998; Von Breymann, Brumsack, & Emeis, 1992; see Figure 2.7.3(c)). Biogeochemical reactions in subseafloor sediments play an important role in the global carbon, sulfur, phosphorus, silica, and metal cycles, by regulating their oceanic reservoirs through transformation, deposition, and burial of aqueous species (e.g., sulfate), and particulate lithogenic and biogenic components in the form of more stable diagenetic products (e.g., authigenic pyrite, carbonate, and apatite phases). Aside from dissolution–precipitation reactions, cation exchange on mineral surfaces is central for the distribution of pore water ions, and thus the availability of reactants for further authigenic mineral formation in the sediment. For instance, the adsorption of ammonium, produced during organic carbon oxidation, onto clay minerals can play an important role for the concentrations and isotopic composition of calcium and magnesium in marine pore waters (Figure 2.7.4; Mavromatis, Meister, & Oelkers, 2014; Ockert, Gussone, Kaufhold, & Teichert, 2013; Teichert, Gussone, & Torres, 2009; Von Breymann, Collier, & Suess, 1990; Von Breymann, Ungerer, & Suess, 1988).

One of the most fundamental reactions of the subseafloor microbial carbon cycle is the formation of diagenetic carbonate phases, such as magnesian calcite and dolomite (Baker & Kastner, 1981; Kelts & McKenzie, 1982; Pisciotto & Mahoney, 1981). This process ultimately controls the amount of DIC that is released back into the ocean and the fraction of carbon that is buried in the sedimentary record over longer timescales (Falkowsi et al., 2000; Higgins, Fischer, & Schrag, 2009; Milliman, 1993; Schrag, Higgins, Macdonald, & Johnston, 2013). It is estimated that diagenetic carbonates may account for at least 10% of global carbonate accumulation at present and may thus play an important role in the removal of carbon from the Earth's surface (Sun & Turchyn, 2014). Additionally, changes in the volume of the global authigenic carbon reservoir over time may have significantly affected the global carbon cycle in the past. This carbon sink should therefore be included in global carbon mass balance estimates throughout Earth's history (Canfield & Kump, 2013; Higgins et al., 2009; Schrag et al., 2013).

Organoclastic sulfate reduction has been proposed as a driver for diagenetic carbonate formation in marine subseafloor sediments (e.g., Burns, Baker, & Showers, 1988; Mazzullo, 2000). This process is dependent on the availability of reactive iron (oxyhydr)oxide phases to form iron sulfides and the concomitant build-up of hydrogen sulfide in the pore water (Morse & McKenzie, 1990;

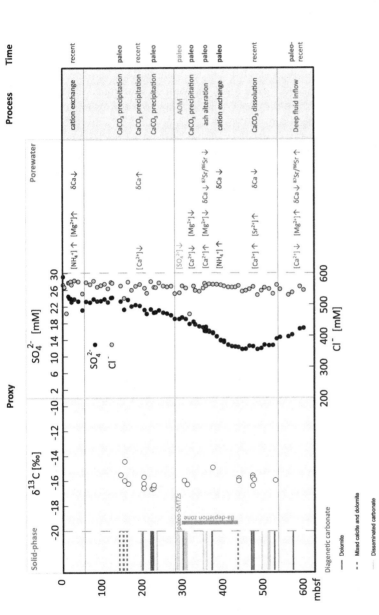

FIGURE 2.7.4 Example of linking biogeochemical processes to solid phase carbonate chemistry (distribution of diagenetic carbonate phases and carbon isotopic composition ($\delta^{13}C$) of carbonates) and interstitial water profiles (sulfate (SO_4^{2-}) and chloride (Cl^-)) at IODP Expedition 323 Site U1341 (Bowers Ridge, Bering Sea; after Wehrmann et al., in press). The present (recent) and past (paleo) biogeochemical processes and their effects on pore water concentrations and isotope values with depth are summarized. Brackets denote concentration, a delta sign the isotopic ratio. Up- and downward arrows indicate that the concentration or isotope ratio is increasing or decreasing at this depth.

Walter & Burton, 1990). In both advective fluid flow systems and diffusive systems, sulfate reduction coupled with reductive iron oxide dissolution and precipitation of pyrite represents an important source of carbonate alkalinity for the marine environment.

As written in Table 2.7.1(b), both organoclastic and methanotrophic sulfate reduction yield equivalent amounts of alkalinity per sulfate reduced. Significant net alkalinity production may result from deep subseafloor sulfate reduction activity (D'Hondt et al., 2002) even given the various wide range of estimates of organic carbon burial within ocean sediments and various estimates of net sulfate reduction (Bowles et al., 2014; D'Hondt et al., 2002). In the absence of reactive iron, a shift in the carbonate equilibrium during the initial stages of sulfate consumption leads to carbonate undersaturation and can facilitate carbonate dissolution until elevated DIC concentrations are high enough to reach carbonate saturation (Meister, 2013; Morse & McKenzie, 1990). Excursions in pore water calcium and magnesium concentrations in the sulfate zone of ocean drilling sites, for example, ODP Leg 201 Site 1226 (D'Hondt et al., 2004), indicate ongoing carbonate formation, which typically encompasses the disseminated precipitation of a variety of authigenic carbonates (Aller, 2014), while distinct diagenetic carbonate nodules at these sites are typically absent.

Methanotrophic sulfate reduction is a major contributor to diagenetic carbonate formation in subseafloor sediments (Meister et al., 2007; Raiswell, 1988; Ussler & Paull, 2008). During the anaerobic oxidation of methane coupled to sulfate reduction, DIC is produced while at the same time the pore water pH converges to 7.9, resulting in carbonate oversaturation and the formation of carbonate phases (Soetaert, Hofmann, Middelburg, Meysman, & Greenwood, 2007). The SMT zone has thus been identified as an important zone of carbonate formation, including the precipitation of dolomite (Figure 2.7.3(d); Malone, Claypool, Martin, & Dickens, 2002; Meister et al., 2007; Meister, Bernasconi, Vasconcelos & Mckenzie, 2008; Moore, Murray, Kurtz, & Schrag, 2004; Rodriguez, Paull, & Borowski, 2000). Here, the anaerobic oxidation of ^{13}C-depleted methane (Alperin, Reeburgh, & Whiticar, 1988; Martens, Albert, & Alperin, 1999; Mccorkle, Emerson, & Quay, 1985) leads to the production of strongly ^{13}C-depleted DIC, which is subsequently recorded in the carbon isotopic composition of the precipitating carbonate phases (Meister et al. 2007; Ritger, Carson, & Suess, 1987; Ussler & Paull, 2008; Wehrmann et al., 2011, Wehrmann et al., 2014). Diagenetic carbonates formed in the SMT zone are often observed as distinct light colored, semilithified bands in unconsolidated, siliciclastic sediments and hold a characteristic light carbon isotope composition (e.g., Meister, Mckenzie, Warthmann, & Vasconcelos, 2006; Meister et al., 2007; Pisciotto & Mahoney, 1981).

Several lines of evidence indicate that diagenetic carbonate phases may also form in and below the active methanogenic zone. At several DSDP, ODP, and IODP sites, diagenetic carbonate phases with a ^{13}C-enriched carbon isotopic signal were found, e.g., at DSDP Leg 64 Hole 479 (maximum δ^{13}C-values > 11‰;

Kelts & McKenzie, 1982), at ODP Leg 175 Site 1081 (+7.7‰; Pufahl & Wefer, 2001), at ODP Leg 201 Site 1227 (+11.5‰; Meister et al., 2007), and at IODP Exp. 323 Site U1343 (+11.37‰; Pierre et al., 2014). The strong ^{13}C-enrichment of the diagenetic carbonate phases is explained by carbonate formation in the methanogenic zone and/or the mixing of ^{13}C-depleted DIC formed during organo-clastic and methanogenic sulfate reduction and ^{13}C-enriched DIC produced during methanogenesis, during the downward movement of the SMT zone in the sediment (Meister et al., 2008; Wehrmann et al., 2014; see Figure 2.7.4). Wehrmann et al. (2011) and Pierre et al., 2014 furthermore provided geochemical and mineralogical evidence for the existence of a deep carbonate formation zone, where dolomite is precipitating at low rates, below the depth of the methanogenic zone at several high-productivity sites drilled during IODP Exp. 323.

An important precondition to the formation of diagenetic carbonate in the methanogenic zone may be the contemporaneous transformation of CO_2, produced during methanogenesis, to bicarbonate (HCO_3^-) via marine silicate weathering (Scholz, Hensen, Schmidt, & Geersen, 2013; Wallmann et al., 2008, 2006). During both acetoclastic and hydrogenotrophic methanogenesis CO_2 is produced whose subsequent (partial) dissociation produces protons and lowers the pH (Soaetert et al., 2007) to conditions unfavorable for carbonate formation. During silicate weathering, however, reactive silicates (e.g., plagioclase feldspars, olivine, and pyroxene) react with CO_2 to form clay minerals and bicarbonate (Wallmann et al., 2008). These reactions can be summarized as follows:

Reactive silicates + $CO_2 \rightarrow$ clay minerals + dissolved cations + dissolved silica + HCO_3^-

This process also releases cations such as magnesium, calcium, and iron, previously bound in the silicate matrix, into the pore water, which can subsequently take part in the precipitation of carbonate. Accordingly, Pierre et al., 2014 describe the occurrence of iron-rich carbonates in Expedition 323 sediments below 130–260 mbsf, which presumably formed as a result of methanogenesis coupled to the microbially assisted low-temperature transformation of Fe-rich clay minerals (smectite and chlorite) to illite.

The diagenesis of biogenic silica in subseafloor sediments, potential linkages to microbially mediated processes, and, more directly, microbe–mineral interactions with silicates and clays in the sedimentary subseafloor biosphere remain poorly investigated. Silicate diagenesis is important in the control of alkalinity and pH in the deep sediments, and may play an important role in the subseafloor phosphorus cycle. The alteration of biogenic silica proceeds via a range of reactions (see Aller, 2014; Loucaides, Van Cappellen, Roubeix, Moriceau, & Ragueneau, 2012; for review) that depend on the availability and interaction of different aqueous and particulate sediment components, for example Al^{3+} and Fe^{3+} (Aplin, 1993; Dixit & Van Cappellen, 2002; Dixit, Van Cappellen, & Van Bennekom, 2001; Mackin & Aller, 1989; Michalopoulos & Aller, 2004; Van Bennekom, Fred Jansen, Van Der Gaast, Van Iperen, & Pieters, 1989). An important process in biogenic opal-rich sediments is the conversion of biogenic silica to aluminosilicates via reverse weathering (Mackenzie & Garrels, 1966; Mackin & Aller, 1984, 1989), which occurs at very early stages of silica diagenesis, i.e., shallow sediment depths (Loucaides et al., 2010; Michalopoulos & Aller, 1995; Michalopoulos, Aller, & Reeder, 2000; Wallmann et al., 2008). Meister et al. (2014) directly link the early diagenetic formation of chert at ODP Site 1226 to the formation of authigenic, poorly crystallized iron oxides and illite. The authors propose that these authigenic iron oxide and iron-bearing silicate phases precipitate at a deep, inverted redox front (400 mbsf) where microbes utilize upward diffusing nitrate to oxidize downward diffusing ferrous iron. The availability of lithogenic Al- and Fe-rich components is also likely a main controlling factor for reverse weathering in open-ocean areas such as the Southern Ocean (Van Cappellen & Qiu, 1997), while Scholz et al. (2013) suggested that the availability of CO_2 deriving from methanogenesis regulates the rate of silicate weathering in underlying methanogenic sediments. Overall, both processes strongly alter the primary biogenic silica component of marine sediments, affect the subseafloor carbonate system through CO_2 production and consumption, respectively, and lead to the uptake or release of specific cations and anions (e.g., K, Mg, and Li).

The reaction of hydrogen sulfide, produced during both organoclastic and methanogenic sulfate reduction, with dissolved iron and reactive iron(oxyhydr) oxide to form iron monosulfide, elemental sulfur, and pyrite also strongly affects the composition of marine sediments, the latter being most important over the extended diagenetic timescales for subseafloor sediments. Several reaction pathways are proposed for the formation of pyrite and other associated iron sulfide phases, e.g., mackinawite (FeS) and greigite (Fe_3S_4) (Goldhaber, 2003; Rickard & Luther, 2007; Rickard, Schoonen, & Luther, 1995; Schoonen, 2004; Table 2.7.1(b)). Pyritization is of fundamental importance for the global sulfur cycle, as the burial of solid-phase sulfur compounds represents a major sink for seawater sulfate, in addition to the formation and burial of organic sulfur and the precipitation of calcium sulfates in evaporites (Raiswell & Canfield, 2012; Vairavamurthy, Orr, & Manowitz, 1995).

The interactions between the iron and sulfur cycles in subseafloor sediments are observed not only in varying concentrations of diagenetic mineral phases such as pyrite, but also in the alteration of magnetic susceptibility records. These records are commonly used in oceanographic studies as a proxy for stratigraphic changes in sediment composition that may be linked to palaeoclimate-controlled depositional processes. In this case, the magnetic signal is created by primary (detrital) iron phases including magnetite and metastable iron sulfides, e.g., greigite and pyrrhotite. The dissolution of primary iron oxide phases during reaction with sulfide and the formation of metastable iron sulfide phases during diagenesis can thus significantly deteriorate the original magnetic susceptibility signal (Figure 2.7.5; e.g., Abrajevitch & Kodama, 2011; Fu, Von Dobeneck, Franke, Heslop, & Kasten, 2008; Karlin & Levi, 1983; März, Hoffmann, Bleil, De Lange, & Kasten, 2008; Reitz, Hensen, Kasten, Funk, & De Lange, 2004; Riedinger et al., 2005).

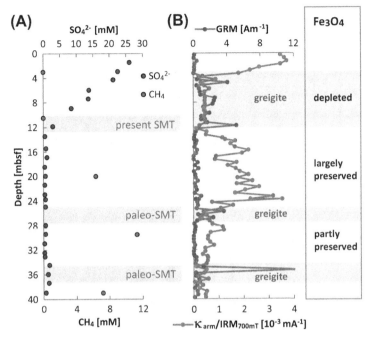

FIGURE 2.7.5 (A) Interstitial water sulfate (SO_4^{2-}) and methane (CH_4) concentrations, (B) distribution of the greigite-sensitive rock magnetic parameter gyroremanent magnetization (GRM) and the $\kappa_{arm}/IRM_{700\,mT}$ parameter, which is sensitive to changes in grain-size variations within magnetite, as well as geochemical interpretation of these records based on Fu et al. (2008).

The dissolution of iron (oxyhydr)oxide by dissimilatory iron reducers or during the reaction with hydrogen sulfide and the mineralization of organic matter release phosphorus into the pore water of marine sediments (Anschutz, Zhong, Sundby, Mucci, & Gobeil, 1998; Filippelli & Delaney, 1996). The phosphorus from these "labile phases" is transformed into an amorphous Ca-phosphate precursor (Gunnars, Blomqvist, & Martinsson, 2004; Van Cappellen & Berner, 1988) and further altered to diagenetically more stable authigenic carbonate fluorapatite over time (Filippelli & Delaney, 1996; Ruttenberg & Berner, 1993). These biogeochemical transformations of phosphorus increase the burial efficiency of labile phosphorus in marine sediments, and distinguish authigenic phosphorus phases in this environment as an important sink in the global phosphorus cycle (Delaney, 1998; Wallmann, 2010). In biosilica-dominated ocean regions, e.g., the Southern Ocean and the North Pacific, biogenic opal-bound phosphorus represents an additional contributor the reactive phosphorus pool (Figure 2.7.6; Latimer, Filippelli, Hendy, & Newkirk, 2006; März, Poulton, Wagner, Schnetger, & Brumsack, 2014). However, the impact of microbial activity on the opal-bound phosphorus pool and its linkage to opal diagenesis in marine sediments remains poorly investigated (Figure 2.7.6; Latimer et al., 2006; März et al., 2014).

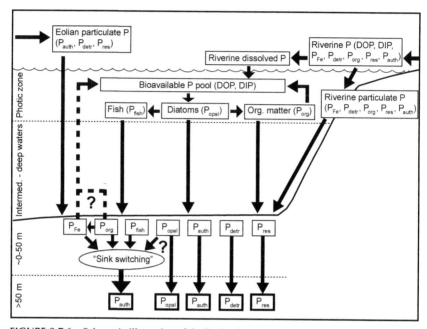

FIGURE 2.7.6 Schematic illustration of the Bering Sea phosphorus cycle including input pathways (eolian, riverine); transformation and recycling processes within surface waters, deep/intermediate waters, surface (<50 m) and deeper (>50 m) marine sediments; and buried phosphorus fractions. *From März et al., 2014.*

At open-ocean sites away from areas of high productivity, the cycling of iron and manganese can play a prevalent role for microbial processes occurring in the subseafloor sediments. Here, extended zones of elevated pore water iron and manganese concentrations have been observed, for example, at ODP Leg 201 Site 1331, and point to the predominance of these TEAPs for carbon mineralization. In particular, if sediments additionally receive increased input of metal oxide phases, such as close to hydrothermal sources in the Eastern Equatorial Pacific (Leg 201 Site 1226; Meister, Bernasconi, Aiello, Vasconcelos, & Mckenzie, 2009) and in the Arctic Ocean (Löwemark, Jakobsson, Mörth, & Backman, 2008; März et al., 2011), the diagenetic Mn (and Fe) cycling facilitates the formation of distinct authigenic mineral phases, including Ca-rhodochrosite, and authigenic Mn-rich layers (Meister et al., 2009; März et al., 2011).

In paleoceanographic studies, sedimentary barite contents, barium to aluminum ratios (Ba/Al), or excess barium concentration over the composition of average crust (Ba_{xs}) are often used as tracers for paleoproductivity, and the strontium isotope composition of deposited barite can give insight into the past seawater strontium composition (e.g., Gingele & Dahmke, 1994; Gingele, Zabel, Kasten, Bonn, & Nürnberg, 1999; Mearon, Paytan, & Bralower, 2003; Paytan, Kastner, & Chavez, 1996; Paytan, Kastner, Martin, Macdougall, & Herbert,

1993s). Microbially mediated diagenetic processes, however, can strongly alter the primary barite and/or barium signal of subseafloor sediments (Brumsack & Gieskes, 1983; McManus et al., 1998; Von Breymann et al., 1992). Specifically, pore water sulfate depletion results in the dissolution of solid-phase barite and the release of barium into the pore water (Figure 2.7.3(c)). Following the subsequent upward diffusion of the released barium, authigenic barite layers form at the interface with downward diffusing sulfate, i.e., immediately above the SMT zone (Dean & Schreiber, 1978; Riedinger, Kasten, Groger, Franke, & Pfeifer, 2006; Torres, Brumsack, Bohrmann, & Emeis, 1996). These barite fronts can persist in marine subseafloor sediments over timescales of several thousands of years, and thus serve as distinct indicators for the past location and shifts of the depth of the SMT zone over time (Arndt, Hetzel, & Brumsack, 2009; Henkel et al., 2012; Riedinger et al., 2006).

2.7.5 TIME AND THE DEEP SUBSEAFLOOR BIOSPHERE

The extended sediment sequences that are retrieved in the framework of IODP not only offer the unique opportunity to study the paleoceanographic and paleoenvironmental history of ocean regions, but these deep-time records also provide the required data to investigate the evolution of biogeochemical processes in seafloor sediments as a function of depth and time. Much of our understanding of deep subseafloor biogeochemistry derives from modeling of concentration–depth profiles, where conditions of steady state, which eliminates any time-dependent variations in input and composition of the sediment can be reasonably assumed (e.g., Berner, 1980; Bowles, Mogollón, Kasten, Zabel, & Hinrichs, 2014; Wang, Spivack, Rutherford, Manor, & D'Hondt, 2008). Nevertheless, non-steady state conditions must be considered at many ODP and IODP sites.

Variations in the amount of the input and the quality of deposited organic matter driving subseafloor biogeochemical processes result from numerous, often co-occurring, factors, including changes in surface water primary productivity, sea ice coverage, the depositional environment, as well as lateral transport processes. Other factors can also influence the magnitude of the input and the quality of organic matter reaching the seafloor, such as water column oxygen levels, the amount and composition of terrigenous ballast material, the community composition of primary producers, and the fraction and degradability of terrestrial organic matter delivered to the ocean (e.g., Arndt et al., 2013; Hedges & Keil, 1995; Hedges et al., 1994; Henson, Sanders, & Madsen, 2012; Hulthe, Hulth, & Hall, 1998; Mollenhauer et al., 2007; Simon, Poulicek, Velimirov, & Mackenzie, 1994; Wilson, Barker, & Ridgwell, 2012; Zonnefeld et al., 2010). Variations of these features are often the result of large-scale changes of the oceanographic regime, e.g., over glacial–interglacial timescales. Important oceanographic changes are modifications in ocean stratification (Gebhardt et al., 2008; Jaccard et al., 2005; Sigman, Jaccard, & Haug, 2004), the location and extent of oxygen minimum zones (e.g., Altabet, Murray, & Prell, 1999), and/or the prevailing current regime (Berger, 1970; Haug & Tiedemann, 1998).

Non-steady state geochemical regimes can furthermore develop as a result of more extreme changes in organic carbon input. This includes periods following the deposition of organic carbon-rich shales and sapropels during oceanic anoxic events, e.g., in the Cretaceous and Neogene (Arthur, Brumsack, Jenkyns, & Schlanger, 1990; Jenkyns, 1980; Wilson & Norris, 2001). Cretaceous black shales, drilled during ODP Leg 207 on Demerara Rise, still imprint the biogeochemical signatures in pore water and the sediment at present; i.e., they are still microbially active "bioreactors," even after almost 100 million years (Arndt, Brumsack, & Wirtz, 2006; Arndt et al., 2009). Similarly, major changes in the environmental setting, such as repeated transitions from freshwater to saltwater conditions like those described for the Baltic Sea and the Black Sea (Björck, 1995; Degens & Ross, 1972; Ryan et al., 1997; Voipio, 1981, p. 418), can drive non-steady state conditions in associated subseafloor sediments.

The occurrence of erosional events is another major cause for non-steady state processes, often linked to changes in the (bottom water) current regime. This process essentially removes the uppermost part of the sediment column, and thus of the pore water profiles. As the geochemical profiles reequilibrate during the downward diffusion of seawater into the sediment column, transient "concave-down" geochemical profiles develop (e.g., Hensen et al., 2003). Erosional events can also lead to the expression of multiple sequences of TEAP in the metabolite profiles, as, for example, observed at IODP Expedition 307 Site U1318 on the Porcupine Seabight continental margin (Figure 2.7.7; Ferdelman, Kano, Williams, Henriet, & the Expedition 307 Scientists, 2006). This submarine weathering of the diagenetically altered deep sediment may have implications for the geochemistry of the ocean and seafloor over geological time frames. On the other hand, the subsequent deposition of eroded sediments downslope can result in the rapid burial of reworked organic matter and fresh iron (oxyhydr)oxides, formed during the oxidation of sedimentary iron monosulfide and pyrite in oxic seawater (Morse, 1991), into deep sediments (Hensen, Zabel, & Schulz, 2000; Hensen et al., 2003; Riedinger et al., 2005; Riedinger, Formolo, Lyons, Henkel, Beck, & Kasten, 2014). Such a depositional environment is, for example, observed along the Argentine Basin, where it led to the burial reactive iron (oxyhydr)oxides below the SMT zone (Hensen et al., 2003; Riedinger et al., 2005). Intriguingly, in this setting the anaerobic oxidation of methane coupled to iron reduction (Beal, House, & Orphan, 2009; Sivan et al., 2011) may play an important role for methane consumption in the deep biosphere (Riedinger et al., 2014).

The long-term evolution of the biogeochemical setting of subseafloor sites is further controlled by plate tectonic movement from the point of original location across the seafloor, for instance, from ocean spreading centers to the continental margins. As a consequence of this drift, the sites cross ocean regions characterized by different extents of surface water productivity, varying inputs

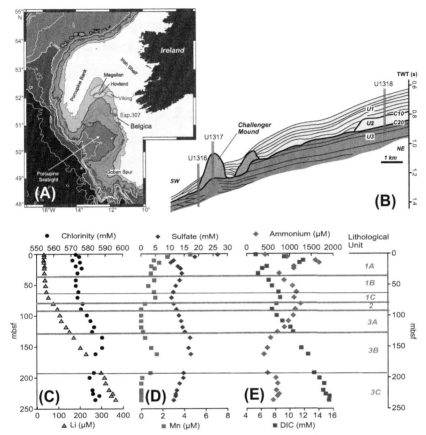

FIGURE 2.7.7 Interstitial water chemistry influenced by periods of rapid sedimentation lying over hiatuses and erosive surfaces at IODP Expedition 307 Site 1318 (Ferdelman et al., 2006). (A) Site 1318 is located upslope from the Porcupine Basin Belgica cold-water coral mound province near SW Ireland. (B) The location and interpreted seismic section of the continental margin sampled during Expedition 307 (after De Mol et al., 2002; Kano, Ferdelman, & Williams, 2010) is shown. The upper seismic unit (U1) corresponds to rapidly buried recent and late Pleistocene sediments from Lithological Unit 1 and Early Pleistocene sediments from Lithological Unit 2. Seismic Units U2 and U3 represent Tertiary (Pliocene and Miocene) deposits ascribed to Lithological Units 3A and 3B & C. The strong seismic reflectors are attributed to erosive unconformities as indicated by the thick blue lines in panels (C)–(E). (C) The distribution of lithium (Li) in the surface 100 m of the sediments indicates rapid burial of seawater Li within the surface sediments that has not come into steady state with respect to a deep source of Li, whereas the chlorinity profile reflects downward propagating changes in ocean salinity. (D, E) Interstitial water profiles of the metabolites sulfate, manganese (Mn), ammonium, and dissolved inorganic carbon (DIC).

of terrigenous and hydrothermal material, and different sedimentation rates. This is often reflected in changes in the amount and quality of deposited organic carbon and in the availability of terminal electron acceptors such as Mn- and Fe-(oxyhydr)oxides. For example, ODP Leg 202 Site 1237 has moved ~3° eastward over the past 6 million years. (Leg 202 Shipboard Scientific Party, 2003a) from a region of higher, to an area of lower temperature, salinity, nutrient concentration, and surface water primary productivity (Wara & Ravelo, 2006). Site 1236, drilled during the same expedition, moved ~20° westward into an area with primary productivity that is ~50% higher than it probably was 25 million years ago at its past location (Leg 202 Shipboard Scientific Party, 2003b).

Ultimately, the result of these important non-steady state diagenetic regimes are changes in metabolic rates and concomitant vertical shifts of biogeochemical zones and diagenetic reaction fronts, including the SMT zone (Arndt et al., 2009; Contreras et al., 2013; Holstein & Wirtz, 2010; Meister et al. 2007; Meister, Liu, Ferdelman, Jørgensen, & Khalili, 2013; Riedinger et al., 2006). Resulting transient geochemical signals remain visible in solid-phase and pore water records over long timescales, e.g., as upward-moving sulfurization fronts (e.g., Eckert et al., 2013; Jørgensen, Böttcher, Lüschen, Neretin, & Volkov, 2004; Neretin et al., 2004) and associated nonlinear geochemical pore water gradients (Contreras et al., 2013; Dickens, 2001; Hensen et al., 2003; Wehrmann, Arndt, März, Ferdelman, & Brunner, 2013). In the sediment these fluctuations are recorded in multiple layers of specific diagenetic phases, e.g., authigenic barite (Arndt et al., 2009; Contreras et al., 2013; Riedinger et al., 2006) and dolomite (Meister et al., 2008; Contreras et al., 2013; Wehrmann et al., 2014), pronounced enrichment of authigenic iron sulfides, and distinct sulfur isotopic excursions of pyrites (Holmkvist, Ferdelman, & Jørgensen, 2011; Kasten, Freudenthal, Gingele, & Schulz, 1998). Furthermore, such changes can be reinforced by variations in sea level and sedimentation rates over time (Kasten et al., 1998; Meister et al., 2008; Riedinger et al., 2005).

Reaction-transport modeling represents an important tool for investigating the biogeochemical history of subseafloor sediments, allowing researchers to infer past and present rates of microbial activity, and helps to reconstruct the evolution of diagenetic processes over geological timescales (Arndt et al., 2006, 2009; Dale, Van Cappellen, Aguilera, & Regnier, 2008; Marquart, Hensen, Piñero, Wallmann, & Haeckel, 2010; Reed, Slomp, & De Lange, 2011). These models have been applied to several sites that have undergone major changes in their diagenetic history due to variations in the supply and quality of organic carbon to the sediment, including the Demerara Rise (ODP Leg 207, Site 1258; Arndt et al., 2009), the Peruvian Shelf (ODP Leg 201, Site 1229; Contreras et al., 2013), the Bering Sea (IODP Exp. 323, Site U1341; Wehrmann et al., 2013), and the Black Sea continental margin (Henkel et al., 2012).

2.7.6 BEYOND INTERSTITIAL WATER AND SOLID PHASE CHEMISTRY?

Our understanding of the biogeochemistry of deeply buried sediments derives primarily from the fitting of reaction-transport models to high-resolution depth distributions of dissolved metabolic reactants and products. Diagenetic modeling provides robust estimates of net reaction rates and fluxes, but may miss processes of biogeochemical importance. For instance, rapid production and consumption of a compound may go unnoticed as the compound's concentration falls below detection limits, or similar rates of consumption and production leave no measurable change in concentration with depth (Fossing, Ferdelman, & Berg, 2000). Experimental rate measurements can lead to additional insight on selected samples and at specific depths. Isotope tracer methods, in particular radiotracer methods, allow the experimental detection of vanishingly small changes in concentration over time. The 10,000-fold increase in the detection limit of sulfate reduction using radiotracers is elaborated upon in the Explanatory Methods of ODP Expedition 201 (D'Hondt, Jørgensen, Miller et al., 2003).

Direct measurement of sulfate turnover using the highly sensitive radiolabel technique affords further insight into sulfur cycling within deep subseafloor sediments. ODP Leg 201 sites along the Peru Continental Margin and Equatorial Pacific have been sampled at high depth resolution from the sediment–water interface down to depths of 300 mbsf for the determination of microbiological sulfate reduction rates (Parkes et al., 2005; see Figure 2.7.8). Based on these studies as well as investigations of other ocean margin sites (Fossing et al., 2000; Holmkvist et al., 2011), direct experimental measurements of sulfate reduction suggest that methane-dependent sulfate reduction appears to consume only a fraction of the total integrated sulfate reduction, and more than 90% of the sulfate turnover (not to be confused with net flux) can be attributed to oxidation of compounds other than methane. Furthermore, sulfate reduction may continue

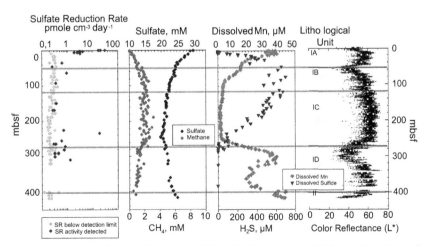

FIGURE 2.7.8 Experimentally determined sulfate reduction rates, interstitial water chemistry, and sediment color reflectance from ODP Leg 201 Site 1226 in the Equatorial Pacific (after Parkes et al., 2005). Sulfate reduction activity was detected using the highly sensitive ^{35}S radiotracer method in a small fraction of the sediments tested, including the darker intervals at depths of 290–320 mbsf that contain Miocene-age diatom and organic-rich sediment. Note the influence of basement seawater flow on the profiles of dissolved sulfate and manganese at depths >380 mbsf. *D'Hondt et al., 2004.*

as an important process even below the sulfate–methane zone (Holmkvist et al., 2011). Alternative sources of sulfate include oxidative or mineral dissolution processes, e.g., dissolution of barite- and gypsum-bearing facies. Other tracer methods used to elucidate key biogeochemical processes include the determination of (radiolabeled) methane, bicarbonate, and acetate turnover (e.g., Parkes et al., 2011; Webster et al., 2009) or hydrogen turnover using tritiated H_2 (Soffientino, Spivack, Smith, Roggenstein, & D'hondt, 2006; Soffientino, Spivack, Smith, & D'hondt, 2009).

Shortcomings of any isotope-experimental approach (radioisotope or stable-isotope probing) are that isotope tracers are limited to a few elements, and that the time and costs of processing such experiments are extensive. It is also important to state that all such isotope rate measurements are experiments whose underlying assumptions and variables must be understood and constrained. An intriguing example is the back flux of tracer from a labeled product pool into the substrate pool under conditions where extreme energy limitation conditions prevail, i.e., conditions typical of the deep subsurface ocean. During the sulfate-dependent oxidation of methane, one of the most endergonic microbial processes known, experimental data show that a transfer of the radiolabeled ^{14}C from $^{14}CO_2$ to methane during the net oxidation of methane to CO_2 occurs (Holler et al., 2011). The total transfer of ^{14}C (or ^{13}C for that matter) becomes decoupled from a straightforward stoichiometric relationship to C reduction or oxidation. Such catabolic back fluxes may occur in the sulfate-reducing system under low-energy conditions as well, with striking consequences for stable sulfur isotope distributions in deeply buried sediments (Sim, Bosak, & Ono, 2011). Tracer experiments nevertheless provide invaluable insight. Such biologically induced back flux or tracer exchange clearly indicates the presence of active, functional enzyme systems, for example, sulfate reduction detected at a depth of approximately 300 mbsf at ODP Site 1226 (Figure 2.7.8). Thus, an important microbial process may be localized that would otherwise remain undiscovered when only evaluating interstitial water chemistry.

These microbial-mediated processes can strongly alter the initial sediment composition of marine sediments during diagenesis within distinct and narrow horizons (as described previously); in turn, sediment lithology influences the composition, population size, and metabolic activity of the local microbial community. For example, Aiello and Bekins (2010) demonstrate that microbial cell counts vary an order of magnitude between organic-rich diatom and nannofossil oozes of a drill site in the eastern equatorial Pacific (ODP Leg 201, Site 1226). The authors suggest that these small-scale variations are ultimately controlled by Milankovitch frequency variations in past oceanographic conditions. Similarly, Picard and Ferdelman (2011) describe changes in microbial heterotrophic activity linked to specific lithologies such as dark clay-rich layers and sandy intervals in oligotrophic sediments of the North Atlantic Gyre. Microbial activity measured with the radiotracer method reveals a tight coupling between sediment lithology and microbial activity in the Equatorial Pacific, along the Peru

Margin (Figure 2.7.8; Parkes et al., 2005), and in the cold-water coral-bearing sediments drilled during IODP Expedition 307 (Webster et al., 2009).

2.7.7 CONNECTING THE PELAGIC OCEAN AND SUBSEAFLOOR SEDIMENTARY OCEAN

Hydrothermal circulation at mid-ocean ridges influences ocean water chemistry through seawater–basement rock chemical reactions. Further off-axis, outcropping of permeable basement rock and seamounts allows conductive flow beneath layers of impermeable sediment accumulating on young ridge flanks (Figure 2.7.1; Fisher et al., 2003; Sclater, 2003). If we consider the sediments as boundaries overlying the hydrologically active crustal environment, then using profiles of bioactive compounds measured within/through the sediment layer is an obvious approach to understanding the role of subseafloor microbial life in the chemistry of the water flowing below the sediment cover. Diagenetic modeling of the overlying sediment has been successfully used to estimate water and heat flow within off-axis, sediment-covered systems such as at the Juan de Fuca Ridge (Elderfield, Wheat, Mottl, Monnin, & Spiro, 1999; Giambalvo, Steefel, Fisher, Rosenberg, & Wheat, 2002; Rudnicki, Elderfield, & Mottl, 2001). Seawater altered by water–rock interactions diffuses into and reacts with the overlying organic-rich sediment. Subsequently, microbial terminal electron acceptor processes such as organoclastic sulfate reduction and metal oxide dissolution; precipitation reactions of carbonates, iron sulfides, and apatite; and silica dissolution alter the chemical composition of the water seeping out at discharge sites. Model estimates and extrapolation to the global scale suggest that these sediment–hydrothermal reactions enhance fluxes of sulfate, ammonium, phosphate, and silica (Giambalvo et al., 2002). When compared to basement hydrothermal and riverine fluxes, these sediment–hydrothermal fluxes, with perhaps the exception of dissolved Si, appear to be insignificant (Giambalvo et al., 2002). However, with the advent of new borehole sampling technologies, and the growing realization that the inflow and outflow paths around basement outcrops are extremely complex (Wheat, Hulme, Fisher, Orcutt, & Becker, 2013), estimates of overall fluxes are under continual revision. Fluxes of trace elements (Wheat, Jannasch, Kastner, Plant, & Decarlo, 2003) and nitrogen (Bouronnais et al., 2012) in and out of hydrothermally influenced off-axis sediments are low, but significant. Wankel et al. (2011) have suggested that subsurface microbial processes may remove more than 50% of the expected flux of H_2 from diffuse venting at the Endeavour hydrothermal field on Juan de Fuca Ridge.

Away from the focused hydrothermal flow fields, subsurface sediment habitats grow and evolve as ocean basins are formed. Off-axis hills and valleys are filled and eventually covered as abyssal deposits accumulate over millions of years of seafloor spreading and subsidence (Figure 2.7.9; Ewing & Ewing, 1967; Tominaga, Lyle, & Mitchell, 2011; Webb & Jordan, 1993). The biogeochemical story of how these sediment basins form and evolve is

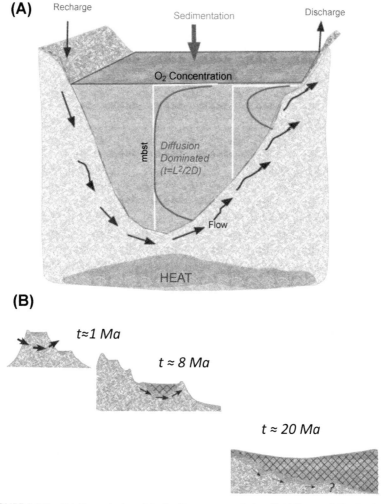

FIGURE 2.7.9   (A) Conceptual model of sediment accumulation, fluid flow, and dissolved oxygen distributions in the middle and nearer to the edges of North Pond, a ponded sediment basin on the flank of the Mid-Atlantic Ridge (figure modified after Ziebis et al., 2012; Sclater, 2003). (B) Conceptual model of the development of an anoxic subsurface biosphere in the open ocean basins, moving from young, about 8 Ma crustal age (e.g., North Pond) to older crust overlain with thick sediment deposits. Arrows indicate decreasing subsediment flow with increasing age, and the hatched area denotes where anaerobic electron acceptor processes dominate (e.g., metal oxide reduction, sulfate reduction, methanogenesis). Note that dissolved oxygen typically penetrates less than a meter below surface in sediments outside of extremely oligotrophic regions.

crucial to understanding the development of a deep sedimentary biosphere. A typical example of younger sediment basin development is the vast area of sea-floor on the western flank of the Mid-Atlantic Ridge, which is characterized by depressions filled with sediment and surrounded by high-relief topography.

The largest depressions are 5–20 km wide and sediment thickness varies but can reach 400 m (Langseth, Becker, Von Herzen, & Schultheiss, 1992). They are believed to overly recharge zones for the venting of fluids that takes place locally through unsedimented young ocean crust.

Deep oxygen profiles obtained from an IODP site survey expedition onboard RV *Maria S. Merian* to North Pond (Figure 2.7.9(a); Ziebis et al., 2012) and the subsequent IODP Expedition 336 provided the first proof of this principal (Orcutt et al., 2013). North Pond is one of the larger ($70 km^2$) and best studied sediment ponds. While surface profiles show typical downward decreasing concentration profiles of dissolved oxygen, indicating oxidation of deposited organic carbon, near the basalt–sediment interface dissolved O_2 concentrations increase (Figure 2.7.9(a)). Degradation of organic carbon within the sediment package creates chemical gradients that drive the diffusive transport of oxidants, from both above and below the sediment. Mean diffusion times scale to the square of length; therefore, in the thicker deposits to be found in the center of the basin, any potential diffusive supply of oxygen, either from above or below, cannot keep up with the low but continual organic carbon remineralization. Ziebis et al. (2012) suggested that the sediment would eventually become anoxic at depth, particularly considering that the sediment package is hundreds of meters deep (Figure 2.7.9). With low to no appreciable rates of aerobic respiration within the sediment, the sediment would remain fully oxic, as is seen in sediments underlying the South Pacific Gyre (D'Hondt et al., 2009). Based on further O_2 measurements during IODP Expedition 336, Orcutt et al. (2013) were able to validate this model and further estimate rates of oxygen consumption in the underlying crust, putatively mediated by microorganisms.

Subsediment, basement flow also functions at much larger scales, for instance, in the equatorial Pacific. The distribution of electron acceptor activities near the basement sediment–water interface is inverted (D'Hondt et al., 2004). Nitrate penetrates 20 m upward from the basalt–sediment interface into the sediment. Sulfate also shows a distinct upward diffusion from seawater-like concentrations from within the basaltic aquifer, and indications are that oxygen penetrates upward from the sediment–water interface. Bekins, Spivack, Davis, and Mayer (2007) provide an elegant mechanism for the convective transport of fluids over these huge distances. They suggest that seawater enters off-axis recharge zones, warms, and allows calcium carbonates to precipitate. As this seawater percolates upward through buried seamounts, the cooling, undersaturated seawater dissolves the overlying biogenic carbonates, thus keeping the discharge zones free. Such a deep thermo-chemo-circulation through the crust may influence sedimentary biogeochemistry over extensive regions of the sedimentary subseafloor ocean. The extent of this flow and sediment alteration is not clear. Initial measurements from IODP Expedition 329 suggest that older basalt crust overlain with deep-sea clays does not show any indication of ongoing basement circulation (Expedition 329 Scientists, 2011). Only near the mid-ocean ridge sediments are there indications for diffusive fluxes of oxygen and nitrate into the sediment.

2.7.8 TOWARD A GLOBAL OCEAN VIEW

Mapping the extent of upward oxidation of reduced sediments in the subseafloor sedimentary ocean and the biogeochemical evolution of sediments will continue to be a major research focus for the coming decade.

The overall effect of this diffuse fluid flow and reaction with overlying sediments on ocean chemistry has not been accurately constrained. More progress has been made with estimating fluxes associated with fluid flow that are advected along fault lines and other permeable horizons, for example, from mud volcanoes at the continental margin of Costa Rica (Hensen & Wallmann, 2005). Boetius and Wenzhöfer (2013) estimate that approximately 0.02 Gt a^{-1} of methane may reach the pelagic ocean from cold venting of methane-containing fluids along continental margins.

Marine sediments are the ultimate and largest reservoir of carbon, bound in both organic carbon and carbonates (DeMarais, 1997). Processes occurring within the deeply buried sediments over Earth's history have controlled the redox poise of the Earth's ocean and atmosphere over most of geological time. In a seminal paper, Hayes and Waldbauer (2006) have examined how the carbon cycle regulates the overall redox state of the planet. In essence, CO_2 that is continuously released from the Earth's mantle is reduced principally through photoautotrophy, thus releasing oxidizing power in the forms of O_2, Fe III (ferric iron), and sulfate. The amount of organic carbon stored in the crust should balance the oxidizing power represented by crustal inventories of Fe^{3+}, SO_4^{2-}, and O_2. A flux of organic carbon from the crust to the mantle, for instance, through burial and subduction along with H_2 loss from the atmosphere, contributes to the accumulation of oxidizing power.

The efficiency of nutrient retention with organic matter buried in marine sediments regulates the overall effectiveness of photosynthesis in the ocean. Lomstein, Langerhuus, D'Hondt, Jørgensen, and Spivack (2012) have shown that amino acids are recycled and retained in so-called necromass in Peru Margin sediments. In the oxic sediments of the South Pacific Gyre, the dominance of aerobic degradation throughout the sediment leads to the situation where there is a flux of nitrate out of the sediments and into the deep waters (D'Hondt et al., 2009). Ocean phosphorus concentrations are strongly regulated by the burial of phosphorus with organic carbon. The efficiency by which phosphorus is removed and leached from organic carbon deposits and made available for photosynthesis will impact the rate of organic carbon formation and burial. Current ongoing studies of phosphorus distributions in sediments obtained from IODP expeditions, in particular in the Pacific Ocean (IODP Expeditions 320, 321, and 329), will better constrain the global phosphorus budget.

The main oxidants released into today's ocean in order of increasing abundance are oxygen, sulfate, and ferric iron. The fluxes of ferric iron and sulfate within deeply buried sediments are key parameters ultimately affecting the overall degradation of organic carbon even at depths below where sulfate reduction and iron reduction expectedly cease (Holmkvist et al., 2011; Riedinger et al., 2005, 2014). In vast swaths of the open ocean, the sediments accumulate slowly and are, more importantly, fully oxic. Data from IODP Expedition 329 (Figure 2.7.2(b); D'Hondt, Inagaki, Alvarez Zarikian et al., 2011) suggest that, although rates are slow, organic carbon decomposition continues over long geological periods of time. Furthermore, the sediment that ultimately becomes subducted may represent a significant sink of oxidants from the global ocean, particularly as Fe(III).

Major insights into biogeochemical processes and implications for sediment diagenetic systems have been gained over the last decade of scientific ocean drilling in the subseafloor sedimentary ocean. The outlines of the consequences for the Earth's climate, ocean chemistry, and biology are broadly apparent. Greater understanding of the subseafloor "plumbing" and transfer of energy and reactants between the deep sediments and the overlying pelagic ocean will certainly progress as data from the most recent scientific drilling investigations of the deep subsurface ocean habitat emerge.

ACKNOWLEDGMENTS

We are so very grateful for opportunities over the past decade to work with the scientists and staff associated with the ODP and IODP programs, both at sea and onshore. We also thank F. Inagaki for his encouragement and patience, as well as the very constructive comments of an anonymous reviewer. We thank C. März for providing Figure 2.7.6. Writing of this manuscript has been supported by the Deutsche Forschungsgemeinschaft (DFG) Schwerpunktprogramm 527 IODP, and the Max Planck Society.

REFERENCES

Abrajevitch, A., & Kodama, K. (2011). Diagenetic sensitivity of paleoenvironmental proxies: a rock magnetic study of Australian continental margin sediments. *Geochemistry, Geophysics, Geosystems, 12*, Q05Z24.

Adkins, J. F., Mcintyre, K., & Schrag, D. P. (2002). The salinity, temperature, and $\delta^{18}O$ of the glacial deep ocean. *Science, 289*, 1769–1773.

Adkins, J. F., & Schrag, D. P. (2001). Pore fluid constraints on deep ocean temperature and salinity during the Last Glacial Maximum. *Geophysical Research Letters, 28*, 771–774.

Aiello, I. W., & Bekins, B. A. (2010). Milankovitch-scale correlations between deeply buried microbial populations and biogenic ooze lithology. *Geology, 38*, 79–82.

Aller, R. C. (2014). Sedimentary diagenesis, depositional environments, and benthic fluxes. In H. D. Holland, & K. K. Turekian (Eds.), *Treatise on geochemistry* (2nd ed.). Oxford: Elsevier.

Alperin, M. J., Reeburgh, W. S., & Whiticar, M. J. (1988). Carbon and hydrogen isotope fractionation resulting from anaerobic methane oxidation. *Global Biogeochemical Cycles, 2*, 278–288.

Altabet, M. A., Murray, D. W., & Prell, W. L. (1999). Climatically linked oscillations in Arabian Sea denitrification over the past 1 m.y.: implications for the marine N cycle. *Paleoceanography, 14*, 732–743.

Anschutz, P., Zhong, S., Sundby, B., Mucci, A., & Gobeil, C. (1998). Burial efficiency of phosphorus and the geochemistry of iron in continental margin sediments. *Limnology and Oceanography*, 53–64.

Aplin, A. C. (1993). The composition of authigenic clay minerals in recent sediments: links to the supply of unstable reactants. In D. A. C. Manning, P. L. Hall, & C. R. Hughes (Eds.), *Geochemistry of clay-pore fluid interactions* (pp. 81–106). London: Chapman and Hall.

Arndt, S., Brumsack, H. J., & Wirtz, K. W. (2006). Cretaceous black shales as active bioreactors: a biogeochemical model for the deep biosphere encountered during ODP Leg 207 (Demerara Rise). *Geochimica et Cosmochimica Acta, 70,* 408–425.

Arndt, S., Hetzel, A., & Brumsack, H.-J. (2009). Evolution of organic matter degradation in Cretaceous black shales inferred from authigenic barite: a reaction-transport model. *Geochimica et Cosmochimica Acta, 73,* 2000–2022.

Arndt, S., Jørgensen, B. B., Larowe, D. E., Middelburg, J. J., Pancost, R. D., & Regnier, P. (2013). Quantifying the degradation of organic matter in marine sediments: a review and synthesis. *Earth-Science Reviews, 123,* 53–86.

Arthur, M., Brumsack, H.-J., Jenkyns, H., & Schlanger, S. (1990). Stratigraphy, geochemistry, and palaeoceanography of organic carbon-rich Cretaceous sequences. In: *Cretaceous resources, events and rhythms* (pp. 75–119). Dordrecht: Kluwer.

Baker, P. A., & Kastner, M. (1981). Constraints on the formation of sedimentary dolomite. *Science, 213,* 214–216.

Beal, E. J., House, C. H., & Orphan, V. J. (2009). Manganese- and iron-dependent marine methane oxidation. *Science, 325,* 184–187.

Bekins, B. A., Spivack, A. J., Davis, E. E., & Mayer, L. A. (2007). Dissolution of biogenic ooze over basement edifices in the equatorial Pacific with implications for hydrothermal ventilation of the oceanic crust. *Geology, 35,* 679–682.

Berger, W. H. (1970). Biogenous deep-sea sediments: fractionation by deep-sea circulation. *Geological Society of America Bulletin, 81,* 1385–1402.

Berner, R. A. (1980). *Early diagenesis.* Princeton: Princeton University Press.

Berner, R. A. (1970). Sedimentary pyrite formation. *American Journal of Science, 268,* 1–23.

Berner, R. A. (1984). Sedimentary pyrite formation - an update. *Geochimica et Cosmochimica Acta, 48,* 605–615.

Björck, S. (1995). A review of the history of the Baltic Sea, 13.0-8.0 ka BP. *Quaternary International, 27,* 19–40.

Boetius, A., & Wenzhöfer, F. (2013). Seafloor oxygen consumption fuelled by methane from cold seeps. *Nature Geosciences, 6,* 725–734.

Bourbannais, A., Juniper, S. K., Butterfield, D. A., Devol, A. H., Kuypers, M. M. M., Lavik, G., et al. (2012). Activity and abundance of denitrifying bacteria in the subsurface biosphere of diffuse hydrothermal vents of the Juan de Fuca Ridge. *Biogeosciences, 9,* 4661–4678.

Bowles, M. W., Mogollón, J. M., Kasten, S., Zabel, M., & Hinrichs, K.-U. (2014). Global rates of marine sulfate reduction and implications for sub–sea-floor metabolic activities. *Science, 344,* 889–891.

Brumsack, H. J., & Gieskes, J. M. (1983). Interstitial water trace-metal chemistry of laminated sediments from the Gulf of California, Mexico. *Marine Chemistry, 14,* 89–106.

Burns, S. J., Baker, P. A., & Showers, W. J. (1988). The factors controlling the formation and chemistry of dolomite in organic-rich sediments: Miocene Drakes Bay formation, California. In V. Shukla, & P. A. Baker (Eds.), *Sedimentology and geochemistry of dolostone* (vol. 43) (pp. 3–10). Tulsa: SEPM Special Publication.

Canfield, D. E., & Kump, L. R. (2013). Carbon cycle makeover. *Science, 339,* 533–534.

Contreras, S., Meister, P., Liu, B., Prieto-Mollar, X., Hinrichs, K.-U., Khalili, A., et al. (2013). Cyclic 100-ka (glacial-interglacial) migration of subseafloor redox zonation on the Peruvian shelf. *Proceedings of the National Academy of Sciences, 110,* 18098–18103.

D'Hondt, S., Inagaki, F., Alvarez Zarikian, C. A., & the IODP Expedition 329 Science Party. (2011). In *Proc. IODP* (vol. 29). Tokyo: Integrated Ocean Drilling Program Management International, Inc. http://dx.doi.org/10.2204/iodp.proc.329.2011.

D'Hondt, S., Inagaki, F., Alvarez Zarikian, C. A., & the IODP Expedition 329 Science Party. (2013). IODP Expedition 329: life and habitability beneath the seafloor of the South Pacific Gyre. *Scientific Drilling, 15,* 4–10. http://dx.doi.org/10.2204/iodp.sd.15.01.2013.

D'Hondt, S., Jørgensen, B. B., Miller, D. J., et al. (2003). In *Proc. ODP, Init. Repts., 201.* College Station, TX: Ocean Drilling Program. http://dx.doi.org/10.2973/odp.proc.ir.201.2003.

D'Hondt, S., Jørgensen, B. B., Miller, D. J., Batzke, A., Blake, R., Cragg, B. A., et al. (2004). Distributions of microbial activities in deep subseafloor sediments. *Science, 306,* 2216–2221.

D'Hondt, S., Rutherford, S., & Spivack, A. J. (2002). Metabolic activity of subsurface life in deep-sea sediments. *Science, 295,* 2067–2070.

D'Hondt, S., Spivack, A. J., Pockalny, R., Ferdelman, T. G., Fischer, J. P., Kallmeyer, J., et al. (2009). Subseafloor sedimentary life in the South Pacific Gyre. *Proceedings of the National Academy of Sciences, 106,* 11651–11656.

Dale, A. W., Van Cappellen, P., Aguilera, D. R., & Regnier, P. (2008). Methane efflux from marine sediments in passive and active margins: estimations from bioenergetic reaction-transport simulations. *Earth and Planetary Science Letters, 265,* 329–344.

Dean, W. E., & Schreiber, B. E. (1978). Authigenic barite, Leg 41 deep sea drilling project. In Y. Lancelot, E. Seibold, et al. (Eds.), *Proc. ODP, Init. Repts.* (vol. 41) (pp. 915–931). Washington, D.C: U.S. Gov. Off.

Degens, E. T., & Ross, D. A. (1972). Chronology of the Black Sea over the last 25,000 years. *Chemical Geology, 10,* 1–16.

Delaney, M. L. (1998). Phosphorus accumulation in marine sediments and the oceanic phosphorus cycle. *Global Biogeochemical Cycles, 12,* 563–572.

Des Marais, D. J. (1997). Isotopic evolution of the biogeochemical carbon cycle during the Proterozoic Eon. *Organic Geochemistry, 27,* 185–193.

De Mol, B., Van Rensbergen, P., Pillen, S., Van Herreweghe, K., Van Rooij, D., Mcdonnell, A., et al. (2002). Large deep-water coral banks in the Porcupine Basin, southwest of Ireland. *Marine Geology, 188,* 193–231.

Dickens, G. R. (2001). Sulfate profiles and barium fronts in sediment on the Blake Ridge: present and past methane fluxes through a large gas hydrate reservoir. *Geochimica et Cosmochimica Acta, 65,* 529–543.

Dixit, S., & Van Cappellen, P. (2002). Surface chemistry and reactivity of biogenic silica. *Geochimica et Cosmochimica Acta, 66,* 2559–2568.

Dixit, S., Van Cappellen, P., & Van Bennekom, A. J. (2001). Processes controlling solubility of biogenic silica and pore water build-up of silicic acid in marine sediments. *Marine Chemistry, 73,* 333–352.

Eckert, S., Brumsack, H.-J., Severmann, S., Schnetger, B., März, C., & Fröllje, H. (2013). Establishment of euxinic conditions in the Holocene Black Sea. *Geology, 41,* 431–434.

Elderfield, H., Wheat, C. G., Mottl, M. J., Monnin, C., & Spiro, B. (1999). Fluid and geochemial transport through oceanic crust: a transect across the eastern flank of the Juan de Fuca Ridge. *Earth and Planetary Science Letters, 172,* 151–165.

Expedition 323 Scientists. (2011). Site U1345. In K. Takahashi, A. C. Ravelo, & C. A. Alvarez Zarikian (Eds.), *Proc. IODP* (vol. 323). Tokyo: Integrated Ocean Drilling Program Management International, Inc. http://dx.doi.org/10.2204/iodp.proc.323.109.2011.

Expedition 329 Scientists. (2011). South Pacific Gyre subseafloor life. *IODP Preliminary Report, 329.* http://dx.doi.org/10.2204/iodp.pr.329.2011.

Ewing, J., & Ewing, M. (1967). Sediment distribution on the mid-ocean ridges with respect to spreading of the sea floor. *Science, 156,* 1590–1592.

Falkowski, P., Scholes, R. J., Boyle, E., Canadell, J., Canfield, D., Elser, J., et al. (2000). The global carbon cycle: a test of our knowledge of earth as a system. *Science, 290,* 291–296.

Ferdelman, T. G., Kano, A., Williams, T., Henriet, J.-P., & the Expedition 307 Scientists. (2006). In: *Proc. IODP* (vol. 307). Washington, DC: Integrated Ocean Drilling Program Management International, Inc. http://dx.doi.org/10.2204/iodp.proc.307.2006.

Filippelli, G. M., & Delaney, M. L. (1996). Phosphorus geochemistry of equatorial Pacific sediments. *Geochimica et Cosmochimica Acta, 60*, 1479–1495.

Fischer, J. P., Ferdelman, T. G., D'Hondt, S., Røy, H., & Wenzhöfer, F. (2009). Oxygen penetration deep into the sediment of the South Pacific gyre. *Biogeosciences, 6*, 1467–1478.

Fisher, A. T., Davis, E. E., Hutnak, M., Spiess, V., Zühlsdorff, L., Cherkaoui, A., et al. (2003). Hydrothermal recharge and discharge across 50 km guided by seamounts on a young ridge flank. *Nature, 421*, 618–621.

Fossing, H., Ferdelman, T. G., & Berg, P. (2000). Sulfate reduction and methane oxidation in continental margin sediments influenced by irrigation (South-East Atlantic off Namibia). *Geochimica et Cosmochimica Acta, 64*, 897–910.

Froelich, P. N., Klinkhammer, G. P., Bender, M. L., Luedtke, N. A., Heath, G. R., Cullen, D., et al. (1979). Early oxidation of organic matter in pelagic sediments of the eastern equatorial Atlantic: suboxic diagenesis. *Geochimica et Cosmochimica Acta, 43*, 1075–1090.

Fu, Y., Von Dobeneck, T., Franke, C., Heslop, D., & Kasten, S. (2008). Rock magnetic identification and geochemical process models of greigite formation in quaternary marine sediments from the Gulf of Mexico (IODP Hole U1319A). *Earth and Planetary Science Letters, 275*, 233–245.

Gebhardt, H., Sarnthein, M., Grootes, P. M., Kiefer, T., Kuehn, H., Schmieder, F., et al. (2008). Paleonutrient and productivity records from the subarctic North Pacific for Pleistocene glacial terminations I to V. *Paleoceanography, 23*.

Giambalvo, E. R., Steefel, C. I., Fisher, A. T., Rosenberg, N. D., & Wheat, C. G. (2002). Effect of fluid-sediment reaction on hydrothermal fluxes of major elements, eastern flank of the Juan de Fuca Ridge. *Geochimica et Cosmochimica Acta, 66*, 1739–1752.

Gingele, F., & Dahmke, A. (1994). Discrete barite particles and barium as tracers of paleoproductivity in south Atlantic sediments. *Paleoceanography, 9*, 151–168.

Gingele, F. X., Zabel, M., Kasten, S., Bonn, W. J., & Nürnberg, C. C. (1999). Biogenic barium as a proxy for paleoproductivity: methods and limitations of application. In G. Fischer, & G. Wefer (Eds.), *Use of proxies in paleoceanography*. Berlin Heidelberg: Springer.

Goldhaber, M. B. (2003). Sulfur-rich sediments. In H. D. Holland, & K. K. Turekian (Eds.), *Treatise on geochemistry*. Oxford: Pergamon.

Gunnars, A., Blomqvist, S., & Martinsson, C. (2004). Inorganic formation of apatite in brackish seawater from the Baltic Sea: an experimental approach. *Marine Chemistry, 91*, 15–26.

Haug, G. H., & Tiedemann, R. (1998). Effect of the formation of the Isthmus of Panama on Atlantic Ocean thermohaline circulation. *Nature, 393*, 673–676.

Hayes, J. M., & Waldbauer, J. R. (2006). The carbon cycle and associated redox processes through time. *Philosophical Transactions of the Royal Society B-Biological Sciences, 361*, 931–950.

Hedges, J. I., Cowie, G. L., Richey, J. E., Quay, P. D., Benner, R., Strom, M., et al. (1994). Origins and processing of organic matter in the Amazon River as indicated by carbohydrates and amino acids. *Limnology and Oceanography, 39*, 743–761.

Hedges, J. I., & Keil, R. G. (1995). Sedimentary organic-matter preservation - an assessment and speculative synthesis. *Marine Chemistry, 49*, 81–115.

Henkel, S., Mogollón, J. M., Nöthen, K., Franke, C., Bogus, K., Robin, E., et al. (2012). Diagenetic barium cycling in Black Sea sediments – a case study for anoxic marine environments. *Geochimica et Cosmochimica Acta, 88*, 88–105.

Henson, S. A., Sanders, R., & Madsen, E. (2012). Global patterns in efficiency of particulate organic carbon export and transfer to the deep ocean. *Global Biogeochemical Cycles, 26*, GB1028.

Hensen, C., & Wallmann, K. (2005). Methane formation at Costa Rica continental margin-constraints for gas hydrate inventories and cross-décollement fluid flow. *Earth and Planetary Science Letters, 236*, 41–60.

Hensen, C., Zabel, M., Pfeifer, K., Schwenk, T., Kasten, S., Riedinger, N., et al. (2003). Control of sulfate pore-water profiles by sedimentary events and the significance of anaerobic oxidation of methane for the burial of sulfur in marine sediments. *Geochimica et Cosmochimica Acta, 67*, 2631–2647.

Hensen, C., Zabel, M., & Schulz, H. D. (2000). A comparison of benthic nutrient fluxes from deep-sea sediments off Namibia and Argentina. *Deep Sea Research Part II: Topical Studies in Oceanography, 47*, 2029–2050.

Higgins, J. A., Fischer, W. W., & Schrag, D. P. (2009). Oxygenation of the ocean and sediments: consequences for the seafloor carbonate factory. *Earth and Planetary Science Letters, 284*, 25–33.

Holler, T., Wegener, G., Niemann, H., Deusner, C., Ferdelman, T. G., Boetius, A., et al. (2011). Carbon and sulfur back flux during anaerobic microbial oxidation of methane and coupled sulfate reduction. *Proceedings of the National Academy of Sciences, 108*, E1484–E1490.

Holmkvist, L., Ferdelman, T. G., & Jørgensen, B. B. (2011). A cryptic sulfur cycle driven by iron in the methane zone of marine sediment (Aarhus Bay, Denmark). *Geochimica et Cosmochimica Acta, 75*, 3581–3599.

Holstein, J. M., & Wirtz, K. W. (2010). Organic matter accumulation and degradation in subsurface coastal sediments: a model-based comparison of rapid sedimentation and aquifer transport. *Biogeosciences, 7*, 3741–3753.

Hulthe, G., Hulth, S., & Hall, P. O. J. (1998). Effect of oxygen on degradation rate of refractory and labile organic matter in continental margin sediments. *Geochimica et Cosmochimica Acta, 62*, 1319–1328.

Inagaki, F., Hinrichs, K.-U., Kubo, Y., & the Expedition 337 Scientists. (2013). In *Proc. IODP* (vol. 337). Tokyo: Integrated Ocean Drilling Program Management International, Inc. http://dx.doi.org/10.2204/iodp.proc.337.2013.

Insua, T. L., Spivack, A. J., Graham, D., D'hondt, S., & Moran, K. (2014). Reconstruction of Pacific Ocean bottom water salinity during the Last Glacial Maximum. *Geophysical Research Letters, 41*, 2014GL059575.

Jaccard, S. L., Haug, G. H., Sigman, D. M., Pedersen, T. F., Thierstein, H. R., & Röhl, U. (2005). Glacial/interglacial changes in subarctic North Pacific stratification. *Science, 308*, 1003–1006.

Jenkyns, H. C. (1980). Cretaceous anoxic events: from continents to oceans. *Journal of the Geological Society, 137*, 171–188.

Jørgensen, B. B. (2006). Bacteria and marine biogeochemistry. Chapter 5. In H. D. Schulz, & M. Zabel (Eds.), *Marine geochemistry* (2nd ed.) (pp. 169–206). Berlin Heidelberg New York: Springer.

Jørgensen, B. B., Böttcher, M. E., Lüschen, H., Neretin, L. N., & Volkov, I. (2004). Anaerobic methane oxidation and a deep H$_2$S sink generate isotopically heavy sulfides in Black Sea sediments. *Geochimica et Cosmochimica Acta, 68*, 2095–2118.

Kano, A., Ferdelman, T. G., & Williams, T. (2010). The Pleistocene cooling built Challenger Mound, a deep-water coral mound in the NE Atlantic: synthesis from IODP Expedition 307. *The Sedimentary Record, 8*, 4–9.

Karlin, R., & Levi, S. (1983). Diagenesis of magnetic minerals in recent haemipelagic sediments. *Nature, 303*, 327–330.

Kasten, S., Freudenthal, T., Gingele, F. X., & Schulz, H. D. (1998). Simultaneous formation of iron-rich layers at different redox boundaries in sediments of the Amazon deep-sea fan. *Geochimica et Cosmochimica Acta, 62*, 2253–2264.

Kastner, M., Elderfield, H., Martin, J. B., Suess, E., Kvenvolden, K. A., & Garrison, R. E. (1990). Diagenesis and interstitial-water chemistry at the Peruvian continental margin—major constituents and strontium isotopes. In E. Suess, R. Von Huene, et al. (Eds.), *Proc. ODP, Sci. Results* (vol. 112) (pp. 413–440). College Station, TX: Ocean Drilling Program.

Kelts, K., & Mckenzie, J. A. (1982). Diagenetic dolomite formation in quaternary anoxic diatomaceous muds of deep-sea drilling project Leg-64, Gulf of California. *Initial Reports of the Deep Sea Drilling Project, 64,* 553–569.

Kennett, J. P. (1982). *Marine geology.* Englewood Cliffs, New Jersey: Prentice Hall, pp.813.

Langseth, M. G., Becker, K., Von Herzen, R. P., & Schultheiss, P. (1992). Heat and fluid flux through sediment on the western flank of the Mid-Atlantic Ridge: a hydrogeological study of North Pond. *Geophysical Research Letters, 19,* 517–520.

Latimer, J. C., Filippelli, G. M., Hendy, I., & Newkirk, D. R. (2006). Opal-associated particulate phosphorus: implications for the marine P cycle. *Geochimica et Cosmochimica Acta, 70,* 3843–3854.

Leg 202 Shipboard Scientific Party. (2003a). Site 1237. In A. C. Mix, R. Tiedemann, P. Blum, et al. (Eds.), *Proc. ODP, Init. Repts.* (vol. 202) (pp. 1–107). College Station, TX: Ocean Drilling Program. http://dx.doi.org/10.2973/odp.proc.ir.202.108.2003.

Leg 202 Shipboard Scientific Party. (2003b). Site 1236. In A. C. Mix, R. Tiedemann, P. Blum, et al. (Eds.), *Proc. ODP, Init. Repts.* (vol. 202) (pp. 1–74). College Station, TX: Ocean Drilling Program. http://dx.doi.org/10.2973/odp.proc.ir.202.107.2003.

Libes, S. M. (2009). *Introduction to marine biogeochemistry* (2nd ed.). California, USA: Academic Press.

Lomstein, B. A., Langerhuus, A. T., D'Hondt, S., Jørgensen, B. B., & Spivack, A. J. (2012). Endospore abundance, microbial growth and necromass turnover in deep sub-seafloor sediment. *Nature, 484,* 101–104.

Loucaides, S., Michalopoulos, P., Presti, M., Koning, E., Behrends, T., & Van Cappellen, P. (2010). Seawater-mediated interactions between diatomaceous silica and terrigenous sediments: results from long-term incubation experiments. *Chemical Geology, 270,* 68–79.

Loucaides, S., Van Cappellen, P., Roubeix, V., Moriceau, B., & Ragueneau, O. (2012). Controls on the recycling and preservation of biogenic silica from Biomineralization to burial. *Silicon, 4,* 7–22.

Löwemark, L., Jakobsson, M., Mörth, M., & Backman, J. (2008). Arctic Ocean manganese contents and sediment colour cycles. *Polar Research, 27,* 105–113.

Lyons, T. W., & Severmann, S. (2006). A critical look at iron paleoredox proxies: new insights from modern euxinic marine basins. *Geochimica et Cosmochimica Acta, 70,* 5698–5722.

Mackenzie, F. T., & Garrels, R. M. (1966). Chemical mass balance between rivers and oceans. *American Journal of Science, 264,* 507–525.

Mackin, J. E., & Aller, R. C. (1984). Ammonium adsorption in marine-sediments. *Limnology and Oceanography, 29,* 250–257.

Mackin, J. E., & Aller, R. C. (1989). The nearshore marine and estuarine chemistry of dissolved Aluminium and rapid authigenic mineral precipitation. *Reviews in Aquatic Sciences, 1,* 537–554.

Malone, M. J., Claypool, G., Martin, J. B., & Dickens, G. R. (2002). Variable methane fluxes in shallow marine systems over geologic time: the composition and origin of pore waters and authigenic carbonates on the New Jersey shelf. *Marine Geology, 189,* 175–196.

Marquardt, M., Hensen, C., Piñero, E., Wallmann, K., & Haeckel, M. (2010). A transfer function for the prediction of gas hydrate inventories in marine sediments. *Biogeosciences, 7,* 2925–2941.

Martens, C. S., Albert, D. B., & Alperin, M. J. (1999). Stable isotope tracing of anaerobic methane oxidation in the gassy sediments of Eckernförde Bay, German Baltic Sea. *American Journal of Science, 299,* 589–610.

März, C., Hoffmann, J., Bleil, U., De Lange, G. J., & Kasten, S. (2008). Diagenetic changes of magnetic and geochemical signals by anaerobic methane oxidation in sediments of the Zambezi deep-sea fan (SW Indian Ocean). *Marine Geology, 255,* 118–130.

März, C., Poulton, S. W., Wagner, T., Schnetger, B., & Brumsack, H. J. (2014). Phosphorus burial and diagenesis in the central Bering Sea (Bowers Ridge, IODP Site U1341): perspectives on the marine P cycle. *Chemical Geology, 363,* 270–282.

März, C., Stratmann, A., Matthiessen, J., Meinhardt, A. K., Eckert, S., Schnetger, B., et al. (2011). Manganese-rich brown layers in Arctic Ocean sediments: composition, formation mechanisms, and diagenetic overprint. *Geochimica et Cosmochimica Acta, 75,* 7668–7687.

Mavromatis, V., Meister, P., & Oelkers, E. H. (2014). Using stable Mg isotopes to distinguish dolomite formation mechanisms: a case study from the Peru Margin. *Chemical Geology, 385,* 84–91.

Mazzullo, S. J. (2000). Organogenic dolomitization in peritidal to deep-sea sediments. *Journal of Sedimentary Research, 70,* 10–23.

Mccorkle, D. C., Emerson, S. R., & Quay, P. D. (1985). Stable carbon isotopes in marine porewaters. *Earth and Planetary Science Letters, 74,* 13–26.

Mcduff, R. W. (1984). The chemistry of interstitial waters, Deep Sea Drilling Project Leg 86. In G. R. Heath, L. H. Burckle, et al. (Eds.), *Init. Repts DSDP* (vol. 86) (pp. 675–687). Washington: US Govt. Printing Office.

McManus, J., Berelson, W. M., Klinkhammer, G. P., Johnson, K. S., Coale, K. H., Anderson, R. F., et al. (1998). Geochemistry of barium in marine sediments: implications for its use as a paleoproxy. *Geochimica et Cosmochimica Acta, 62,* 3453–3473.

Mearon, S., Paytan, A., & Bralower, T. J. (2003). Cretaceous strontium isotope stratigraphy using marine barite. *Geology, 31,* 15–18.

Meister, P. (2013). Two opposing effects of sulfate reduction on carbonate precipitation in normal marine, hypersaline, and alkaline environments. *Geology, 41,* 499–502.

Meister, P., Bernasconi, S. M., Aiello, I. W., Vasconcelos, C., & Mckenzie, J. A. (2009). Depth and controls of Ca-rhodochrosite precipitation in bioturbated sediments of the eastern equatorial Pacific, ODP Leg 201, Site 1226 and DSDP Leg 68, Site 503. *Sedimentology, 56,* 1552–1568.

Meister, P., Bernasconi, S. M., Vasconcelos, C., & Mckenzie, J. A. (2008). Sea level changes control diagenetic dolomite formation in hemipelagic sediments of the Peru margin. *Marine Geology, 252,* 166–173.

Meister, P., Chapligin, B., Picard, A., Meyer, H., Fischer, C., Rettenwander, D., et al. (2014). Early diagenetic quartz formation at a deep iron oxidation front in the eastern equatorial Pacific – a modern analogue for banded iron/chert formations? *Geochimica et Cosmochimica Acta, 137,* 188–207.

Meister, P., Liu, B., Ferdelman, T. G., Jørgensen, B. B., & Khalili, A. (2013). Control of sulphate and methane distributions in marine sediments by organic matter reactivity. *Geochimica et Cosmochimica Acta, 104,* 183–193.

Meister, P., Mckenzie, J. A., Vasconcelos, C., Bernasconi, S., Frank, M., Gutjahr, M., et al. (2007). Dolomite formation in the dynamic deep biosphere: results from the Peru margin. *Sedimentology, 54,* 1007–1032.

Meister, P., Mckenzie, J. A., Warthmann, R., & Vasconcelos, C. (2006). Mineralogy and petrography of diagenetic dolomite, Peru margin, ODP Leg 201. In B. B. Jørgensen, S. L. D'hondt, & D. J. Miller (Eds.), *Proc. ODP, Sci. Results* (vol. 201).

Michalopoulos, P., & Aller, R. C. (2004). Early diagenesis of biogenic silica in the Amazon delta: alteration, authigenic clay formation, and storage. *Geochimica et Cosmochimica Acta, 68,* 1061–1085.

Michalopoulos, P., & Aller, R. C. (1995). Rapid clay mineral formation in Amazon delta sediments: reverse weathering and oceanic elemental cycles. *Science, 270*, 614–617.

Michalopoulos, P., Aller, R. C., & Reeder, R. J. (2000). Conversion of diatoms to clays during early diagenesis in tropical, continental shelf muds. *Geology, 28*, 1095–1098.

Milliman, J. D. (1993). Production and accumulation of calcium-carbonate in the ocean - budget of a nonsteady state. *Global Biogeochemical Cycles, 7*, 927–957.

Mollenhauer, G., Inthorn, M., Vogt, T., Zabel, M., Sinninghe Damsté, J. S., & Eglinton, T. I. (2007). Aging of marine organic matter during cross-shelf lateral transport in the Benguela upwelling system revealed by compound-specific radiocarbon dating. *Geochemistry, Geophysics, Geosystems, 8*, Q09004.

Moore, T. S., Murray, R. W., Kurtz, A. C., & Schrag, D. P. (2004). Anaerobic methane oxidation and the formation of dolomite. *Earth and Planetary Science Letters, 229*, 141–154.

Morse, J. W. (1991). Oxidation kinetics of sedimentary pyrite in seawater. *Geochimica et Cosmochimica Acta, 55*, 3665–3667.

Morse, J. W., & McKenzie, F. T. (1990). *Geochemistry of sedimentary carbonates.* Amsterdam: Elservier.

Murray, R. W., Wigley, R., & Shipboard Scientific Party. (1998). Interstitial water chemistry of deeply buried sediments from the Southwest African margin: a preliminary synthesis of results from Leg 175. In G. Wefer, W. H. Berger, C. Richter, et al. (Eds.), *Proc. ODP, Init. Repts* (vol. 175).

Neretin, L. N., Böttcher, M. E., Jørgensen, B. B., Volkov, I., Lüschen, H., & Hilgenfeldt, K. (2004). Pyritization processes and greigite formation in the advancing sulfidization front in the upper Pleistocene sediments of the Black Sea. *Geochimica et Cosmochimica Acta, 68*, 2081–2093.

Ockert, C., Gussone, N., Kaufhold, S., & Teichert, B. M. A. (2013). Isotope fractionation during Ca exchange on clay minerals in a marine environment. *Geochimica et Cosmochimica Acta, 112*, 374–388.

Orcutt, B. N., Wheat, C. G., Rouxel, O., Hulme, S., Edwards, K. J., & Bach, W. (2013). Oxygen consumption rates in subseafloor basaltic crust derived from a reaction transport model. *Nature Communications, 4*, 2539. http://dx.doi.org/10.1038/ncomms3539.

Parkes, R. J., Linnane, C. D., Webster, G., Sass, H., Weightman, A. J., Hornibrook, E. R. C., et al. (2011). Prokaryotes stimulate mineral H_2 formation for the deep biosphere and subsequent thermogenic activity. *Geology, 39*, 219–222.

Parkes, R. J., Webster, G., Cragg, B. A., Weightman, A. J., Newberry, C. J., Ferdelman, T. G., et al. (2005). Deep sub-seafloor prokaryotes stimulated at interfaces over geological time. *Nature, 436*, 390–394.

Paytan, A., Kastner, M., & Chavez, F. P. (1996). Glacial to interglacial fluctuations in productivity in the equatorial Pacific as indicated by marine barite. *Science, 274*, 1355–1357.

Paytan, A., Kastner, M., Martin, E. E., Macdougall, J. D., & Herbert, T. (1993). Marine barite as a monitor of seawater strontium isotope composition. *Nature, 366*, 445–449.

Picard, A., & Ferdelman, T. G. (2011). Linking microbial heterotrophic activity and sediment lithology in oxic, oligotrophic subseafloor sediments of the North Atlantic Ocean. *Frontiers in Microbiology, 2*, 263. http://dx.doi.org/10.3389/fmicb.2011.00263.

Pierre, C., Blanc-Valleron, M. M., Caquineau, S., März, C., Ravelo, A. C., Takahashi, K., & Alvarez Zarikian, C. (2014). Mineralogical, geochemical and isotopic characterization of authigenic carbonates from the methane-bearing sediments of the Bering Sea continental margin (IODP Expedition 323, Sites U1343–U1345). *Deep Sea Research Part II: Topical Studies in Oceanography*, in press.

Pisciotto, K. A., & Mahoney, J. J. (1981). Isotopic survey of diagenetic carbonates. In R. S. Yeats, B. U. Haq, et al. (Eds.), *Initial Reports of DSDP* (vol. 63) (pp. 595–609). Washington, D.C: US Govt. Printing office.

Pufahl, P. K., & Wefer, G. (2001). Data report: petrographic, cathodoluminescent, and compositional characteristics of organogenic dolomites from the southwest African margin. In G. Wefer, W. H. Berger, & C. Richter (Eds.), *Proc. ODP, Sci. Results* (vol. 175) (pp. 1–17).

Raiswell, R. (1988). Chemical model for the origin of minor limestone-shale cycles by anaerobic methane oxidation. *Geology, 16*, 641–644.

Raiswell, R., Buckley, F., Berner, R. A., & Anderson, T. F. (1988). Degree of pyritization of iron as a paleoenvironmental indicator of bottom-water oxygenation. *Journal of Sedimentary Petrology, 58*, 812–819.

Raiswell, R., & Canfield, D. E. (2012). Section 8. The interaction of iron with other biogeochemical cycles. *Geochemical Perspectives, 1*, 91–114.

Raiswell, R., & Canfield, D. E. (1998). Sources of iron for pyrite formation in marine sediments. *American Journal of Science, 298*, 219–245.

Reed, D. C., Slomp, C. P., & De Lange, G. J. (2011). A quantitative reconstruction of organic matter and nutrient diagenesis in Mediterranean Sea sediments over the Holocene. *Geochimica et Cosmochimica Acta, 75*, 5540–5558.

Reitz, A., Hensen, C., Kasten, S., Funk, J. A., & De Lange, G. J. (2004). A combined geochemical and rock-magnetic investigation of a redox horizon at the last glacial/interglacial transition. *Physics and Chemistry of the Earth, Parts A/B/C, 29*, 921–931.

Rickard, D., & Luther, G. W. (2007). Chemistry of iron sulfides. *Chemical Reviews, 107*, 514–562.

Rickard, D., Schoonen, M. A. A., & Luther, G. W. (1995). Chemistry of iron sulfides in sedimentary environments. In M. A. Vairavanurthy, & M. A. A. Schoonen (Eds.), *Geochemical transformations of sedimentary sulfur* (pp. 168–193). ACS Symposium Series 612, Washington DC: American Chemical Society.

Riedinger, N., Formolo, M. J., Lyons, T. W., Henkel, S., Beck, A., & Kasten, S. (2014). An inorganic geochemical argument for coupled anaerobic oxidation of methane and iron reduction in marine sediments. *Geobiology, 12*, 172–181.

Riedinger, N., Kasten, S., Groger, J., Franke, C., & Pfeifer, K. (2006). Active and buried authigenic barite fronts in sediments from the Eastern Cape Basin. *Earth and Planetary Science Letters, 241*, 876–887.

Riedinger, N., Pfeifer, K., Kasten, S., Garming, J. F. L., Vogt, C., & Hensen, C. (2005). Diagenetic alteration of magnetic signals by anaerobic oxidation of methane related to a change in sedimentation rate. *Geochimica et Cosmochimica Acta, 69*, 4117–4126.

Ritger, S., Carson, B., & Suess, E. (1987). Methane-derived authigenic carbonates formed by subduction induced pore-water expulsion along the Oregon Washington margin. *Geological Society of America Bulletin, 98*, 147–156.

Rodriguez, N. M., Paull, C. K., & Borowski, W. S. (2000). Zonation of authigenic carbonates within gas hydrate-bearing sedimentary sections on the Blake ridge: offshore southeastern North America. In C. K. Paull, et al. (Ed.), *Proc. ODP, Sci. Results* (vol. 164) (pp. 301–312). College Station, Texas: Ocean Drilling Program.

Røy, H., Kallmeyer, J., Adhikari, R. R., Pockalny, R., Jørgensen, B. B., & D'hondt, S. (2012). Aerobic microbial respiration in 86-million-year-old deep-sea red clay. *Science, 336*, 922–925.

Rudnicki, M. D., Elderfield, H., & Mottl, M. J. (2001). Pore fluid advection and reaction in sediments of the eastern flank, Juan de Fuca Ridge, 48°N. *Earth and Planetary Science Letters, 187*, 173–189.

Ruttenberg, K. C., & Berner, R. A. (1993). Authigenic apatite formation and burial in sediments from non-upwelling, continental margin environments. *Geochimica et Cosmochimica Acta, 57*, 991–1007.

Ryan, W. B. F., Pitman, W. C., III, Major, C. O., Shimkus, K., Moskalenko, V., Jones, G. A., et al. (1997). An abrupt drowning of the Black Sea shelf. *Marine Geology, 138*, 119–126.

Scholz, F., Hensen, C., Schmidt, M., & Geersen, J. (2013). Submarine weathering of silicate minerals and the extent of pore water freshening at active continental margins. *Geochimica et Cosmochimica Acta, 100*, 200–216.

Schoonen, M. A. A. (2004). Mechanisms of sedimentary pyrite formation. In J. P. Amend, K. J. Edwards, & T. W. Lyons (Eds.), *Sulfur biogeochemistry - past and present* (vol. 379).

Schrag, D. P., Adkins, J. F., Mcintyre, K., Alexander, J. L., Hodell, D. A., Charles, C. D., et al. (2002). The oxygen isotopic composition of seawater during the Last Glacial Maximum. *Quaternary Science Reviews, 21*, 331–342.

Schrag, D. P., Hampt, G., & Murray, D. W. (1996). Pore fluid constraints on the temperature and oxygen isotopic composition of the glacial ocean. *Science, 272*, 1930–1932.

Schrag, D. P., Higgins, J. A., Macdonald, F. A., & Johnston, D. T. (2013). Authigenic carbonate and the history of the global carbon cycle. *Science, 339*, 540–543.

Sclater, J. G. (2003). Ins and outs on the ocean floor. *Nature, 421*, 590–591.

Sigman, D. M., Jaccard, S. L., & Haug, G. H. (2004). Polar ocean stratification in a cold climate. *Nature, 428*, 59–63.

Sim, M. S., Bosak, T., & Ono, S. (2011). Large sulfur isotope fractionation does not require disproportionation. *Science, 333*, 74–77.

Simon, A., Poulicek, M., Velimirov, B., & Mackenzie, F. T. (1994). Comparison of anaerobic and aerobic biodegradation of mineralized skeletal structures in marine and estuarine conditions. *Biogeochemistry, 25*, 167–195.

Sivan, O., Adler, M., Pearson, A., Gelman, F., Bar-Or, I., John, S. G., et al. (2011). Geochemical evidence for iron-mediated anaerobic oxidation of methane. *Limnology and Oceanography, 56*, 1536–1544.

Soetaert, K., Hofmann, A. F., Middelburg, J. J., Meysman, F. J. R., & Greenwood, J. (2007). The effect of biogeochemical processes on pH. *Marine Chemistry, 105*, 30–51.

Soffientino, B., Spivack, A. J., Smith, D. C., & D'hondt, S. (2009). Hydrogenase activity in deeply buried sediments of the arctic and North Atlantic oceans. *Geomicrobiology Journal, 26*, 537–545.

Soffientino, B., Spivack, A. J., Smith, D. C., Roggenstein, E. B., & D'hondt, S. (2006). A versatile and sensitive tritium-based radioassay for measuring hydrogenase activity in aquatic sediments. *Journal of Microbiological Methods, 66*, 136–146.

Stumm, W., & Morgan, J. J. (1996). *Aquatic chemistry: Chemical equilibria and rates in natural waters* (3rd ed.). New York: Wiley-Interscience.

Sun, X., & Turchyn, A. V. (2014). Significant contribution of authigenic carbonate to marine carbon burial. *Nature Geosciences, 7*, 201–204.

Teichert, B. M. A., Gussone, N., & Torres, M. E. (2009). Controls on calcium isotope fractionation in sedimentary pore waters. *Earth and Planetary Science Letters, 279*, 373–382.

Tominaga, M., Lyle, M., & Mitchell, N. C. (2011). Seismic interpretation of pelagic sedimentation regimes in the 18–53 Ma eastern equatorial Pacific: Basin-scale sedimentation and infilling of abyssal valleys. *Geochemistry, Geophysics, Geosystems, 12*, Q03004.

Torres, M. E., Brumsack, H. J., Bohrmann, G., & Emeis, K. C. (1996). Barite fronts in continental margin sediments: a new look at barium remobilization in the zone of sulfate reduction and formation of heavy barites in diagenetic fronts. *Chemical Geology, 127*, 125–139.

Tromp, T. K., Van Cappellen, P., & Key, R. M. (1995). A global model for the early diagenesis of organic carbon and organic phosphorus in marine sediments. *Geochim. Cosmochim. Acta, 59*, 1259–1284.

Ussler, W., & Paull, C. K. (2008). Rates of anaerobic oxidation of methane and authigenic carbonate mineralization in methane-rich deep-sea sediments inferred from models and geochemical profiles. *Earth and Planetary Science Letters, 266,* 271–287.

Vairavamurthy, M. A., Orr, W. L., & Manowitz, B. (1995). Geochemical transformation of sedimentary sulfur: an introduction. In M. A. Vairavamurthy, & M. A. A. Schoonen (Eds.), *Geochemical transformations of sedimentary sulfur* (pp. 1–17). ACS Symposium Series 612, Washington DC: American Chemical Society.

Van Bennekom, A. J., Fred Jansen, J. H., Van Der Gaast, S. J., Van Iperen, J. M., & Pieters, J. (1989). Aluminium-rich opal: an intermediate in the preservation of biogenic silica in the Zaire (Congo) deep-sea fan. *Deep Sea Research Part A. Oceanographic Research Papers, 36,* 173–190.

Van Cappellen, P., & Berner, R. A. (1988). A mathematical model for the early diagenesis of phosphorus and fluorine in marine sediments; apatite precipitation. *American Journal of Science, 288,* 289–333.

Van Cappellen, P., & Qiu, L. (1997). Biogenic silica dissolution in sediments of the Southern Ocean. I. Solubility. *Deep Sea Research Part II: Topical Studies in Oceanography, 44,* 1109–1128.

Voipio, A. (1981). *The Baltic sea.* Helsinki: Elsevier p. 418.

Von Breymann, M. T., Collier, R., & Suess, E. (1990). Magnesium adsorption and ion exchange in marine sediments: a multi-component model. *Geochimica et Cosmochimica Acta, 54,* 3295–3313.

Von Breymann, M. T., Ungerer, C. A., & Suess, E. (1988). Mg-NH$_4$ exchange on humic acid: a radiotracer technique for conditional exchange constants in a seawater medium. *Chemical Geology, 70,* 349–357.

Von Breymann, M. T., Brumsack, H.-J., & Emeis, K.-C. (1992). Depositional and diagenetic behaviour of barium in the Japan Sea. In *Proc. ODP, Sci. Results* (vol. 127/128) (pp. 651–663).

Wallmann, K. (2010). Phosphorus imbalance in the global ocean? *Global Biogeochemical Cycles, 24,* GB4030.

Wallmann, K., Aloisi, G., Haeckel, M., Obzhirov, A., Pavlova, G., & Tishchenko, P. (2006). Kinetics of organic matter degradation, microbial methane generation, and gas hydrate formation in anoxic marine sediments. *Geochimica et Cosmochimica Acta, 70,* 3905–3927.

Wallmann, K., Aloisi, G., Haeckel, M., Tishchenko, P., Pavlova, G., Greinert, J., et al. (2008). Silicate weathering in anoxic marine sediments. *Geochimica et Cosmochimica Acta, 72,* 2895–2918.

Walter, L. M., & Burton, E. A. (1990). Dissolution of recent platform carbonate sediments in marine pore fluids. *American Journal of Science, 290,* 601–643.

Wang, G., Spivack, A. J., Rutherford, S., Manor, U., & D'Hondt, S. (2008). Quantification of co-occurring reaction rates in deep subseafloor sediments. *Geochimica et Cosmochimica Acta, 72,* 3479–3488.

Wankel, S. D., Germanovich, L. N., Lilley, M. D., Genc, G., Diperna, C. J., Bradley, A. S., et al. (2011). Influence of subsurface biosphere on geochemcial fluxes from diffuse hydrothermal fluids. *Nature Geoscience, 4,* 461–468.

Wara, M. W., & Ravelo, A. C. (2006). Data report: Mg/Ca, Sr/Ca, Mn/Ca, and oxygen and carbon isotope records of Pliocene–Pleistocene foraminifers from ODP Leg 202 site 1237. In R. Tiedemann, A. C. Mix, C. Richter, & W. F. Ruddiman (Eds.), *Proc. ODP, Sci. Results* (vol. 202) (pp. 1–19). College Station, TX: Ocean Drilling Program. http://dx.doi.org/10.2973/odp.proc.sr.202.206.2006.

Webb, H. F., & Jordan, T. H. (1993). Quantifying the distribution and transport of pelagic sediments on young abyssal hills. *Geophysical Research Letters, 20,* 2203–2206.

Webster, G., Blazejak, A., Cragg, B. A., Schippers, A., Sass, H., Rinna, J., et al. (2009). Subsurface microbiology and biogeochemistry of a deep, cold-water carbonate mound from the Porcupine Seabight (IODP Expedition 307). *Environmental Microbiology, 11,* 239–257.

Wehrmann, L. M., Arndt, S., März, C., Ferdelman, T. G., & Brunner, B. (2013). The evolution of early diagenetic signals in Bering Sea subseafloor sediments in response to varying organic carbon deposition over the last 4.3 Ma. *Geochimica et Cosmochimica Acta, 109,* 175–196.

Wehrmann, L.M., Ockert, C., Mix, A. C., Gussone, N., Teichert, B. M. A., & Meister, P. (2014). Repeated occurrences of methanogenic zones, diagenetic dolomite formation and linked silicate alteration in southern Bering Sea sediments (Bowers Ridge, IODP Exp. 323 Site U1341). *Deep Sea Research Part II: Topical Studies in Oceanography,* in press.

Wehrmann, L. M., Risgaard-Petersen, N., Schrum, H. N., Walsh, E. A., Huh, Y., Ikehara, M., et al. (2011). Coupled organic and inorganic carbon cycling in the deep subseafloor sediment of the northeastern Bering Sea Slope (IODP Exp. 323). *Chemical Geology, 284,* 251–261.

Wheat, C. G., Hulme, S. M., Fisher, A. T., Orcutt, B. N., & Becker, K. (2013). Seawater recharge into oceanic crust: IODP Exp 327 Site U 1363 Grizzly Bear outcrop. *Geochemistry, Geophysics, Geosystems, 14,* 1957–1972.

Wheat, C. G., Jannasch, H. W., Kastner, M., Plant, J. N., & Decarlo, E. H. (2003). Seawater transport and reaction in upper oceanic basaltic basement: chemical data from continuous monitoring of sealed boreholes in a ridge flank environment. *Earth and Planetary Science Letters, 216,* 549–564.

Wilson, J. D., Barker, S., & Ridgwell, A. (2012). Assessment of the spatial variability in particulate organic matter and mineral sinking fluxes in the ocean interior: Implications for the ballast hypothesis. *Global Biogeochemical Cycles, 26,* GB4011.

Wilson, P. A., & Norris, R. D. (2001). Warm tropical ocean surface and global anoxia during the mid-Cretaceous period. *Nature, 412,* 425–429.

Ziebis, W., McManus, J., Ferdelman, T., Schmidt-Schierhorn, F., Bach, W., Muratli, J., et al. (2012). Interstitial fluid chemistry of sediments underlying the North Atlantic Gyre and the influence of subsurface fluid flow. *Earth and Planetary Science Letters, 323–324,* 79–91.

Zonneveld, K. A. F., Versteegh, G. J. M., Kasten, S., Eglinton, T. I., Emeis, K. C., Huguet, C., et al. (2010). Selective preservation of organic matter in marine environments; processes and impact on the sedimentary record. *Biogeosciences, 7,* 483–511.

Chapter 3

Environmental Change, Processes and Effects

Chapter 3.1

Introduction: Environmental Change, Processes and Effects—New Insights From Integrated Ocean Drilling Program (2003–2013)

Ruediger Stein

Geosciences Department, Alfred Wegener Institute for Polar and Marine Research,
Bremerhaven, Germany
Email: Ruediger.Stein@awi.de

In general, paleoclimate records document the natural climate, rates of change and variability prior to anthropogenic influence. The instrumental records of temperature, salinity, precipitation, and other environmental observations span only a very short interval (<150 years) of Earth's climate history and provide an inadequate perspective of natural climate variability, as they are biased by an unknown amplitude of anthropogenic forcing. Paleoclimate reconstructions, on the other hand, can be used to assess the sensitivity of the Earth's climate system to changes of different forcing parameters (e.g., CO_2) and to test the reliability of climate models by evaluating their simulations for conditions very different from the modern climate. A precise knowledge of past rates and scales of climate change is the only mode to separate natural and anthropogenic forcings and will enable us to further increase the reliability of prediction of future climate change.

Earth's climate fluctuations occur on very different timescales from 10^0 to 10^6 years: tectonic (>0.5 my); orbital (20–400 ky); oceanic (hundreds to a few thousand years); and anthropogenic (seasonal to millennial). Furthermore, climatic changes may be gradual but also quite rapid and abrupt, and these short-term rapid climatic changes may have occurred under very different boundary conditions. In order to understand the mechanisms controlling natural climate change, detailed continuous paleoclimatic records with/from different temporal and spatial coverage are needed. Such records can only be obtained by scientific drilling such as been carried out by Integrated Ocean Drilling Program (IODP).

Developments in Marine Geology, Volume 7. http://dx.doi.org/10.1016/B978-0-444-62617-2.00010-4

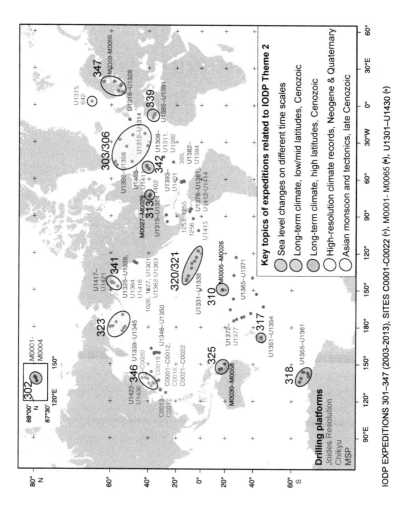

FIGURE 3.1.1 Overview map showing locations of IODP Expeditions 301-347 (2003-2013), SITES C0001-C0022 (◒), Sites U1301-U1430 (USIO/Joides Resolution), C0001-C0022 (CDEX/Chikyu), and M0001-M0065 (ESO/MSP). *Map downloaded from http://iodp.tamu.edu/scienceops/maps.html and supplemented.*

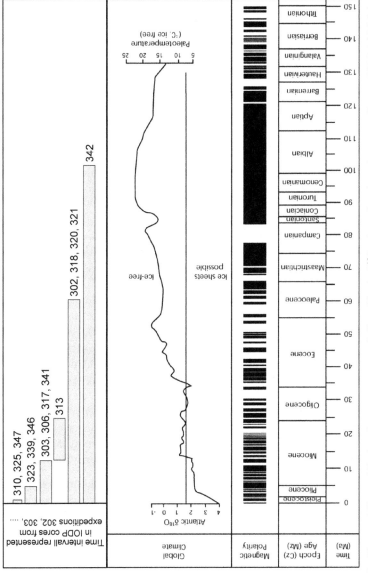

FIGURE 3.1.2   The past 150 my of Earth surface temperatures as indicated by the δ^{18}O values of marine calcareous sediments (IODP Initial Science Plan; http://www.iodp.org/former-program-documents) and time intervals represented in the sedimentary sections recovered during IODP expeditions with paleoclimatic objectives. IODP, Integrated Ocean Drilling Program.

Within the Initial Science Plan of IODP (http://www.iodp.org/former-program-documents), key topics and initiatives related to the overall IODP Theme 2 "Environmental change, processes and effects" as well as key areas and key time intervals have been identified. Between 2003 and 2013, a total of 16 expeditions related to IODP Theme 2 have been carried out (Figures 3.1.1 and 3.1.2). In the following Chapters 3.2 to 3.6, some of the results are described and discussed in more detail:

- Long-term changes in paleoclimatic and paleoceanographic condition during Cenozoic times, representing the change from early Cenozoic Greenhouse conditions to late Cenozoic Icehouse conditions. Here, records from both high latitudes (Arctic Coring Expedition 302 and Wilkes Land Expedition 318—see Chapters 3.2 and 3.3, respectively) and mid-low latitudes (Pacific Equatorial Age Transect Expeditions 320 & 321—see Chapter 3.4) are needed and recovered within IODP. Newfoundland Sediment Drift Expedition 342 even allows to study the paleoclimatic history back to the mid-Cretaceous.
- High-resolution climate records from the North Atlantic (North Atlantic Climate Expeditions 303 and 306; Mediterranean Outflow Expedition 339) (see Chapter 3.5) and the North Pacific (Bering Sea Expedition 323). These records give information on ice sheet–ocean–atmosphere interactions on millennial timescale during the Neogene and Quaternary. Baltic Sea Expedition 347 allows to study in great detail the late Quaternary Eurasian ice sheet and (global) climate history.
- Interaction between climate change and tectonics. Expedition 346 gives the chance to test the hypothesis that the most recent uplift of the Himalaya mountain range and the Tibetan Plateau (beginning about 3.5 Ma) is responsible for the variability of East Asian summer and winter monsoon patterns.
- Reconstruction of the past sea-level change-based continental margin sections (New Jersey Margin Expedition 313 and Canterbury Basin Expedition 317) and tropical coral reef systems (Tahiti Atoll Expedition 310 and Great Barrier Reef Expedition 325). Whereas the former allow to study sea-level cycles during the late Paleocene, Neogene, and Quaternary, the latter give detailed information on amplitude and timing of sea-level change during the last about 30,000 years (see Chapter 3.6).

Some of the objectives and highlights of the other expeditions that are not described in more detail in these chapters are shortly summarized in the appendix (Chapter 5).

Chapter 3.2

Cenozoic Arctic Ocean Climate History: Some Highlights from the Integrated Ocean Drilling Program Arctic Coring Expedition

Ruediger Stein,[1,*] Petra Weller,[1] Jan Backman,[2] Henk Brinkhuis,[3] Kate Moran[4] and Heiko Pälike[5]

[1]*Geosciences Department, Alfred Wegener Institute for Polar and Marine Research, Bremerhaven, Germany;* [2]*Department of Geology and Geochemistry, Stockholm University, Stockholm, Sweden;* [3]*Royal Netherlands Institute for Sea Research NIOZ, Netherlands; and Instituto Andaluz de Ciencias de la Tierra, CSIC-Univ. Granada, Granada, Spain;* [4]*Neptune Ocean Networks Canada, University of Victoria, Victoria, BC, Canada;* [5]*MARUM – Center for Marine Environmental Sciences, University of Bremen, Leobener Strasse, Bremen, Germany*
*Corresponding author E-mail: Ruediger.Stein@awi.de

3.2.1 INTEGRATED OCEAN DRILLING PROGRAM EXPEDITION 302: BACKGROUND AND OBJECTIVES

The permanently to seasonally ice-covered Arctic Ocean is a unique, sensitive, and important component in the Earth's climate system. The melting and freezing of sea ice result in distinct changes in surface albedo, energy balance, and biological processes. Freshwater and sea ice are exported from the Arctic Ocean through Fram Strait into the North Atlantic, and changes in these export rates of freshwater would result in changes of North Atlantic as well as global oceanic circulation patterns. Because factors such as the global thermohaline circulation, sea ice cover, and Earth's albedo have a strong influence on the Earth's climate system, climate change in the Arctic could cause major perturbations in the global environment. Due to complex feedback processes (collectively known as "polar amplification"), the Arctic is both a contributor of climate change and a region that will be most affected by global warming (ACIA, 2004, 2005; IPCC Report Stocker et al., 2013). That means, the Arctic Ocean and surrounding areas are (in real time) and were (over historic and geologic timescales) subject to rapid and dramatic change.

Developments in Marine Geology, Volume 7. http://dx.doi.org/10.1016/B978-0-444-62617-2.00011-6

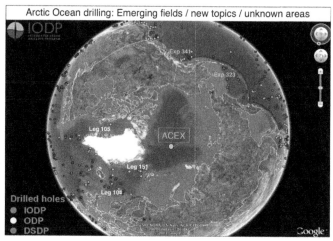

FIGURE 3.2.1 Google map showing the Northern Hemisphere with locations of Deep Sea Drilling Project (DSDP), ODP, and IODP sites.

Despite the importance of the Arctic Ocean in the global climate/earth system, this region is one of the last major physiographic provinces on Earth where the short- and long-term geological history is still poorly known (Figure 3.2.1). This lack in knowledge is mainly due to the major techno-logical/logistical problems in operating within the permanently ice-covered Arctic region, which makes it difficult to retrieve long and undisturbed sedi-ment cores. Prior to 1990, the available samples and geological data from the central Arctic Basins were derived mainly from drifting ice islands such as T-3 (Clark, Whitman, Morgan, & Mackey, 1980) and CESAR (Jackson, Mudie, & Blasco, 1985) (Figure 3.2.2). During the last ~20 years, several international expeditions have greatly advanced our knowledge on central Arctic Ocean paleoenvironment and its variability through Quaternary times (for review see Stein, 2008). Prior to 2004, however, in the central Arctic Ocean piston and gravity coring was mainly restricted to obtaining near-surface sediments, i.e., only the upper 15 m could be sampled. Thus, all studies were restricted to the late Pliocene/Quaternary time interval, with a few exceptions. These include the four short cores obtained by gravity coring from drifting ice floes over the Alpha Ridge, where older pre-Neo-gene organic carbon (OC)-rich muds and laminated biosiliceous oozes were sampled. These were the only samples recording the late Cretaceous/early Cenozoic climate history and depositional environment (Figure 3.2.2). In general, these data suggest a warmer (ice-free) Arctic Ocean with strong sea-sonality and high paleoproductivity, most likely associated with upwelling conditions (Jackson et al., 1985; Clark, Byers, & Pratt, 1986; Firth & Clark, 1998; Jenkyns, Forster, Schouten, & Sinninghe Damsté, 2004; Davies et al., 2009; Davies, Kemp, & Pälike, 2011). Continuous central Arctic Ocean

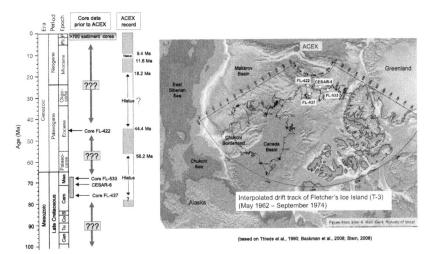

FIGURE 3.2.2   Stratigraphic coverage of existing sediment cores in the central Arctic Ocean prior to IODP-ACEX (based on Thiede, Clark, & Hermann, 1990) and the section recovered during the ACEX drilling expedition (Backman et al., 2006, 2008). The middle Eocene sediments recovered at Core FL-422 are arbitrarily placed at 45 Ma, the three Late Cretaceous (Maastrichtian/Campanian) cores are arbitrarily placed on the time axis as well. Each of these four cores (all recovered on Alpha Ridge) documents a time period of at the most a few hundred thousand years (Backman et al., 2008). The map shows the Amerasian Basin and surrounding continents (shallower water areas in red (darkest gray in print versions), orange (darker gray in print versions), and yellow (gray in print versions) colors, deep-water areas in green (light gray in print versions) and blue (white in print versions) colors and the interpolated drift track of Fletcher's Ice Island (T-3) between May 1962 and September 1974. *Map from John Hall, Geol. Survey of Israel.*

sedimentary records, allowing a development of chronologic sequences of climate and environmental change through Cenozoic times and a comparison with global climate records, however, were missing prior to the Integrated Ocean Drilling Program (IODP) Expedition 302 (Arctic Ocean Coring Expedition (ACEX); Backman et al., 2006; Moran et al., 2006).

In 2004, a new era in Arctic research began with the successful completion of ACEX. For the first time, scientific drilling in the permanently ice-covered central Arctic Ocean was carried out, penetrating 428 m of upper Cretaceous to Quaternary sediments on the crest of Lomonosov Ridge between 87 and 88°N (Figure 3.2.3; Backman, Moran, McInroy, Muyer, & the Expedition 302 Scientists, 2006; Backman & Moran, 2008; Backman & Moran, 2009; Moran et al., 2006). The Lomonosov Ridge was selected as target area as the elevation of the ridge, ~3 km above the surrounding abyssal plains, indicates that sediments on top of the ridge have been isolated from turbidites and are likely of purely pelagic origin (i.e., mainly biogenic, eolian, and/or ice-rafted), an observation borne out by countless shorter cores collected from icebreakers in the past decades. After Heezen and Ewing (1961) recognized that the midocean rift system extended from the North Atlantic into the

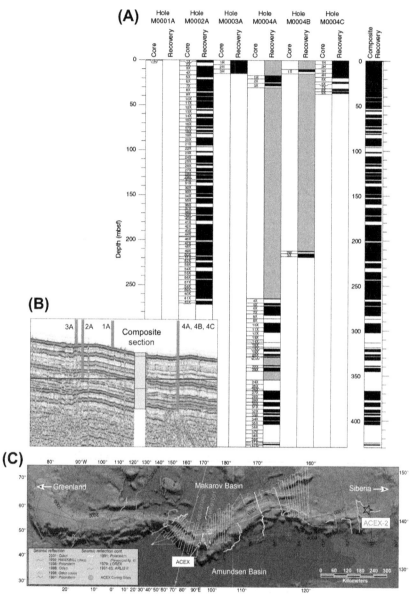

FIGURE 3.2.3   (A) Core recovery summary diagram for all Expedition 302 (ACEX) holes (Backman et al., 2006). Black=qrecovered core, white=no recovery, and shaded (gray in print versions)=washed intervals. Total recovery was 68.4%. (B) Seismic profile AWI 91090 across the Lomonosov Ridge, interpreted as continental crust truncated by a regional unconformity overlain by a continuous sediment sequence (Jokat et al., 1992). The four ACEX sites were positioned on this profile, shown as solid vertical lines. At each of the four sites (1–4), multiple holes (A, B, C) were drilled, and a composite section has been obtained (Backman et al., 2006). (C) The ACEX coring sites have been carefully selected based on comprehensive geophysical data sets, including seismic

Arctic Ocean, it was realized that the 1800-km-long Lomonosov Ridge was originally a continental fragment broken off of the Eurasian continental margin and became separated by seafloor spreading. Regional aeromagnetic data indicated the presence of seafloor spreading anomalies in the basins north and south of the Gakkel Ridge, the active spreading center located in the middle of the Eurasian Basin (Kristoffersen, 1990 and references therein). The interpretation of the magnetic anomalies in terms of seafloor spreading and their correlation with the geomagnetic timescale allowed linking the evolution of the Eurasian Basin to the opening of the Norwegian-Greenland Sea. According to this correlation, seafloor spreading was probably initiated in the Eurasian Basin between chron 24 and 25 in the late Paleocene near 56 Ma (e.g., Kristoffersen, 1990). As the Lomonosov Ridge moved away from the Eurasian plate and subsided, pelagic sedimentation on top of this continental sliver began at that time (e.g., Jokat, 2005; Jokat, Uenzelmann-Neben, Kristoffersen, & Rasmussen, 1992; Jokat, Weigelt, Kristoffersen, & Rasmussen, 1995).

Within ACEX, cores on the Lomonosov Ridge were recovered in five holes across three sites (Holes MSP0002A, MSP0003A, MSP0004A, MSP0004B, and MSP0004C) situated within about 15 km of each other on seismic line AWI-91090 (Jokat et al.,1992) and interpreted as a single site (composite section) because of the internally consistent seismic stratigraphy across that distance (Figure 3.2.3). Key objectives of the ACEX drilling campaign focused on the reconstruction of past (Cenozoic) Arctic ice, temperatures, and climates. Some of the key questions to be answered from ACEX were framed around the evolution of sea ice and ice sheets, the past physical oceanographic structure, Arctic gateways, links between Arctic land and ocean climate, and major changes in depositional environments (Backman et al., 2006, 2008; Backman & Moran, 2009).

ACEX was an outstanding success for two reasons (Backman et al., 2006). First, the biggest technical challenge was to maintain the drillship's location while drilling and coring in heavy sea ice over the Lomonosov Ridge. ACEX has proven that with an intensive ice-management strategy, i.e., a three-ship approach with two powerful icebreakers (*Sovetskiy Soyuz* and *Oden*) protecting the drillship (*Vidar Viking*) by breaking upstream ice floes into small pieces (Figure 3.2.4), successful scientific drilling in the permanently ice-covered central Arctic Ocean is possible. The icebreakers kept the drillship on location in 90% cover of multiyear ice for up to nine consecutive days, a benchmark feat for future drilling in this harsh environment. Second, the outstanding scientific

---

reflection profiles (Jokat, Stein, Rachor, & Schewe, 1999; Jokat et al., 1992, 1995; Kristoffersen, Buravtsev, Jokat, & Poselov, 1997), high-resolution chirp profiles (Jakobsson, 1999), and SCICEX swath bathymetry and sidescan sonar backscatter data (Edwards & Coakley, 2003). In addition to the ACEX site, the potential location of an "ACEX2 site" proposed for future IODP drilling on southern Lomonosov Ridge (Stein, Jokat, et al., 2014b), is indicated.

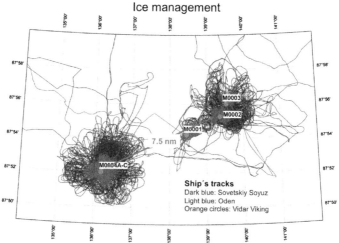

FIGURE 3.2.4 Photographs showing the three ACEX ships and sea-ice operations during the ACEX expedition (Backman et al., 2006). The nuclear icebreaker *Sovetskiy Soyuz* is crashing the large ice floes, to be further cut down by the icebreaker *Oden*. This allowed the drillship *Vidar Viking* to work under crushed-ice conditions and remain for some days on station to carry out the drilling (Backman et al., 2006).

results provide a unique glimpse of the early Arctic Ocean history and its long-term change through Cenozoic times (see below for some results and highlights). To date, the ACEX sites remain the one and only drill location in the central Arctic Ocean (Figure 3.2.2).

3.2.2 MAIN LITHOLOGIES AND STRATIGRAPHIC FRAMEWORK OF THE ACEX SEQUENCE

The almost 430-m-thick sedimentary sequence recovered from Lomonosov Ridge is composed of sediments ranging in age from Campanian, Paleogene, Neogene to Quaternary (Figure 3.2.5). Based on the visual core description and smear slide analysis as well as total OC content and X-ray diffraction measured in core catcher samples, the ACEX sequence was divided into four main lithologic units (Backman et al., 2006; Stein, 2007; Figure 3.2.5); for each unit the ages of the original Backman et al. (2008) stratigraphy are listed:

- Unit 1 (top to 223.6 m of composite depth (mcd); Quaternary to middle Eocene) is dominated by silty clay, ranging from light olive browns at the top, through olives and grays to very dark gray at the bottom; color banding is strong. Millimeter-to-centimeter-scale sandy lenses and isolated pebbles also occur in unit 1. Unit 1 is divided into six subunits. According to Backman et al. (2008), a major hiatus was identified at about 198.7 mcd separating subunits 1/6 and 1/5 and spanning the time interval from about 44.4 to 18.2 Ma; another shorter hiatus lasting 2.2 my occurs within subunit 1/3 in the late Miocene 9.4–11.6 Ma) (Figure 3.2.5). Subunit 1/5 is outstanding due to its very prominent black and white color banding ("Zebra Unit").
- Unit 2 (223.6–313.6 mcd; middle Eocene) is dominated by very dark gray mud-bearing biosiliceous ooze with submillimeter-scale laminations as well as isolated pebbles.
- Unit 3 (about 313.6–404.8 mcd; late Paleocene to early Eocene) is dominated by very dark gray clay with submillimeter-scale laminations. Units 3 and 4 are separated by a second major hiatus representing the time interval of about 56–79 Ma (Backman et al., 2008).
- Unit 4 (424.50–427.63 mbsf; Campanian) is dominated by very dark gray clayey mud and silty sands.

The upper (middle Miocene to Quaternary) part of the ACEX sequence (subunits 1/1 to 1/4) is composed of silty clay with very low OC contents of <0.5%, i.e., values very similar to those known from upper Quarternary records determined in gravity cores from Lomonosov Ridge (Stein, 2008), whereas the Campanian and Paleogene sediments of the ACEX sequence (units 2–4) are characterized by high total OC values of 1 to >5% (Figure 3.2.5). In subunit 1/5 (about 193–199 mcd; late early Miocene) characterized by distinct gray/black color bandings, even OC maxima of 7–14.5% were measured in samples from the black horizons (Stein, 2007).

Backman's et al. (2008) stratigraphic framework of the Cenozoic ACEX sequence ("Age Model 1" in this chapter) is based on biostratigraphic, cosmogenic isotope, magneto- and cyclostratigraphic data, and widely used in the scientific literature so far. Based on age/depth control points, Neogene and Paleogene sedimentation rates reach values of about 1 and 2.4 cm/ky, respectively (Figure 3.2.6). Although this ACEX age model may confirm that the

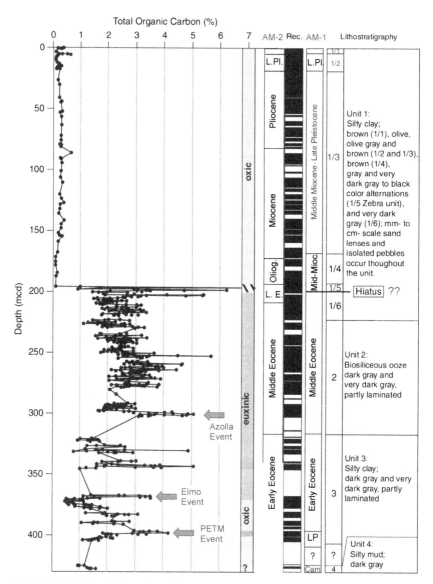

FIGURE 3.2.5 Record of total organic carbon (TOC) contents as determined in the composite ACEX sedimentary sequence (Stein, 2007). *Data on recovery, stratigraphy, and lithological units (1–4) and subunits (1/1 to 1/6) from Backman et al. (2006, 2008).* Cam = Campanian; LP = late Paleocene; Mid Mioc. = middle Miocene; L.Pl. = late Pleistocene. PETM, Elmo (ETM2), and *Azolla* events are indicated. Age models AM-1 (Backman et al., 2008) and AM-2 (Poirier & Hillaire-Marcel, 2011) are shown.

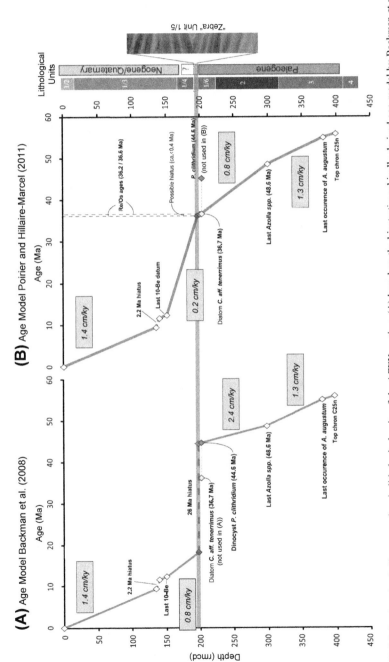

FIGURE 3.2.6 Age-depth diagram and main lithological units of the ACEX section, (A) based on the biostratigraphically derived age model by Backman et al. (2008). In addition, (B) an alternate chronology based on osmium isotopes (Poirier & Hillaire-Marcel, 2011) are shown. Depth of first occurrence of IRD between 240 and 260 mcd are marked as light blue horizontal bar. The different ages of the first occurrence of IRD, obtained by the two age models, AM1 = Backman et al. (2008), AM2 = Poirier and Hillaire-Marcel (2011), are indicated by the large blue arrows. Mean sedimentation rates (centimeters per thousand years) are indicated. *Figure modified from Poirier and Hillaire-Marcel (2011).*

average sedimentation rate mostly is >1 cm/ky (cf., Backman, Jakobsson, Løvlie, Polyak, & Febo, 2004), a highly resolved and robust age model for the ACEX cores is still challenging due to the poor core recovery (about 1/3 of the penetrated section was not recovered), the occurrence of an unexpected major hiatus, the limited availability of biostratigraphic indicators, and the enigmatic preservation of the geomagnetic polarity record (Backman et al., 2008). Recently, Poirier and Hillaire-Marcel (2011) report Rhenium–Osmium (Re–Os) isochron ages and complementary Os–isotope measurements, together with new carbon and nitrogen data, which give information on the redox state of the sediment during deposition, and may allow a better assessment of the timing of events involved. Based on their new data, these authors also challenged the existing age model and the existence of a major hiatus between subunits 1/6 and 1/5. That means, they proposed an improved age model sequence ("Age Model 2" in this chapter) that closes the gap in the ACEX record, resulting in a continuous sedimentary section with three to five times lower sedimentation rates (0.2–0.8 cm/ky) between about 49 and 12 Ma (Figure 3.2.6). The Os ages, of course, would significantly modify the reconstructions of the tectonic evolution of Lomonosov Ridge and the paleoceanographic Arctic Ocean history.

When using the Re-Os and OC records, one should have in mind that these sediments contain significant amounts of reworked Eocene organic matter, on which the Re-Os records are measured. The palynology confirms the presence of late Eocene and even Oligocene elements in subunit 1/5 ("Zebra") (Backman et al., 2008; Sangiorgi, Brumsack, et al., 2008; Sangiorgi, van Soelen, et al., 2008). However, the presumed in situ aquatic palynomorphs are 99% quasi-monotypic assemblages of previously unknown dinocysts that have morphologic similarity with early Miocene ones (Sangiorgi, Brinkhuis, & Damassa, 2009). This points to a restricted marine, or perhaps even freshwater setting, extremely unlikely to have occurred during the Eocene or even Oligocene (Sangiorgi, Brumsack, et al., 2008). Furthermore, there are also massive geochemical breaks in this section that may point to discontinuous sedimentation (Sangiorgi, Brumsack, et al., 2008; Sangiorgi, van Soelen, et al., 2008). More long sedimentary records from the Arctic Ocean are needed to resolve these issues and related paleoenvironmental reconstructions (see outlook below). As no clear and final answer exists which of the two age models is closer to represent the true one, both interpretations are presented in this review article.

3.2.3 HIGHLIGHTS OF ACEX STUDIES

The ACEX Cenozoic record holds a significant number of scientific discoveries that describe previously unknown paleoenvironments (Backman & Moran, 2008, 2009). By studying the unique sedimentary sequence recovered during ACEX in 2004, about 100 papers in highly ranked international peer-reviewed scientific journals have been published throughout the last decade. Key themes discussed in these papers include the following:

- Stratigraphy and chronology (e.g., Backman et al., 2008; Frank et al., 2008; Matthiessen, Brinkhuis, Poulsen, & Smelror, 2009a; Onodera & Takahashi, 2009a; Onodera & Takahashi, 2009b; O'Regan et al., 2008; Poirier & Hillaire-Marcel, 2009, 2011; Sangiorgi et al., 2009);
- Description and analysis of microfossil assemblages (e.g., Onodera & Takahashi, 2009a; Onodera & Takahashi, 2009b; Onodera, Takahashi, & Jordan, 2008; Eynaud et al., 2009; Kaminski, Silye, & Kender, 2009; Matthiessen et al., 2009a; Matthiessen, Knies, Vogt, & Stein, 2009b; Sangiorgi et al., 2009; Setoyama, Kaminski, & Tyszka, 2011; Suto, Jordan, & Watanabe, 2009);
- Subtropical warm conditions during the PETM and the early-mid Eocene (e.g., Ogawa, Takahashi, Yamanaka, & Onodera, 2009; Sluijs et al., 2006, 2008; Weijers, Schouten, Sluijs, Brinkhuis, & Sinninghe Damsté, 2007; Weller & Stein, 2008);
- Arctic Ocean hydrological cycle during the Paleogene (e.g., Brinkhuis et al., 2006; Pagani et al., 2006; Waddell & Moore, 2008);
- Black shales and euxinic conditions in the Eocene Arctic Ocean (e.g., Moran et al., 2006; Stein, Boucsein, & Meyer, 2006; Stein, 2007; Knies, Mann, Popp, Stein, & Brumsack, 2008; Mann et al., 2009);
- Orbital forcing and environmental response during the Paleogene (e.g., Pälike, Spofforth, O'Regan, & Gattacecca, 2008; Sangiorgi, van Soelen, et al., 2008; Spofforth, Pälike, & Green, 2008);
- Early onset of Arctic sea ice and Northern Hemisphere glaciations (NHGs) in the middle Eocene (e.g., Moran et al., 2006; St. John, 2008; Stickley et al., 2009; Immonen, 2013);
- Onset of a phase with perennial sea ice pack in the Arctic Ocean near 14 Ma (Darby, 2008; Kender & Kaminski, 2013; Krylov et al., 2008) (or intermittently, even earlier; Darby, 2014);
- Arctic gateway evolution and circulation changes (e.g., Jakobsson et al., 2007; Kaminski et al., 2009);
- Neogene paleoenvironmental changes (e.g., Cronin, Smith, Eynaud, O'Regan, & King, 2008; Eynaud et al., 2009; Haley, Frank, Spielhagen, & Eisenhauer, 2008);
- Tectonic/subsidence history of Lomonosov Ridge (Moore & the Expedition 302 Scientists, 2006; O'Regan et al., 2008).

Some of the ACEX paleoceanographic highlights with reference to the original literature are summarized here in some more detail (see also Backman & Moran, 2008, 2009; Stein, 2008). A special focus has been drawn on the Eocene history of Arctic sea ice and sea-surface temperatures (SST), a topic for which new and unpublished SST data are presented and discussed.

3.2.3.1 The Paleocene/Eocene Thermal Maximum Event

The Paleocene/Eocene Thermal Maximum (PETM) event, a relatively brief period of widespread, extreme climatic warming (Kennett & Stott, 1991; Röhl, Bralower, Norris, & Wefer, 2000; Tripati & Elderfield, 2005; Zachos, Pagani, Sloan, Thomas, & Billups, 2001; Zachos, Dickens, & Zeebe, 2008), and probably associated with massive atmospheric greenhouse gas input (Dickens, O'Neil, Rea, & Owen, 1995), was identified in the ACEX record by the occurrence of the dinocyst species *Apectodinium augustum*, which is diagnostic of the PETM (Bujak & Brinkhuis, 1998), as well as a distinct negative anomaly in $\delta^{13}C_{org}$ (Pagani et al., 2006; Sluijs et al., 2006, 2008; Stein et al., 2006). During the PETM event, the Arctic Ocean surface-water temperatures reached maximum values around 25 °C as reconstructed from TEX_{86} data (Sluijs et al., 2006, 2008). These are values significantly higher than previous estimates of 10–15 °C (Tripati, Zachos, Marincovich, & Bice, 2001) and model predictions (e.g., Shellito, Sloan, & Huber, 2003; Tindall et al., 2010), indicating a distinctly lower equator-to-pole temperature gradient than previously believed (Sluijs et al., 2006).

Around the PETM event, a drastic change in OC composition is obvious, pointing to a prominent change in the environmental conditions. In the late Paleocene terrigenous OC is predominant, whereas across the PETM the amount of labile OC significantly increased, as indicated by increased hydrogen index values (Figure 3.2.7) as well as increased preservation of algae-type biomarkers (Weller & Stein, 2008). The increased preservation of labile algae-type OC is related to a major change to euxinic conditions, as indicated by a drastic decrease in the organic carbon versus pyritic sulfur (C/S) values (Figure 3.2.8), the occurrence of pyrite framboids (Stein et al., 2006), the absence of benthic foraminiferal linings (Sluijs et al., 2006), and the occurrence of fine lamination (Backman et al., 2006). During the PETM, euxinic conditions expanded even into the photic zone as suggested from the occurrence of the biomarker isorenieratene related to photosynthetic green sulfur bacteria, which requires euxinic conditions to thrive (Sinninghe Damsté, Wakeham, Kohnen, Hayes, & de Leeuw, 1993; Sluijs et al., 2006; Weller & Stein, 2008). Possible cause for the euxinic conditions was a salinity stratification suggested from the high abundance of low-salinity dinocysts (Sluijs et al., 2006). Because these dinocysts were already abundant prior to the PETM event, i.e., during times of oxic water-mass conditions, increased flux of algae-type OC due to some enhanced primary production was probably needed as additional factor causing the change to euxinic conditions (Stein et al., 2006). The increased primary production was probably related to increased fluvial nutrient supply at that time (Pagani et al., 2006).

Toward the end of the PETM event in the earliest Eocene, a gradual return to a more terrestrial influence is obvious and oxic conditions reoccurred as clearly reflected in the C/S diagram (Figure 3.2.8) and the dominance of terrestrial OC (Figure 3.2.7). The termination of the euxinic conditions coincides with

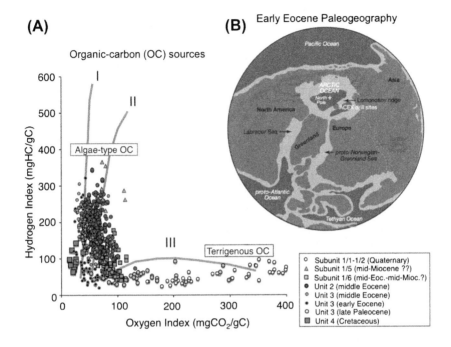

FIGURE 3.2.7 (A) Hydrogen index versus oxygen index ("van Krevelen type") diagrams for Quaternary, Eocene, and Paleocene sediments from the ACEX drill site on Lomonosov Ridge, IODP Expedition 302 (Stein, 2007; modified), and (B) Early Eocene paleogeography (Sickley et al., 2009, supplemented).

increasing surface-water salinities (Pagani et al., 2006) and cooling, suggesting an increased mixing of water masses, which may have caused this change (Sluijs et al., 2006).

Within the early Eocene section of the ACEX sequence (at about 368 mcd), an event with similar characteristics as the PETM event has been identified for the first time by Stein et al. (2006) based on a prominent $\delta^{13}C$ minimum. Dinocyst assemblages confirm the presence of such a second early Eocene hyperthermal, the Eocene Thermal Maximum 2 (ETM2) near 53 Ma (Sluijs et al., 2008). This event may correlate with the global "Elmo Event" (Lourens et al., 2005). According to these authors, the Elmo (or ETM2) event has similar geochemical and biotic characteristics as the PETM event, but of smaller magnitude, and it is coincident with carbon isotope depletion events in world ocean basins, suggesting that it represents a second global thermal maximum. In the ACEX record, this interval is characterized by significantly increased OC contents (Figure 3.2.5) mainly composed of labile algae-type organic matter, as reflected in Rock-Eval data and biomarker composition (Stein et al., 2006; Weller & Stein, 2008). Furthermore, very low C/S ratios <1 point to a euxinic

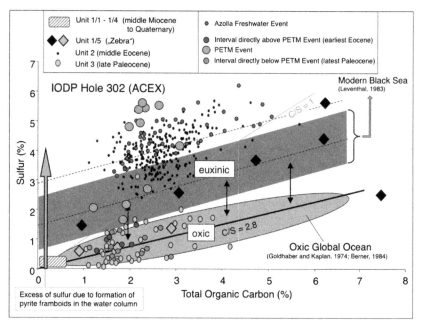

FIGURE 3.2.8 Plot of the total organic carbon versus (pyritic) sulfur (C/S diagram) *(from Stein et al., 2006; supplemented)*. For Quaternary normal (oxic) marine fine-grained detrital sediments shown by the area underline in light gray, a positive correlation between pyritic sulfur and organic carbon with a mean C/S ratio of about 2.8 exists (Berner, 1984; Goldhaber & Kaplan, 1974). In euxinic environments such as the modern Black Sea, however, H_2S already occurs in the water column, and framboidal pyrite can be initially formed, resulting in an excess of sulfur in the C/S diagram (indicated by the open arrow) and very low C/S ratios <1. ACEX data from the different time intervals/units are indicated by different symbols. *Data from the modern euxinic Black Sea are shown by the gray bar (Leventhal, 1983).*

environment causing the preservation of the labile OC (Stein et al., 2006). Of note is the surprising finding of palm pollen, further confirming the overall subtropical conditions during this interval (Sluijs et al., 2008).

3.2.3.2 The *Azolla* Freshwater Episode

The lowermost middle Eocene OC-rich section of the ACEX sequence (about 299–305 mcd; Figure 3.2.5), representing the time interval between about 49 and 48.3 Ma, is composed of microlaminated sediments with extraordinary abundances of microspore clusters (massulae) of the free-floating freshwater fern *Azolla* (Brinkhuis et al., 2006). *Azolla* is typically known from modern freshwater bodies, such as ponds, canals, and flooded rice fields in tropical, subtropical, and warm temperate regions, and cannot tolerate salinities higher than 1–1.6‰ (Rai & Rai, 1998; Arora & Singh, 2003). Based on (1) the presence of mature megaspores with and without attached massulae, (2) single, small groups and large clusters of massulae and probable aborted megaspores of *Azolla*, and (3) support by the relative scarcity of terrestrially derived palynomorphs and extremely low Branched and Isoprenoid Tetraether (BIT) index values of <0.1 (indicating low river-derived terrestrial organic matter), Brinkhuis et al. (2006) favor that *Azolla* grew and reproduced in situ in the Arctic Ocean rather than was brought in by periodic mass transport from freshwater bodies on adjacent continents. That is to say, the *Azolla* event probably represents a distinct episodic freshening of Arctic surface waters lasting about 800 kyrs (Barke et al., 2012; Collinson, Barke, van der Burgh, & van Konijnenburg-van Cittert, 2009; Speelman, Reichart, de Leeuw, Rijpstra, & Sinninghe Damsté, 2009). The freshening of surface waters supports stratification of water masses, causing the euxinic conditions reflected in the C/S ratios that are <1 (Figure 3.2.8).

Over the time interval spanning the *Azolla* (as well as PETM and ETM2) event, the Arctic basin was tectonically isolated from the world's oceans (Figure 3.2.9). This isolation provided the conditions for the development of a brackish water body with a stable fresh water lid. Because the basin served as a fresh water reservoir for the surrounding continental river flow and runoff, it also primed as a "tipping point" to begin freezing of winter sea ice and increasing albedo as the Earth continued on its long-term cooling trend during the Eocene (Backman & Moran, 2009 and further references therein).

The termination of the *Azolla* episode in the Arctic coincides with a local SST rise from about 10 °C to 13 °C (Brinkhuis et al., 2006; Sluijs et al., 2008), which may suggest simultaneously increasing salt and heat supply due to the inflow of waters from adjacent oceans (Brinkhuis et al., 2006). This increased marine influence toward the end of the *Azolla* episode may also explain the occurrence of long-chain alkenones related to marine phytoplankton (Weller & Stein, 2008).

3.2.3.3 From a Euxinic "Lake Stage" to a Fully Ventilated "Ocean Phase"

Black OC-rich biosiliceous silty clays and clayey silts were found throughout the upper early to middle Eocene of the ACEX record, indicating poorly ventilated bottom waters and variable primary production (Moran et al., 2006; Stein et al., 2006). One prerequisite of this extreme paleoenvironmental situation was the paleogeographic boundary setting, i.e., the early Arctic Ocean was isolated from the world ocean in terms of deep-water connection (Figure 3.2.9; Ziegler, 1988; Mutterlose et al., 2003; Backman et al., 2006; Jakobsson et al., 2007). Furthermore, the huge freshwater discharge has favored the development of widespread salinity stratification resulting in a poor ventilation of the subsurface, deeper water masses and causing the high OC preservation rate. A low surface-water salinity (brackish) environment during middle Eocene times is also suggested from the rare and sporadic occurrence of radiolarians (Backman et al., 2006). Runoff-related low salinity might be indicated as well

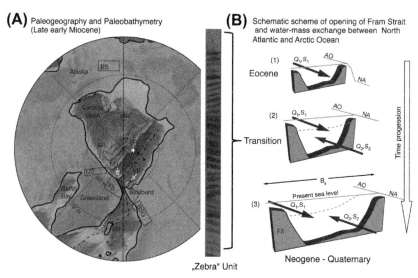

FIGURE 3.2.9 (A) Paleogeographic/paleobathymetric reconstruction for the late early Miocene, based on plate tectonic maps generated from the Ocean Drilling Stratigraphic Network (ODSN) tools available online at http://www.odsn.de *(from Jakobsson et al., 2007; supplemented)*. Physiographic features: AR, Alpha Ridge; BSG, Barents Sea gateway; BS, Bering Strait; FS, Fram Strait; GR, Gakkel Ridge; JM, Jan Mayen microcontinent; LR, Lomonosov Ridge; MR, Mendeleev Ridge; MJ, Morris Jessup rise; NS, Nares Strait; YP, Yermak Plateau. (B) Schematic illustration of the Fram Strait opening and hypothetical water exchange development between the Arctic Ocean and North Atlantic. AO, Arctic Ocean; NA, North Atlantic. (1) A narrow strait resulting in a unidirectional hydraulically controlled outflow from the Arctic. S1 is the salinity of the outflowing flux of water Q1. (2) A wider and deeper strait allowing the establishment of a bidirectional, two-layer flow through the strait due to a compensating inflow (Q2) of saline (S2) North Atlantic water. This phase in the Arctic's paleoceanographic development is analogous to the present Black Sea and proposed for the "Zebra" subunit 1/5. (3) The Fram Strait becomes wide enough that the influence of the Earth's rotation changes the water flow through the strait to a rotationally controlled bidirectional two-layer flow. This opens the possibility of a barotropic current flow through the strait (Jakobsson et al., 2007; modified).

by the abundance of terrestrial palynomorphs and green algae such as *Tasmanites* and *Botryococcus* (Backman et al., 2006). Furthermore, Sr-Nd radiogenic isotope values determined in well-preserved early and middle Eocene ichthyoliths (fish debris) from the ACEX sites are also consistent with a brackish to fresh surface water environment (Gleason, Thomas, Moore, Blum, & Haley, 2007). These data sets even point to lower salinities (5–20 psu) than those indicated by oxygen isotope records from fish bone apatite (21–25 psu) (Gleason et al., 2007). These authors emphasize that the Eocene Arctic Ocean was poorly mixed and characterized by a highly stratified water column with a pervasive "fresh" upper layer and limited, periodic, shallow connections to other oceans.

The remaining questions are how and when the transition from poorly oxygenated to ventilated conditions in the Arctic Ocean occurred. Subunits 1/4 to

1/1 (middle Miocene to Recent) characterized by very low OC values of <0.5% (Figure 3.2.5) and C/S values falling into the "oxic" field (Figure 3.2.8) already represent paleoenvironmental conditions similar to the modern ones, i.e., fully ventilated water masses preventing preservation of high amount of labile algae-type OC. Thus, the transition from euxinic to oxic conditions in the central Arctic Ocean should have occurred within/around "Zebra" subunit 1/5.

Based on a paleogeographic and paleobathymetric reconstruction of the Arctic Ocean, together with a physical oceanographic modeling of the evolving strait and sill conditions in the Fram Strait, Jakobsson et al. (2007) suggest that across subunit 1/5 the Arctic Ocean went from an oxygen-poor "lake stage" to a transitional "estuarine sea" phase with variable ventilation, and finally to the fully ventilated "ocean" phase at 17.5 Ma (Figure 3.2.9). Following Backman et al. (2008) age model, this transition occurred in the late early Miocene between 18.2 and 17.5 Ma. By comparison of the deep-water agglutinated foraminifers in the ACEX record with a similar record from ODP Hole 909C in the Fram Strait (Kaminski, Silye, & Kender, 2005), Kaminski et al. (2009) proposed that the inflow of Atlantic intermediate water began probably already earlier (at least in the early Miocene).

When the Fram Strait opened and deepened through seafloor spreading during Miocene times, the water exchange between the Arctic and North Atlantic must have developed through a series of changes that also influenced the upstream basin circulation and ventilation conditions within the Arctic Ocean (Figure 3.2.9; Jakobsson et al., 2007). During this evolution, the initial opening phase of the Fram Strait was restricted to a unidirectional hydraulically controlled freshwater outflow and reduced ventilation of deep waters via seasonal convection. Later, when the Fram Strait widened and deepened, a compensating inflow of saline North Atlantic water became possible, resulting in a bidirectional, two-layer flow through the strait, similar to that of the modern Black Sea. An inflow of warm near-surface North Atlantic waters in the early Miocene (above the hiatus if using the age model of Backman et al., 2008) may be supported by a distinct increase in TEX_{86}-derived SSTs to ~15–19 °C (Sangiorgi, Brumsack, et al., 2008; Sangiorgi, van Soelen, et al., 2008). At this stage, alternations between euxinic and more oxic conditions (reflected in the black OC-rich and light-gray OC-poor cycles of the "Zebra" subunit 1/5; Figures 3.2.8 and 3.2.9) may have occurred due to oscillations in sea level (Jakobsson et al., 2007). That means sea-level fluctuations could have acted as "on–off switch" for Arctic Ocean circulation, with reduced ventilation during times of lowered sea level and increased ventilation during times of high sea level. According to Miller, Wright, and Browning (2005), sea level may have varied between about 15 and 30 m during late early Miocene times. As the strait deepened further, the sea-level changes were no longer sufficient for a reversal to "Arctic lake" conditions, and oxic conditions similar to the modern ones prevailed (Jakobsson et al., 2007).

The key question when this change from euxinic to well-oxygenated open marine conditions in the Arctic occurred cannot be answered at present. As proposed by Jakobsson et al. (2007) and discussed above, this change was correlated to the tectonically controlled widening of the Fram Strait in the late early Miocene (~17.5 Ma) if using the age model of Backman et al. (2008). The recent Os-Re isotope dates from the cross-banded and underlying Eocene age biosiliceous-rich sediments, however, suggest that the transition from euxinic to well-oxygenated conditions may have occurred much earlier, i.e., already in the late Eocene (Poirier & Hillaire-Marcel, 2009, 2011).

3.2.3.4 Early Onset of Arctic Sea Ice Formation and Cooling of Sea-Surface Temperatures

From sub-Arctic ice-rafted debris (IRD) records in the Norwegian-Greenland, Iceland, and Irminger seas and Fram Strait area it was indirectly inferred that the NHG began no earlier than about 14 Ma (Fronval & Jansen, 1996; Thiede et al., 1998; Winkler, Wolf-Welling, Stattegger, & Thiede, 2002; Wolf & Thiede, 1991). Glaciation of Antarctica, on the other hand, began much earlier, with large ice sheets first appearing near the Eocene/Oligocene boundary at about 34 Ma (Escutia, Brinkhuis, Klaus, & Expedition 318 Scientists, 2010; Kennett & Shackleton, 1976; Lear, Elderfield, & Wilson, 2000; Miller, Wright, & Fairbanks, 1991; Miller, Fairbanks, & Mountain, 1987; Shackleton & Kennett, 1975; Zachos et al., 2001, 2008). The ACEX results, however, push back the date of Northern Hemisphere cooling and onset of sea ice into the Eocene as well. The first occurrence of sea ice-related diatoms, contemporaneously with IRD, was at about 47–46 Ma (when using the ACEX age model of Backman et al., 2008; "Age Model 1") or ~43 Ma* (when using the alternate chronology of Poirier & Hillaire-Marcel, 2011; "Age Model 2," ages marked by an asterisk). Iceberg transport was probably also present in the middle Eocene, as indicated by the occurrence of some isolated pebbles (Moran et al., 2006) and by mechanical surface-texture features on quartz grains from this interval (St. John, 2008; Stickley et al., 2009). Following the discovery of Eocene IRD in the ACEX record (Backman, et al., 2006; Moran et al., 2006), Eldrett, Harding, Wilson, Butler, and Roberts (2007) and Tripati et al. (2008) reexamined the late Eocene–Oligocene sediment record from ODP Site 913 in the Greenland Basin. Together with data on mass accumulation rates of grain size fractions >63, >125, and >250 μm as well as data on the composition of the coarse fractions, Eldrett et al. (2007) report an interval of extensive but highly variable ice rafting in the Greenland Sea between 30 and 38 Ma, related to iceberg transport (rather than sea ice) with East Greenland as the likely source. Tripati et al. (2008) extended the occurrence of IRD at ODP Site 913 to 44 Ma. Furthermore, these authors determined distinct peaks in IRD at 42–40 Ma, 39–37 Ma, and 34–30 Ma, interpreted as phases of extended glaciation on Greenland. These results suggest the potential existence of (at least) isolated glaciers on Greenland at times close to the onset of major glaciations in Antarctica, a hypothesis, however, to be approved by further data.

How are this cooling trend and onset of sea ice formation reflected in the SST, and how do IRD and SST records correlate with each other? Our SST determinations in middle Eocene ACEX sediments are based on the $U_{37}^{K'}$ alkenone approach, following Weller and Stein (2008) who have shown that the alkenones measured in the ACEX section definitely have a marine origin and produced a first low-resolution mid-Eocene SST record. Here, we present a new high-resolution alkenone-based SST record for this time interval. The alkenone-based SST record is discussed together with TEX_{86}-based SST records from different ocean areas (for TEX_{86} background information see Schouten, Hopmans, Schefuß, & Sinninghe Damste´, 2002; Kim et al., 2010; Ho et al., 2014).

In the studied core interval from about 300 to 200 mcd, representing the time interval from about 49 to 44.5 Ma (Age Model 1) or from 48 to 37 Ma (Age Model 2), alkenone based SSTs vary between about 26 and 10 °C (Figure 3.2.10). As the alkenone-producing coccolithophorids need sunlight for photosynthesis, we interpret these SST values as summer temperatures. For the lowermost part of the alkenone-based SST record representing the final phase of the *Azolla* event, TEX_{86}-based SST values are also available (Sluijs et al., 2008). The U_{37}^{K}-based SSTs are still very high and vary between about 20 °C and 25 °C, whereas the TEX_{86}-derived SSTs vary between 8 °C and 13 °C, i.e., they are about 10 °C colder (Figure 3.2.11). To explain these differences, different environmental conditions such as different growth seasons and different water depths were discussed by Weller and Stein (2008).

The overall distinct drop in alkenone-based SST of ACEX record coincides with the global post-Early Eocene Climate Optimum (EECO) cooling trend (Figure 3.2.11; Zachos et al., 2008). Below 263 mcd, prior to about 47 Ma (44 Ma* = "Age Model 2"), warm SST values between 18 and 26 °C were predominant, interrupted by a prominent, short-lived cooling event at 270 mcd, near 47.3 Ma (44.5 Ma*) at which SST dropped down to 5–10 °C (Figure 3.2.10). At about 260 mcd (47 Ma or 44 Ma*), $C_{37:4}$ alkenones occurred in significant amounts (>5% of total C_{37} alkenones). Elevated $C_{37:4}$ concentrations (in combination with other sedimentological and micropaleontological proxies, see below) might be interpreted as decrease in surface-water salinity related to seasonal summer melting of sea ice (cf., Bard, Rostek, Turon, & Gendreau, 2000; Bendle, Rosell-Melé, & Ziveri, 2005; Blanz, Emeis, & Siegel, 2005; McClymont, Rosell-Mele, Haug, & Lloyd, 2008). At about 46.3 Ma (41.5 Ma*), summer SST dropped down to <17 °C (range 8–17 °C), coinciding with a significant increase in IRD (St. John, 2008; Stickley et al., 2009). An absolute SST minimum of 6–8 °C was reached at about 44.8 Ma (37.8 Ma*, coinciding with the Middle Eocene Climate Optimum, Figure 3.2.11), followed by a short but prominent warm phase with an SST around 17 °C near 44.6 Ma (37.5 Ma*). This warming event is characterized by the absence of IRD (Figure 3.2.10), interpreted as sea ice-free time interval. Apart from this warm event, however, widespread sea ice seems to be the more typical phenomenon of the Arctic Ocean

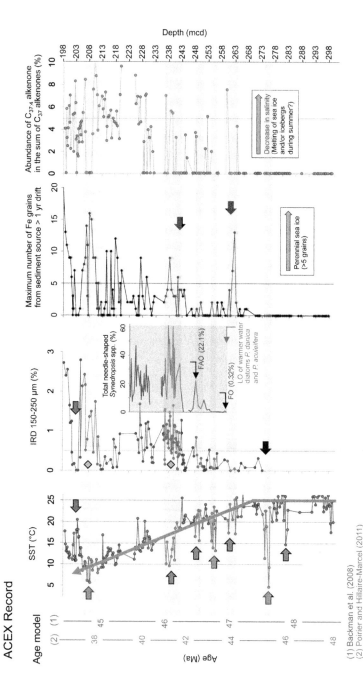

FIGURE 3.2.10 Alkenone-based sea-surface temperature (SST) (red circles; this study), abundance of ice-rafted debris (IRD) (brown circles; St. John, 2008), maximum number of Fe grains from sediment sources with >1 yr drift time and interpreted as indicator for perennial sea ice (Darby, 2014), and abundances of $C_{37:4}$ alkenones (blue circles; this study), determined in the ACEX sequence from 298 to 198 m below seafloor (mcd). Our (summer) SSTs >5 °C do not support perennial sea ice cover during the studied time interval. For the ACEX interval 260–223 mbsf, the abundance of sea-ice diatom species *Synedropsis* spp. is also shown (Stickley et al., 2009). In the diatom record, the first occurrence (FO) and abundant occurrence (FAO) of the sea-ice species and the last occurrence (LO) of warmer water diatoms *Porotheca danica* and *Pterotheca aculeifera* are also shown. The two yellow rhombs in the IRD record indicate occurrence of large-sized single dropstones (Moran et al., 2006). Large black arrows indicate major cooling events; the large red arrow highlights a major warming event near 44.6 Ma. Large black arrows and related ages indicate major steps/increases in sea-ice cover. On the left-hand side, the two different age scales discussed in the text are shown; Age Model 1 (Backman et al., 2008) and Age Model 2 (Poirier & Hillaire-Marcel, 2011).

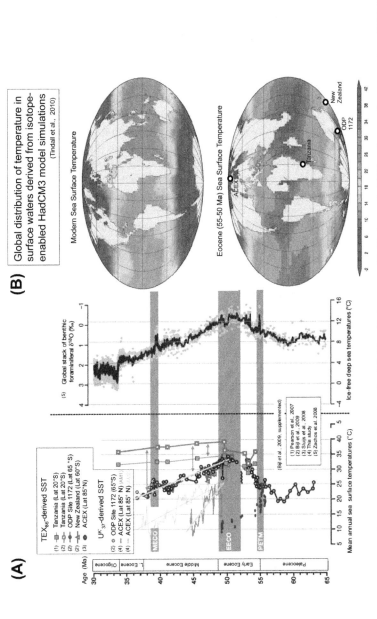

FIGURE 3.2.11 (A) Paleogene deep-sea temperatures (based on the global benthic isotope stack) (Zachos et al., 2008) and sea-surface temperatures (SST) *(main figure from Bijl et al., 2009; supplemented)*. TEX$_{86}$ SST reconstructions (1) from the Tanzania site (based on TEX$_{86}$ calibration of Schouten et al. (2003)) (Pearson et al., 2007); (2) from ODP Site 1172 (East Tasman Plateau), New Zealand, and Tanzania (2), all based on the same calibration of Kim et al. (2008); and (3) from the ACEX record (Sluijs et al., 2008). In addition, alkenone-based SST records from ODP Site 1172 (Bijl et al., 2009) and the ACEX site (this study) are shown. Our ACEX alkenone-based SST record is shown versus both chronologies (AM1 = Age Model 1 of Backman et al., 2008; AM2 = Age Model 2 of Poirier & Hillaire-Marcel, 2011). PETM = Paleocene Eocene Thermal Maximum; EECO = Early Eocene Climate Optimum; MECO = Middle Eocene Climate Optimum. (B) Global distribution of modern and early Eocene sea-surface temperature derived from isotope-enabled "HadCM3 model simulations" *(from Tindall et al., 2010; supplemented)*. Site locations discussed in this chapter are shown.

after about 45.5 Ma (40.5 Ma*), as indicated by high amounts of IRD and high abundances of $C_{37:4}$ alkenones (Figure 3.2.10).

Our interpretation of the biomarker data is supported by data and interpretation of sea ice diatoms determined in the ACEX sequence (Figure 3.2.10) (Stickley et al., 2009). Based on the presence of middle Eocene Arctic sea ice from an extraordinary abundance of a group of sea ice-dependent fossil diatoms (*Synedropsis* spp.) and the abundance of IRD, Stickley et al. (2009) proposed a two-phase establishment of sea ice: an initial phase of episodic formation in marginal shelf areas started near 47.5 Ma (45.5 Ma*), followed 0.5 my later by the onset of seasonally paced sea ice formation in offshore areas of the central Arctic. The latter is marked by the first occurrence of well-preserved sea ice diatoms *Synedropsis* spp., coinciding with the disappearance of a cosmopolitan (warmer water) diatom component characterized by *Porotheca danica* and *Pterotheca aculeifera* (Figure 3.2.10) (Stickley et al., 2009). More or less contemporaneously with the first occurrence of sea ice diatoms, $C_{37:4}$ alkenones indicative for low-saline meltwater supply occurred for the first time (Figure 3.2.10).

Most recently, Darby (2014) postulated ephemeral formation of even perennial sea ice in the Arctic Ocean during the middle Eocene. This statement is based on the occurrence of specific coarse Fe grains (IRD) with a distal source area that has a >1-year drift time to the ACEX core site. The onset of deposition of such Fe grains at the ACEX site is almost contemporaneous with the drop in SST as well as the onset of IRD, sea ice diatoms, and $C_{37:4}$ alkenones (Figure 3.2.10). The latter data, however, are more indicative for a seasonal sea ice cover. Thus, if present, phases of perennial sea ice should have been more the exception in the middle Eocene, whereas seasonal sea ice should have been the rule.

A novel and promising biomarker proxy for reconstruction of Arctic sea ice conditions was developed and is based on the determination of a highly branched isoprenoid with 25 carbons (IP_{25}) (Belt et al., 2007; Müller et al., 2009; for review see; Stein, Fahl, & Müller, 2012; Belt & Müller, 2013). Recent studies by Stein and Fahl (2013) and Stein et al., (2014a) reveal that the IP_{25} biomarker approach has potential for paleo-sea-ice reconstructions covering the entire Quaternary to mid-Pliocene time interval. Testing this approach for ACEX sediments of Eocene age unfortunately remained unsuccessful, i.e., no IP_{25} was found in these sediments (Stein, unpublished data 2010). The absence of this sea ice biomarker, however, may be explained in several ways: (1) only a very limited number of long-time stored ground and nonfrozen sediments have been used (nonoptimum storage for biomarker studies), (2) IP_{25} is not preserved in these very old sediments, and/or (3) Eocene sea ice diatoms did not synthesize IP_{25}. Here, further studies on fresh material are needed.

Overall, the Arctic SST values remain surprisingly high, even in the upper part of our ACEX record. The apparent paradox of coincidence of such a warm SST and sea ice, however, can be explained. Assuming that the alkenone SST represents rather the summer SST and considering the strong seasonal temperature variability of >10 °C during the early-middle Eocene (Basinger, Greenwood, & Sweda, 1994; Wolfe, 1994; Greenwood & Wing, 1995; see Weller & Stein, 2008 for more detailed discussion of ACEX alkenone data and references), favorable conditions for sea ice formation may have occurred during winter time. That means, after about 46.3 Ma (41.5 Ma*) the environmental conditions in part of the Arctic Ocean might have been similar to that observed in the modern Baltic Sea where summer SSTs of >15 °C and winter SSTs <1 °C with sea ice formation are typical (Krause, 1969; Wüst & Brögmus, 1955).

In order to interpret the Arctic SST record in a more global climate view, the new middle Eocene alkenone-based SST record from the Arctic Ocean is compared with TEX_{86} SST records obtained from a stratigraphically continuous sedimentary section from the southwest Pacific Ocean (ODP Leg 189 Site 1172, subpolar East Tasman Plateau, paleolatitude 65°S) (Bijl et al., 2009) and from Tanzania (subtropics, 20°S) (Pearson et al., 2007), respectively (Figure 3.2.11). For the former record, an alkenone-based SST record is also available for the late middle Eocene, supporting the TEX_{86} record. The early to middle Eocene subtropical SST from Tanzania remained quite high and constant with tropical values between 31 and 34 °C (Pearson et al., 2007), whereas the Tasmanian record reached tropical values of about 33 °C during the EECO (53–49 Ma), followed by a continuous cooling trend during the middle Eocene, reaching about 27 °C near 46 Ma and 23 °C near 42 Ma (Figure 3.2.11) (Bijl et al., 2009). Recently, Hollis et al. (2012) recalculated the Site 1172 SST values using the new $TEXL_{86}$ calibration by Kim et al. (2010), resulting in somewhat lower SST values but showing a similar relative SST decrease from about 26 °C during the EECO to about 18 °C near 42 Ma.

The trends of the Arctic ACEX as well as the subpolar Tasman SST records are similar to that in the global compilation of benthic foraminiferal oxygen isotopes (Figure 3.2.11). These records indicate that there was a strongly reduced SST gradient between subequatorial and subpolar/polar regions during the early Eocene. This gradient seems to be significantly lower than predicted by climate models (Figure 3.2.11). During the middle Eocene, on the other hand, the meridional SST gradient distinctly increased (Bijl et al., 2009). If the Eocene cooling was predominantly controlled by a decrease in atmospheric greenhouse gas concentration that cooled both the high and the low latitudes (Huber & Sloan, 2001; Zachos et al., 2008), additional processes are required to explain the relative stability of tropical SSTs and the more significant cooling at higher latitudes (Bijl et al., 2009). That means high-latitude climate feedbacks must have had a more important influence on climate change than previously thought. Differences in cloud/water vapor as well as ice-albedo effects have been mentioned as potential positive-feedback mechanism for the middle Eocene cooling (Bijl et al., 2009). The general coincidence of onset and/or extension of sea ice with major SST cooling steps may support the importance of sea ice formation and related ice-albedo effects for the exceptional drop in Arctic Ocean SST.

The short-term, more drastic, and abrupt cooling events of 5–10 °C probably occurred on Milankovich (orbital) timescales.

A major challenge of the ACEX record remains the age model (see above). If using Age Model 1 (Backman et al., 2008), the ACEX record represents the time interval 49–44 Ma; if using Age Model 2 (Poirier & Hillaire-Marcel, 2011), the ACEX record represents the time interval 48–37 Ma*. That means, the drop in SST seems to be more drastic in the North Polar Region in comparison to the polar southern hemisphere, when using Age Model 1. On the other hand, the middle Eocene drop in SST seems to be more or less parallel for both (sub) polar regions when using Age Model 2. These different SST trends represent very different climatic scenarios to be tested and considered by further (urgently needed) well-dated proxy records and modeling studies.

Despite these uncertainties, the records from ACEX (Backman et al., 2006, 2008; Moran et al., 2006; St. John, 2008) and ODP Site 913 (Eldrett et al., 2007; Tripati et al., 2008) prove an early onset/intensification of NHGs during Eocene times, as already proposed from changes in oxygen-isotope composition across the Eocene/Oligocene boundary and in the late Eocene in records from the tropical Pacific and South Atlantic (Coxall, Wilson, Pälike, Lear, & Backman, 2005; Tripati, Backman, Elderfield, & Ferretti, 2005). Furthermore, the increases in IRD in the ACEX and ODP Site 913 records coincided with major decreases in atmospheric CO_2 concentrations (Lowenstein & Demicco, 2006; Pagani et al., 2006; Pearson & Palmer, 2000). These data suggest that the evolution of Arctic and Antarctic Cenozoic climate is more closely timed, i.e., the Earth's transition from the "greenhouse" to the "icehouse" world was bipolar, which points to greater control of global cooling linked to changes in greenhouse gases in contrast to tectonic forcing, e.g., opening of gateways (Moran et al., 2006). The decline of atmospheric concentrations of CO_2 in the middle Eocene may have driven both poles across the temperature threshold that enabled the nucleation of glaciers on land and partial freezing of the surface Arctic Ocean, especially during times of low insolation (St. John, 2008).

3.2.4 OUTLOOK: NEED FOR FUTURE SCIENTIFIC DRILLING IN THE ARCTIC OCEAN

While the Arctic paleoceanographic and paleoclimate results from ACEX were unprecedented, key questions related to the climate history of the Arctic Ocean on its course from greenhouse to icehouse conditions during early Cenozoic times remain unanswered, in part because of poor core recovery, and in part because of the possible presence of a major mid-Cenozoic hiatus within the ACEX record (Figure 3.2.12; see more detailed discussion above). In addition to elevated atmospheric CO_2 concentrations in the Cenozoic, other boundary conditions such as the freshwater budget, exchange between the Arctic and Pacific/Atlantic oceans as well as the advance and retreat of major circum-Arctic ice sheets have changed dramatically during the mid to late Cenozoic. In

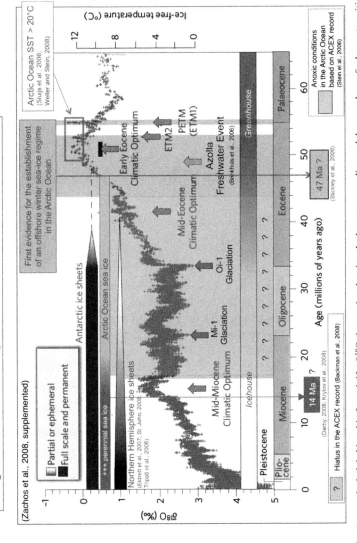

FIGURE 3.2.12 A smoothed global benthic foraminifer δ¹⁸O time series showing the long-term cooling and the greenhouse/icehouse transition through Cenozoic times *(from Zachos et al., 2008; supplemented)*. The occurrence of Cenozoic ice sheets on the Northern and Southern Hemisphere and Arctic sea ice are shown (Eldrett et al., 2007; St. John, 2008; Tripati et al., 2008; Zachos et al., 2008; Stickley et al., 2009). The hiatus in the ACEX record (as based on the age model by Backman, et al., 2008; for alternate age model see Figure 3.2.5) is indicated by a red (light gray in print versions) bar, periods with anoxic water mass conditions are highlighted by gray (dark gray in print versions) bars (Stein et al., 2006). *Figure from Zachos et al. (2008), supplemented by ACEX data.*

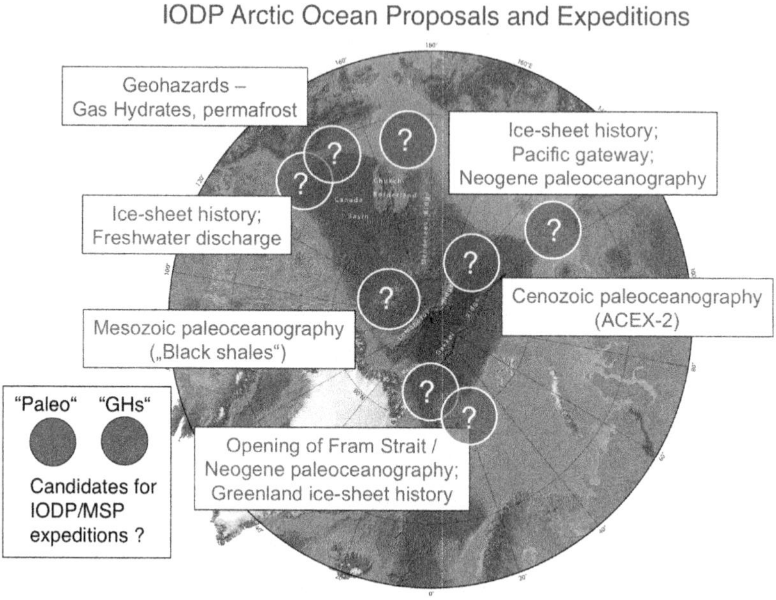

FIGURE 3.2.13 Potential key themes and areas for future IODP drilling campaigns in the Arctic Ocean (cf., Stein, 2008, 2011; Coakley & Stein, 2010).

this context, the development of extensive glaciation and its early onset in the Northern Hemisphere is still a key topic controversially discussed. For example, did extensive glaciations (such as the Oi-1 and Mi-1 glaciations) develop synchronously in both the Northern and Southern Hemispheres (Figure 3.2.12; Zachos et al., 2001, 2008)? Did major East Siberian ice sheets occur during Plio-Pleistocene times? Based on detailed multibeam swath sonar system, sediment echo sounding, and multichannel seismic reflection profiling along the East Siberian continental margin, Niessen et al. (2013) postulated that huge, kilometer-thick marine ice sheets occurred repeatedly off Siberia during Pleistocene glacial intervals, maybe even already since the Pliocene (Hegewald & Jokat, 2013). The existence of such huge ice sheets must have had a significant influence on Earth albedo and oceanic and atmospheric circulation patterns, not considered in climate models so far (Niessen et al., 2013). Furthermore, the history of circum-Arctic ice sheet also controlled the scale and timing of short- and long-term sea-level changes. An understanding of how these boundary conditions have influenced the form, intensity, and permanence of the Arctic sea ice cover can help to improve our understanding of the complex modern ocean–atmosphere–ice system and how it has evolved with global climate (O'Regan, 2011). In order to follow up ACEX and to answer some of the still open key questions, a second scientific drilling on southern Lomonosov Ridge with a focus on the reconstruction of the continuous and complete Cenozoic climate history of the Arctic Ocean has been proposed within the new IODP ("ACEX-2," Stein, Jokat, et al., 2014b; for location see Figures 3.2.3 and 3.2.13). But, of course, all questions dealing with the Arctic Ocean paleoclimatic and tectonic history cannot be answered within one further expedition. Here, more drilling campaigns are needed; some of them hopefully can be carried out within the running phase of the new IODP. In this context, several IODP proposals for future Arctic drilling have already been submitted or are under discussion (Figure 3.2.13; Stein, 2008, 2011).

ACKNOWLEDGMENTS

This research used samples and data provided by IODP. We thank the masters, crew, and Scientific Party of IODP Expedition 302, as well as the IODP-ECORD-ESO.

REFERENCES

ACIA. (2004). *Impacts of a warming arctic: arctic climate impact assessment*. Cambridge: Cambridge University Press pp. 139 http://www.acia.uaf.edu.

ACIA. (2005). *Arctic climate impact assessment*. Cambridge University Press. pp. 1042.

Arora, A., & Singh, P. K. (2003). Comparison of biomass productivity and nitrogen fixing potential of Azolla Spp. *Biomass and Bioenergy, 24*, 175–178.

Backman, J., Jakobsson, M., Frank, M., Sangiorgi, F., Brinkhuis, H., Stickley, C., et al. (2008). Age model and core-seismic integration for the Cenozoic Arctic Coring expedition sediments from the Lomonosov Ridge. *Paleoceanography, 23*, PA1S03.

Backman, J., Jakobsson, M., Løvlie, R., Polyak, L., & Febo, L. A. (2004). Is the central Arctic Ocean a sediment starved basin? *Quaternary Science Reviews, 23*, 1435–1454.

Backman, J., & Moran, K. (2008). Introduction to special section on Cenozoic paleoceanography of the central Arctic ocean. *Paleoceanography, 23*, PA1S01. http://dx.doi.org/10.1029/2 007PA001516.

Backman, J., & Moran, K. (2009). Expanding the Cenozoic paleoceanographic record in the central Arctic ocean: IODP expedition 302 synthesis. *Central European Journal of Geoscience, 1*, 157–175.

Backman, J., Moran, K., McInroy, D. B., Mayer, L. A., & the Expedition 302 Scientists. (2006). *Proceedings IODP, 302. College Station, Texas: Integrated Ocean Drilling Program Management International, Inc.* http://dx.doi.org/10.2204/iodp.proc.302.104.2006.

Bard, E., Rostek, F., Turon, J.-L., & Gendreau, S. (2000). Hydrological impact of heinrich events in the subtropical Northeast Atlantic. *Science, 289*, 1321–1324.

Barke, J., van der Burgh, J., van Konijnenburg-van Cittert, J. H. A., Collinson, M. E., Pearce, M. A., Bujak, J., et al. (2012). Coeval Eocene blooms of the freshwater fern *Azolla* in and around Arctic and Nordic seas. *Palaeogeography, Palaeoclimatology, Palaeoecology, 337–338*, 108–119.

Basinger, J. F., Greenwood, D. G., & Sweda, T. (1994). Early Tertiary vegetation of Arctic Canada and its relevance to paleoclimatic interpretation. In M. C. Boulter, & H. C. Fischer (Eds.), *Cenozoic plants and climates of the high Arctic NATO ASI series 1: Vol. 127.* (pp. 175–198). Heidelberg: Springer Verlag.

Belt, S. T., Massé, G., Rowland, S. J., Poulin, M., Michel, C., & LeBlanc, B. (2007). A novel chemical fossil of palaeo sea ice: IP25. *Organic Geochemistry, 38*, 16–27.

Belt, S. T., & Müller, J. (2013). The Arctic sea ice biomarker IP$_{25}$: a review of current understanding, recommendations for future research and applications in palaeo sea ice reconstructions. *Quaternary Science Reviews, xx*, 1–17.

Bendle, J., Rosell-Melé, A., & Ziveri, P. (2005). Variability of unusual distributions of alkenones in the surface waters of the Nordic seas. *Paleoceanography, 20*. http://dx.doi.org/10.1029/20 04PA001025.

Berner, R. A. (1984). Sedimentary pyrite formation: an update. *Geochim Cosmochim Acta, 48*, 605–615.

Bijl, P. K., Schouten, S., Sluijs, A., Reichart, G.-J., Zachos, J. C., & Brinkhuis, H. (2009). Early Palaeogene temperature evolution of the southwest Pacific ocean. *Nature, 409*, 776–779.

Blanz, T., Emeis, K. C., & Siegel, H. (2005). Controls on alkenone unsaturation ratios along salinity gradient between the open ocean and the Baltic Sea. *Geochimica Cosmochimica Acta, 69*, 3589–3600.

Brinkhuis, H., Schouten, S., Collinson, M. E., Sluijs, A., Damsté, J. S. S., Dickens, G. R., et al. (2006). Episodic fresh surface waters in the Eocene Arctic Ocean. *Nature, 441*, 606–609.

Bujak, J. P., & Brinkhuis, H. (1998). In M.-P. Aubry, S. G. Lucas, & W. A. Berggren (Eds.), *Late Paleocene-early eocene climatic and biotic events in the Marine and terrestrial records* (pp. 277–295). New York: Columbia Univ. Press.

Clark, D. L., Byers, C. W., & Pratt, L. M. (1986). Cretaceous black mud from the central Arctic Ocean. *Paleoceanography, 1*, 265–271.

Clark, D. L., Whitman, R. R., Morgan, K. A., & Mackey, S. D. (1980). Stratigraphy and glacial marine sediments of the Amerasian Basin, central Arctic Ocean. *Geological Society of America Special Paper, 181*, 57 pp.

Coakley, B., & Stein, R. (2010). Arctic ocean scientific drilling: the Next Frontier. *Scientific Drilling, 9*, 45–49.

Collinson, M. E., Barke, J., van der Burgh, J., & van Konijnenburg-van Cittert (2009). A new species of the freshwater fern *Azolla* (Azollaceae) from the Eocene Arctic Ocean. *Review of Palaeobotany and Palynology, 155*, 1–14.

Coxall, H. K., Wilson, P. A., Pälike, H., Lear, C. H., & Backman, J. (2005). Rapid stepwise onset of Antarctic glaciation and deeper calcite compensation in the Pacific Ocean. *Nature, 433*, 53–57.

Cronin, T. M., Smith, S. A., Eynaud, F., O'Regan, M., & King, J. (2008). Quaternary paleoceanography of the central arctic based on integrated ocean drilling program arctic coring expedition 302 foraminiferal assemblages. *Paleoceanography, 23*, PA1S18. http://dx.doi.org/10.1029/2007PA001484.

Darby, D. A. (2008). The Arctic perennial ice cover over the last 14 million years. *Paleoceanography, 23*. http://dx.doi.org/10.1029/2007PA001479.

Darby, D. A. (2014). Ephemeral formation of perennial sea ice in the Arctic Ocean during the middle Eocene. *Nature Geoscience, 7*, 210–213. http://dx.doi.org/10.1038/NGEO2068.

Davies, A., Kemp, A. E. S., & Pike, J. (2009). Late Cretaceous seasonal ocean variability from the Arctic. *Nature, 460*, 254–258.

Davies, A., Kemp, A. E. S., & Pälike, H. (2011). Tropical ocean-atmosphere controls on interannual climate variability in the Cretaceous Arctic. *Geophysical Research Letters, 38.* . http://dx.doi.org/10.1029/2010GL046151.

Dickens, G. R., O'Neil, J. R., Rea, D. K., & Owen, R. M. (1995). Dissociation of oceanic methane hydrate as a cause of the carbon isotope excursion at the end of the Paleocene. *Paleoceanography, 10*, 965–971.

Edwards, M. H., & Coakley, B. J. (2003). SCICEX investigations of the arctic ocean system. *Chemie der Erde, 63*, 281–392.

Eldrett, J. S., Harding, I. C., Wilson, P. A., Butler, E., & Roberts, A. P. (2007). Continental ice in Greenland during the Eocene and Oligocene. *Nature, 446*, 176–179. http://dx.doi.org/10.1038/nature05591.

Escutia, C., Brinkhuis, H., Klaus, A., & Expedition 318 Scientists (2010). Wilkes land glacial history: cenozoic east antarctic ice sheet evolution from Wilkes land margin sediments. In *Proceedings of the integrated ocean drilling program* (Vol. 318) Preliminary Reports.

Eynaud, F., Cronin, T. M., Smith, S. A., Zaragosi, S., Mavel, J., Mary, Y., et al. (2009). Morphological variability of the planktonic foraminifer *Neogloboquadrina pachyderma* in the late Pleistocene of the ACEX cores. *Micropaleontology, 55*, 101–116.

Firth, J. V., & Clark, D. L. (1998). An early Maastrichtian organic-walled phytoplankton cyst assemblage from an organic-walled black mud in Core F1-533, Alpha Ridge: evidence for upwelling conditions in the Cretaceous Arctic Ocean. *Marine Micropaleontology, 34*, 1–27.

Frank, M., Backman, J., Jakobsson, M., Moran, K., O'Regan, M., King, J., et al. (2008). Beryllium isotopes in central Arctic Ocean sediments over the past 12.3 million years: stratigraphic and paleoclimatic implications. *Paleoceanography* 23, http://dx.doi.org/10.1029/2007PA001478.

Fronval, T., & Jansen, E. (1996). Late Neogene paleoclimates and paleoceanography in the Iceland Norweigian Sea: evidence from the Iceland and Vøring Plateaus. In J. Thiede, A. M. Myhre, J. V. Firth, G. L. Johnson, & W. F. Ruddiman (Eds.), *Proc. ODP, Sci. Results, 151* (pp. 455–468). College Station, Texas: Ocean Drilling Program.

Gleason, J. D., Thomas, D. J., Moore, T. C., Jr., Blum, J. D., & Haley, B. A. (2007). Water column structure of the Eocene Arctic Ocean from Nd-Sr isotope proxies in fossil fish debris. *Geochimica et Cosmochimica Acta, 71*, A329.

Goldhaber, M. B., & Kaplan, I. R. (1974). The sulfur cycle. In E. D. Goldberg, et al. (Ed.), *The sea* (Vol. 5) (pp. 569–655). New York: J. Wiley and Sons.

Greenwood, D. G., & Wing, S. L. (1995). Eocene continental climates and latitudinal temperature gradients. *Geology, 23*, 1044–1048.

Haley, B. A., Frank, M., Spielhagen, R. F., & Eisenhauer, A. (2008). Influence of brine formation on Arctic Ocean circulation over the past 15 million years. *Nature Geoscience, 1*, 68–72.

Heezen, B. C., & Ewing, M. (1961). The mid-oceanic Ridge and its extension through the arctic Basin. In G. O. Raasch (Ed.), *Geology of the Arctic* (pp. 622–642). University of Toronto Press.

Hegewald, A., & Jokat, W. (2013). Tectonic and sedimentary structures in the northern Chukchi region. *Arctic Ocean Journal Geophysical Research, 118*. http://dx.doi.org/10.1002/jgrb.50282.

Hollis, C. J., Taylor, K. W. R., Handley, L., Pancost, R. D., Huber, M., Creech, J. B., et al. (2012). Early Paleogene temperature history of the Southwest Pacific Ocean: reconciling proxies and models. *Earth and Planetary Science Letters, 349–350*, 53–66.

Ho, S. L., Mollenhauer, G., Fietz, S., Martinez-Garcia, A., Lamy, F., Rueda, G., et al. (2014). Appraisal of the TEX_{86} and TEX_{86}^L thermometries in the subpolar and polar regions. *Geochimica et Cosmochimica Acta, 131*, 213–226. http://dx.doi.org/10.1016/j.gca.2014.01.001.

Huber, M., & Sloan, L. C. (2001). Heat transport, deep waters, and thermal gradients: coupled simulation of an Eocene 'greenhouse' climate. *Geophysical Research Letters, 28*, 3481–3484.

Immonen, N. (2013). Surface microtextures of ice-rafted quartz grains revealing glacial ice in the Cenozoic Arctic. *Palaeogeography Palaeoclimatology Palaeoecology, 374*, 293–302.

Jackson, H. R., Mudie, P. J., & Blasco, S. M. (1985). *Initial geological report on CESAR -The Canadian expedition to study the alpha Ridge, arctic ocean.* Geol. Surv. Can. Papp. 84–22, 177 pp.

Jakobsson, M. (1999). First high-resolution chirp sonar profiles for the central Arctic Ocean reveal erosion of Lomonosov Ridge sediments. *Marine Geology, 158*, 111–123.

Jakobsson, M., Backman, J., Rudels, B., Nycander, J., Frank, M., Mayer, L., et al. (2007). The early Miocene onset of a ventilated circulation regime in the Arctic Ocean. *Nature, 447*, 986–990.

Jenkyns, H. C., Forster, A., Schouten, S., & Sinninghe Damsté, J. S. (2004). High temperatures in the late Cretaceous Arctic Ocean. *Nature, 432*, 888–892.

Jokat, W. (2005). The sedimentary structure of the Lomonosov Ridge between 88°N and 80°N. *Geophysical Journal International, 163*, 698–726.

Jokat, W., Stein, R., Rachor, E., & Schewe, I. (1999). Expedition gives fresh view of central Arctic geology. *EOS Transactions, 80*(465), 472–473.

Jokat, W., Uenzelmann-Neben, G., Kristoffersen, Y., & Rasmussen, T. M. (1992). Lomonosov Ridge: a double-sided continental margin. *Geology, 20*, 887–890.

Jokat, W., Weigelt, E., Kristoffersen, Y., & Rasmussen, T. M. (1995). New insights into the evolution of Lomonosov Ridge and the Eurasia Basin. *Geophysical Journal International, 122*, 378–392.

Kaminski, M. A., Silye, L., & Kender, S. (2005). Miocene deep-water agglutinated foraminifera from ODP Hole 909C: Implications for the paleoceanography of the Fram Strait area, Greenland Sea. *Micropaleontology, 51*, 373–403.

Kaminski, M. A., Silye, L., & Kender, S. (2009). Miocene deep-water agglutinated foraminifera from the Lomonosov Ridge and the opening of the Fram Strait. *Micropaleontology, 55*, 117–135.

Kender, S., & Kaminski, M. A. (2013). Arctic Ocean benthic foraminiferal faunal change associated with the onset of perennial sea ice in the middle Miocene. *Journal of Foraminiferal Research, 43*, 99–109.

Kennett, J. P., & Shackleton, N. J. (1976). Oxygen isotopic evidence for the development of the psychrosphere 38 Myr ago. *Nature, 260*, 513–515.

Kennett, J. P., & Stott, L. D. (1991). Abrupt deep-sea warming, palaeoceanographic changes and benthic extinctions at the end of the Palaeocene. *Nature, 353*, 225–229.

Kim, J. H., Schouten, S., Hopmans, E. C., Donner, B., & Sinninghe Damsté, J. S. (2008). Global sediment core-top calibration of the TEX86 paleothermometer in the ocean. *Geochim. Cosmochim. Acta, 72*, 1154–1173.

Kim, J. H., van der Meer, J., Schouten, S., Helmke, P., Willmott, V., Sangiorgi, F., et al. (2010). New indices and calibrations derived from the distribution of crenarchaeal isoprenoid tetraether lipids: Implications for past sea surface temperature reconstructions. *Geochim Cosmochim Acta, 74*, 4639–4654.

Knies, J., Mann, U., Popp, B. N., Stein, R., & Brumsack, H.-J. (2008). Surface water productivity and paleoceanographic implications in the Cenozoic Arctic Ocean. *Paleoceanography, 23*, PA1S16. pp. 1–12. http://dx.doi.org/10.1029/2007PA001455.

Krause, G. (1969). Ein Beitrag zum Problem der Erneuerung des Tiefenwassers im Arkona-Becken. *Kieler Meeresforschung, 25*, 268–271.

Kristoffersen, Y. (1990). Eurasian Basin. In A. Grantz, L. Johnson, & J. F. Sweeny (Eds.), *The arctic ocean region, geology of North America* (vol. L) (pp. 365–378). Boulder: Geological Society of America.

Kristoffersen, Y., Buravtsev, V., Jokat, W., & Poselov, V. (1997). *Seismic reflection surveys during Arctic Ocean-96*. Cruise report, In: Polarforskningssekretaariatets arsbok 1995/96, edited by E. Grönlund, pp. 75–77, Polarforskningssekretariatet, Stockholm.

Krylov, A. A., Andreeva, I. A., Vogt, C., Backman, J., Krupskaya, V. V., Grikurov, G. E., et al. (2008). A shift in heavy and Clay Mineral Provenance indicates a middle miocene onset of a perennial sea-ice cover in the arctic ocean. *Paleoceanography, 23*, PA1S06. http://dx.doi.org/10.1029/2007PA001497.

Lear, C. H., Elderfield, P. A., & Wilson, P. A. (2000). Cenozoic deep-sea temperatures and global ice volumes from Mg/Ca in benthic foraminiferal calcite. *Science, 287*, 269–272.

Leventhal, J. S. (1983). An interpretation of carbon and sulfur relationships in Black Sea sediments as indicators of environments of deposition. *Geochim Cosmochim Acta, 47*, 133–137.

Lourens, L. J., Sluijs, A., Kroon, D., Zachos, J. C., Thomas, E., Röhl, U., et al. (2005). Astronomical pacing of late Palaeocene to early Eocene global warming events. *Nature, 435*, 1083–1087.

Lowenstein, T. K., & Demicco, R. V. (2006). Elevated Eocene atmospheric CO_2 and its subsequent decline. *Science, 313*, 1928.

Mann, U., Knies, J., Chand, S., Jokat, W., Stein, R., & Zweigel, J. (2009). Evaluation and modelling of tertiary source rocks in the central Arctic Ocean. *Marine and Petroleum Geology, 26*, 1624–1639. http://dx.doi.org/10.1016/j.marpetgeo.2009.01.008.

Matthiessen, J., Brinkhuis, H., Poulsen, N., & Smelror, M. (2009a). Decahedrella martinheadii Manum 1997 – a stratigraphically and paleoenvironmentally useful Miocene acritarch of the high northern latitudes. *Micropaleontology, 55*, 171–186.

Matthiessen, J., Knies, J., Vogt, C., & Stein, R. (2009b). Pliocene paleoceanography of the Arctic Ocean and subarctic seas. *Philosophical Transactions of the Royal Society A, 367*, 21–48. http://dx.doi.org/10.1098/rsta.2008.0203.

McClymont, E. L., Rosell-Mele, A., Haug, G., & Lloyd, J. (2008). Expansion of subarctic water masses in the north Atlantic and Pacific Oceans and implications for mid-Pleistocene ice-sheet growth. *Paleoceanography, 23*, PA4214. http://dx.doi.org/10.1029/2008PA001622.

Miller, K. G., Fairbanks, R. G., & Mountain, G. S. (1987). Tertiary oxygen isotope synthesis, sea level history, and continental margin erosion. *Paleoceanography, 2*, 1–19.

Miller, K. G., Wright, J. D., & Browning, J. V. (2005). Visions of ice sheets in a greenhouse world. *Marine Geology, 217*, 215–231.

Miller, K. G., Wright, J. D., & Fairbanks, R. G. (1991). Unlocking the ice house: Oligocene-Miocene oxygen isotopes, eustasy, and margin erosion. *Journal of Geophysical Research, 96*(B4), 6829–6849.

Moore, T. C., & the Expedition 302 Scientists (2006). Sedimentation and subsidence history of the Lomonosov Ridge. In J. Backman, K. Moran, D. B. McInroy, L. A. Mayer, & the Expedition 302 Scientists (Eds.), *Proc. IODP, 302: Edinburgh*. Integrated Ocean Drilling Program Management International, Inc. http://dx.doi.org/10.2204/iodp.proc.302.105.2006.

Moran, K., Backman, J., Brinkhuis, H., Clemens, S. C., Cronin, T., Dickens, G. R., et al. (2006). The cenozoic palaeoenvironment of the arctic ocean. *Nature, 441*, 601–605.

Müller, J., Massé, G., Stein, R., & Belt, S. (2009). Extreme variations in sea ice cover for Fram Strait during the past 30 ka. *Nature Geoscience, 2*, 772–776, http://dx.doi.org/10.1038/NGEO665.

Mutterlose, J., Brumsack, H., Flögel, S., Hay, W., Klein, C., Langrock, U., et al. (2003). The Greenland–Norwegian Seaway: a key for understanding late Jurassic to early cretaceous paleoenvironments. *Paleoceanography, 18*, 1010. http://dx.doi.org/10.1029/2001PA000625.

Niessen, F., Hong, J. K., Hegewald, A., Matthiessen, J., Stein, R., Kim, H., et al. (2013). Repeated pleistocene glaciation of the east siberian Continental margin. *Nature Geoscience, 6*, 842–846, http://dx.doi.org/10.1038/NGEO1904.

Ogawa, Y., Takahashi, K., Yamanaka, T., & Onodera, J. (2009). Significance of euxinic condition in the middle Eocene paleo-Arctic basin: a geochemical study on the IODP Arctic Coring Expedition 302 sediments. *Earth and Planetary Science Letters, 285*, 190–197.

Onodera, J., & Takahashi, K. (2009a). Middle Eocene ebridians from the central Arctic Basin. *Micropaleontology, 55*, 187–208.

Onodera, J., & Takahashi, K. (2009b). Taxonomy and biostratigraphy of middle Eocene silicoflagellates in the central Arctic Basin. *Micropaleontology, 55*, 209–248.

Onodera, J., Takahashi, K., & Jordan, R. W. (2008). Eocene silicoflagellate and ebridian paleoceanography in the central Arctic Ocean. *Paleoceanography, 23*, PA1S15. http://dx.doi.org/10.1029/2007PA001474. pp. 1–9.

O'Regan, M. (2011). Late cenozoic paleoceanography of the Central arctic ocean. *IOP Conference Series: Earth and Environmental Science, 14.* http://dx.doi.org/10.1088/1755-1315/14/1/012002.

O'Regan, M., King, J., Backman, J., Jakobsson, M., Pälike, H., Moran, K., et al. (2008). Constraints on the Pleistocene chronology of sediments from the Lomonosov Ridge. *Paleoceanography, 23,* PA1S19.

Pagani, M., Pedentchouk, N., Huber, M., Sluijs, A., Schouten, S., Brinkhuis, H., et al. (2006). The Arctic's hydrologic response to global warming during the Palaeocene-Eocene thermal maximum. *Nature, 442,* 671–675.

Pälike, H., Spofforth, D. J. A., O'Regan, M., & Gattacecca, J. (2008). Orbital scale variations and timescales from the Arctic Ocean. *Paleoceanography, 23,* PA1S10. http://dx.doi.org/10.1029/2007PA001490, pp. 1–13.

Pearson, P. N., & Palmer, M. R. (2000). Atmospheric carbon dioxide concentrations over the past 60 million years. *Nature, 406,* 695–699.

Pearson, P. N., van Dongen, B. E., Nicholas, C. J., Pancost, R. D., Schouten, S., Singano, J. M., et al. (2007). Stable warm tropical climate through the Eocene Epoch. *Geology, 35,* 211–214.

Poirier, A., & Hillaire-Marcel, C. (2009). Os-isotope insights into major environmental changes of the Arctic Ocean during the Cenozoic. *Geophysical Research Letters, 36,* L11602. http://dx.doi.org/10.1029/2009GL037422.

Poirier, A., & Hillaire-Marcel, C. (2011). Improved Os-isotope stratigraphy of the Arctic Ocean. *Geophysical Research Letters, 38.* http://dx.doi.org/10.1029/2011GL047953.

Rai, V., & Rai, A. K. (1998). Growth behaviour of *Azolla pinnata* at various salinity levels and induction of high salt tolerance. *Plant Soil, 206,* 79–84.

Röhl, U., Bralower, T. J., Norris, G., & Wefer, G. (2000). A new chronology for the late Paleocene thermal maximum and its environmental implications. *Geology, 28,* 927–930.

Sangiorgi, F., Brinkhuis, H., & Damassa, S. P. (2009). *Arcticacysta*: a new organic-walled dinoflagellate cyst genus from the early Miocene of the central Arctic Ocean. *Micropaleontology, 55,* 249–258.

Sangiorgi, F., Brumsack, H.-J., Willard, D. A., Schouten, S., Stickley, C., O'Regan, M., et al. (2008). A 26 million year gap in the central Arctic record at the greenhouse-icehouse transition: looking for clues. *Paleoceanography, 23,* PA1S04. http://dx.doi.org/10.1029/2007PA001477, pp. 1–13.

Sangiorgi, F., van Soelen, E. E., Spofforth, D. A. J., Pälike, H., Stickley, C., St John, K., et al. (2008). Cyclicity in the middle Eocene central Arctic Ocean sediment record: orbital forcing and environmental response. *Paleoceanography, 23,* PA1S08. http://dx.doi.org/10.1029/2007PA001487, pp. 1–14.

Schouten, S., Hopmans, E., Forster, A., van Breugel, Y., Kuypers, M. M. M., & Sinninghe Damsteʹ, J. S. (2003). Extremely high sea-surface temperatures at low latitudes during the middle Cretaceous as revealed by archaeal membrane lipids. *Geology, 31,* 1069–1072.

Schouten, S., Hopmans, E. C., Schefuß, E., & Sinninghe Damsteʹ, J. S. (2002). Distributional variations in marine crenarchaeotal membrane lipids: a new tool for reconstructing ancient sea water temperatures? *Earth and Planetary Science Letters, 204,* 265–274.

Setoyama, E., Kaminski, M. A., & Tyszka, J. (2011). Campanian agglutinated foraminifera from the Lomonosov Ridge, IODP Expedition 302, ACEX, in the paleogeographic context of the Arctic Ocean. *Micropaleontology, 57,* 507–530.

Shackleton, N. J., & Kennett, J. P. (1975). *Paleotemperature history of the cenozoic and the initiation of antarctic glaciation: Oxygen and carbon isotope analysis in DSDP sites 277, 279, and 281* (Vol. 29) Initial Rep. Deep Sea Drill. Proj. 743–755.

Shellito, C. J., Sloan, L. C., & Huber, M. (2003). Climate model sensitivity to atmospheric CO_2 levels in the Early-Middle Paleogene. *Palaeogeography Palaeoclimatology Palaeoecology, 193*, 113–123.

Sinninghe Damsté, J. S., Wakeham, S. G., Kohnen, M. E. L., Hayes, J. M., & de Leeuw, J. W. (1993). A 6,000-year sedimentary molecular record of chemocline excursions in the Black Sea. *Nature, 362*, 827–829.

Sluijs, A., Röhl, U., Schouten, S., Brumsack, H.-J., Sangiorgi, F., Sinninghe Damsté, J. S., et al. (2008). Arctic late Paleocene–early eocene paleoenvironments with special emphasis on the Paleocene-Eocene Thermal maximum (Lomonosov Ridge, integrated ocean drilling program expedition 302). *Paleoceanography, 23*. http://dx.doi.org/10.1029/2007PA001495.

Sluijs, A., Schouten, S., Pagani, M., Woltering, M., Brinkhuis, H., Damsté, J. S. S., et al. (2006). Subtropical Arctic Ocean temperatures during the Palaeocene/Eocene thermal maximum. *Nature, 441*, 610–613.

Speelman, E. N., Reichart, G.-J., de Leeuw, J. W., Rijpstra, W. I. C., & Sinninghe Damsté, J. S. (2009). Biomarker lipids of the freshwater fern *Azolla* and its fossil counterpart from the Eocene Arctic Ocean. *Organic Geochemistry, 40*, 628–637.

Spofforth, D. J. A., Pälike, H., & Green, D. (2008). Paleogene record of elemental concentrations from the Arctic Ocean obtained by XRF analyses. *Paleoceanography, 23*, PA1S09. http://dx.doi.org/10.1029/2007PA001489, pp. 1–13.

St John, K. (2008). Cenozoic ice-rafting history of the Central Arctic Ocean: terrigenous sands on the Lomonosov Ridge. *Paleoceanography, 23*. http://dx.doi.org/10.1029/2007PA001483.

Stein, R. (2007). Upper cretaceous/lower tertiary black shales near the North Pole: organic-carbon origin and source-rock potential. *Marine Petroleum Geology, 24*, 67–73.

Stein, R. (2008). Arctic ocean sediments: processes, proxies, and palaeoenvironment. *Developments in marine geology* (Vol. 2)Amsterdam: Elsevier pp. 587.

Stein, R. (2011). The great challenges in Arctic Ocean paleoceanography. *IOP Conference Series: Earth and Environmental Science, 14*. http://dx.doi.org/10.1088/1755-1315/14/1/012001.

Stein, R., Boucsein, B., & Meyer, H. (2006). Anoxia and high primary production in the Paleogene central Arctic Ocean: first detailed records from Lomonosov Ridge. *Geophysical Research Letters, 33*, L18606. http://dx.doi.org/10.1029/2006GL026776.

Stein, R., & Fahl, K. (2013). Biomarker proxy IP_{25} shows potential for studying entire Quaternary Arctic sea-ice history. *Organic Geochemistry, 55*, 98–102. http://dx.doi.org/10.1016/j.orggeochem.2012.11.005.

Stein, R., Fahl, K., & Matthiessen, J. (2014a). Pliocene/Pleistocene changes in Arctic sea-ice cover: biomarker and dinoflagellate records from Fram Strait/Yermak Plateau (ODP Sites 911 and 912). *Geophysical Research Abstracts, 16* EGU2014–6895.

Stein, R., Fahl, K., & Müller, J. (2012). Proxy reconstruction of Arctic Ocean sea ice history: "From IRD to IP_{25}". *Polarforschung, 82*, 37–71.

Stein, R., Jokat, W., Brinkhuis, H., Clarke, L., Coakley, B., Jakobsson, M., et al. (2014b). *Arctic ocean Paleoceanography: towards a continuous cenozoic record from a greenhouse to an icehouse world (ACEX2).* IODP Proposal 708. www.iodp.org/700.

Stickley, C. E., John, K. S., Koc, N., Jordan, R. W., Passchier, S., Pearce, R. B., et al. (2009). Evidence for middle Eocene Arctic sea ice from diatoms and ice-rafted debris. *Nature, 460*, 376–380.

Stocker, T., Dahe, Q., Plattner, G.-K., Tignor, M. M. B., Allen, S. K., Boschung, J., et al. (2013). *Climate change 2013, the physical science basis.* New York: Intergovernmental Panel on Climate Change (IPCC), Cambridge University Press pp. 1535.

Suto, I., Jordan, R. W., & Watanabe, M. (2009). Taxonomy of middle Eocene diatom resting spores and their allied taxa from the central Arctic Basin. *Micropaleontology, 55*, 259–312.

Thiede, J., Clark, D. L., & Hermann, Y. (1990). Late Mesozoic and Cenozoic paleoceanography of the northern polar oceans. In A. Grantz (Ed.), *The Arctic Ocean Region, The geology of North America: Vol. L.* (pp. 427–458). Boulder: Geological Society of America.

Thiede, J., Winkler, A., Wolf-Welling, T., Eldholm, O., Myhre, A., Baumann, K.-H., et al. (1998). Late cenozoic history of the polar North Atlantic: results from ocean drilling. In A. Elverhøi, et al. (Ed.), *Glacial and oceanic history of the polar North Atlantic margins Quat. Sci. Rev.: Vol. 17.* (pp. 185–208).

Tindall, J., Flecker, R., Valdes, P., Schmidt, D. N., Markwick, P., & Harris, J. (2010). Modelling the oxygen isotope distribution of ancient seawater using a coupled ocean-atmosphere GCM: Implications for reconstructing early Eocene climate. *Earth and Planetary Science Letters, 292*, 265–273. Elsevier Amsterdam.

Tripati, A., Backman, J., Elderfield, H., & Ferretti, P. (2005). Eocene bipolar glaciation associated with global carbon cycle changes. *Nature, 436*, 341–346.

Tripati, A. K., Eagle, R. A., Morton, A., Dowdeswell, J. A., Atkinson, K. L., Bahé, Y., et al. (2008). Evidence for glaciation in the Northern Hemisphere back to 44 Ma from ice-rafted debris in the Greenland Sea. *Earth and Planetary Science Letters, 265*, 112–122.

Tripati, A., & Elderfield, H. (2005). Deep-sea temperature and circulation changes at the Paleocene-Eocene thermal maximum. *Science, 308*, 1894–1898.

Tripati, A., Zachos, J., Marincovich, L., Jr., & Bice, K. (2001). Late Paleocene Arctic coastal climate inferred from molluscan stable and radiogenic isotope ratios. *Palaeogeography Palaeoclimatology Palaeoecology, 170*, 101–113.

Waddell, L. M., & Moore, T. C. (2008). Salinity of the Eocene Arctic Ocean from oxygen isotope analysis of fish bone carbonate. *Paleoceanography, 23*, PA1S12. http://dx.doi.org/10.1029/2007PA001451, pp. 1–14.

Weijers, J. W. H., Schouten, S., Sluijs, A., Brinkhuis, H., & Sinninghe Damsté, J. S. (2007). Warm arctic continents during the Palaeocene-Eocene thermal maximum. *Earth and Planetary Science Letters, 261*, 230–238.

Weller, P., & Stein, R. (2008). Paleogene biomarker records from the central Arctic Ocean (IODP Expedition 302): organic-carbon sources, anoxia, and sea-surface temperature. *Paleoceanography, 23*, PA1S17. http://dx.doi.org/10.1029/2007PA001472.

Winkler, A., Wolf-Welling, T. C. W., Stattegger, K., & Thiede, J. (2002). Clay mineral sedimentation in high northern latitude deep-sea basins since the Middle Miocene (ODP Leg 151, NAAG). *International Journal of Earth Sciences, 91*(1), 133–148.

Wolfe, J. A. (1994). Tertiary climatic changes at middle latitudes of western North America. *Palaeogeography Palaeoclimatology Palaeoecology, 108*, 195–205.

Wolf, T. C. W., & Thiede, J. (1991). History of terrigenous sedimentation during the past 10 my in the North Atlantic (ODP Legs 104, 105, and DSDP 81). *Marine Geology, 101*, 83–102.

Wüst, G., & Brögmus, W. (1955). Ozeanographische Ergebnisse einer Untersuchungsfahrt mit Forschungskutter "Südfall" durch die Ostsee Juni/Juli 1954 (anlässlich der totalen Sonnenfinsternis auf Öland). *Kieler Meeresf, 11*, 3–21.

Zachos, J. C., Dickens, G. R., & Zeebe, R. E. (2008). An early Cenozoic perspective on greenhouse warming and carbon-cycle dynamics. *Nature*, *451*, 281–283.

Zachos, J., Pagani, M., Sloan, L., Thomas, E., & Billups, K. (2001). Trends, rhythms, and aberrations in global climate 65 Ma to present. *Science*, *292*, 868–893.

Ziegler, P. A. (1988). Evolution of the Arctic, North Atlantic, and the Western Tethys. *Am. Assoc. Petrol. Geol. Mem*, *43*, 198 pp., Shell Int. Petrol. Maatschapij, The Hague.

Chapter 3.3

From Greenhouse to Icehouse at the Wilkes Land Antarctic Margin: IODP Expedition 318 Synthesis of Results

Carlota Escutia*, Henk Brinkhuis & the Expedition 318 Scientists
Instituto Andaluz de Ciencias de la Tierra, CSIC-Univ. Granada, Granada, Spain
*Corresponding author: E-mail: cescutia@ugr.es

3.3.1 INTRODUCTION

Polar ice is an important component of the modern climate system, affecting, among other, global sea level, ocean circulation and heat transport, marine productivity, air–sea gas exchange, and planetary albedo. The modern ice caps are, geologically speaking, a relatively young phenomenon. Since mid-Permian (~270 Ma) times, parts of Antarctica became reglaciated only ~34 m.y. ago, whereas full scale, permanent Northern Hemisphere continental ice began only ~3 m.y. ago (e.g., Zachos, Dickens, & Zeebe, 2008, Figure 3.3.1). State-of-the-art climate models (e.g., DeConto & Pollard, 2003a, DeConto & Pollard, 2003b; DeConto et al., 2008; DeConto, Pollard, & Harwood, 2007; Huber et al., 2004; Pollard & DeConto, 2005) combined with paleoclimatic proxy data (e.g., Pagani, 2002; Pagani, Zachos, Freeman, Tipple, & Bohaty, 2005; Pearson, Foster, & Wade, 2009; Beerling & Royer, 2011) suggest that the main triggering mechanism for inception and development of the Antarctic ice sheet were the decreasing levels of carbon dioxide (CO_2) and other greenhouse gas concentrations in the atmosphere. This hypothesis opposed a long-held idea that the opening of critical Southern Ocean gateways (i.e., Tasmanian Gateway and Drake Passage), were responsible for Southern Ocean cooling and subsequently the Oligocene Antarctic glaciation (e.g., Kennett, 1977). With current rising atmospheric greenhouse gases resulting in rapidly increasing global temperatures (Intergovernmental Panel on Climate Change [IPCC], 2013, www.ipcc.ch/), studies of climate sensitivity particularly in polar regions are prominent on the research agenda. Understanding changes in Southern Ocean and

Developments in Marine Geology, Volume 7. http://dx.doi.org/10.1016/B978-0-444-62617-2.00012-8

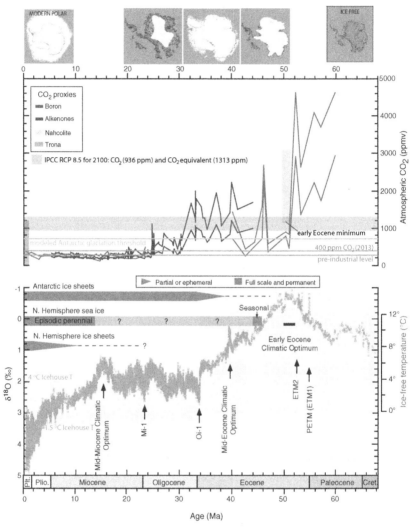

FIGURE 3.3.1   Updated Cenozoic pCO$_2$ and stacked deep-sea benthic foraminifer oxygen isotope curve for 0 to 65 Ma. Updated from Zachos et al. (2008) and converted to the "Gradstein timescale" (Gradstein et al., 2004). Mi-1 = Miocene isotope Event 1, Oi-1 = Oligocene isotope Event 1, ETM2 = Eocene Thermal Maximum 2, PETM = Paleocene/Eocene Thermal Maximum, ETM1 = Eocene Thermal Maximum 1. Antarctic ice sheet evolution (mostly inferred from low-latitude records) under the different CO$_2$ scenarios is shown at the top of the figure. Shaded squares indicate future CO$_2$ (purple (gray in print versions)) and Temperature (blue (light gray in print versions)) scenarios forecasted by the IPCC (2013).

Antarctic ice sheet dynamics and stability is of special relevance because, based on IPCC (2013) forecasts, atmospheric CO$_2$ concentrations ranging from 421 to 936 ppm (1313 ppm when considering the combined greenhouse gases CO$_2$ equivalent) and a global mean surface temperature rise between 1.5 °C and 4 °C are expected by the end of this century (Figure 3.3.1). Paleoclimate records show that the present 400 ppm CO_2 values have not been experienced on Earth since the early Pliocene (5–3.5 m.y.) and the higher values of CO_2 estimates are above the modeled Antarctic glaciation threshold ($2 \times CO_2$ preindustrial values, DeConto & Pollard, 2003a, b) (Figure 3.3.1). With regards to temperatures, the higher values of these estimates have not been experienced on our Planet since around 15 m.y. ago (Beerling & Royer, 2011) (Figure 3.3.1). Considering that the Northern Hemisphere ice caps did not fully develop until 3 m.y. ago, paleoclimate and ice sheet dynamic records from Antarctica and the Southern Ocean are key for providing us with paleoenvironmental reconstructions and models from which to build reference scenarios for future changes.

Despite the critical role of Antarctica and the Southern Ocean in the global system, key geological and geophysical data from this region are lacking. In part, this is because most of the Cenozoic geological record is hidden below Antarctica's massive ice sheets (i.e., 99.7% of Antarctica is ice covered). Although geographically sparse, outcropping terrestrial records from rock exposures around the Antarctic margin provide very valuable snapshots of the state of the ice sheet and regional climate. Coastal and open marine records provide a more continuous and datable window into Antarctic ice sheet behavior and cryosphere–ocean interactions.

Ocean drilling has played a major role in obtaining offshore records of Antarctic Ice Sheet history (Figure 3.3.2). Starting with the Deep Sea Drilling Project (DSDP), Legs 28 and 29 extended the record of Antarctic glaciation back to the Oligocene (Hayes, Frakes et al., 1975) and recorded an increase in ice volume around 14 Ma interpreted to represent the development of a largely stable Antarctic Ice Sheet (Kennett, Houtz et al, 1975; Kennett & Shackleton, 1975). The successor of DSDP, the Ocean Drilling Program (ODP, 1983–2003) drilled four Legs in other areas around the Antarctic margin aiming at determining glacial history from different sectors of the Antarctic Ice Sheet. ODP Leg 113 in the Weddell Sea recovered Paleogene sediments that record a relatively warm climate and then enough cooling during the early Oligocene to allow for the development of land ice and cold deep waters (Barker, Kennett et al., 1988). Drilling by the ODP in Prydz Bay (Leg 119, Barron, Larsen et al., 1989) and the Kerguelen Plateau (Leg 120, Wise, Schlich et al., 1992) established the timing of the first continental Ice Sheet at 34 Ma, around the Eocene–Oligocene transition. ODP Leg 119 in Prydz Bay was the first to drill strata from the east Antarctic continental shelf and determine that a continental scale ice sheet was delivering ice to the shelf edge beyond the limits of the modern ice sheet at least from early Oligocene times (Hambrey, Ehrmann, & Larsen, 1991). In the 1990s, a coordinated community effort to unveil Antarctic glacial history was put forward by the SCAR-ANTOSTRAT (Scientific Committee for Antarctic Research-Antarctic Offshore Seismic Stratigraphy) Program. ANTOSTRAT developed a strategy based on drilling shelf-to-rise transects at a number of key sites around Antarctica to understand terrigenous sediment transport and

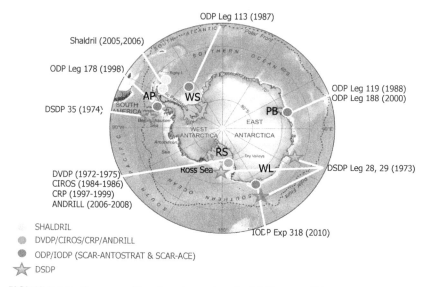

FIGURE 3.3.2 Summary of locations (approximate) of drilling expeditions around Antarctica (south of 60°S) during the last four decades. DSDP: Deep Sea Drilling Project; ODP: Ocean Drilling Program; IODP: Integrated Ocean Drilling Program; DVDP: Dry Valleys Drilling Project; CRP: Cape Roberts Project; ANDRILL (Antarctic Geological Drilling), SHALDRIL (Shallow Scientific Drilling on the Antarctic Continental Margin).

deposition under a glacial regime (Barker et al., 1998). It required direct sampling of the continental shelf prograded wedge of glacial sediments and of the derived more continuous and datable sediments on the continental rise. The former are discontinuous and harder to date because of recovery problems and because of erosion by the advances and retreats of the ice sheet. Data from the two types of records would be combined and used for constraining numerical ice sheet models. Five key regions were identified by ANTOSTRAT (Barker et al., 1998), two of which were drilled by ODP Legs 178 and 188 west of the Antarctic Peninsula and in Prydz Bay, respectively (Figure 3.3.2). ODP Leg 178 obtained a detailed record of glaciation over the past 10 Ma, which showed that, contrary to expectation, the Antarctic Peninsula Ice Sheet stayed large enough to migrate regularly to the shelf edge throughout the past 10 million years of major climate change (Barker, Camerlenghi, Acton, et al., 1999). Leg 178 also recovered a remarkable sequence of diatom ooze and sandy mud that provided a centennial to millenial record of climate over the past several 1000 years (e.g., Domack et al., 2001). Sediments recovered from the continental shelf during ODP Leg 188 to Prydz Bay, constrained the earliest stages of ice sheet development to about 35 Ma and continental rise sediments provided a Miocene to recent record of climate and cryosphere variability (O'Brien, Cooper, Richter, et al., 2001). Proxy measurements from rise sediments suggest dramatic changes at latest Eocene, during the middle Oligocene (24 Ma), and the middle Miocene

(13–15 Ma) but ice proximal records from these intervals are needed to interpret them in terms of east Antarctic ice sheet dynamics.

In summary, since DSDP started operations in 1968 to the end of the Integrated Ocean Drilling Program (IODP 2003–2013), seven expeditions took place in southern high-latitudes (south of 60°S latitude) out of a total of 236 ocean drilling expeditions (Figure 3.3.2). In addition, other programs have recovered coastal and marine sediments from around Antarctica (i.e., the international Cape Roberts Project and ANDRILL Program; and the US Shaldril Program) (Figure 3.3.2). All together, records remain sparse in their local coverage, representing coastal or offshore conditions, but not both, and there are many areas of the Antarctic margin without any records recovered (Figure 3.3.2). For these reasons not much is known yet about the nature, cause, timing, and rate of processes involved in the development and evolution of the Antarctic Ice Sheets. For example, low-latitude proxy records (Coxall, Wilson, Pälike, Lear, & Backman, 2005; Pälike et al., 2006) suggest extreme dynamics in ice volume during the first episode of Antarctic glaciation, the Oligocene. In order for numerical ice sheet models to reproduce these melt events however, they require forcing by extreme climate perturbations (Barker, Camerlenghi, Acton, et al., 1999). Of the two main ice sheets, the West Antarctic Ice Sheet (WAIS) is at present-day mainly marine based and is considered more sensitive to warming (DeConto et al., 2008; Florindo & Siegert, 2009; Naish et al., 2009). The East Antarctic Ice Sheet (EAIS), which overlies continental terrains that are largely above sea level, is considered stable and is believed to respond only slowly to changes in climate (Florindo & Siegert, 2008; DeConto et al., 2008). However, compelling physical evidence for such a stable EAIS is thin, and the deep-sea oxygen isotope records do suggest dynamic ice sheets, including the East Antarctic ice sheet. Moreover, global sea level estimates (e.g., Cramer, Toggweiler, Wright, Katz, & Miller, 2009) imply loss of the WAIS (6 m of sea level equivalent, SLE) and the Greenland ice sheet (6 m of SLE), and some volume loss from the East Antarctic Ice Sheet. Therefore, during episodes of global warmth, with likely elevated atmospheric CO_2 conditions, the EAIS may contribute just as much or more to rising global sea level as the Greenland ice sheet. In the face of rising CO_2 levels (IPCC, 2013), a better understanding of the EAIS dynamics is therefore urgently needed from both an academic as well as a societal point of view.

IODP Expedition 318 (January–March 2010; Wellington, New Zealand, to Hobart, Australia), drilled one unexplored margin of east Antarctica, the eastern Wilkes Land margin (Escutia, Brinkhuis, Klaus, & the Expedition 318 Scientists, 2011). This is a key region for the analysis of the long- and short-term behavior of the EAIS because it is located at the seaward termination of the Wilkes Subglacial Basin (WSB) (Figure 3.3.3(A)). Seismic/Satellite imagery has shown that, even with isostatic rebound of the Eastern Antarctic continent, in the WSB, the ice bed is largely below sea level with steep reversed slopes (Fretwell et al., 2013), which potentially makes the overriding ice sheet extremely vulnerable to climate-change–induced melt. Recent satellite observations indicate significant rates of

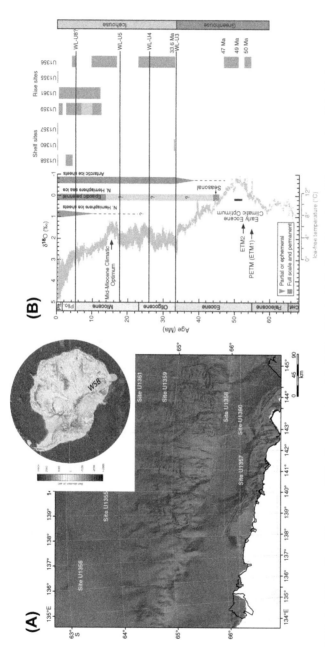

FIGURE 3.3.3 (A) Location of the eastern Wilkes Land margin and the Wilkes Subglacial Basin and locations of the seven sites drilled during IODP Expedition 318 (red (lighter gray in print versions) squares and labels) against a detailed bathymetric map. High-resolution bathymetry in front of the Georges V Land from Beaman, O'Brien, Post, and De Santis (2011), and GEBCO bathymetry for the other areas. (B) Chronostratigraphic and climate framework for the sedimentary sections recovered during Expedition 318. Stacked deep-sea benthic foraminifer oxygen isotope curve for 0–65 Ma (updated from Zachos et al., 2008) converted into Gradstein timescale (Gradstein et al., 2004). ETM2 = Eocene Thermal Maximum 2, PETM = Paleocene/Eocene Thermal Maximum, ETM1 = Eocene Thermal Maximum 1. Modified from Escutia et al. (2011).

change for Wilkes Land catchments, including thinning at their seaward margins (Pritchard et al., 2012) and rapid basal melt rates under their flanking ice shelves (Rignot et al., 2011). The IODP Wilkes Land drilling plan was designed to explore the poorly known terrains of this potentially sensitive sector of the East Antarctic Ice sheet, date the regional unconformities interpreted from the seismic profiles, and reveal and reconstruct past dynamics of the EAIS (Escutia et al. 2011). Of particular interest was testing the sensitivity of the EAIS to episodes of global warming and detailed analysis of critical periods in Earth's climate history, such as the Eocene–Oligocene (i.e., greenhouse–icehouse transition), late Oligocene warmth and early Miocene glaciation (Mi-1 event), late Miocene, the warm Pliocene, and the Holocene deglaciation. During these times, the Antarctic cryosphere evolved in a step-wise fashion to ultimately assume its present-day configuration, thought to be characterized by a relatively stable EAIS.

3.3.2 EXPEDITION 318 SUMMARY OF RESULTS

IODP Expedition 318 occupied seven sites across the Wilkes Land Margin at water depths between ~400 and 4000 mbsl (Figure 3.3.3(A)). In general, the strategy was to core and analyze sedimentary records along inshore–offshore transects to constrain the age, nature, and environments of deposition, until then only inferred from seismic surveys (e.g., Escutia et al. 2002, 2005, Escutia, Eittreim, & Cooper, 1997; De Santis, Brancolini, & Donda, 2003; Donda, Brancolini, De Santis, & Trincardi, 2003; Eittreim, Cooper, & Wannesson, 1995), shallow sediment cores (e.g., Domack, 1982; Domack & Anderson, 1983; Hampton, Eittreim, Richmond, 1987; Escutia et al., 2003; Macrì, Sagnotti, Dinares-Turell, & Caburlotto, 2005; Damiani, Giorgetti, & Memmi Turbanti, 2006; Caburlotto, Lucchi, De Santis, Macrì, & Tolotti, 2010; Presti et al., 2011), and indirect correlations to DSDP Site 269 (Hayes, Frakes, et al., 1975). Sites U1355, U1356, and U1360 targeted the greenhouse–icehouse transition and ice sheet evolution during the icehouse. Sites U1358, U1359, and U1361, targeted high-resolution records of Neogene ice sheet variability. This record included sequences from the Pleistocene and Pliocene that formed during interglacial intervals of exceptional warmth, periods that may provide valuable information about Antarctica's response to warming predicted in the centuries ahead. In addition, the Expedition targeted an ultra-high sediment section from a deep inner shelf basin at Site U1357 to document the Holocene deglaciation and subsequent climate and sedimentological variability over the past 10,000 years. Together, ~2000 m of high-quality upper Eocene–Quaternary sedimentary cores were retrieved (Figure 3.3.3(B)). The cores span ~54 m.y. and reveal the history of the Wilkes Land Antarctic margin from an ice-free "greenhouse Antarctica," to the first cooling, to the onset and erosional consequences of the first glaciation and the subsequent dynamics of the waxing and waning ice sheets (Figure 3.3.3(B)). Furthermore, it also captured a record of the last (Holocene) deglaciation in terms of thick, unprecedented records with annual to seasonal resolution taken in the Adélie Trough (Site U1357) (Figure 3.3.3).

3.3.2.1 Tectonic Evolution

The tectonic evolution of Antarctica involves the separation of Australia from Antarctica and the evolving ocean basin in between these continents. Break-up propagated west to east starting at 83–75 Ma in the central Bight region; at about 65 Ma the western Bight; and likely as late as about 50 Ma in the Térre Adélie-Otway region (Colwell, Stagg, Direen, Bernander, & Borisova, 2006; Close, Watts, & Stagg, 2009; Direen et al., 2013 and references therein). Rifting was slow in the early stages, and generally SE–NW oriented (Hill & Exon, 2004). This direction of motion maintained a continental connection (i.e., the Tasmanian sill) between Antarctica and Australia, at least until 64 Ma, based on biogeographical distribution of Godwana marsupials (Woodbourne & Case, 1996). During middle Eocene times, continental blocks contained in the Tasmanian sill started to drown gradually (Röhl et al., 2004). The rift between Australia and Antarctica changed toward an N–S orientation (Cande & Stock, 2004; Hill & Exon, 2004), and ocean spreading rates markedly increased (Cande & Stock, 2004). Previous Ocean drilling around Tasmania revealed a phase of rapid deepening of the Tasmanian Gateway occurred ~35.5 Ma, with the speedy submersion of the South Australian Margins, the East Tasman Plateau and the South Tasman Rise (Close, Watts, & Stagg, 2009; Stickley et al., 2004), and the Australian Bight (Houben, 2012). The timing of the deepening, allowing through flow of deep (3000 km) water masses, precedes the timing of glaciation (e.g., Coxall et al., 2005; Lear, Bailey, Pearson, Coxall, & Rosenthal, 2008), which indeed precludes a direct link between rifting and the onset of glaciation. Moreover, numerical model evidence suggests that the Eocene Southern Ocean (with closed ocean gateways) featured large clockwise gyres, which prevented the influence of warm, low-latitude surface waters to reach the Antarctic shoreline (Huber, 2001; Huber et al., 2004; Sijp, England, & Huber, 2011). This physics-based oceanographic reconstruction is consistent with biogeographic patterns of organic-walled dinoflagellate cysts (Bijl et al., 2011; Huber et al., 2004). The presence of gyres in the Southern Ocean and the inconsistency in timing of opening of the Tasmanian Gateway and the onset of Antarctic glaciation argues against the hypothesis ascribing a pivotal role for the fundamentally different oceanography to explain Eocene warmth. The question has remained, however, as to what exactly the role was of the opening of Southern Ocean Gateways in regional climate change, and the redistribution of heat, moisture and nutrients. If not a primary cause for glaciation, what was the role of the changed oceanography on, for example, the sensitivity of Antarctica to pCO_2-forced cooling (DeConto & Pollard, 2003a; Sijp et al., 2011)? Can field evidence provide us with evidence in support of the modeling experiments suggesting some effect on ice sheet sensitivity for different Gateway scenarios? The climatic consequences of the different phases of Tasmanian Gateway opening and subsequent deepening on Antarctic climate have been difficult to assess due to the lack of continental archives from the Antarctic margin itself. While a well constrained time scale for the deepening of the Tasmanian Gateway had been established through IODP Leg 189 (Stickley et al., 2004), a comprehensive reconstruction of earlier phases of opening of the Tasmanian Gateway largely hinges on seismic interpretations. A proper assessment was hampered by a paucity of field data from the Australo-Antarctic Gulf (AAG).

Cores retrieved during Expedition 318 reveal details of the tectonic history of the AAG (at 54 Ma), the onset of the second phase of rifting between Australia and Antarctica (Close et al., 2009; Colwell, Stagg, Direen, Bernander, & Borisova, 2006), ever-subsiding margins and deepening, to the present ocean/ continent configuration (Figure 3.3.4(A)). Tectonic and climatic change turned the early-middle Eocene shallow, broad subtropical Antarctic Wilkes Land off- shore shelf into a deeply subsided basin with a narrow ice-infested margin by the earliest Oligocene (by the Oi-1 event at 33.6 Ma) (Figure 3.3.4(A)). This rapid deepening of the margin took place within the ~13 m.y. hiatus between ~47 and 33.6 Ma (Escutia et al., 2011; Tauxe et al., 2012; Bjil, Sluijs, et al., 2013). Following this deepening, the Australian continent accelerated its northward rift during the Oligocene and Miocene, toward its present-day position (Cande & Stock, 2004). As it will be shown below, drilling in the Wilkes Land margin has also contributed key information that open ocean gateways may however have been an indirect, yet important prerequisite for Antarctic continental ice to develop as it did, by reshaping regional ocean circulation patterns, chang- ing regional temperature and moisture/precipitation patterns (Bijl et al., 2013; Houben, 2012; Sijp et al., 2011).

3.3.2.2 The Eocene Hothouse

Prior to the expedition, a prominent regional unconformity, WL-U3, had been interpreted to separate pre-glacial (Eocene) strata from (Oligocene) gla- cial-influenced deposits (De Santis et al., 2003; Escutia et al., 2005; Escutia, Eittreim, & Cooper, 1997). Drilling and dating of WL-U3 at continental rise Site U1356 (Figures 3.3.3 and 3.3.5(A)) confirmed that this surface represents major erosion related to the onset of glaciation at ~34 Ma (early Oligocene), with immediately overlying deposits dated as 33.6 Ma. Below unconformity WL-U3 at Site U1356, we recovered a record that surprisingly turned out to be late early to early middle Eocene in age (Escutia et al., 2011; Bijl, Sluijs, et al., 2013; Tauxe et al., 2012) (Figure 3.3.5(A)), an interval that includes peak green- house conditions, and likely some of the early Eocene so-called hyperthermals (Zachos, Dickens, & Zeebe, 2008) (Figure 3.3.1). Shipboard analyses based on dinocysts, pollen and spores, and the chemical index of alteration, among other indicators, allowed the inference of warm, shallow-water depositional environ- ments for this section (Escutia et al., 2011).

Post cruise, further analyses of pollen and spores preserved in these records revealed an insight into the prevailing vegetation biome on this Antarctic coastal

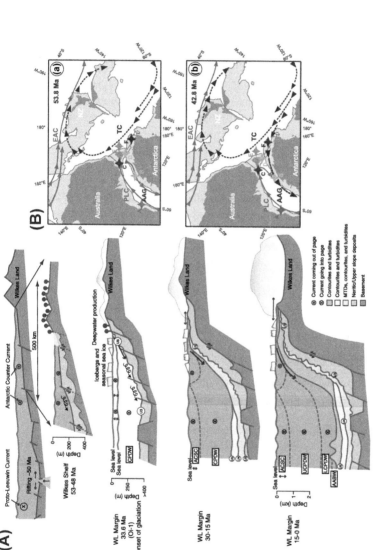

FIGURE 3.3.4   (A) Conceptual illustration of tectonic, geological, sedimentological, and climatic evolution of the Wilkes Land margin since the middle early Eocene (~54 Ma). U3, U4, and U5 refer to seismic unconformities WL-U3, WL-U4, and WL-U5. Oi-1 = Oligocene isotope Event 1, CPDW = Circumpolar Deep Water, ACSC = Antarctic Circumpolar Surface Water, UCPDW = Upper Circumpolar Deep Water, LCPDW = Lower Circumpolar Deep Water, AABW = Antarctic Bottom Water. MTD = mass transport debris flows. Modified from Escutia et al. (2011). (B) Tectonic evolution of the Tasmanian Gateway from Bijl et al. (2013). Eocene continental configurations of the Australian sector of the Southern Ocean for the (A) early Eocene and (B) middle Eocene. Lables A to E show the positions of the study sites. Dark gray areas reflect present day shorelines and light gray areas are submerged continental blocks above 3000 m water depth. The maps are overlain by surface currents as interpreted from modeling experiments (Huber et al., 2004; Sijp et al., 2011). The Tasmanian Gateway was open to shallow circulation from ~50 Ma onward, allowing westward leakage of the Antarctic Counter Current. AAG, Autralo-Antarctic Gulf; EAC, East Australian Current; NZ, New Zealand; PLC,

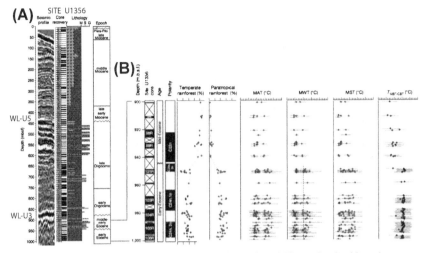

FIGURE 3.3.5 (A) Lithostratigraphic summary from Site 1356 correlated with main unconformities defined in multichannel seismic reflection profile across the Site (Escutia et al., 2011); (B) early and middle Eocene climate reconstruction for the east Antarctic Wilkes Land margin derived from Site U1356 after Pross et al., (2012). Columns show (from left to right): 1) Core recovery; 2) Geological age (Escutia et al., 2011); 3) Relative abundances of sporomorphs representing the temperate and paratropical rainforest biomes; 4) Estimates of Mean Annual Temperatures (MAT), Mean Winter Temperatures (MWT) and Mean Summer Temperatures (MST) for the temperate (blue (dark gray in print versions)) and paratropical (red (light gray in print versions)) rainforest biomes (Pross et al., 2012). Error bars represent the minimum and maximum estimates determined using that method. The vertical dashed line marks the minimum requirements of Bombacoideae for the mean temperature of the coldest month. 5) Temperatures derived from the MBT/CBT index, with horizontal error bars indicating the calibration standard error (65 uC). This error refers to absolute temperature estimates across all environmental settings of the modern calibration; thus, the error of the within-record variation is much smaller.

region during the early Eocene hothouse (Pross et al., 2012; Contreras et al., 2013) (Figure 3.3.5(B)). Applying a nearest living relative approach on pollen and spore assemblages provided quantitative, seasonal temperature reconstructions for the early Eocene greenhouse world on Antarctica. The temperatures indicated by the palynomorphs are similar to those obtained with independent organic geochemical paleotemperature proxies (i.e., branched tetraether lipids, TEX_{86} and methylation of branched tetraether [MBT]–cyclization of branched tetraether [CBT] ratios). Lowland climates along the Wilkes Land coast (at a palaeolatitude of about 70° south) supported the growth of highly diverse, near-tropical forests characterized by mesothermal to megathermal floral elements including palms and Bombacoideae (Pross et al., 2012; Contreras et al., 2013). Notably, winters were extremely mild (frost-free, in some intervals the coldest month mean temperatures were warmer than 10°C) (Figure 3.3.5(B)) despite polar darkness during winter, which provides a critical new constraint for the validation of climate models and for understanding the response of high-latitude terrestrial ecosystems to increased carbon dioxide forcing (Pross et al., 2012). The palynological contents of the mid-Eocene section (49–47 Ma), above a 2 m.y.r hiatus, reflects notably cooler, temperate, conditions; pollen and spore assemblages lacking the megathermal flora were dominated by Nothofagus. The 2–3 °C cooler temperatures for both winter and summer indicated by the palynomorphs is confirmed by similar cooling as reconstructed with the MBT–CBT biomarker proxy (Pross et al., 2012).

The Eocene succession from Site U1356 provides a hitherto missing piece of information to reconstruct biogeographic patterns around the southern opening of the Tasmanian Gateway (Bijl et al., 2013). Dinoflagellate cyst assemblages are particularly useful for assessing through-flow of surface waters across the Tasmanian Gateway as (i) the assemblages in the Eocene are distinctly different on either side of the Tasmanian Gateway and (ii) these different dinocyst assemblages are strongly affiliated with the prevailing surface water currents. Dinocyst analyses from Site U1356 and ODP Leg 189, the Australian Bight (ODP Leg 182) and sections in the southeast Australian Margin suggest earliest through-flow of South Pacific Antarctic waters through the Tasmanian Gateway (Figure 3.3.4(B)) to be coeval with the shift in rifting direction from SE–NW to S–N (Cande & Stock, 2004) and the onset of a gradual deepening of the South Tasman Rise (Hill & Exon, 2004). Moreover, the onset of through-flow coincides with the earliest signs of cooling following the Early Eocene Climatic Optimum (EECO) (Bijl et al., 2013; Pross et al., 2012). The tectonic opening of the Tasmanian Gateway provides a plausible explanation of southern high latitude cooling following the EECO in the absence of significant equatorial cooling in the middle Eocene (Bijl et al., 2013, 2009).

3.3.2.3 From Greenhouse to Icehouse: the Latest Eocene and Early Oligocene

At Site U1356, the upper middle Eocene to the latest Eocene is conspicuously missing in a ~13 m.y. hiatus (~47–33.6 Ma) associated with the unconformity WL-U3, based on dinocyst and paleomagnetic evidence (Escutia et al., 2011; Tauxe et al., 2012). Despite ongoing tectonic reorganizations, it appears likely that the erosive nature of unconformity WL-U3 is probably related to the early stages of EAIS formation. The impact of ice sheet growth, including crustal and sea level response, and major erosion by the ice sheets is proposed as the principal mechanism that formed unconformity WL-U3 (Escutia et al., 2011; Stocchi et al., 2013). This coincides with an abrupt increase in benthic foraminiferal $\delta^{18}O$ values and coeval sea level change globally recorded in complete marine successions (Oligocene isotope event Oi-1; e.g., Miller et al., 1985; Coxall et al., 2005).

Progressive tectonic subsidence, the large accommodation space created by erosion in the margin (300–600 m of missing strata on the shelf; Eittreim et al., 1995), and partial eustatic recovery (Stocchi et al., 2013) allowed sediments of early Oligocene age to accumulate above unconformity WL-U3. The dating of the sediments directly overlying WL-U3 at Site U1356 is based on three key stratigraphic tools: (1) the FO of dinocyst species *Malvinia escutiana* (Houben, Bijl, Guerstein, Sluijs, & Brinkhuis, 2011), indicating a stratigraphic position between 31.5 and 33.5 Ma (Houben et al., 2013), (2) the calcareous nannofossil assemblage suggesting an age older than 31.5 Ma (Escutia et al., 2011; Tauxe et al., 2012), and (3) a normal polarity interval in the lowermost cores indicating the recovery of the interval corresponding to Chron C13n (Tauxe et al., 2012). With the use of the novel stratigraphic marker for the Oi-1, *M. escutiana*, the lowermost sediments recovered from shelf Site U1360 were assigned an age of earliest Oligocene (Houben et al., 2013). In the same way, stratigraphies of cores from previous drilling expeditions (Cape Roberts, Prydz Bay, Weddell Sea) could be improved (Houben et al., 2013).

Geophysical modeling efforts have recently resulted in a concept shift in thinking about past sea-level variability. Particularly during icehouse climate regimes, the tendency has long been that sea level changed uniformly over the globe; the so-called eustatic sea level change. By integrating Glacial Isostatic Adjustment (GIA) geophysical model exercises with field data there is now support for the idea that sea level changed far from uniformly in the past. In contrast, any change in sea level at a certain place is dependent on the magnitude and locus of the sea level forcing (with respect to ice sheet forcing: where is the ice sheet growing/decaying?), the time scales involved and the properties of the lithosphere and resulting gravitational anomalies. The *relative* sea level changes recorded around Antarctica, including the signals in the records from Expedition 318, have recently confirmed this notion. Postcruise studies show that the enormous volume of the Antarctic ice sheet installed by the Oi-1 induced crustal deformation and gravitational perturbations around the continent (Stocchi et al., 2013) (Figure 3.3.6(A)). By forcing a GIA model with the Antarctic ice-sheet model output of DeConto et al. (2008), the crustal response to the progressive continental ice growth on Antarctica, results in a distinct spatial pattern of relative sea-level change that diverges from the expected eustatic signal (Figure 3.3.6(A) and (B)). Sea level rose proximal to the ice sheet despite an overall reduction in the volume of the ocean caused by the transfer of water to the ice sheet. The model output suggests that shelf areas around East Antarctica first shoaled as upper mantle material upwelled and a peripheral forebulge developed (Stocchi et al., 2013). The inner shelf subsequently subsided as lithospheric flexure extended outwards from the ice-sheet margins. Consequently the coasts experienced a progressive relative sea-level rise (Figure 3.3.6(A), (B)). The modeled rise in sea level is supported by lithologic and micropaleontological evidence from Wilkes Land Sites U1360 and U1356, combined with those from Prydz Bay (ODP Site 1166 and 739) and Cape Roberts Project-3 record (Figure 3.3.6(C)). Proximal sites experience first a regression followed by a deepening (sea-level rise), which provides with accommodation space for Oi-1 events to accumulate and be preserved (Figure 3.3.6(C)). In the meantime, distal

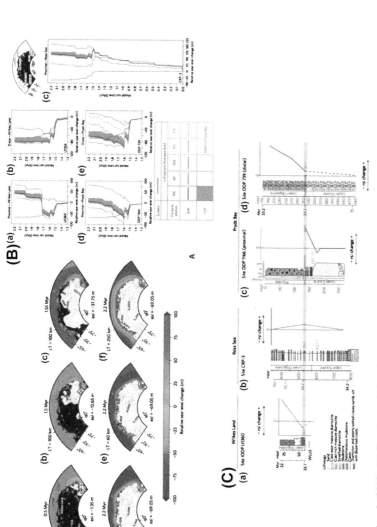

FIGURE 3.3.6   After Stochi et al. (2013). (A) East Antarctic ice-sheet evolution and resulting rsl changes at four model run times and relative to the preglacial state. White: ice-sheet extent and Black: subaerially exposed land. The eustatic values (m) are reported as equivalent sea level (esl). a–d, Rsl model predictions according to Viscosity Profile (RVP)-100km-LT. a, Model run times are 0.5 m.y.r., 1.5 m.y.r., 1.55 m.y.r. (which correlates to the end of the first d¹⁸O step3), and 2.2 Myr., respectively. e–f, Rsl change at 2.2 m.y.r. for RVP-60km-LT. and for RVP-250km-LT, respectively. TAM: *Trans*-Antarctic Mountains; VL: Victoria Land (B) Rsl predictions according to different Earth models and relative to the preglacial state. The rsl change is shown as a function of model run time. a–e, Blue (darkest gray in print versions) rls curve: eustatic sea-level change and showing two main pulses, respectively, at 1.5 and 1.6 myr; Black rsl curve: RVP-100km-LT simulation; Green (darker gray in print versions) rsl curve: RVP-60km-LT simulation; Red (dark gray in print versions) rsl curve: RVP-250km-LT simulation; Dark gray band: EVP-100km-LT simulations. a, Wilkes Land, IODP Site U1360. b, Wilkes Land, IODP Site U1356. c, Ross Sea, Cape Roberts CRP-3. d, Prydz Bay, ODP Site 1166. e, Prydz Bay, ODP Site 739. Here the effects of an extended ice sheet on the continental shelf are included. Solid cyan curve: RVP-100km-LT simulation; Dashed

sites experienced a continuous regression. These results are consistent with the suggestion that near-field processes such as local sea-level change influence the equilibrium state obtained by an ice-sheet grounding line. These results are also consistent with two reported tectonic deformation phases of shelf strata below the WL-U3 on the George V Land coast (De Santis, Brancolini, Donda, & O'Brien, 2010). A first extensional phase resulted in WNW-ESE grabens consistent with the separation between Antarctica and Australia. A second transpressional phase resulted in the uplift and inversion of previously rifted structures and folding in a narrow east–west oriented region near the coast. Sampling of coastal Eocene sediments is needed to test if this second transpressional phase is coeval with forebulge development.

Microfossils, sedimentology, and geochemistry of the early Oligocene sediments from Site U1356, at present occupying a distal setting (i.e., lowermost rise-abyssal plain) and immediately above unconformity WL-U3, unequivocally reflect icehouse environments with evidence of iceberg activity (dropstones), a shift in clay mineralogy suggestive of much colder/drier weathering regimes, and at least seasonal sea ice cover (Escutia et al., 2011) (Figure 3.3.7). The sediments, characterized by hemipelagic sedimentation (including occasional carbonate deposition) with bottom current and gravity flow influences, as well as biota all indicate deeper water settings relative to the underlying middle Eocene environments. These findings imply significant crustal stretching, subsidence of the margin, and deepening of the Tasman Rise and the Adélie Rift Block between 47 and 33.6 Ma (Escutia et al., 2011; Figure 3.3.4(A)). The continuous presence of reworked middle-late Eocene dinocyst species in the Oligocene sediments suggests unabated submarine erosion of the Antarctic shelf (Houben et al., 2013) in agreement with reported erosion along shelf troughs of 300–600 m of strata (Eitreim et al., 1995).

The circum-Antarctic Southern Ocean is an important region for global marine food webs and carbon cycling because of sea-ice formation and its unique plankton ecosystem. However, the mechanisms underlying the appearance of this distinct ecosystem and the geological timing of its development remained unknown. Modern Antarctic phytoplankton communities are adapted to seasonal sea-ice conditions, either as plankton in the upper surface waters, ice dwelling extremophiles occupying brine channels and pockets, or attached subice communities (Houben et al., 2013). During the spring sea-ice melt, accumulated nutrients are released from the ice into the surrounding melt zone, resulting in a nutrient-rich shallow mixed layer that stimulates intensive and relatively short-lived blooms of primary producers. The Oligocene dinoflagellate cyst records from U1356 and U1360, combined with those from other locations across the Antarctic Margin, suggest a major restructuring of the Southern Ocean plankton ecosystem which occurred abruptly and concomitant with the first major Antarctic glaciation in the earliest Oligocene (Houben et al., 2013) (Figure 3.3.7). By analogy to the modern ecosystem, and considering an abrupt regional increase in siliceous sedimentation at Southern Ocean sites across the

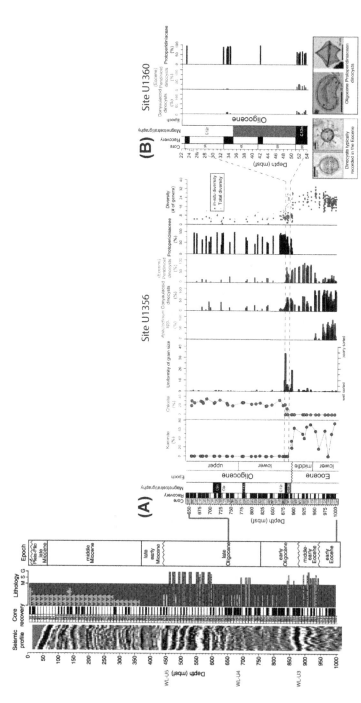

FIGURE 3.3.7 Eocene–Oligocene section recovered in IODP Sites U1356 (A) and U1360 (B), after Houben et al. (2013). A shift in clay mineralogy in the lower-most Oligocene of site U1356A indicates the initiation of a physical (i.e., glacial) weathering regime (Escutia et al., 2011). The uniformity coefficient of grain size—a measure of the sorting of the sediments, with large numbers (>4) indicating poorly sorted sediments—increases episodically in the lower Oligocene, suggesting glacial activity. The magnetostratigraphic interpretations from Tauxe et al. (2012) with gray areas indicating uncertain magnetic polarity. Lower to middle Eocene dinocyst assemblages are diverse (expressed by number of genera), with subtropical components (Apectodinium) and autotrophic (gonyaulacoid taxa) and Eocene heterotrophic (peridinioid dinocysts) representatives. The majority of nonprotoperidiniacean dinocysts in the lower Oligocene succession are reworked (translucent colored bars). Photomicrographs of typical Eocene and Oligocene dinocyst taxa are depicted in the lower right corner. Houben et al. (2013).

Oi-1 glaciation of Antarctica we found that seasonal blooms of phytoplankton initiated in the circum-Antarctic seas around this time (Figure 3.3.7). The proposed scenario involves an abrupt shift to high seasonal primary productivity associated with the development of seasonal sea ice. It provides the most parsimonious explanation for the abundant appearance of protoperidiniacean dinocysts at Oi-1 times (Figure 3.3.7). Moreover, it is in agreement with numerical climate models simulations indicating that sea-ice formation along Antarctic margins may have followed full-scale Antarctic glaciation (DeConto, Pollard, & Harwood, 2007).

3.3.2.4 The Icehouse: Oligocene to Pleistocene Records of EAIS Variability

Icehouse records of EAIS variability and related paleoceanographic changes were recovered from continental rise Sites U1356, U1359, and U1361, and continental shelf Site U1358 (Figure 3.3.3(A) and (B)). Drilling at continental rise Site U1356 recovered a thick section of Oligocene to upper Miocene sediments (Figures 3.3.3, 3.3.5, and 3.3.8) indicative of relatively deep water, sea ice—influenced setting (Escutia et al., 2011; Houben et al., 2013). In general, sediments at this site record a long-term cooling trend from the early Eocene (54–52 Ma) to the middle Miocene (15-13 Ma) on the order of 8 °C (Passchier et al., 2013). Within this trend, Oligocene to upper Miocene sediments are indicative of episodically reduced oxygen conditions either at the seafloor or within the upper sediments prior to ~17 Ma. One of the major climate transitions during this period is the Oligocene–Miocene transition leading to the Mi-1 event (Figure 3.3.1). This climate transition coincides with one of the major regional unconformities in the Wilkes Land margin, unconformity WL-U5 (Escutia, Eittreim, & Cooper, 1997,. 2005). The cooling trend leading to the Mi-1 event glaciation is characterized by mass transport processes (Salabarnada, Escutia, Nelson, Damuth, & Brinkhuis, 2014). Following the Mi-1 event, sedimentation is characterized by hemipelagic, turbidity-, and bottom-current deposition (Escutia, Donda, Lobo, & Tanahashi, 2007; Escutia et al., 2002; 2011; Salabarnada et al., 2014). From the late early Miocene (~17 Ma) onward, progressive deepening and possible intensification of deep-water flow and circulation lead to a transition from a poorly oxygenated low-silica system (present from the early to early middle Eocene to early Miocene) to a well-ventilated silica-enriched system akin to the modern Southern Ocean (Escutia et al., 2011).

A complete record with good recovery (80–100%) of late Miocene to Pleistocene deposits was obtained from continental rise Sites U1359 and U1361 (Figures 3.3.3, 3.3.8(A) and (B)). These Sites were drilled on the top of the east levee of the Jussieau Channel on a proximal (3009 mbsl) and distal (3454 mbsl) position, respectively (Escutia et al., 2011). The records from both Sites should be therefore complimentary. Shipboard, relatively high amplitude variations in sedimentological, wireline logging, and magnetic susceptibility data indicated a

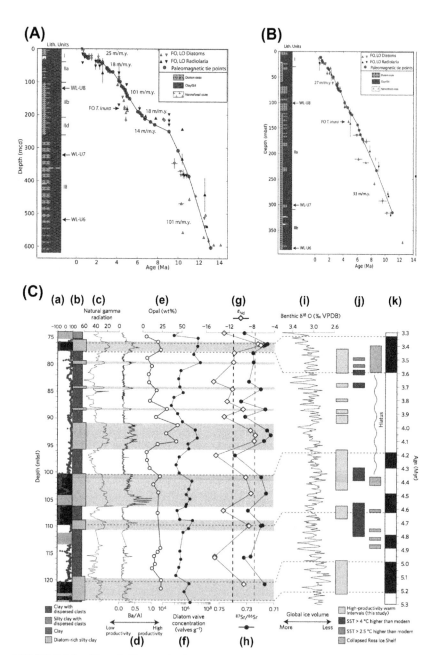

FIGURE 3.3.8 (A) and (B) Age-depth plot for Sites U1359 and U1361, respectively, after (Tauxe et al., 2012). Lithostratigraphic summary and biostratigraphic constraints from Escutia et al. (2011). Last occurrence (LO) and first occurrence (FO) include last common and last abundant occurrences (LCO and LAO, respectively). Same for FCO and FAO in FO is used for the diatom datums. t=top, b=bottom, o=oldest, y=youngest. Blue (light gray in print versions)

strong potential for this record to reveal EAIS dynamics down to orbital times-cales (105 and 41 k.y. cyclicity) (Escutia et al., 2011). This cyclicity documents the successive advances and retreats of the ice sheet and sea ice cover, as well as the varying intensity of cold saline density flows related to bottom water production at the Wilkes Land margin (e.g., high-salinity shelf water flowing from the shelf into the deep ocean to form Antarctic Bottom Water, AABW) (Escutia et al., 2011). In general, typical Southern Ocean open cold-water taxa, with variable abundances of sea ice–associated diatoms were recovered, indicating a high-nutrient, high-productivity sea ice–influenced setting throughout the Neogene. Combined sedimentological and microfossil information indicates the ever-increasing influence of typical Antarctic Counter Current surface waters and intensifying AABW flow. Furthermore, the preservation of calcareous microfossils in several intervals indicates times when bottom waters were favorable to the preservation of calcium carbonate. These observations pointed to a very dynamic ice sheet/sea ice regime during the late Miocene through the Pleistocene (Escutia et al., 2011).

A main focus of postcruise research to date has been the characterization of Pliocene and Pleistocene glacial–interglacial cycles, warm intervals, and transitions. Sediment records from Site U1361 deposited between 5.3 and 3.3 million years ago consist of alternating diatom-rich silty clay layers and diatom-poor clay layers with silt laminations (Figure 3.3.8(C)). Diatom-rich sediments show a correlation with higher diatom valve and bulk sediment biogenic opal concentrations, and distinctively lower signals in natural gamma radiation (Figure 3.3.8(C)), indicating lower clay content. The diatom-rich units also correlate with higher Ba/Al ratios (Figure 3.3.8(C)), suggesting multiple extended periods of increased biological productivity related to less sea ice, and warmer spring and summer sea surface temperatures. This is supported by work on other Antarctic margins that report increases in Southern Ocean surface water productivity and sea ice loss, associated with elevated circum-Antarctic temperatures (e.g., White-head & Bohaty, 2003; Escutia et al., 2009; Naish et al., 2009, Figure 3.3.8(C)).

---

line=best-fit sedimentation rate. Uncertainties in age and position of biostratigraphic datums indicated by horizontal and vertical bars respectively. (C) Pliocene records from IODP Site U1361 in comparison to other circum-Antarctic and global records from Cook et al. (2013). a, Palaeomagnetic chron boundaries based on inclination measurements; gray shading indicates intervals with no data. b, Lithostratigraphy (Escutia et al., 2011). c–h, Expedition 318 shipboard record of natural gamma radiation, and new records of Ba/Al, opal wt%, diatom valve concentrations, and Nd and Sr isotopic compositions; pink (light gray in print versions) shading, high-productivity intervals based on natural gamma radiation; vertical black stippled lines, Holocene Nd and Sr isotopic compositions (core-tops). i, Global benthic oxygen isotope stack (LR04, Lisiecki & Raymo, 2005). j, Circum-Antarctic indicators for warm temperatures; pink (light gray in print versions), Pliocene high-productivity intervals at IODP Site U1361; dark blue (darker gray in print versions), diatom and silicoflagellate assemblages from the Kerguelen Plateau (Bohaty & Harwood, 1998) and Prydz Bay (Escutia et al., 2009); light blue (dark gray in print versions), silicoflagellate assemblages from Prydz Bay (Whitehead & Bohaty, 2003); lilac, diatomite deposits from ANDRILL cores in the Ross Sea (Naish et al., 2009). k, Palaeomagnetic timescale.

In addition, the geochemical provenance of detrital material deposited during these warm intervals also suggests that during the warm periods there is active erosion of continental bedrock from within the WSB, an area today buried beneath the East Antarctic Ice Sheet. These data, as well as the maximum modeled erosion for the northern part of the WSB (Jamieson, Sugden, & Hulton, 2010) are in agreement with retreat of the ice margin several 100 km inland (Cook et al., 2013). Such retreat could have contributed between 3 and 10 m of global sea level rise from the East Antarctic Ice Sheet, providing a new and crucial target for future ice sheet modeling. Irrespective of the extent of ice retreat, Cook et al. (2013) document a dynamic response of the East Antarctic Ice Sheet to varying Pliocene climatic conditions, revealing that low-lying areas of Antarctica's ice sheets are vulnerable to change under warmer than modern conditions, with important implications for the future behavior and sensitivity of the East Antarctic Ice Sheet. Orbital scale time-series of ice-rafted debris from Site U1361 reveal that during the early Pliocene warm conditions mean annual isolation paced by obliquity had more influence on Antarctic ice volume than summer insolation intensity modulated by the precession cycle (Patterson et al., 2014). A transition to precession dominance after 3.5 Ma reflects a declining influence of oceanic forcing as high-latitude southern ocean cooled and perennial summer sea-ice field developed (Patterson et al., 2014).

The drilling of shelf Site U1358 (Figure 3.3.3) was aimed at providing a proximal record of grounding line advances and retreats during Miocene–Pliocene times (Escutia et al., 2011). Visual core descriptions, particle size distribution, and major and trace element ratios indicate that the lower Pliocene strata formed by intermittent glaciomarine sedimentation with open-marine conditions and extensive glacial advances to the outer shelf (Orejola, Passchier, & IODP Expedition 318 Scientists, 2014). In addition, a shift in provenance is recorded by heavy mineral analyses. Sand-sized detritus in the lower Pliocene strata is sourced from local intermediate to high-grade metamorphic rocks near Mertz Glacier (Orejola et al., 2014). In contrast to Pleistocene diamictons sourced from a prehnite–pumpellyite green schist facies terrane suggesting supply via iceberg rafting from northern Victoria Land (Orejola et al., 2014). This sedimentological evidence, is postulated as a shift from a dynamic EAIS margin in the early Pliocene to possible stabilization in the Pleistocene (Orejola et al., 2014).

East Antarctic Ice sheet stability during the Pleistocene is the focus of several ongoing studies. Of these, provenance studies from the heavy mineral fraction from sediments younger than 2.5 Ma recovered from Site U1359, show magmatic and metamorphic affinities that point to two sediment sources (Pant et al., 2013). A high-grade metamorphic source is consistent with the high-grade gneisses and orthogeneisses along the marginal zone of Terre Adèlie. The basaltic component in these sediments is interpreted as to be derived from the Jurassic tholeiites in the Transantarctic Mountains (Pant et al., 2013). Study of sediments from late Miocene and early Pliocene warm events have reported deposition of Ice Rafted Debris (IRD) sourced from the Ross Sea region (Williams et al., 2010). This suggests that the presence of basaltic component at Site U1359, could record changes in ocean circulation during Pleistocene warm events linked to cryosphere loss (Pant et al., 2013).

3.3.2.5 Ultrahigh Resolution Holocene Record of Climate Variability

Coring at Site U1357 (Figures 3.3.3, 3.3.9(A) and (B)) yielded a 185.6 m section of Holocene continuously laminated diatom ooze as well as a portion of an underlying Last Glacial Maximum diamict (Escutia et al., 2011). The site was chosen based on a 40 m long piston core (MD03-2601) previously collected from the Adélie Basin containing laminated diatom oozes with a sedimentation rate of 2 cm/year (Crosta, Debret, Denis, Courty, & Ther, 2007; Denis et al., 2009). However this core did not reach the Last Glacial Maximum (LGM) facies. Site U1357 was triple cored, providing overlapping sequences to aid in the construction of a composite stratigraphy since the LGM. LGM deposits at the bottom of the Hole are characterized by a massive sandy clast-rich diamict

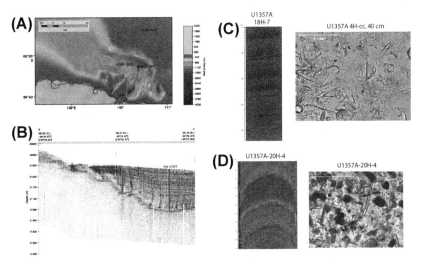

FIGURE 3.3.9 (A) Bathymetric map of eastern Wilkes Land margin showing location of Site U1357 (red (lighter gray in print versions) circle) and areas north and west of Adélie Basin. (B) ODEC 2000 single-channel seismic line across Site U1357. Red (light gray in print versions) bar = approximate section cored in Holes U1357A–U1357C. (C) Core image of typical dark–light laminations (interval U1357A-18H-7, 35–59 cm) and photomicrograph of well-preserved diatoms representative of the upper portion of the Adélie site sedimentary section (Sample 318-U1357A-4H-CC, 40 cm). (D) Core image close-up of lowermost sediments at Site U1357 with distinctive multicolored laminations (interval 318-U1357A-20H-4, 51–61 cm) and photomicrograph of silt-rich layer in Core 318-U1357A-20H-4. Most of the equant silt-sized grains are quartz. Several pieces of broken sponge spicule are also visible.

(Escutia et al., 2011). Overlying the diamict are 15 m of clay-bearing to clay-rich diatom oozes, with well-defined repeating couplets of diatom oozes and mudstone-bearing diatom oozes (Figure 3.3.9(C)). Bulk sediment ^{14}C dates on in this interval range between 13 and 40 cal Ka BP, suggesting there is contamination by old reworked carbon (Mackintosh et al., 2013). Overlying the 15 m interval are 170 m of diatom ooze (averaging 90% of diatom abundance) with well-defined light to dark laminae (Figure 3.3.9(C)), interpreted as representing seasonal variations in diatom deposition (Escutia et al., 2011). The paired light–dark laminae sets range in thickness from ~1 to 3 cm. They lack bioturbation, and exhibit excellent preservation of organic matter as well as calcareous, opaline, phosphatic, and organic fossils. This unit is interpreted as being deposited in seasonally open water conditions, following ice retreat at this site. Based on the calibrated ages, sedimentation rates for the upper 170 m of thick diatom ooze were 1.6 cm/yr with the base of this unit dated at 10,359–11,382 (Hole U1357A) and 10,571–11,756 cal yr BP (Hole U1357B) (Mackintosh et al., 2013). This suggests the retreat of grounded ice from most of the inner continental shelf possibly occurred as early as 11.5 cal Ka BP (Mackintosh et al., 2013). The age model is being finalized at the time of this writing and, as this is the first varved sedimentary sequence extending through the Holocene recovered from the Southern Ocean, it will provide the basis for examining decadal to subdecadal variability in sea ice, temperature, and wind linked to the Southern Annual Mode, Pacific Decadal Variability, and possibly El Niño Southern Oscillation in the near future. It will also allow to address questions regarding rates of change during the hypsithermal Holocene neoglacial events and the time immediately following the first lift-off and pull-back of ice at the end of the last glacial interval. In addition, it will provide an excellent opportunity for ultrahigh resolution correlation to the nearby Law Dome Ice Core (Etheridge, Pearman, & de Silva, 1988), one of the most important Holocene ice cores in Antarctica.

3.3.3 DISCUSSION OF RESULTS

Although much of the work in the nearly 2000 m of core recovered during IODP Expedition 318 is still ongoing, the sedimentary record has already provided a significant number of critical scientific discoveries related to the Cenozoic east Antarctic paleoenvironments. To date, contributions are related to: 1) understanding environmental conditions during two past scenarios of significantly elevated CO_2 concentrations, the Eocene greenhouse world, before the continental ice sheet formed in Antarctica; and the warm early Pliocene, before permanent Northern Hemisphere ice sheets developed; and 2) the physical processes involved in the development of a continental size ice sheet 34 m.y. ago.

Early Eocene continental temperature reconstructions at the Wilkes Land margin, as inferred from the prevailing paratropical rainforest biomes (i.e., with cold month mean temperatures >10 °C) and from organic geochemical proxies (i.e., considering the coldest values of 24–27 °C) (Figure 3.3.5(B)), are much higher than previously thought and provide insight into paleoenvironments in this sector of Antarctica under >1000 ppm CO_2 concentrations. These temperatures provide new constraints for investigations of the value of these data for future climate sensitivity (i.e., Caballero & Huber, 2013). Paleogeographic and paleobathymetric reconstructions indicate that during the early Eocene, Site U1356 was located on a broad continental shelf bathed by the warm Australo-Antarctic Gyre (Escutia et al., 2011; Pross et al., 2012; Bijl et al., 2013). Additional Early Eocene records of peak greenhouse conditions from drilling coastal/ice proximal sedimentary sections from other Antarctic margins with different paleoceanographic and paleogeographic settings (i.e., ANDRILL Coulman High Project) are needed to determine local vs. regional paleoenvironmental conditions.

Climate deterioration from the early Eocene peak greenhouse conditions to the middle Eocene is recorded by the change to a temperate rainforest biome dominated by Notofagus and Araucaria. This biome shift implies a decline of around 2–3 °C from the early Eocene and MBT-CBT paleothermometry provide estimates for the coldest values of 17–20°C (Pross et al., 2012; Contreras et al., 2013) (Figure 3.3.5(B)). A compilation of data from ocean drilling sites across the Tasman Gateway shows the cooling coincides with Ross Sea Gyre cold water flowing through the southern Tasman Gateway (Figure 3.3.4(B)) providing a mechanism for cooling along the eastern Wilkes Land margin following the early Eocene Climate Optimum (Bijl et al., 2013). Similar paleoenvironmental conditions are reported for other Antarctic and Southern Ocean sites from the middle Eocene sediments in southwest Australia and in the Pacific sector of the Southern Ocean (Bijl et al., 2009; Contreras et al., 2014; Hollis et al., 2012, 2009). Palynological work on middle to late Eocene sediments recovered during ODP Leg 188 from the Prydz Bay continental shelf (Site 1166) also record the existence of Nothofagus rainforest (Cooper, O'Brien, Richter, 2004). In the Antarctic Peninsula sector, Seymour Island floras suggests that by the middle Eocene (47–44 Ma) climates were considerably cooler than those of the southwest Pacific and Australo-Antarctic Gulf (Douglas, et al. 2014) and ice may have been present on the Antarctic continent, at least during winter months (Francis, 2000). Several Eocene macrofloras dominated by Nothofagus from King George Island, South Shetland Islands, suggest mean annual temperatures of about 10–11 °C (seasonality unknown) (e.g., Birkenmajer & Zastawniak, 1989; Hunt & Poole, 2003), and the highest latitude Eocene floras, from Minn a Bluff in the McMurdo Sound region, are suggestive of cool temperate climates (Francis, 2000). Further drilling around Antarctica is needed to constrain regional temperature differences and continental gradients.

A hiatus of 13 Myr prevents paleoeviromental reconstructions from the late Eocene (targeted at Sites U1356 and U1360) to the establishment of the continental size Antarctic Ice Sheet (Figure 3.3.3(B)). Sediments below and above the unconformity however, provide evidence for the many changes that occurred in this margin during this missing 13 m.y., namely, rapid subsidence (bringing Site U1356 from a continental shelf to a deep sea environmental setting), and eventually the relatively rapid growth of the Antarctic Ice Sheet (Figure 3.3.4(A)) and the establishment of high seasonal primary productivity associated with the development of seasonal sea ice (Houben et al., 2013; Figure 3.3.7). However, despite major erosion related to this unconformity, we were surprised to find earliest Oligocene sediments dated 33.6 Ma (Oi-1 event) preserved at both Sites. Interestingly, this was not only true for the Wilkes Land Margin but also for other Antarctic margins that have different tectonic and glacial settings. Integration of GIA models with geological data from three Antarctic margins (Wilkes Land, Prydz Bay, and the Ross Sea) shows that the growth the continental-size Antarctic ice sheet by the Oi-1 induced crustal deformation and gravitational perturbations around the continent (Stocchi et al., 2013), which resulted in a complex spatial pattern of relative sea-level change different from the expected 'simple' eustatic signal (Stocchi et al., 2013; Figure 3.3.6). These results point to the erosion forming unconformity WL-U3 to be related with the formation and migration of the forebulge that forced first a regression and then a transgression at Site U1360, in addition to tectonic subsidence. This mechanism is likely responsible for the development of other regional unconformities reported from other Antarctic margins around the Eocene–Oligocene transition and below early Oligocene sediments (33.6 Ma). These results also highlight the relevance of local sea-level change in influencing the equilibrium state of an ice-sheet grounding line.

During the icehouse, warm intervals within the Pliocene Epoch (5.33–2.58 million years ago) were characterized by global temperatures comparable to those predicted for this century (IPCC, 2013) and atmospheric CO_2 concentrations similar to today (Pagani, Liu, LaRiviere, & Ravelo, 2010). Sediment cores collected by the ANDRILL Program (AND-1B) underneath the Ross Ice Shelf show advances and retreats of the grounding line across the site (Naish et al., 2009) and modeling points to repeated times of WAIS collapse (Pollard & DeConto, 2009). Estimates for global sea level highstands during these times (e.g., Miller et al., 2012) imply however, possible retreat of the East Antarctic Ice Sheet. Continental rise sediments from the east Antarctic margin off Prydz Bay (Leg 188) record extended periods of open water conditions and elevated sea surface of >5 °C (Escutia et al., 2009; Whitehead & Bohaty, 2003), but ice-proximal evidence from the Antarctic margin is scarce. Early Pliocene (5-3 Ma) sediments collected from the Wilkes Land continental rise exhibit cyclicity between warm interglacials (i.e., diatom-rich, Ba enriched) and cold glacial (diatom poor) intervals (Figure 3.3.8(C)). In addition, the geochemical provenance of detrital material deposited during warm intervals suggests repeated erosion of continental bedrock several hundreds of kilometres inland of the WLSB. Ice proximal early Pliocene sediments from Site U1358, record intermittent glaciomarine sedimentation with open-marine conditions and extensive glacial advances to the outer shelf (Orejola et al., 2014). All this evidence points to grounding line advances and retreats of the EAIS during the early Pliocene with times of EAIS retreat within the WLSB and suggest the EAIS was sensitive to climatic warmth during the Pliocene (Cook et al., 2013) (Figure 3.3.8(C)). These results imply that at times of elevated temperatures and atmospheric CO_2 concentrations, similar to those predicted within this century, the EAIS can retreat contributing to rising global sea levels. These results support recent early Pliocene modeling simulations that show that, in addition to the collapse of the WAIS (Naish et al., 2009; Pollard & DeConto, 2009), basal melting and increased calving also result in moderate to substantial retreat into the WLSB and around much of the Wilkes Land margin (DeConto & Pollard, 2013).

3.3.4 CONCLUDING REMARKS

Scientific discoveries from IODP Expedition 318 will continue as work shifts to other time windows of past warmth (e.g., late Oligocene), climate transitions (e.g., Mi-1 event), and the ultra-high resolution Holocene climate record. All this work will contribute to the better understanding of Cenozoic Antarctic cryosphere evolution in the context of the Earth's climate system. These palaeoenvironmental records are key for reconstructing equator-to-pole temperature gradients through time, the key factor to determine the strength of "polar amplification" and climate sensitivity in general. These records are also directly relevant to ice sheet and sea level change, because the magnitude of accelerating ice sheet loss, both in Greenland and Antarctica, suggests that ice sheets will be the dominant contributors to sea level rise in coming decades, and their contribution will likely exceed current IPCC projections for the twenty-first century (Rignot et al., 2011). However, to fulfill these contributions, ice proximal records (i.e., records of grounding line retreat and advance) are needed from the eastern sector of the Wilkes Land Margin. Also needed are records from other areas around the margin that have been drilled but are either missing the proximal and/or distal records to advance our current poor understanding of linkages between the ice sheet, ocean circulation, and deep/bottom water processes through time. Ideally, specific drainage basins are of interest to account for different areas having different histories. In addition, there are many areas around the Antarctic margin that are largely unexplored.

The SCAR Scientific Research Program PAIS (Past Antarctic Ice Sheet dynamics) coordinates efforts by the community to obtain sedimentary records from continent-to-abyss transects along specific drainage systems (Figure 3.3.10). PAIS aims to improve understanding of the sensitivity of East, West, and Antarctic Peninsula Ice Sheets to a broad range of time scales (i.e., past "greenhouse" climates warmer than today, and times of more recent warming and ice sheet retreat during glacial terminations) and of climatic and oceanic conditions.

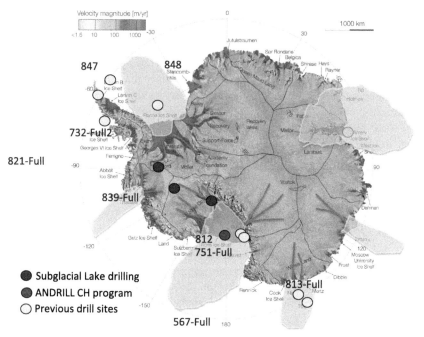

FIGURE 3.3.10   SCAR-PAIS integrated strategy and types of records from ice sheet-ice shelf-offshore transects along specific drainage systems (green (lightest gray in print versions) ellipses) to reconstruct ice sheet dynamics and palaeoclimate from vulnerable areas of the Antarctic margin and the Southern Ocean (Escutia et al., PAIS Proposal, http://www.scar.org/researchgroups/progplanning/#PAIS). Yellow (light gray in print versions) dots indicate locations with previous drilling (e.g., DSDP, ODP, IODP, ANDRILL and SHALDRIL); Maroon (darker gray in print versions) dot indicates the position of proposed Coulman High drilling by the ANDRILL Program; Blue (darkest gray in print versions) dots indicated areas with proposed or ongoing subglacial drilling. Blue (black in print versions) numbers relate to existing proposed IODP drilling programs at different stages of consideration by the IODP FAcility Boards/Panels (http://www.iodp.org/active-proposals)

The IODP Expedition 318 Scientists

C. Escutia[1] (Chief Scientist), H. Brinkhuis[2] (Chief Scientist), A. Klaus[3] (Staff Scientist), J. A. P. Bendle[4], P. K. Bijl[5], S. M. Bohaty[6], S. A. Carr[7], R. B. Dunbar[8], J.A. Flores[9] J. J. González[10], A. Fehr[11], T. G. Hayden[12], M. Iwai[13], F. J. Jimemez-Espejo[14], K. Katsuki[15], G. S. Kong[16], R. M. McKay[17], M. Nakai[18], M. P. Olney[19], S. Passchier[20], S. F. Pekar[21], J. Pross[22], C. Riesselman[23], U. Röhl[24], T. Sakai[25], P. K. Shrivastava[26], C. E. Stickley[27], S. Sugisaki[28], L. Tauxe[29], S. Tuo[30], T. van de Flierdt[31], K. Welsh[32], T. Williams[33], M. Yamane[34].

[1] Instituto Andaluz de Ciencias de la Tierra, CSIC-UGR, Avda de Las Palmeras 4, 18100 Armilla, Spain.
[2] Royal Netherlands Institute for Sea Research (NIOZ), PO Box 591790 AB Den Burg, Texel, The Netherlands.
[3] United States Implementing Organization, Integrated Ocean Drilling Program, Texas A&M University, 1000 Discovery Drive, College Station, Texas 77845, USA.

4 School of Geographical, Earth and Environmental Sciences, AstonWebb Building, University of Birmingham, Edgbaston, B15 2TT, UK.
5 Department of Earth Sciences, Faculty of Geosciences, Utrecht University, Laboratory of Palaeobotany and Palynology, Budapestlaan 4, 3584CD, Utrecht, The Netherlands.
6 Ocean and Earth Science, National Oceanography Center Southampton, University of Southampton, EuropeanWay, SO14 3 ZH, Southampton, UK.
7 Department of Chemistry and Geochemistry, Colorado School of Mines, 1500 Illinois Street, Golden, Colorado 80401, USA.
8 Environmental Earth System Science, Stanford University, Stanford, California 94305-2115, USA.
9 Department of Geology, Universidad de Salamanca, 37008, Salamanca, Spain.
10 Instituto Andaluz de Ciencias de la Tierra, CSIC-UGR, Avda de Las Palmeras 4, 18100 Armilla, Spain.
11 RWTH Aachen University, Institute for Applied Geophysics and Geothermal Energy, Mathieustrasse 6, D-52074 Aachen, Germany.
12 Department of Geology, Western Michigan University, 1187 Rood Hall, 1903West Michigan Avenue, Kalamazoo, Michigan 49008, USA.
13 Department of Natural Science, Kochi University, 2-5-1 Akebono-cho, Kochi 780-8520, Japan.
14 Department of Biogeochemistry, Japan Agency for Marine-Earth Science and Technology (JAMSTEC).
15 Geological Research Division, Korea Institute of Geoscience and Mineral Resources, 124 Gwahang-no, Yuseong-gu, Daejeon 305-350, Korea.
16 Petroleum and Marine Research Division, Korea Institute of Geoscience and Mineral Resources, 30 Gajeong-dong, Yuseong-gu, Daejeon 305-350, Korea.
17 Antarctic Research Center, Victoria University of Wellington, PO Box 600, Wellington 6140, New Zealand.
18 Education Department, Daito Bunka University, 1-9-1 Takashima-daira, Itabashi-ku, Tokyo 175-8571, Japan.
19 Department of Geology, University of South Florida, Tampa, 4202 East Fowler Avenue, SCA 528, Tampa, Florida 33620, USA.
20 Earth and Environmental Studies, Montclair State University, 252 Mallory Hall, 1 Normal Avenue, Montclair, New Jersey 07043, USA.
21 School of Earth and Environmental Sciences, Queens College, 65-30 Kissena Boulevard, Flushing, New York 11367, USA.
22 Paleoenvironmental Dynamics Group Institute of Earth Sciences Heidelberg University Im Neuenheimer Feld 234 D-69120 Heidelberg.
23 Department of Geology, University of Otago, PO Box 56, Dunedin 9054, New Zealand.
24 MARUM—Center for Marine Environmental Sciences, University of Bremen, Leobener Straße, 28359 Bremen, Germany.
25 Department of Geology, Utsunomiya University, 350 Mine-Machi, Utsunomiya 321-8505, Japan.
26 Polar Studies Division, Geological Survey of India, NH5P, NIT, Faridabad 121001, Haryana, India.
27 Evolution Applied Limited, Cornwall, UK.
28 Department of Earth and Planetary Science University of Tokyo 7-3-1 Hongo Bunkyo-ku Tokyo 113-0033 Japan.
29 Scripps Institution of Oceanography, University of California, San Diego, La Jolla, California 92093-0220, USA.
30 State Key Laboratory of Marine Geology, Tongji University, 1239 Spring Road, Shanghai 200092, China.
31 Department of Earth Science and Engineering, Imperial College London, South Kensington Campus, Prince Consort Road, London SW7 2AZ, UK.
32 School of Earth Sciences, University of Queensland, St Lucia, Brisbane, Queensland 4072, Australia.
33 Lamont Doherty Earth Observatory of Columbia University, PO Box 1000, 61 Route 9W, Palisades, New York 10964, USA.
34 Earth and Planetary Science, University of Tokyo, 7-3-1 Hongo, Bunkyo-ku, Tokyo 113-0033, Japan.

ACKNOWLEDGMENTS

We thank the captain and crew of the JOIDES Resolution, the IODP 318 operation super-intendent, ice pilot, weatherman, all technicians, and videographer who were instrumental in the success of Expedition 318. They allowed and assisted us in drilling, documenting, sampling and onboard sample analyses during Expedition 318. Numerous people at IODP-TAMU and the USIO provided their dedicated effort supporting us, including preparation of the expedition and publication of the proceedings. The USIO curatorial team provided us with their able support during the sampling party at College Station. We also thank the numerous shore-based scientists working on samples from this IODP Expedition and who have contributed to this work. We also thank Laura De Santis and Hans-Christian Larsen who provided valuable insights for the improvement of the paper. We acknowledge the Italian Research Program (PNRA), the Australian National Antarctic Research Expedition (ANARE), and the Antarctic Seismic data Library System (SDLS) for providing the seismic reflection data that supported the drilling. Material presented here was based in part upon work supported by the U.S. National Science Foundation. In addition, C. Escutia thanks the support by the Spanish Ministry (Grant CTM2011-24079).

REFERENCES

Barker, P. F., Barrett, P., Camerlenghi, A., Cooper, A. K., Davey, F., Domack, E., et al. (1998). Ice sheet history from Antarctic continental margin sediments: the ANTOSTRAT approach. *Terra Antarctica, 5,* 737–760.

Barker, P. F., Camerlenghi, A., Acton, G. D., et al. (1999). In: *Proc. ODP, Init. Repts.,* (Vol. 178), College Station, TX A&M University: (Ocean Drilling Program).

Barker, P. E., Kennett, J. P., et al. (1988). *Proc. ODP, Init. Repts* (113). College Station: TX. http://dx.doi.org/10.2973/odp.proc.ir.113.1988 (Ocean Drilling Program).

Barron, J., Larsen, B., & the ODP Expedition 119 Scientists (1989). In: *"Proceeding of the ODP scientific results leg 119".* College Station, TX A&M University: (Ocean Drilling Program).

Beaman, R. J., O'Brien, P. E., Post, A. L., & De Santis, L. (2011). A new high-resolution bathymetry model for the Terre Adélie and George V continental margin, East Antarctica. *Antarctic Science, 23,* 95–103.

Beerling, D. J., & Royer, D. L. (2011). Convergent cenozoic CO_2 history. *Nature Geoscience, 7,* 418–420.

Bijl, P. K., Bendle, J. A., Bohaty, S. M., Pross, J., Schouten, S., Tauxe, L., et al. (2013). Onset of Eocene Antarctic cooling linked to early opening of the Tasmanian Gateway. *Proceedings of the National Academy of Sciences of the United States of America, 110*(24), 9645–9650.

Bijl, P. K., Pross, J., Warnaar, J., Stickley, C. E., Huber, M., Guerstein, R., et al. (2011). Environmental forcings of Paleogene Southern Ocean dinoflagellate biogeography. *Paleoceanography, 26,* PA1202.

Bijl, P. K., Schouten, S., Brinkhuis, H., Sluijs, A., Reichart, G., & Zachos, J. C. (2009). Early Palaeogene temperature evolution of the southwest Pacific ocean. *Nature, 461,* 776–779.

Bijl, P. K., Sluijs, A., & Brinkhuis, H. (2013). A magneto- chemo- stratigraphically calibrated dinoflagellate cyst zonation of the early Paleogene South Pacific Ocean. *Earth-Science Reviews, 124,* 1–31.

Birkenmajer, K., & Zastawniak, E. (1989). Late Cretaceous-early Tertiary floras of King George Island, West Antarctica: their stratigraphic distribution and palaeoclimatic significance, Origins and Evolution of the Antarctic Biota. *Geological Society of London, Special Publication, 147,* 227–240.

Bohaty, S. M., & Harwood, D. M. (1998). Southern Ocean Pliocene paleotemperature variation from high-resolution silicoflagellate biostratigraphy. *Marine Micropaleontology, 33,* 241–272.

Caballero, R., & Huber, M. (2013). State-dependent climate sensitivity in past warm climates and its implications for future climate projections. *Proceedings of the National Academy of Sciences of the United States of America, 110*(35), 14162–14167.

Cande, S. C., & Stock, J. M. (2004). Cenozoic reconstructions of the Australia-New Zealand-South Pacific sector of Antarctica. In N. F. Exon, J. P. Kennett, & M. Malone (Eds.), *The cenozoic southern Ocean: Tectonics, sedimentation and climate change between Australia and Antarctica Geophysical monograph series:* (pp. 5–18). American Geophysical Union.

Caburlotto, A., Lucchi, R. G., De Santis, L., Macrì, P., & Tolotti, R. (2010). Sedimentary processes on the Wilkes Land continental rise reflect changes in glacial dynamic and bottom water flow. *International Journal of Earth Sciences (Geol Rundsch), 99*(4), 909–926.

Close, D. I., Watts, A. B., & Stagg, H. M. J. (2009). A marine geophysical study of the Wilkes Land rifted continental margin, Antarctica. *Geophysical Journal International, 177*(2), 430–450.

Colwell, J. B., Stagg, H. M. J., Direen, N. G., Bernander, G., & Borisova, I. (2006). The structure of the continental margin off wilkes land and terre Adelie coast, east Antarctica. In Futterer, Damaslie, Kleinschmidt, Miller, & Tessensohn (Eds.), *Antarctica contributions to global earth sciences* (pp. 327–340). Berlin Heidelberg New York: Springer-Verlag.

Contreras, L., Pross, J., Bijl, P. K., Koutsodendris, A., Raine, J. I., van de Schootbrugge, B., et al. (2013). Early to middle eocene vegetation dynamics at the wilkes land margin (Antarctica). *Review of Palaeobotany and Palynology, 197*(2013), 119–142.

Contreras, L., Pross, J., & Bijl, P. K., O'Hara, B., Raine, J. I., Sluijs, A., Brinkhuis, H. (2014). Southern high-latitude terrestrial climate change during the Paleocene–Eocene derived from a marine pollen record (ODP Site 1172, East Tasman Plateau). Climate of the Past, *10*, 1–20.

Cook, C. P., van de Flierdt, T., Williams, T. J., Hemming, S. R., Iwai, M., Kobayashi, M., et al. (2013). Dynamic behaviour of the east Antarctic ice sheet during pliocene warmth. *Nature Geosciences, 6*(9), 765–769.

Cooper, A. K., O'Brien, P. E., & Richter, C. (Eds.). (2004). *Proc. ODP, Sci. Results, 188.* College Station, TX A&M University: (Ocean Drilling Program). (2004).

Coxall, H. K., Wilson, P. A., Pälike, H., Lear, C. H., & Backman, J. (2005). Rapid stepwise onset of Antarctic glaciation and deeper calcite compensation in the Pacific Ocean. *Nature (London, U. K.), 433*(7021), 53–57.

Cramer, B. S., Toggweiler, J. R., Wright, J. D., Katz, M. E., & Miller, K. G. (2009). Ocean Overturning since the Late Cretaceous: Inferences from a new benthic foraminiferal isotope compilation. *Paleoceanography, 24*, PA4216.

Crosta, X., Debret, M., Denis, D., Courty, M. A., & Ther, O. (2007). Holocene long- and short-term climate changes off Adé lie Land, East Antarctica. *Geochemistry, Geophysics, Geosystems, 8.*

Damiani, D., Giorgetti, G., & Memmi Turbanti, I. (2006). Clay mineral fluctuations and surface textural analysis of quartz grains in Pliocene–Quaternary marine sediments from Wilkes Land continental rise (East-Antarctica): paleoenvironmental significance. *Marine Geology, 226*, 281–295.

DeConto, R. D., & Pollard, D. (2013). A smaller Antarctic Ice Sheet in the Pliocene and in the future. *Geophysical Research Abstracts, 15* EGU2013-13279-2, 2013, EGU General Assembly 2013.

DeConto, R. M., & Pollard, D. (2003a). A coupled climate–ice sheet modeling approach to the early Cenozoic history of the Antarctic ice sheet. *Palaeogeography, Palaeoclimatology, Palaeoecology, 198*(1–2), 39–52.

DeConto, R. M., & Pollard, D. (2003b). Rapid Cenozoic glaciation of Antarctica induced by declining atmospheric CO_2. *Nature (London, U. K.), 421*(6920), 245–249.

DeConto, R., Pollard, D., & Harwood, D. (2007). Sea ice feedback and Cenozoic evolution of Antarctic climate and ice sheets. *Palaeoceanography, 22*(3), PA3214.

DeConto, R. M., Pollard, D., Wilson, P. A., Pälike, H., Lear, C. H., & Pagani, M. (2008). Thresholds for Cenozoic bipolar glaciation. *Nature, 455*, 652–657.

Denis, D., Crosta, X., Zaragosi, S., Romero, O., Martin, B., & Mas, V. (2009). Seasonal and subseasonal climate changes in laminated diatom ooze sediments, Adélie Land, East Antarctica. *Holocene, 16*(8), 1137–1147.

De Santis, L., Brancolini, G., & Donda, F. (2003). Seismo-stratigraphic analysis of the Wilkes Land Continental Margin (East Antarctica): influence of glacially driven processes on the Cenozoic deposition. *Deep-Sea Research, Part II, 50*(8–9), 1563–1594.

De Santis, L., Brancolini, G., Donda, F., & O'Brien, P. (2010). Cenozoic deformation in the George V Land continental margin (East Antarctica). *Marine Geology, 269*, 1–17.

Direen, N. G., Stagg, H. M. J., Symonds, P. A., & Norton, I. O. (2013). Variations in rift symmetry: cautionary examples from the Southern Rift System (Australia-Antarctica). *Geological Society, London, Special Publication, 369*, 453–475.

Domack, E. W. (1982). Sedimentology of glacial and glacial marine deposits on the George V Adelie continental shelf, East Antarctica. *Boreas, 11*, 79–97.

Domack, E. W., & Anderson, J. B. (1983). Marine Geology of the George V continental margin: combined results of Deep Freeze 79 and the 1911-14 Australasian expedition. In R. L. Oliver, P. R. James, & J. B. Jago (Eds.), *Antarctic earth science* (pp. 402–406). New York: Cambridge University Press.

Domack, E., Leventer, A., Dunbar, R., Taylor, F., Brachfeld, S., Sjunneskog, C., & The leg 178 scientific party. (2001). Chronology of the Palmer Deep Site, Antarctic Peninsula: a Holocene paleoenvironmental reference for the circum-Antarctic. *The Holocene, 11*, 1–9.

Donda, F., Brancolini, G., De Santis, L., & Trincardi, F. (2003). Seismic facies and sedimentary processes on the continental rise off Wilkes Land (East Antarctica): evidence of bottom current activity. *Deep-Sea Res., Part II, 50*(8–9), 1509–1527.

Douglas, P. M. J., Affek, H. P., Ivany, L. C., Houben, A. J. P., Sijp, W. P., Sluijs, A., et al. (2014). Pronounced zonal heterogeneity in Eocene southern high-latitude sea surface temperatures. *PNAS, 111*(18), 6582–6587.

Eittreim, S. L., Cooper, A. K., & Wannesson, J. (1995). Seismic stratigraphic evidence of ice-sheet advances on the Wilkes Land margin of Antarctica. *Sedimentary Geology, 96*(1–2), 131–156.

Escutia, C., Donda, F., Lobo, F. J., & Tanahashi, M. (2007). *Extensive debris flow deposits on the eastern Wilkes land margin: A key to changing glacial regimes.* USGS Open File-2007-1047, Short Research Paper 026. U.S. Geological Survey and The National Academies. http://dx.doi.org/10.3133/of2007-1047.srp026.

Escutia, C., Eittreim, S. L., & Cooper, A. K. (1997). Cenozoic sedimentation on the Wilkes Land continental rise, Antarctica. In C. A. Ricci (Ed.), *The Antarctic region: geological evolution and processes. Proceedings of the International Symposium on Earth Sciences 7.* (pp. 791–795). Terra Antartica.

Escutia, C., Nelson, C. H., Acton, G. D., Eittreim, S. L., Cooper, A. K., Warnke, D. A., et al. (2002). Current controlled deposition on the Wilkes Land continental rise, Antarctica. In D. A. V. Stow, C. J. Pudsey, J. A. Howe, J.-C. Faugeres, & A. R. Viana (Eds.), *Deep-water Contourite Systems: Modern Drifts and Ancient series, seismic and sedimentary Characteristics.* Mem. Geol. Soc. London, 22(1), 373–384. Geological Society of London.

Escutia, C., De Santis, L., Donda, F., Dunbar, R. B., Cooper, A. K., Brancolini, G., et al. (2005). Cenozoic ice sheet history from East Antarctic Wilkes Land continental margin sediments. *Global and Planetary Change, 45*(1–3), 51–81.

Escutia, C., Bárcena, M. A., Lucchi, R. G., Romero, O., Ballegeer, M., Gonzalez, J. J., et al. (2009). D. Circum-Antarctic warming events between 4 and 3.5 ma recorded in sediments from the prydz Bay (ODP leg 188) and the antarctic Peninsula (ODP leg 178) margins. *Global and Planetary Change, 69*, 170–184.

Escutia, C., Brinkhuis, H., Klaus, A., & the Expedition 318 Scientists. (2011). Wilkes Land Glacial History:. Cenozoic East Antarctic Ice Sheet evolution from Wilkes Land margin sediments *IODP Proceedings Volume* (318). Tokyo: Integrated Ocean Drilling Program Management International, Inc. http://dx.doi.org/10.2204/ iodp.proc.318.2011.

Escutia, C., Warnke, D., Acton, G. D., Barcena, A., Burckle, L., Canals, M., et al. (2003). Sediment distribution and sedimentary processes across the Antarctic Wilkes Land margin during the Quaternary. *Deep-Sea Research II, 50,* 1481–1508.

Etheridge, D. M., Pearman, G. I., & de Silva, F. (1988). Atmospheric trace-gas variations as revealed by air trapped in an ice core from Law Dome, Antarctica. *Annals of Glaciology, 10,* 28–33.

Florindo, F., & Siegert, M. J. (2008). Recent changes in Antarctuca and future research. In F. Florindo, & M. J. Siegert (Eds.), *Antarctic Climate Evolution* (pp. 571–576). Elsevier.

Florindo, F., & Siegert, M. (2009). Antarctic climate evolution developments in earth and environmental sciences(8).

Francis, J. E. (2000). Fossil wood from Eocene high latitude forests, McMurdo Sound, Antarctica. In J. D. Stilwell, & R. M. Feldmann (Eds.), *Paleobiology and palaeoenvironments of eocene rocks, McMurdo sound, east Antarctica Antarctic research series: 76.* (pp. 253–260). Washington, D.C: American Geophysical Union.

Fretwell, P., Pritchard, H. D., Vaughan, D. G., Bamber, J. L., Barrand, N. E., Bell, R., et al. (2013). Bedmap2: Improved ice bed, surface and thickness datasets for Antarctica. *Cryosphere, 7,* 375–393.

Gradstein, F. M., Ogg, J. G., Smith, A. G., Agterberg, F. P., Bleeker, W., Cooper, R. A., et al. (2004). *Geologic Time Scale 2004.* Cambridge University Press. 589 pp.

Hambrey, M. J., Ehrmann, W. U., & Larsen, B. (1991). Cenozoic glacial record of the Prydz Bay continental shelf, East Antarctica. In J. Barron, B. Larsen, et al. (Eds.), *Proc. ODP, Sci. Results, 119.* College Station, TX A&M University: (Ocean Drilling Program) (pp. 77–132).

Hampton, M. A., Eittreim, S. L., & Richmond, B. M. (1987). Geology of sediment cores from the George V continental margin. In S. L. Eittreim, & M. A. Hampton (Eds.), *The antarctic Continental Margin: Geology and geophysics of offshore wilkes Land: Circum-Pacific Council for Energy and Mineral resources earth sciences series* (Vol. 5A) (pp. 75–88). Houston.

Hayes, D. E., Frakes, L. A., et al. (1975). *Init. Repts. DSDP, 28.* Washington, DC: U.S. Govt. Printing Office.

Hill, P. J., & Exon, N. F. (2004). Tectonics and Basin development of the offshore tasmanian area; Incorporating results from deep ocean drilling. In N. F. Exon, J. P. Kennett, & M. Malone (Eds.), *The cenozoic southern ocean; tectonics, sedimentation and climate change between Australia and Antarctica, geophysical Monograph series* (Vol. 151) (pp. 19–42). Washington, D.C., U.S.A: American Geophysical Union.

Hollis, C. J., Crouch, E. M., Morgans, H. E. G., Handley, L., Baker, J. A., Creech, J., et al. (2009). Tropical sea temperatures in the high latitude South Pacific during the Eocene. *Geology, 37,* 99–102.

Hollis, C. J., Taylor, K. W. R., Handley, L., Pancost, R. D., Huber, M., Creech, J. B., et al. (2012). Early Paleogene temperature history of the Southwest Pacific Ocean: Reconciling proxies and models. *Earth and Planetary Science Letters, 349–350,* 53–66.

Houben, A. J. P. (2012). *Triggers and Consequences of glacial expansion across the Eocene-Oligocene transition.* LLP Contribution Series 39. http://www.narcis.nl/publication/RecordID/oai%3Adspace.library.uu.nl%3A1874%2F256955/coll/person/id/3.

Houben, A. J. P., Bijl, P. K., Guerstein, G. R., Sluijs, A., & Brinkhuis, H. (2011). Malvinia escutiana, a new biostratigraphically important Oligocene dinoflagellate cyst from the Southern Ocean. *Review of Palaeobotany and Palynology, 165,* 175.

Houben, A. J. P., Bijl, P. K., Pross, J., Bohaty, S. M., Stckley, C. E., Passchier, S., et al. (2013). Modern Southern Ocean plankton ecosystems arose at the onset of Antarctic glaciation. *Science, 340*(6130), 341–344.

Huber, M. (2001). *Modeling early Paleogene climate: From the top of the atmosphere to the bottom of the ocean.* PhD dissertation, Univ. California, Santa Cruz.

Huber, M., Brinkhuis, H., Stickley, C. E., Döös, K., Sluijs, A., Warnaar, J., et al. (2004). Eocene circulation of the Southern Ocean: was Antarctica kept warm by subtropical waters? *Paleoceanography, 19*(4), PA4026.

Hunt, R. J., & Poole, I. (2003). Paleogene West Antarctic climate and vegetation history in light of new data from King George Island. In S. L. Wing, P. D. Gingerich, B. Schmitz, & E. Thomas (Eds.), *Causes and consequences of globally warm climates in the early Paleogene* (369) (pp. 395–412). Boulder, Colorado: Geological Society of America.

Intergovernmental Panel on Climate Change (IPCC), report 2013: http://www.ipcc.ch/.

Jamieson, S. S. R., Sugden, D. E., & Hulton, N. R. J. (2010). The evolution of the subglacial landscape of Antarctica. *Earth and Planetary Science Letters, 293*, 1–27.

Kennett, J. P. (1977). Cenozoic evolution of Antarctic glaciation, the circum-Antarctic Ocean, and their impact on global paleoceanography. *Journal of Geophysical Research, 82*(27), 3843–3860.

Kennett, J. P., Houtz, R. E., et al. (1975). *Initial reports of the deep sea drilling project* (Vol. 29)Washington: U.S. Government Printing Office.

Kennett, J. P., & Shackleton, N. J. (1975). Laurentide ice sheet meltwater recorded in gulf of Mexico deep-sea cores. *Science, 188*, 147–150.

Lear, C. H., Bailey, T. R., Pearson, P. N., Coxall, H. K., & Rosenthal, Y. (2008). Cooling and ice growth across the Eocene-Oligocene transition. *Geology, 36*, 251–254.

Lisiecki, L. E., & Raymo, M. E. (2005). A Pliocene-Pleistocene stack of 57 globally distributed benthic $d^{18}O$ records. *Paleoceanography, 20*, PA1003.

Mackintosh, A. N., Verleyen, E., O'Brien, P. E., White, D. A., Jones, R. S., McKay, R., et al. (2013). Retreat history of the east antarctic ice sheet since the last glacial maximum. *Quaternary Science Reviews.* 100: 1–21 (Open Access).

Macrì, P., Sagnotti, L., Dinares-Turell, & Caburlotto, A. (2005). A composite record of Late Pleistocene relative geomagnetic paleointensity from the Wilkes Land Basin (Antarctica). *Physics of the Earth and Planetary Interiors, 151*, 223–242.

Miller, K. G., Aubry, M. P., Kahn, M. J., Melillo, A. J., Kent, D. V., & Berggren, W. A. (1985). Oligocene–Miocene biostratigraphy, magnetostratigraphy, and isotopic stratigraphy of the western North Atlantic. *Geology, 13*(4), 257–261.

Miller, K. G., Wright, J. D., Browning, J. V., Kulpecz, A., Kominz, M., Naish, T. R., et al. (2012). High tide of the warm Pliocene: Implications of global sea level for Antarctic deglaciation. *Geology, 40*, 407–410.

Naish, T., Powell, R., Levy, R., Wilson, G., Scherer, R., Talarico, F., et al. (2009). Obliquity-paced Pliocene West Antarctic ice sheet oscillations. *Nature, 458*, 322–328.

O'Brien, P. E., Cooper, A. K., Richter, C., et al. (2001). *Proc. ODP, Init. Repts (188).* College Station: TX. http://dx.doi.org/10.2973/odp.proc.ir.188.2001 (Ocean Drilling Program).

Orejola, N., Passchier, S., & IODP Expedition 318 Scientists (2014). Sedimentology of lower Pliocene to upper Pleistocence diamictons from IODP Site 1358, Wilkes Land margin, and implications for East Antarctic Ice Sheet dynamics. *Antarctic Science, 26*, 183–192.

Pagani, M. (2002). The alkenone-CO_2 proxy and ancient atmospheric carbon dioxide. *Philosophical Transactions: Mathematical, Physical and Engineering Sciences, 360*, 609–632 Royal Society of London A.

Pagani, M., Liu, Z., LaRiviere, J., & Ravelo, A. C. (2010). High Earth-system climate sensitivity determined from Pliocene carbon dioxide concentrations. *Nature Geoscience, 3,* 27–30.

Pagani, M., Zachos, J. C., Freeman, K. H., Tipple, B., & Bohaty, S. (2005). Marked decline in atmospheric carbon dioxide concentrations during the Paleogene. *Science, 309,* 600–603.

Pälike, H., Norris, R. D., Herrle, J. O., Wilson, P. A., Coxall, H. K., Lear, C. H., et al. (2006). The Heartbeat of the oligocene climate system. *Science, 314,* 1894–1898.

Pant, N. C., Biswas, P., Shrivastava, P. K., Bhattacharya, S., Verma, K., Pandey, M., et al. (2013). Provenance of Pleistocene sediments from Site U1359 of the Wilkes land IODP Leg 318 – evidence for multiple sourcing from the East Antarctic Craton and Ross Orogen. *Geological Society, London, Special Publications, 381,* 277–297.

Passchier, S., Bohaty, S. M., Jiménez-Espejo, F., Pross, J., Röhl, U., van de Flierdt, T., et al. (2013). Early Eocene – to – middle Miocene cooling and aridification of East Antarctica. *Geochemistry, Geophysics, Geosystems (G3), 14*(5), 1399–1410.

Patterson M. O., McKay R., Naish T., Escutia C., Passchier S., Williams T., et al. (2014). The response of the East Antarctic Ice Sheet to orbital forcing during the Pliocene and Early Pleistocene. *Nat. Geosciences online.* http://dx.doi.org/10.1038/NGEO2273

Pearson, P. N., Foster, G. L., & Wade, B. S. (2009). Atmospheric carbon dioxide through the Eocene-Oligocene climate transition. *Nature, 461,* 1110–1113.

Pollard, D., & DeConto, R. M. (2009). Modelling West Antarctic ice sheet growth and collapse through the past five million years. *Nature (London, U. K.), 458*(7236), 329–332.

Pollard, D., & DeConto, R. M. (2005). Hysteresis in Cenozoic Antarctic ice-sheet variations. *Global and Planetary Change, 45,* 9–21.

Pritchard, H. D., et al. (2012). Antarctic ice-sheet loss driven by basal melting of ice shelves. *Nature, 484,* 502–505.

Presti, M., Barbara, L., Denis, D., Schmidt, S., De Santis, L., & Crosta, X. (2011). Sediment delivery and depositional patterns off AdélieLand (East Antarctica) in relation to late Quaternary climatic cycles. *Marine Geology, 284*(1–4), 96–113.

Pross, J., Contreras, L., Bijl, P. K., Greenwood, D. R., Bohaty, S. M., Schouten, S., et al. (2012) Persistent near-tropical warmth on the Antarctic continent during the early Eocene epoch. *Nature Vol. 488,* No. 7409: 73–77.

Röhl, U., Brinkhuis, H., Stickley, C. E., Fuller, M., Schellenberg, S. A., Wefer, G., et al. (2004). Sea level and astronomically induced environmental changes in middle and Late Eocene sediments from the East Tasman Plateau. *Geophysical Monograph Series, 151,* 127–151.

Rignot, E., et al. (2011). Acceleration of the contribution of the Greenland and Antarctic ice sheets to sea level rise. *Geophysical Research Letters.*

Salabarnada, A., Escutia, C., Nelson, C. H., Damuth, J., & Brinkhuis, H. (2014). Varying depositional environments across the Oligocene-Miocene boundary and their relevance for East Antarctic ice sheet history: IODP Site U1356, Wilkes Land margin. *Geophysical Research Abstracts, Vol. 16,* EGU2014–13705.

Sijp, W. P., England, M. H., & Huber, M. (2011). Effect of the deepening of the Tasman Gateway on the global ocean. *Paleoceanography, 26,* PA4207.

Stickley, C. E., Brinkhuis, H., Schellenberg, S. A., Sluijs, A., Röhl, U., Fuller, M., et al. (2004). Timing and nature of the deepening of the Tasmanian Gateway. *Palaeoceanography, 19,* PA4027.

Stocchi, P., Escutia, C., Houben, A. J. P., Bijl, P. K., Brinkhuis, H., DeConto, R., & expedition 318 scientists., et al. (2013). Relative sea level rise around East Antarctica during Oligocene glaciation. *Nature Geosciences, 6,* 380–384.

Tauxe, L., Stickley, C. E., Sugisaki, S., Bijl, P. K., Bohaty, S., Brinkhuis, H., et al. (2012). Chronostratigraphic framework for the IODP Expedition 318 cores from the Wilkes Land Margin: constraints for paleoceanographic reconstruction. *Paleoceanography, 27*, PA2214.

Whitehead, J. M., & Bohaty, S. M. (2003). Pliocene summer sea surface temperature reconstruction using silicoflagellates from Southern Ocean ODP Site 1165. *Paleoceanography, 18*, 20–21.

Williams, T., van de Flierdt, T., Hemming, S. R., Chung, E., Roy, M., & Goldstein, S. L. (2010). Evidence for iceberg armadas from East Antarctica in the Southern Ocean during the late Miocene and early Pliocene. *Earth and Planetary Science Letters, 290*, 351–361.

Wise, S. W., Jr, Schlich, R., et al. (1992). In *Proc. ODP, Sci. Results, 120*. College Station, TX A&M University: (Ocean Drilling Program).

Woodbourne, M. O., & Case, J. A. (1996). Dispersal, variance and the late Cretaceous to early tertiary land mammal biogeography from South America to Australia. *Journal of Mammal Evolution, 3*, 121–161.

Zachos, J. C., Dickens, G. R., & Zeebe, R. E. (2008). An early Cenozoic perspective on greenhouse warming and carbon-cycle dynamics. *Nature (London, U. K.), 451*(7176), 279–283.

Chapter 3.4

The Pacific Equatorial Age Transect: Cenozoic Ocean and Climate History (Integrated Ocean Drilling Program Expeditions 320 & 321)

Heiko Pälike,[1],* Mitchell W. Lyle,[2] Hiroshi Nishi[3] and Isabella Raffi[4]

[1]*MARUM – Center for Marine Environmental Sciences, University of Bremen, Leobener Strasse, Bremen, Germany; [2]College of Geosciences, Texas A&M University, TX, USA; [3]The Center for Academic Resources and Archives, Tohoku University Museum, Tohoku University, Sendai, Japan; [4]Dipartimento di Ingegneria e Geologia (InGeo)–CeRSGeo, Università "G. d'Annunzio" di Chieti-Pescara, Chieti Scalo, Italy*
Corresponding author: E-mail: hpaelike@marum.de

3.4.1 INTEGRATED OCEAN DRILLING PROGRAM EXPEDITIONS 320 & 321 INTRODUCTION: BACKGROUND, OBJECTIVES, AND DRILLING STRATEGY

3.4.1.1 Background

The Pacific Equatorial Age Transect ("PEAT", Integrated Ocean Drilling Program (IODP) Expeditions 320 & 321) was drilled in 2009 to decipher the Cenozoic evolution of paleoequatorial Pacific Ocean and how it interacted with global climatic processes (Lyle et al., 2010; Pälike et al., 2010). The new drill sites (Figure 3.4.1) targeted the period from 52 million years before present (Ma) to today, and are integrated with previous results from Ocean Drilling Program (ODP) Legs 138 (Neogene: Pisias, Mayer, & Mix, 1995) and 199 (Paleogene: Lyle & Wilson, 2006; Lyle et al., 2008) to accomplish a Cenozoic "megasplice" that covers the time interval from 56 Ma to today (Figures 3.4.2 and 3.4.3).

The Pacific Ocean is closely linked to major developments of the global climate cycle during the Cenozoic. As the largest present-day ocean, it occupies approximately 50% of the total ocean surface today, while in the earliest Cenozoic (66 Ma) it spanned nearly two-thirds of the global oceans. The Pacific Ocean is thus a major cog in the Earth's climate system, also owing to the

Developments in Marine Geology, Volume 7. http://dx.doi.org/10.1016/B978-0-444-62617-2.00013-X

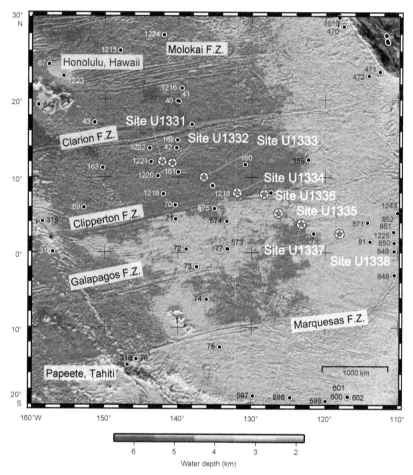

FIGURE 3.4.1   Present locations of Pacific Equatorial Age Transect (PEAT) drill sites (red (gray in print versions) stars) on a bathymetric map of the central Pacific. Also shown are locations of previous DSDP and ODP drilling (solid black circles), as well as Honolulu, Hawaii, and Papeete, Tahiti (open red (light gray in print versions) circles). F.Z. = fraction zone.

large equatorial upwelling zone, and hosting the El Niño–Southern Oscillation pattern with important global teleconnections.

The equatorial Pacific is a major area for trapping incoming solar radiation (Bryden & Brady, 1985), a major region of high primary productivity (Chavez & Barber, 1987; Westberry, Behrenfeld, Siegel, & Boss, 2008), and a primary region for carbon dioxide exchange from the deep ocean to the atmosphere (Takahashi et al., 1997). Using the sediment archive from the paleoequatorial Pacific provides new insights into the operation of the climate and carbon system, for productivity changes across the zone of divergence, time-dependent calcium

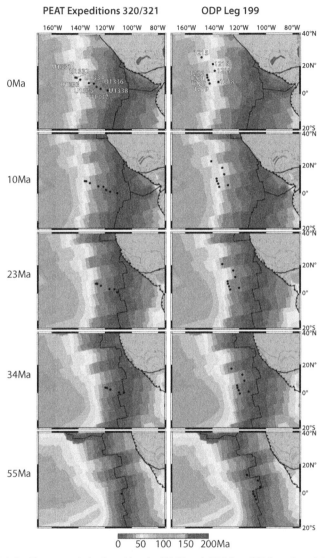

FIGURE 3.4.2 Plate tectonic backtracking for PEAT and ODP Leg 199 for selected time slices. Plate tectonic reconstruction courtesy Mike Gurnis, performed using gplates.org software.

carbonate dissolution and preservation, an integrated astronomically age-calibrated bio- and magnetostratigraphy, the location of the intertropical convergence zone (ITCZ), and evolutionary patterns for times of markedly changing boundary conditions. Integration of results from ocean drilling and detailed regional seismic surveys can now delineate the paleoequatorial position on the Pacific plate and variations in sediment thickness from 144 to 110°W present-day longitude.

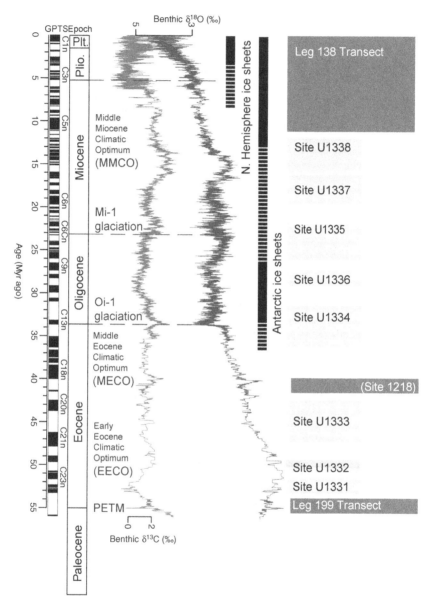

FIGURE 3.4.3   Evolution of oxygen ($\delta^{18}O$) and carbon ($\delta^{13}C$) stable isotopes through the Cenozoic and related major phases of climate change (modified from Zachos, Pagani, Sloan, Thomas, & Billups, 2001; Zachos, Dickens, & Zeebe, 2008). Yellow (lighter gray in print versions) boxes = time slices of interest for the PEAT program, blue (gray in print versions) boxes = ODP legs and sites previously drilled in the equatorial Pacific region. These additional sites will be used with the PEAT sites to obtain a nearly continuous Cenozoic record of the equatorial Pacific region. MMCO = middle Miocene climatic optimum, Mi-1 = Miocene isotopic event 1, Oi-1 = Oligocene isotopic event 1, EECO = early Eocene climatic optimum, PETM = Paleocene/Eocene thermal maximum.

The ocean circulation at the paleoequatorial surface ocean is intimately linked to the trade wind systems. The equatorial Pacific hosts a narrow zone of Coriolis driven divergence, giving rise to strong vertical and horizontal currents providing nutrients boosting biological productivity. Global climate affects the strength of the trade winds, which in turn couple to the biological productivity that is archived in the sediment record. Thus, variations in global climate, interhemispheric differences in temperature gradients, and marked changes in the ocean boundaries are all imprinted on the biogenic rich sediments that are accumulating in the equatorial zone and recovered as part of the PEAT drilling program.

For most of the Paleogene, the carbonate compensation depth (CCD) was around 0.5–1.5 km shallower than at the present day, making it more difficult to obtain sediments with calcareous microfossils that are well preserved and not strongly affected by carbonate dissolution. This difficulty is compounded by initial rapid thermal subsidence of the ridge crest. Nevertheless, the careful coring and site location strategy of the PEAT program allowed drilling of the most promising sites to obtain a unique sedimentary biogenic sediment archive for time periods just after the Paleocene–Eocene boundary event, the Eocene cooling, the Eocene–Oligocene transition, the "one cold pole" Oligocene, the Oligocene–Miocene transition, and the Miocene/Pliocene boundary.

3.4.1.2 Scientific Objectives of PEAT IODP Expeditions 320 & 321

The PEAT drill sites were located (Figures 3.4.1 and 3.4.2) to achieve an age transect of eastern Pacific sediments within the equatorial region (±2° of the Equator) on the Pacific plate at the time of deposition. The age of sediments within the equatorial transect span from the early Eocene through the Pliocene, with Paleocene/Eocene and late Miocene to recent intervals being covered by previous ODP Legs 138 and 199. Drill sites target specific time intervals of interest (Figure 3.4.3) at locations that provide optimum preservation of calcareous sediments and allow a detailed reconstruction of the CCD in the paleoequatorial Pacific (Figure 3.4.4). The overall aim was to obtain a continuous well-preserved equatorial Pacific sediment section that addresses the following primary scientific objectives:

1. To detail the nature and changes of the CCD over the Cenozoic in the paleoequatorial Pacific
2. To determine the evolution of paleoproductivity of the equatorial Pacific over the Cenozoic
3. To validate and extend the astronomical calibration of the geological timescale for the Cenozoic, using orbitally forced variations in sediment composition known to occur in the equatorial Pacific, and to provide a fully integrated and astronomically calibrated bio-, chemo-, and magnetostratigraphy at the equator
4. To determine temperature (sea surface and bottom water), nutrient profiles, and upper water column gradients

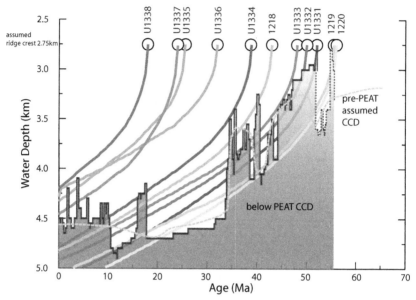

FIGURE 3.4.4   PEAT drill sites in relation to calcium carbonate compensation depth (CCD) history (van Andel, Heath, & Moore, 1975: green (lighter gray in print versions), Pälike et al., 2012: black with blue (darkest gray in print versions) shading), with additional data from Leg 199. Subsidence curves use a subsidence parameter calculated from estimated basement ages of PEAT sites and their present-day depth. Subsidence modeling sediment unloading. The assumed ridge-crest paleodepth is 2.75 km.

5. To better constrain Pacific plate tectonic motion and better locate the Cenozoic equatorial region in plate reconstructions, primarily via paleo-magnetic methods

6. To make use of the high level of correlation between tropical sedimen-tary sections and existing seismic stratigraphy to develop a more complete model of equatorial circulation and sedimentation

7. To provide information about rapid rates of taxonomic evolution during times of climatic stress

8. To improve our knowledge of the reorganization of water masses as a function of depth and time, as the PEAT drilling strategy also implies a paleodepth transect

9. To develop a limited N–S transect across the paleoequator, caused by the northward offset of the proposed sites by Pacific plate motion, providing addi-tional information about N–S hydrographic and biogeochemical gradients

10. To obtain a transect of mid-ocean ridge basalt samples from a fixed loca-tion in the absolute mantle reference frame, and to use a transect of basalt samples along the direction of seafloor spreading that have been erupted in similar formation-water environments to study low-temperature alteration processes by seawater circulation

3.4.1.3 Drilling Design and Strategy

Throughout the Cenozoic the Pacific plate motion has had a northward component, thus moving the thick sediment bulge of biogenic-rich deposits slowly away from the currently narrow equatorial upwelling. The Pacific plate has moved with a northward latitudinal component of around 0.25° per million years since 43 Ma, and moved slightly faster to the north prior to then (Koppers, Morgan, Morgan, & Staudigel, 2001). The northwest movement of the Pacific plate transports the equatorial sediments gradually out from under the zone of highest sediment delivery at the Equator, resulting in a broad mound of biogenic sediments. Thus, older sections are not deeply buried and can be recovered by drilling without extensive diagenesis. The lateral transport of ocean crust away from the equatorial zone of rapid sedimentation into regions with low sedimentation rates keeps older equatorial sediment sections from being buried deeply beneath younger sediments.

The chief design priority for the location of PEAT drilling locations was to backtrack a series of progressively older positions on the Pacific plate to a region about ±2° around the paleoequator in order to recover carbonate-bearing sediments. This transect can then be assembled into an equatorial Pacific "megasplice" covering the interval from 0 to 56 Ma. The sedimentary records from "off-splice" latitudes are not ignored; they give important insight into the strength of winds, currents, upwelling, productivity, and changes in CCD once detailed chronostratigraphy is calibrated and applied (Lyle, 2003) and furthermore allows the application of a depth-transect approach for intermediate ages where multiple drill sites cover the same time interval but at different paleodepths due to thermal subsidence of the ocean crust on which sediment accumulates. The off-equatorial sediments are also important to delineate the position of the ITCZ, and to quantify latitudinal trends across the paleoequatorial upwelling zone.

However, the tectonic movement (Figure 3.4.2) requires that a complete environmental record of the equatorial region be spliced together from different drill sites. Assembling a complete equatorial environmental record across important climatic transitions (Figure 3.4.3) requires periodic shifts to new drill site locations still within the equatorial zone, when equatorial sediments of the appropriate age are being deposited (Figure 3.4.4). In addition, because of the shallow overburden, most of the sediment column can be cored by the advanced piston coring (APC) technique to recover sediments with minimal drilling disturbance, and indeed the enhanced capabilities of the upgraded drilling vessel *JOIDES Resolution* resulted in a set of new APC depth records during IODP Expedition 321 and subsequent cruises, with improved sediment recovery and core quality compared to prior efforts. The northward rate of tectonic displacement in the Pacific Ocean is not so large that a traverse of the equatorial zone (within 2° latitude of the equator) was too rapid to record a reasonable period of equatorial ocean history. Typically, drill sites remain within the equatorial zone for durations of 10–20 Ma before passing beyond the northern edge of high biogenic sedimentation. In summary, the drilling strategy pursued for PEAT resulted not only in a depth transect but also in a limited latitudinal transect.

The PEAT program was designed around a "flow line" strategy, with drill holes located semiparallel to the trajectory of seafloor spreading. The PEAT drilling program pursued a "flow line" that complements "time line" strategies used by previous ODP drilling efforts (ODP Legs 138, 199), for two reasons. A latitudinal transect (time line) best resolves the structure of the equatorial current system, but for only a limited time window of a few million years. Ocean crust cools and sinks as it ages, and the seafloor on which the sediments are deposited approaches the zone of initial carbonate dissolution and CCD within a few million years, especially during the Paleogene when the CCD was >1.5 km shallower than today. Thus, the best preserved part of the sections recovered in such "time line" transects is restricted by the depth at which carbonate dissolution significantly increases, as well as by the northward movement of sediment sections out of the region of high equatorial productivity. This limitation was exemplified by the results from ODP Leg 199, which recovered only limited amounts of carbonate prior to the Eocene/Oligocene boundary (e.g., at ODP site 1218 on 42 Ma aged crust, Coxall, Wilson, Pälike, Lear, & Backman, 2005; Coxall & Wilson, 2011), but allowing the construction of a first initial high-resolution megasplice segment for the Oligocene (Pälike et al., 2006).

Many important paleoceanographic proxies are measured on carbonates, so only a few million years at a time can be studied in detail via the time line approach. It would be time prohibitive to drill the number of time line transects needed to complete a Cenozoic history of the equatorial Pacific. Fortunately, the coherent response of the equatorial Pacific to climate events covers vast areas, so that one site placed near the equator can be used to understand changes in much of the region over a horizontal length scale approaching 1000 km (Pälike et al., 2005; Westerhold et al., 2012a). We were also able to make use of the high level of correlation between tropical sediment sections and seismic stratigraphy to develop a more complete model of equatorial productivity and sedimentation.

When the PEAT flow line strategy is linked to previous drilling and regional seismic stratigraphy, a synoptic view of the Pacific can be developed. The most recent ODP Legs 199 and 138 were drilled along a line of equal oceanic crustal age, thus obtaining an approximate north–south transect across the major east–west currents during time intervals of particular interest. For PEAT, we planned a flow line strategy to collect equatorial sediment sections through the Cenozoic with carbonate preserved. The drilled intervals span the extremely warm times of the early Eocene, through the cooling of the late Eocene through Oligocene, the early Miocene time of relatively warm climates (or low ice volume), and into sections deposited during the development of the major southern and northern hemisphere ice sheets. In this way, we were able to track the paleoceanographic conditions at the paleoequator, in the best preserved sediments obtainable.

3.4.2 MAIN SEDIMENT SEQUENCE

Detailed descriptions of PEAT drilling results can be found in the Expedition 320&321 Proceedings (Pälike et al., 2010), and in Lyle et al. (2010) from which we abstract the following descriptions. Eight sites (Sites U1331–U1338) were drilled, whose basement ages spanned from 52 to 18 Ma. PEAT shipboard science has determined that the sediments recovered fill in gaps from previous drilling and can be used to create a high-resolution megasplice of equatorial Pacific sedimentation. Cross-calibration of magneto-, bio-, and ultimately orbital stratigraphy significantly improve chronological estimates of sedimentation and ages of important paleoceanographic events.

The PEAT program recovered sediments similar in lithology to previous DSDP and ODP expeditions to the central equatorial Pacific region (Lyle et al., 2008). The lithostratigraphy of the northwest–southeast transect of sites drilled during Expedition 320/321 together with the sedimentary sequence from ODP Site 1218, which is also included in the PEAT flow line strategy, is summarized in Figure 3.4.5. As expected of the younging crust toward the southeast, the Eocene sequence (green shading) thins from northwest to southeast, pinching out east of Site U1334, the last drill site on Eocene crust. In contrast, the Miocene sequence (yellow shading) thickens substantially from northwest to southeast. The Miocene section is thickest at Site U1337, targeted on crust of latest Oligocene age and thus the drill site that spent the most time within the Miocene equatorial zone. The Oligocene sequence (blue shading) is thickest in the middle of the PEAT transect (U1334 and U1336) and thins in both directions, marking the Oligocene equatorial zone. The thickness of sediments from different parts of the age transect is compatible with that expected from our drilling strategy. Basement ages for sites that penetrated the ocean crust are similar to expected ages based on plate reconstructions to within ±2 Ma.

Not only is the long-term environmental record of importance, but also key climatic events over the Cenozoic need to be studied in order to understand the role of the equatorial Pacific in these key climate changes. The combined results of both ODP Leg 199 and the PEAT program provide the ability to study these important intervals of climate change within the equatorial Pacific, and significant postcruise research is aimed at these intervals. Important climate intervals (Figure 3.4.3) include the early Eocene climatic optimum (Zachos, Pagani et al., 2001; 2008; Sites U1331 and U1332), the middle Eocene climatic optimum (Bohaty & Zachos, 2003, Bohaty, Zachos, Florindo, & Delaney, 2009, Edgar et al., 2010, Spofforth et al., 2010, Site U1333), the middle through late Eocene carbonate accumulation events (CAE) (Lyle et al., 2005, Pälike et al., 2012, Sites U1333 and U1334), the Eocene–Oligocene transition (Coxall

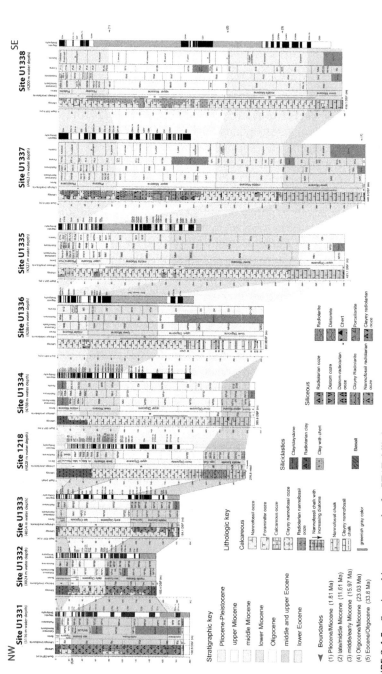

FIGURE 3.4.5 Stratigraphic summary plot for PEAT drill sites and ODP Site 1218. For each site the stratigraphic column and lithology are plotted against drilled depth, together with the magnetostratigraphy and biostratigraphic zonation schemes. The Eocene intervals are shaded in green (gray in print versions), Oligocene in light blue (white in print versions), Miocene in dark and light shades of yellow (dark and light gray in print versions), and Pliocene-Pleistocene in pink (lighter gray in print versions). *Figure reproduced under Creative Commons License from Lyle et al. (2010).*

et al., 2005; Site U1334), the late Oligocene warming (suppl. Material in Pälike et al., 2006, Site U1336), the Oligocene–Miocene transition (Zachos, Shackleton, Revenaugh, Pälike, & Flower, 2001; Pälike, Frazier, & Zachos, 2006; Sites U1335, U1336, and U1337), and the middle Miocene glaciation intensification event (Holbourn, Kuhnt, Schulz, & Erlenkeuser, 2005, Raffi et al., 2006, Sites U1337 and U1338).

3.4.3 RESULTS FROM POSTCRUISE INVESTIGATIONS

3.4.3.1 Integration with Seismic Stratigraphy

The site survey conducted in preparation for the PEAT program (cruise AMAT-03, onboard the *Roger Revelle*, Lyle, Pälike, Moore, Mitchell, & Backman, 2006) was designed to allow a basin-wide seismic correlation, connecting with previous seismic surveys in support for ODP Legs 138 (eastern equatorial Pacific), ODP Leg 199 (cruise Ewing0907), and previous surveys from DSDP Legs 9 and 85. While the PEAT siting strategy results in a restricted north–south transect and recovery of sediments from different paleolatitudes (separated by several degrees) at identical times in the past, the combined regional seismic study provides the opportunity to integrate data from older drill sites with PEAT (Figure 3.4.6) over an even wider area. The PEAT survey crossed 10–53 Ma seafloor and extends over the area between 150°W and 110°W and 5°S and 15°N. The seismic survey lines also cross the following drill sites for ground truth:

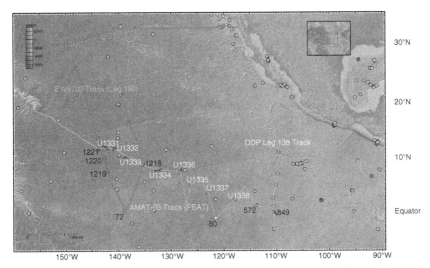

FIGURE 3.4.6 Map of available regional seismic lines (ODP Leg 138, ODP Leg 199, PEAT expeditions). The PEAT seismic survey was conducted onboard the R/V *Roger Revelle* (Lyle et al., 2006), and ties with ODP Site 849 to the east with ODP Leg 199 Sites ODP 1221 and 1218 to the west.

ODP Site 849, DSDP 572 and 80, and IODP Sites U1331–1338. An integrated effort to exploit the traceability of distinct seismic packages across the paleo-equatorial Pacific was presented by Tominaga, Lyle, and Mitchell (2011), and allowed the identification of several distinct reflectors, including the apparent expression of the CCD. Further insights into sediment deposition and transport near the PEAT sites based on the AMAT-03 survey data has been documented by Moore, Mitchell, Lyle, Backman, and Pälike (2007); Dubois and Mitchell (2012), and Mitchell and Huthnance (2013).

3.4.3.2 Integrated Bio-Magneto-Cyclostratigraphy and Sedimentation Rates

The study of paleoceanographic processes and the variations and evolution over time of mass accumulation rates across the PEAT transect depend on a detailed knowledge of sedimentation rates, which result from integrated bio- and magnetostratigraphies, put into a coherent stratigraphic framework. The integrated stratigraphy is the starting point to fully exploit and understand the complex interplay of productivity, dissolution, and spatial biogenic sedimentation patterns. Depending on the crustal subsidence, crustal age, and the time of crossing of the equatorial region, the sedimentation rates vary from site to site over time. Here we provide an overview of shipboard and postcruise results.

3.4.3.2.1 Paleomagnetic Studies

One of the achievements of the PEAT program has been the recovery of sediments recording environmental rock magnetic properties and aiding the development of an integrated magnetostratigraphy. The detailed shipboard results (Lyle et al., 2010; Pälike et al., 2010) already showed that the PEAT sites allow the development of an integrated magnetostratigraphy throughout the recovered sediment column. The identification of magnetic reversals in sediments from a paleoequatorial setting is challenging, but essentially complete for sites U1331–U1333, located on older basement basaltic crust, and characterized by only trace amounts of organic carbon, aiding preservation of paleomagnetic signals. Results from U1334–U1338 proved more challenging due to the effects of redox and diagenetic reactions, resulting in a weaker magnetic signal in parts of the sediment column. Shipboard results indicated the identification of magnetochrons (Cande & Kent, 1995) C1n through C20n, when spliced together from the PEAT sites. Postcruise studies refined the shipboard results significantly (e.g., Channell, Ohneiser, Yamamoto, & Kesler, 2013; Guidry et al., 2012, and Westerhold et al., 2014). A significant finding (Channell et al., 2013; Yamazaki, 2012; Yamamoto, 2013) has been that the magnetic mineral assemblage consists of a dominant biogenic component and a minor terrigenous component, where the carrier of the biogenic component are magnetofossils. The mineral composition of sediments also allows insights into the paleoposition of the ITCZ

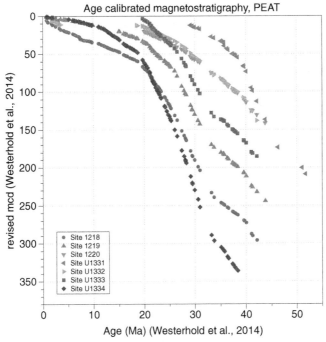

FIGURE 3.4.7 Summary of astronomically age-calibrated magnetic reversals for PEAT Sites U1331–U1334 and ODP Sites 1218–1220, color coded by site, and plotted according to the revised corrected composite depth of Westerhold et al., (2012a). Revised ages for the Paleogene are from Westerhold et al. (2014), younger ages follow the PEAT shipboard age model (Pälike et al., 2010).

(Yamazaki, 2012). One particular postcruise effort has been the development of paleointensity records from select PEAT sites (Channell & Lanci, 2014; Ohneiser et al., 2013; Yamamoto et al., 2013; Yamazaki, Yamamoto, Acton, Guidry, & Richter, 2013), establishing that assembling of paleointensity records from pre-Quaternary times is a distinct possibility. An additional cruise objective was to obtain and analyze a suite of basaltic basement rocks at the PEAT drill sites and to analyze these for magnetic and chemical alteration properties. Basement rocks were obtained at all PEAT sites bar U1336, and initial attempts have been made to study temporal variation in natural remanent magnetization from these samples (Yamamoto, 2013). With particular focus on the Paleogene component of the PEAT megasplice, Figure 3.4.7 illustrates the detail of magnetic reversals for PEAT Sites U1331–U1334 and selected ODP Leg 199 sites, after detailed correlation and age calibration (Westerhold et al., 2014).

3.4.3.2.2 Biostratigraphic Studies

Another achievement of the PEAT expeditions has been the recovery of a well-integrated set of biostratigraphies, spanning all major siliceous and calcareous microfossil groups (diatoms, radiolarians, planktic and benthic foraminifera, calcareous nannofossils). The results from the PEAT expeditions, building on knowledge gained from previous ODP legs, allow a major update in the determination of relative positions of particularly radiolarian, diatom, calcareous nannofossil, and magnetic stratigraphies. Figure 3.4.8 illustrates the level of detail achievable with shipboard determinations of bioevents. Interestingly, at least for the Paleogene, the biostratigraphic resolution achievable with biostratigraphic data from radiolarians matches or exceeds that obtained from calcareous nannofossils. A number of postcruise studies have begun to exploit this new record of evolutionary behavior and change. For siliceous microfossils, a detailed update has been provided for diatoms (Baldauf, 2013; Romero, 2013) and radiolarians (Kamikuri et al., 2012; Moore & Kamikuri, 2012; Moore, Mayer, & Lyle, 2012). Several studies have described the overall distribution of calcareous nannofossils for the Paleogene (Bown & Jones, 2012), supplemented by several detailed species-specific records across different time slices or for specific drill sites (Ciummelli & Raffi, 2013; Toffanin, Agnini, Rio, Acton, & Westerhold, 2013). Backman, Raffi, Ciummelli, and Baldauf (2013) specifically address the species-specific response of a late Miocene coccolithophore to enhanced

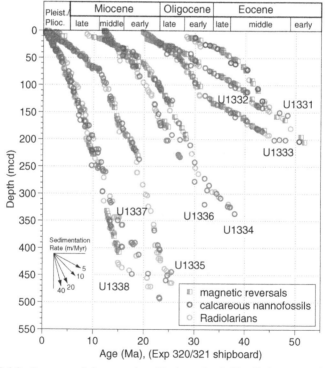

FIGURE 3.4.8 Summary of chronostratigraphic datum levels identified at PEAT sites plotted versus corrected composite depth, color coded by stratigraphy type (magnetic reversals, calcareous nannofossils, radiolarians). A steeper slope corresponds to a higher sedimentation rate.

biosilica productivity, and Beltran, Rousselle, Backman, Wade, and Sicre (2014) investigate environmental conditions responsible for the occurrence of monospecific coccolithophore dominance intervals (acmes). This study shows that the east equatorial Pacific thermocline remained deep and stable during a 100-ka-long acme, and that local surface stratification state fails to explain this acme, therefore contradicting the model-based hypothesis of a Southern Ocean high-latitude nutrient control of the surface waters in the east equatorial Pacific.

A major postcruise effort has been undertaken to describe the detailed biostratigraphic record from benthic foraminifera. Several studies focused on the Paleogene component of PEAT (Nomura, Takata, & Tsujimoto, 2013; Takata, Nomura, Tsujimoto, & Khim, 2012), investigating the faunal response of benthic foraminifera to paleoceanographic changes in the eastern equatorial Pacific. Additional work focused on the Neogene component (Tsujimoto, Nomura, Takata, & Kimoto, 2013) in relation to productivity changes during the constriction and closure of the Central American Seaway. The PEAT record of planktic foraminifera has variable preservation caused by the preservation and dissolution of carbonate, and is thus generally better following the Eocene–Oligocene transition, when the CCD of the equatorial Pacific deepened by more than 1.3 km (Pälike et al., 2012). For the Neogene PEAT section, particularly the detailed Miocene interval, studies on the distribution and evolutionary changes of planktic foraminifera were conducted (Fox & Wade, 2013; Hayashi et al., 2013; Takata, Nomura, Tsujimoto, Khim, & Chung, 2013).

3.4.3.2.3 Stratigraphic Correlation, Cyclostratigraphic and Downhole Logging Studies

Developing a Cenozoic megasplice and the required integrated stratigraphies and age models requires detailed stratigraphic correlation and cyclostratigraphic studies, supported where necessary by additional downhole-logging data. This has been one of the major objectives for PEAT. For the Paleogene section, a detailed effort was undertaken to align all cores using physical property, color line scan, and X-ray fluorescence core scanning data (XRF) as the basis for further cyclostratigraphic work. This correlation for Sites U1331–U1334 includes additional ODP Leg 199 data (Sites 1218, 1219, 1220), which complement the PEAT megasplice. Results were published by Westerhold et al., (2012a). For the Neogene part (chiefly U1337–U1338), similar work was published by Wilkens et al. (2013), employing a novel line-scan image splicing and interpolation algorithm to significantly improve upon initial splicing results. An example of the application of this algorithm to PEAT Sites U1331–U1334, combining age models with spliced image composites, is shown in Figure 3.4.9. The first application of a cyclostratigraphic analysis to generate highly resolved age models has been provided by Westerhold et al. (2014), spanning approximately magnetochrons C12n in the Oligocene to C20n in the middle Eocene (Figure 3.4.10). Additional work is in progress to complete the tuning for the Oligocene, and to finish initial efforts for the Neogene part (Holbourn et al.,

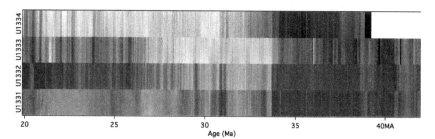

FIGURE 3.4.9 PEAT core image composite record plotted against shipboard age model for PEAT Sites U1331–U1334, courtesy software by Roy Wilkens. Core images are spliced, and interpolated against age.

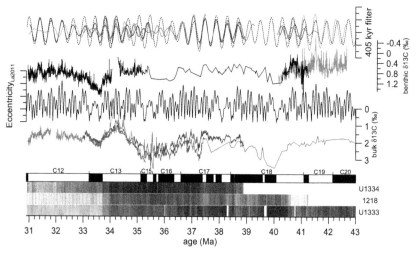

FIGURE 3.4.10 Tuned bulk and benthic stable carbon isotope data plotted against La2011 orbital solution for eccentricity (Laskar, Gastineau, Delisle, Farrés, & Fienga, 2011). Bulk data: 1218 (gray; Pälike et al., 2006), U1333 (green (lightest gray in print versions); Westerhold et al., 2014); Wilson et al., unpublished; Leon-Rodriguez & Dickens, 2013), U1334 (red (lighter gray in print versions); Westerhold et al., 2014; Wilson et al., unpublished). Benthic data: 1218 (black; Lear, Rosenthal, Coxall, & Wilson, 2004; Coxall et al., 2005; Tripati, Backman, Elderfield, & Ferretti, 2005; Coxall & Wilson, 2011), 1260 (blue (darker gray in print versions); Edgar, Wilson, Sexton, & Suganuma, 2007). Long eccentricity (405 ka) band pass filters of the detrended, tuned records are plotted on top La2011 (dashed line). *Figure reproduced under Creative Commons License from Westerhold et al. (2014).*

2013; Tian, Yang, Lyle, Wilkens, & Shackford, 2013). A new method of using Monte Carlo analysis of downhole logging data to achieve core-log integration was provided by Malinverno (2013).

3.4.3.2.4 Sedimentation Rates

The results from the PEAT expeditions reveal the change of linear sedimentation rate (LSR) in both the latitudinal and age transect components of the

PEAT program (Figure 3.4.8). The intercomparison between sites reveals that the highest sedimentation rates occur within the Oligocene and Miocene equatorial zone (Sites U1334 to U1338), with a sedimentation pattern similar to the modern equatorial region (highest deposition at the equator). However, sedimentation rates within the Eocene Equator were not significantly higher than sedimentation rates outside of the Equator. Time-dependent changes in sediment production and preservation strongly affected the Eocene sedimentary record. The linear sedimentation rates of the middle Eocene are high for the pelagic realm, frequently over 10 m/Ma and with a maximum of 18 m/Ma at Site U1331. Rates at Sites U1332 and 1333 are similar (8–6 m/Ma). The sedimentation rates of late Eocene time decrease to 3.5–6 m/Ma at Sites U1331 through U1333. Sedimentation rates are highest (>20 m/Ma) during the early to late Oligocene section at Sites U1333 and 1334, and in the early and middle Miocene at Sites U1337 and U1338. All sites have either a hiatus or reduced sedimentation rates for the youngest sediments, because they have moved out of the Neogene equatorial zone and have low modern deposition rates. The data from the PEAT sites, when combined with available data from ODP Leg 138 for the 0–10 Ma range, and ODP Leg 199 for the intervals between 32 and 42 Ma (Site 1218) and >52 Ma (Sites 1219–1221) produce a continuous history of sedimentation rates in the equatorial Pacific region for the past 56 Ma. A detailed knowledge of sedimentation rates for different sites, ideally in a high-resolution cyclostratigraphic framework, also supports additional studies that are concerned with mass fluxes. One example of such work was provided by Lyle (2003), and was essential to study the detailed variation of biogenic and nonbiogenic mass fluxes to the ocean floor, allowing an improved reconstruction of the Pacific CCD (Pälike et al., 2012). In general, if the underlying age models are highly resolved and accurate, derived flux records (gram per cm^2 per ka) are preferable for paleoceanographic studies as they minimize dissolution effects that are highly nonlinear for closed-sum records such as weight $\%CaCO_3$ (Lyle, 2003).

3.4.3.3 Progress on the Cenozoic Megasplice

Here we provide a short overview of the current status toward the PEAT Cenozoic megasplice in terms of different measured time series. The individual components to achieve this aim initially consist of geochemical XRF measurements, determination of magnetic reversals (described in 3.2.1) and an integrated biostratigraphy (3.2.2), and measurements of $\delta^{18}O$ and $\delta^{13}C$ on bulk sediment and benthic foraminiferal calcite.

3.4.3.3.1 XRF Splice

One of the major achievements of the PEAT program has been the detailed characterization of major and minor element concentrations by XRF core scanning, calibrated to absolute values using independent geochemical measurements. All PEAT sites and ODP Sites 1218, 1219, and 1220 were scanned to facilitate the development of a Cenozoic megasplice. For the Neogene, results were presented in Lyle et al., (2012), Lyle and Backman (2013), and Shackford, Lyle, Wilkens, and Tian (in press). These publications provide a full XRF data set for splices from Sites U1335–U1338. Results for Sites U1333–U1334 and ODP Site 1218 have been published for the interval C12n and older in Westerhold et al. (2014), and together with previous low-resolution studies of CCD changes (Pälike et al., 2012) provide a new high-resolution view on the pacing of carbonate events in the Eocene (Westerhold et al., 2012b). The upper components of Sites U1333 and U1334 have been scanned (Liebrand, unpublished data). Sites U1331 and U1332 have been completely scanned (Westerhold, unpublished data).

For the Sites U1335–U1338, XRF-derived major element concentrations allow the detailed characterization of weight percent $CaCO_3$ variations for the Neogene (Figure 3.4.11). These records track a number of new as well as previously identified $CaCO_3$ percent excursion intervals in much greater detail. Several of these horizons correspond to major regional seismic reflectors (e.g., "Green" ~4.1 Ma and "Lavender" ~16.9 Ma horizons, Mayer, Shipley, Theyer, Wilkens, & Winterer, 1985, Lyle, Liberty, Moore, & Rea, 2002, Tominaga et al., 2011). These horizons are supplemented by additional events such as the carbonate crash (12–9 Ma) and biogenic bloom events (~7–4 Ma) previously described in the equatorial Pacific (Farrell et al., 1995), Atlantic (Diester-Haass, Meyers, & Bickert, 2004), and the mid-Miocene climatic optimum (e.g., Holbourn et al., 2013). As an important contribution from the PEAT results, XRF scanning provides bulk chemical data (Ca, Si, Ti, Ba concentrations) to distinguish between pure $CaCO_3$ dissolution intervals, and zones of low $CaCO_3$ due to biogenic silica/diatom dilution (Shackford et al., in press). Changes in planktonic foraminifera assemblage indicate, e.g., that the "carbonate crash" beginning around 11 Ma and peaking at 9.8 Ma is associated with major ecosystem reorganization. An interpretation for these two different sources of $CaCO_3$ variation (Figure 3.4.12) is that dissolution intervals mark changes in bottom water source and/or reorganization of the carbon cycle while productivity intervals mark changes in shallow/intermediate circulation (Lyle et al., 2013).

3.4.3.3.2 Sediment Diagenesis and Stable Isotope Measurements

A major effort is underway to measure and compile stable isotope measurements from bulk sediment, and benthic and planktonic foraminifera. Much of this work is still in progress, and an important question for the interpretation of benthic isotope records from deep paleodepths in the equatorial Pacific concerns the preservation and fidelity of such stable isotopic measurements. Edgar, Pälike, and Wilson (2013) compared coeval records across the PEAT depth transect and established that different sites show evidence for recrystallization as a function of paleodepth, but that the diagenetic replacement of calcite must have occurred early during the burial history of sites, and therefore has negligible impact on isotopic offsets from these sites. Additional insights

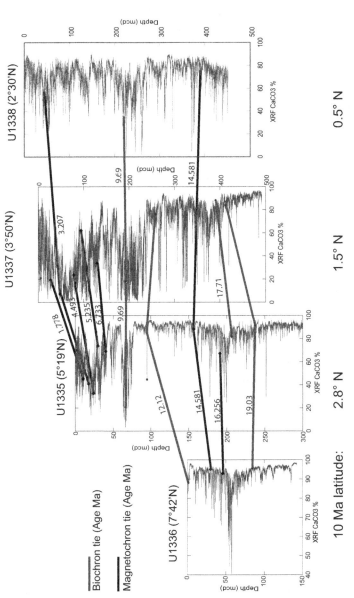

FIGURE 3.4.11 XRF-derived calibrated percent CaCO$_3$ values and stratigraphic correlation between Sites U1335, U1336, U1337, and U1338. Black diamonds and lines indicate magnetostratigrapic correlation points, brown (light gray in print versions) circles and lines indicate nannofossil biostratigraphic correlation points. Ages, in Ma, are shown in blue above the stratigraphic tie lines. Also shown are approximate site latitudes at 10 Ma. *Figure modified from Shackford et al. (in press).*

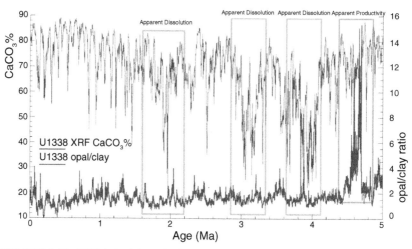

FIGURE 3.4.12 XRF-derived determination of carbonate dissolution and diatom dilution-related lows in percent CaCO₃ (Lyle et al., 2013).

into the diagenetic history of carbonates using Sr isotopes were obtained by Voigt et al. (submitted for publication), documenting $\delta^{88/86}$Sr fractionation during recrystallization of deep-sea carbonates. Results show that the bulk carbonates of Site U1336 are extensively altered compared to the other PEAT sites investigated, and also show a late phase of recrystallization at this more reactive site.

Significant achievements to date include the generation of high-resolution benthic time series from the Miocene for Sites U1337 (Tian et al., 2013) and U1338 (Holbourn et al., 2013) (Figure 3.4.13). These two records were astronomically age calibrated independently, and therefore provide a realistic estimate of uncertainties in ages derived by spectral methods with independent data and tuning approaches. Both records resolve orbital cycles, and exhibit a nicely resolved expansion of Antarctic ice sheets at around 13.9 Ma, with an associated excursion in the benthic δ^{13}C record previously interpreted as reflecting carbon burial feedbacks and coupled atmospheric pCO$_2$ changes concomitant with ice-sheet expansion (Holbourn et al., 2005) during the "Monterey Excursion."

Drury et al. (unpublished) generated a 3- to 4-ka-resolution benthic δ^{13}C and δ^{18}O record from ~3.5–8 Ma from Site U1338, as well as a mixed-layer planktic record from ~4.4–8 Ma (16 ka resolution), supplemented by a Mg/Ca temperature reconstruction from 5–6 Ma. The record has been successfully correlated to existing data from the North Atlantic ODP Site 982, increasing age control and allowing interbasin comparisons. The oxygen record is dominated by an orbital obliquity signal, as well as eccentricity.

The Miocene–Oligocene transition is being analyzed for benthic isotopes by Beddow-Twigg, Wade, Liebrand, Sluijs, and Lourens (2013), and for the Oligocene, a major effort is underway to generate a high-resolution benthic foram record (Zirkel, Herrle, Pälike, & Liebrand, 2013). In the Eocene, multiple

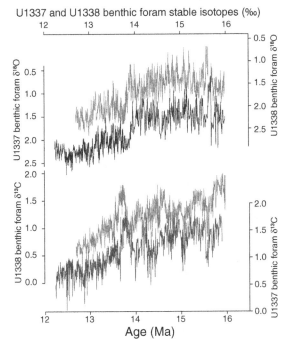

FIGURE 3.4.13   Miocene stable isotope records for Sites U1337 (Tian et al., 2013) and U1338 (Holbourn et al., 2013).

benthic records from Sites U1333 and U1334 are being produced by Kordesch, Pälike, Edgar, Bohaty, and Wilson (2013), Egan et al. (2013), and Wilson et al. (unpublished). A particular focus has been the Eocene–Oligocene transition (Figure 3.4.10), where additional bulk stable isotope measurements have resulted in the construction of an orbital-based age model (Westerhold et al., 2014, Wilson et al. unpublished). Leon-Rodriguez and Dickens (2013) characterized the bulk stable isotope composition of carbonates for Sites U1331, U1332, and U1333.

3.4.3.4 Paleoceanographic, Sedimentological, Geochemical, and Microbiology Results

One of the PEAT objectives was the reconstruction of paleoceanographic and paleoclimatological conditions of the equatorial Pacific Ocean during the Cenozoic. A number of significant studies have been conducted. Erhardt, Pälike, and Paytan (2013) used the Ba proxy to derive a high-resolution record of export production across the Eocene–Oligocene transition. The PEAT effort targeted this transition as one of the science objectives (Figure 3.4.10), and the new high-resolution PEAT records shed new light on the detailed stratigraphy and sequence of events across this transition (Moore, 2013; Moore et al., submitted for publication). One of the major outcomes of the PEAT studies so far has been the reconstruction of the detailed behavior of the CCD in the equatorial Pacific over the Cenozoic (Pälike et al., 2012). Two studies provide alternative interpretations for the reason of the dynamic Eocene CCD behavior, and instead invoke seafloor sedimentological processes (Dubois & Mitchell, 2012; Mitchell & Huthnance, 2013). Further answers to the understanding of geochemical cycles have been obtained by using novel 44/40 Ca isotope measurements on PEAT sediments (Pabich, Gussone, Rabe, Vollmer, & Teichert, 2014). The Ca isotope ratio of paleoseawater that is recorded in biominerals like foraminifers is an ideal tool to study changes in the Ca budget of the ocean, and results suggest considerable differences in the Ca isotope record of benthic foraminifers during the Eocene and the Oligocene, related to the dynamics of the CCD.

The location, sediment characteristics and burial history make the PEAT sediments also an interesting study target for microbiology investigations (Kallmeyer, 2013). Expedition 320/321 sites revealed cell counts that deviate from the usual cell count records found in many other ODP and IODP drill sites. Some of these deviations have been related to lithologic changes caused by the varying productivity regimes in the overlying water. However, several changes cannot be related to lithology and may be caused by geochemical changes, for example, redox fronts. The extent of suboxic diagenesis can partly be assessed using nitrate measurements (Kordesch & Delaney, 2013).

The sediment record of PEAT lends itself also as an archive for other geological events: Kuroda and Westerhold (2013) analyzed glass shards from Site U1333 to derive a likely source area. They regard Central American arc volcanism as the most likely source for these shards.

Two studies used organic and inorganic geochemical proxies to infer sea surface conditions during the Neogene component of the PEAT section. Rousselle, Beltran, Sicre, Raffi, and De Rafélis (2013) study an interval from the early Pliocene (~3.6–4.4 Ma at Site U1338) using alkenone-derived sea surface temperature data, and $\delta^{18}O$ measurements on coccolithophore species-specific small-size fractions to demonstrate the establishment of a cold tongue during the early Pliocene (4.4–3.6 Ma). The data suggest a shallowing of the thermocline in the eastern equatorial Pacific between 6.8 and 6 Ma, and shoaling between 4.8 and 4.0 Ma. In addition, Zhang, Pagani, and Liu (2014) use Site U1338 in combination with other ODP Sites to reconstruct a 12-million-year-long temperature history of the tropical Pacific Ocean. New biomarker sea surface temperature records indicate that the Pacific warm pool was ~4 °C warmer 12 million years ago and that both warm pool and cold tongue slowly cooled toward more modern conditions, while maintaining a ~3 °C zonal temperature gradient. Importantly, then, this result argues against a permanent El Niño-like state.

3.4.4 OUTLOOK

The PEAT sites allowed to address all of the original science objectives, and provided material for additional studies. This overview can only provide a snapshot of currently published results, and the complete Cenozoic megasplice must still await the completion of ongoing stable isotope (particularly for the Paleogene) and astronomical age calibration studies (for selected parts of the Neogene). All the material is in hand, and should be completed soon. Additional studies relate to the reconstruction of ocean circulation patterns and the position of the ITCZ, and are also currently prepared for publication.

ACKNOWLEDGMENTS

This research used samples and data provided by IODP. We thank the masters, crew, and Scientific Parties of IODP Expeditions 320 and 321, as well as the IODP-USIO.

REFERENCES

Backman, J., Raffi, I., Ciummelli, M., & Baldauf, J. (2013). Species-specific responses of late *Miocene Discoaster* spp. to enhanced biosilica productivity conditions in the equatorial Pacific and the Mediterranean. *Geo-Marine Letters, 33*(4), 285–298. http://dx.doi.org/10.1007/s00367-013-0328-0.

Baldauf, J. G. (2013). Data report: diatoms from sites U1334 and U1338, Expedition 320/321. In H. Pälike, M. Lyle, H. Nishi, I. Raffi, K. Gamage, A. Klaus, et al. (Eds.), *Proc. IODP, 320/321*. Tokyo: Integrated Ocean Drilling Program Management International, Inc. http://dx.doi.org/10.2204/iodp.proc.320321.215.2013.

Beddow-Twigg, H., Wade, B., Liebrand, D., Sluijs, A., & Lourens, L. (2013). Orbital forcing and climate response at the Oligocene-Miocene boundary: stable isotope records from the eastern equatorial Pacific. In: *Poster presented at the 11th International conference on Paleoceanography, Sitges.*

Beltran, C., Rousselle, G., Backman, J., Wade, B. S., & Sicre, M. A. (2014). Paleoenvironmental conditions for the development of calcareous nannofossil acme during the late Miocene in the eastern equatorial Pacific. *Paleoceanography, 29*(3), 210–222. http://dx.doi.org/10.1002/2013PA002506.

Bohaty, S. M., & Zachos, J. C. (2003). Significant Southern Ocean warming event in the late middle Eocene. *Geology, 31*(11), 1017–1020. http://dx.doi.org/10.1130/G19800.1.

Bohaty, S. M., Zachos, J. C., Florindo, F., & Delaney, M. L. (2009). Coupled greenhouse warming and deep-sea acidification in the middle Eocene. *Paleoceanography, 24*, PA2207. http://dx.doi.org/10.1029/2008PA001676.

Bown, P. R., & Jones, T. D. (2012). Calcareous nannofossils from the paleogene equatorial Pacific (IODP Expedition 320 Sites U1331–U1334). *Journal of Nannoplankton Research, 32*(2), 3–51.

Bryden, H. L., & Brady, E. C. (1985). Diagnostic model of the three-dimensional circulation in the upper equatorial Pacific ocean. *Journal Physical Oceanography, 15*, 1255–1273. http://dx.doi.org/10.1175/1520-0485(1985)015<1255:DMOTTD>2.0.CO;2.

Cande, S. C., & Kent, D. V. (1995). Revised calibration of the geomagnetic polarity time scale for the Late Cretaceous and Cenozoic. *Journal of Geophysical Research, 100*, 6093–6095. http://dx.doi.org/10.1029/94JB03098.

Channell, J. E. T., & Lanci, L. (2014). Oligocene–Miocene relative (geomagnetic) paleointensity correlated from the equatorial Pacific (IODP Site U1334 and ODP Site 1218) to the South Atlantic (ODP Site 1090). *Earth and Planetary Science Letters, 387*, 77–88. http://dx.doi.org/10.1016/j.epsl.2013.11.028.

Channell, J. E. T., Ohneiser, C., Yamamoto, Y., & Kesler, M. S. (2013). Oligocene-Miocene magnetic stratigraphy carried by biogenic magnetite at sites U1334 and U1335 (equatorial Pacific Ocean). *Geochemistry Geophysics Geosystems, 14.* (2). http://dx.doi.org/10.1029/2012GC004429.

Chavez, F. P., & Barber, R. T. (1987). An estimate of new production in the equatorial Pacific. *Deep Sea Research, 34,* 1229–1243. http://dx.doi.org/10.1016/0198-0149(87)90073-2.

Ciummelli, M., & Raffi, I. (2013). New data on the stratigraphic distribution of the nannofossil genus Catinaster and on evolutionary relationships among its species. *Micropaleontology, 32,* 197–205. http://dx.doi.org/10.1144/jmpaleo2013-002.

Coxall, H. K., & Wilson, P. A. (2011). Early Oligocene glaciation and productivity in the eastern equatorial Pacific: insights into global carbon cycling. *Paleoceanography, 26*(2), PA2221. http://dx.doi.org/10.1029/2010PA002021.

Coxall, H. K., Wilson, P. A., Pälike, H., Lear, C., & Backman, J. (2005). Rapid stepwise onset of Antarctic glaciation and deeper calcite compensation in the Pacific Ocean. *Nature, 433,* 53–57. http://dx.doi.org/10.1038/nature03135.

Diester-Haass, L., Meyers, P. A., & Bickert, T. (2004). Carbonate crash and biogenic bloom in the late Miocene: evidence from ODP sites 1085, 1086, and 1087 in the Cape Basin, southeast Atlantic Ocean. *Paleoceanography, 19*(1), PA1007. http://dx.doi.org/10.1029/2003PA000933.

Dubois, N., & Mitchell, N. C. (2012). Large-scale sediment redistribution on the equatorial Pacific seafloor. *Deep Sea Research Part I: Oceanographic Research Papers, 69,* 51–61. http://dx.doi.org/10.1016/j.dsr.2012.07.006.

Edgar, K. M., Wilson, P. A., Sexton, P. F., & Suganuma, Y. (2007). No extreme bipolar glaciation during the main Eocene calcite compensation shift. *Nature, 448,* 908–911. http://dx.doi.org/10.1038/nature06053.

Edgar, K. M., Wilson, P. A., Sexton, P. F., Gibbs, S. J., Roberts, A. P., Norris, R. D. (2010). New biostratigraphic, magnetostratigraphic and isotopic insights into the Middle Eocene climatic optimum in low latitudes. *Palaeogeography Palaeoclimatology Palaeoecology, 297,* 670–682. http://dx.doi.org/10.1016/j.palaeo.2010.09.016.

Edgar, K. M., Pälike, H., & Wilson, P. A. (2013). Testing the impact of diagenesis on the $\delta^{18}O$ and $\delta^{13}C$ of benthic foraminiferal calcite from a sediment burial depth transect in the equatorial Pacific. *Paleoceanography, 28*(3), 468–480. http://dx.doi.org/10.1002/palo.20045.

Egan, K., Edgar, K., Spencer, M., Pälike, H., Westerhold, T., Bohaty, S., et al. (2013). Late Eocene climate variability and forcing in the eastern equatorial Pacific. In: *Poster presented at the 11th International conference on Paleoceanography, Sitges.*

Erhardt, A. M., Pälike, H., & Paytan, A. (2013). High-resolution record of export production in the eastern equatorial Pacific across the Eocene–Oligocene transition and relationships to global climatic records. *Paleoceanography, 29*(1), 130–142. http://dx.doi.org/10.1029/2012PA002347.

Farrell, J. W., Raffi, I., Janecek, T. R., Murray, D. W., Levitan, M., Dadey, K. A., et al. (1995). Late Neogene sedimentation patterns in the eastern equatorial Pacific. *Proceeding of Ocean Drilling Program Scientific Results, 138,* 717–756. http://dx.doi.org/10.2973/odp.proc.sr.138.143.1995.

Fox, L. R., & Wade, B. S. (2013). Systematic taxonomy of early–middle Miocene planktonic foraminifera from the equatorial Pacific Ocean: Integrated Ocean Drilling Program, site U1338. *Journal of Foraminiferal Research, 43*(4), 374–405. http://dx.doi.org/10.2113/gsjfr.43.4.374.

Guidry, E. P., Richter, C., Acton, G. D., Channell, J. E. T., Evans, H. F., Ohneiser, C., et al. (2012). Oligocene–Miocene magnetostratigraphy of deep-sea sediments from the equatorial Pacific (IODP site U1333). *Geological Society London Special Publication, 373,* 13–27. http://dx.doi.org/10.1144/SP373.7.

Hayashi, H., Idemitsu, K., Wade, B. S., Idehara, Y., Kimoto, K., Nishi, H., et al. (2013). Middle Miocene to Pleistocene planktonic foraminiferal biostratigraphy in the eastern equatorial Pacific Ocean. *Paleontological Research, 17*(1), 91–109. http://dx.doi.org/10.2517/1342-8144-17.1.91.

Holbourn, A., Kuhnt, W., Schulz, M., & Erlenkeuser, H. (2005). Impacts of orbital forcing and atmospheric carbon dioxide on Miocene ice-sheet expansion. *Nature, 438*(7067), 483–487. http://dx.doi.org/10.1038/nature04123.

Holbourn, A., Kuhnt, W., Lyle, M., Schneider, L., Romero, O., & Anderson, N. (2013). Middle Miocene climate cooling linked to intensification of eastern equatorial Pacific upwelling. *Geology, 42*(1), 19–22. http://dx.doi.org/10.1130/G34890.1.

Kallmeyer, J. (2013). Data report: microbial abundance in subseafloor sediments of the equatorial Pacific Ocean, Expedition 320/321. In H. Pälike, M. Lyle, H. Nishi, I. Raffi, K. Gamage, A. Klaus, et al. (Eds.), *Proc. IODP, 320/321*. Tokyo: Integrated Ocean Drilling Program Management International, Inc. http://dx.doi.org/10.2204/iodp.proc.320321.214.2013.

Kamikuri, S., Moore, T. C., Ogane, K., Suzuki, N., Pälike, H., & Nishi, H. (2012). Data report: early to middle Eocene radiolarian biostratigraphy, IODP Expedition 320 site U1331, eastern equatorial Pacific. In H. Pälike, M. Lyle, H. Nishi, I. Raffi, K. Gamage, A. Klaus, et al. (Eds.), *Proc. IODP, 320/321*. Tokyo: Integrated Ocean Drilling Program Management International, Inc. http://dx.doi.org/10.2204/iodp.proc.320321.202.2012.

Koppers, A. A. P., Morgan, J. P., Morgan, J. W., & Staudigel, H. (2001). Testing the fixed hotspot hypothesis using Ar-40/Ar-39 age progressions along seamount trails. *Earth and Planetary Science Letters, 185*(3-4), 237–252. http://dx.doi.org/10.1016/S0012-821X(00)00387-3.

Kordesch, W. E. C., & Delaney, M. L. (2013). Data report: pore water nitrate and silicate concentrations for Expedition 320/321 Pacific Equatorial Age Transect. In H. Pälike, M. Lyle, H. Nishi, I. Raffi, K. Gamage, A. Klaus, et al. (Eds.), *Proc. IODP, 320/321*. Tokyo: Integrated Ocean Drilling Program Management International, Inc. http://dx.doi.org/10.2204/iodp.proc.320321.210.2013.

Kordesch, W., Pälike, H., Edgar, K., Bohaty, S., & Wilson, P. (2013). Middle Eocene climate instability in the equatorial oceans. In: *Poster presented at the 11th International conference on Paleoceanography, Sitges*.

Kuroda, J., & Westerhold, T. (2013). Data report: volcanic glass shards from the Eocene–Oligocene transition interval at site U1333. In H. Pälike, M. Lyle, H. Nishi, I. Raffi, K. Gamage, A. Klaus, et al. (Eds.), *Proc. IODP, 320/321*. Tokyo: Integrated Ocean Drilling Program Management International, Inc. http://dx.doi.org/10.2204/iodp.proc.320321.211.2013.

Laskar, J., Gastineau, M., Delisle, J. B., Farrés, A., & Fienga, A. (2011). Strong chaos induced by close encounters with Ceres and Vesta. *Astronomy and Astrophysics, 532*, L4. http://dx.doi.org/10.1051/0004-6361/201117504.

Lear, C. H., Rosenthal, Y., Coxall, H. K., & Wilson, P. A. (2004). Late Eocene to early Miocene ice sheet dynamics and the global carbon cycle. *Paleoceanography, 19*, PA4015. http://dx.doi.org/10.1029/2004PA001039.

Leon-Rodriguez, L., & Dickens, G. R. (2013). Data report: stable isotope composition of Eocene bulk carbonate at sites U1331, U1332, and U1333. In H. Pälike, M. Lyle, H. Nishi, I. Raffi, K. Gamage, A. Klaus, et al. (Eds.), *Proc. IODP, 320/321*. Tokyo: Integrated Ocean Drilling Program Management International, Inc. http://dx.doi.org/10.2204/iodp.proc.320321.208.2013.

Lyle, M., & Backman, J. (2013). Data report: calibration of XRF-estimated $CaCO_3$ along the site U1338 splice. In H. Pälike, M. Lyle, H. Nishi, I. Raffi, K. Gamage, A. Klaus, et al. (Eds.), *Proc. IODP, 320/321*. Tokyo: Integrated Ocean Drilling Program Management International, Inc. http://dx.doi.org/10.2204/iodp.proc.320321.205.2013.

Lyle, M., & Wilson, P. A. (2006). Leg 199 synthesis: evolution of the equatorial Pacific in the early Cenozoic. In P. A. Wilson, M. Lyle, & J. V. Firth (Eds.), *Proc. ODP, Sci. Results* (199). http://dx.doi.org/10.2973/odp.proc.sr.199.201.2006.

Lyle, M., Liberty, L., Moore, T. C., Jr., & Rea, D. K. (2002). Development of a seismic stratigraphy for the Paleogene sedimentary section, central tropical Pacific Ocean. In M. Lyle, P. A. Wilson, T. R. Janecek, et al. (Eds.), *Proc. ODP, Init. Repts.* (199) (pp. 1–21). http://dx.doi.org/10.2973/odp.proc.ir.199.104.2002.

Lyle, M. W., Olivarez-Lyle, A., Backman, J., Tripati, A. K. (2005). Biogenic sedimentation in the Eocene equatorial Pacific: the stuttering greenhouse and Eocene carbonate compensation depth. In M. Lyle, P. Wilson, T. R. Janecek, & J. Firth (Eds.), *Proc. ODP, Sci. Results* (199). College Station, Texas: Ocean Drilling Program. http://dx.doi.org/10.2973/odp.proc.sr.199.219.2005.

Lyle, M. W., Pälike, H., Moore, T. C., Mitchell, N., & Backman, J. (2006). *Summary report of R/V Roger Revelle site survey AMAT03 to the IODP Environmental Protection and Safety Panel (EPSP) in support for proposal IODP626.* Southampton, UK: University of Southampton. 144pp, Available in electronic form http://eprints.soton.ac.uk/45921/. [Cited 02.06.14].

Lyle, M., Barron, J., Bralower, T. J., Huber, M., Olivarez Lyle, A., Ravelo, A. C., et al. (2008). The Pacific Ocean and the Cenozoic evolution of climate. *Reviews of Geophysics, 46.* (2). http://dx.doi.org/10.1029/2005RG000190 2005RG000190.

Lyle, M., Pälike, H., Nishi, H., Raffi, I., Gamage, K., Klaus, A., et al. (2010). The Pacific equatorial age transect, IODP Expeditions 320 and 321: building a 50-million-year-long environmental record of the equatorial Pacific Ocean. *Scientific Drilling, 9,* 4–15. http://dx.doi.org/10.2204/iodp.sd.9.01.2010.

Lyle, M., Olivarez Lyle, A., Gorgas, T., Holbourn, A., Westerhold, T., Hathorne, E., et al. (2012). Data report: raw and normalized elemental data along the site U1338 splice from X-ray fluorescence scanning. In H. Pälike, M. Lyle, H. Nishi, I. Raffi, K. Gamage, A. Klaus, et al. (Eds.), *Proc. IODP, 320/321.* Tokyo: Integrated Ocean Drilling Program Management International, Inc. http://dx.doi.org/10.2204/iodp.proc.320321.203.2012.

Lyle, M. W., Shackford, J. K., Holbourn, A. E., Tian, J., Raffi, I., Pälike, H., et al. (2013). *The Neogene Equatorial Pacific: A view from 2009 IODP drilling on Expedition 320/321.* Abstract PP34A-02 presented at 2013 Fall Meeting. San Francisco, Calif.: AGU. 9–13 Dec.

Lyle, M. (2003). Neogene carbonate burial in the Pacific Ocean. *Paleoceanography, 18*(3), 1059. http://dx.doi.org/10.1029/2002PA000777.

Malinverno, A. (2013). Data report: Monte Carlo correlation of sediment records from core and downhole log measurements at sites U1337 and U1338 (IODP Expedition 321). In H. Pälike, M. Lyle, H. Nishi, I. Raffi, K. Gamage, A. Klaus, et al. (Eds.), *Proc. IODP, 320/321.* Tokyo: Integrated Ocean Drilling Program Management International, Inc. http://dx.doi.org/10.2204/iodp.proc.320321.207.2013.

Mayer, L. A., Shipley, T. H., Theyer, F., Wilkens, R. H., & Winterer, E. L. (1985). Seismic modeling and paleoceanography at deep sea drilling project site 574. In L. Mayer, F. Theyer, E. Thomas, et al. (Eds.), *Init. Repts. DSDP* (85) (pp. 947–970). http://dx.doi.org/10.2973/dsdp.proc.85.132.1985.

Mitchell, N. C., & Huthnance, J. M. (2013). Geomorphological and geochemical evidence (^{230}Th anomalies) for cross-equatorial currents in the central Pacific. *Deep Sea Research Part 1: Oceanographic Research Papers, 78,* 24–41. http://dx.doi.org/10.1016/j.dsr.2013.04.003.

Moore, T. C., & Kamikuri, S. (2012). Data report: radiolarian stratigraphy across the Eocene/Oligocene boundary in the equatorial Pacific, sites 1218, U1333, and U1334. In H. Pälike, M. Lyle, H. Nishi, I. Raffi, K. Gamage, A. Klaus, et al. (Eds.), *Proc. IODP, 320/321.* Tokyo: Integrated Ocean Drilling Program Management International, Inc. http://dx.doi.org/10.2204/iodp.proc.320321.204.2012.

Moore, T. C., Mitchell, N., Lyle, M., Backman, J., & Pälike, H. (2007). Hydrothermal Pits in the biogenic sediments of the equatorial Pacific ocean. *Geochemistry, Geophysics, Geosystems, 8*(3), Q03015. http://dx.doi.org/10.1029/2006GC001501.

Moore, T. C., Jr., Mayer, L. A., & Lyle, M. (2012). Sediment mixing in the tropical Pacific and radiolarian stratigraphy. *Geochemistry, Geophysics Geosystems, 13,* Q08006. http://dx.doi.org/10.1029/2012GC004198.

Moore T. C., Wade B., Westerhold T., Erhardt A., Baldauf J., Coxall H., Wagner M. Equatorial Pacific Productivity Changes near the Eocene–Oligocene boundary, *Paleoceanography* (submitted for publication).

Moore, T. C., Jr. (2013). Erosion and reworking of Pacific sediments near the Eocene-Oligocene boundary. *Paleoceanography, 28*(2), 263–273. http://dx.doi.org/10.1002/palo.20027.

Nomura, R., Takata, H., & Tsujimoto, A. (2013). Data report: early to middle Eocene benthic foraminifers at sites U1331 and U1333, equatorial central Pacific Ocean, Expedition 320/321. In H. Pälike, M. Lyle, H. Nishi, I. Raffi, K. Gamage, A. Klaus, et al. (Eds.), *Proc. IODP, 320/321*. Tokyo: Integrated Ocean Drilling Program Management International, Inc. http://dx.doi. org/10.2204/iodp.proc.320321.212.2013.

Ohneiser, C., Acton, G., Channell, J. E. T., Wilson, G. S., Yamamoto, Y., & Yamazaki, T. (2013). A Middle Miocene relative paleointensity record from the equatorial Pacific. *Earth and Planetary Science Letters, 374,* 227–238. http://dx.doi.org/10.1016/j.epsl.2013.04.038.

Pabich S., Gussone N., Rabe K., Vollmer C., Teichert B.M.A. (2014). $^{44/40}$Ca isotope fluctuations of the Cenozoic calcium isotope budget and investigations of benthic foraminifer test preservation, Abstract submitted to German IODP/ICDP Kolloquium, Erlangen.

Pälike, H., Moore, T., Backman, J., Raffi, I., Lanci, L., Parés, J. M., et al. (2005). Integrated stratigraphic correlation and improved composite depth scales for ODP sites 1218 and 1219. In P. A. Wilson, M. Lyle, & J. V. Firth (Eds.), *Proc. ODP, Sci. Results* (199) (pp. 1–41). College Station, TX: Ocean Drilling Program. http://dx.doi.org/10.2973/odp.proc. sr.199.213.2005.

Pälike, H., Norris, D., Herrle, J. O., Wilson, P. A., Coxall, H. K., Lear, C. H., et al. (2006). The Heartbeat of the Oligocene climate system. *Science, 314,* 1894–1898. http://dx.doi. org/10.1126/science.1133822.

Pälike, H., Frazier, J., & Zachos, J. C. (2006). Extended orbitally forced palaeoclimatic records from the equatorial Atlantic Ceara Rise. *Quaternary Science Review, 25,* 3138–3149. http:// dx.doi.org/10.1016/j. quascirev.2006.02.011.

Pälike, H., Lyle, M., Nishi, H., Raffi, I., Gamage, K., Klaus, A., et al. (2010). In *Proc. IODP, 320/321*. Tokyo: Integrated Ocean Drilling Program Management International, Inc. http:// dx.doi.org/10.2204/iodp.proc.320321.2010.

Pälike, H., Lyle, M. W., Nishi, H., Raffi, I., Ridgwell, A., Gamage, K., et al. (2012). A Cenozoic record of the equatorial Pacific carbonate compensation depth. *Nature, 488*(7409), 609–614. http://dx.doi.org/10.1038/nature11360.

Pisias, N. G., Mayer, L. A., & Mix, A. C. (1995). Paleoceanography of the eastern equatorial Pacific during the Neogene: synthesis of Leg 138 drilling results. In N. G. Pisias, L. A. Mayer, T. R. Janecek, A. Palmer-Julson, & T. H. van Andel (Eds.), *Proc. ODP, Sci. Results* (138) (pp. 5–21). College Station, TX: Ocean Drilling Program. http://dx.doi.org/10.2973/odp.proc. sr.138.101.1995.

Raffi, I., Backman, J., Fornaciari, E., Pälike, H., Rio, D., Lourens, L., et al. (2006). A review of calcareous nannofossil astrobiochronology encompassing the past 25 million years. *Quaternary Science Reviews, 25,* 3113–3137. http://dx.doi.org/10.1016/j.quascirev. 2006.07.007.

Romero, O. E. (2013). Data report: biogenic silica deposition in the eastern equatorial Pacific. In H. Pälike, M. Lyle, H. Nishi, I. Raffi, K. Gamage, A. Klaus, et al. (Eds.), *Proc. IODP, 320/321.* Tokyo: Integrated Ocean Drilling Program Management International, Inc. http://dx.doi. org/10.2204/iodp.proc.320321.206.2013.

Rousselle, G., Beltran, C., Sicre, M.-A., Raffi, I., & De Rafélis, M. (2013). Changes in sea-surface conditions in the equatorial Pacific during the middle Miocene–Pliocene as inferred from coccolith geochemistry. *Earth and Planetary Science Letters, 361,* 412–421. http://dx.doi. org/10.1016/j.epsl.2012.11.003.

Shackford, J. K., Lyle, M., Wilkens, R., Tian, J. Data report: raw and normalized elemental data along the site U1335, U1336, and U1337 splices from X-ray fluorescence scanning. In H. Pälike, M. Lyle, H. Nishi, I. Raffi, K. Gamage, A. Klaus, and the Expedition 320/321 Scientists (Eds), *Proc. IODP, 320/321: Tokyo.* Integrated Ocean Drilling Program Management International, Inc. (in press).

Spofforth, D. J. A., Agnini, C., Pälike, H., Rio, D., Fornaciari, E., Giusberti, L., et al. (2010). Organic carbon burial following the middle Eocene climatic optimum in the central western Tethys. *Paleoceanography, 25.* http://dx.doi.org/10.1029/2009PA001738.

Takahashi, T., Feely, R. A., Weiss, R. F., Wanninkhof, R. H., Chipman, D. W., Sutherland, S. C., et al. (1997). Global air-sea flux of CO_2: an estimate based on measurements of sea-air pCO_2 difference. *Proceeding of the National Academy of Sciences, 94,* 8292–8299. http://dx.doi. org/10.1073/pnas.94.16.8292.

Takata, H., Nomura, R., Tsujimoto, A., & Khim, B.-K. (2012). Late early Oligocene deep-sea benthic foraminifera and their faunal response to paleoceanographic changes in the eastern equatorial Pacific. *Marine Micropaleontology, 96–97,* 123–132. http://dx.doi.org/10.1016/j. marmicro.2012.09.002.

Takata, H., Nomura, R., Tsujimoto, A., Khim, B.-K., & Chung, I. K. (2013). Abyssal benthic fora-minifera in the eastern equatorial Pacific (IODP Exp. 320) during the middle Miocene. *Journal of Paleontology, 87*(6), 1160–1185. http://dx.doi.org/10.1666/12–107.

Tian, J., Yang, M., Lyle, M. W., Wilkens, R., & Shackford, J. K. (2013). Obliquity and long eccentricity pacing of the middle Miocene climate transition. *Geochemistry, Geophysics, Geosystems, 14*(6), 1740–1755. http://dx.doi.org/10.1002/ggge.20108.

Toffanin, F., Agnini, C., Rio, D., Acton, G., & Westerhold, T. (2013). Middle Eocene to early Oligocene calcareous nannofossil biostratigraphy at IODP Site U1333 (equatorial Pacific). *Micropaleontology, 59*(1), 1–14.

Tominaga, M., Lyle, M., & Mitchell, N. (2011). Seismic interpretation of pelagic sedimentation regimes in the 18–53 Ma eastern equatorial Pacific: basin-scale sedimentation and infilling of abyssal valleys. *Geochemistry, Geophysics, Geosystems, 12,* Q03004. http://dx.doi.org/10.102 9/2010GC003347.

Tripati, A., Backman, J., Elderfield, H., & Ferretti, P. (2005). Eocene bipolar glaciation associated with global carbon cycle changes. *Nature, 436,* 341–346. http://dx.doi.org/10.1038/nature03874.

Tsujimoto, A., Nomura, R., Takata, H., & Kimoto, K. (2013). A deep-sea benthic foraminiferal record of surface productivity changes during the constriction and closure of the Central American seaway: IODP hole U1338B, eastern equatorial Pacific. *Journal of Foraminiferal Research, 43*(4), 361–373. http://dx.doi.org/10.2113/gsjfr.43.4.361.

van Andel, T. H., Heath, G. R., & Moore, T. C. (1975). *Cenozoic history and Paleoceanography of the Central Equatorial Pacific Ocean. A regional synthesis of deep sea drilling project data.* The Geological Society of America, Memoir 143 p. 134.

Voigt J., Hathorne E.C., Frank M., Vollstaedt H., & Eisenhauer A. Variability of carbonate diagene-sis in equatorial Pacific sediments deduced from radiogenic and stable Sr isotopes. *Geochimica et Cosmochimica Acta* (submitted for publication). GCA-D-14–00091R1

Westberry, T. M., Behrenfeld, J., Siegel, D. A., & Boss, E. (2008). Carbon-based primary productivity modeling with vertically resolved photo-acclimation. *Global Biogeochemical Cycles, 22*, GB2024. http://dx.doi.org/10.1029/2007GB003078.

Westerhold, T., Röhl, U., Wilkens, R., Pälike, H., Lyle, M., Jones, T. D., et al. (2012a). Revised composite depth scales and integration of IODP Sites U1331–U1334 and ODP Sites 1218–1220. In H. Pälike, M. Lyle, H. Nishi, I. Raffi, K. Gamage, A. Klaus, et al. (Eds.), *Proc. IODP, 320/321*. Tokyo: Integrated Ocean Drilling Program Management International, Inc. http://dx.doi.org/10.2204/iodp.proc.320321.201.2012.

Westerhold, T., Lyle, M. W., Pälike, H., Röhl, U., & Wilkens, R. (2012b). Middle to late Eocene carbonate accumulation events in the equatorial Pacific – new geochemical records from IODP Exp 320/321 and ODP Leg 199. *Geophysical Research Abstracts, 14* EGU2012–11635.

Westerhold, T., Röhl, U., Pälike, H., Wilkens, R., Wilson, P. A., & Acton, G. (2014). Orbitally tuned and astronomical forcing in the middle Eocene to early Oligocene. *Climate of the Past, 10*, 955–973. http://dx.doi.org/10.5194/cp-10-955-2014.

Wilkens, R. H., Dickens, G. R., Tian, J., Backman, J. (2013). and the Expedition 320/321 Scientists, Data report: revised composite depth scales for Sites U1336, U1337, and U1338. In H. Pälike, M. Lyle, H. Nishi, I. Raffi, K. Gamage, A. Klaus. *and the Expedition 320/321 Scientists, Proc. IODP, 320/321:* Tokyo (Integrated Ocean Drilling Program Management International, Inc.). doi:10.2204/iodp.proc.320321.209.2013.

Yamamoto, Y., Yamazaki, T., Acton, G. D., Richter, C., Guidry, E. P., & Ohneiser, C. (2013). Palaeomagnetic study of IODP Sites U1331 and U1332 in the equatorial Pacific—extending relative geomagnetic palaeointensity observations through the Oligocene and into the Eocene. *Geophysical Journal International, 196*(2), 694–711. http://dx.doi.org/10.1093/gji/ggt412.

Yamamoto, Y. (2013). Data report: temporal variation in natural remanent magnetization observed for Pacific plate basement rocks: compilation from legacy data and new paleomagnetism and rock magnetism data from seafloor basalts cored during Expedition 320/321. In H. Pälike, M. Lyle, H. Nishi, I. Raffi, K. Gamage, A. Klaus, et al. (Eds.), *Proc. IODP, 320/321*. Tokyo: Integrated Ocean Drilling Program Management International, Inc. http://dx.doi.org/10.2204/iodp.proc.320321.213.2013.

Yamazaki, T., Yamamoto, Y., Acton, G., Guidry, E. P., & Richter, C. (2013). Rock-magnetic artifacts on long-term relative paleointensity variations in sediments. *Geochemistry, Geophysics, Geosystems, 14*(1), 29–43. http://dx.doi.org/10.1002/ggge.20064.

Yamazaki, T. (2012). Paleoposition of the intertropical convergence zone in the eastern Pacific inferred from glacial-interglacial changes in terrigenous and biogenic mineral fractions. *Geology, 40*(2), 151–154. http://dx.doi.org/10.1130/G32646.1.

Zachos, J. C., Pagani, M., Sloan, L., Thomas, E., & Billups, K. (2001). Trends, rhythms, and aberrations in global climate 65 Ma to present. *Science, 292*, 686–693. http://dx.doi.org/10.1126/science.1059412.

Zachos, J. C., Shackleton, N. J., Revenaugh, J. S., Pälike, H., & Flower, B. P. (2001). Climate response to orbital forcing across the Oligocene-Miocene boundary. *Science, 292*(5515), 274–278. http://dx.doi.org/10.1126/science.1058288.

Zachos, J. C., Dickens, G. R., & Zeebe, R. E. (2008). An early Cenozoic perspective on greenhouse warming and carbon-cycle dynamics. *Nature, 451*, 279–283. http://dx.doi.org/10.1038/nature06588.

Zhang, Y. G., Pagani, M., & Liu, Z. (2014). A 12-million-year temperature history of the tropical Pacific Ocean. *Science, 344*(6179), 84–87. http://dx.doi.org/10.1126/science.1246172.

Zirkel, J., Herrle, J. O., Pälike, H., & Liebrand, D. (2013). Mid-Oligocene high-resolution benthic foraminiferal productivity records from the central eastern Pacific Ocean (CEPO). In: *Poster presented at the 11th International conference on Paleoceanography, Sitges.*

Chapter 3.5

North Atlantic Paleoceanography from Integrated Ocean Drilling Program Expeditions (2003–2013)

James E.T. Channell[1,*] and David A. Hodell[2]
[1]*Department of Geological Sciences, University of Florida, Gainesville, FL, USA; [2]Godwin Laboratory for Palaeoclimate Research, Department of Earth Sciences, University of Cambridge, Cambridge, UK*
Corresponding author: E-mail: jetc@ufl.edu

3.5.1 INTRODUCTION

The North Atlantic is one of the most climatically sensitive regions on Earth because the ocean–atmosphere–cryosphere system is prone to mode jumps that are triggered by changes in freshwater delivery to source areas of deepwater formation (e.g., Bond et al., 1993; Bond et al., 1999; Clark et al., 2006). During the last glaciation, these abrupt jumps in climate state are manifested by Dansgaard/Oeschger (D/O) cycles and Heinrich Events in ice and marine sediment cores, respectively (e.g., Bond et al., 1993, 1999; Broecker, Bond, McManus, Klas, & Clark, 1992; Dansgaard et al., 1993; Heinrich, 1988). Given the paramount importance of the North Atlantic as a driver of global climate change, the Integrated Ocean Drilling Program (IODP) has recognized the importance of recovering North Atlantic deep-sea sediments suitable for the study of Quaternary climate at millennial-scale resolution.

What is the rationale for studying millennial-scale variability in the North Atlantic over the last few million years, rather than just the last glacial cycle (recoverable by conventional piston cores)? Determining the *long-term evolution* of millennial-scale variability in surface temperature, ice-sheet dynamics and thermohaline circulation can provide clues to the mechanisms responsible for abrupt climate change. For example, the average climate state evolved toward generally colder conditions with larger ice sheets during the Plio-Pleistocene. This shift was accompanied by a change in the spectral character of climate

Developments in Marine Geology, Volume 7. http://dx.doi.org/10.1016/B978-0-444-62617-2.00014-1
359

proxies, from dominantly 41- to 100-ky periods, between ~950 and ~650 ka (e.g., Elderfield et al., 2012). Among the relevant questions are the following: When did "Heinrich Events," and analogous detrital layers characterized by ice-rafted debris (IRD), first appear in the sedimentary record of the North Atlantic, what was their provenance, and are they restricted to the "100-ky world" when ice volume was substantially increased?

Through the advanced piston corer (APC) aboard the RV *Joides Resolution*, IODP has the only available tool for acquiring long, continuous cores with the high core-quality (lack of drilling disturbance) necessary for high-resolution stratigraphy. One of the strategies has been to record distinct components of the millennial-scale climate system (e.g., Heinrich Events) at a wide distribution of sites within the North Atlantic IRD belt of Ruddiman (1977). In addition, the quest for high-resolution Quaternary records has led to the drilling of sediment drifts, both inside and outside the IRD belt, with elevated mean sedimentation rates in the 10–20 cm/ky range. Prior to Ocean Drilling Program (ODP) Leg 162 (in 1995), it was not known whether North Atlantic sediment drifts, such as the Feni, Gardar, and Bjorn Drifts of the Iceland Basin, would be useful sources of Quaternary paleoceanographic data, particularly beyond the last glacial cycle, or whether accumulation was episodic resulting in reworked sequences riddled with unconformities. The apparent continuity of deposition in Iceland Basin sediment drifts, at least on timescales in excess of a few thousand years, as well as the ability to recover stable isotope data and other facets of stratigraphy, led to further exploitation of sediment drifts (particularly, the Eirik and Gardar Drifts) during IODP Expedition 303/306 (2004/2005). It is important to realize that elevated sedimentation rates in sediment drifts are often due to lateral sediment supply from diverse sources, and hence sedimentary components have mixed age (e.g., McCave, 2002). Together with ODP Leg 162, and ODP Leg 172 to the Blake Plateau and Bermuda Rise in 1997, results from IODP Exp. 303/306 have provided Quaternary sedimentary archives that contribute to reconstruction of the history of surface-water characteristics (e.g., surface-water temperature, productivity, and salinity), Laurentide ice-sheet (LIS) dynamics, and North Atlantic thermohaline circulation, on millennial timescales.

IODP Expedition 339 (Nov. 2011–Jan. 2012) extended the theme of drilling sediment drifts to the SW Iberian Margin, which has been a focus for North Atlantic paleoceanography since it was demonstrated that, during the last glacial interval, planktic and benthic $\delta^{18}O$ have signals that can be matched to isotopic data from the Greenland and Antarctic ice cores, respectively (Shackleton, Fairbanks, Chiu, & Parrenin, 2004; Shackleton, Hall, & Vincent, 2000). The region appears to be particularly sensitive to fluctuations of the polar front, as reflected in planktic $\delta^{18}O$ and alkenone sea-surface temperature (SST) data, and to changes in temperature/chemistry related to incursions of Antarctic-sourced bottom waters. IODP Expedition 339 had two overall objectives: (1) to monitor the post-Miocene history of Mediterranean Outflow Water (MOW) through the Strait of Gibraltar and its influence on global circulation and climate, and to integrate this record with the seismic stratigraphy and (2) to recover an extended (Quaternary) sedimentary record at the "Shackleton" site (Site U1385).

Intense interest in early Cenozoic paleoceanography can be attributed to its relevance in understanding the ocean's response to elevated levels of atmospheric CO_2, and hence to our future/present climate dilemma. The IODP/ODP cruises to the equatorial Pacific (ODP Leg 199 and IODP Expedition 320/321) and South Atlantic (ODP Legs 207 and 208) have laid the foundation for these studies. Important contributions to the understanding of this critical time in Earth history are expected through IODP Expedition 342 (June/July, 2012) that recovered Cretaceous to early Miocene sequences from sediment drifts off Newfoundland.

3.5.2 IODP EXPEDITION 303/306 (NORTH ATLANTIC CLIMATE)

IODP Expedition 303/306 was based on two separate proposals (572-Full3 and 543-Full2/543-ADD) entitled "Ice sheet–ocean atmosphere interactions on millennial timescales during the late Neogene–Quaternary using a paleointensity-assisted chronology (PAC) for the North Atlantic" and "Installation of a Cork in Hole 642E to document and monitor bottom water temperature variations through time." The objective of Proposal 572-Full3 was to place late Neogene–Quaternary climate proxies in the North Atlantic into a chronology based on a combination of relative paleointensity (RPI), stable isotope ($\delta^{18}O$), and detrital layer stratigraphies. Primary sites were located off Orphan Knoll (Newfoundland), on the Eirik Drift (southeast Greenland), on the southern Gardar Drift, and in the central Atlantic IRD belt. Mean sedimentation rates were estimated to be in the range of 5–20 cm/ky at these sites. The targeting of high pelagic sedimentation rates, partially on sediment drifts, was designed to yield high-resolution reference records of North Atlantic paleoceanography for the last few million years. The emphasis placed on improving stratigraphic resolution through use of both RPI and $\delta^{18}O$ has meant that the Exp. 303/306 has gone a long way toward testing and ratifying the use of RPI in marine stratigraphy, and providing coupled high-resolution records of the two signals. The establishment of tandem stacks of RPI and $\delta^{18}O$ from globally distributed sites (Channell, Xuan, & Hodell, 2009) was a step toward proving that the RPI signal is suitable for global correlation on millennial scales, an argument consistent with knowledge of the recent geomagnetic field (e.g., Lhuillier, Fournier, Hulot, & Aubert, 2011). The objectives of Exp. 303/306 are both paleoceanographic and geophysical in that paleomagnetic records from high-sedimentation-rate archives lend themselves to comparison with numerical simulations of the geomagnetic field, thereby providing insights into the workings of the geodynamo. Documentation of past characteristics of the geomagnetic field including secular variation, RPI, magnetic excursions, and polarity transitions (e.g., Mazaud, Channell, Xuan, & Stoner, 2008) constrain the range of viable numerical simulations. These geophysical objectives go hand-in-hand with the quest for improved marine stratigraphy, to rival the stratigraphic precision afforded by ice cores, which remains a major challenge in paleoceanography.

Drilling locations for Exp. 303/306 were known, either from previous ODP and Deep Sea Drilling Project (DSDP) drilling, or from conventional piston cores, to: (1) contain distinct records of millennial-scale environmental variability (in terms of ice sheet–ocean interactions, deep circulation changes, or sea-surface conditions), (2) provide the requirements for developing a millennial-scale stratigraphy (through RPI, oxygen isotopes, and regional environmental patterns), and (3) document details of geomagnetic field behavior. Three sites (U1308, U1312, and U1313) constituted reoccupations of classic paleoceanographic sites drilled in 1983 during DSDP Leg 94 (Ruddiman, McIntyre, & Raymo, 1986; Ruddiman, Raymo,Martinson, Clement, & Backman, 1989).

3.5.2.1 Sites U1302/03, U1308, U1312, and U1313 (North Atlantic IRD Belt)

Heinrich layers in the North Atlantic are centimeter-to-decimeter-scale detrital layers dominated by IRD deposited by icebergs shed off surrounding continental ice sheets (Broecker et al., 1992; Heinrich, 1988). Heinrich layers deposited in the last 60 ky have been studied in detail (for review, see Hemming, 2004), and have been associated with surging of the LIS through Hudson Strait. They provide key information on changes in surface- and deepwater conditions associated with these instabilities, have been correlated to cold stadials in Greenland ice cores, and have been implicated in changes in oceanic thermohaline circulation (e.g., Bond et al., 1993, 1999). The characterization and correlation of "Heinrich-like" detrital layers further back in time, from Orphan Knoll (Site U1302/03, Figure 3.5.1) and other North Atlantic (Sites U1308 and U1313, Figure 3.5.1), are important for determination of the tempo of instability of the LIS, the contribution of North Atlantic-bordering ice sheets other than LIS to detrital layer stratigraphy, and the effects of such IRD episodes on surface- and deepwater circulation. For the last glacial cycle, Heinrich layers have been identified in piston cores from throughout the North Atlantic using a combination of high IRD content often comprising detrital carbonate, minima in planktic foraminiferal accumulation rates, and predominance of *Neogloboquadrina pachyderma* (sin.) in the foraminiferal assemblage (e.g., Elliot et al., 1998; Hillaire-Marcel, De Vernal, Bilodeau, & Wu, 1994; Hiscott, Aksu, Mudie, & Parsons, 2001; Van Kreveld, Knappertsbusch, Ottens, Ganssen, & van Hinte, 1996). Magnetic properties of Heinrich layers include high magnetic susceptibility (e.g., Robinson, Maslin, & McCave, 1995) and a coarsening of the magnetic grain size parameter, namely the ratio of anhysteretic susceptibility (κ_{ARM}) to susceptibility (κ) (e.g., Stoner, Channell, & Hillaire-Marcel, 1995; Stoner, Channell, & Hillaire-Marcel, 1996; Stoner, Channell, & Hillaire-Marcel, 1998, Stoner, Laj, Channell, & Kissel, 2000).

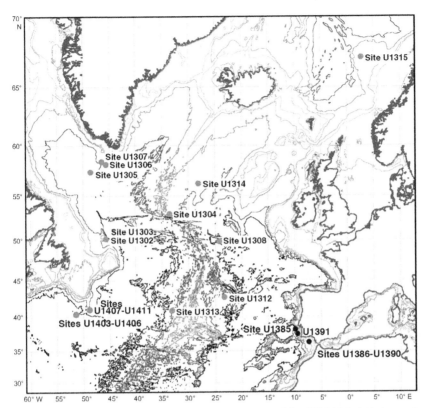

FIGURE 3.5.1 Location of sites occupied during IODP Expedition 303/306 (red (dark gray in print versions)), Expedition 339 (black), and Expedition 342 (purple (light gray in print versions)).

Several new methods have been brought to bear on the problem of identifying and correlating detrital layers at Sites U1302/03, U1308, and U1313. For example, at Site U1308, a Fourier Transform Infrared Spectrophotometer was used to identify dolomite that is characteristic of Ca-rich detrital layers (Ji, Ge, Balsam, Damuth, & Chen, 2009). At Site U1302/03, the lead isotope ratio ^{206}Pb/^{204}Pb has been used to monitor variations in the intensity of continental weathering over the last 37 ky, when marked increases in weathering were associated with Termination I, the Younger Dryas, and H0-H3 (Crocket, Vance, Foster, Richards, & Tranter, 2012). X-ray diffraction (XRD) has been used at Site U1313 to monitor dolomite, quartz, and feldspar components of IRD in detrital layers (Naafs, Hefter, Grützner, & Stein, 2013; Naafs, Hefter, & Stein, 2013; Naafs, Hefter, Ferretti, Stein, & Haug, 2011; Stein, Hefter, Grützner, Voelker, & Naafs, 2009). Alkenone-based SST data and abundance of $C_{37:4}$ alkenones (indicative of cold high-latitude waters) and C28(S) C-ring monoaromatic steroids (indicative of ancient organic matter) have also been used to mark detrital events at Site U1313

(Naafs, Hefter, Grützner et al., 2013, Naafs, Hefter, & Stein, 2013; Naafs et al., 2011). X-ray fluorescence (XRF) core scanning at 1-cm spacing was used at Sites U1302/03 and U1308 as a nondestructive means of detecting detrital layers through the Ca/Sr and Si/Sr ratios, the former being sensitive to detrital carbonate and the latter being sensitive to Si-rich detrital layers (Channell et al., 2012; Hodell, Channell, Curtis, Romero, & Rohl, 2008). Bulk carbonate $\delta^{18}O$ has also been found to be useful for the detection of detrital layers because of more negative $\delta^{18}O$ values for detrital versus biogenic carbonate (Hodell & Curtis, 2008). At Sites U1302/03 and U1308, XRF data were combined with gamma-ray attenuation density, bulk carbonate $\delta^{18}O$, and magnetic ratios sensitive to magnetite grain size (κ_{ARM}/κ, M_r/M_s, H_{cr}/H_c), and magnetic parameters sensitive to magnetic concentration such as susceptibility (Channell & Hodell, 2013; Channell et al., 2012; Hodell et al., 2008).

The first leg of Exp. 303/306 (Exp. 303) occupied two locations at Orphan Knoll, off the continental margin of Newfoundland (Figure 3.5.1), that are separated by 5.7 km, and were combined into a single site owing to their very similar lithostratigraphies (Channell et al., 2006). The combined Site U1302-U1303 is hereafter referred to as Site U1302/03. The location was known from a number of conventional piston cores collected by the R/V *Hudson* and *Marion Dufresne* in the 1990s (e.g., Hillaire-Marcel et al., 1994; Stoner et al., 1995, 1996, 1998) to preserve a detailed record over the last glacial cycle of "Heinrich-like" detrital layers. A debris flow close to the base of the Brunhes Chronozone at Site U1302 interrupted drilling and recovery, leading to the occupation of the second location (Site U1303) in an unsuccessful attempt to avoid the debris flow (Channell et al., 2006). Although the debris flow was encountered at both locations and prevented further drilling, the combined Site (U1302/03) provided an ~105-m-thick section of clay and silty clay with nannofossil ooze to the base of the Brunhes Chron, with the Matuyama-Brunhes boundary being recorded just below an ~10-m-thick debris flow (Channell et al., 2012).

Site U1308 (Figure 3.5.1) is a redrill of a site (Site 609) occupied in 1983 during DSDP Leg 94. Site 609 provided a classic record of Heinrich layers for the distal North Atlantic, and is the origin of the correlation of Heinrich events to cold stadials in Greenland ice cores during marine isotope stage (MIS) 3 (Bond et al., 1993, Bond et al.,1999; Bond & Lotti, 1995), although uncertainty in the continuity of the record has stymied similar studies further back in time. The composite record at Site U1308, comprising mainly nannofossil/foraminiferal oozes with silty clay, reaches a depth of ~250 m, providing a seemingly continuous record back to ~3 Ma (Channell et al., 2006), although the record is now considered to be incomplete in the vicinity of MIS 100 (Bailey et al., 2010).

Site U1313 (Figure 3.5.1), a reoccupation of DSDP Site 607, yielded an apparently complete composite section comprising ~300 m of late Miocene to Quaternary nannofossil ooze with silts and clays reaching back to ~7 Ma (Channell et al., 2006). The site is strategically located at the southern rim of the IRD belt of Ruddiman (1977).

The correlation of detrital layers from Site U1302/03 (Orphan Knoll) into the central Atlantic (Sites U1308 and U1313) presents a stratigraphic challenge because of the short recurrence time of detrital layers, at least in MIS 3. Owing to the scarcity of benthic foraminifera, a planktic oxygen isotope record based on *N. pachyderma* (sin.) was generated at Site U1302/03 (Hillaire-Marcel, de Vernal, & McKay, 2011). At Site U1308, Hodell et al. (2008) generated a benthic oxygen isotope record back to 1.5 Ma that was the basis for an age model through correlation to the LR04 $\delta^{18}O$ stack (Lisiecki & Raymo, 2005). The U1308 age model was used to calibrate an RPI record (Channell, Hodell, Xuan, Mazaud, & Stoner, 2008) that, in turn, was used as the basis for the "PISO-1500" RPI and $\delta^{18}O$ stacks for the last 1.5 my (Channell et al., 2009). These stacks (Figure 3.5.2) were generated by *tandem* correlation, to Site U1308, of $\delta^{18}O$ and RPI records from 12 globally distributed sites (mainly from the Atlantic realm) that had both $\delta^{18}O$ and RPI data for the same sites, although few of the 12 records extended back over the entire 1.5 my. The correlations were achieved using an adaptation of the "Match" protocol (Lisiecki & Lisiecki, 2002) that allows tandem correlation of two (ostensibly) independent global signals ($\delta^{18}O$ and RPI). "Match" correlations are based on set criteria and have the advantage of repeatability over purely visual correlations, particularly for tandem correlations. The PISO RPI stack (Figure 3.5.2) has been used as the calibrated RPI template, together with LR04 for $\delta^{18}O$, to provide age models for certain Exp. 303/306 sites (Channell et al., 2012; Channell, Wright, Mazaud, & Stoner, 2014).

In order to correlate the detrital layer stratigraphy from Site U1302/03 to Site U1308, Channell et al. (2012) used an age model for Site U1302/03 based on tandem correlation of planktic $\delta^{18}O$ and RPI to $\delta^{18}O$/RPI-calibrated templates (LR04 and PISO). The tandem correlation again utilized an adaption of the "Match" protocol (Lisiecki & Lisiecki, 2002) to match the Site 1302/03 $\delta^{18}O$ and RPI records to LR04 and PISO (Channell et al., 2012). Note that the Site U1308 age model (Hodell et al., 2008) provides the chronology for the PISO RPI template (Channell et al., 2009) so that using the PISO stack as the template for Site U1302/03 ties the Site U1302/03 RPI record to that from Site U1308. The use of RPI, together with $\delta^{18}O$, strengthens stratigraphic correlation between the two sites, and the resulting correlation is apparently consistent with the correlation of the diverse parameters that have been used to detect detrital layers over the last glacial cycle (Figure 3.5.3).

Ca/Sr and Si/Sr ratios determined by XRF core scanning at Site U1308 were used to detect Ca-rich and Si-rich IRD in detrital layers, respectively (Hodell et al., 2008). The onset of detrital carbonate layers at this site occurred at ~650 ka within MIS 16 and detrital carbonate layers are present in all subsequent glacial stages (as well as Terminations) other than MIS 6 and 14 (red symbols/lines in Figure 3.5.4(A)) (Hodell et al., 2008). Si-rich detrital layers are observed back to the base of the studied section (MIS 42) at Site U1308, are associated with hematite and magnetic grain sizes indicative of IRD deposition, and are synchronous with fluctuations in benthic $\delta^{13}C$ (Figure 3.5.5) (Channell & Hodell,

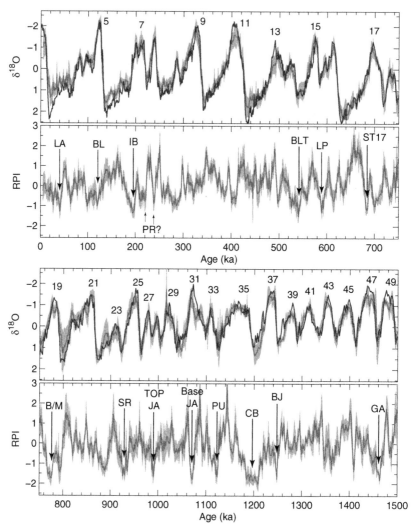

FIGURE 3.5.2 The PISO stacks generated using the Site U1308 age model (modified from Channell et al., 2009). Oxygen isotope and relative paleointensity (RPI) PISO stacks (red (dark gray in print versions)). Half-width of the error envelope in both cases is 2σ (2×standard error) computed using the bootstrap method for 1 million samplings. The oxygen isotope stack (red (dark gray in print versions)) is compared with the LR04 benthic isotope stack (blue (black in print versions)) (Lisiecki & Raymo, 2005). Paleointensity minima in the stack correspond to established ages of magnetic excursions (see Laj & Channell, 2007, chap. 10), and chron/subchron boundaries: LA—Laschamp, BL—Blake, IB—Iceland Basin, PR—Pringle Falls, BLT—Big Lost, LP—La Palma, ST17—Stage 17, JA—Jaramillo Subchron, SR—Santa Rosa, PU—Punaruu, CM—Cobb Mt. Subchron, BJ—Bjorn, GA—Gardar, B/M—Brunhes-Matuyama boundary.

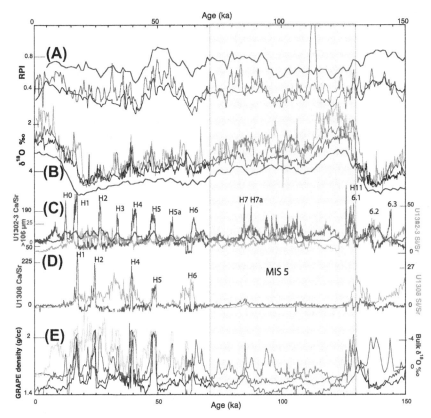

FIGURE 3.5.3 Correlation of IODP Site 1302/03 and U1308 stratigraphies over the last 150 ky. (A) Relative paleointensity proxies for Site U1302/03 (red (dark gray in print versions)) and Site U1308 (black) compared with RPI reference stack (PISO-1500) of Channell et al. (2009) (blue (light black in print versions)). (B) Site U1302/03 planktic δ18O (red (dark gray in print versions)), Site U1308 benthic and planktic δ18O (black and brown (dark black in print versions)) with LR04 (blue (light black in print versions)). (C) Site U1302/03 XRF scanning Ca/Sr (blue (light black in print versions)) and Si/Sr (green (light gray in print versions)), wt% >106 μm grain size fraction (red (dark gray in print versions)). (D) Site U1308 XRF scanning Ca/Sr (blue (light black in print versions)) and Si/Sr (green (light gray in print versions)). (E) Site U1302/03 GRAPE density (red (dark gray in print versions)) and Site U1308 GRAPE density (black), with bulk carbonate δ18O for Site U1302/03 for 0–62 ky (light blue (light gray in print versions)) and at Site U1308 (dark blue (dark black in print versions)). *Data from Channell et al. (2012).*

2013; Hodell et al., 2008). The detrital carbonate layers at Site U1302/03 are far more numerous than at Site U1308 (Figure 3.5.4), presumably reflecting the more proximal location of the site relative to the LIS, and occur within both glacial and interglacial isotopic stages (Figure 3.5.4(A)). Channell et al. (2012) noted a relationship between Ca-rich detrital layers at Site U1302/03 and weak monsoon events in the Chinese speleothem record (Cheng et al., 2009; Dykoski et al., 2005; Wang et al., 2001; Wang et al., 2008) with the orbital-scale variations removed (Barker et al., 2011), implying a link between ice-sheet instability and Asian monsoon precipitation (Figure 3.5.4(B)).

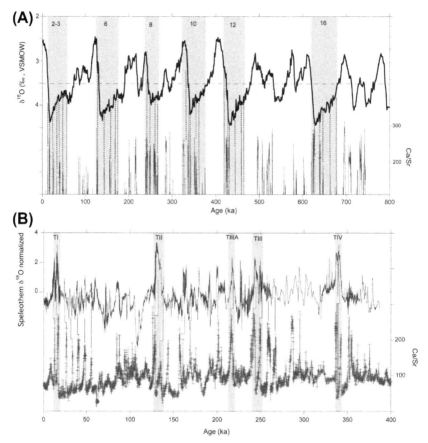

FIGURE 3.5.4 (A) XRF Ca/Sr for Site U1302/03 (blue (dark gray in print versions)) and Site U1308 (red (light gray in print versions)) compared with the LR04 benthic oxygen isotope stack (black, Lisiecki & Raymo, 2005). Dashed horizontal line indicates 3.5‰ threshold value for $\delta^{18}O$ (McManus et al., 1999). Prominent glacial intervals are marked by blue shading (pale gray in print versions). (B) The Chinese speleothem record (Cheng et al., 2009; Dykoski et al., 2005; Wang et al., 2001; Wang et al., 2008) with the orbital-scale variations removed (Barker et al., 2011) in blue (dark gray in print versions), compared with the XRF Ca/Sr records (red (light gray in print versions)) from Site U1302/03. Heinrich-like detrital layers can be tentatively correlated to the speleothem timescale. *Modified after Channell et al. (2012).*

Planktic and benthic oxygen/carbon isotope data and IRD records from Site U1313 have been combined with data from other sites to reconstruct hydrographic conditions in the North Atlantic during the MIS 9-14 interval (Stein et al., 2009; Voelker et al., 2010). Detrital layers in the MIS 9-16 interval at Site U1313 have also been identified through organic geochemistry; XRD evidence for high concentrations of dolomite, quartz, and alkali feldspar; and SST minima associated with meltwater (Stein et al., 2009). Alkenone-based SST, organic biomarkers, and XRD dolomite/calcite and quartz/calcite ratios, have been used to document

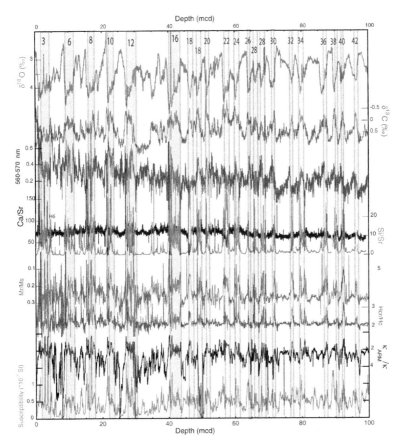

FIGURE 3.5.5 IODP Site U1308, for 0–100 meters composite depth (mcd). From top: benthic $\delta^{18}O$ (brown (pale gray in print versions)) and $\delta^{13}C$ (red (dark gray in print versions)); change in reflectance in the 560–570 nm wavelength band (blue (pale black in print versions)); Ca/Sr counts (black); and Si/Sr (green (light gray in print versions)) from XRF core scanning (Hodell et al., 2008); magnetic hysteresis ratios implying magnetic grain size changes M_r/M_s (red (dark gray in print versions)) and H_{cr}/H_c (blue (pale black in print versions)); an additional magnetic grain size parameter κ_{ARM}/κ (black), and the volume susceptibility magnetic concentration parameter (orange (very light gray in print versions)). M_r/M_s, H_{cr}/H_c, and κ_{ARM}/κ are plotted to indicate larger magnetite grain size up-page. Certain glacial MIS are shaded and numbered back to MIS 42, with yellow (very pale grey in print versions) shading for stages back to MIS 16. *After Channell & Hodell (2013).*

detrital layers at Site U1313 (Figure 3.5.6) (Naafs, Hefter, Ferretti, Stein, and Haug (2011), Naafs, Hefter, and Stein (2013). The first appearance of IRD with high dolomite/calcite ratio is found in MIS 16, consistent with the result from Site U1308, and SST proxies do not indicate anomalously low SST during Heinrich stadials (Figure 3.5.6), implying that the first arrival of Ca-rich detrital layers at Site U1313 was not controlled by iceberg survivability (Naafs et al., 2011). Note that detrital layers characterized by high quartz/calcite XRD ratios predate MIS 16 at Site U1313 (Figure 3.5.6), at least back to MIS 24 (Naafs et al., 2011) and

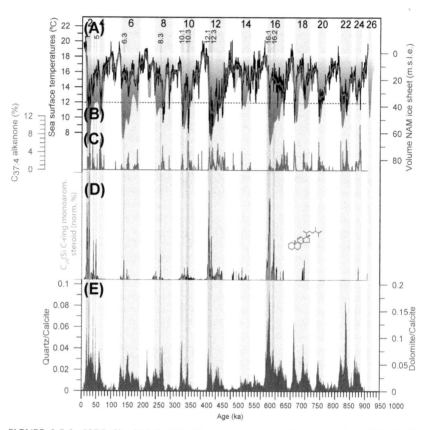

FIGURE 3.5.6 IODP Site U1313: (A) Alkenone-based sea-surface temperature (SST, black) and 10-ky moving average (red (light gray in print versions)) (Naafs et al., 2011) superimposed on (B) modeled size of North American ice sheets based on benthic $\delta^{18}O$ (blue shade (pale gray in print versions)) (Bintanja & van de Wal, 2008), (C) Abundance of $C_{37:4}$ alkenones, indicative of high-latitude waters, (D) Abundance of $C_{28}(S)$ c-ring monoaromatic steroides indicative of ancient organic matter, (E) Abundance of dolomite (red (light gray in print versions)) indicative of IRD from Hudson Bay area superimposed on abundance of quartz (blue (dark gray in print versions)) indicative of circum-Atlantic ice sheets. Light blue (very light gray in print versions) bars indicate Heinrich-like events. Gray shading highlights glacial isotopic stages. *After Naafs, Hefter, & Stein (2013).*

apparently back to ~3 Ma (Naafs, Hefter, & Stein, 2013), presumably equivalent to the Si-rich detrital layers observed through XRF core scanning at Site U1308 where they are observed back to MIS 42 at ~1.4 Ma (Figure 3.5.5) (Channell & Hodell, 2013; Hodell et al., 2008). Over the last 70 ky at Site U1313, Naafs, Hefter, Grützner et al. (2013) observed alkenone-based SST warming, by 2°–4 °C, associated with high dolomite/calcite XRD ratios denoting Ca-rich IRD. They interpret these observations in terms of northward expansion of the subtropical gyre and an antiphase (seesaw) SST pattern, during Heinrich Events, involving midlatitude SST warming accompanying a general North Atlantic SST cooling.

Several conclusions can be drawn from the characterization and correlation of detrital layers at Site U1302/03 (Orphan Knoll) and Sites U1308/U1313 (central Atlantic).

1. At Site U1308, detrital layers rich in detrital carbonate IRD, and therefore supposed to be of Laurentide origin, are present in most glacial isotopic stages and Terminations, other than MIS 6 and 14, back to and including MIS 16 (Hodell et al., 2008). Heinrich layers H3 and H6 are not characterized by high detrital carbonate (high Ca/Sr) at Site U1308, although H3 (but not H6) at Site U1302/03 does correspond to high Ca/Sr values denoting detrital carbonate (Figure 3.5.3). At Site U1313, as at Site U1308, carbonate-rich detrital layers were observed from MIS 16, and their first appearance was not controlled by iceberg survivability through SST (Naafs et al., 2011; Stein et al., 2009). It appears that all Ca-rich detrital layers at Site U1308 have analogues at Site U1302/03 (Figure 3.5.4(A)) and departure from this correspondence at the MIS 9/10 boundary can be attributed to an age model problem caused by the low-resolution $\delta^{18}O$ at Site U1308 (Channell et al., 2012; Hodell et al., 2008). The correspondence of Ca-rich detrital layers from Site U1308 to Site U1302/03 confirms their Laurentide origin; however, at Site U1302/03, Ca-rich detrital layers occur in all glacial and interglacial stages back to MIS 17, other than MIS 11 and MIS 14 implying a mode of deposition not restricted to IRD (Channell et al., 2012). Hodell et al. (2008) suggested that ice volume surpassed a critical threshold at 650 ka (MIS 16) that permitted ice sheets to survive boreal summer insolation maxima, thereby increasing ice volume and thickness, lengthening glacial cycles, and activating the dynamical processes responsible for LIS instability in the region of Hudson Strait (i.e., Heinrich Events).

2. Prior to MIS 16 and back to at least MIS 42 at Site U1308 (~1.35 Ma), and MIS 25 (~950 ka) at Site U1313 (Channell & Hodell, 2013; Naafs et al., 2011; Naafs, Hefter, & Stein, 2013), all detrital layers are apparently Si-rich. They were partly delivered as IRD based on magnetic grain size at Site U1308, contain high-coercivity magnetic minerals (hematite), and caused perturbations in benthic $\delta^{13}C$ implying changes in deepwater circulation associated with IRD delivery (Channell & Hodell, 2013; Hodell et al., 2008). The lack of Heinrich Events containing detrital carbonate and more frequent occurrence of silicate-rich IRD events prior to ~650 ka may indicate the presence of "low-slung, slippery ice sheets" that flowed more readily than their post-Mid-Pleistocene Transition (MPT) counterparts (Bailey et al., 2010).

Mg/Ca ratios from Sites U1313 and U1308 have been used to reconstruct spring/summer SST associated with dinoflagellate cyst assemblages over the last 3.4 my (De Schepper, Fischer, Groeneveld, Head, and Matthiessen (2011). Extant species were found to have comparable spring/summer SST ranges in the past and today, supporting the methodology used to determine SST ranges for extinct species. Sites U1308 and U1313 have yielded abundant and moderate-to-well-preserved calcareous nannofossils and planktic foraminifera over the last 3 my that builds on data from DSDP Sites 609/607, and enhances Sites U1308 and U1313 as well-calibrated North Atlantic reference sections for the late Neogene (Hagino & Kulhanek, 2009; Sato, Chiyonobu, & Hodell, 2008; Sierro, Hernández -Almeida, Alonso-Garcia, & Flores, 2009).

Ferretti, Crowhurst, Hall, and Cacho (2010) have generated planktic and benthic stable isotope data in the MIS 23 to MIS 20 interval of the MPT at Site U1313. Cooling events in the planktic δ^{18}O record of MIS 21 are associated with δ^{13}C excursions, suggesting a link between surface temperature and deepwater circulation, and are thought to indicate climate instability associated with the MPT. They found significant concentrations of spectral power at harmonics of the precession cycle (10.7 and 6 ky) and suggested they originated from low-latitude precession variations. Site U1313 has also been a focus for study of the onset of Northern Hemisphere Glaciation (NHG). A marked increase of eolian plant leaf wax (n-alkanes and n-alkanl-1-ols) at Site U1313 coincides with the intensification of NHG at ~2.7 Ma, with 30 times higher concentrations during glacials versus interglacials since that time, reflecting an expansion of the North American ice sheets (Naafs, Hefter, Acton et al., 2012). Alkenone-based SST and productivity records for the 3.68–2.45 Ma interval of Site U1313 indicate SSTs up to 6 °C higher than today, with low surface-water productivity, prior to 3.1 Ma. This was progressively replaced, after 3.1 Ma, by decreasing SSTs during glacials and higher productivity as the shallow North Atlantic Current (NAC) weakened at the site (Naafs, Hefter, Acton et al., 2012; Naafs, Hefter, & Stein, 2012; Naafs et al., 2010). Bolton et al. (2010) generated a benthic oxygen isotope record for 2.4–3.4 Ma at Site U1313, and planktic δ^{18}O and IRD records for the MIS 96-MIS 102 (2.4–2.6 Ma) interval. In this interval, millennial-scale variability, regardless of glacial/interglacial state, has been observed in both planktic δ^{18}O (Bolton et al., 2010) and Mg/Ca SST data (Friedrich, Wilson, Bolton, Beer, & Schiebel, 2013). At Site U1308, Bailey et al. (2010) and Bailey et al. (2012) studied the provenance of IRD in two glacial MIS (G4 at ~2.64 Ma and MIS 100 at ~2.52 Ma) and found evidence for suborbital ice-rafting events and proposed that ice sheets during this time surged at a lower δ^{18}O threshold (by ~0.45‰) than during the late Pleistocene (McManus, Oppo, & Cullen, 1999), and attributed this to "low-slung slippery" late Pliocene ice sheets.

Further back in time, Mg/Ca ratios, δ18O, IRD proxies, and dinoflagellate cyst assemblages have been used to study the interval around the prominent glacial MIS M2 (~3.30 Ma) at Sites U1308 and U1313, and DSDP Sites 603 and 610 (De Schepper, Head, and Groeneveld (2009); De Schepper et al. (2013). SST data indicate a cooling of approximately 2–3 °C, and the dinoflagellate assemblage was interpreted in terms of reduction in northward heat transport associated with changes in the NAC, ~40 ky prior to the MIS M2 glacial maximum. The lack of an IRD record excludes freshwater input via icebergs as a cause for the observed slackening of the NAC that may be attributed to a brief reopening of the Central American Seaway (CAS) at this time (~3.30 Ma), steepening of the SST gradient in the North Atlantic and thermal isolation of the high latitudes. Sea-level drop during the MIS M2 glaciation then halted the inflow, and terminated the glaciation (De Schepper et al., 2013). This interpretation contrasts with the CAS *closure* often invoked to explain intensification of NHG ~0.5 my later at ~2.75 Ma (e.g., Driscoll and Haug, 1998). The role of the closure of the CAS in paleoceanography, and in faunal interchange between the Americas, remains controversial (e.g., Stone, 2013). Modeling experiments have invoked declining atmospheric CO_2 as the main driver of intensification of NHG (Lunt, Foster, Haywood, & Stone, 2008).

The principal objective at Site U1312 (Figure 3.5.1), a redrill of DSDP Site 608 on the southern flank of the King's Trough, was to recover the Upper Miocene section at 140–260 meters below seafloor (mbsf). Site U1312 was the first site of the second cruise of Exp. 303/306 (Exp. 306). The site was occupied during a severe deterioration of weather conditions, and was eventually abandoned, allowing the RV *Joides Resolution* to ride out the storm on the lee side of the Azores. The relatively poor quality of cores recovered at both holes at Site U1312, and the lack of a composite section in the target interval, has meant that the sediments recovered at the site have not been extensively utilized post cruise. The exception, however, is a study of the paleomagnetic record over the last ~2 my (Kanamatsu, Acton, Evans, Guyodo, & Ohno, 2010) that indicates the presence of a high-fidelity paleomagnetic record, including the potential for an RPI record albeit interrupted by intervals of poor core quality, which would have contributed to the importance of the site had weather conditions allowed the drilling objectives to be achieved.

3.5.2.2 Sites U1304 and U1314 (Gardar Drift)

Two sites (Sites U1304 and U1314, Figure 3.5.1) were collected from the Gardar Drift during Expedition 303/306. At Site U1314, the sediments comprise nannofossil and clay-rich sediments with foraminifera and diatoms deposited at mean sedimentation rates in the 7–11 cm/ky range extending back into the Gauss Chron at ~3 Ma. In contrast, the drilling at Site U1304 recovered a 270-m section of nannofossil and diatom oozes reaching the Olduvai subchronozone at ~1.8 Ma, with mean sedimentation rates in the 12–18 cm/ky range. The high mean sedimentation rates at Site U1304 are accounted for by laminated and massive diatom mats on scales of several tens of meters to several centimeters that are sporadically present throughout the entire section back to ~1.8 Ma. The only analogous documented occurrence of diatom mats in the North Atlantic is in MIS 5e of conventional piston core EW9303-17, located 1115 km NW of Site U1304 (Bodén & Backman, 1996). Shimada, Sato, Toyoshima, Yamasaki, and Tanimura (2008) argued, based on two principal components in the diatom assemblage, that the switching between warm- and cold-water diatom assemblages occurred concurrently with diatom mat deposition implying that diatom productivity was related to transits of the Subarctic Convergence over the site, analogous to the interpretation of Bodén and Backman (1996). Diatom occurrence at Site U1304 is not simply related to glacial–interglacial cycles (Shimada et al., 2008) and, during the last interglacial (LIG), productivity apparently peaked at the MIS 5 d/5e boundary (Romero et al., 2011). According to Romero et al. (2011), surface waters alone would not have provided the necessary nutrients to sustain the observed diatom productivity at Site U1304. In the modern Subarctic Atlantic, diatom production is limited by low concentrations of available silica. The rather unique occurrence of diatom mats throughout the Quaternary section at Site U1304 implies incursion of silicate- and nitrate-rich Subantarctic Mode Water, as a result of slowdown of Iceland–Scotland Overflow Water (ISOW), and sporadic convective mixing of the water column (Romero et al., 2011).

Hodell et al. (2009) studied the LIG period (MIS 5e) and Termination II at Site U1304. They found that deepwater circulation on Gardar Drift was relatively weak during the earliest part (128–124.5 ka) of the LIG, when planktic $\delta^{18}O$ was at a minimum, reflecting warming and/or reduced salinity. The presence of low-$\delta^{13}C$ water and slow current speed on Gardar Drift during the early part of the LIG may have been caused by increased meltwater flux to the Nordic Seas during peak boreal summer insolation, which decreased the flux and/or density of overflow to the North Atlantic. The resumption of the typical interglacial pattern of strong, well-ventilated ISOW was delayed. Similar reductions in NADW during the peak of the LIG have also been inferred from sediments recovered on Eirik Drift (Galaasen et al., 2014). Kuhs, Austin, Abbott, and Hodell, (2014) used geochemical analyses to identify Icelandic tephra within IRD of MIS 6 at Site U1304, indicating that the Icelandic Ice Sheet had calving margins during late MIS 6.

Grützner and Higgins (2010) correlated "blue" reflectance (450–500 nm) and magnetic susceptibility data from Site U1314 to ODP Site 983 in order to generate an age model for Site U1314 for the last 1.8 my. The K/Ti ratio determined from XRF core scanning was used to track changes in sediment provenance, with low (high) K/Ti being typical of interglacials (glacials) and high (low) Icelandic basalt input related to the vigor of ISOW (Grützner & Higgins, 2010). Millennial-scale fluctuations in K/Ti are associated with glacial periods when $\delta^{18}O$ values exceed 4.1‰ (corrected to *Uvigerina*, which is equivalent to ~3.5‰ for *Cibicidoides* as originally proposed by McManus et al., 1999), implying millennial-scale changes in sediment provenance/delivery. Benthic and planktic stable isotope data, foraminiferal abundance data, and carbonate content, for the 779–1069 ka (MIS 19-31) interval at Site U1314 have been interpreted in terms of paleoceanographic change during the MPT (Hernández-Almeida, Sierro, Flores, Cacho, & Filippelli, 2013). The planktic foraminiferal assemblage in the Mid-Pleistocene (400–800 ka) interval at Site U1314 has been used to monitor the migration of the Arctic Front (Alonso-Garcia, Sierro, & Flores, 2011; Alonso-Garcia, Sierro, Kucera et al., 2011).

The magnetic stratigraphy for the 2.1–2.75 Ma interval at Site U1314 has been resolved from the Reunion Subchronozone into the Gauss Chronozone, together with an RPI record and evidence for eight magnetic excursions from this interval of the Matuyama Chron (Ohno et al., 2012). Hayashi et al. (2010) used magnetic susceptibility and natural gamma radiation (NGR) to generate an age model for the 2.1–2.75 Ma interval at Site U1314 and have inferred the presence of millennial-scale iceberg surges, denoted by evidence for IRD, at times when benthic $\delta^{18}O$ exceeded a threshold of 2.8‰. This threshold is 0.7‰ lower than the 3.5‰ threshold proposed by McManus et al. (1999) for the last 0.5 my, and 0.2‰ lower than the early Pleistocene threshold proposed by Bailey et al. (2010).

At Site U1314, a change in the orientation of the maximum axis of the susceptibility ellipsoid at ~0.9 Ma from E–W to NNE–SSW that was attributed to a change in the ISOW at the site at this time associated with the MPT Kanamatsu, Ohno, Acton, Evans, and Guyodo (2009). Nannofossil and ostracode assemblages at Site U1314 have been studied by Hagino and Kulhanek (2009) and Alvarez Zarikian, Stepanova, and Grutzner (2009), respectively.

3.5.2.3 Sites U1305, U1306, and U1307 (Eirik Drift)

The sites drilled on the Eirik Drift during Exp. 303/306 (Sites U1305, U1306, and U1307; Figure 3.5.1) were all occupied during the first phase of the Expedition (Exp. 303). The 4-month time interval between Exp. 303 and Exp. 306 was utilized to build on the results from Exp. 303 and optimize positioning of further Eirik Drift sites for the Exp. 306. Unfortunately, weather/swell conditions off SW Greenland during Exp. 306 (March–April, 2005) prevented the ship from reaching/occupying the planned Eirik Drift sites. Nonetheless, results of drilling during Exp. 303 (Sites U1305, U1306, and U1307) have provided important constraints on the structure and evolution of the Eirik Drift. Building on the results of Arthur, Srivastava, Kaminski, Jarrard, and Osler (1989), the shipboard stratigraphies from Sites U1305-U1307 have been linked to high-resolution seismic reflection data collected in 2009 from the RV *Maria S. Merian* Müller-Michaelis, Uenzelmann-Neben, and Stein (2013). The results indicated early Miocene onset of drift sedimentation with mid-to-late Miocene changes in the deep current system having important implications for the evolution of the Deep Western Boundary Current (DWBC). The DWBC is a major component of Atlantic Meridional Overturning Circulation (AMOC) that, along with detrital flux from Greenland, dictates the sedimentation pattern on Eirik Drift.

The modern configuration of erosion and deposition on Eirik Drift has been inferred from hydrographic data and 3.5/5.1 kHz profiler lines, from which the acoustic character of surface sediment has been interpreted in terms of erosion/deposition, and can be related to the role of the DWBC (Hunter et al., 2007; Stanford, Rohling, Bacon, & Holliday, 2011). Although the characteristics of the DWBC may vary on decadal or millennial timescales, the modern pattern is dominated by erosion on the east side of the drift, in the water-depth range of approximately 2000–3000 m where the DWBC is most active, and by deposition at the toe of the drift in the vicinity of Site U1305 (Figure 3.5.1). The DWBC probably shoaled and slowed during glacial intervals (e.g., Hall & Becker, 2007) thereby reducing detrital deposition during glacials relative to interglacials at Site U1305 (Hillaire-Marcel et al., 1994, 2011).

Sites U1305 (57.48°N, 48.53°W) and U1306 (58.24°N, 45.64°W) are located in water depths of 3460 and 2272 m, respectively (Figure 3.5.1), and the recovered sediments are characterized by silty clays with nannofossils, deposited over the last ~2 my. The sites are suitably located to monitor depositional variability through the Quaternary at the distal toe (Site U1305) and proximal lee-side crest of the drift (Site U1306). The sites were chosen based on interpretation of seismic stratigraphy, partly acquired during cruise KN166-14 in summer 2002, that implied relatively expanded Quaternary sections at both sites (see Channell et al., 2006). At Sites U1305 and U1306, the last ~2 my of drift sedimentation was recovered at mean sedimentation rates in excess of 15 cm/ky. At Site U1305, the shoaling and slowdown of the DWBC during glacial intervals led to reduced detrital deposition during glacials, relative to interglacials, back to ~1.2 Ma (Hillaire-Marcel et al., 2011), consistent with the pattern observed over the last glacial cycle from nearby conventional piston cores (Hillaire-Marcel & Bilodeau, 2000; Hillaire-Marcel et al., 1994). The converse sedimentation pattern is observed at the shallower site (Site U1306) with expanded glacial periods back to ~1.5 Ma (Channell, Wright, Mazaud, & Stoner, 2014). The third Eirik Drift site (Site U1307) was chosen at a location where the older (2–4 Ma) record occurs beneath a relatively condensed younger sediment pile, thereby making it accessible by APC coring (Channell et al., 2006).

Planktic $\delta^{18}O$ data from *N. pachyderma* (sin.) have been generated for most of the Quaternary sequence at Site U1305 since MIS 32 at ~1.1 Ma (Hillaire-Marcel et al., 2011). Prior to MIS 20 at ~820 ka, $\delta^{18}O$ have lower values by ~1‰ during glacials, and ~0.5‰ during interglacials, relative to values since MIS 20. This pattern was attributed to increased production of Denmark Strait Overflow Water and a more vigorous DWBC during interglacial stages leading to high interglacial (vs glacial) sedimentation rates since MIS 20 (Hillaire-Marcel et al., 2011). At Site U1302/03, at the southern edge of the Labrador Sea (Figure 3.5.1), an analogous change in amplitude and values of planktic *N. pachyderma* $\delta^{18}O$ is observed later, during MIS12/13 (Hillaire-Marcel et al., 2011). The magnetic stratigraphy and RPI record for the last 1.2 my at Site U1305 has high fidelity, and the RPI record can be matched to the PISO RPI stack (Mazaud, Channell, & Stoner, 2012). Kawamura, Ishikawa, and Torii (2012) have provided a thorough rock magnetic study of the three Eirik Drift sites and recognized magnetite as the principal magnetic mineral, with maghemite in the surface-sediment oxic zone.

Nicholl et al. (2012) have documented a decimeter-scale red layer containing Laurentide-derived detrital carbonate at ~126 ka within MIS 5e both at Site U1305 and U1302/03. They attribute this layer to a glacial-outburst flood event that postdates Termination II, analogous to the 8.2-ka cooling event that postdates Termination I and has been attributed to final drainage of Lake Agassiz (see Hillaire-Marcel, De Vernal, & Piper, 2007).

Planktic oxygen isotope ($\delta^{18}O$) and RPI data were used to generate an age model for the last 1 my from Site U1306 (Figure 3.5.7) by tandem correlation of RPI to PISO, and $\delta^{18}O$ to LR04 (Channell et al., 2014). For the 1–1.5 Ma interval, the age model is based on RPI alone due to insufficient foraminifera for isotope analyses. Utilizing RPI and $\delta^{18}O$ in tandem allows recognition of

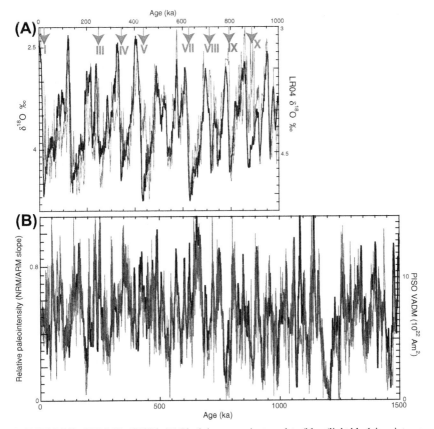

FIGURE 3.5.7 IODP Site U1306: (A) Planktic oxygen isotope data (blue (light black in print versions)) compared to LR04 (black; Lisiecki & Raymo, 2005) with green arrows (dark gray in print versions) marking apparent meltwater effects denoted by low values of $\delta^{18}O$ preceeding glacial Terminations I, III, IV, V, VII, VIII, IX, and X. (B) Relative paleointensity proxies (red (light gray in print versions)) compared to the PISO relative paleointensity stack (black; Channell et al., 2009). *Site U1306 data from Channell et al. (2014).*

low-δ^{18}O "events" prior to glacial Terminations I, III, IV, V, VII, VIII, IX, and X (Figure 3.5.7), that are independently supported by radiocarbon dates through the last deglaciation, and are attributed to local or regional surface-water effects. At Site U1306, Quaternary sedimentation rates (mean ~15 cm/ky) are elevated during peak glacials and glacial onsets, and are reduced during interglacials, in contrast to the pattern at Site U1305 (Channell et al., 2014). The slackening and/or shoaling (due to lowered salinity) of the DWBC during glacial intervals coincided with greater sediment supply to Site U1306, whereas the deepening, and possibly increased vigor, of the DWBC during interglacial intervals boosted sediment supply to Site U1305. The contrasting glacial/interglacial depositional pattern at the two sites appears to have persisted for the last ~1.5 my, spanning the MPT. The Quaternary architecture of the Eirik Drift is intimately related to the variable activity of the DWBC as a principal component of AMOC.

At Site U1307, Sarnthein et al. (2009) found evidence at ~2.8 Ma for changes in IRD flux, SST, and salinity that they associate with the *closure* of the CAS and poleward flux of atmospheric moisture. This, combined with increased flow of low-salinity surface waters from the North Pacific through the Arctic Ocean, is interpreted to have led to increased sea-ice cover and albedo, and onset of NHG at ~2.8 Ma (MIS G10). IRD provenance based on Pb-isotope analyses of feldspars in the central Atlantic at DSDP Site 611 (Bailey et al., 2013) has been interpreted to indicate Laurentide (North American) IRD sources from ~2.6 Ma (MIS G2), with Greenland/Scandinavian sources predominating prior to that, from ~2.7 Ma (MIS G6).

3.5.2.4 Site U1315 (CORK in ODP Hole 642E)

The objective of proposal 543-Full2 was to place a Circulation Obviation Retro-fit Kit (CORK) at ODP Site 642 (Site U1315, Figure 3.5.1) on the Vøring Plateau to document and monitor bottom-water temperature variations through time in a water depth of ~1280 m. Hole U1315A was drilled to a depth of 179 mbsf. A 120-m casing with reentry cone was jetted into the sediments, followed by installation of a thermistor string and a CORK, thereby sealing the borehole from the overlying ocean. The instrumentation at Hole U1315A includes two pressure cases as well as the thermistor string, allowing high-precision temperature measurements as a function of both depth and time. By analyzing sub-bottom temperature perturbations, there is the potential for reconstruction, for the first time, of bottom-water temperatures over at least the last ~100 years, going back in time far beyond directly measured temperatures. This record will contribute to the ongoing discussion about causes and consequences of oceanographic and climatic changes observed in the North Atlantic over the last few decades. The present-day bottom-water temperature and salinity variations are continuously monitored with instrumentation in the water column attached to the reentry cone. After setting of the CORK in the newly drilled Hole U1315A, temperature measurements were made in Hole 642E (Harris et al., 2006) to compare with those acquired 20 years ago when the hole was originally drilled (see Eldholm, Theide, & Taylor, 1987). The comparison indicates fluid flowing from the basement, at an estimated up-hole flow rate of 6–11 m/h, and an estimated permeability of 10^{-13} m^2 reflecting relatively impermeable sediments overlying a relatively permeable basement (Harris & Higgins, 2008).

3.5.3 IODP EXPEDITION 339 (MEDITERRANEAN OUTFLOW)

Expedition 339 represents the marriage of two projects: IODP Proposal 644-Full2 and Ancillary Program Letter (APL)-763. Consequently, Expedition 339 had two broad objectives. The first was to study the contourite depositional system (CDS) in the Gulf of Cádiz and off Portugal, among the best-known examples of contourites in the world. The second was to recover sediments deposited at very high sedimentation rates (several decimeters/ky) suitable for studies of millennial-scale climate variability. During IODP Expedition 339, five sites were drilled in the Gulf of Cádiz and two off the West Iberian margin (Figure 3.5.1) (Expedition 339 Scientists, 2013; Hernández-Molina, Stow, Alvarez-Zarikian, & Expedition IODP 339 Scientists, 2013, 2014). All but one site was drilled at intermediate water depths (560–1073 m) to investigate the history of MOW through the Strait of Gibraltar and its influence on global circulation and climate (Hernández-Molina et al., 2013, 2014). The one deeper site at 2578 m water depth is the so-called "Shackleton" site (Site U1385) on the southwestern Iberian Margin, which was drilled to obtain a marine reference section for marine-terrestrial-ice core correlations and the study of orbital- and millennial-scale variability during the Quaternary (Hodell, Lourens, et al., 2013), at the site of a 29.5-m-long core collected from the RV *Marion Dufresne* in 2001 (Core MD01-2444) that has been extensively studied for its stable isotope records, cyclostratigraphy, pollen, marine biostratigraphy, and paleomagnetism (e.g., Channell et al., 2013; De Abreu et al., 2005; Hodell, Crowhurst, et al., 2013; Margari et al., 2010; Skinner, Elderfield, & Hall, 2007; Vautravers & Shackleton, 2006). Whereas Core MD01-2444 extended back to 200 ky, the recovered section at Site U1385 extends back to ~1.45 Ma or MIS 47, with a mean sedimentation rate of 10.8 cm/ky (Hodell, Lourens, et al., 2013).

A primary objective of Expedition 339 was to determine the timing of onset of contourite deposition in the Pliocene following the opening of the Gibraltar Gateway after the Messinian salinity crisis (Figure 3.5.8). Preliminary results show contourite deposition from about 4.2 to 4.5 Ma onward, suggesting MOW circulation into the Gulf of Cadiz (Hernández-Molina, et al., 2014). Pliocene-Quaternary contourite deposition is punctuated by a series of hiatuses, which are regionally pervasive. Three major hiatuses in sediment deposition (3.2–3.0 Ma, 2.4–2.1 Ma, and 0.9–0.7 Ma), of variable duration, are interpreted as a signal of intensified MOW coupled with flow confinement. Both tectonic and climatic processes affected contourite deposition. Neotectonic activity strongly influenced margin development and sediment deposition both along-slope

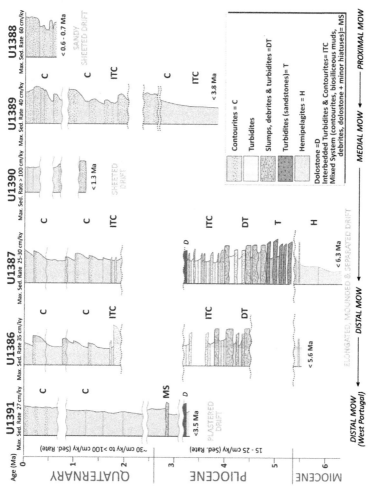

FIGURE 3.5.8 Lithologic summary of IODP Expedition 339 sites designed to monitor Mediterranean Outflow Water (MOW). *From Hernández-Molina et al. (2014).*

(contourite deposition) and down-slope. Climate control on changes in MOW and related bottom-current activity is apparently manifested through orbital- and millennial-scale variability in the sedimentary record. Bulk sediment properties of contourites at several sites can be correlated to physical properties, and these same variations are also recognized in physical properties at the deeper Shackleton site (Site U1385). Sedimentation rates in the contourite-bearing sequences range from moderate (~20 cm/ky) to extremely high (>100 cm/ky), providing high-resolution sequences to study past variations in surface circulation and MOW.

At the Shackleton site (Site U1385), five holes were drilled using the APC system to a maximum depth of ~155.9 mbsf. The sediment lithology consists of uniform nannofossil muds and clays, with varying proportions of biogenic carbonate and terrigenous sediment (Expedition 342 Scientists, 2012). Following the cruise, split cores from all holes were analyzed by core scanning XRF to obtain semiquantitative elemental concentrations at 1-cm spatial resolution. These data were used to precisely correlate among holes and construct a revised spliced stratigraphic section, containing no notable gaps or disturbed intervals to 166.5 revised meters composite depth (rmcd). The high sampling demand required the use of all holes from Site U1385; therefore, a corrected rmcd (crmcd) was defined by aligning features in the Ca/Ti XRF record from all holes to the revised spliced composite section using Analyseries (Paillard, Labeyrie, & Yiou, 1996). This method accommodates stretching and squeezing of cored intervals and corrects for distortion within individual cores, permitting the integration of data from all holes. A low-resolution (20 cm) benthic oxygen isotope record confirms that Site U1385 contains an apparently complete record from the Holocene to ~1.45 Ma (MIS 47). Site U1385 will serve as an important reference section for marine-terrestrial-ice core correlations and the study of orbital- and millennial-scale variability across the MPT. Based on results from Core MD01-2444, at the same location as Site U1385, Hodell, Lourens, et al. (2013); Hodell, Crowhurst, et al. (2013) demonstrated that Ca/Ti is correlated with planktic $\delta^{18}O$ and alkenone SST, resembling the Greenland ice core $\delta^{18}O$ record for the last glacial period (Figure 3.5.9). Cold stadial events are marked by decreases in Ca/Ti, whereas warmer interstadials coincide with Ca/Ti increases. Assuming this relationship held for older glacial periods, the Ca/Ti record will provide a record of millennial variability during glacial periods since 1.45 Ma.

Many years of research will be required to fully decode the climatic signal from the contourites and from the Shackleton site. A highly coordinated sampling effort and detailed shore-based collaboration have been established to produce the widest possible range of proxy measurements on the same set of samples. Building on the success of Site U1385, and given the seminal importance of the Iberian Margin for paleoclimatology and marine-terrestrial-ice core correlations, additional drilling in this region has been proposed (IODP Proposal 771-Full).

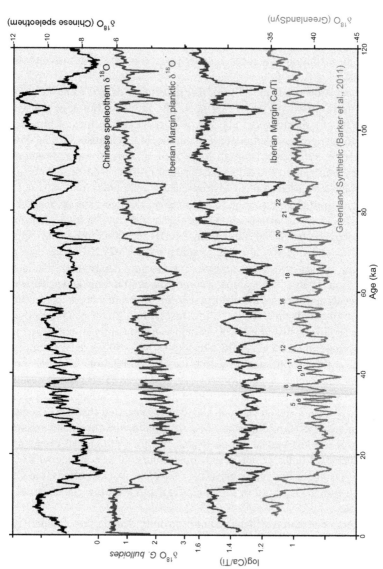

FIGURE 3.5.9 Ca/Ti (blue (light black in print versions)) and planktic δ^{18}O (green (dark gray in print versions)) from piston core MD01-2444 (same location as IODP Site U1385, IODP Exp. 339) compared with the Greenland synthetic δ^{18}O record (Barker et al., 2011) and a composite of Chinese speleothem record (Cheng et al., 2009; Dykoski et al., 2005; Wang et al., 2001; Wang et al., 2008). The similarity of the millennial-scale events offers the opportunity of synchronizing the records and transferring the U-Th chronology of the speleothem record to both the ice core and marine sediment archives. *After Hodell, Lourens, et al. (2013); Hodell, Crowhurst, et al. (2013).*

3.5.4 IODP EXPEDITION 342 (PALEOGENE NEWFOUNDLAND SEDIMENT DRIFTS)

Nine IODP sites were occupied off the Grand Banks of Newfoundland in June–July 2012 during IODP Expedition 342 (Expedition 342 Scientists, 2012; Norris et al., 2014). Four of the sites (Sites U1403-U1406) are located on the J-Anomaly Ridge close to DSDP Site 384 drilled in 1975 and five sites (Sites U1407-U1411) are located on the nearby SE Newfoundland Ridge (Figure 3.5.1). The expedition was designed to monitor, at elevated sedimentation rates in drift depositional settings, Paleogene calcium compensation depths (CCD) for comparison with CCD determinations in the equatorial Pacific (Pälike et al., 2012) and South Atlantic (Zachos et al., 2005). For example, whereas large-scale (~1 km) deepening of the CCD has been postulated for the earliest Oligocene in the equatorial Pacific, apparently synchronous with the expansion of Antarctic glaciation (Coxall, Wilson, Pälike, Lear, & Backman, 2005), a much smaller CCD deepening (~200 m) apparently occurred at the same time in the South Atlantic (Zachos, Kroon, Blum et al., 2004; Zachos et al., 2005). The magnitude and timing of CCD fluctuations provide estimates of ocean under-saturation and hence transient change in CO_2 storage during extreme climate events such as the Paleocene-Eocene Thermal Maximum (PETM), that provide analogues for future greenhouse conditions.

Paleodepths at ~50 Ma for IODP Exp. 342 sites were projected to vary in the 4.5–2.5 km range, and a depth transect was designed to monitor changes in CCD. A deep North Atlantic CCD in the Cretaceous, Paleocene and early-middle Eocene is implied by sporadic calcareous preservation at inferred paleodepths of ~4.5 km. On the other hand, the CCD appears to have been relatively shallow during the Early Eocene Climate Optimum, late Eocene, and middle Oligocene. Carbonate deposition events postdate the K/T (Cretaceous-Tertiary) boundary, the PETM, and the Eocene–Oligocene boundary, possibly indicating a rebalancing of ocean alkalinity after these abrupt climatic/extinction events (Expedition 342 Scientists, 2012; Norris et al., 2014).

Apart from mean sedimentation rates that exceeded 10 cm/ky at Site U1405 in the late Oligocene, sedimentation rates of <1–2 cm/ky were estimated for sites that penetrated Paleogene and Upper Cretaceous sediments (Figure 3.5.10). In general, nannofossil and biosiliceous oozes are associated with higher sedimentation rate intervals (~2 cm/ky) and clays were associated with low sedimentation rate intervals (<0.1 cm/ky). Important boundaries/contacts were recovered including the K/T boundary, PETM, as well as the Eocene–Oligocene and Cenomanian–Turonian boundaries, and oceanic anoxic event (OAE) 2.

Expedition 342 drilling results combined with available seismic stratigraphy will result in a more complete knowledge of the sedimentation patterns of drifts on the J-Anomaly Ridge and SE Newfoundland Ridge. The Exp. 342 drilling transect lies in the path of the DWBC, thereby providing sedimentary sections

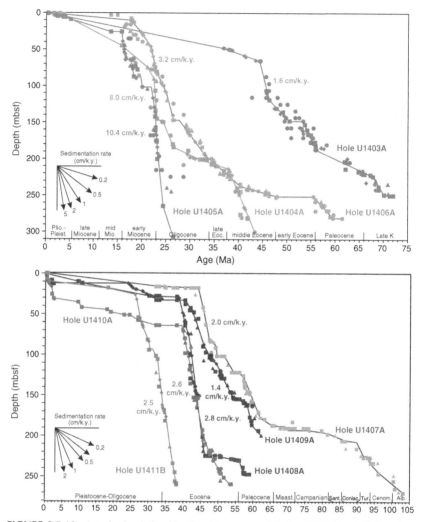

FIGURE 3.5.10 Age-depth relationships for sites recovered during IODP Exp. 342, based on shipboard biostratigraphic and paleomagnetic data. *From Norris et al. (2014).*

that can be used to monitor chemistry, temperature, and flow velocity of a major southward conduit of northern source waters during the Paleogene.

The achievement of Exp. 342 objectives will depend on the chronological control provided by calcareous and siliceous micropaleontology and magnetic stratigraphy. In general, microfossil preservation is moderate to excellent with enhanced preservation in clay-rich sediments. The preservation of calcareous nannofossils at some sites, as indicated by the presence of minute coccoliths, is exceptionally good even in calcareous intervals with higher sedimentation rate. Cycles in L* reflectance (proxy for $CaCO_3$), NGR, and magnetic susceptibility, possibly associated with orbital eccentricity, have been observed in parts of the sedimentary sections recovered during Exp. 342. Astrochronologies based on these cycles may be forthcoming, and would obviously impact the understanding of events recorded by Exp. 342 sediments. The wider impact of such an astrochronology, in terms of calibration of timescales, will depend on the fidelity of the magnetic stratigraphy. Calibration of the geomagnetic polarity timescale (GPTS) is central to the calibration of the Paleogene timescale. Current calibration of the Paleogene timescale (GPTS) still relies on assumptions of uniformity in seafloor spreading rates, rather than precise sedimentary astrochronology (e.g., Ogg, 2012).

3.5.5 SUMMARY

Three IODP expeditions, with focus on North Atlantic paleoceanography, took place in the 2003–2013 decade. The first of these was a two-phase expedition (Expedition 303/306) that sailed during September–November 2004 (Exp. 303) and March–April 2005 (Exp. 306). The central premise was that elevated Quaternary sedimentation rates in the western North Atlantic (Eirik Drift), southern Labrador Sea (Orphan Knoll), and central Atlantic IRD belt provide high-resolution records of Quaternary environmental change. Stratigraphic control was to be a critical aspect of the Expedition with traditional oxygen isotope stratigraphy being used in conjunction with magnetostratigraphy, not only utilizing polarity reversals but also through RPI stratigraphy. Expedition 303/306 was an APC-only expedition, based on the principle that high-resolution stratigraphy is not feasible in cores disturbed by rotary coring techniques, thereby restricting drilling penetration to ~300 mbsf.

The second IODP expedition of the decade, dealing with North Atlantic paleoceanography, took place between November 2011 and January 2012 (Expedition 339) in the Gulf of Cádiz and off the SW Iberian margin. The objectives were to document the CDS as a monitor of MOW through the Strait of Gibraltar gateway over the last 5 my. In addition, a secondary objective was to recover sediments deposited at very high sedimentation rate (reaching several decimeters/ky) that are suitable for studies of millennial-scale climate variability. Several intermediate water-depth sites were located in MOW with one deeper site on the SW Iberian Margin. The so-called "Shackleton" site on the SW Iberian Margin (Site U1385) was occupied to generate a marine reference section for millennial-scale climate variability and changes in surface- and deepwater circulation over the past ~1.5 my (Hodell, Lourens, et al., 2013).

The third IODP expedition of the decade, with a focus on North Atlantic paleoceanography, took place in June–July 2012 (Expedition 342). The target of the expedition was Paleogene sediment drifts off Newfoundland. The objectives are to reconstruct the history of carbonate saturation through Paleogene episodes of abrupt warming, document the flow history of the DWBC, obtain records of the Eocene–Oligocene transition at the onset of major glaciations, and improve Paleogene timescales through astrochronology.

Research on sediments recovered during these expeditions is still in progress, even for Expedition 303/306 that concluded its shipboard-phase over 9 years ago. This was to be expected because generating high-resolution proxy records is particularly time consuming, and these recovered sections are destined to remain important reference sections for North Atlantic paleoceanography. Here we provide an account of the research that has been published to date, as well as initial results from shipboard investigations from Expeditions 339 and 342. The account is heavily weighted toward Exp. 303/306 because postcruise science associated with Expeditions 339 and 342 is largely yet to come.

One general theme binds the three expeditions. They all endeavored to recover high-resolution (high sedimentation rate) paleoceanographic records of critical episodes in North Atlantic Cenozoic evolution: (1) Quaternary NHG (Exp. 303/306, 339), (2) post-Miocene opening of the Gibraltar Strait at the end of the Messinian salinity crisis (Exp. 339), and (3) the greenhouse world of the Paleocene-Eocene and evolution into the glacial world of the Oligocene (Exp. 342).

ACKNOWLEDGMENTS

We would like to acknowledge the diligence of the crew and scientists aboard the RV *Joides Resolution*, the IODP support staff, and the staff of the Bremen Core Repository where the North Atlantic cores are stored. We appreciate thorough reviews of an earlier version of this manuscript by David Naafs and Ian Bailey. The research reported here was partially supported by the US National Science Foundation and the UK National Environmental Research Council.

REFERENCES

Alonso-Garcia, M., Sierro, F. J., & Flores, J.-A. (2011). Arctic front shifts in the subpolar North Atlantic during the Mid-Pleistocene (800-400 ka) and their implications for ocean circulation. *Palaeogeography, Palaeoclimatology, Palaeoecology, 311*, 268–280.

Alonso-Garcia, M., Sierro, F. J., Kucera, M., Flores, J. A., Cacho, I., & Andersen, N. (2011). Ocean circulation, ice sheet growth and interhemispheric coupling of millennial climate variability during the mid-Pleistocene (ca 800-400 ka). *Quaternary Science Reviews, 30*, 3234–3247.

Alvarez Zarikian, C. A., Stepanova, A. Y., & Grutzner, J. (2009). Glacial-interglacial variability in deep sea ostracod assemblage composition at IODP Site U1314 in the subpolar North Atlantic. *Marine Geology, 258*, 69–87. http://dx.doi.org/10.1016/j.margeo.2008.11.009.

Arthur, M. A., Srivastava, S. P., Kaminski, M., Jarrard, R., & Osler, J. (1989). Seismic stratigraphy and history of deep circulation and sediment drift development in Baffin Bay and the Labrador Sea. In S. P. Srivastava, M. A. Arthur, B. Clement, et al. (Eds.), *Proc. ODP, Sci. Results, 105* (pp. 957–988). College Station, TX: Ocean Drilling Program.

Bailey, I., Bolton, C. T., DeConto, R. M., Pollard, D., Schiebel, R., & Wilson, P. A. (2010). A low threshold for the North Atlantic ice rafting from "low-slung slippery" late Pliocene ice sheets. *Paleoceanography, 25*, PA1212. http://dx.doi.org/10.1029/2009PA001736.

Bailey, I., Foster, G. L., Wilson, P. A., Jovane, L., Storey, C. D., Trueman, C. N., et al. (2012). Flux and provenance of ice-rafted debris in the earliest Pleistocene sub-polar North Atlantic Ocean comparable to the last glacial maximum. *Earth and Planetary Science Letters, 341–344*, 222–233.

Bailey, I., Hole, G. M., Foster, G. L., Wilson, P. A., Storey, C. D., Trueman, C. N., et al. (2013). An alternative suggestion for the Pliocene onset of major northern hemisphere glaciation based the geochemical provenance of North Atlantic Ocean ice-rafted debris. *Quaternary Science Reviews, 75C,* 181–194. http://dx.doi.org/10.1016/j.quascirev.2013.06.004.

Barker, S., Knorr, G., Edwards, R. L., Rarrenin, F., Putnam, A. E., Skinner, L. C., et al. (2011). 800,000 years of abrupt climate variability. *Science, 334,* 347–351.

Bintanja, R., & van de Wal, R. S. W. (2008). North American ice-sheet dynamics and the onset of 100,000-year glacial cycles. *Nature, 454,* 869–872.

Bodén, P., & Backman, J. (1996). A laminated sediment sequence from northern North Atlantic Ocean and its climatic record. *Geology, 24,* 507–510.

Bolton, C. T., Wilson, P. A., Bailey, I., Friedrich, O., Beer, C. J., Becker, J., et al. (2010). Millennial-scale climate variability in the subpolar North Atlantic Ocean during the late Pliocene. *Paleoceanography, 25,* PA4218. http://dx.doi.org/10.1029/2010PA001951.

Bond, G., Broecker, W., Johnsen, S., McManus, J., Labeyrie, L., Jouzel, J., et al. (1993). Correlations between climate records from the North Atlantic sediments and Greenland ice. *Nature, 365,* 143–147.

Bond, G. C., & Lotti, R. (1995). Iceberg discharges into the north Atlantic on millennial time scales during the last glaciation. *Science, 267,* 1005–1010.

Bond, G., Showers, W., Elliot, M., Evans, M., Lotti, R., Hajdas, I., et al. (1999). The North Atlantic's 1-2 kyr climate rhythm: relation to Heinrich events, Dansgaard/Oeschger cycles and the little ice age. In: mechanisms of millennial-scale global climate change. (Webb et al., ed.). *AGU Geophysical Monograph, 112,* 35–58.

Broecker, W. S., Bond, G., McManus, J., Klas, M., & Clark, E. (1992). Origin of the Northern Atlantic's Heinrich events. *Climate Dynamics, 6,* 265–273.

Channell, J. E. T., & Hodell, D. A. (2013). Magnetic signatures of Heinrich-like detrital layers in the Quaternary of the North Atlantic. *Earth and Planetary Science Letters, 369–370,* 260–270.

Channell, J. E. T., Hodell, D. A., Margari, V., Skinner, L. C., Tzedakis, P. C., & Kesler, M. S. (2013). Biogenic magnetite, detrital hematite, and relative paleointensity in sediments from the South-west Iberian Margin. *Earth and Planetary Science Letters, 376,* 99–109.

Channell, J. E. T., Hodell, D. A., Romero, O., Hillaire-Marcel, C., de Vernal, A., Stoner, J. S., et al. (2012). A 750-kyr detrital-layer stratigraphy for the North Atlantic (IODP site U1302-U1303, orphan knoll, labrador sea). *Earth and Planetary Science Letters, 317–318,* 218–230.

Channell, J. E. T., Hodell, D. A., Xuan, C., Mazaud, A., & Stoner, J. S. (2008). Age calibrated relative paleointensity for the last 1.5 Myr from IODP Site U1308 (North Atlantic). *Earth and Planetary Science Letters, 274,* 59–71.

Channell, J. E. T., Kanamatsu, T., Sato, T., Stein, R., Alvarez Zarikian, C. A., Malone, M. J., & the Expedition 303/306 Scientists. (2006). *North Atlantic climate, expeditions 303 and 306 of the riserless drilling platform from St. John's, Newfoundland, to Ponta Delgada, Azores (Portugal), Sites U1302–U1308, 25 September–17 November 2004 and from Ponta Delgada, Azores (Portugal) to Dublin, Ireland, Sites U1312–U1315, 2 March–26 April 2005 Integrated Ocean Drilling Program Management International, Inc. for the Integrated Ocean Drilling Program.* Available at http://iodp.tamu.edu/publications/exp303_306/30306title.htm.

Channell, J. E. T., Wright, J. D., Mazaud, A., & Stoner, J. S. (2014). Age through tandem correlation of Quaternary relative paleointensity (RPI) and oxygen isotope data at IODP Site U1306 (Eirik drift, SW Greenland). *Quaternary Science Reviews, 88,* 135–146.

Channell, J. E. T., Xuan, C., & Hodell, D. A. (2009). Stacking paleointensity and oxygen isotope data for the last 1.5 Myrs (PISO-1500). *Earth and Planetary Science Letters, 283,* 14–23.

Cheng, H., Edwards, R. L., Broecker, W. S., Denton, G. H., Kong, X., Wang, Y., et al. (2009). Ice age terminations. *Science, 326,* 248–252.

Clark, P. U., Archer, D., Pollard, D., Blum, J. D., Rial, J. A., Brovkin, V., et al. (2006). The middle Pleistocene transition: characteristics, mechanisms, and implications for long-term changes in atmospheric CO_2. *Quaternary Science Reviews, 25*, 3150–3184.

Coxall, H. K., Wilson, P. A., Pälike, H., Lear, C. H., & Backman, J. (2005). Rapid stepwise onset of Antarctic glaciation and deeper calcite compensation in the Pacific Ocean. *Nature, 433*(7021), 53–57. http://dx.doi.org/10.1038/nature03135 (London, U. K.).

Crocket, K. C., Vance, D., Foster, G. L., Richards, D. A., & Tranter, M. (2012). Continental weathering fluxes during the last glacial/interglacial cycle: insights from marine sedimentary Pb isotope records at orphan Knoll, NW Atlantic. *Quatenary Science Reviews, 38*, 28–41.

Dansgaard, W., Johnsen, S., Clausen, H. B., Dahl-Jensen, D., Gundestrup, N. S., Hammer, C. U., et al. (1993). Evidence for general instability of past climate from a 250-kyr ice core record. *Nature, 364*, 218–220.

De Abreu, L., Abrantes, F. F., Shackleton, N. J., Tzedakis, P. C., McManus, J. F., & Oppo, D. W. (2005). Ocean climate variability in the eastern North Atlantic during interglacial marine isotope stage 11: a partial analogue to the Holocene? *Paleoceanography, 20*, PA3009. http://dx.doi.org/10.1029/2004/PA001091.

De Schepper, S., Head, M. J., & Groeneveld, J. (2009). North Atlantic current variability through marine isotope stage M2 (circa 3.3 Ma) during the mid-Pliocene. *Paleoceanography, 24*, PA4206. http://dx.doi.org/10.1029/2008/PA001725.

De Schepper, S., Fischer, E. I., Groeneveld, J., Head, M. J., & Matthiessen, J. (2011). Deciphering yhe palaeoecology of the late Pliocene and early Pleistocene dinoflagellate cysts. *Palaeogeography, Palaeoclimatology, Palaeoecology, 309*, 17–32.

De Schepper, S., Groeneveld, J., Naafs, B. D. A., Van Renterghem, C., Hennissen, J., Head, M. J., et al. (2013). Northern hemisphere glaciation during the globally warm early late Pliocene. *PLoS One, 8*(12), e815081–15.

Driscoll, N. W., & Haug, G. H. (2008). A short circuit in thermohaline circulation: a cause for Northern Hemisphere glaciation? *Science, 282*, 436–438.

Dykoski, C. A., Edwards, R. L., Cheng, H., Yuan, D., Cai, Y., Zhang, M., et al. (2005). A high-resolution, absolute-dated Holocene and deglacial Asian monsoon record from Dongge Cave, China. *Earth and Planetary Science Letters, 233*, 71–86.

Elderfield, H., Ferretti, P., Greaves, M., Crowhurst, S., McCave, I. N., Hodell, D. A., et al. (2012). Evolution of ocean temperature and ice volume through the Mid-Pleistocene Climate Transition. *Science, 337*, 704–709.

Eldholm, O., Theide, J., & Taylor, E. (1987). Evolution of the Norwegian continental margin: background and objectives. In O. Eldholm, J. Theide, & E. Taylor (Eds.), *Proc. ODP Init. Rep* (Vol. 104) (pp. 5–26).

Elliot, M., Labeyrie, L., Bond, G., Cortijo, E., Turon, J. L., Tisnerat, N., et al. (1998). Millennial-scale iceberg discharges in the Irminger Basin during the last glacial period: relationship with the Heinrich events and environmental settings. *Paleoceanography, 13*, 433–446.

Expedition 339 Scientists. (2013). Expedition 339 summary. In D. A. V. Stow, F. J. Hernández-Molina, C. A. Alvarez Zarikian, & the Expedition 339 Scientists (Eds.), *Proc. IODP, 339*. Tokyo: Integrated Ocean Drilling Program Management International, Inc. http://dx.doi.org/10.2204/iodp.proc.339.101.2013.

Expedition 342 Scientists. (2012). Paleogene Newfoundland sediment drifts. *IODP Preliminary Report, 342*. http://dx.doi.org/10.2204/iodp.pr.342.2012.

Ferretti, P., Crowhurst, S. J., Hall, M. A., & Cacho, I. (2010). North Atlantic millennial-scale climate variability 910–790 ka and the role of the equatorial isolation forcing. *Earth and Planetary Science Letters, 293*, 28–41.

Friedrich, O., Wilson, P. A., Bolton, C. T., Beer, C. J., & Schiebel, R. (2013). Late Pliocene to early Pleistocene changes in the North Atlantic current and suborbital sea-surface temperature variability. *Paleoceanography*, *28*(2), 274–282.

Galaasen, E. V., Ninnemann, U. S., Irvali, N., Kleiven, H. F., Rosenthal, Y., Kissel, C., et al. (2014). Rapid reductions in North Atlantic deep water during the peak of the last interglacial period. *Science*, *343*, 1129–1132.

Grützner, J., & Higgins, S. M. (2010). Threshold behavior of millennial scale variability in deep water hydrography inferred from a 1.1 Ma long record of sediment provenance at the southern Gardar drift. *Paleoceanography*, *25*, PA4204. http://dx.doi.org/10.1029/2009PA001873.

Hagino, K., & Kulhanek, D. K. (2009). Data report: calcareous nannofossils from the upper Pliocene and Pleistocene, exp. 306, sites U1313 and U1314. In J. E. T. Channell, T. Kanamatsu, T. Sato, R. Stein, M. J. Malone, & the Expedition 303/306 Scientists (Eds.), *Proc. IODP 303/306*. College Station TX: Integrated Ocean Drilling Program Management International, Inc.

Hall, I. R., & Becker, J. (2007). Deep western boundary current variability in the subtropical northwest Atlantic Ocean during marine isotope stages 12-10. *Geochemistry, Geophysics, Geosystems*, *8*. (6). http://dx.doi.org/10.1029/2006GC001518.

Harris, R. N., & Higgins, S. M. (2008). A permeability estimate in 56 Ma crust at ODP Hole 642E, Vøring Plateau Norwegian Sea. *Earth and Planetary Science Letters*, *267*, 378–385.

Harris, R. N., & IODP Expedition 306 Scientists. (2006). Borehole observatory installations on IODP Expedition 306 reconstruct bottom-water temperature changes in the Norwegian Sea. *Scientific Drilling*, *2*, 28–31. http://dx.doi.org/10.2204/iodp.sd.2.03.2006.

Hayashi, T., Ohno, M., Acton, G., Guyodo, Y., Evans, H. F., Kanamatsu, T., et al. (2010). Millennial-scale iceberg surges after intensification of Northern Hemisphere glaciation. *Geochemistry, Geophysics, Geosystems*, *11*(9), Q09Z20. http://dx.doi.org/10.1029/2010GC003132.

Heinrich, H. (1988). Origin and consequences of cyclic ice rafting in the Northeast Atlantic Ocean during the past 130,000 years. *Quaternary Research*, *29*, 142–152. http://dx.doi.org/10.1016/0033-5894(88)90057-9.

Hemming, S. R. (2004). Heinrich events: massive late Pleistocene detritus layers of the North Atlantic and their global climate imprint. *Reviews of Geophysics*, *42*, RG1005. http://dx.doi.org/10.1029/2003RG000128.

Hernández -Almeida, I., Sierro, F. J., Flores, J.-A., Cacho, I., & Filippelli, G. A. (2013). Palaeoceanographic changes in the North Atlantic during the Mid-Pleistocene Transition (MIS 31-19) as inferred from planktonic foraminiferal and calcium carbonate records. *Boreas*, *42*, 140–159. http://dx.doi.org/10.1111/j.1502-3885.2012.00283.

Hernández-Molina, F. J., Stow, D. A. V., Alvarez-Zarikian, C., & Expedition IODP 339 Scientists. (2013). IODP Expedition 339 in the Gulf of Cadiz and off West Iberia: decoding the environmental significance of the Mediterranean outflow water and its global influence. *Scientific Drilling*, *16*, 1–11.

Hernández-Molina, F. J., Stow, D. A. V., Alvarez-Zarikian, C., & Expedition IODP 339 Scientists. (2014). Onset of Mediterranean outflow into the North Atlantic. *Science*, *344*, 1244–1250.

Hillaire-Marcel, C., & Bilodeau, G. (2000). Instabilities in the Labrador sea water mass structure during the last glacial cycle. *Canadian Journal of Earth Sciences*, *37*, 795–809.

Hillaire-Marcel, C., De Vernal, A., Bilodeau, G., & Wu, G. (1994). Isotope stratigraphy, sedimentation rates, deep circulation, and carbonate events in the Labrador Sea during the last ~200 ka. *Canadian Journal of Earth Sciences*, *31*, 63–89.

Hillaire-Marcel, C., de Vernal, A., & McKay, J. (2011). Foraminifer isotope study of the Pleistocene Labrador Sea, northwest North Atlantic (IODP Sites 1302/03 and 1305), with emphasis on paleoceanographical differences between its "inner" and "outer" basins. *Marine Geology*, *279*, 188–198. http://dx.doi.org/10.1016/j.margeo.2010.11.001.

Hillaire-Marcel, C., De Vernal, A., & Piper, D. J. W. (2007). Lake Agassiz final drainage event in the northwest North Atlantic. *Geophysics Research Letters, 34,* L15601.

Hiscott, R. N., Aksu, A. E., Mudie, P. J., & Parsons, D. F. (2001). A 340,000 year record of ice rafting, paleoclimatic fluctuations, and shelf crossing glacial advances in the southwestern Labrador Sea. *Global and Planetary Change, 28,* 227–240.

Hodell, D. A., Channell, J. E. T., Curtis, J. H., Romero, O. E., & Rohl, U. (2008). Onset of "Hudson Strait" Heinrich events in the eastern North Atlantic at the end of the middle Pleistocene transition (~640 ka)? *Paleoceanography, 23,* PA4218. http://dx.doi.org/10.1029/200 8PA001591.

Hodell, D. A., Crowhurst, S., Skinner, L., Tzedakis, P. C., Margari, V., Channell, J. E. T., et al. (2013). Response of Iberian Margin sediments to orbital and suborbital forcing over the past 420 ka. *Paleoceanography, 28,* 185–199. http://dx.doi.org/10.1002/paleo.20017.

Hodell, D. A., & Curtis, J. H. (2008). Oxygen and carbon isotopes of detrital carbonate in North Atlantic Heinrich events. *Marine Geology, 256,* 30–35.

Hodell, D. A., Lourens, L., Stow, D. A. V., Hernández-Molina, J., Alvarez Zarikian, C. A., & the Shackleton Site Project Members. (2013). The "Shackleton site" (IODP site U1385) on the iberian margin. *Scientific Drilling, 16,* 13–19.

Hodell, D. A., Minth, E. K., Curtis, J. H., Hall, I. R., Channell, J. E. T., & Xuan, C. (2009). Surface and deep water hydrography on Gardar drift (Iceland Basin) during the last interglacial period. *Earth and Planetary Science Letters, 288,* 10–19.

Hunter, S. E., Wilkinson, D., Louarn, E., McCave, I. N., Rohling, E., Stow, D. A. V., et al. (2007). Deep western boundary current dynamics and associated sedimentation on the Eirik Drift, Southern Greenland margin. *Deep-Sea Research, 54,* 2036–2066.

Ji, J., Ge, Y., Balsam, W., Damuth, J. E., & Chen, J. (2009). Rapid identification of dolomite using a Fourier transform infrared spectrophotometer (FTIR): a fast method for identifying Heinrich events in IODP Site U1308. *Marine Geology, 258,* 60–68.

Kanamatsu, T., Acton, G., Evans, H., Guyodo, Y., & Ohno, M. (2010). Data Report: paleomagnetic and rock magnetic records since the late Pliocene from IODP Site U1312 on the southern flank of the King's Trough. In J. E. T. Channell, T. Kanamatsu, T. Sato, R. Stein, M. J. Malone, & the Expedition 303/306 Scientists (Eds.), *Proc. IODP 303/306.* College Station TX: Integrated Ocean Drilling Program Management International, Inc.

Kanamatsu, T., Ohno, M., Acton, G., Evans, H., & Guyodo, Y. (2009). Rock magnetic properties of the Gardar Drift sedimentary sequence, Site IODP U1314, North Atlantic: Implications for bottom water current change through the mid-Pleistocene. *Marine Geol, 265,* 31–39.

Kawamura, N., Ishikawa, N., & Torii, M. (2012). Diagenetic alteration of magnetic minerals in Labrador Sea sediments (IODP Sites U1305, U1306, and U1307). *Geochemistry, Geophysics, Geosystems, 13*(8), Q08013. http://dx.doi.org/10.1029/2012GC004213.

Kuhs, M., Austin, W. E. N., Abbott, P. M., & Hodell, D. A. (2014). Iceberg-rafted tephra as a potential tool for the reconstruction of ice-sheet processes and ocean surface circulation in the glacial North Atlantic. In W. E. N. Austin, et al. (Ed.), *Marine tephrochronology* (398). Geological Society, London, Special Publication. http:/dx/doi.org/10.1144/SP398.8.

Laj, C., & Channell, J. E. T. (2007). Geomagnetic excursions. In M. Kono (Ed.), *Treatise on Geophysics Geomagnetism: Vol. 5.* (pp. 373–416). Amsterdam: Elsevier. Chapter 10.

Lhuillier, F., Fournier, A., Hulot, G., & Aubert, J. (2011). The geomagnetic secular-variation timescale in observations and numerical dynamo models. *Geophysical Research Letters, 38,* L09306. http://dx.doi.org/10.129/2011GL047356.

Lisiecki, L. E., & Lisiecki, P. A. (2002). Application of dynamic programming to the correlation of paleoclimate records. *Paleoceanography, 17,* 1049. http://dx.doi.org/10.1029/2001PA000733.

Lisiecki, L. E., & Raymo, M. E. (2005). A Pliocene-Pleistocene stack of 57 globally distributed benthic δ^{18}O records. *Paleoceanography, 20*, PA1003. http://dx.doi.org/10.1029/2004PA001071.

Lunt, D. J., Foster, G., Haywood, A. M., & Stone, E. (2008). Late Pliocene Greenland glaciation controlled by a decline in atmospheric CO_2 levels. *Nature, 454*, 1102–1106.

Margari, V., Skinner, L. C., Tzedakis, P. C., Ganopolski, A., Vautravers, M., & Shackleton, N. J. (2010). The nature of millennial-scale climate variability during the pas two glacial periods. *Nature Geoscience, 3*, 127–133.

Mazaud, A., Channell, J. E. T., & Stoner, J. S. (2012). Relative paleointensity and environmental magnetism since 1.2 Ma at IODP Site U1305 (Eirik Drift, NW Atlantic). *Earth and Planetary Science Letters, 357–358*, 137–144.

Mazaud, A., Channell, J. E. T., Xuan, C., & Stoner, J. S. (2008). Upper and lower Jaramillo polarity transitions recorded in IODP Expedition 303 North Atlantic sediments: implications for transitional field geometry. *Physics of the Earth and Planetary Interiors, 172*, 131–140.

McCave, I. N. (2002). A poisoned chalice? *Science, 298*, 1186–1187.

McManus, J. F., Oppo, D. W., & Cullen, J. L. (1999). A 0.5-million-year record of millennial scale climate variability in the North Atlantic. *Science, 283*, 971–975.

Müller-Michaelis, A., Uenzelmann-Neben, G., & Stein, R. (2013). A revised early miocene age for the instigation of the Eirik drift, offshore southern Greenland: evidence from high-resolution seismic reflection data. *Marine Geology, 340*, 1–15.

Naafs, B. D. A., Hefter, J., Acton, G., Haug, G. H., Martínez-Garcia, A., Pancost, R., et al. (2012). Strengthening of North American dust sources during the late Pliocene (2.7 Ma). *Earth and Planetary Science Letters, 317–318*, 8–19. http://dx.doi.org/10.1016/j.epsl.2011.11.026.

Naafs, B. D. A., Hefter, J., Ferretti, P., Stein, R., & Haug, G. (2011). Sea surface temperatures did not controlthe first occurrence of Hudson Strait Heinrich Events during MIS 16. *Paleoceanography, 26*, PA4201. http://dx.doi.org/10.1029/2011PA002135.

Naafs, B. D. A., Hefter, J., Grützner, J., & Stein, R. (2013). Warming of surface waters in the mid-latitude North Atlantic during Heinrich Events. *Paleoceanography, 28*. http://dx.doi.org/10.1029/2012PA002354.

Naafs, B. D. A., Hefter, J., & Stein, R. (2012). Application of the long chain diol index (LDI) paleothermometer to the early Pleistocene (MIS 96). *Organic Geochemistry, 49*, 83–85.

Naafs, B. D. A., Hefter, J., & Stein, R. (2013). Millennial-scale ice rafting events and Hudson Strait Heinrich Events during the late Pliocene and Pleistocene: a review. *Quatenary Science Reviews, 80*, 1–28.

Naafs, D. A., Stein, R., Hefter, J., Khelifi, N., De Schepper, S., & Haug, G. H. (2010). Late Pliocene changes in the North Atlantic current. *Earth and Planetary Science Letters, 298*, 434–442.

Nicholl, J. A. L., Hodell, D. A., Naafs, B. D. A., Hillaire-Marcel, C., Channell, J. E. T., Romero, O. E., et al. (2012). Laurentide outburst flooding event during the last interglacial Period. *Nature Geoscience, 5*, 901–904. http://dx.doi.org/10.1038/NGEO1622.

Norris, R. D., Wilson, P. A., Blum, P., & the expedition 342 Scientists (2014). Proc. IODP, 342: College Station, TX (Integrated Ocean Drilling Program). http://dx.doi.org/10.2204/iodp.proc.342.101.2014.

Ogg, J. G. (2012). The geomagnetic polarity timescale. In F. Gradstein, J. G. Ogg, M. Schmitz, & G. Ogg (Eds.), *The geologic time scale 2012* (pp. 85–128). Amsterdam: Elsevier.

Ohno, M., Hayashi, T., Komatsu, F., Murakami, F., Zhao, M., Guyodo, Y., et al. (2012). A detailed paleomagnetic record between 2.1 and 2.75 Ma at IODP Site U1314 in the North Atlantic: geomagnetic excursions and the Gauss-Matuyama transition. *Geochemistry, Geophysics, Geosystems, 13*(1), Q12Z39. http://dx.doi.org/10.1029/2012GC004080.

Paillard, D., Labeyrie, L., & Yiou, P. (1996). Macintosh program performs time-series analysis. *Eos, Transations American Geophysical Union, 77*. http://dx.doi.org/10.1029/96EO00259.

Pälike, H., Lyle, M.W., Nishi, H., Raffi, I., Ridgwell, A., Gamage, K., et al. (2012). A cenozoic record of the equatorial Pacific carbonate compensation depth. *Nature, 488*, 609–614. http://dx.doi.org/10.1038/nature11360.

Robinson, S. G., Maslin, M. A., & McCave, I. N. (1995). Magnetic susceptibility variations in Upper Pleistocene deep-sea sediments of the NE Atlantic: implications for ice rafting and paleocirculation at the last glacial maximum. *Paleoceanography, 10*, 221–250.

Romero, O. E., Swann, G. E. A., Hodell, D. A., Helmke, P., Rey, D., & Rubio, B. (2011). A highly productive Subarctic Atlantic during the last interglacial and the role of diatoms. *Geology, 39*(11), 1015–1018.

Ruddiman, W. F. (1977). Late quaternary deposition of ice-rafted sand in the subpolar North Atlantic (lat 40 to 65 N). *Geological Society of American Bulletin, 88*, 1813–1827.

Ruddiman, W. F., McIntyre, A., & Raymo, M. (1986). Matuyama 41,000 year cycles: North Atlantic ocean and Northern hemisphere ice sheets. *Earth and Planetary Science Letters, 80*, 117–129.

Ruddiman, W. F., Raymo, M. E., Martinson, D. G., Clement, B. M., & Backman, J. (1989). Pleistocene evolution: northern hemisphere ice sheet and north Atlantic ocean. *Paleoceanography, 4*, 353–412.

Sarnthein, M., Bartoli, G., Prange, M., Schmittner, A., Schneider, B., Weinelt, M., et al. (2009). Mid-Pliocene shifts in ocean overturning circulation and the onset of Quaternary-style climates. *Climate of the Past, 5*, 269–283.

Sato, T., Chiyonobu, S., & Hodell, D. A. (2008). Data Report: Quaternary calcareous nannofossil datums and biochronology in the North Atlantic Ocean, IODP U1308. In J. E. T. Channell, T. Kanamatsu, T. Sato, R. Stein, M. J. Malone, & the Expedition 303/306 Scientists (Eds.), *Proc. IODP 303/306*. College Station TX: Integrated Ocean Drilling Program Management International, Inc.

Shackleton, N. J., Fairbanks, R. G., Chiu, T.-C., & Parrenin, F. (2004). Absolute calibration of the Greenland time scale: implications for Antarctic time scales and for Δ^{14}C. *Quaternary Science Reviews, 23*, 1513–1522.

Shackleton, N. J., Hall, M. A., & Vincent, E. (2000). Phase relationships between millennial-scale events 64,000–24,000 years ago. *Paleoceanography, 15*, 565–569.

Shimada, C., Sato, T., Toyoshima, S., Yamasaki, M., & Tanimura, Y. (2008). Paleoecological significance of laminated diatomaceous oozes during the middle-to-late Pleistocene, North Atlantic Ocean (IODP site U1304). *Marine Micropaleontology, 69*, 139–150.

Sierro, F. J., Hernández -Almeida, I., Alonso-Garcia, M., & Flores, J. A. (2009). Data report: Pliocene-Pleistocene planktonic foraminifer bioevents at IODP site U1313. In J. E. T. Channell, T. Kanamatsu, T. Sato, R. Stein, M. J. Malone, & the Expedition 303/306 Scientists (Eds.), *Proc. IODP 303/306*. College Station TX: Integrated Ocean Drilling Program Management International, Inc.

Skinner, L. C., Elderfield, H., & Hall, M. (2007). Phasing of millennial climate events and northeast Atlantic deep-water temperature change since 50 ka BP, in ocean circulation: mechanisms and Impacts, AGU. *Monograph, 173*, 197–208.

Stanford, J. D., Rohling, E. J., Bacon, S., & Holliday, N. P. (2011). A review of the deep and surface currents around Eirik drift, south of Greenland: comparison of the past with the present. *Global and Planetary Change, 79*, 244–254.

Stein, R., Hefter, J., Grützner, J., Voelker, A., & Naafs, D. A. (2009). Variability of surface-water characteristics and Heinrich events in the Pleistocene mid-latitude North Atlantic Ocean: biomarker and XRD records from IODP Site U1313 (MIS 16–9). *Paleoceanography, 24*, PA2203. http://dx.doi.org/10.1029/2008PA001639.

Stone, R. (2013). Battle for the Americas. *Science, 341*, 230–233.

Stoner, J. S., Channell, J. E. T., & Hillaire-Marcel, C. (1995). Late Pleistocene relative geomagnetic paleointensity from the deep Labrador Sea: regional and global correlations. *Earth and Planetary Science Letters, 134,* 237–252.

Stoner, J. S., Channell, J. E. T., & Hillaire-Marcel, C. (1996). The magnetic signature of rapidly deposited detrital layers from the deep Labrador Sea: relationship to North Atlantic Heinrich layers. *Paleoceanography, 11,* 309–325.

Stoner, J. S., Channell, J. E. T., & Hillaire-Marcel, C. (1998). A 200 kyr geomagnetic chronostratigraphy for the Labrador Sea: indirect correlation of the sediment record to SPECMAP. *Earth and Planetary Science Letters, 159,* 165–181.

Stoner, J. S., Laj, C., Channell, J. E. T., & Kissel, C. (2000). South Atlantic (SAPIS) and North Atlantic (NAPIS) geomagnetic paleointensity stacks (0-80 ka): implications for inter-hemispheric correlation. *Quaternary Science Reviews, 21,* 1141–1151.

Van Kreveld, S. A., Knappertsbusch, M., Ottens, J., Ganssen, G., & van Hinte, J. (1996). Biogenic carbonate and ice-rafted debris (Heinrich layer) accumulation in deep-sea sediments from a Northeast Atlantic piston core. *Marine Geology, 131,* 21–46.

Vautravers, M. J., & Shackleton, N. J. (2006). Centennial-scale surface hydrology off Portugal during marine isotope stage 3: insights from planktonic foraminiferal fauna variability. *Paleoceanography, 21,* PA3004. http://dx.doi.org/10.1029/2005PA001144.

Voelker, A. H. L., Rodrigues, T., Billups, K., Oppo, D., McManus, J., Stein, R., et al. (2010). Variations in mid-latitude North Atlantic surface water properties during the mid-Brunhes (MIS 9-14) and their implications for the thermohaline circulation. *Climate of the Past, 6,* 531–552.

Wang, Y. J., Cheng, H., Edwards, R. L., An, Z. S., Wu, J. Y., Shen, C.-C., et al. (2001). A high-resolution absolute-dated Late Pleistocene monsoon record from Hulu Cave, China. *Science, 294,* 2345–2348.

Wang, Y. J., Cheng, H., Edwards, R. L., Kong, X., Shao, X., Chen, S., et al. (2008). Millennial- and orbital-scale changes in the East Asian monsoon over the past 224,000 years. *Nature, 451,* 1090–1093.

Zachos, J. C., Kroon, D., Blum, P., et al. (2004). In *Proc. ODP, Init. Repts., 208.* College Station, TX: Ocean Drilling Program. http://dx.doi.org/10.2973/odp.proc.ir.208.2004.

Zachos, J. C., Röhl, U., Schellenberg, S. A., Sluijs, A., Hodell, D. A., Kelly, D. C., et al. (2005). Rapid acidification of the ocean during the Paleocene–Eocene Thermal Maximum. *Science, 308*(5728), 1611–1615. http://dx.doi.org/10.1126/science.1109004.

Coral Reefs and Sea-Level Change

Gilbert Camoin[1,*] and Jody Webster[2]
[1]CEREGE, UMR 7330 CNRS, Europôle Méditerranéen de l'Arbois, Aix-en-Provence, France;
[2]Geocoastal Research Group, School of Geosciences, University of Sydney, NSW, Australia
*Corresponding author: E-mail: camoin@cerege.fr

3.6.1 INTRODUCTION/RATIONALE

One of the most societally relevant objectives of Earth Sciences is to success-fully meet the challenges of recent and projected changes in Earth's surface environment, including sea-level rise related to ice sheet instability, changes in hydrological cycle, and increasing atmospheric pCO_2 that is expected to be the main driving force for future climatic and ocean ecological changes.

In the last decade, most of the measured sea-level rise has resulted from thermal expansion of the ocean, but the Greenland and Antarctic ice sheets pro-vide the greatest potential risk for future sea-level rise because of their huge volume, equivalent to ~65 m of sea level. Satellite-based mass balance esti-mates show that the ice sheets are losing mass and that the rate is accelerating. Currently, they are contributing about half of the current global mean sea-level rise (now ~3.4 mm/year), and in coming decades they will be by far the largest source (Bertler & Barrett, 2010). However, uncertainties in sea-level projec-tions remain large (0.5–2.0 m by 2100) because the vulnerability of Greenland and Antarctica ice sheets to ongoing warming and related discharge feedbacks in response to a warming Earth are still poorly understood (Bamber, Riva, Ver-meersen, & LeBrocq, 2009; Milne, Gehrels, Hughes, & Tamisiea, 2009; Pfeffer, Harper, & O'Neel, 2008). The improvement of past and future sea-level change models will rely on the quantification of the relative contributions of Northern and Southern Hemisphere ice sheets, especially based on a better understanding of their instability during recent natural events.

The instrumental record of sea-level and climate variability extends back only about 150 years, a period when sea level has risen only 0.2 m and which does not reflect the conditions that are predicted for the future. Exploiting geological archives is therefore the only way to understand instrumental records of recent

Developments in Marine Geology, Volume 7. http://dx.doi.org/10.1016/B978-0-444-62617-2.00015-3
395

environmental and climate change in the larger context of natural variability, and to examine the response of the earth system to a large dynamic range of boundary conditions, on timescales ranging from annual through geological. Furthermore, geological archives give access to past periods of Earth's environmental history which more closely resemble warm conditions predicted for the next few centuries and beyond, thus providing critical tests to improve climate and Earth System models.

The full extent of sea-level variability can be constrained only by sea-level records from the geological past which provide the most direct estimates of changes in ice volume, including warm periods characterized by sea levels that are meters to tens of meters higher than today, and abrupt climate changes when sea level increased rapidly and dramatically as a consequence of large-amplitude and rapid discharges of freshwater following the collapse of continental ice sheets. Over the past ~800,000 years, the cyclic growth and decay of northern ice sheets induced rapid sea-level change at intervals of ~100,000 years, with maximum amplitudes of 120–140 m (e.g., Lambeck, Yokoyama, & Purcell, 2002; Milne & Mitrovica, 2008; Milne, Mitrovica, & Schrag, 2002; Waelbrock et al., 2002) (Figure 3.6.1). Current discrepancies between models and sea-level reconstructions of past changes are due largely to poor constraints on the timing, rates, and

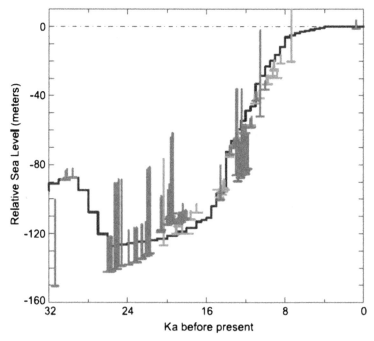

FIGURE 3.6.1 Eustatic sea-level curve for the version of the ICE-5G (VM2) model and relative sea levels (RSL) observations from Barbados with individual data points show as meters below present sea level and corrected for a rate of vertical tectonic uplift of 0.34 mm/year. *From Peltier and Fairbanks (2006).*

relative contributions of the various ice sheets. However, although the correlation between ice and ocean volumes is incontrovertible, the causal link is commonly obscured. Local effects that can obscure the true picture include tectonics, isostatic and hydroisostatic responses, and equatorial ocean siphoning (Mitrovica & Peltier, 1991). The wide regional variation in geophysical processes that affect local relative sea levels (RSLs) implies that sea-level curves are of regional significance only and precludes their use as direct indicators of either ice volume or mean sea-level change. A comparison between interpreted sea-level curves and data from different regions of the planet helps in understanding the interplay of climatic, oceanographic, and geophysical processes invoked to explain relative sea-level positions, in refining the parameters constraining the rheology of the Earth and aspects of the history of past ice sheets, thus providing a useful constraint on geophysical models.

Signals of relative sea-level change are not uniformly distributed and are a function not only of the change in ice volume but also of the planet's rheology through glacial isostatic adjustments (GIA), the changing gravitational potential of the ice sheets due to ice-sheet unloading and the subsequent redistribution of water masses in the global ocean, as well as a combination of local processes (Lambeck, 1993; Lambeck et al., 2006; Lambeck, Purcell, Johnston, Nakada, & Yokoyama, 2003; Milne et al., 2009; Milne & Mitrovica, 2008; Peltier, 1994; Yokoyama, Esat, & Lambeck, 2001a,b). A number of attempts have been made to model both global hydroisostatic adjustments (Bassett, Milne, Bentley, & Huybrechts, 2007; Bassett, Milne, Mitrovica, & Clark, 2005; Clark, Mitrovica, Milne, & Tamisiea, 2002; Lambeck, 1993; Peltier, 1994; Peltier, 1999) and equatorial ocean siphoning (Mitrovica & Milne, 2002; Mitrovica & Peltier, 1991), in order to simulate the lithospheric response to particular deglaciation histories and predict the general shape of local sea-level curves. However, aspects of these models remain controversial.

Geophysical models show that global hydroisostatic adjustment provides the most geographically widespread mechanism to explain local relative sea-level histories (Bassett et al., 2005, 2007; Clark et al., 2002; Milne & Mitrovica, 2008; Stirling, Esat, Lambeck, & McCulloch, 1998). They require to accurately record sea-level changes on a regional scale, at various latitudes, in different tectonic settings, and at variable distances from former glaciated regions ("near-field" vs "far-field" of Peltier (1991) and Mitrovica and Peltier (1991)). At sites distant from former ice sheets (far-field sites), like in tropical coral reefs, the influence of glacio-isostatic rebound is minimized, and sea-level data are therefore more useful in constraining the total volume of land-based ice by using geophysical/inverse modeling techniques, as well as in fingerprinting the source of meltwater contributions (Clark et al., 2002; Mix, Bard, & Schneider, 2001). In contrast, sea-level data from sites close to past ice sheets will provide crucial information regarding the melting history of each drainage basin and complement results from floating ice platforms. Ocean drilling is therefore uniquely positioned to obtain records of sea-level change at a range of latitudes, in different tectonic and sedimentary settings, and at variable distances from former glaciated regions.

The processes leading to the collapse of past ice sheets and the timing and the volume of meltwater released under varying thermal regimes can be better constrained by studying the geologic record of Quaternary glacial–interglacial transitions known as terminations (e.g., the last and penultimate deglaciations, respectively ~23 and ~130 kyr BP). The rapid reduction of the global ice budget during Quaternary glacial–interglacial transitions also affected atmospheric and oceanic circulation through the rapid decrease in ice topography and the large increase in freshwater flux to the ocean. The study of the geologic record of terminations can therefore provide the opportunity to better understand the processes associated with ice sheet instability. Reconstruction of rates and magnitudes of sea-level rise during several terminations, which can only be recovered by scientific ocean drilling, will provide constraints for modeling ice sheet dynamics, clarifying the mechanisms of catastrophic ice sheet collapses, and determining the timing and the volume of meltwater released under varying thermal regimes.

The Last Deglaciation (23–6 kyr BP; Figure 3.6.1) is generally seen as a potential recent analog for the environmental changes that our planet may face in the near future as a consequence of ocean thermal expansion and the melting of polar ice sheets related to a warming Earth. The reconstruction of the magnitude of eustatic changes during the last glacial period may help in constraining the volumes of ice that had accumulated on the continents.

3.6.2 CORAL REEFS: ARCHIVES OF PAST SEA-LEVEL AND ENVIRONMENTAL CHANGES

Carbonate sediments are excellent sea-level markers and contain unique and important components of the records of sea-level timing, amplitude, and stratigraphic response. The relationship of these systems to the carbon cycle allows direct correlation of climatic and eustatic signals. Multiple dating techniques are available for carbonates (including ^{14}C, U/Th, $^{87/86}Sr$, U/Pb, biostratigraphy, and magnetostratigraphy), thus enabling the examination of a wide range of frequencies and amplitudes of sea-level change, from millennial scale to tens of millions of years. While continental margin transects have the advantage that their stratigraphic architectures are well constrained by seismic data (Fulthorpe et al., 2008), coral reef systems provide the most reliable geological estimates of RSL as reef biological communities live in a sufficiently narrow or specific depth range to be useful as absolute sea-level indicators.

The accurate dating of tropical reef corals by mass spectrometry is of prime importance to clarify the mechanisms that drive glacial–interglacial cycles during Quaternary times, to attempt to resolve the rates of millennial-scale changes in sea level, and to constrain geophysical models. Tropical coral reefs are also highly sensitive to variations in water chemistry and physical factors, and are valuable recorders of past climatic and environmental changes. High-resolution records of past global changes (especially changes in sea surface temperatures—SSTs and sea surface salinities—SSSs) are stored in the geo-chemical and physical parameters of coral skeletons and reef sequences and can be used to examine ocean/atmosphere variability and interactions. Changes in other environmental parameters such as light conditions, water energy, and nutrient levels are usually reflected in variations in the composition of reef communities, as reef-dwelling organisms are sensitive to subtle ecological changes affecting their environment. This explains why coral reefs now play a pivotal role in Quaternary paleoclimatic reconstructions.

After Broecker et al. (1968), most coral reef records have concerned sea-level highstands corresponding to the Last Interglacial period approximately 125 kyr BP, and/or to the penultimate Interglacial (isotopic stage 7) from coral reef terraces exposed on the Huon Peninsula, Papua New Guinea (Bloom, Broecker, Chappell, Matthews, & Mesolella, 1974; Chappell, 1974, 2002; Chappell & Veeh, 1978; Chappell et al., 1996; Esat & Yokoyama, 2006; Stein et al., 1993), Barbados (Bard, Hamelin, & Fairbanks, 1990; Edwards, Chen, & Wasserburg, 1987; Gallup, Edwards, & Johnson, 1994; Mesolella, Matthews, Broecker, & Thurber, 1969; Potter & Lambeck, 2004), Sumba (Bard et al., 1996; Pirazzoli et al., 1993), the Red Sea (Dullo, 1990; Gvirtzman & Friedman, 1977; Strasser, Strohmenger, Davaud, & Bach, 1992), and Mexico (Blanchon, Eisenhauer, Fietzke, & Liebetrau, 2009). These generally coincide with major sea-level highstands (i.e., interglacial periods) predicted by the astronomical theory of climate change (Milankovitch, 1941). However, most of these studies concerned uplifted and presently emerged parts of reefs and reef terraces in active subduction zones where relative sea-level records may be biased by variations in rates of tectonic uplift and/or abrupt coseismic vertical motions.

Because the amplitude of Quaternary sea-level changes was in the order of 120 m, the relevant reef and sediment archives, especially recording the glacial stages and the glacial–interglacial transitions, are mostly recorded on modern fore-reef slopes where they have been barely investigated by dredging (e.g., Cabioch et al., 2008; Camoin et al., 2006; Rougerie, Wauthy, & Rancher, 1992) and submersible sampling (e.g., Brachert, 1994; Brachert & Dullo, 1991, 1994; Dullo et al., 1998; Grammer et al., 1993; James & Ginsburg, 1979; Land & Moore, 1980; Macintyre et al., 1991; Webster, Wallace et al., 2004; Webster, Clague et al., 2004). These data are typically fragmentary but have brought to light valuable information regarding the interpretation of morphological features, both accretionary (e.g., terraces, relict reefs) and erosional (e.g., cliffs, notches) in relation to sea-level changes.

The knowledge regarding the coral reef records of glacio-eustatic sea level has been improved by the development of drilling capabilities and radiometric dating techniques over the last 30 years. However, mostly highstand units were recorded in vertical reef drilling. Furthermore, in most cases, U/Th chronology was limited due to the scarcity of datable material, reflecting diagenetic alteration and postdepositional migration of U and Th isotopes. The scarcity of coral reef sea-level records and related data therefore challenges our ability to unravel the rate and timing of the Quaternary sea-level changes, as well as the nature of the coeval climate and environmental changes.

Vertically, deeply cored reef sequences were recovered from barrier reefs located on continental margins, e.g., on the Australian Great Barrier Reef (e.g., Braithwaite & Montaggioni, 2009; Braithwaite et al., 2004; Richards & Hill, 1942; Webster & Davies, 2003), in New Caledonia (Coudray, 1976; Montaggioni et al., 2011), and in Belize (Gischler & Hudson, 2004; Gischler, Lomando, Hudson, & Holmes, 2000; Gischler, Ginsburg, Herrlez, & Prasad, 2010; Purdy, Gischler, & Lomando, 2003).

Limestone columns beneath mid-oceanic atolls contain excellent records of past sea-level change, as the thermal subsidence due to gradual cooling of the oceanic lithosphere is the primary driving component of atoll subsidence and tends toward linearity with time (e.g., Detrick & Crough, 1978; Parsons & Sclater, 1977). Accordingly, mid-oceanic atolls have frequently been referred to as "dipsticks" (e.g., Wheeler & Aharon, 1991). Carbonate sequences of up to a few hundreds of meters, including both sea-level highstands and lowstands, were extracted from a number of atolls (e.g., Funafuti: Bonney, 1904; Cullis, 1904; Ohde et al., 2002; Bikini: Johnson, Todd, Post, Cole, & Wells, 1954; Enewetak: Ladd & Schlanger, 1960; Tracey & Ladd, 1974; Ludwig, Halley, Simmons, & Peterman, 1988; Quinn & Matthews, 1990; Quinn, 1991). However, like continental margins, there have been few drilling opportunities to document a well-defined chronology of sea-level highstand and lowstand reef units. The exception concerns Moruroa, French Polynesia, where highstand (Holocene, stages 5, 7, and 9) and lowstand (stages 2, 4, and 8) reef units were documented and accurately dated based in deviated 300-m long drill holes (Braithwaite & Camoin, 2011; Camoin, Ebren, Eisenhauer, Bard, & Faure, 2001).

The study of coral reef records of the last deglacial events are of prime importance to constrain the timing and amplitude of rapid sea-level changes and to unravel the reef response to dramatic environmental perturbations. Before the Integrated Ocean Drilling Program (IODP) expeditions 310 and 325, only four accurately dated reef sequences that have been attributed to the times reflecting the Holocene-Pleistocene boundary were investigated by drilling, i.e., Barbados (26–7 kyr BP; Fairbanks, 1989; Bard et al., 1990; Fairbanks et al., 2005; Peltier & Fairbanks, 2006), Papua New Guinea (13–6 kyr BP; Chappell & Polach, 1991; Edwards et al., 1993), onshore Tahiti (13.85–2.38 kyr BP; Bard et al., 1996, 2010), and Vanuatu (23–6 kyr BP; Cabioch et al., 2003). Additional fragmentary information concerning the 9–20 kyr BP time span was recorded in Florida (Locker, Hine, Tedesco, & Shinn, 1996), Moruroa (Camoin et al., 2001), Papua New Guinea (Webster, Wallace et al., 2004), Hawaii (Webster, Clague et al., 2004), and the Marquesas Islands (Cabioch et al., 2008). However, uncertainties concerning the general pattern of the last deglacial sea-level rise remain because the apparent sea-level record may not be free of tectonic or isostatic complications. Three of the four major coral reef records of the last deglacial sea-level rise (Barbados, Papua New Guinea, and Vanuatu) are located in active subduction zones where tectonic movements can be large and discontinuous. The reconstructed sea levels may be therefore biased by variations in the rate of tectonic uplift and/or abrupt coseismic vertical motions. Also, Barbados is under the influence of GIA because of the waxing and waning of the North American ice sheet (Lambeck et al., 2002; Milne et al., 2009). Hence, there is a clear need to study past sea-level changes in tectonically stable regions or in areas where vertical crustal deformation is slow and/or regular, located far away from former ice-covered regions (far-field). Furthermore, the abrupt and significant environmental changes that accompanied the deglacial sea-level rise have been barely investigated in these studies, so the accurate reconstruction of the event was obscured.

The only coral reef record that encompasses the whole last deglacial period was that of Barbados which suggested that the global sea-level rise, resulting from melting glaciers following the Last Glacial Maximum (LGM), did not occur uniformly, but was characterized by several centuries of extremely rapid sea-level rise of about 20 m (40 mm/year on average; Fairbanks, 1989; Bard et al., 1990; Fairbanks et al., 2005; Peltier & Fairbanks, 2006), at a time when there was 70% more grounded ice on Earth. These short-term events, thought to be related to massive and rapid discharges of freshwater from continental ice sheets and referred to as meltwater pulses (MWP-1A and MWP-1B, 14.08–13.61 kyr BP and 11.4–11.1 kyr BP, respectively; Figure 3.6.1), probably disturbed oceanic thermohaline circulation and global climate during the last deglacial period (Manabe & Stouffer, 1995; Weaver, Saenko, Clark, & Mitrovica, 2003). Their understanding is of utmost importance when considering the potential collapse of large ice sheets in response to recent climate change. However, the exact timing, origin (Northern Hemisphere ice sheets—NHIS vs Antarctic ice sheet—AIS; see discussion in Deschamps et al. (2012) and Gregoire, Payne, and Valdes (2012)), and consequences of these ice-sheet melting episodes were unclear and have been the subject of a considerable debate (Weaver et al., 2003; Stanford et al., 2006; see also; Deschamps et al., 2012). GIA models disagree on the hemispherical origin of the MWP-1A (Clark et al., 2002; Tarasov, Dyke, Neal, & Peltier, 2012). The duration and amplitude of the maximum lowstand during the LGM (Fleming et al., 1998; Yokoyama et al., 2001a,b; Yokoyama, Esat, Lambeck, & Fifield, 2000; Cutler et al., 2003; Peltier & Fairbanks, 2006), and the timing and the nature of the events following the LGM (Clark, McCabe, Mix, & Weaver, 2004; De Deckker & Yokoyama, 2009; Hanebuth, Stattegger, & Bojanowski, 2009; Yokoyama et al., 2001a,b) are additional controversial topics.

3.6.3 THE LAST DEGLACIAL SEA-LEVEL RISE IN THE SOUTH PACIFIC

The IODP proposal 519 (Camoin, Bard, Hamelin, & Davies, 2002) aimed to reconstruct the last deglacial events by: (1) establishing the course of the last deglacial sea-level rise, (2) analyzing the reef response to sea-level and coeval environmental changes, and (3) defining the climatic variability during that period. Regarding the course of the last deglacial sea-level rise, the objectives were specifically to: (1) establish the minimum sea level during the LGM, (2) assess the validity, the timing, and amplitude of meltwater pulses and thereby identify the exact sources of the ice responsible for these extremely rapid sea-level rises, and (3) test predictions based on different ice and rheological models.

The two drilling areas, Tahiti and the Australian Great Barrier Reef, were selected based on their tectonic setting and their location at a considerable distance from the main former ice sheets ("far-field" site), to minimize the potential tectonic and hydrostatic parameters in the reconstruction of last deglacial sea-level changes. The effects of hydroisostatic processes reflect the geodynamic context: for small islands, the addition of meltwater produces a small differential response between the island and the seafloor, whereas the meltwater load produces significant differential vertical movement between larger islands or continental margins and the seafloor (Lambeck, 1993). There was, therefore, a need to establish the relative magnitudes of hydroisostatic effects at two ideal sites located at a considerable distance from the major former ice sheets, one on an oceanic island (i.e., Tahiti) and another on a continental margin (i.e., the Australian Great Barrier Reef). Furthermore, these sites are located far away from glaciated regions ("far-field") and can therefore provide basic information regarding the melting history of continental ice sheets and the rheological structure of Earth.

The expeditions linked to IODP Proposal 519, Expedition 310 "Tahiti Sea Level" (2005; Camoin, Iryu, McInroy, & The Expedition 310 Scientists, 2007a,b) and the IODP Expedition 325 "Great Barrier Reef Environmental Changes" (2010; Webster, Yokoyama, Cotterill, & The Expedition 325 Scientists, 2011; Yokoyama et al., 2011) aimed to provide the most comprehensive deglaciation curves from tectonically stable regions by conducting offshore drilling of fossil coral reefs now preserved at 40–130 m below present sea level. They were the two first expeditions of the successive drilling programs (Deep Sea Drilling Program—DSDP; Ocean Drilling Program—ODP; and Integrated Ocean Drilling Program—IODP) to drill deglacial reefs.

3.6.4 EXPEDITION 310 "TAHITI SEA LEVEL"

3.6.4.1 Introduction

Tahiti is located in French Polynesia at 17°50'S and 149°20'W, in the central tropical South Pacific and belongs to the Society Archipelago which corresponds to a volcanic linear chain exhibiting an NW–SE general direction and which formed during the last 5 my as the result of the WNW-ward motion of the Pacific Plate over the Society Hotspot. Tahiti is a high volcanic intraplate island (2241 m maximum altitude) that was formed within the last 1.5 my and now lies near the southeastern end of the chain (Hildenbrand, Gillot, & Le Roy, 2004; Le Roy, 1994). Volcanic activity on this island ceased approximately 200 kyrs ago and the currently active volcanic center lies approximately 50 km southeast of

Tahiti on the flanks of Mehetia Island and surrounding seamounts. The volcanic complexes of Tahiti are composed primarily of porphyritic basaltic lavas with alkaline affinities. Toward the end of the main volcanic phase on Tahiti (i.e., at 650–850 ka), the northern and southern flanks of the volcanic edifice collapsed catastrophically and spewed thousands of cubic meters of sediment into the surrounding sea (Hildenbrandt et al., 2004). Today, the island is heavily eroded and dissected by deeply incised radial valleys resulting from its exposure to persistent rainfall punctuated by strong storms.

The subsidence rates of the island were previously estimated to range from 0.15 to 0.4 mm/year by Le Roy (1994) and Montaggioni (1988), respectively. An average subsidence rate of 0.25 mm/year has been deduced from the ages of the subaerial lavas underlying the Pleistocene reef sequence in the Papeete drill cores (Bard et al. 1996). Data obtained on pre-LGM Tahiti corals are consistent with this estimate of 0.25 mm/year. Two corals collected at 147 meters below sea level (mbsl) with U-Th ages of 153 kyr BP (MIS6) indicate an upper limit of 0.4 mm/year for the subsidence of the island, assuming a MIS6 sea-level lowstand of <90 m (Thomas et al., 2009). Weighted average subsidence rates of 0.15±0.15 mm/year were recently deduced from direct measurements by GPS and other geodetic instruments (Fadil, Sichoix, Barriot, Ortèga, & Willis, 2011).

Tahiti lies in the well-ventilated South Pacific gyre. The climate is typically tropical, with two distinct seasons: a warm rainy season from November to April (austral summer), bringing occasional cyclones, with maximum SSTs ranging from 28 to 29 °C, and a cooler and drier season from May to October (austral winter), with lower SSTs averaging 24–25 °C (Delesalle, Galzin, & Salvat, 1985). Rainfall on Tahiti is variable by location; the north side of the island is the driest and receives an average of 1500 mm/year, and the southern and eastern sides may receive up to 3000 mm/year (Crossland, 1928a,b; Williams, 1933). There are marked variations in rain intensity throughout the year, with minimum monthly values in winter less than 50 mm, and maximum monthly values in January and February up to 400 mm. Dominant winds blow from the northeast or the southeast and create strong swells of more than 2 m in amplitude on the eastern side of the island. Tides are semidiurnal and their amplitude averages 0.5 m; reef flats emerge at spring tides and waves commonly break on the central areas of reef flats at high tide.

Tahiti is surrounded by discontinuous fringing reefs that grade locally into a discontinuous chain of barrier reefs. Along the south and west coasts, the reef flat is wide and separated from the fringing reef by only a very shallow lagoon, and on the east coast the reef flat is very narrow and separated by a wide lagoon reaching depths of 35 m (Williams, 1933). The outer reef slope is made up of spurs and groves sloping seaward at 20°. The overall morphology of the reef foreslopes, as established from submersible observations (Salvat, Sibuet, & Laubier, 1985), bathymetric data, and seismic profiling (Camoin et al., 2006, 2012), shows consistent features around Tahiti, although their relative prominence differs from site to site, suggesting a complex history of reef growth and drowning. Two major terraces have been mapped and imaged, at 50–60 and 90–100 mbsl, respectively and, locally, a third narrower terrace at 75–80 mbsl. All terraces are characterized by the occurrence of submerged reefs which developed during the last deglacial sea-level rise (Camoin et al., 2006, 2012).

The extensive terrace at 50–60 mbsl is inclined gently seaward down to 90 m. On the eastern side of the island, it is bounded upslope by a series of pinnacles up to 150 m in diameter that rise from the seafloor to about 20 mbsl. The reef sequence deposited on this terrace forms a sedimentary wedge that pinches out at a depth of 90 m in the Tiarei area (Camoin et al., 2006, 2012). The prominent terrace at 90–100 mbsl bears abundant build-ups that are interpreted as relict reefs based on the study of dredged samples. In the Tiarei area, these range in height from 30 m (base at 100 m, top at 70 mbsl) up to 45 m (base at 90 m, top at 45 mbsl). There is a clear topographic break at 90–100 m deep where the slope steepens sharply.

The Tiarei foreslopes and, to a less extent the Faaa foreslopes, display a distinctive morphology characterized by the occurrence of two prominent ridges seaward of the living barrier reef, which were previously recognized as reef features exhibiting an original irregular morphology consisting of isolated or fused pinnacles (Camoin et al., 2006; Seard, Borgomano, Granjeon, & Camoin, 2013). The last deglacial reef sequence forms the top of the ridges and therefore displays a discontinuous distribution. The outer ridge coincides with a marked break in slope and its top is located at 90–100 mbsl, whereas the inner ridge is located on an extensive terrace and its top occurs at ~60 mbsl. Similar relict reef features have been reported on many modern reef slopes around the world (see reviews in Dullo et al. 1998; Camoin et al., 2006; Beaman, Webster, & Wust, 2008; Abbey, Webster, & Beaman, 2011a). Such reef features have been drilled previously only in Barbados (Fairbanks, 1989) where they have been interpreted as reflecting reef-drowning events during meltwater pulses (Blanchon & Shaw, 1995). However, unlike Tahiti and the Great Barrier Reef (GBR), the lack of available high-resolution multibeam data in Barbados, does not allow the reconstruction of the morphology and the lateral extent of these features. In contrast to the Tiarei and Faaa areas, the Maraa area (south of Tahiti) is characterized by relatively smooth, regular, and more gentle slopes between 45 and 97 m water depth (Camoin et al., 2006). Accordingly, the last deglacial reef sequence forms a continuous sedimentary package from 42 to 122 mbsl in this area.

3.6.4.2 Operational Results

During the IODP Expedition 310 "Tahiti Sea Level" (Camoin et al., 2007a,b), the objective was to recover the entire last deglacial reef sequence by drilling several vertical boreholes along transects on the successive reef terraces that occur seaward of the living barrier reef. Based on the results of previous scientific drilling and bathymetric and seismic data, transects of holes have been drilled in three areas around Tahiti: offshore Faaa, Tiarei, and Maraa (Figure 3.6.2). Water depths

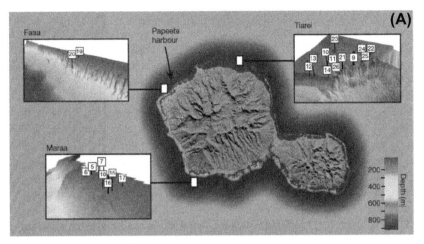

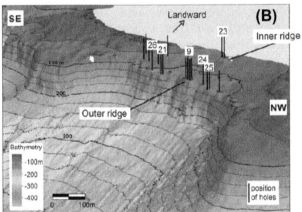

FIGURE 3.6.2 (A). Landsat image of the Tahiti island *(From Deschamps et al. (2012).)*. The three areas (Tiarei, Maraa, and Faaa) drilled during IODP Expedition 310 as well as Papeete harbor where onshore holes were drilled previously are shown. Insets show the bathymetry for each site, with the location of the different drilled holes. (B). 3D bathymetric reconstruction of the Tiarei area and position of holes located on the inner (Site M0023) and outer (Sites M0009, M0021, M0024, and M0025) ridges *(From Seard et al. (2011).)*.

at these locations ranged from 41.65 to 117.54 mbsl. The exact location of the drill holes was determined during the cruise by checking the nature and morphology of the seafloor with a through-pipe underwater camera.

Drill cores were recovered from 37 holes at 22 sites (M0005–M0026) from the vessel *DV/DP Hunter*. Water depths at the sites ranged from 41.6 to 117.5 m, and cores were recovered from 41.6 to 161.8 mbsl. During drilling, the core barrel was advanced in 1.5 m increments, and core depths were measured with ±0.1 m accuracy. The initial strategy of coring along profiles has been applied, although the locations of some proposed drill sites were slightly changed because of difficulties locating and operating at those sites. In the Tia-rei area, drill cores were obtained both from the inner (Site M0023) and outer (Sites M009, M0021, M0024 and M0025) ridges (Figure 3.6.2). Two transects were drilled in the Maraa area: an eastern transect including the Sites M0017, M0015, M0018, and M0016 at increasing water depths, and a western tran-sect hosting Sites M0007 and M0005 (Figure 3.6.2). Two drill cores have been obtained in the Faaa area (Sites M0019 and M0020; Figure 3.6.2). More than 600 m of reef cores displaying an exceptional recovery (>90%, Inwood, Brewer, Braaksma, & Pezard, 2008) and quality were retrieved and, combined with the high-resolution downhole measurement data, correspond therefore to unique archives to resolve in unprecedented detail the reef response to last deglacial sea-level and coeval environmental changes. During the operations, the oppor-tunity was taken to drill deeper beneath the postglacial reef sequence into older reef units, principally to ensure that the entire postglacial sequence was recov-ered but also to provide exciting new data on past sea levels and reef develop-ment around the time of the penultimate deglaciation (i.e., Termination II) (e.g., Fujita, Omori, Yokoyama, Sakai, & Iryu, 2010; Iryu et al., 2010; Ménabréaz, Thouveny, Camoin, & Lund, 2010; Thomas et al., 2009).

The set of borehole geophysical instruments deployed was constrained by the scientific objectives and the geological setting of the expedition. A suite of downhole geophysical methods was chosen to obtain high-resolution images of the borehole wall (OBI40 and ABI40 televiewer tools), to characterize the fluid nature in the borehole (IDRONAUT tool), to measure borehole size (CAL tool), and to measure or derive petrophysical and geochemical properties of the reef units such as porosity, electrical resistivity (DIL 45 tool), acoustic veloci-ties (SONIC tool), and natural gamma radioactivity (ASGR tool). A total of 10 boreholes were prepared for downhole geophysical measurements which were performed under open borehole conditions (no casing) with the exception of a few of spectral gamma-ray logs. Nearly complete downhole coverage of the postglacial reef sequence has been obtained from 72 to 122 mbsl and from 41.65 to 102 mbsl at the Tiarei sites and at the Maraa sites, respectively. Partial down-hole coverage of the underlying older Pleistocene carbonate sequence has been acquired at those sites.

The Tahiti drill cores were also analyzed for evidence of modern microbial activity in the subsurface of the fore-reef slopes. Onboard adenosine triphos-phate activity measurements have shown a certain degree of microbial activ-ity directly attached to rock surfaces; cultivation and microscopic observations were also carried out onboard.

3.6.4.3 Scientific Results

A first reconstruction of sea-level rise and reef development encompassing the last 13.92 kyrs was based on the study of an expanded (85.5–92.5 m thick), con-tinuous, reef sequence recovered in a series of vertical (P6 and P7) and inclined

(P8, P9, and P10; from 30° to 33° by reference to the vertical) drill holes carried out in 1995 through the barrier reef tract off Papeete (Bard et al., 1996, 2010; Cabioch, Camoin, & Montaggioni, 1999; Camoin, Gautret, Montaggioni, & Cabioch, 1999; Montaggioni et al., 1997). The record was continuous from 13.9 kyr BP to present, but did not reach the critical MWP-1A period. The first evidence of reef growth during a period encompassing the MWP-1A came from material dredged on the modern foreslopes (15 kyr BP in situ coral, Camoin et al., 2006).

A specific target of the IODP Expedition 310 was therefore the extension of the previous Tahiti sea-level record to cover earlier portions of the last deglacial sea-level rise.

The IODP Expedition 310 "Tahiti Sea Level" (Camoin et al., 2007a,b) has provided a very accurate and continuous reef record of the last and penultimate deglaciations and brought a wealth of new information in various scientific fields.

3.6.4.3.1 Composition of the Last Deglacial Reef Sequence

The drilled coral reef systems around Tahiti are composed of two major chronological and lithological sequences which are attributed to the last deglaciation and to older Pleistocene time windows (Camoin et al., 2007a,b). The contact between those sequences is characterized by the occurrence of an irregular unconformity caused by the diagenetic alteration and karstification of the older Pleistocene sequence during sea-level lowstand(s). It ranges in depth from ~122 mbsl (Tiarei outer ridge, deep Maraa sites, and Faaa) to 94 mbsl on the Tiarei inner ridge, and 85 mbsl at shallow Maraa sites. The chronological and sedimentological data of the older Pleistocene reef sequences have been detailed in Thomas et al. (2009), Iryu et al. (2010), Fujita et al. (2010), and Ménabréaz et al. (2010).

At all drill sites, the last deglacial reef sequence is mostly composed of reef frameworks comprising three major components—corals, algae, and microbialites—whose proportions vary largely throughout the last deglacial reef sequence (Figures 3.6.3 and 3.6.4). Two major biological communities have been described (Camoin et al., 2007a,b; Seard et al., 2011; Camoin et al., 2012):

1. The coralgal communities include seven distinctive assemblages characterized by various growth forms that form the initial frameworks. The dominant coral morphologies (branching, robust branching, massive, tabular, foliaceous, and encrusting) and the abundance of associated builders and encrusters determine distinctive frameworks displaying a wide range of internal structures, from loose to dense frameworks (Figure 3.6.3). The coral assemblages are described later in this section.
2. Microbialites represent a major structural and volumetric component of the recovered frameworks in which they may locally form up to 80% of the rocks. The microbial communities developed in primary cavities of the coralgal

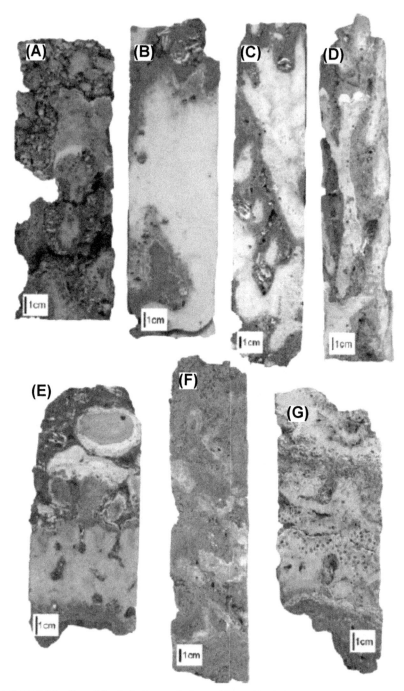

FIGURE 3.6.3 Core slabs displaying coralgal assemblages. (A). Robust-branching *Pocillopora*—encrusting/massive *Montipora* assemblage (PM) showing a highly bioeroded *Pocillopora* in growth position encrusted at its top by an encrusting *Montipora*. Sample 24A15R 1W 19-35, depth ~118.6 m.

frameworks, where they heavily encrusted the coralgal assemblages to form microbialite crusts ranging in thickness from a few centimeters to 20 cm. The microbialite crusts display a wide range of growth forms: laminated crusts, digitate microbialites, and structureless to massive micritic masses (Seard et al., 2011) (Figure 3.6.4). The most widespread microbialite development has been reported in coral frameworks dominated by branching, thin encrusting, tabular and robust branching corals (i.e., PPM, PP, tA, and rbA coral assemblages; see below) which built open frameworks typified by high (i.e., more than 50%) initial porosity values (Seard et al., 2011). The lipid biomarkers and their isotopic patterns (MAGEs and branched fatty acids: 10-Me-C16:0 and iso- and anteiso-C15:0 and -C17:0; Heindel et al., 2012; Heindel, Birgel, Peckmann, Kuhnert, & Westphal, 2010.) indicate that the formation of microbialites was related to the activity of bacterial communities dominated by sulfate-reducing bacteria, thus confirming previous interpretations (Camoin et al., 1999). An accurate chronology has been obtained through the C14-AMS dating of numerous triplets of contiguous corals, coralline algal crusts, and microbialites. It was demonstrated that the microbialite crusts developed a few hundred years (i.e., approximately 100–500 years) after the coralgal communities in cryptic cavities, 1.5–6 m below the living reef surface, as a "filling front" which shortly followed the overall accretion of the coralgal frameworks during the last deglacial sea-level rise (Seard et al., 2011). This implies that there was no direct competition between living corals and microbialites.

In primary cavities of the reef frameworks, microbialites are locally associated or interlayered with skeletal limestone; loose skeletal sediments (rubble, sand, and silt) rich in fragments of corals, coralline, and green algae (*Halimeda*); and, to a less extent, echinoids, molluscs, and foraminifers (mostly *Amphistegina* and *Heterostegina*). The amounts of volcaniclastic sediments (e.g., silt- to cobble-sized lithic volcanic clasts, crystal fragments, clays) are highly variable, from mere sand and silt impurities in the carbonate rock units to minor components (<50 vol%) in carbonate sand units to major components (50 vol%) in sand/silt (or sandstone/siltstone) interbedded with carbonate beds. The last deglacial sequence at Tiarei has a greater volcaniclastic component than the ones at Maraa and Faaa.

(B). Massive *Porites* (mP) assemblage with a mP in growth position encrusted at its top by thin coralline algae and then by microbialites. Sample 24A10R 2W 1-22, depth ~109.4 m. (C). Branching *Porites* and robust-branching *Pocillopora* assemblage (PP) with a robust-branching *Pocillopora* colony in growth position encrusted by microbialites. Sample 23A11R 1W 8-37, depth ~85.4 m. (D). Branching *Porites* and robust-branching *Pocillopora* assemblage (PP) with branching *Porites* in growth position encrusted by thin coralline algae and then by microbialites. Sample 24A9R 2W 83-111, depth ~106.8 m. (E). Tabular *Acropora* (tA) assemblage with a tA colony in growth position encrusted by thick coralline algae. Sample 9B13R 1W 4-18, depth ~117 m. (F). Branching *Porites* and encrusting *Porites* and *Montipora* assemblage with encrusting and branching *Porites* colonies in growth position encrusted by microbialites. Sample 24A5R 2W 26-52, depth ~98.5 m. (G). Encrusting agaricid and faviid assemblage with highly bioeroded encrusting *Leptastrea* colonies in growth position. Sample 21B1R 1W 66-83, depth ~83 m. *From Seard et al. (2013).*

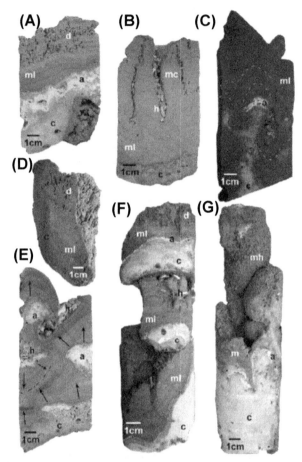

FIGURE 3.6.4 Core slabs displaying the various morphologies of microbialite crusts: m: micro-bialites; ml: laminated microbialites; d: digitate microbialites; mc: columnar microbialites; mh: hemispheroid microbialites; c: coral; a: coralline algae. These pictures show the typical biological successions recorded in deglacial reefs from Tahiti. (A). In situ coral (tabular *Acropora*) encrusted successively by coralline algae and by a microbial compound crust in which digitate microbialites form the last stage of encrustation over laminated crusts. Hole 15B, core 19R, 89.62 m (15B 19R1 57-67). (B). Encrusting coral encrusted by laminated microbialite forming columns with sedimentary infillings composed of Halimeda plates between them. Hole 5C, core 8R, 73.40 m (5C 8R1 60-72). (C). Branching coral (*Pocillopora*) in growth position encrusted successively by coralline algae and by laminated microbialites whose laminations mimic the shape of the coral branches. Hole 23B, core 8R, 83.09 m (23B 8R2 63-77). (D). Branching coral encrusted successively by laminated and digitate microbialites. Hole 18A, core 6R, 89.97 m (18A 6R1 43-54). (E). Tabular coral colonies (*Acropora*) and thick encrusting coralline algae encrusted by laminated microbialites growing in all space directions with a prevalent upward growth typified by the occurrence of convex-upward layered structures. The arrows show the growth direction of laminated microbialites. Hole 7A, core 30R, 78.61 m (7A 30R1 80-102). (F). Microbial compound crusts filling cavities between submassive coral colonies (*Porites*). Hole 24A, core 11R, 111.65 m (24A 11R1 73-91). (G). Hemispheroid microbialites developing around columnar coral colonies (*Porites*). Hole 24A, core 10R, 108.56 m (24A 10R1 49-72). *From Seard et al. (2011).*

At all sites, the top of the last deglacial carbonate sequence is characterized by the widespread development of thin coralline algal crusts indicating deeper-water environments. Extensive bioerosion, black yellow reddish to brown staining (manganese and iron) of the rock surface, and hardgrounds are common within the top 2–3 m of the sequence.

Twenty-six species from 12 genera in 7 scleractinian families were identified from the Tiarei and Maraa reef sequences, and 28 species of coralline algae have been identified in all the studied cores (see Abbey, Webster, Braga et al. (2011)).

Seven distinctive coral assemblages have been defined based on the taxonomy and the morphology of dominant and secondary coral colonies (branching, robust branching, massive, tabular, foliaceous, and encrusting), and associated biota (e.g., coralline algae, vermetid gastropods—*Serpulorbis annulatus* and *Dendropoma maxima*; encrusting foraminifers—*Homotrema rubrum, Carpenteria* cf. *monticularis*, and *Acervulina inhaerens*) (Camoin et al., 2007a,b; Camoin et al., 2012; Seard et al., 2011; Abbey, Webster, Braga et al., 2011). Those assemblages are indicative of a range of modern reef environments, from the reef crest to the reef slope and form a continuum in which most of them are intergradational both vertically and laterally, implying an overlap of their depth ranges (Camoin et al., 2012). Based on our understanding of the modern analogs of the coralgal assemblages, community transitions within a core may represent the response of reef growth to changing paleoenvironmental conditions, but can be also produced by ecological succession and vertical reef accretion, or lateral growth during sea-level still stands (Abbey, Webster, Braga et al., 2011). The Tahiti record allows for an unprecedented investigation of both the stratigraphic and small-scale (meters) to large-scale (island-wide) spatial variations in coralgal assemblages during the last deglacial sea-level rise.

Paleoenvironmental interpretations, especially the paleowater depth estimates, of the various coralgal assemblages are based on comparisons with analogous modern and fossil Indo-Pacific reef communities (Abbey, Webster, Braga et al., 2011). The depth distribution of coralgal assemblages displays some variability throughout the drilled areas. This is mostly due to the influence of local environmental conditions, such as the occurrence of significant siliciclastic inputs in the Tiarei area while siliciclastic deposits are not reported in the Maraa area.

The shallowest coralgal assemblages, at depths less than 10 m, include the robust branching *Acropora* (rbA) and the robust branching *Pocillopora*/massive *Montipora* (PM) assemblages that are characterized by a coralline algal association dominated by *Hydrolithon onkodes* and *Mastophora pacifica*. Shallower depths are considered when vermetid gastropods are associated (see Cabioch, Camoin et al. (1999), Cabioch, Montaggioni, Faure, and Laurenti (1999); Camoin et al. (1999); Montaggioni et al. (1997)). These assemblages characterize the modern reef crest and uppermost reef slope exposed to strong wave action in Tahiti and Moorea. The rbA assemblage (*Acropora* of the *robusta-danai* group) which dominates frameworks younger than 12 kyr BP both in Papeete and Maraa areas was not recovered in the Tiarei cores, probably in relation with local environmental conditions, especially high turbidity, which prevailed prior to that time.

The other end member of the coralgal assemblage continuum is comprised of encrusting agaricids and faviids (AFM) and coralline algae dominated by *Mesophyllum funafutiense* and *Lithoporella*, indicating depths greater than 20 m.

The massive *Porites* (mP) and the tabular *Acropora* (tA) assemblages generally characterize the 5–15 m depth range which can be restricted to the 5–10 m depth range when thick *H. onkodes* crusts are associated.

The branching *Porites/Pocillopora* (PP) assemblage developed at depths ranging from 5 to 15 m, but more frequently between 5 and 10 m, as indicated by the occurrence of thick crusts of *H. onkodes*.

The branching *Porites*/encrusting *Porites* and *Montipora* (PPM) assemblage probably developed in a large depth range (5–25 m), as indicated by the associated coralline algae characterizing either shallow-water, i.e., less than 10 m (*H. onkodes* and *M. pacifica*), or deeper-water environments, i.e., between 15 and 25 m (*Lithophyllum prototypum, Mesophyllum erubescens, Lithothamnion prolifer* assemblage).

3.6.4.3.2 Chronological Frame

The outstanding preservation of the coral samples recovered in the Tahiti IODP cores and the lack of any *postmortem* diagenetic alteration of their aragonite skeleton are typified by their very low calcite content and by initial $(^{234}U/^{238}U)_0$ values averaging 1.1458 ± 0.0020 (2σ), which fall within the most recent determinations of modern seawater and corals (Cutler et al., 2004; Delanghe, Bard, & Hamelin, 2002; Deschamps et al., 2012; Robinson, Belshaw, & Henderson, 2004). Only samples that contain <1% calcite were considered for U-Th and ^{14}C dating as diagenetic calcite may alter original ages. The precision of $^{230}Th/U$ ages generally falls within the 1.5–4‰ range, i.e., ±20–50 years for postglacial samples.

A total of 77 coral samples recovered from 23 cores drilled at 14 different sites were cross-dated by U-Th and C14-AMS. They provided reliable and stratigraphically consistent calendar ages ranging from 10 to 36 kyr BP, including 70 dates for the deglaciation period and 7 for the glacial period with ages from 29 to 37 kyr BP (Deschamps et al., 2012; Durand et al., 2013) (Figure 3.6.5).

In the Tiarei area, the last deglacial reef sequence encompasses the 16.1–10 kyr BP time window (Camoin et al., 2012; Deschamps et al., 2012). Its thickness averages 30 m on the inner ridge (Site M0023) and 40 m on the outer ridge (Sites M0009, M0021, M0024, and M0025; Figure 3.6.5). In the Maraa area, it encompasses the 14.6–9 kyr BP time window (Deschamps et al., 2012) and its thickness ranges from 25 to 39 m in the eastern transect (Sites M0015 through M0018) and from 27.5 to 38.5 m in the western transect (Sites M0005 and M0007) (Figure 3.6.5). In the Faaa area, it encompasses the 14.7–10 kyr BP time window (Deschamps et al., 2012) and its thickness is of 38 m at Site M0020.

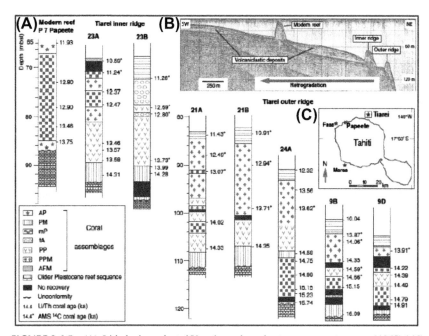

FIGURE 3.6.5 (A). Lithologies, selected U-series and accelerator mass spectrometry (AMS) 14C ages, and distribution of coralgal assemblages at Papeete and Tiarei drill sites (mbsl, meters below sea level). (B). Seismic line across Tiarei area displaying two ridges drilled during Integrated Ocean Drilling Program (IODP) Expedition 310 and modern reef. (C). Offshore (Tiarei, Maraa, and Faaa) and onshore (Papeete P cores) drill sites at Tahiti. Coral assemblages: AP—branching *Acropora* and *Pocillopora*; PM—branching *Pocillopora* and massive *Montipora*; mP—massive *Porites*; tA—tabular *Acropora*; PP—branching *Porites* and *Pocillopora*; PPM—branching *Porites* and encrusting *Porites* and *Montipora*; AFM—encrusting agaricids and faviid. *From Camoin et al. (2012).*

The new data obtained from the IODP 310 expedition extend the previous Tahiti record of the last deglaciation and complement the results obtained on the Papeete drill cores that encompass the 13.9–3 kyr BP time window (Bard et al., 1996; Bard, Hamelin, & Delanghe-Sabatier, 2010; Cabioch, Camoin et al., 1999; Camoin et al., 1999; Montaggioni et al., 1997). They provide an accurate coral reef record of reef development and sea-level change that cover a 10–16 kyr BP time window (Camoin et al., 2012; Deschamps et al., 2012; Seard et al., 2013) and encompass the MWP-1A and MWP-1B events reported at 14.08–13.61 and 11.4–11.1 kyr BP, respectively in the Barbados record (Peltier & Fairbanks, 2006).

3.6.4.3.3 Reef Growth and Sea-Level Change
The reconstruction of the last deglacial reef development and sea-level rise relies mostly on the combination of accurate radiometric U-series and C14-AMS dating results of pristine in situ coral samples and the interpretation of coralgal assemblages that can be considered as reliable depth indicators. In situ

corals were distinguished from drilling disturbance or allochthonous rubble using a suite of criteria established by previous drilling studies (Camoin et al., 1997, 2001; Camoin, Montaggioni, & Braithwaite, 2004; Lighty, Macintyre, & Stuckenrath, 1982; Montaggioni et al., 1997; Webster & Davies, 2003). The reliability of these criteria can vary with growth form, but include (1) orientation of well-preserved corallites, (2) orientation of acroporid, pocilloporid, and poritid branches, (3) coral colonies capped by thick (few centimeters) coralline algal crusts, and (4) presence of macroscopic and microscopic sediment geopetals in cavities and mollusk chambers and valves.

The architecture and geometry of reef systems indicate that their development and growth modes have been controlled by the progressive flooding of the Tahiti slopes and the coeval increase in accommodation space. However, slight differences in reef development patterns are related to local environmental conditions. In the Tiarei area, the formation of the successive ridges is seemingly related to local topographic and substrate conditions provided by the older Pleistocene carbonate sequence in a region dominated by volcaniclastic sediments. No similar ridges were reported in other drilled areas around Tahiti, especially in Maraa area where the last deglacial reef sequence forms a continuous sedimentary body beneath the modern reef slope.

At all drill sites, chronological and sedimentological data do not support any unconformity in the cored reef sequences, thus implying that reefs accreted continuously and mostly through aggradational processes between 16 and 10 kyr BP, and that there was no major break in reef development during this time window (Figure 3.6.6). This suggests that environmental conditions in Tahiti were optimal for reef development and that no significant long-term environmental changes occurred during that period. This precludes any catastrophic impact on reef development such as the temporary cessation of reef growth during the meltwater pulses. At each individual drill site, the last deglacial reef sequence displays a general deepening upward trend, indicating that reef growth gradually lagged behind sea-level rise.

Changes in the composition of coralgal assemblages coincide with abrupt variations in reef growth rates and characterize the response of the upward-growing reef pile to a nonmonotonous sea-level rise and coeval environmental changes, including especially water depth and energy, light conditions, terrigenous fluxes, and nutrient concentrations (Figure 3.6.6).

The overall reef aggradation rate in the Tiarei area averages 7.5 mm/year on the outer ridge and 6.5 mm/year on the inner ridge, for the 16–10 and 14.3–10.6 kyr BP time windows, respectively. In the Maraa area, the aggradation rates average 6–7 mm/year. However, the major part of the Tahiti reef accretion curve expresses maximum growth rates exceeding 10 mm/year.

The oldest period of the last deglacial reef record at Tahiti corresponds to the base of the reef sequence cored on the outer ridge of Tiarei where robust branching Pocillopora collected at the interface of the underlying Pleistocene unit were dated at 15.74 ± 0.03 and 16.09 ± 0.042 kyr BP at 118.6 and 121 mbsl,

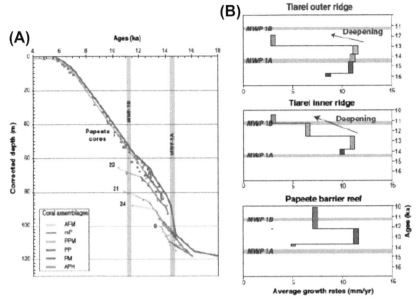

FIGURE 3.6.6 (A). Reef accretion curves based on data obtained from Tiarei offshore drill sites M0009, M0021, M0023, and M0024, and Papeete onshore drill sites. MWP—meltwater pulses. (B). Evolution of average growth rates of successive coral assemblages during 16–10 kyr BP time window for Tiarei and Papeete drill sites. Timing of MWP-1A and MWP-1B based on data from Deschamps et al. (2012) and Peltier and Fairbanks (2006), respectively. Abbreviations of coral assemblages as in Figures 3.6.3 and 3.6.5.

respectively (Figures 3.6.5–3.6.7), thus indicating an RSL of 105–115 m during that time window (Deschamps et al., 2012).

The pre-MWP-1A period (16–14.65 kyr BP) is characterized by a moderate rise in sea level with a magnitude of ~10 m, implying an average rate of 7.4 mm/ year. The PM assemblage represents the first assemblage to colonize the Pleistocene carbonate substrate regardless of the timing of the substrate flooding during the 16.09 to c.15.5 kyr BP time window. The subsequent development of the mP assemblage, composed of turbidity-tolerant corals, occurred during an ~500-year period (i.e., from 15.23 ± 0.03 kyr BP to 14.75 ± 0.03 kyr BP) at the deepest sites of the Tiarei outer margin typifies a shift to quieter and probably more turbid environmental conditions, within the same depth range than that of the PM assemblage. A moderate rise in sea level of approximately 10 m is therefore evidenced during the 16.0–14.65 kyr BP time span. Reef growth rates range from 6 to 9 mm/year during the pre-MWP-1A period, and imply that reefs kept pace with the rising sea level.

The occurrence of an accelerated rise in sea level between 14.65 and 14.3 ky BP, corresponding to the MWP-1A, is typified by a major discontinuity in the upper envelope of the data points in the Tahiti RSL record (Figure 3.6.7). This implies that the MWP-1A occurred at least 500 years earlier than previously proposed

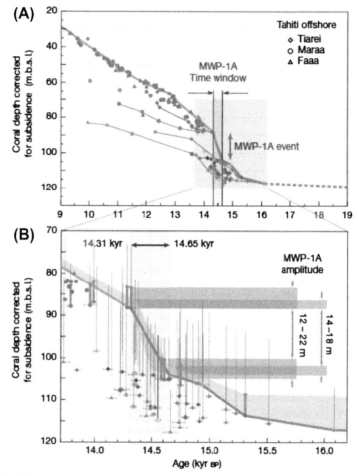

FIGURE 3.6.7 The deglacial Tahiti sea-level curve. (A). Sea level reconstructed from U-Th dated corals recovered in long holes drilled onshore and offshore Tahiti island. Coral depths are expressed in meters below present sea level (mbsl) and are corrected for a constant subsidence rate of 0.25 mm/ year. Gray and colored (gray in print versions) symbols show coral samples collected in onshore holes and in offshore holes (IODP Expedition 310), respectively. Thick blue (pale black in print versions) line shows the lower estimate of the Tahiti RSL curve; the occurrence of a rapid sea-level rise related to the MWP-1A event is indicated by the orange (dark gray in print versions) arrow. The shaded time window and black arrows highlight the tight chronological constraints derived for MWP-1A from the Tahiti record. (B). Magnified view of the MWP-1A time window. The vertical gray bars reported for each coral sample correspond to their optimal bathymetric habitat range inferred from the coralgal assemblage identification and thick orange bars indicate samples associated with vermetid gastropods that are indicative of a shallow environment (0–5 mwd.). The shaded gray band illustrates our estimate of the most likely range of the Tahiti RSL over the last deglaciation. The ranges of uncertainty estimated from the bathymetric range of coralgal assemblages for the pre- and post-MWP-1A sea-level positions are illustrated by the horizontal green bands. The resulting extreme bounds for the MWP-1A amplitude (12 and 22 m) are also indicated (green (light gray in print versions) bands and arrows). *From Deschamps et al. (2012).*

based on the Barbados record (14.082±0.056–13.632±0.032 ky BP, Peltier & Fairbanks, 2006). The amplitude of the inferred sea-level jump is of 16±2 m, in less than 350 years. An average relative sea-level rise of 46±6 mm/year can be deduced for the MWP-1A, based on the 16 m amplitude (Deschamps et al., 2012). The newly defined timing of the MWP-1A implies that the MWP-1A coincided with the inception of the Bølling warming period (14.64 ky BP in the updated GICC05 Greenland ice core chronology) and cannot be considered anymore as a trigger of the Older Dryas cooling event that terminated the Bølling period, as proposed previously (see discussion in Deschamps et al. (2012)). In agreement with Clark et al. (2002) predictions, the sea-level reconstructions during the MWP-1A event at Tahiti seem to preclude a sole Laurentide Ice Sheet contribution to MWP-1A and indicate that a significant, if not the major part, Antarctic contribution to the last deglaciation occurred during the MWP-1A spell (Deschamps et al., 2012), as predicted through a GIA approach (Bassett et al., 2005).

At Tiarei, the reef response to an accelerated sea-level rise at the inception of the MWP-1A is characterized by the replacement of the mP assemblage by fast-growing corals (i.e., branching PP assemblage) which formed loose frameworks with a high initial porosity (averaging 50%; Seard et al., 2011) and aggraded at average vertical rates of 10–13 mm/year, and up to 22.6 mm/year during short time windows (Figure 3.6.5). The PP assemblage exhibits a diachronous development from ~14.6 to ~13.9 kyr BP from the outer ridge to the inner ridge, characterizing a backstepping of the shallow-water assemblages at a rate of 220 mm/year during that time window, indicating that the reef growth, at average vertical rates <10 mm/year, was insufficient to balance the sea-level rise which ultimately induced a gradual deepening and an incipient reef drowning (Camoin et al., 2012); however, no cessation of reef growth, even temporary, has been evidenced during this period (Figures 3.6.5 and 3.6.6).

The rise in sea level during the MWP-1A was sufficient to displace the PM assemblage out of its 10 m or less habitat zone and to induce its relocation upslope, on the inner part of the outer ridge (Site M0021), and then on the inner ridge (Site M0023), implying that a retrogradation of this shallow-water assemblage at average rates >700 mm/year occurred during that time window. The inferred gradual vertical deepening can be estimated to a magnitude of 10–20 m in total, implying an average rate of sea-level rise ranging from 14 to 28 mm/year during that period, in agreement with results of reef growth modeling indicating a 16 m rise in sea level between 14.6 and 14 kyr BP (Seard et al., 2013). On the Tiarei outer ridge, this deepening is characterized by the replacement of the PP assemblage by the PPM assemblage that typifies a quieter and deeper-water (within the 15–25 m range) depositional environment from ~14.35 to ~13.9 kyr BP. A similar evolution of coralgal assemblages has been recorded in the Maraa deep sites with an offset of a few centuries compared to the Tiarei sites (14 kyr BP vs 14.6 kyr BP at Tiarei) and the occurrence of a later incipient drowning. This is probably due to distinctive depositional environment conditions, especially the prevalence of turbid waters in Tiarei in relation to the occurrence of the paleo Papenoo River which probably induced a shrinkage of the depth distribution of coralgal communities. Reef colonization by the PM shallow-water coralgal assemblage occurred at 14.6 kyr BP at Maraa shallow sites and was followed by the successive development of PP and PPM assemblages, indicating a progressive deepening of depositional environments between ~14 and ~11 kyr BP in the present 114–86 mbsl depth range. At all sites, the accelerated rise in sea level did not induce any cessation of reef development but a progressive deepening of depositional environmental conditions. The end of this deepening episode coincides with the coral colonization of the Pleistocene carbonate substrate below the Papeete modern reef (Figure 3.6.5).

At the end of MWP-1A, c.14.3 kyr BP, the Tiarei reefs exhibited a clear lateral zonation from the inner ridge characterized by the growth of the PM shallow-water assemblage to the outer ridge characterized by the development of deeper-water coral assemblages (PP and PPM) (Figure 3.6.5). The reef sequence from the Tiarei inner ridge then records a continuous reef growth from 14.3 to 11.2 kyr BP, characterized by a gradual deepening through time typified by a vertical succession involving the PM and the PP assemblages for >1500 years, followed by the development of tA, mP, and PPM assemblages, while the coeval outer ridge sequences are mostly composed of PPM and AFM assemblages. The transition from mP to AFM coral assemblages may characterize a slight deepening (i.e., a few meters) c.11.2 kyr BP, within the timing of MWP-1B (11.4–11.1 kyr BP; Peltier & Fairbanks, 2006). In contrast, the Maraa shallow sites (Sites M0015, M0017, M0005, and M0007) are characterized by the continuous development of a tA assemblage, locally associated with robust branching *Pocillopora* in the present 82–72 mbsl depth range, between 13 and 10 kyr BP. The offshore data are therefore in good agreement with the results obtained on the Papeete cores where a continuous development of shallow-water coral assemblages (i.e., at depths <6 m) at nearly constant reef accretion rates has been recorded during the same time window (Bard et al., 1996, 2010; Cabioch, Camoin et al., 1999; Camoin et al., 1999; Montaggioni et al., 1997). Thus, the results obtained at all Tahiti drill sites do not support the occurrence of an abrupt reef-drowning event related to a sea-level pulse of ~15 m at an apparent rise of 40 mm/year, coinciding with the MWP-1B centered at 11.5 kyr BP, as it was deduced from the Barbados record (Blanchon & Shaw, 1995). MWPs were originally detected as hiatuses between three separate submerged reef features in the Barbados record. Each of these segments is offset from the next and coincides with the interpreted reef-drowning events, thus hampering the accurate reconstruction of the reef response to sea-level change. It is therefore assumed that the MWP-1B-drowning event may correspond to a drilling bias.

At all drill sites, the upper part of the reef sequence is characterized by the diachronous development of slowly growing (mean growth rate: 3 mm/year) deep-water coralgal assemblages (AFM assemblage at depths generally >20 m) typifying the progressive backstepping of shallow-water assemblages, ranging in age from 12.32 ± 0.03 kyr BP to 10.59 ± 0.10 kyr BP from the Tiarei outer to inner ridge. The top 2–3 m of the sequence exhibit classic platform-drowning signatures characterized by a suite of biological, sedimentary, and diagenetic features, including extensive bioerosion, manganese and iron staining of the rock surface, and submarine hardgrounds. This condensed deep-water sequence developed when the reef features dropped to a depth where carbonate production was limited as a consequence of continued sea-level rise and coeval environmental changes (e.g., light availability, water quality) during Holocene time (see also Camoin et al. (2006)).

3.6.5 EXPEDITION 325 (GBR ENVIRONMENTAL CHANGES)

3.6.5.1 Introduction

The GBR represents the largest extant coral reef system, spanning over 2300 km between 9°30′S in the north and 24°30′S in the south, and includes more than 3600 individual reefs (Davies, 2011; Hopley, Smithers, & Parnell, 2007) (Figure 3.6.8). The NE passive continental margin is comprised of several distinct marginal plateaus and rift troughs: the Eastern; Queensland and Marion Plateaus; the Bligh; Pandora; Osprey Embayment Queensland; and Townsville troughs, and is underlain by "modified continental crust formed as a result of fragmentation of an NE extension of the Tasman Fold Belt" (Davies, Symonds, Feary, & Pigram, 1991). The margin evolved as a result of rifting and seafloor spreading in the Coral Sea Basin (Weissel & Watts, 1979). Rifting probably initiated in the Late Cretaceous (80 Ma) and ceased by the earliest Eocene (56 Ma), at which time the region's major structural and physiographic features had formed (Weissel & Watts, 1979). Davies et al. (1991) concluded that the structural setting of the northeast Australia region is controlled by the gross architecture and morphology of the high-standing structural elements on which the major carbonate platforms in the region, the Queensland Plateau, Marion Plateau, and including GBR, have evolved. In addition to the architecture of margin,

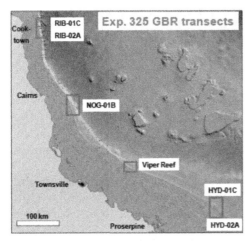

FIGURE 3.6.8 Map showing the location of the IODP Exp. 325 transects. The previously drilled 210 m Ribbon Reef 5 core (see International Consortium (2001)) is located in the red (dark gray in print versions) box adjacent to transect RIB-01C.

a complex ensemble of major factors (e.g., northward drift and collision, sea-level and climate oscillations, and subsidence; see Davies, 2011 for a recent summary) has influenced coral reef development on the margin since the early Miocene, culminating with the late Pleistocene establishment of the modern GBR.

The climate of the GBR is marked by persistent SE trade winds and marked seasonality. During the winter dry season, a high-pressure ridge caused by anticyclones crossing Australia, causes persistent SE winds up to 30 knots. In summer, low-pressure systems are common, resulting in northerly winds up to about 15 knots (Hopley et al., 2007). The period from November through May is also influenced by tropical cyclones generated mainly in the Coral Sea at an average of 2.8 per year (Puotinen, 1997). Rainfall also follows this marked seasonal pattern, mostly falling between December and March associated with the summer Monsoon. Spatially, the distribution of rainfall is highly variable with mean averages ranging from ~900 to 3500 mm/year, the wettest regions being between Daintree and Ingham (Hopley et al., 2007).

The waters of the GBR derive their characteristics from both the influence of the land to the west and the deeper waters of the Coral Sea to the east. Oceanographically, Wolanski (1994) divided the GBR into four main regions: Torres Strait, Northern, Central, and Southern Regions. These waters impact directly upon the biology of the reefs, through temperature, salinity, the provision of nutrients, dispersal of planktonic algae, response to tidal ranges, and the affects of current/wave-energy conditions (Hopley, 1982; Hopley et al., 2007). Average annual SSTs in the GBR range from ~28 °C in the north to about ~25.5 °C in the south (Locarnini et al., 2009). ENSO also has a strong influence on GBR's SSTs, with strong El-Nino events producing 1–2 °C SSTs above the late summer averages, sometimes causing significant coral reef bleaching. SSS also varies seasonally between wet and dry season and ranges between 34.5 and 35.5 psu in the north and south, respectively. Tides vary significantly spatially but are generally mixed between diurnal and semidiurnal modes, with average amplitudes between 4 and 2.5 on most of the outer shelf (Hopley et al., 2007).

Geomorphically, GBR varies considerably from north to south (Figure 3.6.8). In the northern GBR, the reefs on the shelf edge east of off Cooktown form a semicontinuous outer barrier. Here the reef is narrow with ribbon reefs on its eastern edge and extensive coastal fringing reefs and patch reefs. In the south, the GBR broadens, with patch reefs separated by open water or narrow channels. On the outer shelf, east-northeast of Townsville, modern reefs form a line of pinnacles seaward of the main reef edge. South of 15°30′S, the reefs are generally ≥30 km offshore and reach 100 km offshore at 22°30′S. Farther south, the shelf widens again to >200 km. East of Mackay (Australia), the modern reefs form a complex series of flood-tide deltaic reefs (i.e., Pompey Complex; Hopley, 2006).

The origin of the modern GBR remains poorly constrained, but the available evidence suggests ages of <500 kyr BP for the initiation of the northern GBR system (Braithwaite et al., 2004; Davies & Peerdeman, 1998; International

Consortium, 2001; McKenzie & Davies, 1993) and ages between 670 and 560 kyr BP for southern GBR (Dubois, Kindler, Spezzaferri, & Coric, 2008). The most direct evidence comes from the drilling at Ribbon Reef 5 (core depth 210 mbsl) using a reef-mounted jack-up platform (Figure 3.6.8). This demonstrated that the northern GBR is composed of ~100 m thick reef section, resting on a subreef, subtropical coralline algal-dominated section that, in turn, overlies a deepwater temperate grainstone section (Davies & Peerdeman, 1998; International Consortium, 2001). As a result of diagenetic alteration, U/Th series dating failed to accurately date the base of the main reef section (Braithwaite & Montaggioni, 2009). However, strontium isotope and magnetostratigraphic data confirm that the entire core and Pleistocene coral reef sequence is >780 kyr BP (International Consortium, 2001). Further, Webster and Davies (2003), considering all other available data, estimated a tentative turn-on time of 365–452 kyr BP, likely coincident with a very warm MIS11. Detailed stratigraphic and sedimentary facies analysis of the Ribbon Reef 5 drill core shows that reef section is composed of cycles of transgressive cool-water coralline-dominated carbonates topped by at least 6 shallow-water highstand coral reef units (Braga & Aguirre, 2004; Webster & Davies, 2003). These findings confirm that reef growth in the GBR is strongly coupled to major environmental changes. Therefore, combined with its tectonic stability and far-field location, the GBR represents a key site for the reconstruction of last deglacial sea-level and climate changes that are preserved within the drowned shelf-edge reefs (SERs) (Webster et al., 2011).

Previous geophysical studies along the GBR margin have identified a succession of seabed morphologic structures between 120 and 35 mbsl. Over the past 20 years, interpretations of these features have ranged from erosional notches and terraces to constructional paleoshorelines and submerged reefs (Beaman et al., 2008; Carter & Johnson, 1986; Harris & Davies, 1989; Hopley, 2006; Hopley, Graham, & Rasmussen, 1997). However, it is not until more systematic, high-resolution multibeam, seismic, and dredging investigations were conducted in 2007, as part of the Exp. 325 site survey (Webster et al., 2008), that the full extent and nature of the GBR SER systems was revealed. Abbey, Webster, and Beaman (2011) performed a detailed surficial geomorphic interpretation of the GBR SERs focused at the following four shelf and slope areas off: (1) Cooktown (Ribbon Reef), (2) Cairns (Noggin Pass), (3) Townsville shelf (Viper Reef) and Mackay (Hydrographer's Passage) (Figure 3.6.8). They identified and characterized a diverse suite of terraces (~70 to 110 m), pinnacles (~45 to 130 m), and paleo-barrier reefs (~35 to 75 m) and argued that gross shelf geomorphology and oceanographic factors, in conjunction with Mid-Late Pleistocene sea-level oscillations, influenced variability within and between sites.

The Ribbon Reef fore-reef slopes are the narrowest and steepest along the GBR margin (Abbey, Webster, & Beaman, 2011). A modern reef-front talus zone extends to 44 mbsl before merging into a well-developed submerged reef at 50 mbsl. Below this, a gently sloping terrace ~330 m wide at 80–55 mbsl connects with a submerged terrace at 80 and 100 mbsl before descending sharply into an upper slope that is deeply incised by submarine canyons that extend into the Queensland Trough (Puga-Bernabéu, Webster, Beaman, & Guilbaud, 2011). In contrast, the shelf edge to the south at Noggin Pass and is wider and more gently sloping, and forms a double-fronted submerged barrier reef system separated by an 80-m-wide paleo-lagoon. Here the inner and outer barrier reefs occur at 44 and 55 mbsl and represent the prominent 50 mbsl reef feature. Below this, a gently sloping terrace is observed characterized by prominent terraces at 80, 100, and 110 mbsl near the shelf break before descending again to the upper continental slope. The Hydrographer's Passage shelf is significantly wider and more gently sloping again and is characterized by inner and outer barrier reefs at 50 mbsl separated by a paleo-lagoon 2 km wide and as deep as 70 m (Figure 3.6.9). The seaward expression of the outer barrier is characterized by a steeply sloping 500-m-wide terrace with a sharp break in slope at 80 m. Another 1-km-wide paleo-lagoon and reef terrace system occurs at 80 and 90 mbsl, then another 700-m-wide paleo-lagoon that grades into complex system of reef pinnacles and terraces (100–110 mbsl) down to the major break in slope at 110 m that defines the shelf edge. In the north of Hydrographer's Passage, the seafloor grades into a gentle upper slope characterized by fore-reef slope sediments, while in the southern region another prominent 70-m-wide ridge or terrace is observed at 130 mbsl (Abbey, Webster, & Beaman, 2011).

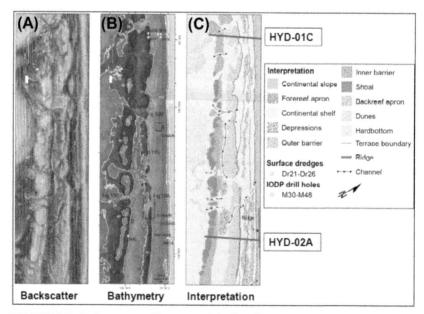

FIGURE 3.6.9 Surface geomorphic context of the Exp. 325 HYD-01C and 02A transects, drill sites (M0030-48), and surface dredge samples collected during the site survey. *After Abbey et al. (2011).*

3.6.5.2 Operation Results

During IODP Expedition 325 (GBR Environmental Changes), 34 holes were cored from 17 sites (M0030-M0058) at three locations (Hydrographer's Passage, Noggin Pass, and Ribbon Reef) in water depths between 42 and 157 mbsl from the vessel RV *Great Ship Maya* (Webster et al., 2011) (Figure 3.6.8). The locations are distributed along the margin to assess the impact of regional variations in oceanographic conditions (SST, sediment input), shelf-edge morphology (width and slope), and glacio-hydroisostatic behavior on reef response (Webster et al., 2011). The drilling strategy focused on recovering fossil coral reef deposits from the LGM to 10 kyr BP. This was achieved by drilling transects of holes through the most prominent fossil barrier reef structures between 40 and 50 mbsl and the series of well-developed reef terraces between 80 and 130 mbsl. Two transects were drilled at Hydrographer's Passage (Figure 3.6.9 and 3.6.10): a northern transect (HYD-01C) that includes Sites M0030-39, and a southern transect (HYD-02A) encompassing Sites M0041-48. In the Ribbon Reef region only four sites (M0049-51) from the southern RIB-02A transect were drilled. However, at Noggin Pass another transect (NOG-01B) consisting of M0052-58—including a hole (M0058A) on the upper continental slope—was completed. Challenging drilling conditions (i.e., serious technical and weather difficulties, unconsolidated sediments, cavities etc.) meant that average percent recoveries (27.2%; Yokoyama et al., 2011) were lower than Exp. 310 (57.5%; Camoin et al., 2012), according to standard IODP calculations. However, several strategies were employed to maximize core recovery and quality including: shorter cores runs (1 m), drilling closely spaced (within 10 m) replicate holes to generate composite cores, and, most importantly, the successful implementation of the HQ drilling string at Sites M0054-57 that saw average recoveries increase to >40–50%. Like Tahiti, efforts were made at several sites (i.e., M0042A, 55A, 56A, 57A) to drill deeper into the older, Pleistocene deposits to understand the nature of the pre-LGM substrate as well as proved new information about reef development, diagenetic environments and sea level prior to this period (i.e., Gischler et al., 2013).

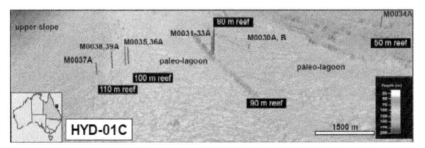

FIGURE 3.6.10 Close-up 3D bathymetric view of Exp. 325 transect HYD-01C and sites (M0030A-38A). *After Yokoyama et al. (2011) and Abbey et al. (2011).*

Borehole geophysical wireline logging was conducted at four holes (M0031A, 36A, 42A, 54B). The suite of slimline borehole logging probes was similar to that used for Exp. 310 and designed to yield a variety of data including high-resolution borehole images (optical (OBI40) and acoustic (ABI40) borehole televiewers); borehole fluid characterization (IDRONAUT); borehole diameter (CAL3); and a variety of petrophysical measurements such as electrical conductivity (DIL45), acoustic velocity (SONIC), spectral natural gamma radiation (ASGR), and magnetic susceptibility (EM51) (Webster et al., 2011). The majority of measurements were made in the open hole but through-pipe spectral gamma was also successfully employed. This provided coverage of both the MIS2 to deglacial reef sequence (Holes M0031A, 36A, 54B) and the older Pleistocene sequence (Hole M0042A). Microbiological sampling and analysis was focused primarily on fine sediment cores collected from transect NOG-01B, Hole M0058A for cell enumeration and phylogenetic analysis.

3.6.5.3 Initial Science Results

At the time of writing this chapter, the majority of Exp. 325 postcruise analyses were still in progress. However, a synthesis of the initial results from the Proceedings and published site survey data confirm that Exp. 325 will provide an important new record of sea level, environmental changes, and reef response over the last 30 kyrs.

3.6.5.3.1 SER Chronostratigraphy

A total of 68 coral samples were dated by U-Th and ^{14}C-AMS from representative cores from the top, middle, and bottom of most Exp. 325 sites (see Figure 3.6.11 for HYD-01C results). This was undertaken to provide a basic chronostratigraphy for the GBR SER system so as to better guide the sampling party and postcruise analyses (see Webster et al. (2011) for details). These preliminary data suggest that the SERs are composed of two basic chronostratigraphic sequences: basal >MIS3 (~30 kyr BP) deposits and the overlying MIS2 to last deglacial coral reef deposits. Below the inner barrier at HYD-01C (Hole M0034A) and NOG-01B (Hole M0057A) and inner terrace at NOG-01B (Hole M0057A), the >MIS3 deposits are clearly reefal and diagenetic evidence (e.g., dissolution, brownish staining) suggests they have been subaerially exposed (Gischler et al. 2013) prior to reflooding reef initiation and growth during the last deglacial. In contrast, the >MIS3 deposits below the deeper terraces (90–110 mbls) (e.g., Holes M0031-39A; 43A; M0055A, 53A, 54A, B) are composed of dark, grainstones and packstones characterized by shells, coral, coralline algae, *Halimeda*, and abundant larger benthic foraminifera representing lower shelf/slope settings (Webster et al., 2011). The contact between the two sequences represents a major unconformity surface and has also been recognized in the Exp. 325 downhole and sample petrophysical data (Webster et al., 2011; Yokoyama et al., 2011), and mapped regionally as well-defined seismic reflectors (see Hinestrosa, Webster, Beaman, and Anderson (2014) for a summary).

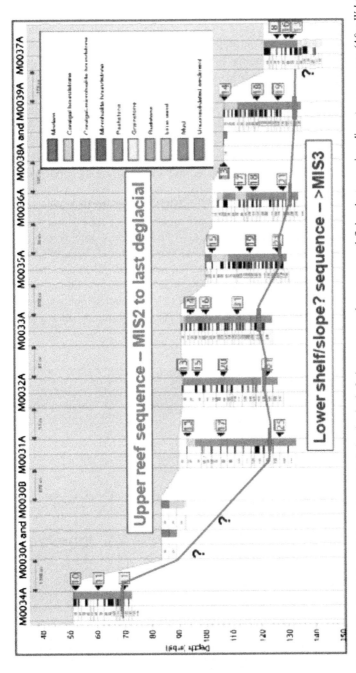

FIGURE 3.6.11 IODP Exp. 325 transect HYD-01C showing basic facies patterns and age structure defining the two main sedimentary sequences *(After Webster et al. (2011))*. The numbers in boxes to the right of the stratigraphic columns represent preliminary core catcher U/Th (red (dark gray in print versions)) and C14-AMS *(Ages From Webster et al. (2011))*.

3.6.5.3.2 Composition of the MIS2 to Deglacial Reef Sequence

The MIS2 to deglacial coral reef deposits are composed of mainly coral reef frameworks and detrital sedimentary facies (Figure 3.6.12). Three bound-stone facies are defined based on their varying proportions of corals, coral-line algae, and microbial sediments. In these framework facies, coral growth forms include massive, robust branching, branching, tabular, encrusting, and foliaceous, and they are commonly encrusted by thick centimeter-scale layers of coralline algae, encrusting foraminifera and associated vermetid gastropods (Webster et al., 2011). While not as ubiquitous as Exp. 310 (Seard et al., 2011), some intervals within the deeper terraces, particularly at NOG-01B (Holes M0053A, 54A, B), are dominated by abundant micro-bialite crusts exhibiting complex laminated and thrombolitic morphologies. The coralgal- to microbialite-dominated boundstones are also associated with abundant consolidated and unconsolidated sediments that are com-posed of mollusks, benthic foraminifera, red algae, and bryozoans that occur locally as internal sediments or as thick (1–19 m) intervals underlying the boundstone facies (e.g., Holes M0031-33A; Figure 3.6.11). Figure 3.6.13 illustrates a representative facies succession from Hole M0033A drilled through the 90 mbsl reef terrace. Here a 4-m interval of unconsolidated sedi-ments is observed directly overlying the >MIS3 lower shelf/slope sequence characterized by a dark packstone facies. These sediments then grade

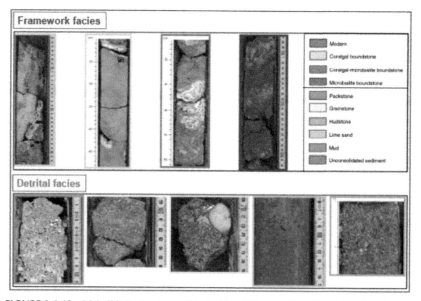

FIGURE 3.6.12 Main lithologies observed in the Exp. 325 cores. The colored (gray in print ver-sions) borders around each facies type correspond to those illustrated in the figure legend. *After Webster et al. (2011).*

into a 25-m-thick framework-dominated interval composed of coralgal-microbialite to coralgal boundstones deposits forming the MIS2 to deglacial reef sequence.

In the Exp. 325 cores, coral assemblages are dominated by massive *Isopora*, rbA, and branching *Seriatopora*, but mP and Faviidae, and encrusting *Porites*, *Montipora* with foliaceous Agariciids are locally abundant. *H. onkodes* is the most abundant coralline algae, together with *L. prototypum* and *Neogoniolithon fosliei* (Webster et al., 2011; Yokoyama et al., 2011) (e.g., Figure 3.6.12 and 3.6.13). Comparison with their modern environments in the GBR suggests that these assemblages are characteristic of shallow reef crest to deeper reef slopes, and are consistent with the reconstructed Exp. 310 environments (Abbey, Webster, Braga et al., 2011; Camoin et al., 2012) and other studies of Indo-Pacific reef systems (Montaggioni, 2005). Preliminary observations confirm that deepening upward coralgal successions are common in the top 3–5 m of the cores (e.g., Figure 3.6.11 and 3.6.13). Combined with other sedimentological characteristics (i.e., intense bioerosion, manganese and iron staining) (Webster et al., 2011), this represents a classic reef-drowning signature also observed at the top of the Exp. 310 deglacial reef (Abbey, Webster, & Beaman, 2011; Camoin et al., 2012), and the adjacent GBR dredge samples (Figure 3.6.10). For example, Abbey et al. (2013) conducted sedimentologic, paleoenvironmental, and chronologic studies of dredged coral, algae, and bryozoan specimens from the tops of the GBR

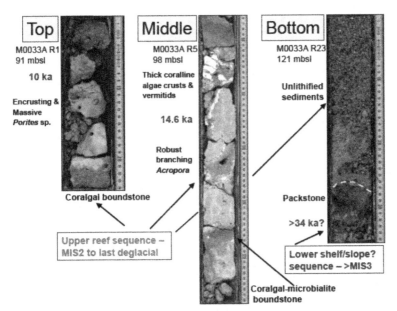

FIGURE 3.6.13 Representative succession of sediment facies and coralgal assemblages from M0033A. *Ages from Webster et al. (2011).*

SERs. Two distinct generations of fossil mesophotic coral community development are observed between 13–10 kyr BP and 8 kyr BP that have been influenced by widespread, massive flux of siliciclastic sediments associated with the flooding of the GBR shelf during deglacial sea-level rise.

3.6.5.3.3 Potential for Reconstructing Reef Growth, Sea-Level, and Paleoclimate Change

IODP Exp. 325 recovered fossil coral reef deposits from 46 to 145 mbsl, with a preliminary age range of between 9 kyr BP to older than 30 kyr BP. Figure 3.6.14 shows the distribution of the core catcher ages and their relationship to previously published sea-level and paleoclimate data since the LGM (see Yokoyama et al. (2011)). This figure highlights the excellent chronologic coverage of the key paleoenvironmental intervals (LGM, Bölling-Alleröd, Younger Dryas, and MWP events) provided by Exp. 325 cores, particularly in the context of the +500 new U/Th and C14-AMS measurements on corals and algae that are in progress. Combined with firm paleowater depth estimates provided by facies and coralgal analysis and GIA modeling, these ages will allow the reconstruction of a robust, new sea-level curve from MIS2 to 10 kyr BP. Numerous massive coral colonies suitable for paleoclimate studies were also recovered that will help define SST and SSS variations during this period in SW Pacific. For example, based on new stable isotope and Sr/Ca data from the Exp. 325 corals, Felis et al. (2014) reported that SST's in the GBR were significantly cooler than previously assumed and that a larger than expected north-south temperature gradient existed 20–13 kyrs ago. Finally, once the precise, stratigraphic, and chronologic sea-level framework has been established, 3D numerical reef modeling will allow the investigation of the response of the GBR to major environmental perturbations over the last 30 kyrs.

3.6.6 CONCLUSIONS

Drilling of multiple sites around the globe, spanning several glacial–interglacial transitions, is needed to better constrain the timing and amplitude of sea-level change that resulted in the disintegration of large ice sheets, and to evaluate the response of ocean meridional overturning circulation to freshwater inputs. A drilling strategy from pole to pole using International Ocean Discovery Program drilling platforms has proposed in the IODP Science Plan (Figure 3.6.15) to collect records linking climate, ice sheet, and sea-level histories on geologic timescales, and to ground truth and test the performance of numerical ice-atmosphere ocean models, in order to improve their ability to project future sea-level rise.

The IODP Expeditions 310 and 325 have demonstrated the pivotal importance of coral reef records to reconstruct the amplitude and timing of sea-level rise (~120 m at mean rates of ~8 mm/year) caused by the melting of land ice after the LGM ~23,000 years ago. In particular, short periods of dramatic sea-level rise indicating varying rates of melting imply the presence of thresholds

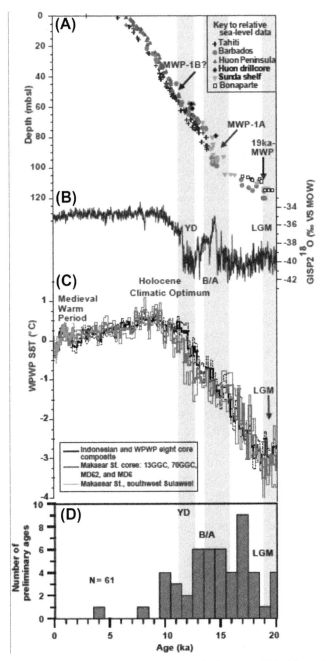

FIGURE 3.6.14 Comparison of preliminary dating results from Exp. 325 with previously published sea level and paleoclimate data. *After Yokoyama et al. (2011) and see this reference for all original data sources.*

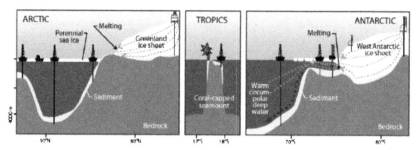

FIGURE 3.6.15 Proposed drilling strategy from pole to pole using International Ocean Discovery Program drilling platforms to collect records linking climate, ice sheet, and sea-level histories on geologic timescales. Ice and coral records are best preserved over the last 100,000 years, Antarctic ice cores go back 850,000 years, and sediment records extend back tens of millions of years. Red (pale gray in print versions) arrows represent warm water flow beneath floating ice, recently recognized as a key factor in accelerating ice loss from West Antarctica. *From the IODP Science Plan 2013-2023.*

within the dynamic behavior of ice sheets and provide a challenging test for ice sheet models used to predict future sea-level rise.

The relative magnitude of isostatic and gravitational processes that impact the position of the sea surface relative to the land will have to be assessed by studying multiple, temporally overlapping sea-level records recovered from a range of latitudes, in different tectonic and sedimentary settings and at varying distances from formerly glaciated regions. Ultimately, the combination of these data with modeling techniques can be used to "fingerprint" the relative contributions of different ice sheets to past sea-level change, providing more realistic scenarios for testing predictive models and a better understanding of ice sheet behavior in a changing world.

Among the several reef drilling proposals which followed the submission of the IODP proposal 519 (Camoin et al., 2002) and the implementation of the expeditions 310 and 325, coral reef drilling on rapidly subsiding margins (e.g., Hawaii, Huon Gulf, Sabine Bank, New Hebrides) has the potential to unlock a unique and largely unexploited archive of sea-level and climate changes. Unlike their mainly transgressive, highstand counterparts that developed in stable (e.g., GBR, Florida Margin) and uplifted settings (e.g., Huon Peninsula, Barbados), these submerged reefs developed in response to rapid subsidence. Depending on the relationship between eustatic sea-level changes and reef growth, the rapid subsidence ensures that these settings have the unique potential to continually create accommodation space, thus generating greatly expanded stratigraphic sections compared to reefs from stable and uplifting margins. Moreover, these drowned reefs evolved mainly during different periods of Earth's sea-level and climate cycles (i.e., glacial periods) that are not well sampled by reefs at stable and uplifting margins (Webster et al., 2009) (see Figure 3.6.16).

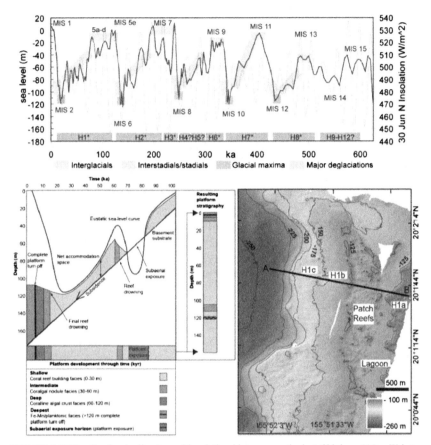

FIGURE 3.6.16   Sea-level, climate, and insolation history over the last 600 kyrs *(After Webster et al. (2007, 2009).)*. Observations, conceptual and numerical modelling data indicate IODP drilling through the succession of drowned Hawaiian reefs (H1-H12) will provide an unprecedented new record of sea-level, climate, and reef response during different interglacial, interstadial/stadial, glacial maxima, and deglacial intervals over the last 600 kyrs.

Hawaii represents the ideal study location to address these major scientific problems because of four main reasons (Figure 3.6.16). First, the rapid subsidence—and the greatly expanded reef sections up to 150 m thick—is caused by flexure of the oceanic lithosphere due to loading of the growing volcanoes, which in turn is controlled by rheology of the lithosphere and mantle, instead of uplift or subsidence due to co-seismic fault displacement, as occurs at convergent margins where most existing fossil reef records are derived (e.g., Barbados, PNG). Short-term fluctuations in volcanic loading are averaged out and the resultant subsidence has been nearly constant at 2.5–2.6 m/ky for the last 500 kyrs (Ludwig, Szabo, Moore, & Simmons, 1991; Moore & Campbell, 1987). Second, the observational, conceptual, and numerical modeling data suggest that the succession of drowned Hawaiian reefs (H1-H12) likely preserve a comprehensive record different interglacial, interstadial/stadial, glacial maxima, and deglacial intervals as the system evolved over the last 600 kyrs. Third, Hawaii's location in the central Pacific Ocean is well away from the confounding influence of large ice sheets and boundary ocean currents that might obscure the sea level and paleoclimate records. Four, an extensive database of bathymetric, submersible, ROV observations, sedimentary and radiometric data collected over the last 30 years is available for these reefs (see Webster et al. (2007, 2009) for a detailed review). These data confirm that the drowned reefs are highly sensitive to abrupt changes in sea level and climate and that there is a wealth of site survey information to successfully plan scientific drilling operations. In summary, IODP proposal 716-Full2 is well positioned to directly address several key IODP Challenges identified in the New Science Plan by investigating the mechanisms that control rapid sea-level and climate change as well as the relationship between changes in mean climate state and high-frequency (seasonal–decadal) climate variability.

REFERENCES

Abbey, E., Webster, J. M., & Beaman, R. J. (2011). Geomorphology of submerged reefs on the shelf edge of the Great Barrier Reef: the influence of oscillating Pleistocene sea-levels. *Marine Geology, 288*, 61–78.

Abbey, E., Webster, J. M., Braga, J. C., Jacobsen, G. E., Thorogood, G., Thomas, A. L., Camoin, G., Reimer, P. J., & Potts, D. C. (2013). Deglacial mesophotic reef demise on the Great Barrier Reef. *Palaeogeogr., Palaeoclim., Palaeoecol., 392*, 473–494.

Abbey, E., Webster, J., Braga, J. C., Sugihara, K., Wallace, C., Iryu, Y., et al. (2011). Variation in deglacial coralgal assemblages and their paleoenvironmental significance: IODP Expedition 310, Tahiti Sea Level. *Global and Planetary Change, 76*, 1–15.

Bamber, J. L., Riva, R. E. M., Vermeersen, B. L. A., & LeBrocq, A. M. (2009). Reassessment of the potential sea-level rise from a collapse of the west antarctic ice sheet. *Science, 324*, 901–903.

Bard, E., Hamelin, B., & Delanghe-Sabatier, D. (2010). Deglacial meltwater pulse 1B and younger dryas sea levels revisited with boreholes at tahiti. *Science, 327*, 1235–1237.

Bard, E., Hamelin, B., & Fairbanks, R. G. (1990). U/Th ages obtained by mass spectometry in corals from Barbados. Sea level during the past 130 000 years. *Nature, 346*, 456–458.

Bard, E., Hamelin, B., Arnold, M., Montaggioni, L. F., Cabioch, G., Faure, G., et al. (1996). Deglacial sea-level record from Tahiti corals and the timing of global meltwater discharge. *Nature, 382*, 241–244.

Bassett, S. E., Milne, G. A., Bentley, M. J., & Huybrechts, P. (2007). Modelling Antarctic sea-level data to explore the possibility of a dominant Antarctic contribution to meltwater pulse IA. *Quaternary Science Reviews, 26*, 2113–2127.

Bassett, S. E., Milne, G. A., Mitrovica, J. X., & Clark, P. U. (2005). Ice sheet and solid earth influences on far-field sea-level histories. *Science, 309*, 925–928.

Beaman, R. J., Webster, J. M., & Wust, R. A. J. (2008). New evidence for drowned shelf edge reefs in the Great Barrier Reef, Australia. *Marine Geology, 247*, 17–34.

Bertler, N. A. N., & Barrett, P. J. (2010). Vanishing polar ice sheets. In J. Dodson (Ed.), *Changing climates, earth systems and society* (pp. 49–84). New York: Springer.

Blanchon, P., Eisenhauer, A., Fietzke, J., & Liebetrau, V. (2009). Rapid sea-level rise and reef back-stepping at the close of the last interglacial highstand. *Nature, 458*, 881–884.

Blanchon, P., & Shaw, J. (1995). Reef-drowning during the last deglaciation: evidence for catastrophic sea level rise and ice-sheet collapse. *Geology, 23*, 4–8.

Bloom, A. L., Broecker, W. S., Chappell, J. M. A., Matthews, R. K., & Mesolella, K. J. (1974). Quaternary sea-level fluctuations on a tectonic coast. *Quaternary Research, 4*, 185–205.

Bonney, T. G. (Ed.). (1904). *The atoll of Funafuti: Borings into a coral reef and the results* (p. 428). London: Being the Report of the Coral Reef committee of the Royal Society.

Brachert, T. C. (1994). Palaeocecology of enigmatic tube microfossils forming "cryptalgal" fabrics (Late Quaternary, Red Sea). *Paläontologische Zeitschrift, 68*, 299–312.

Brachert, T. C., & Dullo, W. C. (1991). Laminar micrite crusts and associated foreslope processes, Red Sea. *Journal of Sedimentary Petrology, 61*, 354–363.

Brachert, T. C., & Dullo, W. C. (1994). Micrite crusts on Ladinian foreslopes of the Dolomites seen in the light of a modern scenario from the Red Sea. *Jahrbuch Geologie BA Wie, 50*, 57–68.

Braga, J. C., & Aguirre, J. (2004). Coralline algae indicate Pleistocene evolution from deep, open platform to outer barrier reef environments in the northern Great Barrier Reef margin. *Coral Reefs, 23*, 547–558.

Braithwaite, C. J. R., & Montaggioni, L. F. (2009). The Great Barrier Reef: a 700 000 year diagenetic history. *Sedimentology, 56*, 1591–1622.

Braithwaite, C. J. R., Dalmasso, H., Gilmour, M. A., Harkness, D. D., Henderson, G. M., Kay, R. L. F., et al. (2004). The Great Barrier Reef: the chronological record from a new borehole. *Journal of Sedimentary Research, 74*, 298–310.

Braithwaite, C. J. R., & Camoin, G. (2011). Diagenesis and sealevel change: lessons from moruroa, French Polynesia. *Sedimentology, 58*, 259–284.

Broecker, W. S., Thurber, D. L., Goddard, J., Ku, T. L., Matthews, R. K., & Mesolella, K. J. (1968). Milankovitch hypothesis supported by precise dating of coral reef and deep sea sediments. *Science, 159*, 297–300.

Cabioch, G., Banks-Cutler, K., Beck, W. J., Burr, G. S., Corrège, T., Edwards, R. L., et al. (2003). Continuous reef growth during the last 23 ka in a tectonically active zone (Vanuatu, SouthWest Pacific). *Quaternary Science Reviews, 22*, 1771–1786.

Cabioch, G., Camoin, G., & Montaggioni, L. F. (1999). Post glacial growth history of a French Polynesian barrier reef tract, Tahiti, central Pacific. *Sedimentology, 46*, 985–1000.

Cabioch, G., Montaggioni, L., Frank, N., Seard, C., Sallé, E., Payri, C., et al. (2008). Successive reef depositional events along the Marquesas foreslopes (French Polynesia) during the sea level changes of the last 26,000 years. *Marine Geology, 254*, 18–34.

Cabioch, G., Montaggioni, L. F., Faure, G., & Laurenti, A. (1999). Reef coralgal assemblages as recorders of paleobathymetry and sea level changes in the Indo-Pacific province. *Quaternary Science Reviews, 18*, 1681–1695.

Camoin, G. F., Bard, E., Hamelin, B., & Davies, P. J. (2002). The last deglacial sea level rise in the south Pacific: offshore drilling in Tahiti (French Polynesia). *IODP Proposal, 519*.

Camoin, G., Cabioch, G., Eisenhauer, A., Braga, J.-C., Hamelin, B., & Lericolais, G. (2006). Environmental significance of microbialites in reef environments during the Last Deglaciation. *Sedimentary Geology, 185*, 277–295.

Camoin, G. F., Colonna, M., Montaggioni, L. F., Casanova, J., Faure, G., & Thomassin, B. A. (1997). Holocene sea level changes and reef development in southwestern Indian Ocean. *Coral Reefs, 16*, 247–259.

Camoin, G. F., Ebren, Ph, Eisenhauer, A., Bard, E., & Faure, G. (2001). A 300,000-yr coral reef record of sea-level changes, Mururoa Atoll (Tuamotu archipelago French Polynesia). *Palaeogeography, Palaeoclimatology, Palaeoecology, 175*, 325–341.

Camoin, G. F., Gautret, P., Montaggioni, L. F., & Cabioch, G. (1999). Nature and environmental significance of microbialites in Quaternary reefs: the Tahiti paradox. *Sedimentary Geology, 126*, 271–304.

Camoin, G. F., Iryu, Y., McInroy, D., & The Expedition 310 scientists. (2007a). *Proc. IODP, 310.* College Station TX: Integrated Ocean Drilling Program Management International, Inc.. http://dx.doi.org/10.2204/iodp.proc.310.101.2007.

Camoin, G. F., Iryu, Y., McInroy, D., & The Expedition 310 scientists. (2007b). IODP Expedition 310 reconstructs sea-Level, climatic and environmental changes in the South Pacific during the Last Deglaciation. *Scientific Drilling, 5*, 4–12. http://dx.doi.org/10.2204/iodp.sd.5.01.2007.

Camoin, G. F., Montaggioni, L. F., & Braithwaite, C. J. R. (2004). Late glacial to post glacial sea levels in the Western Indian Ocean. *Marine Geology, 206*, 119–146.

Camoin, G. F., Seard, C., Deschamps, P., Webster, J. M., Abbey, E., Braga, J. C., et al. (2012). Reef response to sea-level and environmental changes during the last deglaciation: Integrated Ocean Drilling Program Expedition 310, Tahiti Sea Level. *Geology, 40*, 643–646.

Carter, R. M., & Johnson, D. P. (1986). Sea level controls on the post glacial development of the Great Barrier Reef, Queensland. *Marine Geology, 71*, 137–164.

Chappell, J. (1974). Relationships between sealevels, [18]O variations and orbital pertubations during the past 250,000 years. *Nature, 252*, 199–202.

Chappell, J. (2002). Sea level changes forced ice breakouts in the Last Glacial cycle: new results from coral terraces. *Quaternary Science Reviews, 21*, 1229–1240.

Chappell, J., Omura, A., Esat, T., McCulloch, M., Pandolfi, J., Ota, Y., et al. (1996). Reconciliation of late Quaternary sea levels derived from coral terraces at Huon Peninsula with deep sea oxygen isotope records. *Earth Planetary Science Letters, 141*, 227–236.

Chappell, J., & Polach, H. A. (1991). Post-glacial sea level rise from a coral record at Huon Peninsula, Papua New Guinea. *Nature, 349*, 147–149.

Chappell, J., & Veeh, H. H. (1978). [230]Th/[234]U age support of an interstadial sea level of -40 m at 30,000 yr B.P. *Nature, 276*, 602–604.

Clark, P. U., Mitrovica, J. X., Milne, G. A., & Tamisiea, M. E. (2002). Sea-level fingerprinting as a direct test for the source of global meltwater pulse IA. *Science, 295*, 2438–2441.

Clark, P. U., McCabe, A. M., Mix, A. C., & Weaver, A. J. (2004). Rapid rise of sea-level 19,000 years ago and its global implications. *Science, 304*, 1141–1144.

Coudray, J. (1976). Recherches sur le Néogène et le Quaternaire marins de la Nouvelle Calédonie. Contribution de l'étude sédimentologique à la connaissance de l'histoire géologique post-éocène de la Nouvelle-Calédonie. *Expédition Française sur les Récifs Coralliens de Nouvelle Calédonie* (Vol. 8)Paris: Fondation Singer-Polignac. 276.

Crossland, C. (1928a). The island of Tahiti. *Geographical Journal, 71*, 561–583.

Crossland, C. (1928b). Coral reefs of tahiti, Moorea, and Rarotonga. *Zoological Journal of the Linnean Society, 36*, 577–620.

Cullis, G. C. (1904). *The mineralogical changes observed in the cores of the Funafuti borings.* Royal Society of London. 392–420.

Cutler, K. B., Edwards, R. L., Taylor, F. W., Cheng, H., Adkins, J., Gallup, C. D., et al. (2003). Rapid sea-level fall and deep-ocean temperature change since the last interglacial period. *Earth Planetary Science Letters, 206*, 253–271.

Cutler, K. B., Gray, S. C., Burr, G. S., Edwards, R. L., Taylor, F. W., Cabioch, G., et al. (2004). Radiocarbon calibration and comparison to 50 kyr BP with paired C-14 and Th-230 dating of corals from Vanuatu and Papua New Guinea. *Radiocarbon, 463*, 1127–1160.

Davies, P. J. (2011). Great Barrier Reef: origin, evolution, and modern development. In D. Hopley (Ed.), *Encyclopedia of modern coral Reefs: Structure, form and Process* (pp. 504–534). Springer.

Davies, P. J., & Peerdeman, F. M. (1998). The origin of the Great Barrier Reef-the impact of Leg 133 drilling. In G. Camoin, & P. J. Davies (Eds.), *Reefs and carbonate platforms of the Pacific and Indian Ocean* (25) (pp. 23–38). International Association of Sedimentologists. Special Publication.

Davies, P. J., Symonds, P. A., Feary, D. A., & Pigram, C. J. (1991). The evolution of the carbonate platforms of Northwestern Australia. In M. A. J. Williams, P. de-Deckker, & A. P. Kershaw (Eds.), *The Cainozoic in Australia; a re-appraisal of the evidence* (Vol. 18) (pp. 44–78). Special Publication – Geological Society of Australia.

De Deckker, P., & Yokoyama, Y. (2009). Micropalaeontological evidence for late Quaternary sea-level changes in Bonaparte Gulf, Australia. *Global and Planetary Change, 66*, 85–92.

Delanghe, D., Bard, E., & Hamelin, B. (2002). New TIMS constraints on the uranium-238 and uranium-234 in seawaters from the main ocean basins and the Mediterranean Sea. *Marine Chemistry, 80*, 79–93.

Delesalle, B., Galzin, R., & Salvat, B. (1985). French Polynesian coral reefs. In *Proceed. 5th International Coral Reef Congress* (1) (pp. 1–554). Papeete.

Deschamps, P., Durand, N., Bard, E., Hamelin, B., Camoin, G., Thomas, A., et al. (2012). Ice sheet collapse and sea-level rise at the Bølling warming 14,600 yr ago. *Nature, 483*, 559–564.

Detrick, R. S., & Crough, S. T. (1978). Island subsidence, hot spots, and lithospheric thinning. *Journal of Geophysical Research, 83*, 1236–1244.

Dubois, N., Kindler, P., Spezzaferri, S., & Coric, S. (2008). The initiation of the southern central Great Barrier Reef: new multiproxy data from Pleistocene distal sediments from the Marion Plateau (NE Australia). *Marine Geology, 250*, 223–233.

Dullo, W. C. (1990). Facies, fossil record, and age of Pleistocene reefs from the Red Sea (Saudi Arabia). *Facies, 22*, 1–46.

Dullo, W-Ch, Camoin, G. F., Blomeier, D., Casanova, J., Colonna, M., Eisenhauer, A., et al. (1998). Sediments and sea level changes of the foreslopes of Mayotte, Comoro islands: direct observations from submersible. In G. Camoin, & P. J. Davies (Eds.), *Reefs and carbonate platforms of the Pacific and Indian Ocean* (25) (pp. 219–236). International Association of Sedimentologists. Special Publication.

Durand, N., Deschamps, P., Bard, E., Hamelin, B., Camoin, G., Thomas, A., et al. (2013). Comparison of ^{14}C and U-Th ages in corals from IODP #310 cores offshore Tahiti. *Radiocarbon, 55*, 1947–1974.

Edwards, R. L., Beck, J. W., Burr, G. S., Donahue, D. J., Chappell, J. M. A., Bloom, A. L., et al. (1993). A large drop in atmospheric ^{14}C/^{12}C and reduced melting in the Younger Dryas, documented with ^{230}Th ages of corals. *Science, 260*, 962–968.

Edwards, R. L., Chen, J. H., & Wasserburg, G. J. (1987). ^{238}U-^{234}U-^{230}Th-^{232}Th systematics and the precise measurement of time over the past 500,000 years. *Earth Planetary Science Letters, 81*, 175–192.

Esat, T. M., & Yokoyama, Y. (2006). Growth patterns of the last ice age coral terraces at Huon Peninsula. *Global and Planetary Change, 54*, 216–224.

Fadil, A., Sichoix, L., Barriot, J.-P., Ortèga, P., & Willis, P. (2011). Evidence for a slow subsidence of the Tahiti Island from GPS, DORIS, and combined satellite altimetry and tide gauge sea level records. *Comptes Rendus Geoscience, 343*, 331–341.

Fairbanks, R. G. (1989). A 17,000-year glacio-eustatic sea level record: influence of glacial melting rates on the Younger Dryas event and deep-ocean circulation. *Nature, 342*, 637–642.

Fairbanks, R. G., Mortlock, R. A., Chiu, T., Cao, L., Kaplan, A., Guilderson, T. P., et al. (2005). Radiocarbon calibration curve spanning 0 to 50,000 years BP based on paired ^{230}Th/^{234}U/^{238}U and ^{14}C dates on pristine corals. *Quaternary Science Reviews, 24*, 1781–1796.

Felis, T., McGregor, H. V., Linsley, B. K., Tudhope, A. W. T., Gagan, M. K., Suzuki, A., et al. (2014). Intensification of the meridional temperature gradient in the Great Barrier Reef following the Last Glacial Maximum. *Nature Communications*, http://dx.doi.org/10.1038/ncomms5102.

Fleming, K., Johnston, P., Zwartz, D., Yokoyama, Y., Lambeck, K., & Chappell, J. (1998). Refining the eustatic sea-level curve since the last glacial maximum using far- and intermediate field sites. *Earth Planetary Science Letters, 163*, 327–342.

Fujita, K., Omori, A., Yokoyama, Y., Sakai, S., & Iryu, Y. (2010). Sea-level rise during Termination II inferred from large benthic foraminifers: IODP Expedition 310, Tahiti Sea Level. *Marine Geology, 271*, 149–155.

Fulthorpe, C. S., Miller, K. G., Droxler, A. W., Hesselbo, S. P., Camoin, G. F., & Kominz, M. A. (2008). *Drilling to Decipher long-term sea-level changes and effects—A Joint Consortium for Ocean Leadership, ICDP, IODP, DOSECC, and Chevron Workshop. Scientific Drilling, 6.* http://dx.doi.org/10.2204/iodp.sd.6.02.2008.

Gallup, C. D., Edwards, R. L., & Johnson, R. G. (1994). The timing of high sea levels over the past 200,000 years. *Science, 263*, 796–800.

Gischler, E., & Hudson, J. H. (2004). Holocene development of the Belize barrier reef. *Sedimentary Geology, 164*, 223–236.

Gischler, E., Ginsburg, R. N., Herrlez, J. O., & Prasad, S. (2010). Mixed carbonates and siliciclastics in the Quaternary of southern Belize: Pleistocene turning points in reef development controlled by sea-level change. *Sedimentology, 57*, 1049–1068.

Gischler, E., Lomando, A. J., Hudson, J. H., & Holmes, C. W. (2000). Last interglacial reef growth beneath Belize barrier and isolated platform reefs. *Geology, 28*, 387–390.

Gischler, E., Thomas, A. L., Droxler, A. W., Webster, J. M., Yokoyama, Y., Schöne, B. R., et al. (2013). Microfacies and diagenesis of older Pleistocene (pre-last glacial maximum) reef deposits, Great Barrier Reef, Australia (IODP Expedition 325): a quantitative approach. *Sedimentology, 60*, 1432–1466.

Grammer, G. M., Ginsburg, R. N., Swart, P. K., McNeill, D. F., Jull, A. J. T., & Prezbindowski, D. R. (1993). Rapid growth rates of syndepositional marine aragonite cements in steep marginal slope deposits, Bahamas and Belize. *Journal of Sedimentary Petrology, 63*, 983–989.

Gregoire, L. J., Payne, A. J., & Valdes, P. J. (2012). Deglacial rapid sea level rises caused by ice-sheet saddle collapses. *Nature, 487*, 217–222.

Gvirtzman, G., & Friedman, G. M. (1977). Sequence of progressive diagenesis in coral reefs. In S. H. Frost, M. P. Weiss, & J. B. Saunders (Eds.), *Reefs and related carbonates – Ecology and sedimentology Studies in Geology: 4.* (pp. 357–380). American Association of Petroleum Geologists.

Hanebuth, T. J. J., Stattegger, K., & Bojanowski, A. (2009). Termination of the Last Glacial Maximum sea-level lowstand: the Sunda-Shelf data revisited. *Global and Planetary Change, 66*, 76–84.

Harris, P. T., & Davies, P. J. (1989). Submerged reefs and terraces on the shelf edge of the Great Barrier Reef Australia. *Coral Reefs, 8*, 87–98.

Heindel, K., Birgel, D., Brunner, B., Thiel, V., Westphal, H., Gischler, E., et al. (2012). Post-glacial microbialite formation in coral reefs of the Pacific, Atlantic, and Indian Oceans. *Chemical Geology, 304-305*, 117–130.

Heindel, K., Birgel, D., Peckmann, J., Kuhnert, H., & Westphal, H. (2010). Formation of deglacial microbialites in coral reefs off Tahiti (IODP 310) involving sulfate reducing bacteria. *Palaios, 25*, 618–635.

Hildenbrand, A., Gillot, P. Y., & Le Roy, I. (2004). Volcano-tectonic and geochemical evolution of an oceanic intra-plate volcano: Tahiti-Nui (French Polynesia). *Earth Planetary Science Letters, 217*, 349–365.

Hinestrosa, G., Webster, J. M., Beaman, R. J., & Anderson, L. M. (2014). Seismic stratigraphy and development of the shelf-edge reefs of the Great Barrier Reef, Australia. *Marine Geology, 353*, 1–20.

Hopley, D. (1982). *The geomorphology of the great barrier Reef: Quaternary development of coral reefs*. New York: Wiley-Interscience. 320.

Hopley, D. (2006). Coral reef growth on the shelf margin of the Great Barrier Reef with special reference to the Pompey Complex. *Journal of Coastal Research, 22*, 150–158.

Hopley, D., Graham, T. L., & Rasmussen, C. E. (1997). Submerged shelf-edge reefs, coral reefs, Great Barrier Reef Australia. In *Proceedings PACON 96* (pp. 305–315). Honolulu, U.S.A: Recent Advances in Marine Science and Technology.

Hopley, D., Smithers, S. G., & Parnell, K. E. (2007). *The geomorphology of the great barrier reef*. Cambridge. 532.

International Consortium, G. B. R. D. (2001). New constraints on the origin of the Australian Great Barrier Reef from an international project of deep coring. *Geology, 29*, 483–486.

Inwood, J., Brewer, T., Braaksma, H., & Pezard, P. (2008). Integration of core, logging and drilling data in modern reefal carbonates to improve core location and recovery estimates (IODP Expedition 310). *Journal of the Geological Society, 65*, 585–596 London.

Iryu, Y., Takanashi, Y., Fujita, K., Camoin, G., Cabioch, G., Matsuda, H., et al. (2010). Sea-level history recorded in the Pleistocene carbonate sequence in IODP Hole 310-M0005D, off Tahiti. *Island Arc, 19*, 690–706.

James, N. P., & Ginsburg, R. N. (1979). *The seaward margin of Belize barrier and atoll reefs* (3International Association of Sedimentologists. Special Publication, 191.

Johnson, J. H., Todd, R. M., Post, R., Cole, W. S., & Wells, J. W. (1954). *Bikini and Nearby atolls: Part 4: Paleontology*. U.S. Geol. Surv. Prof. Pap. 260-M, N, O and P.

Ladd, H. S., & Schlanger, S. O. (1960). Drilling operations on Eniwetok atoll. *Prof. Pap.-Geol. Surv. (U.S.), 260-Y*, 863–899.

Lambeck, K. (1993). Glacial rebound and sea-level change: an example of a relationship between mantle and surface processes. *Tectonophysics, 223*, 15.

Lambeck, K., Purcell, A., Funder, S., Kjær, K., Larsen, E., & Möller, P. (2006). Constraints on the Late Saalian to early Middle Weichselian ice sheet of Eurasia from field data and rebound modelling. *Boreas, 35*, 539–575.

Lambeck, K., Purcell, A., Johnston, P., Nakada, M., & Yokoyama, Y. (2003). Water-load definition in the glacio-hydro-isostatic sea-level equation. *Quaternary Science Reviews, 22*, 309–318.

Lambeck, K., Yokoyama, Y., & Purcell, A. (2002). Into and out of the Last Glacial Maximum: sea level change during Oxygen Isotope Stages 3 and 2. *Quaternary Science Reviews, 21*, 343–360.

Land, L. S., & Moore, C. H. (1980). Lithification, micritization and syndepositional diagenesis of biolithites on the Jamaican island slope. *Journal of Sedimentary Petrology, 50*, 357–370.

Le Roy, I., (1994). *Evolution des volcans en système de point chaud: Île de tahiti, archipel de la Société (Polynésie Française)* (Thèse Doctorat). Orsay: Université Paris-Sud, 271 pp.

Lighty, R. G., Macintyre, I. G., & Stuckenrath, R. (1982). *Acropora palmata* reef Framework: a reliable indicator of sea level in the western atlantic for the past 10,000 years. *Coral Reefs*, 125–130.

Locarnini, R. A., Mishonov, A. V., Antonov, J. I., Boyer, T. P., Garcia, H. E., Baranova, O. K., et al. (2009). World ocean Atlas 2009. *Temperature* (Vol. 1)NOAA Atlas: NOAA Atlas NESDIS.

Locker, S. D., Hine, A. C., Tedesco, L. P., & Shinn, E. A. (1996). Magnitude and timing of episodic sea-level rise during the last deglaciation. *Geology, 24*, 827–830.

Ludwig, K. R., Halley, R. B., Simmons, K. R., & Peterman, Z. E. (1988). Strontium-isotope stratigraphy of Enewetak Atoll. *Geology, 16*, 173–177.

Ludwig, K. R., Szabo, B. J., Moore, J. G., & Simmons, K. R. (1991). Crustal subsidence rate off Hawaii determined from $^{234}U/^{238}U$ ages of drowned coral reefs. *Geology, 19,* 171–174.

Macintyre, I. G., Rutzler, K., Norris, J. N., Smith, K. P., Cairns, S. D., Bucher, K. E., et al. (1991). An early Holocene reef in the western Atlantic: submersible investigations of a deep relict reef off the west coast of Barbados, W.I. *Coral Reefs, 10,* 167–174.

Manabe, S., & Stouffer, R. J. (1995). Simulation of abrupt climate-change induced by fresh-water input to the North-Atlantic ocean. *Nature, 378,* 165–167.

McKenzie, J. A., & Davies, P. J. (1993). Cenozoic evolution of carbonate platforms on the northeastern Australia margin : synthesis of leg 133 drilling results. In J. A. McKenzie, P. J. Davies, A. Palmer-Julson, et al. (Eds.), *Proceedings of the ocean drilling program, scientific results* (Vol. 133). College Station: TX Ocean Drilling Program. http://dx.doi.org/10.2973/odp.proc.sr.2133.1993.

Ménabréaz, L., Thouveny, N., Camoin, G., & Lund, S. P. (2010). Paleomagnetic record of the Late Pleistocene reef sequence of Tahiti (French Polynesia): a contribution to the chronology of the deposits. *Earth and Planetary Science Letters, 294,* 58–68.

Mesolella, K. J., Matthews, R. K., Broecker, W. S., & Thurber, D. L. (1969). The astronomical theory of climate change: Barbados data. *Journal of Geology, 77,* 250–274.

Milankovitch, M. (1941). Kanon der erdbestrahlung und seine anwendung auf des eiszeitenproblem*König. Serb. Akad. Sond* (Vol. 33). 133, Mathem. und Naturwiss.

Milne, G. A., Gehrels, W. R., Hughes, C. W., & Tamisiea, M. E. (2009). Identifying the causes of sea-level change. *Nature Geoscience, 2,* 471–478. http://dx.doi.org/10.1038/ngeo544.

Milne, G. A., & Mitrovica, J. X. (2008). Searching for eustasy in deglacial sea-level histories. *Quaternary Science Reviews, 27,* 2292–2302.

Milne, G. A., Mitrovica, J. X., & Schrag, D. P. (2002). Estimating past continental ice volume from sea-level data. *Quaternary Science Reviews, 21,* 361–376.

Mitrovica, J. X., & Milne, G. (2002). On the origin of late Holocene sea-level highstands within equatorial ocean basins. *Quaternary Science Reviews, 21,* 2179–2190.

Mitrovica, J. X., & Peltier, W. R. (1991). On postglacial geoid subsidence over the equatorial oceans. *Journal of Geophysical Research, 96,* 20,053–20,071.

Mix, A. C., Bard, E., & Schneider, R. (2001). Environmental processes of the ice age: land, ocean, glaciers (EPILOG). *Quaternary Science Reviews, 20,* 627–657.

Montaggioni, L. F. (1988). Holocene reef growth history in mid plate high volcanic islands. In *Proceed. 6th International Coral Reef Symposium* (3) (pp. 455–460). Australia.

Montaggioni, L. F. (2005). History of Indo-Pacific coral reef systems since the last glaciation: development patterns and controlling factors. *Earth-Science Reviews, 71*(1–2), 1.

Montaggioni, L. F., Cabioch, G., Camoin, G. F., Bard, E., Ribaud-Laurenti, A., Faure, G., et al. (1997). Continuous record of reef growth over the past 14 k.y. on the mid-Pacific island of Tahiti. *Geology, 25,* 555–558.

Montaggioni, L. F., Cabioch, G., Thouveny, N., Frank, N., Sato, T., & Sémah, A. M. (2011). Revisiting the Quaternary development history of the western New Caledonian shelf system: from ramp to barrier reef. *Marine Geology, 280,* 57–75.

Moore, J. G., & Campbell, J. F. (1987). Age of tilted reefs, Hawaii. *Journal of Geophysical Research, 92,* 2641–2646.

Ohde, S., Greaves, M., Masuzawa, T., Buckley, H. A., VanWoesik, R., Wilson, P. A., et al. (2002). The chronology of Funafuti Atoll: revisiting an old friend. *Proceedings Royal Society of London, A,* 1–17.

Parsons, B., & Sclater, J. G. (1977). An analysis of the variation of the ocean floor bathymetry and heat flow with age. *Journal of Geophysical Research, 82,* 803–827.

Peltier, W. R. (1991). The ICE-3G model of late Pleistocene deglaciation: construction, verification, and applications. In R. Sabadini (Ed.), *Isostasy, Sea-level and mantle rheology NATO ASI, C: 334.* (pp. 95–119). Amsterdam: Kluwer.

Peltier, W. R. (1994). Ice-age paleotopography. *Science, 265,* 195–201.

Peltier, W. R. (1999). Global sea level rise and glacial isostatic adjustment. *Global planet. Change, 20,* 93–123.

Peltier, W. R., & Fairbanks, R. G. (2006). Global glacial ice volume and Last Glacial Maximum duration from an extended Barbados sea level record. *Quaternary Science Reviews, 25,* 3322–3337.

Pfeffer, W. T., Harper, J. T., & O'Neel, S. (2008). Kinematic constraints on glacier contributions to 21st-century sea-level rise. *Science, 321,* 1340–1343.

Pirazzoli, P., Ratdke, U., Hantoro, W. S., Jouannic, C., Hoang, C. T., Causse, C., et al. (1993). A one million-year-long sequence of marine terraces on Sumba island. *Marine Geology, 109,* 221–236.

Potter, E.-K., & Lambeck, K. (2004). Reconciliation of sea-level observations in the Western North Atlantic during the last glacial cycle. *Earth Planetary Science Letters, 217,* 171–181.

Puga-Bernabéu, Á., Webster, J. M., Beaman, R. J., & Guilbaud, V. (2011). Morphology and controls on the evolution of a mixed carbonate–siliciclastic submarine canyon system, Great Barrier Reef margin, north-eastern Australia. *Marine Geology, 289,* 100–116.

Puotinen, M. L., Done, T. J., & Skelly, W. C. (1997). *An Atlas of tropical cyclones in the great barrier reef region, 1969–1997.*

Purdy, E. G., Gischler, E., & Lomando, A. J. (2003). The Belize margin revisited. 2. Origin of Holocene antecedent topography. *International Journal of Earth Sciences, 92,* 552–572.

Quinn, T. M. (1991). Meteoric diagenesis of Plio-Pleistocene limestones at enewetak atoll. *Journal of Sedimentary Petrology, 61,* 681–703.

Quinn, T. M., & Matthews, R. K. (1990). Post-Miocene diagenetic and eustatic history of Enewetak Atoll: model and data comparison. *Geology, 18,* 942–945.

Richards, H. S., & Hill, D. (1942). Great Barrier Reef bores, 1926 and 1937. Descriptions, analyses and interpretations. *Report of the Great Barrier Reef Communication, 5,* 1–122.

Robinson, L. F., Belshaw, N. S., & Henderson, G. M. (2004). U and Th concentrations and isotope ratios in modern carbonates and waters from the Bahamas. *Geochimica Cosmochimica Acta, 68,* 1777–1789.

Rougerie, F., Wauthy, B., & Rancher, J. (1992). Le récif barrière ennoyé des Iles Marquises et l'effet d'île par endo-upwelling. *Comptes Rendus Académie des Sciences, Paris II, 315,* 677–682.

Salvat, B., Sibuet, M., & Laubier, L. (1985). Benthic megafauna observed from the submersible "Cyana" on the fore-reef slope of Tahiti (French Polynesia) between 70 and 100 m. In *Proceed. 5th International Coral Reef Congress* (2) (p. 338). Papeete.

Seard, C., Borgomano, J., Granjeon, D., & Camoin, G. (2013). Impact of environmental parameters on coral reef development and drowning: Forward modelling of last deglacial reefs from Tahiti (French Polynesia). *Sedimentology, 60,* 1357–1388.

Seard, C., Camoin, G., Yokoyama, Y., Matsuzaki, H., Durand, N., Bard, E., et al. (2011). Microbialite development patterns in the last deglacial reefs from Tahiti (French Polynesia; IODP Expedition #310). *Implications on reef framework architecture: Marine Geology, 279,* 63–86.

Stanford, J. D., Rohling, E. J., Hunter, S. E., Roberts, A. P., Rasmussen, S. O., Bard, E., et al. (2006). Timing of meltwater pulse 1a and climate responses to meltwater injections. *Paleoceanography, 21.* http://dx.doi.org/10.1029/2006PA001340.

Stein, M., Wasserburg, G. J., Aharon, P., Chen, J. H., Zhu, Z. R., Bloom, A., et al. (1993). TIMS U-series dating and stable isotopes of the last interglacial event in Papua New Guinea. *Geochimica Cosmochimica Acta, 57,* 2541–2554.

Stirling, C. H., Esat, T. M., Lambeck, K., & McCulloch, M. T. (1998). Timing and duration of the Last Interglacial: evidence for a restricted interval of widespread coral reef growth. *Earth Planetary Science Letter, 160*, 745–762.

Strasser, A., Strohmenger, C., Davaud, E., & Bach, A. (1992). Sequential evolution and diagenesis of Pleistocene coral reefs (south Sinai, Egypt). *Sedimentary Geology, 78*, 59–79.

Tarasov, L., Dyke, A. S., Neal, R. M., & Peltier, W. R. A. (2012). Data-calibrated distribution of deglacial chronologies for the North American ice complex from glaciological modeling. *Earth Planetary Science Letters, 315–316*, 30–40.

Thomas, A. L., Henderson, G. M., Deschamps, P., Yokoyama, Y., Mason, A. J., Bard, E., et al. (2009). Penultimate deglacial sea-level timing from uranium/thorium dating of Tahitian corals. *Science, 324*, 1186–1189.

Tracey, J. I., & Ladd, H. S. (1974). Quaternary history of Eniwetok and Bikini atolls, Marshall Islands. In *Proceedings Second International Coral Reef Symposium Brisbane* (2) (pp. 537–550).

Waelbroeck, C., Labeyrie, L., Michel, E., Duplessy, J. C., McManus, J. F., Lambeck, K., et al. (2002). Sea-level and deep water temperature changes derived from benthic foraminifera isotopic records. *Quaternary Science Reviews, 21*, 295–305.

Weaver, A. J., Saenko, O. A., Clark, P. U., & Mitrovica, J. X. (2003). Meltwater pulse 1A from Antarctica as a trigger of the bolling-allerod warm interval. *Science, 299*, 1709–1713.

Webster, J. M., Braga, J. C., Clague, D. A., Gallup, C., Hein, J. R., Potts, D. C., et al. (2009). Coral reef evolution on rapidly subsiding margins. *Global and Planetary Change, 66*, 129–148.

Webster, J. M., Beaman, R. J., Bridge, T. C. L., Davies, P. J., Byrne, M., Williams, S., et al. (2008). From corals to canyons: the great barrier reef margin. *EOS, Transactions, American Geophysical Union, 89*(24), 217–218.

Webster, J. M., Clague, D. A., Coleman-Riker, K., Gallup, C., Braga, J. C., Potts, D., et al. (2004). Drowning of the 150 m reef off Hawaii: a casualty of global meltwater pulse 1A? *Geology, 32*, 49–252.

Webster, J. M., Wallace, L. M., Clague, D. A., & Braga, J. C. (2007). Numerical modeling of the growth and drowning of Hawaiian coral reefs during the last two glacial cycles (0–250 kyr). *Geochem. Geophys. Geosyst., 8*, Q03011, http://dx.doi.org/10.1029/2006GC001415.

Webster, J. M., & Davies, P. J. (2003). Coral variation in two deep drill cores: significance for the Pleistocene development of the Great Barrier Reef. *Sedimentary Geology, 159*, 61–80.

Webster, J. M., Wallace, L., Silver, A. E., Potts, D., Braga, J. C., et al. (2004). Coralgal composition of drowned carbonate platforms in the Huon Gulf, Papua New Guinea; implications for lowstand reef development and drowning. *Marine Geology, 204*, 59–89.

Webster, J. M., Yokoyama, Y., Cotterill, C., & Expedition 325 Scientists. (2011). *Proceedings of the integrated ocean drilling program volume 325 expedition Reports great barrier reef environmental changes*. Integrated Ocean Drilling Program Management International, Inc. for the Integrated Ocean Drilling Program.

Weissel, J. K., & Watts, A. B. (1979). Tectonic evolution of the Coral Sea basin. *Journal of Geophysical Research, 84*, 4572–4582.

Wheeler, C. W., & Aharon, P. (1991). Mid-oceanic carbonate platforms as oceanic dipsticks: examples from the Pacific. *Coral Reefs, 10*, 101–114.

Williams, H. (1933). *Geology of tahiti, Moorea, and Maiao. B.P. Bishop Museum Bulletin 105* Honolulu.

Wolanski, E. (1994). *Physical oceanographic processes of the great barrier reef*. Boca Raton, FL: CRC Press. 376 pp.

Yokoyama, Y., Esat, T. M., & Lambeck, K. (2001a). Coupled climate and sea-level changes deduced from Huon Peninsula coral terraces of the last ice age. *Earth Planetary Science Letters, 193*, 579–587.

Yokoyama, Y., Esat, T. M., & Lambeck, K. (2001b). Last glacial sea-level change deduced from uplifted coral terraces of Huon Penninsula, Papua New Guinea. *Quaternary International, 83–85*, 275–283.

Yokoyama, Y., Esat, T. M., Lambeck, K., & Fifield, L. K. (2000). Last ice age millennial scale climate changes recorded in Huon Peninsula corals. *Radiocarbon, 42*, 383–401.

Yokoyama, Y., Webster, J. M., Cotterill, C., Braga, J. C., Jovane, L., Mills, H., & Expedition 325 Scientists., et al. (2011). IODP Expedition 325: Great Barrier Reefs reveals past sea-level, climate and environmental changes during the end of the last Ice age. *Scientific Drilling, 12*, 32–45.

Solid Earth Cycles and Geodynamics

Chapter 4.1

Introduction

Donna K. Blackman
Scripps Institution of Oceanography, University of California San Diego, La Jolla, CA, USA
E-mail: dblackman@ucsd.edu

Integrated Ocean Drilling Program (IODP) coring and logging documented in situ composition and material properties that tested predrilling hypotheses and that continue to guide new assessments of solid Earth cycling and geodynamic processes. While the pace and scope of advances made from 2003 to 2013 did not meet all the ambitious goals laid out in the Initial Science Plan, due to factors outlined in Chapter 1 of this volume, a number of striking results were obtained, and these have already impacted thinking in the field. In combination with regional data, the core analyses and borehole characterizations performed on data acquired by IODP have been, and will continue to be, a determining factor for structural interpretations, geochemical inferences, insights into physical–chemical interplay, and ongoing modeling efforts. Below, examples of IODP findings that brought new perspectives on crust or mantle processes are noted—these are just a few of the intriguing results presented in each of the papers of this chapter, but they illustrate the pivotal role that drilling has in geoscience research.

Both site-specific findings and general insights into crust and mantle processes were obtained by IODP at each of two to three oceanic spreading centers, convergent margins, and long-lived intraplate volcanic systems. Processes that occur within or near the axial zone of an oceanic spreading center were targeted by six IODP expeditions (Chapter 4.2.1). Three other expeditions had major emphasis on ridge flank processes (Chapter 4.2.2), and additional brief borehole/observatory work was accomplished three times at these sites. Implications of intraplate volcanism in the ocean basins were explored via two IODP expeditions (Chapter 5.3). Ultimately, all but one *Chikyu* IODP expedition (11) and two *JOIDES Resolution* expeditions documented convergent margin geology (Chapters 4.4.1–3).

IODP results anchored structural interpretations that shaped inferences about the formation of oceanic lithosphere. The occurrence of a lava pond many tens of meters thick at Site 1256 on the East Pacific Rise flank indicated that flow volumes of many cubic kilometers can interrupt the pattern of typically smaller seafloor lava flows at a fast spreading ridge. Tilting of the underlying sheeted

Developments in Marine Geology, Volume 7. http://dx.doi.org/10.1016/B978-0-444-62617-2.00016-5

dike complex was modest, indicating that little tectonic disruption occurred after initial accretion of the upper crust. The low ratio of dike: lava thickness at this site contrasts with that exposed in Hess Deep, at IODP Site U1415, despite both being formed at a fast spreading rate. Ildefonse et al. (Chapter 4.2.1) thus infer that tectonic windows may document interplay between axial magmatism and faulting that is distinct from typical intermediate-fast spread oceanic crust. This may be important for gauging which results from tectonic windows should be incorporated in general models of crustal structure. Deep drilling at an oceanic core complex on the young flank of the Mid-Atlantic Ridge discerned extreme strain localization associated with detachment faulting. While the recovered core records deformation in only a few narrow zones, paleomagnetic data indicate that the unroofed intrusive crust rotated several tens of degrees; this is the strongest support yet of the "rolling hinge" model of core complex formation. Drilling and seafloor samples at Atlantis Massif indicate that the detachment fault channeled significant seawater flow, which increased the cooling rate of this unroofed core complex. Hydrothermal circulation in fast spread crust was shown to be dominated by cool temperatures within the lava section and to rapidly transition to higher temperatures across both the lava-dike and the dike–gabbro transitions, the latter sampled in situ for the first time at Site 1256. On the flank of the Juan de Fuca ridge, active flow experiments show that upper crustal permeability varies considerably and that hydrologic connectivity can be limited in this interval (Fisher et al., Chapter 4.2.2).

The structure of mature ocean lithosphere as it enters a subduction zone has been found to fundamentally impact deformation of the sedimentary wedge at subduction zones, with IODP results providing key new insights into Nankai and Costa Rica margins. Together with seismic imaging, drilling results document pulses of thrust faulting, and shifts in the deformation locus from the inner to the outer wedge, sometimes altering trench sediment supply if a forearc basin rapidly subsided, sequestering deposits (Tobin et al., Chapter 4.4.1). Sediment supply at the trench is inferred to control whether a subduction zone is accretionary (high trench sediment supply) or erosive (low), which in turn affects the stress field within the wedge. While regional plate driving forces dominate overall, in situ stress can differ locally in response to splay fault displacements (Kinoshita et al., Chapter 4.4.2), and broader areas can undergo periods of extension, as documented in Costa Rica (Chapter 4.4.1).

Petrology and geochemistry results point to complex crustal magma interactions and document the nature of long-lived melting associated with mantle plumes. Chemical reactivity between rock types traditionally thought of as distinct crustal "layers," sometimes enhanced by magmatic or hydrothermal fluids, has been documented by IODP. Interaction near the dike–gabbro transition was evidenced by contact metamorphism with complete recrystallization and a complex history of cross intrusions likely to have occurred at the past location of an axial melt lens (Site 1256; Chapter 4.2.1). Interaction of mantle melts with lower crustal rocks is strongly suggested by geochemical analyses of core from both the Mid-Atlantic ridge (Site U1309) and Hess Deep (Site U1415) tectonically exposed sections (Chapter 4.2.1). These new findings challenge the simple fractional melting model of midocean ridge basalt genesis, at least locally. Work along the Louisville volcanic chain documents consistent chemistry of magmas erupted from this long-lived mantle melt anomaly. This is unique, globally, raising the question of whether consistent thickness of the lithosphere overlying the plume might have played a role (Koppers and Sager, Chapter 5.3).

The interplay between chemical and physical processes in the ocean lithosphere has been illuminated by IODP data. The extent of alteration recorded in the crust was found to depend directly on lithology. In the upper crust, alteration halos track given lava units, and likely flow-edge permeability paths, rather than the extent of alteration decreasing steadily as a function of depth below seafloor (Chapter 4.2.2). For intrusive crust, olivine-rich intervals can be highly serpentinized while adjacent gabbroic intervals are only modestly altered, as found at Atlantis Massif core complex (Chapter 4.2.1). Local weakening related to alteration that proceeds to talc formation plays an important role in strain localization that characterizes detachment faulting and oceanic core complex formation. While lower rate flow and alteration continue as oceanic plates age, with basement structure impacting local rates and hydrologic directivity (Chapter 4.2.2), it is at the convergent margins where the next round of major physical–chemical interplay strongly impacts geologic processes. Kastner et al. (Chapter 4.4.3) elucidate the key role of thermal structure in sediment dehydration, and other geochemical reactions that fingerprint fluid source region, and determine that fluid pathways depend on whether a subduction zone is accretionary or erosive. In the former, the plate boundary decollement channels much of the flow, whereas in the latter, conduits through the sediment prism are also important.

Finally, IODP 2003–2013 findings have helped advance modeling of solid Earth processes. Recent models of oceanic crust hydrology are beginning to narrow the field of possible flow patterns and heat advection in both volcanic and intrusive crust (Chapter 4.2.2), with measurements of porosity and temperature to constrain numerical experiments. While IODP did not explicitly tackle rifted margins, a new understanding of the nontrivial role of magmatism in the development of long-lived detachments and unroofing of lithospheric blocks (core complexes, Chapter 4.2.1) constrain recent dynamic models of rifting. Thermal models of convergent margins are significantly improved by constraints that IODP fluid chemistry, core alteration, and borehole measurements provide (Chapters 4.3.2 and 4.3.3). Both theoretical and numerical modeling of earthquake processes have been a major focus in the broader geophysical community in 2004–2014, but predictions are highly dependent on assumed material properties within the fault zone. By sampling in situ material (Chapter 4.4.1) and testing its frictional behavior under variable slip velocity (Chapter 4.4.2), IODP has contributed data that are crucial to new understanding of subduction zone fault behavior, including at previously unexpected very shallow subseafloor depths. In a remarkable feat of rapid planning and implementation, in very challenging deep-water conditions, IODP obtained unique samples from a fault displaying evidence of coseismic slip associated with the devastating 2011 Tohoku earthquake and tsunami.

Almost all the authors for this Chapter played a lead role in designing and carrying out a central component of IODP work on the topic presented in each paper. Here, they provide an overview of the scientific context and major findings of that research. For investigators interested in delving more deeply, these summaries can serve as an entry point into the wealth of underlying literature, as well as to the rich set of drilling, logging, and site survey data, many of which are far from being fully tapped and all of which are openly available for future study.

Chapter 4.2.1

Formation and Evolution of Oceanic Lithosphere: New Insights on Crustal Structure and Igneous Geochemistry from ODP/IODP Sites 1256, U1309, and U1415

Benoit Ildefonse,[1,*] Natsue Abe,[2] Marguerite Godard,[1] Antony Morris,[3] Damon A.H. Teagle[4] and Susumu Umino[5]

[1]Géosciences Montpellier, Université Montpellier, Montpellier, France; [2]Institute for Research on Earth Evolution, Japan Agency for Marine-Earth Science and, Technology (JAMSTEC), Yokosuka, Japan; [3]School of Geography, Earth and Environmental Sciences, Plymouth University, Plymouth, UK; [4]National Oceanography Centre Southampton, University of Southampton, Southampton, UK; [5]Department of Earth Sciences, Kanazawa University, Kanazawa, Japan
*Corresponding author: E-mail: ildefonse@um2.fr

4.2.1.1 INTRODUCTION

In March–April 1961, the drilling barge CUSS1 undertook the first scientific ocean drilling operation off Guadalupe Island, ~240 km west of Baja California (Mexico). This expedition, beautifully reported in LIFE magazine by the novelist John Steinbeck and the renowned science photographer Fritz Goro, was the first (and eventually only) concrete manifestation of Project Mohole[1]. Project Mohole was an ambitious endeavor proposed initially in the late 1950s by the American Miscellaneous Society (AMSOC), an informal group of notable U.S. scientists, mostly geophysicists and oceanographers associated with the Office of Naval Research, including Harry Hess and Walter Munk. The principal aim was to drill through the oceanic crust, through the Mohorovičić Discontinuity, and to retrieve samples from Earth's mantle (Teagle & Ildefonse, 2011). In his

---

1. http://www.nationalacademies.org/mohole.html.

Developments in Marine Geology, Volume 7. http://dx.doi.org/10.1016/B978-0-444-62617-2.00017-7

book "A Hole in the Bottom of the Sea" (Bascom, 1961), Willard Bascom, Director of Project Mohole, notes that probably the first written suggestion for deep penetration down into the mantle was given by Frank Estabrook, an astrophysicist from the Basic Research branch of the U.S. Army in Pasadena (Estabrook, 1956). In addition to accomplishing a major technological break-through (including the invention of dynamic positioning, the drilling guide horn, and deepwater drill hole reentry), the Project Mohole drilling expedition in 1961 cored into oceanic basement for the first time, and demonstrated that the uppermost ocean crust consisted of basaltic lavas. This achievement received a personal letter of congratulations from U.S. President Kennedy[2].

More than 50 years later, and since the launch of DSDP in 1968, oceanic basement has been drilled in a range of geodynamic settings. The first inten-tional drilling efforts in the oceanic crust were DSDP Leg 34 in 1973–1974, on the Nazca Plate in the Eastern Pacific Ocean (Yeats et al., 1976), and DSDP Leg 37 in 1974 on the western flank of the Mid-Atlantic Ridge, south of the Azores Plateau (Aumento et al., 1977). These Legs recovered for the first time substan-tial cores of basalt from the upper oceanic crust. In the Atlantic, cores from Site 334 also recovered small amounts of gabbros and serpentinized peridotites for the first time at shallow subseafloor depths, demonstrating, together with future seafloor geology, marine geophysics, and drilling along the Mid-Atlantic Ridge, that the layered "Penrose" model from the early 1970s (Anonymous, 1972) is not applicable worldwide. Since then, scientific ocean drilling has significantly contributed to our understanding of the variability of the architecture of oceanic lithosphere. The style of accretion critically depends on the balance between magma production, hydrothermal cooling, and tectonics, which is on the first order related to spreading rate. Seismic, bathymetric, and marine geological observations indicate that ocean crust formed at fast spreading rates (full rate >80 mm per year) is much less variable than crust formed at slow spreading rates (<40 mm per year) and is closer to the ideal "Penrose" layered model developed from ophiolites (Anonymous, 1972). Yet, numerous basic, direct observations regarding the architecture of in situ, present-day ocean crust, including rock types, deformation, or geochemistry, are yet to be made. A compilation of holes into the ocean crust cored by scientific ocean drilling since the beginning of DSDP is presented in Table 4.2.1.1 and Figures 4.2.1.1 and 4.2.1.2. Only 37 holes deeper than 100 m have been cored in oceanic crust since DSDP Leg 37 in 1974 (Figure 4.2.1.2). The recovered material represents <2% of the cores recovered to date by DSDP, ODP, and IODP. In spite of this cursory sampling, scientific drilling has contributed significantly to advancing knowledge of ocean crust architecture, mid-ocean-ridge accretion, and hydrothermal processes (e.g., Alt et al., 1996; Blackman et al., 2011; Dick et al., 2000; Dick, Natland, & Ildefonse, 2006; Ildefonse et al., 2007a,b; Ildefonse, Rona, & Blackman, 2007c; Teagle, Alt, & Halliday, 1998; Wilson et al., 2006).

---

2. http://nationalacademies.org/includes/KennedyTelegram.pdf.

TABLE 4.2.1.1 Drill Holes into Oceanic Basement in Intact Crust and Tectonically Exposed Lower Crust and Upper Mantle. This Compilation Does Not Include Other "Hard Rock" Drill Holes in Oceanic Plateaus, Arc Basement, Hydrothermal Mounds, or Passive Margins.

Leg/ Expedition	Hole	Location	Ocean	Water Depth (m)	Age (my)	Sediment Thickness (m)	Basement Penetration (m)	Recovery (%)	Spreading Rate	Lithology	Comments
24	238	11°09.21'S 70°31.56'E	Indian	2844.5	30	506	81	50	S/I	Basaltic lavas	Projection of Chagos-Laccadive Plateau
26	257	30°59.16'S 108°20.99'E	Indian	5278	120	262	65	50	S/I	Basalt and breccia	Wharton Basin off Perth, Australia
34	319A	13°01.04'S 101°31.46'W	Pacific	4296	16	98	59	25	F	Basaltic lavas	Bauer Deep, 13°S East Pacific Rise
37	332A	36°52.72'N 33°38.46'W	Atlantic	1851	3.5	104	331	10	S	Basalt, basalt breccia, and interlayered sediments	Mid-Atlantic Ridge 36°–37°N
37	332B	36°52.72'N 33°38.46'W	Atlantic	1983	3.5	149	583	21	S	Basalt and basalt breccia	Mid-Atlantic Ridge 36°–37°N
37	333A	36°50.45'N 33°40.05'W	Atlantic	1665.8	3.5	218	311	8	S	Basalt and basalt breccia	Mid-Atlantic Ridge 36°–37°N
37	335	37°17.74'N 35°11.92'W	Atlantic	3188	15	454	108	38	S	Basaltic lavas	Mid-Atlantic Ridge 36°–37°N
45	395A	22°45.35'N 46°04.90'W	Atlantic	4485	7.3	92	571	18	S	Basaltic lavas and breccia	Mid-Atlantic Ridge 23°N

Continued

TABLE 4.2.1.1 Drill Holes into Oceanic Basement in Intact Crust and Tectonically Exposed Lower Crust and Upper Mantle. This Compilation Does Not Include Other "Hard Rock" Drill Holes in Oceanic Plateaus, Arc Basement, Hydrothermal Mounds, or Passive Margins.—cont'd

Leg/ Expedition	Hole	Location	Ocean	Water Depth (m)	Age (my)	Sediment Thickness (m)	Basement Penetration (m)	Recovery (%)	Spreading Rate	Lithology	Comments
45	396	22°58.88′N 43°30.95′W	Atlantic	4450	9	126	96	33	S	Basaltic lavas	Mid-Atlantic Ridge 23°N
46	396B	22°59.14′N 43°30.90′W	Atlantic	4459	13	151	255	23	S	Basalt and breccia	Mid-Atlantic Ridge 23°N
49	410A	45°30.53′N 29°28.56′W	Atlantic	2987	9	331	49	38	S	Basaltic lavas	Mid-Atlantic Ridge 45°N
49	412A	36°33.74′N 33°09.96′W	Atlantic	2626	1.6	163	131	18	S	Basalt flows and intercalating limestone	Mid-Atlantic Ridge 33°N
51–53	417A	25°06.63′N 68°02.48′W	Atlantic	5478.2	110	208	209	61	S	Basaltic lavas	Western Atlantic
51–53	417D	25°06.69′N 68°02.81′W	Atlantic	5489	110	343	366	70	S	Basaltic lavas	Western Atlantic
51–53	418A	25°02.10′N 68°03.44′W	Atlantic	5519	110	324	544	72	S	Basaltic lavas	Western Atlantic
54	428A	09°02.77′N 105°26.14′W	Pacific	3358.5	2.3	63	53	39	F	Basaltic lavas	9°N East Pacific Rise
63	469	32°37.00′N 120°32.90′W	Pacific	3802.5	17	391	63	34	I	Basaltic lavas	Off California Coast

63	470A	28°54.46'N 117°31.11'W	Pacific	3554.5	15	167	49	33	—	Basaltic lavas	Off California Coast
65	482B	22°47.38'N 107°59.60'W	Pacific	3015	0.5	137	93	54	—	Massive basalt and interlayered sediment	Off Gulf of California
65	482D	22°47.31'N 107°59.51'W	Pacific	3015	0.5	138	50	50	—	Massive basalt and interlayered sediment	Off Gulf of California
65	483	22°53.00'N 108°44.90'W	Pacific	3084	2	110	95	40	—	Massive basalt and pillow basalt with interlayered sediments	Off Gulf of California
65	483B	22°52.99'N 108°44.84'W	Pacific	3084	2	110	157	47	—	Massive basalt and pillow basalt with interlayered sediments	Off Gulf of California
65	485A	22°44.92'N 107°54.23'W	Pacific	2996.5	1.2	153.5	178	51	—	Massive basalt and interlayered sediments	Off Gulf of California

Continued

TABLE 4.2.1.1 Drill Holes into Oceanic Basement in Intact Crust and Tectonically Exposed Lower Crust and Upper Mantle. This Compilation Does Not Include Other "Hard Rock" Drill Holes in Oceanic Plateaus, Arc Basement, Hydrothermal Mounds, or Passive Margins.—cont'd

Leg/ Expedition	Hole	Location	Ocean	Water Depth (m)	Age (my)	Sediment Thickness (m)	Basement Penetration (m)	Recovery (%)	Spreading Rate	Lithology	Comments
68	501	1°13.63′N 83°44.06′W	Pacific	3466.9	6.6	264	73	60	I	Basaltic lavas	South Flank of Costa Rica Rift
69/70/83/111/ 137/140/ 148	504B	1°13.611′N 83°43.818′W	Pacific	3474	6.6	270	1841	20	I	Basalt, stockwork, and diabase	South Flank of Costa Rica Rift
82	559	35°07.45′N 40°55.00′W	Atlantic	3754	35	238	63	37	S	Basaltic lavas	West Flank of Mid-Atlantic Ridge 35°N
82	562	33°08.49′N 41°40.76′W	Atlantic	3172	12	240	90	45	S	Pillow basalt and massive basalt	West Flank of Mid-Atlantic Ridge 33°N
82	564	33°44.36′N 43°46.03′W	Atlantic	3820	35	284	81	43	S	Pillow basalt and minor massive basalt	West Flank of Mid-Atlantic Ridge 34°N
91	595B	23°49.34′S 165°31.61′W	Pacific	5615	80	70	54	28	F	Vesicular aphyric basalt	Central South Pacific
92	597C	18°48.43′S 129°46.22′W	Pacific	4164	30	53	91	53	F	Massive basalt flows	West Flank South East Pacific Rise 18°S

106/109	648B	22°55.32'N 44°56.825'W	Atlantic	3326	0	0	50	12	S	Pillow basalt	Mid-Atlantic Ridge 23°N
109	670A	23°9.996'N 45°1.932'W	Atlantic	3625	0	0	77	6	S	Serpentinized peridotite	Mid-Atlantic Ridge, MARK, 23°N
118/176	735B	32°43.395'S 57°15.959'E	Indian	720	11.8	0	1508	86	S	Gabbro	Atlantis Bank, Southwest Indian Ridge
123	765D	15°58.56'S 117°34.51'E	Indian	5713.8	140	928	267	31	F	North-east mid-ocean-ridge basaltic lavas	Argo Abyssal Plain
129/185	801C	18°38.538'N 156°21.59'E	Pacific	5674	170	462	414	47	F	Pillow basalt, basalt flows, and breccias	Western North Pacific
129	802A	12°5.778'N 153°12.63'E	Pacific	5980	120	509	51	33	F	Basaltic lavas	Western North Pacific
136	843B	19°20.54'N 159°5.68'W	Pacific	4418	95	243	71	37	F	Basaltic lavas	West of Hawaii
147	894E	2°18.059'N 101°31.524'W	Pacific	3014	1	0	29	11	F	Gabbro	Hess Deep
147	894F	2°17.976'N 101°31.554'W	Pacific	3025	1	0	26	7	F	Gabbro	Hess Deep
147	894G	2°17.976'N 101°31.554'W	Pacific	3023	1	0	127.5	35	F	Gabbro	Hess Deep

Continued

TABLE 4.2.1.1 Drill Holes into Oceanic Basement in Intact Crust and Tectonically Exposed Lower Crust and Upper Mantle. This Compilation Does Not Include Other "Hard Rock" Drill Holes in Oceanic Plateaus, Arc Basement, Hydrothermal Mounds, or Passive Margins.—cont'd

Leg/Expedition	Hole	Location	Ocean	Water Depth (m)	Age (my)	Sediment Thickness (m)	Basement Penetration (m)	Recovery (%)	Spreading Rate	Lithology	Comments
147	895A	2°16.638'N 101°26.766'W	Pacific	3821	1	0	17	14	F	Serpentinized peridotite	Hess Deep
147	895B	2°16.638'N 101°26.760'W	Pacific	3821	1	0	10	10	F	Serpentinized peridotite	Hess Deep
147	895C	2°16.632'N 101°26.772'W	Pacific	3820	1	0	38	15	F	Serpentinized peridotite	Hess Deep
147	895D	2°16.638'N 101°26.778'W	Pacific	3821	1	0	94	20	F	Serpentinized peridotite	Hess Deep
147	895E	2°16.788'N 101°26.790'W	Pacific	3753	1	0	88	37	F	Serpentinized peridotite	Hess Deep
147	895F	2°16.902'N 101°26.790'W	Pacific	3693	1	0	26	8	F	Serpentinized peridotite	Hess Deep
148	896A	1°13.006'N 83°43.392'W	Pacific	3459	6.6	179	290	27	I	Basaltic lavas	South Flank of Costa Rica Rift
153	920B	23°20.310'N 45°1.038'W	Atlantic	3339	<1	0	126.4	35.3	S	Serpentinized peridotite	Mid-Atlantic Ridge, MARK, 23°N
153	920D	23°20.322'N 45°1.044'W	Atlantic	3338	<1	0	200.8	47.3	S	Serpentinized peridotite	Mid-Atlantic Ridge, MARK, 23°N
153	921A	23°32.460'N 45°1.866'W	Atlantic	2488	<1	0	17.1	18.1	S	Gabbro	Mid-Atlantic Ridge, MARK, 23°N

Continued

153	921B	23°32.478'N 45°1.842'W	Atlantic	2490	<1	0	44.1	19.4	S	Gabbro	Mid-Atlantic Ridge, MARK, 23°N
153	921C	23°32.472'N 45°1.830'W	Atlantic	2495	<1	0	53.4	11.4	S	Gabbro	Mid-Atlantic Ridge, MARK, 23°N
153	921D	23°32.442'N 45°1.830'W	Atlantic	2514	<1	0	48.6	12.7	S	Gabbro	Mid-Atlantic Ridge, MARK, 23°N
153	921E	23°32.328'N 45°1.878'W	Atlantic	2456	<1	0	82.6	21.4	S	Gabbro	Mid-Atlantic Ridge, MARK, 23°N
153	922A	23°33.162'N 45°1.926'W	Atlantic	2612	<1	0	14.6	63.2	S	Gabbro	Mid-Atlantic Ridge, MARK, 23°N
153	922B	23°31.368'N 45°1.926'W	Atlantic	2612	<1	0	37.4	25.6	S	Gabbro	Mid-Atlantic Ridge, MARK, 23°N
153	923A	23°32.556'N 45°1.896'W	Atlantic	2440	<1	0	70	57.2	S	Gabbro	Mid-Atlantic Ridge, MARK, 23°N
153	924B	23°32.460'N 45°0.858'W	Atlantic	3170	<1	0	30.8	8.7	S	Gabbro	Mid-Atlantic Ridge, MARK, 23°N
153	924C	23°32.496'N 45°0.864'W	Atlantic	3177	<1	0	48.5	23.1	S	Gabbro	Mid-Atlantic Ridge, MARK, 23°N
168	1025C	47°53.250'N 128°38.880'W	Pacific	2602	1.237	106	41	37	I	Basalt	Juan de Fuca Flank
168	1026B	47°45.759'N 127°45.552'W	Pacific	2658	3.511	256	39	5	I	Basalt	Juan de Fuca Flank
168	1026C	47°46.261'N 127°45.186'W	Pacific	2669	3.516	229	19	3.5	I	Basalt	Juan de Fuca Flank

TABLE 4.2.1.1 Drill Holes into Oceanic Basement in Intact Crust and Tectonically Exposed Lower Crust and Upper Mantle. This Compilation Does Not Include Other "Hard Rock" Drill Holes in Oceanic Plateaus, Arc Basement, Hydrothermal Mounds, or Passive Margins.—cont'd

Leg/Expedition	Hole	Location	Ocean	Water Depth (m)	Age (my)	Sediment Thickness (m)	Basement Penetration (m)	Recovery (%)	Spreading Rate	Lithology	Comments
168	1032A	47°46.776'N 128°7.320'W	Pacific	2645	2.621	290	48	6.5	I	Basalt	Juan de Fuca Flank
179	1105A	32°43.135'S 57°16.652'E	Indian	714	11.8	0	158	75	S	Gabbro	Atlantis Bank, Southwest Indian Ridge
185	1149D	31°18.79'N 143°24.03'E	Pacific	5818	133	307	133	17	F	Pillow basalt, basalt flows, and breccias	Western North Pacific
187	1162B	44°37.9'S 129°11.3''E	Indian	5464	18	333	59	17	I	Basaltic lavas and breccia	Australian-Antarctic Discordance
187	1163A	44°25.5'S 126°54.5'E	Indian	4354	17	161	47	33	I	Basaltic lavas	Australian-Antarctic Discordance
187	1164B	43°45.0'S 127°44.8'E	Indian	4798	18.5	150	66	16	I	Basaltic lavas	Australian-Antarctic Discordance
191	1179D	41°04.8'N 159°57.8'E	Pacific	5563.9	129	377	98	44	F	Basaltic lavas	Western North Pacific
200	1224F	27°53.36'N 141°58.77'W	Pacific	4967.1	46	28	147	26	F	Basaltic lavas	Central Pacific
203	1243B	5°18.07'N 110°04.58'W	Pacific	3868	11	110	87	25	F	Basaltic lavas	Western Flank East Pacific Rise 5°N

206	1256C	6°44.18′N 91°56.06′W	Pacific	3634.7	15	251	89	61	F	Basaltic lavas	Cocos plate eastern Flank East Pacific Rise
206/ 309/312	1256D	6°44.16′N 91°56.06′W	Pacific	3634.7	15	250	1257.1	37.1	F	Basaltic lavas, sheeted dike, and varitextured gabbro	Cocos plate Eastern Flank East Pacific Rise
209	1268A	14°50.755′N 45°4.641′W	Atlantic	3007		0	147.6	53.3	S	Serpentinized peridotite	Mid-Atlantic Ridge 15°20′N
209	1270A	14°43.342′N 44°53.321′W	Atlantic	1951		0	26.9	12.2	S	Gabbro	Mid-Atlantic Ridge 15°20′N
209	1270B	14°43.265′N 44°53.225′W	Atlantic	1909		0	45.9	37.4	S	Gabbro	Mid-Atlantic Ridge 15°20′N
209	1270C	14°43.284′N 44°53.091′W	Atlantic	1822		0	18.6	10.6	S	Gabbro	Mid-Atlantic Ridge 15°20′N
209	1270D	14°43.270′N 44°53.084′W	Atlantic	1817		0	57.3	13.4	S	Gabbro	Mid-Atlantic Ridge 15°20′N
209	1271A	15°2.222′N 44°56.887′W	Atlantic	3612		0	44.8	12.9	S	Serpentinized peridotite	Mid-Atlantic Ridge 15°20′N
209	1271B	15°2.189′N 44°56.912′W	Atlantic	3585		0	103.8	15.3	S	Serpentinized peridotite	Mid-Atlantic Ridge 15°20′N
209	1272A	15°5.666′N 44°58.300′W	Atlantic	2560		0	131	28.6	S	Serpentinized peridotite	Mid-Atlantic Ridge 15°20′N

Continued

TABLE 4.2.1.1 Drill Holes into Oceanic Basement in Intact Crust and Tectonically Exposed Lower Crust and Upper Mantle. This Compilation Does Not Include Other "Hard Rock" Drill Holes in Oceanic Plateaus, Arc Basement, Hydrothermal Mounds, or Passive Margins.—cont'd

Leg/Expedition	Hole	Location	Ocean	Water Depth (m)	Age (my)	Sediment Thickness (m)	Basement Penetration (m)	Recovery (%)	Spreading Rate	Lithology	Comments
209	1274A	15°38.867'N 46°40.582'W	Atlantic	3940		0	155.8	22.2	S	Serpentinized peridotite	Mid-Atlantic Ridge 15°20'N
209	1275B	15°44.486'N 46°54.208'W	Atlantic	1562		0	108.7	43.1	S	Gabbro	Mid-Atlantic Ridge 15°20'N
209	1275D	15°44.440'N 46°54.217'W	Atlantic	1554		0	209	50	S	Gabbro	Mid-Atlantic Ridge 15°20'N
301	U1301A	47°45.209'N 127°45.833'W	Pacific	2656	3.5	262	108	0	I	Basaltic lavas	Juan de Fuca Ridge Flank; no coring (CORK)
301	U1301B	47°45.229'N 127°45.826'W	Pacific	2655	3.5	265	318	12.9	I	Basaltic lavas	Juan de Fuca Ridge Flank; recovery is only for the 232 m of cored basement
304	U1309B	30°10.108'N 42°7.110'W	Atlantic	1642	2	2	99.8	45.9	S	Gabbro	Mid-Atlantic Ridge 30°N
304/305	U1309D	30°10.120'N 42°7.113'W	Atlantic	1645	2	2	1413.3	74.8	S	Gabbro	Mid-Atlantic Ridge 30°N

327	U1362A	47°45.663'N 127°45.672'W	Pacific	2661	3.5	236	292	29.6	I	Basaltic lavas	Juan de Fuca Ridge Flank; recovery is only for the 150 m of cored basement
336	U1383C	22°48.124'N 46°03.166'W	Atlantic	4414	8	38.3	293.2	26.5	S	Basaltic lavas	North Pond. Corked Hole. Upper 31.2 m of basement not cored
345	U14515I	2°15.155'N 101°32.662'W	Pacific	4841.8	1	2	35.2	20	F	Gabbro	Hess Deep lower crust
345	U1415J	2°15.160'N 101°32.662'W	Pacific	4838.8	1	2	111.8	15.6	F	Gabbro	Hess Deep lower crust. Upper 15 m of basement not cored
345	U1415P	2°15.149'N 101°32.610'W	Pacific	4852.7	1	2	107.9	32	F	Gabbro	Hess Deep lower crust. Upper 12.5 m of basement not cored

S = Slow (<40 mm per year); I = Intermediate; F = Fast (>80 mm per year); MARK = Mid-Atlantic Ridge Kane Fracture Zone; CORK = Circulation Obviation Retrofit Kit.

Holes Drilled in Ocean Crust (1974-2013)

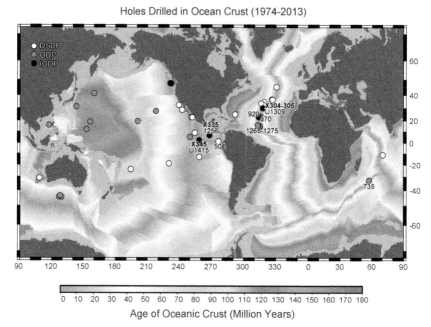

Age of Oceanic Crust (Million Years)

FIGURE 4.2.1.1 DSDP, ODP, and IODP holes drilled in ocean crust >100 mbsf from 1974 to 2013. Sites mentioned in the text are labeled. Also indicated are the locations of the four IODP expeditions reviewed herein. The seafloor age is based on age grid by Müller et al. (2008, revised version 3; www.earthbyte.org/). This map does not include "hard rock" drill holes in oceanic plateaus, arc basement, hydrothermal mounds, or passive margins.

Hole 504B, located on 6.9 my crust formed at an intermediate rate at the Costa Rica Rift (Figure 4.2.1.1), remains the second deepest hole (2111 meters below seafloor (mbsf)) in all of scientific ocean drilling and the deepest into oceanic basement (1836 m subbasement; Alt et al., 1996). This site was host to drilling and other experiments over eight DSDP and ODP cruises, and was the first hole to penetrate completely through the volcanic lava sequences and ~1 km into sheeted dikes in intact, layered oceanic crust. It remains a reference hole for hydrothermal alteration of the upper ocean crust (e.g., Alt, Honnorez, Laverne, & Emmermann, 1986; Alt, Muehlenbachs, & Honnorez, 1986), and for the geological significance of seismic Layers 2A and 2B (e.g., Carlson, 2011), thought to represent a porous lava flow layer and the underlying sheeted dike complex, respectively (e.g., Herron, 1982). In fast-spread crust, the Layer 2/3 boundary is classically assumed to be the boundary between the upper crust (sheeted dikes and lavas) and the underlying plutonic gabbroic crust. This transition between typical layer 2 and layer 3 seismic velocity gradients with depth has been sampled in situ Hole 504B, where it lies in the sheeted dike complex, and appears related to changes in alteration and porosity structure (Carlson, 2010; Detrick, Collins, Stephen, & Swift, 1994).

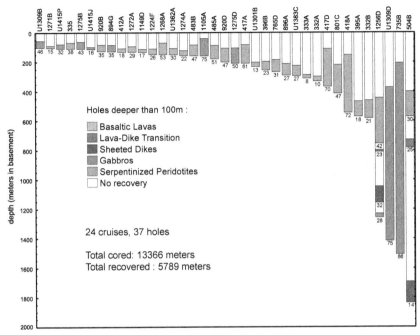

FIGURE 4.2.1.2   Compilation chart showing holes drilled >100 m in intact crust and tectonically exposed lower crust and upper mantle from 1974 to 2013 (drill hole locations in Figure 4.2.1.1). For each hole are indicated the hole number and the recovery (in percent) for each lithology. This compilation does not include "hard rock" drill holes in oceanic plateaus, arc basement, hydrothermal mounds, or passive margins.

A complete, continuous section of intact, layered fast-spread crust down to the cumulate gabbro layers has yet to be drilled and remains a first-order scientific target of ocean drilling for the ocean crust research community (e.g., Chikyu+10 Steering Committee, 2013; Dick & Mével, 1996; Ildefonse et al., 2010a; Ildefonse et al., 2010b; Ildefonse, Christie, & the Mission Moho Workshop Steering Committee, 2007b; IODP Science Plan 2013–2023, 2011; Ravelo et al., 2010; Teagle, Wilson, & Acton, 2004; Teagle et al., 2009). Over the last 10 years, Hole 1256D (ODP Leg 206, Integrated Ocean Drilling Program (IODP) Expeditions 309, 312, and 335; Figure 4.2.1.1) penetrated to the base of the sheeted dike complex and the uppermost gabbro in superfast-spread crust (i.e., >130 mm per year; Lonsdale, 1977); this was the first sampling of the transition to plutonic rocks in intact ocean crust (Teagle et al., 2006; Teagle, Ildefonse, Blum, & the Expedition 335 Scientists, 2012; Wilson et al., 2003; Wilson et al., 2006).

An alternative to drilling deep through intact, fast-spread crust is to access the deepest part of the crust in tectonic windows. This strategy was recently used at Hess Deep, in the eastern equatorial Pacific just off the East Pacific Rise (EPR), during IODP Expedition 345 to access lower crustal gabbroic rocks (Gillis, Snow, Klaus, & the Expedition 345 Scientists, 2014; Gillis et al., 2014a; Figure 4.2.1.1).

In 1986, serpentinized mantle peridotites were intentionally drilled at a mid-ocean ridge for the first time during ODP Leg 109 (Site 670; Detrick et al., 1988) on the west wall of the Mid-Atlantic Ridge median valley near 23°10′N. In the same area, south of the Kane Fracture Zone, 95 m of serpentinized peridotites were recovered from a 200-m-deep hole during ODP Leg 153 (Site 920; Cannat et al., 1995). Ten years later, ODP Leg 209 (Sites 1268–1275) returned to drill in the peridotite-rich area around the 15°20′N fracture zone on the Mid-Atlantic Ridge (Kelemen, Kikawa, Miller, & Shipboard Scientific Party, 2007). Thirteen holes at six sites along the spreading axis penetrated mantle peridotite and gabbroic rocks in proportions that were roughly 70/30, similar to what was previously sampled by dredging and submersible on the seafloor in the same area (compiled data and references in Kelemen et al., 2004). These findings, together with geophysics and seafloor geology, contributed to the understanding that slow-spreading oceanic crust can have a very variable architecture, locally consisting of relatively small gabbro bodies intruded into serpentinized mantle peridotites, and that spreading in nonlayered, magma-poor crust is largely accommodated by detachment faulting (e.g., Cannat, 1993; Dick, 1989; Escartín et al., 2008; Escartín & Canales, 2011; Kelemen et al., 2007; Lagabrielle, Bideau, Cannat, Karson, & Mével, 1998). In this type of crust, the spatially and temporally variable interplay of magmatic and tectonic processes is locally expressed in the development of Oceanic Core Complexes (OCCs). OCCs are domal features that result from the exhumation of lower crustal and upper mantle rocks through long-lived (typically 1–3 my) detachment faulting (e.g., Cann et al., 1997; Dick, Tivey, & Tucholke, 2008; Escartín & Canales, 2011; Escartín, Mével, MacLeod, & McCaig, 2003; MacLeod et al., 2009; Smith, Escartín, Schouten, & Cann, 2008; Tucholke & Lin, 1994; Tucholke, Lin, & Kleinrock, 1998). The first deep hole drilled in an OCC was the 1508-m-deep ODP Hole 735B in gabbroic rocks of the Atlantis Bank on the Southwest Indian Ridge (Dick et al., 2000). More recently, IODP successfully sampled another OCC, with Hole U1309D penetrating 1415 m of gabbroic rocks at the Atlantis Massif on the Mid-Atlantic Ridge during Expeditions 304 and 305 (Blackman et al., 2011; Ildefonse et al., 2007a).

This chapter draws from results of 10 years of IODP ocean drilling (2003–2013) in igneous oceanic crust, focusing on the two deep holes in slow-spread (Hole U1309D) and fast-spread (Hole 1256D) crust, and briefly summarizing the outcome of the recent IODP Expedition 345 at Hess Deep. These drilling expeditions have improved our understanding of crustal architecture and accretion processes at mid-ocean ridges.

4.2.1.2 DEEP DRILLING IN SLOW-SPREAD CRUST: THE ATLANTIS MASSIF

4.2.1.2.1 Background

IODP Expeditions 304 and 305, from November 2004 to March 2005, targeted one of the emblematic OCCs, the Atlantis Massif, at 30°N on the Mid-Atlantic

Ridge (Blackman, Cann, Janssen, & Smith, 1998; Blackman et al., 2002; Cann et al., 1997; Schroeder & John, 2004). The Atlantis Massif is a young (0.5–2 my) OCC, situated at the inside corner of the intersection between the Mid-Atlantic Ridge and the Atlantis Fracture zone at 30°N. Seafloor lithologies observed by dredging and submersible sampling include serpentinites, gabbros, and basaltic rocks (Figure 4.2.1.3; Cann et al., 1997; Blackman et al., 2002). The active, serpentinite-hosted Lost City hydrothermal vent field (Früh-Green et al., 2003; Kelley et al., 2001) is located immediately south of the summit of the massif. Localized deformation and fluid flow associated with detachment faulting are documented from seafloor sampling on the southern wall of the massif (Boschi, Früh-Green, Delacour, Karson, & Kelley, 2006; Karson et al., 2006; Schroeder & John, 2004). Two deep holes were drilled at Site U1309 during IODP Expeditions 304 and 305, in the footwall of the detachment fault (Figure 4.2.1.3; Blackman et al., 2006, 2011). In

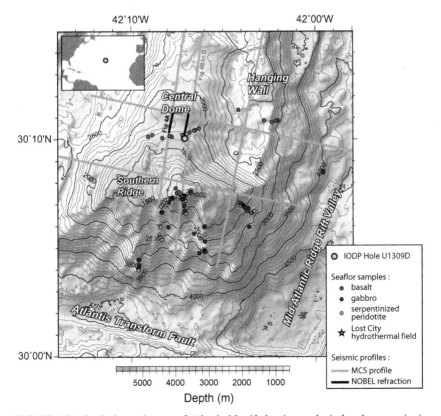

FIGURE 4.2.1.3 Bathymetric map of Atlantis Massif showing geological and some seismic (reflection and refraction) data coverage prior to drilling, and the location of Site U1309 (yellow symbol (pale gray in print versions)), modified from Blackman et al. (2006). The WNW-ESE trending corrugations on the domal core mark the exposed detachment fault; hanging wall block(s) are volcanic.

addition, IODP Expedition 340T revisited this site in 2012 to conduct 5 days of borehole logging in Hole U1309D, to complete the logging program that was hampered by bad weather conditions at the end of IODP expedition 305 (Blackman et al., 2006). The program included completing sonic logging to the bottom of the hole, and acquiring new seismic imaging data from a Vertical Seismic Profile (VSP) experiment (Expedition 340T Scientists, 2012).

4.2.1.2.2 Revisiting the Geophysical Signature of the Atlantis Massif

Based on the dominant occurrence of serpentinized mantle peridotites sampled on the south flank of the southern ridge (Figure 4.2.1.3; e.g., Blackman et al., 2002; Boschi et al., 2006), and on the predrilling interpretation of available seismic (Figure 4.2.1.3; Collins & Detrick, 1998; Collins, Tucholke, & Canales, 2002; Canales, Tucholke, & Collins, 2004) and gravimetric (Blackman, Cann, Janssen, & Smith, 1998; Nooner, Sasagawa, Blackman, & Zumberge, 2003) data, one of the goals of IODP Expeditions 304/305 was to drill and sample the core of an OCC that was inferred to be mostly made of mantle peridotites. It was hoped that fresh mantle peridotite was present at reasonably shallow depths (~800 mbsf) accessible by conventional nonriser drilling and hence, could be sampled for the first time (see Blackman et al., 2011; for a summary of the predrilling rationale). The recovery of a 1415-m-long gabbroic section demonstrated that geological sampling by drilling is essential to better constrain the interpretation of geophysical data in ocean crust, particularly in slow-spread crust, which can be lithologically heterogeneous on small scales. Drill cores, together with wireline geophysical measurements provided critical observations for reexamining the precruise geophysical interpretations and for proposing a revised analysis of the lithological architecture of the Atlantis Massif and other OCCs (Blackman, Karner, & Searle, 2008; Blackman, Canales, & Harding, 2009; Blackman and Collins, 2010; Canales, Tucholke, Xu, Collins, & DuBois, 2008; Collins, Blackman, Harris, & Carlson, 2009; Henig, Blackman, Harding, Canales, & Kent, 2012). The positive 30–40 mGal residual gravity anomaly centered on the core of the OCC is now considered to reflect a gabbroic core with a density of 2900 kg/m^3, the adjacent basaltic block is significantly fractured with an average density of 2600 kg/m^3, and the portion of the lithosphere deeper than 1500 mbsf that has density lower than mantle rock (gabbroic, and/or significantly altered peridotite) is ~3-km thick within the domal core (Blackman, Karner, & Searle, 2008). Postdrilling seafloor refraction modeling showed that shallow mantle velocities are not required (Figure 4.2.1.4; Collins et al., 2009) and velocities in the upper several hundred meters are typical of mafic rock (≤6.5 km/s). Tomographic inversion of refracted arrivals recorded by a 6-km multichannel seismic (MCS) streamer (Canales et al., 2008) produced a similar velocity structure to 1.0–1.5 km depth across the Central Dome (Figure 4.2.1.4(D)). Further detailed processing of the MCS data, continued downward using the "synthetic ocean bottom experiment" technique (Henig et al., 2012),

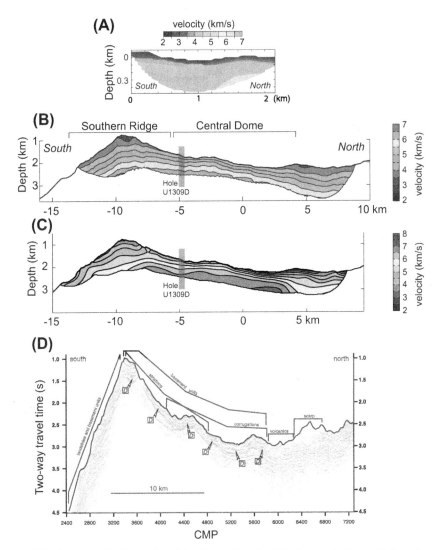

FIGURE 4.2.1.4 Seismic imaging of the Atlantis Massif. (A) Preferred velocity model along the
NOBEL refraction Line 9 on the central dome (location in Figure 4.2.1.3), determined from forward
modeling by Collins et al. (2009). (B) Velocity model based on inversion of multichannel seismic
(MCS) refractions for Line 4 (location in Figure 4.2.1.3) by Canales et al. (2008). (C) Velocity
model based on inversion of MCS refractions for Line 4 by Henig et al. (2012). (D) Time-migrated
section for MCS Line 4 (Canales et al., 2004); D: D-reflector; CMP: Common-Midpoint.

reveals significant spatial (both along-axis and along-flow line) variability of the
P-wave velocity structure at the scale of the Atlantis Massif. This variability is
generally consistent with lithologically heterogeneous crust, made of gabbroic
plutons intruded into a dominantly peridotitic (serpentinized) crust. The western
two-thirds of the southern ridge, to date unsampled by drilling, is dominated by

low-velocity material that is likely to be serpentinized peridotites (at least down to ~1 km below seafloor), whereas the eastern part displays a higher velocity body that is interpreted as being a gabbro pluton. Analysis of a 40-km-long EW air gun refraction profile across the central dome provided constraints on velocities to depths as great as 7 km (Blackman and Collins, 2010). Traveltime tomography suggests that significant volumes of rock with mantle-like velocity (>7.5 km/s) only occur more than ~5 km below seafloor within the dome, and only below 6 km in the axial valley.

Initial results from IODP Expedition 340T seem to indicate that combined lithological changes and local increases in alteration could produce an impedance contrast strong enough to account for the reflectivity in the upper part of the section. In particular, the sonic log confirms that the borehole velocity of altered olivine-rich troctolite intervals in Hole U1309D may be sufficiently distinct from that of surrounding rocks (P-wave velocity (V_P) ~0.5 km/s slower) to produce an MCS reflection (Expedition 340T Scientists, 2012). However, the source(s) of the impedance contrast that gives rise to the strong reflection imaged throughout much of the dome (the "D-reflector"; Canales et al., 2004) is yet to be identified.

4.2.1.2.3 The Interplay between Magmatism and Tectonics Controls the Development of OCCs

Hole U1309D sampled a 1415-m-long section of gabbroic rocks in the core of the Atlantis Massif OCC, with 75% recovery (Figure 4.2.1.5). This is the second longest core drilled in slow-spread crust, after ODP Hole 735B in the Southwest Indian Ocean (Dick et al., 2000). The borehole lithology at Site U1309D contrasts remarkably with the seafloor lithology of the Atlantis Massif. Although seafloor sampling dominantly recovered serpentinized peridotites on the south face of the Southern Ridge (approximately 70% of the recovered samples; Blackman et al., 2002; Karson et al., 2006; Schroeder & John, 2004), only three thin (<1 m) intervals of ultramafic rocks interpreted as residual mantle peridotites, intercalated within gabbros in the upper 225 m of the section (Tamura, Arai, Ishimaru, & Andal, 2008), were recovered in Holes U1309B and U1309D on the Central Dome, 5 km to the north. It is worth noting that including the completion of IODP Expeditions 304/305, ODP and IODP has drilled 16 holes in four different OCCs, and that all holes had recovered exclusively gabbroic sections (Ildefonse et al., 2007a). Hence OCCs are associated with significant magmatism. In contrast with early models (e.g., Escartin et al., 2003; Tucholke & Lin, 1994; Tucholke et al., 1998), this basic observation indicates that OCCs are not the magma-starved endmember of slow-spreading ridges. The revised model proposed that, in heterogeneous crust composed of gabbroic plutons intruded into variably serpentinized mantle peridotites (resulting from episodic magmatic activity), the development of an OCC is associated with, or results from a period of enhanced magmatism at depth. If this model is correct, the

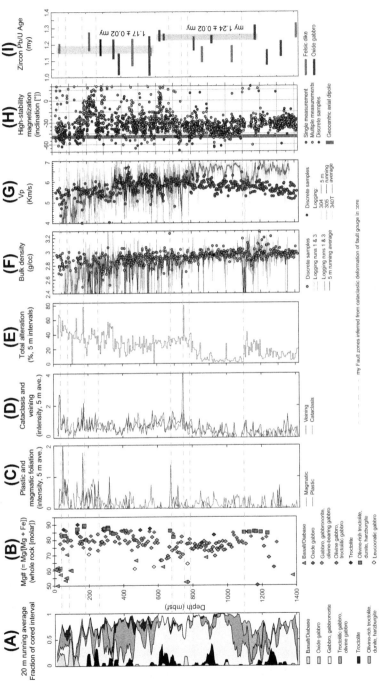

FIGURE 4.2.1.5 Downhole plots of data for IODP Hole U1309D (*modified from Ildefonse et al., 2006*). (A) Lithology (20 m running average, white=no recovery); (B) Mg# (shipboard data); (C) plastic and magmatic foliation intensity (shipboard macroscopic description); (D) cataclasis and veining intensity (shipboard macroscopic description); (E) total alteration (shipboard macroscopic description); (F) bulk density (from borehole and discrete sample shipboard measurements); (G) compressional velocity (from borehole and discrete sample shipboard measurements); (H) inclination of stable remanent magnetization (shipboard measurements); (I) ages obtained for core samples from zircon (postcruise data, Grimes et al., 2008. The gray vertical bars are interpretative coherent age blocks, above and below 600 mbsf, with the average ages indicated aside).

domal morphology of OCCs would essentially be the consequence of having larger gabbroic plutons being exhumed and unroofed by the associated detachment fault (Blackman et al., 2011; Ildefonse et al., 2007a). The model implies that serpentinized fault zones envelop gabbro bodies, explaining why serpentinites can be the dominant lithology on the top or flanks of OCCs, while drilling to date suggests that their core is dominantly gabbroic. The inference that OCCs are associated with significant magma emplacement in the crust is consistent with a series of recent numerical models, which suggest that the optimal magmatic input for the development of long-living detachment faults and OCCs corresponds to about half of the total spreading (Behn & Ito, 2008; Buck, Lavier, & Poliakov, 2005; Olive, Behn, & Tucholke, 2010; Tucholke, Behn, Buck, & Lin, 2008). It is also consistent with observations from the southwest Indian Ocean (Cannat et al., 2006), where the true amagmatic stage of spreading at the Southwest Indian Ridge corresponds to the development of a smooth topography through conjugate, long-lived detachment faulting, marked by broad ridges and the absence of fault scarps and volcanics. OCCs there are identified through their corrugated topography, and are interpreted to represent the intermediate level of magmatic input between the fully magmatic volcanic seafloor and the amagmatic smooth seafloor.

4.2.1.2.4 Paleomagnetic Data and Borehole Imaging Constrain Tectonic Rotation in OCCs

Competing models for the development of OCCs through long-lived detachment faulting involve either displacement on planar, low-angle normal faults (e.g., Wernicke, 1995) or progressive shallowing by rotation of initially steeply dipping (>45°) faults as a result of flexural unloading (i.e., the "rolling-hinge" model; e.g., Lavier, Buck, & Poliakov, 1999). Magmatic rocks acquire a magnetization in the direction of the ancient geomagnetic field at the time of their cooling and crystallization that may be used as a marker for subsequent tectonic rotation. The different models of OCC development can be tested by comparing the orientation of paleomagnetic remanence vectors in recovered magmatic rocks to the expected orientation of the Earth's magnetic field at the sampling location. However, a fundamental difficulty inherent in scientific ocean drilling is that core samples lack azimuthal orientation in the geographical reference frame. Hence, documenting the amount and style of tectonic rotation of OCC footwall sections using paleomagnetic data becomes a problem with too many unknowns and multiple solutions. The way around this problem is to combine paleomagnetic data with structural observations in the core and borehole wireline imaging (e.g., Haggas, Brewer, Harvey, & MacLeod, 2005; MacLeod, Parson, Sager, & the ODP Leg 135 Scientific Party, 1992). In the absence of borehole imaging, Garcés and Gee (2007) had to make an assumption about the orientation of the rotation axis for the proposed tectonic rotations inferred from paleomagnetic measurements in serpentinite and gabbro sections exhumed by detachment faulting in the 15°20′N fracture zone region of the Mid-Atlantic Ridge, sampled during ODP Leg 209 (Kelemen et al., 2007). For the first time, Hole U1309D offered complete, high-quality Formation MicroScanner (FMS) imaging of the borehole coupled with high core recovery, enabling a comprehensive core-log integration study (Figure 4.2.1.6). Using paleomagnetic remanence data and FMS images to reconstruct the true orientation of core pieces, Morris et al. (2009) demonstrated that the upper 400 m of the footwall of the detachment fault in the Atlantis Massif experienced an ~45° anticlockwise rotation around a ridge-parallel horizontal axis trending NNE (010°), providing unequivocal validation of the rolling-hinge model (e.g., Lavier et al., 1999) for the development of OCCs. Analysis of the high-quality FMS imagery from Hole U1309D also allowed the distribution of structures within the Atlantis Massif OCC footwall to be comprehensively documented (Pressling, Morris, John, & MacLeod, 2012), revealing details of the internal structure of an OCC

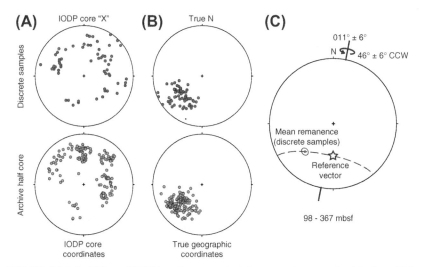

FIGURE 4.2.1.6 Effect of FMS-based core reorientation on paleomagnetic data from Hole U1309D, and resulting constraints on footwall rotation in the Atlantis Massif Oceanic Core Complex (OCC) *(modified after Morris et al., 2009)*. (A) Paleomagnetic remanence directions from discrete samples (red symbols (dark gray in print versions)) and archive half core sections (green symbols (light gray in print versions)) in the unoriented IODP core reference frame have a scattered distribution as a result of the drilling process. (B) The same data after individual core pieces have been reoriented to a true geographic reference frame by matching structures observed in the core with the corresponding structures in the borehole wall (the orientation of which may be determined from the FMS imagery). The independently reoriented paleomagnetic data now display a clustered distribution from which a reliable mean remanence vector may be calculated. (C) Comparison of the full mean remanence vector from discrete samples with the reference geocentric axial dipole vector for Atlantis Massif allows the orientation of the axis of footwall rotation (tilting) and angle of rotation to be determined, indicating substantial rotation around a Mid-Atlantic Ridge (MAR)-parallel axis consistent with rolling-hinge models for OCC development.

for the first time. A zone of brittle fracturing dominates the upper 385 m of the hole, and structures indicating detachment-related damage zones extend down to 750 mbsf into the OCC footwall (See also Michibayashi et al., 2008).

4.2.1.2.5 Protracted Construction of the Lower Crust

Cores recovered in the upper 120 m at Site U1309 contain 35–45% of diabase/basalt in the form of subhorizontal, 5- to 10-m-thick intrusive bodies (Blackman et al., 2011) with chilled margins against cataclastic gabbros, amphibole-rich breccias, and undeformed diabase. They are interpreted as dikes intruded into the detachment fault prior to tectonic rotation (McCaig & Harris, 2012). The composition of all diabase recovered at Site U1309D falls within the range of basalt compositions for the Mid-Atlantic Ridge in the 30°N region (Godard et al., 2009). The dominantly gabbroic sequence in Hole U1309D comprises hundreds of individual lithologic units ranging in composition from oxide-gabbro to olivine-rich troctolite (Figures 4.2.1.5 and 4.2.1.7), identified on the basis of modal composition, grain-size changes, and/or textural changes. Units vary in thickness, from centimeters to tens of meters (John et al., 2009). Many intrusive contacts were recovered and indicate a general relationship of more evolved rock types intruded into more primitive rock types, although the converse sense of intrusion is also observed. Leucocratic intrusions (centimeter-scale thickness) cut the sequence and, together with oxide-gabbro intervals, host zircon grains that have been used to obtain crystallization ages of the recovered crustal section. Ages range from 1.08 ± 0.07 my to 1.28 ± 0.05 my, which indicates that the magmatic accretion of the multiple units described above has occurred over a minimum of approximately 100–200 ky (Figure 4.2.1.5; Grimes, John, Cheadle, & Wooden, 2008). The distribution of ages is inconsistent with models requiring constant depth melt intrusion beneath the detachment fault. Instead, data suggest a protracted construction model (Grimes et al., 2008), in which gabbroic rocks intrude at various depths below the ridge axis over a depth range of at least 1.4 km, before being exhumed by the detachment fault.

   Modal mineralogy and bulk rock geochemistry (e.g., Figure 4.2.1.7(A) and (B)) of the gabbroic sequence are typical of a cumulate series crystallizing from a mid-ocean ridge basalt (MORB) source. However, the bulk composition of Hole U1309D does not represent the complement to basalts recently erupted at the nearby spreading axis (Godard et al., 2009). The composition of the recovered rocks spans a large spectrum (Figure 4.2.1.7), with some troctolites and gabbros, respectively having some of the highest and lowest MgO contents of oceanic crustal rocks sampled to date. The bulk trace element and major element chemistry are consistent with most or all of the section being crystallized from a MORB melt at depth. When all lithologies, including leucocratic intervals, are accounted for, an extra basaltic counterpart to this section is not required (Figure 4.2.1.8; see detailed discussion in Godard et al., 2009).

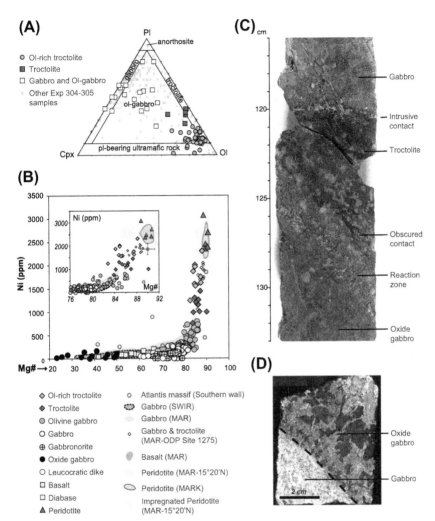

FIGURE 4.2.1.7 Petrology, geochemistry, and photographs of Site U1309 core samples *(From Blackman et al., 2011)*. (A) Compositions of representative suite of samples from site U1309 are shown by small gray symbols. Other symbols indicate samples from within the intervals that are dominantly olivine-rich troctolite that were studied in detail by Drouin et al. (2009). (B) Bulk rock chemistry for Site U1309 and comparison with other areas (Godard et al., 2009). Note how troctolites and olivine-rich troctolites fill in previously sparsely populated Ni-Mg# space. (C) Photo of Core 304-U1309D-69R-1 shows intrusive contact between gabbro and troctolite and lower contact that is a reaction zone between the troctolite and a later oxide gabbro injection. (D) Oxide gabbro dike intrudes gabbro (Hole U1309D, 386 mbsf) with a sharp lower contact with gabbro (Grimes et al., 2008).

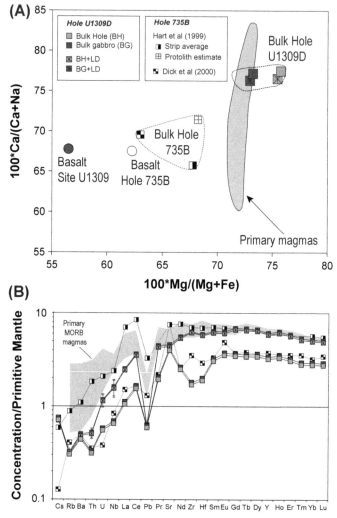

FIGURE 4.2.1.8   Bulk composition of Hole U1309D compared to the compositions of bulk gabbroic crust at Hole 735B and of primary mid-ocean ridge basalt (MORB) magma illustrated on (A) Ca# versus Mg# and (B) extended trace element diagrams *(modified from Godard et al., 2009)*. The bulk composition of Hole U1309D is calculated taking into account all gabbroic rocks (BH) and without the olivine-rich troctolites (BG). The estimate of the bulk composition of Hole U1309 (BH) is characterized by high Mg# compared to primary magma compositions. These high Mg# may indicate a cumulate composition complementary to that of MORB, but also an overestimation of the mass of the most primitive endmember of the gabbroic suite, or an underestimation of the mass of its most evolved endmembers in our calculations. These hypotheses are tested using two models: (1) BH+LD: a 95/5% mix of BH and leucocratic dyke (LD); (2) BG+LD: a 95/5% mix of BG and LD. Estimates of the bulk composition of Hole 735B (Hart et al., 1999; Dick et al., 2000) as well as the mean compositions of basalts at Site U1309 and Hole 735B are shown for comparison. Hole 735B "protolith estimate" was determined on the basis of analyses conducted on a small suite of fresh and unaltered samples representative of the "characteristic lithologies"

Relatively low-pressure crystallization depths (≤200 MPa, ~6 km) are inferred based on the modal relationships and chemistry of cumulus and intercumulus phases (Godard et al., 2009; Suhr, Hellebrand, Johnson, & Brunelli, 2008). This conclusion is supported by a combination of rock ages, assumptions about fault geometry, and isotherm depth that constrain crystallization depths at 5–7 km (Grimes, Cheadle, John, Reiners, & Wooden, 2011).

4.2.1.2.6 Olivine-Rich Troctolites: Mantle Contribution to the Igneous Lower Crust

Olivine-rich troctolites (>70% olivine) are present as relatively thin (approximately 1–12 m) units within two main intervals in Hole U1309D (approximately 310–350 mbsf and 1090–1235 mbsf; Figure 4.2.1.5). Bulk rock geochemical signatures of these troctolites are more primitive (high MgO) than other mafic samples from the ocean crust (Figure 4.2.1.7; Drouin, Godard, Ildefonse, Bruguier, & Garrido, 2009; Godard et al., 2009). They display classical cumulate textures (Figure 4.2.1.9; Blackman et al., 2006). However, several lines of evidence suggest that these rocks are not simply the first crystallized product of a closed, fractionating magma body. Trace element patterns calculated for melts in equilibrium with measured in situ plagioclase, clinopyroxene, and olivine compositions (Figure 4.2.1.9(A)) illustrate that the olivine is in disequilibrium with the MORB melts that crystallized the plagioclase and clinopyroxene (Drouin et al., 2009). Mineral chemistry, textural relations, and crystallographic preferred orientations indicate that the olivine-rich troctolites are the end product of melt–rock interaction between an initial olivine-bearing rock and MORB melt (Drouin, Godard, Ildefonse, Bruguier, & Garrido, 2009; Drouin, Ildefonse, & Godard, 2010). The observed, relatively stronger concentrations of olivine [001] axes in some of these rocks are interpreted as resulting from melt–rock interaction. A first stage of dunitization weakens the olivine crystallographic preferred orientation (e.g., Tommasi, Godard, Coromina, Dautria, & Barsczus, 2004). It is followed by the disaggregation of mantle peridotite

---

of Hole 735B; the "strip average" estimate was obtained using a strip-sampling technique (see Hart et al., 1999) to take into account alteration, small-scale lithologic variability, and patchiness of veins, and thus obtain a possibly more representative sampling. The gray field indicates values of primary MORB melts which represent the most primitive basaltic compositions expected at a spreading centre *(data from Kinzler and Grove (1993) in (A) and from Workman and Hart (2005) in (B))*. Back stripping the contribution of olivine at Hole U1309 gives a composition for the bulk gabbroic crust at Hole U1309 consistent with estimates of the Mg#–Ca# composition of primary MORB magmas (BG). However, this estimate is characterized by trace element signatures indicative of plagioclase accumulation. Only models taking into account the contribution of leucocratic dykes in the bulk hole estimates give results consistent with primary magma compositions for both major and trace elements.

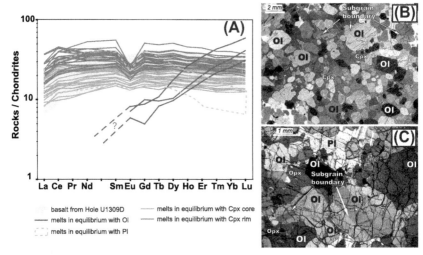

FIGURE 4.2.1.9 Sample characteristics from olivine-rich troctolite intervals *(modified from Blackman et al., 2011)*. (A) Computed rare earth element compositions of melts in equilibrium with the measured mineral chemistry (Drouin et al., 2009). (B) Rounded and subhedral olivine crystals (Ol) included in large clinopyroxene (Cpx) and plagioclase poikiloblasts forming poikilitic texture in olivine-rich troctolite (sample 345_U1309D_247R3_16–18). (C) Coarse–grained subhedral olivine crystals with smooth edges and well-defined and widely spaced subgrain boundaries and poikiloblastic plagioclase (Pl) and orthopyroxene (Opx) (sample 345_U1309D_248R2_22–24). (B) and (C) are cross-polarized photomicrographs from Drouin et al. (2010).

(Figure 4.2.1.10), which, if not heavily fluxed with melt, would be expected to display [100] preferred alignment related to crystal–plastic deformation (Drouin et al., 2010). Disruption of preexisting high-temperature crystallographic preferred orientations during melt influx is also supported by the common occurrence of adjacent grains with similar crystallographic orientations (Drouin et al., 2010; Suhr et al., 2008). Although not unequivocally demonstrated, a mantle origin for the olivine-rich troctolites is favored (Figure 4.2.1.10; Drouin et al., 2009, 2010; Suhr et al., 2008). This proposed scenario suggests a new mechanism for the formation of the lowermost oceanic crust in which small volumes (approximately 1–10-m-thick intervals in the core) of mantle peridotite are impregnated by, and react with, basaltic melts until they become isolated and incorporated into the crust during protracted periods of construction by igneous intrusion. Whatever the initial lithology, impregnation by large volumes of melt has strongly modified the original composition and microstructure of the host rock, and a melt/rock ratio of 3:1 is estimated, based on a fractionation model to match the Mg, Ni, and Cr contents measured in olivine grains (Suhr et al., 2008). The few recovered intervals of mantle peridotites (see above) also show petrographic and geochemical evidence of melt–rock reactions (Tamura et al., 2008; Godard et al., 2009; von der Handt & Hellebrand, 2010).

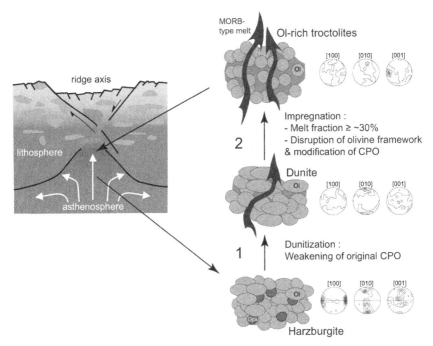

FIGURE 4.2.1.10   Sketch of the proposed scenario for the formation of Hole U1309D olivine-rich troctolites (Drouin et al., 2010). Percolation of basaltic melts through depleted mantle peridotite at near-solidus temperature results in dunitization (orthopyroxene dissolution, precipitation of olivine, and formation of dunite), with weakening of olivine crystallographic preferred orientation (CPO) (indicated by 1), and impregnation when increasing melt fraction (≥30%) locally disrupts the olivine framework and further modifies the olivine CPO (indicated by 2). The dunitization and impregnation are viewed as successive stages of a continuous process. As temperature decreases, melt crystallizes plagioclase and clinopyroxene within the assimilated olivine matrix to form olivine–rich troctolite. The schematic cross-section of the ridge axis is modified from Cannat (1996). The examples of original harzburgite and dunite CPOs are from Polynesian xenoliths (RPA18A and UAH280X, respectively; Tommasi et al., 2004). Note the absence of structural reference frame (no visible foliation in samples) for the dunite and olivine-rich troctolite CPOs.

4.2.1.3 DEEP DRILLING OF INTACT OCEAN CRUST FORMED AT A SUPERFAST SPREADING RATE: HOLE 1256D

4.2.1.3.1 Background

The vast majority (~70%) of magma derived from the mantle is brought into the Earth's crust at mid-ocean ridges, and approximately two-thirds of that magma cools and crystallizes in the lower portion of the oceanic crust. Although <20% of modern ridges are spreading at fast rates, nearly 50% of present-day ocean crust (i.e., ~30% of the Earth's surface) was formed at fast-spreading ridges. Because fast-spread crust is more continuous and regularly layered than slow-spread crust, extrapolating fast-spreading accretion processes from a few sites might reasonably describe a significant portion of the Earth's surface. Additionally, the great majority of crust subducting into the mantle over the past ~200 my formed at fast-spreading ridges (e.g., Müller et al., 2008), making characterizing this style of crust most relevant for understanding the recycling of crustal- and ocean-derived components back into the mantle. Drilling a complete in situ section of ocean crust has been an unfulfilled ambition of Earth scientists for many decades, which provided the impetus for the conception of scientific deep ocean drilling (e.g., Teagle & Ildefonse, 2011); ODP Hole 1256D is the latest, and to date most successful attempt to reach this goal.

The deep drilling campaign to Site 1256 was designed to understand the formation, architecture, and evolution of ocean crust formed at fast spreading rates, and has been the focus of four scientific ocean drilling cruises (ODP 206, Wilson et al., 2003; IODP Expedition 309/312, Teagle et al., 2006; IODP Expedition 335, Teagle et al., 2012). Hole 1256D is located in 3635 m of water in the Guatemala Basin (6°44.2′N, 91°56.1′W), on the Cocos plate in the eastern equatorial Pacific Ocean. Ocean crust here formed 15 my ago during a sustained episode of superfast ocean ridge spreading (>200 mm per year; Wilson, 1996; Figures 4.2.1.1 and 4.2.1.11) at the EPR. The site formed on a ridge segment is at least 400 km in length, ~100 km north of the ridge–ridge–ridge triple junction between the Cocos, Pacific, and Nazca plates. Ocean crust formed at a superfast

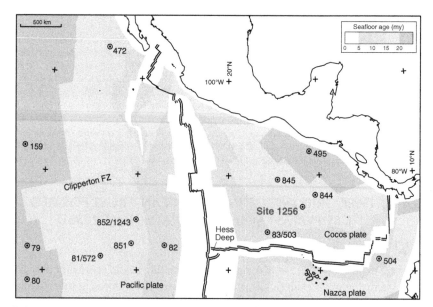

FIGURE 4.2.1.11 Age map of the Cocos plate and corresponding regions of the Pacific and Nazca plates (Wilson et al., 2003). Isochrons at 5 my intervals have been converted from magnetic anomaly identifications according to the timescale of Cande and Kent (1995). Selected DSDP and ODP sites that reached basement are indicated by circles. The wide spacing of the 10–20 my isochrons to the south reflects the extremely fast (200–220 mm per year) full spreading rate.

spreading rate was deliberately targeted because there is strong evidence from mid-ocean ridge seismic experiments that gabbros occur at shallower depths in intact ocean crust with higher spreading rates (Carbotte, Mutter, Mutter, & Ponce-Correa, 1998; Purdy, Kong, Christeson, & Solomon, 1992; Wilson et al., 2006). Consequently, the often difficult-to-drill upper ocean crust should be relatively thin. Site 1256 has a seismic structure reminiscent of typical Pacific off-axis seafloor (Figure 4.2.1.12(B)). Upper Layer 2 velocities are 4.5–5 km/s, and the Layer 2–3 transition is between ~1200 and 1500 m subbasement. The total crustal thickness at Site 1256 is estimated at approximately 5–5.5 km. Site 1256 sits atop a region of smooth seafloor and basement topography (<10 m relief; Figure 4.2.1.12). Unpublished processing (A.J. Harding, personal communication; Teagle et al., 2006) reveals discernible variation in the average seismic velocity (approximately 4.54–4.88 km/s) of the uppermost (~100 m) basement. Two principal features are apparent (Figure 4.2.1.12(A)): a 5–10-km-wide zone of relatively high upper basement velocities (>4.82 km/s) that can be traced ~20 km to the edge of data coverage southeast of Site 1256, and a bull's eye of relatively low velocity (4.66–4.54 km/s) centered around the crossing point of two seismic Lines (21 and 25; Figure 4.2.1.12). The only serious drawback to the Site 1256 area as a crustal reference section for fast spreading rates is its low paleolatitude, making the determination of magnetic polarity from azimuthally unoriented core samples very challenging. In addition, the nearly north–south ridge orientation at the time of accretion makes the magnetic inclination relatively insensitive to tilting.

4.2.1.3.2 Summary of Upper Crustal Lithology at Site 1256

Hole 1256D was the first scientific borehole prepared for deep drilling in ocean crust by installing a large reentry cone secured with almost 270 m of 16-inch casing through the 250-m-thick sedimentary overburden and cemented into the uppermost basement. It was then deepened through an ~810-m-thick sequence of lavas and a thin (~346 m) sheeted dike complex, the lower 60 m of which is strongly contact metamorphosed to granoblastic textures. The first gabbroic rocks were encountered at 1407 mbsf (IODP Expedition 312), where the hole entered a complex dike–gabbro transition zone that includes two 20–50-m-thick gabbro lenses intruded into dikes with granoblastic textures (Figures 4.2.1.13 and 4.2.1.14). At the end of IODP Expedition 312, Hole 1256D had a total depth of 1507 mbsf and was open to its full depth. Five and a half years later, IODP Expedition 335 aimed to deepen Hole 1256D several hundred meters into the cumulate gabbroic rocks of intact lower oceanic crust to address fundamental scientific questions about lower crustal accretion and cooling mechanisms. Unfortunately Hole 1256D was deepened only minimally (to 1521 mbsf), as a number of significant engineering challenges were encountered and overcome during the cruise, each unique but not unexpected in a deep, uncased marine borehole into igneous rocks (see Operations section of the Expedition Summary in Teagle et al., 2012).

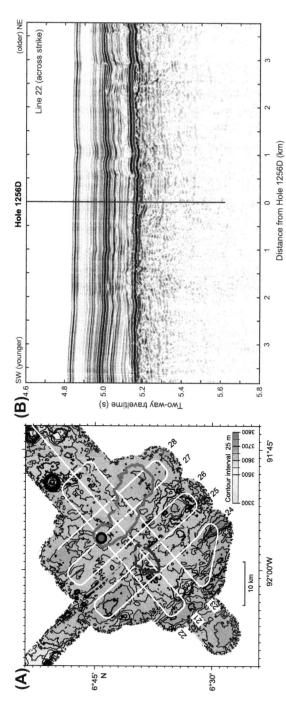

FIGURE 4.2.1.12   Characteristics of the Site 1256 area *(modified from Teagle et al., 2012)*. (A) Bathymetry; abyssal hill relief of as much as 100 m is apparent in the southwest part of the area; relief to the northeast is lower and less organized. Numbers 21–28 identify multichannel seismic (MCS) lines (white) from the site survey. Red (light gray in print versions) and blue (dark gray in print versions) contours are selected top-of-basement seismic velocity contours; the red contour encloses velocity >4.82 km/s, which is interpret as a plausible proxy for the presence of thick ponded lava flows, as sampled in Hole 1256D (Teagle et al., 2012; Wilson et al., 2006); the blue contour encloses velocities <4.60 km/s, possibly reflecting a greater portion of pillow lavas than elsewhere in the region. (B) Site survey MCS profile (Hallenborg et al., 2003) from line 22 that crosses at Site 1256, with penetration as of the end of Expedition 335 scaled approximately to traveltime. Horizontal reflectors in the upper basement in traveltimes of 5.5 s appear to result from contrasts between lava flow sequences, corresponding to depths of at least 800 m in basement.

Formation and Evolution of Oceanic Lithosphere Chapter | 4.2.1 **481**

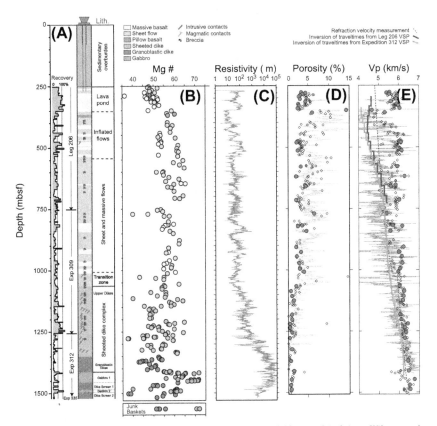

FIGURE 4.2.1.13 Downhole plots of data for ODP Hole 1256D *(modified from Wilson et al., 2006)*. (A) Summary lithostratigraphic column of the basement drilled at Site 1256 showing recovery, and major lithologies. (B) Mg# obtained from shipboard and published whole rock measurements (Neo, Yamazaki, Miyashita, & the Expedition 309/312 Scientists, 2009; Teagle et al., 2006, 2012; Wilson et al., 2003; Yamasaki et al., 2009). (C) Electrical resistivity logging (deep) data (Teagle et al., 2012). (D) Porosity (Gilbert & Salisbury, 2011). Small black symbols are shipboard measurements on discrete samples, red (dark gray in print versions) symbols are postcruise laboratory measurements, and the light blue (pale gray in print versions) curve is calculated from the DLL (Dual Laterolog tool) deep electrical resistivity log. (E) Compressional seismic velocity *(modified from Gilbert & Salisbury, 2011)*. Small black symbols are shipboard measurements on discrete samples corrected for lithostatic pressure, red (dark gray in print versions) symbols are postcruise laboratory measurements under lithostatic pressure, the blue (light gray in print versions) curve is the sonic log data, as reprocessed by Guérin, Goldberg, and Iturrino (2008), the dashed red (dark gray in print versions) line is derived from seismic refraction data (Wilson et al., 2003), the bold red (dark gray in print versions) and orange (light gray in print versions) lines show the inversion of traveltimes from Leg 206 and Expedition 312 vertical seismic profiles (VSPs), respectively (Swift et al., 2008).

However, IODP Expedition 335, returned a unique collection of samples from the granoblastic dikes (Figures 4.2.1.13 and 4.2.1.14), obtained from the many fishing, reaming, and cleaning runs with junk baskets. Some of these samples were much larger than normal core pieces, up to ~20 cm in equivalent diameter.

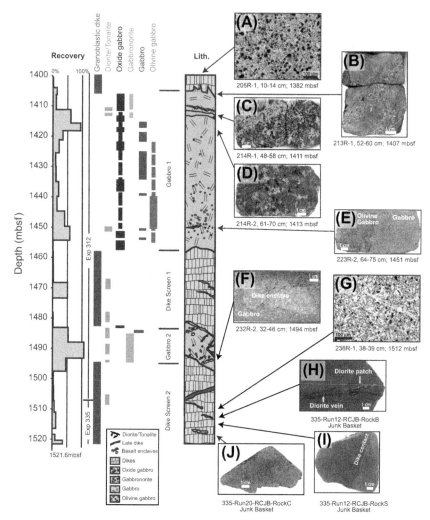

FIGURE 4.2.1.14 Plutonic section from the lower portion of Hole 1256D with representative photomicrographs of key samples *(modified from Wilson et al., 2006; Teagle et al., 2012).* The distribution of rock types is expanded proportionally in zones of incomplete recovery. (A) Photomicrograph of a dike completely recrystallized to a granoblastic association of equant secondary plagioclase, clinopyroxene, magnetite, and ilmenite (plane-polarized light). (B) Uppermost dike/gabbro boundary; medium-grained oxide gabbro is intruded into a granoblastically recrystallized dike along an irregular, moderately dipping contact. (C) Quartz-rich oxide diorite strongly altered to actinolitic hornblende, secondary plagioclase, epidote, and chlorite. (D) Disseminated oxide gabbro with patchy texture and centimeter-scale dark ophitically intergrown clinopyroxene and plagioclase patches separated by irregular, more highly altered leucocratic zones. (E) Sharp modal contact between a medium-grained olivine-rich gabbro (top in the core) and a gabbro. (F) Clast of partially resorbed granoblastic dike within gabbro. (G) Photomicrograph of a granoblastic basalt (plane-polarized light) showing a relict plagioclase microphenocryst with pyroxenes and oxide inclusions in its core (1) and a metamorphic, orthopyroxene vein (2). (H) Dioritic vein and

The volcanic sequences at Site 1256 consist of massive flows and thinner sheet flows with subordinate pillow basalt, hyaloclastite, and breccia (Figure 4.2.1.13). The uppermost crust at Site 1256 comprises an ~100-m-thick sequence of lava dominated by a single massive lava flow >75-m thick, requiring at least this much seafloor relief to pond the lava (Tartarotti & Crispini, 2006; Wilson et al., 2003). On modern fast-spreading ridges, such topography does not commonly develop until 5–10 km from the axis (Macdonald, Fox, Alexander, Pockalny, & Gente, 1996). Although this lava flow cooled off-axis it may have originated at the ridge axis before flowing on to the ridge flanks, as is observed for very large lava flows on modern ocean floor (e.g., Macdonald, Haymon, & Shor, 1989). This flow is relatively unfractured at the scale of the drill hole (Crispini, Tartarotti, & Umino, 2006), with shipboard physical property measurements on discrete samples indicating $V_P > 5.5$ km/s (Wilson et al., 2003). Hence, the area of relatively high uppermost basement seismic velocities likely represents the regional extent of this massive flow, although this hypothesis has not been tested by drilling. Assuming an average thickness of 40 m, the regional survey results would conservatively suggest an eruption volume in excess of 3 km³, plausibly >10 km³. Lava flows this voluminous are not described on-axis at fast-spreading ridges. An off-axis origin is suggested by the presence of a strongly welded and plastically deformed agglutinate in the base of the thick lava in Hole 1256C, that indicates proximity to the source vent for the lava field (Crispini et al., 2006). An off-axis volcanic flow field of comparable size is described on the EPR at 14°S (Geshi et al., 2007). This is extremely large when compared to the size of mid-ocean ridge axial low-velocity zones that are thought to be high-level melt lenses, which typically have volumes approximately 0.05–0.15 km³ per kilometer of ridge axis, and generally appear to be only partially molten (Singh, Kent, Collier, Harding, & Orcutt, 1998). This poses questions regarding the melt sources of these very large submarine flow fields that require either the draining of a relatively large magma reservoir so far not imaged at fast-spreading mid-ocean ridges (Teagle et al., 2012), or a magma system that is being recharged during the course of slow (several decades long), ongoing eruptions, such as described in Iceland (e.g., Eason & Sinton, 2009).

The lava sequence below the lava pond includes sheet and massive flows and minor pillow flows. Subvertical, elongate flow-top fractures filled with quenched glass and hyaloclastite in these lavas indicate flow lobe inflation requiring cooling on a subhorizontal surface off-axis (Umino, Lipman, & Obata, 2000). It is estimated that a total thickness of 284 m of lava flowed and cooled off-axis. Sheet

---

patch within granoblastic basalt. (I) Sharp, curviplanar dike/dike contact between a light-colored, fine-grained basalt (right) and a darker, coarse-grained basalt (left). (J) Medium-grained, orthopyroxene-bearing olivine gabbro.

flows and massive lava flows erupted at the ridge axis make up the remaining extrusive section to 1004 mbsf. This contrasts slightly with the volcanic stratigraphy for Hole 1256D developed from analysis of wireline geophysical imaging (Tominaga & Umino, 2010; Tominaga, Teagle, Alt, & Umino, 2009) that suggests <50% of the lava drilled crystallized within 1000 m of the axis but that the majority of the lava pile had formed within 3000 m of the ridge crest.

From 1004 to 1061 mbsf is the Transition Zone between the extrusive rocks and the sheeted dike complex, marked by the occurrence of subvertical intrusive contacts, a silica-sulfide mineralized hyaloclastic breccia zone, and a step change in hydrothermal alteration from low temperature to greenschist facies assemblages. Below this transition, numerous subvertical intrusive contacts were observed, indicating the start of a relatively thin, ~350-m-thick sheeted dike complex dominated by massive basalt. Some basalts have doleritic textures, and many are crosscut by subvertical dikes with strongly brecciated chilled margins that are mineralized by hydrothermal fluids. There is little evidence from core or geophysical wireline logs for major tilting of the dikes, consistent with subhorizontal seismic reflectors in the lower extrusive rocks that are continuous for several kilometers across the site (Hallenborg, Harding, Kent, & Wilson, 2003). Measurements of dike orientations using FMS images are in close agreement with direct measurements on recovered cores and indicate that the dikes at Site 1256 are tilted ~15° away from the paleospreading axis around a ridge-parallel, subhorizontal axis (Tominaga et al., 2009; Violay, Pezard, Ildefonse, Célérier, & Deleau, 2012).

In the bottommost ~60 m of the sheeted dikes (1348–1407 mbsf), basalts are partially to completely recrystallized to distinctive granoblastic textures characterized by granular secondary clinopyroxene and lesser orthopyroxene, resulting from contact metamorphism by underlying gabbroic intrusions (Figures 4.2.1.13 and 4.2.1.14). Textural changes in oxide minerals indicate that the zone of metamorphic recrystallization may extend for >90 m above the upper contact with the first gabbro interval (Coggon, Alt, & Teagle, 2008). There is unequivocal evidence for hydrothermal alteration before and after the formation of the granoblastic textures and potentially small, irregular patches of partial melting (Alt et al., 2010; France, Ildefonse, & Koepke, 2009; Koepke et al., 2008; Teagle et al., 2006). Simple thermal calculations suggest that the two gabbro bodies so far intersected in Hole 1256D have insufficient thermal mass to be responsible for a 60–90-m-thick high-temperature (600–900 °C) contact-metamorphic halo if the intrusions are simple subhorizontal bodies (Coggon et al., 2008; Koepke et al., 2008). The amount of heat needed to produce this metamorphism requires either a much larger intrusion nearby, or significant topography on the dike/gabbro boundary (Teagle et al., 2006).

Gabbro and trondhjemite intrusions into sheeted dikes at 1407 mbsf mark the top of the plutonic section (Figures 4.2.1.13 and 4.2.1.14) and the start of a complex dike–gabbro transition zone. This first occurrence of plutonic rocks from intact ocean crust occurred within the depth range predicted from the relationship between spreading rate and depth to axial low-velocity zones (Teagle et al., 2006). Two major bodies of gabbro were penetrated beneath this contact, with the 52-m-thick upper gabbro (Gabbro 1) separated from the 24-m-thick lower gabbro (Gabbro 2) by a 24-m screen of granoblastic basalts (Dike Screen 1). The textures and rock types observed in Hole 1256D gabbro intervals are similar to those of varitextured gabbros thought to represent a frozen melt lens between the sheeted dike complex and the underlying cumulate gabbros in ophiolites (e.g., MacLeod and Yaouancq, 2000; France et al., 2009). Gabbro 1 is mineralogically and texturally heterogeneous, comprising olivine gabbros, gabbros, oxide gabbros, quartz-rich oxide diorites, and small trondhjemite dikelets. Oxide abundance decreases irregularly downhole, and olivine is present in significant amounts in the lower part of Gabbro 1. The intervening Dike Screen 1 is interpreted as an interval of sheeted dikes captured between the two intrusions of gabbros. It consists of fine-grained *meta*-basalt similar to the granoblastic dikes overlying Gabbro 1, and is cut by a number of small quartz gabbro and tonalite dikelets of variable thickness (1–10 cm), grain size, and composition. Gabbro 2 comprises gabbro, oxide gabbro, and subordinate orthopyroxene-bearing gabbro and trondhjemite that are altered similarly to Gabbro 1, and has clear intrusive contacts with the overlying granoblastic Dike Screen 1. Partially resorbed, stoped dike clasts are entrained within both the upper and lower margins of Gabbro 2 (Figure 4.2.1.14). Gabbro 2 is characterized by the absence of fresh olivine, high but variable orthopyroxene contents (5–25%), and considerable centimeter- to decimeter-scale heterogeneity. Oxide abundance generally diminishes downhole. The predominant rock type is orthopyroxene-bearing gabbro with gabbronorite in the marginal units. Dike Screen 2, at the bottom of the hole, was extensively sampled with junk baskets during IODP Expedition 335 (Teagle et al., 2012). It is dominantly made of fine-grained granoblastic aphyric basalts, most of which are strongly recrystallized (Figure 4.2.1.14). Some samples preserve dike-to-dike contacts, and many preserve cores of former plagioclase (micro)phenocrysts (Figure 4.2.1.14), indicating that the granoblastic rocks are contact-metamorphosed sheeted dike diabase. Approximately half of the granoblastic basalt samples contain small, irregular patches ($<5 \times 3$ cm), veins (approximately 1–2 mm wide), and dikelets (<1.5 cm wide) of evolved plutonic rocks (oxide gabbro, diorite, and tonalite) (Figures 4.2.1.14 and 4.2.1.15). The veins are observed to be offshoots of the igneous patches, and diffuse patches are issued from the dikelets. Hence these features form part of the same network, marking a single generation of intrusion of melts into the granoblastic basalts. Their magmatic textures (subhedral to euhedral shapes of several phases, along with poikilitic textures) contrast strongly with the granoblastic textures of the host rocks, demonstrating that intrusion occurred after the high-temperature recrystallization of the host rock. The common occurrence of primary magmatic amphibole in the veins and patches suggests high water activities during their formation, hence possibly the partial

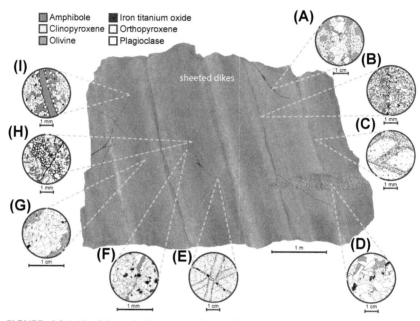

FIGURE 4.2.1.15 Schematic "outcrop" of the dike–gabbro transition zone, showing the micro-textures and contact relationships observed in the cores and junk basket samples recovered during Expedition 335 (Teagle et al., 2012). (A) Diorite dikelet, comprised predominantly of primary igneous amphibole and plagioclase, crosscutting granoblastic basalt (Sample 335-1256D-Run19-RCJB-Rock C). (B) Fine-grained dike chilled against coarse-grained dike (Sample 335-1256D-Run12-RCJB-Rock S). The entire sample is recrystallized to granoblastic assemblages of plagioclase, orthopyroxene, clinopyroxene, magnetite, and ilmenite. Later fine postcontact-metamorphic hydrothermal amphibole veins (not shown) cut across the contact and both dikes. (C) Chilled and brecciated dike margin recrystallized to granoblastic assemblages (Sample 335-1256D-Run14-EXJB-Foliated). Angular clasts consist of granoblastic plagioclase with minor orthopyroxene, clinopyroxene, ilmenite, and magnetite and are recrystallized from chilled dike margin breccia protolith. Clast matrix is orthopyroxene-rich with minor clinopyroxene, plagioclase, and oxides; it is recrystallized from hydrothermal minerals (amphibole and chlorite) that veined and cemented the breccia protolith. (D) Subophitic texture in gabbro (Sample 335-1256D-Run11-EXJB). (E) Diorite vein crosscutting a conjugate set of metamorphic orthopyroxene veins (Sample 335-1256D-Run19-RCJB-Rock B). (F) Postcontact-metamorphic hydrothermal hornblende vein cutting granoblastic basalt (Sample 335-1256D-238R-1, 2–4 cm). (G) Granoblastic basalt with a diorite patch (Sample 335-1256D-Run11-EXJB-Rock). (H) Granoblastic orthopyroxene vein, recrystallized from precursor hydrothermal vein, in granoblastic dike (Sample 335-1256D-238R-1, 2–4 cm). Orthopyroxene vein is cut by small postcontact-metamorphic hydrothermal amphibole vein. (I) Postcontact-metamorphic hydrothermal amphibole vein cutting granoblastic basalt, with 1-mm-wide amphibole-rich alteration halo where pyroxenes are replaced by amphibole (Sample 335-1256D-Run12-RCJB-Rock B).

melting of previously hydrothermally altered rocks. Moreover, the presence of quartz, as well as accessory apatite and zircon, implies that patches, veins, and dikelets crystallized from highly evolved melts. Only a small number of gabbroic rocks (~3 wt%) were recovered during IODP Expedition 335

(Figure 4.2.1.14) and these rocks range in composition from disseminated oxide gabbro to orthopyroxene-bearing olivine gabbro. The IODP Expedition 335 gabbroic rocks are texturally less variable than those recovered during IODP Expedition 312 from the gabbro 1 and 2 intervals. Although the contact relationships with the granoblastic basalts were not recovered, it is likely that at least some of the gabbroic rocks occur intercalated with the granoblastic basalt, perhaps forming small intrusions. The unique sampling with fishing tools achieved during IODP Expedition 335 allowed the documentation of a variety of intrusive structural/textural relationships, and overprinting/cross-cutting hydrothermal alteration and metamorphic paragenetic sequences that had not previously been observed in situ, owing to the one-dimensional nature of drill cores and the very low rates of recovery of the granoblastic material in the sheeted dike/gabbro transition zone. Overall, the bottom of Hole 1256D documents a section of metamorphosed, granoblastic sheeted dikes that underwent small-scale intrusion by both gabbroic and evolved plutonic rocks (Figure 4.2.1.15).

4.2.1.3.3 A Thin upper Crust at Superfast Spreading Rate

Relative to other well-studied upper ocean crust sections (e.g., Karson, 2002), Site 1256 is unique as it shows a thick lava sequence and a thin dike sequence (Umino et al., 2008). Steady-state thermal models require that the conductive lid separating magma from rapidly circulating seawater gets thinner as spreading rate increases, indicating that the thin dike sequence is a direct consequence of the high spreading rate. A thick flow sequence, with many massive individual flows and few pillow lavas, is a consequence of the elevated magma budget (Umino et al., 2008) and/or of the short vertical transport distance from the magma chamber, and is similar to observations from the midsegments of the fast-spreading southern EPR (White, Macdonald, & Haymon, 2000). The high lava/dike thickness ratio appears to be characteristic of superfast spreading, which is in direct contrast to spreading models developed from observations of tectonically disrupted fast-spread crust exposed in Hess Deep (Karson, 2002) that suggest regions of high magma supply should have thin lavas and thick dikes. Similarly, there is little evidence for tilting (at most ~15°) in Hole 1256D and no evidence for significant faulting. In contrast, the upper crust exposed at Hess Deep (fast spreading) and in the Blanco Fracture Zone (intermediate spreading) shows significant faulting and rotation within the dike complex (Karson, 2002), indicating that observations from those tectonic windows may be site specific and not widely applicable to intact ocean crust. The massive lava flow at the top of the Site 1256 basement indicates that faults with throws of approximately 50–100 m must exist in superfast spreading rate crust relatively near to the ridge axis (<10 km) in order to provide the relief necessary to pond the lava (as described at the southern EPR; e.g., Auzende et al., 1996; Macdonald, 1998; Sinton et al., 2002).

4.2.1.3.4 Where Is the Layer 2/3 Transition at Site 1256?

Marine seismologists traditionally subdivided the ocean crust into seismic layers: Layer 1 comprises low-velocity sediments ($V_P < 3$ km/s); Layer 2 has low velocity and a high velocity gradient, with V_P typically ranging from ~3.5 to ~6.7 km/s; and Layer 3 has high velocity and a low velocity gradient (V_P ranges from 6.7 to 7.1 km/s); the step-up to $V_P > 8$ km/s marking the Mohorovičić Discontinuity and is interpreted to be the boundary with ultramafic rocks of the uppermost mantle. In casual language "Layer 3" is often mistakenly used as a synonym for gabbro, although previous drilling in Hole 504B has penetrated Layer 3 but not yet reached gabbroic rocks (Alt et al., 1996; Detrick et al., 1994). From regional seismic refraction data, the transition from seismic Layer 2 to Layer 3 at Site 1256 occurs between 1200 and 1500 m into basement (Wilson et al., 2003). Shipboard determinations of seismic velocities of discrete samples, once corrected for lithostatic pressure, are in close agreement with postcruise laboratory measurements of discrete samples under lithostatic pressure, and with in situ measurements by wireline tools (Figure 4.2.1.13; Gilbert & Salisbury, 2011). Contrary to expectation, porosity increases and P-wave velocities decrease stepwise downward from the lowermost dikes into the uppermost gabbro in Hole 1256D, as the result of the contact metamorphism of the granoblastic dikes and the strong hydrothermal alteration of the uppermost gabbros (Figure 4.2.1.13). Porosity and velocity then increase downhole in the gabbro. Wireline velocity measurements end at the top of gabbro, because additional velocity data closer to the bottom of the hole could not be acquired at the end of IODP Expedition 335 (Teagle et al., 2012). The examination of shipboard and postcruise discrete sample measurements, wireline logging, and VSP data suggests that the base of Hole 1256D has just entered, or is close to the transition from Layer 2 to Layer 3 (Gilbert & Salisbury, 2011; Swift, Reichow, Tikku, Tominaga, & Gilbert, 2008). Encountering gabbro at a depth within Layer 2 reinforces previous inferences that factors such as porosity and hydrothermal alteration (Alt et al., 1996; Carlson, 2010; Detrick et al., 1994) are more important than rock type or grain size in controlling the location of the Layer 2–3 transition. Drilling deeper in Hole 1256D is required to characterize the Layer 2–3 transition at Site 1256.

4.2.1.3.5 Geochemical Characteristics of the Upper Crust at Site 1256

Flows and dikes from Hole 1256D are N-MORB but show a wide range of magmatic fractionation (Figures 4.2.1.13 and 4.2.1.16). The lavas and dikes are on average relatively fractionated compared to global MORB data compilations as anticipated for fast-spreading EPR MORB (e.g., Rubin & Sinton, 2007). The lava pond, cored in both Holes 1256C and D, is more fractionated (Mg# 50 ± 2) than the rocks from the inflated flows, sheet and massive flows, and the sheeted

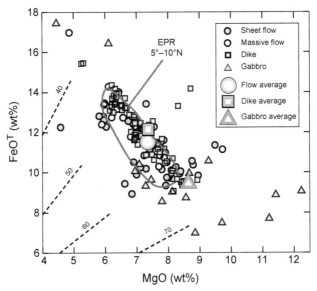

FIGURE 4.2.1.16 FeO_T (Total Fe expressed as FeO) versus MgO for the basement at Site 1256 (Shipboard data from Leg 206 and Expeditions 309, 312, and 335; Teagle et al., 2006, 2012). Data are compared with analyses of northern East Pacific Rise (purple contour; Langmuir, Bender, & Batiza, 1986). Dashed lines show constant Mg#. Possible primary mantle melt compositions should have Mg# of 70–78 and MgO of 9–14 wt%. All flows and dikes, and most gabbros are too evolved to be candidates for primary magmas.

dike complex that all show a similar range of compositions (Mg# 56 ± 6). The more differentiated nature of the lava pond lavas is also displayed by relatively high incompatible element (e.g., Zr, TiO_2, Y, and V) and low compatible element (Cr and Ni) concentrations (Teagle et al., 2006; Wilson et al., 2003). An interval within the lava pond cored in Hole 1256C (294–306 mbsf) shows anomalously high K_2O concentrations (~0.74 wt%) that are difficult to explain by magmatic fractionation and are not a result of hydrothermal alteration by seawater.

There are variations in the basalt chemistry downhole, with a step in igneous trends at 750 mbsf possibly indicating a cycle of fractionation, replenishment, and, perhaps, assimilation (Figure 4.2.1.13). Downhole geochemical compositions within the dike section are variable and do not define trends; primary and evolved compositions are closely juxtaposed, as would be expected for vertically intruded magmas (Figure 4.2.1.13). The range of major element compositions in the dikes is similar to that of the overlying rocks, and the average composition of the dikes is indistinguishable from the average composition of the lavas (Figures 4.2.1.13 and 4.2.1.16).

The compositional ranges of fresh lava and dike samples correspond to typical values for N-MORB for most major elements and many trace elements (Gale, Dalton, Langmuir, Su, & Schilling, 2013; Rubin & Sinton, 2007) and are similar to those observed for the northern EPR (Figure 4.2.1.16). When trace elements are compared to EPR MORB, they are within one standard deviation of average, albeit on the relatively trace element-depleted side of MORB. For example, compared with first-order mid-ocean-ridge segments along the EPR, basalts from Site 1256 have low Zr/TiO_2 and Zr/Y. This decrease in trace element ratios would require ~30% more melting at the superfast ridge, but this appears unlikely because of normal crustal thickness (~5.5 km) at Site 1256. More likely, the relatively depleted trace element signature at ODP Site 1256 results from the upwelling of previously depleted Galápagos plume mantle at the paleo-Site 1256 ridge axis (Geldmacher et al., 2013; Park, MacLennan, Teagle, & Hauff, 2008).

The gabbro compositions, as documented by shipboard geochemical analyses, span a range similar to the flows and dikes but have on average higher MgO and lower FeO, albeit still within the range of EPR basalts (Figure 4.2.1.16). The uppermost gabbros have geochemical characteristics similar to the overlying dikes with MORB chemistries (MgO approximately 7–8 wt%; Zr approximately 47–65 ppm). Deeper, in Gabbro 2, the rocks are less fractionated and there are general downhole trends of increasing MgO, CaO, and Ni and decreasing FeO, Zr, and Y. Decreasing concentrations of FeO and TiO_2 downhole suggest that Gabbro 2 is fractionated similarly to Gabbro 1. The intrusive nature of both gabbro bodies and the chemical variations within them suggest that they intruded into the base of the sheeted dikes and underwent minor internal fractionation in situ, resulting in the observed general geochemical stratification. Partially resorbed xenoliths of granoblastic dikes within Gabbro 2 (Figure 4.2.1.14) indicate that stoping of the intruded dikes may have contaminated the gabbro compositions (e.g., France et al., 2009).

There are linear trends in the gabbro units between MgO and TiO_2, FeO, CaO, Na_2O, and Zr, most likely resulting from fractional crystallization of a gabbroic magma, with the fractionating assemblage consisting of clinopyroxene, plagioclase, and subordinate olivine, as expected for relatively evolved basaltic magmas. Olivines within the gabbros have Forsterite composition Fo<81 (Yamazaki, Neo, & Miyashita, 2009) indicating fractionation. Similarly, even the most primitive dikes and lavas at Site 1256 (Mg# ~66) would be in equilibrium with olivine more evolved than mantle olivine, and hence must have undergone fractionation. Simple mass balance calculations indicate that the average basalt in Hole 1256D has lost >30% of its original liquid mass as solid gabbro, implying the presence of at least 300 m of cumulate gabbro in the crust below the present base of the hole (Teagle et al., 2006). Therefore, the residue removed from primary magma to produce the observed gabbro and basalt compositions must occur below the present base of Hole 1256D.

The lower part of Hole 1256D, below the dike–Gabbro 1 boundary at ~1406 mbsf, is characterized by large chemical variations with, for example, Mg# ranging from 42 to 72 (Figure 4.2.1.13), and Zr from 23 to 117 ppm (Teagle et al., 2012). This variability mainly reflects changes in rock types from the low-Mg#, trace element-rich sheeted and granoblastic dikes and dike screens, to the higher Mg# and trace element-depleted gabbroic rocks of Gabbro 1 and Gabbro 2. There are significant depth-dependent trace element variations in the granoblastic basalts, with a general downhole trend of decreasing incompatible element contents (i.e., Zr and, to a lesser extent, Y). Details of these variations and their possible significance are presented in Teagle et al. (2012).

4.2.1.4 SHALLOW DRILLING IN FAST-SPREAD LOWER CRUST AT HESS DEEP

The primary goal of the "superfast campaign" at Site 1256 was to reach gabbroic rocks of the lower crust. In spite of the operational difficulties encountered during IODP Expedition 335, Hole 1256D has been thoroughly cleaned from debris and cuttings, and is still open. Additional expeditions to Hole 1256D will be required to deepen the hole into the cumulate rocks of the lower crust. An alternative to deep drilling into intact, fast-spread oceanic crust (as at Site 1256) is to access the lower crust in the rare places where it crops out on the seafloor, such as in tectonic windows. This was the goal of the recent IODP Expedition 345, which targeted lower crust gabbros at Hess Deep in the equatorial eastern Pacific (Figure 4.2.1.1; Table 4.2.1.1; Gillis, Snow, Klaus, & the Expedition 345 Scientists, 2014a; Gillis et al., 2014b). The scientific objectives were to test endmember models of magmatic accretion and estimate the intensity of hydrothermal cooling at depth. The Hess Deep Rift was selected to exploit tectonic exposures of young EPR plutonic crust, building upon results from Ocean Drilling Program Leg 147 (Gillis et al., 1993) and submersible surveys (e.g., Francheteau et al., 1990; Hékinian, Bideau, Francheteau, Lonsdale, & Blum, 1993; Lissenberg, MacLeod, Howard, & Godard, 2013). Low to moderate recovery (15–30%) was achieved in three holes (35–110 mbsf). Olivine gabbro and troctolite represent 84% of the plutonic rock types recovered at Site U1415. Additional recovered lithologies are minor gabbro, clinopyroxene oikocryst-bearing troctolite, clinopyroxene oikocryst-bearing gabbro, and gabbronorite. All of these lithologies have Mg# between 78 and 90. Spectacular modal and/or grain-size layering, reminiscent of that classically described in ophiolites (e.g., Boudier, Nicolas, & Ildefonse, 1996; Kelemen, Koga, & Shimizu, 1997; Pallister & Hopson, 1981), was observed in more than half of the recovered material (Figure 4.2.1.17). Magmatic foliation is commonly moderate to strong, particularly in layered lithologies. Together with the compositions of gabbros from the upper part of the igneous crust (known from ODP Hole 894 samples; Gillis et al., 1993), and of the upper crust (known from numerous Alvin and Nautile dive samples; Coogan, Gillis, MacLeod, Thompson, & Hekinian, 2002, and references therein), Gillis et al. (2014b) calculated an estimate of the bulk composition of the Hess Deep crust, the first estimate for fast-spread crust, with an Mg# of 74. This average bulk composition can then be used as that of a parental melt for a MORB fractional crystallization model, which is consistent with the measured sample compositions (Gillis et al., 2014b). However, an unexpected finding, which

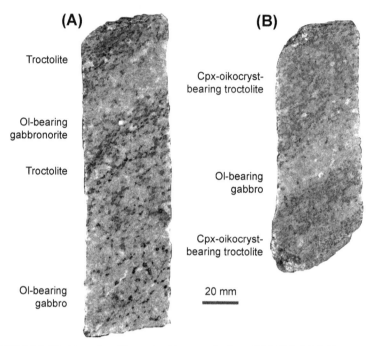

FIGURE 4.2.1.17   Examples of layered gabbro samples from IODP Hole U1415J at Hess Deep (Gillis et al., 2014a). (A) Core 345-U1415J-5R-2, piece 1. (B) 345-U1415J-8R-2, piece 9.

does not fit a simple fractional crystallization trend, is the common abundance (up to 5%) of orthopyroxene as an early cumulus phase in the rocks recovered during Expedition 345. This observation calls for a more complex petrogenetic history of the Hess Deep lower crust, which possibly involves reaction of ascending MORB melts with the upper mantle (hence delivery of melts to the lower crust that are not undersaturated in orthopyroxene; Coogan et al., 2002), and/or the contamination of parental melts by water (Boudier, Godard, & Armbruster, 2000; Nonnotte, Ceuleneer, & Benoit, 2005). The common occurrence of delicate skeletal olivine grains suggests that at this location, the lower crust was not significantly deformed after crystallization. IODP Expedition 345 provides a new, unique, reference section for fast-spreading lower crust (Gillis et al., 2014b), which complements the previous seafloor sampling in this area (Hékinian et al., 1993; Coogan et al., 2002; Lissenberg et al., 2013). Paleomagnetic demagnetization experiments conducted during IODP Expedition 345 on discrete samples of gabbros and troctolites yielded stable magnetization directions with a variety of remanence structures. In all samples, components that unblock close to the magnetite Curie temperature are considered to represent primary thermoremanent magnetizations acquired during crustal accretion. Downhole variations in the inclination of this stable, high-unblocking temperature magnetic component indicate that at least two blocks have been sampled in each of the two deepest holes (U1415J and U1415P; Gillis, Snow,

Klaus, & the Expedition 345 Scientists, 2014a). The data provide evidence of relative displacements of individual, internally coherent blocks (tens of meters in size), consistent with emplacement by very large-scale mass wasting on the southern slope of the Hess Deep intrarift ridge (Ferrini et al., 2013).

4.2.1.5 CONCLUSION

During the last decade, IODP has made substantial progress in sampling of the ocean crust, by doubling the amount of deep (>1 km below seafloor) holes. Two deep holes from sections formed at extremely different spreading rates have shed light on very important unknown and/or poorly understood aspects of the architecture and accretion of modern ocean crust.

In slow-spread crust, Hole U1309D was the second deep hole in an OCC following historic deep drilling at ODP Hole 735B (that remains the deepest hole in slow-spread crust, and returned the highest recovery in hard rock drilling). The spectacular igneous section recovered in the core of the Atlantis Massif is key to demonstrating the important role of magmatism in the development of OCCs. The interplay between magmatism and tectonics is still to be fully documented, described, and quantified, but Hole U1309D is a significant step forward. Olivine-rich troctolites, which represent about 5% of the material recovered in the hole, and which are by far the least altered olivine-rich material (locally <1% serpentinized) ever recovered by scientific ocean drilling, represent an exceptional record of the complex igneous accretion history, involving both fractional crystallization and melt–rock reactions, and possibly also assimilation of mantle peridotites. Another first, resulting from the combination of high recovery and high-quality borehole imaging (FMS) data, was the core-log integration that allowed reorientation of individual core pieces and associated paleomagnetic data and the provision of unequivocal constraints on the tectonic rotation associated with the development of the OCC.

Hole 1256D, in Pacific superfast-spread crust, has for the first time reached the base of the sheeted dikes and penetrated the first gabbroic rocks ever sampled in intact, in situ layered fast-spread crust. The upper crustal section recovered at Site 1256 complements Hole 504B in intermediate-spread crust, which has been for a long time the single available reference section for the modern oceanic upper crust. The sampling of the root zone of the sheeted dike complex, and underlying varitextured gabbros and recrystallized granoblastic basalts, documents for the first time the transition zone between the present-day upper and lower crust formed at fast-spreading ridges. From this unique sampling emerges a vision of a complex, dynamic thermal boundary layer zone. This region of the crust between the principally hydrothermal domain of the upper crust and the intrusive magmatic domain of the lower crust is one of evolving geological conditions. The resulting intimate coupling among temporally and spatially intercalated magmatic, hydrothermal, partial melting, intrusive, metamorphic, and retrograde processes is recorded in the recovered samples.

In addition to the two deep holes in slow- and ultrafast-spread crust, another step forward was recently achieved by sampling the first significant sections of layered gabbros of "Penrose"-style ocean crust (Gillis et al., 2014b), assumed to constitute most of the lower crust based on ophiolite studies since the early 1970s. The drawback of accessing the deeper crustal levels in a tectonic window like Hess Deep is that the recovered sections are tectonically out of place, and hence cannot be used to fully document the depth variations of compositions and structures in the way required to test crustal accretion models (e.g., Teagle et al., 2012). Yet, this unique sampling of the lower, layered crust brings fundamental new constraints on the average composition of fast-spread crust, on the petrological processes involved in building the layered gabbros, and in transporting melt from these deeper levels to the upper crust. Hence, critical samples and data recovered by scientific drilling over the past 10 years provide significant additional constraints on understanding lower oceanic crustal processes. However, we are still far from having a complete set of direct observations regarding the accretion occurring beneath the dike layer at fast-spreading ridges, and we need to achieve deep drilling of intact, fast-spread ocean crust.

Deep drilling into ocean crust has posed, and will continue to present major technical and programmatic challenges to scientific ocean drilling. Only four holes, DSDP Hole 504B, ODP Holes 735B and 1256D, and IODP Hole U1309D (Figures 4.2.1.1 and 4.2.1.2; Table 4.2.1.1), have been cored deeper than 1 km into oceanic basement, and these penetrations are arguably some of the greatest technical achievements of scientific ocean drilling. All were "hard won" multi-expedition experiments. From the experiences of drilling these holes, there are important lessons to be learned for the siting, planning, and implementation of future deep drilling of the oceanic basement, which are described in detail in the IODP Expedition 335 Proceedings (Teagle et al., 2012). Establishing the ideal location for drilling is only part of the challenge of successfully drilling moderately deep holes (2–3 km) to recover the samples and data necessary to address long-standing primary goals of scientific ocean drilling. Experience from Holes 504B and 1256D indicates that such experiments require multiple expeditions to achieve their target depths. A total of ~500 m penetration per expedition is an upper limit for coring in the upper crust, with lesser advances and more frequent drilling challenges as these holes get deeper, and rocks metamorphosed at higher pressures and temperatures are encountered. Penetration and core recovery rates have been low to very low in the two sheeted dike complex sections drilled to date. Average rates of recovery and penetration in the dike section of Hole 1256D are 32% and 0.8 m/h, respectively. The average rate of recovery in the sheeted dike complex of Hole 504B was a miserly 11%. In contrast, experience to date suggests that gabbroic rocks can be cored relatively rapidly at high rates of recovery (e.g., Hole U1309D: penetration rate = 2 m/h; recovery ≥75%), so when the dike–gabbro transition zone is breached in Hole 1256D, solid progress through the plutonic section may be anticipated.

Scientific ocean drilling has little experience in casing long sections (hundreds of meters) of oceanic basement and a poor armory of underreaming tools for opening hard rock basement holes to the diameters required for the insertion of a casing. Casing hundreds of meters of a deep borehole in igneous basement would be a high risk, costly, and ship-time consuming operation that would produce no new scientific output until completed and drilling was resumed. However, it would greatly improve the stability and hydrodynamics of deep basement boreholes. A regular drilling-then-casing approach to investigate the lower oceanic fast-spread crust (target depth ~3 km) will require a long-term commitment by the scientific ocean drilling community to a particular site and experiment, and possibly as many as 10 expeditions to complete. The possibility that even such a highly engineered approach could still fail to reach its target would have to be acknowledged and accepted. Such is the capriciousness of hard rock coring that scientific ocean drilling may have to consider new approaches if it is to ever successfully address some of the major science questions that remain unanswered after more than 50 years. There are unlikely to ever be "quick wins" with targets that require multi-expedition deep boreholes.

Holes U1309D and Hole 1256D are open, stable, and can be deepened further. Drilling down to the bottom of Hole U1309D was achieved without major difficulty, and the hole condition is excellent. During IODP Expedition 335, Hole 1256D was stabilized, thoroughly cleared of cuttings (several hundred kilograms) and debris. Both holes are ready for deepening in the near future, and fundamental observations concerning the formation and evolution of the ocean crust await.

ACKNOWLEDGMENTS

This research used samples and data provided by the Integrated Ocean Drilling Program (IODP). The manuscript benefited from thorough and helpful reviews from Gail Christeson and John Sinton. We thank the USIO teams and JOIDES Resolution crews for their invaluable assistance and outstanding work during IODP Expeditions 304, 305, 309, 312, 335, and 345.

REFERENCES

Anonymous. (1972). Penrose field conference on ophiolites. *Geotimes, 17*, 24–25.

Alt, J. C., Honnorez, J., Laverne, C., & Emmermann, R. (1986). Hydrothermal alteration of a 1 km section through the upper oceanic crust, Deep Sea Drilling Project Hole 504B: mineralogy, chemistry, and evolution of seawater–basalt interactions. *Journal of Geophysical Research: Solid Earth, 91*(B10), 10309–10335. doi:10.1029/JB091iB10p10309.

Alt, J. C., Laverne, C., Coggon, R. M., Teagle, D. A. H., Banerjee, N. R., Morgan, S., et al. (2010). Subsurface structure of a submarine hydrothermal system in ocean crust formed at the East Pacific Rise, ODP/IODP Site 1256. *Geochemistry Geophysics Geosystems, 11*(10), Q10010. doi:10.1029/2010GC003144.

Alt, J. C., Laverne, C., Vanko, D. A., Tartarotti, P., Teagle, D. A. H., Bach, W., et al. (1996). Hydrothermal alteration of a section of upper oceanic crust in the eastern equatorial Pacific: a synthesis of results from Site 504 (DSDP Legs 69, 70, and 83, and ODP Legs 111, 137, 140, and 148.). In J. C. Alt, H. Kinoshita, L. B. Stokking, & P. J. Michael (Eds.), *Proc. ODP, Sci. Results, 148*. College Station, TX: Ocean Drilling Program. 417–434. doi:10.2973/odp.proc.sr.148.159.1996.

Alt, J. C., Muehlenbachs, K., & Honnorez, J. (1986). An oxygen isotopic profile through the upper kilometer of the oceanic crust, DSDP Hole 504B. *Earth and Planetary Science Letters, 80*(3–4), 217–229. doi:10.1016/0012-821X(86)90106-8.

Aumento, F., Melson, W. G., & DSDP Leg 37 Scientific Party. (1977). *Initial Reports of the deep sea drilling project* (Vol. 37) Washington: U.S. Government Printing Office. doi:10.2973/dsdp. proc.37.1977.

Auzende, J. M., Ballu, V., Batiza, R., Bideau, D., Charlou, J. L., Cormier, M. H., et al. (1996). Recent tectonic, magmatic, and hydrothermal activity on the East Pacific Rise between 17°S and 19°S: submersible observations. *Journal of Geophysical Research, 101*, 17995. doi:10.1029/96JB01209.

Bascom, W. (1961). *A hole in the bottom of the sea: The Story of the mohole project*. Garden City, NY: Doubleday.

Behn, M. D., & Ito, G. (2008). Magmatic and tectonic extension at mid-ocean ridges: 1. Controls on fault characteristics. *Geochemistry Geophysics Geosystems, 9*, Q08010. doi:10.1029/2008GC001965.

Blackman, D. K., Canales, J. P., & Harding, A. (2009). Geophysical signatures of oceanic core complexes. *Geophysical Journal International, 178*(2), 593–613. doi:10.1111/j.1365-246X.2009.04184.x.

Blackman, D. K., Cann, J. R., Janssen, B., & Smith, D. K. (1998). Origin of extensional core complexes: evidence from the Mid-Atlantic Ridge at Atlantis Fracture Zone. *Journal of Geophysical Research, 103*, 21315–21321 333. doi:10.1029/98JB01756.

Blackman, D. K., & Collins, J. A. (2010). Lower crustal variability and the crust/mantle transition at the Atlantis Massif oceanic core complex. *Geophys. Res. Lett, 37*, L24303. http://dx.doi.org/10.1029/2010GL045165.

Blackman, D. K., Ildefonse, B., John, B. E., Ohara, Y., Miller, D. J., MacLeod, C. J., et al. (2006). *Proc. IODP, 304/305*. College Station TX: Integrated Ocean Drilling Program Management International, Inc. doi:10.2204/iodp.proc.304305.2006.

Blackman, D. K., Ildefonse, B., John, B. E., Ohara, Y., Miller, D. J., Abe, N., et al. (2011). Drilling constraints on lithospheric accretion and evolution at Atlantis Massif, Mid-Atlantic Ridge 30°N. *Journal of Geophysical Research, 116*, B07103–B07129. doi:10.1029/2010JB007931.

Blackman, D. K., Karner, G. D., & Searle, R. C. (2008). Three-dimensional structure of oceanic core complexes: effects on gravity signature and ridge flank morphology, Mid-Atlantic Ridge, 30°N. *Geochemistry Geophysics Geosystems, 9*(6), Q06007. doi:10.1029/2008GC001951.

Blackman, D. K., Karson, J. A., Kelley, D. S., Cann, J. R., Früh-Green, G. L., Gee, J. S., et al. (2002). Geology of the Atlantis Massif (Mid-Atlantic Ridge, 30 N): implications for the evolution of an ultramafic oceanic core complex. *Marine Geophysical Research, 23*, 443–469. doi:10.1023/B: MARI.0000018232.14085.75.

Boschi, C., Früh-Green, G. L., Delacour, A., Karson, J. A., & Kelley, D. S. (2006). Mass transfer and fluid flow during detachment faulting and development of an oceanic core complex, Atlantis Massif (MAR 30°N). *Geochemistry Geophysics Geosystems, 1*, Q701004. doi:10.1029/2005GC001074.

Boudier, F., Godard, M., & Armbruster, C. (2000). Significance of gabbronorite occurrence in the crustal section of the Semail ophiolite. *Marine Geophysical Research, 21*, 307–326. doi:10.10 23/A:1026726232402.

Boudier, F., Nicolas, A., & Ildefonse, B. (1996). Magma chambers in the Oman ophiolite: fed from the top and the bottom. *Earth and Planetary Science Letters, 144*, 239–250. doi:10.1016/0012-821X(96)00167-7.

Buck, W. R., Lavier, L. L., & Poliakov, A. N. B. (2005). Modes of faulting at mid-ocean ridges. *Nature, 434*, 719–723. doi:10.1038/nature03358.

Canales, J. P., Tucholke, B. E., & Collins, J. A. (2004). Seismic reflection imaging of an oceanic detachment fault: Atlantis megamullion (Mid-Atlantic Ridge, 30°10′N). *Earth and Planetary Science Letters, 222*(2), 543–560. doi:10.1016/j.epsl.2004.02.023.

Canales, J. P., Tucholke, B. E., Xu, M., Collins, J. A., & DuBois, D. L. (2008). Seismic evidence for large-scale compositional heterogeneity of oceanic core complexes. *Geochemistry Geophysics Geosystems, 9*, Q08002. doi:10.1029/2008GC002009.

Cande, S. C., & Kent, D. V. (1995). Revised calibration of the geomagnetic polarity timescale for the Late Cretaceous and Cenozoic. *Journal of Geophysical Research, 100*(B4), 6093–6095. doi:10.1029/94JB03098.

Cann, J. R., Blackman, D. K., Smith, D. K., McAllister, E., Janssen, B., Mello, S., et al. (1997). Corrugated slip surfaces formed at ridge-transform intersections on the Mid-Atlantic Ridge. *Nature, 385*, 329–332. doi:10.1038/385329a0.

Cannat, M. (1993). Emplacement of mantle rocks in the seafloor at mid-ocean ridges. *Journal of Geophysical Research, 98*, 4163–4172. doi:10.1029/92JB02221.

Cannat, M. (1996). How thick is the magmatic crust at slow spreading oceanic ridges? *Journal of Geophysical Research, 101*(B2), 2847–2857. doi:10.1029/95JB03116.

Cannat, M., Sauter, D., Mendel, V., Ruellan, É., Okino, K., Escartín, J., et al. (2006). Modes of seafloor generation at a melt-poor ultraslow-spreading ridge. *Geology, 34*, 605–608. http://dx.doi.org/10.1130/G22486.1.

Cannat, M., Karson, J. A., Miller, D. J., & ODP Leg 153 Scientific Party. (1995). *Proc. ODP, Init. Repts, 153.* College Station, TX: Ocean Drilling Program. doi:10.2973/odp.proc.ir.153.1995.

Carbotte, S., Mutter, C., Mutter, J., & Ponce-Correa, G. (1998). Influence of magma supply and spreading rate on crustal magma bodies and emplacement of the extrusive layer: insights from the East Pacific Rise at lat 16°N. *Geology, 26*, 455–458. doi:10.1130/0091-7613(1998)026<0455:IOMSAS>2.3.CO;2.

Carlson, R. L. (2010). How crack porosity and shape control seismic velocities in the upper oceanic crust: modeling downhole logs from Holes 504B and 1256D. *Geochemistry Geophysics Geosystems, 11*, Q04007. doi:10.1029/2009GC002955.

Carlson, R. L. (2011). The effect of hydrothermal alteration on the seismic structure of the upper oceanic crust: evidence from Holes 504B and 1256D. *Geochemistry Geophysics Geosystems, 12*(9), Q09013. doi:10.1029/2011GC003624.

Chikyu+10 Steering Committee. (2013). *CHIKYU+10 International Workshop Report.* Jamstec & IODP. Available at http://www.jamstec.go.jp/chikyu+10/.

Coggon, R. M., Alt, J. C., & Teagle, D. A. (2008). Thermal history of ODP Hole 1256D lower sheeted dikes: petrology, chemistry and geothermometry of the granoblastic dikes. *Eos, Transactions American Geophysical Union, 89*(53) (Suppl.):V44B-08. (Abstract).

Collins, J. A., Blackman, D. K., Harris, A., & Carlson, R. L. (2009). Seismic and drilling constraints on velocity structure and reflectivity near IODP Hole U1309D on the Central Dome of Atlantis Massif, Mid-Atlantic Ridge 30°N. *Geochemistry Geophysics Geosystems, 10*, Q01010. doi:10.1029/2008GC002121.

Collins, J. A., & Detrick, R. S. (1998). Seismic structure of the Atlantis Fracture Zone megamullion, a serpentinized ultramafic massif. *Eos, Transactions American Geophysical Union, 79*(45), 800 (Abstract).

Collins, J. A., Tucholke, B. E., & Canales, J. P. (2002). Seismic velocity structure of Mid-Atlantic Ridge core complexes. *Eos, Transactions American Geophysical Union, 83*(47) EAE03-A-10390. (Abstract).

Coogan, L. A., Gillis, K. M., MacLeod, C. J., Thompson, G. M., & Hekinian, R. (2002). Petrology and geochemistry of the lower ocean crust formed at the East Pacific Rise and exposed at Hess Deep: a synthesis and new results. *Geochemistry Geophysics Geosystems*, *3*, 1–30. doi:10.10 29/2001GC000230.

Crispini, L., Tartarotti, P., & Umino, S. (2006). Microstructural features of a subaqueous lava from basaltic crust off the East Pacific Rise (ODP Site 1256, Cocos Plate). *Ofioliti*, *31*, 117–127. doi:10.4454/ofioliti.v31i2.334.

Detrick, R., Collins, J., Stephen, R., & Swift, S. (1994). In situ evidence for the nature of the seismic Layer 2/3 boundary in oceanic crust. *Nature (London, UK)*, *370*(6487), 288–290. doi:10.1038/370288a0.

Detrick, R., Honnorez, J., Bryan, W. B., Juteau, T., & ODP Legs 106/109 Scientific Parties. (1988). *Proc. ODP, Init. Repts., 106/109*. College Station, TX: Ocean Drilling Program. doi:10.2973/odp.proc.ir.106109.1988.

Dick, H. J. B. (1989). Abyssal peridotites, very slow spreading ridges and ocean ridge magmatism. In A. D. Saunders, & M. J. Norry (Eds.), *Magmatism in the ocean basins* (pp. 71–105). Geol. Soc. Spec. Publ. 42(1):71-105. doi:10.1144/GSL.SP.1989.042.01.06.

Dick, H. J. B., Natland, J. H., Alt, J. C., Bach, W., Bideau, D., Gee, J. S., et al. (2000). A long in situ section of the lower ocean crust: results of ODP Leg 176 drilling at the Southwest Indian Ridge. *Earth and Planetary Science Letters*, *179*(1), 31–51. doi:10.1016/S0012-821X(00)00102-3.

Dick, H. J. B., Natland, J. H., & Ildefonse, B. (2006). Past and future impact of deep drilling in the oceanic crust and mantle. *Oceanography*, *19*(4), 72–80. doi:10.5670/oceanog.2006.06.

Dick, H. J. B., & Mével, C. (1996). *The oceanic lithosphere and scientific drilling into the 21st Century*. MA: Woods Hole (JOI/USSSP) http://www.odplegacy.org/PDF/Admin/Workshops/1996_05_Ocean_Lithosphere.pdf.

Dick, H. J. B., Tivey, M. A., & Tucholke, B. E. (2008). Plutonic foundation of a slow-spreading ridge segment: oceanic core complex at Kane Megamullion, 23°30′N, 45°20′W. *Geochemistry Geophysics Geosystems*, *9*, Q05014. doi:10.1029/2007GC001645.

Drouin, M., Godard, M., Ildefonse, B., Bruguier, O., & Garrido, C. J. (2009). Geochemical and petrographic evidence for magmatic impregnation in the oceanic lithosphere at Atlantis Massif, Mid-Atlantic Ridge (IODP Hole U1309D, 30°N). *Chemical Geology*, *264*(1–4), 71–88. doi:10.1016/j.chemgeo.2009.02.013.

Drouin, M., Ildefonse, B., & Godard, M. (2010). A microstructural imprint of melt impregnation in slow spreading lithosphere: olivine-rich troctolites from the Atlantis Massif, Mid-Atlantic Ridge, 30°N, IODP Hole U1309D. *Geochemistry Geophysics Geosystems*, *11*(6), Q06003. doi: 10.1029/2009GC002995.

Eason, D. E., & Sinton, J. M. (2009). Lava shields and fissure eruptions of the Western Volcanic Zone, Iceland: evidence for magma chambers and crustal interaction. *Journal of Volcanology and Geothermal Research*, *186*, 331–348. doi:10.1016/j.jvolgeores.2009.06.009.

Escartín, J., & Canales, J. P. (2011). Detachments in oceanic lithosphere: deformation, magmatism, fluid flow, and ecosystems. *Eos, Transactions American Geophysical Union*, *92*, 31. doi:10.10 29/2011EO040003.

Escartín, J., Mével, C., MacLeod, C. J., & McCaig, A. M. (2003). Constraints on deformation conditions and the origin of oceanic detachments: the Mid-Atlantic Ridge core complex at 15°45′N. *Geochemistry Geophysics Geosystems*, *4*, 8. doi:10.1029/2002GC000472.

Escartín, J., Smith, D. K., Cann, J., Scouten, H., Langmuir, C. H., & Escrig, S. (2008). Central role of detachment faults in accretion of slow-spreading oceanic lithosphere. *Nature*, *455*, 790–794. doi:10.1038/nature07333.

Estabrook, F. B. (1956). Geophysical research shaft. *Science, 124*(3324), 686. doi:10.1126/science.124.3224.686.

Expedition 340T Scientists. (2012). *Atlantis massif oceanic core complex: Velocity, porosity, and impedance contrasts within the domal core of atlantis massif: Faults and hydration of lithosphere during core complex evolution*. IODP Prel. Rept., 340T. doi:10.2204/iodp.pr.340T.2012.

Ferrini, V. L., Shillington, D. J., Gillis, K., MacLeod, C. J., Teagle, D. A. H., Morris, A., & the JC21 Scientific Party., et al. (2013). Marine geology. *Marine Geology, 339*, 13–21. doi:10.1016/j.margeo.2013.03.006.

France, L., Ildefonse, B., & Koepke, J. (2009). Interactions between magma and the hydrothermal system in the Oman ophiolite and in IODP Hole 1256D: fossilisation of dynamic melt lens at fast spreading ridges. *Geochemistry Geophysics Geosystems, 10*(10), Q10O19. doi:10.1029/2009GC002652.

Francheteau, J., Armijo, R., Cheminée, J. L., Hekinian, R., Lonsdale, P., & Blum, N. (1990). 1 Ma East Pacific Rise oceanic crust and uppermost mantle exposed by rifting in Hess Deep (equatorial Pacific Ocean). *Earth and Planetary Science Letters, 101*, 281–295. doi:10.1016/0012-821X(90)90160-Y.

Früh-Green, G. L., Kelley, D. S., Bernasconi, S. M., Karson, J. A., Ludwig, K. A., Butterfield, D. A., et al. (2003). 30,000 years of hydrothermal activity at the Lost City vent field. *Science, 301*(5632), 495–498. doi:10.1126/science.1085582.

Gale, A., Dalton, C. A., Langmuir, C. H., Su, Y., & Schilling, J.-G. (2013). The mean composition of ocean ridge basalts. *Geochemistry Geophysics Geosystems, 14*, 489–518. doi:10.1029/2012GC004334.

Garcés, M., & Gee, J. S. (2007). Paleomagnetic evidence of large footwall rotations associated with low-angle faults at the Mid-Atlantic Ridge. *Geology, 35*, 279. doi:10.1130/G23165A.1.

Geldmacher, J., Hofig, T. W., Hauff, F., Hoernle, K., Garbe-Schönberg, D., & Wilson, D. S. (2013). Influence of the Galapagos hotspot on the East Pacific Rise during Miocene superfast spreading. *Geology, 41*, 183–186. doi:10.1130/G33533.1.

Geshi, N., Umino, S., Kumagai, H., Sinton, J. M., White, S. M., Kisimoto, K., et al. (2007). Discrete plumbing systems and heterogeneous magma sources of a 24 km³ off-axis lava field on the western flank of East Pacific Rise, 14°S. *Earth and Planetary Science Letters, 258*, 61–72. doi:10.1016/j.epsl.2007.03.019.

Gilbert, L. A., & Salisbury, M. H. (2011). Oceanic crustal velocities from laboratory and logging measurements of Integrated Ocean Drilling Program Hole 1256D. *Geochemistry Geophysics Geosystems, 12*, Q09001. doi:10.1029/2011GC003750.

Gillis, K., Mével, C., Allan, J., & ODP Leg 147 Scientific Party. (1993). *Proc. ODP, Init. Repts., 147*. College Station, TX: Ocean Drilling Program. doi:10.2973/odp.proc.ir.147.1993.

Gillis, K. M., Snow, J. E., Klaus, A., & the Expedition 345 Scientists. (2014a). *Proc. IODP, 345*. College Station, TX: Integrated Ocean Drilling Program. doi:10.2204/iodp.proc.345.2014.

Gillis, K. M., Snow, J. E., Klaus, A., Abe, N., Akizawa, N., de brito Adrião, A., et al. (2014b). Primitive layered gabbros from fast-spreading lower oceanic crust. *Nature, 505*, 204–207. doi:10.1038/nature12778.

Godard, M., Awaji, S., Hansen, H., Hellebrand, E., Brunelli, D., Johnson, K., et al. (2009). Geochemistry of a long in-situ section of intrusive slow-spread oceanic lithosphere: results from IODP Site U1309 (Atlantis Massif, 30°N Mid-Atlantic-Ridge). *Earth and Planetary Science Letters, 279*, 110–122. doi:10.1016/j.epsl.2008.12.034.

Grimes, C. B., Cheadle, M. J., John, B. E., Reiners, P. W., & Wooden, J. L. (2011). Cooling rates and the depth of detachment faulting at oceanic core complexes: evidence from zircon Pb/U and (U-Th)/He ages. *Geochemistry Geophysics Geosystems, 12*, Q0AG01. doi:10.1029/2010GC003391.

Grimes, C. B., John, B. E., Cheadle, M. J., & Wooden, J. L. (2008). Protracted construction of gabbroic crust at a slow spreading ridge: constraints from 206Pb/238U zircon ages from Atlantis Massif and IODP Hole U1309D (30°N, MAR). *Geochemistry Geophysics Geosystems, 9*, Q08012. doi:10.1029/2008GC002063.

Guérin, G., Goldberg, D. S., & Iturrino, G. J. (2008). Velocity and attenuation in young oceanic crust: new downhole log results from DSDP/ODP/IODP Holes 504B and 1256D. *Geochemistry Geophysics Geosystems, 9*(12), Q12014. doi:10.1029/2008GC002203.

Haggas, S., Brewer, T. S., Harvey, P. K., & MacLeod, C. J. (2005). Integration of electrical and optical images for structural analysis: a case study from ODP Hole 1105A. In P. K. Harvey, T. S. Brewer, P. A. Pezard, & V. A. Petrov, et al. (Ed.), *Petrophysical properties of crystalline rocks* (Vol. 240). Geological Society of London Special Publication. 165–177. doi:10.1144/GSL.SP.2005.240.01.13.

Hallenborg, E., Harding, A. J., Kent, G. M., & Wilson, D. S. (2003). Seismic structure of 15 Ma oceanic crust formed at an ultrafast spreading East Pacific Rise: evidence for kilometer-scale fracturing from dipping reflectors. *Journal of Geophysical Research, 108*(B11), 2532. doi:10.1029/2003JB002400.

Hart, S. R., Blusztajn, J., Dick, H. J. B., Meyer, P. S., & Muehlenbachs, K. (1999). The fingerprint of seawater circulation in a 500-meter section of ocean crust gabbros. *Geochimica et Cosmochimica Acta, 63*, 4059–4080. http://dx.doi.org/10.1016/S0016-7037(99)00309-9.

Hékinian, R., Bideau, D., Francheteau, J., Lonsdale, P., & Blum, N. (1993). Petrology of the East Pacific Rise crust and upper mantle exposed in the Hess Deep (eastern equatorial Pacific). *Journal of Geophysical Research, 98*, 8069–8094. doi:10.1029/92JB02072.

Henig, A. S., Blackman, D. K., Harding, A. J., Canales, J. P., & Kent, G. M. (2012). Downward continued multichannel seismic refraction analysis of Atlantis Massif oceanic core complex, 30°N, Mid-Atlantic Ridge. *Geochemistry Geophysics Geosystems, 13*, Q0AG07. doi:10.1029/2012GC004059.

Herron, T. J. (1982). Lava flow layer - east Pacific rise. *Geophysical Research Letters, 9*, 17–20. doi:10.1029/GL009i001p00017.

Ildefonse, B., Abe, N., Blackman, D. K., Canales, J. P., Isozaki, Y., Kodaira, S., et al. (2010a). *The MoHole: A crustal journey and mantle Quest. Workshop Report.* http://campanian.iodp.org/MoHole/MoHoleWS2010_Report.pdf.

Ildefonse, B., Abe, N., Blackman, D. K., Canales, J. P., Isozaki, Y., Kodaira, S., et al. (2010b). The MoHole: a crustal journey and mantle quest, workshop in Kanazawa, Japan, 3–5 June 2010. *Scientific Drilling, 10*, 56–62. doi:10.2204/iodp.sd.10.07.2010.

Ildefonse, B., Blackman, D. K., John, B. E., Ohara, Y., Miller, D. J., MacLeod, C. J., & the IODP expedition 304 305 scientists. (2006). IODP Expeditions 304 & 305 characterize the lithology, structure, and alteration of an oceanic core complex. *Scientific Drilling, 3*, 4–11. doi:10.2204/iodp.sd.3.01.2006.

Ildefonse, B., Blackman, D. K., John, B. E., Ohara, Y., Miller, D. J., MacLeod, C. J., & Integrated Ocean Drilling Program Expeditions 304/305 Science Party. (2007a). Oceanic core complexes and crustal accretion at slow-spreading ridges. *Geology, 35*(7), 623–626. doi:10.1130/G23531A.1.

Ildefonse, B., Christie, D. M., & the Mission Moho Workshop Steering Committee. (2007b). Mission Moho workshop: drilling through the oceanic crust to the mantle. *Scientific Drilling, 4*, 11–18. doi:10.2204/iodp.sd.4.02.2007.

Ildefonse, B., Rona, P. A., & Blackman, D. K. (2007c). Drilling the crust at mid-ocean ridges: an "in depth" perspective. *Oceanography, 20*(1), 66–77. doi:10.5670/oceanog.2007.81.

IODP Science Plan 2013–2023. (2011). *Illuminating Earth's past, present, and future. The International Ocean Discovery program exploring the earth under the sea.* Available online http://www.iodp.org/doc_download/3885-new-science-plan.

John, B. E., Cheadle, M. J., Gee, J. S., Grimes, C. B., Morris, A., Pressling, N., & the Expedition 304/305 Scientists. (2009). Data report: spatial and temporal evolution of slow spread oceanic crust—graphic sections of core recovered from IODP Hole U1309D, Atlantis Massif, 30°N, MAR (including Pb/U zircon geochronology and magnetic remanence data). In D. K. Blackman, B. Ildefonse, B. E. John, Y. Ohara, D. J. Miller, & C. J. MacLeod (Eds.), *Proc. IODP, 304/305*. College Station, TX: Integrated Ocean Drilling Program Management International, Inc. doi:10.2204/iodp.proc.304305.205.2009.

Karson, J. A. (2002). Geologic structure of the uppermost oceanic crust created at fast- to intermediate-rate spreading centers. *Annual Review of Earth and Planetary Sciences, 30*, 347–384. doi:10.1146/annurev.earth.30.091201.141132.

Karson, J. A., Früh-Green, G. L., Kelley, D. S., Williams, E. A., Yoerger, D. R., & Jakuba, M. (2006). Detachment shear zone of the Atlantis Massif core complex, Mid-Atlantic Ridge, 30°N. *Geochemistry Geophysics Geosystems, 7*(6), Q06016. doi:10.1029/2005GC001109.

Kelemen, P. B., Kikawa, E., Miller, D. J., & ODP Leg 209 Scientific Party. (2004). *Proc. ODP, Init. Repts., 209*. College Station, TX: Ocean Drilling Program. doi:10.2973/odp.proc.ir.209.2004.

Kelemen, P. B., Kikawa, E., Miller, D. J., & Shipboard Scientific Party (2007). Leg 209 summary: processes in a 20-km-thick conductive boundary layer beneath the Mid-Atlantic Ridge, 14°–16°N. In P. B. Kelemen, E. Kikawa, & D. J. Miller (Eds.), *Proc. ODP, Sci. Results, 209*. College Station, TX: Ocean Drilling Program. 1–33. doi:10.2973/odp.proc.sr.209.001.2007.

Kelemen, P. B., Koga, K., & Shimizu, N. (1997). Geochemistry of gabbro sills in the crust-mantle transition zone of the Oman ophiolite: implications for the origin of the oceanic lower crust. *Earth and Planetary Science Letters, 146*, 475–488. doi:10.1016/S0012-821X(96)00235-X.

Kelley, D. S., Karson, J. A., Blackman, D. K., Früh-Green, G. L., Butterfield, D. A., Lilley, M. D., & the AT3-60 Shipboard Party., et al. (2001). An off-axis hydrothermal vent field near the Mid-Atlantic Ridge at 30°N. *Nature, 412*(6843), 145–149. doi:10.1038/35084000.

Kinzler, R. J., & Grove, T. L. (1993). Corrections and further discussion of the primary magmas of mid-ocean ridge basalts, 1 and 2. *Journal of Geophysical Research, 98*, 22339–22347. doi:10.1029/93JB02164.

Koepke, J., Christie, D. M., Dziony, W., Holtz, F., Lattard, D., Maclennan, J., et al. (2008). Petrography of the dike–gabbro transition at IODP Site 1256 (equatorial Pacific): the evolution of the granoblastic dikes. *Geochemistry Geophysics Geosystems, 9*(7), Q07009. doi:10.1029/2008GC001939.

Lagabrielle, Y., Bideau, D., Cannat, M., Karson, J. A., & Mével, C. (1998). Ultramafic-mafic plutonic rock suites exposed along the Mid-Atlantic Ridge (10°N–30°N). Symmetrical–Asymmetrical distribution and implications for seafloor spreading processes. In *Faulting and magatism at mid-ocean ridges* (Vol. 106) Geophys. Monogr. Ser. 153–176. doi:10.1029/GM106p0153.

Langmuir, C. H., Bender, J. F., & Batiza, R. (1986). Petrological and tectonic segmentation of the East Pacific Rise, 5°30'N–14°30'N. *Nature, 322*(6078), 422–429. doi:10.1038/322422a0.

Lavier, L. L., Buck, W. R., & Poliakov, A. N. B. (1999). Self-consistent rolling-hinge model for the evolution of large-offset low-angle normal faults. *Geology, 27*, 1127–1130. doi:10.1130/0091-7613(1999)027 <1127:SCRHMF>2.3.CO;2.

Lissenberg, C. J., MacLeod, C. J., Howard, K. A., & Godard, M. (2013). Pervasive reactive melt migration through fast-spreading lower oceanic crust (Hess Deep, equatorial Pacific Ocean). *Earth and Planetary Science Letters, 361*, 436–447. doi:10.1016/j.epsl.2012.11.012.

Lonsdale, P. (1977). Deep-tow observations at the mounds abyssal hydrothermal field, Galapagos Rift. *Earth and Planetary Science Letters, 36*, 92–110. doi:10.1016/0012-821X(77)90191-1.

Macdonald, K. C. (1998). Linkages between faulting, volcanism, hydrothermal activity and segmentation on fast spreading centers. In *Faulting and magatism at mid-ocean ridges* (Vol. 106) Geophys. Monogr. Ser. 27–58. doi:10.1029/GM106p0027.

Macdonald, K. C., Fox, P. J., Alexander, R. T., Pockalny, R., & Gente, P. (1996). Volcanic growth faults and the origin of Pacific abyssal hills. *Nature, 380*(6570), 125–129. doi:10.1038/380125a0.

Macdonald, K. C., Haymon, R., & Shor, A. (1989). A 220 km² recently erupted lava field on the East Pacific Rise near lat 8°S. *Geology, 17*(3), 212–216. doi:10.1130/0091-7613(1989)017<0212:AKRELF>2.3.CO;2.

MacLeod, C. J., Parson, L. M., Sager, W. W., & the ODP Leg 135 Scientific Party. (1992). Identification of tectonic rotations in boreholes by the integration of core information with Formation MicroScanner and Borehole Televiewer images. In A. Hurst, C. M. Griffiths, & P. F. Worthington (Eds.), *Geological applications of wireline logs ii. 65*). London: Spec. Publ. Geol. Soc. 235–246. doi:10.1144/GSL.SP.1992.065.01.18.

MacLeod, C. J., Searle, R. C., Murton, B. J., Casey, J. F., Mallows, C., Unsworth, S. C., et al. (2009). Life cycle of oceanic core complexes. *Earth and Planetary Science Letters, 287*, 333–344. doi:10.1016/j.epsl.2009.08.016.

MacLeod, C. J., & Yaouancq, G. (2000). A fossil melt lens in the Oman ophiolite: Implications for magma chamber processes at fast spreading ridges. *Earth and Planetary Science Letters, 176*, 357–373. http://dx.doi.org/10.1016/S0012-821X(00)00020-0.

McCaig, A. M., & Harris, M. (2012). Hydrothermal circulation and the dike-gabbro transition in the detachment mode of slow seafloor spreading. *Geology, 40*, 367–370. doi:10.1130/G32789.1.

Michibayashi, K., Hirose, T., Nozaka, T., Harigane, Y., Escartín, J., Delius, H., et al. (2008). Hydration due to high-T brittle failure within in situ oceanic crust, 30°N Mid-Atlantic Ridge. *Earth and Planetary Science Letters, 275*, 348–354. http://dx.doi.org/10.1016/j.epsl.2008.08.033.

Müller, R. D., Sdrolias, M., Gaina, C., & Roest, W. R. (2008). Age, spreading rates, and spreading asymmetry of the world's ocean crust. *Geochem. Geophys. Geosyst, 9*, Q04006. http://dx.doi.org/10.1029/2007GC001743.

Morris, A., Gee, J. S., Pressling, N., John, B. E., MacLeod, C. J., Grimes, C. B., et al. (2009). Footwall rotation in an oceanic core complex quantified using reoriented Integrated Ocean Drilling Program core samples. *Earth and Planetary Science Letters, 287*, 217–228. doi:10.1016/j.epsl.2009.08.007.

Neo, N., Yamazaki, S., Miyashita, S., & the Expedition 309/312 Scientists. (2009). Data report: bulk rock compositions of samples from the IODP Expedition 309/312 sample pool, ODP Hole 1256D. In D. A. H. Teagle, J. C. Alt, S. Umino, S. Miyashita, N. R. Banerjee, & D. S. Wilson (Eds.), *Proc. IODP, 309/312*. Washington, DC: Integrated Ocean Drilling Program Management International, Inc. doi:10.2204/iodp.proc.309312.204.2009.

Nonnotte, P., Ceuleneer, G., & Benoit, M. (2005). Genesis of andesitic–boninitic magmas at mid-ocean ridges by melting of hydrated peridotites: geochemical evidence from DSDP Site 334 gabbronorites. *Earth and Planetary Science Letters, 236*, 632–653. doi:10.1016/j.epsl.2005.05.026.

Nooner, S. L., Sasagawa, G. S., Blackman, D. K., & Zumberge, M. A. (2003). Constraints on crustal structure at the Mid-Atlantic Ridge from seafloor gravity measurements made at the Atlantis Massif. *Geophysical Research Letters, 30*, 1446. doi:10.1029/2003GL017126.

Olive, J.-A., Behn, M. D., & Tucholke, B. E. (2010). The structure of oceanic core complexes controlled by the depth distribution of magma emplacement. *Nature Geoscience, 3*, 491–495. doi:10.1038/ngeo888.

Pallister, J. S., & Hopson, C. A. (1981). Samail ophiolite plutonic suite: field relations, phase variation, cryptic variation and layering, and a model of a spreading ridge magma chamber. *Journal of Geophysical Research, 86*, 2593–2644. doi:10.1029/JB086iB04p02593.

Park, S., MacLennan, J., Teagle, D., & Hauff, F. (2008). Did the Galápagos plume influence the ancient EPR? a geochemical study of basaltic rocks from Hole 1256D. *Eos, Transactions American Geophysical Union, 89* Fall Meeting Supplement, abstract V51F-2097.

Pressling, N., Morris, A., John, B. E., & MacLeod, C. J. (2012). The internal structure of an oceanic core complex: an integrated analysis of oriented borehole imagery from IODP Hole U1309D (Atlantis Massif). *Geochemistry Geophysics Geosystems, 13*, Q04G10. doi:10.1029/201 2GC004061.

Purdy, G. M., Kong, L. S. L., Christeson, G. L., & Solomon, S. C. (1992). Relationship between spreading rate and the seismic structure of mid-ocean ridges. *Nature, 355*(6363), 815–817. doi:10.1038/355815a0.

Ravelo, C., Bach, W., Behrmann, J., Camoin, G., Duncan, R., Edwards, K., et al. (2010). *INVEST report: IODP new ventures in exploring scientific targets—Defining new goals of an international drilling program*. Tokyo: IODP-MI. http://www.ecord.org/rep/INVEST-Report.pdf.

Rubin, K. H., & Sinton, J. M. (2007). Inferences on mid-ocean ridge thermal and magmatic structure from MORB compositions. *Earth and Planetary Science Letters, 260*, 257–276. doi:10.1016/j. epsl.2007.05.035.

Schroeder, T., & John, B. E. (2004). Strain localization on an oceanic detachment fault system, Atlantis Massif, 30°N, Mid-Atlantic Ridge. *Geochemistry Geophysics Geosystems, 5*, Q11007. doi:10.1029/2004GC000728.

Singh, S. C., Kent, G. M., Collier, J. S., Harding, A. J., & Orcutt, J. A. (1998). Melt to mush variations in crustal magma properties along the ridge crest at the southern East Pacific Rise. *Nature, 394*, 874–878. doi:10.1038/29740.

Sinton, J., Bergmanis, E., Rubin, K., Batiza, R., Gregg, T. K. P., Grönvold, K., et al. (2002). Volcanic eruptions on mid-ocean ridges: new evidence from the superfast spreading East Pacific Rise, 17°–19°S. *Journal of Geophysical Research, 107*, 2115. doi:10.1029/2000JB000090.

Smith, D. K., Escartín, J., Schouten, H., & Cann, J. R. (2008). Fault rotation and core complex formation: significant processes in seafloor formation at slow-spreading mid-ocean ridges (Mid-Atlantic Ridge, 13°–15°N). *Geochemistry Geophysics Geosystems, 9*, Q03003. doi:10.1029/2 007GC001699.

Suhr, G., Hellebrand, E., Johnson, K., & Brunelli, D. (2008). Stacked gabbro units and intervening mantle: a detailed look at a section of IODP Leg 305, Hole 1309D. *Geochemistry Geophysics Geosystems, 9*, Q10007. doi:10.1029/2008GC002012.

Swift, S., Reichow, M., Tikku, A., Tominaga, M., & Gilbert, L. (2008). Velocity structure of upper ocean crust at Ocean Drilling Program Site 1256. *Geochemistry Geophysics Geosystems, 9*(10), Q10013. doi:10.1029/2008GC002188.

Tamura, A., Arai, S., Ishimaru, S., & Andal, E. S. (2008). Petrology and geochemistry of peridotites from IODP Site U1309 at Atlantis Massif, MAR 30°N: micro- and macro-scale melt penetrations into peridotites. *Contributions to Mineralogy and Petrology, 155*(4), 491–509. doi:10.1007/s00410-007-0254-0.

Tartarotti, P., & Crispini, L. (2006). ODP-IODP Site 1256 (East Pacific Rise): an in-situ section of upper ocean crust formed at a superfast spreading rate. *Ofioliti, 31*, 107–116. doi:10.4454/ ofioliti.v31i2.333.

Teagle, D. A. H., Alt, J. C., & Halliday, A. N. (1998). Tracing the evolution of hydrothermal fluids in the upper oceanic crust: Sr-isotopic constraints from DSDP/ODP Holes 504B and 896A. In K. Harrison, & R. A. Mills (Eds.), *Modern ocean floor processes and the geological record*. Geol. Soc. Spec. Publ. 148(1):81–97. doi:10.1144/GSL.SP.1998.148.01.06.

Teagle, D. A. H., Alt, J. C., Umino, S., Miyashita, S., Banerjee, N. R., Wilson, D. S., & the Expedition 309/312 Scientists. (2006). *Proc. IODP, 309/312*. Washington, DC: Integrated Ocean Drilling Program Management International, Inc. doi:10.2204/iodp.proc.309312.2006.

Teagle, D., & Ildefonse, B. (2011). Journey to the mantle of the Earth. *Nature, 471*, 437–439. doi:10.1038/471437a.

Teagle, D., Ildefonse, B., Blackman, D., Edwards, K., Bach, W., Abe, N., et al. (2009). *Melting, magma, fluids and life—challenges for the next generation of scientific ocean drilling into the oceanic lithosphere: workshop report.* Southampton: Univ. Southampton.

Teagle, D. A. H., Ildefonse, B., Blum, P., & the Expedition 335 Scientists. (2012). *Proc. IODP, 335.* Tokyo: Integrated Ocean Drilling Program Management International, Inc. doi:10.2204/iodp. proc.335.2012.

Teagle, D. A. H., Wilson, D. S., & Acton, G. D. (2004). The "road to the MoHole" four decades on: deep drilling at Site 1256. *Eos, Transactions American Geophysical Union, 85*(49), 521. doi:1 0.1029/2004EO490002.

Tominaga, M., Teagle, D. A. H., Alt, J. C., & Umino, S. (2009). Determination of volcanostratigraphy of the oceanic crust formed at superfast spreading ridge: electrofacies analyses of ODP/IODP Hole 1256D. *Geochemistry Geophysics Geosystems, 10*(1), Q01003. doi:10.1029/2008GC002143.

Tominaga, M., & Umino, S. (2010). Lava deposition history in ODP Hole 1256D: insights from log-based volcanostratigraphy. *Geochemistry Geophysics Geosystems, 11*(5), Q05003. doi:10 .1029/2009GC002933.

Tommasi, A., Godard, M., Coromina, G., Dautria, J.-M., & Barsczus, H. (2004). Seismic anisotropy and compositionally induced velocity anomalies in the lithosphere above mantle plumes: a petrological and microstructural study of mantle xenoliths from French Polynesia. *Earth and Planetary Science Letters, 227*, 539–556. doi:10.1016/j.epsl.2004.09.019.

Tucholke, B. E., Behn, M. D., Buck, W. R., & Lin, J. (2008). Role of melt supply in oceanic detachment faulting and formation of megamullions. *Geology, 36*, 455. doi:10.1130/G24639A.1.

Tucholke, B. E., Lin, J., & Kleinrock, M. C. (1998). Megamullions and mullion structure defining oceanic metamorphic core complexes on the Mid-Atlantic Ridge. *Journal of Geophysical Research, 103*, 9857–9866. doi:10.1029/98JB00167.

Tucholke, B. E., & Lin, J. (1994). A geological model for the structure of ridge segments in slow spreading ocean crust. *Journal of Geophysical Research, 99*, 11937. doi:10.1029/94JB00338.

Umino, S., Lipman, P. W., & Obata, S. (2000). Subaqueous lava flow lobes, observed on ROV Kaiko dives off Hawaii. *Geology, 28*(6), 503–506. doi:10.1130/0091-7613(2000)28 <503:SLFLOO>2.0.CO;2.

Umino, S., Crispini, L., Tartarotti, P., Teagle, D. A. H., Alt, J. C., Miyashita, S., et al. (2008). Origin of the sheeted dike complex at superfast spread East Pacific Rise revealed by deep ocean crust drilling at Ocean Drilling Program Hole 1256D. *Geochemistry Geophysics Geosystems, 9*, Q06008. doi:10.1029/2007GC001760.

Violay, M., Pezard, P. A., Ildefonse, B., Célérier, B., & Deleau, A. (2012). Structure of the hydrothermal root zone of the sheeted dikes in fats-spread oceanic crust: a core-log integration study of ODP Hole 1256D, Eastern Equatorial Pacific. *Ofioliti, 37*, 1–11. doi:10.4454/ofioliti.v37i1.402.

von der Handt, A., & Hellebrand, E. (2010). *Transformation of mantle to lower crust: Melt-rock reaction processes in peridotites from Atlantis Massif, 30°N Mid–Atlantic Ridge.* Abstract V11A–2244 presented at 2010 Fall Meeting. San Francisco, Calif: AGU. 13–17 Dec.

Wernicke, B. (1995). Low-angle normal fault seismicity: a review. *Journal of Geophysical Research, 100*, 20159–20174. doi:10.1029/95JB01911.

White, S. M., Macdonald, K. C., & Haymon, R. M. (2000). Basaltic lava domes, lava lakes, and volcanic segmentation on the southern East Pacific Rise. *Journal of Geophysical Research, 105*, 23519–23536. doi:10.1029/2000JB900248.

Wilson, D. S. (1996). Fastest known spreading on the Miocene Cocos–Pacific plate boundary. *Geophysical Research Letters, 23*(21), 3003–3006. doi:10.1029/96GL02893.

Wilson, D. S., Teagle, D. A. H., Acton, G. D., & ODP Leg 206 Scientific Party. (2003). *Proc. ODP, Init. Repts., 206.* College Station: TX (Ocean Drilling Program). doi:10.2973/odp.proc.ir.206.2003.

Wilson, D. S., Teagle, D. A. H., Alt, J. C., Banerjee, N. R., Umino, S., Miyashita, S., et al. (2006). Drilling to gabbro in intact ocean crust. *Science*, *312*(5776), 1016–1020. doi:10.1126/science.1126090.

Workman, R., & Hart, S. R. (2005). Major and trace element composition of the depleted MORB mantle (DMM). *Earth and Planetary Science Letters*, *231*, 53–72. doi:10.1016/j.epsl.2004.12.005.

Yamazaki, S., Neo, N., & Miyashita, S. (2009). Data report: whole-rock major and trace elements and mineral compositions of the sheeted dike–gabbro transition in ODP Hole 1256D. In D. A. H. Teagle, J. C. Alt, S. Umino, S. Miyashita, N. R. Banerjee, D. S. Wilson, et al. (Eds.), *Proc. IODP, 309/312*. Washington, DC: Integrated Ocean Drilling Program Management International, Inc. doi:10.2204/iodp.proc.309312.203.2009.

Yeats, R. S., Hart, S. R., & DSDP Leg 34 Scientific Party. (1976). *Initial reports of the deep sea drilling project* (Vol. 34)Washington: U.S. Government Printing Office. doi:10.2973/dsdp.proc.34.1976.

Chapter 4.2.2

Hydrogeologic Properties, Processes, and Alteration in the Igneous Ocean Crust

Andrew T. Fisher,[1,*] Jeffrey Alt[2] and Wolfgang Bach[3]

[1]*Earth and Planetary Sciences Department, University of California, Santa Cruz, CA, USA;*
[2]*Department of Earth and Environmental Sciences, University of Michigan, Ann Arbor, MI, USA;*
[3]*Department of Geosciences, Center for Marine Environmental Sciences (MARUM), University of Bremen, Bremen, Germany*
Corresponding author: E-mail: afisher@ucsc.edu

4.2.2.1 INTRODUCTION

4.2.2.1.1 Motivation

Fluid flow within volcanic ocean crust influences: the thermal and chemical evolution of oceanic lithosphere and seawater; subseafloor microbial ecosystems; diagenetic, seismic, and magmatic activity along plate-boundary faults; and the creation of ore deposits on and below the seafloor (e.g., Coggon, Teagle, Smith-Duque, Alt, & Cooper, 2010; Huber, Butterfield, Johnson, & Baross, 2006; Parsons & Sclater, 1977). The global hydrothermal fluid mass flux through the upper oceanic crust rivals the global riverine fluid flux to the ocean, passing the volume of the oceans through the crust every 10^5–10^6 year (e.g., Johnson & Pruis, 2003; Mottl, 2003; Wheat, McManus, Mottl, & Giambalvo, 2003). Much of this flow occurs at relatively low temperatures, far from volcanically active seafloor spreading centers where new ocean floor is created. This "ridge-flank" circulation can be influenced by off-axis volcanic or tectonic activity, and by exothermic reactions that occur within the crust during fluid transport, but most of the flow is driven by lithospheric heating from below the crust.

Fluid flow in the volcanic oceanic crust appears in several sections of the Initial Science Plan for IODP, including those focusing on solid earth cycles and alteration, ocean chemistry, and the deep biosphere. Many of the expeditions completed in the first decade of IODP operations included components of volcanic crustal hydrogeology and alteration, and several focused specifically on this topic. It is commonly assumed that time-integrated crustal hydrogeology is expressed in the rock record by alteration, as seen in cores and borehole measurements, and that the most recent phase(s) of water–rock interactions should be consistent with the present-day hydrogeologic state of the crust. We assess whether this assumption is valid, based on IODP results from site surveys, borehole sampling, wireline logging, monitoring and active experiments, and associated analyses of geological and biological materials. In the next section, we present a brief overview of operational and borehole measurement methods. Following sections summarize key results from selected expeditions and sites. We end with a synthesis of observational results that seeks to link hydrogeologic data to observations of crustal alteration.

4.2.2.1.2   Drilling, Coring, and Measurement Methods for the Igneous Ocean Crust

Most IODP crustal holes were drilled with a coring bit, but some were drilled without coring to improve hole stability and ease installation of casing. Some crustal holes were logged with wireline geophysical tools, to assess lithostratigraphy, hydrogeology, and rock alteration. Both core samples and geophysical logs can elucidate the physical and chemical states of the formation surrounding a borehole, which results from primary crustal construction overlain by tectonic, magmatic, and alteration (especially, hydrothermal) processes that occur as the lithosphere ages.

The most common logging tools deployed in the ocean crust include caliper (hole diameter), bulk density, porosity, resistivity, seismic velocity, and borehole temperature (we use generic logging tool descriptions throughout this chapter, rather than specific tool names and acronyms). Some basement holes have also been logged with tools that create higher-resolution images of the borehole wall using resistivity and sonic data. In general, wireline geophysical logs are the most accurate, and tools are least likely to get stuck, in holes that are "to gauge" (having a relatively uniform diameter that is slightly larger than the drill bit diameter). Good hole conditions also can contribute to higher core recovery; however, many zones in the crust, inferred to be of greatest hydrogeologic interest, are also where rocks are fractured, porous, and weak, resulting in poor core recovery and poor logging tool response.

Hydrogeologic tests of individual crustal holes drilled during IODP have been conducted using a drillstring packer system (e.g., Becker, 1986; Becker & Davis, 2004; Becker & Fisher, 2000). The IODP packer system is incorporated into the bottom-hole assembly of a drillstring. An inflateable packer element is used to hydraulically isolate part of a borehole so that pressure and fluid flow conditions can be perturbed and the formation response can be monitored. The IODP packer has been inflated in casing to test the entire open-hole interval below, and in open hole where the hole diameter is small enough and the formation is sufficiently massive to hold the inflated element and form a seal against the borehole wall. Interpretation of single-hole packer tests is based on fitting pressure-time observations to an equation of the form: $\Delta P = f(Q,t,T)$, where Q is fluid pumping rate, t is time, and T is formation transmissivity (dimensions of L^2/T). Transmissivity is the product of aquifer thickness and hydraulic conductivity, K, within a horizontal, tabular aquifer, where the latter is related to permeability, k, as $k = K\mu/\varrho g$, μ being dynamic viscosity and ϱg being a unit weight of fluid.

Temperature logs in cased, crustal boreholes have quantified downflow and upflow conditions, and researchers have used these observations and estimated or measured differential pressures to infer near-borehole hydrogeologic properties (e.g., Becker & Davis, 2003; Becker, Langseth, Von Herzen, & Anderson, 1983; Fisher, Becker, & Davis, 1997; Winslow, Fisher, & Becker, 2013). Flow down crustal boreholes is initiated during drilling by pumping cold (dense) seawater in the borehole adjacent to warm (less dense) formation fluid. This creates a positive differential pressure that can drive borehole fluid into the formation, even if the formation is naturally overpressured, and this flow can continue for days to years (e.g., Becker, Bartetzko, & Davis, 2001; Gable, Morin, & Becker, 1992). Downflow can be recognized from curvature of borehole temperature–depth logs and modeled based on heat balance considerations to estimate the flow rate (e.g., Becker et al., 1983; Lesem, Greytok, Marotta, & MecKetta, 1957; Winslow et al., 2013). If a natural formation overpressure exceeds the differential pressure created during downflow, the flow direction can reverse (Becker & Davis, 2004; Fisher et al., 1997). Rising borehole fluids lose heat to the formation surrounding casing, resulting in a borehole thermal profile with curvature opposite to that caused by downward flow; these flows can also be modeled to estimate the flow rate and formation properties.

Long-term borehole observatory systems (CORKs) have contributed additional data and samples related to crustal hydrogeology and alteration (e.g., Becker & Davis, 2005; Davis & Becker, 2004; Davis, Becker, Perrigrew, Carson, & MacDonald, 1992; Davis, Chapman, et al., 1992; Wheat et al., 2011). CORKs (1) seal one or more depth intervals of a borehole so that thermal, pressure, chemical, and microbiological conditions can equilibrate following the dissipation of drilling and other operational disturbances; (2) facilitate collection of fluid and microbiological samples and temperature and pressure data using autonomous samplers and data logging systems; and (3) allow long-term monitoring and large-scale active testing, including the formation response to perturbation experiments.

The CORKs developed for hydrogeologic monitoring and experiments in basement during IODP (deployed on and after Expeditions 301, 327, and 336) share features with systems deployed during earlier drilling expeditions, but have several notable differences. IODP basement CORKs are built around concentric casing strings, with the innermost casing including one or more sets of inflatable packer elements. Later IODP expeditions (327 and 336) were fitted with additional packer elements designed to expand over time through chemical interaction with seawater (Edwards et al., 2012; Fisher, Wheat, et al., 2011). Pressure measurement in these CORKs is accomplished with high-resolution sensors and loggers, deployed on the wellhead and connected at depth using small diameter stainless steel tubing. Data from these systems are accessible using an underwater mateable connector with a submersible or remotely operated vehicle (ROV); pressure data from one IODP CORK (in Hole 1026B) are available through the Ocean Networks Canada (ONC) cabled network (http://www.oceannetworks.ca). Temperatures in most CORKs are recorded at multiple depths using autonomous sensor-logger systems that are attached to a cable suspended inside the inner CORK casing. These instruments must be recovered from the CORKs in order to download data. The CORK in Hole 1026B also provides downhole temperature data in real time through ONC.

IODP CORKs were the first designed to collect relatively pristine samples of basement fluids and microbial materials, using nonreactive casing, tubing, fittings, and coatings. Continuous Osmosampling systems were deployed with these CORKs both on the outside of seafloor wellheads (drawing samples up from below) and at depth within isolated basement intervals. Osmosampling systems were designed specifically for gas sampling (using copper tubing), microbiological incubation, tracer injection, and acid addition (Wheat et al., 2011).

In addition to providing individual monitoring and sampling points in the ocean crust, CORKs provide opportunities for active experiments and monitoring between boreholes. The experiments associated with IODP work on the eastern flank of the Juan de Fuca Ridge were designed with this as a primary project goal, and cross-hole pressure and geochemical perturbations are being used to assess crustal-scale properties and processes, including differences with depth and direction(s) of flow. Several modeling studies have also been constrained and tested using samples and data from IODP boreholes.

4.2.2.1.3 IODP Sites and Results Discussed

The locations of IODP sites discussed in this section are shown in Figure 4.2.2.1, and characteristics of selected holes are listed in Table 4.2.2.1. In several cases, boreholes started during Deep Sea Drilling Project (DSDP) or Ocean Drilling Program (ODP) operations were reoccupied (drilled, cored, and/or configured as observatories) during IODP. Figure 4.2.2.2 shows penetration depths and general lithologies of basement holes drilled during IODP, and holes drilled during DSDP and ODP having ≥100 m of basement penetration. Most sites discussed are located on the flanks of mid-ocean ridges, beyond the direct magmatic and tectonic influence of seafloor creation. Some sites are located where additional characteristics of subseafloor hydrogeology are apparent, including an exposed section of the deep crust and upper mantle near an active spreading center, crust that is about to undergo subduction, and extrusive and intrusive material associated with mid-plate volcanism. In the presentation that follows, results are introduced following the sequence of IODP *Expedition Reports*: borehole operations/configuration, lithostratigraphy, alteration, and downhole/observatory

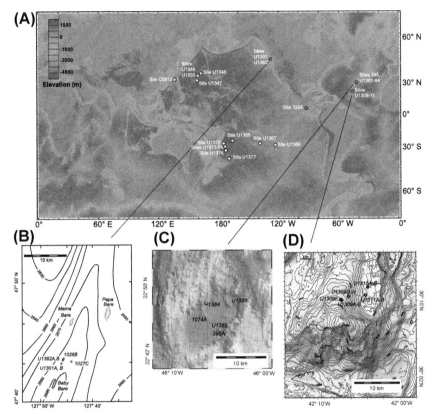

FIGURE 4.2.2.1   Black and White version: Map showing locations of drilling sites at which bore-holes discussed in this chapter were created or occupied during IODP. One or more holes were drilled at each of the sites shown. Table 4.2.2.1 provides details on hole depths, lithologies encountered, and measurements made. Continental topography and ocean bathymetry are shown for reference (Ryan et al., 2009). Sites marked with open circles drilled, sampled, and/or measured basement rocks (on IODP Expeditions 322, Nankai Trough; 324, Shatsky Rise; 329, South Pacific Gyre; and 330, Louisville Seamounts), but are not discussed in this chapter. Color version: Maps showing locations of drilling sites at which boreholes discussed in this chapter were created or occupied during IODP. (A) Global map of sites with significant IODP basement penetration and analyses of hydrogeology and/or alteration. Table 4.2.2.1 provides details on hole depths, lithologies encountered, and measurements made. Continental topography and ocean bathymetry are shown for reference (Ryan et al., 2009). Sites marked with open circles drilled, sampled, and/or measured basement rocks (on IODP Expeditions 322, Nankai Trough; 324, Shatsky Rise; 329, South Pacific Gyre; and 330, Louisville Seamounts), but are not discussed in this chapter. (B) Detail map of Expedition 301/327 field area on eastern flank of Juan de Fuca Ridge *(modified from Fisher et al. (2008))*. (C) Detail map of Expedition 336 Field area in North Pond *(modified from Wheat et al. (2012))*. (D) Detail map of Expedition 304/305 field area on the Atlantis Massif *(modified from Expedition 304/305 Scientists, 2006a)*.

TABLE 4.2.2.1 Summary of Characteristics for IODP Holes with Significant Basement Penetration

Hole	Exp.	TD[a] (mbsf)	TD[a] (msb)	Core top[b] (msb)	Core bot[b] (msb)	Log top[c] (msb)	Log bot[c] (msb)	Setting[d]	Age[e] (my)
U1301A	301/321T	369.7	107.5	–	–			RF	3.5
U1301B	301/321T	582.8	317.6	86.0	317.6	86	314	RF	3.5
U1362A	327	528.0	292.0	110.0	260.0	28	271	RF	3.5
U1362B	327	359.0	117.0	–	–			RF	3.5
U1309B	304/305	101.8	99.8	0.0	99.8	20	94	RF-OCC	1.1–1.3
U1309D[f]	304/305/340T	1415.5	1413.5	18.5	1413.5	19	1413	RF-OCC	
U1310B	304/305	23.0	23.0	0.0	23.0	–	–	RF-OCC	1.1–1.3
U1311A	304/305	12.0	12.0	0.0	12.0	–	–	RF-OCC	1.1–1.3
1256D[g]	309/312/335	1521.6	1271.6	0.0	1271.6	0	1270	RF	15.0
C0012A[h]	322	576.0	38.0	0.0	38.0	–	–	RF-S	20
C0012C[h]	322	630.5	104.8	0.0	104.8	–	–	RF-S	20
U1346A[h]	324	191.8	52.6	0.0	52.6	0	58	LIP	140.0
U1347A[h]	324	317.5	159.9	0.0	159.9	0	156	LIP	145.0
U1349A[h]	324	250.4	85.3	0.0	85.3	0	84.9	LIP	140–142
U1350A[h]	324	315.8	172.7	0.0	172.7	–	–	LIP	140–142

U1365E[h]	329	124.5	53.5	0.0	53.5	–	–	RF	100.0
U1367F[h]	329	55.5	38.5	0.0	37.0	–	–	RF	33.5
U1368F[h]	329	115.1	103.3	0.0	103.3	24.2	80.2	RF	13.5
U1372A[h]	330	232.9	187.3	0.0	187.3	–	–	SM	75–77
U1373A[h]	330	65.7	31.8	0.0	31.8	–	–	SM	72–73
U1374A[h]	330	522.0	505.3	0.0	505.3	108.3	488.3	SM	72–73
U1376A[h]	330	182.8	140.9	0.0	140.9	38.1	128.1	SM	64
U1377A[h]	330	53.3	38.2	0.0	38.2	–	–	SM	50.1
U1377B[h]	330	37.0	27.9	0.0	27.9	–	–	SM	50.1
395A	336	664.0	571.0	–	–	19.0	507.0	RF	7–8
U1382A	336	210.0	120.0	20.0	120.0	8.0	120.0	RF	7–8
U1383C	336	331.5	293.2	31.2	293.2	31.2	293.2	RF	7–8

[a] TD = total depth of hole, mbsf = meters below seafloor, msb = meters subbasement, below the sediment–basement interface.
[b] Core top = depth of bit when first coring began, Core bot = depth of bit when final core ended.
[c] Log top = approximate depth of shallowest open-hole log data collected in basement, Log bot = approximate depth of deepest open hole.
[d] RF = Ridge flank, RF–OCC = Ridge flank, oceanic core complex, RF–S = Ridge flank at subduction zone, LIP = Large igneous province, SM = Seamount.
[e] Age as cited in expedition publications, generally based on magnetostratigraphy, radiometric dating, or paleontological analysis of basal sediments.
[f] Hole U1309D was drilled on IODP Expeditions 304 and 305, then revisited for additional logging on Expedition 340T.
[g] Hole U1256D was first drilled, cored, and logged on ODP Leg 206.
[h] Results from drilling on IODP Expeditions 322 (Nankai Trough), 324 (Shatsky Rise), 329 (South Pacific Gyre), and 330 (Louisville Seamounts) are not discussed in this volume, although these expeditions had significant site surveys, basement penetration, coring and downhole measurements related to crustal hydrogeology and alteration.

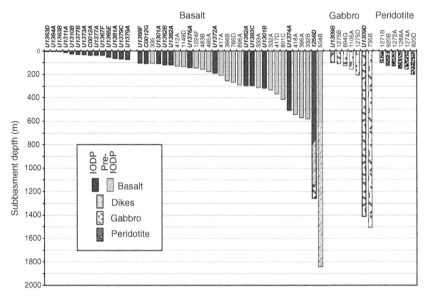

FIGURE 4.2.2.2 Plot showing total depth and general lithology for basement holes drilled during IODP (hole labels in bold italic, lithology with dark shading), and holes having ≥100 m of basement penetration from DSDP and ODP for comparison *(figure modified from Expedition 309/312 Scientists, (2006a).)*. Drilling, coring, and logging intervals for most IODP basement holes are listed in Table 4.2.2.1.

measurements. Drilling and measurement depths are referenced in meters below seafloor (mbsf) and meters subbasement (msb, below the sediment-basement interface).

4.2.2.2 CRUSTAL HYDROGEOLOGY AND ALTERATION

4.2.2.2.1 Eastern Flank of the Juan De Fuca Ridge, Northeastern Pacific Ocean, ~3.5-my-old Upper Crust (IODP Expeditions 301 and 327)

4.2.2.2.1.1 Background and Context

IODP Expeditions 301 and 327 worked on the eastern end of a transect established during ODP Leg 168, on 0.9–3.6 my-old seafloor on the eastern flank of the Juan de Fuca Ridge, generated at a half-spreading rate of ~29 mm per year (Figure 4.2.2.1(B)). IODP project goals were to: resolve linked hydrogeologic, lithologic, biogeochemical, and microbiological properties and processes through analysis of sediment, rock, and fluid samples; determine the thermal, geochemical, and hydrogeologic conditions in basement; and install CORKs in the upper crust (Fisher, Tsuji, Petronotis, & the Expedition 327 Scientists, 2011; Fisher, Wheat, et al., 2011; Fisher, Urabe, Klaus, & the Expedition 301 Scientists, 2005; Fisher, Wheat, et al., 2005; Shipboard Scientific Party, 1997a). Leg

168 conducted limited basement drilling and installed two single-level CORKs in Holes 1026B and 1027C. IODP Expedition 301 researchers drilled deeper into basement at Site U1301, 1 km south of Hole 1026B; sampled sediment, basalt, and microbiological materials; logged upper basement and conducted hydrogeologic tests in Hole U1301B; replaced the borehole observatory in Hole 1026B; and established two new CORKs in Holes U1301A and U1301B. IODP Expedition 327 added two basement holes at Site U1362, between Hole 1026B and Site U1301 on the same buried basement ridge; collected sediment, rock, and microbial samples; logged upper basement and conducted hydrogeologic tests; installed two more CORKs; and recovered/replaced part of a CORK instrument string in Hole U1301B. At of the end of Expedition 327, there were six CORKs operating in this area, two completed across multiple basement depths (Holes U1301B, U1362A). Additional postdrilling expeditions have used a submersible or ROV to download data, manipulate valves, collect fluid samples, and install and recover instruments and samplers from the CORK wellheads and boreholes.

4.2.2.2.1.2 Crustal Petrology and Alteration

Basement was cored below approximately 250–265 m of sediment on Expeditions 301 and 327 in Holes U1301B and U1362A, both penetrating ~300 m into the upper volcanic crust (Table 4.2.2.1) (Expedition 301 Scientists, 2005b; Expedition 327 Scientists, 2011c). Other holes at these sites were drilled to establish borehole observatories but were not cored (U1301A and U1362B), and small amounts of basement were also recovered at Site U1363, adjacent to a seamount through which regional hydrothermal recharge occurs (Expedition 327 Scientists, 2011d). The uppermost ~100 m of basement was not cored at Sites U1301 and U1362 (Expedition 301 Scientists, 2005b; Expedition 327 Scientists, 2011c), and recovery was poor (5%) in the upper ~40 m of basement cored in nearby Hole 1026B (Shipboard Scientific Party, 1997b). Variations in drilling rates at Sites U1301 and U1362 (Becker, Fisher, & Tsuji, 2013; Expedition 301 Scientists, 2005a; Expedition 327 Scientists, 2011a) may indicate massive basalt at 55–60 msb, and breccias or highly altered pillow basalts at 65–80 msb.

Basement cores from Holes U1301B and U1362A consist mainly of pillow basalts with lesser flows, hyaloclastite breccia, and cataclastite (Expedition 301 Scientists, 2005b; Expedition 327 Scientists, 2011c). Some lava flows in both holes are >10 m thick. The basalts are normal depleted mid-ocean-ridge basalt (N-MORB), with similar trace element ratios suggesting a common magmatic source, and are aphyric to highly phyric (with olivine, clinopyroxene, and/or plagioclase phenocrysts). These basalts exhibit low-temperature alteration effects typical of upper oceanic basement (e.g., Alt, 2004), being slightly to highly altered, including a pervasive background alteration characterized by saponite (smectite clay) replacing olivine and mesostasis (Expedition 301 Scientists, 2005b; Lever et al., 2013; Ono, Keller, Rouxel, & Alt, 2012). Alteration intensities in Hole U1301B are typically 5–25%, but can be up to 60% in brecciated intervals (Figure 4.2.2.3). Alteration is more extensive in some Hole U1362A basalts, indicative of more intensive water–rock interaction. Typical alteration features, which are more common in pillows than flows, include halos along veins and fractures, and filled veins and vesicles (with clay, celadonite, carbonate, pyrite, anhydrite and/or Fe-oxyhydroxide). The matrix of hyaloclastite and cataclastite units tends to be more highly altered than are clasts within these materials.

Oxygen isotope analyses of Hole U1301B carbonate minerals suggest temperatures of 30–40 °C during formation (Coggon et al., 2010). Secondary

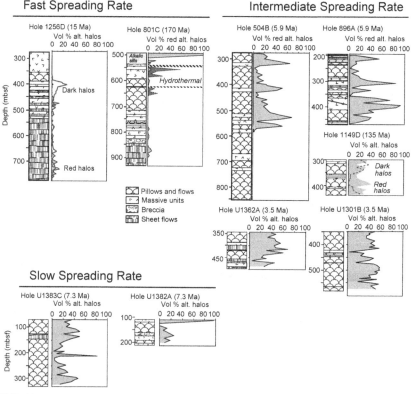

FIGURE 4.2.2.3 Lithology and proportion of alteration halos along veins versus depth for volcanic oceanic basement holes in crust grouped by fast, intermediate, and slow spreading rates. More abundant sheet flows and much lower proportion of alteration halos in crust generated at fast spreading rates may indicate control of fluid flow and low-temperature alteration by lithology and permeability (Alt et al., 2010). Hole 801C contains features not typical of the upper (generally, basaltic) oceanic crust, including a low-temperature hydrothermal deposit with associated alteration, and late alkalic sills that contain abundant alteration halos at top of basement (Alt & Teagle, 2003). Sheet flows may be more abundant than indicated (Tominaga et al., 2009). Figure modified from Alt et al. (2010), with additional data from IODP Sites U1301 (Expedition 301 Scientists, 2005b), U1362 (Expedition 327 Scientists, 2011c), and U1382 and U1383 (Expedition 336 Scientists, 2012c,d).

pyrite and bulk rocks have $\delta^{34}S$ values as low as $-72‰$, indicating the influence of microbial sulfate reduction in basement (Ono et al., 2012; Lever et al., 2013). These rocks also contain organic carbon with low $\delta^{13}C$ values and yield functional genes for methanogens, methane oxidizers, and sulfate reducing bacteria, providing further evidence for microbial activity in basement (Lever et al., 2013).

4.2.2.2.1.3 Geophysical Measurements and Short-term Hydrogeologic Experiments

Holes U1301B and U1362A were geophysically logged to assess borehole diameter and formation, electrical resistivity, bulk density, porosity, and sonic velocity (Expedition 301 Scientists, 2005b; Expedition 327 Scientists, 2011c). Caliper logs show oversized intervals with thicknesses of tens of meters, particularly in the upper 100–200 msb, and more massive formation conditions at greater depth. In general, oversized zones correspond to lithologies of pillow basalt, breccia, and hyaloclastite, but core recovery within these intervals is typically low, so correlating lithology and alteration with geophysical response in these intervals is challenging.

Bulk density logs include zones 1–10-m thick having values of 2900–3000 kg/m^3, consistent with measurements made on core samples (Bartetzko & Fisher, 2008; Becker et al., 2013). There are also thin zones within both boreholes that have bulk density of 1800–2200 kg/m^3 and are not oversized, suggesting more fractured and/or porous conditions. In general, pillow basalts tend to have more of these low-density intervals than do other lithologies.

Hydrogeologic tests were run in Holes U1301B and U1362A with a drill-string packer (Becker & Fisher, 2008; Becker et al., 2013), with the packer set where basement rocks were relatively massive and the hole diameter could support the inflated packer element. The deepest packer setting depth in Hole U1301B corresponds to an abrupt change in the character of basement geophysical logs (near 470 mbsf, 205 msb), with more variable borehole diameter and physical properties (bulk density, resistivity, sonic velocity) at shallower depths. Below this depth, logs indicate more massive conditions, but there are thin zones of lower density that could be brecciated or fractured (Bartetzko & Fisher, 2008; Expedition 301 Scientists, 2005b).

Packer tests of the interval below 472 mbsf (207 msb) in Hole U1301B indicate bulk permeability of $k = 1.7 \times 10^{-12}$ m^2, whereas tests with the packer at 442 and 417 mbsf (177 and 152 msb, respectively) indicate higher permeability, $k = 3.2 \times 10^{-12}$ m^2. If the properties determined at shallower setting depths are applied to the uppermost ~200 m of basement, permeability within this interval is closer to $k = 5 \times 10^{-12}$ m^2. The consistency of properties determined with two shallow setting depths suggests that most of the formation transmissivity may be concentrated between ~150 and 180 msb; if so, the bulk permeability of this 30-m-thick zone would be $k = 2 \times 10^{-11}$ m^2 (Becker & Fisher, 2008).

The single-packer setting depth in Hole U1362A was near 190 msb, within a 20-*m*-thick massive zone that separates two oversized zones (Becker et al., 2013). Results from these tests indicate bulk permeability of $k = 1-2 \times 10^{-12}$ m^2, essentially the same as determined at the deepest setting depth in Hole U1301B, 825 m to the south. Bulk permeability values deeper than ~150 msb in both holes are higher by roughly an order of magnitude than those obtained by Becker and Fisher (2000) in the shallowest basement sections in nearby Holes 1026B and 1027C. This suggests considerable heterogeneity and/or that the highest upper crustal permeabilities in this area are not located immediately adjacent to the sediment–basement interface, but are deeper in the extrusive section (Becker & Fisher, 2008; Becker et al., 2013; Fisher, Davis, & Becker, 2008).

4.2.2.2.1.4 Long-term Observatory Measurements and Samples

Single-level CORKs were installed in Holes 1026B, U1301A and U1362B, and two-level CORKs were installed in Holes U1301B and U1362A (Fisher, Wheat, et al., 2011; Fisher, Wheat, et al., 2005). The CORKs installed in Holes U1301A and U1301B were not sealed as intended during deployment. Incomplete sealing and the imposition of a cold column of water in the boreholes during drilling, casing, and other operations led to self-sustained flow of cold ocean-bottom water down these holes and into the formation. The Hole U1301B CORK was partly sealed in Summer 2007 with cement deployed by submersible, and was fully sealed in Summer 2009 by cementing with the drillship, stopping the downflow of cold bottom water after 5 years (Expedition 327 Scientists, 2011b; Fisher, 2010). The Hole 1301A CORK was also cemented with the drillship in Summer 2009, but much of that cement drained away within a few hours to days, so that CORK has remained unsealed. However, the downflow into Hole 1301A that began in 2004 when the hole was drilled, varied in rate and slowed over several years, then reversed abruptly in September 2007 (Wheat et al., 2010). Since this time, Hole U1301A has been a site of warm (~64 °C) hydrothermal discharge. Interestingly, the geochemistry of fluids discharging from Hole U1301A showed little evidence for recharge of bottom water in nearby Hole U1301B, only 36 m away, even before Hole U1301B was cemented. This observation provides evidence for separation (heterogeneity) of flow systems within the upper oceanic crust.

The CORK in Hole 1027C was supposed to be replaced during Expedition 327, but the old CORK could not be removed. Instead, the Hole 1027C CORK was modified a year later using an ROV by recovering the old data logger and installing a pressure monitoring manifold and a new data logger (Fisher et al., 2012). The other five CORKs installed in this area are installed above a buried basement high, along a transect 1-km-long that is oriented parallel to the active spreading center and the trend of abyssal hill topography. Hole 1027C is located about 2.4 km to the east, above a thicker sediment section where the top of basement is >300 m deeper below the seafloor.

Pressure records recovered from these CORKs show that upper basement is overpressured at Sites 1301 and 1362 by tens of kilopascals with respect to an ambient hydrostatic column, and the upper basement interval monitored in Hole 1027C is underpressured by tens of kilopascals, consistent with earlier measurements (Davis & Becker, 2004; Davis, LaBonte, He, Becker, & Fisher, 2010). Pressure data recovered from Hole 1362A, which monitors two distinct basement intervals, show that the overpressure is greater with greater depth in the hole, consistent with net upward transport of fluid below the buried basement high, as inferred from coupled fluid-heat modeling (Spinelli & Fisher, 2004).

Temperature loggers were recovered from near the top of basement in CORKs in Holes U1301A and U1301B in 2009 (by submersible) and 2010 (by drillship), respectively, following 4–5 years of deployment (Expedition 327 Scientists, 2011b; Wheat et al., 2010). These loggers recorded thermal conditions during a long period of fluid downflow. These thermal data and additional information (borehole geometry, completion details, and basement stratigraphy) were used to assess upper basement hydrogeologic properties with linked analytical models and a statistical analysis using a range of borehole parameters (Winslow et al., 2013). These analyses suggest that the initial flow rate down Hole U1301A (before flow reversal) was ~2 L/s, whereas ~20 L/s flowed down Hole U1301B. Flow may have been more rapid down Hole U1301B, compared to Hole U1301A, because Hole U1301B extends deeper into basement (which imposes a taller column of cold, dense bottom water), and because the formation permeability around Hole U1301B appears to be somewhat greater (Winslow et al., 2013).

Wheat et al. (2010) evaluated the thermal and geochemical response of Hole 1301A before, during, and after the reversal from downflow to upflow. In addition to a thermal response associated with changes in flow rate and direction, two single-hole tracer experiments elucidate the geochemical composition and nature of mixing of borehole and formation fluids. Osmosamplers deployed within slotted casing near the base of the CORK, surrounded by basaltic upper crust, collected fluids throughout the deployment. Major ion chemistry shows relatively consistent concentrations of solutes such as Mg, close to bottom water values, until the flow reversal occurred. After the flow reversal, when reacted crustal fluid flowed rapidly from the formation into and up the borehole, the sampled fluid composition was similar to that seen in fluids sampled from nearby Baby Bare outcrop and ODP Hole 1026B (Shipboard Scientific Party, 1997b; Wheat et al., 2010, 2004). One of the Osmosampler packages deployed in Hole U1301A also injected a tracer solution (containing Cs, Yb, and Tm). Evaluation of tracer concentrations in sampled fluids suggests that the rates of mixing and flow resulted in a volume exchange rate for the sampled interval of ~60% per week during the initially rapid period of downflow, 1–15% per week during the slowdown in downflow that preceded reversal, and ~99% per week after upflow of formation fluids began in 2007 (Wheat et al., 2010).

A large-scale assessment of basement hydrogeologic properties was made from the long-term pressure perturbation observed in Hole 1027B that resulted from leakage of cold bottom seawater into the crust around Hole U1301B following Expedition 301 (Fisher et al., 2008). Both IODP Expedition 301 basement operations and subsequent downflow into the crust influenced pressures in sealed Hole 1027C. Although basement operations in Hole U1301B caused a direct pressure response in Hole 1027C, operations in nearby Holes U1301A and 1026B had little or no influence. The packer experiments in Hole U1301B caused the greatest immediate perturbation, despite modest pumping rates, because fluids injected during this test were forced to enter basement below the seal by the packer element, rather than being allowed to flow back up the borehole to the overlying ocean. These observations suggest that shallow basement surrounding Holes U1301A and 1026B may be less well connected hydrologically to the uppermost oceanic crust around Hole 1027C than is deeper basement in Hole U1301B.

A comparison of pressure conditions in these holes, and correlation with drillship and later borehole operations, suggest a flow rate into Hole U1301B during the 13 months following Expedition 301 of $Q = 2–5$ L/s (Fisher et al., 2008). This flow rate and the observed 13-month pressure change in Hole 1027C (~1.5 kPa in total) suggest a bulk basement permeability of $k = 0.7–2 \times 10^{-12}$ m^2 (Figure 4.2.2.4). This inadvertent cross-hole experiment provided a direct measurement of the storativity of the upper volcanic crust, a term that depends on a combination of crustal and fluid compressibility, and cannot be determined with single-hole pumping experiments. Upper crustal compressibility was calculated to be $\alpha = \sim3$ to 9×10^{-10} Pa^{-1}, close to or somewhat greater than that of seawater under ambient thermal and pressure conditions. Davis et al. (2010) completed additional analyses of the cross-hole response between Sites 1301 and 1027C, and explored the properties required to sustain downflow into an open basement hole. Their analytical and numerical calculations indicate a minimum basement permeability of $k = 3–4 \times 10^{-13}$ m^2, consistent with earlier analytical calculations. Modeling also showed how the time required for a flow reversal (if one occurs) depends on formation permeability and the depth of the flowing borehole.

These CORKs were most recently serviced in Summer 2014, with downhole samplers and data loggers recovered from Holes 1026B, U1362A, and U1362B, and data and samples collected from wellhead instruments in these CORKs and those in Holes 1301A and 1027C. Data and samples analyzed to date show clear evidence for cross-hole pressure perturbations associated with the long-term flow experiment, and the arrival of tracers in multiple holes distant from that used for tracer injection on Expedition 327, Hole U1362B.

4.2.2.2.1.5 Hydrogeologic Modeling

Numerical models were used to calculate hydrologic properties consistent with inferred rates of fluid circulation between Grizzly Bare and Baby Bare outcrops, which are recharge and discharge sites separated by 50 km (south and north of the Expedition 301/327 work area, respectively) (Hutnak et al., 2006).

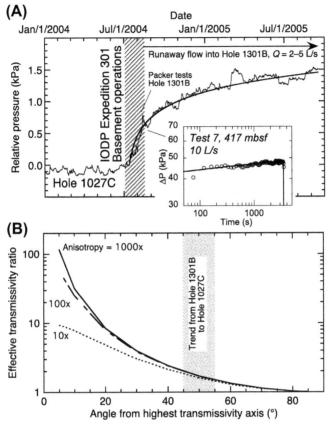

FIGURE 4.2.2.4 Data from hydrogeologic experiments in the IODP Expedition 301/327 field area *(plots modified from Becker & Fisher (2008) and Fisher et al. (2008))*. A. Observations and calculations from cross-hole test. (A) Filtered pressure-time record from Hole 1027C, beginning 6 months before and ending 13 months after Expedition 301. Striped vertical band indicates period of basement drilling, coring, casing, and testing operations during Expedition 301. Smooth curve shows least squares best fit of observations to analytical calculations for the pressure response in Hole 1027C, 2.4 km away, to long-term flow into Hole U1301B. The fit of this curve indicates basement permeability of $k = 0.7$–2×10^{-12} m^2. Inset: Similar fit to short-term (1-h) packer experiment in Hole 1301B, illustrating fit of the same model, but using a basement permeability of 3×10^{-12} m^2. Note relatively large change in pressure during this test (up to 50 kPa) versus that seen from the cross-hole response (~1.5 kPa). (B) Calculations of the effective transmissivity ratio (apparent transmissivity/highest transmissivity) as a function of the angle of measurement between the dominant transmissivity direction and the direction of measurement. Vertical band is orientation of the Site U1301 to Site 1027 experiment, assuming that the direction of highest transmissivity is N20°E (subparallel to the crustal fabric) and the direction of lowest transmissivity is perpendicular to this, N110°E. For essentially any transmissivity anisotropy ratio, the value measured between Sites U1301 and 1027C will be close to that for the lowest transmissivity direction.

This work followed analytical calculations of basement permeability needed to maintain a self-sustaining, hydrothermal siphon (Fisher, Davis, et al., 2003). Two-dimensional, transient numerical models suggest that outcrop-to-outcrop circulation can be sustained across this distance when basement permeability between outcrops is $\geq 10^{-12}$ m^2 (Hutnak et al., 2006). At lower permeabilities, too much energy is lost during lateral fluid transport for circulation to continue without forcing, given the limited driving pressure difference at the base of recharging and discharging fluid columns in the crust.

These models also showed that fluid temperatures in upper basement are highly sensitive to modeled permeability. Observed upper basement temperatures in this area are generally 60–65 °C (Davis, Chapman, et al., 1992; Expedition 301 Scientists, 2005b; Hutnak et al., 2006; Shipboard Scientific Party, 1997b), about the same as would result from fully conductive conditions. This means that, whatever the flow rate between Grizzly Bare and Baby Bare outcrops (and across Sites U1301 and U1362), there is little advective heat extraction on a regional basis. Modeling showed that when crustal permeability is too high ($\geq 10^{-10}$ m^2), fluid circulation between outcrops is so rapid that basement is chilled to temperatures well below those observed (modeled values of 20–50 °C). A better match to observed upper basement temperatures of 60–65 °C is achieved when lateral basement permeability is ~10^{-11} m^2 (Hutnak et al., 2006).

Additional models were crafted to assess the significance of "background" heat flow around the Expedition 301/327 work area. Thermal data collected during regional studies showed that heat flow along a 100-km-long swath of 3.4–3.6-my-old seafloor is lower by 15–20% than predicted by conductive lithospheric cooling models, even after correcting for rapid sedimentation rates (an additional 12–18% correction) (Davis et al., 1999; Hutnak et al., 2006; Zühlsdorff et al., 2005). Observations show no regional recharge-to-discharge trend in heat flow (e.g., Langseth & Herman, 1981; Stein & Fisher, 2003), and no low-heat flow "moat" around Grizzly Bare outcrop (Hutnak et al., 2006; Zühlsdorff et al., 2005), as might be expected if present-day fluid flow were responsible for the regional anomaly.

Calculations show that the Expedition 301/327 work area (and much of the surrounding region) could be undergoing "thermal rebound" following the cessation of a long period of more efficient, advective heat extraction from the crust (Hutnak & Fisher, 2007). Maps of basement relief and sediment thickness in this area show numerous shallowly buried basement highs (Hutnak et al., 2006; Zühlsdorff et al., 2005); basement in many of these areas would have been exposed at the seafloor prior to the last several hundred thousand years of rapid sedimentation, and areas of current basement exposure (e.g., Baby Bare outcrop) would have been larger and better connected to the ocean (Hutnak & Fisher, 2007). Larger areas of basement exposure, and the greater spatial distribution of these areas, would have been permitted more efficient regional advective heat loss, as is currently seen at the western end of the Leg 168 transect (Davis, Chapman, et al., 1992; Hutnak et al., 2006), where measured heat flow is ~20% of lithospheric predictions, and on other ridge flanks where basement outcrops are more common (e.g., Hutnak et al., 2008; Lucazeau et al., 2006; Villinger, Grevemeyer, Kaul, Hauschild, & Pfender, 2002).

4.2.2.2.2 Western Flank of the Mid-Atlantic Ridge, Northern Atlantic Ocean, 7.3-my-old Upper Crust (IODP Expedition 336)

4.2.2.2.2.1 Background and Context

IODP Expedition 336 was designed to explore the subseafloor biosphere, geology, geochemistry, and hydrogeology at a young ridge-flank site known as North Pond (Figure 4.2.2.1(C)). The ocean crust in this area was created at the Mid-Atlantic Ridge (MAR) at a half-spreading rate of 14 mm per year. In contrast to the Expedition 301/327 work area, the seafloor around North Pond is characterized by patchy sediment cover and extensive basement exposure. Rapid fluid flow is thought to occur in the volcanic crust below North Pond, limiting upper basement temperatures to ≤ 20 °C (Langseth, Becker, Von Herzen, & Schultheiss, 1992; Langseth, Hyndman, Becker, Hickman, & Salisbury, 1984). Site 395 was first drilled during DSDP Leg 45 (Shipboard Scientific Party, 1979) and has been revisited multiple times for logging, hydrogeological studies, and other survey work (mapping, seismics, shallow coring, and heat flow) (e.g., Becker et al., 2001; Becker, Langseth, & Hyndman, 1984; Morin, Hess, & Becker, 1992; Morin, Moos, & Hess, 1992). Processes and conditions at North Pond are likely to be typical of ridge-flank hydrothermal circulation through young crust in many settings: rapid flow of cool fluids having limited opportunity to react with basement rocks and overlying sediments before being discharged to the overlying ocean. Samples and data collected during a site survey prior to Exp. 336 are consistent with this interpretation (Ziebis et al., 2012).

The principal objectives of IODP Expedition 336 to North Pond were addressed through coring and analyses of recovered materials, borehole geophysical and fluorescence measurements, and use of CORKs for monitoring and sampling after drilling was complete (Expedition 336 Scientists, 2012a). Basement was cored in two holes on Expedition 336 discussed in this section: Hole U1382A and Hole U1383C (Table 4.2.2.1).

4.2.2.2.2.2 Crustal Petrology and Alteration

Basement coring at Hole 395A recovered mainly pillows and massive basalt units (Shipboard Scientific Party, 1979). The pillow units are typically tens of meters thick and separated by a sedimentary breccia unit that contains cobbles of gabbro and serpentinized peridotite derived from the surrounding basement peaks (Bartetzko, Pezard, Goldberg, Sun, & Becker, 2001; Matthews, Salisbury, & Hyndman, 1984). Nearby Hole 395 also contains a peridotite–gabbro complex that is several meters thick with brecciated contacts (Shipboard Scientific

Party, 1979). Oxygen isotope data for carbonate and clay veins in the volcanic basement from these holes are consistent with low temperatures, around 30 °C for phyllosilicates and 0–15 °C for carbonates (Lawrence & Gieskes, 1981).

Basement cores recovered from Hole U1382A suggest that pillow lavas are slightly more abundant than massive lavas (32% recovery) (Expedition 336 Scientists, 2012c). Fragments of hyaloclastite are present in a few pillow units, and there is sedimentary breccia that contains clasts of gabbroic rocks and weakly serpentinized harzburgite and lherzolite. This breccia likely represents a rockslide deposit and was also encountered in Holes 395 and 395A, within 80 m of Hole U1382A. The full lateral extent of this lithological unit and its influence on the upper basement transmissivity are not known. The basalts are aphyric to highly plagioclase-olivine phyric, are all N-MORB, but have variable immobile trace element ratios (Zr/Y and Ti/Zr) indicative of distinct parental magmas.

Hole U1382A basalts show evidence for low-temperature alteration by seawater: pervasive clay background alteration and common halos around veins, comprising approximately 15–20% of the recovered core (Expedition 336 Scientists, 2012c). Phyllosilicates (smectite > celadonite) are the most abundant secondary phases, followed by Fe-oxyhydroxides and minor zeolites and carbonates. Veins are abundant (13–20 veins/m) but narrow (usually <0.2 mm in width). Common carbonate-filled vein networks are associated with intense oxidative alteration of olivine to clay, oxide, and carbonate (Expedition 336 Scientists, 2012c).

Basalt samples recovered from Hole U1383C feature a greater proportion of thin flows with glassy margins, leading to more extensive palagonitization and greater vein density (33 veins/m) (Expedition 336 Scientists, 2012d). Background alteration of the basalts is ≤40% in the shallower sections and decreases downhole (Figure 4.2.2.3). Vesicle fills of clay, phillipsite, calcium carbonate, and Fe-oxyhydroxide are common. Around 20% of the recovered core consists of brownish alteration halos that flank veins. Veins fills are dominantly clay and Fe-oxyhydroxide with only minor carbonate in the uppermost 150 m of basement. The lowermost 100 m of Hole U1383C feature mixed zeolite/carbonate veins with variable, but usually small, proportions of clay. Veins are more common in this deeper section, but background alteration is less pronounced than in uppermost basement (Expedition 336 Scientists, 2012d).

4.2.2.2.2.3 Geophysical Measurements

Wireline logging in Hole 395A during Expedition 336 included natural gamma ray, temperature, and induced fluorescence to identify microbial material ("DEBI-t" tool) (Expedition 336 Scientists, 2012b). Natural gamma ray and resistivity data are consistent with results from earlier logging (e.g., Bartetzko et al., 2001; Matthews et al., 1984; Moos, 1990), defining individual eruptive and flow units. The upper 300 m of basement comprises lithologic units having thicknesses of approximately 10–70 m, within which there is generally a bottom-to-top decrease in electrical resistivity, bulk density, and compressional velocity, and an increase in natural gamma ray emissions. Some of these units are separated by thinner

(1–10 m) layers of breccia, altered flows, or massive basalt. Fluorescence data from the DEBI-t tool suggests elevated photon counts at several wavelengths, particularly at 455 nm, in the upper 100 msb. The excitation source (224 nm) is intended to give peak fluorescence responses near 300 nm for spores and 320 nm for bacteria, but variations in depth with these wavelengths are not as clear as those at 455 nm (Expedition 336 Scientists, 2012b).

The upper 100–300 msb were logged in Holes U1382A and 1383C with caliper, natural gamma ray, bulk density, and resistivity tools, revealing lithologic layering at a scale of 5–20 m, but without the systematic delineation of individual eruptive units apparent in Hole 395A (Expedition 336 Scientists, 2012c; d). Geophysical logging units defined on the basis of log response are finer than those identified from recovered core ($\leq$30%) (Expedition 336 Scientists, 2012d). Also in contrast to results from Hole 395, logging in these holes revealed no trends in downhole fluorescence (Expedition 336 Scientists, 2012c).

4.2.2.2.2.4 Long-term Observatory Measurements and Samples

Expedition 336 CORKs included multiple nested casing strings and packer systems (inflateable and swellable), casing seals, and nonreactive components deployed at depth (Edwards et al., 2012). Expedition 336 CORKs were the first to use fiberglass inner casing, along with resin-coated steel casing and collars introduced in earlier CORKs, in an effort to allow collection of near-pristine samples of basement fluids and microbial materials. Downhole pressure, temperature, and dissolved oxygen are being monitored for short-term and long-term variability, and downhole and wellhead sampling systems are collecting fluids at a range of rates in different storage media, and providing substrate for microbial incubation. A single-level CORK was deployed in Hole U1382A to monitor conditions in the upper 120 m of volcanic crust (Expedition 336 Scientists, 2012c), and three-level CORKs were deployed in Holes 395A (after removal of an old borehole observatory) and U1383C (Expedition 336 Scientists, 2012b,d). Hole U1382B was drilled and cased during Expedition 336, then left open for later instrumentation using a "CORK-lite" system deployed with a remotely operated vehicle (Expedition 336 Scientists, 2012c; Wheat et al., 2012). Data and samples from these systems will inform understanding of hydrogeologic, geochemical, and microbiological conditions in basement below and around North Pond in coming years.

4.2.2.2.3 Eastern Flank of the East Pacific Rise, Eastern Pacific Ocean, ~15-my-old Upper to Middle Crust (IODP Expeditions 309, 312, and 335)

4.2.2.2.3.1 Background and Context

Site 1256 was drilled to test models for the structure and origin of seismic layering of oceanic crust and the origin of melt lenses at fast-spreading ridges, in 15-my-old crust that formed at a superfast spreading rate (half rate

~100 mm per year) at the East Pacific Rise (EPR) (Expedition 309/312 Scientists, 2006b; Expedition 335 Scientists, 2012; Shipboard Scientific Party, 2003). Drilling, coring, and measurements were initiated on ODP Leg 206 and continued on IODP Expeditions 309, 312, and 335. Coring in Hole 1256D started at 276 mbsf (26 msb) and ended at 1522 mbsf (1246 msb) (Expedition 309/312 Scientists, 2006b; Expedition 335 Scientists, 2012; Shipboard Scientific Party, 2003). Hole 1256D is currently the fourth deepest penetration into oceanic basement, extending through lavas, dikes, and into gabbros (Figure 4.2.2.2).

4.2.2.2.3.2 Crustal Petrology and Alteration

The uppermost basement in Hole 1256D comprises a ~100-m-thick sequence dominated by a ponded flow >75-m thick; the same massive flow is only 32-m thick in nearby Hole 1256C (Expedition 309/312 Scientists, 2006b). The immediately underlying lavas include sheet and massive flows and minor pillow flows. The amount of basement relief indicated by variation in the thickness of the ponded flow, and the presence of flow lobe inflation features, indicate that the upper 284 m of the volcanic section crystallized off-axis (Expedition 309/312 Scientists, 2006b; Tominaga, Teagle, Alt, & Umino, 2009). Sheet flows and massive lavas erupted at the ridge axis, and two thin (1–2 m) hyaloclastite intervals (at 397 and 595 mbsf) make up the remaining extrusive section (Figure 4.2.2.5). The beginning of a lithologic transition at 1004 mbsf is marked by subvertical intrusive dike contacts and sulfide-mineralized breccias. Below 1061 mbsf, subvertical intrusive contacts indicate a ~350-*m*-thick sheeted dike complex, with the lowermost ~60 m characterized by recrystallization to granoblastic textures by contact metamorphism.

The plutonic complex begins at 1407 mbsf (Figure 4.2.2.5), and consists of a 52-m-thick upper gabbro, a 24-m-thick interval of recrystallized granoblastic dikes with minor gabbroic and felsic veins, and a 24-m-thick lower gabbro. Contacts of the gabbros with the intervening dike interval are intrusive, with partly resorbed, stoped dike clasts within the gabbros. The plutonic section has been interpreted as either two separate gabbro units intrusive into the sheeted dikes, with an intervening screen of dikes (Expedition 309/312 Scientists, 2006b; Koepke et al., 2008); or as a single lens of gabbro containing stoped dike clasts, with the poorly recovered intermediate dike interval representing stoped dike fragments within gabbro (France, Ildefonse, & Koepke, 2009; Koepke et al., 2011).

Flows and dikes are aphyic to sparsely phyric and variably fractionated, with MgO contents of approximately 10–4.5 wt% (Expedition 309/312 Scientists, 2006b; Shipboard Scientific Party, 2003). The ranges of most major and many minor element concentrations are similar to those of the northern EPR, suggesting processes analogous to those along the modern EPR. Gabbro compositions are similar to the flows and dikes, but are on average more primitive, with higher MgO numbers (Expedition 309/312 Scientists, 2006b; Koepke et al., 2011).

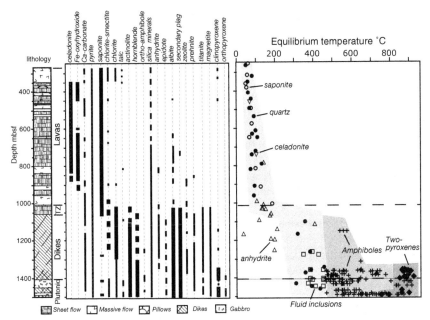

FIGURE 4.2.2.5 Basement lithology, distribution of secondary minerals, and estimated alteration temperatures for Hole 1256D *(modified from Alt et al. (2010))*. Figure illustrates stepped temperature gradient with depth. Secondary minerals and temperature estimates indicate low-temperature seawater fluids in the volcanic section and upwelling high-temperature hydrothermal fluids in the dikes. Stepwise temperature increase at lavas–dikes transition results from mixing of these fluids at this lithologic boundary. Stepwise increase at ~1350 mbsf indicates heating and contact metamorphism of lowermost dikes from underlying gabbroic intrusion. Range of temperatures in plutonic and dike sections results from retrograde reaction during cooling. Temperatures estimated from oxygen isotope analyses of secondary minerals, mineral equilibria, and fluid inclusions.

Hole 1256D sampled the transition between low-temperature and high-temperature hydrothermal alteration in a continuous section of oceanic crust (Alt et al., 2010) (Figure 4.2.2.5). Extrusives exhibit typical low-temperature alteration, with a pervasive saponitic (clay) background alteration, with dark and brownish alteration halos that contain celadonite (mica) and Fe-oxyhydroxide along veins. This vein-related alteration is concentrated in two zones, at 350–450 and 635–750 mbsf (Figure 4.2.2.3). Veins average 30/m, and vein minerals include saponite, Fe-oxyhydroxide, celadonite minor pyrite, and rare carbonate. Alteration temperatures in the lavas, estimated from oxygen isotope analyses of secondary minerals, were approximately 50–125 °C, generally increasing downward (Alt et al., 2010). Sulfur isotope data indicate widespread effects of microbial sulfate reduction in the volcanic sequence (Alt and Shanks, 2011), as in other oceanic basement sections (Lever et al., 2013; Ono et al., 2012; Rouxel, Shanks, Bach, & Edwards, 2008). The basal lava section features intervals with pyrite-rich alteration halos, mixed-layer chlorite-smectite, and anhydrite, and oxygen isotope analyses of secondary minerals indicate higher temperatures (150°–200 °C) than shallower in the crust (Alt et al., 2010).

The appearance of chlorite, albite, actinolite, anhydrite, epidote, and laumontite at around 1000 mbsf and the presence of an intensely altered and mineralized hyaloclastite breccia at 1022 mbsf indicate a stepwise increase in alteration temperatures downward across the top of the transition from lavas to dikes, and reaction with high-temperature hydrothermal fluids (Figure 4.2.2.5). Rocks are typically more intensely altered in centimeter-scale halos along veins in the dike margin breccias. Circulating hydrothermal fluids had temperatures (approximately 320–450 °C), salinities, and oxygen isotope compositions similar to black smoker fluids (Alt et al., 2010). Chlorite is the most common vein filling in the dikes, but quartz, sulfide, actinolite, prehnite, laumontite, and calcite are also present, and anhydrite veins are common to ~1200 mbsf. Hydrothermal veins composed of quartz, epidote, and sulfide postdate the chlorite-dominated veins, and are crosscut by vein assemblages of anhydrite, prehnite, laumontite, and calcite (formed <250 °C). Alteration mineralogy below ~1350 mbsf indicates temperatures >480 °C. In the lower portion of the sheeted dikes (1370–1397 mbsf), the rocks are recrystallized in patches to granoblastic textures, with secondary clinopyroxene, orthopyroxene, actinolitic hornblende, plagioclase, and blebs of oxide (ilmenite and magnetite). The pyroxene-rich assemblage and the granoblastic textures indicate recrystallization of previously hydrothermally altered rocks at temperatures >800 °C, related to underlying gabbroic intrusions (Alt et al., 2010; Koepke et al., 2008).

The plutonic section comprises highly altered gabbro and felsic veins with amphibole, chlorite, plagioclase, titanite, and minor laumontite and epidote, with alteration temperatures and hydrothermal fluid compositions similar to those in the overlying lower dikes. The dike screen separating the two gabbro units is recrystallized to granoblastic textures at temperatures similar to those in the basal granoblastic sheeted dikes. The intensity of gabbro alteration is variable, with intrusive margins and dike screen contacts being most extensively hydrothermal altered (Teagle & Wilson, 2007).

The overall alteration scenario involves cooler seawater fluids circulating in the volcanic section, and high-temperature hydrothermal fluids in the underlying dikes and gabbros. The stepwise increase in temperature downward across the top of the lava–dike transition (Figure 4.2.2.5) and the presence of a mineralized breccia indicate a mixing zone between these two fluid and alteration regimes (Alt et al., 2010). Evidence for black smoker-like fluid compositions and the presence of mineralized dike margins at greater depths indicate upwelling hydrothermal fluids in the dike section. Variations in profiles of oxygen and lithium isotopes result from these thermal regimes and compositional variations within the dikes (Gao et al., 2012). Contact metamorphism of the lowermost hydrothermally altered dikes resulted in a second thermal step at this depth, which decayed as the section cooled back to hydrothermal conditions (Alt et al., 2010).

4.2.2.2.3.3 Geophysical Measurements

Borehole geophysical logs collected in Hole 1256D during IODP Expeditions 309, 312, and 335 included natural gamma ray, bulk density, porosity, sonic velocity, electrical resistivity, temperature, and borehole imaging instruments (Expedition 309/312 Scientists, 2006b; Expedition 335 Scientists, 2012).

Wireline logs from the upper crust show variable borehole diameter, particularly within the interval 350–460 mbsf (100–210 msb), with corresponding decreases in electrical resistivity, bulk density, and sonic velocity, and increases in neutron porosity. These correlations could indicate that local anomalies result in part from poor borehole conditions, but the elevated natural gamma ray values suggest that there may be more extensive alteration in the inflated flows of this interval. The overlying, more massive, lava pond interval generally shows lower natural gamma radioactivity, and higher resistivity, bulk density, and sonic velocity, despite the relatively large borehole diameter. There are additional zones of elevated natural gamma ray emissions within the basalt sheets and massive flows between 780 and 900 mbsf (530–650 msb), despite having a borehole diameter that is mostly to gauge. Borehole imaging provides clear views of formation lithologic changes, for example, showing thin flow and pillow shapes in the upper volcanic crust, and allows assessment of fracture occurrence and orientation (Expedition 309/312 Scientists, 2006b; Expedition 335 Scientists, 2012). Core samples from Hole 1256D down to ~1000 msb yielded P-wave velocity values that are consistent with both wireline borehole logs and a vertical seismic profile experiment, indicating a lack of large-scale porosity (fractures) (Swift, Reichow, Tikku, Tominaga, & Gilbert, 2008). Below this depth, wireline logs indicate velocities higher than those measured with core samples, which may indicate a stress response associated with unloading of samples during coring and recovery (Expedition 309/312 Scientists, 2006b).

Thermal measurements indicate that borehole conditions were conductive prior to drilling and that heat flow through the basement section is essentially the same as that determined for the overlying sediments (Expedition 309/312 Scientists, 2006b; Shipboard Scientific Party, 2003). The lack of curvature in the temperature profile through the cased sedimentary section suggests that there is little or no fluid flow down the borehole, despite imposition of a cold column of fluid during drilling and other operations. Heat flow determined during drilling of the sedimentary section at this site, 113 mW/m^2, is close to the 120–130 mW/m^2 predicted by standard lithospheric cooling curves for seafloor that is 15-my old. This is unusual in comparison to the global heat flow data set for seafloor of this age, which typically shows evidence for advective extraction of ~40% of lithospheric heat (e.g., Stein, Stein, & Pelayo, 1995). Collectively these results suggest that the basement rocks around Hole U1256D may not currently be as hydrothermally active as is common for crust of this age.

4.2.2.2.4 Adjacent to the MAR and Atlantis Transform Fault, Northern Atlantic Ocean, Approximately 1.1–1.3-my-old Upper and Lower Crust (IODP Expeditions 304 and 305)

4.2.2.2.4.1 Background and Context

Expeditions 304 and 305 explored an oceanic core complex at the Atlantis Massif on the young western flank of the MAR at 30°N (Expedition 304/305 Scientists, 2006a) (Figure 4.2.2.1(D)). Exposure of gabbroic and mantle rocks along slow-spreading ridges is common (Escartín et al., 1998), and Exp. 304/305 researchers sought to understand mechanisms of lithospheric accretion by detachment faulting, at a site where the spreading half rate is ~12 mm per year. The "corrugated" central part of the Atlantis Massif is inferred to have been formed by detachment faulting during or soon after crustal formation (e.g., Blackman, Cann, Janssen, & Smith, 1998; Cann, Blackman, Smith, & McAllister, 1997). There is an adjacent basaltic block (to the east) interpreted to be the hanging wall of the detachment fault, and seafloor is covered by a thin drape of variably lithified sediment, volcanic deposits, and rubble. Atlantis Massif also hosts the Lost City Hydrothermal Field (LCHF), which discharges alkaline fluids rich in hydrogen and methane derived from serpentinization reactions (Kelley et al., 2005).

4.2.2.2.4.2 Petrology and Alteration

Hole U1309D was drilled 14–15 km west of the MAR axis and 5 km north of the LCHF, penetrating the Atlantis massif to 1415.5 mbsf, and recovering 75% of the cored interval (Expedition 304/305 Scientists, 2006b). The basement is divided into 770 units comprising dominantly crustal rock types, including dike-/sill-related basalt and dolerite (~3%) and gabbroic rocks (~91%). Interlayered with these are several olivine-rich rock types (~5%; dunites, wehrlites, troctolites). A few thin mantle peridotite intervals are also present in the upper 180 msb (Blackman et al., 2011; Expedition 304/305 Scientists, 2006b).

The gabbbroic rocks are mainly gabbros, including gabbronorite and orthopyroxene-bearing gabbro, making up 55.7% of the core. Olivine-bearing to troctolitic gabbro (25.5%) is the second most abundant rock type, followed by troctolite and other olivine-rich rocks (8%) and oxide gabbro (7%). Contact relations suggest that gabbro is generally intrusive into the more olivine-rich rocks (olivine gabbro and troctolite), and that the gabbroic rocks are intruded by felsic dikes and oxide gabbro. Subhorizontal sheets or sills of diabase intruded other rocks at several depths late in the intrusive history of the site. The abundance of dikes in the uppermost 100 m of Hole U1309D may imply that the detachment fault rooted in the dike–gabbro transition zone (McCaig & Harris, 2012). The gabbroic rocks are primitive, having Mg numbers of 67–87, and may be cumulates related through crystal fractionation to the diabases, which are tholeiitic basalt and minor basaltic andesite.

The fault zone of the detachment is represented by fragments of brecciated talc-tremolite fault schist and fractured metadiabase as deep as 100 mbsf in Holes U1309B and U1309D (Expedition 304/305 Scientists, 2006b). Paleomagnetic data indicate flexural clockwise rotation of the footwall <1 Ma along a ridge parallel horizontal axis (Morris et al., 2009). Logging data, low core recovery, and cataclasis at 108–126 mbsf and 685–785 mbsf indicate fault zones. The core is divided into three structural units based on these and another zone of cataclasis and crystal-plastic deformation.

Site U1310 lies in the hanging wall ~10 km west of the rift valley axis and ~600 m east of the termination of the detachment fault exposed on the central dome (Expedition 304/305 Scientists, 2006c). Hole U1310B recovered olivine + plagioclase phyric, pillow basalt fragments from the uppermost 13.5 m of basement having N-MORB compositions. Site U1311 is located near the termination of the detachment fault and may lie either in the hanging wall or in a klippe atop the footwall (Expedition 304/305 Scientists, 2006d). Drilling in Hole U1311A penetrated into fresh, moderately plagioclase-olivine phyric pillow basalt (with 13% recovery), which is similar in composition to rocks recovered from Site U1310.

Seawater–rock interactions manifest in core from Hole U1309D range from granulite to zeolite facies and alteration assemblages and vein fillings record the unroofing and uplift of the Atlantis Massif (Blackman et al., 2011; Nozaka, Fryer, & Andreani, 2008). Alteration intensity is highly variable, but generally decreases downhole. Commonly, sections of core reveal retrograde overprinting of earlier high-temperature metamorphism by later low-grade conditions. Alteration intensity is a function of time-integrated fluid flow, but also depends on the nature of the rocks. For instance, olivine gabbros and troctolites are more reactive than gabbro, because of the strong contrast in chemical potentials between olivine and plagioclase (e.g., Frost, Beard, McCaig, & Condliffe, 2008). Given the lithological heterogeneity of the basement at Site 1309, it is difficult to link the extent of alteration to the intensity of flow of seawater-derived fluids.

The history of alteration began with the dynamic recrystallization of olivine, clinopyroxene, plagioclase, and brown hornblende under granulite to upper amphibolite-facies conditions during mylonitic deformation (Blackman et al., 2011). Breakdown of clinopyroxene to hornblende under amphibolite-facies conditions is localized in and around sparse mylonitic deformation zones. The continued inflow of seawater-derived fluids led to increasing background alteration facilitated by the generation of microcracks, which helped to distribute fluids across a large volume of rock. This static background alteration is most pronounced in the uppermost 300 m of the basement and decreases in intensity downhole. Actinolitic hornblende and secondary plagioclase (±epidote) formed in the gabbros and oxide gabbros, whereas olivine-rich lithologies grew chlorite-tremolite(±talc) coronas along olivine-plagioclase grain boundaries. When the system had cooled to lower greenschist-facies temperatures, olivine underwent serpentinization, driving alteration of plagioclase to prehnite and hydrogrossular in olivine-rich lithologies (Frost et al., 2008). Chlorite-tremolite ± talc rocks are developed in the uppermost 25 m of basement, where strain localization took place along the detachment. Late-stage prehnite to zeolite-facies veins have saponite + zeolite, zeolite, carbonate, and occasional anhydrite. These veins do not show a systematic relation to high-temperature deformation or lithology. They are likely related to the recent and rapid uplift of Atlantis Massif (e.g.,Nozaka & Fryer, 2011; Nozaka et al., 2008). Thermochrolonological data (Schoolmeesters et al., 2012) indicate that the entire section in Hole U1309D cooled to 780 °C around the same time (0.8 Ma), and then the uppermost 600 m of basement cooled faster than the deeper section, probably because of seawater convection in the damage zone of the footwall.

4.2.2.2.4.3 Geophysical Measurements

The upper 94 m of basement penetration was logged in Hole U1309B, and measurements extended to >1400 msb in Hole U1309D (Expedition 304/305 Scientists, 2006b). Hole U1309D was revisited for additional logging during Expedition 340T (Expedition 340T Scientists, 2012). Wireline tools deployed in these holes included natural gamma ray emission, bulk density, electrical resistivity, neutron porosity, sonic velocity, borehole temperature, and formation imaging. In addition, detailed vertical seismic profile experiments were completed to assist with correlation between regional and borehole geophysical data.

Hole conditions were generally very good to excellent throughout the cored interval, and core recovery was high, allowing direct interpretation of differences in log response in terms of primary or secondary lithology and structure (Expedition 304/305 Scientists, 2006b). Natural gamma ray emissions are generally low in the rocks of Site U1309, but there are thin zones of elevated values. One of these natural gamma anomalies (near 750 mbsf) corresponds to an abrupt increase in formation electrical resistivity, and an increase (and reduction of variability) in bulk density. That change in geophysical properties is associated with an interval containing up to 20% of basalt diabase, in contrast to overlying and underlying intervals comprising mainly gabbro and gabbronorite. Formation resistivity decreases again at 1100 mbsf. Sonic velocities generally increase with depth into the crust, although there are local excursions where lower velocities generally correlate with lower bulk density and electrical resistivity.

Borehole imaging data (electrical and sonic) suggest that small intervals of borehole enlargement often correspond to open faults and fractures. Comparison of core and geophysical logging data shows a strong correlation between deviations in logging parameters (e.g., lower bulk density and sonic velocity) and the intensity and pervasiveness of alteration. There are also good correlations between physical properties determined in the borehole, on recovered core, and inferred from seismic reflection studies.

Logging data collected in Hole U1309D during Expedition 340T are generally consistent with data collected during earlier expeditions, except for the temperature log (Blackman, Slagle, Guerin, & Harding, 2014). Much of the deepest half of the hole had warmed toward a predrilling state, but there was gentle downward curvature in the gradient from the upper 750 mbsf, and small local excursions in downhole temperature near 750 and 1100 mbsf, corresponding to abrupt changes in geophysical properties. Larger excursions in temperature at these depths were also apparent in Expedition 305 data (Expedition 304/305 Scientists, 2006b), but these were superimposed on an overall profile indicative of borehole cooling. These zones could have been more intensively invaded by cool drilling fluids when the holes were cored, and thus would have taken longer to recover. There may be slow fluid convection within the upper 750 msb in this crustal section today, helping to steepen the geothermal gradient in the upper half of Hole 1309D (Blackman et al., 2014). There is no rapid flow down Hole 1309D today, as this would suppress the borehole thermal gradient much more than observed (e.g., Winslow et al., 2013), but slow flow of water down the hole and into thin zones at depth (fractures, faults), could also explain the observed curvature.

4.2.2.3 SYNTHESIS: METHOD AND SITE COMPARISONS AND TRENDS

4.2.2.3.1 Hydrogeologic Properties of the Ocean Crust

A compilation of direct hydrogeologic measurements made during DSDP, ODP, and IODP (Figure 4.2.2.6) illustrates both consistent trends and notable data gaps. Data and interpretations from packer measurements and analyses of thermal logs were added to the global data set during IODP from 3.5-my-old crust on the eastern flank of the Juan de Fuca Ridge, in Holes U1301A, U1301B, and U1362A (Becker & Fisher, 2000; 2008; Becker et al., 2013; Winslow et al., 2013). Collectively these data suggest that the highest permeability within basaltic upper crust is found within the upper 300 msb, particularly within lithologically defined zones that are tens of meters thick (as opposed to the uppermost crust as a whole). Data from 0.9 to 3.6 my-old seafloor in this area suggest that the highest permeabilities are found in upper crust of the youngest sites (Becker & Davis, 2003), but there is considerable variability associated with the scale (test duration), type, and depth range of measurements. Permeability data are sparse below the upper 300 m of the crust, and there is a notable lack of data from a crust older than 8 my (Figure 4.2.2.6). Direct measurements of permeability in gabbroic rocks characteristic of the middle to lower crust have been made during scientific drilling only in Hole 735B (Becker, 1991), and only one set of measurements has been made in fast-spread basaltic crust, in Hole 801C, the oldest site in the data set (Larson, Fisher, Jarrard, & Becker, 1993) (Figure 4.2.2.6).

Permeability calculated from the cross-hole response observed in Hole 1027C to long-term flow down Hole U1301B (Figure 4.2.2.4(A)) is at the lower end of estimates based on single-hole packer experiments (Becker & Fisher, 2000; Becker et al., 2013; Fisher et al., 2008) (Figure 4.2.2.6). This is surprising

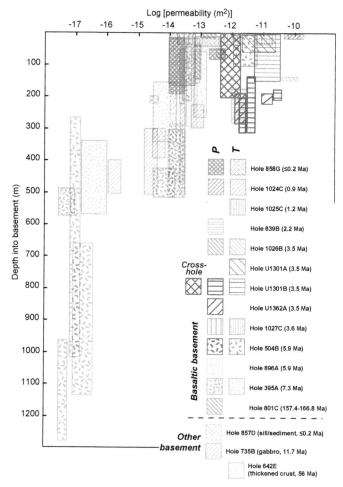

FIGURE 4.2.2.6   Compilation of borehole measurements of permeability in the volcanic ocean crust *(based on summary presented in Becker and Fisher (2008), with additional data from Fisher et al. (2008), Becker et al. (2013), and Winslow et al. (2013)).* Most data are from packer experiments (P) and borehole thermal logs (T), but there was a single cross-hole test, as discussed in the text. Black and white version: Data collected during DSDP and ODP are shown in gray with patterns. Color version: Data collected during DSDP, ODP, and IODP are coded by color and pattern, with DSDP and ODP data being partly transparent.

at first because the long-term cross-hole response should be influenced by a much larger rock volume, extending perhaps 10–30 km from the borehole, and larger test volumes generally correspond to higher apparent permeability in heterogeneous/fractured rock systems such as the ocean crust (e.g., Clauser, 1992; Fisher, 1998; Guéguen, Gavrilenko, & Le Ravalec, 1996). Basement permeability estimated from this cross-hole response is also one to three orders of magnitude lower than estimates based on numerical modeling and calculations based on tidal responses and drainage following tectonic strain events, which evaluated similar crustal volumes (Davis & Becker, 2002; 2004; Davis et al., 2004, 1997; Wang, He, & Davis, 1997).

One possible explanation for the differences in inferred properties based on these methods is that permeability in the crust around Sites U1301 and 1027 is azimuthally anisotropic. Anisotropy in the seismic properties of oceanic basement rocks is thought to result from preferential orientation of cracks, faults, and fractures (i.e., the crustal "fabric") (e.g., Sohn, Webb, Hildebrand, & Cornuelle, 1997; Stephen, 1981). The dominant crustal fabric is generally thought to be subparallel to the orientation of the mid-ocean ridge where the crust was created. This fabric may favor fluid flow in the crust in the "along-strike" direction (Delaney, Robigou, McDuff, & Tivey, 1992; Haymon et al., 1991; Wilcock & Fisher, 2004), an interpretation consistent with geochemical and thermal data from the Expedition 301/327 field area (Fisher, Davis, et al., 2003; Hutnak et al., 2006; Walker, McCarthy, Fisher, & Guilderson, 2007; Wheat, Elderfield, Mottl, & Monnin, 2000). Azimuthal anisotropy could influence the permeability apparent from a cross-hole experiment involving a single observation borehole (Figure 4.2.2.4(B)). If the angle of measurement is oblique relative to the direction of greatest permeability (as between Holes U1301B and 1027C), the measured value will be very close to that in the lowest-permeability direction, even for a large anisotropy ratio. Confirmation that the crust in this area is azimuthally anisotropic with respect to permeability requires simultaneous monitoring of pressure changes in response to pumping or free flow using two or more observation boreholes at significantly different azimuths from the perturbation borehole.

Along-strike consistency of crustal hydrologic properties in this area, at a kilometer scale, is indicated from comparison of packer and wireline logging responses, and rates of drilling penetration, from Holes U1301B and U1362A (Becker et al., 2013; Expedition 301 Scientists, 2005a; Expedition 327 Scientists, 2011a). Similar consistency is not apparent in crustal layering on the basis of core descriptions, but this may result from local heterogeneity at the hand-sample scale and/or limited recovery from the upper crust. New analyses of borehole thermal logs suggest surprising consistency in bulk permeability of the uppermost crust (Winslow et al., 2013), but the global data set remains sparse.

4.2.2.3.2 The Integrated Record of Fluid–Rock Interactions in the Ocean Crust

The distribution and style of low-temperature alteration seen in volcanic sections of oceanic crust cored by IODP are generally similar to those sampled by ODP and DSDP, including the presence of secondary minerals and their distribution in veins, alteration halos, and "background" alteration (e.g., Alt, 2004; Bach, Humphris, & Fisher, 2004). IODP drilling provides valuable new insights regarding lithological controls on alteration in upper oceanic basement and the relation between alteration and crustal spreading rate.

Prior to drilling Hole 1256D it was believed that oxidation effects (as evidenced by reddish alteration halos) should generally decrease with depth in the volcanic section of oceanic crust (e.g., Figure 4.2.2.3, Hole 504B). Hole 1256D does not show this trend with depth. Instead, alteration halos are focused in distinct zones, indicating a strong control on fluid flow by lithology and permeability, and halos are much less abundant in this hole than in crust formed at intermediate spreading rates (Figure 4.2.2.3). In some cases, oxidation and alteration effects are concentrated at the tops of lava flows or eruptive units, which can be rubbly and permeable, allowing focused fluid flow. IODP drilling and logging of cores by shipboard scientists has extended this comparison to crust generated at slow-spreading ridges, showing that alteration halos there are comparable in abundance to those in crust formed at intermediate spreading rates (Figure 4.2.2.3).

Analysis of samples and data from ODP/IODP Site 1256 has also shown that sheet flows are more abundant throughout the volcanic section in crust formed at fast spreading rates, whereas pillows are more abundant at intermediate and slow rates (Figure 4.2.2.3) (Expedition 309/312 Scientists, 2006b; Teagle & Wilson, 2007). This interpretation is consistent with observations made of seafloor lava flows (e.g., Carbotte & Scheirer, 2004), and may help to explain the extreme efficiency of crustal cooling where the crust was formed at a fast-spreading ridge (Fisher, Stein, et al., 2003; Hutnak et al., 2008). At the other extreme, basaltic upper crust may be thinner or absent on slow- and ultraslow-spreading seafloor (e.g., Dick, Lin, & Schouten, 2003; Ildefonse et al., 2007).

A thorough understanding of the distribution of rock types found in oceanic crustal sections is limited by biased and incomplete recovery of core, but careful integration with geophysical logs can put core samples in context, as demonstrated by Tominaga et al. (2009) (Figure 4.2.2.7). Their analysis of wireline logs from Hole 1256D, with an emphasis on resistivity and natural gamma ray responses, suggests that breccias and fragmented flows comprise more than 50% of the borehole interval. In contrast, the vast majority (>90%) of recovered core was classified as sheet and massive basalt flows.

A comparison of age trends in physical properties from multiple DSDP, ODP, and IODP holes using core and wireline logging data illustrates how measurements made at different scales may be related, and how the crust ages (Bartetzko & Fisher, 2008) (Figure 4.2.2.8). Sonic velocity, electrical resistivity, bulk density, and total gamma ray measured with wireline tools in open boreholes in the upper basaltic crust tend to increase with basement age, whereas sonic velocity and bulk density measured on core samples tend to decrease. Properties determined from wireline data and core data tend to converge with increasing crustal age. One explanation for these observations is that borehole-scale measurements record the infilling of small pore spaces by alteration products of water–rock interaction. In contrast, hand samples tend to become less dense

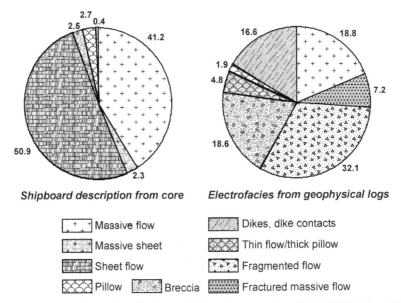

Shipboard description from core   *Electrofacies from geophysical logs*

+·+ Massive flow	▨ Dikes, dike contacts
+·+ Massive sheet	⬡ Thin flow/thick pillow
Sheet flow	▵ Fragmented flow
Pillow ▦ Breccia	▨ Fractured massive flow

FIGURE 4.2.2.7   Comparison of rock types in the volcanic section of Hole 1256D indicated by shipboard core descriptions (left) and analysis of electrofacies (geophysical logs) to identify igneous stratigraphy (right) *(modified from Tominaga et al. (2009))*. Pie slices and numbers indicate the percent of each rock type identified. Shipboard lithostratigraphic data were tabulated from visual core descriptions from Leg 206, Expedition 309 and 312 (Expedition 309/312 Scientists, 2006b; Shipboard Scientific Party, 2003) by calculating a total of recovered core length of each rock type divided by a total of recovered core length. Recovery was highly variable, but was particularly low in zones containing breccia and heavily fractured rock.

and have lower sonic velocities as a result of the same alteration processes, as expressed in the mostly massive samples that are recovered during coring.

Results of hydrogeologic experiments in basement can be inconsistent with alteration patterns in recovered crustal rocks. For example, in Holes U1301B and U1362A, the highest transmissivity occurs at 150–180 mbsf (Figure 4.2.2.6), but alteration halos are abundant throughout the basement sections (Figure 4.2.2.3). Upper basement temperature in this area is ~65 °C, but carbonates in veins from these holes formed at temperatures of 30–40 °C (Coggon, Teagle, Cooper, & Vanko, 2004; Coggon et al., 2010). The difference between present and past temperatures likely results from a relatively recent reduction in the rate of fluid circulation in this setting, as local basement outcrops were systematically buried during the Pleistocene, and basement temperatures warmed in response to reduced heat extraction (Hutnak & Fisher, 2007).

The oceanic crust is a significant global sink for carbon, and carbonate is more abundant in the volcanic section of crust formed at slow spreading rates than in upper crust formed at intermediate and fast spreading rates (Alt & Teagle, 1999; Gillis & Coogan, 2011). Vein carbonates recovered from core samples of

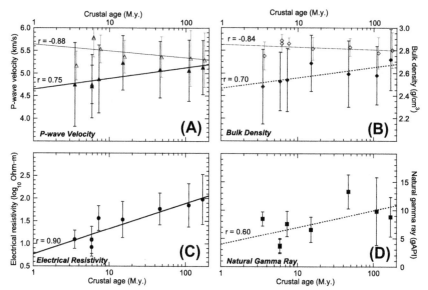

FIGURE 4.2.2.8 Comparison of geophysical logging (borehole) and physical properties (hand sample) values from the upper ocean crustal sections as a function of basement age *(updated from Bartetzko & Fisher (2008))*, based on selected data from DSDP, ODP, and IODP boreholes. Data are from Holes U1301, 504B, 896A, 395A, 1256D, 1224F, 418A, and 801C, with means of borehole logging data plotted using filled symbols and means of core data plotted using open symbols. Lines are least squares best fits based on a semilogarithmic trend (but note $\log_{10}$ scale for electrical resistivity values) with correlation coefficients shown. Solid lines are significant with >99% confidence (Bartetzko & Fisher, 2008).

the oceanic crust are valuable recorders of fluid chemical evolution as the seafloor ages, indicating variations in both crustal and seawater composition and temperature (e.g., Expedition 301 carbonates, Coggon et al., 2004, 2010; Gillis & Coogan, 2011; Rausch, Böhm, Bach, Klügel, & Eisenhauer, 2013).

The temperature of water–rock interaction also has an important influence on the composition of crustal fluids, as shown with a global compilation of data from modern upper basement fluids (Fisher & Wheat, 2010) (Figure 4.2.2.9), including analyses of sedimentary pore fluids recovered from adjacent to the sediment–basement contact (e.g., Elderfield, Wheat, Mottl, Monnin, & Spiro et al., 1999; Mottl, 1989; Orcutt et al., 2013). For major ions involved in mainly inorganic exchange (e.g., magnesium), there is relatively little change in fluid composition at temperatures below about 20 °C, but solutes involved in biogeochemical cycling (e.g., phosphate) show evidence of reaction at essentially all fluid temperatures (Figure 4.2.2.9). In some cases, reactions may occur in both basement rocks and in overlying sediments, with diffusional exchange between crustal and sedimentary layers.

Alteration in core samples from Hole 1256D indicates an abrupt increase in the thermal gradient with depth (Figure 4.2.2.5), corresponding to an increase in alteration temperature across the transition from lavas to dikes, similar to

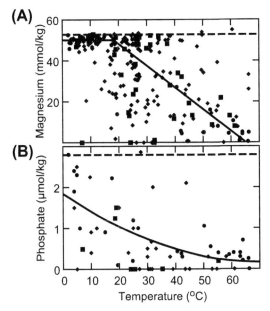

FIGURE 4.2.2.9 Concentrations of magnesium and phosphate in ridge-flank fluids from the upper portion of basaltic basement plotted versus measured and inferred temperatures at the sediment–basement contact *(figure modified from Fisher & Wheat (2010))*. These plots are based on a global compilation of DSDP, ODP, and IODP data, and analyses of samples collected with traditional gravity and piston corers, and from seafloor seeps and springs. Data are classified to indicate quality. Black and white version: circle = highest quality; square = moderate quality; diamond = fair quality. Horizontal dashed lines indicate typical concentrations in bottom seawater. Solid lines/ curves show global trends based on the highest quality data. Color version: red circle = highest quality; blue square = moderate quality; black diamond = fair quality. Green dashed lines indicate typical concentrations in bottom seawater. Pink solid lines/curves show global trends based on the highest quality data.

that documented in DSDP/ODP Hole 504B and at Pito Deep (Alt et al., 2010; Alt et al., 1996; Heft, Gillis, Pollock, Karson, & Klein, 2008). This observation indicates different styles of fluid circulation at depth, controlled by lithology and permeability, with higher-temperature hydrothermal fluids in the dikes and cooler seawater fluids in the volcanic section at the spreading axis, and continued low-temperature circulation in the lavas on the ridge flank. Cores from Hole 1256D also reveal a step in the thermal gradient at the dike-plutonic boundary, corresponding to contact metamorphism of the lowermost dikes driven by intrusion of underlying gabbros near the spreading axis (Alt et al., 2010; Koepke et al., 2008).

Studies of lower crustal rocks from Hole 735B, the deepest seafloor penetration into plutonic rocks, showed that metamorphism, fracturing, and fluid penetration occurred over a wide range of conditions (Bach et al., 2001; Dick et al., 2000). Results from gabbroic rocks from Hole 1309D generally show a similar range of alteration styles, from granulite and amphibolite-facies reactions down to low-temperature smectite formation (Blackman et al., 2011). However, the extent of granulite facies alteration is greater and Sr isotope compositions are less altered in the uppermost 400 m of Hole 735B (McCaig, Delacour, Fallick, Castelain, & Früh-Green, 2010), suggesting less cumulative seawater–rock interaction in comparison to Hole 1309D. In addition, cores from Hole U1309D have a greater abundance of olivine-rich rocks, making serpentinization reactions especially important, and they lack evidence for subgreenschist conditions. Nozaka et al. (2008) interpret this to indicate rapid uplift of the Atlantis Massif (Hole U1309D) in comparison to Atlantis Bank (Hole 735B), but there could also be different fracturing histories during cooling. This would be consistent with packer testing results from Hole 735B, which suggested considerable permeability at present (Becker, 1991), but temperature logs from Hole U1309D suggest limited fluid circulation (Blackman et al., 2014).

4.2.2.3.3 Future Needs and Frontiers

The observations and interpretations presented in this chapter remain limited by the numbers and locations of boreholes that penetrate into the igneous oceanic crust (Figure 4.2.2.2). There are consistent trends in hydrogeologic properties and crustal structure and alteration at many sites (for example, heterogeneity in properties associated with crustal layering). Although scientific ocean drilling is filling gaps across the global map, resolving dependencies on seafloor age and spreading rate will require additional opportunities to sample, analyze, and test crustal conditions and quantify histories of water–rock interaction. Researchers also need access to multiple basement holes separated by a range of distances, to assess the consistency and continuity of properties. This is often not done during scientific ocean drilling because of the cost and time required to establish, drill, sample, test, and instrument basement holes; results from individual holes raise questions about whether these sample points are representative, and how properties and processes may vary laterally, with age, and with proximity to features such as fracture zones and seamounts. These questions can be addressed, in part, by drilling age transects to assess the evolution of ocean crust as a result of fluid–rock interactions.

There are ongoing experiments, results of which are currently incomplete, that were designed to test the nature of permeability and advective transport in upper basement on the eastern flank of the Juan de Fuca Ridge. In 2011, after the new CORKs installed during Expedition 327 were permitted to equilibrate for a year following drilling, a large-diameter ball valve was opened on the wellhead of the CORK in Hole U1362B. Downhole temperature loggers and a flowmeter attached to the discharging wellhead were deployed to assess the rate at which fluids discharged from the overpressured formation for the next 2 years, and the long-term pressure response from surrounding CORKs will help to determine both crustal-scale properties and the extent of permeability anisotropy (vertical, azimuthal). In addition, about 500 m^3 of fluid and tracers (gas, solute, particles) were injected into the formation around Hole U1362B in 24 h as part of a cross-hole tracer experiment initiated during IODP Expedition 327. Multiyear records of tracer recovery are being developed from fluids collected with Osmosamplers and active pumping systems deployed on multiple CORK wellheads before and after tracer injection. Additional records from Osmosamplers and temperature loggers were recovered in Summer 2014.

Drilling deeper basement holes remains important for exploration of hydrothermal processes and effects in the crust. Further drilling in Hole 1256D (or elsewhere) through the uppermost gabbros will allow testing of petrological and geophysical models for melt lenses at mid-ocean ridges. Such drilling will also enable testing of how hydrothermal metamorphism, fluid fluxes, and heat transport in high-temperature axial hydrothermal systems are related throughout the crust, and help reconcile budgets for hydrothermal heat, fluid fluxes, and chemical fluxes. Drilling in nearby ("pilot") holes can be helpful for assessing lithologic and alteration heterogeneity, and could facilitate cross-hole tests that measure rock properties and fluid flow at the scale of crustal fluid flow and in multiple directions. Reconciling present-day hydrologic conditions and the integrated history of water–rock interaction through the crust requires the collection of physical and chemical core samples and borehole data from the same locations, across a range of spatial scales, through long crustal sections. Colocated multidisciplinary studies have long been a strength of scientific ocean drilling, and will continue to be essential in the future, as new tools, methods, and opportunities are developed.

ACKNOWLEDGMENTS

This research used data provided by the Ocean Drilling Program (ODP) and the Integrated Ocean Drilling Program (IODP) and was supported by National Science Foundation grants OCE 1031808, OCE-1131210, OCE-1260408 (AF), and OCE-1129631 and OCE-1334758 (JA), and Consortium for Ocean Leadership Projects T327A7 and T327B7 (ATF). This manuscript benefitted from thoughtful reviews by D. Blackman, A. McCaig, and R. Harris. This is C-DEBI contribution 228.

REFERENCES

Alt, J. C. (2004). Alteration of the upper oceanic crust: mineralogy, chemistry, and processes. In E. E. Davis, & H. Elderfield (Eds.), *Hydrogeology of the oceanic lithosphere* (pp. 456–488). Cambridge, UK: Cambridge University Press.

Alt, J. C., Laverne, C., Coggon, R. M., Teagle, D. A. H., Banerjee, N. R., Morgan, S., et al. (2010). Subsurface structure of a submarine hydrothermal system in ocean crust formed at the East Pacific Rise. *Geochemistry, Geophysics, Geosystems, 11*(10), Q10010. http://dx.doi.org/10.1029/2010GC003144.

Alt, J. C., & Shanks, W. C., III (2011). Microbial sulfate reduction and the sulfur budget for a complete section of altered oceanic basalts, IODP Hole 1256D (eastern Pacific). *Earth Planet. Sci. Lett, 310*, 73–83. http://dx.doi.org/10.1016/j.epsl.2011.1007.1027.

Alt, J. C., & Teagle, A. H. (1999). The uptake of carbon during alteration of ocean crust. *Geochimica Cosmochimica et Acta, 63*, 1527–1535.

Alt, J. C., & Teagle, D. A. H. (2003). Hydrothermal alteration of upper oceanic crust formed at a fast spreading ridge: mineral, chemical, and isotopic evidence from ODP Site 801. *Chemical Geology, 201*(3–4), 191–211.

Alt, J. C., Teagle, D. A. H., Laverne, C., Vanko, D. A., Bach, W., Honnorez, J., et al. (1996). Ridge-flank alteration of upper oceanic crust in the Eastern Pacific: synthesis of results for volcanic rocks of Holes 504B and 896A. In J. C. Alt, H. Kinoshita, & L. Stokking (Eds.), *Proceedings of the ocean drilling program, science results 148* (pp. 435–452). College Station, TX: Ocean Drilling Program.

Bach, W., Alt, J. C., Niu, Y., Humphris, S. E., Erzinger, J., & Dick, H. J. B. (2001). The geochemical consequences of late-stage low-grade alteration of lower ocean crust at the SW Indian Ridge: results from ODP Hole 735B (Leg 176). *Geochimica Cosmochimica et Acta, 65*, 3,267–263,287.

Bach, W., Humphris, S. E., & Fisher, A. T. (2004). Fluid flow and fluid-rock interaction within oceanic crust: reconciling geochemical, geological and geophysical observations. In W. S. D. Wilcock, E. Delong, D. Kelley, J. A. Baross, & S. C. Cary (Eds.), *Subseafloor biosphere at mid-ocean ridges* (pp. 99–117). Washington, D. C: AGU Geophysical Monograph 144, American Geophysical Union. http://dx.doi.org/10.1029/144GM07.

Bartetzko, A., & Fisher, A. T. (2008). Physical properties of young (3.5 m.y.) oceanic crust from the eastern flank of Juan de Fuca Ridge: comparison of wireline and core measurements with global data. *Journal of Geophysical Research*, B05105. http://dx.doi.org/10.1029/2007JB005268.

Bartetzko, A., Pezard, P., Goldberg, D., Sun, Y.-F., & Becker, K. (2001). Volcanic stratigraphy of DSDP/ODP Hole 395A: an interpretation using well-logging data. *Marine Geophysical Researches, 22*, 111–127.

Becker, K. (1986). Special report: development and use of packers in ODP. *JOIDES Journal, 12*, 51–57.

Becker, K. (1991). In-situ bulk permeability of oceanic gabbros in Hole 735B, ODP Leg 118. In R. P. Von Herzen, & P. T. Robinson (Eds.), *Proceedings of the ocean drilling program, science results 118* (pp. 333–347). College Station, TX: Ocean Drilling Program.

Becker, K., Bartetzko, A., & Davis, E. E. (2001). Leg 174B synopsis: revisiting Hole 395A for logging and long-term monitoring of off-axis hydrothermal processes in young oceanic crust. In K. Becker, & M. J. Malone (Eds.), *Proceedings of the ocean drilling program, science results 174B* (pp. 1–12). College Station, TX: Ocean Drilling Program.

Becker, K., & Davis, E. (2003). New evidence for age variation and scale effects of permeabilities of young oceanic crust from borehole thermal and pressure measuements. *Earth and Planetary Science Letters, 201*(3–4), 499–508.

Becker, K., & Davis, E. (2004). In situ determinations of the permeability of the igneous oceanic crust. In E. E. Davis, & H. Elderfield (Eds.), *Hydrogeology of the oceanic lithosphere* (pp. 189–224). Cambridge, UK: Cambridge University Press.

Becker, K., & Davis, E. E. (2005). A review of CORK designs and operations during the Ocean Drilling Program. In A. T. Fisher, T. Urabe, A. Klaus, & the Expedition 327 Scientists (Eds.), *Proc. IODP 301*. Tokyo: Integrated Ocean Drilling Program Management International, Inc. http://dx.doi.org/10.2204/iodp.proc.301.104.

Becker, K., & Fisher, A. T. (2000). Permeability of upper oceanic basement on the eastern flank of the Endeavor Ridge determined with drill-string packer experiments. *Journal of Geophysical Research, 105*(B1), 897–912.

Becker, K., & Fisher, A. T. (2008). Borehole tests at multiple depths resolve distinct hydrologic intervals in 3.5-Ma upper oceanic crust on the eastern flank of the Juan de Fuca Ridge. *Journal of Geophysical Research, 113*, B07105. http://dx.doi.org/10.1029/2007JB005446.

Becker, K., Fisher, A. T., & Tsuji, T. (2013). New packer experiments and borehole logs in upper oceanic crust: evidence for ridge-parallel consistency in crustal hydrogeologic properties. *Geochemistry, Geophysics, Geosystems, 14*. (8). http://dx.doi.org/10.1002/ggge.20201.

Becker, K., Langseth, M., Von Herzen, R. P., & Anderson, R. (1983). Deep crustal geothermal measurements, hole 504B, Costa Rica rift. *Journal of Geophysical Research, 88*, 3447–3457.

Becker, K., Langseth, M. G., & Hyndman, R. D. (1984). Temperature measurements in hole 395A, leg 78B. In R. Hyndman, & M. Salisbury (Eds.), *Init. Repts., DSDP* (pp. 689–698). Washington, D. C: 78B, U. S. Govt. Printing Office.

Blackman, D. K., Cann, J. R., Janssen, B., & Smith, D. K. (1998). Origin of extensional core complexes: evidence from the Mid-Atlantic Ridge at Atlantis Fracture Zone. *Journal of Geophysical Research, 103*(B9), 21,315–321,333.

Blackman, D. K., Ildefonse, B., John, B. E., Ohara, Y., Miller, D. J., Abe, N, et al. (2011). Drilling constraints on lithospheric accretion and evolution at Atlantis Massif, Mid-Atlantic Ridge 30°N. *Journal of Geophysical Research: Solid Earth, 116*(B7), B07103. http://dx.doi.org/10.1029/2010jb007931.

Blackman, D. K., Slagle, A., Guerin, G., & Harding, A. (2014). Geophysical signatures of pasrt and present hydration within a young oceanic core complex. *Geophysical Research Letters, 41*, 1179–1186.

Cann, J. R., Blackman, D. K., Smith, D. K., & McAllister, E. (1997). Corrugated slip surfaces formed at ridge-transform intersections on the mid-atlantic ridge. *Nature, 385*, 329–331.

Carbotte, S., & Scheirer, D. S. (2004). Variability of ocean crustal structure created along the global mid-ocean ridge. In E. E. Davis, & H. Elderfield (Eds.), *Hydrogeology of the oceanic lithosphere* (pp. 59–107). Cambridge, UK: Cambridge University Press.

Clauser, C. (1992). Permeability of crystalline rocks, Transactions. *American Geophysical Union, 73*(233), 237–238.

Coggon, R. M., Teagle, D. A. H., Cooper, M. J., & Vanko, D. A. (2004). Linking basement carbonate vein composition to porewater geochemistry across the eastern flank of the Juan de Fuca Ridge, ODP Leg 168. *Earth and Planetary Science Letters, 219*, 111–128.

Coggon, R. M., Teagle, D. A. H., Smith-Duque, C. E., Alt, J. C., & Cooper, M. J. (2010). Reconstructing past seawater Mg/Ca and Sr/Ca from mid-ocean ridge flank calcium carbonate veins. *Science, 327*(5969), 1114–1117.

Davis, E. E., & Becker, K. (2002). Observations of natural-state fluid pressures and temperatures in young oceanic crust and inferrences regarding hydrothermal circulation. *Earth and Planetary Science Letters, 204*, 231–248.

Davis, E. E., & Becker, K. (2004). Observations of temperature and pressure: constraints on ocean crustal hydrologic state, properties, and flow. In E. E. Davis, & H. Elderfield (Eds.), *Hydrogeology of the oceanic lithosphere* (pp. 225–271). Cambridge, UK: Cambridge University Press.

Davis, E. E., Becker, K., Dziak, R. P., Cassidy, J. F., Wang, K., & Lilley, M. (2004). Hydrologic response to a seafloor spreading episode on the Juan de Fuca Ridge: evidence for co-seismic net crustal dilation. *Nature, 430*, 335–338.

Davis, E. E., Becker, K., Pettigrew, T., Carson, B., & MacDonald, R. (1992). CORK: a hydrologic seal and downhole observatory for deep-ocean boreholes. In E. E. Davis, M. Mottl, & A. T. Fisher (Eds.), *Proceedings of the ocean drilling program, initial reports 139* (pp. 43–53). College Station, TX: Ocean Drilling Program.

Davis, E. E., Chapman, D. S., Mottl, M. J., Bentkowski, W. J., Dadey, K., Forster, C., et al. (1992). FlankFlux: an experiment to study the nature of hydrothermal circulation in young oceanic crust. *Canadian Journal of Earth Sciences, 29*(5), 925–952.

Davis, E. E., Chapman, D. S., Wang, K., Villinger, H., Fisher, A. T., Robinson, S. W., et al. (1999). Regional heat-flow variations across the sedimented Juan de Fuca Ridge eastern flank: constraints on lithospheric cooling and lateral hydrothermal heat transport. *Journal of Geophysical Research, 104*(B8), 17,675–617,688.

Davis, E. E., LaBonte, A., He, J., Becker, K., & Fisher, A. T. (2010). Thermally stimulated "runaway" downhole flow in a super-hydrostatic ocean-crustal borehole: observations, simulations, and inferences regarding crustal permeability. *Journal of Geophysical Research, 115*, B07102. http://dx.doi.org/10.1029/2009JB006986.

Davis, E. E., Wang, K., He, J., Chapman, D. S., Villinger, H., & Rosenberger, A. (1997). An unequivocal case for high Nusselt-number hydrothermal convection in sediment-buried igneous oceanic crust. *Earth and Planetary Science Letters, 146*, 137–150.

Delaney, J. R., Robigou, V., McDuff, R., & Tivey, M. (1992). Geology of a vigorous hydrothermal system on the Endeavor segment, Juan de Fuca Ridge. *Journal of Geophysical Research, 97*, 19,663–619,682.

Dick, H. J. B., Lin, J., & Schouten, H. (2003). An ultraslow-spreading class of ocean ridge. *Nature, 426*(6965), 405–412.

Dick, H. J. B., Natland, J. H., Alt, J. C., Bach, W., Bideau, D., Gee, J. S., et al. (2000). A long in situ section of the lower ocean crust; results of ODP Leg 176 drilling at the Southwest Indian Ridge. *Earth and Planetary Science Letters, 179*, 31–51.

Edwards, K. J., Wheat, C. G., Orcutt, B. N., Hulme, S., Becker, K., Jannasch, H., et al. (2012). Design and deployment of borehole observatories and experiments during IODP Expedition 336, Mid-Atlantic Ridge flank at North Pond. In K. J. Edwards, W. Bach, A. Klaus, & the Expedition 336 Scientists (Eds.), *Proceedings of the integrated ocean drilling program 336*. Tokyo: Integrated Ocean Drilling Program Management International, Inc. http://dx.doi.org/10.2204/iodp.proc.336.109.2012.

Elderfield, H., Wheat, C. G., Mottl, M. J., Monnin, C., & Spiro, B. (1999). Fluid and geochemical transport through oceanic crust: a transect across the eastern flank of the Juan de Fuca Ridge. *Earth and Planetary Science Letters, 172*, 151–165.

Escartín, J., Smith, D. K., Cann, J., Schouten, H., Langmuir, C. H., & Escrig, S. (1998). Central role of detachment faults in accretion of slow-spreading oceanic lithosphere. *Nature, 455*, 790–795.

Expedition 301 Scientists. (2005a). Expedition 301 summary. In A. T. Fisher, T. Urabe, A. Klaus, & the Expedition 301 Scientists (Eds.), *Proceedings of the integrated ocean drilling program 301*. Tokyo: Integrated Ocean Drilling Program Management International, Inc. http://dx.doi.org/10.2204/iodp.proc.301.101.2005.

Expedition 301 Scientists. (2005b). Site U1301. In A. T. Fisher, T. Urabe, A. Klaus, & the Expedition 301 Scientists (Eds.), *Proceedings of the integrated ocean drilling program 301*. Tokyo: IIntegrated Ocean Drilling Program Management International, Inc. http://dx.doi.org/10.2204/iodp.proc.301.106.2005.

Expedition 304/305 Scientists. (2006a). Expedition 304/305 summary. In D. K. Blackman, B. Ildefonse, B. E. John, Y. Ohara, D. J. Miller, C. J. MacLeod, et al. (Eds.), *Proceedings of the integrated ocean drilling program 304/305*. Tokyo: Integrated Ocean Drilling Program Management International, Inc. http://dx.doi.org/10.2204/iodp.proc.304305.101.2006.

Expedition 304/305 Scientists. (2006b). Site U1309. In D. K. Blackman, B. Ildefonse, B. E. John, Y. Ohara, D. J. Miller, C. J. MacLeod, et al. (Eds.), *Proceedings of the integrated ocean drilling program 304/305*. Tokyo: Integrated Ocean Drilling Program Management International, Inc. http://dx.doi.org/10.2204/iodp.proc.304305.103.2006.

Expedition 304/305 Scientists. (2006c). Site U1310. In D. K. Blackman, B. Ildefonse, B. E. John, Y. Ohara, D. J. Miller, C. J. MacLeod, et al. (Eds.), *Proceedings of the integrated ocean drilling program 304/305*. Tokyo: Integrated Ocean Drilling Program Management International, Inc. http://dx.doi.org/10.2204/iodp.proc.304305.104.2006.

Expedition 304/305 Scientists. (2006d). Site U1311. In D. K. Blackman, B. Ildefonse, B. E. John, Y. Ohara, D. J. Miller, C. J. MacLeod, et al. (Eds.), *Proceedings of the integrated ocean drilling program 304/305*. Tokyo: Integrated Ocean Drilling Program Management International, Inc. http://dx.doi.org/10.2204/iodp.proc.304305.105.2006.

Expedition 309/312 Scientists. (2006a). Expedition 309/312 summary. In D. A. H. Teagle, J. C. Alt, S. Umino, S. Miyashita, N. R. Banerjee, D. S. Wilson, et al. (Eds.), *Proceedings of the integrated ocean drilling program 309/312*. Tokyo: Integrated Ocean Drilling Program Management International, Inc. http://dx.doi.org/10.2204/iodp.proc.309312.101.2006.

Expedition 309/312 Scientists. (2006b). Site 1256. In D. A. H. Teagle, J. C. Alt, S. Umino, S. Miyashita, N. R. Banerjee, D. S. Wilson, et al. (Eds.), *Proceedings of the integrated ocean drilling program 309/312*. Tokyo: Integrated Ocean Drilling Program Management International, Inc. http://dx.doi.org/10.2204/iodp.proc.309312.103.2006.

Expedition 327 Scientists. (2011a). Expedition 327 summary. In A. T. Fisher, T. Tsuji, K. Petronotis, & the Expedition 327 Scientists (Eds.), *Proceedings of the integrated ocean drilling program 327*. Tokyo: Integrated Ocean Drilling Program Management International, Inc. http://dx.doi.org/10.2204/iodp.proc.327.101.2011.

Expedition 327 Scientists. (2011b). Site U1301. In A. T. Fisher, T. Tsuji, K. Petronotis, & the Expedition 327 Scientists (Eds.), *Proceedings of the integrated ocean drilling program 327*. Tokyo: Integrated Ocean Drilling Program Management International, Inc. http://dx.doi.org/10.2204/iodp.proc.327.104.2011.

Expedition 327 Scientists. (2011c). Site U1362. In A. T. Fisher, T. Tsuji, K. Petronotis, & the Expedition 327 Scientists (Eds.), *Proceedings of the integrated ocean drilling program 327*. Tokyo: Integrated Ocean Drilling Program Management International, Inc. http://dx.doi.org/10.2204/iodp.proc.327.103.2011.

Expedition 327 Scientists. (2011d). Site U1363. In A. T. Fisher, T. Tsuji, K. Petronotis, & the Expedition 327 Scientists (Eds.), *Proceedings of the integrated ocean drilling program 327*. Tokyo: Integrated Ocean Drilling Program Management International, Inc. http://dx.doi.org/10.2204/iodp.proc.327.106.2011.

Expedition 335 Scientists. (2012). Site 1256. In D. A. H. Teagle, B. Ildefonse, P. Blum, & the Expedition 335 Scientists (Eds.), *Proceedings of the integrated ocean drilling program 335*. Tokyo: Integrated Ocean Drilling Program Management International, Inc. http://dx.doi.org/10.2204/iodp.proc.335.103.2012.

Expedition 336 Scientists. (2012a). Expedition 336 summary. In K. J. Edwards, W. Bach, A. Klaus, & the Expedition 336 Scientists (Eds.), *Proceedings of the integrated ocean drilling program 336*. Tokyo: Integrated Ocean Drilling Program Management International, Inc. http://dx.doi.org/10.2204/iodp.proc.336.101.2012.

Expedition 336 Scientists. (2012b). Site 395. In K. J. Edwards, W. Bach, A. Klaus, & the Expedition 336 Scientists (Eds.), *Proceedings of the integrated ocean drilling program 336*. Tokyo: Integrated Ocean Drilling Program Management International, Inc. http://dx.doi.org/10.2204/iodp.proc.336.103.2012.

Expedition 336 Scientists. (2012c). Site U1382. In K. J. Edwards, W. Bach, A. Klaus, & the Expedition 336 Scientists (Eds.), *Proceedings of the integrated ocean drilling program 336*. Tokyo: Integrated Ocean Drilling Program Management International, Inc. http://dx.doi.org/10.2204/iodp.proc.336.104.2012.

Expedition 336 Scientists. (2012d). Site U1383. In K. J. Edwards, W. Bach, A. Klaus, & the Expedi
tion 336 Scientists (Eds.), *Proceedings of the integrated ocean drilling program 336*. Tokyo:
Integrated Ocean Drilling Program Management International, Inc. http://dx.doi.org/10.2204/
iodp.proc.336.105.2012.

Expedition 340T Scientists. (2012). *Atlantis massif oceanic core complex: Velocity, porosity, and
impedance contrasts within the domal core of atlantis massif: Faults and hydration of lithosphere during core complex evolution, IODP Preliminary Report 340T*. College Station, TX:
Integrated Ocean Drilling Program Management International, Inc. http://dx.doi.org/10.2204/
iodp.pr.340T.2012.

Fisher, A. T. (1998). Permeability within basaltic oceanic crust. *Reviews of Geophysics, 36*(2),
143–182.

Fisher, A. T. (2010). IODP Expedition 321T: cementing operations at Hole U1301A and U1301B,
eastern flank of the Juan de Fuca Ridge Sci. *Drilling, 9*. http://dx.doi.org/10.2204/iodp.
sd.9.02.2010.

Fisher, A. T., Becker, K., & Davis, E. E. (1997). The permeability of young oceanic crust east of
Juan de Fuca Ridge determined using borehole thermal measurements. *Geophysical Research
Letters, 24*, 1311–1314.

Fisher, A. T., Davis, E. E., & Becker, K. (2008). Borehole-to-borehole hydrologic response across
2.4 km in the upper oceanic crust: implications for crustal-scale properties. *Journal of Geophysical Research, 113*, B07106. http://dx.doi.org/10.1029/2007JB005447.

Fisher, A. T., Davis, E. E., Hutnak, M., Spiess, V., Zühlsdorff, L., Cherkaoui, A., et al. (2003).
Hydrothermal recharge and discharge across 50 km guided by seamounts on a young ridge
flank. *Nature, 421*, 618–621.

Fisher, A. T., Stein, C. A., Harris, R. N., Wang, K., Silver, E. A., Pfender, M., et al. (2003). Abrupt
thermal transition reveals hydrothermal boundary and role of seamounts within the Cocos
Plate. *Geophysical Research Letters, 30*(11), 1550. http://dx.doi.org/10.1029/2002GL016766.

Fisher, A. T., Tsuji, T., Petronotis, K., & the Expedition 327 Scientists (2011). *Proceedings of the
integrated ocean drilling program expedition reports 327*. Tokyo: Integrated Ocean Drilling
Program Management International, Inc. http://dx.doi.org/10.2204/iodp.proc.327.2011.

Fisher, A. T., Tsuki, T., Petronotis, K., Wheat, C. G., Becker, K., Clark, J. F., & IODP Expedition
327 Scientific Party, & Atlantis Expedition AT18-07 Scientific Party., et al. (2012). Installing
and servicing borehole crustal observatories to run three-dimensional cross-hole perturbation
and monitoring experiments on the eastern flank of the Juan de Fuca Ridge: IODP Expedition
327 and Atlantis Expedition AT18-07. *Scientific Drilling, 13*. http://dx.doi.org/10.2204/iodp.
sd.13.01.2011.

Fisher, A. T., Urabe, T., Klaus, A., & the Expedition 301 Scientists (2005). *Proceedings of the integrated ocean drilling program 301*. Tokyo: Integrated Ocean Drilling Program Management
International, Inc. http://dx.doi.org/10.2204/iodp.proc.301.2005.

Fisher, A. T., & Wheat, C. G. (2010). Seamounts as conduits for massive fluid, heat, and solute
fluxes on ridge flanks. *Oceanography, 23*(1), 74–87.

Fisher, A. T., Wheat, C. G., Becker, K., Cowen, J., Orcutt, B., Hulme, S., et al. (2011). Design,
deployment, and status of borehole observatory systems used for single-hole and cross-hole
experiments, IODP Expedition 327, eastern flank of the Juan de Fuca Ridge. In A. T. Fisher,
T. Tsuji, K. Petronotis, & the Expedition 327 Scientists (Eds.), *Proceedings of the integrated
ocean drilling program 327*. Tokyo: Integrated Ocean Drilling Program-Management International, Inc. http://dx.doi.org/10.2204/iodp.proc.327.107.2011.

Fisher, A. T., Wheat, C. G., Becker, K., Davis, E. E., Jannasch, H., Schroeder, D., et al. (2005). Scientific and technical design and deployment of long-term, subseafloor observatories for hydro-

geologic and related experiments, IODP Expedition 301, eastern flank of Juan de Fuca Ridge. In A. T. Fisher, T. Urabe, A. Klaus, & the Expedition 301 Scientists (Eds.), *Proceedings of the integrated ocean drilling program 301*. Tokyo: Integrated Ocean Drilling Program Management International, Inc. http://dx.doi.org/10.2204/iodp.proc.301.103.2005.

France, L., Ildefonse, B., & Koepke, J. (2009). Interactions between magma and hydrothermal system in Oman ophiolite and in IODP Hole 1256D: Fossilization of a dynamic melt lens at fast spreading ridges. *Geochemistry, Geophysics, Geosystems, 10*, Q10O19. http://dx.doi.org/10.1029/2009GC002652.

Frost, B. R., Beard, J. S., McCaig, A. M., & Condliffe, E. (2008). The formation of micro-rodingites from IODP Hole U1309D: key to understanding th process of serpentinization. *JP, 49*(9), 1579–1588.

Gable, R., Morin, R. H., & Becker, K. (1992). Geothermal state of DSDP Holes 333A, 395A and 534A: results from the dianaut program. *Geophysical Research Letters, 19*, 505–508.

Gao, Y., Vils, F., Cooper, K. M., Banerjee, N., Harris, M., Hoefs, J., et al. (2012). Downhole variation of lithium and oxygen isotopic compositions of oceanic crust at East Pacific Rise, ODP Site 1256. *Geochemistry, Geophysics, Geosystems, 13*(10), Q10001. http://dx.doi.org/10.1029/2012GC004207.

Gillis, K., & Coogan, L. A. (2011). Secular variation in carbon uptake into the ocean crust. *Earth and Planetary Science Letters, 302*, 385–392.

Guéguen, Y., Gavrilenko, P., & Le Ravalec, M. (1996). Scales of rock permeability. *Surveys in Geophysics, 17*, 245–263.

Haymon, R. M., Fornari, D. J., Edwards, M. H., Carbotte, S., Wright, D., & Macdonald, K. C. (1991). Hydrothermal vent distribution along the East Pacific Rise crest (9° 09'-54' N) and its relationship to magmatic and tectonic processes on fast-spreading mid-ocean ridges. *Earth and Planetary Science Letters, 104*, 513–534.

Heft, K. L., Gillis, K. M., Pollock, M. A., Karson, J. A., & Klein, E. M. (2008). Role of upwelling hydrothermal fluids in the development of alteration patterns at fast spreading ridges: evidence from the sheeted dike complex at Pito Deep. *Geochemistry, Geophysics, Geosystems, 9*, Q05O07. http://dx.doi.org/10.1029/2007GC001926.

Huber, J. A., Butterfield, D. A., Johnson, H. P., & Baross, J. A. (2006). Microbial life in ridge flank crustal fluids. *Environmental Microbiology, 88*, 88–99.

Hutnak, M., & Fisher, A. T. (2007). The influence of sedimentation, local and regional hydrothermal circulation, and thermal rebound on measurements of heat flux from young seafloor. *Journal of Geophysical Research, 112*, B12101. http://dx.doi.org/10.1029/2007JB005022.

Hutnak, M., Fisher, A. T., Harris, R., Stein, C., Wang, K., Spinelli, G., et al. (2008). Large heat and fluid fluxes driven through mid-plate outcrops on ocean crust. *Nature Geoscience, 1*. http://dx.doi.org/10.1038/ngeo264.

Hutnak, M., Fisher, A. T., Zühlsdorff, L., Spiess, V., Stauffer, P., & Gable, C. W. (2006). Hydrothermal recharge and discharge guided by basement outcrops on 0.7-3.6 Ma seafloor east of the Juan de Fuca Ridge: observations and numerical models. *Geochemistry, Geophysics, Geosystems, 7*. http://dx.doi.org/10.1029/2006GC001242.

Ildefonse, B., Blackman, D. K., John, B. E., Ohara, Y., Miller, D. J., & MacLeod, C. J. (2007). Oceanic core complexes and crustal accretion at slow-spreading ridges. *Geology, 35*(7), 623–626.

Johnson, H. P., & Pruis, M. J. (2003). Fluxes of fluid and heat from the oceanic crustal reservoir. *Earth and Planetary Science Letters, 216*, 565–574.

Kelley, D. S., Karson, J. A., Fruh-Green, G. L., Yoerger, D. R., Shank, T. M., Butterfield, D. A., et al. (2005). A serpentinite-hosted ecosystem: the lost city hydrothermal field. *Science, 307*(5714), 1428–1434.

Koepke, J., Christie, D. M., Dziony, W., Holtz, F., Lattard, D., Maclennan, J., et al. (2008). Petrography of the dike-gabbro transition at IODP Site 1256 (equatorial Pacific): the evolution of the granoblastic dikes. *Geochemistry, Geophysics, Geosystems, 9*, Q07O09. http://dx.doi.org/10.1029/2008GC001939.

Koepke, J., France, L., Müller, T., Faure, F., Goetze, N., Dziony, W., et al. (2011). Gabbros from IODP Site 1256, equatorial Pacific: Insight into axial magma chamber processes at fast spreading ocean ridges. *Geochemistry, Geophysics, Geosystems, 12*(9), Q09014. http://dx.doi.org/10.1029/2011GC003655.

Langseth, M. G., Becker, K., Von Herzen, R. P., & Schultheiss, P. (1992). Heat and fluid flux through sediment on the western flank of the Mid-Atlantic Ridge: a hydrogeological study of North Pond. *Geophysical Research Letters, 19*, 517–520.

Langseth, M. G., & Herman, B. (1981). Heat transfer in the oceanic crust of the Brazil Basin. *Journal of Geophysical Research, 86*, 10805–10819.

Langseth, M. G., Hyndman, R. D., Becker, K., Hickman, S. H., & Salisbury, M. H. (1984). The hydrogeological regime of isolated sediment ponds in mid-oceanic ridges. In R. H. Hyndman, & M. H. Salisbury (Eds.), *Init. Repts., DSDP* (pp. 825–837). Washington, D.C.: 78B, U. S. Govt. Printing Office.

Larson, R. L., Fisher, A. T., Jarrard, R., & Becker, K. (1993). Highly permeable and layered Jurassic oceanic crust in the western Pacific. *Earth and Planetary Science Letters, 119*, 71–83.

Lawrence, J. R., & Gieskes, J. M. (1981). Constraints on water transport and alteration in the oceanic crust from the isotopic composition of pore water. *Journal of Geophysical Research, 86*, 7924–7934.

Lesem, L. B., Greytok, F., Marotta, F., & McKetta, J. J., Jr. (1957). A method of calculating the distribution of temperature in flowing gas wells. *Petroleum Transactions of AIME, 210*, 169–176.

Lever, M. A., Rouxel, O., Alt, J. C., Shimizu, N., Ono, S., Coogon, R. M., et al. (2013). Evidence for microbial carbon and sulfur cycling in deeply buried ridge flank basalt. *Science, 339*. (6125). http://dx.doi.org/10.1126/science.1229240.

Lucazeau, F., Bonneville, A., Escartin, J., Von Herzen, R. P., Gouze, P., Carton, H., et al. (2006). Heat flow variations on a slowly accreting ridge: constraints on the hydrothermal and conductive cooling for the Lucky Strike segment (Mid-Atlantic Ridge, 37 N). *Geochemistry, Geophysics, Geosystems, 7*. http://dx.doi.org/10.1029/2005GC001178.

Matthews, M., Salisbury, M., & Hyndman, R. (1984). Basement logging on the Mid-Atlantic Ridge, deep sea drilling project hole 395A. In R. D. Hyndman, & M. H. Salisbury (Eds.), *Init. Repts., DSDP* (pp. 717–730). Washington, D.C.: 78B, U. S. Govt. Printing Office.

McCaig, A. M., Delacour, A. G., Fallick, A. E., Castelain, T., & Früh-Green, G. L. (2010). Detachment fault control on hydrothermal circulation systems: Interpreting the subsurface beneath the TAG hydrothermal field using the isotopic and geological evolution of oceanic core complexes in the Atlantic. In P. A. Rona, C. W. Devey, J. Dyment, & B. Murton (Eds.), *Diversity of hydrothermal systems on slow spreading ocean ridges* (pp. 207–240). Washington DC: AGU Geophysical Monograph 108, American Geophysical Union.

McCaig, A. M., & Harris, A. C. (2012). Hydrothermal circulation and the dike-gabbro transition in the detachment mode of slow seafloor spreading. *Geology, 40*, 367–370.

Moos, D. (1990). Petrophysical results from logging in DSDP Hole 395A, ODP Leg 109. In R. Detrick, & J. Honnorez (Eds.), *Proceedings of the ocean drilling program, scientific results 106/109* (pp. 237–253). College Station, TX: Ocean Drilling Program.

Morin, R. H., Hess, A. E., & Becker, K. (1992). In situ measurements of fluid flow in DSDP Holes 395A and 534A: results from the Dianaut program. *Geophysical Research Letters, 19*, 509–512.

Morin, R. H., Moos, D., & Hess, A. E. (1992). Analysis of the borehole televiewer log from DSDP 395A: results from the Dianaut program. *Geophysical Research Letters, 19*, 501–504.

Morris, A., Gee, J. S., Pressling, N., John, B. E., MacLeod, C. J., Grimes, C. B., et al. (2009). Footwall rotation in an oceanic core complex quantified using reoriented Integrated Ocean Drilling Program core samples. *Earth and Planetary Science Letters, 287*, 217–228.

Mottl, M. J. (1989). Hydrothermal convection, reaction and diffusion in sediments on the Costa Rica Rift flank, pore water evidence from ODP Sites 677 and 678. In K. Becker, & H. Sakai (Eds.), *Proceedings of the ocean drilling program, scientific results 111* (pp. 195–214). College Station, TX: Ocean Drilling Program.

Mottl, M. J. (2003). Partitioning of energy and mass fluxes between mid-ocean ridge axes and flanks at high and low temperature. In P. Halbach, V. Tunnicliffe, & J. Hein (Eds.), *Energy and mass transfer in submarine hydrothermal systems* (pp. 271–286). Berlin, Germany: DWR 89, Dahlem University Press.

Nozaka, T., & Fryer, P. (2011). Alteration of the oceanic lower crust at a slow-spreading axis: Insight from vein-related zoned halos in olivine gabbro from Atlantis Massif, Mid-Atlantic Ridge. *Journal of Petroleum, 52*, 643–664. http://dx.doi.org/10.1093/petrology/egq098.

Nozaka, T., Fryer, P., & Andreani, M. (2008). Formation of clay minerals and exhumation of lower-crustal rocks at Atlantis Massif, Mid-Atlantic Ridge. *Geochemistry, Geophysics, Geosystems, 9*(11), Q11005. http://dx.doi.org/10.1029/2008GC002207.

Ono, S., Keller, N., Rouxel, O., & Alt, J. C. (2012). Sulfur-33 constraints on the origin of secondary pyrite in altered ocean basement rocks. *Geochimica Cosmochimica et Acta, 87*, 323–340.

Orcutt, B. N., Wheat, C. G., Rouxel, O., Hulme, S., Edwards, K. J., & Bach, W. (2013). Oxygen consumption rates in subseafloor basaltic crust derived from a reaction transport model. *Nature Communications, 4*. http://dx.doi.org/10.1038/ncomms3539.

Parsons, B., & Sclater, J. G. (1977). An analysis of the variation of ocean floor bathymetry and heat flow with age. *Journal of Geophysical Research, 82*, 803–829.

Rausch, S., Böhm, F., Bach, W., Klügel, A., & Eisenhauer, A. (2013). Calcium carbonate veins in ocean crust record increases in seawater Mg/Ca and Sr/Ca within the past 30 million years. *Earth and Planetary Science Letters, 362*, 215–224. http://dx.doi.org/10.1016/j.epsl.2012.12.005.

Rouxel, O., Shanks, W. C. I., Bach, W., & Edwards, K. J. (2008). Integrated Fe and S isotope study of seafloor hydrothermal vents at East Pacific Rise 9-10°N. *Chem Geol, 252*, 214–227.

Ryan, W. B. F., Carbotte, S. M., Coplan, J. O., O'Hara, S., Melkonian, A., Arko, R., et al. (2009). Global multi-resolution topography synthesis. *Geochemistry, Geophysics, Geosystems, 10*(3), Q03014. http://dx.doi.org/10.1029/2008GC002332.

Schoolmeesters, N., Cheadle, M. J., John, B. E., Reiners, P. W., Gee, J., & Grimes, C. B. (2012). The cooling history and the depth of the detachment faulting at the Atlantis Massif oceanic core complex. *Geochemistry, Geophysics, Geosystems, 13*, 1–19.

Shipboard Scientific Party. (1979). Site 395. In W. G. Melson, & P. D. Rabinowitz (Eds.), *Deep sea drilling project, initial reports 45* (pp. 131–264). Washington, D.C: U. S. Govt. Printing Office.

Shipboard Scientific Party. (1997a). Introduction and summary: hydrothermal circulation in the oceanic crust and its concequences on the eastern flank of the Juan de Fuca Ridge. In E. E. Davis, A. T. Fisher, & J. Firth (Eds.), *Proceedings of the ocean drilling program, initial reports 168* (pp. 7–21). College Station, TX: Ocean Drilling Program.

Shipboard Scientific Party. (1997b). Rough basement transect (Sites 1026 and 1027). In E. E. Davis, A. T. Fisher, & J. Firth (Eds.), *Proceedings of the ocean drilling program, initial reports 168* (pp. 101–160). College Station, TX: Ocean Drilling Program.

Shipboard Scientific Party. (2003). Site 1256. In D. S. Wilson, D. A. H. Teagle, & G. D. Acton (Eds.), *Proceedings of the ocean drilling program, initial reports 206*. College Station, TX, USA: Ocean Drilling Program. http://dx.doi.org/10.2973/odp.proc.ir.206.103.2003.

Sohn, R. A., Webb, S. C., Hildebrand, J. A., & Cornuelle, B. C. (1997). Three-dimensional tomographic velocity structure of upper crust, CoAxial segment, Juan de Fuca Ridge: implications for on-axis evolution and hydrothermal circulation. *Journal of Geophysical Research, 102*, 17,679–617,695.

Spinelli, G. A., & Fisher, A. T. (2004). Hydrothermal circulation within rough basement on the Juan de Fuca Ridge flank. *Geochemistry, Geophysics, Geosystems, 5*(2), Q02001. http://dx.doi.org/10.1029/2003GC000616.

Stein, C. A., Stein, S., & Pelayo, A. M. (1995). Heat flow and hydrothermal circulation. In S. E. Humphris, R. A. Zierenberg, L. S. Mullineaux, & R. E. Thompson (Eds.), *Seafloor hydrothermal Systems: Physical, chemical, biological and geological interactions Geophysical monograph series: Vol. 91.* (pp. 425–445). Washington, D.C.: American Geophysical Union.

Stein, J. S., & Fisher, A. T. (2003). Observations and models of lateral hydrothermal circulation on a young ridge flank: numerical evaluation of thermal and chemical constraints. *Geochemistry, Geophysics, Geosystems, 4*. http://dx.doi.org/10.1029/2002GC000415.

Stephen, R. (1981). Seismic anisotropy observed upper oceanic crust. *Geophysical Research Letters, 8*, 865–868.

Swift, S. A., Reichow, M., Tikku, A., Tominaga, M., & Gilbert, L. (2008). Velocity structure of upper ocean crust at Ocean Drilling Program Site 1256. *Geochemistry, Geophysics, Geosystems, 9*(10), Q10O13. http://dx.doi.org/10.1029/2008GC002188.

Teagle, D. A. H., & Wilson, D. S. (2007). Leg 206 Synthesis: Initiation of drilling an intact section of upper oceanic crust formed at a superfast spreading rate at Site 1256 in the eastern equatorial Pacific. In D. S. Wilson, D. A. H. Teagle, G. D. Acton, & D. A. Vanko (Eds.), *Proceedings of the ocean drilling program, initial reports 206* (pp. 1–7). College Station, TX, USA: Ocean Drilling Program. http://dx.doi.org/10.2973/odp.proc.sr.206.001.2007.

Tominaga, M., Teagle, D. A. H., Alt, J. C., & Umino, S. (2009). Determination of the volcanostratigraphy of oceanic crust formed at superfast spreading ridge: electrofacies analyses of ODP/IODP Hole 1256D. *Geochemistry, Geophysics, Geosystems, 10.* (9). http://dx.doi.org/10.102 9/2008GC002143.

Villinger, H., Grevemeyer, I., Kaul, N., Hauschild, J., & Pfender, M. (2002). Hydrothermal heat flux through aged oceanic crust: where does the heat escape? *Earth and Planetary Science Letters, 202*(1), 159–170.

Walker, B. D., McCarthy, M. D., Fisher, A. T., & Guilderson, T. P. (2007). Dissolved inorganic carbon isotopic composition of low-temperature axial and ridge-flank hydrothermal fluids of the Juan de Fuca Ridge. *Marine Chemistry, 108*. http://dx.doi.org/10.1016/j.marchem.2007.11.002.

Wang, K., He, J., & Davis, E. E. (1997). Influence of basement topography on hydrothermal circulation in sediment-buried oceanic crust. *Earth and Planetary Science Letters, 146*, 151–164.

Wheat, C. G., Edwards, K. J., Pettigrew, T., Jannasch, H. W., Becker, K., Davis, E. E., et al. (2012). CORK-Lite: Bringing legacy boreholes back to life. *Scientific Drilling, 14*. http://dx.doi.org/10.2204/iodp.sd.2214.2205.2012.

Wheat, C. G., Elderfield, H., Mottl, M. J., & Monnin, C. (2000). Chemical composition of basement fluids within an oceanic ridge flank: implications for along-strike and across-strike hydrothermal circulation. *Journal of Geophysical Research, 105*(B6), 13437–13447.

Wheat, C. G., Jannasch, H., Fisher, A. T., Becker, K., Sharkey, J., & Hulme, S. M. (2010). Subsea-floor seawater-basalt-microbe reactions: continuous sampling of borehole fluids in a ridge flank environment. *Geochemistry, Geophysics, Geosystems*, Q07011. http://dx.doi.org/10.1029/2010GC00305.

Wheat, C. G., Jannasch, H., Kastner, M., Hulme, S., Cowen, J., Edwards, K., et al. (2011). Fluid sampling from oceanic borehole observatories: design and methods for CORK activities (1990-2010). In A. T. Fisher, T. Tsuji, K. Petronotis, & the Expedition 327 Scientists (Eds.), *Proceedings of the integrated ocean drilling program 327*. Tokyo: Integrated Ocean Drilling Program Management International, Inc. http://dx.doi.org/10.2204/iodp.proc.327.109.2011.

Wheat, C. G., McManus, J., Mottl, M., & Giambalvo, E. G. (2003). Oceanic phosphorus imbalance: magnitude of the mid-ocean ridge flank hydrothermal sink. *Geophysical Research Letters, 30*. http://dx.doi.org/10.1029/2003GL017318.

Wheat, C. G., Mottl, M. J., Fisher, A. T., Kadko, D., Davis, E. E., & Baker, E. (2004). Heat and fluid flow through a basaltic outcrop on a ridge flank. *Geochemistry, Geophysics, Geosystems, 5*. (12). http://dx.doi.org/10.1029/2004GC000700.

Wilcock, W. S. D., & Fisher, A. T. (2004). Geophysical constraints on the sub-seafloor environment near mid-ocean ridges. In W. S. D. Wilcock, E. Delong, D. Kelley, J. A. Baross, & S. C. Cary (Eds.), *Subseafloor biosphere at mid-ocean ridges* (pp. 51–74). Washington, D.C: AGU Geophysical Monograph 144, American Geophysical Union. http://dx.doi.org/10.1029/144GM05.

Winslow, D. M., Fisher, A. T., & Becker, K. (2013). Characterizing borehole fluid flow and formation permeability in the ocean crust using linked analytic models and Markov Chain Monte Carlo analysis. *Geochemistry, Geophysics, Geosystems, 14*. (9). http://dx.doi.org/10.1002/ggge.20241.

Ziebis, W., McManus, J., Ferdelman, T., Schmidt-Schierhorn, F., Bach, W., Muratli, J., et al. (2012). Interstitial fluid chemistry of sediments underlying the North Atlantic gyre and the influence of subsurface fluid flow. *Earth and Planetary Science Letters, 323-324*, 79–91.

Zühlsdorff, L., Hutnak, M., Fisher, A. T., Spiess, V., Davis, E. E., Nedimovic, M., et al. (2005). Site Surveys related to IODP Expedition 301: ImageFlux (SO149) and RetroFlux (TN116) expeditions and earlier studies. In A. T. Fisher, T. Urabe, & A. Klaus (Eds.), *Proceedings of the integrated ocean drilling program 301*. Tokyo: Integrated Ocean Drilling Program Management International, Inc. http://dx.doi.org/10.2204/iodp.proc.301.102.2005.

Chapter 4.3

Large-Scale and Long-Term Volcanism on Oceanic Lithosphere

Anthony A.P. Koppers[1],* and William W. Sager[2]
[1]*College of Earth, Ocean & Atmospheric Sciences, Oregon State University, Corvallis, OR, USA;*
[2]*Department of Earth and Atmospheric Sciences, University of Houston, Houston, TX, USA*
Corresponding author: E-mail: akoppers@coas.oregonstate.edu

4.3.1 INTRODUCTION

Intraplate basaltic volcanism includes all volcanic activity away from tectonic plate boundaries both on the continents and in the oceans. This type of volcanism stands apart from volcanism that forms oceanic crust senso stricto and from volcanism that results in volcanic arcs above subduction zones. It may include volcanism at midocean spreading centers, if magma production exceeds that of the normal plate formation processes, such as observed in Iceland. It may also include highly magmatic divergent margins upon the opening of large ocean basins. Below the sea surface, hundreds of thousands of seamounts and a small number of larger oceanic plateaus reside on the ocean floor (Wessel, Sandwell, & Kim, 2010). Together, these volcanic structures formed on top of the oceanic lithosphere and cover a surface area equivalent in size to Europe and Asia combined. Considering the large number of intraplate volcanoes and their globally scattered distribution beneath the oceans, it is no surprise that our knowledge about their evolution remains rudimentary.

The concept of intraplate volcanism was introduced as an adjunct to the theory of plate tectonics. These volcanoes require other explanation(s) than seafloor spreading and subduction, such as volcanism resulting from upwelling mantle plumes (Koppers & Watts, 2010) or leakage through lithosphere at extensional zones and along fractures (Foulger, 2007; Natland & Winterer, 2005). Geochronological studies of the volcanic islands of Hawaii provided the first clues of what may cause intraplate volcanism. K/Ar dating of the basaltic lava flows (McDougall, 1964) indicated a systematic (linear) increase in the age of

Developments in Marine Geology, Volume 7. http://dx.doi.org/10.1016/B978-0-444-62617-2.00019-0

the volcanism in a northwesterly direction away from the active volcanoes on the Big Island of Hawaii, evidently a locus or "hotspot" of active intraplate volcanism (Wilson, 1963, Wilson, 1965). This kind of age-progressive volcanism has become the hallmark of hotspot seamount trails and was subsequently used to model the past motions of tectonic plates with respect to a reference frame of deep-seated, stationary mantle plumes (Morgan, 1972, Morgan, 1981). This model was extended to include the formation of large igneous provinces (LIPs) from the head of a nascent mantle plume emerging beneath the lithospheric lid, causing "catastrophic" intraplate volcanism following large amounts of partial melting over relatively short geological intervals (Coffin & Eldholm, 1994; Duncan & Richards, 1991). In this model, it also follows that the later arrival of the mantle plume "tails" then result in the formation of narrow seamount trails (e.g., Campbell, 2006; Campbell & Kerr, 2007).

Although earlier work on intraplate volcanism in the 1950–1980s focused on subaerial portions of a few volcanic island groups, later advances in seafloor mapping (through the introduction of multibeam sonar) and satellite altimetry allowed scientists to grasp both the global scale and high volume of intraplate volcanism from the expansive presence of seamounts and LIPs on the ocean floor (Figure 4.3.1). The vast majority of these seafloor features are of a volcanic origin (Staudigel & Clague, 2010), with seamounts alone combining to an estimated total volume of $4.4 \times 10^7 \, \text{km}^3$ (Figure 4.3.2). They span in age from historical and currently active to only about 170 million years ago (Ma), because the older seamounts and oceanic LIPs have been subducted and are no longer part of the geological record. Interestingly, the oldest seamount dated today is Look seamount in the Western Pacific ocean at 140 Ma (Koppers, Staudigel, Pringle, & Wijbrans, 2003), and the oldest LIP is Shatsky Rise dated at 146 Ma (Mahoney, Duncan, Tejada, Sager, & Bralower, 2005), meaning we have not sampled and studied the oldest 25 million years of intraplate volcanism yet.

During the past two decades of scientific ocean drilling, a better understanding of the overall history of intraplate volcanism has been gained, from increased mapping and dredging campaigns and 13 scientific ocean drilling expeditions. The improved mapping of the ocean floor has allowed us to recognize that some seamount trails were split onto two or more tectonic plates (e.g., Easter and Salas y Gómez Ridge, Walvis Ridge, and Rio Grande Rise) and that many seamount trails are in fact relatively short lived, not representing more than approximately 30 million years of volcanism (Koppers et al., 2003). Moreover, it has been recognized that seafloor volcanism is uneven in time, with focused periods of more intense intraplate volcanism, in particular, around 120 Ma, and more recently with a high abundance of active hotspot trails (0–30 Ma) throughout the oceans (Figure 4.3.2). A few seamount trails, however, are exceptionally long lived and have well-developed morphologies, and two of those unique "primary" hotspot systems, Hawaii and Louisville, have received the most attention during scientific ocean drilling. On the other hand, Ontong Java Plateau, the Kerguelen Plateau, and the Shatsky Rise have been the main focus of drilling into oceanic LIPs.

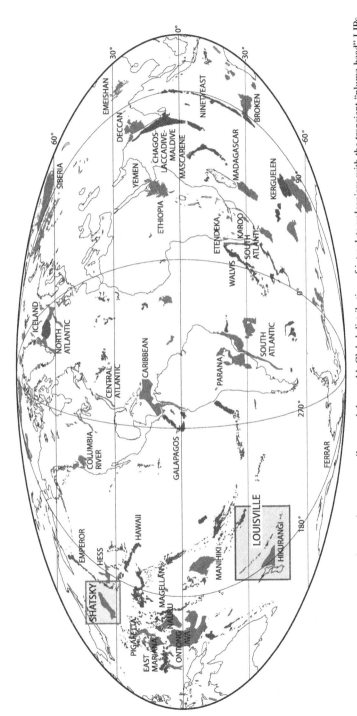

FIGURE 4.3.1 Large-Igneous provinces and seamount trails around the world. Global distribution of intraplate volcanism with the transient "plume head" LIPs and persistent "plume tail" seamount trails indicated in red (light gray in print versions) and blue (dark gray in print versions), respectively. Shatsky Rise in the NW Pacific Ocean and the Louisville seamount trail in the SW Pacific are highlighted. *Figure from Coffin et al. (2006).*

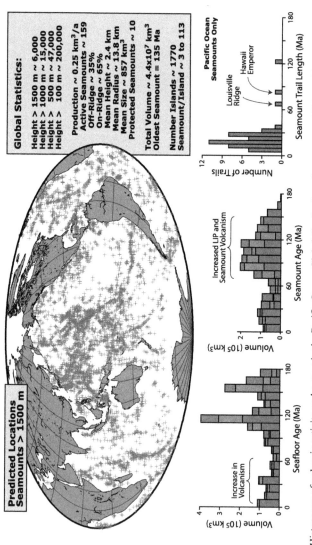

FIGURE 4.3.2 Histograms of volcanic activity and output in the Pacific Ocean. Global distribution of seamounts >1500 m with some general seamount statistics (based on data from Koppers et al., 2003; Watts, Sandwell, Smith, & Wessel, 2006; Hillier, 2007; Hillier & Watts, 2007) but excluding oceanic plateaus.

The Integrated Ocean Drilling Program (IODP) Initial Science Plan (IODP, 2003–2013) laid out a broad vision for a greatly upgraded program of scientific drilling building on the successes of the preceding Deep Sea Drilling Program (DSDP) and Ocean Drilling Program (ODP). It envisioned a three-platform program including the traditional nonriser US drilling vessel, D/V *JOIDES Resolution*, as well as new capabilities offered by the Japan riser drilling vessel, D/V *Chikyu*, and European mission specific platforms for environments inaccessible to the two larger drilling ships. Intraplate volcanism was included within the "Solid Earth Cycles and Geodynamics" theme, which sought continued progress in understanding the cycling of magma and energy from the Earth's mantle to surface. Intraplate volcanism generating both seamounts and LIPs, was recognized as a signal of poorly understood "dynamic" mantle processes that coexist with the steady-state plate tectonic process of lithosphere creation and destruction. Further, it was recognized that large-scale magmatism, especially in LIPs, may profoundly affect the Earth's environment and surface geological processes, yet the details of this coupling with the mantle are unclear.

During IODP, only two seagoing programs investigated fundamental questions about the nature and behavior of intraplate volcanism as occurring on the oceanic portions of Earth's tectonic plates. Expedition 324 examined the formation of the Shatsky Rise oceanic plateau in the western Pacific and was envisioned as a direct test of whether it was formed by a mantle plume or surface processes (Sager, Sano, Geldmacher, & Expedition 324 Scientists, 2010; Sager, Sano, & Geldmacher, 2011). Expedition 330 cored the Louisville Seamount Trail, one of the two primary hotspot trails on the Pacific plate, mainly as a paleomagnetic test of mantle plume motions and also to better understand the formation of hotspot trails (Koppers, Yamazaki, Geldmacher, & Expedition 330 Scientists, 2012; Koppers, Yamazaki, Geldmacher, & Expedition 330 Scientists, 2013). Given that these two expeditions were the primary focus of intraplate volcanism research during IODP, we focus here on the results from these two efforts. By doing so, we provide a history and review of our advancements in understanding "marine" intraplate volcanism and the mantle processes governing their causes, notwithstanding the significant progress in our understanding of continental flood volcanism (e.g., Bryan et al., 2010; Jerram & Widdowson, 2005).

4.3.2 HISTORY OF DRILLING LIPs AND HOTSPOT TRAILS DURING DSDP AND ODP

4.3.2.1 DSDP

The drilling of submarine volcanic features has evolved with scientific drilling since 1969. During the DSDP (1968–1983) coring of seamounts and oceanic plateaus was at first not overtly hypothesis driven but rather exploratory. Especially during the early years of the program, when geoscientists were examining the fundamental nature of seafloor features, the D/V *Glomar*

Challenger probed many seafloor features to find out what they were made of. Submarine highs were popular targets, and many seamounts and oceanic plateaus were cored, although with minimum penetration depths into the basaltic basement. A typical DSDP "basement" hole might touch the igneous surface and recover a small amount of basalt or related rocks, or core a few meters into the igneous pile (Table 4.3.1). In this manner, it was discovered that seamounts and oceanic plateaus (rises) mostly have foundations of basaltic rock. The first coring of seamount rocks occurred on DSDP Leg 6 at Site 57 on the Caroline Ridge in the western Pacific (Fischer & Heezen, 1971), where igneous basement was cored for approximately 8 m and in total 5.5 m of rock was recovered (Shipboard Scientific Party, 1971). With a similar approach and results, Legs 16, 17, and 19 cored across the Pacific basin, sampling hotspot ridges on the Cocos plate (Leg 16, van Andel & Heath, 1973), Horizon Guyot at the north end of the Line Islands seamount trail (Leg 17, Winterer, 1973), and Meiji Guyot at the northern terminus of the Emperor seamount trail (Leg 19, Scholl & Creager, 1973).

Attempts to drill into igneous basement during the early years of DSDP were often frustrated by (still) primitive drilling technologies, which frequently failed when trying to drill through thick or strongly indurated sedimentary units lying atop the basement. A large percentage of holes drilled on the top of seamounts and plateaus thus sampled only sediments. An example of such frustration was the coring of Shatsky Rise on DSDP Leg 32, where deep holes were drilled at Sites 305 and 306, penetrating 641 and 381 m into the sediment cap, respectively. Both failed to reach the igneous basement because of bit failure brought on by having to drill through hard chert layers and having no way to reenter the hole after replacing the bit (Shipboard Scientific Party, 1975a,b).

As DSDP progressed, projects became more hypothesis driven, and linear hotspot trails and ridges were a particular target in the wake of Morgan's (1971) mantle plume hypothesis. Ninetyeast Ridge was cored on Leg 22, the Emperor Seamounts on Leg 55, and Walvis Ridge on Legs 74 and 75. With these expeditions, the substantial technical improvement during DSDP drilling is evident. Leg 22 drilled only 14 and 15 m into the igneous basement at two sites, whereas both Legs 55 and 74 cored several hundreds of meters into the basaltic basements of these seamounts (Jackson et al., 1980; Moore, Rabinowitz, Borella, Shackleton, & Boersma, 1984). Leg 55 produced the notable result of coring

TABLE 4.3.1 Large-Igneous Provinces, Seamount Trails, and Magmatic Divergent Margins Drilled During DSDP, ODP, and IODP

All drill sites are compiled, and the total cored section are given for DSDP and ODP, whereas the total cored section into igneous basement is given for the IODP Expedition 324 to Shatsky Rise and Expedition 330 to the Louisville seamount trail.

385 m into the volcanic pedestal at Site 433 on Suiko Guyot, recovering 65 lava flows and producing the most precise paleomagnetic paleolatitude from ocean drilling at the time or since (Kono, 1980).

Oceanic plateaus were cored in several settings during DSDP, but not in any systematic fashion. Most holes were single penetrations into a particular feature, and sometimes, this occurred because of the aforementioned problem with penetrating through thick sedimentary caps and reaching into basement. During DSDP, Magellan Rise was cored on Leg 17 (Site 167, Winterer, 1973), Ontong Java Plateau on Leg 30 (Site 289, Shipboard Scientific Party, 1975c), Benham Rise on Leg 31 (Site 292, Shipboard Scientific Party, 1975d), Manihiki Plateau on Leg 33 (Site 317, Jackson & Schlanger, 1976), and Rio Grande Rise on Leg 72 (Site 516, Barker, 1983). Geochemical data from these holes demonstrated the similarity of plateau basalts to midocean ridge basalts and thus contrasting the ocean island basalts (OIBs; Bass, Moberly, Rhodes, Shih, & Church, 1973; Houghton, 1979; Jackson, Bargar, Fabbi, & Heropoulis, 1976; Kirkpatrick, Clague, & Freisen, 1980; Richardson, Erlank, Reid, & Duncan, 1984; Weaver, Marsh, & Tarney, 1983).

By the end of the DSDP, the igneous basement of approximately two dozen seamounts and plateaus had been cored, of which approximately 60% were located in the Pacific Ocean. Notably, neither seamounts nor oceanic plateaus were targeted in the final 22 legs (>3 years) of the program, reflecting a need for new hypotheses and rationales for targeting these features as scientific drilling progressed from exploratory to hypothesis-driven research. With the transition to ODP (1985–2003), many of the same igneous features would be cored, but with improved technology and greater focus, while addressing new paradigms.

4.3.2.2 ODP

During ODP, seamount and plateau coring began with a continuation of programs (Table 4.3.1) similar to those of the DSDP. Leg 115 (Mascarene Plateau-Chagos-Laccadive Ridge) and Leg 122 (Ninetyeast Ridge) cored Indian Ocean hotspot ridges, testing the hotspot hypothesis in another ocean than the Pacific and finding the expected age progressions (Duncan, 1990; Royer, Peirce, & Weissel, 1991). Later in the program, the focus on hotspot trails would shift, with evidence that hotspots were not fixed in the mantle, as suggested by Morgan (1971). Basalt samples drilled during Leg 145 from Detroit Guyot (Rea et al., 1995) near the north end of the Emperor seamount trail, showed a larger departure from the current latitude of the Hawaiian hotspot than those from Suiko Guyot, cored during DSDP Leg 55 (Tarduno & Cottrell, 1997). Although the Suiko Guyot datum showed a distinct departure from the Hawaiian hotspot latitude, the idea of "fixed" hotspots was entrenched, so its effect was muted. By the time the Detroit Guyot data publication, evidence from independent studies suggested hotspot mobility, so the stage was set to test this new view.

Leg 197 cored three guyots of the Emperor seamount trail, including a 453-m penetration into Detroit Guyot (Shipboard Scientific Party, 2002a), showing conclusively that the Hawaiian hotspot moved southward rapidly during the Late Cretaceous and early Cenozoic (Tarduno et al., 2003). These results were instrumental in dislodging the prior "fixist" view, leading to wider acceptance of mobile hotspots.

ODP coring also evolved the next steps in oceanic plateau research. Early in the program, expeditions were planned to address the origin and evolution of the two largest oceanic plateaus Kerguelen Plateau (Legs 119 and 120; Shipboard Scientific Party, 1989a,b) and Ontong Java Plateau (Leg 130, Berger, Kroenke, Mayer, & Shipboard Scientific Party, 1991). In both cases, two to three sites penetrated into igneous basement for several hundreds of meters, reflecting the recognition that more sites and greater penetration were required compared to those obtained during DSDP. During the early 1990s, a new hypothesis for oceanic plateau formation was generating excitement. It posited that the eruption of oceanic plateau basalts were cousins to continental flood basalts and the result of the impact on the lithosphere by the head of a nascent mantle plume that had risen from the core–mantle boundary (Coffin & Eldholm, 1994; Duncan & Richards, 1991; Richards, Duncan, & Courtillot, 1989). This hypothesis seemed an elegant explanation for the massive eruptions that must have formed oceanic plateaus, so it was readily accepted and tested on two subsequent ODP cruises, Leg 183 to Kerguelen Plateau (Shipboard Scientific Party, 2000) and Leg 192 to Ontong Java Plateau (Shipboard Scientific Party, 2001). Both expeditions gave important insights into oceanic plateau development but also yielded complications that did not fit the simple plume head theory. Evidence of effusive eruptions similar to continental flood basalts were recovered in the Kerguelen Plateau that is characterized by thick sequences of subaerial compound pahoehoe flows, whereas the Ontong Java Plateau is characterized by more massive sheet flows (Self, Thordardon, & Keszthelyi, 1997) up to tens of meters in thickness. However, the Kerguelen Plateau exhibited a long period of peak output of approximately 25 Myr (Coffin et al., 2002) instead of the expected rapid outpouring of 1–2 Myr (Coffin & Eldholm, 1994). Moreover, rocks with continental affinity were cored at Site 1137 (Shipboard Scientific Party, 2000), an unexpected result implying interaction with lithosphere from microcontinents. Although samples cored from Ontong Java Plateau fit the plume head hypothesis better by providing evidence for an overall short eruption period, high degrees (~30%) of partial melting and the occurrence of some units with high-MgO lava flows, a confounding finding was that almost all lavas were erupted in submarine environments, precluding the large uplift and associated expansive subaerial volcanism expected from a thermal plume (Fitton & Godard 2004). As a result, investigators started to entertain more complex plume models (e.g., Korenaga, 2005) or to seek out other mechanisms, including bolide impacts (e.g., Ingle and Coffin, 2004).

Near the end of the ODP, a paleoceanography-focused cruise took time to core igneous basement for the first time on Shatsky Rise at Site 1213 (Shipboard Scientific Party, 2002b), setting the foundation for the next step of coring oceanic plateaus during IODP. The Site 1213 cores produced a radiometric date of 144.6 ± 0.8 Ma (Mahoney et al., 2005) for the southwestern flank of Tamu Massif, the oldest edifice of Shatsky Rise, which proved that it formed in a near-ridge environment at a triple junction. Further, isotopic characteristics showed that the cored basalts were similar to midocean ridge basalts (Mahoney et al., 2005) and unlike the expected lower mantle material that was expected to be entrained by a rising plume. As a result, Shatsky Rise has the characteristics of both a mantle plume and plate-edge midocean ridge magmatism (Sager, 2005), setting up an argument that Shatsky Rise is a good location to test these two opposing models for the formation of oceanic plateaus in general.

ODP also included two expeditions dedicated to coring seamounts, Legs 143 and 144. Dubbed "Atolls and Guyots" these back-to-back expeditions set out to test the hypothesis that guyot sediment caps record a history of sea level change and to learn why these former atolls drowned when neighboring atolls maintained themselves at sea level. Leg 143 focused on two guyots of the Mid-Pacific Mountains, whereas Leg 144 visited five guyots in the Marshall Islands and other seamount groups farther northwest. Both expeditions demonstrated that Cretaceous atolls had a different and shallower structure than do present-day atolls and showed that the record of sea level recorded in carbonate banks was complex with periods of exposure as well as inundation (Haggerty & Premoli Silva, 1995; Winterer & Sager, 1995). In addition, Leg 144 results implicated passage of these western Pacific atolls beneath the equatorial high-productivity zone inhibited coral growth, allowing the banks to drown (Haggerty & Premoli Silva, 1995). Legs 143 and 144 also added substantially to studies of seamount geochemistry, geochronology, and tectonics (Baker, Castillo, & Condliffe, 1995; Christie, Dieu, & Gee, 1995; Janney & Baker, 1995; Koppers, Staudigel, Christie, Dieu, & Pringle, 1995; Nakanishi and Gee, 1995; Pringle & Duncan, 1995a,b; Tarduno & Sager, 1995).

4.3.3 IODP EXPEDITION 324 TO THE SHATSKY RISE

4.3.3.1 Rationale and Objectives: A Shift in Focus

After assimilating results from drilling oceanic plateaus during ODP, the focus shifted with IODP. During ODP, the Kerguelen and Ontong Java Plateaus were each cored by two to three cruises with the goal of testing the plume head hypothesis (Coffin & Eldholm, 1994; Coffin et al., 2006). By the time of IODP, debate was ongoing about the existence of plumes and their role in mantle convection (Anderson, 2001; Anderson, Tanimoto, & Zhang, 1992), and an alternative mechanism was proposed for plateau formation. It was suggested that excess volcanism from the mantle could also be caused by decompression melting of fertile mantle, which is material with a lower melting point than normal mantle (Foulger, 2007). A crack or plate boundary was required to cause the decompression, suggesting a link to lithospheric extension found at plate edges or boundaries such as midocean spreading centers and triple junctions.

As many oceanic plateaus, including the giants Kerguelen and Ontong Java, were formed during the Cretaceous Normal Superchron (CNS; a.k.a. Cretaceous Quiet Period, ~125–83 Ma) when there were no magnetic reversals that could be recorded by seafloor spreading, it is difficult to determine whether spreading ridges played any role in their formation. Shatsky Rise, a large oceanic plateau that formed at a spreading-ridge triple junction during the Late Jurassic and Early Cretaceous (Sager, Handschumacher, Hilde, & Bracey, 1988), a time with magnetic reversals, provided the opportunity to make this connection (Figure 4.3.3). What is more, Shatsky Rise displayed characteristics that are supportive of both the plume and plate-edge hypotheses. Based on morphology, it appeared that initial Shatsky Rise eruptions rapidly formed Tamu Massif, the largest volcanic edifice of the plateau in a short period of time (Sager & Han, 1993) followed by the expected transition from a plume head (Tamu Massif) to a plume tail as the edifices (Ori Massif, Shirshov Massif, and Papanin Ridge) become smaller with time (Sager, Kim, Klaus, Nakanishi, & Khankishieva, 1999). Despite the morphologic support for the plume head model, other aspects did not fit. Tamu Massif formed close to a triple junction, and the odds of this coincidence are small, if there is no connection. This coincidence is seen with other Pacific oceanic plateaus as well (Sager, 2005). Shatsky Rise igneous rocks also had isotopic characteristics similar to Mid-Ocean Ridge Basalt (MORB) rather than signatures that could be traced to the lower mantle as expected for a deep plume (Mahoney et al., 2005). Thus, Expedition 324 to the Shatsky Rise was conceived as a test between plume and plate-edge mechanisms for oceanic plateau formation.

The primary objective of Expedition 324 was to core the igneous basement of the three main Shatsky Rise massifs. Although dredging had previously obtained samples from Shatsky Rise, most of these rocks are highly altered by long exposure to seawater, making geochronology and geochemical studies difficult. Prior drilling had penetrated igneous basement at only one spot, Site 1213, cored during ODP Leg 198 (Shipboard Scientific Party, 2002b), so both the amount and distribution of igneous samples were limited. The vision for Expedition 324 was to core at multiple locations, recovering sections of basal sediments and igneous rock to gain an improved understanding of plateau evolution. Igneous rocks were to be studied for geochemical and isotopic characteristics and geochronology, the former seeking to understand where in the mantle and in what conditions the volcanism arose and the latter to determine the timing of Shatsky Rise formation. Studies of physical properties and volcanologic characteristics of core samples along with downhole logging data would give a picture of the structure of the volcanic edifices. Basal sediments would tell of the water depth and environment of the eruptions.

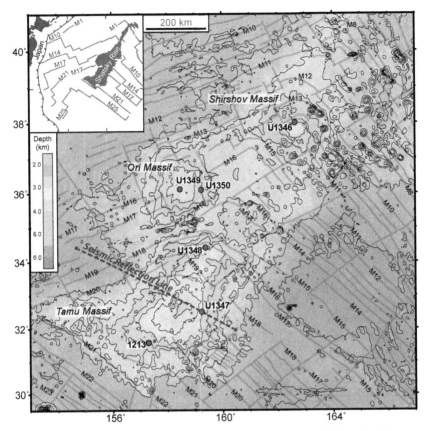

FIGURE 4.3.3   Map of Shatsky rise oceanic plateau. Bathymetric map of the Jurassic–Cretaceous Shatsky Rise showing three massif plateaus, Tamu, Ori, and Shirshov, which progressively become younger toward the northeast. ODP Site 1213 and IODP Sites U1346–1350 are indicated, as well as the M-series magnetic anomalies from Nakanishi, Sager, and Klaus (1999) and the seismic reflection line (dashed) shown in Figure 4.3.7 from Sager et al. (2013).

4.3.3.2 Outcomes

4.3.3.2.1 Volcanism

Expedition 324 cored at five sites on Shatsky Rise (Figure 4.3.3; Table 4.3.1): two each on Tamu Massif (Sites U1347, U1348) and Ori Massif (Sites U1349, U1350) and one on Shirshov Massif (Site U1346). Basaltic lava flows were cored at Sites U1346, U1347, U1349, and U1350, but at Site U1348, only highly altered volcaniclastic material was recovered (Figure 4.3.4). The recovered lava flows were of two types, pillow lavas and massive flows (Sager et al., 2010; 2011). Pillow flows represent common, low-effusion rate submarine volcanism (Ballard, Holcomb, & Van Andel, 1979), whereas the massive flows indicate effusive volcanism (Self et al., 1997). These massive flows are likely

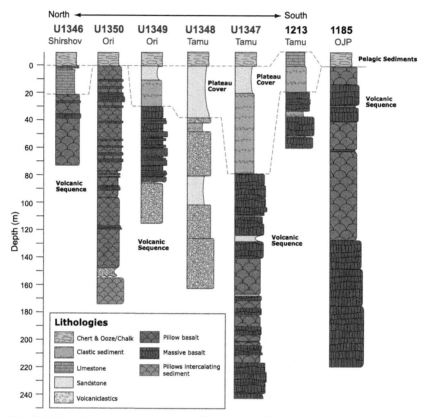

FIGURE 4.3.4 Overview of stratigraphy of Shatsky Rise drill sites. Lithostratigraphy of Shatsky Rise drill sites and comparison with ODP Site 1185 from the Ontong Java Plateau (OJP). Sites are arranged in order from northeast (left) to southwest (right), and depths have been shifted to align the tops of the basement sections. Data from Sites U1346–U1350 are from IODP Expedition 324 (Sager et al., 2010), whereas data from Site 1213 are from ODP Leg 198 (Shipboard Scientific Party, 2002b; Koppers et al., 2010a) and Site 1185 from ODP Leg 192 (Shipboard Scientific Party, 2001). Site locations are shown in Figure 4.3.3. The most notable features of cores from Tamu Massif are massive lava flows, which imply effusive volcanism.

analogous to the massive sheet flows of Self et al. (1997) and are similar to thick flows found in continental flood basalts (Bryan et al., 2010; Jerram & Widdowson, 2005) and on Ontong Java Plateau (Shipboard Scientific Party, 2001). Within the small collection of massive flows cored at Site U1347, the greatest thickness is approximately 23 m, an extraordinary value for a flow on the flank of a volcano. Expedition 324 scientists redescribed the Site 1213 section (Figure 4.3.5) because Leg 198 was a paleoceanographic study whose scientists had little experience with igneous cores. The three "diabase sills" cored at the site (Shipboard Scientific Party, 2002b) were reclassified as massive flows similar to those at Site U1347 (Koppers et al., 2010a). These results showed

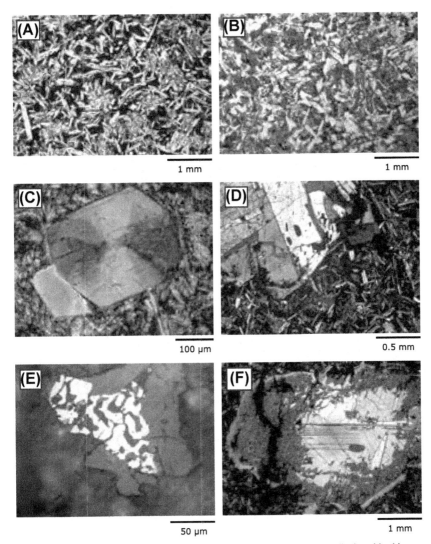

FIGURE 4.3.5 Image gallery of Shatsky Rise rocks. Lithological variation displayed in thin section images from Shatsky Rise at Site 1213. (A + B) Standard texture photomicrographs showing the development of an interlocking network of plagioclase–clinopyroxene crystal clumps with increasing grain size (decreasing cooling rate). (C) Photomicrograph in crosspolarized light of a small euhedral clinopyroxene microphenocryst with sector zoning. (D) Photomicrograph of the edge of a large skeletal plagioclase xenocryst with marginal overgrowths. (E) Photomicrograph in reflected light of a skeletal grain of titanomagnetite that is intergrown with plagioclase. (F) Photomicrograph of a partially altered plagioclase phenocryst, with the inner part remaining relatively fresh, but with a rim that has been completely replaced, likely by feldspar enriched in K and/or Na and associated with some secondary sericite. *Figure after Koppers et al. (2010a).*

that Tamu Massif formed by effusive eruptions of lavas that formed thick, massive flows on the flanks of the volcano. Their similarity to massive flows cored by Leg 192 from Ontong Java Plateau (Figure 4.3.4) demonstrates the similar physical volcanology of these two plateaus. Further, their similarity to thick flows of continental flood basalts implies a connection between the largest subaerial and submarine LIPs (Duncan & Richards, 1991; Richards et al., 1989).

The distribution of lava flow types on Shatsky Rise fits the trend of waning volcanism with time inferred from morphology (Sager et al., 1999, 2010). Thick massive flows occur only on Tamu Massif, making up approximately 67% of the entire section and approximately 100% of two contiguous sections separated by an interval of pillow flows. At Ori Massif, thin massive flows occur at Site U1349 on the summit, but make up only 14% of the Site U1350 section from the flank. Site U1346 recovered only two thin massive flows but otherwise consists of pillow lavas (90%). Although this finding is based on a small number of holes with limited penetration, it suggests that Tamu Massif eruptions were effusive, fluid, and thick, whereas Ori and Shirshov massifs experienced smaller, less effusive eruptions, more like the formation of large (deep) submarine seamounts (Staudigel & Clague, 2010; Staudigel and Koppers, 2015).

4.3.3.2.2 Geochronology

Age data are still preliminary at the time of writing. Prior to Expedition 324 studies, only one modern radiometric date was available for Shatsky Rise igneous rock, an age of 144.6 ± 0.8 Ma from an average of samples from two Site 1213 flows (Mahoney et al., 2005). Being nearly coincident with the age of magnetic anomaly M19, which transects the north flank of Tamu Massif, this datum showed that Tamu Massif formed close to the triple junction on young lithosphere. New age data from Site U1347 on the upper east flank of Tamu Massif give nearly the same age, whereas the ages for Site 1213 on its lower southwestern flank are readjusted to approximately 142 Ma (Geldmacher, van den Bogaard, Heydolph, & Hoernle, 2014; Heaton & Koppers, 2014). Ori and Shirshov massifs are approximately 5–10 Myr younger in accordance with their underlying magnetic anomalies (Heaton & Koppers, 2014). Most interesting, Tamu Massif appears to have been active for a longer period than expected. Although the deepest cored units at Site U1347 are close in age to magnetic anomaly M19, the youngest volcanic units are a few million years younger and coincident in eruption age with younger flows dated for Site 1213. The large and shallow Toronto Ridge, appearing on Tamu's summit (Figure 4.3.6), is even younger than the main shield volcano and apparently formed approximately 10 Myr later by small volume postshield eruptions that are found on many large volcanoes (e.g., Clague and Dalrymple, 1989; Heaton & Koppers, 2014). This finding was expected because it was inferred that this ridge would have experienced subaerial erosion (which is not observed) had it formed atop the shield summit prior to subsidence (Sager et al., 2013).

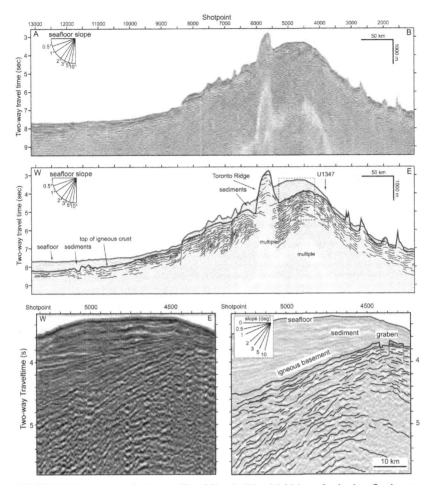

FIGURE 4.3.6 Seismic reflection profile of Shatsky Rise. Multichannel seismic reflection profile over Tamu Massif (see Figure 4.3.3 for the location) demonstrating that the edifice is a huge, single shield volcano (Sager et al., 2013). Darker image (top) shows original data, whereas grayed image (middle) shows an interpretation with dark lines highlighting intrabasement reflectors and colored (dark gray and light gray in print versions) lines indicating the water–sediment interface, basement boundary and faults. The two seismic profiles on the bottom show enlarged area from the overall profile around the ancient summit of Tamu Massif. Notably, many subparallel reflectors are observed within the basement, trending downslope from the summit at shotpoint 4500. Secondary volcanism (such as Toronto Ridge) appears to tower over the main shield because of the vertical exaggeration and the low slopes of the shield. The enlargement shows the piecewise continuous character of intrabasement reflectors as well as a summit graben (likely a caldera). *Figure after Sager et al. (2013).*

An unexpected result, however, is that the upper massive flows at Site U1347 may be millions of years younger than the flows at the base of the hole. It is unclear how such massive flows can occur after a significant interval of quiescence in the volcano's shield-building stage, but the presence of interbedded sediments is in agreement with extended intervals between eruptions (Sager et al., 2010). An implication is that some magmas are available for a long time period, perhaps from communication with magma sources beneath the younger north flank of Tamu Massif or Ori Massif. Further, paleomagnetic data show that the magnetic polarity of samples from Sites 1213 and U1347 are different (Sager et al., in press), implying that the bulk of the volcano formed over more than one magnetic polarity interval, in contrast to an earlier inference that Tamu Massif contains only one polarity (Sager & Han, 1993). The ages and magnetic polarity systematics found on Tamu Massif thus show a complex history in the building of this gigantic edifice, with the deepest massive lava flows at Site U1347 providing evidence for rapid shield building, practically on top of a spreading center, and with the youngest lava flows at Sites U1347 and 1213 and on Toronto Ridge providing ammunition for an extended, low volume eruptive period occurring later on.

4.3.3.2.3 Geochemistry and Isotopic Chemistry

Expedition 324 geochemical and isotopic studies have sought to identify a signature in Shatsky Rise rocks that could be tied to a particular part of the mantle. For example, a deep, primitive source would favor the mantle plume hypothesis, whereas a shallow source would favor the plate edge-driven melting hypothesis. Recent results have provided important insights into the source of Shatsky Rise igneous rocks, but implications with regard to plume or plate-edge mechanisms are still uncertain, mainly because geochemists are still debating the structure of isotopic reservoirs of the mantle. Similar to prior results from Site 1213 (Mahoney et al., 2005), new geochemical data confirm that Shatsky Rise basalts are tholeiites with compositions similar to MORB, but with slight enrichment of incompatible elements (Sano et al., 2012). Most recovered basalts, including all those recovered at Tamu Massif Sites 1213 and U1347 (except for one flow) and Shirshov Massif Site U1346, plot wholly or partly on top of the N-MORB (normal midocean ridge basalt) field in plots of oxides versus MgO (Sano et al., 2012). Nearly half of Ori Massif samples were similar, but other samples with different compositions were also found, including primitive, depleted basalts at Site U1349, and samples with low-Ti or high-Nb contents at Site U1350. These differences indicate source heterogeneity at the least, and it is also possible to suggest that the increased variability at Ori Massif indicate an evolution of the source with time (Heydolph et al., 2014; Sano et al., 2012).

Despite obvious similarities to MORB compositions, geochemical data indicate that Shatsky Rise basalts were produced under different conditions but with similarities to samples from Ontong Java Plateau. Shatsky Rise magmas formed at greater depths and degrees of partial melting than normal MORB. From trace element ratios, Sano et al. (2012) estimated a degree of partial melting of approximately 15% in the garnet stability field, which is on the upper end of expected MORB melt percentages and at deeper depths. Husen et al. (2013) interpreted fractionation-corrected Na_2O concentrations and CaO/Al_2O_3 ratios to indicate 20–23% partial melting, greater than MORB, but less than the approximately 30% inferred for Ontong Java Plateau (Fitton & Godard, 2004). Systematic differences in sample MgO contents indicate that lavas (before their eruption) resided within a steady-state magma chamber at shallow depths (<6 km) that was episodically filled from a deeper reservoir, 18–24 km deep, near the base of the thick oceanic plateau crust (Husen et al., 2013). Husen et al. (2013) also noted that Na_2O and FeO contents for Shatsky Rise rocks fall on the evolutionary trend defined by Ontong Java Plateau samples, suggesting similar melting conditions. Additional similarity to Ontong Java Plateau is demonstrated by the isotopic compositions of the high-Nb basalts from Site U1350 (Ori Massif), which overlap those of the Kwaimbata and Kroenke-type basalts from Ontong Java Plateau (Heydolph et al., 2014). These results underscore that despite obvious similarity to MORB, an anomalous higher melt production is required to create the large bulk of the Shatsky Rise oceanic plateau and that these anomalous conditions are similar to those of Ontong Java Plateau and likely other oceanic plateaus.

Isotopic evidence indicates enrichment of Shatsky Rise lavas, either by plume components (e.g., Campbell, 2007) or by fusible, recycled oceanic crust (e.g., Korenaga, 2005; Foulger, 2007). High-Nb basalts from Site U1350 plot (Figure 4.3.7a) on the Iceland sample array for Nb/Y versus Zr/Y (Sano et al., 2012) and these samples have incompatible trace element and isotopic compositions characteristic of OIBs (Heydolph et al., 2014). Although OIBs are commonly attributed to a plume source (Hoffman & White, 1982), this view is not universally accepted (Foulger, 2007). $^3He/^4He$ ratios from Site U1350 rocks show a limited range (5.5–5.9 Ra) that is systematically lower than MORB, implying a different source but not primitive lower mantle material (Hanyu et al., 2014a). Instead, both isotopic and noble gas evidence implies enrichment from subducted slab material (Heydolph et al., 2014; Hanyu et al., 2014a). Left unanswered is how the enriched component was delivered. One pathway is through recycling slab material carried into the mantle transition zone or deep mantle by subduction (e.g., Jackson & Jellinek, 2013) and then delivered to the surface by narrow plume conduits (e.g., Campbell, 2007) or by the upwelling of broad-scale mantle "updrafts" (Anderson & Natland, 2014). Another possible pathway is for this material to be trapped in the upper mantle where it is sampled by chance as a plate boundary or extension zone passes over it (Foulger, 2007; Machida, Hirano, & Kimura, 2009; Niu, Waggoner, Sinton, & Mahoney, 1996).

4.3.3.2.4 Seismic Evidence

Although data from cored samples are crucial for understanding the origin and evolution of Shatsky Rise, these results are constrained by the fact that only a very small number and the shallow parts of the oceanic plateau were sampled.

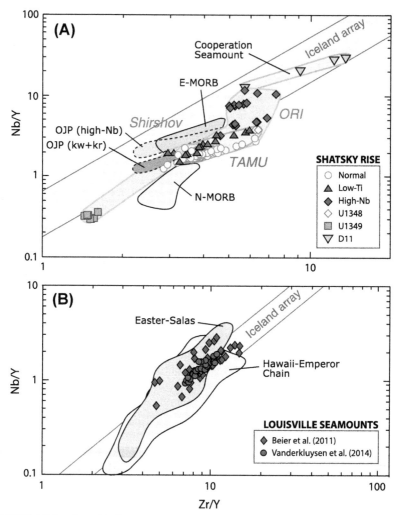

FIGURE 4.3.7 Fitton Plot for Shatsky Rise and Louisville Seamount trail. Plot of Nb/Y versus Zr/Y of basalts measured for Shatsky Rise (A) and the Louisville Seamount Trail (B) with reference to the Icelandic Array. Reference fields for MORB and the Ontong Java Plateau are given to contrast Shatsky Rise, whereas Easter-Salas and the Hawaii-Emperor Chain are given to contrast with the Louisville Seamount Trail basalts. *Figure A after Sano et al. (2012) and Figure B after Vanderkluysen et al. (2014).*

Data from recent seismic refraction and reflection (Korenaga & Sager, 2012; Sager et al., 2013) provide an important context by elucidating the large-scale structure of the Shatsky Rise oceanic plateau (Figure 4.3.6). Seismic refraction data reached the Moho beneath Tamu Massif for the first time, demonstrating that the crust is approximately 30 km thick beneath the shallowest part of the massif (Korenaga & Sager, 2012). This result was expected, considering that other oceanic plateaus have thick crusts (e.g., Ontong Java Plateau has a crust ~32–33 km thick, Gladczenko, Coffin, & Eldholm, 1997; Miura et al., 2004) and that Shatsky Rise is completely Airy compensated (Sandwell & MacKenzie, 1989). The seismic data, however, display a peculiar negative correlation between lower crustal thickness and seismic velocity. Vigorous upwelling and melting at higher than normal mantle temperatures, as expected for a plume source, should create thick crust with higher seismic velocities (Korenaga, Kelemen, & Holbrook, 2002), yet this negative correlation implies that chemically anomalous mantle may have been tapped in the formation of Tamu Massif keeping the observed seismic velocities relatively low (Korenaga & Sager, 2012).

Multichannel seismic profiles across Tamu Massif show that the upper structure of the volcanic mountain consists of subparallel reflecting horizons within the volcanic basement. Correlation with core and wire-line log data from Expedition 324 indicates that these reflectors result from groups of lava flows, especially the contrast between thick massive flows and pillow basalts, and between lava flows and thick intraflow sediment layers (Zhang et al., in press). These intrabasement reflectors demonstrate that Tamu Massif is a single, enormous shield volcano with anomalously low flank slopes, typically <1° (Sager et al., 2013). On every seismic profile across the massive volcano, the same picture is seen: reflecting horizons subparallel to the surface descend from the summit of Tamu Massif and no large secondary eruption centers are observed (Figure 4.3.6). Tamu Massif is thus a single volcano far larger than any other. This finding and the similarity of morphology with other oceanic plateaus imply that these features are a special class of volcanic edifice created by voluminous, effusive eruptions that produce massive lava flows and low flank slopes that likely result from eruption of copious quantities of fluid magma that travel long distances.

4.3.4 IODP EXPEDITION 330 TO THE LOUISVILLE SEAMOUNT TRAIL

4.3.4.1 Rationale and Objectives: Mirroring ODP Leg 197

The 6000-km-long Hawaiian-Emperor seamount trail (Figure 4.3.8) is our epitome of a hotspot and the most famous example used in many textbooks. The 4300-km-long Louisville seamount trail is its South Pacific counterpart (Figure 4.3.9) because it also has a striking linear morphology, lacks cross-cutting seamount trails, and exhibits long-lived age-progressive volcanism for almost 80 Myr (Koppers, Duncan, & Steinberger, 2004; Lonsdale, 1988). Both hotspots are considered *primary* plumes with a deep mantle origin (Courtillot, Davaille, Besse, & Stock, 2003) of which there exist only a handful worldwide. They stand apart from *secondary* and *tertiary* hotspots that form shorter seamount trails, potentially above short-lived shallow mantle plumelets and extensional regions or fractures in the oceanic lithosphere

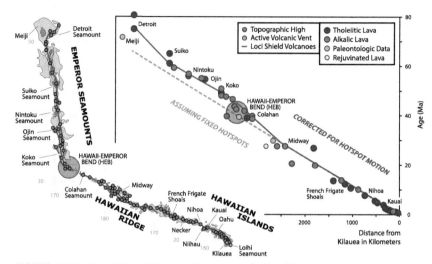

FIGURE 4.3.8   Map of Hawaii-Emperor Seamount trail. Simplified map and age–distance relationship for the Hawaiian-Emperor seamount trail. A consistent decreasing of seamount age from the old end of the trail toward the active end is apparent, whereby all ages were determined with modern-day $^{40}Ar/^{39}Ar$ incremental heating techniques and following careful sample selection and acid cleaning. Data shown fit better to APM models for the Pacific plate that are incorporating the motion of mantle plumes, such as the approximately 15° southward shift observed for the Hawaiian plume between approximately 80 and 50 Ma (Tarduno et al., 2003, 2009). *Figure from Staudigel and Koppers (2015).*

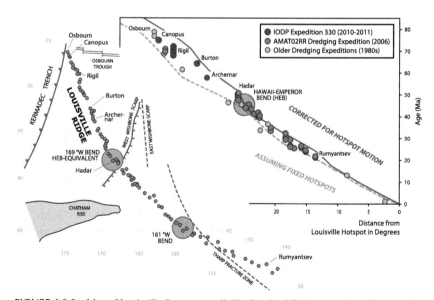

FIGURE 4.3.9   Map of Louisville Seamount trail. Similar simplified map and age–distance relationship for the Louisville Seamount trail. Also, here the age data better fit the APM model corrected for hotspot motions, especially for seamounts that are >50 Ma and that were drilled during IODP Expedition 330. *Figure from Staudigel and Koppers (2015).*

(Koppers, 2011). The Louisville seamount trail includes at least 65 volcanic seamounts, ranging in age from approximately 79 Ma at its northwestern end to 1.11 Ma at its southeastern end (Koppers et al., 2004).

It has long been envisioned that the Hawaii and Louisville hotspots (and all other recognized hotspots) remained fixed in the mantle. However, DSDP Leg 55 and ODP Legs 145 and 197 to the Emperor Seamounts in the Hawaii hotspot trail documented an approximately 15° southward drift of this hotspot between 80 and 50 Ma (Kono, 1980; Tarduno, 2007; Tarduno & Cottrell, 1997; Tarduno et al., 2003, Tarduno, Bunge, Sleep, & Hansen, 2009) upsetting the status quo. This hotspot shift was reproduced in numerical models of large-scale mantle flow, which indicate that plume conduits may become tilted in the so-called "mantle wind," which then results in an absolute motion of hotspots along the Earth's surface (e.g., Steinberger & O'Connell, 1998). Compared to Hawaii, the models predict essentially *no* southward motion for the Louisville hotspot but instead a strong west-to-east drift over the last 80 Myr (Steinberger & Antretter, 2006; Steinberger & Calderwood, 2006; Steinberger, Sutherland, & O'Connell, 2004). The primary objective of IODP Expedition 330 was to determine how much the Louisville hotspot has moved in latitudinal sense during the 80–50 Ma time frame and whether it moved independently from the southward-moving Hawaiian hotspot as predicted from these mantle flow models.

IODP Expedition 330 to the Louisville Seamount Trail thus centered on testing the fixity of the Louisville hotspot and on comparing it directly to previous DSDP and ODP work for the Hawaii hotspot that allowed for significant mobility of (all) mantle plumes. In fact, the drill plan was to mirror the Hawaiian experiment and to drill seamounts of approximately the same age in the Louisville seamount trail (Koppers et al., 2010b). This would provide direct insights into the Hawaiian–Louisville interplume drift, as well as a test of the validity of a (moving) reference frame for modeling absolute plate motion (APM) in the Pacific. The first two primary objectives of Expedition 330 therefore required a large number of measurements of paleomagnetic inclination onboard the D/V *JOIDES Resolution* to determine the paleolatitude of the Louisville hotspot between 80 and 50 Ma, as well as sampling of unaltered basalt flows for onshore $^{40}Ar/^{39}Ar$ geochronology to establish an accurate time frame. Based on results from previous drilling legs during DSDP and ODP, it was estimated that at least 40 different lava flow units needed to be cored for each seamount to average out paleosecular variation in the Earth's magnetic field and to achieve paleolatitude estimates with 4°–5° precisions or better. To do so successfully entailed that igneous basement of the Louisville seamounts needed to be cored for at least 350 m, which was much deeper than the average penetration depths into seamounts as achieved during DSDP and ODP.

The third primary objective of Expedition 330 addressed the remarkable mantle plume source homogeneity that characterizes Louisville hotspot, which for as long as 80 Myr erupted alkali basalts with only a very narrow range in isotopic and trace element compositions (Beier, Vanderkluysen, Regelous, Mahoney, & Garbe-Schönberg, 2011; Cheng et al., 1987; Hawkins, Lonsdale, & Batiza, 1987). This required high-quality shipboard geochemistry and onshore isotopic analyses of recovered lava flows in order to establish a complete picture of the geochemical evolution of the Louisville seamounts and how those compare to the evolution of archetype Hawaiian volcanoes, which progress typically from a main tholeiitic shield stage into an alkali basalt postshield or rejuvenated stage over several millions of years (Clague et al., 1989). The same analyses also would provide key insights into the mantle source region of the Louisville seamounts in comparison to Hawaii and all other hotspot systems worldwide.

Secondary objectives included studying the Louisville plume–lithosphere interactions and testing whether the Ontong Java Plateau formed from a nascent Louisville mantle plume head reaching the lithospheric lid around 120 Ma. Finally, Expedition 330 sampled a wide range of lithologies for geomicrobiological studies to study living and extant microbial residents within the 80- to 50-Ma seamount basement rocks.

4.3.4.2 Outcomes

4.3.4.2.1 Volcanism

Six sites were drilled during Expedition 330 along the old end of the Louisville Seamounts (Figure 4.3.9) on Canopus Guyot (U1372), Rigil Guyot (U1373, U1374), Burton Guyot (U1367), Achernar Guyot (U1375), and Hadar Guyot (U1377). Coring reached 522.0 m below the seafloor (mbsf) at Site U1374 and 232.9, 65.7, 11.5, 182.8, and 53.3 mbsf at Sites U1372, U1373, U1375, U1376, and U1377 (Table 4.3.1). Average recovery rates were high at 72.4%, and igneous basement was reached at four of the drilling sites, yet at none >40 individual cooling or lava flow units were encountered before time ran out for drilling. Even on Rigil Guyot, where two sites were drilled about 10 km apart on the same summit platform, only 35 in situ lava flows were recovered after a total penetration of 537.1 m into igneous basement. Routine deep coring into seamounts remains problematic, specifically because during Expedition 330 abundant loose volcanic breccias and poorly consolidated hyaloclastites provided difficult drilling conditions, necessitating early abandonment of some drill sites.

All seamounts targeted are guyots with flat, tablemount-like summits, indicating they once were volcanic islands whose tops were eroded away by wave action. On these guyots, the sequences drilled are characterized by (1) an absent or thin layer of recent pelagic ooze on top; underlain by (2) an older, heavily eroded sedimentary cover of volcanic breccias, conglomerates, and sandstones deposited in a subaerial to shallow submarine environment and interspersed with a small number of (condensed) limestones and/or late-stage lava flows; (3) an upper volcanic sequence dominated by mostly massive lava flows that are interlayered with volcaniclastic sediments; and (4) a lower volcanic sequence characterized by mainly volcaniclastics and hyaloclastites, which are indicative of a submarine environment, interspersed with in situ massive lava flows, smaller lava pods, pillow lavas, and subvertical dike intrusions (Figures 4.3.10 and 11). Based on shipboard geochemistry and thin section descriptions, it is evident that all Louisville lavas cored are alkalic basalts that progressed from (shallow) submarine eruptive environments deeper in the volcanic sequences to subaerial at the top of the basement. Volatile chemistry (H_2O, S, and CO_2) from volcanic glasses found in pillow basalts, intrusive sheets and abundant volcaniclastics from the Louisville Seamounts show an almost complete degassing of these materials. Assuming

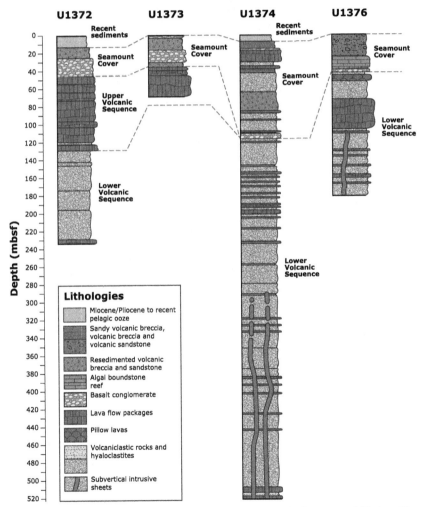

FIGURE 4.3.10 Overview of the lithostratigraphy of Louisville Seamounts drill sites. Sites are arranged in order from old (left) to young (right) and different evolutionary stages in the seamount construction have been indicated. Site locations are shown in Figure 4.3.5. *Figure is after Koppers et al. (2013).*

FIGURE 4.3.11 Image gallery of Louisville Seamount rocks. Lithological variation displayed in core images from the Louisville Seamounts. (A) Monolithic bioturbated brown coarse tuff (top) and heterolithic multicolor volcanic sandstone (bottom) from Site U1376 deposited as turbidites in a hemipelagic or pelagic environment. (B) Algal boundstone with a branching growth form of algae from Site U1376 interpreted as a reef. (C) Multicolor coarse (layered) volcanic sandstone/breccia at Site U1374 interpreted to have been emplaced on a shallow-marine slope as a hyperconcentrated flow. (D) Multicolor (light gray and dark gray in print versions) basalt conglomerate emplaced under hemipelagic conditions at Site U1375 with interpebble spaces composed of foraminiferal limestone, finer-grained volcaniclastic sediment, and carbonate cement. (E) Angular aphyric basalt clasts in hyaloclastite matrix at Site U1374. (F) Example of magma–sediment (peperitic) interaction observed at Site U1374. (G) Flow top at the upper (peperitic) boundary of a massive (~23 m thick) aphyric basalt flow at Site U1373. (H) Aphyric subvertical dike intrusion (top) into hyaloclastite breccia sequence (bottom) at Site U1376. *Figure after Koppers et al. (2013).*

the volcanic glasses were saturated with H_2O before eruption, this may indicate that all glasses quenched at eruption depths shallower than 200 m below sea level (mbsl) and most of them shallower than 20 mbsl or subaerially (Nichols et al., 2014). It is expected that syneruptive transport over short distances was required to deposit the quenched glasses and other "hot" volcaniclastic materials, but only onto slightly deeper seamount flanks. Volcanic breccias, which are the most abundant type of volcaniclastics in the Louisville Seamounts, commonly have clasts that are lobate with intricate margins, suggesting in situ cooling and only limited transport (Koppers, Yamazaki, Geldmacher, & Expedition 330 Scientists, 2012). Syneruptional talus deposits forming below shallow submarine eruptions may explain the above observations as may flow-foot breccias forming where active subaerial lava flows were entering the sea (Nichols et al., 2014).

In essence, the volcanology of Louisville differs distinctly from that observed in the Emperor Seamounts, where the latter is dominated by subaerial massive lava flows (Duncan, 2004) and the former by volcaniclastics only interspersed with thin lava flows or pods (Koppers, Yamazaki, Geldmacher, & Expedition 330 Scientists, 2012). The principal difference, however, simply may be a function of seamount size. Structures in the Emperor Seamounts are larger and taller than those in the Louisville seamount trail, allowing for a thicker subaerial (upper) volcanic sequence to build up during the construction of the Emperor volcanoes. These thick sequences may not have been eroded deeply enough by the time the volcanic islands started drowning. In fact, the limited approximately 280-m depth of basement penetration during ODP Leg 197 might not have been deep enough to drill through the upper volcanic units in the Emperor Seamounts. On the other hand, the subaerial summits of the Louisville guyots were much smaller and more easily (and thus almost completely) removed by wave action and other types of island erosion (Koppers, 2009). This explains why only a thin upper sequence of massive flows remained in the larger of the drilled Louisville Seamounts, as observed at Site U1372 on Canopus Guyot and Site U1373 on Rigil Guyot. In the smaller edifice of Burton Guyot, the subaerial sequence is lacking at Site U1376, likely because sea surface erosion on this former island incised deeper into the volcanic structure and removed the entire subaerial summit.

4.3.4.2.2 Geochronology

Establishing time frames for the construction of single Louisville seamounts as well as the age progression along the Louisville seamount trail are both crucial for interpreting changes in absolute motion of the Pacific Plate and potential movements of the underlying mantle plume. A seamount can take several millions of years to finish its construction, and there is a wide variety in construction times between seamount trails worldwide. This means that the time that a seamount volcano resides over a hotspot varies from region to region and principally depends on the rate of plate motion. For example, Hawaiian volcanoes finish shield building relatively quickly, in only hundreds of thousands of years (MacDonald and Abbott, 1970; Clague et al., 1989), consistent with rapid Pacific Plate motion, yet they also experience late-stage volcanism that may occur up to several millions of years following the onset of erosion (Clague et al., 1989, 2009). Construction of volcanoes in the Canary Islands takes much longer, because plate motion is much slower in the Atlantic, with volcanic activity on these volcanic islands lingering up to 20 millions of years (Geldmacher et al., 2005). For the Louisville seamounts, we only have sparse indications by comparing dredge ages from neighboring seamounts that can span intervals up to six millions of years (Koppers et al., 2011), but these timings are uncertain as it is unclear to which evolutionary stages the dredge samples belong. Results from ongoing ^{40}Ar/^{39}Ar geochronology studies on the Expedition 330 cores show that the upper and lower volcanic sequences in the Louisville Seamounts formed over short time intervals, likely within less than one million of years at Burton Guyot and less than three million of years at Rigil Guyot (Heaton & Koppers, 2014). No direct age evidence has been found (yet) in the drill cores that suggest late-stage volcanism, even though multibeam mapping revealed posterosional parasitic volcanic cones on the summit planes of many Louisville guyots (Koppers et al., 2011). As in Hawaii, the relatively short duration of volcano construction allows us to positively co-locate Louisville seamounts with its hotspot and establish an accurate age progression.

If one assumes that "fixed" hotspots are underpinned by a global network of stationary mantle plumes, then coeval seamounts trails formed on the Pacific Plate should all have comparable age progressions. Thus, comparing the age progressions observed for the Hawaiian and Louisville hotspots and other Pacific hotspots provides a direct test of the "fixed" hotspot hypothesis (Koppers, Morgan, Morgan, & Staudigel, 2001). By ^{40}Ar/^{39}Ar age dating of in situ lava flows, pillow basalt units and subvertical dike intrusions from the deepest units cored in the Louisville seamounts (Koppers, Yamazaki, Geldmacher, Gee, et al., 2012) and by combining those with dredge ages (Koppers et al., 2004; 2011), we can confirm that the age progression for Louisville becomes systematically older from the youngest 1.11-Ma seamount to the oldest, 78-Ma Osbourn Guyot, similar to the age progression for Hawaii (Figure 4.3.8). However, the Louisville ages clearly deviate from APM models that assume "fixed" hotspots. In contrast, models that take into account the southward motion of the Hawaiian plume between 80 and 50 Ma provide a better fit (Figure 4.3.9). This result indicates that the mantle plumes underpinning the Hawaiian and Louisville hotspots cannot have been fixed, either with respect to each other or with the Earth's spin axis.

The ages measured for seamounts within the Hawaiian and Louisville seamount trails also can be used to directly measure the separation (in kilometers) of both hotspots over geological time. If the hotspots were "fixed" over long periods of time, it is expected that this separation remains the same. Conversely, if the separation increases or decreases, it means that the hotspots are, respectively, drifting away from or toward each other. Presently, the distance between the

Hawaiian hotspot (located beneath Loihi seamount) and the Louisville hotspot is approximately 8200 km (Figure 4.3.12). Because of the high number of modern ^{40}Ar/^{39}Ar ages available for both the Hawaiian and Louisville seamount trials, we can now measure that distance for all seamounts of coeval age and record their "separation" history. This exercise is illuminating as it shows that there is a period with hardly any increasing or decreasing separation, from approximately 55 Ma to the present day, but prior to that period, there is a clear increase in the measured interhotspot distance up to approximately 8480 km, showing that both plumes were drifting apart at rates up to 10 km/Myr (O'Connor et al., 2013). This result confirms prior suggestions that the hotspots have remained fixed since the time of the Hawaiian-Emperor Bend (HEB), while demonstrating a clear divergent motion during the formation of the Emperor Seamounts (Koivisto, Andrews, & Gordon, 2013; Sager, Lamarche, & Kopp, 2005).

4.3.4.2.3 Paleolatitudes and Paleomagnetic Intensities

Expedition 330 aimed to compare observable shifts in paleolatitude for the Louisville hotspot with that of the large approximately 15° southern shift of the Hawaiian hotspot between 80 and 50 Ma. A large quantity of paleomagnetic inclination data was required from independent lava flow units to average out secular variation and to achieve high precisions. By the end of Expedition 330, inclinations were measured at 9235 intervals every 2 cm in the archive halves of the drill cores, and in total, 493 discrete samples were analyzed (Koppers, Yamazaki, Geldmacher, Gee, et al., 2012). Natural remanent magnetization was retained very well in these cores with most lava flows having relatively high magnetic coercivities and a minimal drilling-induced magnetic overprinting.

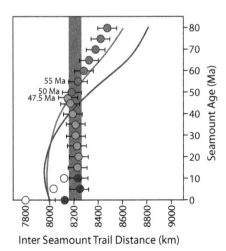

FIGURE 4.3.12 Louisville versus Hawaii Interhotspot Motions Based on ^{40}Ar/^{39}Ar Geochronology. Interseamount trail separation between coeval locations on the Hawaiian-Emperor and Louisville seamount trails estimated in 5-Ma increments and assuming linear age–distance relations on both trails. Between 0 and 55 Ma, this separation is rather constant around 8200 km, but prior to 55 km, the separation was up to 334 km larger, which can be translated in 10 km/Myr of interhotspot motion. Horizontal error bars simply reflect the average trail width between Hawaii and Louisville of approximately 75 km. Also shown are predictions from two past large-scale mantle flow models (blue (light gray in print versions) and red (dark gray in print versions) curves) after Steinberger et al. (2004). *Figure after O'Connor et al. (2013).*

At Sites U1372 and U1374, magnetic polarity reversals were observed in the seamount sediment covers, yet the underlying igneous basements are all characterized by single polarities, emphasizing the relatively fast construction of the Louisville volcanoes (Koppers, Yamazaki, Geldmacher, Gee, et al., 2012).

Inclinations measured for the igneous basement units (Figure 4.3.13) are generally steep for the Louisville drilled sites (Koppers, Yamazaki, Geldmacher, Gee, et al., 2012). For example, at Site U1372 on Canopus Guyot, the mean inclination determined from discrete samples from nine in situ lava flow units is $-61.7° \pm 7.2°/8.8°$ (2σ, $n=9$). On Rigil Guyot, two sites were drilled at opposite edges of the same summit platform, with Sites U1373 and U1374 yielding mean inclinations of $-55.2° \pm 10.6°$ (2σ, $n=9$) and $-68.7° \pm 8.4°$ ($n=19$). Because lava flows at both sites are different in age by less than 1 million years, the large inclination difference shows the clear effects of secular variation. The most reliable inclination estimate for Rigil Guyot therefore is the combined inclination of $-65.0° \pm 4.6°/7.3°$ (2σ, $n=28$) and is within error with the geocentric axial dipole inclination ($\pm68°$) for the present-day Louisville hotspot location at approximately 51°S (Koppers, Yamazaki, Geldmacher, Gee, et al., 2012). Sites U1376 and U1377 at Burton and Hadar Guyots are similarly consistent with today's Louisville hotspot inclination at $-67.1° \pm 3.5°/3.5°$ (2σ, $n=8$) and $-68.9° \pm 14.6°/13.6°$ (2σ, $n=3$). Statistical analyses of the inclination data shows that the Louisville hotspot experienced a limited north-to-south motion of 3°–4° between 80 and 50 Ma, much less than that observed for the Hawaiian hotspot, but consistent with the 10 km/Myr interhotspot motion based on interseamount distances and $^{40}Ar/^{39}Ar$ ages. What is more, the observed limited latitudinal motion for the Louisville hotspot is consistent with predictions of whole mantle flow modeling (Steinberger & Antretter, 2006; Steinberger & Calderwood, 2006; Steinberger et al., 2004). For the Louisville hotspot, this may mean that, if its mantle plume moves at similar rates as the Hawaiian plume, it is not doing so from a north to south but rather in a west-to-east direction. Subduction of the Pacific Plate in the Tonga-Kermadec trench is likely the cause of a pronounced return flow at midmantle depths that governed such a large eastern movement of the Louisville plume (Koppers, Yamazaki, Geldmacher, Gee, et al., 2012).

The Louisville Seamounts drilled during Expedition 330 all formed just after the end of the CNS during which the strength of geomagnetic field was consistently high (Tauxe, 2006; Tarduno et al., 2006). Paleointensities measured for the Louisville Seamounts thus potentially are important for understanding the behavior of the geodynamo while it recovered and returned to "normal" intensity values after the CNS. An average virtual axial dipole moment (VADM) of 3.75 ± 1.52 ($\times10^{22}$ Am^2, $n=16$) was obtained at Site U1372 on Canopus Guyot (~74 Ma) based on 16 independent measurements. Similar VADMs of 3.79 ± 1.40 ($n=16$) were obtained for Sites U1373 and U1374 on Rigil Guyot (~70 Ma) and 3.70 ± 1.37 ($n=8$) for Site U1376 on Burton Guyot (~65 Ma). These values are similar to the long-term VADM average (Tauxe et al., 2013) for the last 200 million years and close to the mean of the

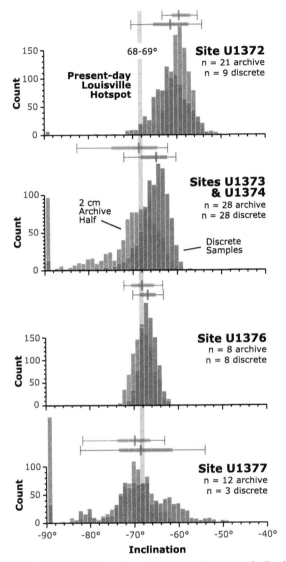

FIGURE 4.3.13 Limited Latitudinal motion of the Louisville hotspot. Inclination distributions for four Louisville seamounts are shown as measured for in situ lava flows drilled during IODP Expedition 330. Bootstrap results of 1000 resamplings are shown for both 2-cm archive half (red (light gray in print versions)) and discrete sample (blue (dark gray in print versions)) flow means. Resulting inclination averages and 1σ (filled rectangles) and 2σ uncertainties (lines) are statistically overlapping (for three out of four seamounts) with the expected geocentric axial dipole inclination of 68°–69° (vertical gray line) at the present-day hotspot location (50.9–52.4°S), which indicates that only a limited amount of latitudinal hotspot motion occurred between approximately 80 and 50 Ma. *Figure after Koppers, Yamazaki, Geldmacher, Gee, et al. (2012).*

last 5 million years (Yamamoto and Tsunakawa, 2005). Therefore, the geomagnetic field must have recovered relatively quickly and completely following the CNS and by the time that the oldest drilled Louisville seamount formed around 74 Ma.

4.3.4.2.4 Geochemistry and Isotopic Chemistry

Another principal difference between the Louisville and Emperor seamounts is that not a single tholeiitic basalt was cored for Louisville during Expedition 330 (Koppers, Yamazaki, Geldmacher, Gee, Expeditions 330 Scientists, 2012). Neither were tholeiites encountered in dredge hauls on as many as 40 different Louisville seamounts (Cheng et al., 1987; Hawkins et al., 1987; Beier et al., 2011; Vanderkluysen et al., 2014). Microprobe analyses of olivine, clinopyroxene, and plagioclase phenocrysts in Expedition 330 basalts confirm their alkalic nature (Dorais, submitted). Analyses of volcanic glass from chilled margins of pillow basalts and dike intrusions and abundant hyaloclastite glass shards show the same picture, again with all glasses being alkalic in nature and plotting in the basalt, trachybasalt, and basanite index fields (Nichols et al., 2014). The likelihood of (still) finding tholeiitic shield stage lavas beneath the deepest Expedition 330 drill holes has become small, and we may have to accept that Louisville shield volcanoes are produced by massive amounts of alkali basalt, rather than the tholeiites expected of most large Hawaiian seamounts.

The homogeneity in basalt petrology is underscored by the fact that Louisville geochemistry is likewise remarkably homogeneous, with major and trace element compositions and Sr-Nd-Pb-Hf isotopic signatures showing little variation downhole, across the six Expedition 330 drill sites, and throughout all dredges available for the entire Louisville seamount trail. With exception of volcanic glass in the youngest volcaniclastic sediments on top of Rigil Guyot and Burton Guyot, which are slightly enriched, typical downhole trace element ratios of, for example, La/Yb vary only between 8 and 15 and Nb/Zr between 0.12 and 0.18 (Nichols et al., 2014). In a similar fashion, downhole $^{87}Sr/^{86}Sr$ isotope ratios in plagioclases phenocrysts are identical (with overlapping 2σ error bars) over a depth range of almost 500 m in Hole U1374A on Rigil Guyot (Dorais, submitted). Interestingly, the observed $^{87}Sr/^{86}Sr$ ratios altogether have a range that is smaller than that of the "Focal Zone" (FOZO) mantle end member, the mantle end member common to source regions for intraplate volcanic systems and some midocean spreading centers (Hart et al., 1992; Stracke et al., 2005; Jackson et al., 2007). Moreover, measured trace elemental as well as isotope ratios do not vary much along the 4300-km-long Louisville seamount trail, which represents close to 80 Myr of intraplate volcanism (Cheng et al., 1987; Hawkins et al., 1987; Beier et al., 2011; Nichols et al., 2014; Vanderkluysen et al., 2014). In fact, isotopic variability is about 10× smaller than in Hawaii (for $^{143}Nd/^{144}Nd$ for example) and Louisville's chemical signatures are much like FOZO. In a Zr/Y versus Nb/Y diagram, Louisville data define a narrow and short array parallel to Icelandic lavas, indicating that there is little difference between incompatible elements Zr and Nb (Figure 4.3.7b). Therefore, melting conditions cannot have varied much, and modeling shows that small variations (2–7%) in the degree of partial melting of a uniform peridotitic source residing beneath a lithosphere 90–110 km thick will produce an array as that shown for Louisville (Vanderkluysen et al., 2014).

The overall homogeneous character of the Louisville Seamounts puts this hotspot in its own class, compared to Hawaii and any other intraplate volcanic system. Louisville sample homogeneity indicates a remarkable lack of variation in melting degree or a lack of mantle source heterogeneities. A plain perioditic mantle plume source is sufficient to explain most of the observed geochemical characteristics for Louisville (Vanderkluysen et al., 2014). In fact, melting conditions have remained constant during the construction of each individual Louisville seamount and during the approximately 80 millions of years while the Louisville seamount trail was formed. One aspect that may help in this matter is the fact that the lithospheric thickness along the seamount trail also has been nearly constant, with the exception perhaps of the oldest seamounts (Vanderkluysen et al., 2014) and seamounts close to the anomalous Wishbone Scarp (Beier et al., 2011). Although we are measuring the "same" geochemical characteristics again-and-again in Louisville samples, this is not a "boring" intraplate system, rather it occupies a unique niche in the spectrum of hotspots worldwide.

Finally, $^3He/^4He$ measurements in olivine phenocrysts and volcanic glasses yield ratios in the 6.5–10.6 R_A range (Hanyu, 2014) with some higher than characteristic approximately 8 R_A in MORB (Graham, 2002). In combination with elevated $^{20}Ne/^{22}Ne$ and $^{21}Ne/^{22}Ne$ ratios that are higher than the atmospheric values, this may indicate a deep mantle origin for the Louisville mantle plume (Hanyu 2014). This places Louisville more firmly in the small group of *primary* hotspots that potentially have plumes originating deep in the Earth's mantle (Courtillot et al., 2003; Koppers, 2011).

4.3.5 OCEANIC PLATEAUS: PLUMES OR PLATE BOUNDARIES?

Scientific drilling of Shatsky Rise addressed the question of where does oceanic plateau volcanism come from? The problem was billed as a test between the plume head and the plate-edge extension hypotheses (Sager et al., 2010). The essential difference between the two mechanisms is that the plume head arises from a depth in the mantle and is driven by thermal uplift (Campbell & Griffiths, 1990; Griffiths & Campbell, 1991; Farnetani & Richards, 1994; Campbell, 2006, 2007), whereas plate extension is a shallow phenomenon that occurs because of anomalous upper mantle conditions and composition (Anderson et al., 1992; Foulger, 2007). The plume head hypothesis predicts massive, rapid eruptions when the plume head impinges on the base of the lithosphere, a result of high degrees of melting over the upper mantle caused by elevated temperatures (Duncan & Richards, 1991; Campbell, 2006, 2007). In laboratory experiments and models, the plume head is followed by a tail, so the massive eruption of the plume head should be followed by lesser volcanism from the tail (Campbell, 2006, p. 2007). A shift in eruption conditions may occur as the plume head transitions to the tail, with lesser melting and lower volume magma emplacements. From a plume head, one would expect to find indications of higher temperatures and high partial melting, primitive rocks such as picritic basalts, dynamic uplift on the order of 1 km owing to thermal buoyancy, and the chemical and isotopic characteristics of entrained deep mantle material (Campbell, 2007; Duncan & Richards, 1991). In contrast, the plate-edge extension hypothesis predicts only shallow melting at normal mantle temperatures in relation to a tear in the lithosphere or a divergent plate boundary (Foulger, 2007). This model is not yet well developed and is unspecific about how large quantities of magma are created or about a particular eruption geometry, timing, or evolution. Presumably, these factors are dictated by geometry, timing, and evolution of the lithospheric breach or plate boundary and the geometry of the fertile patch of mantle beneath.

Results from Shatsky Rise investigations can be interpreted to support either plume or plate hypotheses because there is no clear "smoking gun" that supports either. Shatsky Rise eruptions began with Tamu Massif, building an enormous shield volcano underlain by thick crust and resulting in massive outpourings of lavas similar to flood basalts (Sager et al., 2013). Subsequently, the volume and style of volcanism shifted to lesser emplacements of smaller volume pillow flows (Sager et al., 2010). This sequence of events fits the plume head-to-tail transition expected from a plume head. While there are many indications that Shatsky Rise magmas are distinct from MORB and presumably from a different or modified source, the location and characteristics of the source are unclear. Heydolph et al. (2014) suggest that in aggregate, a number of observations tilt the balance toward a plume source for Shatsky Rise. They note that basalt samples show enriched trace element and isotope patterns (Sano et al., 2012) and that $^3He/^4He$ ratios are not like that of MORB (Heydolph et al., 2014). These helium isotope ratios are lower than MORB values, indicating that the magma was not from a primitive, undergassed source (Hanyu et al., 2014a). However, the $^3He/^4He$ ratios can be explained by a source containing subducted oceanic crust (Heydolph et al., 2014; Hanyu et al., 2014a), but whether this is a plume signature or not depends on one's view of mantle convection and mixing. One view is that the crustal component is recycled from the lower mantle or transition zone by mantle plumes (Heydolph et al., 2014), but it is also possible that the crustal component comes from subducted slabs lingering in the transition zone or even in the upper mantle (Hanyu et al., 2014a), consistent with the plate-edge extension hypothesis. Heydolph et al. (2014) also cite as support for the plume hypothesis that Shatsky Rise can be reconstructed near the edge of the Pacific large low shear velocity province, a potential source of mantle plumes (Torsvik, Smethhurst, Burke, & Steinberger, 2006; Torsvik, Muller,

Van der Voo, Steinberger, & Gaina, 2008). This is not a strong proof because reconstruction of the paleoposition of the Pacific plate prior to approximately 80 Ma is highly uncertain (e.g., Wessel, Harada, & Kroenke, 2006). The increasing heterogeneity of the Shatsky Rise source with time is also interpreted as a plume characteristic (Heydolph et al., 2014) based on certain plume models (Farnetani, Legras, & Tackley, 2002). Again, this is not a definitive proof because such models assume initial heterogeneity and show that it can be preserved despite upwelling through the mantle, not that it is an essential characteristic of a mantle plume. In addition, the difference in vanadium isotope signature between Shatsky Rise samples and MORB (Prytulak et al., 2013) is inferred as further evidence favoring a plume origin (Heydolph et al., 2014). Studies of vanadium isotope differences are in their infancy, so although this observation implies a component of non-MORB source, it is not necessarily evidence that the source for Shatsky Rise magmas was a plume.

If evidence for a plume is lacking, should the plate-edge extension hypothesis be preferred? There is certainly strong evidence to link oceanic plateaus and plate extension zones, particularly near triple junctions. Virtually every western Pacific oceanic plateau occurs along or near the trace of a triple junction (Sager, 2005) and many others (e.g., Azores; Rio Grande Rise-Walvis Ridge, Ninetyeast Ridge) can be linked to spreading ridges or triple junctions. Shatsky Rise clearly erupted along the path of a triple junction with geochronology data implying that the Tamu, Ori, and Shirshov volcanic edifices formed virtually atop the junction (Heaton & Koppers, 2014; Mahoney et al., 2005). This association is unlikely to have occurred by chance (Sager, 2005). In addition, the plate-edge extension hypothesis could explain the apparent lack of dynamic uplift that is implied by the dearth of evidence for significant subaerial exposure on Shatsky Rise summits (Korenaga, 2005; Sager et al., 2010, 2013). Models of uplift caused by the thermal buoyancy of hotter-than-normal mantle predict greater elevations of 0.5–4.0 km (Farnetani & Richards, 1994; Griffiths & Campbell, 1991; Monnereau, Rabinowicz, & Arquis, 1993), yet Shatsky Rise drill cores provided evidence of shallow water, but not significant subaerial exposure (Sager et al., 2010). Further, seismic data also show no evidence of subaerial erosion (Sager et al., 2013), whereas subsidence data and geochemistry give no compelling evidence for elevated mantle temperatures in the upper mantle beneath Shatsky Rise (e.g., Shimizu, Shimizu, Sano, Matsubara, & Sager, 2013; Heydolph et al., 2014).

Although these discrepancies favor the plate extension hypothesis, it can be argued that the path of the triple junction was determined by heat and uplift of the plume (Sager et al., 1999) instead of the other way around. Further, it is unclear where fertile subducted crust in the mantle beneath Shatsky Rise would have come from, since the Central Pacific has not been the site of subduction during the past several hundred million years (Seton et al., 2012; Torsvik et al., 2012). It is also unclear whether melting of fertile mantle in plate extension zones can produce the high degrees of partial melting and extract copious magmas to build a huge edifice such as Shatsky Rise (Foulger, 2007). As things stand, we cannot definitively accept either the plume or the plate hypothesis.

One could say that Shatsky Rise studies have been ineffectual because the plume versus plate-edge debate is still unsettled. However, coring from Expedition 324 recovered a large amount of basaltic material that has been the nucleus for a new generation of studies about the (isotope) geochemistry, geochronology, and mantle geodynamics in relation to oceanic plateau formation. This research has spawned new proposals, new cruises to sample and survey Shatsky Rise, and continued research about the mantle sources of LIPs. The debate has moved forward since the mid-1990s when the focus was on simple oceanic plateau geology, simple thermal plumes, and simple mantle convection. In 2014, our view of all those subjects is more complex, partly as a result of unexpected findings that came from drilling oceanic plateaus during DSDP, ODP, and IODP. Indeed, the present results suggest that the plume versus plate-edge dichotomy may be overly simplistic and artificial. Recent modeling suggests that plumes can have a variety of types (both thermal and thermochemical) and sizes and shapes (e.g., Farnetani & Samuel, 2005), and thus, multiple plume types may exist (Koppers, 2011). These complications require more complete sampling of oceanic plateaus to understand the variations in timing, structure, and chemistry.

4.3.6 LARGE-SCALE MANTLE MOVEMENTS TRACED BY SEAMOUNT TRAILS

Sir Arthur Holmes in his 1945 geology textbook (Holmes, 1945) predicted that akin to the atmosphere and hydrosphere, Earth's mantle must be convecting—albeit at a much slower pace. The nature of mantle convection remains an enigma, because we have little means of studying the deep Earth directly. By now, plate tectonics is commonplace, and scientists accept the notion of lithospheric plates moving across Earth's surface, which ultimately results in their subduction. To maintain a (mass) balance, that also means that the subducted plates need to be redistributed in the mantle by large-scale mantle movements (e.g., Anderson & Natland, 2014). The downgoing mass transport is well imaged using mantle tomography (Van der Hilst, Widiyantoro, & Engdahl, 1997), yet mantle upwelling, especially that of narrow mantle plumes, cannot easily be seen using similar techniques (Montelli et al., 2004; Wolfe et al., 2009). This leaves the study of *primary* hotspot trails, such as Hawaii and Louisville, as one of the few means to understand upwelling mantle plumes and how these behave within a convecting Earth. Scientific ocean drilling has been instrumental in tracing large-scale mantle movements by deciphering the paleolatitude histories of mantle plumes that are recorded in seamount trails.

ODP Leg 197 and IODP Expedition 330 provided compelling evidence for motions of the Hawaiian and Louisville mantle plumes, but it turns out that these motions were not identical between approximately 80 and 50 Ma. While Hawaii showed a large southern shift of approximately 15° in its paleolatitude

(Tarduno et al., 2003), the Louisville plume exhibited only a limited motion up to 3°–5° (Koppers, Yamazaki, Geldmacher, Gee, et al., 2012). Important conclusions can be drawn from these results, namely, that mantle plumes are not fixed with respect to the deep mantle and that they move independently from each other. Computer simulations of large-scale mantle flow (Koppers et al., 2004; Steinberger, 2000; Steinberger et al., 2004) corroborate these plume behaviors, indicating that mantle plume motions are likely governed by the geometry of subduction zones and their shifts over time and the location of large-scale upwellings, in particular those of superplumes. In the northern hemisphere, subduction of the Pacific Plate occurred toward the north around 80 Ma as at that time the Izu–Bonin–Mariana subduction zone was not yet active. With the Pacific Superplume located to the south of Hawaii, such a configuration would lead to a strong north-to-south "mantle wind" that affected the Hawaiian plume (Tarduno et al., 2003). Alternatively, the presence of a midocean spreading center close by might have caused the Hawaiian plume to be captured in an upper mantle upwelling and causing it to be deflected more strongly (Tarduno et al., 2009). In the southern hemisphere, however, subduction around 80 Ma was toward the west and into the precursor of the Kermadec–Tonga Trench, thus leading to a strong west-to-east mantle flow that may have caused a more pronounced eastern motion for the Louisville plume, with a minimal component of north–south motion (Koppers, Yamazaki, Geldmacher, Gee, et al., 2012).

Interestingly, the motion of mantle plumes also has a direct effect on the formation of seamount trails, in terms of the morphology, direction and age progressions recorded in them. Seamount trails that formed on the same plate should have spatial and age patterns that match plate motion parameters (Duncan & Clague, 1985; Koppers et al., 2001; Wessel & Kroenke, 2008; Wessel et al., 2006) assuming that their associated hotspots remained fixed in the mantle. For example, the sharp "bend" in the Hawaiian-Emperor seamount trail could reflect a drastic change in the direction of Pacific plate motion, and it also could result from a large amount of plume motion (Tarduno et al., 2009) or a combination of both (Duncan & Keller, 2004). If plate motion is the main cause of the HEB, such a sharp bend is expected to be present in other Pacific seamount trails and also around the same time. However, there is no evidence for coeval bends across the Pacific (Koppers & Staudigel, 2005) and the Louisville seamount trail lacks a sharp bend altogether, whereby the 'faint' bend that may reflect the HEB is roughly 3 Myr older (Koppers et al., 2011). Considering the paleolatitude evidence, it is becoming more and more likely that the HEB in Hawaii is simply caused by the cessation of the extraordinarily large north-to-south motion between 80 and 50 Ma.

4.3.7 CONCLUSIONS AND FUTURE WORK

The first phases of scientific ocean drilling barely scratched oceanic plateaus, typically coring 50–250 m into the upper surface at about half a dozen sites.

Unsurprisingly, the recovered cores have posed as many questions as they have answered because oceanic plateau geology is more complex than simple models can predict. Scientific drilling remains the only feasible way to obtain the samples required to understand oceanic plateau formation, structure, and evolution. Nevertheless, because of the extremely large size and thickness of oceanic plateaus, drilling must be complemented by state-of-the art geophysical investigations, particularly seismic imaging. The way forward for studying of oceanic plateaus therefore will not be easy or inexpensive. This likely precludes drilling deeply into the middle of an oceanic plateau, even though the technology exists, in the form of the drillship *Chikyu*, for deep-penetration riser drilling. A profitable approach is to drill at the lower edges of oceanic plateaus, in an effort to recover the longest-traveled flows over a long period of time, in order to gain a more complete picture of oceanic plateau history (Neal et al., 2008). Moreover, drilling through sections of sediments, located proximal in nearby ocean basins may document its entire history (Neal et al., 2008). Both approaches are nevertheless challenging because they require deep drilling (in terms of water depth and penetration) to recover complete sedimentary sections that record the entire history of a plateau and that also can be cored with good recovery (which has proven difficult in Mesozoic submarine sedimentary sections). Interest in oceanic plateau formation and evolution is still high because of the possible clues these features can give about mantle convection, geochemistry, environmental impacts, and the flux of material from the interior of the Earth to its surface.

In a similar way, scientific ocean drilling has targeted only a few seamount trails, whereby most emphasis has been laid on two *primary* hotspots, Hawaii and Louisville, which are not necessarily representative of most of the hundreds of thousands of seamounts found in the ocean basins. These expeditions have shown that hotspots are good recorders of past large-scale mantle flow, with ODP Leg 197 defining the motion of the Hawaiian hotspot in Earth's mantle, and IODP Expedition 330 showing the independent motion between the Louisville and Hawaiian hotspots, thus defining mantle flow within an ocean basin. Additional drilling of hotspots in other ocean basins is essential to understand how these motions relate to global mantle convection patterns. Future seamount drilling will also test the hotspot reference frame that has been used to provide an absolute framework for studying geographic climate responses into geologic time, whether true polar wander has occurred during Earth's history, what the nature of nonhotspot seamount trails may be, and it will provide a better understanding of mantle end members for the study of mantle geochemistry. Ideal targets are the long-lived Walvis Ridge hotspot trail in the Atlantic Ocean and the Réunion hotspot in the Indian Ocean that connects the Deccan Traps with the Chagos-Laccadive Ridge and the southern part of the Mascarene Plateau. Although much has been done and achieved by exploration of seamounts and oceanic plateaus during DSDP, ODP, and IODP, our knowledge of these features has become more complex and nuanced, leading to new hypotheses and questions, requiring future ocean drilling to seek the answers.

REFERENCES

van Andel, T. H., & Heath, G. R. (1973). Geological results of Leg 16: the central equatorial Pacific west of the East Pacific Rise. *Initial Reports DSDP, 16*, 937–949. http://dx.doi.org/10.2973/proc.dsdp.16.137.1973.

Anderson, D. L. (2001). Top down tectonics? *Science, 293*, 2016–2018.

Anderson, D. L., & Natland, J. H. (2014). Mantle updrafts and mechanisms of oceanic volcanism. *Proceedings of National Academy of Sciences, 111*(41), E4298–E4304 http://dx.doi.org/10.1073/pnas.1410229111.

Anderson, D. L., Tanimoto, T., & Zhang, Y.-S. (1992). Plate tectonics and hotspots: the third dimension. *Science, 256*, 1645–1651.

Baker, P. E., Castillo, P. R., & Condliffe, E. (1995). Petrology and geochemistry of igneous rocks from Allison and Resolution guyots, Sites 865 and 866. *Proceedings of ODP, Scientific Results, 143*, 245–261. http://dx.doi.org/10.2973/odp.proc.sr.143.216.1995.

Ballard, R. D., Holcomb, R. T., & Van Andel, Tj H. (1979). The Galapagos Rift at 86°W: 3. Sheet flows, collapse pits, and lava lakes of the rift valley. *Journal of Geophysical Research, 84*, 5407–5422.

Barker, P. F. (1983). Tectonic evolution and subsidence history of the Rio Grande Rise. *Initial Reports DSDP, 72*, 953–976. http://dx.doi.org/10.2973/dsdp.proc.72.151.1983.

Bass, M. N., Moberly, R., Rhodes, J. M., Shih, C.-Y., & Church, S. E. (1973). Volcanic rocks cored in the central Pacific, Leg 17. *Initial Reports DSDP, 17*, 429–503. http://dx.doi.org/10.2973/dsdp.proc.17.114.1973.

Beier, C., Vanderkluysen, L., Regelous, M., Mahoney, J. J., & Garbe-Schönberg, D. (2011). Lithospheric control on geochemical composition along the Louisville Seamount Chain. *Geochemistry, Geophysics, Geosystems, 12*, Q0AM01. http://dx.doi.org/10.1029/2011GC003690.

Berger, W. H., Kroenke, L. W., Mayer, L. A., & Shipboard Scientific Party. (1991). Ontong java plateau, leg 130: synopsis of major drilling results. *Proceedings of ODP, Initial Reports, 130*, 497–537 odp.proc.ir.130.110.1991.

Bryan, S. E., Peate, I. U., Peate, D. W., Self, S., Jerram, D. A., Mawby, M. R., et al. (2010). The largest volcanic eruptions on Earth. *Earth Science Reviews, 102*, 207–229.

Campbell, I. H. (2007). Testing the plume theory. *Chemical Geology, 241*, 153–176.

Campbell, I. H., & Griffiths, R. W. (1990). Implications of mantle plume structure for the evolution of flood basalts. *Earth and Planetary Science Letters, 99*, 79–93.

Campbell, I. H., & Kerr, A. C. (2007). The Great plume debate: testing the plume theory. *Chemical Geology, 241*(3–4), 149–152.

Campbell, I. H. (2006). Large igneous provinces and the mantle plume hypothesis. *Elements, 1*, 265–269.

Cheng, Q., Park, K.-H., MacDougall, J. D., Zindler, A., Lugmair, G. W., Hawkins, J., et al. (1987). Isotopic evidence for a hot spot origin of the Louisville seamount chain. In B. H. Keating, P. Fryer, R. Batiza, & G. W. Boehlert (Eds.), *Seamounts, Islands and Atolls. Geophys. Monogr.*, (pp. 283–296).

Christie, D. M., Dieu, J. J., & Gee, J. S. (1995). Petrologic studies of basement lavas from northwest Pacific guyots. *Proceedings of ODP, Scientific Results, 144*, 495–512. http://dx.doi.org/10.2973/odp.proc.sr.144.028.1995.

Clague, D. A., Dalrymple, G. B., Wright, T. L., Klein, F. W., Koyanagi, R. Y., Decker, R. W., et al. (1989). The Hawaiian-Emperor chain. In E. L. Winterer, D. M. Hussong, & R. W. Decker (Eds.), *The Geology of North America: The Eastern Pacific Ocean and Hawaii.* (pp. 187–237). Boulder, CO: Geol. Soc. Amer.

Clague, D. A., & Calvert, A. T. (2009). Postshield stage transitional volcanism on Mahukona Volcano, Hawaii. *Bulletin of Volcanology, 71*(5), 533–539.

Coffin, M. F., & Eldholm, O. (1994). Large igneous provinces: crustal structure, dimensions, and external consequences. *Reviews of Geophysics, 32*, 1–36.

Coffin, M. F., Duncan, R. A., Eldholm, O., Fitton, J. G., Frey, F. A., Larsen, H. C., et al. (2006). Large igneous provinces and scientific ocean drilling—status quo and a look ahead. *Oceanography, 19*, 150–160.

Coffin, M. F., Pringle, M. S., Duncan, R. A., Gladczenko, T. P., Storey, M., Müller, R. D., et al. (2002). Kerguelen hotspot magma output since 130 Ma. *Journal of Petrology, 43*, 1121–1139.

Courtillot, V., Davaille, A., Besse, J., & Stock, J. (2003). Three distinct types of hotspots in the Earth's mantle. *Earth and Planetary Science Letters, 205*(3–4), 295–308.

Dorais, M. Exploring the Mineralogical heterogenities of the Louisville seamount trail. Geochemistry, Geophysics, Geosystems, submitted for publication.

Duncan, R. A. (1990). The volcanic record of the Réunion hotspot. *Procedings of ODP, Scientific Results, 115*, 3–10. http://dx.doi.org/10.2973/odp.proc.sr.115.206.1990.

Duncan, R. A., & Richards, M. A. (1991). Hotspots, mantle plumes, flood basalts, and true polar wander. *Reviews of Geophysics, 29*, 31–50.

Duncan, R. A., & Clague, D. A. (1985). Pacific plate motion recorded by linear volcanic chains. In A. E. A. Nairn, F. L. Stehli, & S. Uyeda (Eds.), *The ocean basins and margins* (pp. 89–121). New York: Plenum Press.

Duncan, R. A. (2004). Radiometric ages for basement rocks from the Emperor Seamounts, ODP Leg 197. *Geochemistry, Geophysics, Geosystems, 5*(8). http://dx.doi.org/10.1029/2004GC000704.

Duncan, R. A., & Keller, R. (2004). Radiometric ages for basement rocks from the Emperor Seamounts, ODP Leg 197. *Geochemistry Geophysics Geosystems, 5*(8).

Farnetani, C. G., Legras, B., & Tackley, P. J. (2002). Mixing and deformations in mantle plumes. *Earth and Planetary Science Letters, 196*, 1–15.

Farnetani, C. G., & Richards, M. A. (1994). Numerical investigations of the mantle plume initiation model for flood basalt events. *Journal of Geophysical Research, 99*, 13,813–13,833.

Farnetani, C. G., & Samuel, H. (2005). Beyond the thermal plume paradigm. *Geophysical Research Letters, 32*. http://dx.doi.org/10.1029/2005GL022360.

Fischer, A. G., & Heezen, B. C. (1971). Introduction. *Initial Reports DSDP, 6*, 3–16. http://dx.doi.org/10.2973/dsdp.proc.6.101.1971.

Fitton, J. G., & Godard, M. (2004). Origin and evolution of magmas of the Ontong Java Plateau. In J. G. Fitton, J. J. Mahoney, P. J. Wallace, & A. D. Saunders (Eds.), *Origin and evolution of the Ontong Java plateau Spec. Pub., the Geol. Soc.: Vol. 229.* (pp. 151–178). London.

Foulger, G. R. (2007). The "plate" model for the genesis of melting anomalies. In G. R. Foulger, & D. M. Jurdy (Eds.), *Plates, plumes, and Planetary processes Spec. Paper, Geol. Soc. Amer.: Vol. 430.* (pp. 1–28). Boulder, CO.

Geldmacher, J., Hoernle, K., Bogaard, P. v. d., Duggen, S., & Werner, R. (2005). New 40Ar/39Ar age and geochemical data from seamounts in the Canary and Madeira volcanic provinces: Support for the mantle plume hypothesis. *Earth and Planetary Science Letters, 237*(1-2), 85–101.

Geldmacher, J., van den Bogaard, P., Heydolph, K., & Hoernle, K., (2014). The age of the Earth's largest volcano: Tamu Massif on Shatsky Rise (northwest Pacific Ocean). *International Journal of Earth Science*. http://dx.doi.org/10.1007/s00531-014-1078-6.

Gladczenko, T., Coffin, M. F., & Eldholm, O. (1997). Crustal structure of the Ontong Java Plateau: modeling of new gravity and existing seismic data. *Journal of Geophysical Research, 102*, 22,711–22,729.

Griffiths, R. W., & Campbell, I. H. (1991). Interaction of mantle plume heads with the Earth's surface and onset of small-scale convection. *Journal of Geophysical Research, 96*, 18,295–18,310.

Graham, D. W. (2002). Noble gas isotope geochemistry of mid-ocean ridge and ocean island basalts: characterization of mantle source reservoirs. In D. Porcelli, C. J. Ballentine, & R. Wieler (Eds.), *Noble gases in geochemistry and cosmoshemistry. Reviews in mineralogy & geochemistry* (pp. 247–317).

Haggerty, J. A., & Premoli Silva, I. (1995). Comparison of the origin and evolution of northwest Pacific guyots drilled during Leg 144. *Proceedings of ODP, Scientific Results, 144*, 935–949. http://dx.doi.org/10.2973/odp.proc.sr.144.2074.1995.

Hanyu, T., Shimizu, K., & Sano, T. (2014a). Noble gas evidence for the presence of recycled material in magma sources of Shatsky Rise. In C. Neal, T. Sano, E. Erba, & W. Sager (Eds.), *The origin, evolution, and environmental evolution of oceanic large igneous provinces, Special Paper*. Geological Society of America, Boulder, CO, in press.

Hanyu, T. (2014). Deep plume origin of the Louisville hotspot: Noble gas evidence. Geochemistry, Geophysics. *Geosystems, 15*(3), 565–576.

Hart, S. R., Hauri, E. H., Oschmann, L. A., & Whitehead, J. A. (1992). Mantle Plumes and Entrainment - Isotopic Evidence. *Science, 256*(5056), 517–520.

Hawkins, J. W., Lonsdale, P. F., & Batiza, R. (1987). Petrologic evolution of the Louisville seamount chain. In B. H. Keating, P. Fryer, R. Batiza, & G. W. Boehlert (Eds.), *Seamounts, Islands and Atolls. Geophys. Monogr.*, (pp. 235–254). Geophysical Union Washington, DC.

Heaton, D. E., & Koppers, A. A. P. (2014). *Constraining the rapid construction of Tamu Massif at a 146 Myr old triple junction, Shatsky Rise* Abstract 4093, 2014 Goldschmidt Conference, Sacramento, CA, 8-13 June 2014.

Heydolph, K., Murphy, D. T., Geldmacher, J., Romanova, I. V., Greene, A., Hoernle, K., et al. (2014). Plume versus plate origin for the Shatsky Rise oceanic plateau (NW Pacific): Insights from Nd, Pb, and Hf isotopes. *Lithos, 200–201*, 49–63.

Hillier, J. K., & Watts, A. B. (2007). Global distribution of seamounts from ship-track bathymetry data. *Geophysical Research Letters, 34*(13).

Hillier, J. K. (2007). Pacific seamount volcanism in space and time. *Geophysical Journal International, 168*(2), 877–889.

Hoffman, A. W., & White, W. M. (1982). Mantle plumes from ancient oceanic crust. *Earth and Planetary Science Letters, 57*, 421–436.

Holmes, A. (1945). Principles of Physical Geology (Edinburgh: Thomas Nelson and Sons, 1944 and New York: Ronald Press).

Houghton, R. L. (1979). Petrology and geochemistry of basaltic rocks recovered on Leg 43 of the Deep Sea Drilling Project. *Initial Reports DSDP, 43*, 721–738. http://dx.doi.org/10.2973/dsdp.proc.43.133.1979.

Husen, A., Almeev, R. R., Holtz, F., Koepke, J., Sano, T., & Mengel, K. (2013). Geothermobarometry of basaltic glasses from the Tamu Massif, Shatsky Rise oceanic plateau. *Geochemistry, Geophysics, Geosystems, 14*. http://dx.doi.org/10.1002/ggge.20231.

Ingle, S., & Coffin, M. F. (2004). Impact origin for the greater Ontong Java Plateau? *Earth and Planetary Science Letters, 218*(1-2), 123–134.

Integrated Ocean Drilling Program Initial Science Plan. (2003–2013). Earth, Oceans and Life: Scientific Investigation of the Earth System using Multiple Drilling Platforms and New Technologies. International Working Group Support Office, Washington, DC 20036, USA, 110 pages.

Jackson, E. D., & Schlanger, S. O. (1976). *Initial Reports DSDP*, Vol. 33, 915–927. http://dx.doi.org/10.2973/dsdp.proc.33.136.1976.

Jackson, E. D., Bargar, K. E., Fabbi, B. P., & Heropoulis, C. (1976). Petrology of the basaltic rocks drilled on Leg 33 of the Deep Sea Drilling Project. *Initial Reports DSDP, 33*, 571–630. http://dx.doi.org/10.2973/dsdp.proc.33.120.1976.

Jackson, E. D., Koizumi, I., Dalrymple, G. B., Clague, D. A., Kirkpatrick, R. J., & Greene, H. G. (1980). Introduction and summary of results from DSDP Leg 55, the Hawaiian-Emperor hot-spot experiment. *Initial Reports DSDP*, *55*, 5–31. http://dx.doi.org/10.2973/dsdp.proc.55.101.1980.

Jackson, M. G., & Jellinek, A. M. (2013). Major and trace element composition of the high ^{3}He/^{4}He mantle: Implications for the composition of a nonchrondritic Earth. *Geochemistry, Geophysics, Geosystems*, *14*, 2954–2976. http://dx.doi.org/10.1002/ggge.20188.

Jackson, M., Kurz, M., Hart, S., & Workman, R. (2007). New Samoan lavas from Ofu Island reveal a hemispherically heterogeneous high 3He/4He mantle. *Earth and Planetary Science Letters*, *264*(3-4), 360–374.

Janney, P. R., & Baker, P. E. (1995). Petrology and geochemistry of basaltic clasts and hyaloclastites from volcaniclastic sediments at Site 869. *Proceedings of ODP, Scientific Results*, *143*, 263–276. http://dx.doi.org/10.2973/odp.proc.sr.143.217.1995.

Jerram, D. A., & Widdowson, M. A. (2005). The anatomy of continental flood basalt provinces: geological constraints on the processes and products of flood volcanism. *Lithos*, *79*, 385–405.

Kirkpatrick, R. J., Clague, D. A., & Freisen, W. (1980). Petrology and geochemistry of volcanic rocks, DSDP Leg 55, Emperor Seamount chain. *Initial Reports DSDP*, *55*, 509–557. http://dx.doi.org/10.2973/dsdp.proc.55.120.1980.

Koivisto, E. A., Andrews, D. L., & Gordon, R. G. (2013). Tests of fixity of the Indo-Atlantic hot spots relative to Pacific hot spots. *Journal of Geophysical Research*, *119*. http://dx.doi.org/10.1002/2013JB010413.

Kono, M. (1980). Paleomagnetism of DSDP Leg 55, Honolulu, Hawaii to Yokohama, Japan, basalts and implications for the tectonics of the Pacific plate. In E. D. Jackson, I. Koisumi, et al. (Eds.), *Initial Reports DSDP*, *55*, 737–752.

Koppers, A. A. P., Sano, T., Natland, J. H., Widdowson, M., Almeev, R., Greene, A. R., et al. (2010a). Massive basalt flows on the southern flank of Tamu Massif, Shatsky Rise: A reappraisal of ODP Site 1213 basement. *Proc. IODP, Integrated Ocean Drilling Program – Management International, Tokyo*. http://dx.doi.org/10.2204/ iodp.proc.324.109.2010.

Koppers, A. A. P., Staudigel, H., Christie, D. M., Dieu, J. J., & Pringle, M. S. (1995). Sr–Nd–Pb isotope geochemistry of Leg 144 west Pacific guyots: Implications for the evolution of the "SOPITA" mantle anomaly. *Proceedings of ODP, Scientific Results*, *144*, 535–545. http://dx.doi.org/10.2973/odp.proc.sr.144.031.1995.

Koppers, A. A. P., & Staudigel, H. (2005). Asynchronous bends in Pacific seamount trails: a case for extensional volcanism? *Science*, *307*(5711), 904–907.

Koppers, A. A. P., & Watts, A. B. (2010). Intraplate Seamounts as a window into deep earth processes. *Oceanography*, *23*(1), 42–57.

Koppers, A. A. P. (2011). Mantle plumes persevere. *Nature Geoscience*, *4*(12), 816–817.

Koppers, A. A. P., Duncan, R. A., & Steinberger, B. (2004). Implications of a non-linear ^{40}Ar/^{39}Ar age progression along the Louisville seamount trail for models of fixed and moving hotspots. *Geochemistry, Geophysics, Geosystems*, *5*(6), Q06L02.

Koppers, A. A. P., Gowen, M. D., Colwell, L. E., Gee, J. S., Lonsdale, P. F., Mahoney, J. J., et al. (2011). New ^{40}Ar/^{39}Ar age progression for the Louisville hot spot trail and implications for inter–hot spot motion. *Geochemistry, Geophysics, Geosystems*, *12*.

Koppers, A. A. P., Morgan, J. P., Morgan, J. W., & Staudigel, H. (2001). Testing the fixed hotspot hypothesis using Ar-40/Ar-39 age progressions along seamount trails. *Earth and Planetary Science Letters*, *185*(3–4), 237–252.

Koppers, A. A. P., Staudigel, H., Pringle, M. S., & Wijbrans, J. R. (2003). Short-lived and discontinuous intraplate volcanism in the South Pacific: hot spots or extensional volcanism? *Geochemistry, Geophysics, Geosystems*, *4*(10).

Koppers, A. A. P., Yamazaki, T., & Geldmacher, J. (2010b). Louisville Seamount Trail: implications for geodynamic mantle flow models and the geochemical evolution of primary hotspots. *IODP Sci Prospectus, 330*, 85.

Koppers, A. A. P., Yamazaki, T., Geldmacher, J., & Expedition 330 Scientists. (2013). IODP expedition 330: drilling the Louisville seamount trail in the SW Pacific. *Scientific Drilling, 15*, 11–22.

Koppers, A. A. P., Yamazaki, T., Geldmacher, J., & Expedition 330 Scientists. (2012). In *Proceedings IODP expedition 330: Tokyo*. Integrated Ocean Drilling Program Management International, Inc. http://dx.doi.org/10.2204/iodp.proc.330.2012.

Koppers, A. A. P., Yamazaki, T., Geldmacher, J., Gee, J. S., Pressling, N., Hiroyuki, H., et al. (2012). Limited latitudinal mantle plume motion for the Louisville hotspot. *Nature Geoscience*. http://dx.doi.org/10.1038/NGEO1638.

Koppers, A. A. P. (2009). Pacific Region. In G. R. Gillespie, & D. A. Clague (Eds.), *Encyclopedia of Islands* (pp. 702–715). Berkeley: University of California Press.

Korenaga, J. (2005). Why did not the Ontong Java Plateau form subaerially? *Earth and Planetery Science Letters, 234*, 385–399.

Korenaga, J., & Sager, W. W. (2012). Seismic tomography of Shatsky Rise by adaptive importance sampling. *Journal of Geophysical Research, 117*. http://dx.doi.org/10.1029/2012JB009248.

Korenaga, J., Kelemen, P. B., & Holbrook, W. S. (2002). Methods for resolving the origin of large igneous provinces from crustal seismology. *Journal of Geophysical Research, 107*, 2178. http://dx.doi.org/10.1029/2001JB001030.

Lonsdale, P. (1988). Geography and history of the Louisville hotspot chain in the southwest Pacific. *Journal of Geophysical Research Solid Earth and Planets, 93*(B4), 3078–3104.

MacDonald, G. A., & Abbott, A. T. (1970). *Volcanoes in the Sea*. Honolulu: University of Hawaii press. 441 pp.

Machida, S., Hirano, N., & Kimura, J.-I. (2009). Evidence for recycled plate material in Pacific upper mantle unrelated to plumes. *Geochimica et Cosmochimoca Acta, 73*, 3028–3037.

Mahoney, J. J., Duncan, R. A., Tejada, M. L. G., Sager, W. W., & Bralower, T. J. (2005). Jurassic-Cretaceous boundary age and mid-ocean-ridge-type mantle source for Shatsky Rise. *Geology, 33*(3), 185–188.

McDougall, I. (1964). Potassium–argon ages from lavas of the Hawaiian Islands. *Geological Society of America Bulletin, 75*(2), 107–128.

Miura, S., Suyehiro, K., Shinohara, M., Takahashi, N., Araki, E., & Taira, A. (2004). Seismological structure and implications of collision between the Ontong Java Plateau and Solomon Island Arc from ocean bottom seismometer–airgun data. *Tectonophysics, 389*(3), 191–220. http://dx.doi.org/10.1016/j.tecto.2003.09.029.

Monnereau, M., Rabinowicz, M., & Arquis, E. (1993). Mechanical erosion and reheating of the lithosphere: a numerical model for hotspot swells. *Journal of Geophysical Research, 98*, 809–823.

Montelli, R., Nolet, G., Dahlen, F. A., Masters, G., Engdahl, E. R., & Hung, S. H. (2004). Finite-frequency tomography reveals a variety of plumes in the mantle. *Science, 303*(5656), 338–343.

Moore, T. C., Jr., Rabinowitz, P. D., Borella, P. E., Shackleton, N. J., & Boersma, A. (1984). History of the walvis Ridge. *Initial Reports DSDP, 74*, 873–894. http://dx.doi.org/10.2973/dsdp.proc.74.131.1984.

Morgan, W. J. (1971). Convection plumes in the lower mantle. *Nature, 230*(5288), 42–43.

Morgan, W. J. (1972). Deep mantle convection plumes and plate motions. *American Association of Petroleum Geologists Bulletin, 56*, 42–43.

Morgan, W. J. (1981). Hotspot tracks and the opening of the Atlantic and Indian Oceans. In C. Emiliani (Ed.), *The sea* (pp. 443–487). New York: Wiley & Sons.

Nakanishi, M., & Gee, J. S. (1995). Paleomagnetic investigations of volcanic rocks: paleolatitudes of the norhtwestern Pacific guyots. In J. A. Haggerty, I. Premoli Silva, F. Rack, & M. K. McNutt (Eds.), Proc. ODP, Sci. Results, 144: College Station, TX (Ocean Drilling Program), (pp. 585–604).

Nakanishi, M., Sager, W. W., & Klaus, A. (1999). Magnetic lineations within Shatsky Rise, northwest Pacific Ocean: Implications for hot spot-triple junction interaction and oceanic plateau formation. *Journal of Geophysical Research, Solid Earth, 104*(B4), 7539–7556.

Natland, J., & Winterer, E. L. (2005). Fissure control on volcanic action in the Pacific. In G. R. Foulger, J. Natland, D. Presnall, & D. L. Anderson (Eds.), *Plumes, plates and paradigms* (pp. 687–710). Geological Society of America Book.

Neal, C. R., Coffin, M. F., Arndt, N. T., Duncan, R. A., Eldholm, O., Erba, E., et al. (2008). Investigating large igneous province formation and associated paleoenvironmental events: a white paper for scientific drilling. *Scientific Drilling, 6*, 4–18. http://dx.doi.org/10.2204/iodp.sd.6.01.2008.

Nichols, A. R. L., Beier, C., Brandl, P. A., Buchs, D. M., & Krumm, S. H. (2014). Geochemistry of volcanic glasses from the Louisville Seamount Trail (IODP Expedition 330): Implications for eruption environments and mantle melting. Geochemistry, Geophysics. *Geosystems, 15*(5), 1718–1738.

Niu, Y. L., Waggoner, D. G., Sinton, J. M., & Mahoney, J. J. (1996). Mantle source heterogeneity and melting processes beneath seafloor spreading centers: the East Pacific Rise, 18°–19°S. *Journal of Geophysical Research, 101*, 27711–27733.

O'Connor, J. M., Steinberger, B., Regelous, M., Koppers, A. A. P., Wijbrans, J. R., Haase, K. M., et al. (2013). Constraints on past plate and mantle motion from new ages for the Hawaiian-Emperor Seamount Chain. *Geochemistry, Geophysics, Geosystems, 14*(10), 4564–4584. http://dx.doi.org/10.1002/ggge.20267.

Pringle, M. S., & Duncan, R. A. (1995a). Radiometric ages of basement lavas recovered at Loen, Wodejabato, MIT, and Taikuyo-Daisan guyots, northwestern Pacific ocean. *Proceedings of ODP, Science Results, 143*, 547–557. http://dx.doi.org/10.2973/odp.proc.sr.143.033.1995.

Pringle, M. S., & Duncan, R. A. (1995b). Radiometric ages of basaltic lavas recovered at Sites 865, 866, and 869. *Proceedings of ODP, Science Results, 144*, 277–283. http://dx.doi.org/10.2973/odp.proc.sr.144.218.1995.

Prytulak, J., Nielsen, S. G., Ionov, D. A., Halliday, A. N., Harvey, J., Kelley, K. A., et al. (2013). The stable vanadium isotope composition of the mantle and mafic lavas. *Earth and Planetary Science Letters, 365*, 177–189.

Rea, D.K., Basov, I.A., Scholl, D.W., & Allan, J.F. (Eds.), 1995. Proc. ODP, Sci. Results, 145: College Station TX, (Ocean Drilling Program).

Richards, M. A., Duncan, R. A., & Courtillot, V. E. (1989). Flood basalts and hot-spot tracks: plume heads and tails. *Science, 246*, 103–107.

Richardson, S. H., Erlank, A. J., Reid, D. L., & Duncan, A. R. (1984). Major and trace elements and Nd and Sr isotope geochemistry of basalts from the Deep Sea Drilling Project Leg 74 Walvis Ridge transect. *Initial Reports DSDP, 74*, 739–754. http://dx.doi.org/10.2973/dsdp.proc.74.125.1984.

Royer, J.-Y., Peirce, J. W., & Weissel, J. K. (1991). Tectonic constraints on the hotspot formation of Ninetyeast Ridge. *Proceedings of ODP, Scientific Results, 121*, 763–776. http://dx.doi.org/10.2973/odp.proc.sr.121.122.1991.

Sager, W. W. (2005). What built Shatsky Rise, a mantle plume or ridge tectonics? In G. R. Foulger, J. H. Natland, D. C. Presnall, & D. L. Anderson (Eds.), *Plates, plumes, and paradigms Spec. Paper, Geol. Soc. Amer.: Vol. 388*. (pp. 721–733). Boulder, CO.

Sager, W. W., & Han, H.-C. (1993). Rapid formation of Shatsky rise oceanic plateau inferred from its magnetic anomaly. *Nature, 364*, 610–613.

Sager, W. W., Handschumacher, D. W., Hilde, T. W. C., & Bracey, D. R. (1988). Tectonic evolution of the northern Pacific plate and Pacific-Farallon-Izanagi triple junction in the Late Jurassic and Early Cretaceous (M21-M10). *Tectonophysics, 155*, 345–364.

Sager, W. W., Kim, J., Klaus, A., Nakanishi, M., & Khankishieva, L. M. (1999). Bathymetry of Shatsky Rise, northwest Pacific Ocean: Implications for oceanic plateau development at a triple junction. *Journal of Geophysical Research, 104*, 7557–7576.

Sager, W. W., Lamarche, A. J., & Kopp, C. (2005). Paleomagnetic modeling of seamounts near the Hawaiian-Emperor Bend. *Tectonophysics, 405,* 121–140.

Sager, W. W., Pueringer, M., Carvallo, C., Ooga, M., Housen, B., & Tominaga, M. (2014). Paleomagnetism of igneous rocks from Shatsky Rise and implications for oceanic plateau volcanism and paleolatitude. In C. Neal, T. Sano, E. Erba, and W. Sager (Eds.), *The origin, evolution, and environmental evolution of oceanic large igneous provinces,* Special Paper, Geological Society of America, Boulder, CO, in press.

Sager, W. W., Sano, T., & Geldmacher, J. (2011). How do oceanic plateaus form? Clues from drilling Shatsky Rise. *EOS, Trans AGU, 92,* 37–44.

Sager, W. W., Sano, T., Geldmacher, J., & Expedition 324 Scientists. (2010). *Proc IODP, 324, Integrated Ocean Drilling Program-Management International, Tokyo.* http://dx.doi.org/10.2204/iodp.proc.324.2010.

Sager, W. W., Zhang, J., Korenaga, J., Sano, T., Koppers, A. A. P., Widdowson, M., et al. (2013). An immense shield volcano within the Shatsky Rise oceanic plateau, northwest Pacific Ocean. *Nature Geoscience, 6,* 976–981.

Sandwell, D. T., & MacKenzie, K. R. (1989). Geoid height versus topography for oceanic plateaus and swells. *Journal of Geophysical Research, 94,* 7403–7418.

Sano, T., Shimizu, K., Ishikawa, A., Senda, R., Chang, Q., Kimura, J.-I., et al. (2012). Variety an origin of magmas on Shatsky Rise, northwest Pacific Ocean. *Geochemistry, Geophysics, Geosystems, 13.* http://dx.doi.org/10.1029/2012GC004235.

Scholl, D. S., & Creager, J. S. (1973). Geologic synthesis of Leg 19 (DSDP): far north Pacific, and Aleutian Ridge, and Bering Sea. *Initial Reports DSDP, 19,* 897–913. http://dx.doi.org/10.2973/dsdp.proc.19.137.1973.

Self, S., Thordardon, T., & Keszthelyi, L. (1997). Emplacement of continental flood basalt lava flows. In J. J. Mahoney, & M. Coffin (Eds.), *Large igneous provinces: Continental, oceanic, and planetary flood volcanism Geophys. Mon. Ser., Amer. Geophys. Union: Vol. 100.* (pp. 381–410). Washington, DC.

Seton, M., Müller, R. D., Zahirovic, S., Gaina, C., Torsvik, T., Shephard, G., et al. (2012). Global continental and ocean basin reconstructions since 200 Ma. *Earth-Science Reviews, 113,* 212–270.

Shimizu, K., Shimizu, N., Sano, T., Matsubara, N., & Sager, W. W. (2013). Paleo-elevation and subsidence of ~145 Ma Shatsky Rise inferred from CO_2 and H_2O in fresh volcanic glass. *Earth and Planetary Science Letters, 383,* 37–44.

Shipboard Scientific Party. (1971). Site 57. *Initial Reports DSDP, 6,* 493–537. http://dx.doi.org/10.2973/dsdp.proc.6.115.1971 158.

Shipboard Scientific Party. (1975a). Site 305: Shatsky Rise. *Initial Reports DSDP, 32,* 75–158. http://dx.doi.org/10.2973/dsdp.proc.32.104.1975.

Shipboard Scientific Party. (1975b). Site 306: Shatsky Rise. *Initial Reports DSDP, 32,* 159–191. http://dx.doi.org/10.2973/dsdp.proc.32.105.1975.

Shipboard Scientific Party. (1975c). Site 289. *Initial Reports DSDP, 30,* 231–398. http://dx.doi.org/10.2973/dsdp.proc.30.107.1975.

Shipboard Scientific Party. (1975d). Site 292. *Initial Reports DSDP, 31,* 67–129. http://dx.doi.org/10.2973/dsdp.proc.31.104.1975.

Shipboard Scientific Party. (1989a). Principle results. *Proceedings of ODP, Initial Reports, 119,* 5–14. http://dx.doi.org/10.2973/odp.proc.ir.119.1989.

Shipboard Scientific Party. (1989b). Principle results. *Proceedings of ODP, Initial Reports, 120,* 73–85. http://dx.doi.org/10.2973/odp.proc.ir.120.108.1989.

Shipboard Scientific Party. (2000). Leg 183 summary: Kerguelen plateau-Broken Ridge— a large igneous province. *Proceedings of ODP, Initial Reports, 183,* 1–101, http://dx.doi.org/odp.proc.ir.183.101.2000.

Shipboard Scientific Party. (2001). Leg 192 summary. *Proceedings of ODP, Initial Reports, 192,* 1–75, http://dx.doi.org/odp.proc.ir.192.101.2001.

Shipboard Scientific Party. (2002a). Leg 197 summary. *Proceedings of ODP, Initial Reports, 197,* 1–92, http://dx.doi.org/odp.proc.ir.197.101.2002.

Shipboard Scientific Party. (2002b). Site 1213. *Proceedings of ODP, Initial Reports, 198,* 1–110, http://dx.doi.org/odp.proc.ir.198.109.2002.

Staudigel, H., & Clague, D. A. (2010). The geological history of deep-sea volcanoes: biosphere, hydrosphere, and lithosphere interactions. *Oceanography, 23*(1), 58–71.

Staudigel, H., & Koppers, A. A. P. (2015). Seamounts and island building. In: H. Sigurdsson, B, Houghton, S, McNutt, H, Rymer, & J, Stix (Eds.), *The encyclopedia of volcanoes, chapter 22. (In review).* Oxford, UK: Elsevier.

Steinberger, B., & Antretter, M. (2006). Conduit diameter and buoyant rising speed of mantle plumes: Implications for the motion of hot spots and shape of plume conduits. *Geochemistry, Geophysics, Geosystems, 7*(11), Q11018. http://dx.doi.org/10.1029/2006GC001409.

Steinberger, B., & Calderwood, A. R. (2006). Models of large-scale viscous flow in the Earth's mantle with constraints from mineral physics and surface observations. *Geophysical Journal International, 167*(3), 1461–1481.

Steinberger, B., & O'Connell, R. J. (1998). Advection of plumes in mantle flow: Implications for hotspot motion, mantle viscosity and plume distribution. *Geophysical Journal International, 132*(2), 412–434.

Steinberger, B. (2000). Plumes in a convecting mantle: models and observations for individual hotspots. *Journal of Geophysical Research B: Solid Earth, 105*(B5), 11127–11152.

Steinberger, B., Sutherland, R., & O'Connell, R. J. (2004). Prediction of Emperor-Hawaii seamount locations from a revised model of global plate motion and mantle flow. *Nature, 430*(6996), 167–173.

Stracke, A., Hofmann, A. W., & Hart, S. R. (2005). FOZO, HIMU, and the rest of the mantle zoo. *Geochemistry, Geophysics, Geosystems, 6*(5).

Tarduno, J. A., & Sager, W. W. (1995). Polar standstill of the Mid-Cretaceous Pacific plate and its geodynamic consequences. *Science, 269,* 956–959.

Tarduno, J., Bunge, H.-P., Sleep, N., & Hansen, U. (2009). The Bent Hawaiian-Emperor hotspot Track: Inheriting the mantle wind. *Science, 324*(5923), 50–53.

Tarduno, J. A., & Cottrell, R. D. (1997). Paleomagnetic evidence for motion of the Hawaiian hotspot during formation of the Emperor seamounts. *Earth and Planetary Science Letters, 153*(3–4), 171–180.

Tarduno, J. A., Cottrell, R. D., & Smirnov, A. V. (2006). The paleomagnetism of single silicate crystals: Recording geomagnetic field strength during mixed polarity intervals, superchrons, and inner core growth. *Reviews of Geophysics, 44*(1).

Tarduno, J. A. (2007). On the motion of Hawaii and other mantle plumes. *Chemical Geology, 241*(3–4), 234–247.

Tarduno, J. A., Duncan, R. A., Scholl, D. W., Cottrell, R. D., Steinberger, B., Thordarson, T., et al. (2003). The Emperor Seamounts: southward motion of the Hawaiian hotspot plume in earth's mantle. *Science, 301*(5636), 1064–1069.

Tauxe, L. (2006). Long-term trends in paleointensity: The contribution of DSDP/ODP submarine basaltic glass collections. *Physics of the Earth and Planetary Interiors, 156*(3-4), 223–241.

Tauxe, L., Gee, J. S., Steiner, M. B., & Staudigel, H. (2013). Paleointensity results from the Jurassic : New constraints from submarine basaltic glasses of ODP Site 801C. *Geochemistry, Geophysics, Geosystems 8.* http://dx.doi.org/10.1002/2013GC004704.

Torskvik, T. H., Müller, R. D., Van der Voo, R., Steinberger, B., & Gaina, C. (2008). Global plate motion frames: toward a unified model. *Reviews of Geophysics, 46.* http://dx.doi.org/10.1029 /2007RG000227.

Torsvik, T. H., Smethhurst, M. A., Burke, K., & Steinberger, B. (2006). Large igneous provinces generated from the margins of the large low-velocity provinces in the deep mantle. *Geophysical Journal International, 167,* 1447–1460.

Torsvik, T. H., Van der Voo, R., Preeden, U., Niocaill, C. M., Steinberger, B., Doubrovine, P. V., et al. (2012). Phaerozoic polar wander, palaeogeography and dynamics. *Earth-Science Reviews, 114,* 325–368.

Van der Hilst, R. D., Widiyantoro, S., & Engdahl, R. L. (1997). Evidence for deep mantle circulation from global tomography. *Nature, 386,* 578–584.

Vanderkluysen, L., Mahoney, J. J., Koppers, A. A. P., Beier, C., Regelous, M., Gee, J. S., & Lonsdale, P. F. (2014). Louisville Seamount Chain: Petrogenetic processes and geochemical evolution of the mantle source. *Geochemistry, Geophysics, Geosystems.* http://dx.doi.org/10.1002/2014gc005288.

Watts, A. B., Sandwell, D. T., Smith, W. H. F., & Wessel, P. (2006). Global gravity, bathymetry, and the distribution of submarine volcanism through space and time. *Journal of Geophysical Research B: Solid Earth, 111*(8).

Weaver, B. L., Marsh, N. G., & Tarney, J. (1983). Trace element geochemistry of basaltic rocks recovered at site 516, Rio Grande rise, deep sea drilling project leg 72. *Initial Reports DSDP, 72,* 451–456. http://dx.doi.org/10.2973/dsdp.proc.72.114.1983.

Wessel, P., & Kroenke, L. W. (2008). Pacific absolute plate motion since 145 Ma: an assessment of the fixed hot spot hypothesis. *Journal of Geophysical Research-Solid Earth, 113*(B6), B06101. http://dx.doi.org/10.1029/2007JB005499.

Wessel, P., Harada, Y., & Kroenke, L. W. (2006). Toward a self-consistent, high-resolution absolute plate motion model of the Pacific. *Geochemistry, Geophysics, Geosystems, 7.* http://dx.doi.org /10.10029/2005GC001000.

Wessel, P., Sandwell, D. T., & Kim, S. S. (2010). The global seamount census. *Oceanography, 23*(1), 24–33.

Wilson, J. T. (1963). A possible origin of the Hawaiian Islands. *Canadian Journal of Physics, 41,* 863–870.

Wilson, J. T. (1965). Evidence from oceanic islands suggesting movement in the Earth. *Philosophical Transactions of the Royal Society of London, Mathematical Physical and Engineering Sciences, A 258,* 145–167.

Winterer, E. L. (1973). Regional problems. *Initial Reports DSDP, 17,* 911–922. http://dx.doi. org/10.2973/dsdp.proc.17.129.1973.

Winterer, E. L., & Sager, W. W. (1995). Synthesis of drilling results from the Mid-Pacific Mountains: regional context and implications. *Proceedings of ODP, Science Results, 143,* 497–535. http://dx.doi.org/10.2973/odp.proc.sr.143.245.1995.

Wolfe, C. J., Solomon, S. C., Laske, G., Collins, J. A., Detrick, R. S., Orcutt, J. A., et al. (2009). Mantle shear-wave velocity structure beneath the hawaiian hot spot. *Science, 326*(5958), 1388–1390.

Yamamoto, Y., & Tsunakawa, H. (2005). Geomagnetic field intensity during the last 5 Myr: LTD-DHT Shaw paleointensities from volcanic rocks of the Society Islands, French Polynesia. *Geophys. J. Int., 162,* 79–114.

Zhang, J., Sager, W.W., and Korenaga, J. (2014). Shatsky Rise oceanic plateau structure from 2D multichannel seismic reflection profiles and implications for oceanic plateau formation. In C. Neal, T. Sano, E. Erba, and W. Sager (Eds.), *The origin, evolution, and environmental evolution of oceanic large igneous provinces,* Special Paper, Geological Society of America, Boulder, CO, in press.

Chapter 4.4.1

Subduction Zones: Structure and Deformation History

Harold Tobin,[1,*] Pierre Henry,[2] Paola Vannucchi[3] and
Elizabeth Screaton[4]

*[1]Department of Geoscience, University of Wisconsin–Madison, Madison, WI, USA; [2]CEREGE,
Aix-Marseille Université, Marseille, France; [3]Department of Earth Sciences, University
of London, London, UK; [4]Department of Geological Science, University of Florida,
Gainesville, FL, USA*
*Corresponding author: E-mail: htobin@wisc.edu

4.4.1.1 INTRODUCTION

The transition from the Ocean Drilling Program (ODP) to the Integrated
Ocean Drilling Program (IODP) marked a substantial change in the focus of
subduction zone drilling investigations. Whereas the Deep Sea Drilling Project (DSDP) and ODP drilling (1972–2002) primarily focused on deformation
and fluid flow processes near the trench, the main driver for IODP investigations has been the understanding of large earthquakes and tsunamis. Between
2003 and 2013, IODP drilling expeditions in the Nankai Trough Kumano transect, the Costa Rica Osa Peninsula transect, and the Japan Trench have provided great insight into deformation processes in subduction zone forearcs.
In pursuit of IODP's objective of investigating the onset of seismogenic locking and rupture, the overarching aim of all these transects was to drill into
plate boundary fault systems, both at shallow, presumed aseismic levels and
at greater, presumed seismogenic, depths to study faults and their immediate
environment. Closely allied objectives included determination of ages of tectonic and stratigraphic events in marginal wedges, elucidation of hydrogeologic and diagenetic processes in these active margins, and monitoring stress
conditions and seismic activity. The bulk of 2003–2013 scientific drilling has
taken place as part of the Nankai Trough Seismogenic Zone Experiment (Nan-
TroSEIZE) and the Costa Rica Seismogenesis Project (CRISP). Both projects
have the ambitious goal of reaching the updip portion of the seismogenic zone
through drilling with the capabilities of *Chikyu*. As of 2014, drilling has not

Developments in Marine Geology, Volume 7. http://dx.doi.org/10.1016/B978-0-444-62617-2.00020-7

yet reached the plate boundary faults at the target depths of 5–6 km below the seafloor. Nonetheless, scientific ocean drilling has provided a wealth of information about the history and current state of the Nankai and Costa Rica margins. In addition, drilling of the Japan Trench (JFAST project) in response to the March 2011 Tohoku earthquake provided invaluable information on subduction zone deformation and coseismic fault friction during slip, by sampling a plate boundary fault zone in a place of known, large, and very recent displacement. In this chapter, we review the results of IODP drilling at all three margins. The largest amount of drilling activity has taken place at Nankai, which correspondingly is presented here in the most detail.

IODP drilling during this phase has examined both dominantly accretionary (Nankai Trough; Figures 4.4.1.1 and 4.4.1.2) and dominantly erosive (Costa Rica; Figures 4.4.1.3 to 4.4.1.5; and Japan Trench; Figure 4.4.1.6) subduction zones. Globally, in late Cenozoic time approximately half of convergent margins are accretionary, in which sediment is scraped off the downgoing plate, and half are nonaccretionary or undergoing subduction erosion, where material is removed from the upper plate and subducted (von Huene & Scholl, 1991).

Accretionary processes tend to dominate where incoming sediment thickness exceeds 1 km (±0.5 km) (Clift & Vannucchi, 2004). At the deformation

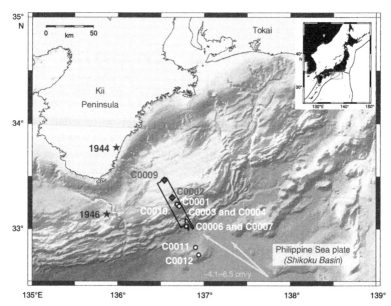

FIGURE 4.4.1.1 NanTroSEIZE transect location map with primary drill sites indicated (some sites left off for clarity). Black outline box is area of 2006 3D seismic survey. Circles are riserless drilled sites; red (gray in print versions) diamonds are riser-drilled site. Stars with dates indicate the epicentral location of the last megathrust M8+ events; in the 1944 Tonankai earthquake, rupture propagated up dip and leftward to fill the area under the large forearc basin. *Modified from Tobin et al. (2014).*

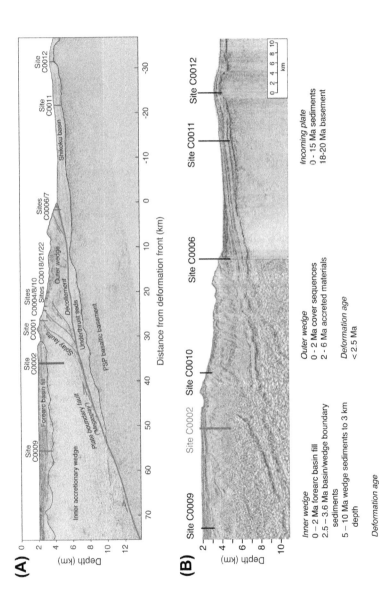

FIGURE 4.4.1.2 NanTroSEIZE composite depth seismic sections, extracted and spliced from several Kumano and Shikoku 2D and 3D seismic surveys, with drill sites shown. (A) Regional 2D seismic line, interpreted after Park et al. (2002) and Moore et al. (2007), with modifications based on Tobin et al. (2014). Annotations show the age of sediments and rocks, and of incorporation into the accretionary wedge. At riser Site C0002, drilling through Expedition 348 in early 2014 is shown in blue line; planned completion to ~5200 mbsf in upcoming drilling is dashed black line. (B) Lower panel: spliced line through most NanTroSEIZE drill sites, extracted from Kumano and Shikoku 3D surveys. Inset illustrates details of inner wedge, forearc basin, and megasplay domains, with sequential casing shown schematically for Site C0002. *After Tobin et al. (2014).*

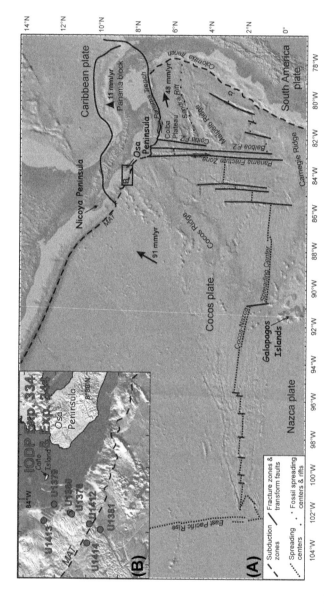

FIGURE 4.4.1.3 Tectonic setting of the Central America subduction zone (A) and location map for CRISP drilling project (B). The CRISP area is located where the NW flank of the Cocos Ridge is subducting. This setting is responsible for the shallow location of the seismogenic zone, as well as for the presence of enhanced subduction erosion.

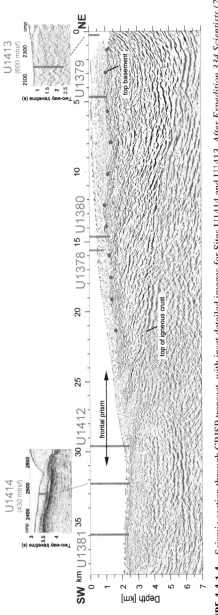

FIGURE 4.4.1.4 Seismic section through CRISP transect, with inset detailed images for Sites U1414 and U1413. *After Expedition 334 Scientists (2012).*

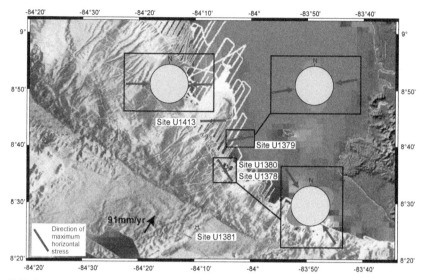

FIGURE 4.4.1.5 Bathymetric map of CRISP transect area with borehole-based maximum horizontal principal stress directions. *Modified from Vannucchi, Ujiie, et al. (2013).*

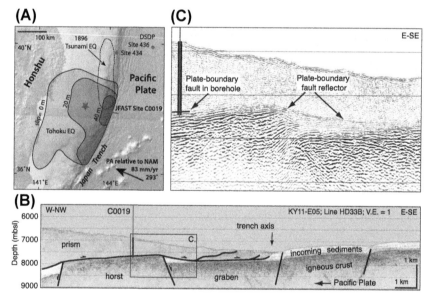

FIGURE 4.4.1.6 Summary diagram for the Expedition 343 (JFAST) site. *(After Chester et al. (2013).)* (A) Regional map with selected total slip contours during the March 11, 2011 M9 earthquake, and DSDP and IODP drill site locations. (B) Interpreted seismic line of drill site area. (C) Detail of seismic reflection image and drill location for Hole C0019; interpreted plate boundary reflector is indicated (but see also Strasser et al., 2013 for an alternative interpretation). *Copyright AAAS.*

front, the incoming sediment section is partitioned across a décollement into an offscraped, imbricate-thrust-dominated accretionary prism and an underlying package of underthrust sediment transported beneath the advancing wedge. The fold-and-thrust belt is generally formed by progressive forward development of each successive thrust by frontal imbrication ("in-sequence thrusting"), but this accretionary prism commonly also is internally deformed by younger "out-of-sequence thrusts" (OOST); (Boyer & Elliott, 1982). Further back from the deformation front, additional material may be added to the upper plate by underplating, in which the plate boundary fault steps downward into previously underthrust sediments. Large OOST systems, termed "megasplays" when they form the boundary between major tectonic domains in the subduction complex, may cut across the upper plate of accretionary subduction zones to the seafloor (e.g., Moore et al., 2007; Park, Tsuru, Kodaira, Cummins, & Kaneda, 2002). Megasplay faults both create and mark the boundary between the more consolidated, internally inactive inner wedge and the less consolidated actively deforming outer wedge (e.g., Kimura et al., 2007). These megasplay fault systems have received renewed attention because coseismic rupture of large earthquakes could branch along them and enhance vertical seafloor motion leading to tsunami generation—not a new idea (see e.g., Pflaker, 1969) but one which has gained renewed currency (e.g., Park et al., 2002).

In nonaccretionary systems, the entire incoming sediment column is subducted and additional material from the upper plate may be tectonically eroded. Subduction erosion acts through a combination of frontal and basal erosion. Frontal erosion removes slope sediment that has been transported downslope by small- and large-scale slumping and turbidity currents. In contrast, basal subduction erosion acts to remove material from the base of the overriding forearc. Subduction erosion is the dominant mechanism for recycling continental crust and is responsible for one third of the ~5.25 km^3 of continental crust consumed annually (Stern, 2011). The transfer of material from the upper plate to the subducting plate is indicated by subsidence of convergent margins, with local studies indicating total subsidence of several km and inferred subsidence rates exceeding several 100 m/Myr (Vannucchi et al., 2003). Extensional faulting is commonly observed in the subsiding region, but can only account for a small proportion of the inferred total subsidence (Vannucchi et al., 2003).

The occurrence of subduction erosion implies that the upper plate must have a region that is weaker than the plate boundary. One category of proposed mechanisms for subduction erosion is the abrasion model, invoking high friction along the plate boundary due to subducting basement topography on large and small scales. An alternate mechanism calls on fluid overpressure to hydrofracture and weaken the upper plate (von Huene, Ranero, & Vannucchi, 2004; Lallemand, Schnurle, & Malavieille, 1994; Murauchi and Ludwig, 1980). Onland, exposures suggest that hydrofracturing may be a common process in plate boundary shear zones (Vannucchi, Remitti, & Bettelli, 2008). Even if dominantly one or the other, margins clearly commonly pass through periods of both accretionary wedge construction and also of tectonic erosion, episodically. As discussed later in this chapter, sediment supply and distribution likely dominates this alternation.

Coulomb wedge or critical taper theory (Davis, Suppe, & Dahlen, 1983) provides one conceptual framework for understanding deformation at convergent margins. At a critical taper angle, a wedge that is not accreting sediment will slide without internal deformation, and a wedge that is accreting sediment will internally deform to maintain a self-similar geometry through time. Important parameters controlling critical taper include the frictional properties of the décollement zone and the wedge material and pore fluid pressures. Variations in sedimentation and/or topography on the surface of the downgoing plate can cause changes in accretion versus nonaccretion (or erosion) as mechanisms of critical taper adjustment.

Taper angles and style of deformation can vary significantly from seaward to landward regions of a marginal wedge (Kimura et al., 2007; Wang & Hu, 2006). The outer wedge generally consists of folded and thrust-faulted sediments offscraped from the incoming plate or margin rocks. The transition between the inner and outer wedge is marked by a sharp break in slope (Figure 4.4.1.2, for example), and may be understood as the limit where a seafloor slope toward the trench is no longer required by mechanical stability criteria as a consequence of either the progressive bending of the downgoing plate as it goes into subduction, sedimentation within a forearc basin, and/or variations in rheology (Fuller, Willett, & Brandon, 2006). In detail, a sharp slope break and presence of discontinuities (such as splay faults) in the transition from the inner to the outer wedge causes deviations from the self-similar Coulomb wedge model, strongly affecting the stress state in the shallow parts of the margin (Conin, Henry, Godard, & Bourlange, 2012). In the case of an accretionary margin, active compression and contractional deformation appears to be concentrated within the outer wedge, and in the case of a nonaccreting margin, the outer wedge may either be stable or under critical extension (Wang & Hu, 2006).

Furthermore, the inner and outer wedges have been hypothesized to overlie the velocity weakening (seismogenic) and velocity strengthening (or aseismic) portion of the plate boundary décollement zone, respectively (see e.g., Fuller et al., 2006; Wang & Hu, 2006). However, this hypothesis has been partially called into question by seismological and geodetic observations before and after Tohoku 2011 earthquake (Fujiwara et al., 2011; Ide, Baltay, & Beroza, 2011; Ito et al., 2012; Ito et al., 2011; Yue & Lay, 2011). As described later, drilling through main thrust planes at shallow levels during IODP expeditions on the Nankai margin and in the Japan Trench brought further evidence that coseismic slip can and does propagate to the seafloor, or at least to shallow depths beneath it. New insight was also obtained on the mechanical properties of fault zones and stress state within the margin. In turn, the distribution of materials of different strength, the geometry of fault zones, and the seafloor slope will affect the seaward propagation of seismic slip as well as the volume of water displaced to cause tsunamis (Ryan, von Heune, Scholl, & Kirby, 2012). Moreover, deformation and fluid flow are linked through pore fluid pressures which impact strength of fault zones and surrounding rock, both at the time scale of coseismic slip (e.g., by thermal pressurization and clay dehydration) and at the time scale of tectonic stress build up during the earthquake cycle. If fluid migration to the seafloor cannot keep pace with pressure generation through applied tectonic and gravitational stresses and by mineral dehydration, pore pressures will build, and sediments will remain weak and unconsolidated. As a result, hydrogeology can constitute an important control on margin geometry (e.g., Saffer & Bekins, 2002; also see review in Saffer & Tobin, 2011).

In this chapter, we highlight some of the contributions provided by scientific ocean drilling to the understanding of these complex tectonic systems. Drilling has allowed dating of forearc basin sediments, of slope sediments overridden by splay and frontal thrust faults, and of the material accreted from the incoming plate to the deforming wedge. By integrating these data with seismic reflection data, the large-scale history of the investigated margins has been largely unraveled. Borehole imaging has provided information on the structural geometry and current stress state in the hanging wall of the main plate boundary and across subsidiary faults. Extensive laboratory tests on samples of the incoming sediment and basaltic crust have provided information on frictional and other geotechnical properties. Finally, fault zone samples from shallow portions of fault zones have challenged our previous conceptual models concerning the development and seismic behavior of fault zones.

4.4.1.2 IODP DRILLING AT THREE SUBDUCTION ZONES: TARGETS AND OBJECTIVES

4.4.1.2.1 Nankai Trough Seismogenic Zone Experiment

The Nankai Trough is formed by subduction of the Philippine Sea plate (PSP) to the northwest beneath the Eurasian plate at a rate of 4.1–6.5 cm/yr (Figure 4.4.1.1) (DeMets, Gordon, & Argus, 2010; Miyazaki & Heki, 2001; Seno, Stein, & Gripp, 1993). With its ~1300 year historic record of great (Mw > 8.0) earthquakes that are typically tsunamigenic (Ando, 1975; Hori, Kato, Hirahara, Baba, & Kaneda, 2004), this margin was deemed especially favorable for seismogenic zone drilling study because the entire downdip width of the seismogenic zone ruptures in each event (Tobin & Kinoshita, 2006). Additionally, it is apparently a mostly locked plate interface (Henry, Mazzotti, & Le Pichon, 2001; Mazzotti, Le Pichon, Henry, & Miyazaki, 2000; Miyazaki & Heki, 2001). The rupture area and zone of tsunami generation for the 1944 Mw 8.1 Tonankai earthquake (where the drilling transect is located) are reasonably well understood (Baba, Cummins, Hori, & Kaneda, 2005; Ichinose, Thio, Somerville, Sato, & Ishii, 2003). Similarly, the microseismic activity observed near the updip limits of the 1940s earthquakes could result from stress built up and strain at the edge of the locked zone (Obana, Kodaira, Mochizuki, & Shinohara, 2001). How-ever, recent observations of very low frequency (VLF) earthquakes and shallow tremor within or just below the accretionary prism in the drilling area (Obana & Kodaira, 2009; Obara & Ito, 2005; Sugioka et al., 2012) demonstrate that strain release along the megathrust is not restricted solely to slow interseismic strain accumulation punctuated by recurring great earthquakes.

The portion of the PSP subducting in the region of Nankai Trough is termed the Shikoku Basin, where widespread abyssal turbiditic and hemipelagic sedi-ments have blanketed the PSP crust. In the Shikoku Basin, sediment deposition was controlled by large-scale tectonic history and locally by basement topog-raphy (Pickering et al., 2013), resulting in thickness variations from <500 to >1000 m even before trench sedimentation, as well as lithological variations that have been documented by drilling along three transects (Ike et al., 2008a, 2008b; Pickering et al., 2013; Taira, 2001; Underwood & Moore, 2012; Watts & Weissel, 1975). DSDP and ODP boreholes were drilled along the Ashizuri and Muroto transects, whereas IODP NanTroSEIZE boreholes lie along the Kumano Basin transect off the Kii peninsula, several hundred kilometers to the northeast of the former regions (Figure 4.4.1.2). At the Muroto transect, incoming sedi-ments are monotonous hemipelagic muds overlain by turbidites deposited near the deformation front and trench. In contrast, the Ashizuri and Kumano incom-ing sediments contain substantial turbidite deposits within the Shikoku Basin facies. Turbidite thickness is generally much less on basement highs than adja-cent regions. Variable but up to ≥1 km thicknesses of coarse clastic sediments accumulate in and near the trench overlying the Shikoku basin section (e.g., Moore et al., 2009). These lateral variations in sedimentation along the trench are potentially an important factor for understanding large-scale spatial varia-tions of the seismogenic behavior of the plate interface.

The NanTroSEIZE Complex Drilling Project has been the largest undertak-ing of a single transect in the history of scientific ocean drilling, begun in 2007 and still ongoing at the time of this writing. It has included 10 IODP expe-ditions with more than 180 scientists directly involved, and both 2D and 3D seismic reflection imaging efforts. A 3D seismic reflection volume imaged the transect region in detail (Figures 4.4.1.1 and 4.4.1.2; Bangs et al., 2009; Moore et al., 2007; Park et al., 2010) providing detailed structural interpretation and reflector targets for drilling. Imaging (Figure 4.4.1.2) reveals that the margin can be subdivided into several tectono-structural domains: an *incoming plate* domain outboard of the deformation front; an *outer wedge* dominated by imbri-cate, mostly in-sequence thrusting, and an *inner wedge* domain encompassing an older, dominantly inactive imbricate thrust wedge with an overlying thick forearc basin sequence, displaying progressive tilting and minor deformation, termed the Kumano Basin. A megasplay marks the boundary between the inner and outer wedge domains (Tobin & Kinoshita, 2006; Moore et al., 2007, 2009).

NanTroSEIZE drilling has sampled incoming sediments and basement, the frontal thrust region, the megasplay fault region, the Kumano Basin sediments, and the inner wedge domain underlying that basin. The objectives were to document the materials (sediments, rocks, and fluids) being delivered to the subduction plate boundary system, the properties and effective stress conditions in the shallow thrusts at the accretionary toe and the tip of the megasplay system, and the properties of corresponding thrusts at seismogenic depth in order to understand fault evolution across the aseismic–seismogenic transition (Tobin & Kinoshita, 2006). Closely related goals were to instrument the shallow and deep fault systems and surrounding host rocks with an array of seismic, geodetic, hydrologic, and thermal sensors to obtain in situ conditions and document evolution during the seismic cycle. Although this complex project is still ongoing and the plate boundary fault zone at depth had not yet been reached by drilling by 2014, the main riser hole at Site C0002 has been drilled to more than 3000 meters-below-sea-floor (mbsf) and cased to 2922 mbsf, making it the deepest hole in scientific ocean drilling to date. Remaining objectives for 2015 and beyond include (1) extending this hole to 5200 mbsf across the plate boundary reflector (Figure 4.4.1.2), (2) installing long-term observatory instruments in that hole, and (3) installing 1 to 2 additional long-term observatories at other sites.

4.4.1.2.2 Costa Rica Seismogenesis Project

Costa Rica is a seismically active area with a history of Mw > 7 earthquakes, most recently a Mw 7.6 that occurred in 2012 (Yue et al., 2013). It exhibits along-strike contrasts in the nature of the seismogenic zone (Newman et al., 2002; Protti, Gündel, & McNally, 1994). Offshore of Costa Rica, the Middle America Trench (MAT) forms where the oceanic Cocos plate subducts under the Caribbean plate at rates approaching 9 cm/yr (DeMets, 2001). The Cocos plate consists of crust formed at two ridges, with generally smoother crust formed at East Pacific Rise to the north and generally rougher crust of the Cocos-Nazca spreading center to the south (Figure 4.4.1.3).

In 2011 and 2012, the *JOIDES Resolution* carried out IODP Expeditions 334 and 344, respectively, completing the CRISP riserless program (CRISP A) in preparation for proposed future deep drilling through the aseismic and seismic portion of the Cocos-Caribbean plate boundary (Harris, Sakaguchi, Petronotis, & the Expedition 344 Scientists, 2013; Vannucchi, Ujiie, Stroncik, & the Expedition 334 Scientists, 2012) (Figure 4.4.1.3). The area chosen for the experiment lies at the southeastern end of the Costa Rica subduction zone offshore the Osa Peninsula where the Cocos Ridge is subducting beneath the Caribbean plate. Although subduction erosion is the main tectonic process shaping the MAT forearc from Costa Rica to Guatemala (Ranero & von Huene, 2000; Vannucchi, Galeotti, Clift, Ranero, & von Huene, 2004; Vannucchi et al., 2003; Vannucchi, Scholl, Meschede, & McDougall-Reid, 2001), the Cocos ridge has caused subduction erosion to be particularly effective offshore Osa Peninsula (Vannucchi, Sak, et al., 2013). CRISP A results constrained the beginning of subduction of the Cocos Ridge at 2.5–2.3 Ma, using as a proxy the dramatic progression of events defining uplift and subsidence of the forearc. Subduction of the ridge also caused upward migration of the plate boundary so that its shallower seismogenic portion is at ~6 km depth. This relatively shallow depth makes the area offshore southern Costa Rica suitable for a deep drilling experiment aiming at the direct sampling of the plate boundary material of an erosive margin as planned for the second part of the CRISP project, CRISP B.

Through CRISP A, two transects were drilled from shelf to trench, located about 12 km apart along the northeast flank of the subducting Cocos Ridge (Figure 4.4.1.3). In addition to coring, acquisition of logging while drilling (LWD) data in holes on the shelf and slope was successfully achieved. CRISP A drill site locations are shown on Figure 4.4.1.3 and focused on the incoming sediment section, the frontal thrust region, and the upper plate and forearc basin. Incoming sediments consist of pelagic carbonate-rich sediments overlain by hemipelagic mud. Drilled sediment thickness ranges from ~400 m (Sites 1039 and U1414) to <100 m on the flanks of the Cocos Ridge (Site U1381). In complementary investigations, 3D seismic reflection data were collected in 2011 and results are currently emerging (e.g., Kluesner et al., 2013; Bangs et al., in press). The results of CRISP A, summarized later in this chapter, help to shape a new view of erosive margins, their sedimentary budget, their stress regime and ultimately the rock input into the seismogenic portion of the plate boundary.

CRISP B is proposed to investigate the processes that become active within the plate boundary at the onset of seismogenesis. From reflection seismic investigations, this area corresponds to the downdip transition from a highly reflective to a weakly reflective plate boundary, interpreted as the transition from fluid-rich/poorly drained to well-drained conditions (Bangs et al., in press; Ranero et al., 2008) and to a temperature range from 100 °C to 150 °C (Harris et al., 2010; Ranero et al., 2008). CRISP B has four major goals: (1) Quantify effective stress and plate boundary migration via focused investigation of fluid pressure gradients and fluid advection across the erosional plate boundary; (2) Determine the structure and fault mechanics of an erosional convergent margin and identify the processes that control the updip limit of seismicity; (3) Constrain fluid–rock interaction above and within the updip limit of the seismogenic zone by studying fluid chemistry and residence time, basement alteration, diagenesis, and low-grade metamorphism; and (4) Obtain physical properties of a 3-D volume that spans the seismogenic zone.

4.4.1.2.3 Japan Trench Fast Drilling Project

At the Japan Trench, Cretaceous-age Pacific plate subducts below Honshu Island at ~8.3 cm/yr (e.g., Apel et al., 2006; Argus, Gordon, & DeMets, 2011; DeMets et al., 2010). During 2012, the Japan Trench Fast Drilling Project (JFAST) investigated the shallow, active slip zone of the 2011 Mw = 9.0 Tohoku-Oki earthquake just landward of the Japan Trench (Chester et al., 2012, 2013). In addition to sampling the slip zone, a borehole observatory consisting of an array of temperature sensors hung inside well casing was established to monitor the thermal profile across the fault, allowing detection of residual heat from rapid fault slip and thereby inference of frictional shear stress during the 2011 event (Fulton et al., 2013).

Multiple lines of seismologic, geodetic, and bathymetric evidence have shown that 50 m or more of rapid coseismic fault displacement extended all the way to the trench during that earthquake (Fujiwara et al., 2011; Ito et al., 2011; Lay, Ammon, Kanamori, Xue, & Kim, 2011), displacing the seafloor substantially and shaping the destructive tsunami. JFAST was conceived specifically to access the fault zone within ~1 year after the earthquake to measure temperature at high spatial resolution across the fault in an attempt to detect the residual signature of frictional heating directly. Additional objectives were to sample the freshly slipped fault material and surrounding wall rocks (or sediments) to characterize the material properties and structure of the shallow plate boundary fault zone (Chester et al., 2013). Drilling was carried out at Site C0019, where the fault zone was encountered at about 820 mbsf in a very clay-rich pelagic sedimentary unit. Previous drilling at DSDP Site 436 characterized the incoming sediments, which consist of hemipelagic silty clay overlying pelagic clay and chert (Shipboard Scientific Party, 1980). On the Japan Trench slope, investigations were conducted by DSDP legs 56–57 and 87, and ODP leg 186. DSDP Sites 438 and 439 located at the boundary between the upper slope and the shelf provided the first data on forearc subsidence which characterized the margin as erosive.

4.4.1.3 HIGHLIGHTS OF SCIENTIFIC RESULTS FROM IODP SUBDUCTION ZONE DRILLING

In the first decade of IODP (2003–2013), drilling expeditions in subduction zone forearcs have been carried out as part of NanTroSEIZE, CRISP, and JFAST on 13 individual expeditions of varying duration and objectives. Some key themes have emerged from all of this work: (1) that the complex tectonic evolution results in part from the interplay of sedimentation and plate movements and the record of that evolution in the basinal sections; (2) that elevated pore fluid pressure plays a strong role in fault and wedge strength at depth, but strong overpressure may not be present in upper and mid-wedge settings away from active fault systems; (3) that the mechanical properties of fault zone materials, particularly frictional characteristics, are strongly rate-dependent and those faults contain evidence for past co-seismic slip; and (4) that principal stress orientations are governed by tectonic convergence but relative stress magnitudes exhibit a more complex relationship with tectonics than anticipated. Each of these themes are developed in the sections that follow.

4.4.1.3.1 Tectonic History and Evolution

4.4.1.3.1.1 Nankai: Punctuated Activity of Thrust Fault Systems in an Accretionary Margin

The NantroSEIZE drilling results have documented the ages of the incoming plate section, the inner and outer accretionary wedges (both sediment age and timing of wedge deformation), and the slope and forearc basins in some detail (Figure 4.4.1.2). The incoming plate age is 18–20 Ma, and the Shikoku basin sediments span depositional ages from ~15 Ma to Quaternary (Pickering et al., 2013; Underwood, Saito, Kubo, & the Expedition 322 Scientists, 2010); they are additionally overlain by Neogene trench fill sediments that are restricted to the proximal trench environment. The outer accretionary wedge fold and thrust belt incorporates sediments of both the Shikoku basin and trench fill systems. The deformational age of this entire ~30 km wide thrust belt is 0–2 Ma, dated via slope basin sediments that overlie and in some cases are partially overridden by wedge thrusts (Strasser et al., 2009; Underwood & Moore, 2012), and by age inversions across drilled thrust faults (Screaton et al., 2009). Seismic interpretation and drilling results show that the entire outer wedge domain has hosted a complex interplay of frontal imbrication, out-of-sequence thrusting, and slope deposition throughout this time.

By contrast, the inner wedge domain is made of similar Miocene-age sediments but thrust deformation at Site C0002 ended between 5 and 3.6 Ma, no younger than mid-Pliocene. Results of drilling document punctuated and at times rapid sedimentation on top of the inner wedge domain. The formation of the Kumano Basin was initiated by uplift at its outer margin due to megasplay fault activity that began at ~1.95 Ma (Strasser et al., 2009). Following basin creation, rapid turbidite deposition (400–800 m/Myr) is documented by drilling to have initiated 1.67–1.6 Ma (Underwood & Moore, 2012). The lag between the basin development and the rapid deposition may reflect the time needed for development of through-going sediment delivery systems (Underwood & Moore, 2012). A progressive landward tilting and concurrent shift of the center of forearc basin deposition through time has been attributed to megasplay activity (Gulick et al., 2010), although this large-scale tilt occurred at a time when the slip on the drilled branch of the megasplay system slowed down. The tilting may also reflect a progressive change of geometry of the main plate interface at a larger scale as a large packet of sediments is subducted (see Figure 4.4.1.2) (Bangs et al., 2009).

In NanTroSEIZE, drilling results combined with 3D seismic interpretation have provided a detailed understanding of timing of fault displacements on geologic time scales, highlighting the episodic nature of deformation in the active outer wedge. Biostratigraphic dating of sediments that overlie material caught up in prism deformation (Figure 4.4.1.2) has elucidated periods of activity and inactivity of those thrusts. The frontal thrust system was active beginning ~0.436 to 0.78 Ma, resulting in ~6 km of total throw (Screaton et al., 2009).

Assuming 0.436–0.78 Myr of activity and estimated plate convergence rates of 4–6 cm/yr, Screaton et al. (2009) estimated that the ~6 km displacement on the frontal thrust accounted for only 13–34% of the total tectonic convergence. Out of sequence faults or other diffuse smaller faults must accommodate the majority of total wedge displacement.

Strasser et al. (2009) found that the OOST fault system (a.k.a. the shallow part of the megasplay) drilled at Sites C0004 and associated cover sediments from Site C0008 began as a frontal accretion fault at a time between 2.5 and 2.0 Ma and built a thrust wedge "prism" of strongly deformed sediments, then was reactivated as an OOST at 1.95 Ma or later, overriding mass wasting deposits that accumulated at the toe of this wedge. After a burst of slip lasting until ~1.8 Ma, OOST displacement slowed. A second pulse of activity between 1.6 and 1.24 Ma was also identified based on accumulation rate and calcite occurrence in the slope sediments of Site C0008. Displacement largely ceased along this branch after 1.24 Ma, but likely branched to other, undrilled faults in higher structural positions (Strasser et al., 2009). The overall implication of this work is alternations of activity between in-sequence and OOST on a ~100 kyr time scale. In a related work, Kimura et al. (2011) found spatial variability in periods of high activity of this same OOST splay along strike within the 3D area (only 11 km in width). Kington (2012, p. 125) suggested the differences in fault activity were due to subduction of lower plate topography (buried basement high, or "seamount," collision), affecting the eastern but not the western portion of the 3D volume.

Finally, drilling results at Site C0002 and C0009 in the main Kumano forearc basin and underlying strongly deformed inner accretionary wedge sediments constrain the timing of megasplay OOST activity at a broader scale. Expedition 315 Scientists (2009) and Expedition 319 Scientists (2011) found at Site C0002 and C0009 that a late Miocene to Pliocene age buried accretionary wedge is overlain by Pleistocene age Kumano basin turbiditic sediments spanning 1.67 Ma to <0.5 Ma. From then until the locus of deposition shifted landward with progressive tilting of the basin fill at about 0.5 Ma, accumulation rates reached 400–800 m/Myr, via some combination of evolving accommodation space and sediment availability, as megasplay OOST deformation modified the margin (Underwood & Moore, 2012).

4.4.1.3.1.2 Costa Rica: Extension in an Erosive Convergent Plate Boundary Setting

In Costa Rica, away from the deformation front, seismic imaging shows distinct normal displacement along reflectors concentrated in or cutting through the slope sediments. Other reflectors are present as landward dipping reflectors limited to the upper plate framework (Figure 4.4.1.3). The chemistry of the abundant and active fluid venting suggests deeply sourced fluids possibly migrating from the plate boundary (Hensen & Wallmann, 2005; Ranero et al., 2008). CRISP A targeted the termination of a high-amplitude seismic reflector cutting through the forearc framework, but not propagating into the slope sediments. Drilling results show the presence of geochemical anomalies suggestive of deeply sourced fluids coincident with intervals of the cores where normal faults are particularly concentrated (Harris et al., 2013; Vannucchi et al., 2012).

Extension of the forearc has been observed in seismic reflection imaging (e.g., Bangs et al., in press; McIntosh, Silver, & Shipley, 1993; Park et al., 2002; Ranero et al., 2008; Sacks et al., 2012) and, under certain conditions, has also been predicted from standard or dynamic Coulomb wedge theory (Davis et al., 1983; Wang & Hu, 2006). Whether this extension is localized into a few large-displacement normal faults or is broadly distributed depends on specific conditions of plate interaction, sediment distribution, and the mechanical properties of the rock units involved. In erosive margins, the removal of upper plate material triggers subsidence that can be accommodated by extension of the forearc (Vannucchi et al., 2003). In the case of the CRISP transect, the absence of age reversals below the base of slope sediments and through the termination of the high-amplitude reflector crossed at Site U1380 (Figure 4.4.1.3) supports the interpretation of extension as the main factor responsible for the development of the forearc basin. Further ongoing analysis on the material recovered during CRISP A is also suggestive that normal faults are the main control over forearc tectonics: sediment provenance, the development of differential subsidence (more efficient in the shelf than in the middle slope), and the evolution of a forearc basin depocenter rather than a constant thickness sequence carpeting the forearc.

4.4.1.3.1.3 Significance of Local Extension on Nankai Margin

In contrast, the NanTroSEIZE region does not exhibit any evidence of wholesale extension or tectonic erosion, but it nonetheless contains prominent normal faulting in both the Kumano forearc basin and in shallow locations on the outer wedge (Figure 4.4.1.2) (Byrne et al., 2009; Moore et al., 2007; Park et al., 2002; Sacks, Saffer, & Fisher, 2013). Analysis of the Kumano basin normal faults suggests only minor total extension, linked to either gravitational effects due to basin tilting or sediment underthrusting beneath this zone (Bangs et al., 2009; Gulick et al., 2010). In the outer wedge, normal faults occur and are even locally dominant in core-scale structures and in seismic imaging of regions shallower than ~1 km (Expedition 316 Scientists, 2009). Additionally, anelastic strain recovery (ASR) data from a pair of core samples from Sites C0001 and C0007 suggest a normal-faulting stress state in those cores, which led Byrne et al. (2009) to propose that the outer wedge is dominated by horizontal extension rather than compression. However, broader structural analysis indicates that the core samples come from specific areas in the near surface of the hanging wall of major thrusts and normal faulting evidence in the seismic data is restricted to these specific areas (e.g., Kington, 2012, p. 125; Moore et al., 2009). Overall, the amount of extension is quite minor relative to the total convergent shortening documented in the thrust wedge and, as discussed later in this article, the normal faults may result from stress transients related to the earthquake cycle. For the entire margin transect, there is no evidence of large-scale subsidence (e.g., Underwood & Moore, 2012).

4.4.1.3.2 Feedbacks with Sediment Supply and Role of Forearc Basins

Clift and Vannucchi (2004) suggested that subduction erosion and/or accretion are controlled, among other parameters, by the amount of sediment present in the trench. From their global compilation of the present situation at subduction zones, erosion is favored if the sediments in the trench are less than 1 km thick. Accordingly, they described erosive margins as sediment-poor systems. Erosive margin drilling, such as ODP Leg 170 off Nicoya Peninsula in northern Costa Rica, supported this model with slope sedimentation rates on the order of ~40 m/Myr. However, results from CRISP A challenge this paradigm. During CRISP A, the 960 m of sediments cored in the shelf at IODP Site U1379 and the 800 m cored in the middle slope at IODP Site U1380 are in both cases younger than 2.5 Ma (Harris et al., 2013; Vannucchi et al., 2012). The sedimentation rates along the CRISP transects is of order ~350–400 m/Myr, on average, but a more detailed dating of the sediment sequence reveals intervals where the sedimentation rates can reach ~1 km/Myr (Vannucchi, Sak et al., 2013). Therefore sediment supply in the forearc is actually quite high.

Drilling has demonstrated an extraordinary variation in the amount of sediments being delivered to the forearc and deposited on the slope along the 260 km that separate the Leg 170 and the CRISP A transects. The trench itself has variable amounts of fill, but generally less than 1 km, with only 400 m offshore Nicoya Peninsula to <100 m offshore Osa Peninsula. Along the CRISP transect, the forearc has developed a basin efficient in sequestering all the sediments, which are therefore not reaching the trench despite their abundance. The storage capacity of the forearc basin is especially high in the portion of the MAT where the Cocos Ridge is subducting, causing extreme subsidence driven by subduction erosion. The paucity of sediment in the trench is, therefore, the result of the efficient sequestration of sediment in the forearc, which, in turn, is the effect of subduction erosion-driven subsidence. In this case subduction erosion is the cause of little sediment fill in the trench, which in turn reinforces the tendency toward subduction erosion.

In contrast to the CRISP transect, trench sedimentation rates remain high for the NanTroSEIZE transect. This trench sedimentation can be attributed to submarine canyon systems on either side of Kumano Basin (Tenryu and Shionomisaki) that funnel sediments from Honshu to the trench (Underwood & Moore, 2012). Moreover, a large part of the sediment infill in the Nankai Trough originates from the Izu collision zone and Mount Fuji area and is transported laterally along the trough by the Nankai axial channel (e.g., Pickering et al., 2013; Pickering, Underwood, & Taira, 1992).

In addition to affecting sediment supply at the trench, development of forearc basins has mechanical implications for the stability of the inner wedge (e.g., Fuller et al., 2006). Sedimentation in the forearc basin increases vertical stress more than horizontal effective stress, resulting in a supercritical state that potentially stabilizes the wedge beneath the basin and transfers displacement outward (Simpson, 2010). As a result, variations in sediment supply or routing could drive transitions from accretionary to erosive conditions.

Interaction between sedimentary and tectonic processes thus involves reinforcing feedbacks that could result in self-sustained regimes:

1. Fold and thrust tectonics in the outer part of an active accretionary wedge favor trapping of sediment in shallow basins on the inner part of the margin and deposition of surplus sediment in the trench. This deposition pattern stabilizes the inner wedge (Fuller et al., 2006) and promotes accretionary wedge growth as long as sufficient sediment input at the trench is maintained.
2. Subduction erosion appears to be associated with deposition on the outer slope at the expense of deposition in the trench, which limits the thickness of sediment available for frontal accretion. Moreover, feedback of slope sedimentation on stress state (increase of vertical stress relative to the horizontal stresses) contributes to maintenance of the outer wedge under stable (subcritical) or extensional stress condition. However, slope sedimentation and instability can cause compression downslope of the deposition area, which may contribute to the formation of a small frontal prism near the trench composed of slope sediments.

4.4.1.3.3 Impact of Basement Topography on Deformation Near the Trench

All three drilling regions illustrate that basement topography plays an important role in deformation, particularly near the trench. The frontal thrust region of the Kumano Basin transect is unusual, in that it appears that normal frontal imbrication may have been interrupted due to passage of a basement high (Bangs et al., 2009; Moore et al., 2009), now beneath the accretionary wedge. Another yet to be subducted basement high, perhaps analogous, is present in the Shikoku Basin on Kumamo transect and was drilled during IODP Expedition 333. As stated earlier, Screaton et al. (2009) estimated that the frontal thrust began activity ~0.436 to 0.78 Ma and accommodated 13–34% of the plate convergence. In contrast to the Kumano Basin frontal thrust, the Muroto and Ashizuri frontal thrusts of the Nankai prism have steeper dip angles ($\sim25°–30°$) and are only displaced by ~1–2 km. Development of a proto-thrust zone seaward of the frontal thrust suggests that a new décollement is propagating outward at a deeper level again and the system is returning to more typical frontal imbrication (Moore et al., 2009).

The forearc toe along the CRISP transects shows a small (3–5 km-wide) accretionary prism just landward of the trench. The frontal prism of erosive margins is often a transient feature rapidly disrupted by subducting relief and re-grown. A common characteristic is its formation by material accumulated at the toe by mass wasting processes, such as debris flows (von Huene et al., 2004). This is, however, not the case in the area offshore Osa peninsula, where IODP Site U1412 recovered pelagic and subordinate hemipelagic sediment of the Cocos plate forming this prism. The age of this sediment would support the idea of a re-growth of the accretionary prism after its disruption caused by Cocos Ridge impingement. Accretion at the toe also suggests that the wedge equilibrium size and shape could be recovered only by transferring material from the incoming plate, implying that not enough material is transported from the slope to reconstruct the prism toe. Seismic images show very little deformation in the incoming section beneath the frontal prism, suggesting pronounced decoupling across the basal décollement.

Similar to Costa Rica, the Japan Trench forearc is a dominantly erosive margin, but includes a frontal accretionary prism of a few 10s of km width (Kodaira et al., 2012). The architecture and the length of the frontal prism is strongly influenced by the incoming plate structure, which is characterized by high relief horsts and grabens (Figure 4.4.1.6). Although the sediment thickness in the incoming plate is about 1 km irrespective of the horst and graben geometry, the steep edges of the structures interacting with the leading edge of the overriding wedge may cause gravitational instability associated with thrusting. Because the shallow décollement largely follows a specific lithological unit of pelagic clay (Chester et al., 2013) with extremely low friction coefficient (Ujiie, et al., 2013), it has an irregular geometry as it rides over the horst and graben structure. Interpretation of seismic profiles intersecting the J-FAST site before and after the Tohoku 2011 earthquake (Figure 4.4.1.6; Chester et al., 2013; Kodaira et al., 2012) suggest that coseismic slip on the frontal thrust occurred along a downward propagating décollement (labeled "plate boundary fault reflector" in Figure 4.4.1.6(C)). As interpreted, this fault actually has the kinematics of a normal fault (rather than climbing up-section toward the seafloor, it moves the hanging-wall downward). However, an alternative—and perhaps more structurally plausible—interpretation of the seismic and bathymetric images is that a large rotational slump block caused by megathrust slip is present at the toe of the wedge, such that the "plate boundary fault reflector" is actually its basal rotational plane (see Figure 4.4.1.4 of Strasser et al., 2013 for this interpretation).

4.4.1.3.4 Fluid Pressures and Deformation

Fluid pressures are closely tied to deformation through control on effective fault friction (as reviewed in Saffer & Tobin, 2011). In turn, fluid pressures can be increased by rapid burial of the footwall of thrust faults, by deformation caused by faulting, by chemical reactions that increase volume, and transiently by thermal pressurization. During the ODP era, monitoring of shallow fault zones within the Barbados, Costa Rica, and Nankai wedges yielded low to moderate excess pore pressures (e.g., Bekins & Screaton, 2007; Screaton, 2010). Elevated excess pore pressures were inferred from high porosities, low seismic velocities, and consolidation testing within and beneath the décollement zone at the Muroto transect (e.g., Screaton, Saffer, Henry, Hunze, & Leg 190 Shipboard Scientific Party, 2002; Tobin & Saffer, 2009) and Costa Rica's Nicoya transect (Saffer, 2003). Interpretation of consolidation tests can be complicated by cementation and diagenetic effects, which can cause overestimation of pore pressures from porosity (e.g., Ask & Morgan, 2010; Hüpers & Kopf, 2009; Morgan & Ask, 2004; Morgan, Sunderland, & Ask, 2007); however expected effects of cementation on seismic velocity were included in velocity–porosity transforms (Hoffman & Tobin, 2004).

During IODP efforts at the NanTroSEIZE transect, direct downhole pressure monitoring within the megasplay fault at ~410 mbsf at Site C0010 documented pore pressure slightly in excess of hydrostatic (Hammerschmidt, Davis, & Kopf, 2013). This result is consistent with shipboard porosity data which show no indication of undercompaction beneath the shallow megasplay (Screaton et al., 2009). Furthermore, comparison of logging data and physical properties measurement on core samples does not indicate overpressuring in any of the shallow boreholes, except perhaps below 450 m at Site C0001 (Conin, Henry, Bourlange, Raimbourg, & Reuschlé, 2011). In contrast, annular pressures measured while drilling from Site C0001 suggest lithostatic excess pore pressures between 500 and 1000 mbsf, although these may have been increased by drilling operations (Moore, Barrett, & Thu, 2012).

Pore pressures within the Kumano Basin sediments have been measured by downhole testing. Measured pore pressures were near hydrostatic despite high sedimentation rates in the basin (Saffer et al., 2013). Based on calculations, Saffer et al. (2013) suggested that these conditions were consistent with measured permeabilities ranging from $\sim 6.5 \times 10^{-17}$ to $\sim 4 \times 10^{-14} \, m^2$. At greater depths, elevated fluid pressures along the plate boundary have been inferred from low-velocity zones in seismic profiles of the Kumano transect (Bangs et al., 2009; Kitajima & Saffer, 2012; Park et al., 2010), Costa Rica subduction zone (Ranero et al., 2008), and the Japan Trench (Kimura et al., 2012). However, these depths have not yet been reached by drilling. Because direct measurements at depths >1 km will only rarely be possible, indirect estimates requires ground-truthing of velocity–porosity relationships to quantify pore pressures (e.g., Kitajima & Saffer, 2012; Tobin & Saffer, 2009; Tsuji, Tokuyama, Pisani, & Moore, 2008) and inferences from numerical modeling. This calibration is challenging because smectite-rich sediments, which are common in the shallow portions of subduction zones, may not fit previous velocity–porosity relationships (Tudge & Tobin, 2013), and the effects of cementation and clay mineral alterations during progressive diagenesis on velocity–porosity relations remain a subject for further research.

Drilling has also provided information on fluid sources and permeabilities needed to simulate pore pressures and interactions with deformation. Samples from ODP and IODP drilling have provided information on permeability evolution during early compaction that has been extrapolated to greater depths (e.g., Daigle & Screaton, 2014; Gamage, Screaton, Bekins, & Aeillo, 2011). Permeable trench sediments in the Nankai Trough and buried turbidite layers in the underthrusting Shikoku Basin sediments may aid fluid drainage at the toe of the prism (Rowe, Screaton, & Ge, 2012). Similar drainage along permeable layers has been simulated for the Ashizuri transect of the Nankai Trough (Saffer, 2010). Despite the potential for drainage provided by turbiditic sediments, coupled deformation and fluid flow modeling suggest that excess fluid pressures can rapidly build in the underthrust sediments of the Kumano Basin transect and may contribute to décollement downstepping (Rowe et al., 2012). Numerical modeling has also been applied to understand low seismic velocities observed beneath the megasplay faults (Screaton & Ge, 2012). Results suggest that differences in permeability between accreted trench sediments (higher permeability) and fine-grained hemipelagic sediments greatly affect simulated pore pressure distributions.

Pore pressures have not yet been measured in the CRISP region. Backscatter and bathymetric mapping suggests a high density of potential fluid seepage sites on the shelf and slope (Kluesner et al., 2013). These sites correspond to folds and faults identified within the underlying wedge, and the shape of the shelf suggests passage of a seamount or ridge in the CRISP region (Kluesner et al., 2013), which may contribute to fracturing that allows seepage. The observed high density of seepage may aid drainage of the wedge and subducting sediments.

4.4.1.3.5 Fault Zone Development and Mechanical Properties

Core samples from IODP drilling have provided new insight on fault zone deformation processes. Several observations and studies strongly suggest that fault slip has been remarkably localized and at least sometimes occurred very rapidly, indicating coseismic slip rates, even to the toe region of marginal wedges.

A recent review by Ujiie and Kimura (2014) summarizes many of the key observations from fault zone drilling during NanTroSEIZE in detail. At both the accretionary wedge toe/frontal thrust region (Site C0007) and the shallow portion of the mega-splay fault system (Site C0004), dark, very fine grained fault gouges several mm to ~1 cm in thickness were found, embedded in wider zones of fractured and brecciated rocks, defined as zones of cm to sub-mm size fragments bounded by fractures, the surfaces of which are commonly polished and slickenlined. In the frontal thrust, two zones of foliated gouge occur, bounded by anastomosing slip surfaces. The dark gouge zone at 438 mbsf at the frontal thrust (Site C0007) (Figure 4.4.1.7) is 2 mm thick, bounded by sharp planar surfaces from the surrounding brecciated zone, and marks a 1.67 million year

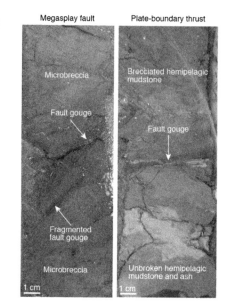

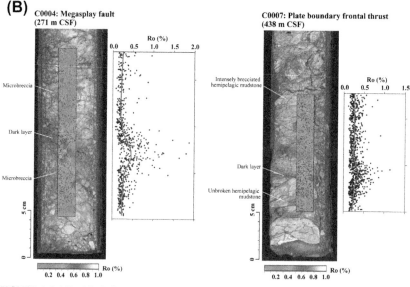

FIGURE 4.4.1.7 (A) Split faces of NanTroSEIZE core samples from Site C0004 in the shallow branch of the megasplay fault system (left) and Site C0007 in the main frontal thrust fault (right) with labeled main structural features. *(After Expedition 316 Scientists (2009) and Ujiie and Kimura (2014).)* (B) Vitrinite reflectance mapping of core samples from the shallow branch of the megasplay fault system drilled at Site C0004 (left) and frontal thrust zone at Site C0007 (right). Superimposed on photograph of slabbed core sample is map of Reflectance (R0) value, also plotted to the left of each image. *Sakaguchi et al. (2011); copyright GSA.*

age reversal based on biostratigraphy, all indicative that this is the main plate boundary thrust and marks a strong localization of faulting over protracted time (Ujiie & Kimura, 2014).

Several workers have addressed the evidence for frictional shear heating during past slip in these gouge zones. Sakaguchi et al. (2011) found anomalies in vitrinite reflectance values from which they estimated that the fault gouges had experienced 330 °C to 390 ± 50 °C temperatures, compared to ambient background of only ~20 °C, consistent with frictional heating of the 2–10 mm wide gouge zones. However, the quantitative estimation of peak paleotemperature from this proxy has been questioned (Furuichi, Ujiie, Sakaguchi, & Tsutsumi, 2013; Kitamura, Mukoyoshi, Fulton, & Hirose, 2012; Ujiie & Kimura, 2014). Nonetheless, the vitrinite observations do seem to require substantial frictional heating only possible during rapid slip (Fulton & Harris, 2012).

This result was corroborated by XRF mapping of the split core across the gouge zones coupled with XRD, which together revealed geochemical variation consistent with the preferential occurrence of illite in the fault gouge mixed clays as compared to the surrounding breccia (Yamaguchi et al., 2011), a possible result of frictional heating. By contrast, Hirono et al. (2009) investigated several geochemical proxies for paleo-temperature in the C0004 splay fault gouge, including fluid-mobile trace element concentrations, Sr isotopes, magnetic minerals, inorganic carbon content, and Raman spectra of carbonaceous material, and concluded that there was no evidence for elevated paleotemperature of greater than 300 °C, and also no evidence for past seismic slip velocities (~1 m/s) on this fault. These two results may be reconcilable, given uncertainty, if the fault zone experienced heating of up to about 300 °C, but not significantly more. A plausible scenario for this observation would be rapid coseismic slip under low effective friction, which is consistent with frictional experiment data discussed in the next section.

Taken together, these detailed studies of the gouge deformational fabrics, strong localization, and geochemical proxies for frictional heating suggest that past earthquake rupture did propagate all the way to the trench, at least in one or more events, and also all the way to the sea floor at the studied branch of the megasplay fault.

It was unexpected to find that this large total displacement slip must have occurred via extreme localization onto ~mm wide fault slip surfaces, despite the implied large displacements. The total displacement of the frontal thrust is ~6 km, which would require ~1000 individual megathrust earthquakes to have propagated through the cored interval, if *all* slip is coseismic. At present, we know of no way to determine whether this is the case, or what fractions of total displacement happen in rapid versus slower slip events. Nevertheless, these studies support the hypothesis that at Nankai, large-displacement coseismic rapid slip can propagate to the sea floor at least in some events, contrary to common previous assumptions about the "aseismic zone" (see e.g., Hyndman, 1997; Tobin & Kinoshita, 2006) that rapid slip could not be supported through this zone. The documented occurrence of tens of meters of fault slip reaching to the trench during the 2011 Tohoku earthquake (Fujiwara et al., 2011; Ito et al., 2011) dramatically confirmed that this can occur and can produce measurable frictional heating (Fulton et al., 2013). Taken together, the Tohoku and Nankai observations raise the question whether the large shallow slip at Tohoku was in fact anomalous, or is instead a common event. This question remains open for future work at these and other subduction zones.

4.4.1.3.6 Frictional Properties and Implications for Deformation

An important contribution of drilling in all three regions has been the recovery of core samples for testing of frictional properties. Frictional properties of subduction zone materials may play a key role in seismogenesis, but also control the overall structure of the margin through their influence on critical taper.

Large reduction of intrinsic friction coefficient when slip velocity is increased to coseismic rates ≥ ~1 m/s) has been demonstrated in a variety of rocks since the first tests performed on gabbro (Tsutsumi & Shimamoto, 1997). Moreover, clay-rich fault zones are prone to thermal pressurization (Wibberley & Shimamoto, 2005) and concomitant extreme weakening can occur when slip propagates from depth into impermeable clayey fault rocks that would ordinarily be velocity strengthening (Faulkner, Mitchell, Behnsen, Hirose, & Shimamoto, 2011). Experiments on IODP samples summarized in this section shows subduction megathrusts fault rocks sampled at shallow depth display large friction reduction at coseismic slip rates and thus perhaps do not constitute a barrier against rupture propagation to the seafloor during large earthquakes.

Paralleling investigations in other fault zones, such as the San Andreas drilling project, frictional studies (shear experiments) have revealed coefficients of friction significantly lower than the ~0.4 to 0.6 that was generally anticipated for clay-rich fault rocks (Byerlee, 1978). For example, samples from the San Andreas fault zone have yielded coefficients of friction of 0.15 and 0.21 (Carpenter, Marone, & Saffer, 2011; Lockner, Morrow, Moore, & Hickman, 2011). For the Nankai Trough, results range from 0.30 to 0.45 along the frontal thrust and megasplay fault of the NanTroSEIZE Kumano transect, and from 0.20 to 0.28 within the Muroto décollement zone in low (0.1–10 μm/s) to intermediate velocity (0.01–3 mm/s) experiments (Ikari & Saffer, 2011; Tsutsumi, Fabbri, Karpoff, Ujiie, & Tsujimoto, 2011). Most samples generally displayed velocity strengthening behavior (a–b>0, in rate-state friction terms; see e.g., Scholz, 1998) in the laboratory tests, with the exception that neutral or velocity-weakening behavior appeared in some at low slip rates of 1–3 μm/s. By contrast, at high slip velocities (>10 mm/s), wet samples of megasplay fault gouge exhibited strong velocity weakening, with friction coefficients of only 0.1–0.2 at 0.1–1 m/s (Ujiie & Tsutsumi, 2010).

Clay-rich samples from the Nicoya transect of the Costa Rica subduction zone yielded coefficients of friction values (0.22–0.35 in Ikari, Niemeijer, Spiers, Kopf, & Saffer, 2013; 0.19 in Kopf, 2013) similar to those of hemipelagic clays from Nankai. In contrast, pelagic carbonate frictional coefficients were considerably higher (0.56 peak and 0.43 residual in Kopf, 2013; 0.71 to 0.88 in Ikari et al., 2013). Ikari et al. (2013) suggested that the contrast in properties may yield variations in fault-zone behavior as basement topography causes patches of carbonate to occur along the plate boundary fault.

CRISP collected samples from two possible sources of input material to the plate boundary: the incoming plate sediments and the forearc sediments (Harris et al., 2013; Sakaguchi, et al., 2013; Vannucchi, Ujiie, Stroncik, & the IODP Expedition 334 Science Party, 2013; Vannucchi et al., 2012). Preliminary results on silicic to calcareous ooze collected at the incoming plate Site U1381 reveal that friction values at slow slip velocities (v < ~30 mm/s) are about ~0.7, comparable to the typical friction values for rocks. Velocity-weakening behavior occurred at v < 0.3 mm/s and neutral to velocity-strengthening at 0.3 < v < ~3 mm/s. At higher velocities (v > ~30 mm/s), steady state friction decreases dramatically. For example, at a velocity of 260 mm/s, the friction coefficient show a gradual decrease with a large weakening displacement toward the establishment of a nearly constant level of friction at ~0.1 (Namiki, Tsutsumi, Ujiie, & Kameda, 2014).

Drilling of the single site in the Japan Trench found that the shallow portion of the plate boundary fault is localized within a thin layer (<15 m at the drill site) of pelagic clay with high smectite content (Chester et al., 2013; Ujiie et al., 2013). Frictional testing of the clay found low apparent friction (<0.2) during high-velocity slip, with friction becoming vanishingly small under impermeable conditions (Ujiie et al., 2013). The thermal anomaly observed with the borehole observatory one to two years after the 2011 Tohoku earthquake constrains frictional heating to a modest value with a calculated effective coefficient of friction of 0.08 (Fulton et al., 2013), much lower than static values for these rocks, but consistent with high velocity friction under impermeable conditions.

4.4.1.3.7 Stress State: Observations and Modeling

Past stress state has been inferred from structures in cores, geophysical logs, and seismic profiles and from measurements of anisotropy of magnetic susceptibility. Details of stress-related observations can be found in the accompanying chapter by Kinoshita et al. (this volume), and are only summarized here. Borehole breakouts provide an assessment of current orientation of horizontal stresses (Chang et al., 2010; Lin et al., 2013, 2010, 2011; Malinverno & Saito, 2013; Vannucchi et al., 2012) and have been complemented by measurements of anelastic recovery of strain conducted on recovered cores (Byrne et al., 2009; Yamamoto et al., 2011) and seismic velocity anisotropy studies (Tsuji et al., 2011). The first IODP applications of a Modular Formation Dynamics Tester tool were conducted in the hanging wall of the Kumano Basin transect (Ito et al., 2013; Saffer et al., 2013). This tool allows in situ hydraulic fracturing tests to determine stress magnitude.

In Nankai, the present-day orientation of the maximum horizontal stress is roughly perpendicular to the trench (Byrne et al., 2009; Tobin et al., 2009) in the outer wedge but more oblique, nearly parallel to the subduction velocity vector at Site C0009 in the forearc Basin (Lin et al., 2010). More surprising is the presence of a swath of trench-normal extension (minimum principal stress is trench-normal) above the megasplay system that has been documented to persist down to at least 2 km below seafloor at Site C0002. This extension is expressed tectonically as an array of normal faults affecting the seaward edge of the Kumano forearc basin (Sacks, Saffer, & Fisher, 2013) that cannot be explained by partitioning of the subduction oblique component along the megasplay system (Martin et al., 2010). The inception of trench normal extension is recent (younger than 0.44 Ma; Sacks et al., 2013) and thus post-dates the pause in frontal accretion observed along the Kumano transect (Screaton et al., 2009). The progressive slowing of thrusting on the megasplay (Strasser et al., 2009), and the subsequent development of the extensional stress regime (vertical maximum principal stress) above it could both result from a change of the stress regime within the outer wedge from critical (compressive) to subcritical (stable) (Conin et al., 2011). Alternatively, extension along the edge of the forearc basin could result at least in part from plate bending, alone or in combination with underthrusting (Byrne et al., 2009).

Analysis of borehole stability, ASR, and fault kinematics indicate an extensional (vertical maximum stress) to strike-slip (vertical intermediate stress) stress state at drilled depths in the wedge (Byrne et al., 2009; Chang et al., 2010), implying a subcritical state of stress today, but as argued earlier, that may only apply to shallow and localized positions in the wedge. Both active out-of-sequence thrusting and strike-slip faulting are observed within the accretionary wedge (Kinoshita, Moore, & Kido, 2011), and VLF earthquake focal mechanisms suggested a transpressive (horizontal least principal stress) regime (Ito, Asano, & Obara, 2009). However, detailed analysis of VLF activity shows these events are dominated by low-angle thrusting occurring along or near the basal décollement surface (Sugioka et al., 2012) and thus are not necessarily indicative of the state of stress within the overlying accretionary wedge. Taken as a whole, the various observations suggest that the overall stress state for NanTroSEIZE is thrust to transpressive, whereas extensional states are quite localized and shallow, or possibly transient during a seismic cycle.

Variations of the stress field during the seismic cycle may cause deviations between the stress state snapshot inferred by geomechanical analysis of borehole stability or geophysical methods on one hand, and the tectonic stress state inferred from tectonically active structures and tectonic strain on the other. In Nankai, such evidence of stress cycling remain subtle (e.g. Sacks et al., 2013) but in the Japan Trench, the Tohoku 2011 earthquake appears to have caused a major change in stress conditions (Lin et al., 2013). There, the maximum horizontal compressive stress remains oriented parallel to fault slip but the maximum stress is now vertical indicating an extensional state of stress. The state of stress calculated before the earthquake taking into account the updip increase of slip toward the trench would be compressive, consistent with focal mechanisms recorded before and after the earthquake that also indicate a change from compressive to extensional conditions (Asano et al., 2011; Ide et al., 2011). The nearly complete stress release observed after Tohoku earthquake may be the consequence of strong velocity weakening on the subduction plane (see frictional properties section).

In the CRISP transect, borehole breakouts were imaged at sites on the middle and upper slope. The middle slope, 15 km landward of the deformation front, shows a maximum principal horizontal stress with average azimuth NNW-SSE (Figure 4.4.1.5). At the upper slope/shelf, about 25 km landward of the deformation front, the in situ maximum principal horizontal stresses are diverse over ~90° from WSW-ENE to E–W (Malinverno & Saito, 2013). These directions are approximately parallel and perpendicular to NNE-directed GPS deformation vectors on land (LaFemina et al., 2009). The relationships with the relative plate motion direction and the Cocos Ridge trend are more difficult to reconcile. Stress and strain analyses using ASR on core samples help to reconstruct the 3D orientation of stress. ASR indicates that a normal-fault stress regime characterizes the upper slope, while the stress state at the middle slope is typical of a strike-slip regime with σ_2 oriented vertically (Yamamoto et al., 2011). For the middle region, it has been suggested that in the outer forearc the maximum horizontal stress is related to the Cocos Ridge working as an indenter, whereas the inner forearc region instead, shows trench-parallel extension.

Stress estimation by drilling provides constraint for the mechanical modeling of the seismic cycle, but material properties remain an important yet imperfectly known component influencing shallow fault slip to near the seafloor during megathrust earthquakes. Recent numerical modeling of dynamic rupture with realistic variations of elastic parameters toward the trench suggests that large slip can propagate to a subduction trench even through tens of kilometers of velocity-strengthening materials with ordinary friction values of ~0.6 (Kozdon & Dunham, 2013). This implies that velocity strengthening (as found for most clay-rich wedge materials at moderate laboratory velocity) and coseismic rupture are not incompatible. Furthermore, if high velocity weakening in clays is as fundamental a process as the laboratory studies suggest, then rupture propagation to the trench, or to the seafloor along splay faults, should be a common occurrence. The Tohoku 2011 earthquake provides clear evidence that this can happen, but available data are not yet adequate to know whether megathrust rupture propagation to the trench, or up megasplays to the surface, is or is not as unusual as was previously thought.

Tectonic stress state as elucidated by drilling may also be relevant to understanding the conditions favoring rupture propagation to the trench, its branching into a splay fault, or its arrest at depth. It has been found that static force balance largely determines long-term variations of related splay fault and outer décollement activity and thus should influence coseismic rupture branching

(Conin et al., 2012). Additionally, quasi-static stress modeling shows that stress variations before and after a subduction earthquake can cause important variations in the stress distribution within the margin and bring portions of the margin to compressional or extensional failure during or immediately after the earthquake (e.g., Wang & Hu, 2006). For instance, some authors have recently proposed that large-displacement forearc normal faults are evidence for very efficient dynamic weakening along the megathrust, typifying megathrusts with high tsunamigenic potential (Conin et al., 2012; Cubas, Avouac, Leroy, & Pons, 2013; Tsuji et al., 2011). On the other hand, the case of the Maule earthquake suggests that rupture propagation does not occur below portions of an active margin that are under compressively critical stress in the sense of Coulomb wedge theory (Cubas, Avouac, Souloumiac, & Leroy, 2013). Moreover, generic numerical studies have shown that the minimum stress level for self-sustained rupture propagation depends on the proximity of the initial stress state to the yield surface (Dunham, Belanger, Cong, & Kozdon, 2011) and this could imply that rupture propagation to the trench is less likely below a wedge poised near compressive failure than in other cases. The influence of the initial stress state (and of failure outside the main fault plane) on rupture propagation and branching has yet to be fully taken into account in dynamic models of subduction zone earthquakes.

4.4.1.4 FUTURE DIRECTIONS

The science plan for the 2013–2023 International Ocean Discovery Program (*IODP*) prioritizes several aspects of subduction zone research. Given the past decade's history of destructive megathrust earthquakes and tsunami, understanding the mechanisms that control the occurrence of destructive earthquakes, landslides, and tsunami will certainly continue to drive future subduction zone science. Completion of the work started via NanTroSEIZE and CRISP would help achieve goals of understanding fault zone processes. In particular, realizing the goal of borehole sampling of the deep (>2–3 km) plate boundary fault zones remains high in priority (CHIKYU+10 Steering Committee, 2013; Science Plan Writing Committee, INVEST Steering Committee, & Additional Contributors, 2011). At the time of this writing, plans are active for advancing NanTroSEIZE Site C0002 from 3000 mbsf to ~5200 mbsf, as a flagship project for Chikyu, with coring and LWD logging of the main fault zone (Figure 4.4.1.8), and casing that main plate boundary fault for installation of a long-term borehole monitoring package (see e.g., Tobin, Hirose, Saffer, & Expedition 348 Shipboard Scientific Party, 2014). CRISP B has also been proposed as one of the flagship projects for the *Chikyu* platform, but as of 2014, a decision whether or not drilling will go ahead and when has not yet been reached.

Because the seismogenic cycle in subduction zones is decades to centuries in duration, a key strategy for future efforts will be to investigate different plate boundaries that are at different points in the megathrust event seismic cycle (see,

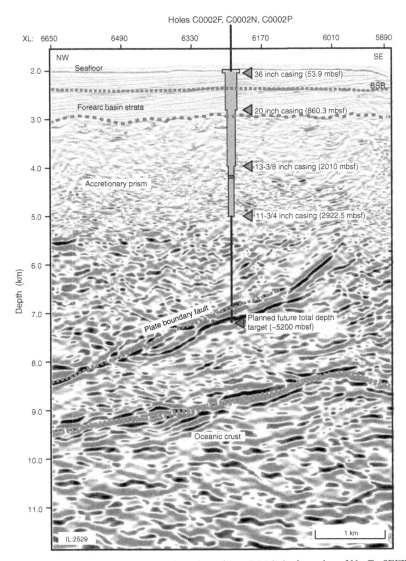

FIGURE 4.4.1.8 Detail of seismic depth section (Figure 4.4.1.2) in the region of NanTroSEIZE Deep riser Site C0002, illustrating casing installed in the borehole on Expeditions 326, 338, and 348, and planned future drilling target across the main high-amplitude plate boundary (so called megasplay) reflector and into its footwall. The NanTroSEIZE science team proposes to drill to ~5200 m below sea floor, first in a logging-while-drilling hole, then in a second sidetrack hole to collect cores from above, within, and below the main fault zone as interpreted from this reflection.

e.g., Wang, Hu, & He, 2012). Further investigations of the Japan Trench would address, in the near-term postseismic period, questions on the mechanisms of the surprising and devastating 2011 Tohoku earthquake. Another expedition, now scheduled for 2016, would examine the inputs to the Sumatra 2004 M9 nucleation region, also in the early postseismic phase. Both of these complement the Nankai and Costa Rica efforts, which have focused on systems that are at mid- to late stages in the presumed seismic cycle.

Recent advances in broadband seismology and continuous GPS geodetic monitoring have revolutionized our understanding of fault displacement, showing that it can occur across a much wider spectrum of time scales than previously appreciated (e.g., Beroza & Ide, 2011; Obara, 2002; Schwartz & Rokosky, 2007; Wallace & Beavan, 2010; Wallace, Beavan, Bannister, & Williams, 2012, and many others). Slow slip and VLF earthquake events (Ito & Obara, 2006; Sugioka et al., 2012) in particular, those known to occur at shallow depths offshore in several subduction zones including the Hikurangi, Mexico, and Nankai margins, may elucidate fault behavior in the transitional state between genuinely aseismic and fully seismogenic conditions, and are promising targets for future drilling projects, for example, as proposed for the north Hikurangi margin of New Zealand.

Additional high priority scientific goals in the new *IODP* Science Plan include understanding how subduction zones initiate and how new continental crust is generated. This will be investigated through drilling of regions in incipient or initial stages of subduction, examination of ash as a recorder of arc volcanism, and through drilling transects across regions of ancient and active subduction. Another subduction zone target is the outer trench environment, where normal faulting is suspected to allow fluid influx and mantle serpentinization.

4.4.1.5 SUMMARY AND CONCLUSIONS

IODP drilling from 2006 to 2013 took place during a period when understanding of active subduction-related forearc processes, particularly seismicity and tsunamigenesis, was evolving rapidly. The 2004 Sumatra megathrust earthquake was the first magnitude 9 class event in 40 years, and the first in the modern era of broadband seismic and geodetic monitoring. It was accompanied by a devastating tsunami which focused the world's attention on megathrust processes and contributed to the rapid development of marine tools to monitor and mitigate hazards. Meanwhile, those same types of observations revealed slow slip, tremor, and VLF events, all illustrating that the concept of aseismic versus seismogenic faulting, separated by a clear transition, are overly simplistic. The series of large events since 2004, especially the 2011 Tohoku tsunamigenic earthquake, made it clear that a better understanding of shallow (upper 5–10 km) fault slip is critical if we wish to better understand subduction zone fault behavior and evaluate potential hazards. Investigation of the Nankai, Costa Rica, and Tohoku margins by IODP drilling has become a key component of this new and more fully realized picture.

Key results to date include the following:

- Rapid and localized fault slip can extend all the way to the trench, even where hosted in porous, poorly consolidated marine sediments. The best current explanation for these observations is that dynamic frictional processes, perhaps coupled to preexisting stress conditions in the outer wedge, control the likelihood of shallow propagation of rapid coseismic slip.

• Although seismic reflection data and numerical modeling suggest that excess pore pressures may be high at depths <1 km, drilling and monitoring observation have not found strong evidence for high pore pressures in the shallow regions of the NanTroSEIZE and JFAST transects. The presence of a strong seismic reflection from a low-angle thrust plane permits but does not require that the fault is a localized fluid channeling system. The lack of evidence for any identified geochemical anomaly at either the frontal thrust or the megasplay system in the NanTroSEIZE transect, or from the Tohoku borehole (Expedition 343 Scientists, 2013), suggests that the presence of a fault-hosted fluid regime is specific to certain margins and perhaps to certain lithologies.

• It is now well documented that the tectonics of subduction are strongly punctuated in time by events at both dominantly accretionary and erosional margins. Evidence from both Costa Rica and Nankai suggest that bathymetric features in the incoming plate likely have large impact on the deforming wedge, kicking off out-of-sequence thrusting, modifying frontal accretion, causing episodes of tectonic erosion, causing variations in stress regime from compressional to extensional and transform, among other impacts. Along the CRISP transect a rapid pulse of subduction erosion has strongly shaped the present-day forearc. Similar pulses may punctuate the evolution of many margins, contributing disproportionately to the recycling of continental material at subduction zones. Along the NanTroSEIZE transect, subducting basement topography has apparently flipped the locus of wedge deformation from frontal to out-of-sequence several times as well as controlled pulses of major forearc basin formation and filling, and the underthrusting of a thick package of Shikoku basin sediments. The NanTroSEIZE transect also shows clear evidence for episodic activation of OOST, both the megasplay and other faults, simultaneously with or alternating with frontal thrusting. All of these modifications at both these example margins cause variation in the criticality of the deforming wedges, which may cycle from stable to critical on a range of time scales.

ACKNOWLEDGMENTS

First and foremost, we acknowledge all of the many participating scientists, both shipboard and shore-based, in IODP Expeditions 314, 315, 316, 319, 322, 332, 333, 334, 338, 343, 344, and 348 and the personnel onboard the Chikyu and JOIDES Resolution. It is their work that we only review in this chapter. We also thank the editors of this volume for their patient support, as well as reviewers Julia Morgan and Roland von Huene for their insightful comments which improved the manuscript.

REFERENCES

Ando, M. (1975). Source mechanisms and tectonic significance of historical earthquakes along the Nankai Trough, Japan. *Tectonophysics, 27*(2), 119–140. http://dx.doi.org/10.1016/0040-1951(75)90102-X.

Apel, E. V., Bürgmann, R., Steblov, G., Vasilenko, N., King, R., & Prytkov, A. (2006). Independent active microplate tectonics of northeast Asia from GPS velocities and block modeling. *Geophysical Research Letters, 33*(11), L11303. http://dx.doi.org/10.1029/2006GL026077.

Argus, D. F., Gordon, R. G., & DeMets, C. (2011). Geologically current motion of 56 plates relative to the no-net-rotation reference frame. *Geochemistry Geophysics Geosystems, 12*(11), Q11001. http://dx.doi.org/10.1029/2011GC003751.

Asano, Y., Saito, T., Ito, Y., Shiomi, K., Hirose, H., Matsumoto, T., et al. (2011). Spatial distribution and focal mechanisms of aftershocks of the 2011 off the Pacific Coast of Tohoku Earthquake. *Earth Planets and Space, 63*, 669–673. http://dx.doi.org/10.5047/eps.2011.06.016.

Ask, M. V. S., & Morgan, J. K. (2010). Projection of mechanical properties from shallow to greater depths seaward of the Nankai accretionary prism. *Tectonophysics, 482*, 50–64. http://dx.doi.org/10.1016/j.tecto.2009.08.023.

Baba, T., Cummins, P. R., Hori, T., & Kaneda, Y. (2005). High precision slip distribution of the 1944 Tonankai earthquake inferred from tsunami waveforms: possible slip on a splay fault. *Tectonophysics, 426*, 119–134.

Bangs, N. L., McIntosh, K. D., Silver, E. A., Kluesner, J. W., Ranero, C. R. (in press). Fluid accumulation along the Costa Rica subduction thrust development of the seismogenic zone. *Journal of Geophysical Research.*

Bangs, N. L. B., Moore, G. F., Gulick, S. P. S., Pangborn, E. M., Tobin, H. J., Kuramoto, S., et al. (2009). Broad, weak regions of the Nankai megathrust and implications for shallow coseismic slip. *Earth and Planetary Science Letters, 284*(1–2), 44–49. http://dx.doi.org/10.1016/j.epsl.2009.04.026.

Bekins, B., & Screaton, E. (2007). Pore pressure and fluid flow in the northern Barbados accretionary complex. In T. Dixon, & J. Casey Moore (Eds.), *The seismogenic zone of subduction thrust faults* (pp. 148–170). Columbia University Press.

Beroza, G., & Ide, S. (2011). Slow earthquakes and non-volcanic tremor. *Annual Review of Earth and Planetary Sciences, 39*, 271–296.

Boyer, S. E., & Elliott, D. (1982). Thrust systems. *Bulletin of the American Association of Petroleum Geologists, 66*(9), 1196–1230.

Byerlee, J. D. (1978). Friction of rocks. *Pure and Applied Geophysics, 116*, 615–626.

Byrne, T. B., Lin, W., Tsutsumi, A., Yamamoto, Y., Lewis, J. C., Kanagawa, K., et al. (2009). Anelastic strain recovery reveals extension across SW Japan subduction zone. *Geophysical Research Letters, 36*, L23310. http://dx.doi.org/10.1029/2009GL040749.

Carpenter, B. M., Marone, C., & Saffer, D. M. (2011). Weakness of the San Andreas fault revealed by samples of the active fault zone. *Nature Geoscience, 4*, 251–254. http://dx.doi.org/10.1038/ngeo1089.

Chang, C., McNeill, L. C., Moore, J. C., Lin, W., Conin, M., & Yamada, Y. (2010). In situ stress state in the Nankai accretionary wedge estimated from borehole wall failures. *Geochemistry Geophysics Geosystems, 11*, Q0AD04. http://dx.doi.org/10.1029/2010GC003261.

Chester, F. M., Mori, J. J., Toczko, S., Eguchi, N., & the Expedition 343/343T Scientists. (2012). Japan trench fast drilling project (JFAST). In *IODP Prel. Rept., 343/343T.* http://dx.doi.org/10.2204/iodp.pr.343343T.2012.

Chester, F. M., Rowe, C., Ujiie, K., Kirkpatrick, J., Regalla, C., Remitti, F., et al. (2013). Structure and composition of the plate-boundary slip zone for the 2011 Tohoku Oki Earthquake. *Science*, *342*(6163), 1208–1211. http://dx.doi.org/10.1126/science.1243719.

CHIKYU+10 Steering Committee. (2013). *Chikyu+10 international Workshop Report*. Available at http://www.jamstec.go.jp/chikyu+10/docs/C+10_report_textbody.pdf.

Clift, P., & Vannucchi, P. (2004). Controls on tectonic accretion versus erosion in subduction zones: implications for the origin and recycling of the continental crust. *Reviews of Geophysics*, *42*, RG2001. http://dx.doi.org/10.1029/2003RG000127.

Conin, M., Henry, P., Bourlange, S., Raimbourg, H., & Reuschlé, T. (2011). Interpretation of porosity and LWD resistivity from the Nankai accretionary wedge in light of clay physicochemical properties: evidence for erosion and local overpressuring. *Geochemistry Geophysics Geosystems*, *12*, Q0AD07. http://dx.doi.org/10.1029/2010GC003381.

Conin, M., Henry, P., Godard, V., & Bourlange, S. (2012). Splay fault slip in a subduction margin, a new model of evolution. *Earth and Planetary Science Letters*, *341–344*, 170–175. http://dx.doi.org/10.1016/j.epsl.2012.06.003.

Cubas, N., Avouac, J. P., Leroy, Y. M., & Pons, A. (2013). Low friction along the high slip patch of the 2011 Mw 9.0 Tohoku-Oki earthquake required from the wedge structure and extensional splay faults. *Geophysical Research Letters*, *40*, 4231–4237. http://dx.doi.org/10.1002/grl.50682.

Cubas, N., Avouac, J.-P., Souloumiac, P., & Leroy, Y. (2013). Megathrust friction determined from mechanical analysis of the forearc in the Maule earthquake area. *Earth and Planetary Science Letters*, *381*, 92–103. http://dx.doi.org/10.1016/j.epsl.2013.07.037.

Daigle, H., & Screaton, E. J. (2014). Evolution of sediment permeability during burial and subduction. *Geofluids*, *15*. http://dx.doi.org/10.1111/gfl.12090.

Davis, D., Suppe, J., & Dahlen, F. A. (1983). Mechanics of fold-and-thrust belts and accretionary wedges. *Journal of Geophysical Research*, *88*, 1153–1172. http://dx.doi.org/10.1029/JB088iB02p01153.

DeMets, C. (2001). A new estimate for present-day Cocos-Caribbean plate motion: implications for slip along the Central American volcanic arc. *Geophysical Research Letters*, *28*(21), 4043–4046. http://dx.doi.org/10.1029/2001GL013518.

DeMets, C., Gordon, R. G., & Argus, D. F. (2010). Geologically current plate motions. *Geophysical Journal International*, *181*(1), 1–80. http://dx.doi.org/10.1111/j.1365-246X.2009.04491.x.

Dunham, E. M., Belanger, D., Cong, L., & Kozdon, J. E. (2011). Earthquake ruptures with strongly rate-weakening friction and off-fault plasticity, 1: planar faults. *Bulletin of the Seismological Society of America*, *101*(5), 2296–2307. http://dx.doi.org/10.1785/0120100075.

Expedition 316 Scientists. (2009). Expedition 316 site C0007. In M. Kinoshita, H. Tobin, J. Ashi, G. Kimura, S. Lallemant, E. J. Screaton, et al. (Eds.), *Proc. IODP, 314/315/316*. Washington, DC: Integrated Ocean Drilling Program Management International, Inc. http://dx.doi.org/10.2204/iodp.proc.314315316.135.2009.

Expedition 334 Scientists. (2012). Expedition 334 summary. In P. Vannucchi, K. Ujiie, N. Stroncik, A. Malinverno, & the Expedition 334 Scientists (Eds.), *Proc. IODP, 334*. Tokyo: Integrated Ocean Drilling Program Management International, Inc. http://dx.doi.org/10.2204/iodp.proc.334.101.2012.

Expedition 343/343T Scientists. (2013). Site C0019. In F. M. Chester, J. Mori, N. Eguchi, S. Toczko, & the Expedition 343/343T Scientists (Eds.), *Proc. IODP, 343/343T*. Tokyo: Integrated Ocean Drilling Program Management International, Inc.. http://dx.doi.org/10.2204/iodp.proc.343343T.103.2013.

Faulkner, D. R., Mitchell, T. M., Behnsen, J., Hirose, T., & Shimamoto, T. (2011). Stuck in the mud? Earthquake nucleation and propagation through accretionary forearcs. *Geophysical Research Letters, 38.* (18). http://dx.doi.org/10.1029/2011GL048552.

Fujiwara, T., Kodaira, S., No, T., Kaiho, Y., Takahashi, N., & Kaneda, Y. (2011). The 2011 Tohoku-Oki earthquake: displacement reaching the trench axis. *Science, 334*(6060), 1240. http://dx.doi.org/10.1126/science.1211554.

Fuller, C. W., Willett, S. D., & Brandon, M. T. (2006). Formation of forearc basins and their influence on subduction zone earthquakes. *Geology, 34*, 65–68. http://dx.doi.org/10.1130/G21828.1.

Fulton, P. M., Brodsky, E. E., Kano, Y., Mori, J., Chester, F., Ishikawa, T., et al. (2013). Low coseismic friction on the Tohoku-Oki fault determined from temperature measurements. *Science, 342*(6163), 1214–1217. http://dx.doi.org/10.1126/science.1243641.

Fulton, P. M., & Harris, R. N. (2012). Thermal considerations in inferring frictional heating from vitrinite reflectance and implications for estimates of shallow coseismic slip within the Nankai subduction zone. *Earth and Planetary Science Letters, 335–336*, 206–215. http://dx.doi.org/10.1016/j.epsl.2012.04.012.

Furuichi, H., Ujiie, K., Sakaguchi, A., & Tsutsumi, A. (2013). What is the factor controlling the increase in vitrinite reflectance along faults? In *Japan Geoscience Union Meeting 2013. Chiba: Makuhari.*

Gamage, K., Screaton, E., Bekins, B., & Aeillo, I. (2011). Permeability-porosity relationships of subduction zone sediments. *Marine Geology, 279*(1–4), 19–36.

Gulick, S. P. S., Bangs, N. L. B., Moore, G. F., Ashi, J., Martin, K. M., Sawyer, D. S., et al. (2010). Rapid forearc basin uplift and megasplay fault development from 3D seismic images of Nankai margin off Kii Peninsula, Japan. *Earth and Planetary Science Letters, 300*(1–2), 55–62. http://dx.doi.org/10.1016/j.epsl.2010.09.034.

Hammerschmidt, S., Davis, E. E., & Kopf, A. (2013). Fluid pressure and temperature transients detected at the Nankai Trough megasplay fault: results from the SmartPlug borehole observatory. *Tectonophysics*, ISSN 0040-1951, *600*, 116–133. http://dx.doi.org/10.1016/j.tecto.2013.02.010.

Harris, R. N., Sakaguchi, A., Petronotis, K., & the Expedition 344 Scientists. (2013). In *Proc. IODP, 344.* College Station, TX: Integrated Ocean Drilling Program. http://dx.doi.org/10.2204/iodp.proc.344.2013.

Harris, R. N., Spinelli, G., Ranero, C. R., Grevemeyer, I., Villinger, H., & Barckhausen, U. (2010). Thermal regime of the Costa Rican convergent margin: 2. Thermal models of the shallow Middle America subduction zone offshore Costa Rica. *Geochemistry Geophysics Geosystems, 11*, Q12S29. http://dx.doi.org/10.1029/2010GC003273.

Henry, P., Mazzotti, S., & Le Pichon, X. (2001). Transient and permanent deformation of central Japan estimated by GPS, 1. Interseismic loading and subduction kinematics. *Earth and Planetary Science Letters, 184*, 443–453.

Hensen, C., & Wallmann, K. (2005). Methane formation at Costa Rica continental margin: constraints for gas hydrate inventory and cross-décollement fluid flow. *Earth and Planetary Science Letters, 236*, 41–60.

Hirono, T., Ujiie, K., Ishikawa, T., Mishima, T., Hamada, Y., Tanimizu, M., et al. (2009). Estimation of temperature rise in a shallow slip zone of the megasplay fault in the Nankai Trough. *Tectonophysics, 478*(3–4), 215–220. http://dx.doi.org/10.1016/j.tecto.2009.08.001.

Hoffman, N., & Tobin, H. (2004). An empirical relationship between velocity and porosity for underthrust sediments in the Nankai Trough accretionary prism. In H. Mikada, G. F. Moore, A. Taira, K. Becker, J. C. Moore, & A. Klaus (Eds.), *Proc. ODP, Sci. Results, 190/196.* http://www-odp.tamu.edu/publications/190196SR/355/355.htm.

Hori, T., Kato, N., Hirahara, K., Baba, T., & Kaneda, Y. (2004). A numerical simulation of earthquake cycles along the Nankai Trough in southwest Japan: lateral variation in frictional property due to the slab geometry controls the nucleation position. *Earth and Planetary Science Letters, 228*(3–4), 215–226. http://dx.doi.org/10.1016/j.epsl.2004.09.033.

von Huene, R., Ranero, C. R., & Vannucchi, P. (2004). Generic model of subduction erosion. *Geology, 32*(10), 913–916. http://dx.doi.org/10.1130/G20563.1.

von Huene, R., & Scholl, D. W. (1991). Observations at convergent margins concerning sediment subduction, subduction erosion, and the growth of continental crust. *Reviews of Geophysics, 29*, 279–316. http://dx.doi.org/10.1029/91RG00969.

Hüpers, A., & Kopf, A. J. (2009). The thermal influence on the consolidation state of underthrust sediments from the Nankai margin and its implications for excess pore pressure. *Earth and Planetary Science Letters, 286*, 324–332.

Hyndman, R., Yamano, M., & Oleskevich, D. (1997). The seismogenic zone of subduction thrust faults. *Island Arc, 6*, 244–260.

Ichinose, G. A., Thio, H. K., Somerville, P. G., Sato, T., & Ishii, T. (2003). Rupture process of the 1944 Tonankai earthquake (Ms 8.1) from the inversion of teleseismic and regional seismograms. *Journal of Geophysical Research, 108*(B10), 2497. http://dx.doi.org/10.1029/2003JB002393.

Ide, S., Baltay, A., & Beroza, G. C. (2011). Shallow dynamic overshoot and energetic deep rupture in the 2011 Mw 9.0 Tohoku-Oki earthquake. *Science, 332*(6036), 1426–1429. http://dx.doi.org/10.1126/science.1207020.

Ikari, M. J., Niemeijer, A. R., Spiers, C. J., Kopf, A. J., & Saffer, D. M. (2013). Experimental evidence linking slip instability with seafloor lithology and topography at the Costa Rica convergent margin. *Geology, 41*(8), 891–894. http://dx.doi.org/10.1130/G33956.1.

Ikari, M. J., & Saffer, D. M. (2011). Comparison of frictional strength and velocity dependence between fault zones in the Nankai accretionary complex. *Geochemistry Geophysics Geosystems, 12*, Q0AD11. http://dx.doi.org/10.1029/2010GC003442.

Ike, T., Moore, G. F., Kuramoto, S., Park, J.-O., Kaneda, Y., & Taira, A. (2008a). Variations in sediment thickness and type along the northern Philippine Sea Plate at the Nankai Trough. *Island Arc, 17*(3), 342–357. http://dx.doi.org/10.1111/j.1440-1738.2008.00624.x.

Ike, T., Moore, G. F., Kuramoto, S., Park, J.-O., Kaneda, Y., & Taira, A. (2008b). Tectonics and sedimentation around Kashinosaki Knoll: a subducting basement high in the eastern Nankai Trough. *Island Arc, 17*, 358–375. http://dx.doi.org/10.1111/j.1440-1738.2008.00625.x.

Ito, Y., Asano, Y., & Obara, K. (2009). Very-low-frequency earthquakes indicate a transpressional stress regime in the Nankai accretionary prim. *Geophysical Research Letters, 36*, L203309. http://dx.doi.org/10.1029/2009GL039332.

Ito, T., Funato, A., Lin, W., Doan, M.-L., Boutt, D. F., Kano, Y., et al. (2013). Determination of stress state in deep subsea formation by combination of hydraulic fracturing in situ test and core analysis: a case study in the IODP Expedition 319. *Journal of Geophysical Research Solid Earth, 118*, 1203–1215. http://dx.doi.org/10.1002/jgrb.50086.

Ito, Y., Hino, R., Kido, M., Fujimoto, H., Osada, Y., Inazu, D., et al. (2012). Episodic slow slip events in the Japan subduction zone before the 2011 Tohoku-Oki earthquake. *Tectonophysics, 600*, 14–26. http://dx.doi.org/10.1016/j.tecto.2012.08.022.

Ito, Y., & Obara, K. (2006). Dynamic deformation of the accretionary prism excites very low frequency earthquakes. *Geophysical Research Letters, 33*(2), LO2311. http://dx.doi.org/10.1029/2005GL025270.

Ito, Y., Tsuji, T., Osada, Y., Kido, M., Inazu, D., Hayashi, Y., et al. (2011). Frontal wedge deformation near the source region of the 2011 Tohoku-Oki earthquake. *Geophysical Research Letters, 38*(15), L00G05. http://dx.doi.org/10.1029/2011GL048355.

Kimura, G., Hina, S., Hamada, Y., Kameda, J., Tsuji, T., Kinoshita, M., & Yamaguchi, A. (2012). Runaway slip to the trench due to rupture of highly pressurized megathrust beneath the middle trench slope: the tsunamigenesis of the 2011 Tohoku earthquake off the east coast of northern Japan. *Earth and Planetary Science Letters*, *339–340*, 32–45. http://dx.doi.org/10.1016/j.epsl.2012.04.002.

Kimura, G., Kitamura, Y., Hashimoto, Y., Yamaguchi, A., Shibata, T., Ujiie, K., et al. (2007). Transition of accretionary wedge structures around the up-dip limit of the seismogenic subduction zone. *Earth and Planetary Science Letters*, *255*(3–4), 471–484. http://dx.doi.org/10.1016/j.epsl.2007.01.005.

Kimura, G., Moore, G. F., Strasser, M., Screaton, E., Curewitz, D., Streiff, C., et al. (2011). Spatial and temporal evolution of the megasplay fault in the Nankai Trough. *Geochemistry Geophysics Geosystems*, *12*, Q0A008. http://dx.doi.org/10.1029/2010GC003335.

Kington, J. (2012). *The structure and kinematics of the Nankai trough accretionary prism, Japan* (Ph.D. Dissertation). Madison: University of Wisconsin.

Kinoshita, M., Moore, G. F., & Kido, Y. N. (2011). Heat flow estimated from BSR and IODP borehole data: Implication of recent uplift and erosion of the imbricate thrust zone in the Nankai Trough off Kumano. *Geochemistry Geophysics Geosystems*, *12*, Q0AD18. http://dx.doi.org/10.1029/2011GC003609.

Kitajima, H., & Saffer, D. M. (2012). Elevated pore pressure and anomalously low stress in regions of low frequency earthquakes along the Nankai Trough subduction megathrust. *Geophysical Research Letters*, *39*, L23301. http://dx.doi.org/10.1029/2012GL053793.

Kitamura, M., Mukoyoshi, H., Fulton, P. M., & Hirose, T. (2012). Coal maturation by frictional heat during rapid fault slip. *Geophysical Research Letters*, *39*, L16302. http://dx.doi.org/10.1029/2012GL052316.

Kluesner, J. W., Silver, E. A., Bangs, N. L., McIntosh, K. D., Gibson, J., Orange, D., et al. (2013). High density of structurally controlled, shallow to deep water fluid seep indicators imaged offshore Costa Rica. *Geochemistry Geophysics Geosystems*, *14*, 519–539. http://dx.doi.org/10.1002/ggge.20058.

Kodaira, S., No, T., Nakamura, Y., Fujiwara, T., Kaiho, Y., Miura, S., et al. (2012). Coseismic fault rupture at the trench axis during the 2011 Tohoku-Oki earthquake. *Nature Geoscience*, *5*(9), 646–650. http://dx.doi.org/10.1038/ngeo1547.

Kopf, A. (2013). Effective strength of incoming sediments and its implications for plate boundary propagation: Nankai and Costa Rica as type examples of accreting vs. erosive convergent margins. *Tectonophysics*, *608*, 958–969. http://dx.doi.org/10.1016/j.tecto.2013.07.023.

Kozdon, J. E., & Dunham, E. M. (2013). Rupture to the trench: dynamic rupture simulations of the 11 March 2011 Tohoku earthquake. *Bulletin of the Seismological Society of America*, *103*(2B), 1275–1289. http://dx.doi.org/10.1785/0120120136.

LaFemina, P., Dixon, T. H., Govers, R., Norabuena, E., Turner, H., Saballos, A., et al. (2009). Forearc motion and Cocos Ridge collision in Central America. *Geochemistry Geophysics Geosystems*, *10*. http://dx.doi.org/10.1029/2008GC002181.

Lallemand, S. E., Schnurle, P., & Malavieille, J. (1994). Coulomb theory applied to accretionary and non-accretionary wedges-possible causes for tectonic erosion and/or frontal accretion. *Journal of Geophysical Research*, *99*(B6), 12033–12055.

Lay, T., Ammon, C., Kanamori, H., Xue, L., & Kim, M. (2011). Possible large near-trench slip during the great 2011 Tohoku (Mw 9.0) earthquake. *Earth Planets and Space*, *63*, 687–692. http://dx.doi.org/10.5047/eps.2011.05.033.

Lin, W., Conin, M., Moore, J. C., Chester, F. M., Nakamura, Y., Mori, J. J., et al. (2013). Stress state in the largest displacement area of the 2011 Tohoku-Oki Earthquake. *Science*, *339*(6120), 687–690. http://dx.doi.org/10.1126/science.1229379.

Lin, W., Doan, M.-L., Moore, J. C., McNeill, L., Byrne, T. B., Ito, T., et al. (2010). Present-day principal horizontal stress orientations in the Kumano forearc basin of the southwest Japan subduction zone determined from IODP NanTroSEIZE drilling site C0009. *Geophysical Research Letters, 37*, L13303. http://dx.doi.org/10.1029/2010GL043158.2010.

Lin, W., Saito, S., Sanada, Y., Yamamoto, Y., Hashimoto, Y., & Kanamatsu, T. (2011). Principal horizontal stress orientations prior to the 2011 Mw 9.0 Tohoku-Oki, Japan, earthquake in its source area. *Geophysical Research Letters, 38*(17), L00G10. http://dx.doi.org/10.1029/201 1GL049097.

Lockner, D. A., Morrow, C., Moore, D., & Hickman, S. (2011). Low strength of deep San Andreas fault gouge from SAFOD core. *Nature, 472*, 82–85. http://dx.doi.org/10.1038/nature09927.

Malinverno, A., & Saito, S. (2013). In *Borehole breakout orientation from LWD data (IODP Exp. 334) and the present stress state in the Costa Rica Seismogenesis Project transect, 2013 Fall AGU Meeting, San Francisco, Calif.* Abstract T34C-07.

Martin, K. M., Gulick, S. P. S., Bangs, N. L. B., Moore, G. F., Ashi, J., Park, J.-O., et al. (2010). Possible strain partitioning structure between the Kumano fore-arc basin and the slope of the Nankai Trough accretionary prism. *Geochemistry Geophysics Geosystems, 11*, Q0AD02. http://dx.doi.org/10.1029/2009GC002668.

Mazzotti, S., Le Pichon, X., Henry, P., & Miyazaki, S. (2000). Full interseismic locking of the Nankai and Japan-West Kurile subduction zones: an analysis of uniform elastic strain accumulation in Japan constrained by permanent GPS. *Journal of Geophysical Research, 105*, 13159–13177.

McIntosh, K., Silver, E., & Shipley, T. (1993). Evidence and mechanisms for fore-arc extension at the accretionary Costa-Rica convergent margin. *Tectonics, 12*(6), 1380–1392.

Miyazaki, S., & Heki, K. (2001). Crustal velocity field of southwest Japan: subduction and arc-arc collision. *Journal of Geophysical Research, 106*. http://dx.doi.org/10.1029/2000JB900312.

Moore, G. F., Bangs, N. L., Taira, A., Kuramoto, S., Pangborn, E., & Tobin, H. J. (2007). Three-dimensional splay fault geometry and implications for tsunami generation. *Science, 318*, 1128–1131. http://dx.doi.org/10.1126/science.1147195.

Moore, J. C., Barrett, M., & Thu, M. K. (2012). High fluid pressures and high fluid flow rates in the megasplay fault zone, NanTroSEIZE Kumano Transect, SW Japan. *Geochemistry Geophysics Geosystems, 13*, Q0AD25. http://dx.doi.org/10.1029/2012GC004181.

Moore, G. F., Park, J.-O., Bangs, N. L., Gulick, S. P., Tobin, H. J., Nakamura, Y., et al. (2009). Structural and seismic stratigraphic framework of the NanTroSEIZE Stage 1 transect. In M. Kinoshita, et al. (Ed.), *Proceedings of Integrated Ocean Drilling Program* (vol. 314/315/316). Ocean Drill. Program, Washington, DC, Integrated Ocean Drilling Program Management International, Inc. http://dx.doi.org/10.2204/iodp.proc.314315316.102.2009.

Morgan, J. K., & Ask, M. V. S. (2004). Consolidation state and strength of underthrust sediments and evolution of the decollement at the Nankai accretionary margin: results of uniaxial reconsolidation experiments. *Journal of Geophysical Research, 109*, B03102. http://dx.doi.org/10.1 029/2002JB002335.

Morgan, J. K., Sunderland, E. B., & Ask, M. V. S. (2007). Deformation and mechanical strength of sediments at the Nankai subduction zone: implications for prism evolution and decollement initiation and propagation, in the Seismogenic Zone of subduction thrust faults. In T. H. Dixon, & J. C. Moore (Eds.), *MARGINS Theoretical and Experimental Earth Science Series* (pp. 210–256). Columbia University Press.

Murauchi, S., & Ludwig, W. J. (1980). Crustal structure of the Japan trench: the effect of subduction of ocean crust. In scientific party (Ed.). *Initial Reports of the Deep Sea Drilling Project: vol. 56–57.* (pp. 463–469). Washington: U.S. Govt. Printing Office. Pt. 1.

Namiki, Y., Tsutsumi, A., Ujiie, K., & Kameda, J. (2014). Frictional properties of sediments entering the Costa Rica subduction zone offshore the Osa Peninsula: implications for fault slip in shallow subduction zones. *Earth Planets and Space*, *66*, 72. http://dx.doi.org/10.1186/1880-5981-66-72.

Newman, A. V., Schwartz, S. Y., Gonzalez, V., DeShon, H. R., Protti, J. M., & Dorman, L. M. (2002). Along-strike variability in the seismogenic zone below Nicoya Peninsula, Costa Rica. *Geophysical Research Letters*, *29*(20), 1977. http://dx.doi.org/10.1029/2002GL015409.

Obana, K., & Kodaira, S. (2009). Low-frequency tremors associated with reverse faults in a shallow accretionary prism. *Earth and Planetary Science Letters*, *287*(1–2), 168–174. http://dx.doi.org/10.1016/j.epsl.2009.08.005.

Obana, K., Kodaira, S., Mochizuki, K., & Shinohara, M. (2001). Micro-seismicity around the seaward updip limit of the 1946 Nankai earthquake dislocation area. *Geophysical Research Letters*, *28*, 2333–2336.

Obara, K. (2002). Nonvolcanic deep tremor associated with subduction in southwest Japan. *Science*, *296*, 1679. http://dx.doi.org/10.1126/science.1070378.

Obara, K., & Ito, Y. (2005). Very low frequency earthquakes excited by the 2004 off the Kii Peninsula earthquakes: a dynamic deformation process in the large accretionary prism. *Earth Planets and Space*, *57*(4), 321–326.

Park, J.-O., Fujie, G., Wijerathne, L., Hori, T., Kodaira, S., Fukao, Y., et al. (2010). A low-velocity zone with weak reflectivity along the Nankai subduction zone. *Geology*, *38*(3), 283–286. http://dx.doi.org/10.1130/G30205.1.

Park, J.-O., Tsuru, T., Kodaira, S., Cummins, P. R., & Kaneda, Y. (2002). Splay fault branching along the Nankai subduction zone. *Science*, *297*, 1157–1160. http://dx.doi.org/10.1126/science.1074111.

Pflaker, G. (1969). *Tectonics of the March 27, 1964 Alaska Earthquake*, U.S. Geological Survey Professional Paper 543–I, p. 74, 2 sheets. http://pubs.usgs.gov/pp/0543i/.

Pickering, K. T., Underwood, M. B., Saito, S., Naruse, H., Kutterolf, S., Scudder, R., et al. (2013). Depositional architecture, provenance, and tectonic/eustatic modulation of Miocene submarine fans in the Shikoku Basin: results from Nankai Trough Seismogenic Zone experiment. *Geochemistry Geophysics Geosystems*, *14*(6), 1722–1739. http://dx.doi.org/10.1002/ggge.20107.

Pickering, K. T., Underwood, M. B., & Taira, A. (1992). Open-ocean to trench turbidity-current flow in the Nankai Trough: flow collapse and reflection. *Geology*, *20*, 1099–1102.

Protti, M., Gündel, F., & McNally, K. (1994). The geometry of the Wadati-Benioff zone under southern Central America and its tectonic significance: results from a high-resolution local seismographic network. *Physics of the Earth and Planetary Interiors*, *84*(1–4), 271–287. http://dx.doi.org/10.1016/0031-9201(94)90046-9.

Ranero, C. R., Grevemeyer, I., Sahling, H., Barckhausen, U., Hensen, C., Wallman, K., et al. (2008). Hydrogeological system of erosional convergent margins and its influence of tectonics and intraplate seismogenesis. *Geochemistry Geophysics Geosystems*, *9*, Q03S04. http://dx.doi.org/10.1029/2007GC001679.

Ranero, C. R., & von Huene, R. (2000). Subduction erosion along the Middle America convergent margin. *Nature*, *404*, 748–752. http://dx.doi.org/10.1038/35008046.

Rowe, K., Screaton, E. J., & Ge, S. (2012). Coupled fluid-flow and deformation modeling of the frontal thrust region of the Kumano basin transect, Japan: implications for fluid pressures and decollement downstepping. *Geochemistry Geophysics Geosystems*, *13*, Q0AD23 18 pp. http://dx.doi.org/10.1029/2011GC003861

Ryan, H., von Heune, R., Scholl, D., & Kirby, S. (2012). Tsunami hazards to U.S. coasts from giant earthquakes in Alaska. *EOS Transactions American Geophysical Union, 93*(19), 185. http://dx.doi.org/10.1029/2012EO190001.

Sacks, A., Saffer, D. M., & Fisher, D. (2013). Analysis of normal fault populations in the Kumano forearc basin, Nankai Trough, Japan: 2. Principal axes of stress and strain from inversion of fault orientations. *Geochemistry Geophysics Geosystems, 14*(6), 1973–1988. http://dx.doi.org/10.1002/ggge.20118.

Saffer, D. M. (2003). Pore pressure development and progressive dewatering in underthrust sediments at the Costa Rica subduction margin: comparison with northern Barbados and Nankai. *Journal of Geophysical Research, 108*(B5), 2261.

Saffer, D. M. (2010). Hydrostratigraphy as a control on subduction zone mechanics through its effects on drainage: an example from the Nankai margin, SW Japan. *Geofluids, 10*(1–2), 114–131.

Saffer, D. M., & Bekins, B. A. (2002). Hydrologic controls on the morphology and mechanics of accretionary wedges. *Geology, 30*, 271–274. http://dx.doi.org/10.1130/0091-7613(2002)030<0271:HCOTMA>2.0.CO;2.

Saffer, D. M., Flemings, P. B., Boutt, D., Doan, M.-L., Ito, T., McNeill, L., et al. (2013). In situ stress and pore pressure in the Kumano Forearc Basin, offshore SW Honshu from downhole measurements during riser drilling. *Geochemistry Geophysics Geosystems, 14*(5), 1454–1470. http://dx.doi.org/10.1002/ggge.20051.

Saffer, D. M., & Tobin, H. (2011). Hydrogeology and mechanics of subduction zone forearcs: fluid flow and pore pressure. *Annual Reviews of Earth and Planetary Sciences, 39*. http://dx.doi.org/10.1146/annurev-earth-040610-133408.

Sakaguchi, A., Chester, F., Curewitz, D., Fabbri, O., Goldsby, D., Kimura, G., et al. (2011). Seismic slip propagation to the updip end of plate boundary subduction interface faults: vitrinite reflectance geothermometry on Integrated Ocean Drilling Program NanTroSEIZE cores. *Geology, 39*(4), 395–398. http://dx.doi.org/10.1130/G31642.1.

Sakaguchi, A., Yamamoto, Y., Hashimoto, Y., Harris, R. N., Vannucchi, P., & Petronotis, K. E. (2013). In *Characteristic magnitude of subduction earthquake and upper plate stiffness. 2013 Fall AGU Meeting, San Francisco, Calif.* Abstract T34C-06.

Scholz, C. H. (1998). Earthquakes and friction laws. *Nature, 391*, 37–42.

Schwartz, S., & Rokosky, J. (2007). Slow slip events and seismic tremor at circum-Pacific subduction zones. *Reviews of Geophysics, 45*, RG3004.

Science Plan Writing Committee, INVEST Steering Committee, & Additional Contributors. (2011). In E. Kappel, & J. Adams (Eds.), *Illuminating Earth's past, present, and future, The International Ocean Discovery Program: Exploring the Earth under the Sea, Science Plan for 2013–2023* Available at www.iodp.org/science-plan-for-2013-2023.

Screaton, E. J. (2010). Recent advances in subseafloor hydrogeology: focus on basement–sediment interactions, subduction zones, and slopes. *Hydrogeology Journal*, 1547–1570. http://dx.doi.org/10.1007/s10040-010-0636-7.

Screaton, E. J., & Ge, S. (2012). The impact of megasplay faulting and permeability contrasts on Nankai Trough subduction zone pore pressures. *Geophysical Research Letters, 39*, L22301. http://dx.doi.org/10.1029/2012GL053595.

Screaton, E., Kimura, G., Curewitz, D., Moore, G., Chester, F., Fabbri, O., et al. (2009). Interactions between deformation and fluid in the frontal thrust region of the NanTroSEIZE transect offshore the Kii Peninsula, Japan; Results from IODP Expedition 316 Sites C0006 and C0007. *Geochemistry Geophysics Geosystems, 10*, Q0AD01. http://dx.doi.org/10.1029/2009GC002713.

Screaton, E., Saffer, D., Henry, P., Hunze, S., & Leg 190 Shipboard Scientific Party. (2002). Porosity loss within the underthrust sediments of the Nankai accretionary complex: Implications for overpressure. *Geology, 30*, 19–22. http://dx.doi.org/10.1130/0091-7613(2002)030<0019:PLWTUS>2.0.CO;2.

Seno, T., Stein, S., & Gripp, A. E. (1993). A model for the motion of the Philippine Sea plate consistent with NUVEL-1 and geological data. *Journal of Geophysical Research, 98*(B10), 17,941–17,948. http://dx.doi.org/10.1029/93JB00782.

Shipboard Scientific Party. (1980). Site 436: Japan trench outer rise, leg 56. In Scientific Party (Ed.). *Init. Repts. DSDP: vol. 56–57.* (pp. 399–446). Washington, DC: U.S. Govt. Printing Office. http://dx.doi.org/10.2973/dsdp.proc.5657.107.F. Pt. 1.

Simpson, G. D. H. (2010). Formation of accretionary prisms influenced by sediment subduction and supplied by sediments from adjacent continents. *Geology, 38*, 131–134. http://dx.doi.org/10.1130/G30461.1.

Stern, C. R. (2011). Subduction erosion: rates, mechanisms, and its role in arc magmatism and the evolution of the continental crust and mantle. *Gondwana Research, 20*, 284–308. http://dx.doi.org/10.1016/j.gr.2011.03.006.

Strasser, M., Kölling, M., dos Santos Ferreira, C., Fink, H. G., Fujiwara, T., Henkel, S., et al. (2013). A slump in the trench: tracking the impact of the 2011 Tohoku-Oki earthquake. *Geology, 41*, 935–938. http://dx.doi.org/10.1130/G34477.1.

Strasser, M., Moore, G. F., Kimura, G., Kitamura, Y., Kopf, A. J., Lallemant, S., et al. (2009). Origin and evolution of a splay fault in the Nankai accretionary wedge. *Nature Geoscience, 2*, 648–652. http://dx.doi.org/10.1038/NGEO609.

Sugioka, H., Okamoto, T., Nakamura, T., Ishihara, Y., Ito, A., Obana, K., et al. (2012). Tsunamigenic potential of the shallow subduction plate boundary inferred from slow seismic slip. *Nature Geoscience, 5*, 414–418. http://dx.doi.org/10.1038/NGEO1466.

Taira, A. (2001). Tectonic evolution of the Japanese island arc system. *Annual Reviews of Earth and Planetary Science Letters, 29*, 109–134.

Tobin, H., Hirose, T., Saffer, D., & Expedition 348 Shipboard Scientific Party. (2014). NanTroSEIZE Stage 3: NanTroSEIZE plate boundary deep riser 3. In *IODP Prel Rept, 348.* http://dx.doi.org/10.2204/iodp.pr.348.2015.

Tobin, H. J., & Kinoshita, M. (2006). NanTroSEIZE: the IODP Nankai trough seismogenic zone experiment. *Scientific Drilling, 2*, 23–27.

Tobin, H., Kinoshita, M., Ashi, J., Lallemant, S., Kimura, G., Screaton, E. J., et al. (2009). NanTroSEIZE Stage 1 expeditions: introduction and synthesis of key results. In M. Kinoshita, H. Tobin, J. Ashi, G. Kimura, S. Lallemant, E. J. Screaton, et al. (Eds.), *Proc. IODP, 314/315/316.* Washington, DC: Integrated Ocean Drilling Program Management International, Inc. http://dx.doi.org/10.2204/iodp.proc.314315316.101.2009.

Tobin, H. J., & Saffer, D. M. (2009). Elevated fluid pressure and extreme mechanical weakness of a plate boundary thrust, Nankai Trough subduction zone. *Geology, 37*(8), 679–682. http://dx.doi.org/10.1130/G25752A.1.

Tsuji, T., Dvorkin, J., Mavko, G., Nakata, N., Matsuoka, T., Nakanishi, A., et al. (2011). VP/VS ratio and shear-wave splitting in the Nankai Trough seismogenic zone: insights into effective stress, pore pressure, and sediment consolidation. *Geophysics, 76*(3), WA71. http://dx.doi.org/10.1190/1.3560018.

Tsuji, T., Tokuyama, H., Pisani, P. C., & Moore, G. (2008). Effective stress and pore pressure in the Nankai accretionary prism off the Muroto Peninsula, southwestern Japan. *Journal of Geophysical Research, 113*, B11401. http://dx.doi.org/10.1029/2007JB005002.

Tsutsumi, A., Fabbri, O., Karpoff, A.-M., Ujiie, K., & Tsujimoto, A. (2011). Friction velocity dependance of clay-rich fault material along a megasplay fault in the Nankai subduction zone at intermediate to high velocities. *Geophysical Research Letters*, *38*, L19301. http://dx.doi.org/10.1029/2011GL049314.

Tsutsumi, A., & Shimamoto, T. (1997). High-velocity frictional properties of gabbro. *Journal of Geophysical Research*, *24*, 699–702.

Tudge, J., & Tobin, H. J. (2013). Velocity–porosity relationships in smectite-rich sediments: Shikoku Basin, Japan. *Geochemistry Geophysics Geosystems*, *14*, 5194–5207. http://dx.doi.org/10.1002/2013GC004974.

Ujiie, K., & Kimura, G. (2014). Earthquake faulting subduction zones: insights fault rocks accretionary prisms. *Progress in Earth and Planetary Science*, *1*(7). http://www.progearthplanetsci.com/content/1/1/7.

Ujiie, K., Tanaka, H., Saito, T., Tsutsumi, A., Mori, J. J., Kameda, J., et al. (2013). Low coseismic shear stress on the Tohoku-Oki megathrust determined from laboratory experiments. *Science*, *342*(6163), 1211–1214. http://dx.doi.org/10.1126/science.1243485.

Ujiie, K., & Tsutsumi, A. (2010). High-velocity frictional properties of clay-rich fault gouge in a megasplay fault zone, Nankai subduction zone. *Geophysical Research Letters*, *37*, L24310. http://dx.doi.org/10.1029/2010GL046002.

Underwood, M. B., & Moore, G. F. (2012). Evolution of sedimentary environments in the subduction zone v of southwest Japan: recent results from the NanTroSEIZE Kumano transect. In C. Busby, & A. Azor (Eds.), *Tectonics of Sedimentary Basins: Recent Advances*. Chichester, UK: John Wiley & Sons, Ltd. http://dx.doi.org/10.1002/9781444347166.ch15.

Underwood, M. B., Saito, S., Kubo, Y., & the Expedition 322 Scientists. (2010). Expedition 322 summary. In S. Saito, M. B. Underwood, Y. Kubo, & the Expedition 322 Scientists (Eds.), *Proc. IODP, 322*. Tokyo: Integrated Ocean Drilling Program Management International, Inc. http://dx.doi.org/10.2204/iodp.proc.322.101.2010.

Vannucchi, P., Galeotti, S., Clift, P. D., Ranero, C. R., & von Huene, R. (2004). Long-term subduction-erosion along the Guatemalan margin of the Middle America Trench. *Geology*, *32*(7), 617–620. http://dx.doi.org/10.1130/G20422.1.

Vannucchi, P., Ranero, C. R., Galeotti, S., Straub, S. M., Scholl, D. W., & McDougall-Reid, K. (2003). Fast rates of subduction erosion along the Costa Rica Pacific margin: implications for nonsteady rates of crustal recycling at subduction zones. *Journal of Geophysical Resarch*, *108*(B11), 2511. http://dx.doi.org/10.1029/2002JB002207.

Vannucchi, P., Remitti, F., & Bettelli, G. (2008). Geologic record of fluid flow and seismogenesis along an erosive subducting plate boundary. *Nature*, *451*, 699–703. http://dx.doi.org/10.1038/nature06486.

Vannucchi, P., Sak, P. B., Morgan, J. P., Ohkushi, K., Ujiie, K., & the IODP Expedition 334 Shipboard Scientists. (2013). Rapid pulses of uplift, subsidence, and subduction erosion offshore Central America: implications for building the rock record of convergent margins. *Geology*, *41*(9), 995. http://dx.doi.org/10.1130/G34355.1.

Vannucchi, P., Scholl, D. W., Meschede, M., & McDougall-Reid, K. (2001). Tectonic erosion and consequent collapse of the Pacific margin of Costa Rica: combined implications from ODP Leg 170, seismic offshore data, and regional geology of the Nicoya Peninsula. *Tectonics*, *20*(5), 649–668. http://dx.doi.org/10.1029/2000TC001223.

Vannucchi, P., Ujiie, K., Stroncik, N., & the Expedition 334 Scientists. (2012). In *Proc. IODP, 334*. Tokyo: Integrated Ocean Drilling Program Management International, Inc. http://dx.doi.org/10.2204/iodp.proc.334.2012.

Vannucchi, P., Ujiie, K., Stroncik, N., & the IODP Expedition 334 Science Party. (2013). IODP Expedition 334: an investigation of the sedimentary record, fluid flow and state of stress on top of the seismogenic zone of an erosive subduction margin. *Scientific Drilling*, *15*, 23–30. http://dx.doi.org/10.2204/iodp.sd.15.03.2013.

Wallace, L. M., & Beavan, J. (2010). Diverse slow slip behavior at the Hikurangi subduction margin, New Zealand. *Journal of Geophysical Research*, *115*, B12402. http://dx.doi.org/10.1029/2010JB007717.

Wallace, L. M., Beavan, J., Bannister, S., & Williams, C. (2012). Simultaneous long-term and short-term slow slip events at the Hikurangi subduction margin, New Zealand: implications for processes that control slow slip event occurrence, duration, and migration. *Journal of Geophysical Research*, *117*. http://dx.doi.org/10.1029/2012JB009489.

Wang, K., & Hu, Y. (2006). Accretionary prisms in subduction earthquake cycles: the theory of dynamic Coulomb wedge. *Journal of Geophysical Research*, *111*, B06410. http://dx.doi.org/10.1029/2005JB004094.

Wang, K., Hu, J., & He, Y. (2012). Deformation cycles of subduction earthquakes in a viscoelastic earth. *Nature*, *484*, 327–332. http://dx.doi.org/10.1038/nature11032.

Watts, A. B., & Weissel, J. K. (1975). Tectonic history of the Shikoku MarginalBasin. *Earth and Planetary Science Letters*, *25*, 239–250.

Wibberley, C., & Shimamoto, T. (2005). Earthquake slip weakening and asperities explained by thermal pressurization. *Nature*, *436*, 689–692. http://dx.doi.org/10.1038/nature03901.

Yamaguchi, A., Sakaguchi, A., Sakamoto, T., Iijima, K., Kameda, J., Kimura, G., et al. (2011). Progressive illitization in fault gouge caused by seismic slip propagation along a megasplay fault in the Nankai Trough. *Geology*, *39*, 995–998. http://dx.doi.org/10.1130/G32038.

Yamamoto, Y., Lin, W., Usui, Y., Kanamatsu, T., Saito, S., Zhao, X., et al. (2011). In *Preliminary results of stress and strain analyses, IODP Expedition 334, Costa Rica Seismogenesis Project (CRISP). American Geophysical Union Fall Meeting, San Francisco, CA, 5–9 December 2011*. Abstract T21B-2360.

Yue, H., & Lay, T. (2011). Inversion of high-rate (1 sps) GPS data for rupture process of the 11 March 2011 Tohoku earthquake (Mw 9.1). *Geophysical Research Letters*, *38*, L00G09. http://dx.doi.org/10.1029/2011GL048700.

Yue, H., Lay, T., Schwartz, S. Y., Rivera, L., Protti, M., Dixon, T. H., et al. (2013). The 5 September 2012 Nicoya, Costa Rica M-w 7.6 earthquake rupture process from joint inversion of high-rate GPS, strong-motion, and teleseismic P wave data and its relationship to adjacent plate boundary interface properties. *Journal of Geophysical Research*, *118*, 5453–5466. http://dx.doi.org/10.1002/jgrb.50379.

Chapter 4.4.2

Seismogenic Processes Revealed Through the Nankai Trough Seismogenic Zone Experiments: Core, Log, Geophysics, and Observatory Measurements

Masataka Kinoshita,[1],* Gaku Kimura[2] and Saneatsu Saito[3]
[1]*Kochi Institute for Core Sample Research, Japan Agency for Marine-Earth Science & Technology (JAMSTEC), Nankoku, Kochi, Japan;* [2]*Department of Earth and Planetary Science, Graduate School of Science, University of Tokyo, Japan;* [3]*R&D Center for Ocean Drilling Sciences, JAMSTEC, Japan*
Corresponding author: E-mail: masa@jamstec.go.jp

4.4.2.1 INTRODUCTION

The Integrated Ocean Drilling Program (IODP) Nankai Trough Seismogenic Zone Experiments (NanTroSEIZE) is the first attempt to drill into, sample, and instrument the seismogenic portion of a plate boundary fault or mega-splay fault within a subduction zone. Access to the interior of active faults where in situ processes can be measured and monitored and pristine fault zone materials can be sampled is of fundamental importance to the understanding of earthquake mechanics. As we learned from recent great earthquakes and tsunamis (e.g., December 2004 Sumatra, March 2011 Tohoku, February 2010 Chile), large subduction earthquakes represent one of the greatest natural hazards on the planet. Accordingly, drilling into and instrumenting an active interplate seismogenic zone was identified as a very high priority in the IODP Initial Science Plan (2000). Within the international scientific community, consensus was reached that the Nankai margin is of the highest priority to be drilled using the riser drilling vessel through a long series of workshops Conference on Cooperative Ocean Riser Drilling (CONCORD), Conference on Multi-Platform Experiments (COMPLEX), and the Seismogenic Zone Detailed Planning Group (Hyndman et al., 1999)).

In 2003, the NanTroSEIZE Complex Drilling Project (CDP) proposal was submitted to the IODP Science Advisory Structure, followed by four independent proposals. All proposals were reviewed and ranked highly by 2004, and the

Developments in Marine Geology, Volume 7. http://dx.doi.org/10.1016/B978-0-444-62617-2.00021-9

Implementation Organizations of the United States (JOIDES Resolution) and Japan (Chikyu) started planning drilling operations through the newly formed "Project Management Team."

The primary goal of NanTroSEIZE is to advance our knowledge of fundamental aseismic and seismic faulting processes and controls on the transition between them, through comprehensive logging, sampling, and in situ monitoring. To reach the seismogenic zone of the plate boundary megathrust at 5–7 km below seafloor, a careful well condition control using the riser pipe is essential. Riser drilling was conducted for the first time in the scientific ocean drilling during the NanTroSEIZE project.

Figure 4.4.2.1 shows the location and seismic section of NanTroSEIZE drill sites (Moore et al., 2007, 2009), and Figure 4.4.2.2 shows the history of Nan-TroSEIZE expeditions and drill sites. So far 14 sites were drilled, including two riser holes at Sites C0002 and C0009 (Expedition 326 Scientists, 2011; Henry, Kanamatsu, Moe, & the Expedition 333 Scientists, 2012; Kinoshita et al., 2009; Kopf, Araki, Toczko, & the Expedition 332 Scientists, 2011; Saffer et al., 2010; Saito, Underwood, Kubo, & the Expedition 322 Scientists, 2010; Screaton, Kimura, Curewitz, & the Expedition 316 Scientists, 2009; Strasser et al., 2014; Tobin et al., 2009).

4.4.2.2 STRESS STATE AND PHYSICAL PROPERTIES IN SHALLOW FORMATIONS

In order to understand the dynamics of the seismogenic process, we rely on the Mohr–Coulomb failure theory for assessment of nucleation/initiation of slip (both typical high-frequency and low-frequency events), and on the rate- and state-dependent friction law for the rupture nucleation, propagation, and arrest behavior. One of the key parameters is the present stress state (stress tensor) at and around the fault.

4.4.2.2.1 Stress State

4.4.2.2.1.1 Orientation of Horizontal Principal Stress

The present and the past stress fields (orientation and magnitude) in the incoming Shikoku Basin, Nankai accretionary prism and forearc, was estimated using logged images and core measurements, and compared with regional-scale stress and strain estimated from seismic data analyses.

In a vertical borehole, the orientation of breakouts is a well-established indicator of the orientation of the horizontal maximum principal stress in the present-day stress field (Zoback et al., 2003). Borehole breakouts were observed through the logging-while-drilling (LWD) resistivity imaging tool at Sites C0001, C0002, C0004, and C0006, C0010, C0011, C0012, C0018, C0021, C0022, and through the wireline resistivity imaging tool at Site C0009 (Figure 4.4.2.3).

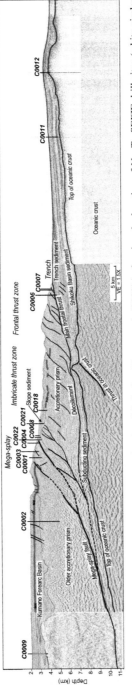

FIGURE 4.4.2.1 Bathymetry and seismic cross-section (*modified after Moore et al. (2009).*), showing the location of NanTroSEIZE drill sites (white circles). Solid stars indicate epicenters of 1944 and 1946 earthquakes. Contours show the seismic slip on the plate interface for the 1944 event (after Kikuchi, Nakamura, & Yoshikawa, 2003). Yellow (light gray in print versions) arrows show the plate convergence vector between Philippine Sea and Eurasia plates (Heki & Miyazaki, 2001; Seno, Stein, & Gripp. 1993).

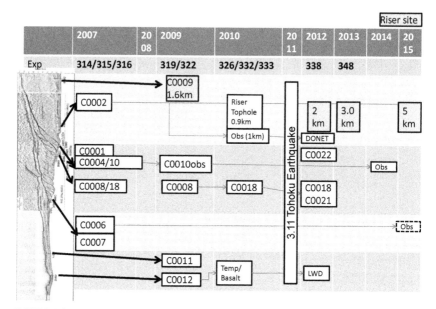

FIGURE 4.4.2.2 History and expected plan of NanTroSEIZE expeditions and the drill sites (as of July 2014). "Obs" indicates the borehole observatory installation.

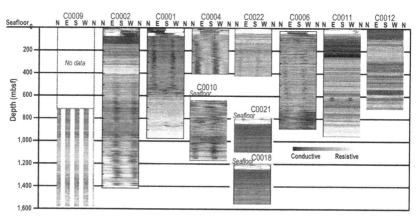

FIGURE 4.4.2.3 Unwrapped borehole resistivity images at NanTroSEIZE drill sites aligned on the depth below seafloor, with a common depth scale (Kinoshita et al., 2009; Kopf et al., 2011; Saito et al., 2010; Strasser et al., 2014). Darker colors are conductive, indicating porous formation or larger calipers. Dark vertical bands are the breakouts. "N" and "S" indicate north and south azimuth, respectively.

At Site C0011 located in the incoming oceanic plate, breakouts are prevalent in the hemipelagic sediments (at ~600 and 650 m below seafloor; mbsf). The orientation of maximum horizontal principal stress (S_{Hmax}) is N25E (Saito et al., 2010), roughly perpendicular to the convergence direction of the Philippine Sea plate.

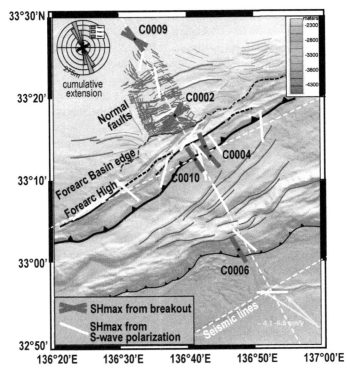

FIGURE 4.4.2.4   Maximum principal horizontal stress (S_{Hmax}) orientations (Byrne et al., 2009; Tsuji et al., 2011b) and fault traces. Bars at the drilling sites show the S_{Hmax} orientations in boreholes (blue (dark gray in print versions) in the accretionary prism and red (dark gray in print versions) in the overlying sediments; Lin et al., 2010). White bars indicate the fast S-wave polarization direction estimated from OBS data along white broken lines (Tsuji et al., 2011b). Dark gray curves are traces of faults mapped in the 3D seismic volume at a depth of 150 mbsf (Sacks et al., 2013). Rose diagram in the upper-left corner shows cumulative extension of these fault traces. Yellow (white in print versions) arrows show the far-field convergence vectors between the Philippine Sea plate and Japan (Heki & Miyazaki, 2001).

Breakout orientations at sites C0001, C004, C0006, C0009, and C0010 indicate approximately northwest–southeast (135–155°) azimuths of S_{Hmax} (Figure 4.4.2.4) (Chang et al., 2010; Lin et al., 2010; Saffer et al., 2010; Yamada, McNeill, Moore, & Nakamura, 2011) (note that the stress magnitude is positive for a compressional state). A slight deviation of the S_{Hmax} orientations from the far-field plate motion (Heki & Miyazaki, 2001) implies a strain partitioning between convergence and strike-slip motions.

At Site C0002 in the Kumano Forearc Basin, by contrast, the orientation of S_{Hmax} is northeast–southwest (N41E) or near-perpendicular to that in the more trenchward sites. This is generally consistent with the stress state inferred from past and present normal fault populations that extend in the southeastern part of the basin and also in the upper 400 m, at least, of the underlying prism

(Gulick et al., 2010; Moore, Barrett, & Thu, 2013; Sacks, Saffer, & Fisher, 2013). From the stress inversion analysis, Sacks et al. (2013) suggest that the horizontal stresses are almost isotropic, reflected by the widely varying fault strikes, and that the tectonic stress fluctuation through the seismic cycle can possibly determine the local stress anisotropy at Site C0002.

S_{Hmax} is normal to the trench at Site C0009 in the center of the Kumano Forearc Basin (Saffer et al., 2013), suggesting that the margin-normal extension near Site C0002 is a local feature in the southeastern part of the basin. The orientation of S_{Hmax} can also be inferred from seismic anisotropy observations. Tsuji et al. (2011a,b) and Saffer et al. (2010) estimated seismic anisotropy from walk-around vertical seismic profiling (VSP) data recorded at Site C0009 and active source seismic data recorded by wide-angle multicomponent ocean bottom seismometers (OBS). Both the fast P-wave velocity direction and S-wave polarization axes are normal to the plate convergence direction (Figure 4.4.2.4), suggesting that the horizontal maximum stress at Site C0009 is parallel to the direction of plate subduction.

Ito, Asano, and Obara (2009) estimated stress orientations and stress ratios by applying a stress tensor inversion to very-low-frequency (VLF) earthquakes. The stress orientations indicate that the region off Kumano is within a transpressional stress regime with trench-normal shortening.

4.4.2.2.1.2 Stress Magnitude

Assessment of the horizontal principal stress magnitude and eventually determination of the full stress tensor is our ultimate goal. Attempts were made through in situ (breakout width analyses and hydrofracturing) and laboratory (the elastic and anelastic strain recovery methods and rock consolidation tests) experiments. Some of them have limited quality because experiments were not conducted under the ideal condition, or because calibration is not yet completed.

Chang et al. (2010) and Olcott and Saffer (2012) estimated S_{Hmax} and S_{hmin} at sites logged during Expedition 314, from the width of breakouts combined with the rock strength (unconfined compressive strength) inferred from an empirical relationship with P-wave velocity. Downhole experiments in a riser hole were carried out during Expedition 319 to estimate in situ S_{hmin} (and potentially S_{Hmax}) using a special hydrofracturing tool "modulation formation dynamic tester (MDT)" for the first time in IODP operations, as well as through the leak-off test for measurement of S_{hmin} (Expedition 319 Scientists, 2010a; Ito et al., 2013; Saffer et al., 2013).

When core samples are cut by a drill bit, stress-induced strain is released instantaneously as elastic rebound. Ito et al. (2013) applied this to estimate the horizontal differential stress by measuring the eccentricity of recovered cores. The recovered core then experiences a gradual time-dependent release, called the anelastic strain recovery (ASR). The ASR versus time curve can provide constraints on the in situ, present state of stress field.

From stress analyses at Sites C0001 and C0002, Chang et al. (2010) suggest that the stress state in the shallow wedge (forearc basin and slope sediment) is in favor of normal faulting and that the stress state in the deeper accretionary prism is in favor of strike-slip or possible reverse faulting. Chang et al. (2010) also suggest that the horizontal differential stress is small (<5 MPa) in the shallow portion of the forearc (Figure 4.4.2.5). This can explain the difference in the orientation of S_{Hmax} between Site C0002 and the other sites (C0009, C0001, C0004, and C0006); at Site C0002, nontectonic forces such as trench-normal gravity-driven extension may be larger than tectonic forces due to plate convergence.

The stress magnitude estimation by Chang et al. (2010) is based on the width of breakouts. However, Moore et al. (2011) noticed that the initial borehole breakout width widens with time after drilling. They attribute this to the apparent strengthening caused by the drilling, and argue that the borehole image logged a few minutes after the bit would underestimate the breakout width, leading to an underestimate of stress magnitude. Thus, the stress magnitude by Chang et al. (2010), derived from LWD data collected shortly after the drilling, may be lower than in situ values, and the faulting regime in the sediment at Sites C0001 and C0002 may favor strike-slip rather than normal faulting.

Olcott and Saffer (2012) suggest that Sites C0004 and C0010 are in a reverse faulting regime both in the hanging wall and footwall of the mega-splay. At Site C0010 LWD was relogged 3 days later, and they found a breakout wider than the first run, indicating that horizontal stresses are even larger.

At Site C0009, a minimum horizontal stress (S_{hmin}) determination was attempted using the MDT hydraulic fracturing test in addition to a conventional leak-off test (Saffer et al., 2013) to assess the present state of stress around the borehole. Ito et al. (2013) combined their elastic rebound method with the in situ hydrofracturing test in order to estimate S_{Hmax} and S_{hmin}. From the analysis of the MDT test, they found that the reopening pressure at 877 mbsf in the early Kumano Forearc Basin (35 MPa) is identical to the shut-in pressure, and concluded that $S_{Hmax} \sim S_{hmin}$. On the other hand, at 1534 mbsf in the older slope deposits and/or accretionary prism sediment, the horizontal stress was determined from shut-in pressure and elastic core deformation, and yields $S_{hmin} = 42$ MPa and $S_{Hmax} = 55$ MPa. Saffer et al. (2013) compiled the stress magnitude estimations and concluded that S_{hmin} is consistently lower in magnitude than the vertical stress and suggest a normal or strike-slip stress regime at Site C0009.

Byrne et al. (2009) showed results from the ASR analysis and compared with other independent data [borehole breakouts, core-scale faults, and Anisotropy of Magnetic Susceptibility (AMS) results]. All samples from C0001, C0002, and C0006 yielded a vertical to nearly vertical direction for the maximum principal stress (σ_1). Byrne et al. (2009) also showed that late-stage faults observed in core samples are dominated by normal faults, and the intermediate axes are consistent with the maximum horizontal stress (hence σ_2) orientation estimated by ASR. This indicates that at least the shallow portion of the Kumano Forearc

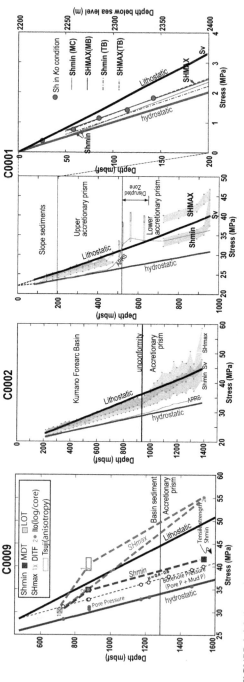

FIGURE 4.4.2.5 In situ stress profile with depth in Sites C0009 (Ito et al., 2013; Saffer et al., 2013; Tsuji, Hino, et al., 2011b), C0002 (Chang et al., 2010), and C0001 (Chang et al., 2010; Song et al., 2011). For stress estimation, pore pressure is assumed hydrostatic throughout. In the shallow depths, the two horizontal principal stresses (S_{Hmax} and S_{hmin}) are generally lower than lithostatic (Sv), suggesting a normal faulting stress regime. At the greater depths, S_{Hmax} appears to be larger than Sv, suggesting that the stress regime there is in favor of strike-slip faulting. MDT: modular dynamic ester, LOT: leak-off test, DITF: drilling-induced tensile fracture, APRS: annular pressure (=borehole pressure).

Basin and forearc slope sediments overlying the prism to the toe are dominated by normal faulting and show little evidence of horizontal shortening.

Another approach for understanding stress and hydrological state is through geotechnical experiments (e.g., Karig, 1990). Song, Saffer, and Flemings (2011) performed a series of mechanical experiments on mudstone core samples from slope sediment at Site C0001 down to 150 mbsf. They show that the stress regime supports uniaxial burial where the effective horizontal stress is ~40% of effective vertical stress, and that the magnitude of S_{hmin} is significantly lower than S_{Hmax}, which is comparable to that expected for sedimentation and uniaxial burial. These suggest that the shallow sedimentary section (<200 mbsf) is subjected to extension subparallel to the trench, and that stresses associated with plate convergence are not effectively transmitted to the shallow sediments.

Regional-scale differential stress magnitude can also be inferred from the analyses of seismic wave propagation. Tsuji et al. (2011b) calculated the anisotropy of P-wave velocity as ~5%, from the azimuthal normal moveout analysis of the circular VSP data around Site C0009. If the anisotropy is caused by stress-induced, selective closure of cracks that were originally oriented randomly, the degree of anisotropy can be used to infer the magnitude of differential stress. Tsuji et al. (2011b) used the laboratory measurements on the relationship between P-wave velocity and effective stress for rock samples (e.g., Hashimoto, Tobin, & Knuth, 2010) at Site C0009 and estimated the magnitude of horizontal differential stress as 2.7–5.5 MPa at Site C0009 (Figure 4.4.2.5).

Through 2-D finite element elastic modeling, Kinoshita and Tobin (2013) showed that the shear stress accumulated during one seismic cycle at the Nankai locked zone does not exceed a few megapascals at the downdip limit of the locked zone along plate interface, which is consistent with the static stress drop at the time of an earthquake. They also mention that the accumulated stress in the accretionary prism is much smaller than a few megapascals. It may be compared to the small values of horizontal differential stress in the shallow formation.

Conin, Henry, Godard, and Bourlange (2012) used a finite element elasto-plastic model to estimate a quasi-static stress and strain of a décollement and splay fault system in the Nankai wedge. They showed that slip partitioning between the splay fault and décollement is controlled by their friction coefficients such that an interface with lower friction has a larger slip, and that an extension localized at a shallow level in the splay hanging wall occurs for a wide range of both frictional coefficients. The trench-normal extension observed at Site C0002 may be explained through this mechanism.

4.4.2.2.2 In Situ Physical Properties

In ocean drilling research, physical properties have served as good proxies for lithology, sediment compaction or lithification, tectonic processes, and for permeability, fluid flow, and possible overpressure. They have been routinely measured either on core samples or through downhole logging. The core sample measurement is a direct measure, but the porosity, for instance, is often biased toward lower values than in situ due to selective core recovery of stiffer material. Core-based porosity should also be corrected for postcoring expansion due to depressurization, which apparently increases the porosity. The log-derived properties do not have this problem, but accuracy is restricted by a nonideal hole condition, limited spatial resolution, etc.

Of the various physical properties, porosity and P-wave velocity are the basic properties most commonly measured/estimated. Here we briefly review results of porosity and P-wave velocity estimation within NanTroSEIZE and their implications on the fault mechanics and forearc deformation.

4.4.2.2.2.1 Porosity

The porosity measured on core samples onboard is larger than in situ values due to unloading upon recovery. Porosity will change under applied effective pressure, P_{eff} (the difference between the lithostatic pressure and the pore pressure; $P_{litho} - P_{pore}$). Raimbourg, Hamano, Saito, Kinoshita, and Kopf (2011) carried out a series of isotropic loading tests for sediment samples at various drill sites along the Muroto (ODP Site 1173) and Kumano transects (IODP Sites C0001, C0002, C0006, and C0007), and derived acoustic (Vp) and mechanical properties for various P_{eff} values. They applied a correction for both elastic and irreversible rebounds to in situ conditions. The amount of this correction depends on site and on depth, and is ~10% at Site C0006 (~200 mbsf) and is up to 4% at other sites. They also suggest that the irreversible shrinkage is dominant at less than 3–4 MPa effective confining stress, or possible at a domain larger than the critical porosity.

Measured porosity may also be affected by clay-bound water, which does not affect hydrological state at shallow depths. Porosity measurements on core samples use oven drying at 105 °C to remove total volume of fluid from samples, including the water contained in hydrous minerals such as smectites, zeolites, and gypsum. In order to calculate interstitial porosity, therefore, the effect of clay-bound water must be estimated and corrected. Hashimoto et al. (2010) adopted the method suggested by Brown and Ransom (1996), and determined that the interstitial (or effective) porosity is reduced by ~15% (Figure 4.4.2.6). Conin, Henry, Bourlange, Raimbourg, and Reuschlé (2011) used cation exchange capacity (CEC) to correct for hydrous mineral water content, and showed that apparent porosity anomalies are significantly reduced by up to 0.1 of fractional porosity. Through comparison between core and log data at Site C0009, Doan et al. (2011) also indicated that the difference between the interstitial (effective) porosity derived from sonic and resistivity logs and the total porosity derived from the lithodensity log and core data reflects differences in the hydrous clay mineral abundance.

Porosities measured on core samples in the prism toe region are anomalously low at shallow depth (48% at 5 and 34 mbsf at Sites C0006 and C0007, respectively), as compared to porosities of fine-grained sediments at Sites C0004 and C0008 that do not decrease below 50% above 200 mbsf (Screaton

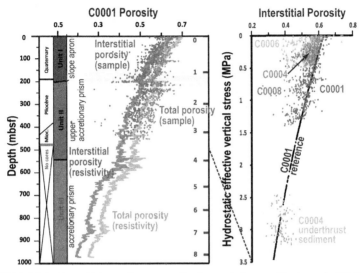

FIGURE 4.4.2.6 Interstitial porosity versus effective vertical stress (Conin et al., 2011). Left: Total and interstitial porosities at Site C0001. Right: Comparison of interstitial porosities for slope sediments at Sites C0001, C0004, C0006, and C0008 with the reference curve defined at Site C0001. Shallow porosity at C0004 (dark green (dark gray in print versions)) is lower than the reference, whereas underthrust sediment porosity (light green (light gray in print versions)) match the reference trend.

et al., 2009). Still, low overall porosity at the toe and in the mega-splay fault area is best explained by erosion of surface sediment (Conin et al., 2011) by 50–100 m thickness. This amount coincides with regional slide scars and deposits observed seaward of the mega-splay fault in 3-D seismic data (Strasser et al., 2011). Conin et al. (2011) pointed out that underthrust sediments below the fault at Site C0004 appear to follow the normal consolidation trend defined from the Site C0001 porosity. They suggest that unloading (and thus overconsolidation) due to erosion is compensated by the overpressure (leading to the undercompacted state) due to thrusting along the splay fault under the undrained condition. The overpressure near the mega-splay fault zone is suggested through a waveform tomography analysis by Tsuji, Kamei, and Pratt (2014).

4.4.2.2.2.2 P-Wave Velocity–Porosity Relationship

Hashimoto et al. (2010) measured the P-wave velocity (Vp) under in situ effective pressure for samples at C0001. The relationship between Vp and porosity (corrected for smectite abundance) was examined for slope versus accreted sediments in the outer forearc. Accreted sediments show a slightly steeper gradient, thus a higher velocity relative to porosity than that of the slope sediment, possibly explained by weak cementation, critical porosity, or differences in loading history. Raimbourg et al. (2011; their Figure 4.4.2.12) examined Vp–porosity relationship for the forearc slope/basin sediments (C0001, C0002, C0004, C0006, and C0007) and showed that P-wave velocities are primarily controlled by porosity for high-porosity sediments. All samples are isotropic except for Site C0001 (accretionary prism unit), which shows a sign of incipient cementation (generally consistent with results from Hashimoto et al., 2010).

Tudge and Tobin (2013) examined the relationship between Vp (corrected for in situ condition using that at nearby sites (Expedition 315 Scientists, 2009); 2.4% offset plus 1% per 100 m depth) and porosity (corrected for smectite abundance; fractional porosity by ~0.1) at Shikoku Basin input sites C0011 and C0012 seaward of Nankai Trough, and proposed a new empirical relationship for the smectite-rich Shikoku Basin sediments.

During Expedition 319, high-quality sonic data were acquired using the Sonic Scanner™ tool (Expedition 319 Scientists, 2010b; Methods). Sonic Scanner data provide accurate radial as well as axial measurements, which are converted to P-wave and S-wave velocities and to their anisotropy. It can also provide data on cement bonding quality. Using the log data, Doan et al. (2011) evaluated the porosity and gas content of the fine-grained forearc basin-filling sediments at Site C0009 and estimated gas saturation as 0–5% from 1050 to ~1200 mbsf within a lithological unit characterized by a high abundance of wood fragments and lignite. Doan et al. (2011) proposed that the gas migrates upward and updip toward the seaward edge of the Kumano Forearc Basin to form strong Bottom-Simulating Reflectors (BSRs) near Site C0002.

4.4.2.2.2.3 Mechanical Properties

Mechanical properties such as consolidation or shear strength change as a result of tectonic and sedimentological processes. Thus, it is essential to understand how such processes lead to the observed consolidation states, conveniently represented in terms of porosity or P-wave velocity, in order to assess processes affecting the margin.

Kitajima, Chester, and Biscontin (2012) conducted triaxial deformation tests for selected sediment samples cored in NanTroSEIZE expeditions. To understand the effect of loading history in the Nankai accretionary subduction zone, several different potential load paths were tested. Adopting a critical state soil mechanics model, in situ stress regime and deformation modes (brittle or ductile) were constrained by the critical state line and yield envelope determined from the consolidation tests. The results suggest that the sediments in the vicinity of the mega-splay fault (Site C0004) and the frontal thrust (C0006) are highly overconsolidated, which may be reconciled as a combined effect of a decrease in effective vertical stress (due to surface erosion or increase in pore pressure) and horizontal contraction caused by the tectonic horizontal stress. Larger values of overconsolidation ratio at Sites C0004 and C0006 suggest brittle deformation rather than ductile.

The study by Kitajima et al. (2012) demonstrates the importance of mechanical and hydraulic property changes as a function of effective stress path during subduction. It has also been useful for interpreting the stress state and hydro-logic/pore pressure conditions through the relationship with seismic attributes (e.g., Kitajima & Saffer, 2012).

Conin, Bourlange, Henry, Boiselet, and Gaillot (2014) investigated the LWD resistivity images and the core X-ray computed tomography images to interpret the stress path, and suggested that sediments in different locations of the wedge followed different stress paths.

Ikari, Hüpers, and Kopf (2013a) conducted laboratory tests measuring the shear strength of sediments of the incoming plate approaching the trench at Sites C0011 and C0012, including major lithological sedimentary units. At both sites, they observe nonlinear increase of shear strength with depth, and a decrease in apparent friction coefficient with depth calculated from the shear strength and effective normal stress. Assuming that the wedge toe region is in a (compres-sionally) critical state, the observed apparent friction of 0.58 and surface slope angle in the prism/wedge toe region (10°) would correspond to a plate bound-ary position on the main frontal thrust (see Figure 4.4.2.1) with its dip of 7.6° assuming a hydrostatic state. Ikari, Hüpers, and kopf (2013a) claim that the higher friction on the frontal thrust would suppress earthquake rupture propaga-tion to the trench compared to adjacent areas on the margin, such as off Muroto.

4.4.2.2.2.4 Permeability and Pore Pressure

Pore pressure anomalies, also referred to as excess pore pressure or overpres-sure, along a fault zone play a key role in the earthquake mechanism. Saffer and Tobin (2011) summarized recent knowledge of hydrology and mechanics of subduction forearcs, and provide insights into the fault slip processes. Pore pres-sure inferred from seismic and borehole data is generally high in wedge faults and underlying sediments, and excess pore pressure in subduction forearcs is dynamically balanced between the geologic forcing for fluid generation and the hydrological state for a fluid escape.

Through laboratory experiments, Ikari, Saffer, and Marone (2009) showed that permeability normal to the gouge layer measured before, during, and after shear ranges from $8.3 \times 10^{-21}\,\mathrm{m}^2$ to $3.6 \times 10^{-16}\,\mathrm{m}^2$ and that permeability decreases dramatically with shearing. Tanikawa et al. (2012) also measured per-meability for the mega-splay (C0004) and frontal thrust (C0007) core samples, and showed that the gouge permeability ($10^{-18}\,\mathrm{m}^2$) decreased after sliding for wet gouge and increased for dry gouge. They attribute such changes to thermal and mechanical pressurization upon shear-induced compaction.

So far there are few direct measurements of in situ pore pressure within a fault zone through borehole observatories. Moore et al. (2013) analyzed the annular pressure data near the bit acquired while drilling (APWD) during IODP Expedition 314. Although APWD data are affected by the cuttings load, pump-ing rate at the rig floor, and viscous resistance to flow up the annulus, two holes (Sites C0001 and C0003) crossing a fracture zone in the hanging wall of the mega-splay showed pressure above lithostatic conditions. Moore et al. (2013)

anticipated that the possibly overpressurized fluid in the fracture zone was injected at 3000 L/m into the annulus at Site C0001, which may reflect a possible residual overpressure at the 1944 M 8.1 earthquake.

During riser drilling at Site C0009 (IODP Expedition 319), in situ stress, pore pressure, and permeability were measured by using the MDT downhole tool (Expedition 319 Scientists, 2010a). Saffer et al. (2013) analyzed the data and concluded that pore fluid pressures are near hydrostatic throughout the forearc basin and underlying prism section despite rapid sedimentation in the Kumano Forearc Basin. In situ permeability was measured using a "single-probe" tool and theoretically estimated using a "dual packer." The determined permeability ranges from 6.5×10^{-17} to 4×10^{-14} m^2 through the basin sediment (Unit II) and the accretionary wedge (Unit IV), but with no systematic variation with depth and unit.

4.4.2.3 FAULT ZONE STATE AND PROPERTIES

During IODP NanTroSEIZE expeditions, the shallow portions of the mega-splay and frontal thrust/décollement faults were drilled and cored (Figure 4.4.2.1). If these faults have velocity-strengthening behavior (e.g., Wang & Hu, 2006), they would not nucleate seismic slip, and if coseismic slip propagates from a nucleation zone deeper on the fault, it would need to overcome these stabilizing properties for rupture to reach the seafloor. To test whether this portion is in a velocity-strengthening state, laboratory experiments to measure effective frictional coefficients were conducted. Geochemical proxies for the past temperature increase were also sought to test whether the shallow faults have had past coseismic slip.

4.4.2.3.1 Friction

One of the most important goals of NanTroSEIZE has been to reveal the frictional behavior of fault rocks across the aseismic–seismogenic transition. While we keep drilling toward the updip limit of seismically locked portion of the fault at Site C0002, we also attempted to recover samples from very shallow portion of faults as an expected aseismic end member.

A convenient parameter is the effective frictional coefficient μ', which links material friction properties to the pore pressure effect. The value of μ' varies with the fault material, effective pressure, change in slip velocity, and shear displacement (e.g., Scholz, 1998). In the last decade or two, intensive advances have been made in rock friction experiments for a wide range of slip velocity from ~1 μm/s to ~1 m/s (e.g., Di Toro et al., 2011). The "seismic" or "aseismic" character of a fault rock is measured whether μ' during a steady state seismic slip is smaller or larger, respectively, than that before the slip occurs. It is referred to as a velocity-weakening or velocity-strengthening behavior using "a-b" values. The other measure is the peak friction or the friction when the slip velocity is suddenly increased.

Using a triaxial testing machine with slip rates of 0.03–100 µm/s, Ikari et al. (2009) and Ikari and Saffer (2011) measured frictional and hydrological properties of fault gouge and surrounding wall rock from the shallow mega-splay fault at Site C0004, the frontal thrust at Site C0007, and at the décollement at Site 1174 off Muroto, and showed that the fault zone is frictionally weak and impermeable, and velocity strengthening in all cases (Figure 4.4.2.7). They also show that, although the samples show velocity-strengthening behavior, an observed minimum in a-b for the velocity range of 1–10 µm/s could host slow slip or very-low-frequency events in the accretionary wedge (e.g., Ito & Obara, 2006). Tsutsumi, Fabbri, Karpoff, Ujiie, and Tsujimoto (2011) and Tsutsumi performed friction experiments on water-saturated samples from the mega-splay fault using a rotary shear testing machine (see also Tsutsumi & Shimamoto, 1997) with slip rates of 0.026–2.6 mm/s. They found velocity-weakening behavior both at slow (<2.6 mm/s) and at high (>26 mm/s) slip rates. Tsutsumi et al. (2011) interpret that the velocity-weakening behavior, apparently inconsistent with results by Ikari et al. (2009), is caused by a development of localized shear at large shear displacements.

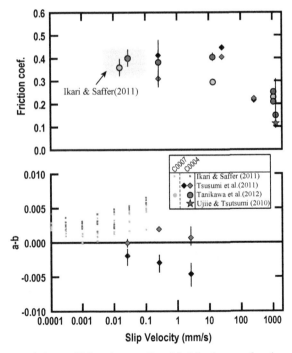

FIGURE 4.4.2.7 Friction coefficient (top panel) and "a-b" value as a function of slip velocity experimentally determined for the fault rock samples the shallow mega-splay (C0004) and frontal thrust (C0007). *Modified from Tanikawa et al. (2012) by adding data of Ikari et al. (2009) and Ujiie and Tsutsumi (2010).*

Ujiie and Tsutsumi (2010) conducted high-velocity friction experiments at 1.27 m/s for clay-rich fault gouge from the mega-splay fault zone at Site C0004 under dry and wet conditions. Based on observations of pore-fluid pressurization via shear-enhanced compaction and frictional heating for wet tests, they propose that the earthquake rupture propagates easily through clay-rich fault gouge by high-velocity slip weakening, if the stabilizing slip properties and initial peak friction can somehow be overcome.

By studying synthetic and natural fault gouge samples, Ikari, Marone, and Saffer (2011) found a systematic relationship between absolute frictional strength and the potential of stick slip for clay gouge samples with a slip velocity of 1–300 μm/s. They indicate that the weakest gouges are uniformly velocity strengthening, whereas the frictionally stronger gouges exhibit both velocity-weakening and velocity-strengthening behavior (Figure 4.4.2.8(A)). Tsutsumi et al. (2011) also show similar results for Nankai samples, and suggest that velocity-weakening, localized shear deformation could occur in higher frictional strength material, whereas velocity-strengthening behavior may be widely distributed along a frictionally weak section. Such a relationship is also recognized for axial displacement tests (10 μm/s) on samples from the old accretionary prism underlying the forearc basin at Site C0002, where clayey mud samples are weaker but experience a long-lasting slow failure, whereas silty mud samples have a higher strength and then experience a rapid failure (Takahashi, Azuma, Uehara, Kanagawa, & Inoue, 2013).

Ikari et al. (2011) found that the friction dependence on slip velocity evolves systematically with accumulated shear strain. Following this, Ikari, Marone, Saffer, and Kopf (2013b) found that the fault rocks exhibit decreasing strength trend with the slip distance over a millimeter scale, which is most enhanced at the slip velocity range ~0.3–3 mm/s (Figure 4.4.2.8(B)). They propose that such a slip-weakening behavior may be a viable mechanism for the generation of slow slip events or VLF earthquakes, without requiring velocity-weakening behavior by the rate- and state-dependent friction law.

4.4.2.3.2 Activity of Shallow Faults

When a fault slips seismically, fault rocks will fracture around the shear zone, the strong ground-shaking motion will disturb the surface sediment, and the slip friction will generate heat, generating thermally induced signals in and around the slip zone. During NanTroSEIZE expeditions some results supporting these signals have been reported.

Surface samples cored from the slope sediment above the hanging wall of mega-splay show repeated occurrences of mud breccia. Since their radioisotope lead (^{210}Pb) and carbon (^{14}C) dating indicates a deposition time consistent with the 1944 Tonankai earthquake, Sakaguchi et al. (2011b) suggest that the mud-breccia layers were possibly caused by seismic shaking during the 1944 event.

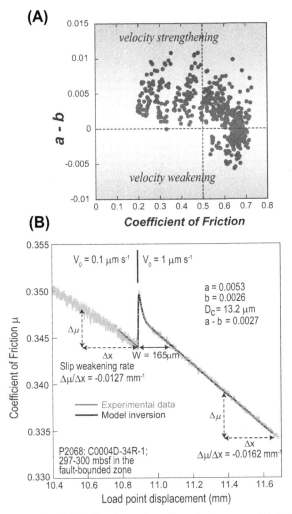

FIGURE 4.4.2.8 Frictional behavior for a slow slip velocity range. (A) Friction rate a-b versus frictional coefficient with a slip velocity of 1–300 μm/s (Ikari et al., 2011). (B) Experimental data to extract the constitutive friction parameters a, b, a-b, and Dc for the fault rock sample from Site C0004 (Ikari et al., 2013b). Slip weakening rates (Δμ/Δx) decrease after the velocity upstepping, exhibiting decreasing strength over millimeter-scale slip distances.

Sakaguchi et al. (2011a) identified another evidence for past seismic slip along the shallow mega-splay (C0004) and the frontal thrust décollement (C0007). They showed a vitrinite reflectance anomaly localized at the fault, revealing a potential past thermal anomaly of up to ~400 °C (Figure 4.4.2.9), and suggested this is evidence of coseismic slip that should have caused huge tsunami. Yamaguchi et al. (2011) analyzed the elemental and mineral abundance in the thin dark fault gouge layers at Site C0004, and found a higher illite content relative to host rock.

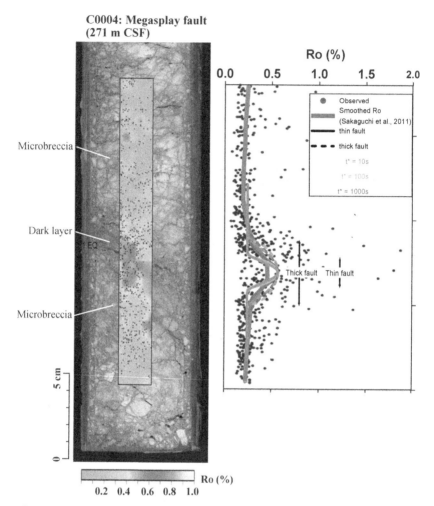

FIGURE 4.4.2.9 Measured vitrinite reflectance (Ro) distribution overlain on photographs of slab core samples from mega-splay fault *(modified from Sakaguchi et al. (2011).)*. Graph to right shows Ro; distance is normalized to dark layer. Curves show a smoothed Ro (Sakaguchi et al., 2011), and simulated Ro values for "thin" and "thick" fault scenarios for a single earthquake (Fulton & Harris, 2012). Different colors reflect different slip duration (t*) scenarios.

They attribute this to thermally enhanced smectite to illite transformation, which represents additional evidence of possible frictional heating.

In contrast, Hirono et al. (2014) performed geochemical analyses of a mega-splay fault rock and reexamined previously reported trace-element and isotope compositions (Hirono et al., 2009), and concluded no specific changes related to temperatures higher than 250 °C. Through their kinetic evaluation, they claim that the illitization process hardly progresses at temperature <250 °C, and that instead the illite content could have increased by other non-thermally induced chemical reactions. They also argue that the large coseismic slip (>5m), which may be supported by the existence of thin gouge layer (Sakaguchi, Chester, et al., 2011a), can be theoretically achieved alongside a temperature less than 250 °C.

Fulton and Harris (2012) reexamined the effects of frictional heat generation on the spatial distribution of vitrinite reflectance, as well as on the potential slip zone width and slip duration scenarios and on the cumulative effects of multiple events. They suggested high vitrinite reflectance values and relatively low temperature increase (<250 °C) can be reconciled as the result of cumulative effects of ~100 coseismic slip events of several tens of meters. However, this requires unrealistically large amounts of total displacement (>1 km) along a single fault zone. Kitamura, Mukoyoshi, Fulton, and Hirose (2012) argue that the coal can mature coseismically (in ~10 s) with a peak temperature from 26 to 266 °C, significantly lower than 300 °C. However, they notice that the measured temperature using a thermocouple sensor only detects temperature at the edge of a slip zone and does not measure the peak temperature at the center, which could have increased up to 570 °C.

As mentioned by Fulton and Harris (2012), our understanding of how vitrinite reflectance is affected by fault slip is still incomplete. In order to better estimate heat generation at a fault during an earthquake, it is necessary to revise the kinetic model of vitrinite maturation that involves the effects of flash temperature and mechanochemical reaction at fast sliding conditions (Kitamura et al., 2012). As suggested by Hirono et al. (2014), accurate assessment of the slip behavior of the mega-splay fault could be obtained by sampling the fault zone at a deeper depth of approximately 1.5 km, where records of high temperatures should be clearly detectable.

4.4.2.4 BOREHOLE OBSERVATORY

Borehole observatories are essential in order to detect and monitor small and low-frequency deformation (e.g., slow slip events, strain rate, pore pressure variation) that continue at and around the plate boundary fault. As stated in the NanTroSEIZE proposal (603-CDP; www.iodp.org/doc_download/359-603-cdp3cover), "Our fundamental goal is the creation of a distributed observatory spanning the up-dip limit of seismogenic and tsunamigenic behavior." Toward the ultimate "deep riser borehole observatory" at the locked portion of the fault (5–7 km below seafloor), a temporary hydrological observatory (SmartPlug) was deployed in 2009 at C0010 (Saffer et al., 2010), which was replaced with another one (GeniusPlug) in 2010 at the same site (Kopf et al. 2011), followed by the shallow (1-km-long) permanent observatory installed at Site C0002 in 2010 (Kopf et al., 2011).

4.4.2.4.1 SmartPlug

The SmartPlug is a stand-alone sensor package, with a capability of self-recording temperature and pressure, attached below a retrievable bridge plug (Saffer et al., 2010; Figure 4.4.2.10). It was designed as a temporary monitoring system until a permanent observatory was ready for installation. Although simple, it

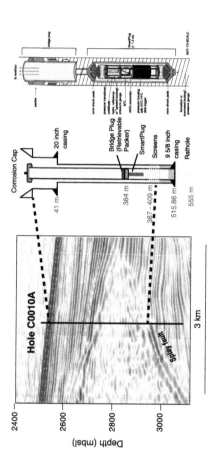

FIGURE 4.4.2.10 Schematic drawing of the SmartPlug installed at Hole C0010A. *Modified from Hammerschmidt et al. (2013).*

represents the first long-term monitoring package installation in NanTroSEIZE. During Expedition 319, the retrievable casing packer was installed at 410 mbsf at Site C0010 above the screen casing joints spanning the mega-splay fault zone, thus the smart plug was able to record temperature and pressure within the fault zone.

The smart plug was retrieved during Expedition 332 (Kopf et al., 2011), after 15 months of recording from August 23, 2009, to November 7, 2010, with 1-min sampling interval. The data detected pressure variations related to local storms, seismic waves, and tsunami waves from the Chile M8.8 earthquake (December 27, 2010) (Hammerschmidt et al., 2013). They also suggest that an increase in formation permeability was induced by regional seismic waves, which led to an abrupt drop (0.2 kPa) in the pressure data.

4.4.2.4.2 GeniusPlug

Although the concept of the GeniusPlug remains the same as for the SmartPlug, it also includes a 30-cm-long extension containing an osmotically driven fluid sampler (OsmoSampler) to collect a time series of pore fluid samples for 2 years (Jannasch et al., 2003) and a flow-through osmocolonization system for examination of in situ microbial growth under controlled conditions (Wheat et al., 2011; Orcutt, Wheat, & Edwards, 2010).

The GeniusPlug was deployed from Chikyu on November 9, 2010, into Hole C0010A (Kopf et al., 2011). Recovery of the GeniusPlug is planned at the time of long-term borehole monitoring system (LTBMS) installation at Site C0010, as part of future NanTroSEIZE operations.

4.4.2.4.3 Long-Term Borehole Monitoring System

The LTBMS is a permanent geodetic and hydrological observatory installed in the southern Kumano Forearc Basin. The sensors were installed in 2010 during the Expedition 332 into the base of the forearc basin sediments (Unit III) and the topmost portion of accretionary prism sediments (Unit IV) (Kopf et al., 2011) (Figure 4.4.2.11).

The suite of sensors for the observatory includes three pore pressure ports, volumetric strainmeter, broadband seismometer, tiltmeter, three-component geophones, three-component accelerometers, and a thermometer array with five thermistors. One pressure port was installed below the strainmeter in Unit IV. Another port was set at the screened joint from 757 to 780 mbsf to sample pore fluid pressure in Unit II. Prior to installation, Kimura et al. (2013) compared the performance of borehole seismometers with those used in the Dense Ocean-floor Network system for Earthquakes and Tsunamis (DONET) (Kaneda et al., 2009). They show less noise level in the high-frequency region, but the noise level in the low-frequency (<0.1 Hz) region was greater.

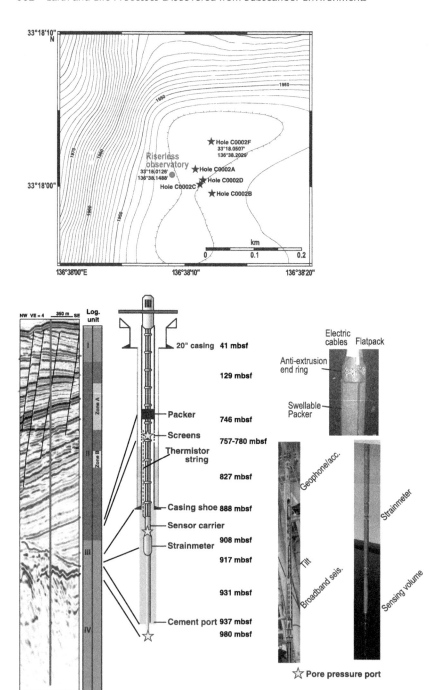

FIGURE 4.4.2.11 (top) Bathymetric map at Site C0002. (Bottom) The long-term borehole monitoring system (LTBMS) deployed at Site C0002. *Modified after Kopf et al. (2011).*

The Kuroshio is among the world's fastest currents, flowing at up to 5 knots in the Nankai Trough area. The vibration to the drill pipe, called vortex-induced vibration (VIV), was induced by the Kuroshio current, which caused severe damage to sensors being lowered into the formation inside the dill pipe (Kitada et al., 2013b; Saffer et al., 2010). Drilling operators and scientists therefore collaborated to prevent damages to the sensors before installations. First, the sensors were assembled and lowered in a low-current area, and the vessel drifted at <1 knot toward the site (Kitada et al., 2013a). Second, attaching a couple of ropes (diameter = 24 mm) and a heavy weight near the bottom of the assembly was extremely useful for reducing vibration.

The LTBMS system was connected to the DONET in January 2013 by using the Japan Agency for Marine-Earth Science and Technology remotely operated vehicle "Hyper Dolphin." Since then, downhole data are transmitted to shore for real-time monitoring (Kitada et al., 2013a). Using the temperature data from LTMBS and the thermal conductivities measured for core samples, Sugihara et al. (2014) determined the temperature and heat flow at 900 mbsf as 37.9 °C and 56 mW/m^2, respectively. Using these values, they evaluated the temperature at the mega-splay fault below Site C0002 (5000 mbsf) as 130–150 °C.

4.4.2.5 SUMMARY AND IMPLICATIONS

Ten expeditions were carried out since the start of NanTroSEIZE operation in 2007. Core and log data were obtained from 14 sites from the Shikoku Basin of the oceanic plate across the accretionary prism to the Kumano Forearc Basin. The maximum drilled depth at any one site is 3059 m below seafloor at riser site C0002. Previous drilling experiences in heavily indurated or fractured formations were known to be very difficult. Based on the lessons learned from NanTroSEIZE and previous experiences, we adopted extensive use of LWD for obtaining geological information including possible fault locations, prior to taking core samples. This has proven effective and is now used in other drill sites (Japan Trench Fast Drilling Project (JFAST); Chester, Mori, Eguchi, Toczko, & the Expedition 343/343T Scientists, 2013) in similar tectonic settings.

1. The stress state in the outer forearc basin or slope sediments is predominantly in favor of normal faulting, whereas the stress in the deeper accretionary prism is probably in the strike-slip or thrust faulting regime. In the thrust-dominated tectonic environment, this is consistent either with trench-normal shortening or with strike-slip and/or normal faulting stress state. Small differential horizontal stress (<5 MPa) is indicated consistently from log image data, core measurements, and seismic data analyses. The stress both in the hanging wall and footwall of the mega-splay faults (Sites C0004 and C0010) are in favor of thrust faulting with the trench-normal shortening, suggesting fault activities in the future as well as in the past.

2. In situ porosity and P-wave velocity were determined from core samples and log data. The porosity measured on core samples onboard is larger than the in situ value by up to 10–15% due to unloading upon recovery and clay dehydration. Correction for clay-bound effects yields an acceptable agreement with the interstitial porosity derived from sonic or resistivity log data. Mechanical and hydraulic properties as a function of effective stress path are useful for modeling a large-scale hydromechanical behavior of shallow subduction zones. Consolidation tests suggest that the sediments near the mega-splay fault and the frontal thrust are overconsolidated. They are caused by the surface erosion inferred from lower porosity (~50%) and the tectonic horizontal contraction. Porosity in the underthrust sediments below the splay fault is larger than the trend extrapolated from overconsolidated shallow slope sediments. It is possibly supported by overpressures inferred from the seismic analysis. The decrease in permeability with shearing measured on fault samples supports a possible thermal pressurization that can promote coseismic slip along shallow faults.

3. Frictional properties were determined from mega-splay and frontal décollement fault samples for a slip velocity ranging from 10^{-7} to 1 m/s. At slower slip rates, velocity-weakening behavior was detected in the frictionally stronger gouges from the shallow portion of the mega-splay and frontal décollement, which is interpreted as shear localization at large slip magnitudes and that these fault sections could host slow slip events. Friction under high-velocity slip experiments show velocity-weakening behavior, suggesting earthquake rupture could propagate to the seafloor if the initial peak friction can be overcome.

4. High vitrinite reflectance values and higher illite content within the localized fault gouge of shallow fault zones indicate potential coseismic slip(s) in the past, accompanied by transient temperature anomaly of up to >300 °C. However, absence of geochemical anomalies related to temperatures higher than 250 °C suggests either nonthermally induced chemical reactions or cumulative effects of ~100 coseismic slip events of several tens of meters. Our understanding of how vitrinite reflectance is affected by fault slip is still incomplete.

5. The first permanent geodetic and hydrological observatory was installed in the southern Kumano Forearc Basin (C0002) and connected to the DONET seafloor cable network in 2013 for real-time monitoring of fault behavior.

ACKNOWLEDGMENTS

This manuscript was motivated through a series of discussions within the Project Management Team for the Integrated Ocean Drilling Program (IODP) Nankai Trough seismogenic zone experiments (NanTroSEIZE). We are grateful to Dr Lisa McNeill and an anonymous reviewer for very productive comments and suggestions on the manuscript. Funding for this study was provided by the Grant-in-Aid for Scientific Research on Innovative Areas (21107006). We are grateful to the crew and technicians of D/V Chikyu and the Center for Deep Earth Exploration of Japan Agency for Marine-Earth Science and technology for the successful operations during IODP NanTroSEIZE Expeditions.

REFERENCES

Brown, K. M., & Ransom, B. (1996). Porosity corrections for smectite-rich sediments: impact on studies of compaction, fluid generation, and tectonic history. *Geology, 24,* 843–846. http://dx.doi.org/10.1130/0091-7613(1996)024.

Byrne, T. B., Lin, W., Tsutsumi, A., Yamamoto, Y., Lewis, J. C., Kanagawa, K., et al. (2009). Anelastic strain recovery reveals extension across SW Japan subduction zone. *Geophysical Research Letters, 36,* L23310. http://dx.doi.org/10.1029/2009GL040749.

Chang, C., McNeill, L. C., Moore, J. C., Lin, W., Conin, M., & Yamada, Y. (2010). In situ stress state in the Nankai accretionary wedge estimated from borehole wall failures. *Geochemistry Geophysics Geosystems, 11,* Q0AD04. http://dx.doi.org/10.1029/2010GC003261.

Chester, F. M., Mori, J., Eguchi, N., Toczko, S., & the Expedition 343/343T Scientists. (2013). *Proc. IODP* (Vol. 343/343T) Tokyo: Integrated Ocean Drilling Program Management International, Inc. http://dx.doi.org/10.2204/?iodp.proc.343343T.2013.

Conin, M., Bourlange, S., Henry, P., Boiselet, A., & Gaillot, P. (2014). Distribution of resistive and conductive structures in Nankai accretionary wedge reveals contrasting stress paths. *Tectonophysics, 611,* 181–191.

Conin, M., Henry, P., Bourlange, S., Raimbourg, H., & Reuschlé, T. (2011). Interpretation of porosity and LWD resistivity from the Nankai accretionary wedge in light of clay physicochemical properties: evidence for erosion and local overpressuring. *Geochemistry Geophysics Geosystems, 12,* Q0AD07. http://dx.doi.org/10.1029/2010GC003381.

Conin, M., Henry, P., Godard, V., & Bourlange, S. (2012). Splay fault slip in a subduction margin, a new model of evolution. *Earth and Planetary Science Letters, 341–344,* 170–175. http://dx.doi.org/10.1016/j.epsl.2012.06.003.

Di Toro, G., Han, R., Hirose, T., De Paola, N., Nielsen, S., Mizoguchi, K., et al. (2011). Fault lubrication during earthquakes. *Nature, 471,* 494–498. http://dx.doi.org/10.1038/nature09838.

Doan, M.-L., Conin, M., Henry, P., Wiersberg, T., Boutt, D., Buchs, D., et al. (2011). Quantification of free gas in the Kumano fore-arc basin detected from borehole physical properties: IODP NanTroSEIZE drilling site C0009. *Geochemistry Geophysics Geosystems, 12,* Q0AD06. http://dx.doi.org/10.1029/2010GC003284.

Expedition 315 Scientists. (2009). Site C0001. In M. Kinoshita, H. Tobin, J. Ashi, G. Kimura, S. Lallemant, E. J. Screaton, et al. (Eds.), *Proc. IODP* (Vol. 314/315/316). Washington, DC: Integrated Ocean Drilling Program Management International, Inc. http://dx.doi.org/10.2204/iodp.proc.314315316.123.2009.

Expedition 319 Scientists. (2010a). Site C0009. In D. Saffer, L. McNeill, T. Byrne, E. Araki, S. Toczko, N. Eguchi, et al. (Eds.), *Proc. IODP* (Vol. 319). Tokyo: Integrated Ocean Drilling Program Management International, Inc. http://dx.doi.org/10.2204/iodp.proc.319.104.2010.

Expedition 319 Scientists. (2010b). Methods. In D. Saffer, L. McNeill, T. Byrne, E. Araki, S. Toczko, N. Eguchi, et al. (Eds.), *Proc. IODP* (Vol. 319). Tokyo: Integrated Ocean Drilling Program Management International, Inc. http://dx.doi.org/10.2204/iodp.proc.319.104.2010.

Expedition 326 Scientists. (2011). *NanTroSEIZE stage 3: Plate boundary deep riser: Top hole engineering* (Vol. 326) IODP. Prel. Rept. http://dx.doi.org/10.2204/iodp.pr.326.2011.

Fulton, P. M., & Harris, R. N. (2012). Thermal considerations in inferring frictional heating from vitrinite reflectance and implications for shallow coseismic slip within the Nankai subduction zone. *Earth and Planetary Science Letters, 335-336*, 206–215. http://dx.doi.org/10.1016/j.epsl.2012.04.012.

Gulick, S. P. S., Bangs, N. L. B., Moore, G. F., Ashi, J., Martin, K. M., Sawyer, D. S., et al. (2010). Rapid forearc basin uplift and megasplay fault development from 3-D seismic images of Nankai margin off Kii Peninsula, Japan. *Earth and Planetary Science Letters, 300*, 55–62. http://dx.doi.org/10.1016/j.epsl.2010.09.034.

Hammerschmidt, S., Davis, E. E., & Kopf, A. (2013). Fluid pressure and temperature transients detected at the Nankai Trough megasplay fault: results from the SmartPlug borehole observatory. *Tectonophysics, 600*, 116–133. http://dx.doi.org/10.1016/j.tecto.2013.02.010.

Hashimoto, Y., Tobin, H. J., & Knuth, M. (2010). Velocity - porosity relationships for slope apron and accreted sediments in the Nankai Trough seismogenic zone experiment, integrated ocean drilling program expedition 315 site C0001. *Geochemistry Geophysics Geosystems, 11*, Q0AD05. http://dx.doi.org/10.1029/2010GC003217.

Heki, K., & Miyazaki, S. (2001). Plate convergence and long-term crustal deformation in central Japan. *Geophysical Research Letters, 28*, 2313–2316. http://dx.doi.org/10.1029/2000GL012537.

Henry, P., Kanamatsu, T., Moe, K., & the Expedition 333 Scientists. (2012). In *Proc. IODP* (Vol. 333) Tokyo: Integrated Ocean Drilling Program Management International, Inc. http://dx.doi.org/10.2204/iodp.proc.333.2012.

Hirono, T., Tanikawa, W., Honda, G., Kameda, J., Fukuda, J., & Ishikawa, T. (2014). Importance of mechanochemical effects on fault slip behavior during earthquakes. *Geophysical Research Letters, 40*, 1–5.

Hirono, T., Ujiie, K., Ishikawa, T., Mishima, T., Hamada, Y., Tanimizu, M., et al. (2009). Estimation of temperature rise in a shallow slip zone of the megasplay fault in the Nankai Trough. *Tectonophysics, 478*, 215–220.

Hyndman, R., Ashi, J., Brown, K., Favali, P., Harjes, P., Huchon, P., et al. (1999). Seismogenic zone deep drilling and measurement report of the detailed planning group. *JOIDES Journal, 25-2*, 20–23.

Ikari, M. J., Hüpers, A., & Kopf, A. J. (2013a). Shear strength of sediments approaching subduction in the Nankai Trough, Japan as constraints on forearc mechanics. *Geochemistry Geophysics Geosystems, 14*, 2716–2730. http://dx.doi.org/10.1002/ggge.20156.

Ikari, M. J., Marone, C., & Saffer, D. M. (2011). On the relation between fault strength and frictional stability. *Geology, 39*, 83–86. http://dx.doi.org/10.1130/G31416.1.

Ikari, M., Marone, C., Saffer, D. M., & Kopf, A. (2013b). Slip weakening as a mechanism for slow earthquakes. *Nature Geoscience, 6*. http://dx.doi.org/10.1038/NGEO1818.

Ikari, M. J., & Saffer, D. M. (2011). Comparison of frictional strength and velocity dependence between fault zones in the Nankai accretionary complex. *Geochemistry Geophysics Geosystems, 12*, Q0AD11. http://dx.doi.org/10.1029/2010GC003442.

Ikari, M. J., Saffer, D. M., & Marone, C. (2009). Frictional and hydrologic properties of a major splay fault system, Nankai subduction zone. *Geophysical Research Letters, 36*, L20313. http://dx.doi.org/10.1029/2009GL040009.

IODP Initial Science Plan. (2000). *International working group support office*. 1755 Massachusetts Ave., NW, Suite 700, Washington, DC 20036–2102, USA.

Ito, Y., Asano, Y., & Obara, K. (2009). Very low-frequency earthquakes indicate a transpressional stress regime in the Nankai accretionary prism. *Geophysical Research Letters, 36*, L20309. http://dx.doi.org/10.1029/2009GL039332.

Ito, T., Funato, A., Lin, W., Doan, M.-L., Boutt, D. F., Kano, Y., et al. (2013). Determination of stress state in deep subsea formation by combination of hydraulic fracturing in situ test and core analysis: a case study in the IODP expedition 319. *Journal of Geophysical Research: Solid Earth, 118*, 1203–1215. http://dx.doi.org/10.1002/jgrb.50086.

Ito, Y., & Obara, K. (2006). Dynamic deformation of the accretionary prism excites very low frequency earthquakes. *Geophysical Research Letters, 33*, L02311. http://dx.doi.org/10.1029/2005GL025270.

Jannasch, H. W., Davis, E. E., Kastner, M., Morris, J. D., Pettigrew, T. L., Plant, J. N., et al. (2003). CORK-II: long-term monitoring of fluid chemistry, fluxes, and hydrology in instrumented boreholes at the Costa Rica subduction zone. In J. D. Morris, H. W. Villinger, & A. Klaus (Eds.), *Proc. ODP, Init. Repts* (Vol. 205) (pp. 1–36). College Station: TX (Ocean Drilling Program). http://dx.doi.org/10.2973/odp.proc.ir.205.102.2003.

Kaneda, Y., Kawaguchi, K., Araki, E., Sakuma, A., Matsumoto, H., Nakamura, T., et al. (2009). Dense ocean floor network for earthquakes and tsunamis (DONET) - development and data application for the mega thrust earthquakes around the Nankai trough. *Eos Transactions AGU, 90*(52), S53A–S1453 Fall Meet. Suppl., Abstract.

Karig, D. E. (1990). Experimental and observational constraints on the mechanical behaviour in the toes of accretionary prisms. In R. J. Knipe, & E. H. Rutter (Eds.), *Deformation mechanisms, rheology, and tectonics* (Vol. 54) (pp. 383–398). Geol. Soc. Spec. Publ. London.

Kikuchi, M., Nakamura, M., & Yoshikawa, K. (2003). Source rupture processes of the 1944 Tonankai earthquake and the 1945 Mikawa earthquake derived from low-gain seismograms. *Earth Planet Space, 55*, 159–172.

Kimura, T., Araki, E., Takayama, H., Kitada, K., Kinoshita, M., Namba, Y., et al. (2013). Development and performance tests of a sensor suite for a long-term borehole monitoring system in seafloor settings in the Nankai Trough, Japan. *IEEE Journal of Oceanic Engineering, 38*, 383–395. http://dx.doi.org/10.1109/JOE.2012.2225293.

Kinoshita, M., & Tobin, H. J. (2013). Interseismic stress accumulation at the locked zone of Nankai Trough seismogenic fault off Kii Peninsula. *Tectonophysics, 600C*, 153–164. http://dx.doi.org/10.1016/j.tecto.2013.03.015.

Kinoshita, M., Tobin, H., Ashi, J., Kimura, G., Lallemant, S., Screaton, E. J., et al. (2009). *Proc. IODP* (Vol. 314/315/316) Washington, DC: Integrated Ocean Drilling Program Management International, Inc. http://dx.doi.org/10.2204/iodp.proc.314315316.2009.

Kitada, K., Araki, E., Kimura, T., Kinoshita, M., Kopf, A., & Saffer, D. M. (2013a). Long-term monitoring at C0002 seafloor borehole in Nankai Trough seismogenic zone. In *Procs. Of underwater technology symposium (UT), 2013 IEEE international* (pp. 1–3). http://dx.doi.org/10.1109/UT.2013.6519882.

Kitada, K., Araki, E., Kimura, T., Mizuguchi, Y., Kyo, M., Saruhashi, T., et al. (2013b). Field experimental study on vortex-induced vibration behavior of the drill pipe for the ocean borehole observatory installation. *IEEE Journal of Oceanic Engineering, 38*, 158–166. http://dx.doi.org/10.1109/JOE.2012.2213973.

Kitajima, H., Chester, F. M., & Biscontin, G. (2012). Mechanical and hydraulic properties of Nankai accretionary prism sediments: effect of stress path. *Geochemistry Geophysics Geosystems, 13*, Q0AD27. http://dx.doi.org/10.1029/2012GC004124.

Kitajima, H., & Saffer, D. M. (2012). Elevated pore pressure and anomalously low stress in regions of low frequency earthquakes along the Nankai Trough subduction megathrust. *Geophysical Research Letters, 39*, L23301. http://dx.doi.org/10.1029/2012GL053793.

Kitamura, M., Mukoyoshi, H., Fulton, P. M., & Hirose, T. (2012). Coal maturation by frictional heat during rapid fault slip. *Geophysical Research Letters, 39*, L16302. http://dx.doi.org/10.1029/2012GL052316.

Kopf, A., Araki, E., Toczko, S., & the Expedition 332 Scientists. (2011). *Proc. IODP* (Vol. 332) Tokyo: Integrated Ocean Drilling Program Management International, Inc. http://dx.doi.org/10.2204/iodp.proc.332.104.2011.

Lin, W., Doan, M.-L., Moore, J. C., McNeill, L., Byrne, T. B., Ito, T., et al. (2010). Present-day principal horizontal stress orientations in the Kumano forearc basin of the southwest Japan subduction zone determined from IODP NanTroSEIZE drilling Site C0009. *Geophysical Research Letters*, *37*, L13303. http://dx.doi.org/10.1029/2010GL043158.

Moore, G. F., Bangs, N. L., Taira, A., Kuramoto, S., Pangborn, E., & Tobin, H. J. (2007). Three-dimensional splay fault geometry and implications for tsunami generation. *Science*, *318*(5853), 1128–1131. http://dx.doi.org/10.1126/science.1147195.

Moore, J. C., Barrett, M., & Thu, M. K. (2013). Fluid pressures and fluid flows from boreholes spanning the NanTroSEIZE transect through the Nankai Trough, SW Japan. *Tectonophysics*, *600*, 108–115. http://dx.doi.org/10.1016/j.tecto.2013.01.026.

Moore, J. C., Chang, C., McNeill, L., Thu, M. K., Yamada, Y., & Huftile, G. (2011). Growth of borehole breakouts with time after drilling: implications for state of stress, NanTroSEIZE transect, SW Japan. *Geochemistry Geophysics Geosystems*, *12*, Q04D09. http://dx.doi.org/10.1029/2010GC003417.

Moore, G. F., Park, J.-O., Bangs, N. L., Gulick, S. P., Tobin, H. J., Nakamura, Y., et al. (2009). Structural and seismic stratigraphic framework of the NanTroSEIZE Stage 1 transect. In *Proc. Integrated Ocean Drill. Program* (Vol. 314/315/316) (pp.1–46). http://dx.doi.org/10.2204/iodp.proc.314315316.102.2009.

Olcott, K., & Saffer, D. (3–7 December 2012). *Constraints on in situ stress across the shallow megasplay fault offshore the Kii Peninsula, SW Japan from borehole breakouts.* Abstract T13A-2576 presented at 2012 Fall Meeting. San Francisco, Calif.: AGU.

Orcutt, B., Wheat, C. G., & Edwards, K. J. (2010). Subseafloor ocean crust microbial observatories: development of FLOCS (flow-through osmo colonization system) and evaluation of borehole construction materials. *Geomicrobiology Journal*, *27*(2), 143–157. http://dx.doi.org/10.1080/01490450903456772.

Raimbourg, H., Hamano, Y., Saito, S., Kinoshita, M., & Kopf, A. (2011). Acoustic and mechanical properties of Nankai accretionary prism core samples. *Geochemistry Geophysics Geosystems*, *12*, Q0AD10. http://dx.doi.org/10.1029/2010GC003169.

Sacks, A., Saffer, D. M., & Fisher, D. (2013). Analysis of normal fault populations in the Kumano forearc basin, Nankai Trough, Japan: 2. Principal axes of stress and strain from inversion of fault orientations. *Geochemistry Geophysics Geosystems*, *14*, 1973–1988. http://dx.doi.org/10.1002/ggge.20118.

Saffer, D. M., Flemings, P. B., Boutt, D., Doan, M.-L., Ito, T., McNeill, L., et al. (2013). In situ stress and pore pressure in the Kumano Forearc Basin, offshore SW Honshu from downhole measurements during riser drilling. *Geochemistry Geophysics Geosystems*, *14*. http://dx.doi.org/10.1002/ggge.20051.

Saffer, D., McNeill, L., Byrne, T., Araki, E., Toczko, S., Eguchi, N., et al. (2010). *Proc. IODP* (Vol. 319)Tokyo: Integrated Ocean Drilling Program Management International, Inc. http://dx.doi.org/10.2204/iodp.proc.319.104.2010.

Saffer, D. M., & Tobin, H. J. (2011). Hydrogeology and mechanics of subduction zone forearcs: fluid flow and pore pressure. *Annual Review of Earth and Planetary Sciences*, *39*, 157–186. http://dx.doi.org/10.1146/annurev-earth-040610-133408.

Saito, S., Underwood, M. B., Kubo, Y., & the Expedition 322 Scientists. (2010). *Proc. IODP* (Vol. 322) Tokyo: Integrated Ocean Drilling Program Management International, Inc. http://dx.doi.org/10.2204/iodp.proc.322.2010.

Sakaguchi, A., Chester, F., Curewitz, D., Fabbri, O., Goldsby, D., Kimura, G., et al. (2011a). Seismic slip propagation to the updip end of plate boundary subduction interface faults: vitrinite reflectance geothermometry on integrated cean drilling program NanTro SEIZE cores. *Geology*, *39*, 395–398. http://dx.doi.org/10.1130/G31642.1.

Sakaguchi, A., Kimura, G., Strasser, M., Screaton, E. J., Curewitz, D., & Murayama, M. (2011b). Episodic seafloor mud brecciation due to great subduction zone earthquakes. *Geology, 39*, 919–922. http://dx.doi.org/10.1130/G32043.1.

Scholz, C. H. (1998). Earthquakes and friction laws. *Nature, 391*, 37–42.

Screaton, E. J., Kimura, G., Curewitz, D., & the Expedition 316 Scientists. (2009). Expedition 316 summary. In M. Kinoshita, H. Tobin, J. Ashi, G. Kimura, S. Lallemant, E. J. Screaton, et al. (Eds.), *Proc. IODP* (Vol. 314/315/316). Washington, DC: Integrated Ocean Drilling Program Management International, Inc. http://dx.doi.org/10.2204/iodp.proc.314315316.131.2009.

Seno, T., Stein, S., & Gripp, A. E. (1993). A model for the motion of the Philippine Sea plate consistent with NUVEL-1 and geological data. *Journal of Geophysical Research, 98*, 17941–17948.

Song, I., D. Saffer, & P. Flemings (2011). Mechanical characterization of slope sediments: constraints on in situ stress and pore pressure near the tip of the megasplay fault in the Nankai accretionary complex. *Geochemistry Geophysics Geosystems, 12*, Q0AD17, doi:10.1029/2011GC003556, in press.

Strasser, M., Dugan, B., Kanagawa, K., Moore, G. F., Toczko, S., Maeda, L., et al. (2014). *Proc. IODP* (Vol. 338) Yokohama: Integrated Ocean Drilling Program 12, Q0AD17. http://dx.doi.org/10.2204/iodp.proc.338.2014.

Strasser, M., Moore, G. F., Kimura, G., Kopf, A. J., Underwood, M. B., Guo, J., et al. (2011). Slumping and mass transport deposition in the Nankai fore arc: evidence from IODP drilling and 3-D reflection seismic data. *Geochemistry Geophysics Geosystems, 12*, Q0AD13. http://dx.doi.org/10.1029/2010GC003431.

Sugihara, T., Kinoshita, M., Araki, E., Kimura, T., Kyo, M., Namba, Y., et al. (2014). Re-evaluation of temperature at the updip limit of locked portion of Nankai megasplay inferred from IODP Site C0002 temperature observatory. *Earth Planet Space, 66*, 107. http://www.earth-planets-space.com/content/66/1/107.

Takahashi, M., Azuma, S., Uehara, S., Kanagawa, K., & Inoue, A. (2013). Contrasting hydrological and mechanical properties of clayey and silty muds cored from the shallow Nankai Trough accretionary prism. *Tectonophysics, 600*, 63–74. http://dx.doi.org/10.1016/j.tecto.2013.01.008.

Tanikawa, W., Mukoyoshi, H., Tadai, O., Hirose, T., Tsutsumi, A., & Lin, W. (2012). Velocity dependence of shear-induced permeability associated with frictional behavior in fault zones of the Nankai subduction zone. *Journal of Geophysical Research, 117*, B05405. http://dx.doi.org/10.1029/2011JB008956.

Tobin, H., Kinoshita, M., Ashi, J., Lallemant, S., Kimura, G., Screaton, E., et al. (2009). NanTroSEIZE stage 1 expeditions 314, 315, and 316: first drilling program of the nankai trough seismogenic zone experiment. *Scientific Drilling, 8*, 4–17. http://dx.doi.org/10.2204/iodp.sd.8.01.2009.

Tsuji, T., Dvorkin, J., Mavko, G., Nakata, N., Matsuoka, T., Nakanishi, A., et al. (2011a). Vp/Vs ratio and shear-wave splitting in the Nankai Trough seismogenic zone: insights into effective stress, pore pressure and sediment consolidation. *Geophysics, 76*, 1–12. http://dx.doi.org/10.1190/1.3560018.

Tsuji, T., Hino, R., Sanada, Y., Yamamoto, K., Park, J.-O., No, T., et al. (2011b). In site stress state from walkaround VSP anisotropy in the Kumano basin southeast of the Kii Peninsula, Japan. *Geochemistry Geophysics Geosystems, 12*, Q0AD19. http://dx.doi.org/10.1029/2011GC003583.

Tsuji, T., Kamei, R., & Pratt, R. G. (2014). Pore pressure distribution of a mega-splay fault system in the Nankai Trough subduction zone: Insight into up-dip extent of the seismogenic zone. *Earth and Planetary Science Letters, 396*, 165–178. http://dx.doi.org/10.1016/j.epsl.2014.04.011.

Tsutsumi, A., Fabbri, O., Karpoff, A. M., Ujiie, K., & Tsujimoto, A. (2011). Friction velocity dependence of clay-rich fault material along a megasplay fault in the Nankai subduction zone at intermediate to high velocities. *Geophysical Research Letters, 38*, L19301. http://dx.doi.org/10.1029/2011GL049314.

Tsutsumi, A., & Shimamoto, T. (1997). High - velocity frictional properties of gabbro. *Geophysical Research Letters, 24*, 699–702. http://dx.doi.org/10.1029/97GL00503.

Tudge, J., & Tobin, H. J. (2013). Velocity-porosity relationships in smectite-rich sediments: Shikoku Basin, Japan. *Geochemistry Geophysics Geosystems, 14*, 5194–5207. http://dx.doi.org/10.1002/2013GC004974.

Ujiie, K., & Tsutsumi, A. (2010). High-velocity frictional properties of clay-rich fault gouge in a megasplay fault zone, Nankai subduction zone. *Geophysical Research Letters, 37*, L24310. http://dx.doi.org/10.1029/2010GL046002.

Wang, K., & Hu, Y. (2006). Accretionary prisms in subduction earthquake cycles: the theory of dynamic Coulomb wedge. *Journal of Geophysical Research, 111*, B06410. http://dx.doi.org/10.1029/2005JB004094.

Wheat, C. G., Jannasch, H. W., Kastner, M., Hulme, S., Cowen, J., Edwards, K. J., et al. (2011). Fluid sampling from oceanic borehole observatories: design and methods for CORK activities (1990–2010). In A. T. Fisher, T. Tsuji, K. Petronotis, & the Expedition 327 Scientists (Eds.), *Proc. IODP* (Vol. 327). Tokyo: Integrated Ocean Drilling Program Management International, Inc. http://dx.doi.org/10.2204/iodp.proc.327.109.2011.

Yamada, Y., McNeill, L., Moore, J. C., & Nakamura, Y. (2011). Structural styles across the Nankai accretionary prism revealed from LWD borehole images and their correlation with seismic profile and core data: results from NanTroSEIZE stage 1 expeditions. *Geochemistry Geophysics Geosystems, 12*, Q0AD15. http://dx.doi.org/10.1029/2010GC003365.

Yamaguchi, A., Sakaguchi, A., Sakamoto, T., Iijima, K., Kameda, J., Kimura, G., et al. (2011). Progressive illitization in fault gouge caused by seismic slip propagation along a megasplay fault in the Nankai Trough. *Geology, 39*, 995–998. http://dx.doi.org/10.1130/G32038.1.

Zoback, M. D., Barton, C. A., Brudy, M., Castillo, D. A., Finkbeiner, T., Grollimund, B. R., et al. (2003). Determination of stress orientation and magnitude in deep wells. *International Journal of rock mechanics and mining sciences, 40*, 1049–1076. http://dx.doi.org/10.1016/j.ijrmms.2003.07.001.

Chapter 4.4.3

Fluid Origins, Thermal Regimes, and Fluid and Solute Fluxes in the Forearc of Subduction Zones

Miriam Kastner,[1,*] Evan A. Solomon,[2] Robert N. Harris[3] and
Marta E. Torres[3]
[1]Scripps Institution of Oceanography, La Jolla, CA, USA; [2]School of Oceanography, University of
Washington, Seattle, WA, USA; [3]Oregon State University, Corvallis, OR, USA
*Corresponding author: E-mail: mkastner@ucsd.edu

4.4.3.1 INTRODUCTION

Fluids play a fundamental role in subduction zone (SZ) processes that include
chemical, isotopic, and material cycling; earthquake and volcanic processes; and
the evolution of physical properties in the forearc. The study of SZ processes were
greatly facilitated by scientific ocean drilling (ODP and IODP). Fluid-rich sediments and hydrated oceanic crust enter SZs where the fluids and rocks are chemically and isotopically altered in response to fluid rock reactions at progressively
higher temperatures and pressures (e.g., Lauer & Saffer, 2012; Moore & Vrolijk,
1992; Ranero et al., 2008; and references therein). A portion of the fluid produced
is expelled back to the ocean from the forearc by sediment compaction, mostly
through focused flow. Fluids that reach the depth of arc magma generation promote melting and recycling of volatiles to the ocean and atmosphere through arc
volcanoes. The residual fluid and chemically modified rocks are released in the
back-arc or are transported to the mantle where they impact the long-term chemical and isotopic evolution of the mantle (e.g., Bebout, 1996; Bekins, McCaffrey,
& Dreiss, 1995; Füri et al., 2010; Hacker, 2008; Hilton, Fischer, & Marty, 2002;
Jarrard, 2003; Kastner, Elderfield, & Martin, 1991; Kerrick & Connolly, 2001;
Moore et al., 2001; Moore & Vrolijk, 1992; Peacock, 1990; Plank & Langmuir,
1993, 1998; Rea & Ruff, 1996; Saffer & Tobin, 2011; Wallmann, 2001; and
references therein).

In addition to chemical cycling, fluids play an important role in the dynamics of SZ processes; the two most important dynamic processes in SZs are
earthquakes and volcanism. The role of fluids in earthquakes remains a key
question (e.g., Rea & Ruff, 1996). Recently, the role of fluids in modulating the

Developments in Marine Geology, Volume 7. http://dx.doi.org/10.1016/B978-0-444-62617-2.00022-0
671

extent of the locked portion of the subducting plate, and thus their role in subduction zone earthquake generation, has been suggested (e.g., Saffer & Tobin, 2011, and references therein), and other recent studies suggest that slow slip events and tremors are linked to high pore fluid pressure (e.g., Audet, Bostock, Christensen, & Peacock, 2009; Audet & Schwartz, 2013; Liu & Rice, 2007; Song et al., 2009; and references therein). In contrast, the role of fluids in arc magmatism has largely been established (e.g., Gill, 1981; Marty & Tolstikhin, 1998), but remaining questions include what are the source and composition of the fluids, and what are the fluxes of fluids, solutes, and isotope ratios in accretionary versus erosive SZs.

Fluid generation, elevated fluid pressure, and fluid flow also have profound implications for the evolution of forearc physical, chemical, and thermal properties (e.g., Carson & Screaton, 1998; Chan & Kastner, 2000; Hyndman & Wang, 1993; Langseth & Silver, 1996; Saffer & Tobin, 2011; Silver et al., 2000; Spinelli & Saffer, 2004). High pore fluid pressures develop due to rapid burial, tectonic loading, and mineral dehydration (Moore & Vrolijk, 1992; Saffer, Silver, Fisher, Tobin, & Moran, 2000), promoting fluid flow that influences the thermal regimes of SZs (Foucher et al., 1990; Grevemeyer et al., 2004; Harris, Grevemeyer, et al., 2010; Le Pichon et al., 1990; Le Pichon, Kobayashi, & Kaiko-Nankai Scientific Crew, 1992; Spinelli & Wang, 2008) that enhance sediment diagenesis and rock metamorphism (Colten-Bradley, 1987; Ernst, 1990; Jarrard, 2003; Kastner et al., 1991; Kerrick & Connolly, 2001; Martin, Kastner, & Elderfield, 1991; Moore & Saffer, 2001; Peacock, 1990, 1993; Peacock & Hyndman, 1999; Solomon & Kastner, 2012; Spivack, Kastner, & Ransom, 2002; Teichert, Torres, Bohrmann, & Eisenhauer, 2005; and references therein) and modulates the C budget (Kvenvolden, 1993).

In comparison to geochemical cycling at hydrothermal systems at ridge crests and flanks, the recycling of fluids at SZs and its impact on seawater chemistry is poorly constrained. Key questions concerning the fluid origins, the volumes, the spatial distribution of advecting fluids, the main fluid-rock reactions, the nature and paths of flow, and the global ocean geochemical consequences of this fluid transport remain. In this paper, we provide an in-depth analysis and synthesis of geochemical cycling in SZs and the associated thermal regimes, with a focus on the forearc. Our analysis and synthesis are based on both previously published and newly acquired chemical and isotopic data from five extensively studied SZs, representing both accreting and erosive margins, with special emphasis on the Nankai Trough and Costa Rica SZs (Figure 4.4.3.1). The characteristics of the five margins are provided in Table 4.4.3.1. Our focus is on H_2O and solute cycling in the forearc and the impact on ocean chemistry and on the forearc shallow thermal structure. Fluids venting at the seafloor are included in our global flux calculations. The data synthesized are from IODP and ODP expeditions, with an emphasis on IODP expeditions at the Costa Rica and Nankai subduction zones. Nankai drilling results include IODP expeditions 322 and 315/316 and ODP Legs 190 and 131. Costa Rica drilling

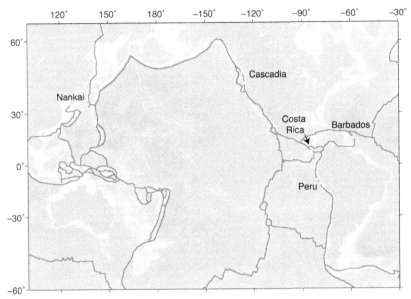

FIGURE 4.4.3.1 Suduction zones discussed and locations of DSDP, ODP, and IODP holes in those subduction zones.

results come from IODP expeditions 334 and 344 and ODP Legs 205 and 170. These results are supplemented with drilling results from Cascadia (expedition 311 and Legs 146 and 204), Barbados (Legs 110 and 156), and Peru (Leg 112).

Research over the past two decades has shown the importance of hydrologic differences between accretionary and erosive margins (Figure 4.4.3.2). We start by highlighting these differences and then discuss thermal and fluid inputs to convergent margins and provide an updated estimate of the global fluid input to subduction zones. Fluids that are not subducted to the depths of magma generation return to the surface with modified composition; we discuss their pathways, geochemistry, and flow rates that lead us to an estimate of the discharge flux through the forearc. We conclude by discussing geochemical signatures, solutes, and isotope ratio fluxes.

The main objective of this study is to provide insight into the fluid budgets in subduction zone forearcs. For this synthesis, primarily ODP and IODP expedition data are used. More specifically, we aim:

- To use geochemistry to determine the source of the fluid and the temperature at the source. This will aid in understanding the interplay between diagenetic dehydration reactions and the thermal structures of SZs, which can be used to evaluate the relationship between the depths of the fluid source and seismicity.
- To integrate the chemical and isotopic compositions and estimates of the output flux of the forearc fluids that return to the ocean in order to evaluate the impact on seawater chemistry.

TABLE 4.4.3.1 Main Geologic Characteristics of the Five Subduction Zones Considered

	Sediment Type	Sediment Thickness (m)	Fate of Incoming Sediments	Convergence Rate (mm/year)	Age of Subducting plate (Ma)	Geothermal Gradient (°C/km)	Heat Flow (mW/m²)	Exp/Hole
Nankai Trough, Muroto and Ashizuri, respectively	Ash-rich silt-sand turbidites, hemipelagic muds	500[a]–1250 500[a]–1000	Most accreted, some subducted	40 40	15 20	183[b] 46[i]	180[b] 63[i]	190/1173 87/582
Costa Rica, Nicoya and Osa respectively	Hemipelagic clays, carbonate ooze	0–380[i] 0–300[i]	No accretion, entirely subducted. Some underplated	87	24 14	9.8[b] 222[d]	9.1[c] 178[d]	170/1039 334/ U1381
Cascadia	Clayey-silt with sand turbidites, hemipelagic mud	2500[e]	Most sediment accreted	35	6–8	68[e]	84[e]	146/888
Barbados	Fine-grained pelagic clays and muds	800[f]	Most accreted, some subducted	20	90	79[g]	92[g]	110/672[g]

Peru	Organic-rich diatomaceous muds and silts	115[h]	Most subducted, some accreted	90	30–40	90[h], 0–55 mbsf 35, 55–115 mbsf	51[h]–64	201/1231

[a]Ike et al. (2008).

[b1]Moore, G.F., Taira, A., Klaus, A., et al. (2001). Proceedings of the Ocean Drilling Program, Initial Reports: Vol. 190.

[b2]Kimura, G., Silver, E.A., Blum, P., et al. (1997). Proceedings of the Ocean Drilling Program, Initial Reports: Vol. 170.

[c]Ruppel, C., & Kinoshita, M. (2000). Fluid, methane, and energy flux in an active margin gas hydrate province, offshore Costa Rica. Earth and Planetary Science Letters, 179(1), 153–165.

[d]Vannucchi, P., Ujiie, K., Stroncik, N., Malinverno, A., & the Expedition 334 Scientists. Proceedings of the Integrated Ocean Drilling Program: Vol. 334.

[e]Westbrook, G.K., Carson, B., Musgrave, R.J., et al. (1994). Proceedings of the Ocean Drilling Program, Initial Reports: Vol. 146 (Part one) Mascle, A., Moore, J.C., et al. (1988). Proceedings of the Ocean Drilling Program, Initial Reports: Vol. 110. College Station, TX (Ocean Drilling Program). doi:10.2973/odp.proc.ir.110.1988.

[f]Mascle, A., et al., 1988, Proceedings of the Ocean Drilling Program, Initial Reports, Part A: Vol. 110. College Station, Tex: Ocean Drilling Program.

[g]Fisher, A., & Hounslow, M. (1990). Transient fluid flow through the toe of the Barbados accretionary complex: constraints from ocean drilling program leg 110 heat row studies and simple models. Journal of Geophysical Research, 95 (B6), 8845–8858.

[h]D'Hondt, S.L, Jørgensen, B.B., Miller, D.J., et al. (2003). Proceedings of the Ocean Drilling Program, Initial Reports: Vol. 201.

[i]Kagami, H., Karig, D. E., Coulbourn, W. T., et al. (1986). Initial Reports. DSDP, 87. Washington: U.S. Govt. Printing Office.

[i]Von Huene, Ranero, Weinrebe, and Hinz (2000)

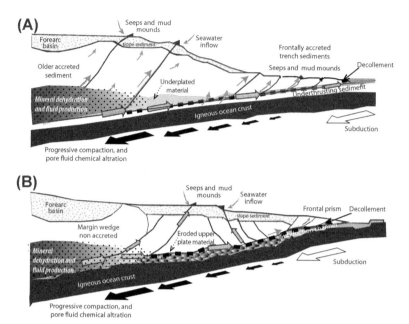

FIGURE 4.4.3.2 Schematic diagram illustrating hydrogeology of (A) accretionary, and (B) erosive margins. In both systems fluid generation increases progressively with distance from the trench, by compaction, and mineral dehydration (solid back arrows). Blue (light gray in print versions) arrows denote the relative magnitude and direction of flow, which in accretionary margins is focused mainly along the décollement (thick dashed line) and underthrust coarser grain stratigraphic horizons. In contrast at nonaccreting and erosive margins, fluids are partitioned between the décollement and focus conduits through the prism.

These data provide an estimate on the residual fluid transported to the arc plus mantle. By using the available estimates on fluid flux through arcs to the atmosphere (e.g., Hilton et al., 2002) an estimate of the flux to the mantle could be obtained. We define subducted fluid flux as the sum of the incoming fluid flux and erosion flux less the accretion flux and underplate flux.

$$(Subducted\ fluid\ flux\ =\ (Incoming\ fluid\ flux + Erosion\ flux)$$
$$-\ (Accretion\ flux + Underplated\ flux)$$

4.4.3.2 ACCRETIONARY AND EROSIVE CONVERGENT MARGINS

Convergent margins may be divided into two end-member types that can be understood in terms of either accretionary or erosive plate boundaries. A fundamental tectonic difference between these styles of margins is the transfer of material between the overriding and down-going plates. At accretionary margins, material is transferred from the down-going plate to the upper plate through either frontal accretion or underplating at the base of the margin prism. At erosive margins, all incoming sediments are subducted and the upper plate is eroded at its front and base such that the upper surface of the margin subsides and retreats (Lallemand, Schnule, & Malavieille, 1994; Von Huene & Scholl, 1991). In this scheme, nonaccretionary margins reside between these end members. In general, accretionary margins are better understood because they were recognized early in plate tectonic history (e.g., Ernst, 1970; Hamilton, 1969) due in part to large accumulations of oceanic and trench sediments fronting these margins. In contrast, erosive margins are less well understood because it takes longer to recognize them (e.g., Von Huene & Lallemand, 1990) and because the forearc is eroded making its study difficult. As this synthesis indicates, our knowledge of accretionary margins, as revealed by drilling and other studies, comes from multiple sites whereas our knowledge of erosive margins largely comes from Costa Rica (Figure 4.4.3.1) (Ranero and von Huene, 2000). It is important to note that convergent margins likely transition between accretionary and erosive as the dynamics of plate interactions change (e.g., Simpson, 2010).

Von Huene and Scholl (1991) and Scholl and von Huene (2007) estimated that of the ~42,000 km of global subduction zone length, presently ~75% of convergent margins are nonaccreting. Clift and Vannucchi (2004) estimated that ~57% of the global convergent margins are erosive. Accretion preferentially occurs where convergence rates are slow (<6 cm/year) and/or trench sediments are thick (>1 km), whereas tectonic erosion is favored in regions with convergence rates >6 ± 1 cm/year (Clift & Vannucchi, 2004; Von Huene & Scholl, 1991). In general, at accreting margins, 60% of the sediments are being accreted and ~40% subducted (Von Huene & Scholl, 1991). In erosional margins, estimates of eroded hydrated forearc basement constitutes ~55% of the crustal material subducted to depth, the remainder being subducted sediments and basement (Von Huene & Lallemand, 1990; Von Huene & Scholl, 1991). Below the overriding plate, sediments will either be transferred to the overriding plate through basal accretion or be removed from the overriding plate through basal erosion.

In general, subducted sediments differ between accretionary and nonaccretionary margins. In accretionary margins the subducted lithology is mostly silty-sand—consisting of primarily terrigenous clastic deposits—whereas in nonaccretionary margins the predominant sediments subducted are clayey-pelagic and hemipelagic (Von Huene & Scholl, 1991). In addition, erosive margins incorporate material from the overriding plate. Both types of margins show a correlation between subducted sediment type and volcanic arc compositions (Plank & Langmuir, 1993, 1998), and there is strong evidence for sediment contribution to arc volcanism, with [10]Be isotopes providing the most profound evidence for this process (Morris & Tera, 1989; Tera, Brown, Morris, & Sacks, 1986).

The difference in sediment composition likely has important implications for the geochemistry and hydrology of these margins through differences in fluid production via dehydration as well as the permeability structure and draining characteristics of these margins. Accretionary margins are typically in compression and the majority of fluids are thought to drain through the décollement. In contrast, high-resolution seismic images at erosive margins show high-angle normal faults (Ranero, Phipps Morgan, Mcintosh, & Reichert, 2003) that may provide permeable pathways. This leads to the hypothesis that at erosive margins most of the fluids are draining through the prism (e.g., Ranero et al., 2003, 2008). For example, a large number of recent studies in the Middle America erosional margins have observed numerous active venting sites associated with high-angle normal faults that penetrate the prism (Hensen, Wallmann, Schmidt, Ranero, & Suess, 2004; Kluesner et al., 2013; Silver et al., 2000; and references therein).

Moore and Saffer (2001) suggested that accretionary margins might be expected to show higher rates of fluid and gas venting because of greater sediment flux. At fluid and gas venting sites at nonaccretionary margins, however, the relative contribution of deeply sourced fluids may be larger than at accretionary margins. This partitioning is likely because at erosive systems most of the compaction fluids drain along the décollement, whereas most of the dehydrating fluids may be migrating up through the prism along conduits. Erosive margins tend to have higher taper angles such that compaction may happen closer to the deformation front (e.g., Ranero et al., 2008). At both types of subduction margins fluid and gas venting sites are primarily associated with faults; folds are also important in directing fluids. Based on presently existing fluid and gas venting data at both accretionary and erosional subduction zones (e.g., Carson, Clarke, & Powell, 2000; Henry et al., 1996; Hensen et al., 2004; Kluesner et al., 2013; Le Pichon et al., 1992; Martin, Kastner, Henry, LePichon, & Lallement, 1996; Suess et al., 1998; Von Rad et al., 2000; and references therein), it is not yet possible to quantitatively test these ideas. Future testing of these hypotheses will illuminate the dynamics, differences, and consequences of fluid flow between the tectonic styles of margins.

4.4.3.3 GLOBAL ESTIMATES OF FLUID SOURCES AND INPUT FLUXES

The fluid inventory entering SZs is comprised of the pore water in void spaces, including fracture porosity and water bound in hydrous minerals, both in the sediments and in the oceanic basement. Subduction zone fluid input fluxes and thermal regimes are highly variable in time and space, which impacts fluid budgets and cycling. Table 4.4.3.2 summarizes estimated global inventories of H_2O inputs into SZs and provides the various assumptions behind the calculations. Sediment input is mostly known from ocean drilling data. Published syntheses of global sediments include Rea and Ruff (1996), Plank and Langmuir (1998), Jarrard (2003), and Spinelli and Underwood (2004). Hydration of oceanic lithosphere begins at the ridge crest and by the time oceanic crust begins to subduct, the percent hydration by alteration is estimated to be from

TABLE 4.4.3.2 Material and Water Input Fluxes

References	Sediment Input (10^{15} g/ year)	H_2O Input* (km^3/ year = 10^{15} g/ year)	Comments
Von Huene and Scholl (1991)	1.9 input total (~1.5 subducted 1.1 in accretion. Margins 0.4 in non-accretion.)	1.0	Assuming 400 m sediments on average, and the H_2O is from pore vol, not from dehydrat reactions or altered underthrust basement. 0.9 km^3 (90%) of the water is lost by compaction (0.1 km^3 from accreted and 0.8 from subducted sed).
Moore and Vrolijk (1992)		2.6	1.8 km^3 in pore V in incoming sed; 0.43 in hydrous minerals; 0.35 in ocean crust (top 1 km). Ocean volume recycled in ~500 my.
Rea and Ruff (1996)	1.43 (61%) (~0.54 subducted)	0.9 (39%)	1.08 Terrig (76%); 0.22 $CaCO_3$ (15%); 0.13 opal (9%) sediments.
McLennan (1988)	1.6–2.3		
Plank and Langmuir (1998)	1.8 ~0.73 subducted	0.86	The water is the H_2O lost at 100 °C. The sed flux is subducted sediment, assuming 50–500 m. (Global subd. Sed.: 78% terigenous; 7% $CaCO_3$; 10% opal; 7% mineral bound H_2O).
Wallmann (2001)		1.4–2.1	Range of H_2O in sediments. Pore water 1.1–1.8; total structurally bound H_2O in sed. (0.1) in upper crust (0.2), gabbro and peridotite.
Jarrard (2003)		1.83	0.76 in sed. Pore water, 0.13 in hydrated seds, 0.92 in igneous crust (0.6 structural and 0.32 pore water). Contribution from seds and crust~equal.
Rüpke et al. (2006)		0.9–1.8	Only ~20% in the forearc, the rest is subducted is subducted beyond the arc.

Continued

TABLE 4.4.3.2 Material and Water Input Fluxes—cont'd

References	Sediment Input (10^{15} g/ year)	H_2O Input[*] (km^3/ year = 10^{15} g/ year)	Comments
Hacker (2008)		2.4	From the 2.4×10^{15} g/year entering the trench 1.1 is in pore fluids, and 0.84 reaches postarc depth.
Kastner et al. (in this volume)		2.9–3.2	Based on Moore and Vrolijk (1992) sediment pore V and hydrated minerals input, and Jarrard (2003) and Hacker (2008) igneous ocean crust input (7 km), respectively. Ocean volume recycled in ~400 my.
Ocean water		1.37×10^{21} kg	
Global arcs		1–6×10^{14} g/year H_2O	Peacock (1990), Wallman (2001), Hilton (2002), Jarrard (2003)
MOR hydrothermal flux		$3.7 \pm 0.5 \times 10^{16}$ g/ year	Mottl (2003)

[*]In some references, the H_2O input only includes the sediment pore fluids that enter the subduction zones, some also include the H_2O in the sediment hydrous minerals, notably in opal and clay minerals. Several of the values also include H_2O in the subducting altered slab, as indicated under comments.

7–8% (Becker et al., 1990; Kastner et al., 1991) to 3.4% and 2% (Schmidt & Poli, 1998; Hacker, 2008) to 1.3% (Alt et al., 1996). Estimates of H_2O input fluxes vary between 0.86 and 1.0×10^{15} g/year (km^3/year) when only pore water and sediment hydrous minerals are considered; they vary between 1.4 and 2.1×10^{15} g/year in estimates that also include the hydrated oceanic basement (e.g., Wallmann, 2001).

We update these estimates by using the average global sediment input of solids estimated by Plank and Langmuir (1998), which are ~76% terrigenous, 7% $CaCO_3$, 10% opal, and 7% mineral-bound H_2O. We estimate that terrigenous sediments on average contain ~7.5% H_2O and Opal-A contains ~11% H_2O. For oceanic basement hydration we used the values by Jarrard (2003) and Hacker (2008) of 300 m upper volcanic, 300 m lower volcanic, 1.4 km dikes, and 5 km gabbro, for a total of 7 km. These values lead to an estimated H_2O input of $2.9\text{-}3.2 \times 10^{15}$ g/year such that the total ocean volume is recycled approximately every 400 my. Our values lead to a shorter ocean volume recycling time than estimated by Moore and Vrolijk (1992) and Wallmann (2001) because of different assumptions used in the thickness and hydration state of the oceanic basement (Table 4.4.3.2).

Most of this water input is recycled in the forearc; estimates of volatile outputs through the global arcs range from $1\text{-}6 \times 10^{14}$ g/year (e.g., Hilton et al., 2002; Jarrard, 2003; Peacock, 1990; Wallmann, 2001), which is ~3–20% of our estimated global input flux of H_2O that does not include serpentine. The residual fluid and chemically modified rocks are released in the back-arc or are transported to the mantle, as discussed above.

4.4.3.3.1 The Role of Serpentine Minerals in Volatile Cycling in the Forearc

Except for the Cascadia forearc, calculated temperatures of the uppermost forearc mantle of 150–250 °C were reported by Hyndman and Peacock (2003); in this thermal regime serpentine minerals are stable; therefore, they are not involved in the fluid and volatile cycling budgets in forearcs considered in this paper. However, we describe the nature and significance of this process in Appendix 1.

4.4.3.4 FOREARC THERMAL REGIMES

Temperature plays a fundamental role in the mechanical evolution of sediments, and the fluid cycle, through its influence on the nature, depth, and kinetics of dehydration reactions (Table 4.4.3.3). After sediment compaction, dehydration reactions provide the largest source of water. Of the sediment entering SZs dehydration of smectite and opal-A are the most important minerals for fluid flux considerations because they contain the largest amounts of bound water (Table 4.4.3.3). In most SZs smectite is more abundant than opal-A; smectite often comprises 40 and up to 50 wt% of the sediment section (Underwood, 2007). At temperatures greater than about 30 °C opal-A dehydration may become noticeable (Kastner, 1981); however, the most important temperature range of smectite and opal-A dehydration is between 60 and 150 °C. Dehydration reactions continue at 150 to >300 °C (Table 4.4.3.3). Of the dehydration reactions, the smectite dehydration and transformation to illite reactions provide the largest fluid source, mostly between ~60 and 150 °C (Bekins et al., 1995, 1994; Moore & Vrolijk, 1992; Saffer, Underwood, & Mckiernan, 2008; Spinelli & Saffer, 2004). Diagenesis and metamorphism thus affect the hydrological behavior and the chemical and physical properties of SZs. Temperature is also the main control on gas hydrate formation, stability, and dissociation in SZ sediments (Sloan & Koh, 2008, 721 pp.).

TABLE 4.4.3.3 Main Fluid-Rock Reactions Impacting Volatile Cycling in SZ Forearcs

Reaction	Comments	References
Opal-A → Opal-CT & Opal-CT → Quartz	The transformation to opal-CT begins at 10–20°C; completion T is not well documented. The reaction to quartz is mostly complete at ~85–100°C; it is a time-T dependent reaction kinetics. Opal A (diatoms and radiolarians) has ~11% structural H_2O, the water released when transformed to quartz comprises ~23% of the volume. Except at hot SZs, the reactions are kinetically controlled, and the activation energies are low.	Ernst and Calvert (1969) Kastner (1981) Pytte and Reynolds (1988) Moore and Vrolijk (1992) Behl and Garrison (1994) Murata, Friedman, and Gleason (1977)
Opal-A → Quartz	The direct transformation to quartz may occur in SZ sediments. Having a low abundance of opal-A, at >10–20°C, and is kinetically controlled by the dissolution rate of Opal-A.	Kastner, Keene, and Gieskes (1977) Kastner (1981)
Pressure solution and quartz cementation	In silicoclastic sediments pressure solution of quartz and quartz cementation occurs at >100–150°C, depending on the pressure.	Fisher and Byrne (1992)
Smectite (S) dehydration and smectite to illite (I)	Smectite in the marine environment contains $2H_2O$ interlayers, ~20 wt% H_2O. The dehydration and transformation reactions are Thermally driven, and slowed-down by P. The S → I transformation does not impact their structural (OH)– that begins to be lost at >350°C but mostly at 500–650°C, therefore not important in the forearc. Most loss of the S interlayer water during the transformation to I occurs at 60–150°C, it continues to 250°C. The transformation occurs stepwise; one step is associated with the formation of an ordered S/I phase (rectorite). The S → I transformation reaction is kinetically controlled. Smectite may just dehydrate and not transform to illite if the K and Al required for illite formation are not available. Pure (100%) smectite is not common, it forms by volcanic glass diagenesis. The majority of detrital smectite that enters SZs is 70–75% mixed-layer S/I.	Ernst and Calvert (1969) Perry and Hower (1970, 1972) Colten-Bradley (1987) Pytte and Reynolds (1988) Bekins et al. (1994) Spinelli and Saffer (2004) Hower, Eslinger, and Hower (1976)

Zeolites formation	Zeolitization of detrital immature sediments occurs mostly at 50–275 °C and some of the zeolites occur as cements. The zeolite facies is also associated with carbonate and chlorite. The most common zeolites at the lower temperature end, at ≤100–250 °C are heulandite, analcite, and laumontite, depending on the chemistry of the sediment. Zeolite diagenesis consumes water that is returned to the system upon their dehydration and transformation to anhydrous silicates (i.e. analcite to albite). The reactions are rather often kinetically controlled.	Coombs and Whetten (1967) Ernst (1990) Bish and Ming Eds (2001)
Oceanic basement hydration and dehydration	Estimated % hydration by alteration of oceanic basement when it enters SZs ranges from 7–8 to 3.4%, 2% and to 1.3 wt%. Hydration of the basement begins at the ridge crest and continues on the ridge flank. Low temperature; alteration in the shallow basement consists mostly of celadonite and Fe-oxy-hydroxides, alteration to saponite follows with some chlorite layers at >100 °C. For dehydration T of saponite, a Mg smectite, and celadonite, a type of illite, see above discussion on S and I. Late phases in fractures and veins are carbonates, and the zeolites heulandite and analcite. Palagonoitizaton of volcanic glass = hydration and alteration to smectite + phillipsite + Fe-oxyhydroxides is widespread in the shallow part (also in volcanic ash in the sediments) at <60 °C. Dehydration of greenschist facies hydrous silicates in the altered basement dehydrate at ≥400 °C, thus volatile contribution to the forearc is insignificant. For example, chlorite is stable in the forearc and persists to 680 °C at 4.2 GPa.	Becker et al. (1990) Kastner et al. (1991) Peacock (1993) Schmidt and Polli (1995) Alt et al. (1996), Alt and Honnorez (1984) Kerrick and Connolly (2001) Jarrard (2003) Hacker (2008)
Decarbonation	Does not occur in the forearc; the Mariana forearc may be an exception.	
Hydrocarbons maturation	Hydrocarbon maturation begins at ~60–70 °C and peaks at ~100–150 °C the maturation involves fluid formation. Although some volatiles, in particular CO_2 and CH_4 are produced at lower-temperatures organic matter remineralization, they are not considered. $>C_3$ hyddrocarbons in pore fluids are important indicators of fluid flow from greater depths.	Claypool and Kaplan (1974) Selley (1998, 470 pp.) Gieskes and Lawrence (1981) Silver et al., 2000 Kastner et al., 2006

The overall thermal structure of SZs is sensitive to the geotherm of the incoming plate, the convergence rate, the subduction geometry, thermal physical rock properties of the forearc, and frictional heating along the plate interface (e.g., Dumitru, 1991; Harris, Spinelli, et al., 2010; Hesseman, Grevemeyer, & Villinger, 2009; Hyndman & Wang, 1993; McCaffrey, 1993, 1997; Molnar & England, 1995). Generally, the convergence rate is well known from plate motion studies and the gross margin geometry and slab dip are estimated from active and passive seismic studies. Core samples and downhole logs from IODP drilling are an important source of information about the thermal physical properties of the rocks making up the forearc. However, the thermal structure of the incoming plate, particularly in the presence of hydrothermal circulation in the upper oceanic crust is less well known.

As a first-order approximation, the thermal structure of the incoming plate can be estimated from conductive plate cooling models (e.g., Harris & Chapman, 2004; Stein & Stein, 1992, 1994). For crustal ages less than about 65 Ma, global heat flow averages are less than conductive values, because hydrothermal circulation extracts heat from the incoming plate. IODP and its predecessors ODP and DSDP have typically drilled convergent margins where subducting oceanic crust is younger than 65 Ma (Table 4.4.3.1), with Barbados being the sole exception. In the presence of hydrothermal circulation, these conductive models may overestimate temperatures as a function of depth and underestimate their variability. Understanding hydrothermal circulation within the oceanic crust under the forearc is important because of its ability to continue altering and hydrating oceanic crust, its ability to influence the position of dehydration reactions, and its ability to transfer heat and solutes back to the ocean. Hydrothermal circulation within subducting oceanic crust may be best exemplified at Nankai and Costa Rica where the young plate and variably thin sediment cover promotes advective heat exchange between the crust and overlying ocean. Below, we briefly review the evidence for hydrothermal circulation at Nankai and Costa Rica.

Heat flow measurements along the Nankai Trough are relatively abundant. Features discussed in previous compilations (Kinoshita & Yamano, 1986; Yamano, Honda, & Uyeda, 1984; Yamano, Kinoshita, Goto, & Matsubayashi, 2003) that indicate the presence of hydrothermal circulation with the subducting crustal aquifer include: (1) relatively large variability between closely spaced heat flow measurements, (2) heat flow higher than conductive predictions within or just seaward of the Nankai Trough, and (3) a large and steep decrease in heat flow just landward of the deformation front. These features are best expressed in the Muroto transect where heat flow declines from values of approximately $210\,mW/m^2$ near the deformation front to values of approximately $60\,mW/m^2$ over a distance of 65 km and are fit by models that include hydrothermal circulation (Harris et al.,

2013; Spinelli & Wang, 2008). The steep landward descent in heat flow is less clear along the Kumano Basin and Ashizuri transect where heat flow values decline from $140\,mW/m^2$ to values of approximately $60\,mW/m^2$ over a distance of $50\,km$.

Offshore the Nicoya Peninsula of Costa Rica, in the region of ODP Legs 170 and 205, heat flow values are generally very low. In this region conductive cooling models predict values of approximately $100\,mW/m^2$ (e.g., Stein & Stein, 1992). Instead very low values near $10-20\,mW/m^2$ are documented near the prism toe and curiously increase landward to slightly higher values of $\sim30\,mW/m^2$ (Langseth & Silver, 1996). The low values and general pattern were confirmed with detailed downhole temperature measurements ODP drilling of Leg 170 (Ruppel & Kinoshita, 2000). Subsequently, the region of very low heat flow was found to extend well seaward of the trench and indicate the efficiency of hydrothermal circulation in cooling oceanic crust (Fisher et al., 2003; Hutnak et al., 2007, 2008). Although Bottom Simulating Reflectors (BSRs) are widespread along the margin, they are absent in the Leg 170 drilling area, because the cool thermal conditions put the base of gas hydrate stability in the oceanic crust (Ruppel & Kinoshita, 2000).

Normally, heat flow decreases landward from the deformation front, consistent with the subducting plate advecting heat back towards the Earth's interior. The heat flow pattern imaged by Langseth & Silver (1996) was most recently interpreted as being due to continuing hydrothermal circulation in the upper oceanic crust and the geotherm relaxing from one that reflected open communication with the ocean to a more restricted flow pattern under the forearc (Harris, Spinelli, et al., 2010). It is interesting to note that this same pattern of heat flow, with anomalously low values at the prism toe that then increase slightly landward, was detected by Yamano and Uyeda (1990) at the Peru margin in support of Leg 112. Similar to Costa Rica, the Peru margin is erosive with relatively thin incoming sediment.

South of the Nicoya Peninsula, on Cocos-Nazca Spreading Center (CNS) generated crust, regional values have a mean that is not only consistent with conductive cooling model but also show substantial variation about the mean (Harris, Spinelli, et al., 2010). These values imply ongoing hydrothermal circulation in this portion of the margin that is much less efficient at removing heat from the crust (Fisher et al., 2003; Harris, Spinelli, et al., 2010). The transition between these two regions appears to be very sharp, $<\sim5\,km$ (Fisher et al., 2003) and may reflect differing styles of hydrothermal circulation between East Pacific Rise (EPR) and CNS crust. Heat flow transects in the region of Expeditions 334 and 344 where CNS crust is subducting show anomalously high heat flow at the toe and a rapid landward decline (Harris, Grevemeyer, et al., 2010). This heat flow pattern is consistent with the rapid advection of heat along the upper basement aquifer from deep within the forearc to the trench (Harris, Spinelli, et al., 2010).

Temperatures along the plate boundary are shown in Figure 4.4.3.3. Accretionary plate boundaries are warmer at the deformation front than erosive plate boundaries. The influence of the thick sediment section at the Cascadia deformation front is apparent and represents the warm end member of these margins. The high temperatures associated with Cascadia imply that many sediment dehydration reactions have occurred by the time the sediments get to the deformation front. In contrast, Barbados is associated with the oldest and therefore coolest oceanic crust and represents the cold end member of the margins. Although the age of the oceanic crust at the Nankai Trough and Japan are similar (Table 4.4.3.1), differences in the plate boundary temperatures can be ascribed to convergence rate and plate geometry. The fast rise in temperature with distance from the trench at Costa Rica is due in part to the larger taper angle.

Figure 4.4.3.4(A) shows a comparison of fluid sources downdip of the trench due to compaction and illitization along the Muroto and Ashizuri transects (Saffer et al., 2008). Sediment thickness along both transects is variable but generally similar (Ike et al., 2008). The higher heat flow and plate boundary temperatures along the Muroto transect leads to dehydration reactions seaward of the trench, whereas the lower heat flow along the Ashizuri transect leads to negligible dehydration seaward of the trench (Saffer

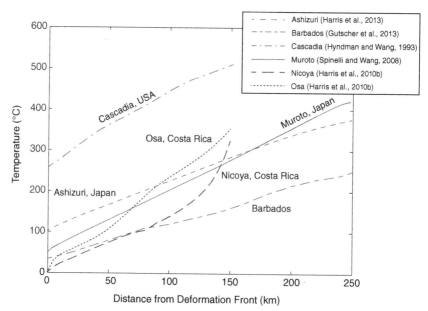

FIGURE 4.4.3.3 Modeled temperature along the plate boundary. Margins that are generally considered accretionary and erosive are plotted in red (light gray in print versions) and blue (dark gray in print versions), respectively.

et al., 2008). The difference in thermal regimes leads to greater dehydration fluid production along the Ashizuri transect compared to the Muroto transect. The difference between estimates using a heat flow value of 90 and 120 mW/m² highlights the sensitivity of peak dehydration to heat flow.

The influence of the regional change in thermal regimes across the plate suture led Spinelli, Saffer, and Underwood (2006) to investigate the differences in position of dehydration reactions (Figure 4.4.3.4(B)). This thermal model is based on the algorithm of Ferguson (1990, 155 pp.), a transient 1-D model that accounts for heat advection and conduction during burial of subducted sediment. These models do not account for ongoing hydrothermal circulation in the subducting oceanic crust and lead to slightly higher temperatures along the subduction thrust than estimated by Harris, Spinelli, et al. (2010). The progress from opal to quartz and smectitie to illite is based on kinetic expressions of Ernst and Calvert (1969) and Pytte and Reynolds (1988). Figure 4.4.3.4(B) contrasts the warm CNS crust with the cool EPR and shows that these diagenetic

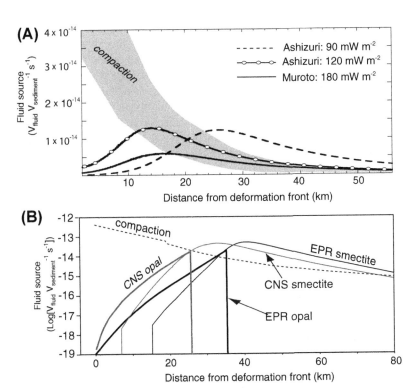

FIGURE 4.4.3.4 Simulated fluid release within the underthrust sediment by smectite transformation, and compaction, reported in units of $V_{fluid}/V_{sediment}$/time for (A) models along the Muroto and Ashizuri transects *(modified from Saffer et al. (2008))* and (B) grey-shaded area in (A) shows the range of fluid sources from sediment compaction *(after Saffer and Bekins (1998))*.

reactions complete at shallower depths for CNS crust than for EPR crust. Importantly, this figure also shows that within 25 km of the trench, compaction-driven fluid flow dominates whereas at distances greater than 25 km from the trench fluids from dehydration reactions dominate. Dehydrating fluid sources peak at ~25 and 37 km from the trench and depths of 7 and 9 km on CNS and EPR crust, respectively. Observed variations in the updip limit of seismicity (Newman et al., 2002) are consistent with these variations in dehydrating sources (Spinelli et al., 2006).

4.4.3.5 FLUID OUTPUTS, FLOW RATES AND FLUXES

4.4.3.5.1 Fluid Geochemistry

The chemistry and isotopic composition of the pore and seep fluids transported from the forearc back to the seafloor are distinct from seawater composition. During initial compaction and pore fluid expulsion from sediment with porosities of ~70–75%, the impact of fluid-rock reactions on the chemistry is relatively small. The expelled pore fluid is impacted by low-temperature diagenetic reactions, primarily adsorption/desorption and ion exchange, volcanic ash alteration, and organic matter diagenesis. The slightly modified seawater in the sediment pore space becomes chemically more exotic with increasing depth and increased control of fluid-rock reactions on the pore fluid composition. Table 4.4.3.4 summarizes the fluid composition as impacted by the most common fluid-rock reactions in SZs. Quantitatively, the most important early diagenetic reactions (Table 4.4.3.3) are the smectite to illite (S/I) and Opal-A to CT to quartz transformations, both involve dehydration, thus producing low Cl fluids. Water is liberated stepwise by these two reactions that are most intense at 50–150 °C, but continue to ~250 °C (Table 4.4.3.3). Based on the chemistry of the fluids sampled in flow horizons within the forearc, temperatures up to 120–150 °C have often been reported (e.g., Chan, Alt, & Teagle, 2002; Kastner, Solomon, Wei, Chan, & Saether, 2006; Elderfield et al., 1990; Martin et al., 1996; Solomon et al., 2009; Spivack et al., 2002; Teichert et al., 2005; Torres, Teichert, Trehu, Borowski, & Tomaru, 2004; Tryon, Wheat, & Hilton, 2010; and references therein). This temperature regime commonly reflects the temperature near the updip limit of the seismogenic zone (e.g., Hyndman & Wang, 1993, Figures 4.4.3.2–4.4.3.5).

These diagenetic reactions also impact several important elements that are either released into the fluid or consumed by the solids; in the smectite to illite reaction the important elements consist of B, K, Li, Rb, Cs, Ba, and Si. In the opal phases B is relatively abundant (Furst, 1981; Kolodny, Taraboulos, & Frieslander, 1980). Thus, the most commonly reported and revealing characteristics about the source of these fluids are: the considerable freshening (best indicated by the Cl concentration), low K concentrations; elevated Si, B, and Li concentrations; presence of thermogenic hydrocarbons; and O

TABLE 4.4.3.4 Pore Fluid Composition as Impacted by Fluid-Rock Reactions[a]

Reaction	Chemical tracers consumed	Chemical tracers released	Temperature	Comments
1. Volcanic ash dissolution		[Ca] [Na] [Al] [Si] [Sr] [Ba] [Li] Nonradiogenic $^{87}Sr/^{86}Sr$	Reaction proceeds at all temperatures. Rate is temperature dependent.	In some instances may induce Si cementation, which will affect porosity and shear strength.
2. Volcanic ash alteration and smectite formation	[Mg] [Si] [Li] [B] $^{6}Li, ^{10}B$ $^{18}O, ^{1}H$ [H_2O]	[Ca] [Sr] [Al] Non-radiogenic $^{87}Sr/^{86}Sr$	Reaction proceeds at all temperatures. Rate is temperature dependent.	Impacts overall smectite budget. All solute concentrations increase in proportion to H_2O uptake, e.g., [Cl].
3. Palagonitization of volcanic ash (= hydration+alteration to mostly smectite, phillipsite, and Fe-oxyhydroxides)	[K] [Mg] [Si] [Li] [B] [Rb] [Cs] $^{18}O, ^{1}H$ $^{6}Li, ^{10}B$ [H_2O]	[Ca] [Al] [Sr] Nonradiogenic $^{87}Sr/^{86}Sr$	Reaction mostly at low temperatures ~2–60°C.	Fe^{3+}/Fe^{2+} ratio increases; all solute concentrations increase in proportion to H_2O uptake, e.g., [Cl].
4. Volcanic ash alteration and zeolite formation	[K] [Na] [Si] at lower temp. [Ca] [Al] at higher temp. Possibly [Cl] and ^{37}Cl ^{18}O ^{1}H [H_2O]	A range of trace elements depending on volcanic ash composition	<80°C, K and Na zeolites >80°C, Ca zeolites	Zeolite cementation likely impacts sediment mechanical properties; all solute concentrations increase in proportion to H_2O uptake, e.g., [Cl].
5. Smectite ion exchange and desorption reactions	Depending on the smectite composition, NH_4	Mostly [Mg] [B] [Li] possibly [K] [Sr] [Rb] [Cs] $^{6}Li, ^{10}B$	Each element likely has a characteristic behavior with respect to partitioning into the fluid with temperature.	

Continued

TABLE 4.4.3.4 Pore Fluid Composition as Impacted by Fluid-Rock Reactions[a]—cont'd

Reaction	Chemical tracers consumed	Chemical tracers released	Temperature	Comments
6. Smectite To illite transformation	[K] [Al] [Cl] [B] [10]B [1]H	[Si] [Mg] [18]O [H$_2$O]	~60–150°C continues to ~250°C	Dehydration, pore pressure increase, induce fluid flow; all solutes will be diluted in proportion to H$_2$O released, e.g., [Cl]; some quartz may precipitate from the Si released, as cement it may change the frictional properties of the rock. The impact on the O and H isotopes depends on the system pore fluid/interlayer water ratio.
7. Hydrous Silicates formation	[Si] possibly [Al] and a range of major and minor elements depending on the mineral [Cl] [37]Cl [18]O [1]H [H$_2$O]		≥200–250°C	All solute concentrations increase in proportion to H$_2$O uptake, e.g., [Cl].
8. Opal A to CT Opal CT to quartz	[Si] [Cl] [10]B [37]Cl [H$_2$O]		20–100°C Mostly complete at >80–100°C	Dehydration, pore pressure increase, induce fluid flow; all solutes will be diluted in proportion to H$_2$O released, e.g., [Cl].

9. Pressure Solution and quartz cementation	[Si]	[K] [Al] [Mg] and other solutes depending on minerals dissolved	Mostly at $\geq$150 °C	Elements released depend on the minerals involved; quartz cementation changes the frictional properties of the rock, leading to a velocity-weakening behavior.
10. Plagioclase albitization	[Na] [Si] ^{18}O	[Ca] [Al]	At >75–100 °C	Carbonate cement may form and impact sediment physical properties.
11. Carbonate diagenesis	[Ca] and sometimes [Mg] [Mn] [Fe]	[Sr], ^{87}Sr/^{86}Sr ^{18}O depending on the temp.	Cement and veining >100 °C	Changes sediment physical properties, velocity-weakening.
12. Thermogenic hydrocarbon alteration		C_3 and higher hydrocarbons	>80 °C Peaks at 100 °C	CH_4 and other gases formed inducing high pore pressure.

[a]Based on too many references to be cited in this table.

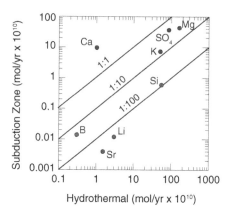

FIGURE 4.4.3.5 Fluxes of key solute concentrations (Ca, Mg, Sr, K, B, Li, Si, and SO_4) in subduction zones versus in ridge-crest hydrothermal systems in mol/year $\times 10^{10}$. Blue (dark gray in print versions) indicates the solute is consumed from seawater in subduction zones and red (light gray in print versions) indicates the solute is released. Some of these solutes are partitioned into the solid phase at low temperatures and released to the fluid at moderate to high temperatures. To account for this behavior, the output fluxes of B, Li, and Sr are based only on the dehydration H_2O flux, and the estimate does not consider uptake of these elements at lower temperatures. While Ca can be released at higher temperatures, it is consumed through authigenic carbonate formation. The flux of Ca is the sink flux considering the total water flux in the forearc (compaction + dehydration + other sources). The fluxes of K, Mg, and SO_4 were computed as for Ca. Ridge crest fluxes are from Elderfield & Schultz (1996).

and D isotopic values that suggest temperatures at the source of ~80–150 °C (Chan and Kastner, 2000; Henry et al., 1996; Hensen et al., 2004; Kopf et al., 2003; Martin et al., 1996; Moore & Vrolijk, 1992; Sahling et al., 2008; Silver et al., 2000). The inverse relations between O and D isotopes (Sheppard and Gilg, 1996) have been used by some to indicate the temperature at the source (e.g., Hensen et al., 2004, and references therein). These characteristics indicate that the fluids are deeply sourced and are related to dehydration reactions. A detailed discussion of the characteristics of these geochemical tracers is provided in Appendix 3.

Other tracers of deep-sourced fluids include distinctive patterns of alkali metal enrichment and characteristic Li, B, and Cl isotope ratios (Table 4.4.3.4). The alkali metals are preferentially released into the fluid phase at temperatures ≥30–40 °C; the exact threshold temperature for the release into the fluid phase is poorly constrained but differs for each alkali metal (Wei, Kastner, Rosenbauer, Chan, & Weinstein, 2010). Furthermore, Li isotope ratios together with concentrations provide important insights on the hydrology and approximate fluid-rock reaction temperatures at the source in subduction zones. Clay minerals (especially those rich in Mg) and also zeolite diagenesis reactions preferentially fractionate ^6Li from low to greenschist facies temperatures, enriching the fluid in ^7Li. The reactivity of Li and fractionation factor of $^7Li/^6Li$ has been studied in laboratory experiments at elevated temperatures of 100–250 °C (e.g. Benton, Ryan, & Savov, 2004; Williams & Hervig, 2005; and references therein). The chemical and isotopic characteristics described here are observed more distinctly at arcward sites, apparently because there is less dilution with in situ pore fluids during transport to the seafloor. Specific chemical signatures, such as the $\delta^{37}Cl$ value of the fluids (Ransom, Spivack, & Kastner, 1995; Spivack et al., 2002; Wei, Kastner, & Spivack, 2008), indicate that in some instances the fluids originate even deeper, where high-temperature (250–400 °C) hydration reactions such as chloritization or talc formation occur (e.g., Bebout, 1996; Hacker, 2008; Peacock, 1990; Peacock & Hyndman, 1999; and references therein).

As the arc and subarc dewatering continues, the slab's higher-temperature hydrous minerals dehydrate and transform to anhydrous minerals, and deserpentinization must occur. As in the forearc, the kinetics of these reactions mostly depends on the thermal regime of the SZ. Various aspects of this overall sequence of events have been described preiously (e.g., Bebout, 1996; Bekins, McCaffrey, & Dreiss, 1994; Moore & Vrolijk, 1992; Saffer & Tobin, 2011; Spinelli & Saffer, 2004; Vrolijk et al., 1991; and references therein).

4.4.3.5.2 Flow Pathways

The fluids that originate from compaction and thermally induced dehydration reactions must exit the system. The most profound and direct evidence for large-scale fluid flow includes widespread channelized fluid venting at the forearc seafloor. These fluids sustain prolific benthic biological communities that form extensive seafloor carbonate crusts and cements (Carson, Suess, & Strasser, 1990; Füri et al., 2010; Henry et al., 1996; Hensen et al., 2004; Kleusner et al., 2013; Kopf, 2002; Kulm et al., 1986; Labaume et al., 1997; Le Pichon et al., (1992); Martin et al., 1996; Sahling et al., 2008; Tryon et al., 2010; and references therein). Geochemical evidence for extensive lateral and vertical fluid flow is well established, for example, by the low Cl concentrations and $\delta^{37}Cl$ anomalies in the pore fluids (e.g., Ransom et al., 1995; Spivack et al., 2002; Wei et al., 2008) and the pervasive observations of mature hydrocarbons ($>C_3$) along faults. These fluids transport dissolved components from depth into the ocean and may be important in global chemical and isotopic budgets.

Compaction dewaters the subducting sediment rather continuously; it is closely followed by sediment hydrous mineral dehydration reactions that occur stepwise, depending on the mineralogy of the hydrous phase. With decreasing porosity and permeability, dehydration reactions increase the pore fluid pressure that becomes more significant the greater the distance from the deformation front and with increasing temperatures and pressures (e.g., Bekins et al., 1994; Saffer & Tobin, 2011; Spinelli & Saffer, 2004). The focused fluid flow that evolves is transient and associated with changes in fluid pressure and fluid flow rates along the décollement as documented by data from in situ monitoring of fluid pressure and flow rates on the seafloor and the décollement in boreholes (Becker et al., 1990; Bekins et al., 1995; Brown, Tryon, Deshon, Dorman, & Schwartz, 2005; Davis & Villinger, 2006; Davis, Heesemann, & Wang, 2011; Solomon et al., 2009).

The dehydration reactions that mostly follow compaction provide a large source of water. Of the sediment entering SZs dehydration of smectite and opal-A are the most important minerals for fluid flux considerations because they contain the largest amounts of bound water (Table 4.4.3.3). In most SZs smectite is more abundant than opal-A; smectite often comprises 40 and up to 50 wt% of the sediment section (Underwood, 2007). At temperatures greater than about 30 °C opal-A dehydration may become noticeable (Kastner 1981); however, the most important temperature range of smectite and opal-A dehydration is between 60 and 150 °C. Dehydration reactions continue at 150 to >300 °C (Table 4.4.3.3). Of the dehydration reactions, the smectite dehydration and transformation to illite reactions provide the largest fluid source, mostly between ~60 and 150 °C (Bekins et al., 1995, 1994; Moore & Vrolijk, 1992; Saffer et al., 2008; Spinelli & Saffer, 2004).

The most important features of fluid flow in the forearc include seafloor venting through mud volcanoes and other seafloor bathymetric features that often sustain chemosynthetic communities; distinct pore fluid chemical anomalies; and diagenetic mineral precipitation as crusts, cements, or as veins. Most of these features are associated with faults. Thermal anomalies centered at faults are well documented (e.g., Carson et al., 2003) and suggest that the focused flow through décollement and other faults, including out of sequence and splay faults, are the most important pathways in forearc fluid recycling (Moore & Vrolijk, 1992). Seeps, mud volcanoes, and mounds that are also important in fluid discharge are mostly associated with faults (Lauer & Saffer, 2012; Ranero et al., 2008; Sahling et al., 2008).

Differences in tectonic style have led to speculation that accretionary and erosional margins differ in their geochemistry and hydrology. Accretionary margins are typically in compression and the majority of fluids are thought to drain through the décollement. Modeling and geochemical data documented the important role the décollement plays in transporting fluids from depth to the trench (Kastner et al., 1991; Henry 2000; Silver et al., 2000; Spinelli et al., 2006) and indicated that 60–70% of the fluid flux is through the décollement (Screaton et al., 1990; Saffer and Bekins, 1998). However, as discussed above, at nonaccretionary margins, the relative contribution of deeply sourced fluids may be larger than at accretionary margins and most fluid migrates to the seafloor through high-angle normal faults and fractures in the upper plate rather than through the décollement. Lauer and Saffer (2012) estimated the fluid flux to the ocean through permeable splay faults (10^{-16} to $10^{-13}\,m^2$) to be 6–35% of the total dewatering flux at Costa Rica. Additional important fluid flow pathways in both erosional and accretionary margins are high permeability sediments, i.e., basal sands of turbidites or volcanic ash layers. Significant lateral flow must exist along stratigraphic horizons, as indicated, for example in Costa Rica, where the pore fluid Cl and concentration profile in the underthrust section is decoupled from the décollement (Saffer et al., 2000; Solomon & Kastner, 2012).

Based on presently existing reports on fluid and gas venting at both accretionary and erosional subduction zones (e.g., Carson et al., 2000; Henry et al., 1996; Hensen et al., 2004; Kluesner et al., 2013; Le Pichon et al., 1992; Martin et al., 1996; Suess et al., 1998; Von Rad et al., 2000; and references therein) as yet it is not possible to conclusively state if the extent of venting differs between them.

Recently, heat flow and geochemical data at both Nankai and Costa Rica have emphasized the role the igneous oceanic crust plays as a permeable conduit (Harris, Spinelli, et al., 2010, Harris et al., 2013; Solomon et al., 2009; Spinelli & Wang, 2008). Quantifying processes associated with these permeable conduits and fluxes is important.

Key questions thus are: what is the fluid flux to the ocean/atmosphere in the forearc and what is the impact of these fluids on seawater chemistry?

4.4.3.5.3 Flow Rates

As discussed above, fluid flow along the décollement and major fault zones in the prism has been well documented in SZs worldwide (e.g., Bekins et al., 1995; Carson et al., 1990; Grevemeyer et al., 2004; Henry, 2000; Hensen et al., 2004; Kastner et al., 1991; Le Pichon et al., 1991, 1992; Moore et al., 1990; Ranero et al., 2008; Sahling et al., 2008; Saffer, 2003; Solomon & Kastner, 2012; Solomon et al., 2009). However, the role of SZ forearcs in global geochemical cycles, and in particular the magnitude of fluid egress and chemical fluxes to the ocean are still poorly constrained.

Focused fluid flux surveys in the late 1980s to early 1990s directly and indirectly measured very high fluid flow rates at seafloor seeps near the toe of the Barbados, Cascadia, and Nankai subduction zones (Table 4.4.3.5). Flow rates measured near the deformation front at these subduction zones ranged from 1–150 m/year (Carson et al., 1990; Davis, Hyndman, & Villinger, 1990; Foucher et al., 1990; Le Pichon et al., 1992, 1990). Based on the number of seafloor seeps in each of these margins, the output flux of fluid from only the toe of the accretionary prism was estimated to be $200\,m^3/$ year/m along strike at Nankai (Le Pichon et al., 1990, 1992), $410–490\,m^3/$ year/m along strike at Cascadia (Carson et al., 1990; Davis et al., 1990), and $\sim400\,m^3/year/m$ along strike at Barbados (Foucher et al., 1990; Fisher and Housnlow, 1990). These fluxes are extraordinary considering that the amount of fluid that could be produced in the accretionary prism through local steady-state compaction dewatering ranges from $\sim5–30\,m^3/year/m$

TABLE 4.4.3.5 Measured and Modeled Fluid Output Fluxes at Individual Subduction Zones

Location	Flow Rate Range	# of Seeps	Individual Seep Area (m²)	Discharge (m³/year)	Comments	References
Eastern Nankai Trough	70–150 m/year	445 clam colonies covering a surface area of 94 m²	ND	200 m³/year/m of margin	Minimum estimate based on toe area deeper than 2000 m. total sediment pore water input of 18 m³/year/m of margin. Local steady-state dewatering of the accretionary wedge produces 5 m³/year/m of margin. The output flux is 40 times the amount produced by local steady-state compaction dewatering. Suggests shallow seawater circulation at seep sites and/or transient fluid flow.	Le Pichon et al. (1991, 1992, 1993)
Cascadia	ND	10–43 over 10 km long survey, ~1–4 seeps/km	ND	ND	Survey only at 1st anticlinal ridge. Seepage indicators are clam shells, tubeworms, carbonates, and bacterial mats.	Moore et al. (1990)
Cascadia	100 m/year	4 seeps over 1.8 km, ~2 seeps/km	9 m²	125–150 l/ m²d, assuming 9 m² seep area 410–493 m³/ year	Based on flux chambers. The expected steady-state flux from compaction in the prism would only be 6–15 m³/year/m. The measured flux is many orders of magnitude greater than fluid that produced through local compaction. Either non-steady-state flow or convection of seawater.	Carson et al. (1990), Davis et al. (1990)

Cascadia	10–250 cm/year (bacterial mats), average of 85 cm/year <10 cm/year out of clam beds	ND	Bacterial mats cover area of 20 m^2	Assuming average flow rate and area of bacterial mats 17 m^3/year at S. Hydrate ridge	These estimates are based on models of pore fluid profiles from short cores, benthic flux chambers, and CAT flow meters.	Torres et al. (2002), Tryon et al. (2002)
Barbados	~2 cm/year	ND	ND	30 m^3/year/m of margin	Based on heat flow at 1st and 2nd anticlinal ridges. Max water from compaction is 30 m^3/year. Suggests all fluid entering must exit near deformation front.	Foucher et al. (1990)
Barbados (ODP Sites 671/672)	10 m/year	ND	ND	400 m^3/year/m of margin	Based on heat flow. Requires a sediment hydraulic conductivity of 10^{-6} to 10^{-7} m/s. Output flux greatly exceeds input flux of ~9–18 m^3/year. Potentially from short-lived non-steady-state flow or seawater circulation.	Foucher et al. (1990), Fisher and Hounslow (1990), Fisher et al. (2003)
Barbados, 12 km seaward of the deformation front	17 m/year	1	0.55 km^2	10^6 m^3/year	Estimate based on thermal and chemical considerations. Amount of water produced is equivalent to steady state compaction from the entire sediment section over 100 km of the margin. Suggests transient flow or seawater convection. Fluid has low Cl and Ca concentrations and high CH$_4$ concentrations. Chemistry suggests 75–115 °C at source of fluid.	Le Pichon et al. (1990), Martin et al. (1996), Henry et al. (1996)

Continued

TABLE 4.4.3.5 Measured and Modeled Fluid Output Fluxes at Individual Subduction Zones—cont'd

Location	Flow Rate Range	# of Seeps	Individual Seep Area (m²)	Discharge (m³/year)	Comments	References
Costa Rica	0.3–1.5 cm/year (background) up to 300 cm/year (focused)	ND	7,900–785,000 m², focused seepage in area of 25 m²	Up to 75 m³/year (focused) 100 m³/year background	They estimate 70 m³/year per 100 m margin of H$_2$O in clays subducted per year. Focused from 5 × 5 m area of one mound equals the total dehydration flux from a distance of 100 m. even the background flow rate is high enough to account for all of the water. Fluid is low in Cl and high in CH$_4$ (thermogenic) and B concentrations. δ^{18}O/δD inversely correlated suggesting T at origin of 85–135 °C. Suggests mixing between seawater and low-Cl fluid.	Hesen et al. (2004)
Costa Rica	ND	161 over 11 × 55 km area, ~14 seeps/km	ND	ND	Highest resolution geophysical survey. They observe 77 indicators of seepage on outer shelf, 19 along upper slope, 49 in the mid-slope region, and 16 along the lower slope.	Kluesner et al. (2013)
Costa Rica	ND	124, ~0.25 seep/km margin	Mounds are up to 1 km in diameter	ND	Their study shows almost all of the seeps are in the mid to upper slope region centered at 28 ± 7 km landward of the deformation front.	Sahling et al. (2008), Ranero et al. (2008)

Costa Rica	2–5 cm/year	ND	ND		Based on 2 mounds (11 & 12), fluids collected forearc for 1 year. Fluid at one (11a) has low Cl, Ca, Mg, K, Na, and high B concentrations; high B/Li reported by Tryon et al. (2010) venting through deep-penetrating faults. $^3He/^4He = 1.04–1.3\ R_A$.	Füri et al. (2010)
Costa Rica	0.3–20 cm/year	ND	ND	ND	Focused fluid flow through faults inferred from geochemical profiles and heat flow. Estimated flow rates in the decollement range from 1.4 to 45 cm/year (Saffer and Screaton, 2003) and measured directly between 0 and 5 cm/year (Solomon et al., 2009).	Lauer and Saffer (2012)

along strike (Le Pichon et al., 1990). Hence, these measured output fluxes are ~15–40 times the amount that can be produced through local steady-state compaction. The discharge of fluid from a single mud volcano seaward of the deformation front at Barbados was estimated to be ~10^6 m³/year (Henry et al., 1996; Martin et al., 1996), equivalent to the amount of fluid produced through steady-state compaction from the entire subducting sediment section over 100 km of the margin.

Clearly there must be another source of fluid feeding these seeps beyond local steady-state compaction. These data suggest that, in addition to compaction, there is some combination of deep-sourced fluid flow from clay dehydration, nonsteady state fluid flow, and circulation of seawater feeding seeps near the deformation front (Carson et al., 1990; Henry et al., 1996; Le Pichon et al., 1990). Other sources of water include water in both fractures and alteration minerals in the subducting igneous basement and circulation of meteoric water at shallow seep sites.

At Southern Hydrate Ridge, offshore Oregon, fluid flow rates of 10–250 cm/year at bacterial mat sites (average of 85 cm/year) and less than 10 cm/year out of clam beds were determined from modeling of solute profiles obtained from shallow cores (Torres et al., 2002). During the same field program, flow rates measured directly with a benthic flux meter were consistent with the results from chemical profiles, ranging from ~0.1–10 m/year in and adjacent to bacterial mats and generally less than 10 cm/year in clam beds (Tryon, Brown, & Torres, 2002). The range of flow rates measured during the Hydrate Ridge study (Torres et al., 2002; Tryon et al., 2002) is about 1–2 orders of magnitude less than previous measurements at the Cascadia subduction zone that were made closer to the deformation front (Carson et al., 1990). The flow rates measured by the flux meters are transient over several different time scales showing that indeed fluid flow at seeps is not at steady-state (Tryon et al., 2002). Furthermore, fluid flow at some of the clam beds and at off-seep locations is mostly into the sediments showing that there can be considerable downward fluid flow (circulation) of seawater into the sediment column (Tryon et al., 2002). These results were corroborated by direct measurements of fluid flow rates using a seafloor fluid flow meter at a seep location in the Gulf of Mexico by Solomon, Kastner, Jannasch, Robertson, and Weinstein (2008). They also measured highly transient flow rates and considerable circulation of seawater at and around the seep site. These results show that complex dynamic processes control fluid expulsion at seafloor seeps at continental margins, and the steady background flow of fluids produced through tectonic compaction and mineral dehydration at subduction zones is overprinted by additional processes that may include gas expulsion-driven pumping and aqueous entrainment in migrating gas (seawater circulation), salinity-driven seawater circulation through the sediment column, and rapid changes in sediment permeability driven by two-phase flow and mineral precipitation (e.g., Henry et al., 1996; Solomon et al., 2008; Tryon et al., 2002). Recirculation of seawater as a result of these processes must be an important component of the overall forearc output flux from seep sites at subduction zones.

The most detailed recent studies of fluid discharge along the forearc were conducted at the erosive Costa Rica convergent margin. These studies show widespread discharge of deep-sourced fluids at the seafloor. During a series of expeditions between 1999 and 2003, a systematic search for methane-rich fluid seeps at the seafloor documented more than 100 seeps offshore southern Nicaragua and northern central Costa Rica, corresponding to, on average, one seep site every 4 km along the continental slope (Sahling et al., 2008). The majority of the seeps are located in the mid to upper slope region centered at 28 ± 7 km landward of the deformation front and occur near or at faults/fractures that reach the seafloor (Ranero et al., 2008). These faults penetrate deep into the slope sediment and some extend into the basement providing channels for fluid migration from depth. Modeling of pore fluid chemical profiles from a TV-guided core recovered from a patch of bacterial mats ($25 \, m^2$) indicated a flow rate of 300 cm/year (Hensen et al., 2004). However, numerous gravity cores and heat flow probe data across mound structures and fault segments suggest the entire seepage-related structures are affected by a less intense, but pervasive, fluid flux (Harris, Grevemeyer, et al., 2010; Ranero et al., 2008). Heat flow data across two mounds indicate that pervasive fluid flow occurs over a $0.5 \, km^2$ area on each of them (Grevemeyer et al., 2004; Schmidt et al., 2005) with estimated fluid flow rates ranging from 0.1–1.0 cm/year for most sites and ~2–5 cm/year at fewer locations (Harris, Grevemeyer, et al., 2010; Ranero et al., 2008).

During a recent geophysical survey at the southern Costa Rica subduction zone offshore Osa Peninsula, Kluesner et al. (2013) documented 161 seep sites over a $605 \, km^2$ area, indicating ~14 seeps per km along the continental slope, which is approximately 56 times larger than estimated earlier by Sahling et al. (2008). Kluesner et al. (2013) also shows a broader distribution of seeps that span the entire width of the forearc with 77 indicators of seepage on the outer shelf, 19 along the upper slope, 49 in the midslope region, and 16 along the lower slope. This shows that fluid flow is much more pervasive at the Costa Rica margin than previously thought with fluids migrating along a wide variety of flow pathways. Using the assumptions outlined in Ranero et al. (2008), each seep is characterized by an average flow rate of 0.5–1 cm/year across the average $0.5 \, km^2$ area of the seep and yields a conservative average discharge of $2.5–5 \times 10^3 \, m^3$/year per seep site. Assuming a coverage of 14 seeps per km margin and that all seeps are active (Kluesner et al. 2013), the potential fluid output flux at the Costa Rica margin is 35×10^3 to $70 \times 10^3 \, m^3$/year per km of margin. Considering that subducting sediment supplies ~$0.9 \times 10^3 \, m^3$/year of mineral-bound water per linear km of trench (Spinelli & Underwood, 2004), the estimated output flux is 44 times the amount of water that is produced through clay mineral dehydration per year. The estimated inventory of pore fluids entering the subduction zone is $21.7 \times 10^3 \, m^3$/year per km along strike (Lauer & Saffer, 2012), thus the conservative estimate of fluid output is roughly 1.6 to 3 times greater than the amount of fluid that could be produced through sediment compaction. This estimate does not consider fluid flow along the décollement, subducting oceanic basement, or through intergranular flow through the forearc prism and is a conservative estimate of the fluid flux out of seafloor seeps associated with faults across the prism. As for the accretionary prisms discussed above, the mismatch in input and output fluxes requires an additional source of water and/or nonsteady state flow. Since the input flux considers the water in sediment pore space and hydrous minerals in the sediment column, this water must come from hydrous minerals in the subducting oceanic basement, through seawater circulation at seeps (e.g., Solomon et al., 2008; Tryon et al., 2002) and/or through submarine discharge of meteoric groundwater at seeps concentrated on the continental shelf.

4.4.3.5.4 Global Fluid Fluxes

Even though there have been efforts to estimate fluid inventories at individual margins, as described above, the global ocean geochemical consequences of fluid flow through the forearc have not yet been documented. Based on the measurements of fluid flow rates at both accretionary and erosive subduction zones made to date, we attempt to make a first-order estimate of the global output flux of water from subduction zones. There are five seafloor regions that are important to consider when estimating the flux of fluid from the forearc: (1) focused seepage areas, (2) diffuse seepage near focused seepage sites, (3) background fluid flow not associated with seep sites, (4) flow along the décollement, and (5) potential flow within the subducting igneous basement.

The region of diffuse seepage associated with seeps (hereafter referred to as background seepage area) is more than 300 times larger than the focused seepage area (Table 4.4.3.6). Thus, if the flow rate at focused seep sites is 25 cm/year, the average background seepage rate only needs to exceed 0.09 cm/year for it to be the dominant component of flow at a seep site. Considering that background seepage operates over a larger area at seafloor seeps (e.g., Ranero et al., 2008), the most important variables in estimating the magnitude of the output flux from seafloor seep sites are the background fluid flow rate and the area of seafloor characterized by this flow rate (the maximum flow rates are far less important because of the small area affected; Table 4.4.3.6). Unfortunately, these parameters are the least well known and there are only a few direct measurements of fluid flow rates at the flanks of mud volcanoes and seeps. Future field studies should focus on constraining these background flow rates and the areas represented by them.

Extrapolation #1 (Table 4.4.3.6) is our most conservative estimate. This estimate assumes the seep distribution observed at the Cascadia subduction zone

(Carson et al., 1990; Moore et al., 1990), and a background flow rate at each seep of 2 cm/year (Füri et al., 2010). We also assume that there is fluid flow along the décollement at 1 cm/year as measured at the Costa Rica subduction zone by Solomon et al. (2009). Based on drilling results from ODP expeditions 190 and 205, it is assumed that the décollement fluid conduit is 30 m thick. With these assumptions, the total output flux per km of trench is 680 m³/year. However, we view this estimate as being nonrepresentative. If this flux is multiplied by the global trench length of 42,000 km, the global output flux of water from the forearc would only be 2% of the global input flux of water in subducting sediment pore space. Compaction-derived fluid release is at a peak within the first 10 km of the forearc, thus the output flux should at least match the input flux of water in the sediment pore space. Given that both mid-slope seeps and the décollement in subduction zones have a strong signature of deep-sourced, clay-derived fluids (e.g., Hensen et al., 2004; Kastner et al., 1991, 2006; Silver et al., 2000; Spivack et al., 2002), the fluid output arguably has to be equal to or larger than the input flux from interstitial water. Even if the background seepage area is quadrupled and intersititial fluid flow in the prism is included in the Extrapolation #2 estimate (Table 4.4.3.6), the global fluid output would be ~80% of the amount of fluids entering subduction zones in sediment pore space per year. Output H_2O flux calculations, based on Extrapolation #1 and #2, indicate much longer ocean recycling times through SZs than the water input values (Table 4.4.3.2).

The most likely estimate for the output flux is Extrapolation #3 (Table 4.4.3.6). This estimate includes focused flow at seeps, uses the seep distribution offshore Osa Peninsula (Kluesner et al., 2013), and fluid flow within the décollement, intergranular flow in the prism, and fluid flow within the subducting oceanic basement (Harris, Spinelli, et al., 2010; Solomon et al., 2009). It is assumed that the basement flow conduit is 100 m thick (Solomon et al., 2009). Using these flow rate estimates, the output flux of water per km of subduction zone is 300×10^3 m³/year. These results are strikingly similar to the earlier estimates of the fluid output fluxes near the deformation front of the Cascadia, Barbados, and Nankai subduction zones (e.g., Le Pichon et al., 1991, 1992; Foucher et al., 1990). This flux is seven times the input flux of water in the sediment pore space and four times the total flux of water into subduction zones that also includes hydrous minerals in the subducting sediment and upper oceanic basement. This calculation suggests that seawater circulation driven by seep processes (e.g., the ebullition of free gas or salinity gradients between seawater and advecting deep-sourced fluids; Henry et al., 1996; O'Hara et al., 1995; Solomon et al., 2008; Tryon et al., 2002) at subduction zones is indeed more important than previously recognized.

Our estimates indicate that when fluid venting at the forearc seafloor is included in the total forearc water flux calculations, fluid recycling through SZs is much faster. Using Extrapolation #3 with a global ocean volume of 1340×10^6 km³, the global ocean residence time in subduction zones is

TABLE 4.4.3.6 A 1st Order Estimate of the Global Output Flux and Water from Subduction Zones

	Seep Distribution (Seeps/km Margin)	Seep Diameter (m)	Focused Seepage Area (m²)	Background Seepage Area (m²)	Focused Seep Flow Rate (cm/year)	Background Seep Flow Rate (cm/year)
Extrapolation#1	4	100	25	7850	50	2
Extrapolation#2	14	200	25	31,400	300	5
Extrapolation#3	14	500	50	196,350	500	10

[a]It is assumed that the prism is 10 km wide.
[b]It is assumed that the basement fluid flow horizon is 100 m thick (Solomon et al., 2009).
[c]Based on a global trench length of 42,000 km (Von Huene and Scholl, 1991).
[d]Based on the global amount of water subducted per year in this paper. First number is based on total pore volume subducted and second is total water subducted (pore volume + hydrous minerals).
[e]Global ocean residence time is based on a global ocean volume of 1340 × 10⁶ km³.

olle-nt v Rate /year)	Prism Flow Rate (cm/ year)[a]	Basement Flow Rate (cm/ year)[b]	Total Flux (m³/year per km Margin)[c]	Input/ Output[d]	τ_R[e]	References
0	0	0	680	0.02 – 0.01	34 Gyr	Seep distribution from Moore et al. (1990). Seep diameter and focused seep area from Hensen et al. (2004). Focused seep flow rate from Torres et al. (2002). Background seep flow rate from Furi et al. (2010). Décollement flow rate and area from Solomon et al. (2009).
0.1	0.1	0	34,500	0.8 – 0.5	924 Myr	Seep distribution from Kluesner et al. (2013). Seep diameter and focused seep flow rate from Hensen et al. (2004). Décollement flow rate from Solomon et al. (2009). Prism flow rate from Lauer and Saffer (2012).
0.1	0.1	12	306,000	7.1 – 4.3	104 Myr	Seep distribution from Kluesner et al. (2013). Seep diameter from Hensen et al. (2004) and Solomon et al. (unpublished data). Focused seep flow rate is minimum from Le Pichon et al. (1991). Background seep flow rate from Torres et al. (2002). Décollement and prism flow rate from Lauer and Saffer (2012). Basement flow rate from Solomon et al. (2009). This total flux is roughly equivalent to the output flux from seeps at the toe of the Nankai, Barbados, and Cascadia subduction zones in ; Le Pichon et al. (19911992), Foucher et al. (1990), and Carson et al. (1990). Using these estimates for the seep flux instead, the total output flux could be as high as 430,000 m³/km year and t_R could be ~75 Myr

estimated to be ~100 Myr. This residence time is roughly 3 times longer than the best estimate of the residence time in the global ridge crest of 20–36 Myr (German & von Damm, 2004; Mottl, 2003). The estimate is about five times faster than the maximum estimate of 500 Myr (Moore & Vrolijk, 1992) as the time to cycle the ocean through the global subduction system. As described previously, fluid-rock reactions in the forearc of subduction zones act as a source and sink for many elements and isotope ratios. Considering the potential enormous fluxes of fluid through the forearc, less but comparable to those at the ridge crest system (Elderfield & Schultz, 1996; German & von Damm, 2004; Mottle, 2003), subduction zone fluid cycling may as well have an impact on the global geochemical cycles of some elements and isotope ratios and seawater chemistry.

For comparison, global estimates of volatile outputs through the arcs range from $1-6 \times 10^{14}$ g/year (e.g., Hilton et al., 2002; Jarrard, 2003; Peacock, 1990; Wallmann, 2001). Our estimates yield a global flux of ~1×10^{16} g/year, that is, two orders of magnitude greater than the volatile output of arcs. It is worth noting that the residual fluid and chemically modified rocks are released in the back-arc or are transported to the mantle where they impact its long-term chemical and isotopic evolution (e.g., Bebout, 1996; Bekins et al., 1995; Füri et al. 2010; Hacker, 2008; Hilton et al., 2002; Jarrard, 2003; Kastner et al., 1991; Moore & Vrolijk, 1992; Peacock, 1990; Plank & Langmuir, 1993, 1998; Rea & Ruff, 1996; Saffer & Tobin, 2011; Wallmann, 2001; and references therein).

4.4.3.5.5 Solute Fluxes

The major players in global element cycling of solutes are rivers and hydrothermal circulation in the ocean. They act as major sources and sinks of some elements and isotopes; in addition, carbonate precipitation, diagenesis in marine sediments, and submarine groundwater discharge contribute to a lesser but in some cases, significant degree. As discussed above, fluid-rock reactions in the forearc of subduction zones act as a source and sink for many elements and isotope ratios. The role of subduction zones in the seawater inventory of most elements and isotope ratios is, however, poorly constrained, but is critical for evaluating the residence time of solutes in seawater and for constraining the relative importance of this seafloor environment in controlling the composition of seawater over time.

Considering the average composition of the end-member fluids (Table A.1) across the five subduction zones reviewed herein (Table 4.4.3.1), the return fluxes of these solutes and isotope ratios from subduction zone forearcs to the ocean are estimated (Table 4.4.3.7). Fluxes in mol/year are calculated as follows:

$$Flux = (C_{SZ} - C_{SW}) \times Q_{SZ} \qquad (4.4.3.1)$$

where C_{SZ} is the average end-member concentration of each solute in deep-sourced fluids sampled in the forearc of subduction zones, C_{SW} is the

TABLE 4.4.3.7 Estimated solute and isotope ratio fluxes within subduction zones

Solute	$C_{SZ\,reflux}$	$C_{seawater}$	Source (+) or Sink (−)	$F_{SZ\,dehydration}$ (mol/year) × 10^{10}	$F_{SZ\,extrapolated}$ (mol/year) × 10^{10}	$F_{hydrothermal}^{a}$ (mol/year) × 10^{10}	F_{river}^{a} (mol/yr) × 10^{10}	$F_{SZ\,dehydration}/F_{hydrothermal}$	$F_{SZ\,extrapolated}/F_{hydrothermal}$	$F_{SZ\,extrapolated}/F_{river}$
Ca	3–19.9 mM	10.5 mM	?	−0.323–0.4	−9.64–12.08	0.9–1.3	1200	0.31–0.44	9.29–10.7	0.01
Mg	20.2 mM	54 mM	−	−1.453	−43.44	−160	530	0.009	0.27	0.08
Sr	190.3 µM	87 µM	+	0.004	0.133	1.5	2.2	0.002	0.089	0.06
K	4.68 mM	10.4 mM	−	−0.246	−7.35	23–69	190	0.01–0.003	0.11–0.32	0.04
Li	310 µM	27 µM	+	0.012	0.364	1.2–3.9	1.4	0.01–0.003	0.09–0.30	0.26
Si	593 µM	125 µM	+	0.02	0.601	43–66	640	0	0.01	0
B	777 µM	450 µM	+	0.014	0.42	0.11–0.45	5.4	0.13–0.03	0.93–3.82	0.08
SO$_4$	0	28.9	−	−1.24	−37.14	−84	370	0.015	0.44	0.1

Isotope	SZ avg	Seawater	Hydrothermal	River	F_{SZ}	$F_{hydrothermal}^{a}$	F_{river}^{a}	$F_{SZ}/F_{hydrothermal}$	F_{SZ}/F_{river}
$^{87}Sr/^{86}Sr$	0.70793	0.7092	0.7035	0.7119	−0.000169	−0.00855	0.00594	0.020	−0.028
$\delta^{7}Li$	18.256	31	8.3	23	−4.639	−27.240	−11.20	0.17	0.41

^{a}Estimates from Elderfield and Schultz (1996). Note that the hydrothermal fluxes are estimated for mid-ocean ridge systems.

contemporary seawater concentration of each solute, and Q_{SZ} is the volumetric discharge of fluid from the forearc (km^3/year). The isotopic fluxes are estimated by:

$$Flux = (R_{SZ} - R_{SW}) \times F_{SZ} \qquad (4.4.3.2)$$

where R_{SZ} is either the isotopic ratio (e.g., $^{87}Sr/^{86}Sr$) in the end-member fluid or the ratio in delta notation (e.g. δ^7Li), R_{SW} is the ratio in contemporary seawater, and F_{SZ} is the solute flux calculated using Eqn (4.4.3.1).

Using this approach, we estimate solute fluxes for two cases: (1) Assuming that only the mineral dehydration-derived fluid discharge is important for the specific solute flux ($\omega_{SZ}=0.43\,km^3$/year) and (2) the solute flux is computed considering the total output flux of fluid estimated from extrapolation of measured fluid flow rates at specific subduction zones ($\omega_{SZ}=12.85\,km^3$/year), which includes fluid derived from compaction, dehydration, and seawater circulation (Extrapolation #3; Table 4.4.3.6).

The reason for providing these two estimates of solute fluxes is that the end-member fluid compositions are measured in deep-sourced fluids sampled along fluid flow horizons (e.g., the décollement) that are largely derived by mineral dehydration and fluid-rock reactions at elevated temperatures. The solutes and isotope ratios that are most affected by deep-sourced fluid-rock reaction are Sr, K, Li, B, $^{87}Sr/^{86}Sr$, and δ^7Li. However, there must be some degree of mixing between the fluid end-member and shallower fluids during migration that produces the composition observed in these flow horizons. Thus, these should be considered minimum estimates. The second flux estimate is likely to be most representative for Ca, Mg, and SO_4 as these solutes are impacted by shallow biogeochemical reactions such as organoclastic sulfate reduction, anaerobic oxidation of methane, and associated authigenic carbonate precipitation. This flux estimate accounts for the excess fluid output observed in subduction zones (Tables 4.4.3.5 and 4.4.3.6). The excess fluid is hypothesized to be the result of shallow circulation of seawater at seafloor seep sites and must be an important control on Ca, Mg, and SO_4 fluxes from subduction zones, which act as a significant sink for these solutes. A more detailed discussion of the chemical components and isotope ratios of the solute fluxes is provided in Appendix 3.

These results show that subduction zones are an important, commonly overlooked source and sink for many elements and isotope ratios. SZs are an important sink of seawater Mg. The output flux estimated for subduction zones is up to 27% of the output flux estimated for hydrothermal circulation at midocean ridges (Elderfield & Schultz, 1996) and accounts for 8% of the input flux from rivers (Table 4.4.3.7). Subduction zones are also an important sink of seawater SO_4 accounting for 10% of the river flux and up to 44% of the hydrothermal flux. These results show that the output flux of Ca at subduction zones is high enough to completely consume the input of Ca from hydrothermal circulation. The flux of B from subduction zones are ~13–100% of the flux from hydrothermal circulation, and the input flux of Li is ~10–30% of the input flux from hydrothermal circulation (Figure 4.4.3.5). Subduction zones are an important source of ^6Li to the global ocean representing 17% of the hydrothermal flux and 41% of the river flux.

4.4.3.6 GLOBAL VOLATILE AND MASS CYCLING IN SZs, HAS IT EVOLVED OR FLUCTUATED THROUGH TIME?

In the past ~2.5 bys "a volume of continental crust ~100% of that existing has been recycled to the mantle", this is ~0.6% of the mantle volume (Scholl & von Huene, 2007). The effect of this recycling on mantle geochemical heterogeneity must have been significant. The impact may be even greater assuming subduction was underway since at least ~3 billion years ago (Carlson et al., 2000) or even as early as 3.5–4.0 by ago (Condie, 2005). The evolution of continental crust and climate has changed through time. Changes in continental weathering have most likely influenced the mass and the average composition of sediments available for subduction with consequences for elemental, H_2O, and other volatile cycling on Earth. Changes in continental weathering likely also influenced margin architecture, hence, the ratio of accretionary versus erosional dominated subduction during high and low weathering rates, respectively.

The biogenic components of marine sediments have also changed through time. The opal in diatoms contains 10–11 wt% H_2O, and it is a significant component of modern sediments beneath upwelling zones (e.g., in western continental margins and equatorial regions). Presently, mostly diatomaceous ooze sediments that consist of silica-opal are being subducted in polar region SZs, such as the Alaska SZ. However, the amount of opal in sediments subducted must have varied greatly over time. The earliest known fossil diatoms date from early Jurassic (~185 Ma ago), and large deposits of fossil diatoms with well-developed silica frustules date back to early Cretaceous. No traces of diatoms have been observed in any siliceous sediments older than Jurassic; the preservation record of diatoms since the Cretaceous may simply be a biomineralization event. Diatoms rapidly adapted to all climatic regions and to diverse ecosystems, including the polar regions as well as fresh and brackish water environments. Radiolarians (and siliceous sponges), having a similar wt% H_2O in their opal structures, were present throughout the Phanerozoic and were abundant before the proliferation of the diatoms. Radiolarians, however, typically occur only in equatorial regions and primarily in the pelagic environment. The evolution of diatoms must have impacted the amount and distribution of opal in pelagic and hemipelagic sediments, hence, the amount of H_2O-bound in opal that became available for subduction. The global opal inventories available for subduction likely increased since the "take-over" of the Si cycle in the ocean by diatoms, probably in the early Cenozoic, and in particular following the Oligocene-Miocene climate cooling, thus affecting the amount of diatom-H_2O-rich siliceous sediments that became available for subduction.

Similarly, deep-sea calcareous sediments were not available for subduction until the successful evolution of the calcareous nannoplankton (coccolithophores) in the Late Triassic and the evolution of the planktic foraminifera in the Jurassic. The Paleozoic benthic foraminifera, the fusulinids, which probably evolved in the Carboniferous, were not deep-water forms; they occur together with reef forming corals, thus seem to have lived in the shallow water photic zone. Shelf carbonates are rarely subducted, except during large-scale collisions. Following the evolution of the coccolithophorides and planktic foraminifera, there was an important shift from shelf to deep-water carbonate deposition in the ocean (e.g., Ridgwell, 2005; Blättler et al., 2012). The shift in $CaCO_3$ deposition from the shallow (continental shelves) to the pelagic deep-water environment in the Late Jurassic and Cretaceous is not a matter of preservation, rather this shift was driven by the evolution of biomineralizing calcareous zoo- and phyto-plankton.

Hence, both the evolution of the calcareous phyto- and zoo-plankton since the Jurassic-Cretaceous and the proliferation of the diatoms, particularly in the Tertiary, must have impacted the composition of the subducting sediments. The subduction of pelagic carbonates likely has influenced the long-term carbon cycling on Earth since 200–175 Ma ago, and the subduction of diatoms must have influenced the water content of subducted sediments in the shallow forearc and the increase in fluxes of Si and B into the ocean since ~34–25 Ma ago. The evolution of calcareous zoo- and phyto-plankton and proliferation of diatoms in the tertiary and their impact on the composition of the subducting slab could be explored by modeling.

4.4.3.7 CONCLUDING REMARKS

In this paper we provided an in-depth analysis and synthesis of H_2O and volatile cycling in SZs and the associated thermal regimes, with a focus on the forearc and on the Costa Rica and Nankai SZs. We have reconsidered the H_2O and volatile fluxes into and out of the forearc, including seawater recycling processes that also affect serpentinization in the outer rise upper mantle and extensive venting at the seafloor. We also extrapolated globally data from a few well-studied representative accretionary and erosive margins to construct global inventories, and note the outstanding questions and discrepancies in these estimates. Subduction zone fluid input fluxes are highly variable in time and space and the thermal regimes vary as well. The most important early diagenetic reactions involved in volatile fluxes were summarized (Table 4.4.3.3).

How much of the H_2O and other volatile budgets reach the arc, and thus the deep mantle is still an important outstanding question. There are rather different estimates: Hacker (2008) for example concluded that the global H_2O flux to the deep mantle is about one ocean mass over the age of Earth. Wallmann (2001), however, concluded that water released into the mantle is $0.3 \times km^3$/year or 10^{15} g/year, which if extrapolated assuming an average constant rate, is three orders of magnitudes higher over the age of Earth.

For clear reasons, there is a much smaller discrepancy between estimates of H_2O released to the atmosphere in the global arc volcanoes. The estimates range is $1-6 \times 10^{14}$ g/year (Hilton at al., 2002; Jarrard, 2003; Peacock, 1990; Wallmann, 2001) (Table 4.4.3.2).

The main characteristics of nonaccreting and accreting margins were analyzed. The nature of fluid venting seems to differ at the two margin types. At nonaccreting margins the fluids from compaction and dehydration that are venting are partitioned between the décollement and the conduits through the prism, respectively. At both types of subduction margins fluid and gas venting sites are primarily associated with faults.

The décollement and other faults, including out of sequence and splay faults, play the most important role in fluid recycling in the forearc. Seeps, mud volcanoes, and mounds that also play an important role are as well mostly associated with faults. The measured fluid output fluxes at seeps are ~15–40 times the amount that can be produced through local steady-state compaction. This suggests that, in addition to compaction and mineral dehydration, there is some combination of nonsteady state fluid flow, shallow circulation of seawater at seeps, from hydrous minerals in the subducting oceanic basement, and submarine discharge of meteoric groundwater at seeps concentrated on the continental shelf.

Using our best fluid output flux estimate in Extrapolation #3 in Table 4.4.3.6 and considering an ocean volume of $1340 \times 106 \, km^3$, the global ocean residence time in subduction zones is ~100 Myr. This residence time is similar to the best estimate of the residence time in the global ridge crest of 90 Myr by Elderfield and Schultz (1996) and roughly three times the residence time of 20–36 Myr estimated by German and von Damm (2004) and Mottl (2003).

Considering the average composition of end-member fluids across the five subduction zones reviewed herein, the return fluxes of these solutes and isotope ratios from subduction zone forearcs to the ocean were estimated. The results show that subduction zones are an important sink and source for several elements and isotope ratios. For example, they are an important sink of seawater Ca, Mg, and SO_4 and an important source of B and Li into seawater.

4.4.3.8 APPENDICES

4.4.3.8.1 Appendix 1. The Role of Serpentine Minerals in Volatile Cycling in the Forearc

Except for the Cascadia forearc, calculated temperatures of the uppermost forearc mantle of 150–250 °C were reported by Hyndman and Peacock (2003); in this thermal regime the serpentine minerals are stable; therefore, they are not involved in the fluid and volatile cycling budgets in forearcs considered in this paper.

In SZs the two processes that are responsible for serpentinization of the uppermost mantle are: (1) at the outer rise where the plate bends, and deep faulting occurs before entering the SZ, the deep faults provide pathways for seawater circulation into the upper mantle causing serpentinization (e.g., Ranero et al., 2005, 2003; Rüpke, Morgan, & Connolly, 2004) and (2) forearc uppermost mantle serpentinization that is linked to hydration of the overlying mantle by dewatering of the H_2O-rich subducting plate, first suggested by Fyfe and McBirney (1975) and manifested, for example, by the serpentine mud volcanoes at the Mariana forearc (e.g., Fryer, 1996, p. 1999; Fryer, Wheat, & Mottl, 1999). Hyndman and Peacock (2003) suggested that such serpentinization of the forearc uppermost mantle is a global phenomenon. Percent serpentinization can be approximated seismically; the seismic velocity decreases and the Poisson's ratio increases relative to unaltered rocks (e.g., Hyndman & Peacock, 2003; Grevemeyer et al., 2007).

Serpentinization of the upper mantle, however, does not require high temperature; serpentine forms under a wide range of temperatures, for example, according to Alt (2004); Bonnatti, Lawrence, and Morandi (1984); Wenner and Taylor (1971) at 120–130 °C. The serpentine minerals (lizardite, chrysotile, and antigorite) contain ~13 wt% H_2O. The serpentine minerals lizardite and chrysotile may transform to antigorite at >350 °C and antigorite is stable to ≥600 °C and 1–6 GPa (Hyndman & Peacock, 2003; Rüpke et al., 2004). Accordingly, serpentine dehydration fluxes should contribute to the deep water cycling in SZs and should be important in cycling of volatiles in arcs and subarcs, but not in the forearc, the focus of this paper; the fluxes involved are as yet unknown (e.g., Hacker, 2008; Rüpke et al., 2004; Ranero at al., 2005).

In addition to deep water cycling in SZs, serpentines are considered as the most important sink of seawater B (Bonnatti et al., 1984; Vils, Pelletier, Kalt, Muntener, & Ludwig, 2008), with concentrations inversely related to the formation temperature (Bonnatti et al., 1984; Wenner & Taylor, 1971), and are important in the Cl isotope cycle of seawater, exhibiting enrichment in ^{37}Cl (Spivack et al., 2002; Wei et al., 2008).

4.4.3.8.2 Appendix 2

Table A1. Composition of fluid end members in the forearc of well studied SEs.

4.4.3.8.3 Appendix 3. The Most Important Chemical and Isotopic Compositions in Pore Fluids Used for Interpreting Fluid-Rock Reactions

Table 4.4.3.3 and 4.4.3.6 provide comprehensive summaries of the discussed reactions.

4.4.3.8.3.1 Cl and Br Concentrations and Cl Isotope Ratios

It is commonly assumed that Cl behaves conservatively, and the observed changes in concentrations simply reflect dilution by mineral dehydration reactions, gas hydrate dissociation, or mixing with a dilute fluid; values higher than seawater indicate water uptake by hydration reactions, such as volcanic ash alteration to smectite and zeolites, by evaporation or by mixing with a brine. Also, during maximum glaciation, seawater Cl concentrations were higher by ~3% (McDuff, 1981; Adkins & Schrag, 2001).

Cl-isotopes provide key information on mineral/fluid exchange reactions (Ransom et al., 1995; Spivack et al., 2002; Wei et al., 2008). Cl has two stable isotopes, ^{35}Cl and ^{37}Cl. The preferred uptake of ^{37}Cl by high-temperature hydrous silicates, such as serpentine, talc, chlorite, and amphiboles (Ransom et al., 1995; Spivack et al., 2002) enriches the pore fluid in ^{35}Cl. Upon dehydration the pore fluids become enriched in ^{37}Cl.

Thus, Cl concentrations, together with Cl isotope ratios, allow us to distinguish between simple clay dehydration and gas hydrate dissociation reactions and dehydration plus formation of high temperature hydrous minerals. Higher-temperature fluid-rock reactions have greater impact on the Cl concentration and isotope ratio in the fluids than moderate to lower temperature reactions. This is because the higher temperature silicates also contain hundreds or more ppm of Cl, whereas, for example, the smectite formed from volcanic ash alteration contains ~100 ppm Cl. Although ash alteration to smectite increases the Cl concentration in the pore fluid and also preferentially consumes ^{37}Cl, thus enriching the residual more saline fluid in ^{35}Cl, the effect on the pore fluid Cl isotopes is minimal because of the low Cl concentration in smectites.

For <2 cm/year advective velocities, pore fluid Cl concentration is a more reliable indicator of advective velocity calculations than the $\delta^{18}O$ value (Wheat & McDuff, 1994), because of the higher precision to which Cl concentration can be analyzed relative to the expected signal ratio of Cl. The Cl isotopes fractionation must as well be evaluated.

The fractionation between Cl and Br concentrations, thus the Cl/Br ratio in pore fluids is another diasgnotic diagenetic proxy; Br concentration (therefore the Cl/Br ratio) is primarily impacted by organic matter diagenesis in the shallow subsurface where Br is being released to the pore fluids. At greater depths the ratio is affected by the exclusion of Br from the high-temperature hydrous silicates, such that hydration reactions decrease the Cl/Br ratio and dehydration reactions increase it. The pore fluid Cl isotope ratios can be used to distinguish between the release of Br from organic matter and the exclusion of Br from high-temperature hydrous minerals.

4.4.3.8.3.2 O and H Isotopes

Oxygen and Hydrogen isotopes are used to constrain the nature of the pore fluid freshening reactions. Enrichment in $\delta^{18}O$ and depletion in Cl concentration indicate either gas hydrates dissociation or hydrous silicates (especially smectite) dehydration as the source reaction (e.g., Kastner, Elderfield, Jenkins, Gieskes, & Gamo, 1993, Tables 4.4.3.3 and 4.4.3.6). To discriminate between the two reactions the δD values need as well to be determined. An inverse relation between δ^{18}O and δD values, especially where fluids are low in Cl, is typical for hydrous silicates dehydration, assuming unaltered seawater pore water isotopic composition. This is well exemplified in fluids from the Costa Rica seeps (Hensen et al., 2004). In gas hydrates the relation is positive, both δ^{18}O and δD values are enriched in the clathrate cage, with δ^{18}O values of ~3.5‰ and δD values of ~21‰ (O'Neil, 1968; Suzuki and Kimura, 1973; Davidson et al., 1978).

4.4.3.8.3.3 Sr Concentrations and $^{87}Sr/^{86}Sr$ Ratios

Volcanic ash alteration only exerts minor influence on Sr concentration and considerable influence on the Sr isotopic value of pore fluids. In contrast, biogenic calcite is a major Sr source to pore waters, commonly used as fluid sources tracer; it is a minor source to the Sr isotopic value (Lawrence et al., 1979).

Oceanic crust $^{87}Sr/^{86}Sr$ value is 0.703 to 0.704, biogenic calcite values range from 0.7092 to 0.7075, depending on the age, and continental detritus values vary between 0.7119 and 0.7133 depending on the drainage terrain.

4.4.3.8.3.4 Silica Concentration

Silica concentration in the fluid provides the best established geothermometer to date. It can be applied to most natural fluids with pH <~8.7 in which the silica monomer is the dominant dissolved silica species. It is particularly useful at temperatures >150 °C, when quartz controls the silica concentration.

The Si geothermometer is based on experimental data and on thermodynamic calculations of the temperature dependence of silica concentrations in the fluid at equilibrium with quartz, chalcedony, cristobalite, and opal-A (Fournier, 1973; Fournier & Rowe, 1996; Von Damm, Bischoff, & Rosenbauer, 1991, and references therein).

4.4.3.8.3.5 Ca and Mg Concentrations

The concentrations of both Ca and Mg vary considerably by authigenic carbonate formation and recrystallization and volcanic ash alteration. Carbonate formation mostly consumes both Ca and Mg and recrystallization via dissolution-precipitation releases Sr into the pore fluid. Volcanic ash alteration effects Ca and Mg concentrations inversely, releasing Ca from silicates weathering and consuming Mg, in a 1.1 to 1.2 ratios (e.g., Lawrence et al.; Lawrence & Gieskes, 1981).

4.4.3.8.3.6 The Alkali Metals and δ^7Li

Alkali metals are highly reactive due in part to their larger atomic radii and low ionization energies. At low temperatures of ≤30–40 °C Li and K are preferentially partitioned into the solid phase, whereas at ≥50–60 °C all alkali metals are preferentially released into the fluid phase. The high temperature reactions, based primarily on recent hydrothermal experiments by Wei (2007, 165 pp.) and Wei et al., (2010) show that the exact threshold temperature for the release into the fluid phase is poorly defined and differs somewhat for each alkali metal. Previous experiments indicated the 'fluid loving' nature of Li and K, in particular at ≥100 °C (Seyfried et al., 1998; You & Gieskes, 2001, and references therein). To date, Li and K concentrations (and their ratios) have been the alkali metals most commonly used as approximate geothermometers.

In addition, Li isotope ratios together with concentrations provide important insights on the hydrology and the approximate fluid-rock reaction temperatures at the fluid source in SZs. Both (concentration and isotopes) are affected by mostly clay minerals (especially those rich in Mg) and also zeolite diagenesis. These reactions preferentially fractionate ^6Li from low to greenschist facies temperatures, enriching the fluid in ^7Li. The reactivity of Li and fractionation factor of $^7Li/^6Li$ have been studied in laboratory experiments at elevated temperatures of 100–250 °C.

At higher temperature fluid-rock reactions the increase in Li concentration in the pore fluids is accompanied by an increase in ^6Li, which is consistent with a clay source.

Other alkali metals used for approximate geothermometry are the Na^+ and K^+ concentrations for >150–200 °C (Fournier, 1979; Truesdell, 1976) and the Na^+ and Li^+ geothermometer, suggested by Fouillac and Michard (1981). The latter provided very similar temperatures obtained by the Mg^{2+} and Li^+ geothermometer (Kharaka & Mariner, 1989) to estimate fluid temperatures at several mud volcanoes at Barbados (Martin and Kastner, 1996).

Based on the recent hydrothermal experiments by Wei (2007) and Wei et al., (2010), a larger matrix of alkali metal ratios, used together with Li isotope ratios, may be established as additional approximate geothermometers for deep-sourced fluids at SZs and ridge flanks.

4.4.3.8.3.7 Boron and $\delta^{11}B$

Boron isotope ratios and concentrations provide evidence for clay alteration and serpentine formation at depth (Deyhle et al., 2003). B is released into the fluid phase at all temperatures. Experiments by You, Spivack, Gieskes, and Rosenbauer (1995), and Williams and Hervig, (2001) show desorption of B to be complete at 100 °C, and at ~150 °C most of the B is in the fluid. However, in retrograde reactions B is readsorbed at <60 °C. The light B isotope ^{10}B is preferentially adsorbed on clay minerals leading to pore fluids enriched in ^{11}B. In clay alteration and transformation reactions, B is released and the isotope ratio of the residual solution decreases. When serpentine forms, the opposite occurs—B uptake is high (e.g., in the Mariana, Vils et al. 2008) and isotope ratios in the residual solution increase (Seyfried, Janecky, & Mottle, 1984; Spivack, Palmer, & Edmond, 1987; You, Castillo, Gieskes, Chan, & Spivack, 1997). Illite is particularly enriched in B and has the most negative B isotope values among the clay minerals; thus the smectite→illite reaction may be recognized using B isotope ratios. The B isotope ratios and B concentration data together with other major and minor element concentration data involved in the smectite-to-illite transformation reaction, such as the K or Rb concentrations that are enriched in illite, provide unequivocal evidence for this reaction.

4.4.3.8.3.8 $^3He/^4He$ Ratios

The noble gas ^3He is a clear indicator of primordial volatile flux from the mantle, and thus provides important insights on the interaction between Earth's interior and exterior reservoirs. Higher than atmospheric $^3He/^4He$ ratios indicate the presence of fluids that have reacted with mantle-derived rocks.

A recent study by Kastner, Hilton, Jenkins, Solomon, and Spivack (in this volume) on the He isotope ratios in the Nankai Trough and Costa Rica subduction zones showed that at the Nankai Trough SZ the $^3He/^4He$ ratios relative to atmospheric ratio (R_A) range from mostly crustal 0.47 R_A to 4.30 R_A, which is ~55% of the MORB value of 8 R_A. Whereas at the Costa Rica SZ, offshore Osa Peninsula, the ratios just range from 0.86 to 1.14 R_A, indicating the dominance of crustal radiogenic ^4He. The difference is attributed to the contrasting nature of fluid cycling in the accretionary versus erosive SZs, respectively.

Focused flow along faults, e.g., the décollement, splay, and out of sequence faults, and fractured and permeable horizons at SZs play a key role in controlling fluid, heat, and mantle He transport.

4.4.3.8.3.9 Hydrocarbons

At low temperatures in shallow subsurface organic-rich sediments ($\geq 0.5\,wt\%$) and sedimentation rates of $>10–20\,cm/year$, methane is produced by microbially mediated organic matter reactions following sulfate reduction and depletion. At elevated temperatures ($>80–100\,°C$) higher hydrocarbons (C_3 to C_6) form. The presence and relative abundance of the higher hydrocarbons and C and H isotopic compositions are used to distinguish between microbial and thermogenic methane and yield information about reaction temperatures and fluid migration (e.g., Claypool & Kaplan, 1974; Schoell, 1980; Whiticar, 1989; Whiticar et al., 1986).

TABLE A.1 Composition of Fluid End Members in the Forearc of Well-Studied Subduction Zones

Expedition	Subduction Zone	Site	Conduit	Cl (mM)	Ca (mM)	Mg (mM)	K (mM)	Na (mM)
131	Muroto	808	Lower Shikoku Basin	457	28.7	2.8	1.34	422
190	Muroto	1174	Lower Shikoku Basin	473	25.7	6.27	1.52	420
190	Muroto	1173	Lower Shikoku Basin	497	43.5	3.91	1.75	420
110	Barbados	671	Décollement	505	37.2	28.1	3	404
110	Barbados	672	Incipient Décollement	505	NS	NS	NS	NS
110	Barbados	674	Fault zone	394	25.6	10.7	1	324
156	Barbados	948	Décollement	461	42.5	18.7	2.8	364
170/205	Nicoya	1040/1254	Décollement	480	27	12.9	3.3	389
205	Nicoya	1253	Basement-CORK data	ND	31.5	24.95	9.88	ND

Continued

TABLE A.1 Composition of Fluid End Members in the Forearc of Well-Studied Subduction Zones—cont'd

Expedition	Subduction Zone	Site	Conduit	Cl (mM)	Ca (mM)	Mg (mM)	K (mM)	Na (mM)
334	Osa	U1378	Shear zone	355	5.58	12.23	5.01	303
334	Osa	U1379	Shear zone	524	11.25	21.06	5.94	ND
334	Osa	U1379	Upper prism	548	34.55	12.95	3.84	ND
344	Osa	U1380	Upper prism	392	31.2	0.401	1.94	313
315	Nantroseize	C0002	Upper accretionary prism	433	20.6	10	1.78	369
322	Nantroseize	C0011	Lower Shikoku Basin turbidites	504	NS	NS	NS	NS
204	Cascadia	1244	Background deep fluid	467	5.5	19.5	5	420
204	Cascadia	1251	Background deep fluid	432	7.5	10.4	4.6	425
311	Cascadia	1327	Background deep fluid	369	3.7	10.2	6	328
195	Marianas	1200	S. Chamarro Seamount	513	0.38	0	19	611
	Costa Rica	Slope	Mud mounds, mound 11	213	4.2	8.3	4.3	189

TABLE A.1 Composition of Fluid End Members in the Forearc of Well-Studied Subduction Zones—cont'd

Li (μM)	Rb (μM)	Sr (μM)	Ba (μM)	B (μM)	Si (μM)	F (μM)	C₂	C₃
725	ND	289	ND	2873	381	2600	4.07 μL/kg	3.31 μL/kg
ND	ND	ND	ND	ND	340	2400	19.8 ppmv	8.22 ppmv
ND	ND	ND	ND	ND	248	2000	3.75 ppmv	2.26 ppmv
316	ND	352	ND	ND	815	ND	ND	ND
NS	NS	NS	NS	NS	NS	NS	NS	NS
ND	ND	ND	ND	ND	285	ND	ND	ND
107	0.157	ND	2.25	ND	257	ND	1.06 ppmv	0.93 ppmv
239.18	0.48	113	13.21	76.68	84	948	781 ppmv	169 ppmv
59.5	2.595	106	0.4558	182	ND	67	ND	ND
56.64	ND	33.81	2.67	192	169	ND	11.2 ppmv	3.0 ppmv
67.37	ND	58.63	3.71	156	214	ND	11.3 ppmv	14.5 ppmv
76.51	ND	36.48	1.019	119	66	ND	6.8 ppmv	0.5 ppmv

Continued

TABLE A.1 Composition of Fluid End Members in the Forearc of Well-Studied Subduction Zones—cont'd

Li (µM)	Rb (µM)	Sr (µM)	Ba (µM)	B (µM)	Si (µM)	F (µM)	C_2	C_3
10.4	ND	63.4	2.82	187	112	ND	10.4	0 ppmv
192	ND	67.8	20.2	325	379	ND	95 ppmv	19.1 ppmv
NS	NS	NS	NS	NS	NS	NS	NS	NS
104	ND	112	40.3	733	ND	BDL	36 ppmv	5.1 ppmv
83.7	ND	208	101.7	1303	ND	ND	11.3 ppmv	NS
84	ND	54.6	73	188	794	ND	313.9 ppmv	12.8 ppmv
0.5	ND	11.5	500	3228	69	81	1432 ppmv	20.5 ppmv
13.2	ND	20.4	ND	1786	599	ND	ND	ND

Continued

TABLE A.1 Composition of Fluid End Members in the Forearc of Well-Studied Subduction Zones—cont'd

C_4	$\delta^{18}O$	δD	$^{87}Sr/^{86}Sr$	$\delta^{11}B$	δ^7Li	$\delta^{37}Cl$	Notes/References
0.615 μL/kg	0.55	−10.78	0.708	28	ND	−7.5	Most data from initial reports, B from You Sci Res, $\delta^{11}B$ from You Scie Results, F from Wei 2008, $\delta^{18}O$ and δD from Kastner 1993, $^{87}Sr/^{86}Sr$ from Kastner 1993, $\delta^{37}Cl$ from Wei 2008
ND	ND	ND	ND	ND	ND	−7.5	Most data from initial reports, Cl isotopes and F from Wei 2008
ND	ND	ND	ND	ND	ND	−6	Data from initial reports, Cl isotope and F data from Wei et al., 2008
ND	−1.2	−1	0.707504	ND	ND	ND	Same from the initial reports, majors and Sr isotopes from Gieskes et al., $\delta^{18}O/\delta D$ from Vrolijk et al.
NS	−2.2	2	NS	ND	ND	ND	Cl from initial reports, $\delta^{18}O/\delta D$ from Vrolijk et al.
ND	−1.2	−0.9	ND	ND	ND	ND	Major chem from Gieskes et al., $\delta^{18}O/\delta D$ from Vrolijk et al.
ND	ND	ND	ND	ND	ND	ND	Most initial reports, Ba and Rb from Zheng and Kastner.
29 ppmv	ND	ND	0.707379	36.1	24.73	−5.6	Majors from initial reports, Ba from Solomon and Kastner, 2012, F, Rb, $^{87}Sr/^{86}Sr$, δ^7Li from Kastner et al., 2006, Cl isotopes from Wei et al., 2008, $\delta^{11}B$ from Kopf et al., 2001.
ND	−0.1	ND	0.70969	ND	25.2	ND	All data from Solomon et al., 2009.

TABLE A.1 Composition of Fluid End Members in the Forearc of Well-Studied Subduction Zones—cont'd

C_4	$\delta^{18}O$	δD	$^{87}Sr/^{86}Sr$	$\delta^{11}B$	δ^7Li	$\delta^{37}Cl$	Notes/References
1.5 ppmv	ND	ND	0.7084	ND	23.61	ND	Majors, Cl, and C_1–C_4 from initial reports. Minors and Si from Torres and Solomon, unpublished data, Sr and Li isotopes from Torres and Solomon, unpublished data.
5.7 ppmv	ND	ND	0.7082	ND	23.73	ND	Majors, Cl, and C_1–C_4 from initial reports. Minors and Si from Torres and Solomon, unpublished data, Sr and Li isotopes from Torres and Solomon, unpublished data.
0 ppmv	ND	ND	0.70775	ND	ND	ND	Majors, Cl, and C_1–C_4 from initial reports. Minors and Si from Torres and Solomon, unpublished data, Sr and Li isotopes from Torres and Solomon, unpublished data.
0 ppmv	ND	ND	ND	ND	ND	ND	Data from initial reports.
ND	−2.37	−12	0.707716	ND	20.39	ND	Data from initial reports except Sr and Li isotopes which are Solomon unpublished data.
NS	NS	NS	NS	ND	ND	ND	Intitial reports.
ND	−0.73	−10.7	ND	ND	ND	ND	Initial reports, Fl from Dickens, $\delta^{18}O/\delta D$ from Tomaru et al.
ND	−0.58	−7.2	ND	ND	ND	ND	Initial reports, $\delta^{18}O/\delta D$ from Tomaru et al.
ND	ND	ND	ND	ND	ND	ND	Initial reports.
0.5 ppmv	ND	ND	ND	ND	ND	ND	
3.5 ppmv	ND	ND	ND	16.1	ND	1.8	Initial reports, F and B isotopes from Wei et al., 2005; Cl isotopes from Wei et al., 2008.
ND	4.8	−1.6	0.708699	ND	29.7	ND	$\delta^{18}O/\delta D$ from Hensen et al., 2004; Li, δ^7Li, $^{87}Sr/^{86}Sr$ from Scholz et al., 2012; the rest from Tryon et al., 2010.

ACKNOWLEDGMENTS

We thank the US National Science Foundation for providing financial support, in particular for awards OCE #1153870 to M.K., NSF OCE-1233587 to E.A.S., and NSF OCE–0637120 and OCE-0652315 to R.N.H., and NSF OCE-1061189 to M.E.T. This analysis/synthesis used samples and data provided by ODP and IODP, sponsored by the US National Science Foundation and participating countries under management of Joint Oceanographic Institutions Inc. Funding was also provided by US Science Support Program grants to all the authors. This manuscript has benefited from helpful comments from two reviewers, Drs Glen A. Spinelli and C. Geoff Wheat. We also thank Dr Ellen Thomas for helpful comments on the evolution of biogenic sediments through time.

REFERENCES

Adkins, J. F., & Schrag, D. P. (2001). Pore fluid constraints on deep ocean temperature and salinity during the last glacial maximum. *Geophysical Research Letters*, *28*(5), 771–774.

Alt, J. C. (2004). Alteration of the upper oceanic crust: mineralogy, chemistry, and processes. In E. Davis, & H. Elderfield (Eds.), *Hydrogeology of the oceanic lithosphere* (pp. 495–533). Cambridge University Press.

Alt, J. C., & Honnorez, J. (1984). Alteration of the upper oceanic crust, deep sea drilling project, site 417: mineralogy and chemistry. *Contributions to Mineralogy and Petrology*, *87*, 149–169.

Alt, J. C., Teagle, D. A. H., Laverne, C., Vanko, D. A., Bach, W., Honnorez, J., et al. (1996). Ridge flank alteration of upper ocean crust in the eastern Pacific: synthesis of results for volcanic rocks of holes 504B and 896A. *Proceedings of the Ocean Drilling Program Scientific Results*, *148*, 435–450.

Audet, P., Bostock, M. G., Christensen, N. I., & Peacock, S. M. (2009). Seismic evidence for over-pressured subducting oceanic crust and megathrust fault sealing. *Nature*, *457*, 76–78.

Audet, P., & Schwartz, S. Y. (2013). Hydrologic control of forearc strength and seismicity in the Costa Rica subduction zone. *Nature Geoscience*, *6*, 852–855.

Bebout, G. E. (1996). Volatile transfer and recycling at convergent margins: mass-balance and insights from high-P/T metamorphic rocks. In G. E. Bebout, D.,W. Scholl, S. H. Kirby, & J. P. Platt (Eds.), *Subduction top to bottom* (Vol. 96) (pp. 179–193). Amer. Geophys. Union.

Becker, K., Sakai, H., Adamson. A. C., Alexandrovich, J., Alt, J. C., Anderson, R. N., et al. (1990). Drilling deep into oceanic crust. Hole 504B. *Costa Rica Rift Reviews of Geophysics*, *27*, 79–102.

Bekins, B. A., McCaffrey, A. M., & Dreiss, S. J. (1994). Influence of kinetics on the smectite to illite transformation in the Barbados accretionary prism. *Journal of Geophysical Research*, *99*(18), 147–158.

Bekins, B. A., McCaffrey, A. M., & Dreiss, S. J. (1995). Episodic and constant flow models for the origin of low-chloride waters in a modern accretionary complex. *Water Resources Research*, *31*, 3,205–3,215.

Benton, L. D., Ryan, J. G., & Savov, I. P. (2004). Lithium abundances and isotopic systematics of forearc serpentinites, Conica Seamount, Mariana forearc: insights into the mechanism of slab-mantle exchange during subduction. *Geochemistry, Geophysics, Geosystems (G3)*, *5*, Q08J12.

Bish, L. D., & Ming, D. W. (Eds.). (2001). *Natural zeolites: occurrence, properties, applications. Reviews in mineralogy and geochemistry* (Vol. 45) (p. 654). Mineral. Soc. Amer.

Blättler, C., Henderson, G. M., & Jenkyns, H. C. (2012). Explaining the Phanerozoic Ca isotope history of seawater. *Geol.*, *40*(9), 843–846.

Bonnatti, E., Lawrence, J., & Morandi, N. (1984). Serpentinization of ocean peridotites: temperature dependence of mineralogy and boron content. *Earth and Planetary Science Letters, 70*, 88–94.

Brown, K. M., Tryon, M., DeShon, H., Dorman, L. M., & Schwartz, S. (2005). Correlated transient fluid pulsing and seismic tremor in the Costa Rica subduction zone. *Earth and Planetary Science Letters, 238*, 189–203.

Carlson, R. W., Boyd, F. R., Shirely, S. B., Janney, P. E., Groves, T. L., Bowring, S. A., et al. (2000). Continental growth, preservation, and modification in southern Africa. *GSA Today, 10*(2), 1–6.

Carson, C., Clarke, G., & Powell, R. (2000). Hydration of eclogite, Pam Peninsula, New Caledonia. *Journal of Metamorphic Geology, 18*(1), 79–90.

Carson, B., Kastner, M., Bartlett, D., Jaeger, J., Jannasch, H., & Weinstein, Y. (2003). Implications of carbon flux from the Cascadia accretionary prism: results from long-term measurements at ODP Site 892B. *Marine Geology, 198*, 159–180.

Carson, B., & Screaton, E. J. (1998). Fluid flow in accretionary prisms: evidence for focused, time-variable discharge. *Reviews of Geophysics, 36*, 329–351.

Carson, B., Suess, E., & Strasser, J. C. (1990). Fluid flow and mass flux determinations at vent sites on the Cascadia margin accretionary prism. *Journal of Geophysical Research, 95*(B6), 8,891–8,897.

Chan, L. H., Alt, J. C., & Teagle, D. A. H. (2002). Lithium and lithium isotope profile through the upper oceanic crust: a study of seawater-basalt exchange at ODP sites 504B and 896A. *Earth and Planetary Science Letters, 201*, 187–201.

Chan, L. H., & Kastner, M. (2000). Lithium isotopic compositions of pore fluids and sediments in the Costa Rica subduction zone: Implications for fluid processes and sediment contribution to the arc volcanoes. *Earth and Planetary Science Letters, 183*, 275–290.

Claypool, G., & Kaplan, I. R. (1974). The origin and distribution of methane in marine sediments. In I. R. Kaplan (Ed.), *Natural gases in marine sediments* (pp. 99–139). New York: Plenum.

Clift, P., & Vannucchi, P. (2004). Controls on tectonic accretion versus erosion in subduction zones: implications for origin and recycling of continental crust. *Reviews of Geophysics, 42*, RG2001.

Colten-Bradley, V. A. (1987). Role of pressure in smectite dehydration; Effects on geopressure and smectite-to-illite transformation. *American Association of Petroleum Geologists Bulletin, 71*, 1414–1427.

Condie, K. C. (2005). *Earth as an evolving planetary system*. Amsterdam: Elsevier Academic Press. 447 pp.

Davidson, D. W., El-Defrawy, M. K., Fuglem, M. O., & Judge, A. S. (1978). Natural gas hydrates in northern Canada. Proc. 3rd Int. Conf. *Permafrost, 3*, 937–943.

Davis, E., Heesemann, M., & Wang, K. (2011). Evidence for episodic aseismic slip across the subduction seismogenic zone off Costa Rica: CORK borehole pressure observations at the subduction prism toe. *Earth and Planetary Science Letters, 306*, 299–305.

Davis, E. E., Hyndman, R. D., & Villinger, H. (1990). Rates of fluid expulsion across the northern Cascadia accretionary prism: constraints from new heat flow and multichannel seismic reflection data. *Journal of Geophysical Research, 95*(B6), 8,869–8,889.

Davis, E. E., & Villinger, H. W. (2006). Transient formation of fluid pressures and temperatures in the Costa Rica forearc prism and subducting oceanic basement: CORK monitoring at ODP Sites1253 and 1255. *Earth and Planetary Science Letters, 245*, 232–244.

Deyhle, A., Kopf, A., & Aloisi, G. (2003). Boron and boron isotopes as a tracer for diagenetic reactions and depth of mobilization, using muds and authigenic carbonates from eastern Mediterranean mud volcanoes. *Geological Society of London Special Publications, 216*, 493–505.

Dumitru, T. A. (1991). Effects of subduction parameters on geothermal gradients in forearcs, with an application to Franciscan subduction in California. *Journal of Geophysical Research, 96*(B1), 621–641.

Elderfield, H., Kastner, M., & Martin, J. B. (1990). Composition and sources of fluids in sediments of the Peru subduction zone. *Journal of Geophysical Research, 95,* 8819–8828.

Elderfield, H., & Schultz, A. (1996). Mid-ocean ridge hydrothermal fluxes and the chemical composition of the ocean. *Annual Review of Earth and Planetary Sciences, 24,* 191–224.

Ernst, W. G. (1970). Tectonic contact between the Franciscan mélange and the Great Valley Sequence – crustal expression of a late Mesozoic Benioff zone. *Journal of Geophysical Research, 75,* 886–901.

Ernst, W. G. (1990). Thermobarometric and fluid expulsion history of subduction zones. *Journal of Geophysical Research, 95,* 9047–9053.

Ernst, W. G., & Calvert, S. E. (1969). An experimental study of the recrystallization of porcelanite and its bearing on the origin of some bedded cherts. *American Journal of Science, 267-A,* 114–133.

Ferguson, I.J. (1990). *Numerical modeling of heat flow and fluid flow in subduction-accretion complexes* (Ph.D. thesis). Birmingham, U.K: Univ. of Birmingham.

Fisher, D., & Byrne, T. (1992). Strain variations in an ancient accretionary complex: Implications for forearc evolution. *Tectonics, 11,* 330–347.

Fisher, A. T., Davis, E. E., Hutnak, M., Speiss, V., Zuehlsdorff, L., Cherkaoui, A., et al. (2003). Hydrothermal recharge and discharge across 50 km guided by seamounts on a young ridge flank. *Nature, 421,* 618–621 *Geochimica et Cosmochimica Acta, 57,* 4377–4389.

Fisher, A. T., & Hunslow, M. W. (1990). Transient fluid fow through the toe of the Barbados accretionary complex: constraints from Ocean Drilling Program Leg 110 heat flow studies and simple models. *J. Geophys, Res., 95,* 8,845–8,858.

Foucher, J. P., Le Pichon, X., Lallemant, S., Hobart, M. A., Henry, P., Benedetti, M., et al. (1990). Heat flow, tectonics and fluid circulation in the toe of the Barbados Ridge accretionary prism. *Journal of Geophysical Research, 95,* 8,859–8,867.

Fouilac, C., & Michard, G. (1981). Sodium/lithium ratio in water applied to geothermometry of geothermal reservoirs. *Geothermics, 10,* 55–70.

Fournier, R. O. (1973). Silica in thermal waters: laboratory and field investigation. In *Proceedings of the International Symposium on Hydrogeochemistry and Biogeochemistry, Japan 1970* (pp. 122–139). Washingotn, D. C: J. W. Clark.

Fournier, R. O. (1979). A revised equation for the Na/K thermometer. *Transactions on Geothermal Resources Council, 3,* 221–224.

Fournier, R. O., & Rowe, J. J. (1966). Estimation of underground temperatures from the silica content of water, hot springs and steam wells. *American Journal of Science, 264,* 685–697.

Fryer, P. (1996). Evolution of Mariana convergent plate margin system. *Reviews of Geophysics, 34,* 89–125.

Fryer, P., Wheat, C. G., & Mottl, M. J. (1999). Mariana blueschist mud volcanism: implications for conditions within the subduction zone. *Geology, 27,* 103–106.

Füri, E., Hilton, D., Tryon, M. D., Brown, K. M., McMurtry, G. M., Brückmann, W., et al. (2010). Carbon release from submarine seeps at the Costa Rica fore arc: Implications for the volatile cycle at the Central America convergent margin. *Geochemistry, Geophysics, Geosystems (G3), 11.* http://dx.doi.org/10.1029/2009GC002810.

Furst, M. J. (1981). Boron in siliceous materials as a paleosalinity indicator. *Geochimica et Cosmochimica Acta, 45,* 1–13.

Fyfe, W. S., & McBirney, A. R. (1975). Subduction and the structure of andesitic volcanic belts. *American Journal of Science, 275A,* 285–297.

German, C., & von Damm, K. (2004). In K. K. Turekian, & D. Holland (Eds.), *Treatise on geochemistry Seawater chemistry: Vol. 6.* (pp. 181–222). Elsevier.

Gieskes, J. M., & Lawrence, J. R. (1981). Alteration of volcanic matter in deep-sea sediments: evidence from the chemical composition of interstitial waters from deep-sea drilling cores. *Geochimica et Cosmochimica Acta, 45,* 1,687–1,703.

Gill, J. B. (1981). *Orogenic andesites and plate tectonics.* Berlin: Springer-Verlag.

Hacker, B. R. (2008). H_2O subduction beyond arcs. *Geochemistry, Geophysics, Geosystems, 9,* QO3001. http://dx.doi.org/10.1029/2007/gc001707.

Hamilton, W. B. (1969). Mesozoic California and the underflow of Pacific mantle. *Geological Society of America Bulletin, 80,* 2409–2430.

Harris, R. N., & Chapman, D. S. (2004). In: E. Davis, & H, Elderfield (Eds.), *Deep-seated oceanic heat flux, heat deficits, and hydrothermal circulation, hydrology of the oceanic lithosphere* (pp. 311–336).

Harris, R. N., Grevemeyer, I., Ranero, C. R., Villinger, H., Barckhausen, U., Henke, T., et al. (2010). Thermal regime of Costa Rica convergent margin: 1. Along-strike variations in heat flow from probe measurements and estimated from bottom-simulating reflectors. *Geochemistry, Geophysics, Geosystems, 11,* Q12S28. http://dx.doi.org/10.1029/2010GC003272.

Harris, R. N., Spinelli, G., Ranero, C. R., Grevemeyer, I., Villinger, H., & Barckhausen, U. (2010). Thermal regime of the Costa Rican convergent margin: 2. Thermal models of the shallow Middle America subduction zone offshore Costa Rica. *Geochemistry, Geophysics, Geosystems, 11*(12), Q12S29.

Harris, R., Yamano, M., Kinoshita, M., Spinelli, G., Hamamoto, H., & Ashi, J. (2013). A synthesis of heat flow determinations and thermal modeling along the Nankai Trough, Japan. *Journal of Geophysical Research Solid Earth, 118.* http://dx.doi.org/10.1002/jgrb.50230.

Heesemann, M., Grevemeyer, I., & Villinger, H. (2009). Thermal constraints on the frictional conditions of the nucleation and rupture area of the 1992 Nicaragua tsunami earthquake. *Geophysical Journal International, 179*(3), 1265–1278.

Henry, P. (2000). Fluid fllow at the toe of the Babrbados accretionary wedge constrained by thermal, chemical, and hydrologic observations and models. *J. Geophys. Res., 105,* 25,855–25,872.

Henry, P., Le Pichon, X., Lallement, S., Lance, S., Marting, J. B., Foucher, J.-P., et al. (1996). Fluid flow in and around a mud volcano field seaward of the Barbados accretionary wedge: results from Manon cruise. *Journal of Geophysical Research, 101*(B9), 20,297–20,323.

Hensen, C., Wallmann, K., Schmidt, M., Ranero, C., & Suess, E. (2004). Fluid expulsion related to mud extrusion off Costa Rica – a window to the subducting slab. *Geology, 32*(3), 201–204.

Hilton, D. R., Fischer, T. P., & Marty, B. (2002). Noble gases and volatile recycling at subduction zones. In J. Masuda (Ed.), *Noble gas geochemistry and cosmochemistry, Rev. Mineral. Geochem.* (Vol. 47) (pp. 319–370). Mineral. Soc. Am.

Hower, J., Eslinger, E. V., & Hower, M. E. (1976). Mechanism of burial metamorphism of argillaceous sediment: mineralogical and chemical evidence. *Geological Society of America Bulletin, 87,* 235–737.

Hutnak, M., Fisher, A., Harris, R., Stein, C., Wang, K., Spinelli, G., et al. (2008). Large heat and fluid fluxes driven through mid-plate outcrops on ocean crust. *Nature Geoscience, 1*(9), 611–614.

Hutnak, M., Fisher, A., Stein, C., Harris, R., Wang, K., Silver, E., et al. (2007). The thermal state of 18-24 Ma upper lithosphere subducting below the Nicoya Peninsula, northern Costa Rica margin. In T. H. Dixon, & C. J. Moore (Eds.), *The seismogenic zone of subduction thrust faults* (pp. 86–122). New York: Columbia University Press.

Hyndman, R. D., & Peacock, S. M. (2003). Sepentinization of the forearc mantle. *Earth and Planetary Science Letters, 212,* 417–432.

Hyndman, R. D., & Wang, K. (1993). Thermal constraints on the zone of major thrust earthquake failure: the Cascadia subduction zone. *Journal of Geophysical Research, 98*(B2), 2039–2060.

Ike, T., Moore, G. F., Kuramoto, S., Park, J. O., Kaneda, Y., & Taira, A. (2008). Variations in sediment thickness and type along the northern Philippine Sea Plate at the Nankai Trough. *Island Arc, 17*(3), 342–357.

Jarrard, R. D. (2003). .Subduction fluxes of water, carbon dioxide, chlorite, and potassium. *Geochemistry, Geophysics, Geosystems (G3), 4*(5), 8905.

Kastner, M. (1981). Authigenic silicates in deep-sea sediments: formation and diagenesis. In C. Emiliani (Ed.), *The oceanic lithosphere The sea: Vol. 7.* (pp. 915–980). Wiley.

Kastner, M., Elderfield, H., Jenkins, W. J., Gieskes, J. M., & Gamo, T. (1993). Geochemical and isotope evidence for fluid flow in the western Nankai subduction zone, Japan. In L. A. Hill, et al. (Ed.), *Proceedings of the ocean drilling program, scientific results* (Vol. 131) (pp. 397–413).

Kastner, M., Elderfield, H., & Martin, J. B. (1991). Fluids in convergent margins: what do we know about their composition, origin, role in diagenesis and importance for oceanic chemical fluxes. *Philosophical Transactions of the Royal Society – London A, 335,* 243–259.

Kastner, M., Hilton, D. R., Jenkins, W. J., Solomon, E. A., & Spivack, A. J. (2013). He isotope ratios in the Nankai Trough and Costa Rica subduction zones – implications for volatile cycling. *Trans. Amer. Geophysical Union, T34C, 2013.*

Kastner, M., Keene, J. B., & Gieskes, J. M. (1977). Diagenesis of siliceous oozes-I. Chemical controls on the rate of opal-A to opal-CT transformation-and experimental study. *Geochimica et Cosmochimica Acta, 41,* 1041–1059.

Kastner, M., Solomon, E., Wei, W., Chan, L. H., & Saether, O. M. (2006). Chemical and isotopic compositions of pore fluids and sediments from across the Middle America Trench, offshore Costa.Rica. J. Morris, H. Villinger, & A. Klaus (Eds.). *Proceedings of the Ocean Drilling Program Scientific Results, 205,* 1–21.

Kerrick, D. M., & Connolly, J. A. D. (2001). Metamorphic devolatilization of subducted marine sediments and the transport of volatiles into the Earth's mantle. *Nature, 411,* 293–296.

Kharaka, Y. K., & Mariner, R. H. (1989). Chemical geothermometers and their applications to formartion waters from sedimentary basins. In N. D. Naesser, & T. H. McCulloch (Eds.), *Thermal history of sedimentary basins: Methods and case histories* (pp. 99–117). New York: Springer-Verlag.

Kinoshita, H., & Yamano, M. (1986). In *The heat flow anomaly in the nankai trough area, initial reports of the deep sea drilling project* (Vol. 87) (pp. 737–743).

Kluesner, J. W., Silver, E. A., Gibson, J., Bangs, N. L., McIntosh, K. W., Orange, D., et al. (2013). High density of structurally controlled, shallow to deep water fluid seep indicators imaged offshore Costa Rica. *Geochemistry, Geophysics, Geosystems (G³), 14*(3), 519–539.

Kolodny, Y., Taraboulos, A., & Frieslander, U. (1980). Participation of fresh water in chert diagenesis: evidence from oxygen isotopes and boron α–track mapping. *Sedimentology, 27,* 305–316.

Kopf, A. J. (2002). Significance of mud volcanism. *Reviews of Geophysics, 40*(2), 1005–1057.

Kopf, A., Mora, G., Deyhle, A., Frape, S., & Hesse, R. (2003). *Fluid geochemistry in the Japan Trench forearc (ODP Leg, 186),* 1–23 College Station TX, Ocean Drilling Program.

Kulm, L. D., Suess, E., Moore, J.C., Carson, B., Lewis, B.T., Ritger S.D., et al. (1986). Oregon subduction zone: Venting, fauna, and carbonates. *Science, 231,* 561–566.

Kvenvolden, K. A. (1993). Gas hydrates: geological perspectives and global change. *Reviews of Geophysics, 31,* 173–181.

Labaume, P., Kastner, M., Trave, A., & Henry, P. (1997). Carbonate veins from the décollement zone at the toe of the northern Barbados accretionary prism: microstructure, mineralogy, geochemistry, and relations with prism structures and fluid regimes. *Proceedings of the Ocean Drilling Program Scientific Results, 156,* 79–96.

Lallemand, S. E., Schnule, P., & Malavieille, J. (1994). Coulomb theory applied to accretionary and nonaccretionary wedges: possible causes for tectonic. *Journal of Geophysical Research, 99*(B6), 12,033–12,055.

Langseth, M., & Silver, E. (1996). The Nicoya convergent margin: a region of exceptionally low heat flow. *Geophysical Research Letters, 23,* 891–894.

Lauer, R. M., & Saffer, D. M. (2012). Fluid budgets of subdution zone forearcs: the contribution of splay faults. *Geophysical Research Letters, 39,* L13604. http://dx.doi.org/10.1029/20 12GL052182.

Lawrence, J. R., Drever, J. L., Anderson, T. F., & Brueckner, H. K. (1979). Importance of volcanic alteration in sediments of Site 323: chemistry, O^{18}/O^{16}, Sr^{87}/Sr^{86}. *Geochimica et Cosmochimica Acta, 43,* 573–588.

Lawrence, J. R., & Gieskes, J. M. (1981). Constraints on water transport and alteration in the oceanic crust from the isotopic composition of pore water. *Journal of Geophysical Research, 86,* 7924–7934.

Le Pichon, X., Foucher, J. P., Boulegue, J., Henry, P., Lallemannt, S., Benedetti, M., et al. (1990). Mud volcano field seaward of the Barbados accretionary complex: a submersible survey. *Journal of Geophysical Research, 95,* 8,931–8,943.

Le Pichon, X., & Henry, P. & the Kaiko Nankai ScientificCrew. (1991). Water budget in the accretionary wedge: a comparison. *Philos. Trans. R. Soc. London A, 335,* 315–330.

Le Pichon, X., Kobayashi, K., & Kaiko-Nankai Scientific Crew (1992). Fluid venting activity within the eastern Nankai Trough accretionary wedge: a summary of the 1989 Kaiko-Nankai results. *Earth and Planetary Science Letters, 109,* 303–313.

Liu, Y., & Rice, J. R. (2007). Spontaneous and triggered seismic deformation transients in a subduction fault model. *Journal of Geophysical Research, 112,* B09404.

Martin, J. B., & Kastner, M. (1996). Chemical and isotopic evidence for sources of fluids in mud volcano field seaward of the Barbados accretionary wedge. *J. Geophys. Res., 101,* 20,325–20,345.

Martin, J. B., Kastner, M., & Elderfield, H. (1991). Lithium: sources in pore fluids of Peru Slope sediments and implications for oceanic fluxes. *Marine Geology, 102,* 281–292.

Martin, J. B., Kastner, M., Henry, P., LePichon, X., & Lallement, S. (1996). Chemical and isotopic evidence for sources of fluids in a mud volcano field seaward of the Barbados accretionary wedge. *Journal of Geophysical Research, 101,* 20,325–20,345.

Marty, B., & Tolstikhin, I. N. (1998). CO$_2$ fluxes from mid-ocean ridges, arcs and plumes. *Chemical Geology, 145,* 233–245.

McCaffrey, R. (1993). On the role of the upper plate in great subduction zone earthquakes. *Journal of Geophysical Research, 98*(B7), 11953–11966.

McCaffrey, R. (1997). Influences of recurrence times and fault zone temperatures on the age-rate dependence of subduction zone seismicity. *Journal of Geophysical Research, 102,* 22.

McLennan, S. M. (1988). Recycling of the continental crust. *Pure and Applied Geophysics, 28,* 683–724.

Molnar, P., & England, P. (1995). Temperatures in zones of steady-state underthrusting of young oceanic lithosphere. *Earth and Planetary Science Letters, 131*(1–2), 57–70.

Moore, G. F., Shipley, T. H., Stoffa, P. L., Karig, D. E., Taira, A., et al. (1990). Structure of the Nankai Trough accretionary zone from multichannel seismic reflection data. *J. Geophys. Res., 95,* 8,753–8,765.

Moore, G. F., Taira, A., Klaus, A., Becker, L., Boeckel, B., Cragg, B.A., et al. (2001). New insights into deformation and fluid flow processes in the Nankai Trough accretionary prism; results of Ocean Drilling Program Leg 190. *Geochemistry, Geophysics, Geosystems, 2*(10). http://dx.doi.org/10.1029/2001GC000166.

Moore, J. C., & Saffer, D. (2001). Updip limit of the seismogenic zone beneath the accretionary prism of southwest Japan: an effect of diagenetic to low-grade metamorphic processes and increasing effective stress. *Geology, 29,* 183–186.

Moore, J. C., & Vrolijk, P. (1992). Fluids in accretionary prisms. *Reviews of Geophysics, 30,* 113–135.

Morris, J., & Tera, F. (1989). [10]Be and [9]Be in mineral separates and whole rocks from volcanic arcs: Implications for sediment subduction. *Geochimica et Cosmochimica Acta, 53*(12), 3197–3206.

Mottl, M. J. (2003). Partitioning of energy and mass fluxes between mid-ocean ridge axes and flanks at high and low temperature. In P. E. Halbach, V. Tunnicliffe, & J. R. Hein (Eds.), *Energy and mass transfer in marine hydrothermal systems* (pp. 271–286). Berlin: Dahlem University Press. 365 pp.

Murata, K. J., Friedman, I., & Gleason, J. D. (1977). Oxygen isotope relations between diagenetic silica minerals in Monterey Shale, Temblor Range, California. *American Journal of Science, 277,* 259–272.

Newman, A., Schwartz, S., Gonzalez, V., DeShon, H., Protti, J., & Dorman, L. (2002). Along-strike variability in the seismogenic zone below Nicoya Peninsula, Costa Rica. *Geophysical Research Letters, 29*(20), 31–38.

O'Neil, J. R. (1968). Hydrogen and oxygen isotope fractionation between ice and water. *J. Phys. Chem., 72,* 3,683–3,684.

O', S. C., Dando, P. R., Schuster, U., Bennis, A., Boyle, J. D., Chui, F. T. W., et al. (1995). Gas seep induced interstitial water circulation: observations and environmental implications. *Continental Shelf Research, 15*(8), 931–948.

Peacock, S. M. (1990). Fluid processes in subduction zones. *Science, 248,* 329–337.

Peacock, S. M. (1993). The importance of blueschist to eclogite dehydration reactions in subducting oceanic crust. *Geological Society of America Bulletin, 105,* 684–694.

Peacock, S. M., & Hyndman, R. D. (1999). Hydrous minerals in the mantle wedge and the maximum depth of subduction thrust earthquakes. *Geophysical Research Letters, 26,* 2517–2531.

Perry, E. A., & Hower, J. (1970). Burial diagenesis in Gulf Coast pelitic sediments. *Clays Clay Minerals, 18,* 165–177.

Perry, E. A., & Hower, J. (1972). Late-stage dehydration in deeply buried pelitic sediments. *Bulletin American Association of Petroleum Geologists, 56,* 2013–2021.

Plank, T., & Langmuir, C. (1993). Tracing trace elements from sediment input to volcanic output at subduction zones. *Nature, 362,* 739–742.

Plank, T., & Langmuir, C. H. (1998). The chemical composition of subducting sediment and its consequences for the crust and mantle. *Chemical Geology, 145,* 325–394.

Pytte, A. M., & Reynolds, R. C. (1988). The thermal transformation of smectite to illite. In N. D. Naeser, & T. M. McCulloh (Eds.), *Thermal histories of sedimentary basins* (pp. 133–140). New York: Springer.

Ranero, C. R., Grevemeyer, I., Sahling, H., Barckhausen, U., Hensen, C., Wallmann, K., et al. (2008). Hydrogeological system of erosional convergent margins and its influence on tectonics and interplate seismogenesis. *Geochemical, Geophysics, Geosystems, 9*(3), Q03S04.

Ranero, C. R., Phipps Morgan, J., Mcintosh, K., & Reichert, C. (2003). Bending-related faulting and mantle serpentinization at the Middle America trench. *Nature, 425,* 367–373.

Ranero, C. R., & von Huene, R. (2000). Subduction erosion along the Middle America convergent margin. *Nature, 404,* 748–752.

Ransom, B., Spivack, A. J., & Kastner, M. (1995). Stable chlorine isotopes in subduction zone pore waters: implications for fluid-rock reactions and the cycling of chlorine. *Geology, 23,* 712–715.

Rea, D. K., & Ruff, L. J. (1996). Composition and mass flux of sediment entering the world's subduction zones: Implications for global sediment budgets, great earthquakes, and volcanism. *Earth and Planetary Science Letters, 140*, 1–12.

Ridgewell, A. (2005). Mid Mesozoic revolution in the regulation of ocean chemistry. *Mar. Geol., 217*, 339–357.

Rüpke, L. H., Morgan, J. P., & Connolly, J. A. D. (2004). Serpentine and the subduction zone water cycle. *Earth and Planetary Science Letters, 223*, 17–34.

Ruppel, C., & Kinoshita, M. (2000). Fluid, methane, and energy flux in an active margin gas hydrate province, offshore Costa Rica. *Earth and Planetary Science Letters, 179*(1), 153–165.

Saffer, D. M. (2003). Pore pressure development and progressive dewatering in underthrust sediments at the Costa Rica subduction margin: comparison with northern Barbados and Nankai. *J Geophys. Res., 108*, 2261.

Saffer, D. M., & Bekins, B. A. (1998). Episodic fluid flow in the Nankai accretionary complex: timescale, geochemistry, and fluid budget. *J. Geophys. Res., 103*, 30,351–30,370.

Saffer, D. M., & Screaton (2003). Fluid flow at the toe of convergent margins: Interpretation of sharp pore-water geochemical gradients. *Earth and Planetary Science Letters, 213*, 261–270.

Saffer, D. M., Silver, E. A., Fisher, A. T., Tobin, H., & Moran, K. (2000). Inferred pore pressures at the Costa Rica subduction zone: Implications for dewatering processes. *Earth and Planetary Science Letters, 177*, 193–207.

Saffer, D. M., & Tobin, H. J. (2011). Hydrogeology and mechanics of subduction zone forearcs: fluid flow and pore pressure. *Annual Reviews of Earth and Planetary Sciences, 39*, 157–186.

Saffer, D. M., Underwood, M. B., & Mckiernan, A. W. (2008). Evaluation of factors controlling smectite transformation and fluid production in subduction zones: applications to Nankai Trough. *Island Arc, 17.* 208–230. http://dx.doi.org/10.1111/j.1440-1738.2008.00614.x.

Sahling, H., Masson, D. G., Ranero, C. R., Hühnerbach, V., Weinrebe, W., Klaucke, I., et al. (2008). Fluid seepage at the continental margin offshore Costa Rica and southern Nicaragua. *Geochemical, Geophysics, Geosystems (G3), 9*(5), Q05S05. http://dx.doi.org/10.1029/200 8GC001978.

Schmidt, M., Hensen, C., Mörz, T., Müller, C., Grevemeyer, I., Wallmann, K. et al. (2005). Methane hydrate accumulation in "Mound 11" mud-volcano, Costa Rica forearc. *Mar. Geol., 216*, 83–100.

Schmidt, M. W., & Poli, S. (1998). Experimentally based water budgets for dehydrating slabs and consequences for arc magma generation. *Earth and Planetary Science Letters, 163*, 361–379.

Schoell, M. (1980). The hydrogen and carbon isotope composition of methane from natural gases of various sources. *Geochimica et Cosmocimica Acta, 44*, 641–661.

Scholl, D. W., & von Huene, R. (2007). Crustal cycling at modern subduction zones applied to the past – issues of growth and preservation of continental basement crust, mantle geochemistry, and supercontinent reconstruction. *Geological Society of America Memoirs, 200*, 9–32.

Screaton, E. J., Wuthrich, D. R., & Dreiss, S. J. (1990). Permeabilities, fluid pressures, and flow rates in the Barbados Ridge Complex. *J.Geophys. Res., 95*, 8,997–9,007.

Selley, R. C. (1998). *Elements of petroleum geology.* Academic Press, 470 pp.

Seyfried, W. E., Chen, X., & Chan, L.-H. (1998). Trace element mobility and lithum exchange during hydrothermal alteration of seafloor weatherd basalt; an experimental study at 350 degrees C, 500 bars. *Ceochim. Cosmochim. Acta, 62*, 949–960.

Seyfried, W. E., Janecky, P. R., & Mottle, M. J. (1984). Alteration of the oceanic crust: Implications for geochemical cycles of lithium and boron. *Geochimica et Cosmochimica Acta, 48*, 557–569.

Sheppard, S. M. E., & Gilg, H. A. (1996). Stable isotope geochemistry of clay minerals. *Clay Minerals, 31,* 1–24.

Silver, E., Kastner, M., Fisher, A., Morris, J., McIntosh, K., & Saffer, D. (2000). Fluid flow paths in the Middle America Trench and Costa Rica margin. *Geology, 28*(8), 679–682.

Simpson, G. D. H. (2010). Formation of accretionary prisms influenced by sediment subduction and supplied by sediments from adjacent continents. *Geology, 38,* 131–134.

Sloan, D. E., & Koh, C. A. (2008). *Clathrate hydrates of natural Gases* (3rd ed.). CRC Press, Taylor & Francis Group, LLC, 721 pp.

Solomon, E. A., & Kastner, M. (2012). Progressive barite dissolution in the Costa Rica forearc – implications for global fluxes of Ba to the volcanic arc and mantle. *Geochimica et Cosmochimica Acta, 83,* 110–124.

Solomon, E. A., Kastner, M., Jannasch, H., Robertson, G., & Weinstein, Y. (2008). Dynamic fluid flow and chemical fluxes associated with a seafloor gas hydrate deposit on the northern Gulf of Mexico slope. *Earth and Planetary Science Letters, 270,* 95–105.

Solomon, E. A., Kastner, M., Wheat, G., Jannasch, H. W., Robertson, G., Davis, E. E., et al. (2009). Long-term hydrogeochemical records in the oceanic basement forearc prism at the Costa Rica subduction zone. *Earth and Planetary Science Letters, 282*(1–4), 240–251. http://dx.doi.org/10.1016/j.epsl.2009.03.022.

Song, T. A., Helmberger, D. V., Brudzinski, M. R., Clayton, R. W., Davis, P., Perez-Campos, X., et al. (2009). Subducting slab ultra-slow velocity layer coincident with silent earthquakes in southern Mexico. *Science, 324,* 502–506. http://dx.doi.org/10.1126/science.1167595.

Spinelli, G. A., & Saffer, D. M. (2004). Along-strike variations in underthrust sediment dewatering on the Nicoya margin, Costa Rica related to the updip limit of seismicity. *Geophysical Research Letters, 31,* L04613. http://dx.doi.org/10.1029/2003GL018863.

Spinelli, G. A., Saffer, D. M., & Underwood, M. B. (2006). Hydrogeologic responses to three-dimensional temperature variability, Costa Rica subduction margin. *Journal of Geophysical Research, 111*(B4), B04403.

Spinelli, G. A., & Underwood, M. B. (2004). Character of sediments entering the Costa Rica subduction zone: Implications for partitioning of water along the plate interface. *Island Arc, 13,* 432–451.

Spinelli, G. A., & Wang, K. (2008). Effects of fluid circulation in subducting crust on Nankai margin seismogenic zone temperatures. *Geology, 36*(11), 887.

Spivack, A. J., Kastner, M., & Ransom, B. (2002). Elemental and isotopic chloride geochemistry and fluid flow in the Nankai Trough. *Geophysical Research Letters, 29.* (14). http://dx.doi.org/10.1029/2001GL014122.

Spivack, A. J., Palmer, M. R., & Edmond, J. M. (1987). The sedimentary cycle of the boron isotopes. *Geochimica et Cosmochimica Acta, 57,* 1033–1047.

Stein, C. A., & Stein, S. (1992). A model for the global variation in oceanic depth and heat flow with lithospheric age. *Nature, 359*(6391), 123–129.

Stein, C. A., & Stein, S. (1994). Constraints on hydrothermal heat flux through the oceanic lithosphere from global heat flow. *Journal of Geophysical Research, 99,* 3081–3095.

Suess, E., Bohrmann, G., von Huene, R., Linke, P., Wallmann, K., Lammers, S., et al. (1998). Fluid venting in the eastern Aleutian subduction zone. *Journal of Geophysical Research, 103*(B2), 2,597–2.614.

Suzuki, T., & Kimura, T. (1973). D/H and $^{18}O/^{16}O$ fractionation in ice-water systems. *Mass Spectroscopy, 21,* 229–233.

Teichert, B. M. A., Torres, M. E., Bohrmann, G., & Eisenhauer, A. (2005). Reactions, fluid sources, fluid pathways and diagenetic reactions across an accretionary prism revealed by Sr and B geochemistry. *Earth and Planetary Science Letters, 239,* 106–121.

Tera, F., Brown, L., Morris, J., & Sacks, I. S. (1986). Sediment incorporation in island-arc magmas: Inferences from [10]Be. *Geochimica et Cosmochimica Acta, 50,* 535–550.

Torres, L. M., McManus, J., Hammond, D. E., de Angelis, M. A., Heeschen, K. U., Colbert, S. L., et al. (2002). Fluid and chemical fluxes in and out of sediments hosting methane hydrate deposits on Hydrate Ridge, OR, I: hydrological provinces. *Earth and Planetary Science Letters, 201,* 525–540.

Torres, M. E., Teichert, B. M. A., Trehu, A. M., Borowski, W. T., & Tomaru, H. (2004). Relationship of pore water freshening to accretionary processes in the Cascadia margin: fluid sources and gas hydrate abundance. *Geophysical Research Letters, 31,* L 22305.

Truesdell, A. H. (1976). Geochemical techniques in exploration, summary of section III. In *Proceedings of the second United Nations Symposium on the development and use of geothermal resources, San Francisco* (pp. 53–58). Berkeley: Univ. of California.

Tryon, M. D., Brown, K. M., & Torres, M. E. (2002). Fluid and chemical flux in and out of, sediments hosting methane hydrate deposits on Hydrate Ridge, Or, II: hydrological processes. *Earth and Planetary Science Letters, 201,* 541–557.

Tryon, M. D., Wheat, C. G., & Hilton, D. R. (2010). Fluid sources and pathways of the Costa Rica erosional convergent margin. *Geochemistry, Geophysics, Geosystems (G3), 11*(4), Q04S22. http://dx.doi.org/10.1029/2009GC002818.

Underwood, M. B. (2007). Sediment inputs to subduction zones: why lithostratigraphy and clay mineralogy matter. In T. H. Dixon, & J. C. Moore (Eds.), *The seismogenic zone of the subduction thrust faults* (pp. 42–85). New York: Columbia Univ. Press.

Vils, F., Pelletier, L., Kalt, A., Muntener, O., & Ludwig, T. (2008). The lithium, boron, beryllium content of serpentinized peridotites from ODP Leg 209 (Sites 1272A and 1274A): Implications for lithium and boron budgets of oceanic lithosphere. *Geochimica et Cosmochimica Acta, 72,* 5,475–5,504.

Von Damm, K. L., Bischoff, J. L., & Rosenbauer, R. J. (1991). Quartz solubility in hydrothermal seawater: an experimental study and equation describing quartz solubility for up to 0.5 M NaCl solutions. *American Journal of Science, 291,* 977–1007.

Von Huene, R., & Lallemand, S. (1990). Tectonic erosion along the Japan and Peru convergent margin. *Geological Society of America Bulletin, 102,* 704–720.

Von Huene, R., Ranero, C. R., Weinrebe, W., & Hinz, K. (2000). Quaternary convergent margin tectonics of Costa Rica, segmentation of the Cocos Plate, and Central American volcanism. *Tectonics, 19*(2), 314–334. http://dx.doi.org/10.1029/1999TC001143.

Von Huene, R., & Scholl, D. W. (1991). Observations at convergent margins concerning sediment subduction, subduction erosion, and the growth of continental crust. *Reviews of Geophysics, 29,* 279–316.

Von Rad, U., Berner, U., Delisle, G., Doose-Rolinski, H., Fechner, N., Linke, P., et al. (2000). Gas and fluid venting at the Makran accretionary wedge off Pakistan. *Geo-Marine Letters, 20,* 10–19.

Vrolijk, P., Fisher, A., & Gieskes, J. (1991). Geochemical and geothermal evidence for fluid migration in the Barbados accretionary prism (ODP Leg 110). *Geophysical Research Letters, 18,* 947–950.

Wallmann, K. (2001). The geological water cycle and the evolution of marine $\delta^{18}O$ values. *Geochimica et Cosmchimica Acta, 65,* 2,469–2,485.

Wei, W. (2007). *Fluid origins, paths, and fluid-rock reactions at convergent margins, using halogens, Cl Stable Isotopes, and alkali metals as geochemical tracers* (Ph.D. thesis). University of California San Diego, 165 pp.

Wei, W., Kastner, M., Rosenbauer, R., Chan, L. H., & Weinstein, Y. (2010). *Alkali elements as geothermometers for ridge flanks and subduction zones.* London: Water-Rock Interaction – Taylor & Francis Publication (ISBN).

Wei, W., Kastner, M., & Spivack, A. (2008). Chlorine stable isotopes and halogen concentrations in convergent margins with implications for the Cl isotopes cycle in the ocean. *Earth and Planetary Science Letters, 266*, 90–104.

Wenner, D. B., & Taylor, H. P. (1971). Temperature of serpentinization of ultramafic rocks based on $^{18}O/^{16}O$ fractionation between coexisting serpentine and magnetite. *Contributions to Mineralogy Petrology, 32*, 165–185.

Wheat, C. G., & McDuff, R. E. (1994). Hydrothermal flow through the Mariana Mounds: dissolution of amorphous silica and degradation of organic matter on a mid-ocean ridge flank. *Geochimica et Cosmochimica Acta, 58*, 2461–2475.

Whiticar, M. J. (1989). A geochemical perspective of natural gas and atmospheric ethane. *Organic Geochemistry, 16*, 531–547.

Whiticar, M. J., Faber, E., & Schoell, M. (1986). Biogenic methane formation in marine and freshwater environments — CO_2 reduction vs acetate fermentation isotope evidence. *Geochimica et Cosmochimica Acta, 50*(5), 693–709.

Williams, L. B., & Hervig, R. L. (2005). Lithium and boron isotopes in illite-smectite: the importance of crystal size. *Geochemical, Geophysics, Geosystems (G^3)* 14(6)ta, 69, 5705–5716.

Williams, L. B., Hervig, R. L., Halloway, J. R., & Hutcheon, I. E. (2001). Boron isotope geochemistry during diagenesis: Part 1. Experimental determination of fractionation during illitization of smectite. *Geochimica et Cosmochimica Acat, 65*, 1,768–1,782.

Yamano, M., Honda, S., & Uyeda, S. (1984). Nankai Trough: a hot trench? *Marine Geophysical Research, 6*(2), 187–203.

Yamano, M., Kinoshita, M., Goto, S., & Matsubayashi, O. (2003). Extremely high heat flow anomaly in the middle part of the Nankai Trough. *Physics and Chemistry of the Earth, Parts A/B/C, 28*(9–11), 487–497.

Yamano, M., & Uyeda, S. (1990). Heat-flow studies in the Peru Trench subduction zone. In E. Suess, et al. (Ed.), *Proceedings of the ocean drilling program, scientific results* (Vol. 112) (pp. 653–661).

You, C.-F., Castillo, P. R., Gieskes, J. M., Chan, L.-H., & Spivack, A. J. (1997). Trace element behavior in hydrothermal experiments: implications for fluid processes at shallow depth in subduction zones. *Earth and Planetary Science Letters, 140*, 41–52.

You, C.-F., & Gieskes (2001). Hydrothermal alteration of hemi-pelagic sediments: experimental evaluation og geochemical processes in shallow subduction zones. *Applied Geochemistry, 16*, 1055–1066.

You, C.-F., Spivack, A. J., Gieskes, J. M., & Rosenbauer, R. (1995). Experimental study of boron geochemistry: Implications for fluid processes in subduction zones. *Geochimica et Cosmochimica Acta, 59*, 2435–2442.

Chapter 5: Appendix

One-Page Summaries of IODP Expeditions 301–348

Jeannette Lezius and Ruediger Stein*

Geosciences Department, Alfred Wegener Institute for Polar and Marine Research, Bremerhaven, Germany
Corresponding author: E-mail: Ruediger.Stein@awi.de

In the preceding main chapters of this book, highlights and syntheses of IODP research carried out over the last decade are presented, ordered by the three major IODP themes *Subseafloor Life and Deep Biosphere*; *Environmental Change, Processes, and Effects*; and *Solid Earth and Geodynamics*. In this appendix, background information about each expedition, i.e., Expedition 301 to Expedition 348 (Table A.1), are shortly presented in one-page summaries (full color PDF versions are available at: http://booksite.elsevier.com/9780444626172/), ordered by expedition numbers. These summaries are all structured consistently as follows.

• Expedition's number and short name with schedule, port calls, site numbers, science operator and drilling platform, co-chief scientists, and staff scientist
• Global overview map with expedition's area
• Direct link to the expedition Website where further details about the expedition can be found
• Full title of expedition
• Summary of objectives
• Highlights of expedition
• Further information and/or selected key figures
• References

Some expeditions are part of a larger program with two, three, or more related expeditions (e.g., North Atlantic Climate 1 & 2; Superfast 1 to 4; NanTroSEIZE, etc.). In these cases, overall objectives are presented at the beginning and repeated for the different individual expeditions.

For the online version of this issue, three different background colors are used for the summary pages, depending on the major IODP theme the expedition is related to: (1) light green (*Subseafloor Life and Deep Biosphere*), (2) light blue (*Environmental Change, Processes, and Effects*), and (3) light red (*Solid Earth and Geodynamics*).

The listed key phrases and statements for the objectives and highlights as well as the figures are extracted from the initial report, the scientific report, the prospectus and/or the scientific drilling issue of each expedition. The references to these sources are listed at the end of the summaries.

Developments in Marine Geology, Volume 7. http://dx.doi.org/10.1016/B978-0-444-62617-2.19001-2

TABLE A.1 Summary Table of IODP Expeditions 301–348, Indicating Short Title, Drilling Platform, and Major IODP Theme of Each Single Expedition

Exp Number	Short Title	Platform	Main IODP Theme
301	Juan de Fuca Hydrogeology	Joides Resolution	1 = Deep Biosphere and Subseafloor Ocean
302	Arctic Coring Expedition (ACEX)	MSP	2 = Environmental Change, Processes and Effects
303	North Atlantic Climate 1	Joides Resolution	2 = Environmental Change, Processes and Effects
304	Oceanic Core Complex Formation, Atlantis Massif 1	Joides Resolution	3 = Solid Earth Cycles and Geodynamic
305	Ocean Core Complex Formation, Atlantis Massif 2	Joides Resolution	3 = Solid Earth Cycles and Geodynamic
306	North Atlantic Climate 2	Joides Resolution	2 = Environmental Change, Processes and Effects
307	Porcupine Basin Carbonate Mounds	Joides Resolution	1 = Deep Biosphere and Subseafloor Ocean
308	Gulf of Mexico Hydrogeology	Joides Resolution	1 = Deep Biosphere and Subseafloor Ocean
309	Superfast Spreading Rate Crust 2	Joides Resolution	3 = Solid Earth Cycles and Geodynamic
310	Tahiti Sea Level	MSP	2 = Environmental Change, Processes and Effects
311	Cascadia Margin Gas Hydrates	Joides Resolution	1 = Deep Biosphere and Subseafloor Ocean
312	Superfast Spreading Rate Crust 3	Joides Resolution	3 = Solid Earth Cycles and Geodynamic
313	New Jersey Shallow Shelf	MSP	2 = Environmental Change, Processes and Effects
314	NanTroSEIZE Stage 1: LWD Transect	Chikyu	3 = Solid Earth Cycles and Geodynamic
315	NanTroSEIZE Stage 1: Megasplay Riser Pilot	Chikyu	3 = Solid Earth Cycles and Geodynamic

316	NanTroSEIZE Stage 1: Shallow Megasplay and Frontal Thrusts	Chikyu	3 = Solid Earth Cycles and Geodynamic
317	Canterbury Basin	Joides Resolution	2 = Environmental Change, Processes and Effects
318	Wilkes Land	Joides Resolution	2 = Environmental Change, Processes and Effects
319	NanTroSEIZE Stage 2: Riser/Riserless Observatory 1	Chikyu	3 = Solid Earth Cycles and Geodynamic
320	Pacific Equatorial Age Transect I	Joides Resolution	2 = Environmental Change, Processes and Effects
321	Pacific Equatorial Age Transect II/Juan de Fuca	Joides Resolution	2 = Environmental Change, Processes and Effects
322	NanTroSEIZE Stage 2: Subduction Input	Chikyu	3 = Solid Earth Cycles and Geodynamic
323	Bering Sea Paleoceanography	Joides Resolution	2 = Environmental Change, Processes and Effects
324	Shatsky Rise	Joides Resolution	3 = Solid Earth Cycles and Geodynamic
325	Great Barrier Reef Environmental Changes	MSP	2 = Environmental Change, Processes and Effects
326	NanTroSEIZE Stage 3: Plate Boundary Deep Riser 1	Chikyu	3 = Solid Earth Cycles and Geodynamic
327	Juan de Fuca Hydrogeology	Joides Resolution	1 = Deep Biosphere and Subseafloor Ocean
328	Cascadia CORK	Joides Resolution	1 = Deep Biosphere and Subseafloor Ocean
329	South Pacific Gyre Microbiology	Joides Resolution	1 = Deep Biosphere and Subseafloor Ocean
330	Louisville Seamount Trail	Joides Resolution	1 = Deep Biosphere and Subseafloor Ocean
331	DEEP HOT BIOSPHERE	Chikyu	1 = Deep Biosphere and Subseafloor Ocean
332	NanTroSEIZE Stage 2: Riserless Observatory	Chikyu	3 = Solid Earth Cycles and Geodynamic
333	NanTroSEIZE Stage 2: Subduction Inputs 2 and Heat Flow	Chikyu	3 = Solid Earth Cycles and Geodynamic

Continued

TABLE A.1 Summary Table of IODP Expeditions 301–348, Indicating Short Title, Drilling Platform, and Major IODP Theme of Each Single Expedition—cont'd

Exp Number	Short Title	Platform	Main IODP Theme
334	Costa Rica Seismogenesis Project (CRISP)	Joides Resolution	1 = Deep Biosphere and Subseafloor Ocean
335	Superfast Spreading Rate Crust 4	Joides Resolution	3 = Solid Earth Cycles and Geodynamic
336	Mid-Atlantic Ridge Microbiology	Joides Resolution	1 = Deep Biosphere and Subseafloor Ocean
337	Shimokita Coal-Bed Biosphere	Chikyu	1 = Deep Biosphere and Subseafloor Ocean
338	NanTroSEIZE Stage 3, Plate Boundary Deep Riser-2	Chikyu	3 = Solid Earth Cycles and Geodynamic
339	Mediterranean Outflow	Joides Resolution	2 = Environmental Change, Processes and Effects
340	Lesser Antilles Volcanism and Landslides	Joides Resolution	3 = Solid Earth Cycles and Geodynamic
340T	Atlantis Massif Oceanic Core Complex	Joides Resolution	3 = Solid Earth Cycles and Geodynamic
341	Southern Alaska Margin Tectonics, Climate & Sedimentation	Joides Resolution	2 = Environmental Change, Processes and Effects
341S	SCIMPI & 858G CORK	Joides Resolution	1 = Deep Biosphere and Subseafloor Ocean
342	Paleogene Newfoundland Sediment Drifts	Joides Resolution	2 = Environmental Change, Processes and Effects
343	Japan Trench Fast Drilling Project	Chikyu	3 = Solid Earth Cycles and Geodynamic
344	Costa Rica Seismogenesis Project 2 (CRISP)	Joides Resolution	1 = Deep Biosphere and Subseafloor Ocean
345	Hess Deep Plutonic Crust	Joides Resolution	3 = Solid Earth Cycles and Geodynamic
346	Asian Monsoon	Joides Resolution	2 = Environmental Change, Processes and Effects
347	Baltic Sea Paleoenvironment	MSP	2 = Environmental Change, Processes and Effects
348	NanTroSEIZE Plate Boundary Deep Riser	Chikyu	3 = Solid Earth Cycles and Geodynamic

IODP Expedition 301
Juan de Fuca Hydrogeology

Dates: 27 June – 21 August 2004
Ports: Astoria, Oregon, to Astoria, Oregon
Sites: 1026, U1301
Operator: USIO, Platform: JR
Co-chief Scientists: Andrew Fisher & Tetsuro Urabe
Staff Scientist: Adam Klaus

Expedition Website: http://iodp.tamu.edu/scienceops/expeditions/exp301.html

IODP Expeditions 301 & 327:
The hydrogeological architecture of basaltic oceanic crust: compartmentalization, anisotropy, microbiology, and crustal–scale properties on the eastern flank of Juan de Fuca Ridge, eastern Pacific Ocean

OBJECTIVES
- First part of a two-expedition, multidisciplinary program (incl. offset seismic experiment, long-term monitoring and cross-hole tests (ROV expeditions extending 6–10 years after IODP Exp. 301)
- Evaluate the formation-scale hydrogeological properties within oceanic crust
- Determine how fluid pathways are distributed within an active hydrothermal system
- Establish linkages between fluid circulation, alteration, and microbiological processes
- Determine relations between seismic and hydrologic anisotropy

HIGHLIGHTS
- Replaced one existing borehole observatory penetrating the upper oceanic crust on the eastern flank of the Juan de Fuca Ridge
- Established two new observatories penetrating to depths as great as 583 m below seafloor or 318 m into basement
- Sampled sediments, basalt, fluids, and microbial samples
- Collected wireline logs
- Conducted hydrogeological tests in two basement holes
- Documentation about microbiological communities life in the crust and how communities cycle carbon, alter rocks, and are influenced by fluid flow paths

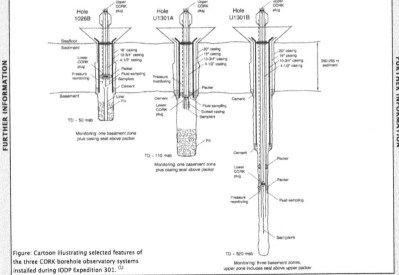

Figure: Cartoon illustrating selected features of the three CORK borehole observatory systems installed during IODP Expedition 301. [1]

REFERENCES (1)• Shipboard Scientific Party, 2004. Juan de Fuca hydrogeology: the hydrogeologic architecture of basaltic oceanic crust: compartmentalization, anisotropy, microbiology, and crustal-scale properties on the eastern flank of Juan de Fuca Ridge, eastern Pacific Ocean. IODP Prel. Rept., 301. doi:10.2204/ iodp.pr.301.2004 (2)• Fisher, A.T., Urabe, T., Klaus, A., and the IODP Expedition 301 Scientists, 2005. IODP Expedition 301 installs three borehole crustal observatories, prepares for three-dimensional, cross-hole experiments in the northeastern Pacific Ocean. Sci. Drill., 1:6–11. doi:10.2204/ iodp.sd.1.01.2005 (3)• Davis, E., and Becker, K., 2007. On the fidelity of "CORK" borehole hydrologic observatory pressure records. Sci. Drill., 5:54–59. doi:10.2204/ iodp.sd.5.09.2007

IODP Expedition 302
Arctic Coring Expedition (ACEX)

Dates: 07 August – 15 September 2004
Ports: Tromsö to Tromsö, Norway
Sites: M0001, M0002, M0003 and M0004
Operator: ESO, Vidar Viking (drillship), Oden and Sovetskiy Soyuz (icebreakers)
Co-chief Scientists: Jan Backman & Kate Moran
Staff Scientist: David McInroy

Expedition Website: http://www.eso.ecord.org/expeditions/302/302.php

Arctic Coring Expedition – ACEX: Paleoceanographic and tectonic evolution of the central Arctic Ocean

OBJECTIVES

- Understand history of ice rafting and sea ice
- Study local versus regional ice-sheet development
- Determine the density structure of Arctic Ocean surface waters, nature of NorthAtlantic conveyor, and onset of Northern Hemisphere glaciation
- Determine the timing and consequences of the opening of the Bering Strait
- Study land-sea links and response of the Arctic to Pliocene warm events
- Investigate the development of the Fram Strait and deep-water exchange between Arctic Ocean, Greenland–Iceland–Norwegian Seas, and world ocean
- Determine the history of biogenic sedimentation
- Investigate the nature and origin of Lomonosov Ridge by sampling oldest rocks below regional unconformity to establish pre-Cenozoic environmental setting of the ridge
- Study history of rifting and the timing of tectonic events that affected the ridge.

HIGHLIGHTS

- Penetrated 428 meters of Quaternary, Neogene, Paleogene and Campanian sediment
- Navigated thick and relentless ice floes
- Research team strategically managed three icebreakers
- Recovered sediment record from Lomonosov Ridge reveals transition from past Arctic greenhouse conditions to present-day ice-house conditions
- High Arctic Ocean surface water temperatures and a hydrologically active climate during the Paleocene Eocene Thermal Maximum (PETM) and large part of the early-middle Eocene
- Distinct freshening of Arctic surface waters in the Eocene
- Ice-rafted debris as old as middle Eocene
- Early-middle Eocene euxinic environment with high preservation rate of organic carbon
- Ventilation of the Arctic Ocean to the North Atlantic through the Fram Strait near the early-middle Miocene boundary

FURTHER INFORMATION

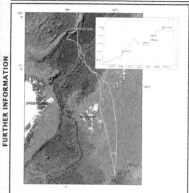

Map: Location of ACEX study area on the Lomonosov Ridge and locations of ACEX sites. [1]

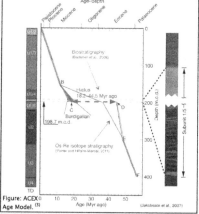

Figure: ACEX Age Model. [3]

REFERENCES: (1)• Expedition 302 Scientists, 2005. Arctic Coring Expedition (ACEX): paleoceanographic and tectonic evolution of the central Arctic Ocean. IODP Prel. Rept., 302. doi:10.2204/iodp.pr.302.2005 •(2) Backman, J., Moran, K., McInroy, D., and the IODP Expedition 302 Scientists, 2005. IODP Expedition 302, Arctic Coring Expedition (ACEX): a first look at the Cenozoic paleoceanography of the central Arctic Ocean. Sci. Drill., 1:12–17. doi:10.2204/iodp.sd.1.02.2005 (3)• Jakobsson, M., Backman, J., Rudels, B., Nycander, J., Frank, M., Mayer, L., Jokat, W., Sangiorgi, F., O'Regan, M., Brinkhuis, H., King, J., and Moran, K., 2007. The early Miocene onset of a ventilated circulation regime in the Arctic Ocean. Nature (London, U. K.), 447(7147):986–990. doi:10.1038/nature05924 (4)• Backman, J., and Moran, K., 2009. Expanding the Cenozoic paleoceanographic record in the Central Arctic Ocean: IODP Expedition 302 synthesis. Cent. Eur. J. Geosci., 1(2):157–175. doi:10.2478/v10085-009-0015-6

IODP Expedition 303
North Atlantic Climate 1

Dates: 25 September–17 November 2004
Ports: St. John's, Newfoundland, to Ponta Delgada, Azores
Sites: U1302–U1308
Operator: USIO, Platform: JR
Co-chief Scientists: James Channell & Tokiyuki Sato
Staff Scientist: Mitchell Malone

IODP Expedition 303

Expedition Website: http://publications.iodp.org/preliminary_report/303/

IODP Expeditions 303 & 306:
Ice sheet–ocean atmosphere interactions on millennial timescales during the late Neogene–Quaternary using a paleointensity-assisted chronology for the North Atlantic

OBJECTIVES

- Capture Miocene–Quaternary millennial-scale climate variability in sensitive regions at the mouth of the Labrador Sea and in the North Atlantic ice-rafted debris (IRD) belt
- Provide sedimentary and paleomagnetic attributes, including adequate sedimentation rates, for constructing high-resolution isotopic and magnetic stratigraphies

- Designed to sample and study climate records, including the composition and structure of surface or bottom waters and detrital layer stratigraphy indicative of ice sheet instability, at strategic sites that record North Atlantic Pliocene - Quaternary climate
- Sites chosen on basis of importance of the climate or paleoceanographic record, adequate sedimentation rates in the 5–20 cm/ky range, and attributes for a stratigraphic template based on relative geomagnetic paleointensity and oxygen isotope data.

HIGHLIGHTS

- Recovery of high-sedimentation-rate sedimentary sections
- Generation of stratigraphic framework based on biostratigraphy, oxygen isotope data and magnetic stratigraphy, including RPI data
- Placed detrital record and other climate proxies into a precise stratigraphic framework
- Detection of basic climate-related lithologic variability throughout the late Miocene–Quaternary

FURTHER INFORMATION

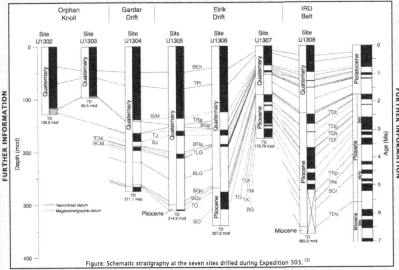

Figure: Schematic stratigraphy at the seven sites drilled during Expedition 303. [1]

REFERENCES: (1)• Shipboard Scientific Party, 2005. North Atlantic climate: ice sheet–ocean atmosphere interactions on millennial timescales during the late Neogene–Quaternary using a paleointensity-assisted chronology for the North Atlantic. IODP Prel. Rept., 303. doi:10.2204/ iodp.pr.303.2005 •(2) Channell, J.E.T., Sato, T., Kanamatsu, T., Stein, R., Malone, M., Alvarez-Zarikian, C., and the IODP Expeditions 303 and 306 Scientists, 2006. IODP Expeditions 303 and 306 monitor Miocene–Quaternary climate in the North Atlantic. Sci. Drill., 2:4–10. doi:10.2204/ iodp.sd.2.01.2006

IODP Expedition 304
Oceanic Core Complex Formation, Atlantis Massif 1

Dates: 17 November 2004–8 January 2005
Ports: Ponta Delgada, Azores, to Ponta Delgada, Azores
Sites: U1309–U1311
Operator: USIO, Platform: JR
Co–chief Scientists: Donna Blackman & Barbara John
Staff Scientist: Jay Miller

Expedition Website: http://iodp.tamu.edu/scienceops/expeditions/exp304.html

IODP Expedition 304

IODP Expeditions 304 & 305:
Oceanic Core Complex Formation, Atlantis Massif, Mid–Atlantic Ridge: drilling into the footwall and hanging wall of a tectonic exposure of deep, young oceanic lithosphere to study deformation, alteration, and melt generation

OBJECTIVES
- Designed to investigate the processes that control formation of oceanic core complexes (OCCs), as well as the exposure of ultramafic rocks in very young oceanic lithosphere
- Understand dominant mechanism(s) of footwall uplift and interactions between tectonics and magmatism in OCCs
- Sample and assess petrogenetic relationships between volcanic rocks in the hanging wall and potential source rocks recovered in the footwall

HIGHLIGHTS
- Penetration of 400 m at Hole U1309D
- Detection of series of interfingered gabbroic intrusions based on variation in olivine content, the presence of intercumulus phases, the extent of late magmatic dikes, and the presence of oxide gabbro
- Lack of widespread deformation suggests that strain is concentrated in a small number of very localized zones
- Establishment of reentry hole for subsequent deep penetration (Exp. 305)

FURTHER INFORMATION

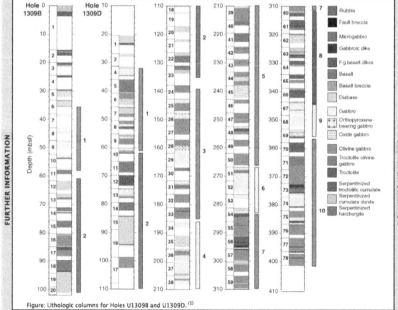

Figure: Lithologic columns for Holes U1309B and U1309D. [1]

REFERENCES: (1)* Expedition Scientific Party, 2005. Oceanic core complex formation, Atlantis Massif—oceanic core complex formation, Atlantis Massif, Mid–Atlantic Ridge: drilling into the footwall and hanging wall of a tectonic exposure of deep, young oceanic lithosphere to study deformation, alteration, and melt generation. IODP Prel. Rept., 304. doi:10.2204/ iodp.pr.304.2005 *(2) Ildefonse, B., Blackman, D., John, B.E., Ohara, Y., Miller, D.J., MacLeod, C.J., and the IODP Expeditions 304–305 Scientists, 2006. IODP Expeditions 304 & 305 characterize the lithology, structure, and alteration of an oceanic core complex. Sci. Drill., 3:4–11. doi:10.2204/ iodp.sd.3.01.2006

IODP Expedition 305
Oceanic Core Complex Formation, Atlantis Massif 2

Dates: 8 January–2 March 2005
Ports: Ponta Delgada, Azores, to Ponta Delgada, Azores
Sites: U1309
Operator: USIO, Platform: JR
Co-chief Scientists: Benoit Ildefonse & Yasuhiko Ohara
Staff Scientist: Jay Miller

Expedition Website: http://iodp.tamu.edu/scienceops/expeditions/exp305.html

IODP Expeditions 304 & 305:
Oceanic Core Complex Formation, Atlantis Massif, Mid–Atlantic Ridge: drilling into the footwall and hanging wall of a tectonic exposure of deep, young oceanic lithosphere to study deformation, alteration, and melt generation

OBJECTIVES

• Designed to investigate the processes that control formation of oceanic core complexes (OCCs), as well as the exposure of ultramafic rocks in very young oceanic lithosphere
• Understand dominant mechanism(s) of footwall uplift and interactions between tectonics and magmatism in OCCs
• Sample and assess petrogenetic relationships between volcanic rocks in the hanging wall and potential source rocks recovered in the footwall

HIGHLIGHTS

• Deepening of Hole U1309D in the footwall of Atlantis Massif to 1415.5 mbs
• High recovery (average = 74.8%) of dominantly gabbroic rocks
• Rocks range from dunitic troctolite, troctolite, (olivine) gabbro, and gabbronorite to evolved oxide gabbro that locally contains abundant zircon and apatite, and diabase
• Footwall composed of uplifted mantle section where serpentinization was responsible for lower densities/seismic velocities in upper few hundred meters
• Support of complex model of complicated lateral heterogeneity in slow-spreading oceanic crust

General outcome of Expeditions 304 & 305:
• Recovered just over 1400 m of little-deformed, gabbroic lower crust from a tectonic window along the slow-spreading Mid–Atlantic Ridge
• Attempted drilling at three sites: one in the footwall through the inferred detachment fault; two in the hanging wall; Site in the footwall of the detachment fault, penetrated 1415.5 m below the seafloor, and recovery averaged 75% but did not succeed in drilling the fractured basalt in the hanging wall

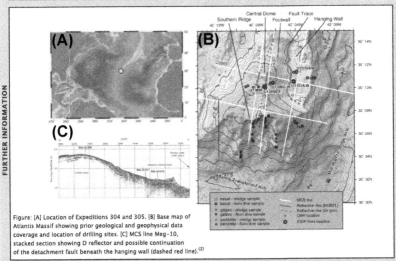

Figure: [A] Location of Expeditions 304 and 305. [B] Base map of Atlantis Massif showing prior geological and geophysical data coverage and location of drilling sites. [C] MCS line Meg-10, stacked section showing D reflector and possible continuation of the detachment fault beneath the hanging wall (dashed red line).[2]

REFERENCES: (1)• Expedition Scientific Party, 2005. Oceanic core complex formation, Atlantis Massif—oceanic core complex formation, Atlantis Massif, Mid–Atlantic Ridge: drilling into the footwall and hanging wall of a tectonic exposure of deep, young oceanic lithosphere to study deformation, alteration, and melt generation. IODP Prel. Rept. 305. doi:10.2204/ iodp.pr.305.2005 •(2) Ildefonse, B., Blackman, D., John, B.E., Ohara, Y., Miller, D.J., MacLeod, C.J., and the IODP Expeditions 304–305 Scientists, 2006. IODP Expeditions 304 & 305 characterize the lithology, structure, and alteration of an oceanic core complex. Sci. Drill., 3:4–11. doi:10.2204/ iodp.sd.3.01.2006

IODP Expedition 306
North Atlantic Climate 2

Dates: 2 March–25 April 2005
Ports: Ponta Delgada, Azores, to Dublin, Ireland
Sites: U1312–U1315
Operator: USIO, Platform: JR
Co–chief Scientists: Toshiya Kanamatsu & Ruediger Stein
Staff Scientist: Carlos Alvarez Zarikian

Expedition Website: http://iodp.tamu.edu/scienceops/expeditions/exp306.html

IODP Expeditions 303 & 306:
Ice sheet–ocean atmosphere interactions on millennial timescales during the late Neogene–Quaternary using a paleointensity–assisted chronology for the North Atlantic

OBJECTIVES

- Capture Miocene-Quaternary millennial-scale climate variability in sensitive regions at the mouth of the Labrador Sea and in the North Atlantic ice-rafted debris (IRD) belt
- Provide sedimentary and paleomagnetic attributes, including adequate sedimentation rates, for constructing high-resolution isotopic and magnetic stratigraphies

303 & 306

Map: Drill sites of Expeditions 303 & 306. (3)

OBJECTIVES

- Place late Neogene-Quaternary climate proxies in the North Atlantic into a chronology based on a combination of geomagnetic paleointensity, stable isotope, and detrital layer stratigraphies
- Generate integrated North Atlantic millennial-scale stratigraphies for the last few million years
- Document and monitor bottom water temperature variations through time: installing a CORK at Site 642, Norwegian-Greenland Sea

HIGHLIGHTS

- Drilling of complete sedimentary sections by multiple advanced piston coring directly south of central Atlantic "ice-rafted debris belt" and on southern Gardar Drift
- Installation of borehole observatory in a new 170 m deep hole close to Ocean Drilling Program Site 642

FURTHER INFORMATION

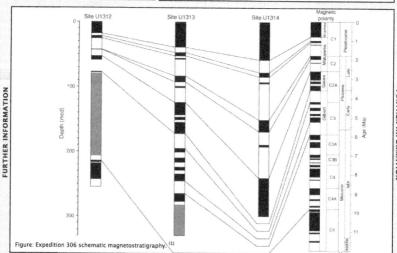

Figure: Expedition 306 schematic magnetostratigraphy. (1)

REFERENCES: (1)• Expedition Scientists, 2005. North Atlantic climate 2. IODP Prel. Rept., 306. doi:10.2204/ iodp.pr.306.2005 •(2) Channell, J.E.T., Sato, T., Kanamatsu, T., Stein, R., Malone, M., Alvarez-Zarikian, C., and the IODP Expeditions 303 and 306 Scientists, 2006. IODP Expeditions 303 and 306 monitor Miocene-Quaternary climate in the North Atlantic. Sci. Drill., 2:4–10. doi:10.2204/ iodp.sd.2.01.2006 •(3) http://iodp.tamu.edu/scienceops/expeditions/ exp306.html

IODP Expedition 307
Modern Carbonate Mounds: Porcupine Drilling

Dates: 25 April–30 May 2005
Ports: Dublin, Ireland, to Mobile, Alabama
Sites: U1316–U1318
Operator: USIO, Platform: JR
Co–chief Scientists: Timothy Ferdelman & Akihiro Kano
Staff Scientist: Trevor Williams

Expedition Website: http://iodp.tamu.edu/scienceops/expeditions/exp307.html

Modern Carbonate Mounds: Porcupine Drilling

OBJECTIVES

- Drilling Cold–Water Coral Mound along the Irish Continental Margin: Challenger Mound, a putative carbonate mound structure covered with dead deepwater coral rubble
- Additional drilling of one site downslope of Challenger Mound and an up–slope site
- Establish whether the mound base rested on a carbonate hardground of microbial origin and whether past geofluid migration events acted as a prime trigger for mound genesis
- Definition of relationship between mound initiation, mound growth phases, and global oceanographic events
- Analyzes of geochemical and microbiological profiles that define the sequence of microbial communities and geomicrobial reactions throughout the drilled sections
- Examination of high–resolution paleoclimatic records from the mound section using a wide range of geochemical and isotopic proxies
- Description of stratigraphic, lithologic, and diagenetic characteristics, including timing of key mound–building phases, for establishing a depositional model of deepwater carbonate mounds and for investigating how they resemble ancient mud mounds.
- Constrain of stratigraphic framework of the slope/mound system
- Identification and correlation of erosional surfaces observed in slope sediment seismics
- Investigation of potential gas accumulation in the sediments underlying the mound.

HIGHLIGHTS

- Gas seeps act as a prime trigger for mound genesis—a case for geosphere-biosphere coupling
- Prominent erosional surfaces reflect global oceanographic events
- The mound may be a high–resolution paleoenvironmental recorder because of its high depositional rate and abundance of micro- and macrofossils
- The Porcupine mounds are present–day analogs for Phanerozoic reef mounds and mud mounds

FURTHER INFORMATION

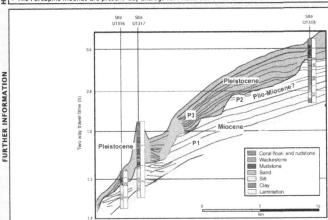

Figure:
Lithostratigraphy of three drilling sites projected on seismic profile of Challenger Mound along a north–northwest to south–southeast transect. [2]

REFERENCES: (1)• Expedition Scientists, 2005. Modern carbonate mounds: Porcupine drilling. IODP Prel. Rept., 307, doi:10.2204/iodp.pr.307.2005 (2)• erdelman, T.G., Kano, A., Williams, T., and the IODP Expedition 307 Scientists, 2006. IODP Expedition 307 drills cold–water coral mound along the Irish continental margin. Sci. Drill., 2:11–16. doi:10.2204/iodp.sd.2.02.2006

IODP Expedition 308
Gulf of Mexico Hydrogeology

IODP Expedition 308

Dates: 30 May-8 July 2005
Ports: Mobile, Alabama, to Cristobal, Panama
Sites: U1319-U1324
Operator: USIO, Platform: JR
Co-chief Scientists: Peter Flemings & Jan Behrmann
Staff Scientist: Cédric John

Expedition Website: http://iodp.tamu.edu/scienceops/expeditions/exp308.html

Overpressure and fluid flow processes in the deepwater Gulf of Mexico: slope stability, seeps, and shallow-water flow

OBJECTIVES
- Study of overpressure and fluid flow on Gulf of Mexico continental slope
- Examine sedimentation, overpressure, fluid flow, and deformation mechanisms and how they are coupled in passive margin settings
- Test of multidimensional flow model by examining how physical properties, pressure, temperature, and pore fluid composition vary within low-permeability mudstones that overlie permeable and overpressured aquifer

HIGHLIGHTS

Ursa region:
- Very rapid Pleistocene sedimentation where overpressure is known to be present:
- First-ever in-situ measurements of how physical properties, pressure, temperature, and pore fluid compositions vary within low-permeability mudstones that overlie a permeable, overpressured aquifer
- Documentation of severe overpressure in the mudstones overlying the aquifer
- Illumination of controls on slope stability, seafloor seeps, and large-scale crustal fluid flow
- Substantial whole-core geotechnical samples taken for later shore-based analysis and to deploy the temperature and dual pressure probe to measure in situ pressure

Brazos-Trinity Basin:
- Reference location where little overpressure is deemed to be present
- Drilling, logging and in-situ measurements

FURTHER INFORMATION

Figure: Stratigraphic cross-section of Brazos-Trinity Basin. [1]

REFERENCES (1)• Expedition 308 Scientists, 2005. Gulf of Mexico hydrogeology—overpressure and fluid flow processes in the deepwater Gulf of Mexico: slope stability, seeps, and shallow-water flow. IODP Prel. Rept., 308. doi:10.2204/ iodp.pr.308.2005 (2)• Behrmann, J.H., Flemings, P.B., John, C.M., and the IODP Expedition 308 Scientists, 2006. Rapid sedimentation, overpressure, and focused fluid flow, Gulf of Mexico continental margin. Sci. Drill., 3:12-17. doi:10.2204/ iodp.sd.3.03.2006

IODP Expedition 309
Superfast Spreading Rate Crust 2

IODP Expedition 309

Dates: 8 July–28 August 2005
Ports: Cristobal, Panama, to Balboa, Panama
Sites: 1256
Operator: USIO, Platform: JR
Co–chief Scientists: Damon Teagle & Susumu Umino
Staff Scientist: Neil Banerjee

Expedition Website: http://iodp.tamu.edu/scienceops/expeditions/exp309.html

The Superfast Spreading Rate Crust mission multi–cruise program (ODP Leg 206, IODP Legs 309, 312, 335):
A complete in–situ section of upper oceanic crust formed at a superfast spreading rate

OBJECTIVES

- First sampling of intact section of upper oceanic basement (upper oceanic crust, through lavas and the sheeted dikes) into uppermost gabbros at Ocean Drilling Program Site 1256 (ODP Leg 206)
- Crust at Site 1256 formed at superfast (>200 mm/y) spreading rate ~15 m.y. ago at the East Pacific Rise
- Exploit inverse relationship between spreading rate and depth to axial low–velocity zones

- Testing of prediction that gabbros representing the crystallized melt lens should be encountered at a depth of 900–1300 m sub–basement at Site 1256 from the correlation of spreading rate with decreasing depth to the axial melt lens
- Determination of lithology and structure of the upper oceanic crust for the superfast spreading end–member
- Correlation and calibration of remote geophysical seismic and magnetic imaging of the structure of the crust with basic geological observations
- Investigation of interactions between magmatic and alteration processes, including the relationships between extrusive volcanic rocks, the feeder sheeted dikes, and the underlying gabbroic rocks

HIGHLIGHTS

- Deepening of Hole 1256D to a total depth of 1255 mbsf (1005 m sub–basement)
- Penetration through >800 m of extrusive normal mid–ocean–ridge basalt
- Entering region dominated by intrusive rocks with numerous sub–vertical chilled dike margins
- Cleaning and preparation of Hole 1256D for deepening during IODP Expedition 312
- Uppermost basement comprises a ~100–m thick sequence of lava dominated by a single flow up to 75 m thick (ponded lava overlying massive, sheet, and minor pillow flows)

FURTHER INFORMATION

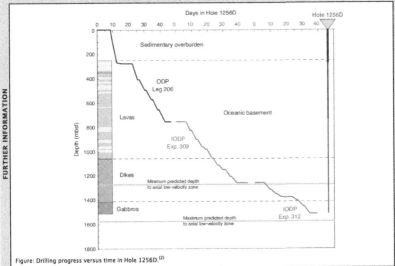

Figure: Drilling progress versus time in Hole 1256D.[2]

REFERENCES: (1)• Expedition 309 Scientists, 2005. Superfast spreading rate crust 2: a complete in situ section of upper oceanic crust formed at a superfast spreading rate. IODP Prel. Rept., 309. doi:10.2204/ iodp.pr.309.2005 •(2) Alt, J.C., Teagle, D.A.H., Umino, S., Miyashita, S., Banerjee, N.R., Wilson, D.S., and the IODP Expeditions 309 and 312 Scientists, and the ODP Leg 206 Scientific Party, 2007. IODP Expeditions 309 and 312 drill and intact section of upper oceanic basement into gabbros. Sci. Drill., 4:4–10. doi:10.2204/ iodp.sd.4.01.2007

IODP Expedition 310
Tahiti Sea Level Expedition

Dates: 6 October–17 November 2005 (coring expeditions only)
Ports: Papeete to Papeete,Tahiti, France
Sites: M0005 to M0026
Operator: ESO, Platform: DP Hunter
Co-chief Scientists: Gilbert Camoin & Yasufumi Iryu
Expedition Project Manager: David McInroy
Expedition Website: http://www.eso.ecord.org/expeditions/310/310.php

IODP Expedition 310

The last deglacial sea level rise in the South Pacific: offshore drilling in Tahiti (French Polynesia)

OBJECTIVES

- Establish the course of postglacial sea level rise at Tahiti
- Define sea–surface temperature (SST) variations for the region over the period 20–10 ka
- Analyze the impact of sea level changes on reef growth and geometry
- Investigate the geomicrobiology processes
- Core postglacial reef sequence, which consists of successive reef terraces seaward of the living barrier reef, from dynamically positioned vessel

HIGHLIGHTS

- Cored total of 37 boreholes across 22 sites
- Water depths from 41.65 to 117.54 m
- Borehole logging operations in 10 boreholes provided continuous geophysical information about drilled strata

FURTHER INFORMATION

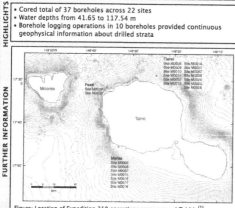
Figure: Location of Expedition 310 operations areas around Tahiti. [1]

Figure: View of corals from the surface of the Tahiti reefs. [2]

Figure: Core displaying coralgal frameworks heavily encrusted with microbialites. [2]

REFERENCES: (1)• Expedition 310 Scientists, 2006. Tahiti Sea Level: the last deglacial sea level rise in the South Pacific: offshore drilling in Tahiti (French Polynesia). IODP Prel. Rept., 310. doi:10.2204/ iodp.pr.310.2006 (2)• Camoin, G.F., Iryu, Y., McInroy, D., and the IODP Expedition 310 Scientists, 2007. IODP Expedition 310 reconstructs sea level, climatic, and environmental changes in the South Pacific during the last deglaciation. Sci. Drill., 5:4–12. doi:10.2204/ iodp.sd.5.01.2007

IODP Expedition 311
Cascadia Margin Gas Hydrates

Dates: 28 August–28 October 2005
Ports: Balboa, Panama, to Victoria, B.C., Canada
Sites: U1325–U1329
Operator: USIO, Platform: JR
Co-chief Scientists: Michael Riedel & Tim Collett
Staff Scientist: Mitch Malone

IODP Expedition 311

Expedition Website: http://iodp.tamu.edu/scienceops/expeditions/exp311.html

OBJECTIVES

- Study formation of natural gas hydrate in marine sediments in accretionary prism of the Cascadia subduction zone
- Test and constrain geological models of gas hydrate formation by upward fluid and methane transport in accretionary prisms
- Determine the mechanism of development, nature, magnitude, and global distribution of gas hydrate reservoirs
- Investigate the gas transport mechanism, and migration pathways through sedimentary structures, from site of origin to reservoir
- Examine the effect of gas hydrate on the physical properties of the enclosing sediments, particularly as it relates to the potential relationship between gas hydrates and slope stability
- Investigate the microbiology and geochemistry associated with gas hydrate formation and dissociation

HIGHLIGHTS

- Drilled transect of four sites across the Northern Cascadia margin, representing different stages in the evolution of gas hydrate across the margin from the earliest occurrence (westernmost first accreted ridge) to its final stage (eastward limit of gas hydrate occurrence on the margin in shallower water)
- Additional fifth site representing a cold vent with active fluid and gas flow
- Logging while drilling/measurement while drilling provided a set of measurements that guided subsequent coring and special tool deployments at all five sites
- Additional wireline logging at each site
- Two vertical seismic profiles
- Total of 1217.76 m of sediment core was recovered: 24 (16 successful) pressure core sampler, 19 Fugro piston corer/HYACE deployments, 4 pressure cores stored under in-situ pressure for subsequent shore-based studies
- Occurrences of unexpectedly high concentrations of gas hydrate at relatively shallow depths, 50–120 meters below the seafloor
- Strong support for sediment porosity as a controlling factor in gas hydrate information:
  - Gas hydrate occurs within coarser-grained turbidite sands and silts
  - Occurrence of gas hydrate appears to be controlled by: local methane solubility linked with pore-water salinity, fluid and gas advection rates, the availability of suitable host material such as coarse-grained sediments

FURTHER INFORMATION

Figure: Multibeam bathymetry map along transect across accretionary prism offshore Vancouver Island established during IODP Expedition 311. [2]

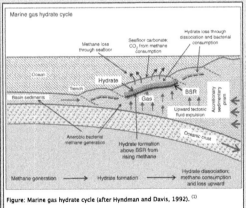

Figure: Marine gas hydrate cycle (after Hyndman and Davis, 1992). [1]

REFERENCES (1)* Expedition 311 Scientists, 2005. Cascadia margin gas hydrates. IODP Prel. Rept., 311. doi:10.2204/iodp.pr.311.2005 (2)* Riedel, M., Collett, T.S., Malone, M.J., and the IODP Expedition 311 Scientists, 2006. Stages of gas-hydrate evolution on the northern Cascadia margin. Sci. Drill., 3:18–24. doi:10.2204/iodp.sd.3.04.2006

IODP Expedition 312
Superfast Spreading Rate Crust 3

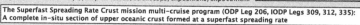

IODP Expedition 312

Dates: 28 October–28 December 2005
Ports: Victoria, B.C., Canada, to Cristobal, Panama
Sites: 1256
Operator: USIO, Platform: JR
Co–chief Scientists: Jeffrey C. Alt & Sumio Miyashita
Staff Scientist: Neil Banerjee

Expedition Website: http://iodp.tamu.edu/scienceops/expeditions/exp312.html

The Superfast Spreading Rate Crust mission multi–cruise program (ODP Leg 206, IODP Legs 309, 312, 335):
A complete in–situ section of upper oceanic crust formed at a superfast spreading rate

OBJECTIVES

- First sampling of intact section of upper oceanic basement (upper oceanic crust, through lavas and the sheeted dikes) into uppermost gabbros at Ocean Drilling Program Site 1256 (ODP Leg 206)
- Crust at Site 1256 formed at superfast (>200 mm/y) spreading rate ~15 m.y. ago at the East Pacific Rise
- Exploit inverse relationship between spreading rate and depth to axial low-velocity zones
- Testing of prediction that gabbros representing the crystallized melt lens should be encountered at a depth of 900–1300 m sub-basement at Site 1256 from the correlation of spreading rate with decreasing depth to the axial melt lens
- Determination of lithology and structure of the upper oceanic crust for the superfast spreading end-member
- Correlation and calibration of remote geophysical seismic and magnetic imaging of the structure of the crust with basic geological observations
- Investigation of interactions between magmatic and alteration processes, including the relationships between extrusive volcanic rocks, the feeder sheeted dikes, and the underlying gabbroic rocks

HIGHLIGHTS

- Deepening of Hole 1256D to a total depth of 1507.1 meters below seafloor (mbsf)
- Hole 1256D now extends through the 345 m sheeted dike complex and 100.5 m into the upper portions of the plutonic complex
- Uppermost basement comprises a ~100–m–thick sequence of lava dominated by a single flow up to 75 m thick (ponded lava overlying massive, sheet, and minor pillow flows)
- Estimated total thickness of 284 m for off–axis
- Sheet flows and massive lavas in remaining extrusive section down to 1004 mbsf
- Lithologic transition below 1061 mbsf: start of (~350–m–thick) sheeted dike complex that is dominated by massive basalts
- Steep thermal gradient in the dikes
- First gabbroic rocks as intrusion of underlying gabbros of upper gabbroic body at 1406.6 mbsf
- Lower gabbroic body at 1483–1507.1 mbsf: complex zone of fractionated gabbros intruded into contact metamorphosed dikes
- Gabbroic rocks are highly altered

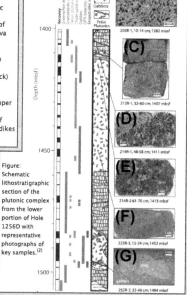

(A)

(B)

(C)

(D)

(E)

(F)

(G)

Figure: Schematic lithostratigraphic section of the plutonic complex from the lower portion of Hole 1256D with representative photographs of key samples.[2]

FURTHER INFORMATION

Figure: Age map of the Cocos plate and corresponding regions of the Pacific plate. [1]

REFERENCES: (1)• Expedition 309 and 312 Scientists, 2006. Superfast spreading rate crust 3: a complete in situ section of upper oceanic crust formed at a superfast spreading rate, IODP Prel. Rept., 312. doi:10.2204/ iodp.pr.312.2006 •(2) Alt, J.C., Teagle, D.A.H., Umino, S., Miyashita, S., Banerjee, N.R., Wilson, D.S., and the IODP Expeditions 309 and 312 Scientists, and the ODP Leg 206 Scientific Party, 2007. IODP Expeditions 309 and 312 drill and intact section of upper oceanic basement into gabbros. Sci. Drill., 4:4–10. doi:10.2204/ iodp.sd.4.01.2007

Expedition 313
New Jersey Shallow Shelf

Dates: 30 April 2009–17 July 2009
Ports: Atlantic City, New Jersey, USA
Sites: MAT–1, MAT–2 and MAT–3
Operator: ESO, Platform: L/B Kayd
Co-chief Scientists: Gregory Mountain & Jean-Noël Proust
Expedition Project Manager: David McInroy

Expedition Website: http://www.eso.ecord.org/expeditions/313/313.php

Shallow-water drilling of the New Jersey continental shelf: global sea level and architecture of passive margin sediments

OBJECTIVES

- Date late Paleogene-Neogene depositional sequences and compare ages of unconformable surfaces that divide these sequences with times of sea level lowerings predicted from the $\delta^{18}O$ glacio-eustatic proxy
- Estimate the corresponding amplitudes, rates, and mechanisms of sea level change
- Evaluate sequence stratigraphic facies models that predict depositional environments, sediment compositions, and stratal geometries in response to sea level change

HIGHLIGHTS

- Drilled at three locations in 35 m of water depth 45–67 km offshore, targeting the topsets, foresets, and toesets of several clinoforms at 180–750 m core depth below seafloor
- 612 cores with 80% recovery totaling 1311 m in length
- Eight lithologic units are recognized
- Nearly continuous composite record of ~1 m.y. sea level cycles (22–12 Ma)

FURTHER INFORMATION

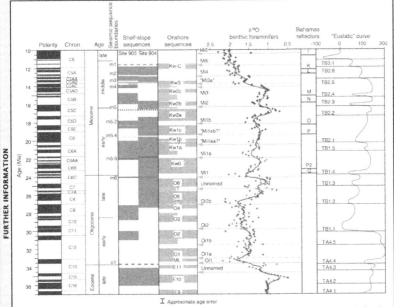

Figure: Comparison of Oligocene-Miocene slope sequences, onshore sequences, oxygen isotopes, Bahamian reflections, and inferred eustatic record. [1]

REFERENCES: (1)• Expedition 313 Scientists, 2010. New Jersey Shallow Shelf: shallow-water drilling of the New Jersey continental shelf: global sea level and architecture of passive margin sediments. IODP Prel. Rept., 313. doi:10.2204/ iodp.pr.313.2010 •(2) Mountain, G., Proust, J.-N., and the Expedition 313 Science Party, 2010. The New Jersey margin scientific drilling project (IODP Expedition 313): untangling the record of global and local sea-level changes. Sci. Drill., 10:26–34. doi:10.2204/ iodp.sd.10.03.2010

IODP Expedition 314
NanTroSEIZE Stage 1: LWD Transect

Dates: 21 September to 15 November 2007
Port: Shingu, Japan
Sites: C0001 – C0006
Operator: CDEX, Platform: Chikyu
Co–chief Scientists: Masataka Kinoshita & Harold J. Tobin
Staff Scientist: Moe Kyaw Thu

Exp. Website: http://www.jamstec.go.jp/chikyu/eng/Expedition/NantroSEIZE/exp314.html

Nankai Trough Seismogenic Zone Experiment Stage 1 – Riserless Drilling Expeditions:
September 2007–February 2008

Expedition Breakdown of Drilling Activities:
• Expedition 314:
Logging-while-drilling (LWD) only transect of the al prism and forearc basin sites
• Expedition 315:
Coring-focused drilling of the mega–splay fault thrust sheet and forearc basin
• Expedition 316:
Coring-focused drilling of the shallow portions of the frontal thrust and mega–splay fault zone

Scientific Targets: • The frontal thrust system at the toe of the accretionary wedge • The mid-wedge megasply fault system

OBJECTIVES

Riserless drilling without coring at all six Stage 1 sites using dedicated Logging-While-Drilling (LWD) technology at unstable formations associated with accretionary prism environment. Results from the variety of logging measurements of this expedition are crucial for optimizing the subsequent Stage 1 expeditions for coring and observatory operations, and for planning future stages.

HIGHLIGHTS

• Investigation of fault mechanics and seismogenesis along subduction megathrusts through direct sampling, in situ measurements, and long–term monitoring
• Obtaining a comprehensive suite of geophysical logs and other downhole measurements at sites along a transect from the incoming plate to the Kumano forearc basin using state-of-the-art LWD technology
• Drilling and logging successfully completed at four sites
• Depth range below the seafloor from 400 to 1400 m
Documentation of:
  • In–situ physical properties
  • Stratigraphic and structural features
  • Sonic to seismic scale velocity data for core–log–seismic integration
  • Stress, pore pressure, and hydrological parameters through both scalar and imaging log measurements

• The megasplay thrust sheet is composed of highly deformed and fractured rocks that are anomalously well indurated relative to their present depth.
• Present-day stress varies markedly along the NanTroSEIZE transect, and stresses in the upper 1.4 km are strongly compressional in the outer, active accretionary prism but extensional in the forearc basin.
• Occurrence of gas hydrate as a cement preferentially located in sandy portions of turbidite beds above a bottom-simulating reflector was quantified.
• Drilling at Sites C0001 and C0002 also provides important pilot hole information that will help prepare for the planned deep-riser sites for later stages of NanTroSEIZE drilling.

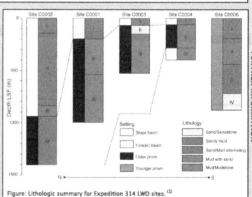

Figure: Lithologic summary for Expedition 314 LWD sites. [1]

REFERENCES: (1)• Kinoshita, M., Tobin, H., Moe, K.T., and the Expedition 314 Scientists, 2008. NanTroSEIZE Stage 1A: NanTroSEIZE LWD transect. IODP Prel. Rept., 314. doi:10.2204/ iodp.pr.314.2008 •(2) Tobin, H., Kinoshita, M., Ashi, J., Lallemant, S., Kimura, G., Screaton, E., Thu, M. K., Masago, H., & Curewitz, D., 2009. NanTroSEIZE stage 1 expeditions 314, 315, and 316; first drilling program of the nankai trough seismogenic zone experiment. Scientific Drilling, 8, 4–17. http://dx.doi.org/10.2204/iodp.sd.8.01.2009

IODP Expedition 315
NanTroSEIZE Stage 1: Megasplay Riser Pilot

Dates: 16 November to 18 December 2007
Ports: Shingu, Japan
Sites: C0001 – C0006
Operator: CDEX, Platform: Chikyu
Co-chief Scientists: Juichiro Ashi & Siegfried Lallemant
Staff Scientist: Hideki Masago

Exp. Website: http://www.jamstec.go.jp/chikyu/eng/Expedition/NantroSEIZE/exp315.html

Nankai Trough Seismogenic Zone Experiment Stage 1 - Riserless Drilling Expeditions:
September 2007-February 2008

Expedition Breakdown of Drilling Activities:
• Expedition 314:
Logging-while-drilling (LWD) only transect of the
al prism and forearc basin sites
• Expedition 315:
Coring-focused drilling of the mega-splay fault
thrust sheet and forearc basin
• Expedition 316:
Coring-focused drilling of the shallow portions
of the frontal thrust and mega-splay fault zone

Scientific Targets: • The frontal thrust system at the toe of the accretionary wedge • The mid-wedge megasply fault system

OBJECTIVES

Pilot study for the future deep riser drilling of the megasplay fault in Stage 2. The primary engineering and scientific objectives of this expedition are to obtain geotechnical information needed for well planning of future riser drilling to 3500 meters below seafloor. This site's location is also critical for understanding the nature of the shallow portions of splay faults. The scientific targets of this expedition are deformation mechanics, fault-related fluid source and migration pathways, and correlations between fault activity and slump deposits on the trench slope.

HIGHLIGHTS
• Revisited same sites of Exp 314 in order to collect cores and other downhole measurements (e.g., temperature) in adjacent holes (typically 25-50 m away)
• Cored at two planned riser drilling sites was conducted:
• C0001: cored to 458 m CSF
• C0002: drilled to 1057 m CSF

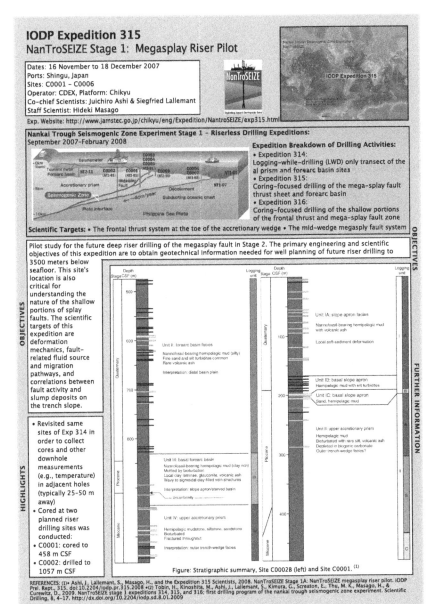

Figure: Stratigraphic summary, Site C0002B (left) and Site C0001. [1]

REFERENCES: (1)• Ashi, J., Lallemant, S., Masago, H., and the Expedition 315 Scientists, 2008. NanTroSEIZE Stage 1A: NanTroSEIZE megasplay riser pilot. IODP Prel. Rept., 315. doi:10.2204/iodp.pr.315.2008 •(2) Tobin, H., Kinoshita, M., Ashi, J., Lallemant, S., Kimura, G., Screaton, E., Thu, M. K., Masago, H., & Curewitz, D., 2009. NanTroSEIZE stage 1 expeditions 314, 315, and 316; first drilling program of the nankai trough seismogenic zone experiment. Scientific Drilling, 8, 4–17. http://dx.doi.org/10.2204/iodp.sd.8.01.2009

IODP Expedition 316
NanTroSEIZE Stage 1: Shallow Megasplay and Frontal Thrusts

Dates: 19 Decemer to 05 February 2008
Port: Shingu, Japan
Sites: C0004, C0006, C0007, C0008
Operator: CDEX, Platform: Chikyu
Co-chief Scientists: Gaku Kimura & Elizabeth Screaton
Staff Scientist: Daniel Curewitz

Exp. Website: http://www.jamstec.go.jp/chikyu/eng/Expedition/NantroSEIZE/exp316.html

Nankai Trough Seismogenic Zone Experiment Stage 1 – Riserless Drilling Expeditions:
September 2007-February 2008

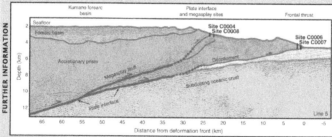

Expedition Breakdown of Drilling Activities:
- Expedition 314:
Logging-while-drilling (LWD) only transect of the
al prism and forearc basin sites
- Expedition 315:
Coring-focused drilling of the mega-splay fault
thrust sheet and forearc basin
- Expedition 316:
Coring-focused drilling of the shallow portions
of the frontal thrust and mega-splay fault zone

Scientific Targets: • The frontal thrust system at the toe of the accretionary wedge • The mid-wedge megasply fault system

OBJECTIVES

- Evaluate: • The deformation • Inferred depth of detachment • Structural partitioning • Fault zone physical characteristics
  and • Fluid flow at the frontal thrust and at the shallow portion of the megasplay system
- Clarify the character and behavior of the shallow portion of the megasplay:
  • Is the megasplay an active blind thrust or an inactive fault?
  • Is there evidence for past deformation mechanisms in the region above the (inferred) unstable seismogenic fault;
- Clarify the slip and deformation mechanisms in the region above the (inferred) unstable seismogenic fault;
- Clarify the relationship between fluid behavior, slip, and deformation along the megathrust
- Clarify the evolutionary development of the splay fault
- Clarify the function of the frontal thrust with respect to large earthquakes:
  • Do great earthquakes trigger slip along this fault plane, and if so, are they tsunamigenic?
  • Does the frontal thrust generate low-frequency events or does it creep during the interseismic period?
- Evaluate the relationship between fluid behavior and slip and deformation.
- Assess the evolution of the frontal thrust from its birth to death.

HIGHLIGHTS

- Drilling conducted at two sites in the megasplay region (one within the fault zone, one in the slope basin seaward of
  the megasplay) and at two sites within the frontal thrust region :
- C0004: Accretionary prism was sampled and the megasplay fault zone was successfully drilled: marine sediments,
  redeposited material from upslope, complex history of deformation of fault zone
- C0008: basin records the history of fault movement, sediment layers within basin provide reference for sediment
  underthrusting Site C0004
- (C0006&) C0007: plate boundary frontal thrust was successfully drilled, thrust fault material ranging from breccia to
  fault gouge was successfully recovered at Site C0007

FURTHER INFORMATION

Figure:
Interpretation of
Line KR0108-5
(after Park et al.,
2002 [see 1])
showing locations
of sites cored
during IODP
Expedition 316. [1]

REFERENCES: (1) Kimura, G., Screaton, E.J., Curewitz, D., and the Expedition 316 Scientists, 2008. NanTroSEIZE Stage 1A: NanTroSEIZE shallow megasplay and
frontal thrusts. IODP Prel. Rept., 316. doi:10.2204/iodp.pr.316.2008 •(2) Tobin, H., Kinoshita, M., Ashi, J., Lallemant, S., Kimura, G., Screaton, E., Thu, M. K.,
Masago, H., & Curewitz, D., 2009. NanTroSEIZE stage 1 expeditions 314, 315, and 316: first drilling program of the nankai trough seismogenic zone
experiment. Scientific Drilling, 8, 4–17. http://dx.doi.org/10.2204/iodp.sd.8.01.2009

IODP Expedition 317
Canterbury Basin Sea Level

Dates: 4 November 2009 to 4 January 2010
Ports: Townsville, Australia to Wellington, New Zealand
Sites: U1351–U1354
Operator: USIO, Platform: JR
Co-chief Scientists: Craig Fulthorpe & Koichi Hoyanagi
Staff Scientist: Peter Blum

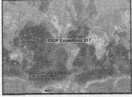

Exp. Website: http://iodp.tamu.edu/scienceops/expeditions/canterbury_basin.html

Global and Local Controls on Continental Margin Stratigraphy: Canterbury Basin, Eastern South Island, New Zealand

OBJECTIVES
- Date clinoform seismic sequence boundaries and sample associated facies to provide information necessary for estimation of eustatic amplitudes using backstripping
- Understand the relative importance of global sea level (eustasy) versus local tectonic and sedimentary processes in controlling continental margin sedimentary cycles
- Estimate the timing and amplitude of global sea level change
- Document sedimentary processes that operate during sequence formation
- Study complex interactions between preserved sedimentary architectures and strong ocean currents
- Evaluate predictions of future changes in global sea level and shoreline location

HIGHLIGHTS
- Neogene sediments, focusing on late Miocene to recent
- High-frequency (0.1–0.5 m.y.) record of depositional cyclicity
- Drilling of multi-target sequences on transect of sites on continental shelf plus on continental slope; transect provides a stratigraphic record of depositional cycles across the shallow-water environment most directly affected by relative sea level change
- Record will be used to estimate the timing and amplitude of global sea level change and to document the sedimentary processes that operate during sequence formation

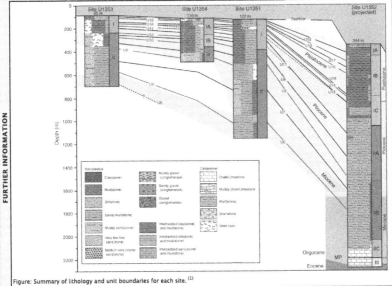

Figure: Summary of lithology and unit boundaries for each site. [1]

REFERENCES: (1)• Expedition 317 Scientists, 2010. Canterbury Basin Sea Level: Global and local controls on continental margin stratigraphy. IODP Prel. Rept., 317. doi:10.2204/iodp.pr.317.2010 •(2) Fulthorpe, C.S., Hoyanagi, K., Blum, P., and the IODP Expedition 317 Scientists, 2011. IODP Expedition 317: exploring the record of sea-level change off New Zealand. Sci. Drill., 12:4–14. doi:10.2204/iodp.sd.12.01.2011

IODP Expedition 318
Wilkes Land Glacial History

Dates: 4 January to 8 March 2010
Ports: Wellington, New Zealand to Hobart, Australia
Sites: U1355–U1361
Operator: USIO, Platform: JR
Co-chief Scientists: Carlota Escutia & Henk Brinkhuis
Staff Scientist: Adam Klaus

Expedition Website: http://iodp.tamu.edu/scienceops/expeditions/wilkes_land.html

Cenozoic East Antarctic Ice Sheet evolution from Wilkes Land margin sediments
From Greenhouse to Icehouse at the Wilkes Land Antarctic Margin: Reveal the history of East Antarctic ice sheet dynamics and climate variations of the Antarctic from Eocene to Quaternary

OBJECTIVES

- Provide a long-term record of the sedimentary archives along an inshore to offshore transect of Cenozoic Antarctic glaciation and its intimate relationships with global climatic and oceanographic change
- Obtain the timing and nature of the first arrival of ice at the Wilkes Land margin inferred to have occurred during the earliest Oligocene
- Obtain the nature and age of the changes in the geometry of the progradational wedge interpreted to correspond with large fluctuations in the extent of the East Antarctic Ice Sheet and possibly coinciding with the transition from a wet-based to a cold-based glacial regime
- Obtain a high-resolution record of Antarctic climate variability during the late Neogene and Quaternary
- Obtain an unprecedented ultrahigh resolution (i.e., annual to decadal) Holocene record of climate variability

HIGHLIGHTS

- 1973m of sediments recovered
- Records span ~53 million years of Antarctic history : Wilkes Land Antarctic margin from an ice-free "greenhouse" Antarctica, to the first cooling, to the onset and erosional consequences of the first glaciation and the subsequent dynamics of the waxing and waning ice sheets, all the way to thick, unprecedented "tree ring style" records with seasonal resolution of the last deglaciation that began ~10,000 years ago
- Revealing details of tectonic history of Australo-Antarctic Gulf

FURTHER INFORMATION

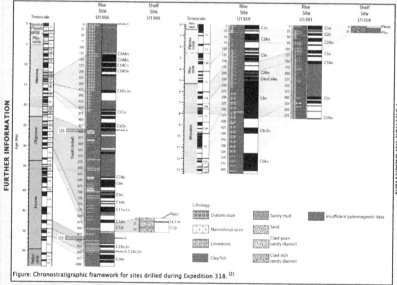

Figure: Chronostratigraphic framework for sites drilled during Expedition 318. [2]

REFERENCES: (1)·Expedition 318 Scientists, 2010. Wilkes Land Glacial History: Cenozoic East Antarctic Ice Sheet evolution from Wilkes Land margin sediments. IODP Prel. Rept., 318. doi:10.2204/ Iodp.pr.318.2010 •(2) Escutia, C., Brinkhuis, H., Klaus, A., and the IODP Expedition 318 Scientists, 2011. IODP Expedition 318: from Greenhouse to Icehouse at the Wilkes Land Antarctic margin. Sci. Drill., 12:15–23. doi:10.2204/ Iodp.sd.12.02.2011

IODP Expedition 319
NanTroSEIZE Stage 2: Riser/Riserless Observatory 1

Dates: 10 May to 31 August 2009
Port: Shingu, Japan
Sites: C0009, C0010
Operator: CDEX, Platform: Chikyu
Co-chief Scientists: Elichiro Araki, Tim Byrne, Lisa McNeill & Demian Saffer
Expedition Project Manager: Nobuhisa Eguchi, Kyoma Takahashi & Sean Toczko

Exp. Website: http://www.jamstec.go.jp/chikyu/nantroseize/e/expedition_319.html

Nankai Trough Seismogenic Zone Experiment Stage 2 – Riser drilling and geophysical logging in the area straight above the seismogenic zone

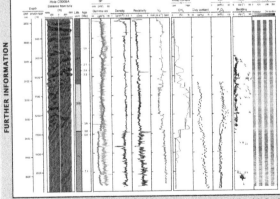

OBJECTIVES

Understanding of:
• In-situ physical conditions and the state of stress within different parts of the subduction system during an earthquake cycle
• The mechanisms controlling the updip aseismic–seismic transition along the plate boundary fault system
• Processes of earthquake and tsunami generation and strain accumulation and release
• The mechanical strength of the plate boundary fault
• The potential role of a major fault system (termed the "megasplay" fault) in accommodating earthquake slip and hence influencing tsunami generation

HIGHLIGHTS

• First riser drilling in IODP history: successfully drilled down to a depth of 1,603.7 meters beneath the sea floor at the site C0009 (at water depth of 2,054 meters)
• Measurements of in situ pore pressure, permeability and minimum principal stress magnitude, real-time mud gas analysis, and laboratory analyses of cuttings throughout the entire riser–drilled depth range (~700-1600 meters below seafloor)
• Conduction of leak–off test at one depth interval
• Deployment of wireline Modular Formation Dynamics Tester 12 times to directly measure in situ stress magnitude, formation pore pressure, and permeability
• Collection of mud gas for geochemical analyses and cuttings samples throughout the entire riser–drilled depth range
• Definition of a single integrated set of lithologic units and comparison with previously drilled IODP Site C0002 to determine the evolutionary history of the forearc basin
• Conduction of a long–offset (30 km) "walkaway" vertical seismic profile (VSP) to image the megasplay and master décollement beneath the borehole, and to evaluate seismic velocity and anisotropy of the forearc basin and accretionary prism sediments around the borehole

FURTHER INFORMATION

Figure: Site summary diagram for Site C0009 within the central forearc basin and hanging wall of the seismogenic plate boundary fault. [2]

REFERENCES: (1)• Saffer, D., McNeill, L., Araki, E., Byrne, T., Eguchi, N., Toczko, S., Takahashi, K., and the Expedition 319 Scientists, 2009. NanTroSEIZE Stage 2: NanTroSEIZE riser/riserless observatory. IODP Prel. Rept., 319. doi:10.2204/ Iodp.pr.319.2009 •(2) McNeill, L., Saffer, D., Byrne, T., Araki, E., Toczko, S., Eguchi, N., Takahashi, K., and the IODP Expedition 319 Scientists, 2010. IODP Expedition 319, NanTroSEIZE Stage 2: first IODP riser drilling operations and observatory installation towards understanding subduction zone seismogenesis. Sci. Drill., 10:4-13. doi:10.2204/ Iodp.sd.10.01.2010

IODP Expedition 320
Pacific Equatorial Age Transect I

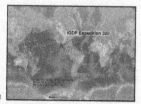

Dates: 5 March to 5 May 2009
Ports: Honolulu, Hawaii to Honolulu, Hawaii
Sites: U1331–U1336
Operator: USIO, Platform: JR
Co-chief Scientists: Heiko Pälike & Hiroshi Nishi
Staff Scientist: Adam Klaus

Exp. Website: http://iodp.tamu.edu/scienceops/expeditions/equatorial_pacific.html

Pacific Equatorial Age Transect (PEAT) – Expeditions 320 & 321:
• Drill an age transect ("flow line") along the position of the paleoequator
• Recover a continuous Cenozoic climate record of the paleoequatorial Pacific

OBJECTIVES

• Provide rare material to validate and extend the astronomical calibration of the geological time scale for the Cenozoic
• Provide information about the detailed nature of calcium carbonate dissolution and changes of the CCD
• Enhance the understanding of bio- and magnetostratigraphic datums at the equator
• Provide information about rapid biological evolution and turnover during times of climatic stress

HIGHLIGHTS

• Cored eight sites from the sediment surface to at or near basement
• Basalt aged between 52 and 18 Ma
• Covered the time period following maximum Cenozoic warmth, through initial major glaciations, to today
• Reconstructed extreme changes of the calcium carbonate compensation depth (CCD) across major geological boundaries during the last 52 Ma
• Recovered a unique sedimentary biogenic sediment archive for time periods just after the Paleocene/Eocene boundary event, the Eocene cooling, the Eocene–Oligocene transition, the "one cold pole" Oligocene, the Oligocene–Miocene transition, and the middle Miocene cooling

FURTHER INFORMATION

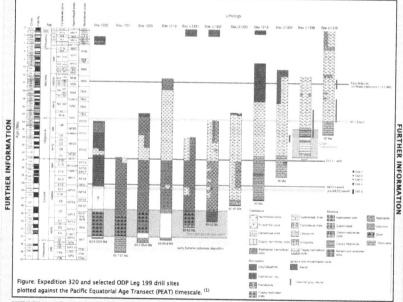

Figure: Expedition 320 and selected ODP Leg 199 drill sites plotted against the Pacific Equatorial Age Transect (PEAT) timescale. [1]

REFERENCES: (1)• Pälike, H., Nishi, H., Lyle, M., Raffi, I., Klaus, A., Gamage, K., and the Expedition 320/321 Scientists, 2009. Pacific Equatorial Transect. IODP Prel. Rept., 320. doi:10.2204/ iodp.pr.320.2009 •(2) Lyle, M., Pälike, H., Nishi, H., Raffi, I., Gamage, K., Klaus, A., and the IODP Expeditions 320/321 Science Party, 2010. The Pacific Equatorial Age Transect, IODP Expeditions 320 and 321: building a 50-million-year-long environmental record of the equatorial Pacific Ocean. Sci. Drill., 9:4–15. doi:10.2204/ iodp.sd.9.01.2010

IODP Expedition 321
Pacific Equatorial Age Transect II

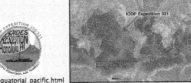

IODP Expedition 321

Dates: 5 May to 22 June 2009
Ports: Honolulu, Hawaii to San Diego, California
Sites: U1337–U1338
Operator: USIO, Platform: JR
Co-chief Scientists: Mitch Lyle & Isabella Raffi
Staff Scientist: Kusali Gamage
Exp. Website: http://iodp.tamu.edu/scienceops/expeditions/equatorial_pacific.html

Pacific Equatorial Age Transect (PEAT) – Expeditions 320 & 321:
• Drill an age transect ("flow line") along the position of the paleoequator
• Recover a continuous Cenozoic climate record of the paleoequatorial Pacific

OBJECTIVES
• Provide rare material to validate and extend the astronomical calibration of the geological time scale for the Cenozoic
• Provide information about the detailed nature of calcium carbonate dissolution and changes of the CCD
• Enhance the understanding of bio- and magnetostratigraphic datums at the equator
• Provide information about rapid biological evolution and turnover during times of climatic stress

HIGHLIGHTS
• Cored eight sites from the sediment surface to at or near basement
• Basalt aged between 52 and 18 Ma
• Covered the time period following maximum Cenozoic warmth, through initial major glaciations, to today
• Reconstructed extreme changes of the calcium carbonate compensation depth (CCD) across major geological boundaries during the last 52 Ma
• Recovered a unique sedimentary biogenic sediment archive for time periods just after the Paleocene/Eocene boundary event, the Eocene cooling, the Eocene-Oligocene transition, the "one cold pole" Oligocene, the Oligocene-Miocene transition, and the middle Miocene cooling

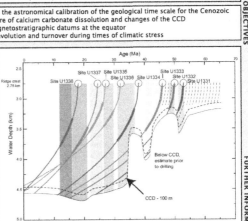

Figure: Targeting drill sites prior to coring based on calcium carbonate compensation depth (CCD) history (van Andel, 1975), with additional data from Leg 199. [2]

FURTHER INFORMATION

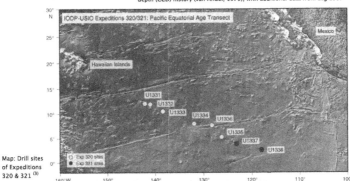

Map: Drill sites of Expeditions 320 & 321 [3]

REFERENCES: (1)• Lyle, M., Raffi, I., Pälike, H., Nishi, H., Gamage, K., Klaus, A., and the Expedition 320/321 Scientists, 2009. Pacific Equatorial Transect. IODP Prel. Rept., 321. doi:10.2204/ iodp.pr.321.2009 •(2) Lyle, M., Pälike, H., Nishi, H., Raffi, I., Gamage, K., Klaus, A., and the IODP Expeditions 320/321 Science Party, 2010. The Pacific Equatorial Age Transect, IODP Expeditions 320 and 321: building a 50–million–year–long environmental record of the equatorial Pacific Ocean. Sci. Drill., 9:4–15. doi:10.2204/ iodp.sd.9.01.2010 •(3) http://iodp.tamu.edu/scienceops/expeditions/equatorial_pacific.html

IODP Expedition 322
NanTroSEIZE Stage 2: Subduction Input

Dates: 01 September to 10 October 2009
Port: Shingu, Japan
Sites: C0011, C0012
Operator: CDEX, Platform: Chikyu
Co-chief Scientists: Michael B. Underwood & Saneatsu Saito
Expedition Project Manager: Yu'suke Kubo
Exp. Website: http://www.jamstec.go.jp/chikyu/nantroseize/e/expedition_322.html

IODP Expedition 322

OBJECTIVES

Nankai Trough Seismogenic Zone Experiment Stage 2 – Characterization of composition, architecture, and stte of pre-subduction sediments transported to the seisogenic zone

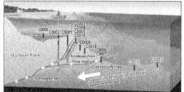

• Sample and log the incoming sedimentary strata and uppermost igneous basement of the Shikoku Basin, seaward of the Nankai Trough to study:
• Determine the initial pre-subduction conditions to assess how various material properties evolve down the dip of the plate interface, and potentially change the fault's behavior from stable sliding to seismogenic slip
• Demonstrate the response of facies character and sedimentation rates to bathymetric architecture

HIGHLIGHTS

Characterizing input sediments and basement rock entering to the seismogenic zone for great earthquakes:
• Volcanic material supply to Shikoku Basin
• Sediment supply from southwestern Japan
• Two types of groundwater running beneath the seafloor
• Identification of sediment /basement interface, and the collection of basement samples
• Downhole geophysical properties by the comparison of cores and LWD data

FURTHER INFORMATION

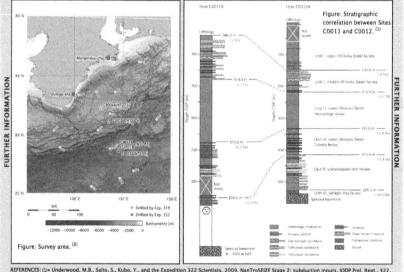

Figure: Survey area. [3]

Figure: Stratigraphic correlation between Sites C0011 and C0012. [1]

REFERENCES: (1)• Underwood, M.B., Saito, S., Kubo, Y., and the Expedition 322 Scientists, 2009. NanTroSEIZE Stage 2: subduction inputs. IODP Prel. Rept., 322. doi:10.2204/?iodp.pr.322.2009 •(2) Underwood, M.B., Saito, S., Kubo, Y., and the IODP Expedition 322 Scientists, 2010. IODP Expedition 322 drills two sites to document inputs to the Nankai Trough Subduction Zone. Sci. Drill., 10:14–25. doi:10.2204/?iodp.sd.10.02.2010 •(3) http://www.jamstec.go.jp/e/about/press_release/20091009/

IODP Expedition 323
Bering Sea Paleoceanography

Dates: 5 July to 4 September 2009
Ports: Victoria, Canada to Yokohama, Japan
Sites: U1339–U1345
Operator: USIO, Platform: JR
Co-chief Scientists: Christina Ravelo & Kozo Takahashi
Staff Scientist: Carlos Alvarez-Zarikian

Expedition Website: http://iodp.tamu.edu/scienceops/expeditions/bering_sea.html

Pliocene–Pleistocene paleoceanography and climate history of the Bering Sea

OBJECTIVES
- Elucidate detailed evolutionary history of climate and surface ocean conditions since the earliest Pliocene in the Bering Sea, where amplified high-resolution changes of climatic signals are recorded
- Shed light on temporal changes in the origin and intensity of North Pacific Intermediate Water and possibly deeper water mass formation in the Bering Sea
- Characterize history of continental glaciation, river discharge, and sea ice formation in order to investigate link between continental and oceanic conditions of Bering Sea and adjacent land areas;
- Investigate linkages through comparison to pelagic records between ocean/climate processes that occur in the more sensitive marginal sea environment and processes that occur in the North Pacific and/or globally
- Evaluate how ocean/climate history of the Bering Strait gateway region may have affected North Pacific and global

HIGHLIGHTS
- Drilling of 5741 m sediment at seven sites (97.4% recovery)
- Four deep holes with depth ranges from 600 to 745 mbsf, spanning 1.9 to 5 Ma
- Water depths between 818 to 3174 m in order to characterize past vertical water mass distribution and circulation.
- Understanding of long-term evolution of surface water mass distribution during the past 5 Ma, including expansion of seasonal sea ice to Bowers Ridge between 3.0 and 2.5 Ma and intensification of seasonal sea ice at both Bowers Ridge and the Bering slope at ~1.0 Ma, the mid-Pleistocene Transition.
- Characterization of intermediate and deep water masses, including evidence from benthic foraminifers and sediment laminations, for episodes of low-oxygen conditions in the Bering Sea throughout the last 5 Ma

FURTHER INFORMATION

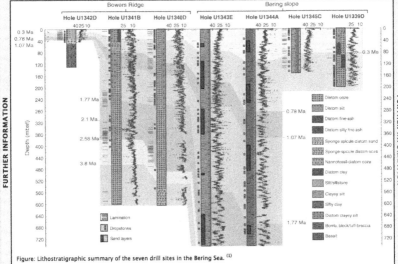

Figure: Lithostratigraphic summary of the seven drill sites in the Bering Sea. (1)

REFERENCES: (1)• Expedition 323 Scientists, 2010. Bering Sea paleoceanography: Pliocene–Pleistocene paleoceanography and climate history of the Bering Sea. IODP Prel. Rept., 323. doi:10.2204/ iodp.pr.323.2010 •(2) Takahashi, K., Ravelo, A.C., and Alvarez Zarikian, C., 2011. IODP Expedition 323—Pliocene and Pleistocene paleoceanographic changes in the Bering Sea. Sci. Drill., 11:4–13. doi:10.2204/iodp.sd.11.01.2011 iodp.sd.9.01.2010

IODP Expedition 324
Shatsky Rise Formation

IODP Expedition 324

Dates: 4 September to 3 November 2009
Ports: Yokohama, Japan to Townsville, Australia
Sites: U1346–U1350
Operator: USIO, Platform: JR
Co-chief Scientists: Will Sager & Takashi Sano
Staff Scientist: Jörg Geldmacher

Exp. Website: http://iodp.tamu.edu/scienceops/expeditions/shatsky_rise.html

Testing plume and plate models of ocean plateau formation at Shatsky Rise, northwest Pacific Ocean

OBJECTIVES

- Study processes of oceanic plateau formation and evolution
- Study the origin and temporal and geochemical evolution of the submarine Shatsky Rise Plateau, a large igneous province (LIP)
- Core igneous rocks from the volcanic massifs of Shatsky Rise and the sediments above to determine the age, sources, and evolution of this plateau and to test hypotheses of a plume head or plate-controlled origin as well as sedimentary history
- Illuminate plate motions and volcanic structure from paleomagnetic data
- Explore links between Shatsky Rise formation and the synchronous Jurassic–Cretaceous boundary through stratigraphic, geochemical, and age data.

HIGHLIGHTS

- Site penetrations from 191.8 m to 324.1 m with coring of 52.6 m to 172.7 m into igneous basement
- Average recovery in basement 38.7%–67.4%
- Igneous sections consist mainly of variably evolved tholeiitic basalts emplaced as pillows or massive flows
- Shatsky Rise began with massive eruptions forming a huge volcano
- Subsequent eruptions waned in intensity, forming volcanoes that are large, but which did not erupt with unusually high effusion rates

FURTHER INFORMATION

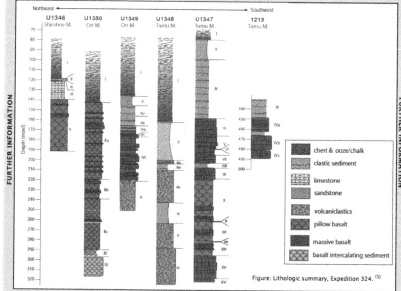

Figure: Lithologic summary, Expedition 324. [1]

REFERENCES: (1)• Expedition 324 Scientists, 2010. Testing plume and plate models of ocean plateau formation at Shatsky Rise, northwest Pacific Ocean. IODP Prel. Rept., 324. doi:10.2204/iodp.pr.324.2010 •(2) Sager, W.W., Sano, T., Geldmacher, J., and the IODP Expedition 324 Scientists, 2011. IODP Expedition 324: ocean drilling at Shatsky Rise gives clues about oceanic plateau formation. Sci. Drill., 12:24–31. doi:10.2204/iodp.sd.12.03.2011

IODP Expedition 325
Great Barrier Reef environmental changes

IODP Expedition 325

Dates: 11 February – 6 April 2010
Ports: Townsville, Queensland, Australia
Sites: Ribbon Reef 5 (RIB–01C) – 5 sites, Ribbon Reef 3 (RIB–02A) – 4 sites
Noggin Pass (NOG–01B) – 8 sites, Hydrographer's Passage (HYD–01C) –
11 sites, Hydrographer's Passage (HYD–02A) – 12 sites
Operator: ESO, Platform: Greatship Maya
Co-chief Scientists: Jody Webster & Yusuke Yokoyama
Expedition Project Manager: Carol Cotterill

Expedition Website: http://www.eso.ecord.org/expeditions/325/325.php

The last deglacial sea level rise in the South Pacific: offshore drilling northeast Australia

OBJECTIVES
- Establish the course of sea-level rise during the last deglaciation
- Reconstruct the nature and magnitude of seasonal–millennial scale climate variability
- Define sea-surface temperature variations
- Determine the biologic and geologic response of the Great Barrier Reef (GBR) to abrupt sea-level and climate changes
- Analyze the impact of environmental changes on reef growth and geometry for the region over the period of 20–10 ka

HIGHLIGHTS
- Cored a succession of fossil reef structures preserved on the shelf edge seaward of the modern barrier reef
- Total of 34 boreholes across 17 sites cored in depths ranging from 42.27 to 167.14 mbsl
- Recovered coral reef deposits from water depth down to 126 m that ranged in age from 9 ka to older than 30 ka
- Reconstructed the course of postglacial sea-level change in the GBR from 20 ka to 10 ka
- Established sea-surface variations in the GBR from 20 ka to 10 ka
- Investigated the response of the GBR to environmental changes caused by sea-level and climate change

FURTHER INFORMATION

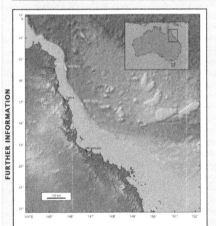

Figure: Regional map showing locations of the five proposed drill sites. [1]

Figure: Representative litholostratigraphic succession from the fossil coral reef terraces (M0033A). [2]

REFERENCES: (1)• Expedition 325 Scientists, 2010. Great Barrier Reef environmental changes: the last deglacial sea level rise in the South Pacific: offshore drilling northeast Australia. IODP Prel. Rept., 325. doi:10.2204/ iodp.pr.325.2010 •(2) Yokoyama, Y., Webster, J.M., Cotterill, C., Braga, J.C., Jovane, L., Mills, H., Morgan, S., Suzuki, A., and the IODP Expedition 325 Scientists, 2011. IODP Expedition 325: the Great Barrier Reef reveals past sea-level, climate, and environmental changes since the last ice Age. Sci. Drill., 12:32–45. doi:10.2204/ iodp.sd.12.04.2011

IODP Expedition 326
NanTroSEIZE Stage 3: Plate Boundary Deep Riser 1

Dates: 15 July to 20 August 2010
Port: Shingu, Japan
Sites: C0002F
Operator: CDEX, Platform: Chikyu
Co-chief Scientists: Harold Tobin & Masataka Kinoshita
Expedition Project Manager: Nobuhisa Eguchi & Simon Nielsen

Exp. Website: http://www.jamstec.go.jp/chikyu/nantroseize/e/expedition_326.html

NanTroSEIZE Stage 3: Plate Boundary Deep Riser: Top Hole Engineering, Prepare the hole for future ultra-deep riser drilling by installing the wellhead and casing

OBJECTIVES

- First preparatory stage of drilling and coring of the main IODP Hole C0002F to the boundary zone between the Philippine Sea and Eurasian plates in the Nankai accretionary margin
- Purely operational objectives, with the main goal of installation of the wellhead assembly and drilling and casing the uppermost 800 m of the planned 7 km deep hole (Thus, no science party was on board during the expedition and no scientific results are reported. Scientific objectives for the uppermost 1400 m at this site were previously fulfilled during NanTroSEIZE Stage 1 Expeditions 314 and 315).

HIGHLIGHTS

- Drilling of Hole C0002F to 872.5 meters below seafloor
- Lining hole with successfully cemented 20 inch casing
- Setting of corrosion cap in preparation for a future return to continue drilling
- Hole C0002F is ready for further deep riser drilling

FURTHER INFORMATION

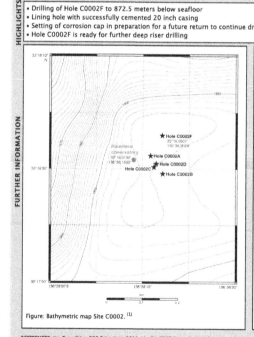

Figure: Bathymetric map Site C0002. [1]

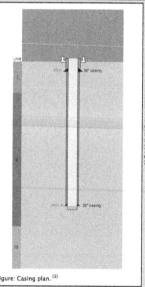

Figure: Casing plan. [1]

REFERENCES: (1)• Expedition 326 Scientists, 2011. NanTroSEIZE Stage 3: plate boundary deep riser: top hole engineering. IODP Prel. Rept., 326. doi:10.2204/iodp.pr.326.2011

IODP Expedition 327
Juan de Fuca Ridge–Flank Hydrogeology

Dates: 5 July–5 September 2010
Ports: Townsville, Queensland, Australia
Sites: U1362, U1363, U1301, 1027
Operator: USIO, Platform: JR
Co–chief Scientists: Andrew Fisher & Takeshi Tsuji
Staff Scientist: Katerina Petronotis

IODP Expedition 327

Exp. Website: http://iodp.tamu.edu/scienceops/expeditions/juan_de_fuca.html

The hydrogeological architecture of basaltic oceanic crust: compartmentalization, anisotropy, microbiology, and crustal-scale properties on the eastern flank of Juan de Fuca Ridge, eastern Pacific Ocean: IODP Expeditions 301 & 327

OBJECTIVES

- Second part of a two-expedition, multidisciplinary program, built on the achievements of IODP Expedition 301 and subsequent ROV expeditions
- Multidimensional, cross-hole experiments in oceanic crust, incl. linked hydrology, microbiological, seismic and tracer components
- Understand fluid-rock interactions in young, upper ocean crust on the eastern flank of the Juan de Fuca Ridge
- Delineate the magnitude and distribution of hydrologic properties, the rates and spatial extent of ridge–flank fluid circulation and links between ridge–flank circulation, crustal alteration, and geomicrobial processes
- Evaluate the extent to which oceanic crust is connected vertically and horizontally; the influence of these connections on fluid, solute, heat, and microbiological processes; and the importance of scaling on hydrologic properties
- Understand the nature of permeable pathways, the depth extent of circulation, the importance of permeability anisotropy, and the significance of hydrogeologic barriers in the crust

HIGHLIGHTS

- Installed subseafloor borehole observatories ("CORKs") in basement holes to allow borehole conditions to recover to a more natural state after the dissipation of disturbances caused by drilling, casing, and other operations
- Cored two basement holes and drilled to 528 meters below seafloor (mbsf) (292 meters subbasement [msb]) and to 359 mbsf (117 msb) respectively
- Provided a long-term monitoring and sampling presence for determining fluid pressure, temperature, composition, and microbiology
- Facilitated completion of active experiments to resolve crustal hydrogeologic conditions and processes

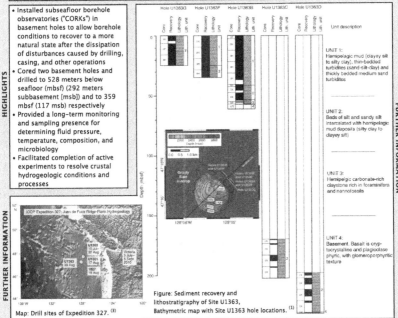

FURTHER INFORMATION

Unit description

UNIT 1:
Hemipelagic mud (clayey silt to silty clay), thin-bedded turbidites (sand-silt-clay) and thickly bedded medium sand turbidites

UNIT 2:
Beds of silt and sandy silt intercalated with hemipelagic mud deposits (silty clay to clayey silt)

UNIT 3:
Hemipelgic carbonate-rich claystone rich in foraminifera and nannofossils

UNIT 4:
Basement. Basalt is cryptocrystalline and plagioclase phyric, with glomeroporphyritic texture

Figure: Sediment recovery and lithostratigraphy of Site U1363, Bathymetric map with Site U1363 hole locations. (1)

Map: Drill sites of Expedition 327. (3)

REFERENCES (1)• Expedition 327 Scientists, 2010. Juan de Fuca Ridge–flank hydrogeology: the hydrogeologic architecture of basaltic oceanic crust: compartmentalization, anisotropy, microbiology, and crustal-scale properties on the eastern flank of Juan de Fuca Ridge, eastern Pacific Ocean. IODP Prel. Rept., 327. doi:10.2204/ iodp.pr.327.2010 (2)• Fisher, A.T., Tsuji, T., Petronotis, K., Wheat, C.G., Becker, K., Clark, J.F., Cowen, J., Edwards, K., Jannasch, H., and the IODP Expedition 327 and Atlantis Expedition AT18–07 Shipboard Parties, 2012. IODP Expedition 327 and Atlantis Expedition AT 18–07: observatories and experiments on the eastern flank of the Juan de Fuca Ridge. Sci. Drill., 13:4–11. doi:10.2204/ iodp.sd.13.01.2011 (3)• http://iodp.tamu.edu/scienceops/ expeditions/juan_de_fuca.html

IODP Expedition 328
Cascadia subduction zone ACORK observatory

Dates: 5–18 September 2010
Ports: Victoria to Victoria, British Columbia
Sites: U1364
Operator: USIO, Platform: JR
Co-chief Scientist: Earl Davis
Staff Scientist: Mitch Malone

IODP Expedition 328

Expedition Website: http://iodp.tamu.edu/scienceops/expeditions/cascadia.html

Cascadia subduction zone ACORK observatory

OBJECTIVES
- Operational objective: installation of a new permanent hydrologic borehole observatory near Ocean Drilling Program Site 889
- Format: Advanced CORK design (Circulation Obviation Retrofit Kit), facilitate pressure monitoring at multiple formation levels on the outside of a 10? inch casing string

HIGHLIGHTS
- Observatory successfully installed
- Documentation of:
  - Average state of pressure in the frontal part of the Cascadia accretionary prism
  - Pressure gradients driving flow from the consolidating sediments
  - Mode of formation of gas hydrates
  - Influence of gas hydrates and free gas on the mechanical properties of their host lithology
  - Response of the material to seismic ground motion
  - Magnitude of deformation at the site caused by secular strain and episodic seismic and aseismic slip in this subduction setting
- Casing sealed: inside available for future installation of additional monitoring instruments, observatory will be connected to the NEPTUNE Canada fiber-optic cable for power and real-time communications from land in near future

FURTHER INFORMATION

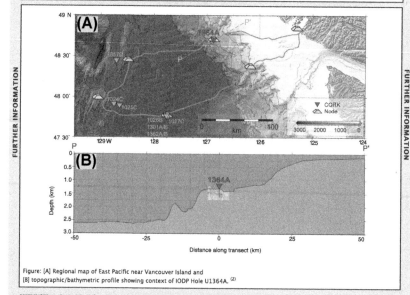

Figure: [A] Regional map of East Pacific near Vancouver Island and
[B] topographic/bathymetric profile showing context of IODP Hole U1364A. [2]

REFERENCES (1)• Davis, E.E., Malone, M.J., and the Expedition 328 Scientists and Engineers, 2010. Cascadia subduction zone ACORK observatory. IODP Prel. Rept., 328. doi:10.2204/ Iodp.pr.328.2010 (2)• Davis, E., Heesemann, M., and the IODP Expedition 328 Scientists and Engineers, 2012. IODP Expedition 328: early results of Cascadia subduction zone ACORK observatory. Sci. Drill., 13:11–18. doi:10.2204/iodp.sd.13.02.2011

IODP Expedition 329
South Pacific Gyre subseafloor life

Dates: 9 October–13 December 2010
Ports: Papeete, Tahiti, to Auckland, New Zealand
Sites: U1365–U1371
Operator: USIO, Platform: JR
Co–chief Scientists: Steven D'Hondt & Fumio Inagaki
Staff Scientist: Carlos Alvarez Zarikian

Exp. Website: http://iodp.tamu.edu/scienceops/expeditions/south_pacific_gyre_microbio.html

OBJECTIVES

- Documentation of many fundamental aspects of subseafloor sedimentary habitats, metabolic activities, and biomass in this very low-activity sedimentary ecosystem
- Improvement of understanding of how oceanographic factors control variation in subseafloor sedimentary habitats, activities, and biomass from gyre center to gyre margin
- Quantification of availability of dissolved hydrogen throughout the sediment column
- Documentation of first-order patterns of basement habitability and potential microbial activities

HIGHLIGHTS

- Reconstruction of glacial seawater characteristics through the South Pacific Gyre (SPG)
- Throughout the SPG dissolved oxygen and nitrate are present the entire sediment sequence
- Sedimentary microbial cell counts are lower than at all previously drilled IODP/Ocean Drilling Program (ODP)/Deep Sea Drilling Program (DSDP) sites
- Detectable oxygen and nitrate in the upwelling zone just south of the gyre are limited to the top and bottom of the sediment column
- Manganese reduction is a prominent electron-accepting process
- Cell concentrations are higher than at the same depths in the SPG sites throughout the sediment column
- Geographic variation in subseafloor profiles of dissolved and solid-phase chemicals are consistent with the magnitude of organic-fueled subseafloor respiration declining from outside the gyre to the gyre center
- Chemical profiles in the sedimentary pore water and secondary mineral distributions in the basaltic basement indicate that basement alteration continues on the timescale of formation fluid replacement, even at the sites with the oldest basement

FURTHER INFORMATION

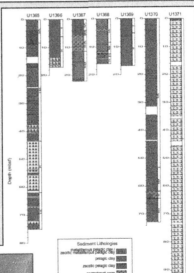

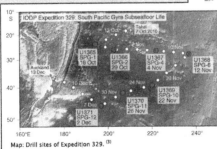

Map: Drill sites of Expedition 329. [3]

Figure: Representative lithologic columns, Sites U1365–U1371. [2]

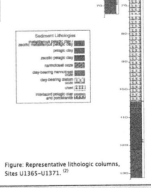

REFERENCES (1)• Expedition 329 Scientists, 2011. South Pacific Gyre subseafloor life, IODP Prel. Rept., 329. doi:10.2204/ iodp.pr.329.2011 (2)• D'Hondt, S., Inagaki, F., Alvarez Zarikian, C.A., and the IODP Expedition 329 Science Party, 2013. IODP Expedition 329: life and habitability beneath the seafloor of the South Pacific Gyre. Sci. Drill., 15:4–10. doi:10.2204/ iodp.sd.15.01.2013 (3)•http://iodp.tamu.edu/scienceops/expeditions/south_pacific_gyre_microbio.html

IODP Expedition 330
Louisville Seamount Trail

Dates: 13 December 2010 to 11 February 2011
Ports: Auckland, New Zealand, to Auckland, New Zealand
Sites: U1372-U1377
Operator: USIO, Platform: JR
Co-chief Scientists: Anthony Koppers & Toshitsugu Yamazaki
Staff Scientist: Jörg Geldmacher

Exp. Website: http://iodp.tamu.edu/scienceops/expeditions/louisville_seamounts.html

Implications for geodynamic mantle flow models and the geochemical evolution of primary hotspots

OBJECTIVES

OBJECTIVES

- Test end-member geodynamic models: is a significant plume motion required to determine whether it moved independently from Hawaii (as predicted from the mantle flow models) or in concert
- Determine the Louisville hotspot movement during the 80 Ma to 50 Ma time interval
- Determine the paleolatitude of the Louisville hotspot at the time of seamount formation
- Sample unaltered basalt flows for onshore 40Ar/39Ar geochronology to establish the time frame for potential changes in the Louisville hotspot paleolatitude
- Determine origin and magmatic evolution of the Louisville volcanoes and hotspot source
- Determine possible plume-lithosphere interaction
- Validity of the old hypothesis that the Ontong Java Plateau formed from the initial activity of the Louisville mantle plume around 120 Ma
- Determine the existence of subsurface biosphere in older volcanic basement rocks

HIGHLIGHTS

- Drilled into the summits of the five Louisville guyots and reached volcanic basement at four of these drilling targets
- Collected of more than sixty microbiology samples from four seamounts up to a maximum depth of 516 meters below seafloor (mbsf)
- Cored 1114 m sediment and igneous basement with an average 72.4% recovery at five seamounts
- Cored materials are relatively unaltered, providing an affluence of well-preserved basaltic samples that contain, for example, pristine olivine crystals and fresh volcanic glass
- Recovered large fractions of mostly submarine hyaloclastites, volcanic sandstone, and basaltic breccia at all drill sites

FURTHER INFORMATION

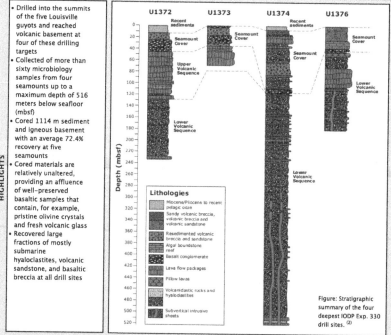

Figure: Stratigraphic summary of the four deepest IODP Exp. 330 drill sites. [2]

REFERENCES (1)• Expedition 330 Scientists, 2011. Louisville Seamount Trail: implications for geodynamic mantle flow models and the geochemical evolution of primary hotspots. IODP Prel. Rept., 330. doi:10.2204/ Iodp.pr.330.2011 (2)• Koppers, A.A.P., Yamazaki, T., Geldmacher, J., and the IODP Expedition 330 Scientific Party, 2013. IODP Expedition 330: drilling the Louisville Seamount Trail in the SW Pacific. Sci. Drill., 15:11–22. doi:10.2204/ Iodp.sd.15.02.2013

IODP Expedition 331
DEEP HOT BIOSPHERE

Dates: 01 September to 03 October 2010
Port: Shingu, Japan
Sites: C0013–C0017
Operator: CDEX, Platform: Chikyu
Co–chief Scientists: Michael Mottl & Ken Takai
Expedition Project Manager: Simon H.H. Nielsen

Expedition Website: http://www.jamstec.go.jp/chikyu/exp331/e/expedition.html

OBJECTIVES
- Test for existence of a functionally active, metabolically diverse subvent biosphere associated with subseafloor hydrothermal activity
- Clarify the architecture, function, and impact of subseafloor microbial eco–systems and their relationship to physical, geochemical, and hydrogeologic variations within the hydrothermal mixing zones around the discharge area
- Establish artificial hydrothermal vents in cased holes from potential subseafloor hydrothermal flows, and to prepare a research platform at each cased hole for later study of fluids tapped from various parts of the hydrothermal system and their associated microbial and macrofaunal communities

FURTHER INFORMATION

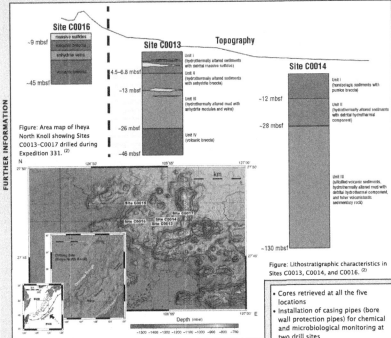

Figure: Area map of Iheya North Knoll showing Sites C0013–C0017 drilled during Expedition 331. [2]

Figure: Lithostratigraphic characteristics in Sites C0013, C0014, and C0016. [2]

HIGHLIGHTS
- Cores retrieved at all the five locations
- Installation of casing pipes (bore wall protection pipes) for chemical and microbiological monitoring at two drill sites
- Discovery of the sub–seafloor hydrothermal fluid structure and hydrothermal alternation zone
- Discovery of the sub–seafloor hydrothermal fluid reservoir
- Discovery of the distribution and mineralogy of hydrothermal sulfide minerals: clues to uncover the hydrothermal ore genesis

REFERENCES (1)•Expedition 331 Scientists, 2010. Deep hot biosphere. IODP Prel. Rept., 331. doi:10.2204/?iodp.pr.331.2010 (2)• Takai, K., Mottl, M.J., Nielsen, S.H.H., and the IODP Expedition 331 Scientists, 2012. IODP Expedition 331: strong and expansive subseafloor hydrothermal activities in the Okinawa Trough. Sci. Drill., 13:19–27. doi:10.2204/?iodp.sd.13.03.2011 (3)• http://www.jamstec.go.jp/chikyu/exp331/e/expedition.html

IODP Expedition 332
NanTroSEIZE Stage 2: Riserless Observatory 2

Dates: 25 October to 12 December 2010
Port: Shingu, Japan
Sites: C0002, C0010
Operator: CDEX, Platform: Chikyu
Co-chief Scientists: Achim Kopf & Eiichiro Araki
Expedition Project Manager: Sean Toczko

Exp. Website: http://www.jamstec.go.jp/chikyu/nantroseize/e/expedition_332.html

NanTroSEIZE Stage 2: Install a permanent riserless CORK observatory and replacing the currently deployed temporary observatory with a newly designed sensor

- Focus mainly on engineering work
- Retrieval of a temporary observatory instrument installed during Expedition 319 at IODP Site C0010, which penetrates the shallow "megasplay" fault in the mid-forearc, and installation of a new suite of temporary sensors
- Deployment of an upgraded temporary observatory at Site C0010
- Installation of a permanent observatory at IODP Site C0002 in the outer Kumano Basin, at the location of planned future deep riser drilling

HIGHLIGHTS
- Installation of permanent riserless long-term borehole observatory at Site C0002
- Recovering the currently deployed temporary observatory and replacing it with a newly designed sensor at Site C0010

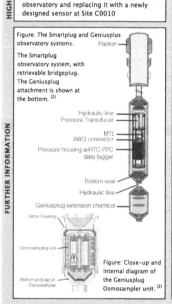

Figure: The Smartplug and Geniusplus observatory systems.

The Smartplug observatory system, with retrievable bridgeplug. The Geniusplug attachment is shown at the bottom. [2]

Figure: Close-up and internal diagram of the Geniusplug Osmosampler unit. [2]

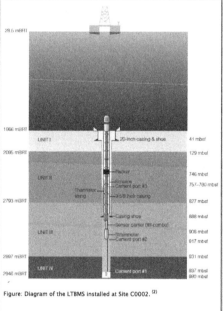

Figure: Diagram of the LTBMS installed at Site C0002. [2]

REFERENCES: (1)• Kopf, A., Araki, E., Toczko, S., and the Expedition 332 Scientists, 2011. NanTroSEIZE Stage 2: riserless observatory. IODP Prel. Rept., 332. doi:10.2204/?iodp.pr.332.2011 (2)• Toczko, S.T., Kopf, A.J., Araki, E., and the IODP Expedition 332 Scientific Party, 2012. The IODP Expedition 332: eyes on the prism, the NanTroSEIZE observatories. Sci. Drill., 14:34-38.

IODP Expedition 333
NanTroSEIZE Stage 2: subduction inputs 2 & heat flow

Dates: 12 December 2010 to 10 January 2011
Port: Shingu, Japan
Sites: C0011, C0012, C0018
Operator: CDEX, Platform: Chikyu
Co-chief Scientists: Pierre Henry & Toshiya Kanamatsu
Expedition Project Manager: Moe Kyaw Thu

Exp. Website: http://www.jamstec.go.jp/chikyu/nantroseize/e/expedition_333.html

NanTroSEIZE Stage 2: Sediment coring with downhole temperature measurements and basement coring

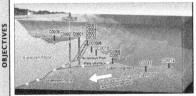

OBJECTIVES
- Drilling and coring of previously unsampled intervals of sediment and basalt at IODP Sites C0011 and C0012 in the Shikoku Basin, together with downhole measurements of temperature
- Drilling and coring at a site near the updip terminus of the megasplay fault, as proposed in Ancillary Project Letter (NanTroSLIDE): Correlating history of submarine landslides along the lower forearc slope to history of slip along the megasplay fault

HIGHLIGHTS
- Determination of heat flow at Sites C0011 and C0012
- Coring through a transition in physical properties within the upper part of the Shikoku hemipelagite
- Repeated coring and sampling of fluid above the sediment/basement interface for shore-based geochemical studies
- Coring into the basalt to 100 m below the sediment/basalt interface, and
- Drilling and sampling a nearly complete slope basin stratigraphic succession comprising six mass transport deposits that record ~1 m.y. submarine landsliding history near the shallow megasplay fault zone area.

FURTHER INFORMATION

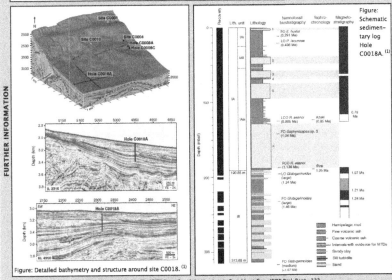

Figure: Detailed bathymetry and structure around site C0018. [1]

Figure: Schematic sedimentary log Hole C0018A. [1]

REFERENCES: (1)•Expedition 333 Scientists, 2011. NanTroSEIZE Stage 2: subduction inputs 2 and heat flow. IODP Prel. Rept., 333. doi:10.2204/?iodp.pr.333.2011 (2)• Henry, P., Kanamatsu, T., Moe, K.T., Strasser, M., and the IODP Expedition 333 Scientific Party, 2012. IODP Expedition 333: return to Nankai Trough subduction inputs sites and coring of mass transport deposits. Sci. Drill., 14:4–17. doi:10.2204/iodp.sd.14.01.2012

IODP Expedition 334
Costa Rica Seismogenesis Project, Program A Stage 1 (CRISP-A1)

IODP Expedition 334

Dates: 13 March to 13 April 2011
Ports: Puntarenas to Puntarenas, Costa Rica
Sites: U1378–U1381
Operator: USIO, Platform: JR
Co-chief Scientists: Paola Vannucchi & Kohtaro Ujiie
Staff Scientist: Nicole Stroncik

Exp. Website: http://iodp.tamu.edu/scienceops/expeditions/costa_rica_seismogenesis_334.html

Costa Rica Seismogenesis Project (CRISP), IODP Expeditions 334 & 344: Sampling and quantifying input to the seismogenic zone and fluid output

OBJECTIVES

- Designed to understand the processes that control nucleation and seismic rupture of large earthquakes at erosional subduction zones
- Involve erosional end-member of convergent margins: relatively thin sediment cover, fast convergence rate, abundant seismicity, subduction erosion, and change in subducting plate relief along strike
- CRISP complements other deep-fault drilling and investigates the first-order seismogenic processes common to most faults and those unique to erosional margins
- CRISP Program A is the first step toward deep riser drilling through the seismogenic zone
- Estimate composition, texture, and physical and frictional properties of the upper plate material
- Quantify subduction channel thickness and the rate of subduction erosion
- Characterize of fluid/rock interaction, the hydrologic system, and the geochemical processes active within the upper plate
- Measure the stress field along the updip limit of the seismogenic zone

CRISP Program A is also considered a standalone project providing data to solve longstanding problems related to tectonics of the region.:

- Cocos Ridge subduction • Evolution of the Central America volcanic arc • Death of a volcanic arc

HIGHLIGHTS

- Coring was conducted at three slope sites and at one site on the Cocos plate
- CRISP is first step toward deep riser drilling through the aseismic and seismic plate boundary
- Logging while drilling (LWD) to document in situ physical properties, stratigraphic and structural features, and stress state
- Continuous core sampling to target depth
- Examination of slope sediments and the underlying upper plate basement
- Provide important information about tectonic, hydrologic, and seismic features along this erosive convergent margin

FURTHER INFORMATION

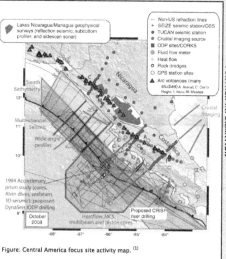

- Non-US refraction lines
- Lakes Nicaragua/Managua geophysical surveys (reflection seismic, subbottom profiler, and sidescan sonar)
► SEIZE seismic station/OBS
● TUCAN seismic station
● Crustal imaging source
■ ODP sites/CORKS
● Fluid flow meter
○ Heat flow
○ Rock dredges
○ GPS station sites
▲ Arc volcanoes (many studies) A. Arenal; C. Cerro Negro; I. Irazu; M. Masaya

Figure: Central America focus site activity map. [1]

Map: Drill Sites of Expedition 334. [3]

REFERENCES (1)• Expedition 334 Scientists, 2011. Costa Rica seismogenesis project (CRISP): sampling and quantifying input to the seismogenic zone and fluid output. IODP Prel. Rept., 334. doi:10.2204/ iodp.pr.334.2011 (2)• Vannucchi, P., Ujiie, K., Stroncik, N., and the IODP Expedition 334 Science Party, 2013. IODP Expedition 334: an investigation of the sedimentary record, fluid flow and state of stress on top of the seismogenic zone of an erosive subduction margin. Sci. Drill., 15:23–30. doi:10.2204/ iodp.sd.15.03.2013 (3)• http://iodp.tamu.edu/scienceops/expeditions/costa_rica_seismogenesis_334.html

IODP Expedition 335
Superfast Spreading Rate Crust 4

IODP Expedition 335

Dates: 13 April to 3 June 2011
Ports: Puntarenas, Costa Rica to Balboa, Panama
Sites: 1256
Operator: USIO, Platform: JR
Co-chief Scientists: Damon Teagle & Benoît Ildefonse
Staff Scientist: Peter Blum

Exp. Website: http://iodp.tamu.edu/scienceops/expeditions/superfast_rate_crust.html

The Superfast Spreading Rate Crust mission multicruise program (ODP Leg 206, IODP Legs 309, 312, 335): A complete in-situ section of upper oceanic crust formed at a superfast spreading rate

OBJECTIVES

- First sampling of intact section of upper oceanic basement (upper oceanic crust, through lavas and the sheeted dikes) into uppermost gabbros at Ocean Drilling Program Site 1256 (ODP Leg 206)
- Crust at Site 1256 formed at superfast (>200 mm/y) spreading rate ~15 m.y. ago at the East Pacific Rise
- Exploit inverse relationship between spreading rate and depth to axial low-velocity zones
- Does the lower crust form by the recrystallization and subsidence of a high-level magma chamber (gabbro glacier), crustal accretion by intrusion of sills throughout the lower crust, or some other mechanism?
- Is the plutonic crust cooled by conduction or hydrothermal circulation?
- What is the geological nature of Layer 3 and the Layer 2/3 boundary at Site 1256?
- What is the magnetic contribution of the lower crust to marine magnetic anomalies?
- Try to deepen the hole >300 m into plutonic rocks, past the transition from dikes to gabbro, and into a region of solely cumulate gabbroic rocks.

HIGHLIGHTS

- Reentered hole
- Stabilized borehole wall
- Opened hole to its full depth of 1521.6 mbsf
- Unique collection of samples recovered, including large cobbles, angular rubble, and fine cuttings of granoblastic basalt with minor gabbroic rocks, some of which intrude previously recrystallized dikes

FURTHER INFORMATION

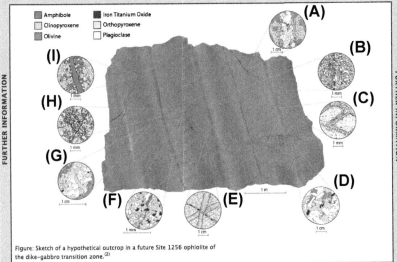

Figure: Sketch of a hypothetical outcrop in a future Site 1256 ophiolite of the dike-gabbro transition zone.[2]

REFERENCES: (1)•Expedition 335 Scientists, 2011. Superfast spreading rate crust 4: drilling gabbro in intact ocean crust formed at a superfast spreading rate. IODP Prel. Rept., 335. doi:10.2204/ iodp.pr.335.2011 •(2) Teagle, D.A.H., Ildefonse, B., Blum, P., and the IODP Expedition 335 Scientists, 2012. IODP Expedition 335: deep sampling in ODP Hole 1256D. Sci. Drill., 13:28–34. doi:10.2204/ iodp.sd.13.04.2011

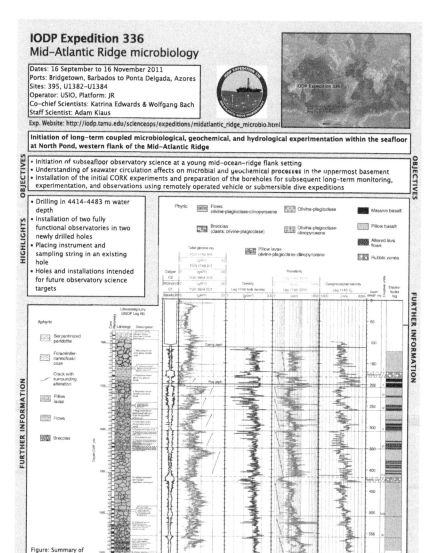

IODP Expedition 336
Mid-Atlantic Ridge microbiology

Dates: 16 September to 16 November 2011
Ports: Bridgetown, Barbados to Ponta Delgada, Azores
Sites: 395, U1382–U1384
Operator: USIO, Platform: JR
Co–chief Scientists: Katrina Edwards & Wolfgang Bach
Staff Scientist: Adam Klaus

Exp. Website: http://iodp.tamu.edu/scienceops/expeditions/midatlantic_ridge_microbio.html

Initiation of long-term coupled microbiological, geochemical, and hydrological experimentation within the seafloor at North Pond, western flank of the Mid-Atlantic Ridge

OBJECTIVES
- Initiation of subseafloor observatory science at a young mid–ocean-ridge flank setting
- Understanding of seawater circulation affects on microbial and geochemical processes in the uppermost basement
- Installation of the initial CORK experiments and preparation of the boreholes for subsequent long-term monitoring, experimentation, and observations using remotely operated vehicle or submersible dive expeditions

HIGHLIGHTS
- Drilling in 4414–4483 m water depth
- Installation of two fully functional observatories in two newly drilled holes
- Placing instrument and sampling string in an existing hole
- Holes and installations intended for future observatory science targets

Figure: Summary of Hole 395A logging results. [1]

REFERENCES (1)• Expedition 336 Scientists, 2012. Mid–Atlantic Ridge microbiology: Initiation of long-term coupled microbiological, geochemical, and hydrological experimentation within the seafloor at North Pond, western flank of the Mid–Atlantic Ridge. IODP Prel. Rept., 336. doi:10.2204/ iodp.pr.336.2012 (2)• Edwards, K., Bach, W., Klaus, A., and the IODP Expedition 336 Scientific Party, 2014. IODP Expedition 336: initiation of long-term coupled microbiological, geochemical, and hydrological experimentation within the seafloor at North Pond, western flank of the Mid–Atlantic Ridge. Sci. Drill., 17:13–18. doi:10.5194/ sd–17–13–2014

IODP Expedition 337
Deep Coalbed Biosphere off Shimokita

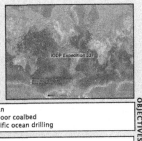

Dates: 25 July to 30 September 2012
Port: Shimizu, Japan
Sites: C0020
Operator: CDEX, Platform: Chikyu
Co-chief Scientists: Kai-Uwe Hinrichs & Fumio Inagaki
Expedition Project Manager: Yu'suke Kubo
Expedition Website: http://www.jamstec.go.jp/chikyu/exp337/e/expedition.html

OBJECTIVES
- Riser-drilling over 2,200 m below the seafloor off Shimokita Peninsula of Japan
- Study the relationship between the deep microbial biosphere and the subseafloor coalbed
- Explore the limits of life in horizons deeper than ever probed before by scientific ocean drilling

HIGHLIGHTS
- Penetration of a 2466 m deep sedimentary sequence with a series of coal layers at ~2 km below the seafloor
- Conduction of downhole fluid analysis and sampling, and logging operations yielded data of unprecedented quality that provide a comprehensive view of sediment properties at Site C0020
- Provide an unprecedented record of dynamically changing depositional environments in the former forearc basin off the Shimokita Peninsula during the late Oligocene and Miocene
- Record comprises a rich diversity of lithologic facies reflecting environments ranging from warm-temperate coastal backswamps to cool-water continental shelf

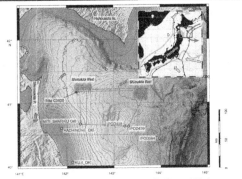

Figure: Index map of Site C0020 off the Shimokita Peninsula of Japan with bathymetry, seismic survey track lines, and locations of existing drill holes. [1]

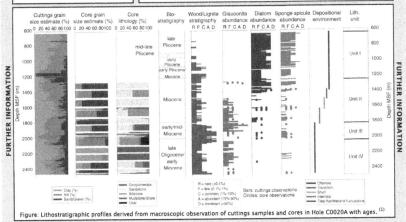

Figure: Lithostratigraphic profiles derived from macroscopic observation of cuttings samples and cores in Hole C0020A with ages. [1]

REFERENCES (1)• Inagaki, F., Hinrichs, K.-U., Kubo, Y., and the Expedition 337 Scientists, 2012. Deep coalbed biosphere off Shimokita: microbial processes and hydrocarbon system associated with deeply buried coalbed in the ocean. IODP Prel. Rept., 337. doi:10.2204/iodp.pr.337.2012

IODP Expedition 338
NanTroSEIZE Stage 3 – Plate Boundary Deep Riser — 2

Dates: 01 October 2012 to 13 January 2013
Port: Shingu, Japan
Sites: C0002
Operator: CDEX, Platform: Chikyu
Co-chief Scientists: Brandon Dugan, Kyuichi Kanagawa, Gregory Moore & Michael Strasser
Expedition Project Manager: Lena Maeda & Sean Toczko

Exp. Website: http://www.jamstec.go.jp/chikyu/nantroseize/e/expedition_338.html

NanTroSEIZE Stage 2: Ultra-deeo drilling toward the megasplay fault (continued)

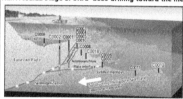

OBJECTIVES

- Access the deep interior of the Miocene accretionary prism
- Riser drilling with continuous cuttings recovery and mud gas analysis from ~856 mbsf to total achieved depth (TD) for this phase of drilling, currently planned for ~ 3600 mbsf
- Coring of 50 meters at each casing TD (approximately 2250–2300 mbsf and 3550–3600 mbsf) in intervals deep within the inner wedge accretionary complex (more coring will be performed if the schedule permits)
- An extensive suite of LWD and wireline logging, downhole stress, pore pressure, permeability tests and a planned zero-offset vertical seismic profile

HIGHLIGHTS

- Riser drilling was conducted in Hole C0002F to 2005.5 mbsf
- Because of riser damage at 2005.5 mbsf, riser drilling operations were suspended for future reoccupation and completion of the NanTroSEIZE project
- Riserless contingency operations were conducted: LDW and coring in Holes C0002H (1100.5–1120 mbsf), C0002J (902–940 mbsf), C0002K (0–286.5 mbsf), C0002L (277–505 mbsf), C0021B (0–194.5 mbsf), and C0022B (0–419.5 mbsf) and LWD in Holes C0012H (0–709 mbsf), C0018B (0–350 mbsf), C0021A (0–294 mbsf), and C0022A (0–420 mbsf)

FURTHER INFORMATION

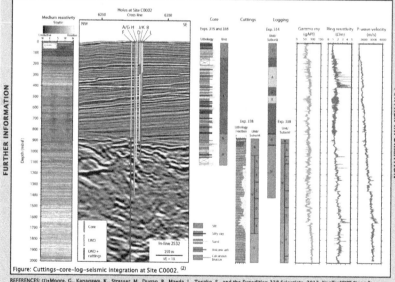

Figure: Cuttings–core–log–seismic integration at Site C0002. [2]

REFERENCES: (1)•Moore, G., Kanagawa, K., Strasser, M., Dugan, B., Maeda, L., Toczko, S., and the Expedition 338 Scientists, 2013. NanTroSEIZE Stage 3: NanTroSEIZE plate boundary deep riser 2. IODP Prel. Rept., 338. doi:10.2204/iodp.pr.338.2013 (1)• Moore, G., Kanagawa, K., Strasser, M., Dugan, B., Maeda, L., and Toczko, S., 2014. IODP Expedition 338: NanTroSEIZE Stage 3: NanTroSEIZE plate boundary deep riser 2. Sci. Drill., 17:1–12. doi:10.5194/ sd-17-1-2014

IODP Expedition 339
Mediterranean Outflow

Dates: 16 November 2011 to 17 January 2012
Ports: Ponta Delgada, Azores to Lisbon, Portugal
Sites: U1385–U1391
Operator: USIO, Platform: JR
Co-chief Scientists: Francisco J. Hernández-Molina & Dorrik Stow
Staff Scientist: Carlos Alvarez–Zarikian

Exp. Website: http://iodp.tamu.edu/scienceops/expeditions/mediterranean_outflow.html

Environmental significance of the Mediterranean Outflow Water (MOW) and its global implications

OBJECTIVES

- Understanding of the opening of the Gibraltar Gateway and onset of MOW
- Determination of MOW paleocirculation and global climate significance
- Identification of external controls on sediment architecture of the Gulf of Cadiz and Iberian margin
- Ascertaining synsedimentary tectonic control on architecture and evolution of the contourite depositional system

HIGHLIGHTS

- Drilled five sites in the Gulf of Cadiz and two off the west Iberian margin
- Recovered 5.5 km of sediment cores (recovery of 86.4 %)
- Penetrated into Miocene at two different sites, into Pliocene at four sites
- Established a strong signal of MOW in the sedimentary record of the Gulf of Cadiz
- Initiation of contourite deposition at 4.2–4.5 Ma, possible onset of MOW?
- Low bottom current activity linked with a weak MOW in Pliocene succession
- Pronounced phase of contourite drift development with two periods of MOW intensification in Quaternary succession – Establishment of contourite depositional system architecture
- Significant climate control on evolution of MOW and bottom–current activity, but even stronger tectonic control on margin development, downslope sediment transport and contourite drift evolution
- High quantity and extensive distribution of clean and well sorted contourite sands as untapped and important exploration target

Figure: Expedition 339 sites in the Gulf of Cádiz and west Iberian margin. (2)

FURTHER INFORMATION

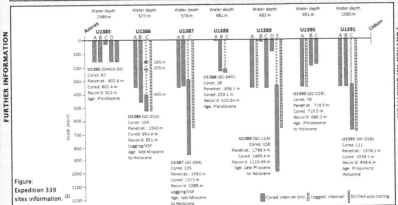

Figure: Expedition 339 sites information. (2)

REFERENCES: (1)• Expedition 339 Scientists, 2012. Mediterranean outflow: environmental significance of the Mediterranean Outflow Water and its global implications. IODP Prel. Rept., 339. doi:10.2204/ iodp.pr.339.2012 •(2) Hernández-Molina, F.J., Stow, D., Alvarez–Zarikian, C., and Expedition IODP 339 Scientists, 2013. IODP Expedition 339 in the Gulf of Cadiz and off West Iberia: decoding the environmental significance of the Mediterranean outflow water and its global influence. Sci. Drill., 16:1–11. doi:10.5194/ sd–16–1–2013

IODP Expedition 340
Lesser Antilles Volcanism and Landslides

Dates: 2 March to 17 April 2012
Ports: San Juan, Puerto Rico to Curacao, Dutch Antilles
Sites: U1393–U1401
Operator: USIO, Platform: JR
Co-chief Scientists: Anne Le Friant & Osamu Ishizuka
Staff Scientist: Adam Klaus

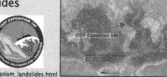

Exp. Website: http://iodp.tamu.edu/scienceops/expeditions/antilles_volcanism_landslides.html

Implications for hazard assessment and long-term magmatic evolution of the arc

OBJECTIVES
- Further understanding of the constructive and destructive processes related to island arc volcanism
- Investigation of magmatic evolution and eruptive activity along the Lesser Antilles arc
- Understanding of the mechanisms involved in both the transport and deposition of volcanic debris avalanche deposits
- Assessing the potential for volcanic hazards associated with avalanches
- Understanding of timing and emplacement processes of potentially tsunamigenic large debris avalanche emplacements
- Documentation of long-term eruptive history of the arc to assess volcano evolution (cycles of construction and destruction) and major volcanic hazards
- Characterization the magmatic cycles and long-term magmatic evolution of the arc
- Documentation dispersal of sediment into the deep ocean
- Determination of processes and element fluxes associated with submarine alteration of volcanic material

HIGHLIGHTS
- Drilling of two holes for each of the nine sites, a total of 2384 m of core recovered
- Identification of mechanisms controlling processes and timing of potentially tsunamigenic large mass transport deposit emplacement
- Characterization of eruptive history, magmatic cycles, and long-term evolution of the arc
- Characterization of sedimentation processes along the deep backarc Grenada Basin

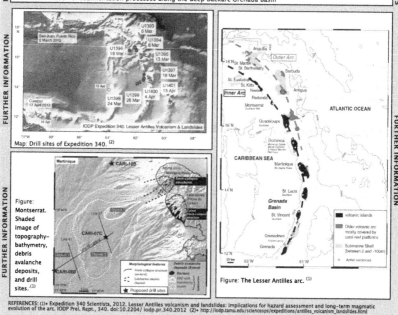

Map: Drill sites of Expedition 340. [2]

Figure: Montserrat. Shaded image of topography-bathymetry, debris avalanche deposits, and drill sites.[1]

Figure: The Lesser Antilles arc. [1]

REFERENCES: (1)• Expedition 340 Scientists, 2012. Lesser Antilles volcanism and landslides: Implications for hazard assessment and long-term magmatic evolution of the arc. IODP Prel. Rept., 340. doi:10.2204/ Iodp.pr.340.2012 (2)• http://iodp.tamu.edu/scienceops/expeditions/antilles_volcanism_landslides.html

IODP Expedition 340T
Atlantis Massif Oceanic Core Complex

Dates: 15 February to 2 March 2012
Ports: Lisbon, Portugal to San Juan, Puerto Rico
Sites: U1309, U1392
Operator: USIO, Platform: JR
Co-chief Scientist: Donna Blackman
Staff Scientist: Angela Slagle

IODP Expedition 340T

Expedition Website: http://iodp.tamu.edu/scienceops/expeditions/atlantis_massif.html

Velocity, porosity, and impedance contrasts within the domal core of Atlantis Massif: faults and hydration of lithosphere during core complex evolution

OBJECTIVES

- Returning to the 1.4–km-deep Hole U1309D at Atlantis Massif to carry out borehole logging including vertical seismic profiling (VSP)
- Testing the hypothesis that highly altered intervals and/or fluid-bearing fault zones at depth might be responsible for these contrasts, thus allowing interpretation of the reflectivity patterns in terms of hydration pathways within young oceanic crust

HIGHLIGHTS

- Obtaining seismic, resistivity, and temperature logs of the 800–1400 mbsf interval of Hole 1309D
- Vertical seismic profile station coverage extends the full length of the hole
- Sampling seafloor features: Recovering fragments of possible cap rock that may provide information on processes within the exposed detachment
- Characterization of boundaries of altered, olivine–rich troctolite intervals within the otherwise dominantly gabbroic sequence
- Confirmation of known faults at 750 mbsf and 1100 mbsf
- Confirmation of interplay between lithology, structure, lithospheric hydration, and core complex evolution
- Percolation of seawater along the fault zone is still active

FURTHER INFORMATION

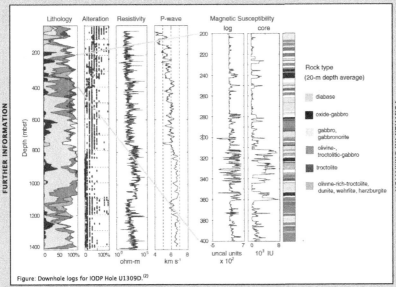

Figure: Downhole logs for IODP Hole U1309D.[2]

REFERENCES: (1)• Expedition 340T Scientists, 2012. Atlantis Massif Oceanic Core Complex: velocity, porosity, and impedance contrasts within the domal core of Atlantis Massif: faults and hydration of lithosphere during core complex evolution. IODP Prel. Rept., 340T. doi:10.2204/iodp.pr.340T.2012 (2)•Blackman, D., Slagle, A., Harding, A., Guerin, G., McCaig, A. (2013): IODP Expedition 340T: Borehole Logging at Atlantis Massif Oceanic Core Complex. – Scientific Drilling, 15, 31–35, 10.2204/iodp.sd.15.04.2013.

IODP Expedition 341
Southern Alaska Margin

IODP Expedition 341

Dates: 29 May to 29 July 2013
Ports: Victoria, Canada to Valdez, Alaska, USA
Sites: U1417–U1421
Operator: USIO, Platform: JR
Co-chief Scientists: John Jaeger & Sean Gulick
Staff Scientist: Leah LeVay

Ex. Website: http://iodp.tamu.edu/scienceops/expeditions/alaska_tectonics_climate.html

Interactions of tectonics, climate, and sedimentation

OBJECTIVES

- Drilling of a cross-margin transect
- Investigation of the northeast Pacific continental margin sedimentary record formed during orogenesis during a time of significant global climatic deterioration in the Pliocene–Pleistocene
- Generation of detailed records of changes in the locus and magnitude of glacial erosion, degree of tectonic shortening, and sediment and freshwater delivery to the coastal ocean and their impact on oceanographic conditions in the Gulf of Alaska
- Documentation of tectonic response of an active orogenic system to late Miocene to recent climate change.
- Establishment of the timing of advance and retreat phases of the Northwestern Cordilleran Ice Sheet (NCIS) to test its relation to dynamics of other global ice sheets.
- Implementation of expanded source-to-sink study of the complex interactions between glacial, tectonic, and oceanographic processes responsible for creation of one of the thickest Neogene–Quaternary high-latitude continental margin sequences.
- Understanding dynamics of productivity, nutrients, freshwater input to the ocean, and surface and subsurface circulation in the Northeast Pacific and their role in the global carbon cycle.
- Document spatial and temporal behavior during the Neogene of the geomagnetic field at extremely high temporal resolution in an undersampled region of the globe.

HIGHLIGHTS

- Recovered 3240 m sedimentary record that extends from the late Pleistocene/Holocene through the middle Miocene
- Discovered substantial sediment volume accumulating on the shelf, slope, and fan since the early Pleistocene intensification of Northern Hemisphere glaciation and more significantly since the mid-Pleistocene transition
- All five sites include the middle Pleistocene to recent and demonstrate exceptional accumulation rates

FURTHER INFORMATION

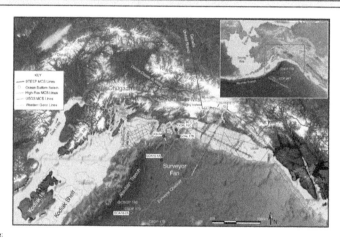

Figure:
The Gulf of Alaska region: geography and location of previous DSDP and ODP drilling locations (see inset) and Expedition 341 drilling sites. [1]

REFERENCES: (1)•Expedition 341 Scientists, 2014. Southern Alaska Margin: interactions of tectonics, climate, and sedimentation. IODP Prel. Rept., 341. doi:10.2204/iodp.pr.341.2014

IODP Expedition 341S
SCIMPI & 858G CORK

Dates: 19 to 29 May 2013
Ports: Victoria to Victoria, Canada
Sites: 898, U1416
Operator: USIO, Platform: JR
Co-chief Scientists: Ian Kulin & Michael Riedel
Staff Scientist: Adam Klaus

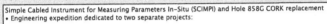

IODP Expedition 341S

Expedition Website: http://iodp.tamu.edu/scienceops/expeditions/scimpi.html

OBJECTIVES

Simple Cabled Instrument for Measuring Parameters In-Situ (SCIMPI) and Hole 858G CORK replacement
• Engineering expedition dedicated to two separate projects:

1) First deployment of the Simple Cabled Instrument for Measuring Parameters In Situ (SCIMPI) on the Cascadia margin
• study dynamic processes in the subseabed based on a simple and low-cost approach
• successfully installed in Hole U1416A
• deployment of a single module in Hole U1416B

2) Replacement of the CORK in Hole 858G for formation pressure monitoring in the Middle Valley axial rift of the Juan de Fuca Ridge
• New CORK was not installed because the old CORK could not be removed from Hole 858G

& OUTCOMES

FURTHER INFORMATION

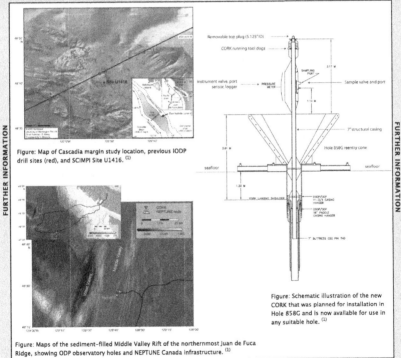

Figure: Map of Cascadia margin study location, previous IODP drill sites (red), and SCIMPI Site U1416. [1]

Figure: Maps of the sediment-filled Middle Valley Rift of the northernmost Juan de Fuca Ridge, showing ODP observatory holes and NEPTUNE Canada infrastructure. [1]

Figure: Schematic illustration of the new CORK that was planned for installation in Hole 858G and is now available for use in any suitable hole. [1]

FURTHER INFORMATION

REFERENCES (1)• Expedition 341S Scientists and Engineers, 2013. Simple Cabled Instrument for Measuring Parameters In Situ (SCIMPI) and Hole 858G CORK replacement. IODP Prel. Rept., 341S. doi:10.2204/iodp.pr.341S.2013

IODP Expedition 342
Paleogene Newfoundland Sediment Drifts

Dates: 2 June to 1 August 2012
Ports: St. George, Bermuda to St. John's, Newfoundland
Sites: U1402 (MDHDS test) and U1403–U1411
Operator: USIO, Platform: JR
Co–chief Scientists: Richard Norris & Paul Wilson
Staff Scientist: Peter Blum

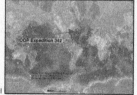

Exp. Website: http://iodp.tamu.edu/scienceops/expeditions/newfoundland_sediment_drifts.html

OBJECTIVES

- Reconstruction of detailed history of the carbonate saturation state of the North Atlantic through numerous episodes of abrupt global warming
- Obtaining a very detailed record of the flow history of the Deep Western Boundary Current issuing from the North Atlantic
- Obtaining a detailed record of the Eocene–Oligocene transition (EOT, ~33.7 Ma) and the onset of major glaciation following the warm climates of the Eocene
- Obtaining cores useful for resolving major uncertainties in Eocene chronostratigraphy that can be used to link the astronomical timescale developed for the last ~40 m.y. to the floating timescale of the early Paleogene developed over a series of IODP and earlier drilling expeditions
- North Atlantic history of carbonate compensation depth change
- North Atlantic history of ocean structure and sediment drift formation
- High–resolution records of climate from rapidly accumulating sediment drifts
- Exceptional preservation of calcareous microfossils in sediment drifts
- Paleogene and Cretaceous hyperthermal and hypoxic events
- Stability of Cenozoic ice sheets and early northern hemisphere glaciation
- Astrochronology and calibration of the Cenozoic timescale

HIGHLIGHTS

- 5724 m cored sediment, average recovery 94% in 25 APC/XCB holes across 10 sites
- Majority of recovery: Sediment of Paleogene age, including records of the Eocene together with Paleocene/Eocene and Eocene/Oligocene boundaries
- Recovery of Cretaceous and Neogene sections: including Cenomanian/Turonian boundary, a deepwater record of the Cretaceous/Paleogene boundary and Oligocene/Miocene boundary
- Most sediments characterized by excellent age control and exquisitely well preserved microfossils
- Detection of rapid ecosystem changes on a warm Earth and on a cold Earth
- Reconstruction of the history of the Deep Western Boundary Current
- High–resolution records of the onset and development of Cenozoic glaciation
- Astronomical calibration of chronostratigraphic markers and the geological timescale
- Test of MDHDS with goal to wash to a depth of 100 m coring depth below seafloor

FURTHER INFORMATION

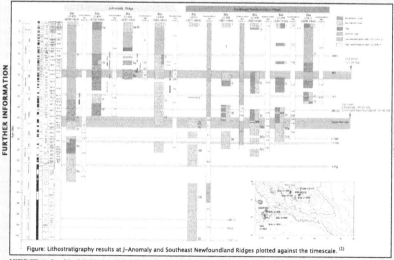

Figure: Lithostratigraphy results at J-Anomaly and Southeast Newfoundland Ridges plotted against the timescale. [1]

REFERENCES: (1)• Expedition 342 Scientists, 2012. Paleogene Newfoundland sediment drifts. IODP Prel. Rept., 342. doi:10.2204/iodp.pr.342.2012

IODP Expedition 343 (&343T)
Japan Trench Fast Drilling Project

Dates: 01 April to 24 May 2012 (& 05 July to 19 July 2012)
Port: Shimizu (& Hachinohe), Japan
Sites: C0019
Operator: CDEX, Platform: Chikyu
Co-chief Scientists: Frederick M. Chester & James J. Mori
Expedition Project Manager: Nobuhisa Eguchi

Expedition Website: http://www.jamstec.go.jp/chikyu/exp343/e/

Rapid Respone to the Tohoku Great Earthquake to Understand the Devasting Tsunami

OBJECTIVES
- Understanding of the physical mechanisms and dynamics of large slip earthquakes
- Investigation of the level of frictional stress during the earthquake rupture and the physical characteristics of the fault zone
- Locating the fault that ruptured during the Tohoku event using logging while drilling (LWD)
- Characterizing the composition, architecture, and fundamental mechanisms of dynamic frictional slip and healing processes along the fault by taking core samples
- Estimating the frictional heat and stress within and around the fault zone by placing a temperature measurement observatory across the fault

HIGHLIGHTS
- Drilling in 7000 m of water depth
- LWD was completed in a borehole drilled to 850.5 mbsf
- Coring hole drilled to 844.5 mbsf
- Single major plate-boundary fault accommodated the large slip of the Tohoku–Oki earthquake rupture, as well as nearly all the cumulative interplate motion at the drill site
- Defined characteristic of the shallow earthquake fault: localization of deformation onto a limited thickness (less than 5 meters) of pelagic clay -- pelagic clay may be a regionally important control on tsunamigenic earthquakes
- Deployment of temperature sensors in July 2012 (Exp 343T)

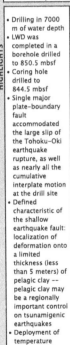

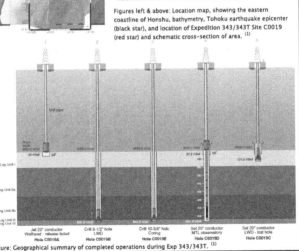

Figures left & above: Location map, showing the eastern coastline of Honshu, bathymetry, Tohoku earthquake epicenter (black star), and location of Expedition 343/343T Site C0019 (red star) and schematic cross-section of area. [1]

FURTHER INFORMATION

Figure: Geographical summary of completed operations during Exp 343/343T. [1]

REFERENCES: (1)• Chester, F.M., Mori, J.J., Toczko, S., Eguchi, N., and the Expedition 343/343T Scientists, 2012. Japan Trench Fast Drilling Project (JFAST). IODP Prel. Rept., 343/343T. doi:10.2204/iodp.pr.3433431.2012 (2)• Chester, F.M., Rowe, C., Ujiie, K., Kirkpatrick, J., Regalla, C., Remitti, F., Moore, J.C., Toy, V., Wolfson-Schwehr, M., Bose, S., Kameda, J., Mori, J.J., Brodsky, E.E., Eguchi, N., Toczko, S., and the Expedition 343 and 343T Scientists, 2013. Structure and composition of the plate-boundary slip zone for the 2011 Tohoku-Oki Earthquake. Science, 342(6163):1208-1211. doi:10.1126/science.1243719

IODP Expedition 344
Costa Rica Seismogenesis Project,
Program A Stage 2 (CRISP–A2)

Dates: 23 October to 11 December 2012
Ports: Balboa, Panama to Puntarenas, Costa Rica
Sites: U1380, U1381, U1412, U1413, U1414
Operator: USIO, Platform: JR
Co-chief Scientists: Robert Harris & Arito Sakaguchi
Staff Scientist: Katerina Petronotis

Exp. Website: http://iodp.tamu.edu/scienceops/expeditions/costa_rica_seismogenesis.html

Costa Rica Seismogenesis Project (CRISP)
Sampling and quantifying input to the seismogenic zone and fluid output (IODP Expeditions 334 & 344)

OBJECTIVES

• Designed to understand the processes that control nucleation and seismic rupture of large earthquakes at erosional subduction zones
• Involve erosional end–member of convergent margins: relatively thin sediment cover, fast convergence rate, abundant seismicity, subduction erosion, and change in subducting plate relief along strike
• CRISP complements other deep–fault drilling and investigates the first-order seismogenic processes common to most faults and those unique to erosional margins
• CRISP Program A is the first step toward deep riser drilling through the seismogenic zone
• Estimate the composition, texture, and physical and frictional properties of the upper plate material
• Quantify the subduction channel thickness and the rate of subduction erosion
• Characterization of fluid/rock interaction, the hydrologic system, and the geochemical processes active within the upper plate
• Measure the stress field along the updip limit of the seismogenic zone
CRISP Program A is also considered a standalone project providing data to solve longstanding problems related to tectonics of the region.:
• Cocos Ridge subduction • Evolution of the Central America volcanic arc • Death of a volcanic arc

HIGHLIGHTS

• Material recovered across Costa Rica erosive convergent margin offshore the Osa Peninsula: incoming Cocos plate, toe of the margin, midslope region, and upper slope region
• Recovery: very good at incoming plate sites, good at midslope and upper slope sites, and poor at the toe site
• Increase understanding of subduction erosion that has led to important new insights

Figure: A. Seismic Line BGR99-7 showing location of Expedition 344 drill sites.

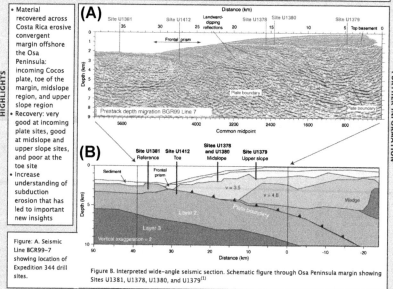

Figure B. Interpreted wide-angle seismic section. Schematic figure through Osa Peninsula margin showing Sites U1381, U1378, U1380, and U1379[1]

REFERENCES (1)• Expedition 344 Scientists, 2013. Costa Rica Seismogenesis Project, Program A Stage 2 (CRISP–A2): sampling and quantifying lithologic inputs and fluid inputs and outputs of the seismogenic zone. IODP Prel. Rept., 344. doi:10.2204/ iodp.pr.344.2013 (2)•Expedition 334 Scientists, 2011. Costa Rica seismogenesis project (CRISP): sampling and quantifying input to the seismogenic zone and fluid output. IODP Prel. Rept., 334. doi:10.2204/ iodp.pr.334.2011

IODP Expedition 345
Hess Deep Plutonic Crust

IODP Expedition 345

Dates: 11 December 2012 to 12 February 2013
Ports: Puntarenas, Costa Rica to Balboa, Panama
Sites: U1415
Operator: USIO, Platform: JR
Co-chief Scientists: Kathryn Gillis & Jonathan Snow
Staff Scientist: Adam Klaus

Expedition Website: http://iodp.tamu.edu/scienceops/expeditions/hess_deep.html

Exploring the plutonic crust at a fast-spreading ridge: new drilling at Hess Deep

OBJECTIVES

- Recovering the first drilled sections of primitive gabbroic rocks formed at a fast-spreading ridge
- Detection of origin and significance of layering
- Determination of transport mechanisms from the mantle through the lower crust
- Evaluation of mechanisms of heat extracted from the lower plutonic crust
- Detection of fluid and geochemical fluxes in the East Pacific Rise (EPR) lower plutonic crust
- Determining the origin of orthopyroxene in primitive lower ocean crust gabbros,
- Determining how plate separation is accommodated in the crust given the evidence against high-strain flow in a partially molten or subsolidus state,
- Reconstructing parental/primary MORB melt compositions using recovered olivine-phyric basalts in Hole U1415N,
- Distinguishing an EPR versus Cocos-Nazca Ridge source for basaltic dikes that intrude the gabbroic lithologies, and
- Evaluating the fluid flux and timing of localized zones of brittle fracturing and cataclasis.

HIGHLIGHTS

- Occupation of 16 holes in water depths ranging from 4675 to 4853 m
- Cored seven holes by rotary core barrel (RCB), recovering 55.2 m of gabbroic rock, additional "ghost core" of 19.8m cored during hole-cleaning operations
- U1415J and U1415JP cored to >100mbsf, other five RCB holes cored to total depth between 12.9 and 37.0 mbsf
- Provided the first confirmation of predictions that fast-spreading lower oceanic crust is layered
- Revealed a diversity of layering whose characteristics have similarities and differences to both layered sequences in ophiolites and layered mafic intrusions
- Revealed significant unexpected mineralogical and textural diversity

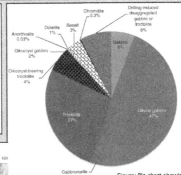

Figure: Pie chart showing the relative proportions of all recovered igneous lithologies except samples from ghost cores, Site 1415.[1]

FURTHER INFORMATION

Figure: Regional bathymetric map of the Hess Deep Rift showing key morphological features and locations of IODP Expedition 345 Site U1415 and ODP Site 895.[1]

REFERENCES: (1)=Expedition 345 Scientists, 2013. Hess Deep plutonic crust: exploring the plutonic crust at a fast-spreading ridge: new drilling at Hess Deep. IODP Prel. Rept., 345. doi:10.2204/iodp.pr.345.2014

IODP Expedition 346
Asian Monsoon

Dates: 29 July to 27 September 2013
Ports: Valdez, Alaska, USA to Busan, Korea
Sites: U1422–U1430
Operator: USIO, Platform: JR
Co–chief Scientists: Ryuji Tada & Rick Murray
Staff Scientist: Carlos Alvarez Zarikian

Exp. Website: http://iodp.tamu.edu/scienceops/expeditions/asian_monsoon.html

OBJECTIVES

Onset and evolution of millennial–scale variability of Asian monsoon and its possible relation with Himalaya and Tibetan Plateau uplift

Test hypothesis that the most recent uplift of the Himalaya mountain range and the Tibetan Plateau (beginning about 3.5 Ma) is responsible for the variability of East Asian summer and winter monsoon patterns (EASM, EAWM)

HIGHLIGHTS

- Drilling of seven sites covering a wide latitudinal range in the Sea of Japan/East Sea and two closely spaced sites in the East China Sea
- 6135.3 m of core with an average recovery of 101%
- Deepest continuous piston core sequence ever recovered in DSDP/ODP/IODP history (490.4 mbsf in IODP Hole U1427A)
- Retrieving of unparalleled record of climate cyclicity in the Miocene epoch
- Address timing of onset of orbital– and millennial–scale variability of the EASM and EAWM and their relation with variability of Westerly Jet circulation
- Reconstruction of orbital– and millennial–scale changes in surface and deepwater circulation and surface productivity in the Sea of Japan/East Sea during at least the last 5 Ma.
- Reconstruction of history of the Yangtze River discharge using cores from the northern end of the East China Sea, as it reflects variation and evolution in the EASM and exerts an impact on the paleoceanography of the Sea of Japan/East Sea
- Examination of interrelationship among the EASM, EAWM, nature and intensity of the influx through the Tsushima Strait, intensity of winter cooling, surface productivity, ventilation, and bottom water oxygenation in the Sea of Japan/East Sea and their changes during the last 5 Ma

FURTHER INFORMATION

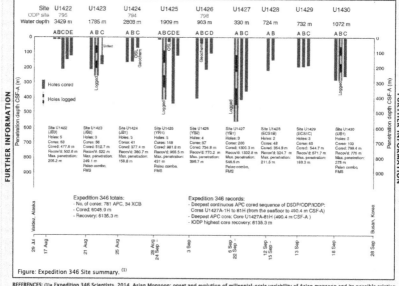

Figure: Expedition 346 Site summary. [1]

REFERENCES: (1)• Expedition 346 Scientists, 2014. Asian Monsoon: onset and evolution of millennial–scale variability of Asian monsoon and its possible relation with Himalaya and Tibetan Plateau uplift. IODP Prel. Rept., 346. doi:10.2204/iodp.pr.346.2014

IODP Expedition 347
Baltic Sea Basin Paleoenvironment

Dates: 12 September to 01 November 2013
Ports: Kiel, Germany
Sites: BSB-1,3,5,7,9,10,11
Operator: ESO, Platform: Greatship Manisha
Co-chief Scientists: Thomas Andrén & Bo Barker Jørgensen
Expedition Project Manager: Carol Cotterill
Expedition Website: http://www.eso.ecord.org/expeditions/347/347.php

Paleoenvironmental evolution of the Baltic Sea Basin through the last glacial cycle

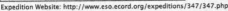

OBJECTIVES

- Coring sediments from different settings in the Baltic Sea Basin (BSB) spanning the last glacial-interglacial cycles
- Climate and sea level dynamics of marine oxygen Isotope Stage (MIS) 5, including onsets and terminations,
- The complexities of the last glacial (MIS 4 – MIS 2),
- Glacial and Holocene climate forcing (MIS 2 – MIS 1)
- Deep biosphere responses to glacial – interglacial cycles

FURTHER INFORMATION

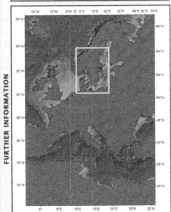

Map: Overview map of Expedition 347 drill sites. [3]

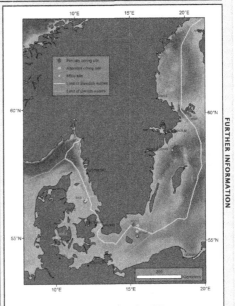

Detailed Site map: Location of proposed Expedition 347 sites. [1]

HIGHLIGHTS

- Coring of 30 holes across eight sites with about 1620 m of high-quality sediment core sediments collected with an average recovery of 91.5% in six different subbasins. These sediments were expected to contain sediment sequences representative of the last ~140,000 years.
The different sub basins are:
  1) Gateway of the Baltic Sea Basin; focus: sediments from MIS 6–5 as well as MIS 2–1 (Site M0060/BSB-1)
  2) Subbasin in the southwestern Baltic Sea (Little Belt) that possibly holds unique MIS 5 record (Site M0059/BSB-3)
  3) and 4) Two subbasins in the south (Bornholm Basin and Hanö Bay) that may hold long complete records from MIS 4–2 (Site M0064/BSB-5 and Site M0065/BSB-7)
  5) 450 m deep subbasin in the central Baltic (Landsort Deep) that promised to contain a thick and continuous record of the last ~14,000 years (Site M0063/BSB-9)
  6) Subbasin in the very north (Ångermanälven River estuary) that contains a uniquely varved (annually deposited) sediment record of the last 10,000 years (Site M0061/BSB-10 and Site M0062/BSB-11)

REFERENCES: (1)• Andrén, T., Jørgensen, B.B., and Cotterill, C., 2012, Baltic Sea Basin Paleoenvironment: paleoenvironmental evolution of the Baltic Sea Basin through the last glacial cycle. IODP Sci. Prosp., 347. doi:10.2204/iodp.sp.347.2012 (2)• Expedition 347 Scientists, 2014. Baltic Sea Basin Paleoenvironment: paleoenvironmental evolution of the Baltic Sea Basin through the last glacial cycle. IODP Prel. Rept., 347. doi:10.2204/iodp.pr.347.2014 (3)• http://www.eso.ecord.org/expeditions/347/347.php

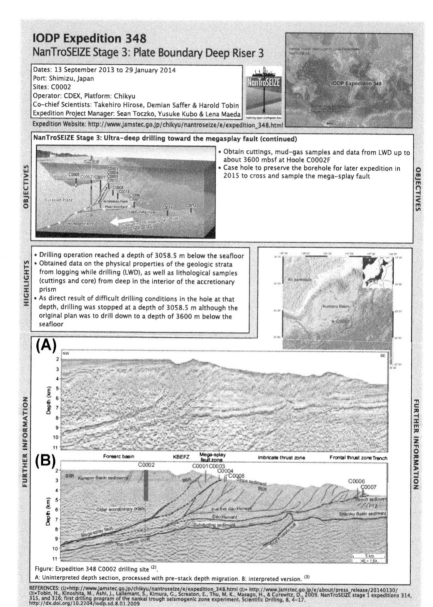

IODP Expedition 348
NanTroSEIZE Stage 3: Plate Boundary Deep Riser 3

Dates: 13 September 2013 to 29 January 2014
Port: Shimizu, Japan
Sites: C0002
Operator: CDEX, Platform: Chikyu
Co–chief Scientists: Takehiro Hirose, Demian Saffer & Harold Tobin
Expedition Project Manager: Sean Toczko, Yusuke Kubo & Lena Maeda
Expedition Website: http://www.jamstec.go.jp/chikyu/nantroseize/e/expedition_348.html

NanTroSEIZE Stage 3: Ultra–deep drilling toward the megasplay fault (continued)

OBJECTIVES
- Obtain cuttings, mud–gas samples and data from LWD up to about 3600 mbsf at Hoole C0002F
- Case hole to preserve the borehole for later expedition in 2015 to cross and sample the mega–splay fault

HIGHLIGHTS
- Drilling operation reached a depth of 3058.5 m below the seafloor
- Obtained data on the physical properties of the geologic strata from logging while drilling (LWD), as well as lithological samples (cuttings and core) from deep in the interior of the accretionary prism
- As direct result of difficult drilling conditions in the hole at that depth, drilling was stopped at a depth of 3058.5 m although the original plan was to drill down to a depth of 3600 m below the seafloor

(A)

(B)

Figure: Expedition 348 C0002 drilling site [2].
A: Uninterpreted depth section, processed with pre–stack depth migration. B: interpreted version. [3]

REFERENCES: (1)•http://www.jamstec.go.jp/chikyu/nantroseize/e/expedition_348.html (2)• http://www.jamstec.go.jp/e/about/press_release/20140130/ (3)•Tobin, H., Kinoshita, M., Ashi, J., Lallemant, S., Kimura, G., Screaton, E., Thu, M. K., Masago, H., & Curewitz, D., 2009. NanTroSEIZE stage 1 expeditions 314, 315, and 316: first drilling program of the nankai trough seismogenic zone experiment. Scientific Drilling, 8, 4–17. http://dx.doi.org/10.2204/iodp.sd.8.01.2009

Index

Printed in the United States
By Bookmasters